COAL AND COALBED GAS

FUELING THE FUTURE

Romeo M. Flores

AMSTERDAM • BOSTON • HEIDELBERG • LONDON • NEW YORK • OXFORD
PARIS • SAN DIEGO • SAN FRANCISCO • SINGAPORE • SYDNEY • TOKYO

Elsevier
225 Wyman Street, Waltham, MA 02451, USA
525 B Street, Suite 1900, San Diego, CA 92101-4495, USA

First edition 2014

Library of Congress Cataloging-in-Publication Data
Flores, Romeo M.
Coal and coalbed gas: fueling the future / Romeo M. Flores. – First edition.
pages cm
Summary: "Coalbed gas has emerged as an important energy source and understanding this resource requires knowledge of elements of coal geology. The diverse heritage of coal and the maturing field of coalbed gas have given rise to definitions, and terminologies that are used throughout this book. The ability to comprehend basic principles is the key to extrapolate from coal science to engineering-applied coalbed gas. Understanding the global coal endowment, occurrence, and distribution related to the dynamic structural changes in the Earth's outer shell through geologic time is essential to the exploitation of coalbed gas. Vast global coal resources of 9-27 trillion metric tons potentially hold coalbed gas-in-place of 67-252 trillion m^3 based on comparisons with producing American fields"– Provided by publisher.
Includes bibliographical references.
ISBN 978-0-12-396972-9 (hardback)
1. Coalbed methane. 2. Natural gas. 3. Coal Geology. I. Title. II. Title: Coal and coalbed gas.
TN844.5.F55 2014
622'.3385–dc23
2013025155

British Library Cataloguing in Publication Data
A catalogue record for this book is available from the British Library

For information on all Elsevier publications
visit our web site at store.elsevier.com

ISBN: 9780123969729

Contents

Author's Biography

ROMEO M. FLORES

Consulting Associate with Cipher Coal Consulting Limited for international companies. Serves in the International Journal of Coal Geology editorial board and author of innumerable publications in coal geology, sedimentology, stratigraphy, hydrostratigraphy, and basin analysis, coal and coalbed gas resource assessments, coal-to-biogenic coalbed gas, and conference organizer and lecturer. Editor of several international special publications. Recipient of awards from the U.S. Department of Interior and U.S. Geological Survey, Geological Society of America, University of the Philippines, and University of Canterbury (New Zealand). Degrees: PhD, Louisiana State University, MS, University of Tulsa, and BS, University of the Philippines.

Foreword

It was a typical Powder River Basin summer day—dry, hotter than a kiln even by 9 am, dust permeating every pore, and a sky so azure that it reminded a thirsty young geologist of exotic tropical drinks and beaches. Despite these distractions, we climbed up the soft steep side of a moderately high hill. The "we" in this, as you might have guessed, is Romey Flores and I. For the previous 2 weeks, I had been in the field gathering data from coal mines and testing my skills at measuring sections, all in the pursuit of my Masters thesis. Romey had come up the night before from Denver and in the morning we had set off up the Tongue River Road to examine some of the Tertiary age coal-bearing sediments which reside there, in abundance. Romey has never been an overly talkative man, and that morning was no exception; thus it was easy to lose oneself in thought while going up that hill and think about, well, think about nothing much at all really. As we approached two-thirds of the way up, I suddenly had a premonition that perhaps I should be paying attention and, as you may have once again guessed, it was at that moment that Romey turned around, sat down on the lip of a beige sandstone ledge, repositioned his toothpick and said, "So, now tell me what you just saw on the way up". I felt like a possum staring into the headlights of a tractor-trailer smack in the middle of the expanse of Interstate-90. Doomed.

Yet, the lesson I learned was to always pay attention, look at the details, and gather data wherever you are; you never know when it will come in handy. In other words, be curious, and of course, think. Who knew 25 years ago that collecting all that data both large and small scale on coal would be so useful for so many things? As it turns out, one of those things has been in the understanding of how gas is formed in coal and how to get it out. By an odd professional convention, anything to do with gas (as with liquid hydrocarbons) usually falls within the domain of petroleum geoscientists and engineers. But coal is not quite like those "other" rocks. And thus, the exploitation of gas from coal (sometimes called coal seam gas, other times called coalbed methane) really does rely on a true overlap of classical coal geology and conventional petroleum geoscience. But where does a professional go to find a summary, a text, and a concise synopsis of coalbed methane—a field that truly encompasses an incredibly wide set of disciplines? Who has the time, or access, to scour through thousands of articles in hundreds of journals? Ergo, enter Romeo Flores and the publication of this book, *Coal and Coalbed Gas: Fueling the Future*.

The breadth of the book you are about to read is large, as it must be. You will be asked to visualize geology on the basin scale, as well as envision processes that happen on a subangstrom level. To many geoscientists, neither scale will be totally unfamiliar but in order to understand coalbed methane, sometimes we must have both scales operating in our heads at once. Similarly, the coal geologists among us will be quite comfortable with peat formation, cleats, and depositional environments, although perhaps a few synapses will need stretching when we enter the realm of submicron porosity, saturation, and van der Waals forces. Switch the analogy around and it works equally well for petroleum geoscientists and engineers. And that is what makes this book exciting.

In order to write such a wide-ranging book on such a big topic, the author needs to be equally expansive. Many of us know Romey as "King of the Powder River Basin". Certainly he has done a copious amount of data gathering there over many decades and published innumerable reports. In the early years of my career, I would return from some far flung area (think Indonesia, New Zealand, remote Northwest China) and roll into Romey's office sure that I had trumped him in seeing some new and exotic geology. But invariably, after a moment of thought, he'd reach under some pile of papers on his desk and say, "As a matter of fact, I was there in 1977..." or 1982 or 1994, whatever the case. After a while, I gave up and just consulted Romey *before* I went somewhere!

Now, what is in the book? I'll leave that for you to explore. But I think the basic tenets and the germane issues are best summed up by a list given in Chapter 7: (1) Coals are not conventional oil and gas reservoirs; (2) coals are thin, compressible, and chemically reactive; (3) coalbed gas production techniques are coal basin specific; and (4) multiple completions are necessary in some basins in order to maximize recovery from multiple coal beds. I'd encourage all coalbed methane scientists, engineers, and managers to re-read these four points. Now go read the details in the book, before Romey sits down on that sandstone ledge again.

Tim A. Moore

Managing Director, Cipher Consulting Ltd. (www.ciphercoal.com), Adjunct Associate Professor, Department of Geological Sciences, University of Canterbury, Member, Royal Society of New Zealand and Member, Australasian Institute of Mining and Metallurgy, Christchurch, New Zealand 2013

Preface

Coal has a long-established tradition as a discipline that utilizes diverse scientific specialties from coal geology, sedimentology, stratigraphy, depositional environments (peat to coal), organic petrology, and inorganic mineralogy to geochemistry. These specialties have been applied to basic research and development, assessment of coal resources and reserves, and processes of hydrocarbon generation and accumulation. The use of coal for heat is over 2000 years old, as documented in East Asia and the Mediterranean. Coal has been an important catalyst to industrialization of countries for the past 50 years. Coal is estimated to be the largest endowment of fossil fuels on Earth and is on a trajectory to replace oil as the primary energy source fueling more proposed efficient power plants worldwide. The high proportion of organic matter in coal has undoubtedly contributed to higher carbon dioxide emissions, which in a carbon-constrained world, has implications for its future as a fuel source. Emissions of greenhouse gases can be alleviated by utilization of virgin coalbed and coal-mine gas as a cleaner energy source. However, the global greenhouse gas emission from coal use belies the full extent of the contribution because its peat precursor, which covers 2–3% of the total Earth's surface, continues to release methane that is a more potent gas than carbon dioxide. Methane is both sourced and stored within the coal, and this fundamental property has generated worldwide interest in commercial production over the past 30 years, even though the presence of gas in coal has been known since the earliest coal mine explosions in the early nineteenth century.

The viability and sustainability of coalbed gas as an energy source has emerged in many developing and developed countries. In Indonesia where the petroleum reserves have been depleted, coalbed gas is being extensively explored as a replacement energy source. In China, coalbed gas has ignited countless coal mine explosions with the resultant loss of lives of thousands of miners. China has now started to recover coalbed gas from active mines as well as ahead of mining and from virgin coals and this gas is taken to the market as liquid and compressed natural gas. Great Britain has produced coalbed gas from abandoned coal mines and delivered the gas to electricity utilities for power generation. About 37 coal-producing countries are exploiting and testing the economic potential of coalbed gas from coal mines for recovery and utilization in order to mitigate greenhouse gas emissions. The Indian government has awarded coalbed gas blocks in several coalfields for exploration and development. The United States, Australia, and Canada aided by technological advances in drilling, stimulation, and completion are the major producers of coalbed gas. This gas is a significant part of their overall portfolio of unconventional gas assets. Despite the groundbreaking successes in duplicating methanogenic processes in gas- and nongas-producing coals to generate biogenic coalbed gas, it remains to be proved that the technology can sustain gas production. Challenges to future application of coal-to-biogenic coalbed gas technology also come from environmental factors, coal reservoir properties, and hydrogeological factors.

More than 70 countries possess coal deposits and even though much of these coal resources may not be economically mineable, they are exploitable for coalbed gas. The world coal resources have been reported to vary from 9 to 27 trillion metric tons with most reserves found in the United States, China, Russia, India, and Australia. Coal is produced in more than 35 countries with potential for coalbed gas recovery in advance of mining. Although huge coal resources do not directly translate to huge volumes of coalbed gas, it is estimated that the global gas-in-place resource is from 67 to 252 trillion m^3. The discrepancy between coal resources and coalbed gas-in-place resources and gas recoverability is mainly a result of geological characteristics of the coal in relationship to engineering properties of the coal reservoir rock as well a general lack of reliable data. In spite of the large coalbed gas-in-place resources, major commercial production is limited to about six countries because of technical and economic limitations and lack of knowledge of coalbed gas as a potential source of energy. Many of the countries interested in exploiting coalbed gas are experiencing much of the same problems that occurred early in coalbed gas development in the United States where government regulators have to navigate through contentious issues such as "who owns coalbed gas: the petroleum or coal mining industry?" and the ever-present management of coproduced water.

The emergence of global exploitation and development of coalbed gas in many countries rich in coal resources have given impetus to the writing of this book. The goal has been for the book to serve as a reference, bridging the gap between scientific knowledge of coal and recently acquired technology of coalbed gas. There are many notable books and scientific journal publications, referenced in this book, which have focused on coal and coalbed gas separately as energy sources. This book strives to integrate the wealth of knowledge of these disciplines and is dedicated to multidisciplinary practitioners. These practitioners include petroleum and reservoir engineers, surface and subsurface hydrologists, geologists, microbiologists, molecular engineers, biogeochemists, environmentalists, land managers, policy makers, and others involved in research and development of coalbed gas and management of related coproduced water. In order to inform and affect the learning of the reader, strategically placed subsections of the book focus on "facts vs fictions" as well as "concepts and misconceptions" of coal and coalbed gas. Coalbed gas has long been produced in the United States, so most of the scientific, technical, and economic information used in this book is derived from this country.

Many people contributed to this book, beginning with those cited in the reference section without whose foundation papers this book would not have been possible. I wish to thank the authors and publishers for kindly providing copyright permissions to complete this book. To colleagues whose work I have used, but have somehow failed to cite, my deepest apologies. I am indebted to many colleagues who provided extremely valuable comments, constructive criticisms, and detailed analysis, which shaped and greatly improved the book. However, I am responsible for any remaining errors or inaccuracies. This book benefited from incisive reviews and insightful suggestions by colleagues from academia, government, and industry: Mike Brogan, Fred Crockett, Frank Ethridge, Mohinudeen Faiz, Song Jin, Dwain McGarry, Ricky Pena, Tim Moore, Dave Rutledge, and Russell Stands-Over-Bull. In particular, Frank Ethridge who patiently reviewed each chapter, recommending clarity, identifying inconsistency and repetition, and consistently, but gently, preaching readability has acted as a friendly gadfly.

I owe debts of gratitude to other colleagues who provided selfless assistance; these people include Melody Bragg, Mick Brownfield, Randy

Caber, Larry Claypool, Joseph Crisostomo, Fred Crockett, Tom Doll, Chris Eisinger, Joan Esterle, Paul Fallgren, Alejandro Flores, Gustavo Navas Guzman, Trinetta Herdy, Scott Kinney, Tim Moore, Dave Mathew, Dwain McGarry, Tracey Mercier, Pierce Norton, Ian Palmer, Cindy Rice, James Rumbaugh, Christopher Scotese, Ernie Slucher, Brianne Spear, Gary Stricker, James Welch, and Sadiq Zarrouk. My learning experience of China's coalbed gas and CMM was made possible by Dr Tian YongDong, Jincheng Lanyan CBM Group, and Dr Hu QianTing, China Coal Technology Group, ChonQing Division, with skilled guidance by Yuewen Xi. Dr Hadiyanto, formerly Assistant Minister of Energy and Mineral Resources, provided my introduction to Indonesian peat, coal, and coalbed gas. Lisa Ramirez Rukstales' advice has been priceless.

The major effort in publishing this book, besides that of the author, comes from the publication staff of Elsevier Science and Technology Books. My appreciation is expressed to Jill Cetel, and Sean Coombs editorial project managers; Sharmila Vadivelan, production manager; Nithya Sambantham, copyrights coordinator; copy editors; and others who all contributed to the book in its final and present form. Finally, thanks to Candice Janco, Acquisitions Editor, and reviewers of the book proposal for believing that the project can be done.

The book is not a research monograph, although I have relied on more than 50 years of research starting from the Allegheny coal mines in eastern Ohio for my PhD thesis to 35 years in the USGS working in the Powder River Basin and other coal basins in the conterminous United States, Alaska, and worldwide. Working in the right place and the right time in the Powder River Basin in a consortium of USGS, BLM, and gas operators from 1999 to 2009 has helped immeasurably in my understanding of coal and coalbed gas. I wish to express deep appreciation to many colleagues who extended kindness and assistance throughout a long career in coal and coalbed gas, especially John Ferm and Ron Stanton, both of them mentors and colleagues, who unfortunately passed away before their time and to Frank, Tim, and Gary, friends and colleagues, for wholehearted support of this grand endeavor.

Most importantly, my appreciation to Lejo, Emily, and Ellie Amihan and the rest of my family and friends; hopefully this book will not only act as a good summary of my career but also provide insight to a father's and grandfather's continuing love of geology.

SI/Metric Units

Approximate Conversions from SI/Metric Units to Standard/Imperial Units*

Known (Symbol)	Multiply By	Obtain (Symbol)
LENGTH		
Centimeter (cm)	0.3937	Inch (in.)
Millimeter (mm)	0.03937	Inch (in.)
Meter (m)	3.281	Foot (ft)
Kilometer (km)	0.6214	Mile (mi)
Kilometer (km)	0.5400	Mile, nautical (nmi)
Meter (m)	1.094	Yard (yd)
AREA		
Square meter (m^2)	0.0002471	Acre
Hectare (ha)	2.471	Acre
Square kilometer (km^2)	247.1	Acre
Square centimeter (cm^2)	0.001076	Square foot (ft^2)
Square meter (m^2)	10.76	Square foot (ft^2)
Square centimeter (cm^2)	0.1550	Square inch (ft^2)
Hectare (ha)	0.003861	Square mile (mi^2)
Square kilometer (km^2)	0.3861	Square mile (mi^2)
VOLUME		
Cubic meter (m^3)	6.290	Barrel (petroleum, 1 barrel = 42 gal)
Liter (l)	33.82	Ounce, fluid (fl. oz)
Liter (l)	2.113	Pint (pt)
Liter (l)	1.057	Quart (qt)
Liter (l)	0.2642	Gallon (gal)

(Continued)

Approximate Conversions from SI/Metric Units to Standard/Imperial Units* (*cont'd*)

Known (Symbol)	Multiply By	Obtain (Symbol)
Cubic meter (m^3)	264.2	Gallon (gal)
Cubic meter (m^3)	0.0002642	Million gallons (Mgal)
Cubic centimeter (cm^3)	0.06102	Cubic inch ($in.^3$)
Liter (l)	61.02	Cubic inch ($in.^3$)
Cubic meter (m^3)	35.31	Cubic foot (ft^3)
Cubic meter (m^3)	1.308	Cubic yard (yd^3)
Cubic kilometer (km^3)	0.2399	Cubic mile (mi^3)
Cubic meter (m^3)	0.0008107	Acre-foot (acre-ft)
Cubic hectometer (hm^3)	810.7	Acre-foot (acre-ft)
Cubic centimeter per gram (cc/gm)	32.037	Cubic feet per ton (cft)
FLOW RATE		
Cubic meter per second (m^3/s)	70.07	Acre-foot per day (acre-ft/day)
Cubic meter per year (m^3/year)	0.000811	Acre-foot per year (acre-ft/year)
Cubic hectometer per year (hm^3/year)	811.03	Acre-foot per year (acre-ft/year)
Meter per second (m/s)	3.281	Foot per second (ft/s)
Meter per minute (m/min)	3.281	Foot per minute (ft/min)
Meter per hour (m/hr)	3.281	Foot per hour (ft/hr)
Meter per day (m/day)	3.281	Foot per day (ft/day)
Meter per year (m/year)	3.281	Foot per year (ft/year)
Cubic meter per second (m^3/s)	35.31	Cubic foot per second (ft^3/s)
Cubic meter per second per square kilometer ($(m^3/s)/km^2$)	91.49	Cubic foot per second per square mile ($(ft^3/s)/mi^2$)
Cubic meter per day (m^3/day)	35.31	Cubic foot per day (ft^3/day)
Liter per second (l/s)	15.85	Gallon per minute (gal/min)
Cubic meter per day (m^3/day)	264.2	Gallon per day (gal/day)
Cubic meter per day per square Kilometer ($(m^3/d)/km^2$)	684.28	Gallon per day per square mile ((gal/day)/mi^2)
Cubic meter per second (m^3/s)	22.83	Million gallons per day (Mgal/day)
Cubic meter per day per square kilometer ($(m^3/day)/km^2$)	0.0006844	Million gallons per day per square mile ((Mgal/day)/mi^2)

Approximate Conversions from SI/Metric Units to Standard/Imperial Units* (*cont'd*)

Known (Symbol)	Multiply By	Obtain (Symbol)
Cubic meter per hour (m^3/h)	39.37	Inch per hour (in./h)
Millimeter per year (mm/year)	0.03937	Inch per year (in./year)
MASS		
Gram (g)	0.03527	Ounce, avoirdupois (oz)
Kilogram (kg)	2.205	Pound avoirdupois (lb)
Megagram (Mg)	1.102	Ton, short (2000 lb)
Megagram (Mg)	0.9842	Ton, long (2240 lb)
Metric ton per day	1.102	Ton per day (ton/day)
Megagram per day (Mg/day)	1.102	Ton per day (ton/day)
Megagram per day per square kilometer ((Mg/day)/km^2)	2.8547	Ton per day per square mile ((ton/day)/mi^2)
Megagram per year (Mg/year)	1.102	Ton per year (ton/year)
Metric ton per year	1.102	Ton per year (ton/yesr)
PRESSURE		
Kilopascal (kPa)	0.009869	Atmosphere, standard (atm)
Kilopascal (kPa)	0.01	Bar
Kilopascal (kPa)	0.2961	Inch of mercury at 60 °F (in Hg)
Kilopascal (kPa)	0.1450	Pound-force per inch (lbf/in.)
Kilopascal (kPa)	20.88	Pound per square foot (lb/ft^2)
Kilopascal (kPa)	0.1450	Pound per square inch (lb/ft^2)
DENSITY		
Kilogram per cubic meter (kg/m^3)	0.06242	Pound per cubic foot (lb/ft^3)
Gram per cubic centimeter (g/cm^3)	62.4220	Pound per cubic foot (lb/ft^3)
ENERGY		
Joule (J)	0.0000002	Kilowatthour (kWh)
SPECIFIC CAPACITY		
Liter per second per meter ((l/s)/m)	4.831	Gallon per minute per foot ((gal/min)/ft)
HYDRAULIC CONDUCTIVITY		
Meter per day (m/day)	3.281	Foot per day (ft/day)

(Continued)

Approximate Conversions from SI/Metric Units to Standard/Imperial Units* (*cont'd*)

Known (Symbol)	Multiply By	Obtain (Symbol)
HYDRAULIC GRADIENT		
Meter per kilometer (m/km)	5.27983	Foot per mile (ft/mi)
TRANSMISSIVITY		
Meter squared per day (m^2/day)	10.76	Foot squared per day (ft^2/day)

* *Modified from U.S. Geological Survey publications.*

Units: one metric tonne = 2204.6 lb; I kilocalorie (kcal) = 4.18 kJ = 3.96 Btu; 1 kJ (kJ) = 0.24 kcal = 0.95 Btu; 1 British thermal init (Btu) = 0.25 kcal = 1.05 kJ; 1 kilowat hour (kWh) = 860 kcal = 3600 kJ = 3412 Btu. Calorific Equivalents: 1 ton of oil equivalent = 10 million kilocalories = 42 gigajoules = 40 million Btu of heat; 1.5 tonnes of hard coal = 3 tonnes of lignite; 12 megawatt hours; One million tones of oil produces 4400 gigawatt hours. Temperature in degrees Celsius (°C) may be converted to degrees Fahrenheit (°F) as follows: F = (1.8 ×°C) + 32. Concentrations of chemical constituents in water are given either in milligrams per liter (mg/l) or micrograms per liter (μg/l). Use of hectare (ha) is for the measurement of small land or water areas. Use of liter (l) is for measurement of liquids and gases. The prefix pf milli is used with liter. Metric ton (Mt) is commercial usage with prefixes used.

CHAPTER

1

Introduction and Principles

OUTLINE

KEY ITEMS

- Coalbed gas has emerged as an important energy source and understanding this resource requires the knowledge of elements of coal geology.
- The diverse heritage of coal and the maturing field of coalbed gas have given rise to definitions, and terminologies that are used throughout this book.
- Ability to comprehend basic principles is the key to extrapolate from coal science to engineering-applied coalbed gas.
- Understanding the global coal endowment, occurrence, and distribution related to the dynamic structural changes in the Earth's outer shell through geologic time is essential to the exploitation of coalbed gas.

http://dx.doi.org/10.1016/B978-0-12-396972-9.00001-X

- Vast global coal resources of 9–27 trillion metric tons (mt) potentially hold coalbed gas-in-place of 67–252 trillion m^3 (Tcm) based on comparisons with producing American fields.
- About 2100 known coal deposits from Late Devonian (380 million years ago or mya) to Pliocene (3 mya) in age are identified resources/reserves in 74 countries and are inferred resources in 27 other countries.
- Peak coal production following M. King Hubbert peak oil theory has led to forecasts of exhaustion of global coal reserves to 90% in 2070, global peak coal production in 2050, and in the United States as early as 2060 and as late as 2105.
- Despite projection of global peak coal production and reserve exhaustion occurring within this century, coal will continue to fuel our foreseeable future with 2.5–3.5% of the global total coal resource having been consumed.
- Potential reduction of coal consumption and CO_2 emissions from coal-fired power plants using more efficient technology (e.g. supercritical, ultra-supercritical power plants) and CO_2 capture and sequestration is believed to be the key to future of coal. Availability of natural gas and environmental challenges for coal use have limited construction of power plants in the United States with 5% of new capacity of electricity for proposed plants to be fueled by coal. However, developing countries will accelerate the construction of new coal plants to generate electricity and are unconstrained by CO_2 emissions.
- Emergence of unconventional gases (e.g. coalbed gas, shale gas) for power generation is a positive alternative in a carbon-constrained world. According to U.S. Energy Information Administration, 80% of the new capacity of electricity for proposed power plants will be fueled by natural gas.

INTRODUCTION

This introductory chapter discusses the philosophy and scope of the book but more importantly, it introduces the reader to the terminology, definition, and fundamentals of coal and coalbed gas. The mixing of the long-established principles of coal geology and maturing precepts of coalbed gas, as an emerging unconventional energy, has introduced entirely rich subjects that blend old and new concepts in this book for a target multidisciplinary audience. A result of the mixture of old and new concepts is the proliferation in the literature of metaphors and misconceptions, which are stumbling blocks to familiarizing, learning, and understanding the subject matter, and this is explained in this chapter. The success in blending coal and coalbed gas as a discipline lies in the understanding of the basic principles of coal discussed in the context of its worldwide geologic and geographic distributions as overprinted by the dynamism of the earth's outer shell and constrained by geologic time. Essential to the discussion of basic principles is the foundation of global coal endowment, which is critical to the assessment of coalbed gas because it addresses how coal is measured, calculated, and classified (e.g. resources and reserves) as practiced by different countries and reported by worldwide agencies. The methods of classification and reporting of coal resources and reserves vary from country to country and knowing this difference is critical to the evaluation and application to coalbed gas. Also important is the consideration of how much coal remains in the ground from the standpoint of historical and production perspectives. The past and future coal production is relevant to the deliberation of potential development of coalbed gas resource. Finally, the ramifications of greenhouse gas emission (e.g. carbon dioxide or CO_2) from coal use to generate electricity directly impact the continued switch of natural gas (e.g. coalbed gas) for generation of power plants in the future.

PHILOSOPHICAL OVERVIEW AND SCOPE

The purpose of this overview is to introduce the reader to the scale and complexity of the subject and scope of the book, with particular emphasis on the trans-disciplinary nature of the state-of-the-art in coal and coalbed gas research and development. More than 30 years ago, coalbed gas emerged globally as a valuable and potential energy resource with a growing role in the exploitation and development of unconventional gas. Unlike other natural gas from conventional reservoirs, coalbed gas is self-sourced with atypical generation, storage, migration, and entrapment of methane within the coal. Another unconventional gas that is self-sourced is shale gas. These reservoir properties and the abundance of coal worldwide at shallow depths make the finding, processing, and development costs of coalbed gas relatively low compared to conventional gas. Increasing global gas supplies for cleaner energy source for electrical generation puts coalbed gas as an important environment-friendly component of the total energy mix (Chapter 2).

The scope of this book is focused toward the study of coalbed gas, which includes: (1) understanding the origin, composition, and physical properties of coal (Chapter 3) and their relationships to coalification, gasification, and gas storage (Chapter 4); (2) the knowledge of coal reservoir characterization in terms of macrolithotype (maceral) composition in relations to coal bed fracture systems, permeability, and porosity as functions of gas flow (Chapter 5); (3) the familiarization of the methodologies of resource assessments of both commodities (Chapter 6); (4) the acquaintance of production advances in relations to drilling and completion technology (Chapter 7); (5) the awareness of implementing environmentally and economically viable disposal of co-produced water to meet regulations (Chapter 8), (6) the knowledge of global development of coalbed gas and potential of coalmine gas recovery and utilization (Chapter 9); and the consideration of the short and long term outlook of coal and coalbed gas, and the future role of sustainability of coalbed gas through microbial generation and the coal-to-biogenic coalbed gas technology (Chapter 10). Thus, the key to the success of improving sustainability of coal and coalbed gas to fuel the foreseeable future and beyond is a better understanding of their economic and environmental impacts in relations to the carbon-constrained world and technological advances.

Philosophically, because of the inherent lithogenetic relationship of coal and coalbed gas, one cannot be separated from the other in order to elevate the level of comprehension of both energies to meet new challenges in their exploration and development. Coal is a physically unique reservoir almost totally composed of organic matter, which is different in characteristics compared to other natural gas reservoir rocks composed of epiclastics such as sandstones, siltstones, shales, and precipitates (Figure 1.1). Early exploitation of coalbed gas relied on fairly well-defined resources from surface and subsurface mapping of coal beds. However, the lateral and vertical behaviors of coal beds are sensitive to depositional environments, which vary greatly from coastal to alluvial settings. The coal environments found in various paleoclimatic settings, in turn, dictate the nature of vegetation and type of peat bog/swamp, which control the organic/inorganic matter

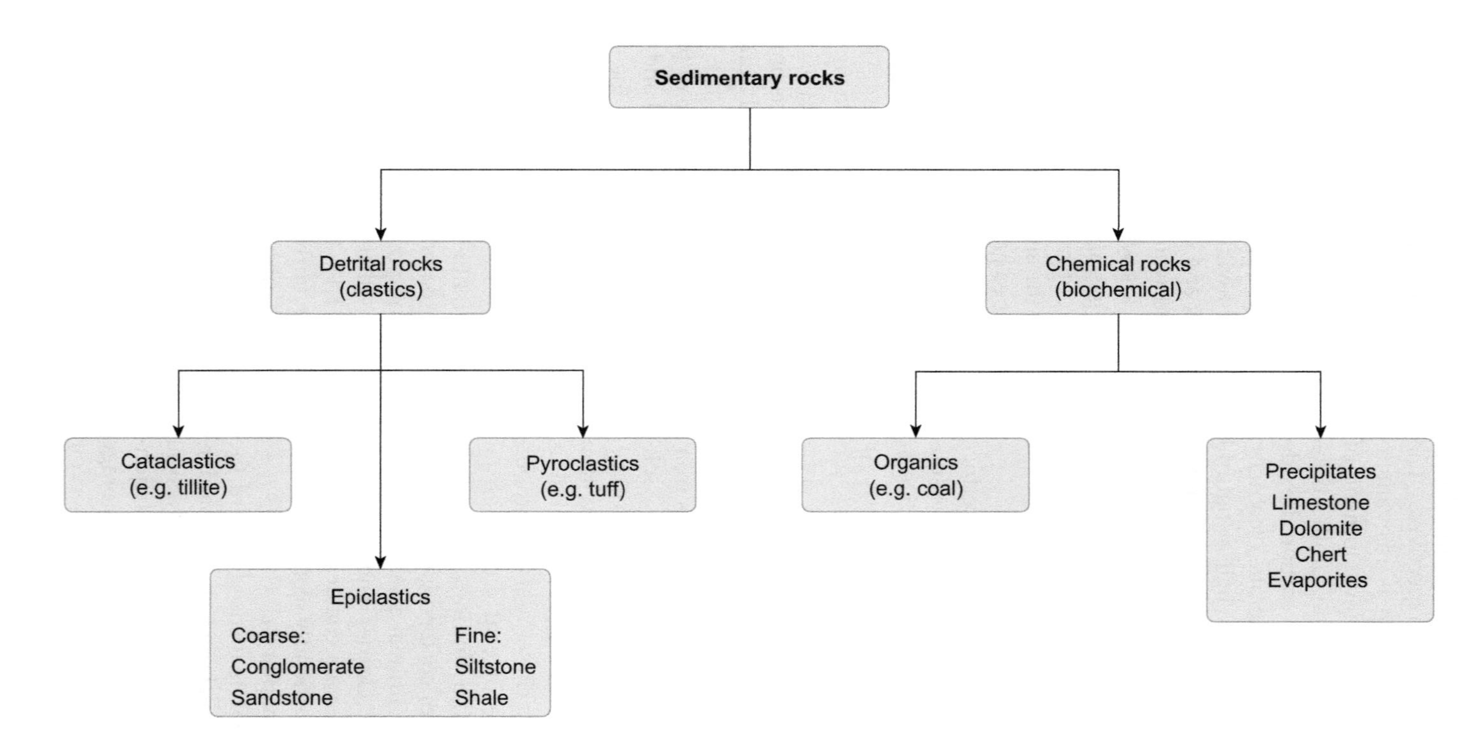

FIGURE 1.1 Flow chart showing the hierarchy of sedimentary rocks divided into detrital and chemical rocks, which in turn, is subdivided into the clastic, organic, and precipitate types.

composition and ultimately potential hydrocarbon byproducts. The process of metamorphism controlled by heat and pressure during burial of the peat precursors, in turn, affects maturation of organic matter and control gasification. Thus, understanding the generation, storage, and entrapment of the gas in the coal plays a major role in predicting economic recoverability and production.

The technical aspect and success of drilling and completion of coalbed gas wells depend on the characterization of the coal reservoir. Often the ease or difficulty of drilling through the reservoir rock depends on the types of maceral composition, bedding planes, and presence or absence of fractures or cleats in coal. "Tight coal" described during slow drilling and slow gas flow is often times caused by the physical and maceral composition of the coal reservoir. Owing to these coal properties, the reservoir is responsive to well completion techniques and stimulation treatments (fluids), which in turn, affects reservoir performance. Another technical aspect is the volume of water coproduced during development, which often is related to the origin of the gas. That is, coalbed gas generated by biogenic activity, which involved introduction of meteoric water into the coal from surface recharge yield more water than normal during production.

Like conventional natural gas plays when coalbed gas prospect areas are exhausted and exploitation expands into potential areas, the most important requirements for data collection and evaluation are: (1) areal distribution of coal, which involve understanding of the process sedimentology and stratigraphy of the coal beds; (2) areal subsurface and surface mapping to determine thickness and extent of coal beds leading to estimation of coal resources and certification of reserves; (3) gas content and desorption rate of the coal to evaluate the gas-in-place, which in turn, assist in forecasting gas recoverability and production; (4) modeling regional hydrology and hydrostratigraphy to establish that gas is stored in the coal by hydrostatic pressure and the volume of water can flow gas to the well and how many wells are needed to optimize gas desorption (well spacing); (5) recognizing factors that control coal permeability and porosity, which affects reservoir stimulation, completion, and production; and (6) awareness that reservoir performance on the whole is a reflection of the physical, compositional, and hydrological conditions of the coal reservoir.

Thus, analyses of coalbed gas plays and developmental challenges such as optimizing gas desorption within a water drawdown area require comprehensive examination of both coal and coalbed gas from their formation to production. In order to execute this unique combination of disciplines, the integration of knowledge of coal beds must utilize interdisciplinary concepts in coal geology, hydrology, geophysics, and engineering. Finally, meeting future advances require amalgamating insights of multiple disciplines from environmental specialists, microbiologists, biochemists, power plant combustion engineers to reservoir modelers (Figure 1.2) on the following: (1) managing coproduced water in an environmentally and technically cost-effective manner; (2) sustaining generation and production of methane from biotechnology by introducing microbes, nutrients or amendments coal hydrology; and (3) sequestering carbon dioxide in coal to enhance the recovery of methane.

LEARNING METAPHORS

A major distraction to learning basic principles, in any discipline, is widely used metaphors and dialects inherent to the rich heritage of coal as applied to the maturing field of coalbed gas. Thus, before further discussions, it is important to explain and clarify metaphors to provide effective understanding and to improve comprehension of terms. Coal-related metaphors such as coal bed vs coal seam, coalbed gas vs coalbed

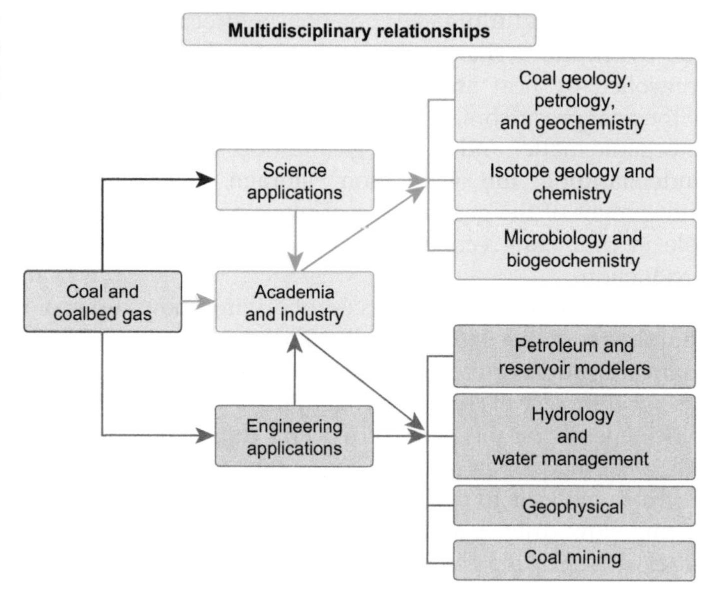

FIGURE 1.2 Flow chart showing relationships of coal and coalbed gas to multidisciplines and applications to research and development in academia and industry, which also may apply to governmental activities.

methane (popularly known as CBM), coal seam gas vs. coalbed natural gas, and coalbed gas are commonly used interchangeably. Other metaphors such as coalmine methane (CMM) vs abandoned mine methane (AMM), and unconventional coalbed gas vs conventional natural gas are attempts to relate the coal hydrocarbons to coal-mine conditions and the petroleum systems, respectively.

Coal Metaphors

The term coal seam originated during the Industrial Revolution, which began in Great Britain during the eighteenth century and subsequently spread to Europe and North America. At this time in Great Britain, where parts of the country lacked or experienced shortage of wood for building fires, natives found and dug or "burrowed" into "outcrops" where coal "layers" or "seams" emerged on the surface and extended along hillsides. When the coal seam was exhausted from being mined attempts were made to find other outcrops during which the coal seams were discovered to occur as several layers. Thus, coal was found to occur as seams separated by other layers of rocks such as sandstones, shales, fireclays, which as a whole was called "coal measures" (Singh, 1997).

In contrast, the term coal bed originated in the United States when anthracite was mined in the Pennsylvanian coal measures in the central Appalachian Basin. Anthracite, originally described by early miners as coal veins, was mined during the late eighteenth and early nineteenth centuries to fuel the Industrial Revolution. Rhone (2011) reported from old coal mine records that miners described the anthracites coal measures as comprising "alternate layers of rock and coal piled upon each like the layers of a jelly cake, in which the thick layers of cake represent rock-strata and the thin layers of jelly the coal "beds". The term bed in this sense is defined and identified from other rock layers or strata according to "thickness", which was eventually recognized as the smallest rock-stratigraphic unit. On the other hand, the term coal bed has been formally recognized by the North American Stratigraphic Code as the smallest lithostratigraphic unit of sedimentary rocks, which may be formally named if economically of importance (North American Commission on Stratigraphic Nomenclature, 2005).

Thus, the terms coal bed and coal seam have similar definition and their descriptions were both established from mining coal measures in different parts of the world. However, the use of the term coal seam remained a tradition in Great Britain, Australia, India, Europe and other countries. It was originally coined to refer to the coal-bearing succession of the coal-bearing rocks of the Carboniferous System in Great Britain (Conybeare & Phillips, 1822).

Coalbed Gas Metaphors

Coalbed gas, CBM, and coalbed natural gas are terms referring to the nature of the gas adsorbed in the coal. The term coalbed gas, used in this book, represents an all-encompassing definition reflecting that more than one gas is adsorbed in the coal. The coalbed gases are composed of molecular and isotopic components with methane as a major component and carbon dioxide and nitrogen as minor components (Stricker et al., 2006). Other hydrocarbon gases such as ethane, propane, butane, and pentane are present in trace amounts as well as isotopic composition of carbon dioxide, ethane, and methane measured in parts per million (Flores, Stricker, Rice, Warden, & Ellis, 2008; Rice, 1993).

In contrast, the term CBM reflects that only methane is the gas adsorbed in the coal. The term, however, correctly emphasized that methane is the major component of the adsorbed gas in the coal. More importantly, it focuses on the "clean" and "pure" component of the coalbed gases. Thus, the relative abundance of methane and as a principal component of the coalbed gas makes it an attractive fuel.

The term coalbed natural gas reflects a definition more in line to that used by the petroleum industry. It is recognized industry-wide that gas from the coal is natural gas, which is composed mainly of methane and minor carbon dioxide, hydrogen sulfide, hydrogen, nitrogen, helium, and argon. However, natural gas is typically higher in heavy hydrocarbons primarily in ethane as well as propane, butane, and pentane than the coalbed gas. More importantly, natural gas is commonly associated with crude oil either dissolved in oil at high pressures or separated as gas cap above the oil in the reservoir. This natural gas is called associated gas compared to nonassociated gas with no oil as exemplified by coalbed gas.

Natural gas may be classified either as conventional or unconventional gas based on: (1) the characteristics of the reservoir rock, (2) the geologic nature of accumulation, and (3) the industry-standard level of technological and investment requirements that take to develop these resources (Kuuskraa et al., 1992; Rogner, 1997; Scott, 2004; Charpentier & Ahlbrandt, 2003; Holditch, Perry, & Lee, 2007; Schmoker & Klett, 2005). Conventional natural gas accumulations are localized in structural or stratigraphic traps where the reservoir and fluid properties permit gas to flow to well bore. These accumulations can be drilled and produced with traditional technologies (Figure 1.3). The unconventional gas resources, which commonly include coal bed, shale, and tight-sand gases in low-permeability reservoirs, may require drilling a horizontal well and hydraulic fracture treatment to stimulate and optimize gas production. Thus, unconventional gas may be relatively more difficult and expensive to develop.

DEFINITIONS AND TERMINOLOGIES

The basic principles and definition of concepts that are fundamental to the understanding of coal and coalbed gas are reviewed in this section. Explanation of coal and coal bed principles will facilitate application and use of the terms and concepts throughout the book. Also, it is

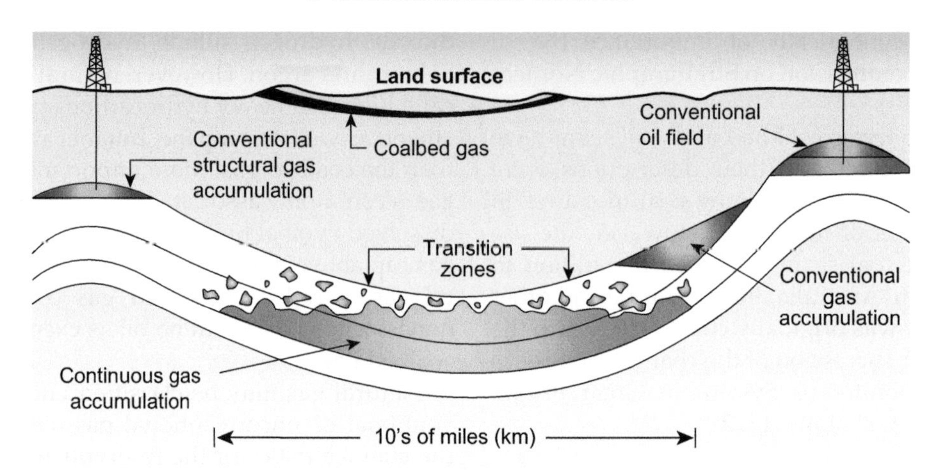

FIGURE 1.3 Schematic geological cross section showing stratigraphical and structural positions of conventional and unconventional oil and gas accumulations. *Source: Modified from Charpentier and Ahlbrandt (2003).*

intended to familiarize a broad spectrum of readers including specialists and experts in coal and coalbed gas researchers and practitioners as well as teachers, students, managers, consultants, and environmentalists. The proliferation of coal principles and terminologies, which are explained and defined in various contexts in both the scientific and nonscientific literature, has led to many viewpoints and confusion.

The complex nature of coal science requires multidisciplinary approach to understand this highly organic substance (Lyons & Alpern, 1989a, 1989b). Thus, the study of coal created multifaceted subdisciplines with each specialist propagating their own terminologies and definitions. The advent of coalbed gas in the past few decades has added more complexity and new principles and terminologies with introduction of hydrocarbons from coal (Law & Rice, 1993). In order to assist in clarification of commonly used concepts and terminologies in this book, text boxes of "facts or fictions" and "concepts and misconceptions" are strategically placed. Hopefully this pedagogy will shed some light on the misunderstandings instilled in the literature.

The most commonly accepted scientific and technical definition of coal by practitioners in the field is "a readily combustible rock containing more than 50% by weight and more than 70% by volume of carbonaceous material formed from compaction of variously altered plant remains similar to peaty deposits" (Schopf, 1956).

Thus, inherent to the Schopf (1956) definition of coal is a mixture of organic and inorganic matters including original moisture similar to modern peaty swamps/bogs (Bates & Jackson, 1987). The American Society for Testing and Materials (ASTM, 1991) and many coal specialists have recognized and accepted this definition; however, the definition of coal varies according to different subdiscipline viewpoints. For example, to a biologist/botanist, coal is a successor of peat, which is predominantly composed of organic material derived from plants that accumulated in certain types of ecosystems (Moore, 1989). To a coal petrologist, coal is a rock derived mainly of plant remains that have undergone peatification and coalification or maturation (Teichmuller, 1989). To a coal sedimentologist and stratigrapher, coal is a product of paleodepositional environments and tectonic settings not too different from modern day analogs (Ferm & Horne, 1979; Ferm & Weisenfluch,

BOX 1.1

NOT ALL COAL IS COMBUSTIBLE—FACT OR FICTION

Coal in any dictionary is generally defined as a "combustible, natural fuel and a solid, brownish-black rock occurring in coal beds or seams formed by degradation of vegetal matter without free access to air, under the influence of moisture, increased pressure and temperature." The fact is not all coal is combustible as a natural fuel for coal-fired power plants to generate electricity. The generation of heat by combusting coal is the result the reactions between the burnable organic matter of coal and oxygen (Stach et al., 1982; Taylor et al., 1998; Thomas, 2002; Schweinfurth, 2002). The burnable material, which discharges heat when oxidized, includes carbon, hydrogen, nitrogen, and sulfur; however, coal is also composed of inorganic material (e.g. mineral matter), which along with oxygen consumes heat during combustion. Thus, the combustibility and calorific value (e.g. heat of combustion) are controlled by the rank and organic composition (e.g. macerals, see Chapter 5 for definition) and inorganic matter (e.g. quartz, pyrite, calcite, clay) content of the coal. For example, certain macerals (e.g. exinite) in low-rank coal (e.g. subbituminous) owing to high hydrogen content has higher calorific value than associated vitrinite maceral. In contrast, both macerals (e.g. exinite and vitrinite) in high-rank coal (e.g. low-volatile bituminous) have about the same calorific values. In addition, coal loses moisture depending on rank, which range from 1% to 2% in higher rank coal (e.g. anthracite–bituminous) to more than 22% in lower rank coal (e.g. subbituminous–lignite). As a result of environmental concerns of coal utilization regulations are imposed to achieve significant reductions in pollutants (e.g. carbon dioxide or CO_2, nitrogen oxides or NOx, sulfur oxides or SOx, particulates) and hazardous trace elemental components (e.g. arsenic, boron, cadmium, chromium, lead, mercury, nickel, selenium, zinc) associated with the burning of coal (Schweinfurth, 2002).

1989). Simply put, to a geochemist, coal is a peat that has undergone chemical and physical transformation during diagenesis, catagenesis, and metagenesis (Mukhopadhyay & Hatcher, 1993; Tissot & Welte, 1984). Finally, to a petroleum geologist coal is a source of hydrocarbons capable of generating and expelling economic quantities of gas (e.g. methane and carbon dioxide) and oil (Boreham & Powell, 1993; Clayton, 1993). Generation of methane and carbon dioxide and other minor volatiles during coalification is mainly related to the organic composition of the coal (Hunt, 1979). The various concepts embodied in these definitions by subdisciplines are a result of the historical evolution of approaching the investigation of coal. This approach has changed and enhanced the fundamental principles in integration and interpretation of data, which should lead to a better understanding of exploitation of coalbed gas.

"Dirty" Coal vs "Clean" Coal

To many people "clean" coal is an oxymoron. For them there is only "dirty" coal that produces pollution when it is burned in power plants. More accurately, the "dirty" aspect of coal is both compositional and technical in nature. Compositionally, the major component of coal is organic material with minor constituent of inorganic material. More precisely, the boundary of "pure" and "impure" coal is 30% of inorganic mineral matter constituent, which is classified as carbonaceous shale (between shale and coal;

Alpern, 1981). The organic matter includes barks, wood, stems, roots, leaves, spores, pollens, cuticles, and grasses, and algae as well as minor fungal, bacterial, and other animal remains. The inorganic matter includes mineral grains (e.g. quartz, clays, volcanic fragments from ash fall), minerals associated with the organic macerals (e.g. pyrite that formed in cavities in the organic matter), minerals formed during or after coalification (e.g. calcite from mineralization along cleats), and trace elements from within the mineral and organic matter (e.g. Brownfield, 2002). A complete "clean" coal is one that contains no inorganic matter.

Technically, the "dirty" aspect of burning coal by power plants comes from the generation of heat during combustion, which is a function of the reaction of the organic matter and oxygen. Organic material, with its carbon, hydrogen, nitrogen, and sulfur, releases heat during combustion, whereas the incombustible inorganic matter or minerals with oxygen consume heat during the combustion as well release trace elements, particulates (e.g. fly ash) and deposit slag in the boiler. The heat of combustion or calorific value of the coal depends on the organic matter composition and rank on one hand and the inorganic material on the other. Thus, the "dirty" components of coal during combustion are mainly from the particulates or ash from inorganic matter and compounds generated with reaction with oxygen such as oxides of sulfur and nitrogen for the organic matter. Coal-fired power plants are equipped at various stages of combustion to control these pollutants, the concentrations of which are regulated to meet health standards. Thus, "clean" and "dirty" coals are defined at the basic compositional level of the coal to its utilization for electric generation by power plants.

The concept of "clean" coal has expanded into "clean coal technology". "Clean coal technology" is defined as a collection of applied engineering and science knowledge employed to improve and increase productivity and efficiency of machines such as power plant equipment in order to reduce environmental impacts of coal utilization (USDOE, 2010). For example, the primary focus to capture and sequester carbon dioxide in deep formations toward reduction of greenhouse gas emissions and sulfur dioxide and particulates to prevent acid rain effects is addressed in "clean coal technology". This technology, in development stage, is exemplified by FutureGen in which "oxy-combustion burns coal with a mixture of oxygen and carbon dioxide (CO_2) instead of air to produce a concentrated CO_2 stream for safe and permanent storage as well as near-zero emissions of mercury, sulfur and nitrogen oxides, and particulates" (USDOE, 2010). Under this technology coal-fired power plants will combine carbon capture and sequestration technology with CO_2 capture (e.g. 90%) and separation from other gases, CO_2 transport to sites of sequestration, and CO_2 injection into subsurface formations for permanent storages (Folger, 2013). Equally importantly to reduction of CO_2 emission is a coal combustion system at high temperatures and pressures attained by using more efficient supercritical and ultra-supercritical power plants, which are emerging technology of utilization in developed and developing countries (IEA, 2006). Additionally, "clean coal technology" has evolved over time to include CO_2 sequestration outside the power stations particularly in coal beds or seams and other geological sinks because of the focus on mitigating greenhouse gas emissions.

BASIC PRINCIPLES

Coal Occurrence in the Geologic Past

To understand coal and petroleum geology we must first have an understanding of the Earth's dynamic changes in the geologic record. Coal is the most abundant mineral energy resource and is widespread around the world but is unevenly distributed in all the continents

BOX 1.2

COAL GREENHOUSE GAS EMISSION—CONCEPTS AND MISCONCEPTIONS

Coal combustion contributes twice as much CO_2 per unit of heat energy, which is mainly from carbon, as natural gas and lesser for crude oil in the United States (Hong & Slatick, 1994). However, the ratio of carbon to heat energy vary with coal rank in which carbon on a dry basis ranges from >60% for lignite and >80% for anthracite (Hong & Slatik, 1994). Carbon dioxide and methane contribute 26% and 6%, respectively, to the greenhouse effect of clear sky radiative forcing (Kiehl & Trenberth, 1997). Carbon dioxide dominates the total forcing with methane (CH_4), nitrous oxide (N_2O), and halocarbon compounds (mainly aerosols particles or chlorofluorocarbons) becoming smaller contributors to the total forcing over time (Butler, 2011). Of these greenhouse gases and particles that contribute 96% to radiative climate forcing, Butler (2011) reported that only CO_2 and N_2O continue to increase at a regular rate and that radiative climate forcing from methane or CH_4 increased during 2007–2010 after remaining nearly constant 1999–2006. Carbon dioxide has accounted for nearly 80% of the increased long-lived greenhouse gases from 1990 to 2010. Thus, CO_2 is the most abundant and has a much longer life span; however, methane short life span is more potent at trapping heat (Blasting, 2013). The global warming potential (GWP) is a measure of how much heat a greenhouse gas traps compared to CO_2 in the atmosphere and is measured over a time period. For example, the 100-year GWP of CH_4 is 25; that is, CH_4 will trap 25 times more heat than CO_2 over the next 100 years (Blasting, 2013). Thus, CH_4 has a greater effect for a shorter period of lifetime in the atmosphere than CO_2 over 100 years.

including Antarctica. The worldwide distribution is linked to the geological factors that control the origin of coal and coalbed gas, which are transformed through the geologic time. The mineral energy resource base of the world, which includes oil, gas, tar, oil shale, coal, and uranium, is dominated by coal (40–49%) and uranium (40–56%) with the hydrocarbons making up the remainder 5–10% according to Moody (1978) and the World Energy Conference (1978). The occurrence of coal and coalbed gas is best described in terms of their geologic age and geographic distribution in the context of their plate tectonic and paleogeologic settings.

Geologic Coal Distribution

Coal distribution through the geologic time, summarized in Figure 1.4, ranges from Late Devonian (about 380 mya) to Pliocene (about 3 mya) for a duration of about 375 mya. The coal occurrence through the geologic time is not continuous but rather episodic and culminated by significant mass extinctions of plant and animal life, which created from marked to very minor coal hiatus at the Permian–Triassic and Cretaceous–Tertiary boundaries (Figure 1.4). Coal accumulations were influenced by the appearance, evolution, dispersion, and ecological adaption of terrestrial plants in ancestral wetlands. Terrestrial plants first colonized the landscape especially wetlands in the Late Paleozoic Era during the Devonian Period as small vascular (e.g. water-transport tubes), tree-like plants, which became increasingly diversified, evolved, and exploded into tall woody, vascular plants (e.g. angiosperms) in the

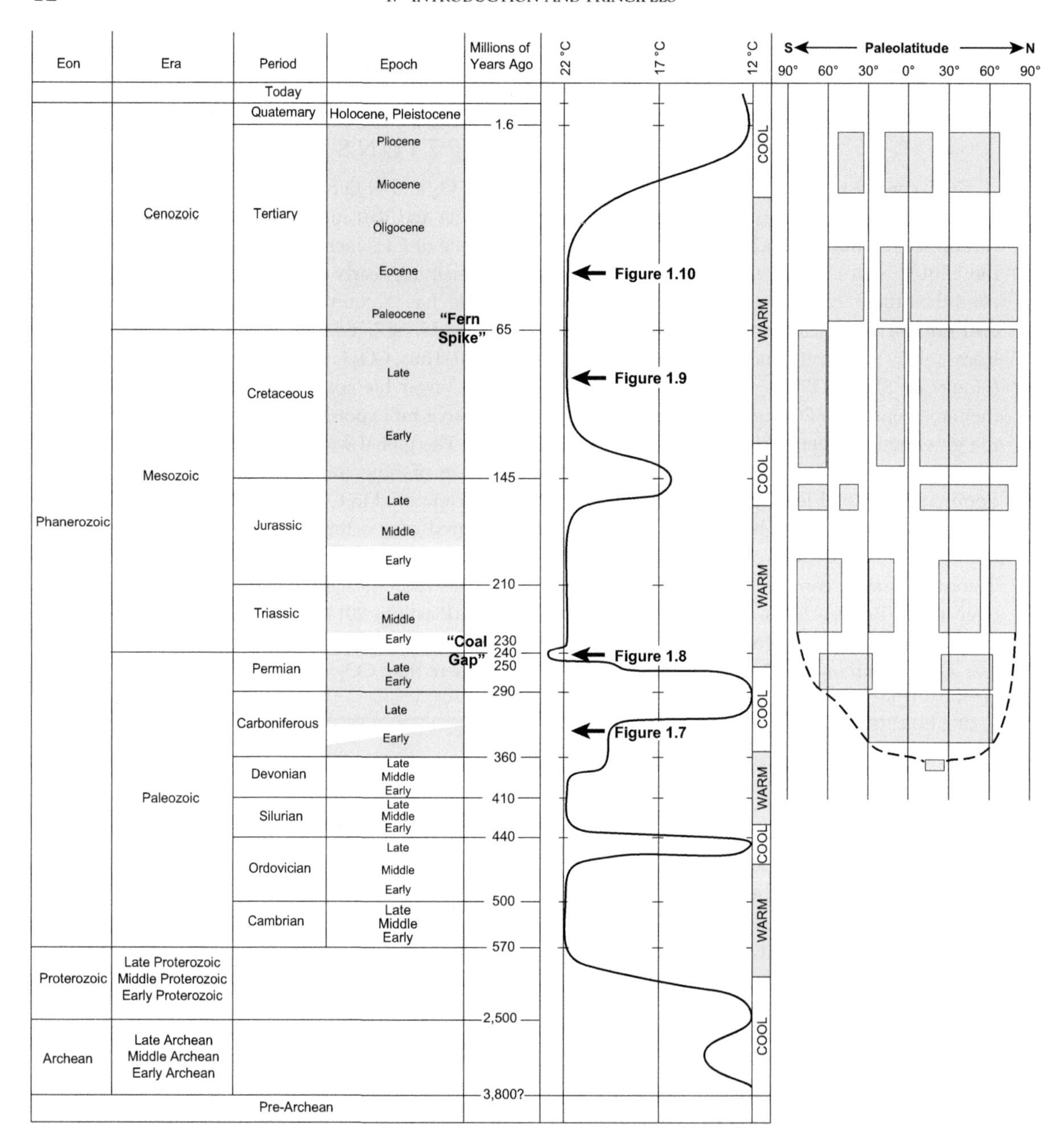

FIGURE 1.4 Geologic time showing breakdown into Eon, Era, Period, Epoch, and millions of years ago (mya) as well as the geologic distribution of coal by paleolatitude. Average global paleotemperature chart shows warming and cooling trends through the geologic record. Dash lines indicate marked paleolatitudinal change in coal accumulation. *(Source: Modified after McCabe (1984), Retallack et al. (1996), Schweinfurth (2002), Greb et al. (2006), and Scotese (2013)).* Arrows designate approximate geologic age for Figures 1.7–1.10.

Carboniferous peaty wetlands (Cross & Phillips, 1990; DiMichele & Phillips, 1996; Greb, DiMichele, & Gastaldo, 2006; Phillips, Peppers, Avcin, & Loughman, 1974; Stewart & Rothwell, 1993). The vascular system led to the extraction of water and minerals from paleosols through root-like systems to grow tall in peaty wetlands along shoreline-coastal areas, river-delta plains, and lakes. The Carboniferous plants passed on into the peaty wetlands of the Permian and Triassic Periods but most became extinct probably due to hotter and drier climate for peat accumulation (Worsely, Moore, Fraticelli, & Scotese, 1994), landmasses were too high to form peaty wetlands (Faure, de Wit, & Willis, 1995), peaty wetlands were overwhelmed by sulfuric acid from massive eruptions of the Siberian Traps (McCartney, Hoffman, & Tredoux, 1990) or by nitric acid from impact of an extraterrestrial bolide at the Permian–Triassic boundary (Zahnle, 1990), and acidic and nutrient-poor peaty wetlands (Retallack, Veevers, & Morante, 1996). This geologic time of marked demise of peat accumulation is called the global "coal hiatus" or "coal gap" (Faure et al., 1995; Retallack et al., 1996).

During the succeeding Mesozoic Era particularly in the Jurassic and Cretaceous Periods, forest vegetation (e.g. conifers) became reestablished, flourished, and became totally different from those of the Permian and Carboniferous landscapes. More importantly, the Cretaceous Period was characterized by the sudden appearance and proliferation of flowering plants and leafy trees, which along with conifers and ferns dominated the vegetation of peaty wetlands in coastal and low-lying continental areas (Bremer, 2000; Cross & Phillips, 1990). The end of the Cretaceous Period like the Permian–Triassic boundary event was marked by the mass extinction of plants and animal life. Also, like the Permian–Triassic extinction event, one of the explanations (e.g. volcanic eruptions) of the Cretaceous–Tertiary boundary extinction is the collision and impact of an asteroid or comet on the Earth's surface (Alvarez, Alvarez, Asaro, & Michel, 1980). The asteroid impact created dust clouds that enveloped the earth, which blocked sunlight and triggered cooling and disappearance of the majority of Cretaceous life forms. However, it is also argued that the Cretaceous extinction was caused by massive volcanic eruptions that created gases (e.g. sulfur dioxide, carbon dioxide) and lava flows from the Deccan Traps in India (Royer, Berner, Montañez, Tabor, & Beerling, 2004). However, unlike the marked Permian–Triassic boundary "coal hiatus or gap" recorded from 250 to 230 mya (Retallack et al., 1996), the Cretaceous–Tertiary boundary indicates continuous deposition of peat/coal except for micro-lithostratigraphic disruptions of spores and pollens in many U.S. coal basins (Fassett & Rigby, 1987; Pillmore & Flores, 1987). For instance, in the Raton Basin in New Mexico, the proportion of fern spores to angiosperm pollen in coal and carbonaceous shale is 15–30% fern spores, which shifts to as much as 99% (e.g. recognized as "fern spike" by Nichols, Jarzen, Orth, and Oliver (1986)) at the top of the iridium-rich boundary clay and then returns to 10–30%, 10–15 cm above the Cretaceous–Tertiary (K/T) boundary (Pillmore & Flores, 1987). This microlaminated parting containing the K/T boundary clay in carbonaceous shale vary in thickness from 0.64 to 2.5 cm and indicates very minor interruption of coal accumulation across the boundary. In the Powder River Basin in Wyoming, 30% of mostly angiosperm pollen suddenly disappeared within the Cretaceous–Tertiary boundary (Nichols, Brown, Attrep, & Orth, 1992). These observations suggest that only small proportion of plant species disappeared during the Cretaceous–Tertiary impact event and those that disappeared *en masse* were able to recover immediately and reestablish as coal-forming vegetation in peaty wetlands of the Paleocene Epoch particularly in coal basins in the U.S. western interior (Flores, 2003).

The Cretaceous flowering plants, which resemble many plants of today, flourished and were dispersed farther into more diversified ecosystems during Cenozoic Era, in general and in the Tertiary Period, in particular (Cross & Phillips, 1990; Stewart & Rothwell, 1993). Dispersal of flowering plants was aided by the presence of and coevolution with insects, which nourished on and pollinated them leading to growth of modern plant species. For example, modern bees and other insects as well as birds, which have as many as six ultraviolet light receptors, focused on flowers from leaves due to ultraviolet patterns (Williams, 2011). In addition, a wide range of grasses evolved during the early Tertiary or Paleogene (e.g. Paleocene and Eocene Epochs) and thrived, spread, and dominated grassland ecosystems in the Neogene (e.g. Miocene, Pliocene). However, although they covered about a third of the earth's surface, grasslands played a minor role in coal formation. Tertiary peaty wetland forests, which formed in low-lying coastal and intermontane areas, were more widespread than in previous coal-forming geologic periods probably controlled by warmer paleoclimate.

Although terrestrial plants dominated the Late Paleozoic, Mesozoic, and Cenozoic Eras, the remains of which produced organic matter for coal formation, only the Carboniferous System (e.g. Pennsylvanian Period), and Permian Period of the Paleozoic Era, Jurassic and Cretaceous Periods in the Mesozoic Era, and Tertiary Period of the Cenozoic Era contained significant coal deposits in the world (Figure 1.4). The global coal reserves estimated by Bestougeff (1980) and Bouska (1981) 30 years ago put majority of the anthracite–bituminous or hard coal in the Carboniferous (24.5%), Permian (31.5%), Triassic (1%), Jurassic (16.5%), and Cretaceous (13%), which are buried deeper and undergone longer coalification (see Chapter 4) than the minor subbituminous–lignite coal in the Tertiary (13.5%) (Figure 1.5). The global proven coal reserves reported by BP (2011) at the end of 2010 shows that the Tertiary subbituminous–lignite coals increased about 53% and the anthracite–bituminous coals decreased about 47%.

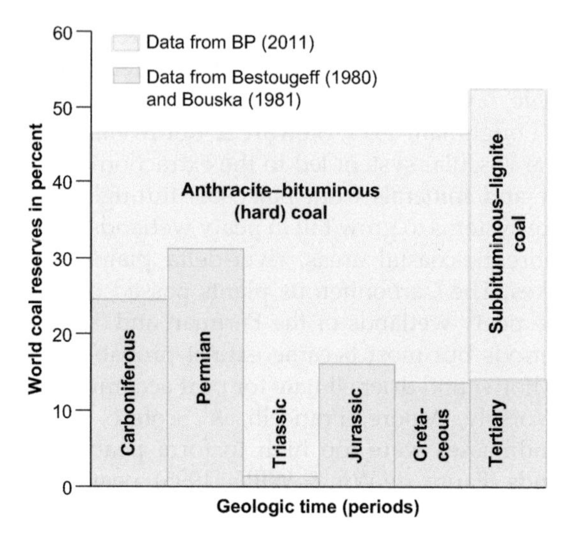

FIGURE 1.5 A comparison of frequency distributions of the world coal reserves from 1980 to 2010 relative to the geologic periods. *Source: Data from Bestogeff (1980), Bouska (1981), and BP (2011).*

Geographic Coal Distribution

The present-day worldwide geographic distribution of coal related to the Paleozoic–Mesozoic age (Carboniferous, Permian, Jurassic, and Cretaceous) and Tertiary age is shown in Figure 1.6. The geographic locations of the Paleozoic- and Mesozoic-age coals are mostly in larger basins (e.g. Euro-Asia and North America) than the Tertiary coal basins (e.g. Southeast Asia and eastern Europe). Furthermore, these coal basins are unevenly distributed across major continents, microcontinents, and islands. The current geographic coal distribution is a result of constant motion of drifting landmasses from the Carboniferous to Tertiary time when coal mainly accumulated.

The Carboniferous coals are found mainly in large areas of Europe, Russia, Far East Asia, and North America, which were in the equatorial latitude and northern hemisphere. In Europe, the Carboniferous coals were formed in the Paralic Coal Basin, which stretched from Ireland through Great Britain, northern France,

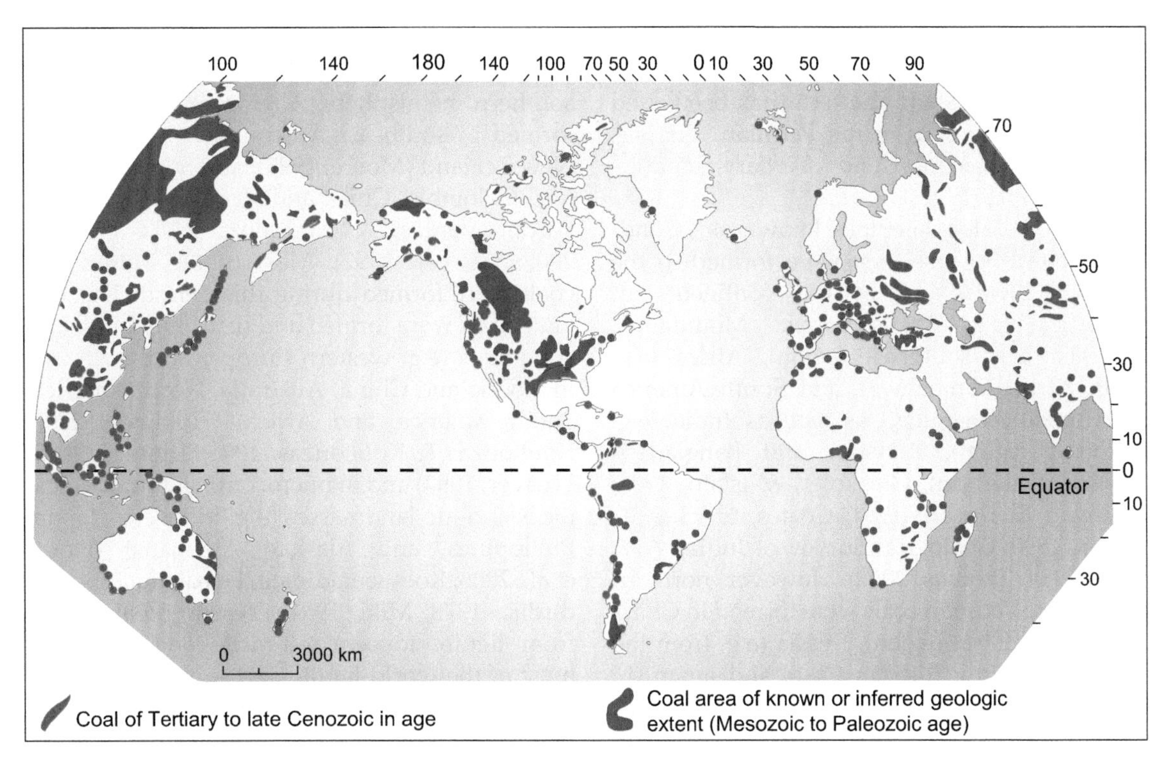

FIGURE 1.6 Present-day coal distribution worldwide. The coal areas represent from known or inferred geological extent (large area) to one or more known occurrences or unknown geological extent (small area). *Source: Modified from Landis and Weaver (1993).*

Belgium, the Netherlands, Ruhr region of Germany to Poland, and Romania (Allen, 1995; Dreesen, Bossiroy, Dusar, Flores, & Verkaeren, 1995; Guion, Fulton, & Jones, 1995; Thomas, 2002; Whateley & Spears, 1995). In Russia, Carboniferous coals formed along the northern hemisphere and the Far East region, which also extended into eastern and southeastern China. Carboniferous coals were also formed in Turkey, Morocco, and Egypt (Lapo & Drozdova, 1989; Liu, 1990; Thomas, 2002). In North America, the Carboniferous (Pennsylvanian) coals were formed along the eastern region, which stretched from Alabama, United States to Nova Scotia and New Brunswick, Canada and in the middle-central United States (Ferm & Horne, 1979). The majority of the world's anthracite and bituminous coals or hard coal resources were formed during the Carboniferous System. Detailed discussions of coal rank and their origin will be discussed in Chapter 4.

The Carboniferous "coal measures" that refer to the rich coal deposits in Great Britain, northern Europe, and Asia formed during the Paleozoic Era. The coal deposits are often interbedded with conglomerates, sandstones, siltstones, shales, and limestones grouped as "coal measures". The Carboniferous in North America is divided into the Mississippian (lower Carboniferous) and Pennsylvanian (upper Carboniferous) Periods to separate the limestone of the former period from the coal-bearing rocks or "coal measures" of the latter period, which are mainly distributed in the midwestern and eastern North America. Although "coal measures" are strictly defined in the context of the Carboniferous System, the term has been used of coal-bearing rocks in the Permian, Triassic, Jurassic, Cretaceous, and Tertiary Periods worldwide.

Permian coals sometimes known as "Gondwana coals" because they were formed in the southern hemisphere in the continents of Antarctica (e.g. Transantarctic Mountains), Australia (e.g. southeast region), Africa (e.g. South Africa, Zimbabwe), and South America (e.g. Brazil, Argentina), as well as India (e.g. northeast region), Pakistan, and Bangladesh (Bostick, Betterhorn, Gluskoter, & Islam, 1991; Casshyap & Tewari, 1984; Coates, Stricker, & Landis, 1990; Geological Survey of India, 1977; Hunt, 1989; Thomas, 2002). However, northern hemisphere Permian coals were formed in China (e.g. south-central region), Russia (e.g. from the northern region to the Far East), and Japan (Li & Gong, 2000; Suping & Flores, 1996; Thomas, 2002). Like Permian coals, Triassic coals were formed mainly in the continents of Antarctica (e.g. Transantarctic Mountains), India, Australia (e.g. east, central, west), South America (e.g. southern region), and South Africa; however, Triassic coals were also formed in Euro-Asia (e.g. China, Russia, Europe, North America) (Farabee, Taylor, & Taylor, 1990; Retallack et al., 1996). Most of the world's anthracite coal resources were formed during the Permian Period.

Jurassic coals were formed mainly in the northern hemisphere in Euro-Asia (e.g. Russia, northwest China, Inner Mongolia, Iran, Afghanistan, and South Korea), Australia (e.g. New South Wales, Queensland) and South America (e.g. Argentina) (Lapo & Drozdova, 1989; Scott, Anderson, Crosdale, Dingwall, & Leblang, 2007; Thomas, 2002; Zhang, Li, & Xiong, 1998). Cretaceous coals were formed mainly in China, United States and Canada (e.g. western interior from the southern to northern Rocky Mountains), and Alaska (e.g. North Slope, central, and southern) (Flores, Stricker, & Kinney, 2004; McCabe & Parrish, 1992). In the southern hemisphere, Cretaceous coals were formed in southeast Australia (Fielding, 1992); New Zealand (Moore, 1995); and South America (e.g. Colombia, Chile, and Venezuela) (Barrero, Pardo, Vargas, & Martinez, 2007; Thomas, 2002; Weaver, 1992). Most of the bituminous coals were formed during this period. The Tertiary coals were formed and distributed in major continents (e.g. western Europe, northeast Asia in Siberia and China, Australia, North America, South America, and Africa) (Holdgate, 2005; Koukouzas & Koukouzas, 1995; Thomas, 2002; Weaver, 1992) and in microcontinents and volcanic island-arc landmasses (e.g. Indonesia, Japan, Philippines, and Alaskan Aleutians) (Flores et al., 2006; Koesoemadinata, Hardjono, & Sumadirdija, 1978; Miki, 1994). Tertiary coals range from lignite to semianthracite and make up most of the world lignite and subbituminous or low-rank brown coal resources/reserves. Tertiary coals are mined in Australia, Indonesia, Canada, Colombia, Germany, Pakistan, Philippines, Poland, New Zealand, Thailand, Turkey, and Venezuela (Barrero et al., 2007, p. 91; Durrani & Warwick, 1992; Flores & Sykes, 1993; Flores et al., 2006; Suggate, 1959; Thomas, 2002; Weaver, 1992; Whateley & Tuncali, 1995). Coals from this period range from anthracite to lignite with the latter coal making up the bulk of the world's reserves. In the United States, the Tertiary coals mined in the Powder River Basin in Wyoming make up more than 40% of the nation's total annual coal production (Flores, Spear, Kinney, Purchase, & Gallagher, 2010). However, the reserves of the 10 coal mines in the Gillette coalfield, which produce most of this coal, have 10–30 years left remaining of coal production (Milici, Flores, & Stricker, 2009).

Role of the Dynamic Earth Outer Shell

The geologic and geographic distributions of coal in present-day continents may be partly explained by the dynamic lithosphere (outer layer or shell) of the earth because of plate tectonics through geologic time. The plate tectonic theory has been generally accepted and described extensively in the literature and textbooks for the past 100 years and it is not the intention of this book to duplicate this exemplary endeavor (e.g. Condie, 1997, 2011, p. 574; Dinwiddie, Lamb, & Reynolds, 2011; Frisch, Meschede, & Blakey, 2011; Keary & Vine, 1996; Kious & Tilling, 1996; Moores & Twiss, 1995; Wegener, 1912; Windley, 1995). More exactly this section of the introduction will concentrate on the process of coal accumulation in the manner of evolution of the continents through movements of the outer shell of the earth through geologic time.

According to the plate tectonic theory, the Early Paleozoic Era (from 540 to 580 mya; see Figure 1.4) indicates that a shot-lived supercontinent Pannotia, which lay in the southern Hemisphere, formed as Laurentia, Baltica, and Siberia then collided and amalgamated the continent of Gondwana followed by fragmentation (Condie, 2011; Dinwiddie et al., 2011; Scotese, 2013). This paleogeographic setting with Pannotia is controversial but it ushered in the geologic periods of major coal accumulations during the Carboniferous and Permian Periods (240–360 mya) of the late Paleozoic Era followed by the Triassic (210–230 mya)—Jurassic Periods (145–185 mya), Cretaceous Period (65–145 mya) of the late Mesozoic Era and culminated in the Tertiary Period (1.6–65 mya) of the Cenozoic Era (Figure 1.4). The paleolatitudinal distribution of major coal (peat swamp) accumulations from Carboniferous to Tertiary times (Figs 1.7–1.10) reconstructed from Habitch (1979), McCabe (1984), Retallack et al. (1996), and Greb et al. (2006) shows a precursor, narrow coal distribution between 20° and 30° north during the Devonian Period. However, this isolated coal distribution evolved into a broader

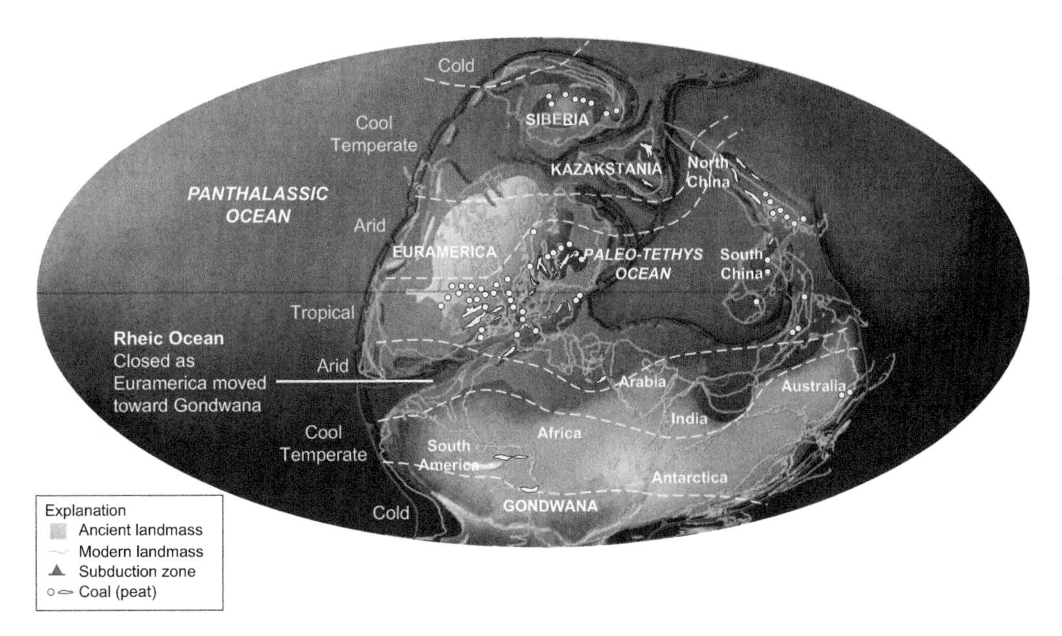

FIGURE 1.7 Paleogeographic distribution of ancestral landmasses (e.g. supercontinents, continents, microcontinents) driven by mechanism of plate tectonics during the Carboniferous time when coals were formed in peaty wetlands. Approximate global paleoclimatic zones are indicated. *(Source: Modified from Greb et al. (2006), Dinwiddie et al. (2011), and Scotese (2013)).* See Figure 1.4 for approximate geologic age. Black dots and elongate areas are coal (peat accumulations).

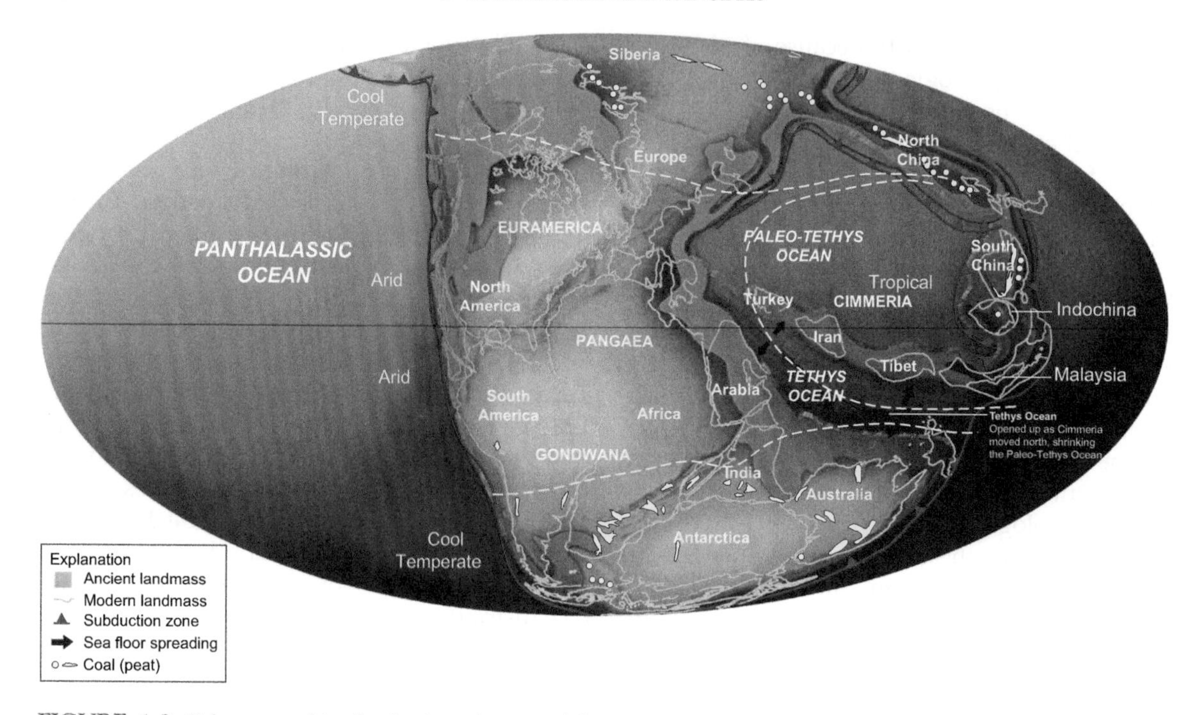

FIGURE 1.8 Paleogeographic distribution of ancestral landmasses (e.g. supercontinents, continents, microcontinents) driven by mechanism of plate tectonics during the Permian–Triassic times when coals were formed in peaty wetlands. Approximate global paleoclimatic zones are indicated. *(Source: Modified from Greb et al. (2006), Dinwiddie et al. (2011), and Scotese (2013)).* See Figure 1.4 for approximate geologic age. Black dots and elongate areas are coal (peat accumulations).

paleolatitudinal extent during the Permian but reached its broadest paleolatitudinal extent during the Permian–Triassic Periods.

The marked change from low to high paleolatitudinal extent covering both hemispheres from Devonian to Carboniferous Periods (Figure 1.4) may be explained by the evolution of tropical decomposers (e.g. fungi and bacteria) to exist in less warmer regions (Briden & Irving, 1964) and evolutionary improvements of lignin or phenolic production and biomass or productivity, which increased effectiveness of peat-forming vegetation to survive landscape disturbances (Greb et al., 2006; Retallack et al., 1996). Coal accumulations during the Cretaceous Period remained nearly extended from pole to pole, which became less extensive during the Tertiary Period (Figure 1.4) when peat-forming bogs formed in much warmer climatic belts and controlled by their depositional environments (Greb et al. 2006; McCabe & Parrish, 1992; Rahmani & Flores, 1984).

Fast forward from Early Paleozoic to Late Paleozoic during the coal (peat) accumulations in the Carboniferous Period (290–360 mya; Figure 1.7) the ancestral landmasses of Laurentia, Baltica, and Avalonia merged by collisions to form a large landmass called Euramerica while Siberia drifted farther north toward Kazakhstan (Condie, 2011; Dinwiddie et al., 2011; Scotese, 2013). Euramerica, in turn, drifted southward toward Gondwana in the southern Hemisphere, which at this time formed an ice cap (Condie, 2011; Dinwiddie et al., 2011; Scotese, 2013). Gondwana comprised the present-day continents of Antarctica, Australia, Africa, and South America and the subcontinents of India and Arabia (Figure 1.7). Through time as the ancestral

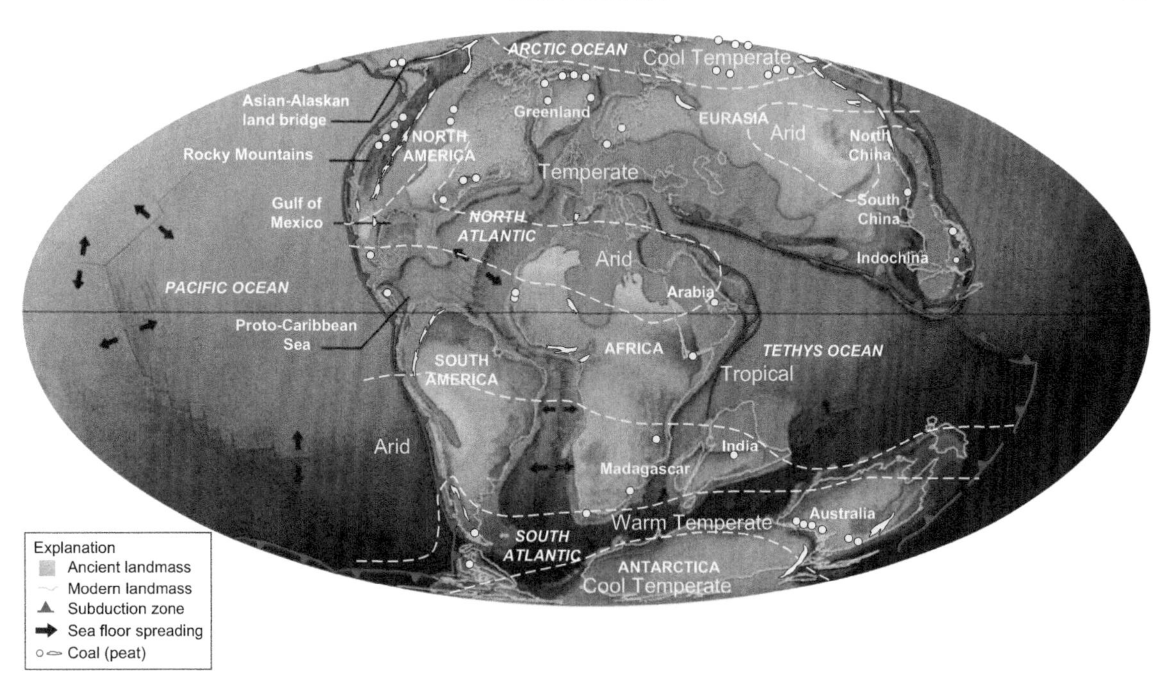

FIGURE 1.9 Paleogeographic distribution of ancestral landmasses (e.g. supercontinents, continents, microcontinents) driven by mechanism of plate tectonics during the Cretaceous time when coals were formed in peaty wetlands. Approximate global paleoclimatic zones are indicated. *(Source: Modified from Greb et al. (2006), Dinwiddie et al. (2011), and Scotese (2013)).* See Figure 1.4 for approximate geologic age. Black dots and elongate areas are coal (peat accumulations).

landmasses drifted toward each other, microcontinents (e.g. Kazakhstani/Siberia and Europe) collided to form the Ural Mountains. About 280 mya all the landmasses merged to form a Pangea supercontinent, which in turn split about 185 mya (Condie, 2011). Collision of larger continents created mountain belts (e.g. Appalachian and Hercynian Mountains) in North America and Europe, respectively (Frisch et al., 2011; Kious & Tilling, 1996). These ancestral crustal uplifts were precursors of subsiding marginal marine, coastal, and low-lying continental environments where widespread forest peaty wetlands (e.g. bogs/mires/swamps; see Chapter 3) formed during the Carboniferous Period. Cyclical transgressions and regressions of the sea on these marginal peaty wetlands permitted coal successions to form, which are mainly distributed in paleolatitudes between 0° and 60° north (Habitch, 1979; McCabe, 1984; Figure 1.4) with minor coal deposits forming in the south of the equator (Figure 1.7). These peaty wetlands were mainly developed in tropical belts although Carboniferous coals were also formed at high paleolatitudes or cool temperate belt (Greb et al., 2006; Retallack et al., 1996). Carboniferous coals formed in Europe (e.g. Spain, Portugal, France, Germany, Netherlands, Poland, Turkey, United Kingdom, and Ireland), Russia, China, and Africa (e.g. Morocco and Egypt).

Carboniferous–Permian coal accumulations controlled by paleogeography, paleoclimate, plate paleotectonic, and paleosealevel fluctuation are best exemplified by the separate North and South China microcontinents (Figure 1.7). In North China microcontinent, coal deposits were deposited in alluvial fans, fluvial floodplain, and paralic mires located in low northern

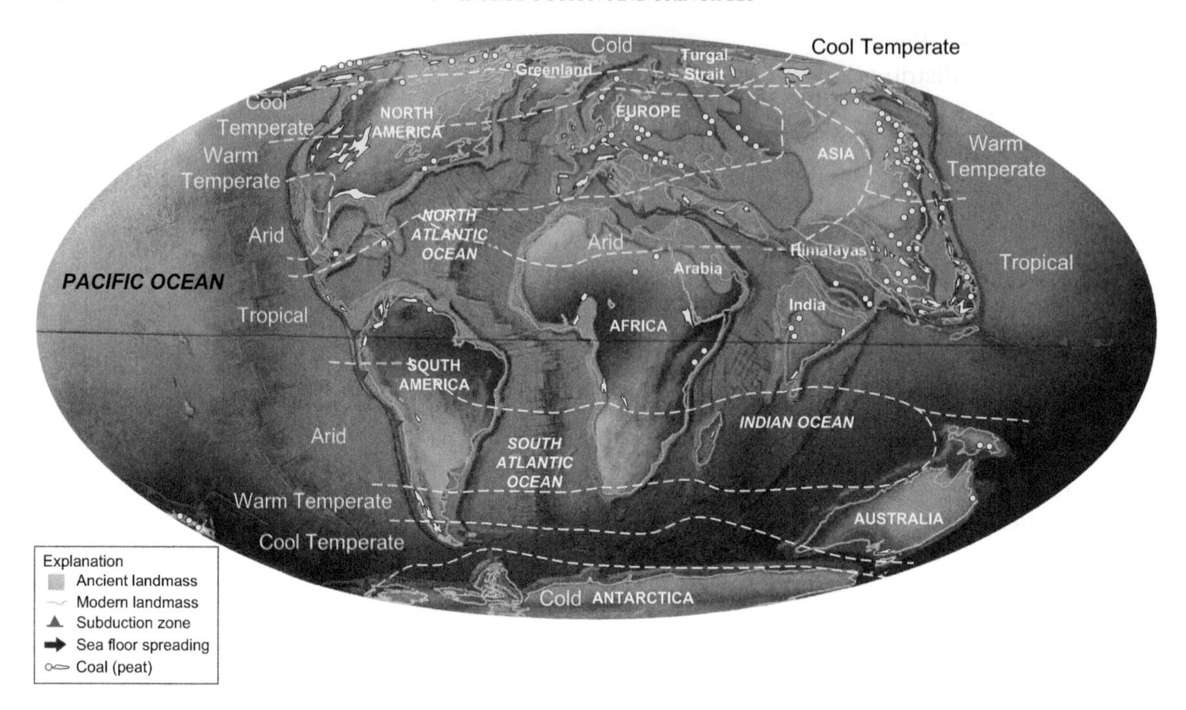

FIGURE 1.10 Paleogeographic distribution of ancestral landmasses (e.g. supercontinents, continents, microcontinents) driven by mechanism of plate tectonics during the Tertiary–Paleogene times when coals were formed in peaty wetlands. Approximate global paleoclimatic zones are indicated. *(Source: Modified from Greb et al. (2006), Dinwiddie et al. (2011), and Scotese (2013)).* See Figure 1.4 for approximate geologic age. Black dots and elongate areas are coal (peat accumulations).

paleolatitudes characterized by moist tropical paleoclimate (Liu, 1990). Thick coals were accumulated in favorable transitional environments between alluvial fans, fluvial floodplains, and paleoshorelines where elevated groundwater levels were fed by rainfall. In Late Permian, coal accumulations shifted to the South China microcontinent with continued northward drift and collision of the North China microcontinent with the Siberian microcontinent (Figure 1.8). In South China, the accumulations of coal were limited to paralic delta-plain mires where oceanic-generated wet paleomicroclimate and elevated groundwater level existed (Suping & Flores, 1996). This paleogeographic setting was succeeded by subduction of the Paleo-Tethys oceanic plate beneath the southern margin of North China resulting in uplift such that the South China deltaic systems kept pace with accommodation space and paleosealevel fluctuations resulting in cyclical deposition of deltaic coals. Liu (1990) indicated that in South China, coals were formed in coastal plains of an epeiric and shelf sea which connected the oceanic basin frequently controlled by plate-marginal tectonic movements related to back-arc extension (Figure 1.8).

The Permian Period (240–290 mya; Figure 1.4) was marked by the reduction of paleoshorelines and coastal wetlands with the almost complete coalescing of Euramerica and Gondwana into the supercontinent Pangaea, which extended from pole to pole by early Triassic Period (230–240 mya; Figures 1.4 and 1.8) (Condie, 2011; Dinwiddie et al., 2011; Kremp, 1977; Scotese, 2013). Gondwana coals (peats) were commonly formed in these wetlands in present-day continents of Antarctica, Australia, India,

South Africa, and South America (Figure 1.8). Paleolatitudinal distribution of the Gondwana coals extended from 30° to 75° south (Habitch, 1979; McCabe, 1984; Figure 1.4). Peaty wetlands were formed in mainly cool temperate paleoclimates at high paleolatitudes (Greb et al., 2006; Retallack et al., 1996). Permian coals (peats) were well developed in east–southeast Africa from Tanzania to South Africa (Karroo system), south Madagascar, southern South America, northeast India, and throughout Australia (Figure 1.8). Depositional settings of Lower Permian coals vary from coastal plain–fluviodeltaic mires in south Brazil, South America probably during precoalescing of Gondwana to braided–meandering fluvial mires in northeast India during postcoalescing of Gondwana (Casshyap & Tewari, 1984; Cazzulo-Klepzig, Guerra-Sommer, Menegat, Simas, & Filho, 2007). Additional Permian coals were commonly formed in ancestral peaty wetlands in tropical paleoclimates (Greb et al., 2006) of present-day continents of Asia (e.g. China, Siberia) and in cool temperate Europe (e.g. Russia) where paleolatitudinal distribution of the coals extended from 30° to 60° north (Habitch, 1979; McCabe, 1984; Figure 1.4). Drying of the interiors of Pangaea, formation of glaciers in the high latitude areas, and mass extinction of terrestrial plants due to inhospitable peaty wetlands as previously discussed during the Late Permian Period, which continued into the Triassic Period may explain the decline and demise of coal deposition. Retallack et al. (1996) suggested that the Triasssic coal gap was caused by extinction of irreplaceable peat-forming vegetation of widespread mires in Antarctica, Africa, Australia, South America, China, and Siberia. However, as shown in Figure 1.4 peat-forming plants reappeared in the Jurassic Period, which initiated the extensive distribution from pole to pole through Cretaceous and Tertiary Periods.

The combined Jurassic–Cretaceous coal-forming period was highlighted by plant diversification as a result of the completed breakup of Gondwana with the separation of the present-day South America and Africa from Antarctica accompanied by the widening of the South Atlantic Ocean from separation of both continents (Condie, 2011; Dinwiddie et al., 2011; Scotese, 2013). Jurassic coals formed in Australia, China, North America, Europe (e.g. Russia), and Asia (e.g. Afghanistan, China, Iran, Korea). The separation of continents continued during the Cretaceous Period (95 mya; Figure 1.9), led to warmer paleoclimatic zones and creation of more paleoshoreline and paleocoastal areas where coals (peats) accumulated (Greb et al., 2006). In addition, widening of the North Atlantic Ocean promoted drift of North America from Eurasia (e.g. Europe and China).

Also, during this time India and Madagascar are now a microcontinent separated from Africa and Antarctica but through time India drifted northward toward Eurasia. Australia remained attached to Antarctica until about 100 mya when it rifted (Condie, 2011) and Greenland to North America. Major subduction zones particularly along the western coast of the North and South American continents where volcanic activity and tectonic uplift occurred led to development of epicontinental seas along the western interior of these landmasses. Paleotectonic movements also lowered the earth's crust epicontinental areas providing accommodation spaces, which were flooded by the sea and whose coastal environments created new wetlands for colonization by plants. The waxing and waning of the epicontinental seas promoted repetitive appearances, on large scales, new wetlands and niches for flowering plants to invade, adapt, and establish worldwide leading to evolution into plants of the present. Coals commonly formed in peaty wetlands of now separated continents during the Cretaceous Period where they were distributed in paleolatitudes from 50° to 80° south and from 40° to 80° north (Habitch, 1979; McCabe, 1984; Figure 1.4). The Cretaceous peaty wetlands were mainly formed in subtropical paleoclimatic belt in the northern hemisphere with average of about 50° paleolatitude (McCabe, 1984; Retallack et al., 1996).

During the Tertiary Period especially from Paleocene to Eocene Epochs (50 mya; Figure 1.10) Greenland has moved for some distance from North America, Australia has moved northward from Antarctica, and India has moved closer to Asia (Condie, 2011; Dinwiddie et al., 2011; Scotese, 2013). At the beginning of the Tertiary Period (65 mya), the shape of the North American continent was nearly similar to that of present day with only marine encroachments along the Atlantic, Gulf, and Pacific coasts. Accompanying coastal wetlands formed coals (peats) only along the Atlantic, Gulf, and Pacific coasts from Paleocene to Miocene Periods (Figure 1.10). However, not shown in Figure 1.10 during the Paleocene is the boreal seaway, which was a remnant of the Cretaceous epicontinental sea, which formed coastal wetlands in North Dakota and Canada prior to retreating northward (Flores, 2003). The culmination of the Laramide orogeny, which structurally partitioned the western interior of the United States, formed intermontane basins that accumulated coals, which differs from other significant coal deposits of the world. At the close of Tertiary Period, the shape of Europe had emerged substantially in its present form with the marine encroachments in Europe being moderately extensive and where coastal areas accumulated coals (peats). Collisions of major continental and oceanic plates created a complex of subduction zones overprinted by volcanic island arcs, forearc, back arc, and rift basins, which formed mainly Eocene–Miocene coal-bearing sequences in southeast Asia (e.g. Indonesia, Philippines) (Sidi & Darman, 1995). Coals commonly formed in ancestral peaty wetlands in these continents during the Tertiary Period were distributed in paleolatitudes from 0° to 60° south and from 0° to 80° north (Habitch, 1979; McCabe, 1984; Figure 1.4). Coal-forming wetlands shifted mainly from subtropical to tropical belts during the Tertiary Period (Greb et al., 2006; Habitch, 1979; Retallack et al., 1996).

GLOBAL COAL ENDOWMENT

An unsubstantiated estimate of the world's original endowment of "mineable coal" is more than 7.6 trillion mt and of this total, only about 2.5% has so far been exploited. Coal reserves are projected to last from a few 100 years to more than 1000 years, depending on the rate of production, consumption, and technological advances. This claim is beset by the following problems: (1) the major coal-bearing rocks, as previously demonstrated, are unequally distributed and mainly concentrated in the Australian, Euro-Asian, Indian, South African, and North American continents; (2) the estimate of coal endowment is based on geological data, which varies in quality and quantity from continent to continent, in general and from nation to nation, in particular; (3) the methodology of estimation of coal resources and reserves differs from nation to nation, (4) the best quality and known sources of coal have been mined at shallow depths and utilized since recorded historical times, and (5) the remaining coal endowment is in less accessible regions and uneconomical at present-day technology. The remaining coal endowment, which is deeper, is more amenable to coalbed gas development.

Concepts of Coal Resources and Reserves

The world's coal endowment is measured and reported in terms of coal resources and reserves; the coal resources, in turn, is used to estimate coalbed gas resources as discussed in Chapter 6. The total world coal resource is difficult to determine; however, Landis and Weaver (1993) reported it is as much as 27 trillion mt and is unevenly distributed in 74 countries in six continents. A more reliable estimate of the world coal endowment, which is assessed from the total coal resource, is the coal reserve or the amount/volume of coal determined by geological and engineering information with reasonable certainty of extraction in the future under existing economic and mining parameters. Eighty two percent of the world recoverable reserves (amount/volume of coal reserve extractable at present day economic and mining conditions) are located in the United States (29%), Russia (19%), China (14%), Europe and Asia (10%), and Australia/New Zealand (9%) (USEIA, 2010b). In 2007, these regions altogether produced 4.4 billion mt of coal representing 71% of the total world coal production (BP, 2010). At the end of 2009, BGR (2010) reported that global total coal reserves are 1001 Gigatonnes or Gt (723 Gt hard coal and 278 Gt lignite).

The world's coal resources and reserves, which are subdivided into discovered and undiscovered (inferred/assumed/speculative quantities) and proved and recoverable, respectively. These categories vary over time as a result of exploration and development activities, which led to depletion of coal supply or reserves and replenishment by conversion of coal resources previously technologically and economically deemed inaccessible. More importantly, these categories are based on geological uncertainty and economic feasibility, which vary worldwide; thus, accurate and reliable reporting of coal resources and reserves has received increasing attention. Stevens (2004) summarized the various current practices of reporting by the United States, Australia, Canada, South Africa, and Great Britain where coal classification systems have been developed for the last 10–30 years and mainly modeled after "Codes for Reporting Mineral Resources and Mineral Reserves". However, mineral and energy resources classification are not without issues (Edens & DiMatteo, 2007). Nevertheless, the objective of the present classification systems is to put into practice a reliable approach for reporting coal resources and reserves using international standard terminology and methodology (Rendu, 2000), which are summarized in the following sections.

International Codes of Reporting Coal Assets

Reporting of estimates of global coal assets in the forms of resources and reserves is guided by codes, which differ regionally and geological as well as economic considerations. Although progress is being done to standardize reporting codes internationally, no unifying guidelines currently exist. Similarly codes have been developed to guide the valuation of mineral prospects, discoveries and operations.

As Practiced by United States

In the United States, the concept of coal categorization into resources and reserves was derived from the classification system for mineral resources designed as a "box" by Vincent McKelvey, former Director of the United States Geological Survey (USGS) (McKelvey, 1972). The McKelvey "box" (Figure 1.11) of Mineral Resources Classification is a two-dimensional scheme that combines increasing geologic assurance (e.g. undiscovered, possible, probable, and

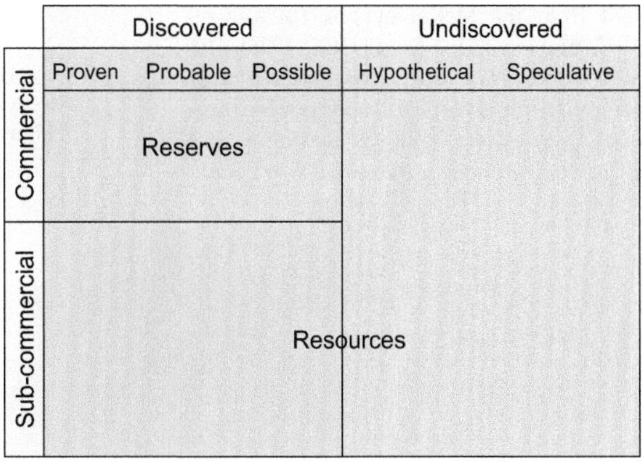

FIGURE 1.11 The original McKelvey box showing classification system of mineral resources and reserves categorized as discovered and undiscovered versus commercial and subcommercial. *Source: Modified from McKelvey (1972).*

proven reserves) and economic feasibility (e.g. commercial and subcommercial resources as compared with commercial reserves depending on price and cost levels and available extraction technologies). The proven reserves are more certain to be recovered or extracted than the probable and possible reserves based on more reliable closely spaced geological data. Discovered resources are confirmed by drilling test wells and undiscovered resources are inferred from geophysical and geological data.

The standards for reporting coal resources and reserves in the United States began with using the framework of the McKelvey "box" giving rise to a classification system published jointly by the United States Bureau of Mines and the USGS (USGS, 1976). This USGS report was modified and followed by the publication of the principles of a resource/reserve classification for metallic minerals, industrial minerals, and coal (USGS, 1980). The USGS subsequently published an updated classification system, which included criteria, guidelines or rules, and methods required for uniform application to classification of coal resources and reserves (Wood, Kehn, Carter, & Culbertson, 1983). This classification system continues to be the definitive reporting system for coal resources and reserves used by the industry and other countries despite attempts to modify the system of apparent shortcomings on economic and technical parameters.

The USGS coal classification system employs a concept based on increasing economic feasibility, the geologic assurance of their existence, and the economic feasibility of their recovery much like the McKelvey "box" (Wood et al., 1983). The geologic assurance is based on the distance of control points (e.g. boreholes, geophysical logs, outcrops) where coal is measured; thickness of coal and overlying rocks; coal rank, quality, correlations, and areal extent; and stratigraphy, structural geology, and depositional environments. Economic feasibility is based on environmental regulations, technical constraints, price of coal, and cost of operations (e.g. equipment, labor, processing, transportation, taxes). Figure 1.12 provides an internationally recognized and accepted classification system of quantifying coal resources and reserves

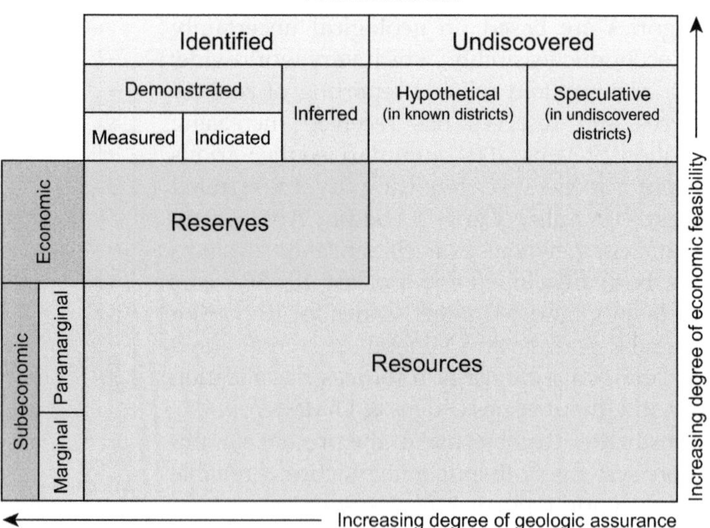

FIGURE 1.12 USGS classification system for coal resources and reserves modified from the McKelvey box (McKelvey, 1972) and expanded from USGS (1976). The system also categorized coal in terms of identified reserves and undiscovered resources based on economic feasibility and increasing geologic assurance (Wood et al., 1983).

from the USGS that has been tested through time (Gluskoter, 2009 Hughes, Klatzel-Mudry, & Nikols, 1988; Richardson, 2010; Ruppert et al., 2002; USGS, 1980; Wood et al., 1983).

The classification system is developed to estimate amounts of coal on the ground before mining (original resources), after mining (remaining resources), that are known (identified resources), and that remains to be discovered (undiscovered resources). More specifically, the classification system quantifies the amounts of coal in terms of measured, indicated inferred, and hypothetical resources from which the coal reserves are estimated from what is identified (e.g. measured, indicated, and inferred resources), and in turn economically classified. The reserve base is identified by the physical and chemical properties of coal such as thickness, rank, quality, and heat value as well as thickness of overburden rocks and distance of control points (Wood et al., 1983). However, the most important part of the coal classification system, which is relevant only to the estimation of coalbed gas-in-place resources, is the coal resources as demonstrated in the next section.

Coal Resource Calculation

The estimate of the tonnage for any coal bed can be performed if the thickness, areal extent, and weight per unit volume (or density) of the coal are known. The density or specific gravity varies with rank, ash content, and the amount of organic content of the coal. The estimate of coal tonnage (volume) or coal mass-in-place, in turn, is used to calculate the coalbed gas resource by multiplying the in situ gas content of the coal, measurement of which will be described in Chapter 6. Calculation of coal resources adopted from Wood et al. (1983) is shown the following equation:

$$A \times B \times C = \text{Tonnage or volume of coal}$$

where

A = Weighted average thickness of coal in inches or feet (centimeters or meters),

B = Weight of coal per appropriate unit volume in short or metric tons, and

C = Area underlain by coal in acres or hectares.

The weight of coal per unit volume or density (specific gravity) varies with rank, ash content, and the amount of coal maceral macrolithotypes (e.g. vitrain, durain, clarain), which is discussed in Chapter 5. Coal density should be determined from average specific gravity of various coal rank as in 1.29 for lignite coal, 1.30 for subbituminous coal, 1.32 for bituminous coal, and 1.47 for anthracite and semianthracite coal. The specific gravity is based on coal samples from United States; thus, it is recommended that the specific gravity, which is variable, be replaced as applied to coal from country to country. This assessment methodology has been widely utilized for estimation of coalbed gas-in-place resources (Richardson, 2010).

As Practiced by Other Countries

In many countries for which coal classification systems have been developed over the past two decades, coal resource and reserve reporting guidelines are generally similar to that for mineral commodities. Significant advances were made to establish a consistent globally recognized code or standard of reporting for mineral resources and reserves, which have been adopted and modified for coal resources and reserves. The international approaches are generally based on guidelines or rules and methods as applied in the Australia, South Africa, China, Great Britain, Europe, Canada, and United States. Although the approaches are mainly similar, they may vary in specific methods and criteria for quantification (e.g. minimum coal bed/seam thickness, maximum parting thickness, maximum drill-hole spacing, depths) from country to country. Some reporting standards may reference supplementary rules to allow for a few additional provisions to address considerations unique to the coal deposits (e.g. coal rank) of the country. For example, the classification system for coal resources and reserves of China follows a strict criteria laid down by the Chinese Ministry of Land and Resources (Qian et al., 1999), which are subject to annual recalculation and review. These criteria include application of minimum coal seam thickness of 0.70 m, exclusion of "dirt" bands (partings) >0.05 m within the coal bed thickness, maximum depth of 1400 m, exclusion of weathered coal, and calculation in relatively small block areas. The 1999 Chinese Mineral Resource/Reserve Classification Standard was based in part on the United Nations Framework Classification of Fossil Energy and Mineral Resources.

Using the USGS and U.B. Bureau of Mines coal resource classification system (McKelvey, 1972; USGS, 1976, 1980; Wood et al., 1983) as templates, other countries have developed their own classification systems and guidelines for mineral resources and reserves adopted for coal such as that by the Joint Ore Reserves Committee Code and Guidelines (2004) recognized in Australia, New Zealand and Asian countries; thus, also called the Australasian Joint Ore Reserves Committee (JORC) Code (Figure 1.13). Professional organizations such as the Australasian Institute of Mining and Metallurgy, Australian Institute of Geoscientists and Minerals Council of Australia played a major role in the development of international standard definitions for these Codes and Guidelines. The new Chinese classification system for resources/reserves of solid fuels and mineral commodities (Qian et al., 1999) is not readily comparable to the classification nomenclature of Australasian JORC Code of guidelines for estimating and reporting of inventory coal, coal resources and coal reserves (Stoker, 2009). The Government of Indonesia, Ministry of Energy and Mineral Resources and the New Energy Development Organization of Japan coal resources and reserves study (2007–2008) also adopted both the Australian and American classification systems (Lucarelli, 2010). The South African code for reporting of exploration results of mineral resources and reserves (SAMREC Code) was adopted for reporting coal resources and reserves, which allows for the reporting of several categories of reserves, including mineable in situ coal reserves (that only include geologic loss factors within a mine plan), run of mine coal reserves (that include mining loss, dilution, and moisture factors), and saleable coal reserves (coal available for sale after beneficiation resulting from coal processing) (SAMREC Working Group, 2009).

In Great Britain and Europe, the JORC Code was modified by the Institute of Materials, Minerals & Mining Working Group (2001) in conjunction with the European Federation of Geologists, the Geological Society of London, and Institute of Geologists of Ireland and created a reporting code for mineral resources

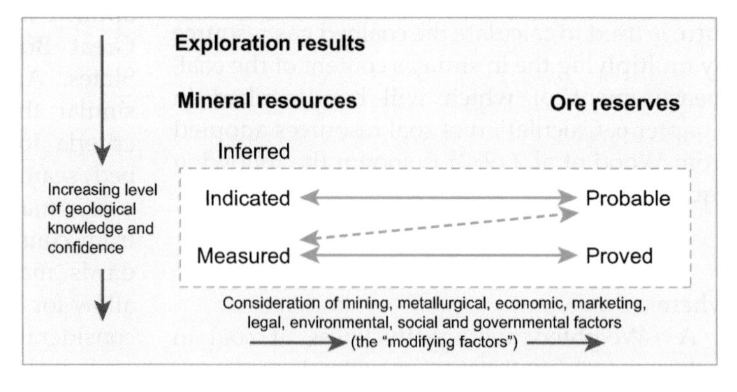

FIGURE 1.13 Classification code developed for mineral resources and reserves for coal by JORC and other countries in Asia. *Source: Adopted from Joint Ore Reserves Committee Code and Guidelines (2004).*

and reserves including coal. The reporting code requires that a clear distinction be made to identify mineable coal reserves (mining losses taken into account) and saleable coal reserves (mining and beneficiation losses taken into account); however, if saleable are reported, the corresponding recoverable coal reserves must also be reported. The Geological Survey of Canada Paper 88–21 (Hughes et al., 1988), a standardized coal resource/reserve reporting system for Canada, reverts to guidelines similar in approach to the USGS classification system. The Russian coal classification system for coal reserves differs from the United States classification system because the categories are based on economic factors identified and classified by the State, which is the owner and operator of the coal (Lawson, 2002). The philosophy behind all these different guidelines is very similar with emphasis on the coal thickness, rank, quality, distance of control points (points of observation), economic (marketability), and technology (mining methods). Thus, the close resemblance of the reporting codes and guidelines of the various countries eventually will lead to a single international reporting code, which will be followed by reporting agencies worldwide.

The United Nations Framework Classification (UNFC, 2010) is closest to a consensus code, which is internationally accepted reporting and evaluation of fossil energy and mineral resources and reserves. This reporting code, which is applicable to coal resources and reserves, is a three-dimensional framework based on the following categories: (1) economic and social viability, (2) field-project status and feasibility, and (3) geological knowledge. The first category assigns the level of favorability of economic and social conditions in determining commercial viability of the project. This category takes into account the market prices and related legal, regulatory, environmental, and contractual circumstances. The second category is the growth process of investigations from early exploration through confirmation of accumulation of deposits to implementation of mining plans, development selling the commodities. The third category assigns the degree of confidence of geological knowledge and potential recoverability of the deposits. The UNFC is a flexible system that permits integration of existing terms and definitions from current reporting codes (e.g. JORC, SAMREC Codes) for application at national, industrial, and institutional levels, as well as for global assessments. However, this approach clearly applies more extractable and marketable energy and mineral reserves than resources; the latter more appropriate to assessment of coalbed gas-in-place.

As Reported by World Information Agencies

Annual status of the world coal reserves, production, and consumption are reported by a few reliable information-gathering agencies such as the World Energy Council (WEC), World Coal Association, International Energy Agency (IEA), British Petroleum (BP) Statistical Review, and United States Energy Information Agency (USEIA). Although these organizations record and report coal statistics worldwide spending considerable efforts to ensure comparability of national estimates, terminologies used by each reporting organizations may or may not be similar.

For example, the WEC classifies reserves according to the following definitions. (1) Proved/proven in-place reserve is the amount or volume of the remaining coal resource (after mining), which is known coal deposits that have been measured and assessed as minable under present and future economic conditions with existing available technology. (2) Proved/proven recoverable reserve is the amount or volume in the proved/proven in-place reserve that is recoverable in the future under present and future economic conditions with existing available technology. (3) Proved/proven estimated reserve is the amount or volume in the proved/proven in-place reserve that is indicated and inferred that is of likely interest additional to the proved/ proven in-place reserve. Also, it includes amount or volume of proved/proven estimated reserve that exist in unexplored areas of known and undiscovered coal deposits identified in coal-bearing areas as well as inferred from the knowledge of favorable geological conditions. The proved/proven estimated reserve, which is recoverable amount or volume from the proved/proven reserve in-place, is that deemed to be recoverable with reasonable certainty in the future based on geological and engineering information.

The proved/proven recoverable reserve as defined by WEC is the same as proved/proven reserve defined by IEA and BP Statistical Review of the World Energy. The USEIA uses the following criteria of coal reserves (USEIA, 1996): (1) demonstrated reserve base is measured and indicated coal determined by the degrees of geological accuracy, which is found at depths and coal bed thickness considered technologically minable at the time of determinations; and (2) estimated recoverable reserve is in the demonstrated reserve base considered recoverable after excluding the coal estimated to be unavailable due to land use restrictions or economically unattractive for mining, and after applying assumed mining recovery rates. This classification is similar to proved reserves according to BP statistics.

The BP Statistical Review of World Energy publishes proven reserves together with production data for each year (BP, 2010, 2011). Each new edition is published with a listing of "proven reserves at end year". However BP just reproduces the data collected by WEC, which collects these data from its member countries from time to time. The latest WEC Survey of Energy Resources was published in 2010 with data as of year-end 2008. This results in the published proven reserves at year-end 2008 in BP Statistical Review of World Energy in reality being those that were reported for the year 2009 (BP, 2010). Also, global annual coal reserves up to 2009 have been estimated by BGR (2010).

FROM PAST TO FUTURE COAL PRODUCTION

Coal production is closely linked to coalbed gas desorption especially in relationship to underground coal mining as discussed in Chapter 2. Gas explosions or outbursts are as old as the coal mines "dug" in the early nineteenth century and remained a major hazard affecting safety and productivity in underground coal mines for more than 200 years (Deul & Kim, 1986; Flores, 1998; Lama, 1995). Extensive gas outbursts and emissions during coal production are a global phenomenon and attest to the rich coalbed gas endowment. Thus, the historical and future perspectives of coal production provide a better insight into the coalbed gas development.

Historical Coal Production

Coal production dates back to the earliest civilizations cited by Aristotle (340) and Agricola (1556) for use in fire combustion during the middle of the fourth century B.C. (Landis & Weaver, 1993). It was thought that the Chinese first commercialized coal production for smelting 2000–3000 years ago mining coal from the Fushun coalfield in Manchuria (Inouye, 1913). Coal replaced wood and charcoal for fuel by the thirteenth century and commercialized for smelting and casting of brass by the seventeenth century in Great Britain (Eavenson, 1939). Coal was exported to other European countries by 1325 as a result of widespread mining with increased coal production from 2.27 to 9.07 million mt during 1700–1800 (Eavenson, 1939). The Industrial Revolution, confined in Great Britain from 1760 to 1830, increasingly utilized coal to make coke for smelting iron and other metals and to fuel coal-fired steam engine for locomotives (Lindbergh & Provorse, 1977).

Coal was known to fire kilns for native pottery in the southwest United States during the eleventh century (Landis & Weaver, 1993). Immigration of Europeans brought the Industrial Revolution to North America and increased commercialization of coal for use in smelting and space heating in the northeast United States. Coal mining of bituminous and anthracite coals in the central Appalachian Basin occurred in the early 1800 with production of both to as much as 2721 mt increasing to 1,814,000 mt prior to the American Civil War and reached to about 516,990,000 mt by 1913 (Moore, 1922). Dramatic increase of coal production in the United States occurred during the First and Second World Wars after which coal was supplanted by a major shift to oil and gas for energy sources. The rest of the world's reliance of coal resources for energy mimicked that of Great Britain, Europe, and North America. In 1913 prior to the First World War the other countries that relied on coal for energy source were China producing about 13,967,800 mt, Australia producing 12,698,000 mt, and Russia producing 32,652,000 mt (Lesher (1916). In 1989, China has increased coal production by about 67%, Australia by about 14%, and Russia by about 22% (USEIA, 1991). In 2007, the largest coal producers, which account for 75% of global production, in descending order, are China, United States (39% of China production), Australia (37% of United States production), India, South Africa, and Russia (BP, 2012).

The increased coal production worldwide through underground mining led to explosions of gas (e.g. methane and carbon dioxide) and coal from the working mine face. Outburst was not a problem when coal was mined from outcrops and shallow shafting in Great Britain and Europe because gas was easily liberated to the atmosphere. As shallow coal resources were slowly exhausted at the end of the eighteenth century and technology was improved to permit construction of deep mines, coalbed gas was recorded in Great Britain, France, and United States. In the early nineteenth and twentieth centuries minor and major outbursts were reported in Australia, Canada, Belgium, China, Germany, Japan, Poland, Russia, and United States (Flores, 1998).

Future Peak Coal Production

Coal is a finite resource contrary to the conventional wisdom that coal is abundant and can be used as energy far into the future. Statistical methods, which have been controversial, have been used to test this conventional wisdom and to forecast near future peak years of coal production and/or exhaustion of reserves. The first method is to use the historical coal production curve fits (ratio of annual production to cumulative production is plotted vs. cumulative production) in the form of Hubbert's (1982) peak oil or modified version to project peak years of coal production (Hook, Zittel, Schindler, & Aleklett, 2010; Mohr & Evans, 2009; Patzek & Croft, 2010; Stricker & Flores, 2009). The second method is to use the original coal reserves plus the annual rate of coal production and estimated recoverable reserves to forecast peak years of production from a curve fit (Milici et al., 2009). The third method does not predict peak years but instead apply dual tests for mature coal regions (declining coal production) using curve fits to historical coal production history rather than reserve estimates and active coal regions (predeclining coal production) using the reserves plus cumulative production as the estimate for long-term production (Rutledge, 2011). The concept of global peak coal has been very contentious, which evoked issues such as imminent decline of coal supply leading to reduction of CO_2 emissions from burning coal and development of more efficient power generation from coal-fired power plants making carbon capture and sequestration unnecessary.

In order to understand the principles behind these statistical methods, it is necessary to review how much global coal is produced against the calculated coal reserves. The coal

BOX 1.3

PEAK COAL—CONCEPTS AND MISCONCEPTIONS

The concept or theory of peak coal like its predecessor peak oil (Hubbert, 1982) is controversial. However, on the one hand it offers reasonable possibilities and on the other presents reasons of when the maximum global coal production may be reached after which production is at a decline. To environmentalists, numerical analytical scientists, climatologists, and other scientists, the peak coal concept proposes the prospect of the future decline of coal production and utilization, which leads to the solution of reduction of carbon dioxide (CO_2) emission from coal-fired power plants (USEIA, 2010a). To energy enablers, economists, scientists, and engineers, it is the potential to redouble efforts to find cleaner coal supply and to build more efficient technologies for power generation (USDOE 2010; 2011). More importantly, to energy entrepreneurs this situation is seen as an opportunity to harvest the clean coal-bed gas from the vast coal-resource endowment.

Perhaps the answer to these extreme viewpoints is the misconception of the focus on the future coal supply and production. The view that environmental regulations and socioeconomic restrictions will curtail coal from being produced and utilized in the near future is a double edge sword because it also leads to reduced supply and conservation due to a predicted higher price of coal from that of the current market. This principle of supply and demand elevates the economics of coal, which will lead to investment in extensive exploration of recoverable coal reserves from previously inaccessible areas resulting to large-scale conversion of coal resources to potential reserves. On the other hand, the economics of coal may depend on the natural gas supply and demand. Finally, the environmental and socioeconomic constraints along with favorable financial investments will lead to technical breakthrough of construction of more efficient coal-fired power plants with CO_2 captured and sequestered in a geological sink (McFarland et al., 2004).

reserves-to-production (R/P) ratio has decreased worldwide during the past decade, which begs the question about the world's reaching coal peak production. The world's R/P ratio is 224 in 1996–1997 compared to 119 in 2009 (BP, 2010; World Energy Council, 2010). Of the two countries with the most coal reserves and production, the R/P ratio of the United States was reduced from 465 in 1997 to 224 in 2009 and China declined from 72 in 1997 to 41 in 2009. Eurasia holds the largest coal reserves and the highest regional R/P ratio of 257 (BP, 2011). The decrease in the R/P ratio can be attributed to infrequent cycles of mapping, assessment, and conversion of coal reserves from resources. Mining companies assess on the short-term conversion of coal reserves from resources of leases immediate to their mine properties. Government-funded geological surveys typically perform larger-scale coal reserve exploration and assessment activity. The USGS has performed coal resource/reserve assessments only twice since 1975 with the latest assessment completed in 1999–2000 or once every 25 years. During that period the coal resources of the conterminous (48 states) United States changed from 3599 billion mt of total coal remaining in the ground in 1974 to 1451 billion mt of coal resources based on over 60 minable coal beds and zones in 1999 (Averitt, 1975; Averitt, 1981;

Ruppert et al., 2002). However, this estimate does not include the largely hypothetical coal resources of Alaska, which was estimated to be as much as 5012 billion mt with 2902 billion mt in the North Slope coal province (Flores, Stricker, & Kinney, 2003, 2004). Thus, including Alaska coal resources the United States has about 6463 billion mt and Alaska surpassed the 1974 total coal resources of the conterminous United States by 40% and only 12.2 billion mt are identified coal resources mainly from the Central Alaska-Nenana and Southern Alaska-Cook Inlet coal provinces, which is a small fraction (0.002%) of the total coal resources of Alaska (Flores et al., 2004). BP (2011) reported proven coal reserves for the United States of 215,227 million mt or about 3.3% of the total coal resources.

The Energy Information Administration (USEIA, 2011) reported proved coal reserves worldwide of 1038 billion mt in 1991, 982 billion mt in 2000, and 825 billion mt in 2008 with the latter estimate reflecting reserves/production ratio of 129 years. The World Coal Association (2010) reported lower proved coal reserves worldwide to 768 billion mt indicating that there is enough coal to continue production (R/P ratio) at the current rates for about 118 years. The key point to raise with the two decades of decreased proved coal reserves worldwide is in when (e.g. year) does peak coal production occur following the permanent decline of production? Jevons (1856) raised this question 155 years ago with increasing coal production of Great Britain, which at that time consumed about 84 million mt of mainly bituminous coal. Hull (1861) followed with the question on how long Great Britain coalfields will last with the increased production. Based on coal consumption at a rate of 3.5% per year extrapolated from previous decades and the price of coal becoming too high as mining progressed from 610 to 1220 m depth, Jevons (1856) predicted more than 100 years for the coal to continue fueling the economy of Great Britain. At 1220 m depth, Hull (1861) estimated coal resource of 72,418 million mt for beds 0.61 m thick and greater. It was not until 132 years later that the peak coal production of Great Britain was determined to have occurred in 1913 as a result of increased mining costs, declined profits, lack of technological advances, and weakened economy coinciding with World Wars I and II (Mitchell, 1988). However, Rutledge (2009, 2011) suggested that the immediate reason of Great Britain's coal peak in 1913 was that 40% of the miners volunteered for service in 1914 but the long-term cause was "geological exhaustion" (David Rutledge, Personal Commun., 2012). Using M. King Hubbert (1982) model and additional data from 1988 to 2005, Stricker and Flores (2009) revised the peak coal production for Great Britain to have been reached in 1923.

The concept of peak coal production is based on M. King Hubbert peak oil theory (Hubbert, 1982) and is defined as the point in time (e.g. year) at which the maximum rate of coal production has occurred or reached after which production permanently declines. For example, Hubbert (1956) estimated that United States coal production would peak in about the year 2150. To determine the potential timing of this presumably irreversible decline, Milici et al. (2009) obtained estimates of original reserves (EOR) by adding cumulative production data from government sources to the estimated recoverable reserves published by the Energy Information Administration, state energy boards, and the USGS. For calculated EOR of three scenarios of 307, 370, and 462 billion mt of reserves, United States peak coal production will be reached in the years 2062, 2074, and 2105, respectively (Figure 1.14). However, these scenarios do not account to the fact that certain rank coals from overmatured United States coal basins have already reached their peak coal production. Stricker and Flores (2009) reported that the bituminous coal from the Illinois Basin has reached coal peak production in 1977 and the anthracite coal from the central Appalachian Basin has reached coal peak production in 1917. It should be noted that 60% of the estimated original coal reserves are

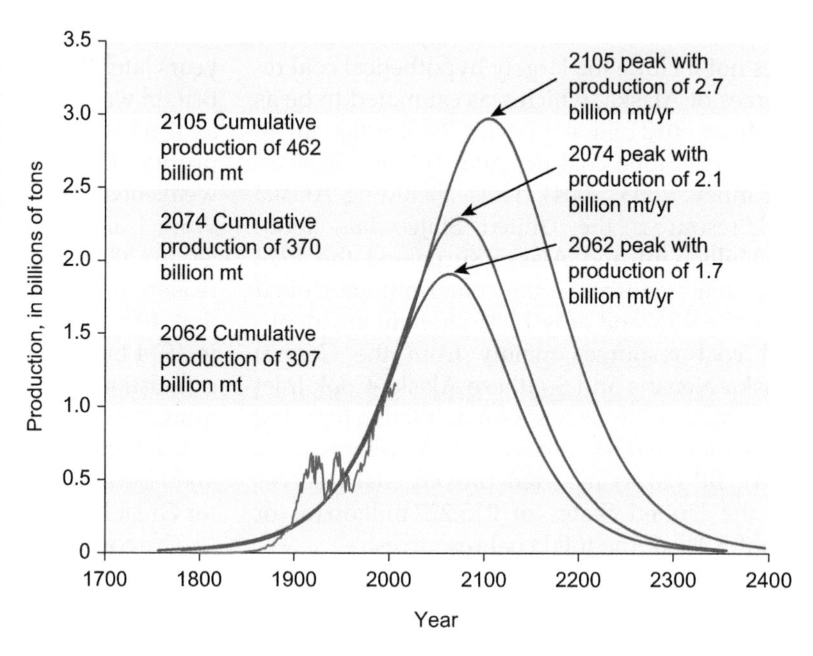

FIGURE 1.14 Cumulative U.S. coal production (green curve) and projected curve fits for calculated estimate of original reserves of 307, 370, and 462 billion mt, which indicate peak coal production in 2062, 2074, and 2105, respectively (Milici et al., 2009). mt = metric tons; yr = year.

concentrated in three states of the United States of which Wyoming produces about 90% of subbituminous coal most from the Powder River Basin. Stricker and Flores (2009) estimated at the rate of coal production from 16 coal mines in the Powder River Basin, which started development at the turn of the century and accounts for more than 40% of the United States total coal production as of 2010, will reach peak coal production in 2015 (Stricker & Flores, 2009).

The trends of peak coal production in Great Britain and United States are not isolated cases as noted by Mohr and Evans (2009), Hook et al. (2010), and Patzek and Croft (2010). The Hubbert (1982) model was applied by Hook et al. (2010) to coal production from 1815 to 2005 indicates peak coal for Germany in 1945, Japan in 1950, and Great Britain in 1924; the latter prediction similar to that of Stricker and Flores (2009). This curve fitting following Hubbert's (1982) approach gives some information on coal resources and reserves and projection of coal peak production based on current technology as shown by Mohr and Evans (2009). Patzek and Croft (2010) estimated that global coal production would decline 50% in 2050. Hook, Zittel, Schindler, and Aleklett (2008) reported global coal production forecasts for six countries, which possess majority of the coal reserves of the world (e.g. United States, China, Russia, Australia, India, and South Africa), and major coal exporting countries (e.g. Canada, Colombia, Indonesia, Kazakhstan, and Poland) based on the upper limit of current reserves relevant to be produced in the future (Figure 1.15). The global coal production is projected to increase over the next decade by about 30%, mainly driven by China, India, Australia, and South Africa. However, flattening of the coal production will occur around 2020 and global peak coal production will be reached and go into decline after 2050. Furthermore, the coal production trends indicate that Russia will be the last major coal producer and China will

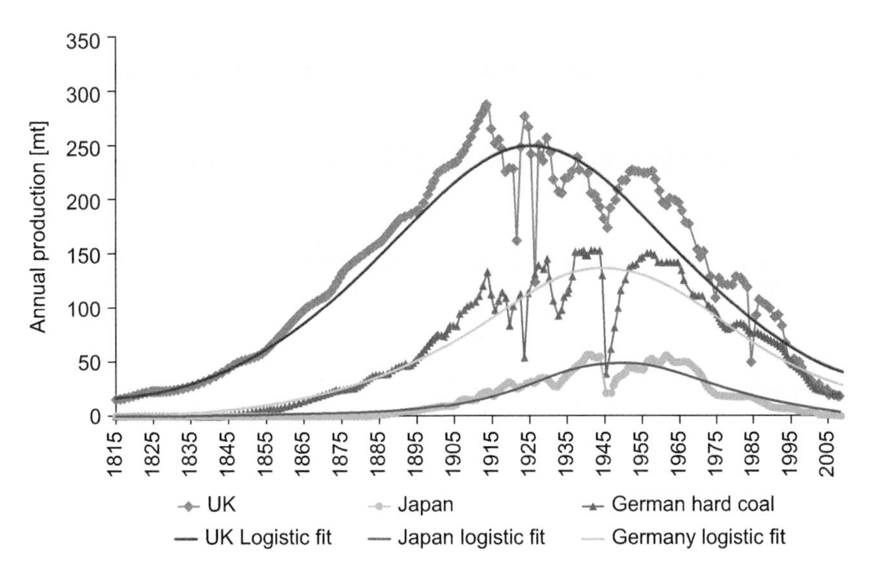

FIGURE 1.15 Peak coal production and curve fits of Japan and Germany compared to Great Britain (UK) from 1815 to 2007. Coal peaks have been reached in 1924 for United Kingdom, 1945 for Germany, and 1948 for Japan. mt = metric tons. *Source: Adopted from Hook et al. (2008).*

rapidly decline after 2020 (Hook et al., 2008). More importantly, the United States coal production will plateau after the peak is reached around 2035, which is about three decades earlier than the earliest coal peak production forecasted by Milici et al. (2009).

Although Mohr and Evans (2009), Hook et al. (2010), and Patzek and Croft (2010) used only coal production curve fits to arrive at various peak years for world coal production, Rutledge (2009, 2011) used long-term coal production and cumulative production plus coal reserves from international surveys (e.g. WEC) to estimate more accurately the exhaustion of world coal production. Additionally, instead of a global curve fit, Rutledge (2009, 2011) used regional fits for better estimates of mature coal regions (e.g. Great Britain, Pennsylvania anthracite, France, Belgium, Japan, and South Korea) compared to active coal regions (e.g. Australia, China, Africa, Europe, Russia, Western United States, Eastern United States, Canada, South Asia and Latin America). In each case for the mature coal regions, reserves are available early in the production cycle, but they were too high, and curve fit to the coal production history gave better accuracy. Based on ultimate production, Rutledge (2011) estimated the year of peak production varying from 43% exhaustion for Great Britain to 78% for France and Belgium. Also, similar patterns for mature coal regions that have reached 99% exhaustion ranging from 7 Gt for France and Belgium to 4 Gt for Japan and South Korea. Rutledge (2011) then tried to estimate ultimate world production on the active regions (e.g. Eastern United States without Pennsylvania anthracite, Western United States, Canada, Russia, Australia, Africa, and Europe) using the reserves plus cumulative long-term production and was able to fit either a logistic or a cumulative normal; thus, making projections ranging from 117 Gt for China to 4 Gt for Canada with 127 for the rest of Europe. These mature and active regions were projected to make up the world ultimate of 676 Gt, 60% of the current WEC reserves plus cumulative

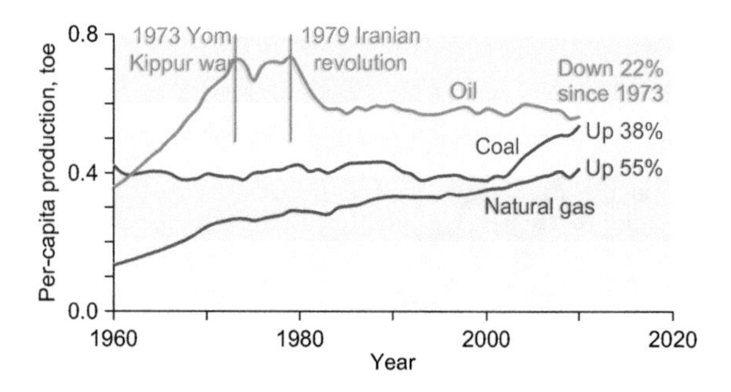

FIGURE 1.16 Historical (1990–2011) and future projection through 2035 of the world energy-related carbon dioxide emission by fuel types. toe = tonnes of oil equivalent. *Source: Adopted from Dave Rutledge Lecture Notes and USEIA (2010a, 2010b).*

production (1130 Gt). Finally, Rutledge (2011) observed that the curve fits also indicate that 90% of the total coal production will be exhausted by 2070. The global trends of per-capita production of coal compared to natural gas and oil for 50 years from 1960 to 2010 are shown in Figure 1.16 (David Rutledge, Personal Commun., 2011). Coal is up 38% compared to natural gas, which is up 55% through 2010. Oil has peaked in 1973 and 1979 during the Middle East conflicts (e.g. Yom Kippur war, Iranian revolution) and has declined and remained flat through 2010.

COAL USE IN A CARBON-CONSTRAINED WORLD

Global Impact

The future of coal use in a carbon-constrained world is directly related to CO_2 emissions from generation of electricity by coal-fired power plants. About 86% of coal consumption mainly for electric generation is by United States, Russia, China, Australia, India, Germany, South Africa, Poland, South Korea, and Japan (BP, 2010). These countries and other developing countries (e.g. Brazil, Indonesia, Turley) have contributed to the rise of coal use in the past 50 years (Cattaneo, 2013). However, of the 10 leading coal users, China showed the fastest rate of coal consumption by more than double from about 454 million mt oil equivalents in 1990 to about 1134 million mt oil equivalent in 2010 (BP, 2010). During the same period the other countries showed stable coal consumption except Russia, which reduced consumption by about 50%. The world coal consumption is 3,403,229 million mt oil equivalent in 1980 increasing to 6,872,683 million mt oil equivalent in 2009 (BP, 2010). BP (2010) reported that for the past decade global coal consumption rise about 50%. USEIA (2008) projected the United States, India, and China to increase consumption through 2030 with China increasing from about 58 to 233 quadrillion kJ/kg compared to the rest of the world (Figure 1.17). Electricity generation fueled by coal is about 80% in China and 68% in India (Pew Center, 2011).

Approximately 85% of the coal consumption by 2030 will be from China and India (IEA, 2007; USEIA, 2010a). Because of the inherently high carbon content of coal, about 20% of global greenhouse emission is mainly from coal-fired power plants in which a 500 MJ/s plant produces about 2.7 million mt of carbon dioxide (CO_2) (Pew Center, 2011). The U.S. Energy Information Administration (USEIA, 2010a) reported that the total CO_2 emission is from energy mix of coal, liquid fuels, and natural gas. Figure 1.18 shows that CO_2 contributions of these different fossil fuels have changed over time since 1990 and will continue to vary through 2035 (USEIA, 2011). The CO_2 emission from liquid fuels is 42% in 1990, 38% in 2007, and 34% in 2035 of the total world emissions. The CO_2 emission from natural gas is less than 19% in 1990 and 20% in 2007 of the total world emissions. However, the CO_2 emission from natural gas is relatively stable from 20% to 22% during the projected period of 2007 through 2035. Of the three fossil fuels that account for the world CO_2 emissions, coal contribution shows the most rapid rate of

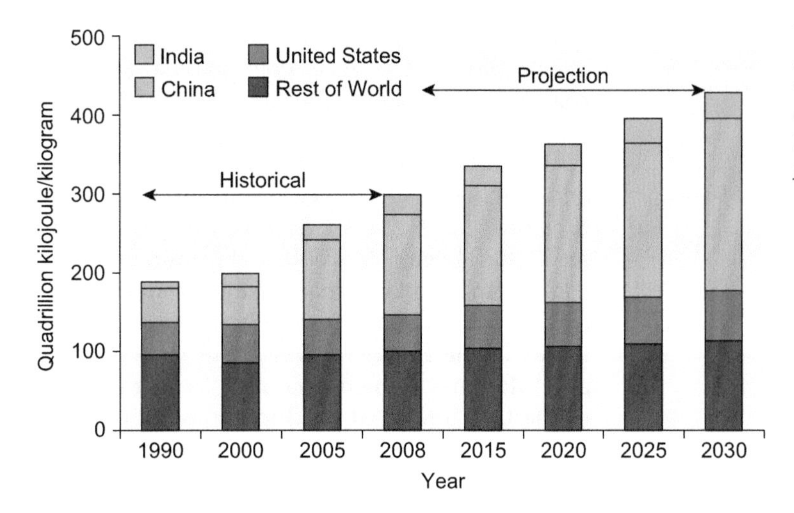

FIGURE 1.17 Historical (1990–2008) and future projection through 2030 of the world coal consumption with emphasis on the growth of consumption in China, India, and United States. *Source: Data from USEIA (2008), BP (2010, 2011, 2012).*

growth (Figure 1.18). The CO_2 emission from coal is 39% in 1990, grew to 42% in 2007, and projected to about 46% in 2035. More specifically, World CO_2 emissions from coal combustion increase by 56%, or 1.6% per year on average, from 12.5 billion mt in 2007 to 19.4 billion mt in 2035 (USEIA, 2010a). Thus, from 1990 to 2035 coal would have contributed to the total world CO_2 emissions by 7%, the natural gas by 2%, and the liquid fuels will have reduced contribution by 8%. These trends are understandable because coal is the most carbon-rich of the fossil fuels and increasingly consumed source of heat energy especially in China and India. USEIA (2011) reported that China and India together accounted for 13% in 1990 and 26% in 2007 of the world CO_2 emissions because of robust economic growth and increase use of coal to provide energy to sustain the economic growth. Of the world total the CO_2 emissions from China and India combined is projected to be 37% in 2035 with China accounting for 31%.

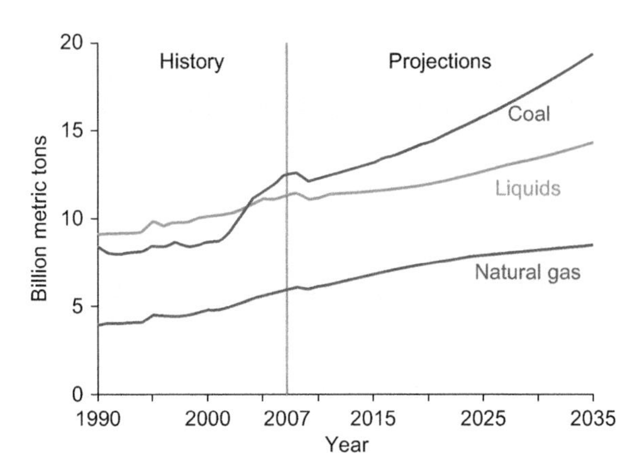

FIGURE 1.18 A graph showing the historical (1990–2007) and future projection through 2035 of the world energy related carbon dioxide emission by fuel types. *Source: Adopted from USEIA (2010a).*

Coal use is projected to not only continue but also increase mainly for the generation of electricity because of the inexpensive cost per Btu/lb (or 2.326 kJ/kg), which is ideal for the 10 developing and developed countries. However, the high carbon content of coal in a carbon-constrained world will continue to exacerbate the CO_2 emission unless coal-fired power plants are installed with very high efficiency carbon capture and sequestration system as well-as utilized, high-temperature combusting supercritical, ultra-supercritical power plants (McFarland

TABLE 1.1 Comparison of emissions from combustion of coal and natural gas. Particulates are fine particles (e.g. fly ash). Emission levels in kilograms per billion kilojoules of energy input

Pollutants	Coal	Natural Gas
Carbon dioxide (CO_2)	483,808	272,142
Carbon monoxide (CO)	483	93
Nitrogen oxides (NOx)	1063	214
Sulfur oxides (SOx)	6027	2
Particulates	6382	16
Mercury (Hg)	0.037	0.000

Source: Data from USEIA (1998).

et al., 2004; IEA, 2006). Another option in favor of coalbed gas is to expand the natural gas-fueled generation, which makes up about one-fifth of all electricity in the United States. A comparison of pollutants from coal versus natural gas combustion (Table 1.1) in the United States shows that carbon dioxide (CO_2), carbon monoxide (CO), nitrogen oxides (NOx), sulfur oxides (SOx), mercury (Hg), and particulates are much lower for natural gas. The percent pollutants from coal are 44% for CO_2 and about 80% for CO and NOx more than natural gas.

Impacts of Coal Use in the United States

In the United States, approximately 50% of the electricity is generated from coal-fired power plants in 2007, declined to about 33% in mid-2012, and recovered or rebounded to about 40% in the early 2013, which produce about 1.8 billion mt of CO_2 per year (Pew Center, 2011; USEIA, 2013). There are 471 coal-fired power plants in the conterminous United States, which consumed more than 0.9 billion mt/yr during the last few years. During the past two centuries approximately 169 billion mt of CO_2 have been emitted from combustion of solid fuels in the United States, most from steam coal-fired power plants (Stricker & Flores, 2009). Bituminous feed coal played a major role in carbon CO_2 emission during the first part of the twentieth century, but bituminous coal production peaked in the Appalachian and Illinois Basins in 1970 and 1977 (Stricker & Flores, 2009), respectively. Subbituminous feed coal began to replace bituminous feed coal during the latter part of the twentieth century as production gradually increased from mines in the Powder River Basin in Wyoming and Montana, which accounts for about 40% of the U.S. total coal production in 2011.

In order to determine the impact of burning low-rank coal, Stricker, Flores, Ellis, and Klein (2008) and Stricker and Flores (2009) investigated CO_2 emissions from present and proposed coal-fired power plants supplied by subbituminous coals from the Powder River Basin, Wyoming and Montana. The present power plants are nearby the basin, which include the Colstrip (Colstrip, Montana), Corette (Billings, Montana), French (Rapid City, South Dakota), Johnston (Glenrock, Wyoming), Laramie River (Wheatland, Wyoming), Neil Simpson 1 and 2 (Gillette, Wyoming), Osage (Weston, Wyoming), Wygen 1, and Wyodak (Gillette, Wyoming). The 2005 gross gigajoules (or megawatts) generated varied from 528,303 for the Neil Simpson 1 plant to 59,514,678 for the Colstrip plant (Table 1.2). In 2005 the total generation for the 10 power plants was 144,811,584 GJ, consuming a total of 24,054,420 mt of subbituminous coal from the Powder River Basin. The volume of CO_2 emission from these coal-fired power plants for the year 2005 varied from 65,296,000 m^3 for the Neil Simpson 1 plant to 8,819,029,000 m^3 for the Colstrip plant. The total CO_2 emission of these 10 coal-fired power plants in 2005 was more than 22 billion m^3 (Bcm), which produced a total of 146,911,624 GJ (Stricker et al., 2008). Traditionally, because of the low carbon (low-energy density) and high-moisture content of low rank coals (e.g. subbituminous and lignite)

TABLE 1.2 Characteristics of 10 coal-fired power plants in and within 161 km of the Powder River Basin (PRB) in Wyoming and Montana, United States

Power Plant	Distance from PRB (km)	Consumption of PRB Subbituminous Coal in 2005 (mt)	Gigajoules Produced	Volume of CO_2 (thousands of m^3)
Colstrip	0	9,677,846	59,514,678	8,819,029
J.E. Corette	24	580,000	3,638,865	581,972
French	120	121,000	548,521	68,098
Johnston	0	3,460,002	20,446,621	3,272,024
Laramie river	64	7,228,448	46,849,955	7,038,052
Neil Simpson 1	0	107,000	528,303	65,296
Neil Simpson	0	439,000	2,333,801	382,263
Osage (BKH)	24	221,000	882,324	114,363
Wygen 1	0	462,000	2,551,003	477,531
Wyodak	0	1,758,124	9,617,513	1,546,674

km = kilometer; m^3 = cubic meter; mt = metric tons.
Source: Adopted from Stricker and Flores (2008).

emission is higher than in power plants, which use high rank coals (e.g. bituminous coal). The CO_2 emission factors for subbituminous and lignite coals, which are mainly mined west of the Mississippi River, are 3–5% lower than bituminous coal, which is mainly mined east of the Mississippi River (Hong & Slatick, 1994). However, the lignite, which makes up about 20% of feed coal for power plants, vary in CO_2 emission factors from 1% mined in Texas to 3% mined in North Dakota and Montana. The environmental concern may be offset by burning low rank coals in mine-mouth power plants close to coal mines exemplified in the Powder River Basin; thus, avoiding CO_2 footprint from inefficient transportation.

NETL (2007) reported 151 proposed and new coal-fired power plants for a total of 90 GJ/s of power were in some stage of planning or permitting before State Commissions. The coal-fired electricity-generating technologies which are proposed for the new power plants include: (1) circulating fluid bed, (2) supercritical pulverized coal (PC), (3) ultra-supercritical PC, and (4) integrated combined cycle gasification. These technologies can include oxygen-combustion and postcombustion CO_2 capture (e.g. chemical solvents such as amine, flue-gas refrigeration), which can be used in new power plant boilers to reduce greenhouse gas emissions per unit of power output. Stricker and Flores (2009) reported that due to the reduced coal reserves and utilization in the mature Appalachian and Illinois coal basins, 30 coal-fired power plants were being planned, developed, or constructed in North Dakota, South Dakota, Montana, Wyoming, Nebraska, Colorado, New Mexico, and Texas (NETL, 2009; Sierra Club, 2009). These power plants, mainly located in the high plains states, will generate 10,060 kJ/s of electricity, utilize 65.7 million mt of "brown" coal (e.g. subbituminous and lignite), and contribute an additional 114 million mt of CO_2 per year. Twenty plants will use Wyoming subbituminous coals—19 of which are from the Powder River Basin and one of which is from the Green River Basin— and five will use lignite coal from North Dakota and Montana, three will use lignite from Texas, and two will use coal from New Mexico and Arizona (Hong & Slatick, 1994; Stricker & Flores, 2009). Sixteen of the coal-fired power plants are planned for construction in Montana, North Dakota, and Wyoming, 11 of which will consume Powder River Basin coal at an estimated 40.3 million mt/yr (based on calculations of Hong and Slatick (1994)).

Carbon dioxide emissions from solid fuels in the United States, including coal, pet coke, metallurgical coal, and wood, are approximately 2.2 billion mt/yr as of 2007 (Marland, Boden, & Andres, 2008; USEIA–DOE, 2008); most is from the combustion of coal in power plants. The historical pattern of CO_2 emissions, from 1900 to 2000, mimics the patterns of coal production of bituminous coal in the Appalachian and Illinois Basins (Stricker & Flores, 2009). During this period, "peak" emissions (as much as 1.5 billion mt) occurred during World Wars I and II, and periods of lesser emissions (as low as 0.8 million mt) occurred during the depression of the 1930s and other economic downturns. During much of this 100-year span, bituminous coals utilized by coal-fired power plants and mines in the Appalachian supplied other industrial users and Illinois coal basins, and the emission volumes from 1965 to the present reflect the increased production and use of subbituminous coal from the Powder River Basin. The overall rise in CO_2 emissions from 1955 to 2007 in the United States can be interpreted to reflect the increasing use of coal to generate electricity by coal-fired power plants. During the last 50 years or so, the emissions increased from 0.9 to 2.4 billion mt/yr, and from 1970 to 2007 Powder River Basin coals have contributed significantly to this increase because of their sheer volume of use (Stricker & Flores, 2009).

Finally, the proposed deployment of new power plants according to estimates of the Energy Information Administration (USEIA, 2010b) will be natural gas-fired electricity generation. USEIA (2010b) indicated that 20% of the 97,000 MJ/s of new capacity of electricity to be

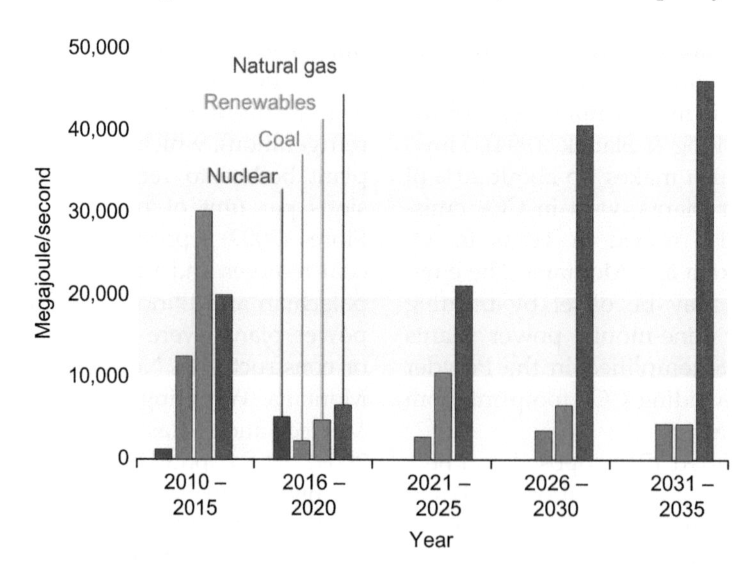

FIGURE 1.19 A graph showing the capacity of electricity generation by fuel type with emphasis on natural gas from 2010 to 2035 in the United States. *Source: Modified from USEIA (2010b).*

BOX 1.4

CARBON DIOXIDE: FROM POWER PLANTS TO VEHICLES—FACT OR FICTION

The carbon dioxide (CO_2) emitted from coal-fired power plants and passenger or personal vehicles is elusive to pin down. However, it is a fact that coal use from combustion in coal-fired power plants is one of many sources of CO_2 emission (e.g. passenger vehicles defined as two-axle four-tire vehicle, gasoline, crude oil, propane, railcars, and homes) (USEPA, 2011; USDOE, 2010, 2011). USEPA (2011) calculated emission of 4,023,304 mt CO_2/power plant for 1 year compared to 5.1 mt CO_2 vehicle per year in the United States. Based on a total of 464 coal-fired power plants that generated at least 95% of electricity, these plants emitted 1,866,813,072 mt in 2007 (USEPA, 2011). Of the 700 million vehicles worldwide, 30% is driven in the United States but personal vehicles account for 314 million mt in 2004, which according to DeCicco and Fung (2006) indicate "that much carbon could fill a coal train 88,514 km long—long enough to circle the Earth twice." DeCicco and Fung (2006) argued that the disproportionate amount of CO_2 emission from personal vehicles in the United States account for half of all the greenhouse gases emitted by vehicles worldwide.

added in the United States would be fueled by natural gas. In addition, USEIA (2010b) projected that 80% of new capacity of electricity in the United States by 2035 will be fueled by natural gas. The following Figure 1.19 shows the future rate of addition of natural gas for electricity capacity from about 7000 MJ/s in 2016–2020 to about 47,000 MJ/s in 2031–2035 or 40,000 MJ/s in the later 19 years. During the same time period USEIA (2010b) projected coal to add of the new capacity of electricity from about 2000 MJ/s in 2016–2020 to about 4000 MJ/s in 2031–2035.

Thus, environmental and technological advances in electricity-generation power plants will place natural gas like coalbed gas as an increasingly significant option with coal for fueling the future.

SUMMARY

This introductory chapter discussed the scope of the book and the importance of definitions, metaphors, and basic principles, which originated from the rich tradition of coal geology, redefined, amalgamated, and adopted by the youthful but maturing discipline of coalbed gas. Central to this coal tradition, which is directly related to understanding coalbed gas, is the fundamental of coal endowment, geologic occurrence, geographic distribution, and the role of the dynamic outer shell of the earth as driven by continental drift and tectonic plate movements on the evolution of coal accumulations in supercontinents and microcontinents. To the extent that the coal endowment is unequally distributed worldwide and the coal resources and reserves are regional in accumulation, their classification system and estimation are as diverse as there are many countries that practice reporting them. In addition, the classification system of coal resources being derived from mineral resources and reserves exacerbates confusion of these concepts. However, much like mineral commodities the concepts of coal resources and reserves are determined by geological assurance and engineering data with reasonable certainty of future extraction under existing economic and mining parameters, and marketability.

The interplay of coal supply, demand, and consumption yield a dynamic debate about coal utilization mainly for electric power generation and ramifications to the environment in a carbon-constrained world concerned about emissions of greenhouse gases. In addition, the vigorous discussion and controversy about the concept of peak coal production, which is modeled from the M. King Hubbert peak oil production theory, and its influence on consumption and potential reduction of CO_2 emissions from coal-fired power plants play an important role in the future of coal use. More importantly it also affects the energy mix such as the unconventional gas portfolio, which includes the emerging coalbed gas, for future power generation. In an effort to prolong the use of coal, proposed new worldwide power plants include more efficient electricity-generating technologies (e.g. supercritical, ultra-supercritical power plants) that can include oxygen-combustion and postcombustion CO_2 capture (e.g. chemical solvents such as amine, flue-gas refrigeration) to be used in new power-plant boilers and sequestered to reduce greenhouse gas emissions per unit of power output. In addition, despite challenges for CO_2 capture and sequestration, the technology provides a viable option to reduce footprint of CO_2 power plant emissions. These technologies may potentially alleviate using abundant low-rank coal for combustion, which have lower CO_2 emission factors (3–5%) than bituminous coal, which is gradually being depleted especially deposits east of the Mississippi River. The trend of deploying coal-fired power plants near coal mines or mine-mouth plants also may reduce the CO_2 footprint from inefficient transportation of coal. In addition, the Energy Information Administration projected that 80% of new capacity of electricity in the United States will be fueled by natural gas and 5% by coal by 2035. Thus, the U.S. Energy Information Administration projection suggests that both coal and natural gas such as coalbed gas will continue to fuel the future.

CHAPTER

2

Coal as Multiple Sources of Energy

OUTLINE

Coal and Coalbed Gas

http://dx.doi.org/10.1016/B978-0-12-396972-9.00002-1

KEY ITEMS

- Coal as multiple sources of energy is mined as a solid fuel exploited for gas-, oil- and condensate-derived hydrocarbons, and converted for its liquid byproducts.
- Worldwide, coal is estimated to be a significant natural gas (coalbed gas) resource with gas-in-place ranging as low as 67 Tcm to as high as 252 Tcm depending on the global total coal resources.
- Coalbed gas is an unconventional gas (e.g. shale gas and tight sand gas), which accounts for about 8% of the total natural gas production of the United States.
- Recoverable coalbed gas resources lie mainly in Russia, United States, China, Australia, Canada, and Indonesia, which total about 24 Tcm with Russia comprising 24% and Indonesia comprising 6%.
- Coalbed gas has evolved from hazardous explosive gas in coal mines to an environmental challenge as methane and carbon dioxide emissions from coal use.
- Coal is mined in more than 50 countries with 40% of coal production from surface mines and 60% from underground mines. China is the largest producer, which is 3.2 billion metric tons as of 2010, compared to the United States the closest producer with less than 1.1 billion metric tons in the same year.
- Large-scale global coal production from underground mining led to deadly coal mine outbursts caused by violent ejection of coalbed gas, coal, and rock during mining operations.
- More than 30,000 coal mine explosions occurred worldwide through 1996 with more than half in China. As of 2009, there are 14,673 coal mines in China of which 3412 or 23% have high-gas content and are outburst prone.
- Fugitive coalbed gas emissions of coalmine methane (CMM) from surface and underground mines and abandoned mine methane (AMM) are commonly released to the atmosphere where CMM and AMM are not exploited for industrial and commercial uses.
- Coal-derived hydrocarbons are explored and developed worldwide as conventional energy. Coal-derived oil and condensate are exploited in Carboniferous, Permian, Jurassic, Cretaceous, and Tertiary coal-bearing rocks.
- Coalbed gas is expelled, trapped, and sealed in adjoining sandstone reservoirs of coal-bearing rocks. Gas-charged sandstones, often neglected coalbed gas prospects, are recognized as conventional gas reservoirs.

OVERVIEW OF RESOURCES

The continuously growing need for energy resources of the world will come from a mix of sustainable and secure supply of fossil and renewable energy (USEIA, 2011). However, there is a conundrum of what energy mix can be added to the dwindling fossil fuel reserves such as oil, gas, and coal. This challenge is exacerbated by growing concerns about climate change from emissions of greenhouse gases with the use of fossil fuels; hence, natural gas has gradually replaced coal to generate electric power during the past decade. However, a new

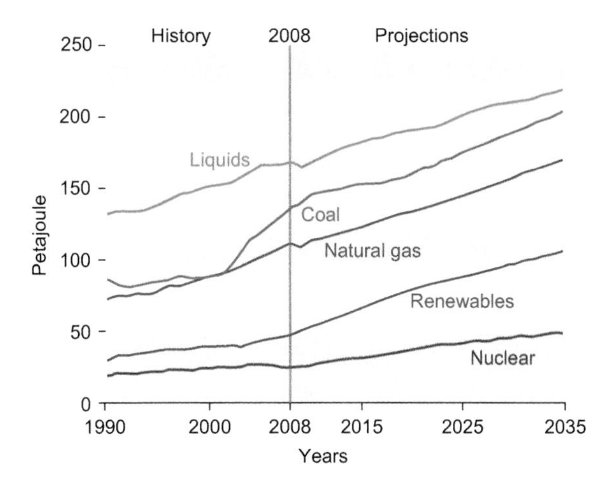

FIGURE 2.1 A graph showing worldwide consumption from 1990 to 2035 of energy mix consisting of nuclear, renewable, natural gas, coal, and liquids (oil and condensates). *Source: Adopted from USEIA (2011).*

and emerging breed of efficient coal-burning power plants can generate electricity and in the process capture carbon dioxide and separate other gases. Thus, combined diverse fossil and renewable energy should be in the mix in order to maintain and secure reliable and affordable energy supply. Despite this scenario, USEIA (2011) reported that coal and natural gas (e.g. coalbed gas) would still supply about two-thirds of world's electricity in 2035. Figure 2.1 shows the worldwide energy consumption through 2035 with coal and gas as major energy suppliers through 2035 (USEIA, 2011). Thus, coal and coalbed gas will continue to be an important part of the energy mix with both fueling the near future and probably beyond with potential renewability and sustainability of biogenic coalbed gas.

BOX 2.1

IS NATURAL GAS CLEANER FUEL THAN COAL FOR ELECTRICITY GENERATION?—FACT VS FICTION

Worldwide and in the United States, more than 40% of the electricity is generated by coal-fired power plants and in the United States about 20% by natural gas. Natural gas proponents suggest that the use of the fuel for electricity generation should be increased because it produces lesser greenhouse gases than coal (e.g. generates less carbon dioxide per energy unit released) and supply is more readily available with the advent of unconventional gas production of coalbed gas, shale gas, and tight gas sands (e.g. 60% of production in 2009 and in 2035 it has been projected that there will be about three times more production than conventional gas) (USEIA, 2011; MIT, 2011). Although these arguments are mainly accurate, the fact according to a USEPA (2007) study is that when gas leakage is considered during extraction, treatment, transportation, and utilization natural gas may not be any cleaner than coal. Scientific American (2011) indicated that based on estimates of the World Bank, gas emissions during production accounts for more than a fifth of the atmosphere's total load of methane in the atmosphere. Purification of natural gas to concentrated methane requires removal of ethane, butane, and pentane byproducts. In addition, USEPA (2007) reported that deliberate gas venting for safe operations and leakage from gas-line infrastructure release gas equivalent to that from 35 million vehicles annually. ProPublica (2011) reported that about 800 of the 1600 gas-fired power plants in the United States operate at low efficiency only improving by about 25% the greenhouse gas emission of a typical coal-fired power plant. The average air emissions from coal-fired generation in comparison to natural gas produce half as much carbon dioxide, less than a third as much nitrogen oxides, and one percent as much sulfur oxides (USEPA, 2007). Thus, natural gas as a cleaner fuel may be attained when the fuel is combusted by more efficient power plants.

COAL AS RESOURCE OF COALBED GAS

Coals have long been recognized as the source of significant volume of natural gas ever since the first documented coal mine explosion caused by methane in the early 1800s (Flores, 1998). However, gas from coals has not become a resource until about the middle of the twentieth century. Before then the first gas-drainage technology to isolate gas from coal mines to prevent explosion was put into practice in Great Britain in about mid-1800 (Bromilow & Jones, 1955). Ventilation via shafts sunk through coal beds by cross-measures has been the early method of draining gas from coal mines until it was replaced by vertical and horizontal boreholes, which were drilled in advance of coal mining in the early 1900s (Darton, 1915). Consequently, gas was liberated from coals to the atmosphere until later it was successfully recovered from vertical wells in front of mining thus becoming a natural gas resource and viable commercial commodity (Dunn, 1995). This was made possible in the United States when the government passed the "Crude Oil Windfall Profit Tax of 1980" (Soot, 1988). This federal tax credit was intended for production of unconventional energy such as: (1) gas, liquid, and solid synthetic fuels from coal, (2) oil shale and tar sands, (3) gas from Devonian shale (shale gas) and geo-pressured brines, (4) gas from biomass, (5) processed wood fuels, and (6) steam from agriculture byproducts. In order to qualify for this federal tax incentive the wells must have been drilled by 1990 and on production by 2001, which qualified for tax credit for $1.34 per million British Thermal Unit (or Joules). The unconventional energy was redefined as unconventional gases (Schmoker and Dyman, 1996; Schmoker, 2002) to include coalbed gas, shale gas, tight gas sands, and gas hydrates with the first three gases increasing natural gas production in the United States from 30% (Holditch, Perry, & Lee, 2007) to 48% (Vidas & Hugman, 2008). The contribution of coalbed gas to the total natural gas production in the United States is about 8% depending on the price of gas.

Coal is the most abundant fossil fuel in the world; thus, the associated gas found in coal or coalbed gas is an increasingly important part of the energy mix of dependable and adequate global supply of natural gas. The evolution, coexistence, and accumulation of coal and coalbed gas coal is an extraordinary occurrence compared to conventional hydrocarbon accumulation. Traditionally, most conventional oil and gas are sourced from organic-rich mudrock (e.g. shale, mudstone) or limestone and migrated into reservoir rocks (e.g. sandstone), in turn trapped by a seal rock (e.g. shale or other impermeable rock). Conventional reservoir rocks such as sandstones hold oil and gas mainly in intergranular pore spaces but coal stores gas in nanometer-scale interstitial (e.g. space between grains) molecular pore structures and in fractures in which small molecules of methane, carbon dioxide, and other gases (e.g. nitrogen) can exist (Ajdukiewicz & Lander, 2010; Levine, 1993). More importantly, the organic-dominated coal beds generated, stored, and trapped coalbed gas during the burial and coalification processes. However, unlike in conventional reservoirs where oil and gas are trapped by overlying seal rock, coalbed gas is mainly held or adsorbed and "trapped" in the coal reservoir by weak Van der Waals force or bond with moderate hydrostatic pressure and contained by seal rock (Flores, 2008; Levine, 1993). The Van der Waals forces are attractions between molecules, surfaces, and other intermolecular forces, which permit sliding and rupture to occur (Parsegian, 2006). The methane molecule (CH_4) and other gas molecules (e.g. CO_2) are held in place on coal surfaces (e.g. fractures and pore systems) by both weak Van der Waals bond and hydrostatic or water pressure, which have permanent dipole molecules. These chemical, physical, and hydrological properties of coal uniquely serve as the source and reservoir of coalbed gas. The capacity of the coal to generate gas is dependent on its organic matter composition, chemical and physical properties, and the degree of coalification, thus, understanding of these properties in relationship to other geological factors is important to successful exploration and development of coalbed gas. Because of variable properties and levels of coalification, not all coal beds contain significant amounts of coalbed gas resources (note discussed in the next section).

Coal Resources vs Size of Gas Resources

Central to the understanding of coalbed gas resources is the vast and widely variable volume of coal resources in the ground, which was estimated from 8757 to 27,210 billion metric tons (Averitt, 1981; BGR, 2010; BP, 2011; Landis & Weaver, 1993; Matveev, 1976; World Energy Conference, 1978). Unfortunately, these coal resource estimates have not been assessed and updated to modern classification standards resulting in large hypothetical resources based on unreliable geologic assurance (Averitt, 1981). Thus, the direct translation of coal resources to coalbed gas resources (e.g. gas-in-place) is complicated by outdated information and wide discrepancy of reporting coal resources. This relationship is also compounded by the variable physical properties, coal rank or maturation, depth, and geological characteristics of coal. Consequently, under these conditions coalbed gas resources are very difficult to determine and present estimations are unreliable but provided here as ideal and imperfect.

Coalbed Gas-in-Place Resources

Coalbed gas resource is generally expressed as the total volume of gas-in-place (GIP) resources regardless of recoverability and irrespective of economical and technical constraints. The GIP is a volume of gas held or stored within a particular amount of a reservoir rock such as coal and interbedded partings (e.g. carbonaceous shale etc.). The GIP is measured by using estimates of the total coal resources (derived from the equation in Chapter 1) and in situ gas content measured from coal desorption. It is calculated by the following equation:

$$\text{Gas-in-place (GIP)} = \text{Coal resources} \times \text{Gas (in situ) content}$$

Thus, calculation of GIP relies on the: (1) area covered by the coal, (2) coal thickness including partings, (3) average coal density, and (4) in situ gas content, which is normalized with ash yield and moisture content of the coal. The first three sets of data are determined from coal mining and/or exploration wells and the last set of data is measured directly from gas desorption of fresh coal core samples, although at times measured from coal cuttings but widely variable and inaccurate. Usually coalbed gas investigation often occurs in areas of existing coal, oil, and gas exploration and development such that coal resources data can be determined although gas desorption data are not available. Thus, this method of GIP calculation is not applicable to frontier areas without drillhole coal cores for gas desorption data. This limitation may be overcome in frontier coalfields by using viable analogs based on similarities in coal rank, coal thickness, thermal maturity, burial history, and overall geologic setting from different coal basins.

The global coalbed GIP estimates by Holditch et al. (2007), which vary from 67 to 241 Tcm, are less than the estimate by Kuuskraa, Boyer, and Kelafant (1992) and Rogner (1997), which is 252 Tcm. The vast size of the global coalbed gas resource is based on using parameters of U.S. coal-reservoir paradigms, and the wide range of the size is attributed to the differences in coal characteristics and geological factors between the U.S. and other countries. The size variability of the global coalbed GIP as reported by Holditch et al. (2007) is shown in Figure 2.2, which highlights the former Soviet Union with

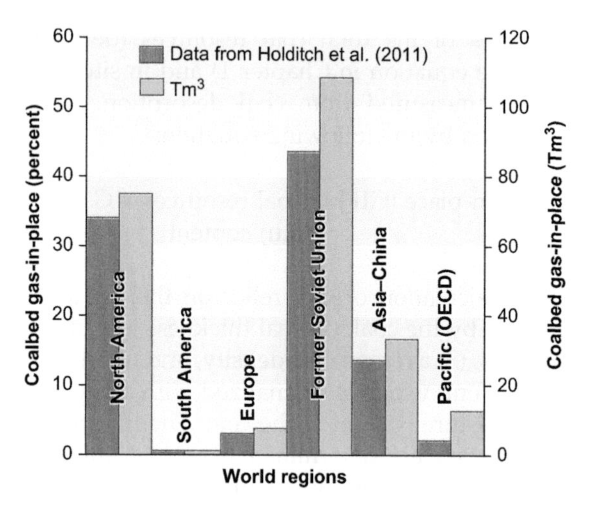

FIGURE 2.2 Variability of the coalbed GIP in percent and volume of various regions in the world. Tm^3, trillion cubic meter. *Source: Data from Holditch et al. (2007).*

the most coalbed GIP of 111 Tcm or about 44% of the total resources compared to the least amount for South America with more than 1 Tcm or about 0.43% of the total resources. Forty seven percent of the total global coalbed GIP is found in North America (e.g. United States and Canada), which is more than 75 Tcm or 33%, and in Asia–China (e.g. China and India), which is more than 35 Tcm or about 14%. The variability of coalbed GIP between these countries is controlled by differences in coal resources, in situ gas content, adsorption isotherms, and coal reservoir properties. In addition, to understand the variability of the volume of coalbed gas one must consider parameters such as gas content and adsorption capacity of the coals, which play a major role in the resource calculation. For example, the gas content and adsorbed or stored gas vary with coal rank, within and between coal beds, stratigraphic positions, depths from the surface, and geologic age (Esterle, Williams, Sliwa, & Malone, 2006; Stricker et al., 2006; Warwick, 2005). Esterle et al. (2006) reported that given rank (e.g. high to low bituminous coals) gas content greatly vary and increases at depth (Figure 2.3) in the Bowen Basin and Hunter

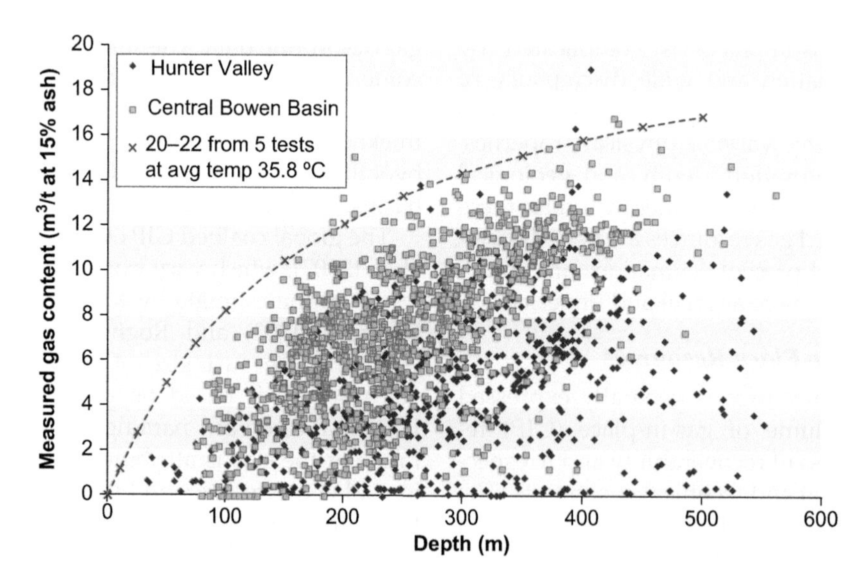

FIGURE 2.3 A diagram showing variability of coalbed gas content related to coal rank, and depth of coal beds in the Bowen Basin and Hunter Valley coalfields in Australia. m^3/t, cubic meter per ton; m, meter; avg, average. *Source: Adopted from Esterle et al. (2006).*

Valley in Australia. In addition, Esterle et al. (2006) concluded that gas content variability is very high at the borehole, mine-site, and basin scales. Stricker et al. (2006) reported that the coal beds vary in average gas content within and between coal beds and between depths of coal beds in the Powder River Basin, Wyoming. The adsorption isotherms of the same coal bed/zone (e.g. Wyodak) vary between subbituminous C coal in the shallow part of the basin and the subbituminous A coal in deepest part of the basin. In comparison, lignite A coal in the Williston Basin, North Dakota, United States displays lower adsorption isotherms than the subbituminous coals in the Powder River Basin. In the Gulf Coast Basin, United States, Cretaceous subbituminous A coal in south Texas contains higher gas content than Paleocene–Eocene lignite–subbituminous coals coal beds in Louisiana (Warwick, 2005).

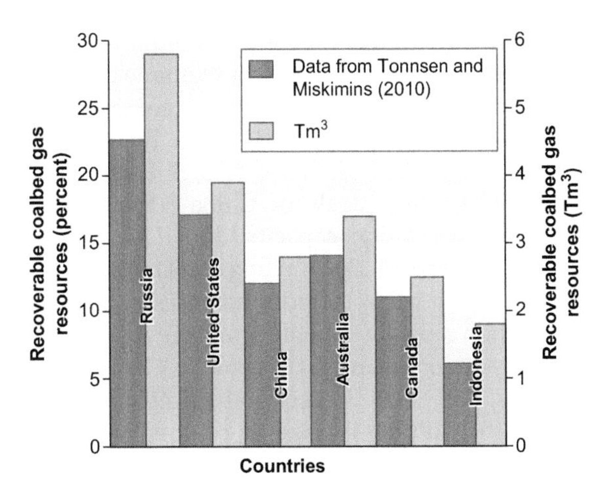

FIGURE 2.4 A frequency diagram of recoverable coalbed gas resources in percent and volume. Tm^3, trillion cubic meter. *Source: Adopted Data from Tonnsen and Miskimins (2010).*

Technologically Recoverable Coalbed Gas Resources

The size of coalbed gas is also reported as recoverable resources (or reserves), which is defined as the amount of gas that is technologically and economically feasible to develop. Tonnsen and Miskimins (2010) reported that the largest coalbed gas recoverable resources occur in Russia, United States, China, Australia (includes New Zealand), Canada, and Indonesia (Figure 2.4). Together these six countries have total recoverable coalbed gas resources of about 24 Tcm of which Russia has the most with 5.8 Tcm (24% of total) and Indonesia with the least of 1.8 Tcm (6% of total). According to Tonnsen and Miskimins (2010) the United States has 3.9 Tcm (17% of total), Australia has 3.4 Tcm (14% of total), and China has 2.8 Tcm (12% of total) of recoverable coalbed gas resources. The recoverable coalbed gas estimates of the United States are lower than reported by the assessments of the Potential Gas Committee for year end 2010 (Curtis, 2011), which is 4.4 Tcm as a "baseline estimate" of what is technically recoverable gas resource potential of the United States. Much of the world's coalbed gas recoverable resources remain unrecovered with about 1% only recovered as of 2007. This is due to unstable gas prices in the United States and the lack of incentive in some countries to fully exploit coalbed gas such as in Russia where conventional natural gas is abundant.

Variability of the size estimates of recoverable coalbed gas is controlled primarily by gas storage capacity and secondarily by the gas content and other coal reservoir properties (e.g. porosity, permeability, pressure or hydrology). Gas storage capacity, which is determined from adsorption isotherms, varies greatly within the same coal rank, between coal beds, depths, within and between coalfields and basins, and much more with extrapolation within and between countries (Flores et al., 2006; Saghafi & Hadiyanto, 2000; Wang, 2011). For example, coalbed methane adsorption isotherms indicate that Philippine coals (e.g. lignite to semianthracite) hold more than U.S. coals in both as-received and dry-ash-free basis (Flores & Stricker, 2010). Also,

Paleogene low rank coals (e.g. subbituminous C and B) of Indonesia have higher methane adsorption capacity compared to the same coal rank and age in the Powder River Basin in the United States (Saghafi & Hadiyanto, 2000; Stricker et al., 2006). The lignite coals of China have higher methane adsorption capacity than U.S. lignite coals (Stricker et al., 2006; Wang, 2011). Additionally, Permian coals in India have less methane adsorption capacity than the same-age Australian coals (Laxminarayana & Crosdale, 2002).

Thus, the use of recoverable resources to report the volume or size of coalbed gas is a less inflated estimate than GIP resources, which presume that all coals equally contain and adsorb gas. Also, it presents a wrong impression about the size of coalbed gas because the GIP estimate is normally much higher than what is technologically and economically recoverable for a coalfield, coal basin, and more importantly for a country. For example, the U.S. Geological Survey (USGS) estimated total coalbed GIP resource for the conterminous United States is about 20 Tcm; however, almost 3 Tcm of the gas or about 14% of the total GIP may be recoverable resources with the use of currently available technology and industry experience in spite of economic and accessibility reasons. That is, the technically recoverable resources account for ultimately technologies such as exploration and production know-how (e.g. drilling, geophysical) and engineering procedures will improve the growth of the recoverable resources.

COAL RESOURCES VS GAS PRODUCTION POTENTIAL

Coalbed gas production, which evolved from coal mine hazard to environmental concern, through past, present, and future contributions is modest compared to the total United States production of unconventional gas (Figure 2.5). In 2009, coalbed gas production was less than one-third of tight gas and less than one-half of shale gas; however, coalbed gas and tight gas will remain almost constant in production trends through 2035 while shale gas production is projected to grow two to five times (USEIA, 2011). The ratios of production for coalbed gas vs tight gas sands and shale gas are impressive considering that according to Kuuskraa and Van Leeuwen (2011) based on the assessment of 100 shale gas, tight gas sands and coalbed gas basins in the United States that the technically recoverable resources are 23 Tcm for shale gas, 17 Tcm for tight gas sands, and 3.5 Tcm for coalbed gas or about five to six times more resources. This may be attributable to favorable economic development of coalbed gas reservoirs due to less expensive completion at shallow depth. Coalbed gas presently provides over 0.45 Tcm of gas production per year in the United States but is underdeveloped worldwide. The limited global development of coalbed gas is partly due to sparse available information, shortage of needed technology transfer and expertise, unfavorable governmental policies, and unstable market circumstances for exploitation in many countries.

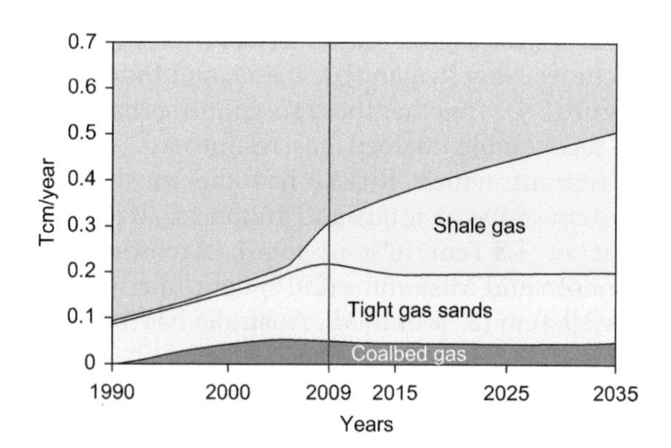

FIGURE 2.5 A chart showing the United States unconventional gas production, which includes coal bed, tight sand, and shale gases, from 1990 projected to 2035. Tcm, trillion cubic meter. *Source: Data modified from USEIA (2011).*

Chronological Developments of Coalbed Gas

Coal mining and the concern about mine safety in relation to coal mine explosions and derivative incidents originally spurred on coalbed gas development. For the past 100 years, efforts on improving technology and safety of coal mines worldwide focused on controlling methane from causing mine explosions. The early goal was to control methane accumulations in coal mines by a mixture with air and ventilating to the surface. The gas drainage system method led to collection of the gas at the surface from active underground mines and in the advance of mining, which in turn led to industrial and commercial uses of coalbed gas. The technology used toward drainage and collection of CMM led to development of gas from virgin coal beds with the original coalbed gas wells. Presently, coalbed gas is emitted from coal mines to the atmosphere as a significant part of the greenhouse gases contributing to potential climate change.

Anecdotal vs Recorded History of Coalbed Gas

Although coalbed gas recovery and collection continue to be utilized for industrial and commercial purposes in the United States, the first effort to isolate and pipe coalbed gas to prevent a coal mine explosion was in the United Kingdom in 1733 (Bromilow & Jones, 1955). The gas that caused the explosion was from a gob or accumulation of fractured rocks and coal that collapsed from roof and refuse rocks from coal mining operations. Gas is desorbed and accumulated in the pile of rocks of the gob and piped the gas through ventilation to the surface prevented gas concentration, which when ignited will eventually explode. Subsequently, the gas recovery from gobs led to production from underground coal mines through surface pipelines, which were extensively used throughout Europe in the 1940s. However, it was not until 1943 that the first successful vertical borehole was drilled to extract gas from a virgin coal bed in the Mansfield Colliery in Ruhr, Germany (Venter & Stassen, 1953).

In the United States, vertical boreholes to drain coalbed gas were drilled in front of coal mining during the early 1900s in the central Appalachian Basin (Darton, 1915; Lawall & Morris, 1934; Ranney, 1941). However, it was in 1952 that dual completions in two separate beds with vertical boreholes were drilled to produce gas from the Pennsylvanian-age Sewickley and Pittsburgh coals at 113 m and 140 m depths, respectively, at a coal mine in the central Appalachian Basin (Diamond, 1993; Spindler & Poundstone, 1960). Coalbed gas was measured for potential production from these coal beds with the Sewickley coal yielding no measurable gas and the Pittsburgh coal producing up to $1.1 \times 103\ m^3$/day. The coalbed gas borehole in the Pittsburgh coal pioneered in the use of precursor gas recovery technology, which is now improved by the industry, such as "downhole water pump and vacuum pump" to draw gas to the surface and "shoot the well with nitroglycerine" to stimulate and increase production (Diamond, 1993). The commercial value of coalbed gas produced from U.S. coal mines in the central Appalachian Basins was first recognized by Lawall and Morris (1934). Both workers observed that coalbed gas from two mines in West Virginia released $0.37 \times 106\ m^3$/day and they estimated the market value of $1300/day based on the price of gas at that time. Although it was estimated by 1940 that U.S. coal mines released $14.2 \times 106\ m^3$/day of coalbed gas causing the death of 275 miners from explosions, no mine-safety drainage system was planned to recover gas. It was thought at the time that coalbed gas recovery and production for commercial use was either "visionary or crazy" (Diamond, 1993). However, economic production of coalbed gas was paved by the adoption of the drilling and completion technologies of the petroleum industry (Price & Headlee, 1943). For example, Halliburton fractured the first coalbed gas well in 1954 to improve degasification in cooperation with the United States Bureau of Mines (USBM). Thus, commercial coalbed gas production evolved from a simple coal-mining technology to improve mine safety to the present production of unmineable coals with vertical to horizontal drilling assisted by specialized stimulations and multibed completions.

Success of Coalbed Gas Forged with Governments

The investigation of the feasibility of coalbed gas as a technically and economically recoverable resource was forged in the United States and exported to other countries. In the United States, the catalyst for this investigation was the government in cooperation with the industry, state, and federal government agencies (e.g. USBM, Department of Energy). The significant role of the United States government through funding incentives enhanced applied research toward the commercial production of coalbed gas. Foremost of this cooperation was the Department of Energy "Coalbed Methane Recovery Projects", which financed researches in 1973 in the San Juan, Illinois, and Black Warrior Basins with state, university, and industry scientists and engineers to desorb coalbed gas and characterized coal reservoirs. The USBM funded coalbed gas projects in 1978 to improve degasification procedures during and preceding mining as well as studied fracturing coals in Pennsylvania, Virginia, West Virginia, Ohio, and Illinois mines. The Department of Energy and Gas Research Institute embarked on joint projects in the Black Warrior Basin, Alabama and studied response of coal beds to fracturing, reservoir engineering, completion improvements, and evaluated methane commercial possibilities. In 1980, Federal tax credit was established as incentive for coalbed methane exploration and development.

The successes on coalbed gas and CMM recovery and production in the United States continued into major coal-producing countries during the middle 1970s to 1980s. In Australia, the first commercial CMM recovery operation started with the 1990s underground mines in central Queensland methane drainage project, which led to CMM being used to fuel on-site generator sets of gas-fired power stations. In China, only 130 coal mines have CMM drainage and recovery systems in place but as of 1994 this number had more than doubled by 2007 (IEA, 2009). About half of the drained gas is used mainly for generating electricity. Russia is the third largest source of CMM but the obstacle to CMM drainage and recovery is the low number of mines employing the system. Although Russia has a long history of mine degasification and despite significant opportunities for development, there are relatively few CMM utilization projects in Russia. Globally, Ukraine ranks ninth in coal production, but it is the fourth largest emitter of CMM. There are many gassy mines in Ukraine, and mine explosions remain a critical safety issue. Nationally, Ukrainian coal mines recover about 12% of all CMM liberated but only utilize about 30% of what is captured.

COAL MINING DEVELOPMENT AND GAS OUTBURSTS

Coal is mined in more than 50 countries and worldwide about 40% of coal production is from surface mines and 60% from underground mines. Coal mines of the major coal-producing countries (e.g. Australia, Canada, China, India, Russia, South Africa, and United States) have suffered either explosions or outbursts. The occurrence of explosive gases in coal mines have been known ever since the first mine explosion in 1733 in Great Britain and 1810 in the United States (Bromilow & Jones, 1955; Flores, 1998). An early attempt to understand the explosive gases in coals was by Darton (1915) who reported, "all coals contain in its pores and crevices some inflammable gas". Darton (1915) observed that gas is "given off when the coal is in the ground during mining and long after the coal has been taken from the mine". All reported modern coal mine outbursts or explosions occurred in underground mines. Presently, underground coal mines account for majority of the global coal production with China leading the major coal-producing countries. However, other major coal-producing countries like Australia account for 80% of production from surface coal mining and the United States, which accounts for 67% of production (World Coal Association, 2011).

Underground Coal-Mining Development

Underground coal mine development began with extraction from shallow entries of horizontal or nearly horizontal tunnels and/or adits dug into coal beds from the surface. Early on, underground coal mines were shallow because of difficulties of hauling coal from the mine face to the surface, roof collapse, and running the risks of potential coal mine outbursts. However, during small-scale coal production in the nineteenth century in Europe and the United States, the common technique of coal development in underground mines was to access from the surface through rocks via roof-controlled, ventilated ramps or declines, and vertical shafts to extract deep coal beds. By the twentieth century this technique was used worldwide for large-scale coal production (Figure 2.6). Mined coal was moved from the mine workings through along graded ramps along mine cars or on belt conveyors. In the early to middle 1900s, underground mine development accounted for about half of all the production of hard coal (e.g. bituminous, anthracite) mined worldwide. By late 1900s the hard coal production ranged from 30% to 95% worldwide with China being the largest producer.

Methodologies of extracting coal in underground mines are discussed in many literature and the reader is directed to these references for detailed discussions (Cantrill, 2012; Peng, 2006; Singh, 2004; USEIA, 1995). There are two principal techniques to extract coal in underground mines: (1) room-and-pillar and (2) longwall mining (Peng, 2006; World Coal Association, 2011). The room-and-pillar mining is named after cutting a network of rooms into the coal bed and leaving behind large pillars of coal spaced at regular intervals or grids during mining to provide support to the ceiling or mine roof. The coal pillars can be up to 40% of the total coal bed; thus, coal production using this technique is not efficient with extraction

FIGURE 2.6 Photographs of underground coal mines from an inclined mine shaft in Colombia (A) to horizontal mine in the Philippines (B). Both are high-gas mines with the Philippines mine experienced an outburst months before the photograph was taken by the author.

only as much as 60% although these coal pillars may be extracted during retreat mining and increasing production to as much as 90%. Room-and-pillar coal mining in the United States is used in 1 m thick coal beds and in South Africa in as much as 6 m thick coal beds in which coal production exceeds 9 million metric tons per year.

In comparison, longwall mining involves the full extraction of a section or face, 100–350 m long, of the coal bed using mechanical, electrically powered shearers equipped with two coal cutting drums on either side of the panel. This methodology achieves a total extraction of large sections or panels of a coal bed, causing the roof rocks to collapse into the mined-out area and creating gob areas where gas may accumulate through time. The mine roof is propped up by self-advancing, hydraulically powered supports, which temporarily hold up the ceiling while the coal is extracted, which is allowed to collapse after coal extraction is completed in the area. More than 75% of the coal bed can be extracted by the longwall panels, which can be extended to 3 km along the coal bed. Longwall panels and mining face have been substantially increased over the past 40 years. In the United States the average lengths of the longwall face increased from 150 to 227 m. In Germany the average length of the longwall face was 270 m with plans of increasing to more than 300 m. In the United Kingdom and Poland, the lengths of the longwall face are as much as 300 m. As the length of longwall face increases, panel lengths, which are controlled by coal bed continuity and faults, may possibly extend to as long as 6.7 km in the United States.

Surface Coal Mining Development

Surface coal mine development began in the sixteenth century but did not became widespread until the twentieth century from 1970s to 1980s, particularly in the United States, Australia, South Africa, and India. Surface coal mines, also known as "opencast or open-pit mines", expose shallow (e.g. 50 to 100 m depths but can be deeper to 450 m as in Germany) coal beds by removing overburden soil and rock with the use of draglines, trucks, shovels, and/or bucket wheel excavators along open cuts or strips; thus, surface mines are sometimes known as "strip mines". As shown in Figure 2.7 surface coal mining can extract from one thick bed, multiple thick beds, to multiple thin to thick beds by building benches in the highwall to lessen the angle of repose and prevent mass movements. Compared to underground coal mining, surface coal mining recovers as much as 90% or more of the mineable coal bed. In the United States shallow lignite to bituminous coals are mined by area, contour, and mountaintop valley methods. Mountaintop and contour mining methods are applied in the Appalachian Basin where the topography or terrain is steep and rugged. Area mining is mainly applied in the western United States where flat terrain is common.

Area mining exposes the coal bed in successive long cuts or strips (e.g. width × length) by first removing the soil and second the overburden (e.g. rock below the soil to top of coal bed) as spoils, which are both deposited on previously mined-out area. The exposed coal bed or block is drilled and filled with explosives, which are in turn blasted to break up the coal for extraction by earth-moving equipment, loaded, and transported by conveyors and/or trucks to preparation areas. After the coal has been extracted the spoils from the subsequent cut are used to fill the previous cut. This mining process is repeated with new and adjoining cuts until an area is stripped empty of coal after which the overburden infill is shaped by earth-moving equipment to its near-original topography or terrain in preparation for reclamation. The soil is then redeposited on the overburden and reclaimed by seeding the soil with native vegetation.

Contour mining removes the soil and overburden rocks above the coal bed by following

FIGURE 2.7 Photographs of strip mines around the world showing extracting from (A) a thick (25 m) coal bed in the United States, and (B) two thick (8–10 m) coal beds in New Zealand. The mine highwalls range from 75 to 100 m in height and the beds are almost flat-lying to dipping about 20°.

the topographic contours around a hill or mountain. Because this method of mining is applied commonly to rolling and steep terrains, spoil disposal areas are very much prone to erosion and landslides. Spoils are commonly deposited on the downslope side of the mining bench; however, the spoil disposal area took up much space disturbed frequently by mass movements. The disposal space and mass wasting problems were resolved by refilling mined out areas with freshly cut overburden spoils. That is, the overburden spoil of the initial cut is set aside downslope of the mine bench or at another mine site and then the spoil from the second cut is used to refill the first cut. This is called a "haul-back" or lateral movement backfilling mined out areas. A perimeter of undisturbed material is left off the mined-out area and used to stabilize reclaimed slopes and to prevent downslope mass wasting of the spoils.

Mountaintop mining combines contour mining in areas with rolling or steep terrain by extracting a coal bed (or multiple beds) at or near the top of a mountain or hill. The top of the mountain is completely removed in a series of parallel cuts following the contour of the terrain, which resulted in a flat plateau. The spoils from overburden rocks are disposed in adjoining valleys and hollows creating "valley infills" and "head-of-hollow-fills", which cover up streams or creeks. Prior to infilling a valley, soil and vegetation are removed and a rock drain or "underdrain" is constructed in the area to be filled where a stream or creek previously existed. This drainage system conveys a water runoff from the upper to lower ends of the fill. Reclamation of the valley fills includes grading and terracing the spoil disposal sites to create stable slopes.

GAS OUTBURSTS VS COAL DEVELOPMENT

Gas outbursts or explosions related to coal development are global occurrences, which vary from one incident every few years to as many as 145 incidences in 1 year during modern times. The phenomenon of gas outbursts particularly in underground mines are centuries old and are always related to high gas emissions from coal during mining operations. In a carbon-constrained world, this phenomenon has become a challenge because it has contributed significantly to greenhouse gases. Considering that more than 30,000 coal mine outbursts have

BOX 2.2

MOUNTAINTOP MINING PRODUCES COAL BUT GENERATES CONTROVERSY—CONCEPTS AND MISCONCEPTIONS

According to the NMA (2009; 2011a), mountaintop mining extracts coal from reserves near the surface in steep and hilly terrains mainly in Central Appalachia of the United States (e.g. eastern Kentucky, southern West Virginia). The mountaintop mining contributed about 10% of the total coal production of the United States, which was 0.95 billion metric tons in 2010. Regionally, coal production from mountaintop mining makes up 40% of the coal mined in Kentucky and West Virginia. Environmental activists label mountaintop mining as the most controversial surface mining because it removes the tops of mountains and piles the mine spoils as high as 244 m along more than 1.6 km streambeds. The Impact Statement by the USEPA (2005a) estimated that valley fills cover about 1165 km of Appalachian streams buried by mine spoils from 1985 to 2001. Also, USEPA (2005a) reported that increased minerals in the water as well as less diverse and more pollutant-tolerant macroinvertebrates and fish species characterize streams in watershed areas where mountaintop mining and valley fills occur. Sulfate, calcium, chloride, magnesium, potassium, sodium, antimony, and selenium have higher concentrations below mining areas and even higher concentrations downstream of valley fills. However, USEPA (2005a) indicates that mining activity does not appear to cause any difference in aluminum, iron, manganese, beryllium, cadmium, copper, manganese, mercury, phosphorous, silver, and zinc. Stream water quality criteria were not exceeded except for selenium, which was exceeded even in unmined watersheds. A USGS study of groundwater in mountaintop mine reclaimed areas indicates that valley fills discharged groundwater is the equivalent to a spring, but residence times are reduced from the original undisturbed aquifer system (Eychaner, 2000). The study also suggests that small, low-flow streams of valleys filled during construction will act as large-capacity groundwater reservoir during drier periods.

occurred since the early nineteenth century, this phenomenon singularly has caused tens of thousands of deaths and large volumes of released methane to the atmosphere.

Historical Survey of Coal Mine Outbursts

The history, causes, prevention, control, and management of underground coal mine outbursts or explosions are best described and discussed in Flores (1998), Lama (1995), Lama and Saghafi (2002), and Taylor and Ozgen (2011). Lama (1995) noted that high-gas emissions resulting in outbursts are worldwide events and some of the underground coal mines in Australia with gas emissions during outbursts of about 100 m^3/t. Lama and Saghafi (2002) reported that in the last 150 years up to 1996 about 30,000 outbursts have been recorded with almost half of this number has occurred in China (Table 2.1). To highlight this phenomenon, China had about 145 accidents caused by gas and coal explosions in 2010, with 623 casualties. Among which, 57 are category three accidents (four to nine casualties) accidents with 299 casualties; 11 are category two accidents (10–29 casualties) with 220 casualties; and the rest is category 4 (less than three casualties) and category 1 (more than 30 casualties).

TABLE 2.1 Incidence of Outbursts Worldwide

Country	Coal–Rock Explosions	Gas Type	Number of Explosions	Largest Quantity of Ejected Materials		Depth (m)
				Coal + Rock (metric tons)	Gas (m^3)	
Australia	Coal	CH_4+CO_2	>669	1000	14,000	95
Belgium	Coal	CH_4	487	1600	34,000	250
Bulgaria	Coal	CH_4	250	350	12,000–19,000	130
Canada	Coal + rock	CH_4+CO_2	411	3500	60,000–140,000	200
China	Coal + rock	CH_4	>14,477	12,780	10,000–1,400,000	600
Colombia	Coal	CH_4	55	ND	ND	ND
Czechoslovakia	Coal + rock	CH_4+CO_2	482	4310	96,000	80
France	Coal + rock	CH_4+CO_2	>6814	330	400,000	270
Germany	Coal + rock	CH_4	359	2500	66,000	1150
Hungary	Coal	CH_4	600	1800	27,000	140
Japan	Coal	CH_4	920	5200	600,000	ND
Kazakhstan	Coal	CH_4	45	ND	ND	400
Philippines	Coal + rock	CH_4+CO_2	9	200	250	60
Poland	Coal	CH_4+CO_2	1738	5000	750,000	80
Romania	Coal	CH_4	24	500	ND	300

(Continued)

TABLE 2.1 Incidence of Outbursts Worldwide (*cont'd*)

Country	Coal–Rock Explosions	Gas Type	Number of Explosions	Largest Quantity of Ejected Materials: Coal + Rock (metric tons)	Largest Quantity of Ejected Materials: Gas (m^3)	Depth (m)
South Africa	Coal	CH_4	5	200	ND	ND
Russia	Coal + rock	CH_4	521	ND	ND	200
Taiwan	Coal	CH_4	60	ND	ND	ND
Turkey	Coal	CH_4	58	700	11,000	300
Ukraine	Coal + rock	CH_4	4694	14,500	600,000	200
United Kingdom	Coal	CH_4	>219	400	60,000	183
United States	Coal + rock	CH_4 + coal dust	479	1814*	980–4816/h*	225

** Data are based on the Monongah coal mine (Nos. 6 and 8) explosion determined to be the worst outburst in the United States in 1907. Volumes of gas determined for No. 8 (980 m^3) and No. 6 (4816 m^3) are test results repeated for several months. Weight of debris (1814 metric tons) only measures the largest concrete blocks in addition to timbers, coal, rock, rail cars, etc.*

CO_2, carbon dioxide; CH_4, methane; ND, no data; m^3, cubic meter, m, meter.

Source: Modified from Bodziony and Lama (1996) and Lama and Saghafi (2002). Data are updated from Philippines Department of Energy, China Coal Mining Research Institute in Chongqing, USMRA (2011), and World Nuclear Association (2011). The Colombian data are from 2004 to 2011 provided by Servicio Geologico Colombiano in Bogota. The data on the United States are from Bass (1907).

Coal beds, which at low-stress levels show high-gas emissions, will always experience outbursts when the strength of the coal is low or stress levels are high (Lama & Saghafi, 2002). Lama (1995) indicated that although gas drainage within coal beds has been generally a successful control to outbursts in high-methane mines, it was not as successful in high-carbon dioxide mines, which resulted in more outbursts. The occurrence of outbursts in countries including the United States related to coal and rock bursts, gas types (e.g. CH_4 and CO_2), number of occurrence of outbursts, largest quantities of materials and volume of gas related to the outbursts, and depth where outbursts occurred are shown in Table 2.1 based on data from Bodziony and Lama (1996) and modified from other sources. NIOSH (2010), formerly the United States Mining Safety and Health Agency, reported that there were 625 coal mine disasters mainly caused by outbursts in the United States. Figure 2.8 shows the relationship between number of coal mine outbursts and related incidents from 1875 to 2010 and number of deaths based on data from USMRA (2011). The number of coal mine outbursts and related incidents decreased significantly from 1926 through 1975, which is related to the widespread use of gas drainage systems and application of control-prevention management of coal mines under the United States Coal Mine Safety and Health Act of 1969.

Settings of Gas Outbursts

The impact of large-scale coal production from underground mining worldwide inevitably has led to common occurrences of mine hazards particularly gas poisoning and outbursts mainly caused by coalbed gas. An underground coal mine outburst is defined as the instantaneous violent ejection of gas, coal, and rock in the

BOX 2.3

DEADLIEST JOB IN CHINA: UNDERGROUND COAL MINING—CONCEPT AND MISCONCEPTION

According to Zhao and Jiang (2004) coal mining has become the most deadly job in China as reported by the State Administration of Work Safety (SAWS). During 2003, China produced about 35% of the world's coal but accounted for 80% of total number of deaths from underground coal mine outbursts mainly due to accidental ignition of methane or coal dust. It was estimated by SAWS that China averaged one death per week from underground coal mine outbursts from 2001 to 2004. SAWS has reported several coal mine accidents causing 200 deaths during one month in 2004. In 2003, a coal miner in China produced an average of 321 tons of coal, which is about 2% of that produced by a coal miner in the United States and 8% in South Africa. However, the death rate per 100 tons of coal is 100 times that in the United States and 30 times in South Africa. Compared to pre-2000 statistics, SAWS reported that the number of coal mine accidents decreased 8% every year and death toll also decreased by about 13%. This statistics is a misconception as hundreds of miners are reported killed by mine explosions in the post 2000. However the Chinese government has promised to improve mine safety by primarily closing thousands of small, illegal, and dangerous coal mines (Tu, 2007).

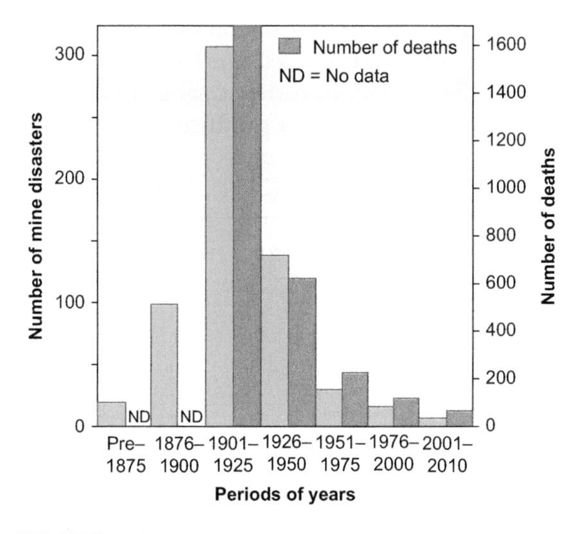

FIGURE 2.8 A frequency diagram showing the relationship of the number of United States coal mine outbursts and related incidents from 1875 to 2010 and number of deaths as first recorded in 1907 through 2010. Data from Dubaniewicz (2009) and MSHA (2011). However, USMRA (2011) reported that United States coal mine explosions, probably from methane and coal dust, have reached to a total of 10,546 from 1909 to 2010.

process of breaking into and developing a freshly exposed coal face during mining operations (Chen, Barron, & Chan, 1995; Hargraves, 1983). Additionally, outbursts commonly result from ignition of the gas that leaked and/or accumulated during mining. The violent failures are prevalent and pronounced in underground coal mines because of the development of markedly high gas emissions and pressures as well as accumulations of large quantities of methane, carbon dioxide or both, which are stored in the coal. For example, the underground coal mine outbursts in Japan and Russia expelled from 5200 to 14,500 tons of coal and associated rock fragments and up to 600,000 m^3 of methane (Deguchi, Yu, & Jiao, 1995; Lama & Bodzinoy, 1998; Lama & Saghafi, 2002) (Table 2.1). Bodziony and Lama (1996) indicated that underground coal mine outbursts occurred between 80 and 1150 m depths. However, coal mine outbursts in China occurred at shallow depths (less than 100 m) and expelling 12,780 tons of coal and rock and 3.5×10^6 m^3 of methane (Table 2.1).

The major factors that control coal mine outbursts are: (1) gaseousness of the coal in relation to the state of free gas in fractures or crevices and sorbed gas in pore spaces; (2) geology/tectonics in relation to the bed discontinuity, structural deformation, and depth of the coal; (3) properties of the coal and rock in relation to degree of coalification, volatile matter, and permeability; and (4) vertical and lateral stresses in the coal and/or related rock in relation to strengths and shear zones (Lama & Bodziony, 1998; Lama & Saghafi, 2002). For example, Lama and Saghafi (2002) suggested that coals with methane content of more than 8 m^3/ton, permeability of less than 2 millidarcies, and mechanical strength less than the lowest principal stress are prone to outbursts.

Gas Drainage Systems in Coal Mines

A byproduct of coal development from underground and surface mining is the emission of gas, which is very common in underground coal mines. In order to prevent outbursts, gas has to be desorbed before, during, and after mining with the use of drainage and ventilation systems. A drainage system to extract gas in a coal leased area before or in advanced of underground and surface mining uses vertical wells, which also can simultaneously extract gas from multiple coal beds. The vertical wells can drain and produce commercial gas as well as reduce and prevent the risk of underground coal mine outbursts. Also, the vertical drainage wells can avoid potential gas desorption from highwalls of surface coal mines, which otherwise is emitted to the atmosphere as part of the greenhouse gases. According to Schultz (2000) one underground coal mine alone in the United States employing predrainage extraction of gas will reduce emissions by more than 453,500 metric tons/year carbon dioxide equivalents or an equivalent of less than 100,000 cars off the road. If gas extraction is applied a few years prior to mining a virgin coal bed, it can potentially recover as much as 90% of the coalbed gas unless the original gas and/or hydrostatic pressures have been reduced in which case recovery is 50–60% or less. Also, gas drainage by vertical wells in front of mining results in recovery of higher quality gas with composition more than 90–100% methane, although typically composition of 95% or greater methane is ideal for injection to pipelines for the market (Schultz, 2000; USEPA, 2005a). An EPA (1999) study indicated that an underground mine with more than 11 m^3/t of gas, which is predrained by vertical wells in order to reduce ventilation quantities will save power costs of about $11 million over 20 years. In the same underground mine, an additional $3 million in ventilation power costs can be saved when horizontal wells are used in tandem with vertical wells.

A drainage system to extract gas from underground coal mines is performed in gob areas, and before or after mining operations (Figure 2.9). Gob areas of collapsed coal and rock from the strata of the mine roof can be drilled by vertical wells from the surface to produce accumulated

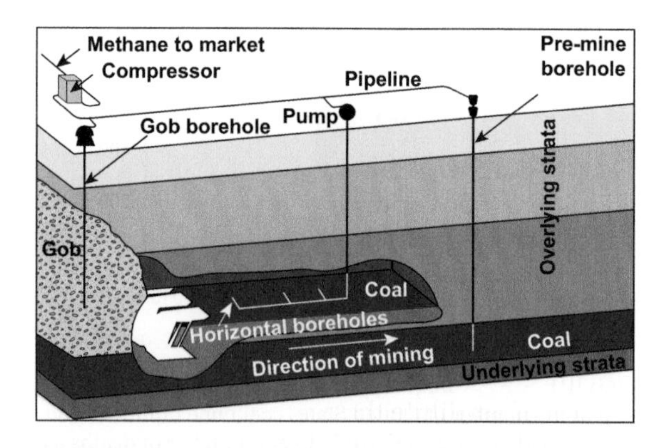

FIGURE 2.9 A diagram of an active underground coal mine showing gob, coal worked face, and vertical and horizontal wells recovering CMM in advance of and after mining. *Source: From Environmental Protection Agency (Schultz, 2000; Franklin, 2004).*

gas that migrated from nearby coal beds. In order to prevent accumulation of gob gas, vertical wells are drilled from the surface into the roof strata to about 3–15 m above the coal being mined before operations pass the mining area. Thus, the vertical wells extract the gas before roof strata collapses, avoiding the release of gas into the mine workings. The gas is desorbed from fractures and pores of the strata, which flows into the vertical wells often assisted by blowers on the surface to stimulate low-gas flow rates caused by the low-density gas natural-head pressures. The methane composition of the gob gas varies from 30% to 90% concentrations (Schultz, 2000).

Gas emitted during mining from the worked coal face is diluted and extracted by a large ventilation system designed to move vast volumes of air through the underground coal mine. The ventilation system dilutes by mixing air with the CMM (and/or carbon dioxide) to less than 1% concentration from the explosive range of 5–15% concentrations. All coal-producing countries are regulated to a maximum methane concentration of 1.0–1.25% at the coal face and within the mine workings (Thakur, 2006). The air-mixed gas is moved by the ventilation system out from the working mine areas through shafts toward the surface. The methane extracted from underground coal mines by this system is known as ventilation air methane (VAM). VAM released through ventilation shafts is either destroyed by converting methane to carbon dioxide (e.g. flare the gas) or captured and compressed for use to generate power for mine equipments or for electricity generation rather than released methane into the atmosphere as part of the greenhouse gases. VAM contains low methane concentration of less than 2% because volumes of air dilute the coalmine gas (Schultz, 2000). The VAM system requires a minimum concentration of 0.2% methane per volume to operate and it accounts for more than 60% of methane emissions from underground coal mines in the United States.

Coalbed Gas in Active Coal Mine Development

Many coal-producing countries have conducted drainage of methane from coal beds and/or surrounding rock strata in addition to ventilation in active or operating and abandoned coal mines (USEPA, 2010). Coalbed gas extraction is performed more often from active underground coal mines than from active surface coal mines. The coalbed gas extracted from active mines is known as CMM. Many abandoned or closed coal mines continue to release low to medium quality methane (30–75%) from mine openings and ventilation pipes, shafts, and boreholes. For example, Japan, which has the gassiest coal mines, has abandoned about 669 coal mines from 1960 to 1998 (Nambo, 2000). The coalbed gas extracted from these abandoned coal mines is known as abandoned mine methane (AMM). A global methane initiative has been successful in reducing fugitive methane emissions as a potent greenhouse gas by recovering methane before, during, and after coal mining, and utilizing methane as a clean energy source (ICMM, 2010; USEPA, 2010). In the United States CMM emissions were reduced to more than 216 million metric tons CO_2 equivalent from 1994 to 2006 (USEPA, 2008a).

Coalbed Gas in Underground Coal Mines

Underground coal mines are the largest source of CMM, which according to Schultz (2000) makes up total gas emissions varying from 408 to 680 million metric tons CO_2 equivalent/year worldwide (Figure 2.10). The United States underground coal mines alone emitted CMM in 1997 of about 17 million metric tons CO_2 equivalent per year (Schultz, 2000). Franklin, Jemelkova, and Somers (2008) reported that the drained gas from the U.S. underground coal mines was 22 million m^3 or 5% from the total CMM emissions in 2006 (Figure 2.11). CMM is emitted from the underground mine workings depending on the methods used in mining

FIGURE 2.10 Gas bubbles indicating methane emissions in an active underground coal mine.

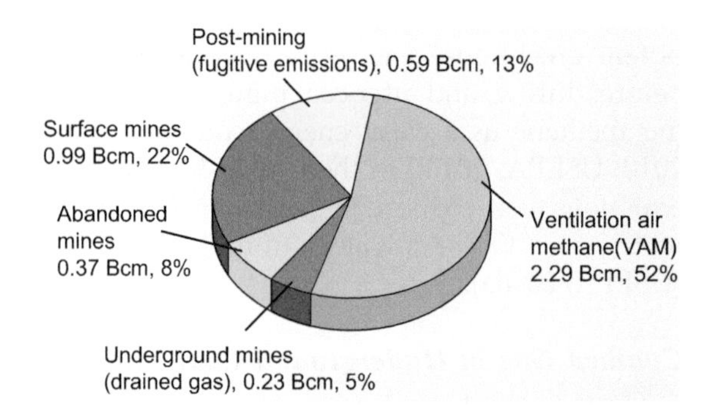

FIGURE 2.11 A pie chart showing the percentage and volume of CMM emissions from coal mining in the United States as of 2006. Bcm, billion cubic meter. *Source: Modified from Environmental Protection Agency (Franklin et al., 2008).*

(USEPA, 2009). CMM is emitted from coal crosscuts and entries during room-and-pillar mining as well as from overlying strata (e.g. rock and coal) of the mine roof and from fracturing of underlying strata below the mine floor by loading/unloading weight of mine equipment. Gob areas formed by roof collapse are another source of CMM emissions. CMM is emitted during longwall mining from the worked coal face and from hauling the coal in mine cars and/or conveyors to the surface. Gas flow is permitted from high gas pressure areas in surrounding coal and rocks into low pressure mine-workings areas. Geological structures and framework in the mine site where coal and rock are deformed and disrupted coal continuity are significant influence to gas flow, thus becoming potential and unexpected sources of CMM emissions (Diamond, 1993; Ulery, 2008). Geological structures such as faults that displaced and fractured coal and rock, structural and compaction folds that contain stressed and strained coal and rock, igneous intrusions through the coal and rock, and geological framework as lithological discontinuities and splits of coal beds and pinchouts of channel-want deposits and related sandstones are very effective in the control of CMM emissions.

An essential component of the coal mine drainage system is the underground infrastructure for CMM collection and transportation

(USEPA, 2009). Below the mine surface, the infrastructure serves to move CMM collected from degasification vertical wells to the surface where gas gathering complexes include gob wells, pipelines, compression and processing facilities (if the methane is to be used commercially), flare wellhead blowers and exhausters (USEPA, 2009). CMM collected from degasification vertical wells comes to the surface via a network of pipelines fitted with safety devices, water separators, monitors and controls, and vacuum pumps. On the surface, CMM is collected from vertical frac wells, surface-drilled horizontal wells, gob wells, and centralized vacuum stations, which collect the CMM produced by in-mine vertical wells (Figure 2.9). An excellent review by Karacan et al. (2011) discussed the technical aspects of CMM capture, factors affecting CMM accumulations in underground coal mines, methods and effective designs of methane capture by boreholes, aspects of removing AMM and sealed/active gas well gobs of operating mines, benefits of capturing and controlling CMM for mine safety, and benefits for energy production and greenhouse gas reduction.

In many drainage systems worldwide, where CMM is not technologically and commercially feasible, the captured gas is emitted to the atmosphere by exhauster and wellhead blower, which in turn, exacerbate the greenhouse gas concentrations. However, many gassy underground coal mines worldwide still do not emphasize CMM capture; thus, no gas recovery and use programs are employed. Nevertheless, under these circumstances USEPA (2000) proposed that CMM emissions be burned under controlled flare system but a byproduct is carbon dioxide, which is 25 times less in global warming potential than methane as greenhouse gas. The concept of maintaining open flares at coal mine sites poses safety concerns such that acceptance is slow by the United States coal-mining industry even though open flare system tested 98% destruction efficiency (USEPA, 2000). However, CMM flare system has been used successfully in United Kingdom and Australia.

Benefits of CMM Utilization

CMM is a valuable and clean energy source primarily composed of methane and secondarily carbon dioxide, nitrogen, ethane, and propane (USEPA, 2009). CMM is composed of pure methane with a calorific value of about 8455 kcal/m^3 but the calorific value is reduced with varying quantities of the other gases (USEPA, 2009). The mixture of CMM with ventilation air with oxygen decreases the calorific value depending on the ratio of methane and air mixture but 1:1 ratio yields a calorific value of about 4450 kcal/m (USEPA, 2009). Different methods of CMM recovery produce varying concentrations of methane at the surface collection points. The benefits of recovering CMM in active underground coal mines are: (1) quality of gas is medium to high, which can be utilized for generation of electricity at mine site, (2) not constrained by mine operations, (3) not limited to old mine infrastructure, (4) accessible and close proximity to demand for the captured gas, and (5) manageable scale, which is smaller than conventional natural gas wells.

A number of possible uses for CMM depend on the methane concentration or heating value (USEPA, 2009). High-quality CMM (e.g. greater than 95% methane) with a calorific value normally greater than 8455 kcal/m^3 is mainly injected into natural gas pipelines for domestic and industrial uses and/or compressed and liquefied for vehicular transportation (Table 2.2). Medium-quality CMM is used for pipeline injection when added with propane and cofiring with coal in industrial boilers. Depending on the CH_4 concentration, medium-quality gas may be used to fuel internal gas combustion engines, blast furnace, microturbines, and cells as well as production of liquefied gas. Low-quality CMM with about 2670 kcal/m^3, which is mixed with air, is used mainly for combustion air in internal combustion engines and/or turbines for

TABLE 2.2 Potential Uses of Recovered CMM and AMM Based on the Quality of the Gases

Btu Quality	Recovery Method(s)	Utilization Options
High-Btu gas (>950 Btu/scf or >35,395 kJ/m^3)	1. Vertical wells 2. Horizontal boreholes	1. Injection into natural gas pipeline fuel (>97% CH_4) 2. Feedstock for production of ammonia, methanol, and acetic acid (>89% CH_4) 3. Compressed and liquefied gas for transportation vehicles
Medium-Btu gas (350–950 Btu/scf or 13,040–35,395 kJ/m^3)*	1. Gob wells 2. Cross-measure boreholes	1. Additive with propane or other gases to increase Btu for pipeline injection 2. Cofiring with coal in utility and industrial boilers 3. Fuel for internal gas combustion engines (>20% CH_4) 4. Enrichment through gas processing 5. Brine water treatment (>50% CH_4) 6. Greenhouse heating 7. Blast furnace fuel as supplement to natural gas 8. Production of liquefied gas (>80% CH_4) 9. Fuel for coal thermal dryers in processing plant 10. Fuel for microturbines (>35% CH_4) 11. Fuel for heating mine facilities 12. Fuel for heating mine intake air 13. Use in fuel cells (>30% CH_4)
Ventilation air	Ventilation air	1. Combustion air in power production (< 1.0% CH_4) 2. Combustion air in internal combustion engines or turbines (< 1.0% CH_4) 3. Conversion to energy using oxidation technologies (< 1.0% CH_4)

* *In the United States, drained gas is above 350 Btu/scf (13,040 kJ/m^3) but is lower in other countries (e.g. China).*
CH_4, methane; Btu/scf, British thermal unit per standard cubic foot; kJ/m^3, kilojoule per cubic meter.
Source: Modified from USEPA (2008b).

electricity generation and combustion air in power production.

USEPA (2005b) reported that in 2001 and 2003 CMM recovery and use in the United States amounted to approximately 1.12 Bcm/year each year; however, recovery and use was up slightly in 2002 to 1.20 Bcm. However, CMM recovery and use increased from 1997, which was estimated as 0.78 Bcm. The success of CMM recovery and use prompted USEPA (2005b) to profile 50

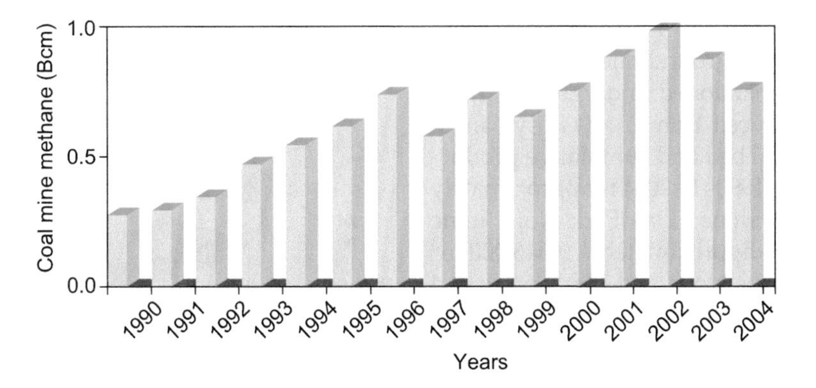

FIGURE 2.12 A frequency diagram showing the annual use of recovered CMM from 1990 to 2004 in the United States. Bcm, billion cubic meter. *Source: Modified from USEPA (2005b).*

United States gassy underground coal mines from 1999 to 2003, which are potential candidates to continue reduction of CMM emissions that account to about 10% of all man-made methane emissions in the United States. USEPA (2005b) reported that from 1990 to 2003 about 49 United States coal mines with CMM recovery and use projects have shown average annual rate of use by approximately 0.6 Bcm (Figure 2.12). In eastern Ohio, the Unionvale Nelms No. 1 and No. 2, and Oak Park No. 7 underground coal mines recovered and used CMM from the early

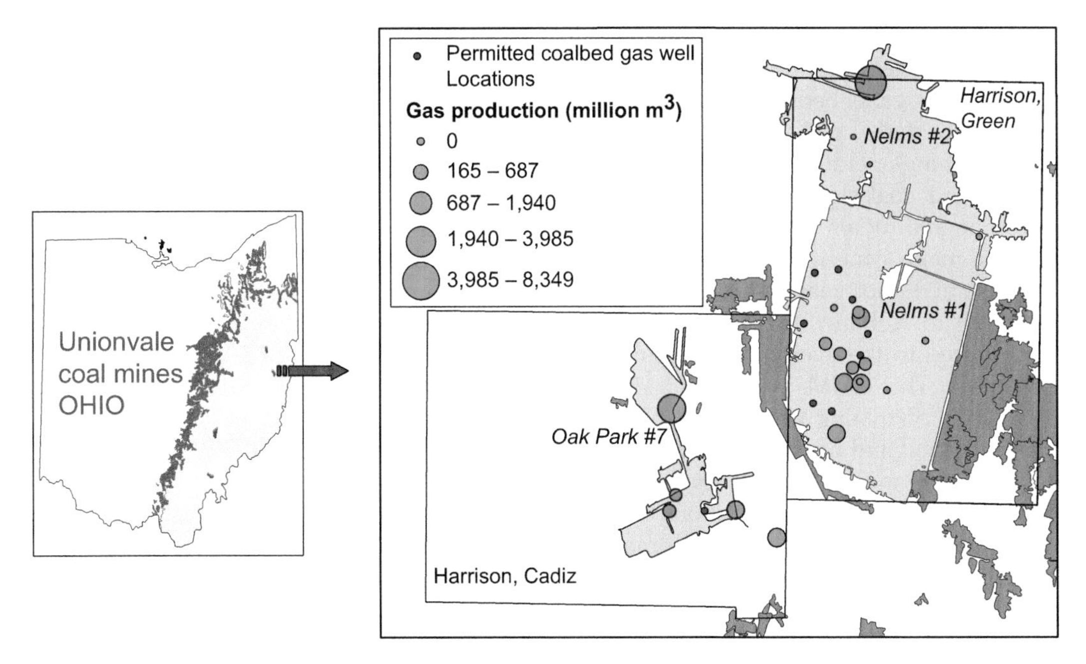

FIGURE 2.13 Underground Unionvale Nelms No. 1 and No. 2, and Oak Park No. 7 coal mines in eastern Ohio, which recovered and used CMM from the early 1990s to 2004. m^3, cubic meter. *Source: From data of Ernie Slucher (USGS).*

1990s to 2004 (Figure 2.13). The Nelms No. 1 and No. 2 are active underground mines and the Oak Park No. 7 is an abandoned underground mine. The CMM is produced from the Pennsylvanian Lower Freeport coal bed and cumulative production from the early 1990s to 2004 is approximately 0.06 Bcm. The abandoned Oak Park No. 7 mine produced CMM approximately 9,646,000 m^3 and the active Nelms No. 1 and No. 2 produced CMM approximately 591,354,000 m^3 in about 10 years. The produced CMM was utilized primarily for electrical generation (generation capacity of 270 MJ/unit) and secondarily fed directly to natural gas pipelines for marketing.

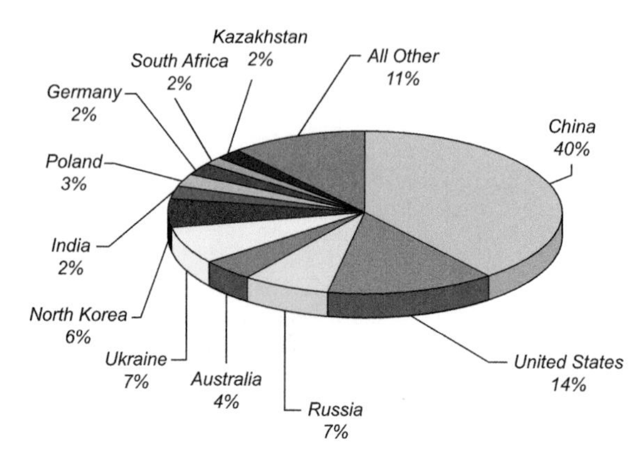

FIGURE 2.14 A pie chart showing the percent CMM emissions in 2000 from 12 major coal-producing countries in the world. *Source: From EPA (2011).*

Coalbed Gas in Surface Coal Mines

Surface mining accounts from 30% to 95% of the world coal production whereas in the United States 67% of coal production is from surface mining. However, the surface coal mining in the United States only accounts for about 16% of the CMM emissions (USEPA, 2008a). The major concerns about CMM associated with surface coal mining are emissions from: (1) the minable coal bed during excavation and processing, (2) the coal and other gassy beds in overlying (overburden) and underlying (underburden) strata exposed by mining operations, and (3) excavated overburden and coal stored as waste piles on mine sites (Saghafi et al., 2005; USEPA, 2008a). Of the three major concerns, the mineable coal bed is presumed (not mandatory to measure gas volume) to contain the most available CMM compared to the overburden and underburden strata. The CMM is released during various phases of mining operations from the removal of the soil and overburden to the excavation, exposure, and blasting of the mineable coal bed. Then these mining operations are followed by the disturbance of earth-moving equipment in preparation to hauling (e.g. trucks, conveyors) the coal from the opencast to the preparation facilities or to offloading areas.

Worldwide, USEPA (2008a) identified 12 major coal-producing countries that account for emitting 89% of the CMM from surface coal mines (Figure 2.14). The CMM emissions from the major coal-producing countries in 2000 vary from 40% for China to 2% for India, Germany, Kazakhstan, and Republic of South Africa. As of 2010, China's surface coal mine production has increased from 29.7 million metric tons in 2001 or about 3% of the total coal production to 435 million metric tons or 15% in 2010 (Data from China Coal Technology Group, Chonqing Division). Still, coal production from surface mines in China is relatively very small compared to production from underground mines considering that China's total coal production in 2010 is 3.2 billion metric tons (BP, 2011). In comparison, in India about 85% of the total coal production of 506 million metric tons in 2009 is from surface coal mines.

Coalbed Gas in Underground Abandoned Coal Mines

Worldwide, there are thousands of underground abandoned coal mines, which continue to emit methane and carbon dioxide after closures contributing to the global greenhouse gas emissions. As of 2005, there were 8000 abandoned underground mines, 400 of which as considered gassy. These gas emissions are called

AMM in contrast to CMM emissions, which are from active underground and surface coal mines. The AMM is emitted mainly from old entrances, shafts, boreholes, ventilation systems, and vent pipes as well as from fractures in the overburden rocks. Underground coal mines are closed and abandoned as coal reserves are depleted or eventually mined out. As a part of the reclamation of abandoned coal mines, the entrances and shafts are sealed by concrete, and boreholes and vent pipes are plugged to prevent AMM emissions. In a number of abandoned coal mines vent pipes and boreholes are left open at the surface in order to provide a controlled release point for AMM to the atmosphere. In some countries and in the United States, abandoned coal mines are required to operate ventilation systems for months or years by local and state regulations in order to control AMM emissions. Additional control of AMM emissions from abandoned coal mines requires a recovery system to collect the gas as shown Figure 2.15.

The sealing of man-made infrastructure in the abandoned underground coal mines does not totally contain the liberation of AMM because of the geological characteristics of the overburden rocks. Fracture systems caused by displacement and deformational stress and strain of the overburden rocks before, during, and after mining cannot be totally accounted by sealing man-made structures. According to Khoudin (1995) underground coal mining operations produce redistribution of rock pressure and different stressed and strained zones. For example, the overburden rocks consist of mainly interbedded sandstone, siltstone, mudstone, and coal beds. During mining, displacement of beds in the overburden strata occurs by bending of layers due to differences in tensile and compressive strengths as well as rigidity of the rock layers. Rock strains are propagated into the rocks away from the working face and into the overburden rocks, which differ in strengths and consist of weak contacts along the bedding planes, which are displaced in the form of folding. During mining, the peak of support is away from and at the ends of the coal face where stress during extraction depends on: (1) the rate of mining, (2) width of the coal extraction face, (3) roof support assistance, (4) thickness of the rock layer immediately above the coal, and (5) presence/absence of higher pressure zones. Thus, the stress and strain conditions of the coal and overburden rocks influence the formation of rock displacements and associated fracture systems. The rock displacements and fracture systems in the overburden strata, which are difficult to predict and recognize, are natural conduits for AMM flow to the surface. AMM emissions may be inhibited by reestablishment of groundwater flow in the abandoned coal mines and overburden strata. The restoration of hydrostatic pressure underground will hold the residual gas on pore and fracture surfaces of the coal and overburden rocks. The restoration of the groundwater system requires flooding the coal mine with water by natural or artificial methods. However, a disadvantage to reestablishment of groundwater system in the abandoned coal mine is the potential generation

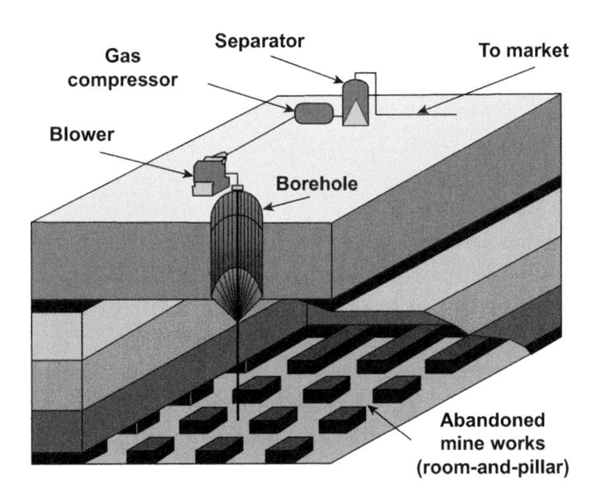

FIGURE 2.15 Diagram of an abandoned coal mine showing infrastructure of AMM recovery after the mine is closed. *Source: From Environmental Protection Agency (Franklin, 2004).*

of biogenic gas or methane by active methanogens through aceticlastic methanogenesis (Beckman et al., 2011; see Chapter 4 for more discussions).

Role of Mine Dewatering to CMM Emissions

The extensive coalbed gas development in the Powder River Basin in the United States since the late 1980s, which so far yielded a cumulative production of more than 0.1 Tcm, may shed light on the CMM emissions in the surface coal mines (Flores, Spear, Kinney, Purchase, & Gallagher, 2010). One of the major problems in the surface coal mines in the Powder River Basin and elsewhere in the world is the drainage of groundwater caused by dewatering of the mineable coal and overburden rocks. The strata overlying the mineable coal bed(s) consist of permeable sandstones interbedded with impermeable mudstones, thus acting as aquifers and aquitards, respectively. The groundwater in the overburden rocks results in residual water impeding mining excavation, combustion hazards in the exposed coal, and instability of highwall and spoil slopes. In order to alleviate these mining problems, mine operators are permitted to pump water from the overburden rocks and underlying mineable coal bed to the surface with as many as 30 wells in a quarter–quarter section. A major impact of dewatering groundwater from the overburden rocks and underlying virgin coal bed to coalbed gas development is the lowering of the hydrostatic pressure, which holds the methane and carbon dioxide molecules in the coal reservoir. Prolonged surface coal mine dewatering since the mid-1970s gradually lowered the groundwater level subsequently decreasing the hydrostatic head to the top of the coal beds, permitting gas desorption and flow to the rocks in the highwall, and releasing of the gas to the atmosphere (Flores, 2004). Figure 2.16 shows a lowering of groundwater level or drawdowns as of 2002 in the vicinity of coal mines in the Gillette coalfield, Wyoming in the United States (BLM, 2008). Long-term pumping of dewatering wells in nearby surface coal mines discharged large volumes of water, which created cone-shaped depressions or drawdowns of about 50 m deep from the premining groundwater level. These groundwater drawdowns acted as low hydrostatic pressure areas permitting gas to flow from the Wyodak coal in adjoining high hydrostatic pressure areas, which eventually released CMM to the atmosphere. Another groundwater level changes shown in Figure 2.17 was developed as of 2002 in the Upper Fort Union resulting in coal mine and coalbed gas pumping in the Gillette coalfield (BLM, 2008). The regional drawdown is about 107 m deep from the premining groundwater level; thus, had about twice the effect of degasification of the Wyodak coal compared to the drawdown north of the coalfield.

The fugitive CMM emissions from the Powder River Basin surface coal mines caused by dewatering wells is exemplified in the southern part of the Gillette coalfield. Flores and McGarry (2004) determined that the premining (1982) hydrostatic head of groundwater level in the overburden rocks and mineable Wyodak coal bed in the vicinity of a group of surface coal mines in this area ranged from 0 m at the outcrop to 55 m several km toward the basin (Figures 2.18(A) and (B)). In 2001, the hydrostatic head changed from 0 m at a more basinward outcrop to 30 m also at a more basinward location; thus a drop of about 23 m of the hydrostatic head (Flores & McGarry, 2004; McGarry, 2003). The 0 m contour line of the hydrostatic head has moved about 5 km farther basinward. Based on the lowering of hydrostatic head of the groundwater, the hydrostatic pressure for the pre- and postmining was estimated (Figures 2.19(A) and (B)) to have decreased to about 22 kPa. Using the hydrostatic pressure gradient for the pre- and postmining, the gas content was estimated to have lost 0.12 cm^3/g in 20 years (Figures 2.20(A) and (B)). This lost gas content can be applied to estimate the lost gas as the result of mining the Wyodak coal across the entire Gillette

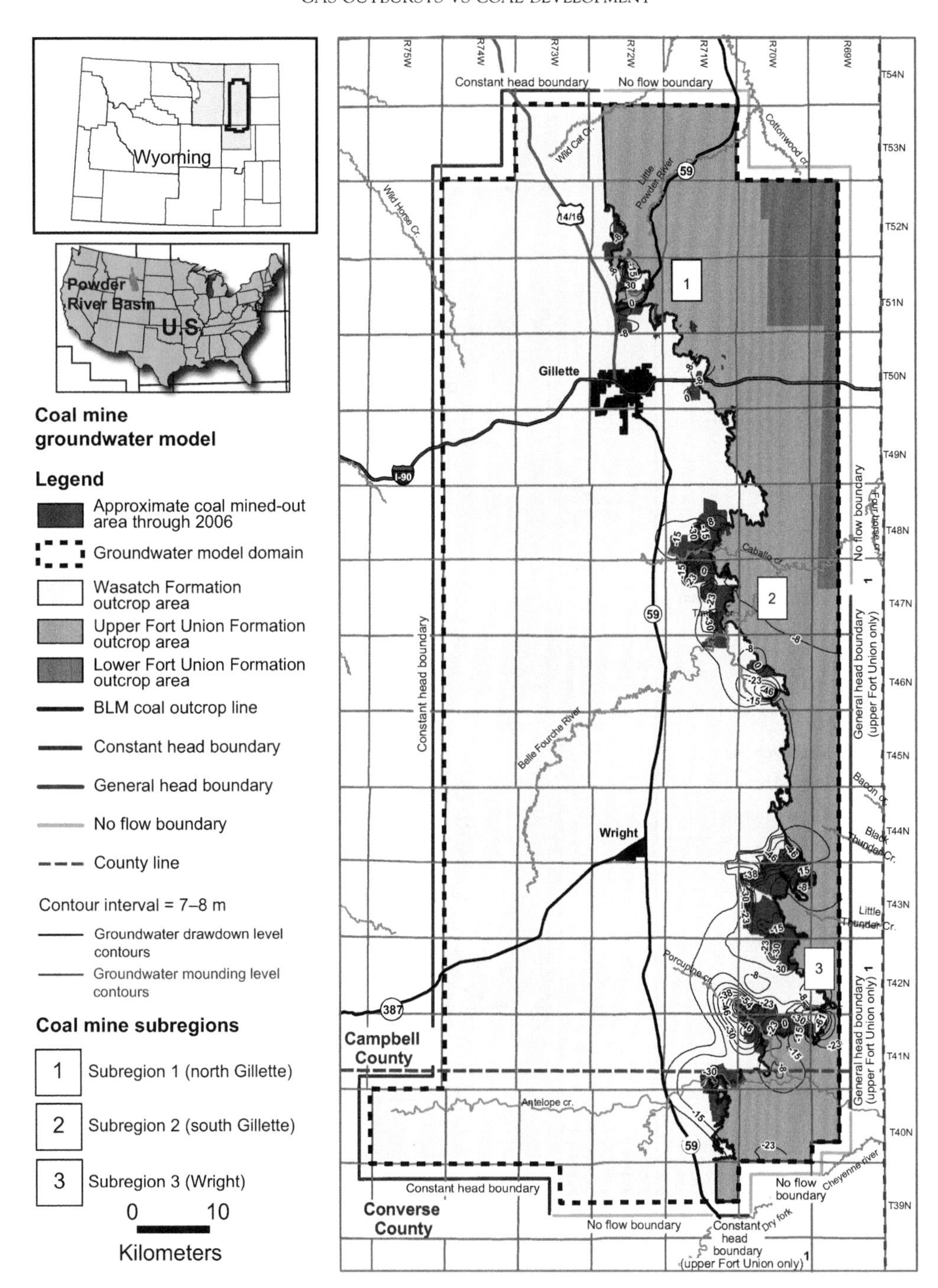

FIGURE 2.16 Modeled groundwater changes as of 2002 in the Upper Fort Formation in the Gillette coalfield, Wyoming. Includes effects of groundwater drawdown of coal mines. *Source: Modified from BLM (2008) and Flores (2004).*

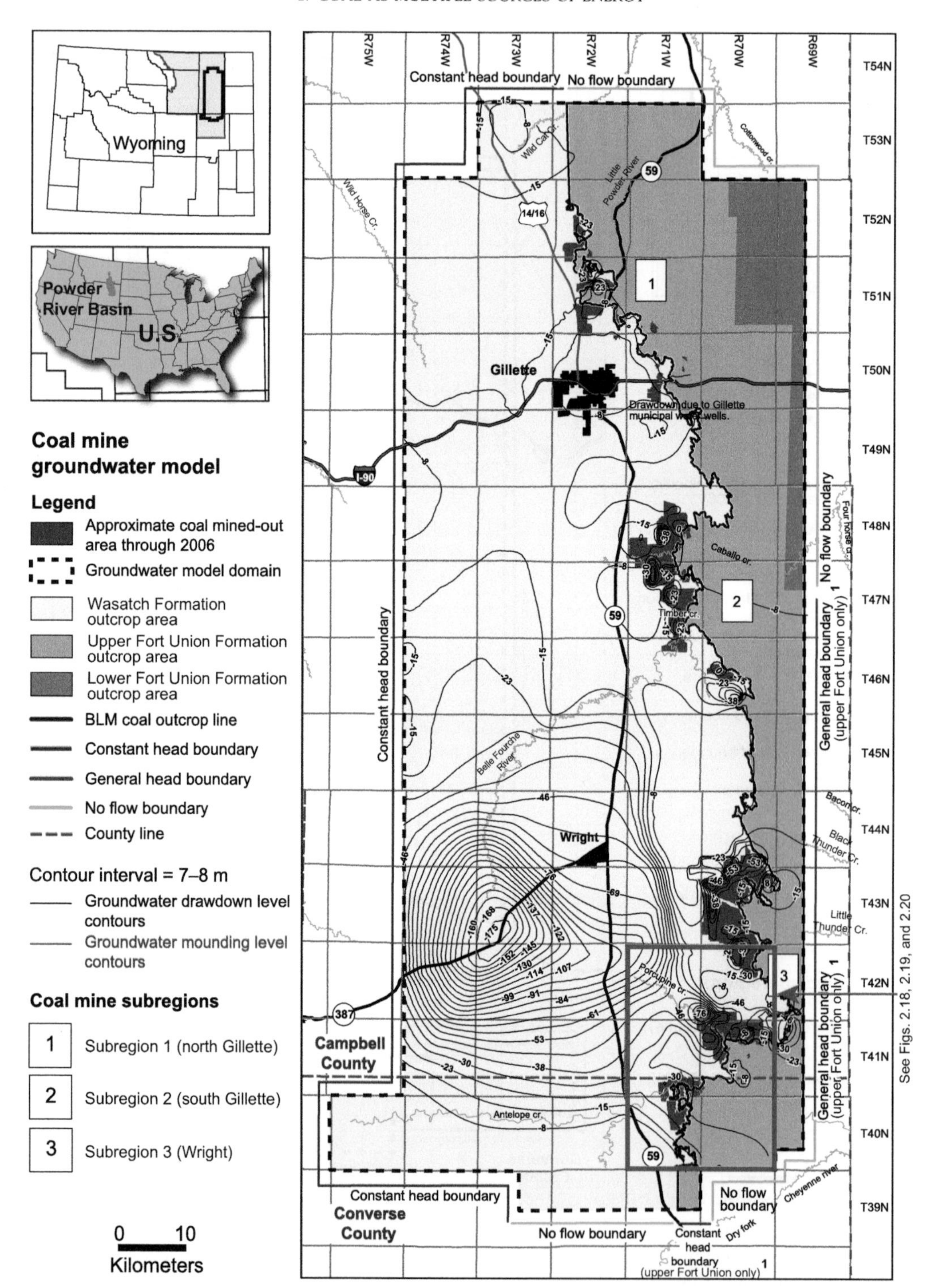

FIGURE 2.17 Modeled groundwater changes as of 2002 in the Upper Fort Union Formation in the Gillette coalfield, Wyoming. Includes effects of coal mines and coalbed gas groundwater pumping. *Source: Modified from BLM (2008) and Flores (2004)*

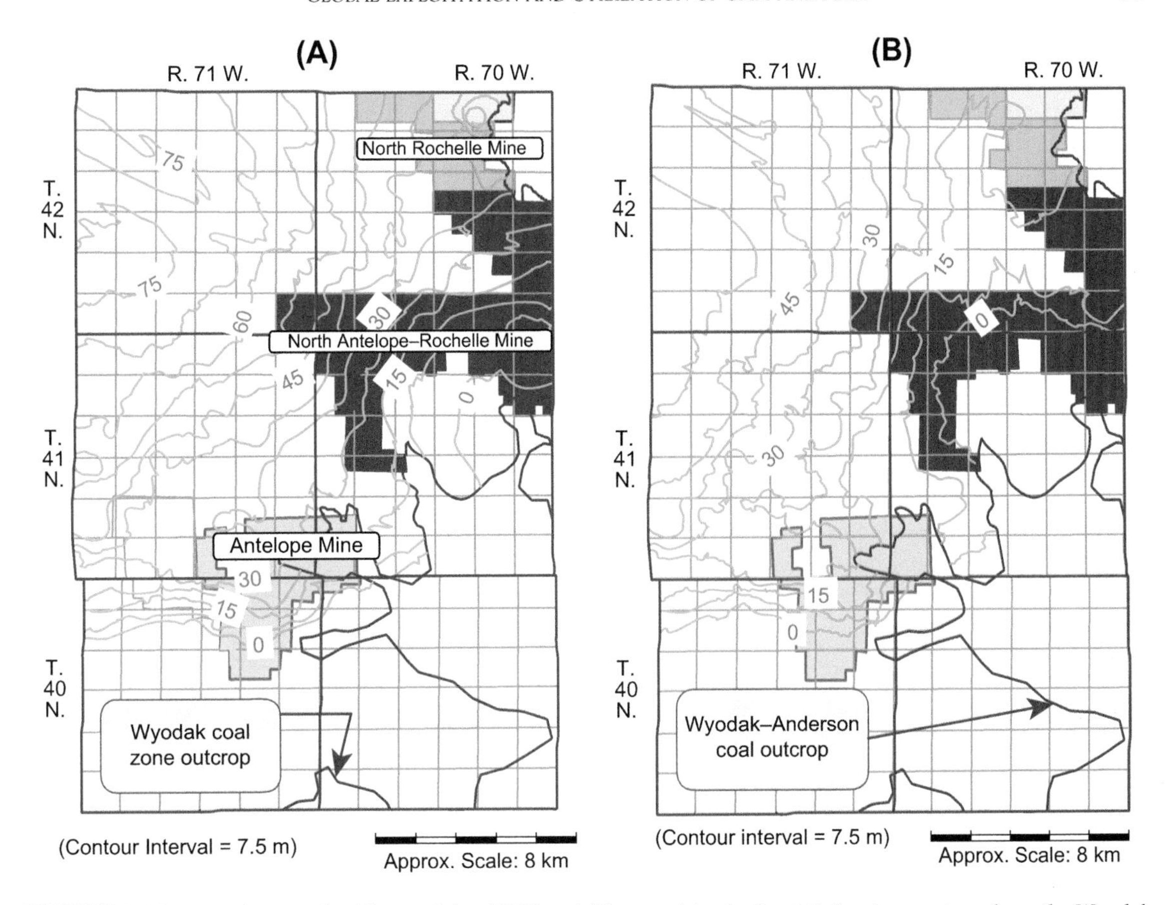

FIGURE 2.18 Map showing the (A) premining (1982) and (B) postmining hydrostatic heads or meters above the Wyodak coal in the vicinity of the surface coal mines in the southernmost part of the Gillette coalfield, Wyoming (see Fig. 2.17 for location). Hydrostatic head calculated from GAGMO groundwater monitoring well data (1982) and geologic structure mapped from publicly available drill-hole data. *Source: Adopted from McGarry (2003) and Flores and McGarry (2004).*

coalfield from 1998 to 2010. Twelve surface coal mines during the period of 13 years produced a total of 4089 million metric tons or average 314 million metric tons, which makes up about 34% of the total coal production of the United States during the same time. If all the surface coal mines practiced coalbed gas drainage system in advance of mining during that period of time, a total of about 490 million m^3 of gas would have been recovered and produced. An evidence of lost gas from surface coal mining is best demonstrated in Figure 2.21 in which coalbed gas wells about a few hundred meters to 6.5 km away from a highwall of one of the mines in the southern part of the Gillette coalfield show decreased production due to mine dewatering operations compared to wells farther basinward.

GLOBAL EXPLOITATION AND UTILIZATION OF CMM AND AMM

Coalbed gas from active underground coal mines extracted by drainage and VAM systems,

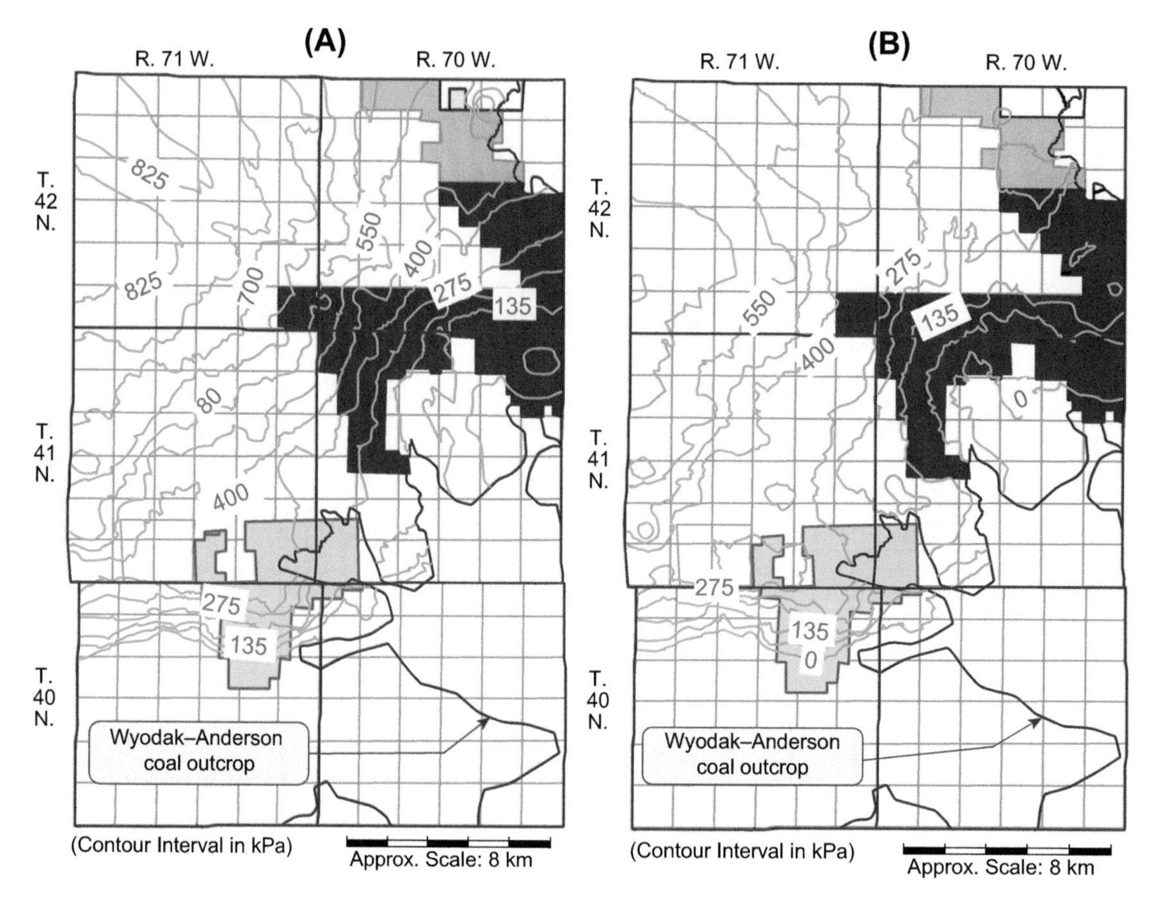

FIGURE 2.19 Map showing the (A) premining (1982) and (B) 2001 postmining hydrostatic pressures in the Wyodak coal in the vicinity of the surface coal mines in the southernmost part of the Gillette coalfield, Wyoming (see Fig. 2.17 for location). Derived by multiplying 1982 hydrostatic grid by about 9 kPa/m pressure gradient. *Source: Adopted from McGarry (2003) and Flores and McGarry (2004).*

referred to as CMM, is exploited in major coal-producing countries for industrial and commercial use. According to USEPA (2010; 2011a) the total CMM emissions of 37 major coal-mining countries amounted to 298 million metric tons CO_2 equivalent in 2005. Nine of these major coal-producing countries with more than 100 million metric tons (about 90%) of total world coal production in 2007 such as Australia, China, India, Indonesia, Germany, Poland, Russia, South Africa, and the United States account for about 87% of the total CMM emissions worldwide. In 2010, coal production expanded in the United States and Asia but contracted in the Europe. Thus, in terms of coal production the nine major coal-producing countries may be grouped into four regional hierarchies: (1) Australasian (Australia, China, India, and Indonesia), (2) United States, (3) Europe (Germany, Poland, and Russia), and (4) South Africa. Of the total world coal production in 2010, Australia–Asia accounts for two thirds, the United States more than one sixth, and the remainder is accounted for by Europe and South Africa.

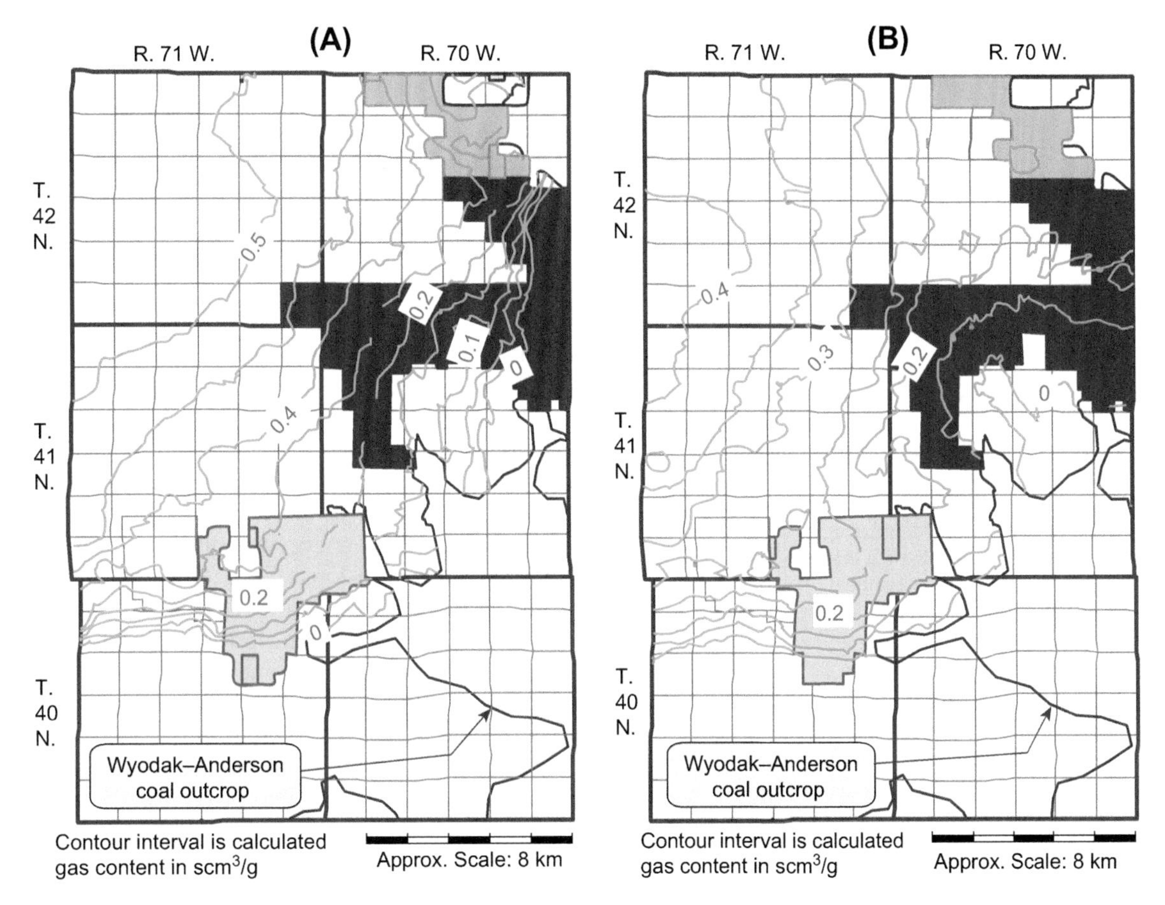

FIGURE 2.20 Maps showing (A) 1982premining and (B) 2001 postmining gas contents in the Wyodak coal in the vicinity of the surface coal mines in the southern part of the Gillette coalfield, Wyoming (see Fig. 2.17 for location). Calculated from 1982 pressure surface and average adsorption isotherm. scm^3/g: standard cubic meter per gram. *Source: Modified from McGarry (2003).*

China, which is by far the largest coal producer in the world, accounted for the largest CMM emissions of 123.1 million metric tons CO_2 equivalent. Indonesia accounts for the smallest CMM emissions of 0.4 million metric tons CO_2 equivalent of the nine major coal-producing countries. The CMM emissions of the United States and Russia are 50.2 and 23.8 million metric tons CO_2 equivalent, respectively. In general, the largest coal-producing countries such as China, United States, Russia, Australia, and India emit more CMM than other countries (see Figure 2.14). This correlation is probably related to countries like China and the United States producing large-scale coal from underground and surface mines, respectively. Underground mining accounted for 95% of China coal production in 1996 but more recently, underground mines produced about 90% of China's coal (Tu, 2007). However, as of 2010 underground coal mine production was about 85% of the total annual coal production of 2.946 billion metric tons. Indonesia maintains only a few underground coal mines. Another factor to consider

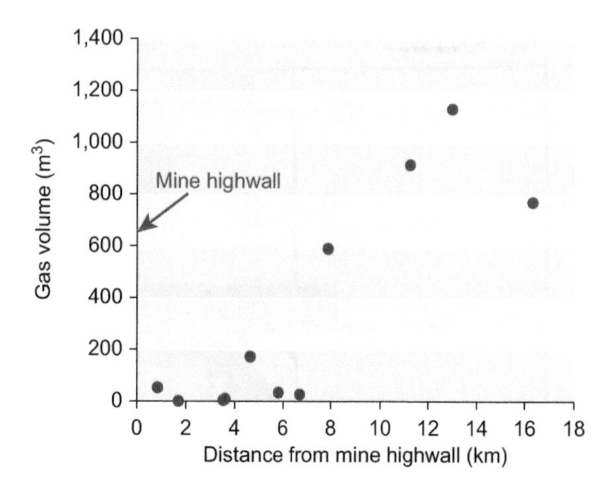

FIGURE 2.21 Chart showing the distance and depleted coalbed gas due to mine dewatering in six producing wells away from the highwall of a surface coal mine in the southern part of Gillette coalfield, Powder River Basin, Wyoming. Gas volume is from January to February, 2001. m^3, cubic meter; km, kilometer.

as demonstrated in Australia is that lower gas emission is primarily attributed to the mining of less gassy coal reserves. Exploitation of coal bed mine gas varies between the hierarchical regions from no CMM recovery and use programs to very active program in China.

Australasian Coalbed Mine Gas Exploitation

IEA (2009) estimated that there were nearly 100,000 coal mines in China during the mid-1990s and majority of these mines belonged to villages and towns. China has implemented a program to closedown underperforming (annual capacity below 136,000 metric tons) or unsafe mines, especially town and village coal mines in the last decade (GLG, 2010). In 2009, there were 14,673 underground coal mines in China among which 1061 mines have the potential for coalbed gas outbursts and 2351 mines have high gas contents. According to Chinese mining regulations, all high gas and gas-outburst prone mines are required to be degassed. Thus, 3412 Chinese coal mines have to be degassed, which is about 23.3% of the total number of coal mines. Underground coal mines in China have increased installation of drainage and VAM systems so that by 2006, more than 300 mines have installed CMM drainage and VAM systems, which collectively extracted more than 3.24 Bcm of CMM (USEPA, 2010). Approximately 80% of the CMM drained in China was from state-owned coal mines (Huang, 2007). China's CMM emissions nearly doubled in 2009, however, the recovered and utilized CMM in the same year tripled the levels of 2005 (Huang, 2010). In 2011, the Chinese government and China Coal Technology Engineering Group established a 12-year plan to develop gob gas from clean energy supply and eliminate coalmine gas explosions.

In China, recovered CMM is used for town gas, electricity generation, industrial boiler fuel feed, vehicle fuel, and thermal applications (e.g. office space heating) as well as for multiple uses (Table 2.2). Many coal mines in China recover low concentration gas (e.g. less than 30% methane concentration) (USEPA, 2010). CMM used for power generation produced more than 900 MW (774×10^6 kcal/h) by the end of 2008.

Additionally, approximately 4000 vehicles operate on CMM as fuel (Huang, 2010). The largest CMM power generation project in the world is at the Sihe Coal Mine in Jincheng, Shanxi Province, which uses Caterpillar engines to generate electricity at a 120-MW (103×10^6 kcal/h capacity power plant (Figures 2.22(A) and (B)). The Sihe project utilizes 180 million m^3 of both coalbed gas recovered in front of mining and CMM recovered from the mine sites (Huang, 2008; USEPA, 2006a). A large part of the recovered CMM is compressed and liquefied to market as well as used to fuel mine vehicles. The Sihe project has prevented the emission of 2.26 million metric tons of CO_2 equivalent annually (M2M Projects, 2010a).

FIGURE 2.22 (A) Photograph of the Jiafeng liquid gas factory of the Shanxi Yigao coal seam gas company where coalbed and coalmine methane are liquefied. (B) Photograph of a transport vehicle containing compressed gas for the domestic market.

The Australian government estimated that VAM is responsible for 64% of the country's CMM emissions. Typical gassy Australian coal mines produced VAM at a rate of 150–300 m^3/s (M2M-Australia, 2005). CMM are recovered in 10 underground and five abandoned Australian coal mines, which involved flaring gas, using gas in the generation of electricity using reciprocating compressor engines, destruction using VAM, and pipeline injection of high-quality CMM for commercial use (M2M-Projects, 2010b).

In India, there is some drainage of CMM but no large-scale commercial recovery or use is planned; however, the government and world organizations (e.g. United Nations Development Program, the Global Environmental Facility) have cooperated through the "Coalbed Methane Recovery & Commercial Utilization" with successful demonstration of the commercial feasibility of utilizing CMM recovered before, during, and after coal extraction. Recovered coalbed gas and CMM are currently utilized to generate electricity to meet local power needs, power converted bi-fuel engines for dump trucks. Future results of the cooperation could lead to expanded application of CMM technologies, such as the production of compressed natural gas fuel for mine vehicles and commercial scale projects.

United States Coalbed Mine Gas Exploitation

In the United States, coal production for the past 10 years has reached record levels with the average of less than 1 billion metric tons, which is second to China's production. Much of the coal production from 2007 to 2011 is from surface coal mines, which has grown 10–12% in contrast to underground coal mines, which has grown only about 2% during the same period (USEIA, 2012). And there are currently a limited number of coal mine gas recovery and CMM-to-power projects in existence or planned coal mines. Many United States coal mining companies have worked toward reduction and prevention of CMM and AMM emissions using recovery and utilization technologies, which provide alternative energy as well as economic and environmental benefits. The 50 coal mines profiled by USEPA (2008b) for CMM recovery and use projects are located in six states most of which are found in the eastern and midwestern regions of the United States. The estimated CMM emissions of these coal mines range from 2 to 680 m^3 per ton. The coal mine (V8) in the Central Appalachian Basin in Virginia with the most estimated specific emissions recovers, uses, and operates the second largest CMM power generation station in the world (88 MW or 75×10^6 kcal/h) but it is run infrequently and only for power peaking (USEPA, 2010). In addition, this coal mine uses about 56,620 m^3/day of recovered CMM for drying coal and piped 80% of the rest of the recovered CMM. The coal mine, which produced 0.3 million metric tons in 2006, employs longwall method of underground mining of steam and metallurgical coals. The coal mine utilizes vertical gob and horizontal and vertical premine boreholes as drainage systems with 83% efficiency. The coal mine is reported to have an estimated daily CMM drained of 0.44 million m^3/day, daily ventilation emission of 0.09 million m^3/day, and a total liberated CMM of 0.5 million m^3/day in 2006.

Flaring gas has been used at closed mines but has not been widely implemented at active mines in the United States. The coal industry has expressed concerns about the safety of flaring due to the potential for the flame to propagate back down to the mine and cause an underground explosion. Implementing flaring at active mines requires greater acceptance by miners, union parties, mine owners, and the Mine Safety and Health Administration. However, some form of flaring occurs in some active mines where drainage gas is used to preheat incoming ventilation air in cold months to condition incoming air and prevent the formation of large icicles at the ventilation shaft opening.

European Coalbed Mine Gas Exploitation

Europe, Germany, and Poland are actively pursuing CMM and AMM exploitations. In Germany, CMM emissions amounted to 336 million m^3 or 4.8 million metric tons carbon dioxide equivalent in 2006 (IEA, 2009). Most of Germany's coal mine gas recovery is AMM and CMM makes up a minor part of the recovery. The recovered AMM and CMM are primarily used for electricity power generation and secondarily for combined heat and power (M2M Projects, 2010a). In Poland, coal mines are the source of 26% of the country's total methane emissions, with the top 20% of underground mines accounting for 90% of the CMM emissions (USEPA, 2005b; USEPA, 2010). All the CMM recovery in Poland is from active underground mines. The recovered CMM is used for boiler fuel, coal drying, combined heat and power, industrial use, and power generation. Poland has extensive CMM recovery implementing mine-site utilization for electricity, heating, and cooling. About 300 million m^3 of CMM has been drained from Poland's coal mines annually as of 1997, with 65–70% of the recovered gas used at the mine sites or sold to outside consumers and the rest vented to the atmosphere (USEPA, 2010). CMM recovery has decreased over the years such that from an estimated 870 million m^3 of CMM emissions in 2006, less than 30% was removed through degasification (IEA, 2009) and in 2008, 269 million m^3 was removed through degasification and of this about 166 million m^3 utilized and 103 million m^3 released into the atmosphere (Skiba, 2009). The number of gassy mines in Poland has been reduced by 48% from 1989 to 2005; however, total gassiness dropped only 19% over this period, which indicates increasing share of emissions from gassy coal mines (USEPA, 2010). This situation has increased the potential of CMM recovery and utilization in Poland (IEA, 2009) with CMM capture forecasted to be 320.5 million m^3 in 2015 and estimated utilization of 1068 GW-hours (918×10^9 kcal/h) (Skiba, 2009).

In Russia, CMM emissions are all from active underground coal mines, which totaled 1.8 Bcm in 2005 (UNFCCC, 2009). The CMM in Russia is primarily from coal mines in three coal basins: Kuzbass, Pechora, and Donetsk (also known as Donbass) most of which is located in Ukraine. The Kuzbass and Pechora Basins account for 78% and 12%, respectively, of CMM emissions. IEA (2009) estimated 1.5 Bcm of CMM from these two coal basins alone as of 2008. CMM drainage was initiated in coal mines in the Kuzbass Basin in 1951 and Pechora Basin in 1956;

however, degasification peaked in 1990 when many of these coal mines had to be shut down. The coal mines in the Pechora Basin vented 289.8 million m^3 of CMM in 1998, which decreased to 42.05 million m^3 in 2000 (Ugle Metan, 2005). In recent years, the rate of CMM recovery in these coal mines has averaged 27–30%, with only 25% of active coal mines utilizing drainage system in 2009 (Ugle Metan, 2010). The total CMM drained from coal mines in the Kuzbass and Pechora Basins was estimated to be 320 million m^3 in 2008 (USEPA, 2010). The CMM is of poor quality (<25% methane) and is recovered mainly for outburst prevention. Currently, only CMM drained from the coal mines in the Pechora Basin is used for boiler fuel (40 million m^3 in 2006) (IEA, 2009).

South African Coalbed Mine Gas Exploitation

South Africa as the world's seventh largest coal producer dominated the coal industry in the African continent in 2010. However, South Africa was one of the world's top five CMM emitters attributable to the high coal production and high gas content similar to Australia Permian coals. In 2008, South Africa was downgraded to ninth in worldwide CMM emissions estimated to be about 10 million metric tons of carbon dioxide equivalent (USEPA, 2008b). Gas emissions in South Africa are from CMM and AMM in underground coal mines and CMM in surface coal mines. In 1990, CMM from underground coal mines was estimated to be 418 million m^3 and from surface mine was about 57 million m^3 (USEPA, 2008b). Although South Africa has a number of abandoned coal mines, which was reduced in half from 1986 to 2004, no appreciable AMM was measured because the percentage of gassy mines was low. It is projected by USEPA (2006b) that the total emission (total liberated—recovered and used) in 2015 will increase by about 25%. The main recovery and use programs in South Africa are in the form of flaring the gas, which is mixed with air before burning. The potential uses of recovered CMM in South Africa include electric power generation, feedstock for the petrochemical industry, and fuel for boilers and transportation.

PETROLEUM DERIVED FROM COAL

Coal as source of oil and condensate has been known for as long as coal is the source of natural gas (Bradfield, 1943; Hedberg, 1968; Hubbard, 1950). It is generally accepted that coal contains total organic matter above 1–1.5% to be a viable source of petroleum liquids that are expelled, migrated, and trapped in economic quantities to nearby sedimentary reservoir rocks (Bertrand, 1984; Boreham, Horsfield, & Schenk, 1999; Boreham & Powell, 1993; Brooks & Smith, 1967; Davis, Noon, & Harington, 2007; Horsfield, Yordy, & Crelling, 1988; Killops, Mills, & Johansen, 2008; Li, Jin, &Lehrmann, 2008; Lu & Kaplan, 1990; Petersen, 2006; Shanmugam, 1985; Smyth, 1983). Proven and potential coal-derived oil reservoirs are mostly associated with coal-bearing sequences deposited in lacustrine, alluvial, deltaic, and tidal environments. For example, the Cretaceous to Paleocene coal-bearing sequences in the Bass and Gippsland Basins, Australia with coals and coaly shales as terrestrial source rocks of the produced oils were deposited in dominantly bogs in aggradational coastal and barrier depositional settings as well as lacustrine environments (Bishop, 2000; Boreham, Blevin, Radlinski, & Trig, 2003; Fielding, 1992). In contrast, the coals of the Eocene coal-bearing sequences, which served as source of oil and condensate in the Taranaki Basin, New Zealand, were deposited in transgressive, tidal-influenced coastal-plain swamps (Flores, 2003; King & Thrasher, 1996). Although coal-derived oil or petroleum liquids make up a small fraction of the commercial petroleum accumulations worldwide, the potential of this mix energy continues to be evaluated (Large, Marshall, Meredith, Snape, & Spiro, 2011; Ogala, 2011).

Oil Potential of Coal

Understanding the oil potential of coal requires mainly knowledge of the organic matter or maceral composition of coal. The maceral composition of coal is directly related to the parent plant material, which in turn, is influenced by the depositional environments of the precursor peat deposits. For example, according to Littke and Haven (1989) the Ruhr Basin coals in Germany are rich in vitrinite macerals in the lower part of the beds and are overlain mainly by inertinite macerals, which they ascribed to be formed in raised peat bogs. However, where the coal bed contains mineral-rich interbeds of mudstones and vitritnite macerals, the peat was likely deposited in areas prone to floods or accumulated in floodplains. These changes of coal maceral facies are dictated by growth and deposition of particular plant parts (e.g. roots, barks, trunks, stems, leaves, etc.), and their degree of alteration, and mineral influx into the peat, which in turn, control the migration of hydrocarbons out of the coals (Littke & Haven, 1989). The basic maceral groups of coal identified by petrology are: (1) vitrinite macerals derived from coalified woody tissue, and contain relatively higher oxygen; (2) liptinite macerals derived from the resinous and waxy parts of plants (e.g. lipids, alginites), which contain relatively higher hydrogen; and (3) inertinite macerals derived from charred and biochemically altered plant cell walls, which relatively contain higher carbon (see more discussion in Chapters 3 and 4). The maceral groups occur at various levels in the form of Types I, II, III, and IV in the van Krevelen diagram (Figure 2.23) based on the ratio of hydrogen and oxygen ratio, and oxygen and hydrogen ratio following different

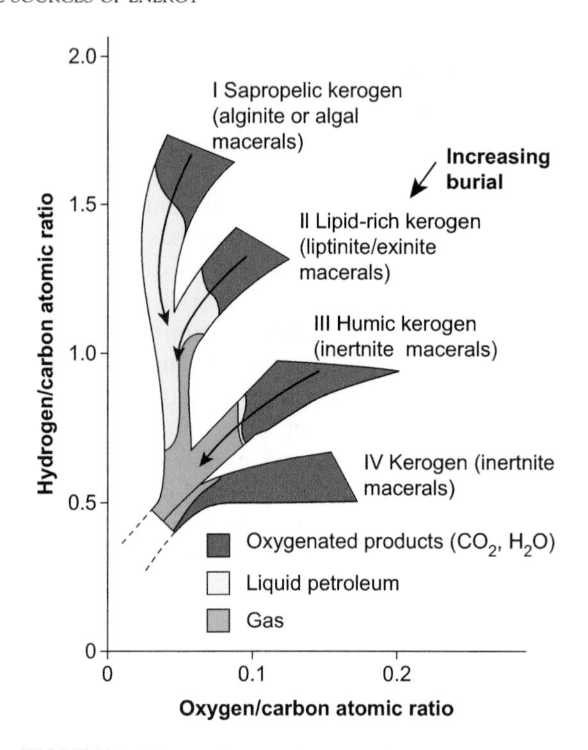

FIGURE 2.23 Coal maceral groups form Types I, II, III, and IV in the van Krevelen (1961) diagram based on the ratio of hydrogen/carbon atomic ratio, and oxygen/carbon atomic ratio ratio. *Source: Adopted from Monash University (2011) website.*

tracks of coalification and thermal maturation (Van Krevelen, 1961).

It is generally acknowledged that liptinite (exinite) macerals, which include lipid-rich kerogen (Type II) and alginite/algal or sapropelic kerogen (Type I), are oil-prone source organic matter commonly comprised hydrogen-rich coals capable of generating oil (Boreham & Powell, 1993). In contrast, coals that are dominantly composed of woody or vitrinite-rich matter, derived principally from vascular plants, are classified as oxygen-rich humic Type III coal (humic kerogen). It has been determined that certain submacerals of vitrinite macerals such as exinites (e.g. resinites and cutinites) and desmocollinite are hydrogen-rich and other terrestrial organic matter are capable of generating and expelling liquid hydrocarbons (Clayton, 1993; Hunt, 1991; Ogala, 2011; Shanmugam, 1985; Snowdon, 1977). Thus, the hydrogen proportions of the organic matter may be used to evaluate the oil potential of coal. Powell (1988) suggested that in order for coal to be the potential source of hydrocarbons, 10–20% of organic matter must be equal to the Type I organic matter and 20–30% of organic matter equal to the Type II organic matter. The three petrographic maceral groups of coal (e.g. Types I, II, and III) generally correlate to the organic matter or kerogen types determined for sedimentary rocks as shown in Figure 2.23 (Tissot, Durand, Espitalie, & Combaz, 1974; Tissot & Welte, 1984). The liptinite-rich macerals correspond to the Types I and II kerogens, the vitrinite-rich macerals correspond to Type III kerogen and the inertinite-rich macerals correspond to Type IV kerogen.

Global Occurrences of Coal-Derived Oil

The occurrences of coal-derived oil and/or condensate have become increasingly important targets of exploration and development in many parts of the world. These hydrocarbons have been investigated in various coal-bearing basins worldwide (e.g. Australia, China, Denmark, India, Indonesia, New Zealand, Nigeria, United States) of Pennsylvanian through Tertiary-age coal deposits (Ahmed et al., 2009; Bagge & Keeley, 1994; Curry, Emmett, & Hunt, 1994; Davis et al., 2007; Johnston, Collier, & Maidment, 1991; Petersen & Brekke, 2001; Wilkins & George, 2002). Clayton (1993) described 23 basins worldwide with potential and proven coal-derived oil and condensate accumulations mainly in Australia, Canada, China, India, Indonesia, Greece, New Zealand, Nigeria, North Korea, and United States. Since then additional potential and proven coal-derived hydrocarbons were investigated in Burma, Denmark (North Sea), Egypt, Germany, Greenland, Hungary, Iran, Kuwait, Norway (North Sea and Barents sea) Russia, Saudi Arabia, Tanzania, Thailand, Turkey, Venezuela, and Vietnam. A list of several of these proven coal-derived hydrocarbons in basins in Australia, China, Denmark, Indonesia, New Zealand, and Nigeria are summarized in Table 2.3. Examples of proven coal-derived petroleum have been reported from the coal-bearing rocks of Eocene–Miocene age in Kutai (or Kutei) Basin in Borneo, Indonesia; Jurassic age in Junggar, Tarim, and Turpan Basins in northwestern China and Pennsylvanian age in Bohai Gulf Basin in northeastern China; Cretaceous–Paleocene age in the Bass and Gippsland Basins in Australia, Cretaceous age in Anambra Basin in Nigeria; Eocene age in Taranaki Basin in New Zealand; and Lower Carboniferous age in the Central Graben in North Sea, Denmark. In all these examples of coal-derived hydrocarbons, understanding of the maceral or organic matter composition, organic geochemistry, coalification and maturation, biological source of organic matter, and depositional environments of coal have been paramount importance and useful criteria for coal as source rock (Boreham & Powell, 1993; Clayton, 1993; Snowdon, 1991).

Geologic Age Distribution of Coal-Derived Hydrocarbons

Clayton (1993) reported that the geologic age distribution of coal-derived oils corresponds to the occurrences of waxy crudes. Coal-derived oils are commonly found worldwide in source-and-reservoir rocks of Cenozoic age. Petersen, Lindström, Nytoft, and Rosenberg (2009) reported that Cenozoic coal and coaly sedimentary rocks potentially generate oil, which is related to the paraffinic composition of the coals. For example, Cenozoic coals from Colombia, Venezuela, Indonesia, and Vietnam are composed mainly of huminite particularly detrohuminite, except for the coals in Vietnam, which are rich in liptinite macerals. The palynology (e.g. spore and pollen) of these Cenozoic coals is interpreted to be mainly from ancestral angiosperm plants and subordinately from ferns, which thrived in subtropical to

TABLE 2.3 Coal-Derived Hydrocarbons are Exemplified in Coal-Bearing Sequences from Carboniferous to Tertiary Age Worldwide

Country	Coal Basin	Age and Rock Formation	Depositional Environments	TOC (wt%)	Hydrogen Index (HI mg HC/g)	Vitrinite Reflectance (% R_o)	Types of Organic Matter	References
Nigeria	Anambra	Upper Cretaceous Mamu Fm	Deltaic mires	0.07–61.42	0.05–332	0.47–0.78	Types III and II	Ogala (2011)
Denmark	Central Graben, North Sea	Lo. Carb. Firth Coal Fm	Coastal plain and fluvial mires	ND	171–219	0.90–0.95	Types III and II	Petersen and Nytoft (2007)
Australia	Gippsland	Upper Cretaceous Latrob Group	Coastal plain mires	61.34–69.57	200–350	0.92–1.0	Types II and III	Bishop (2000)
	Bass	Paleocene–Eocene Eastern View coal sequence	Fluvial–deltaic mires	59.03–78.53	133–405	0.3–4.9	Type III	Boreham et al. (2003); Schimmelmann, Sessions, and Mastalerz (2006)
New Zealand	Taranaki	Paleocene–Eocene Kapuni Group	Fluvial and tidal dominated Coastal plain mires	0.25–45	235–420	0.50–0.71	Types III and II	Sykes, Suggate, and King (1991); Killops et al. (1994); King et al. (1998); Thrasher (1991); Flores (2003)
China	Tarim, Jungar, and Turpan	Jurassic Sangonghe, Badaowan, and Xishanyao Fms	Fluvial and lacustrine delta mires	4.7–81.7	4–232	0.42–1.77	Types I or II and Type III	Hendrix, Brassell, Carroll, and Graham (1995)
	Bohai Gulf	Carboniferous Taiyuan Fm	Coastal plain mires	ND	ND	0.56	Type III	Li et al. (2007) Li et al. (2008)
Indonesia	Kutai (Kutei)	Eocene–Miocene Balikpapan Group	Fluvial and littoral–sublittoral mires	62.9	284	0.24–0.59	Types II and III	Davis et al. (2007)

TOC, Carboniferous Taiyuan Fm; HC, hydrocarbon; g, gram; ND, no data; wt%, weight; percent; %Ro. vitrinite reflectance; Fm(s), Formation(s); lo. Carb., Lower Carboniferous.

tropical bogs or mires. Petersen et al. (2009) argued that coals with the same depositional origin and maturation usually show a wide range in potential generation of hydrocarbons as measured by hydrogen index, which vary from 200 to 300 mg HC/g C total organic carbon (TOC). The variation of hydrogen index may be controlled by the different plant material deposited in the mires and mode of preservation of the organic matter of the precursor peat deposits. The New Zealand Tertiary coals, which are waxy, may be comparable and attributable to angiosperm-dominated mire vegetation (Shearer & Moore, 1994).

The potential for coal-derived oils in Indonesia, which is exemplified by the Kutai also called Kutei Basin, is more widespread than originally determined for Cenozoic (Paleogene–Neogene) coals according to Davis et al. (2007). A study by Davis et al. (2007) of more than 500 coaly terrigenous clastics including 432 coal samples from 14 basins in western Indonesia indicates that the coals are hydrogen rich with average hydrogen index of 250–300 mg HC/g C. The Paleogene (e.g. Eocene) subbituminous high-volatile to low-volatile coals are richer in hydrogen than the Neogene (Oligocene–Pliocene) subbituminous high-volatile to lignite coals in which hydrogen ranges from 224 to 365 mg HC/g C and 138–318 mg HC/g C, respectively. In a larger context, Davis et al. (2007) concluded that the most hydrogen-rich coals or most oil prone are high ash and liptinitic Eocene coals deposited in submerged mires in a transgressive synrift tectonic setting in contrast with equally hydrogen-rich, low ash Oligocene–Miocene coals deposited in raised (ombrotrophic) mires in regressive syntectonic setting. Thus, Tertiary low-rank coals in Indonesia, in particular, and in others parts of the world, in general, have been underestimated for their petroleum potential.

Coal-derived hydrocarbons in older rock units are exemplified in coal-bearing sequences of Carboniferous age in the Bohai Gulf Basin in China and Central Graham in North Sea, Denmark; Permian and Jurassic age in the Cooper Basin in Australia, and Cretaceous age in the Anambra Basin in Nigeria (Table 2.3). Petersen (2004) reported that depositional setting and geologic age mainly govern oil generation from coal. For example, generation potential of Jurassic coal source in the North Sea (Søgne Basin) recorded the generation of gas and condensate in the landward area in contrast to generation of oil in the seaward area independent of age of the coal. Petersen (2004) argued that hydrogen enrichment is directly related to sulfate-reducing bacteria, which in turn, are dependent on the increase sulfur content (Figure 2.24). The direct relationship between oil generation potential and geologic age is demonstrated by the large oil accumulations derived from Cenozoic coals in the Australasian region, which include the Bass and Gippsland Basins in Australia, Taranaki Basin

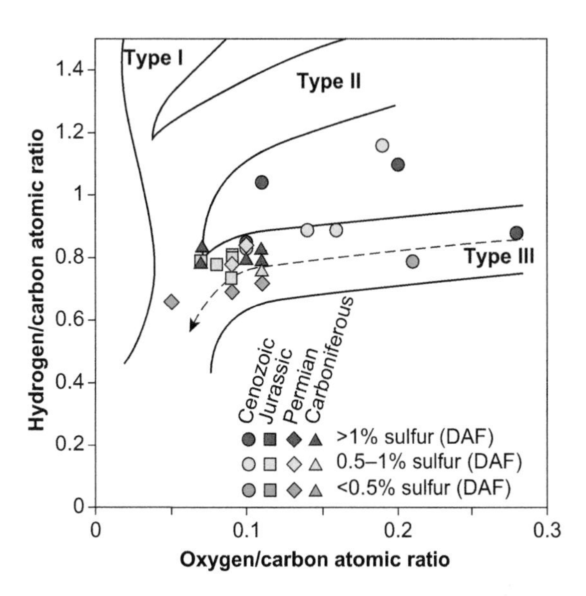

FIGURE 2.24 A van Krevelen (1961) diagram showing the relationship between hydrogen enrichment to sulfate-reducing bacteria as a function of increased sulfur content caused by the coal-forming mire proximal to the marine environment. DAF, dry ash free. *Source: Adopted from Petersen (2004).*

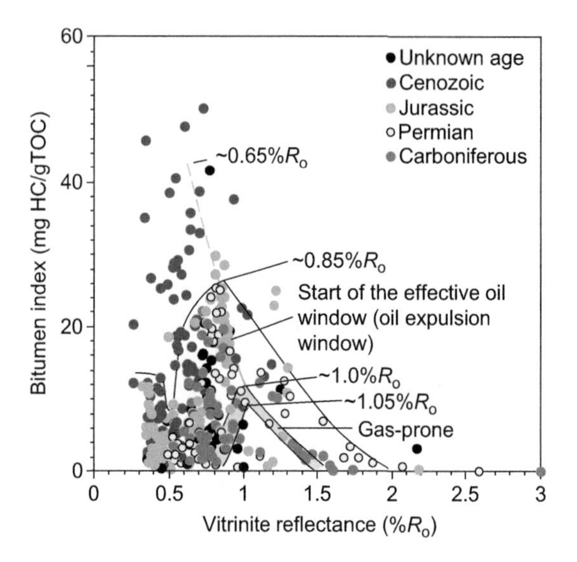

FIGURE 2.25 A diagram showing coals plotted against vitrinite reflectance and bitumen index, which indicates that Cenozoic coals have better potential to generate oil than older coals. mg, milligram; g, gram; HC, hydrocarbon; TOC, total organic content. *Source: Adopted from Petersen (2004).*

in New Zealand, and Kutai (Kutei) Basin in Indonesia. In contrast, Carboniferous coals are associated with large gas accumulations exemplified in the North Sea. The Permian and Jurassic coal-derived oils are typified in the Cooper Basin, Australia and the Danish North Sea, respectively. In general, only minor quantities of oil are related to coals of Permian and Jurassic ages. According to Petersen (2004), the variations in the types of hydrocarbon accumulations sourced from different geologic-age coals may be due to the evolution of more complex and diversified plants (e.g. angiosperms) in the Cenozoic Era, which have better potential to generate oil. As shown in Figure 2.25 Cenozoic coals produce the largest amount of oils, which are expelled at the lowest maturity index (0.65% R_o) and have the broadest effective oil window (0.65–2.0% R_o). The Jurassic, Permian, and Carboniferous coals reach the beginning of effective oil window at a higher maturity index (0.85–0.90% R_o). The Jurassic coals are able to generate and expel hydrocarbons compared to the Carboniferous and Permian coals have limited expulsion efficiency and mainly gas prone. More specifically the effective oil window of Carboniferous coals is actually a gas-condensate window. The Jurassic coal-derived hydrocarbons in the Tarim, Jungar, and Turpan Basins of western China best exemplify the Petersen (2004) model, which generated and expelled both gas and oil. In the Bohai Gulf Basin in eastern China, the results of the study by Li et al. (2007) indicated that the Pennsylvanian coals have not only some oil-generation potential but also high expulsion efficiency and consequently have a large capacity for gas production.

BOX 2.4

LIQUEFIED VERSIONS OF COAL—FACT VS FICTION

The concept of conversion of coal-to-liquids (CTL) commonly known as coal liquefaction is a proven technology and an integral part of the global energy mix in Germany in the 1940s and in South Africa in the 1950s (Hook & Aleklett, 2010; Leckel, 2009; NPC, 2007; Li, 2009; Van Vliet, 2009). In South Africa, Sasol (2011) possesses the only currently operated, commercial-scale CTL plants with proven Fischer–Tropsch (FT) technology with realistic conversion ratios (barrel of oil:ton of coal) of: low (1:1), mean (1.5:1), and high (2:1) (Figure 2.26; Hook & Aleklett, 2010). Eighty-eight CTL projects have gained approval in China with 20–30 more projects pending for approval by the government (AAAS, 2011; Richardson, 2007). Hook and Aleklett (2010) reported that the proposed Chinese CTL projects were initially to produce 9 million metric tons annually of crude oil equivalents by 2010 and projected to increase to 27 million metric tons of crude oil equivalents by 2020 or about 16% of China's crude oil production.

Since the late-twentieth century according to USDOE (2008) the world oil consumption has increased 20% and is expected to increase for additional 30% in 2030. Projected increase of oil consumption is almost entirely accounted, about 95%, by world oil used in transportation. USDOE (2008) forecasted doubling the number of vehicles worldwide through 2030 and during the same period of time is projected to triple in number in developing countries. The United States is the largest oil importer in the world but has about 27% of the world coal supply. Proponents of the CTL technology argue that to solve this imbalance of energy demand and supply is to tap the United States coal reserves for liquefaction (NTIS, 1977). The United States has more than 227 billion metric tons of recoverable U.S. coal reserves, which according to National Mining Association (2011a, b) and USDOE (2008) when converted to oil equivalent will eclipse Saudi Arabia's proven oil reserves by three times. The United States National Coal Council (NCC, 2006) projected CTL production of 429 million liters/day by 2025 using 363 million metric tons for coal consumption. Patzek and Croft (2009) argued that the CTL conversion and inherent low efficiency of the FT process are not the best use of coal as long as it is economical to use natural gas.

The opponents of CTL technology argue that conversion of coal to synthetic fuel (synfuel) will generate a huge volume of greenhouse gases. AAAS (2011) suggests that if the CO_2 extracted from the synthetic fuel during refining is not sequestered, the quantity of CO_2 released is more than twice the amount emitted by conventional hydrocarbons. USDOE (2008) reported that from the environmental standpoint, the CO_2-intensive CTL technology can be resolved by adopting carbon capture and sequestration systems (CSS) of the proposed CTL plants. Although sequestration of anthropogenic CO_2 is possible, the viability is still in question as to potential of leakage and integrity of the geologic sinks for CO_2 sequestration, ecological impacts of sequestration, geologic sinks (e.g. coal, oil and gas fields, saline formations) that make the best storage, safety potential, and commercial feasibility. However, the MIT (2007) study on "The Future of Coal" argued that the proposed CSS to keep from releasing gas emissions into the atmosphere is untested and even if feasible, it would fail to create a clean fuel. Furthermore, large amounts of water are required for the CTL process, which cannot be sustained without disturbing the surface and ground water systems. Currently, the Sasol FT process uses 795 l of water per 159 l of produced synfuel (Patzek and Croft, 2009).

COAL HYDROCARBONS AS PETROLEUM SYSTEMS

The concept of petroleum system introduced by Magoon and Dow (1994) is a "pod" or contiguous volume of active source rock and all related oil and gas including the essential elements and processes required for hydrocarbons occurrence. The elements, which include source, reservoir (e.g. porosity and permeability properties), seal, and overburden rocks are both geological and geochemical in nature. However, coal as source and reservoir rocks of hydrocarbons, knowledge of organic petrology and paleobotany are important elements to consider in the assessment of petroleum system (Abdullah & Abolins,

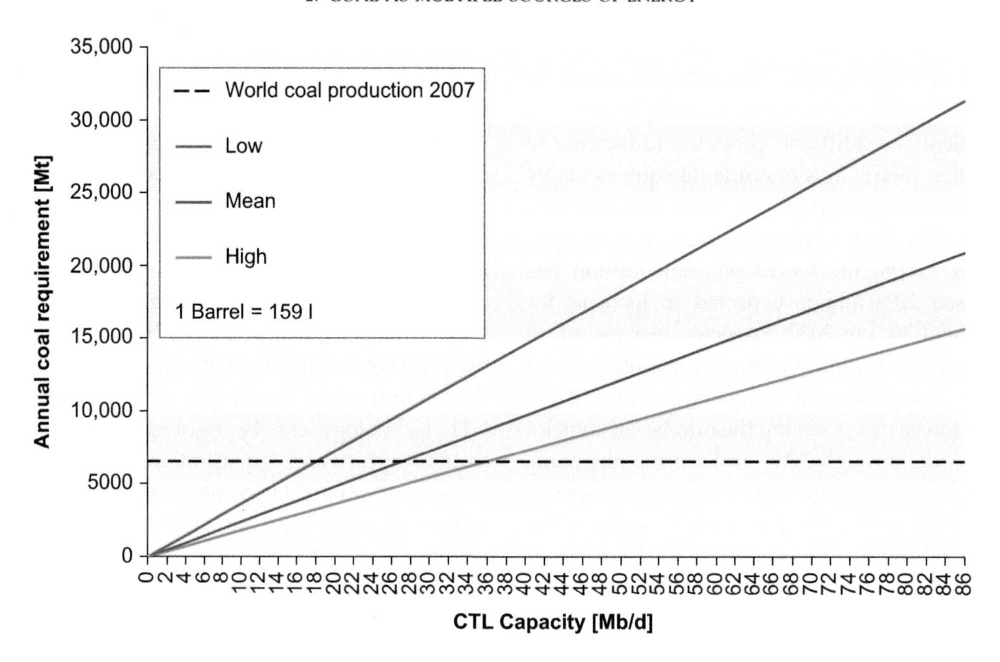

FIGURE 2.26 A diagram showing estimated coal consumption as a function of CTL capacity, as of 2007, for various coal conversion ratios from 1–2 barrels/ton based on the Sasol FT model. Mb/d, million barrels per day; Mt, million tonnes; CTL, coal-to-liquids. *Source: Adopted from Hook and Aleklett (2010).*

2006; Macgregor, 1994; Shearer & Moore, 1994; Sykes & Johansen, 2006). The processes, which include timing of hydrocarbon generation, migration (expulsion), accumulation, and trap formation, are also both geological and geochemical in nature; however, in coal, understanding of depositional environments and processes (e.g. modern and ancient analogs) are key to the an in-depth analysis of the source rock (Abdullah & Abolins, 2006; Sykes & Johansen, 2006). Additionally, a petroleum system is three-dimensional consisting of stratigraphic, geographic, and time components, are important factors in assessment of coal as a source of gas (Flores, 2004).

Assessments of the potential hydrocarbon occurrences in terms of the total petroleum system traditionally have been categorized as conventional and unconventional accumulations. The accepted definition of conventional hydrocarbon accumulation occupies limited, discrete volumes of rock bounded by traps, seals, down-dip water contacts, and depending upon the buoyancy of oil or gas in water for their existence (Schmoker & Klett, 2007). In contrast, the unconventional accumulation is defined as mainly requiring unusual engineering techniques to extract the hydrocarbons, which unlike the conventional accumulation does not take into account the geological elements and processes as well as the three-dimensional components of the petroleum system. That is, unconventional accumulation, which is the traditional category for coalbed gas, requires the hydrocarbon reservoir to be stimulated (e.g. hydraulic fracturing, treatments) in order to increase flow rates to produce economic volumes of gas or oil derived from coal. Thus, under these definitions application of the parameters of petroleum systems toward the assessment of coal as source and reservoir is a conundrum. Furthermore, this problem is exacerbated by coal-derived hydrocarbons as mixture of gas and oil occurring independently as in situ accumulation and/or accumulation expelled from the coal source rock into adjoining reservoir rocks. The gas and oil derived from coal are all genetically related hydrocarbons that can be traced to the source or provenance by using principles of coal geology, petroleum geology, geochemistry, and hydrology. However, generation and expulsion of oil and gas from coal have been controversial for the lack of strong evidence that all coals are sources of oil or only of gas (McCormack et al., 2006). Schimmelmann et al. (2004) highlighted this controversy in a study of terrestrially derived petroleum system in Australia where oils are sourced from multiple and heterogeneous terrestrial sequences and geochemical investigation is central to oil-to-source correlations. However, these problems were overcome in assessing the total petroleum system of technically undiscovered unconventional and conventional accumulations of coalbed gas in the Powder River Basin in Wyoming, United States (Flores, 2004).

Conventional Accumulations of Coal-Derived Oil

The conventional accumulations of hydrocarbons derived from coal are oil and gas, which migrated or expelled from the coal source pod at a critical moment (Magoon & Dow, 1994) and trapped/sealed in adjoining porous and permeable reservoir rocks (e.g. sandstone). Coal-derived oil conventional accumulations are exemplified by the Taranaki Basin in New Zealand (Johnston et al., 1991), Bass Basin in Australia (Boreham et al., 2003), and Kutai (Kutei) Basin in Indonesia (Curaile et al., 2006; Petersen et al., 2009). These entire coal-derived conventional hydrocarbon accumulations in the Australasian region are in Cenozoic deposits ranging from Paleocene to Miocene in age. According to Boreham et al. (2003) age, paleoclimate, depositional environments, and floral evolution mainly governed the geochemical setting of these oil-prone coal sequences. The Australasian region is unique during the Cretaceous–Cenozoic Eras when tropical angiosperm and temperate gymnosperms were common vegetation in coastal plain mires or bogs (Macgregor, 1994; Shearer & Moore, 1994). Shearer and Moore (1994) study of coal banding of Cretaceous and Eocene deposits in the South Island coal basins, New Zealand indicated transformation of mire vegetation from gymnosperms to angiosperms, respectively. Killops, Woolhouse, Weston, and Cook (1994) and Killops et al. (2008) noted that this paleobotanical change is reflected by the geochemical evolution of diterpane-rich older source rocks to triterpane-rich younger sources in the Taranaki Basin. Coals formed during the Cenozoic Period are interpreted to have accumulated in mires/bogs resulted in concentration of the more resistant, hydrogen-rich components of plant remains such as cuticles and resins due to microbial degradation (Powell, 1988).

Analogs of Coal-Derived Oil Conventional Accumulations

TARANAKI BASIN, NEW ZEALAND

The Taranaki Basin is the only area that produces commercial hydrocarbons in New Zealand (Figure 2.27). Production is from sandstone reservoirs in onshore and offshore fields in the coal-bearing Eocene Kapuni Group, which is a proven exploration play for oil, condensate, and gas in the Taranaki Basin, New Zealand (Figure 2.27) (Cook & Gregg, 1997; King et al., 1998; King & Thrasher, 1996). The Kapuni Group contains alternating cycles of stacked coarsening-upward marine-shoreface mudstone, siltstone, and sandstone or parasequence sets laterally interfingering with fluvio-tidal coal, carbonaceous shale, mudstone, siltstone, sandstone, and conglomeratic sandstone (Flores,

2003; see Chapter 3 for definitions of terminology on sequence stratigraphy). Regional erosion surfaces or sequence boundaries within the Kapuni Group reflect maximum-flooding surfaces during sea-level rise. The tidal influenced coastal plains were formed during the intermittent transgressions of paleoshorelines where coal buildup in tidal-influenced coastal marshes was bounded seaward by stacked marine shoreface sandstones and in the hinterland by alluvial-belt mires. These terrestrial depositional environments were controlled by third and fourth order fluctuations of sea level, basin subsidence, and accommodation space. The vertical and lateral stratigraphic distributions of the Kapuni coals which served as source for gas, condensate, and oil are typified by the geophysical and lithological logs of Kapuni-2

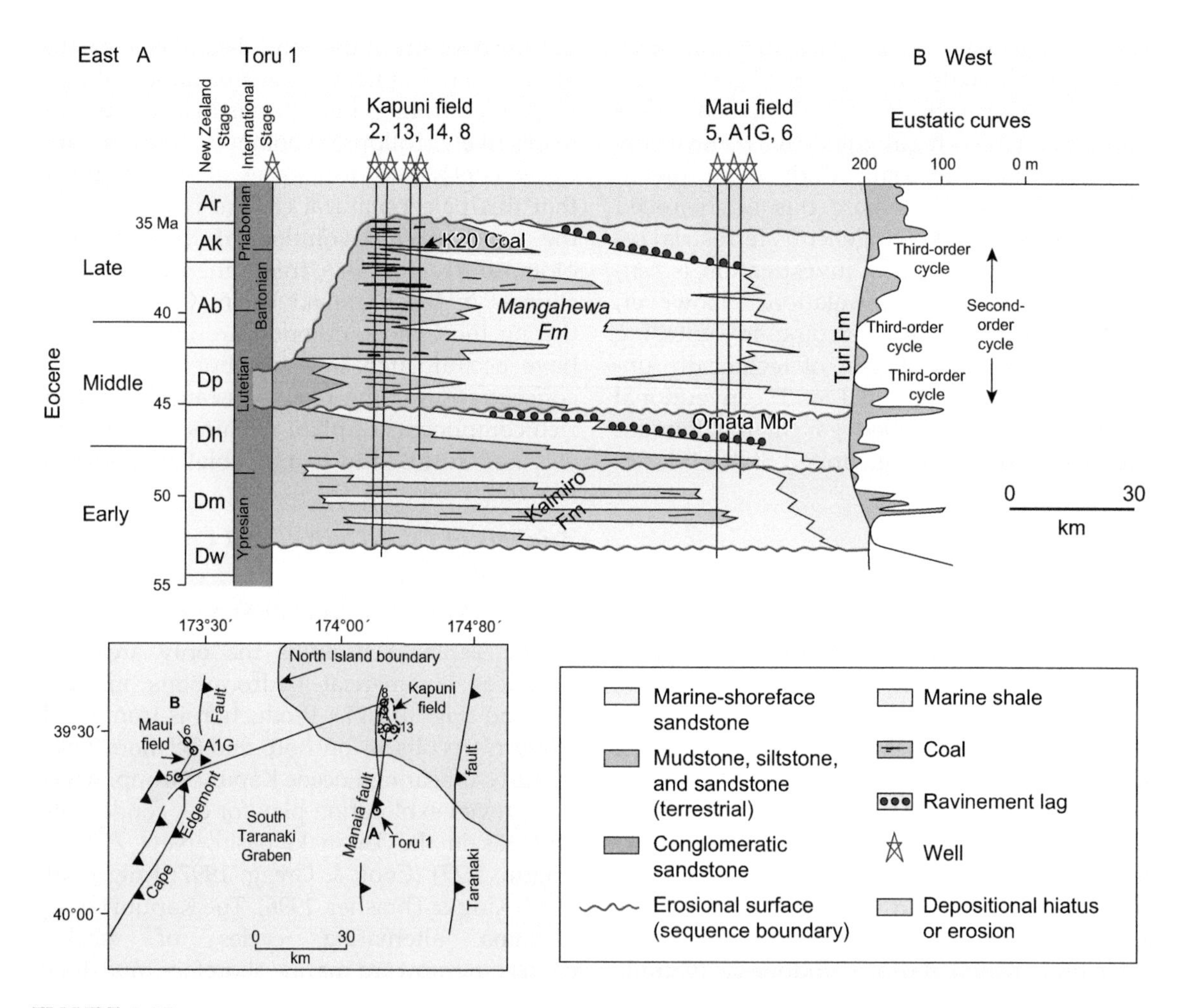

FIGURE 2.27 A generalized cross section showing the vertical and lateral stratigraphic variations of the Eocene Kapuni Group, which contains coal-derived oil, gas, and condensate. The lateral and vertical shifts of facies (coal-bearing vs shoreface marine sandstones) are defined by the landward and seaward migrations of the paleoshorelines during transgressions and regressions. *Source: Adopted from Flores (2003).*

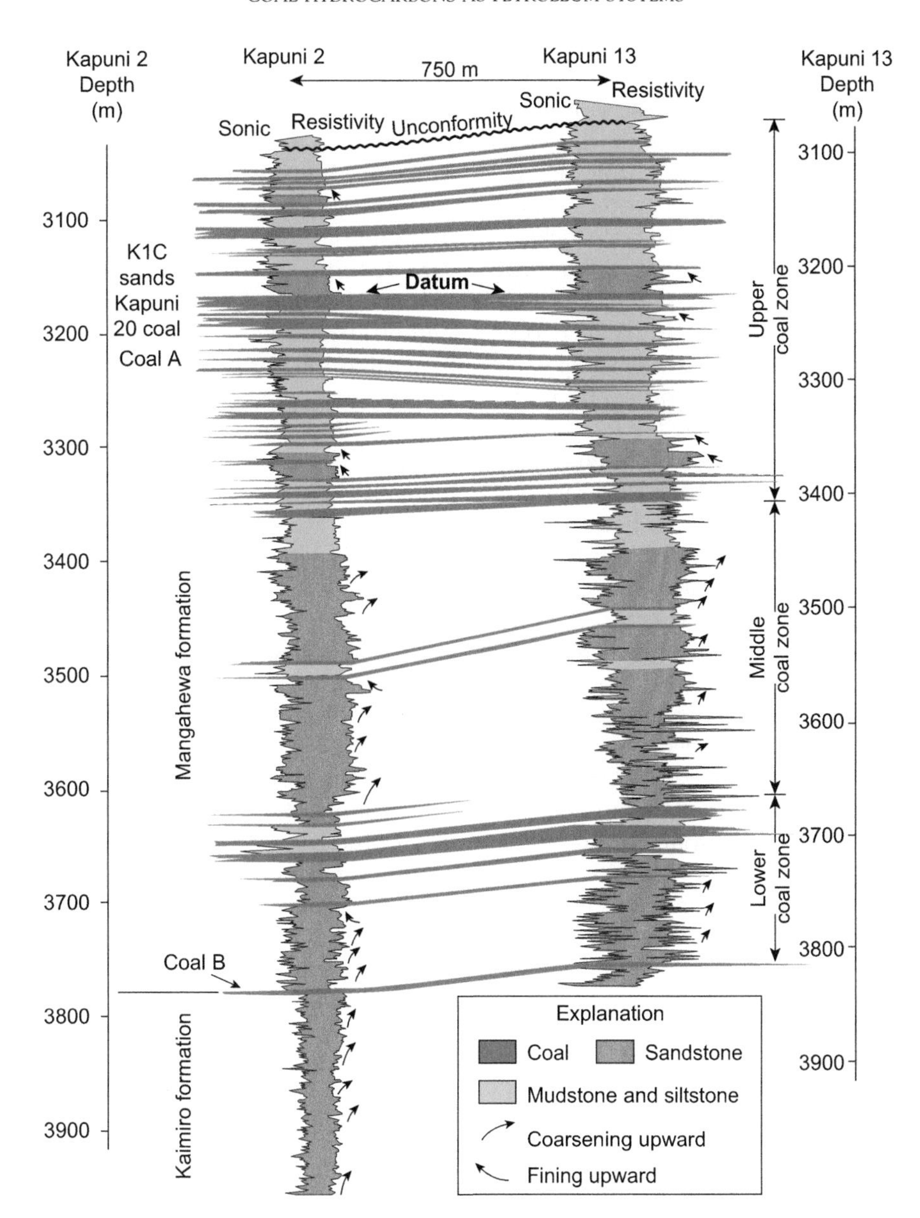

FIGURE 2.28 A cross section showing detailed vertical and lateral variations of the Kapuni Group, which contains mainly more numerous and thickening coal beds upward. See Fig. 2.27 for locations of Kapuni 2 and 13 wells. *Source: Adopted from Flores (2003).*

and -13 wells, which are 0.75 km apart (Figure 2.28). The Kapuni Group consists of the Kamiro Formation overlain by the Mangahewa Formation, which in turn can be divided into three coal zones: lower coal zone (3660–3810 m), middle coal zone (3405–3660 m), and upper coal zone (3080–3405 m). These coal zones are separated by coarsening-upward marine shoreface sandstone-dominated intervals as much as 125 m thick and fining-upward fluvial-tidal channel sandstones commonly serving as reservoir rocks. The lower coal zone consists of four coal beds ranging from 2 to 10 m thick. The middle coal zone contains two to three coal beds ranging from 2 to 5 m thick. The upper coal zone includes 28 to 33 coal beds ranging from 2 to 14 m thick. Thus, the coal buildup in the Kapuni Group coal became well established from Early to Late Eocene time as evidenced by the upward thickening of the coal beds. Each of the lower, middle, and upper coal zones, which are separated by landward marine-shoreface sandstones, represents coal buildup related to the fourth order fluctuation of sea level during a time interval of about 1.5 mya (Flores, 2003). However, within this fourth order cycle coal buildups occur in a number of coal beds that vary in thickness and lateral continuity. This is exemplified by the 14 m thick K20 coal formed in ombrotrophic mire (less than 2% ash yield), which is split by fining-upward fluvial channel sandstones into three beds within 750 m distance (Figure 2.28).

Sykes and Johansen (2006) developed a coal depositional facies model that differentiates rheotrophic (planar) and ombrotrophic (raised or domed) mire coals (see Chapter 3 for more discussions), which can be used to predict gas:oil ratios in terrestrially sourced petroleum systems in Cretaceous–Tertiary coal basins in New Zealand. Rheotrophic mires are low- and flat-lying (hence planar) mires that are prone to detrital floods from sediment-laden waters (e.g. fluvial, deltaic, and tidal channels, marine incursions). Thus, peat formed in this setting due to abundance of mineral nutrients from floods, according to Sykes and Johansen (2006) results in thin coals containing relatively abundant wood tissue within a collodetrinite-rich matrix (see Chapter 5 for definition), together with variable leaf biomass. In contrast, raised mire above flood levels results in thick coals containing abundant cork tissue in a largely cork-derived, vitrodetrinite-rich matrix, and rare leaf material (see Chapter 3). Multivariate analysis by Sykes and Johnson (2006) has shown that nonvolatile paraffinic oil potential is related to the abundance of leaf-derived liptinites (cutinite and liptodetrinite). Thus, rheotrophic coals have a predisposition for better oil potential than ombrotrophic (see Chapter 3 for definition) coals because of more leaf biomass input and/or likely preservation under higher groundwater and/or floodwater levels. Sykes and Johansen (2006) concluded that rheotrophic coals are more oil-prone influenced by high rates of accommodation increase compared to ombrotrophic coals, formed in an ever-wet climate, are gas-prone influenced by low-moderate rates of accommodation (see Chapter 3). Flores (2003) interpreted occurrence of thin coals in the lower and middle parts of the Kapuni Group, which is separated by thick marine-shoreface sandstones, suggests coal buildup was interrupted by rapid landward stepping shoreface events followed by shelfal mudstones, which served as seal rocks. However, the buildup of thick (up to 14 m) and numerous vertically stacked coals in the upper part of the Kapuni Group represents prolonged accumulation of peat mires behind a very thick, vertically stacked marine-shoreface sandstones, which served as platforms for freshwater mires that accumulated thick coals in an aggrading fluvial–tidal influenced coastal plain (see Chapter 3).

Application of the study of Sykes and Johansen (2006) to the stratigraphic–sedimentologic study of Flores (2003) in the Taranaki Basin suggests that the relatively thin coal beds in the lower and middle parts of the Kapuni Group were probably formed in rheotrophic mires in contrast to the relatively thicker coals in the upper part of the Kapuni Group, which probably were formed in ombrotrophic (raised or domed) mires (see Chapter 3). Indeed, if the thin rheotrophic coals are oil-prone according to Sykes and Johansen (2006) then by inference the thin coal beds in the lower-middle part of the Kapuni

Group may have sourced the oil-charged sandstone reservoirs in the upper part of the Kapuni Group. By the same token, the thick ombrotrophic coal beds in the upper part of the Kapuni Group may have sourced gas-charged reservoir sandstones in the upper Kapuni Group. The study of Killops et al. (1994), Killops et al. (2008) on the biomarkers triterpene, hopanes and steranes extracted from coals from selected onshore wells in the deeper part of the Taranaki Basin may shed light into the viability of the coal facies model of Sykes and Johansen (2006). Results of Killops et al. (1994), Killops et al. (2008) indicate that the produced hydrocarbons in the Kapuni Group have thermal maturities comparable to or approaching those of the deepest Kapuni coals. Although the hydrocarbons were generated within the Kapuni Group the hydrocarbons are expelled only from the deepest Kapuni coals or possibly from Paleocene–Upper Cretaceous coals of the underlying Pakawau Formation, thus, exhibiting higher maturity levels. Specific biomarkers in the produced hydrocarbons suggest the possibility of multiple coal source rocks and that the hydrocarbons have migrated to the present shallower sandstone reservoirs. Thus, the coal facies model of Sykes and Johansen (2006) support the idea that expulsion of hydrocarbons probably occurred from deeper and thinner Kapuni coals to stratigraphically higher charged sandstone reservoirs. King et al. (1998) reported that oil produced in the Taranaki Basin is related to source rocks of particular age from biomarkers (e.g. relative angiosperm and gymnosperm component) such that three families of oils in various fields are derived from (1) Late Cretaceous, (2) Paleocene or mixed Late Cretaceous to Eocene, and (3) Eocene sources.

BASS BASIN, AUSTRALIA

The main targets of exploration and development in the Bass Basin, in Australia (Figure 2.29) are mainly Paleocene–Eocene fluvial-deltaic reservoir sandstones (Figure 2.30) (Boreham et al., 2003). Recently, discovered hydrocarbons are reservoired in the Late Paleocene–Middle Eocene age sandstones, with the largest discovered accumulations 12.6–16.8 Bcm GIP and 11,129 million liters oil-in-place (OIP) (Lennon, Suttill, Guthrie, & Waldron, 1999). A well (White Ibis 1 in Figure 2.29) also encountered a gas column in the Paleocene to Early Eocene sandstones, with estimated mean resource of around 2.4 Bcm and 1749 million liters OIP (Lennon et al., 1999). Shale units within the Eastern View sequence act as intraformational seals for gas reservoirs; whereas, both intraformational seals of the Eastern View sequence and the overlying marine Demons Bluff Formation have sealed the liquid hydrocarbons (Figure 2.30).

Boreham et al. (2003) and Trigg and Blevin (2004) interpreted the depositional environments (see Chapter 3 for more discussions) in the Bass Basin initially as fluvial and lacustrine during the Early Cretaceous succeeded mainly by fluvial during the Upper Cretaceous to Middle Eocene. Fluvial-deltaic and lacustrine depositional environments followed during the late Early Eocene–Middle Eocene. These depositional settings, in turn, were succeeded by aggradational accumulation of stacked coaly facies in delta plains and marginal lacustrine environments influenced by slow rate of accommodation at the onset of latest Early Eocene and younger time period. The coals deposited during this time are significantly thicker (e.g. 22 m thick) than early deposits. The Early Eocene coal depositional trend of upward thickening with thicker coals most likely formed in ombrotrophic mires in the Bass Basin, which is very similar to the mires in the Taranaki Basin, New Zealand except the Eocene Kapuni Group was deposited mainly by fluvial-tidal influenced to shoreface–shelf environments in transgressive system tracts (see Chapter 3). The potential targets of hydrocarbon-charged reservoirs are in the Eastern View Coal Measures of Paleocene to Early Eocene in age (Figure 2.30), which are typified by upward-coarsening nearshore and shoreface sandstones, deposited in coastal strandplain

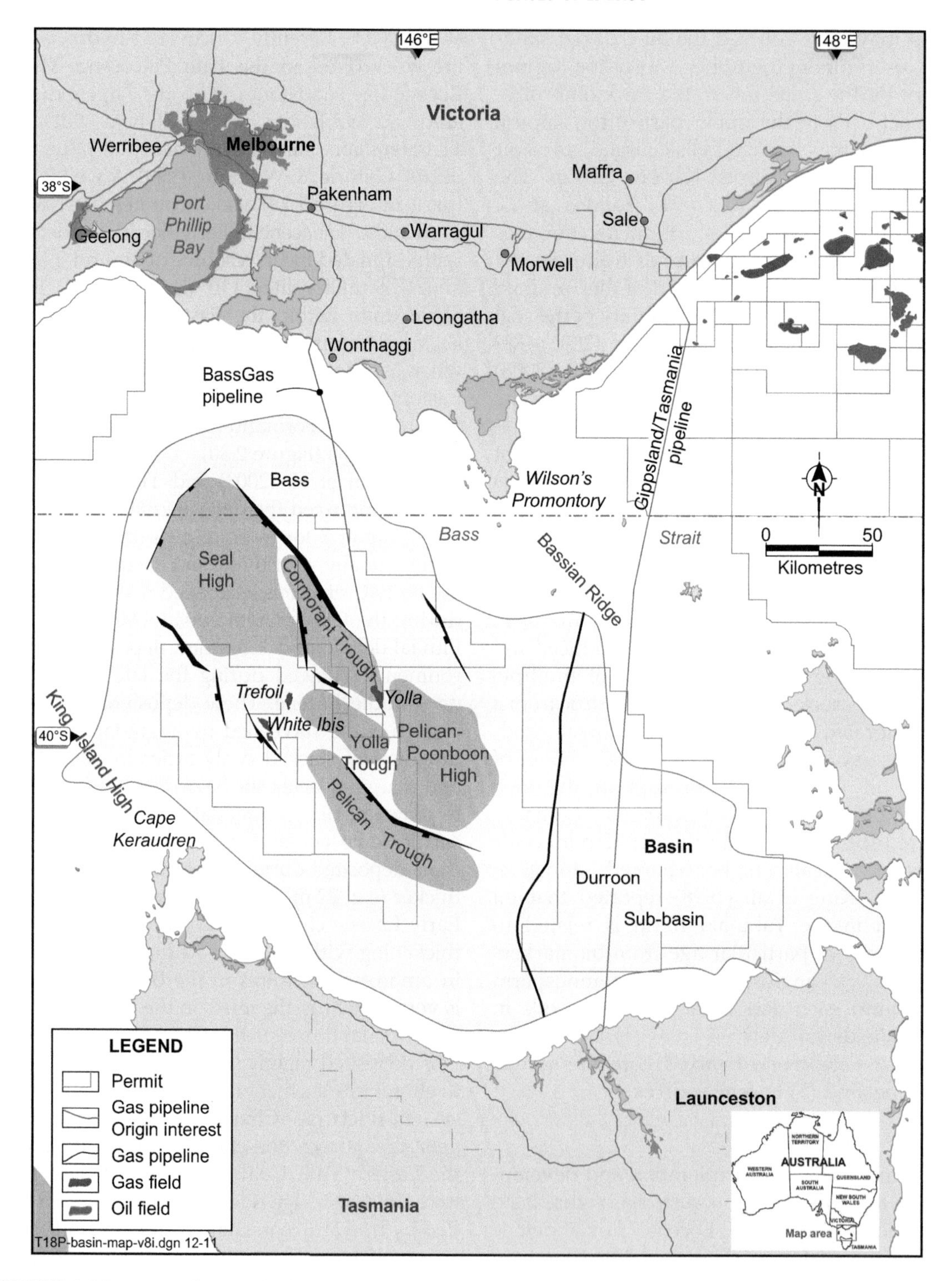

FIGURE 2.29 A map showing the location and leases in the Bass Basin between mainland Australia and Tasmania. See inset map for location of map area. Illustration from Origin Energy, Australia.

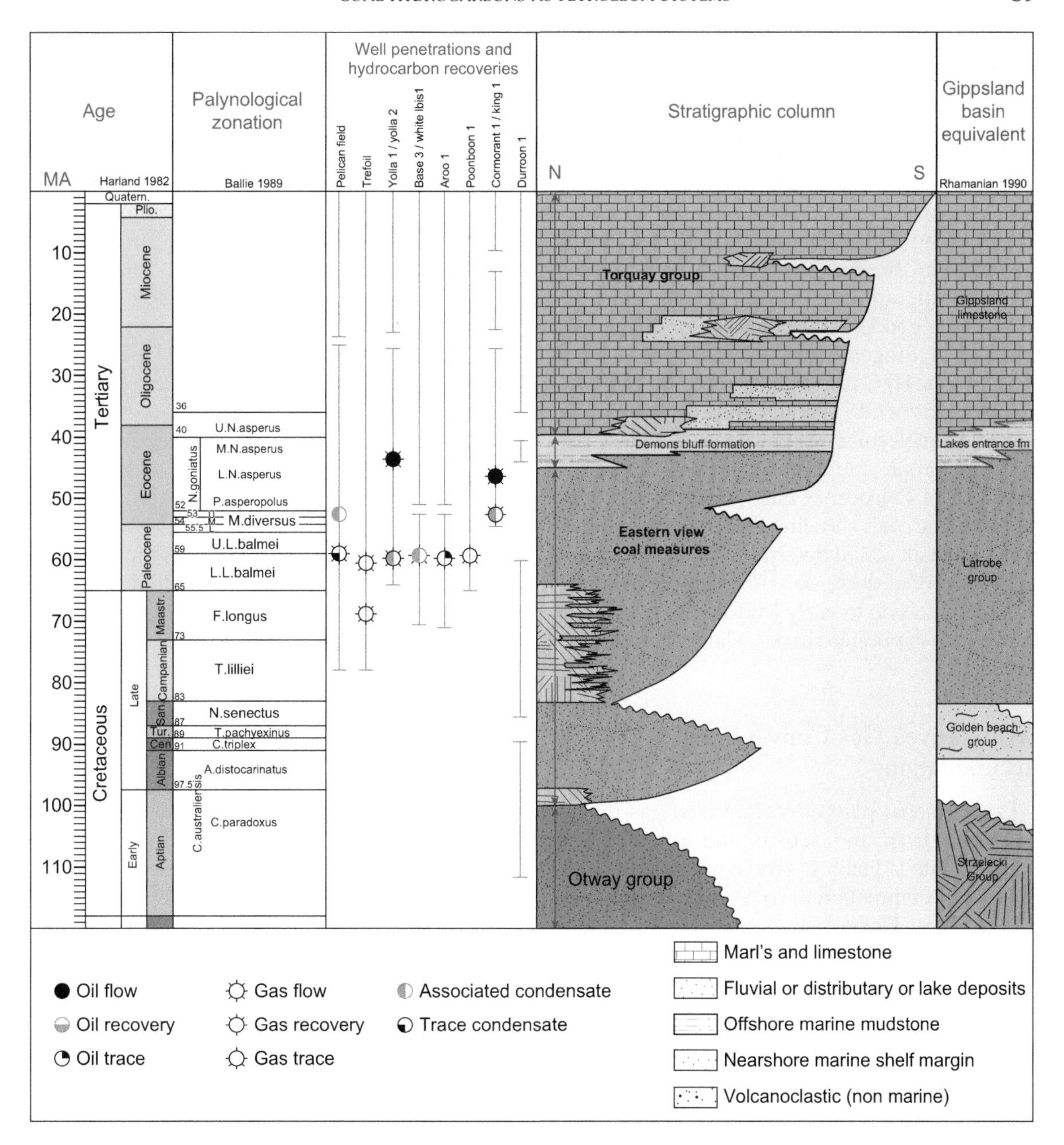

FIGURE 2.30 A generalized stratigraphic column and facies of Cretaceous and Cenozoic rocks in the Bass Basin (see Fig. 2.29 for location). Illustration from Origin Energy, Australia.

environment adjacent to a large lake or restricted sea (Boral Energy, 2011). Specifically, the strandplain environment is interpreted to include lagoon fills, aeolian dunes, beach backshores, tidal channels, and shoreface swash/backwash and beach ridge zones (see Chapter 3 for more discussions).

According to Boreham et al. (2003) coal facies dominate in the Early Eocene to early Middle Eocene stratigraphic interval. The coals in this interval show good source potential for oil generation with hydrogen index (HIs) >300 mg HC/g TOC, atomic H/C ratios of 0.8–1.0, and high pyrolysis yields of C15 + nonvolatile oil with abundant high molecular weight linear alkanes. The most oil prone and volumetrically significant coals are of Paleocene–Early Eocene age with the latter age coals showing the best oil potential. Boreham et al. (2003) determined that coals with liquid potential decrease with younger rocks of late Early Eocene to early Middle Eocene in age and much younger rocks show no liquids potential.

Coal-Derived Gas Conventional Accumulations

Conventional play of coal-derived gas that is generated from the source and migrated or expelled into adjoining sandstone reservoirs is common accumulation in coal-bearing sequences (Flores, 2004; Flores, Stricker, & Kinney, 2004). This play is often dismissed in favor of targeting the unconventional gas in the coal reservoirs because of smaller gas reserves. That is, a cubic meter of coal can contain about 35 times the volume of natural gas that exists in a cubic meter of a conventional sandstone reservoir. In addition, the coal-derived gas is often classified as part of conventional natural gas, sometimes known as coalbed natural gas, being accumulated outside the coal reservoirs and for the lack of evidence of source-to-reservoir correlations. Conventional coalbed gas plays are mainly as stratigraphic and structural traps. Stratigraphic traps are commonly developed in channel sandstones with gas contained by lithologic pinchouts into adjoining finer-grained floodplain mudstones and siltstones, which serve as seal rocks. Structural traps are mainly tectonic folds from horizontal compression and compaction folds. Compaction folds are very common in coal-bearing sequences because of the interbedded sandstones, finer-grained shales, mudstones, and siltstones, and coals and carbonaceous shales. During burial and loading of overlying sedimentary rocks physical compaction develops in which the sands (e.g. fluvial and distributary channel deposits) compact less than surrounding fine-grained sediments. The difference of average initial porosity of sands (about 30%) and muds (about 50–60%) produces differential compaction due to varying porosity. Channel sandstones, which are lenticular or lensoid shape, which pinched out at the margins create fold effects as interbedded fine-grained sediments and coal wrap around the sandstone bodies during differential compaction. Thus, a channel sandstone reservoir form a localized gas field in contrast to the structural folds, which consist of numerous channel sandstone reservoirs, occurring as regional gas fields.

Powder River Basin, Wyoming and Montana, USA

Fluvial channel sandstone reservoirs interbedded with coal beds and finer grained rocks in the Paleocene Fort Union Formation in the Powder River Basin in Wyoming and Montana in the United States are excellent analogs of conventional gas-charged reservoirs (Flores, 2004). The Fort Union coals are well known to contain coalbed gas, which served as the source. These coal beds were interpreted by Flores (2004) as deposited in tropical to subtropical ombrogenous (raised or domed) mires or bogs due to low ash yield (see Chapter 3 for more discussions). The Fort Union fluvial channel sandstones, which are up to 90 m thick, were deposited mainly in streams flanked by fine-grained floodplains and ombrogenous and topogenous peat mire deposits. The fluvial channel sandstones, which make up more than one-third of the total volume of the Fort Union Formation, commonly pinchout into and bounded mainly by floodplain mudstones. The coal beds directly or indirectly underlie and overlie the fluvial channel sandstones, indicating immediate communication or buffered between the reservoir beds and the coalbed gas source rocks. When the channel sandstones directly overlie the coal beds, direct communication is via the basal erosional contact. When barriers of finer grained sediments are between coal beds and fluvial channel sandstones, communication with the reservoirs is via fractures, microfaults, and sandstone dikes. Early formed biogenic coalbed gas (Flores, Stricker, Rice, Warden, & Ellis, 2008) potentially migrated into adjoining sandstone reservoirs during burial and compaction of detrital sediments, which permitted entrapment of the gas by intervening mudstone seal rocks.

Flores (2004) suggested that the most important mechanism of gas expulsion or migration from the coal source and entrapment in the sandstone reservoirs was the lowering of hydrostatic pressure that held the gas in the coal source. Groundwater was originally established in the coal beds by meteoric water recharge at the outcrops about 10 mya during erosion and uplift of the Powder River Basin. Flores (2004) proposed that during the Pleistocene glacial and interglacial periods the repetitive lowering and rising of groundwater levels, respectively, affected the hydrostatic pressures in the source and reservoir rocks. That is, the lowering of groundwater level during the Pleistocene glacial times decreased hydrostatic pressure in the coal beds permitting release or expulsion of gas from the source rock into the fluvial channel sandstones, in turn, entrapped by seal rocks. During the Pleistocene interglacial times when recharge from surface and/or meteoric water ensued at the outcrops along the basin margin, groundwater level rose. Subsequent recharged water flow basinward flushed and brought nutrients to the coal beds, which served as catalysts for bacterial activity and regeneration of biogenic gas in the shallow part of the Powder River Basin (Flores et al., 2008). Permeable and 4.0–2.8 mya old clinkers (burned coal and roof rock) covered about 4144 km^2 of the basin margins and served as groundwater recharge and storage areas, which flowed down dip along coal beds in proximity to potential sandstone reservoirs in the upper part of the Fort Union Formation.

The Fort Union coalbed gas source-to-reservoir correlation was determined by Rice and Flores (1990; 1991) using methane and deuterium isotopes of both coal and sandstone gases. The methane is isotopically light values ($\delta^{13}C1$ values −56.7 to −60.9 per mil) and enriched in deuterium (δD values −307 to −315 per mil). The gases are interpreted as biogenic in origin and generated in the coals by fermentation prior to main phase of thermogenic generation by devolatization (see Chapter 4). Flores et al. (2008) reinterpreted the origin of biogenic gas particularly in the deeper part of the Powder River Basin as mainly generated from coal by carbon dioxide reduction and only selected areas in the shallow basin margin where gas was generated by fermentation. Thermogenic gas was probably generated from deeper Cretaceous coal beds. Commercial quantities of coalbed gas have been mainly produced from shallow (61–122 m below the surface) fluvial channel sandstone reservoirs at the northeast margin of the Powder River Basin since 1987. In the same area, the Chan field produced about 0.04 Bcm from fluvial channel sandstones 91–152 m deep interbedded by Canyon, Cook, and Wall coal beds, which may have served as source rocks. Coalbed gas saturation in the upper parts of these sandstone reservoirs is recognized by upward increasing resistivity and unchanged gamma ray response (Oldham, 1997).

Initial coalbed gas exploration and development in the Powder River Basin focused on the

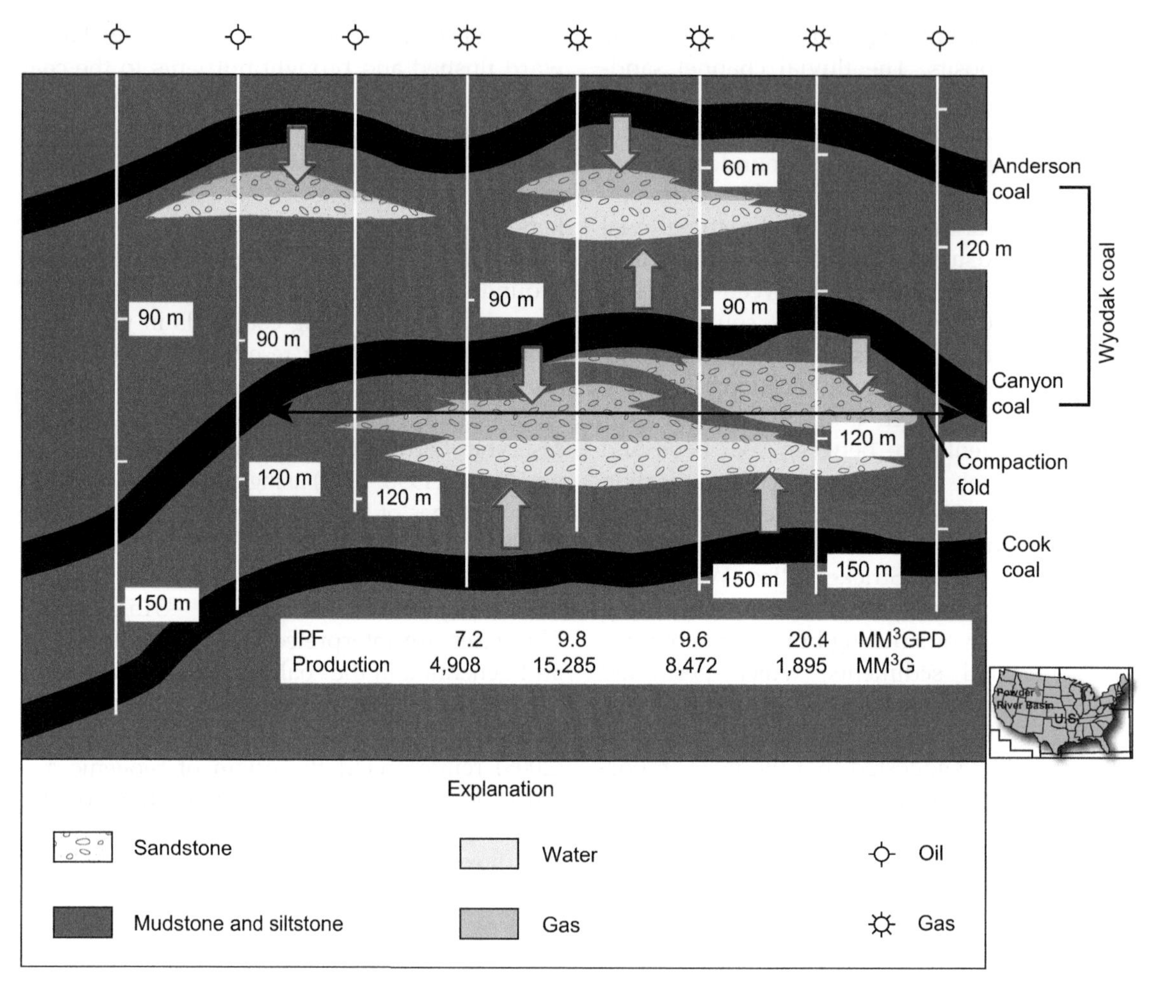

FIGURE 2.31 A generalized cross section showing compaction fold in the Fort Union Formation, which contains fluvial-channel sandstone reservoirs and associated water/gas contact, in coalbed gas fields in the eastern Powder River Basin, Wyoming. Block arrows show likely gas migration from coal sources. IPF; initial production flow; MM^3G, 1000 cubic meter of gas; MM^3GPD, 1000 cubic meter of gas per day. Vertical scale at sea level. No horizontal scale. *Source: Adopted from Flores (2004) as modified from Oldham (1997).*

Fort Union fluvial channel sandstones (Oldham, 1997; Randall, 1989). Commercial quantities of coalbed gas have been produced from shallow (60–150 m below the surface) fluvial channel sandstone reservoirs at the northeast margin of the Powder River Basin from 1987 to the 1990s. In the same part of the basin, Oldham (1997) reported large coalbed gas accumulation ($CH_4 = 94.1\%$; $N_2 = 5.8\%$; $CO_2 = 0.1\%$) and production (0.05 Bcm) from a depth of 104 m from fluvial channel sandstones between the thick coal beds (Figure 2.31). The gas accumulation in the coalbed gas fields occurs in differential compaction folds, which resemble an anticline as shown in Figure 2.32 (Oldham, 1997). An earlier study by Randall (1989) reported a

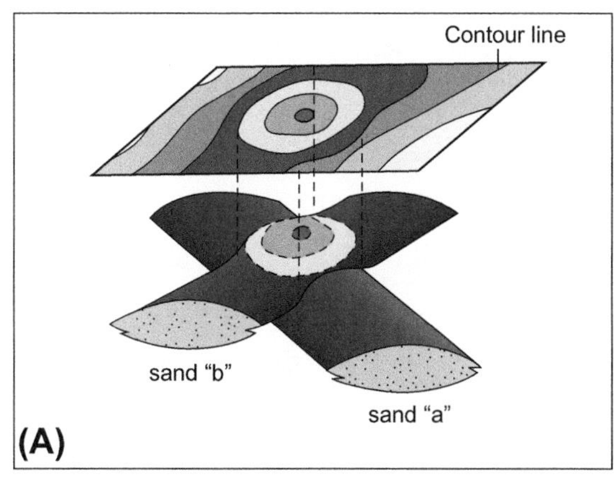

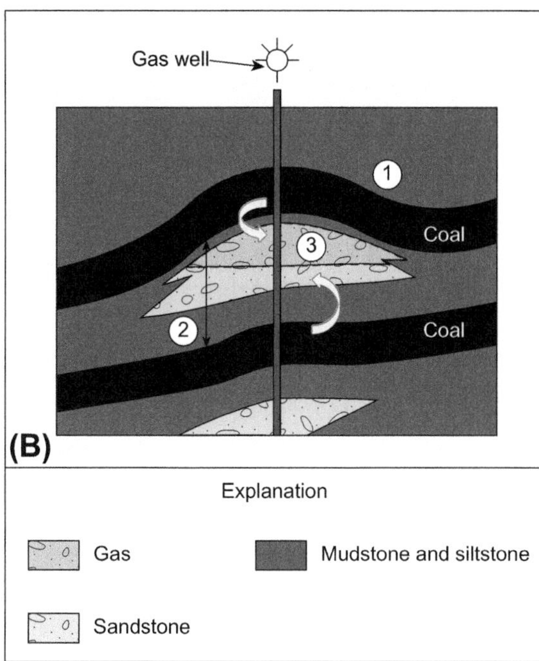

FIGURE 2.32 Diagrams (A) and (B) showing Fort Union sandstone reservoirs between coal beds in the Powder River Basin coalbed gas field (see Fig. 2.31). Diagram (A) shows enhanced closure on a compaction trap in the stacked sandstone reservoirs (sands "a" and "b"). Not shown between the sandstone reservoirs is interbedded coal and fine-grained sediments. Diagram (B) shows (1) a compaction fold with a structural closure, (2) increased noncoal interval (arrow) below the upper coal bed, and (3) gas-saturated top of the sandstone. Curve arrows show likely gas migration from coal sources. *Source: Adopted from Flores (2004) as modified from Oldham (1997).*

significant relationship of thick, stacked sandstone bodies with compaction-induced closures and increased cumulative coalbed gas production in the gas fields (Figures 2.32 (A) and (B)). The coalbed gas sandstone reservoirs have high self-potential and very little water production; however, some sandstone reservoirs have gas/water contacts. The best producing wells in the coalbed gas fields are found at the crest of structures (e.g. anticline) and structural lows with the thickest sandstone pays. The anticlines, which exhibit closures at the top of overlying coal beds where coalbed gas is trapped, resemble compaction folds described by Law, Rice, and Flores (1991).

Cook Inlet Basin, Alaska, USA

The Oligocene–Pliocene (Neogene) coal-bearing rocks on the Cook Inlet Basin in Alaska typify another example of conventional gas accumulations (Flores et al., 2004). Coal beds identified in lithologic logs of the Neogene formations are directly associated with the high-gas shows indicated on the mud logs. However, the coal beds are low rank (e.g. subbituminous) with vitrinite reflectance from 0.32% to 0.66%. Thus, these coal beds are thermally immature and incapable of generating thermogenic gas. However, the isotopic markers, much like those in the Powder River Basin, indicate biogenic origin (Flores et al., 2004). The biogenic origin of the gas may be formed by the influx of fresh melt water that brought in nutrients and bacteria into the coal-bed outcrops during time of glaciation. The gas fields along the northern and eastern parts of the Cook Inlet may have been impacted by glacial drift, which infiltrated and recharged the groundwater promoting biogenic activity along the basin margins much like in the Powder River Basin. The high-latitude location of the Cook Inlet Basin where topographically low-lying glaciers advanced and retreated during the Pleistocene time may have played a major role in establishment and change of

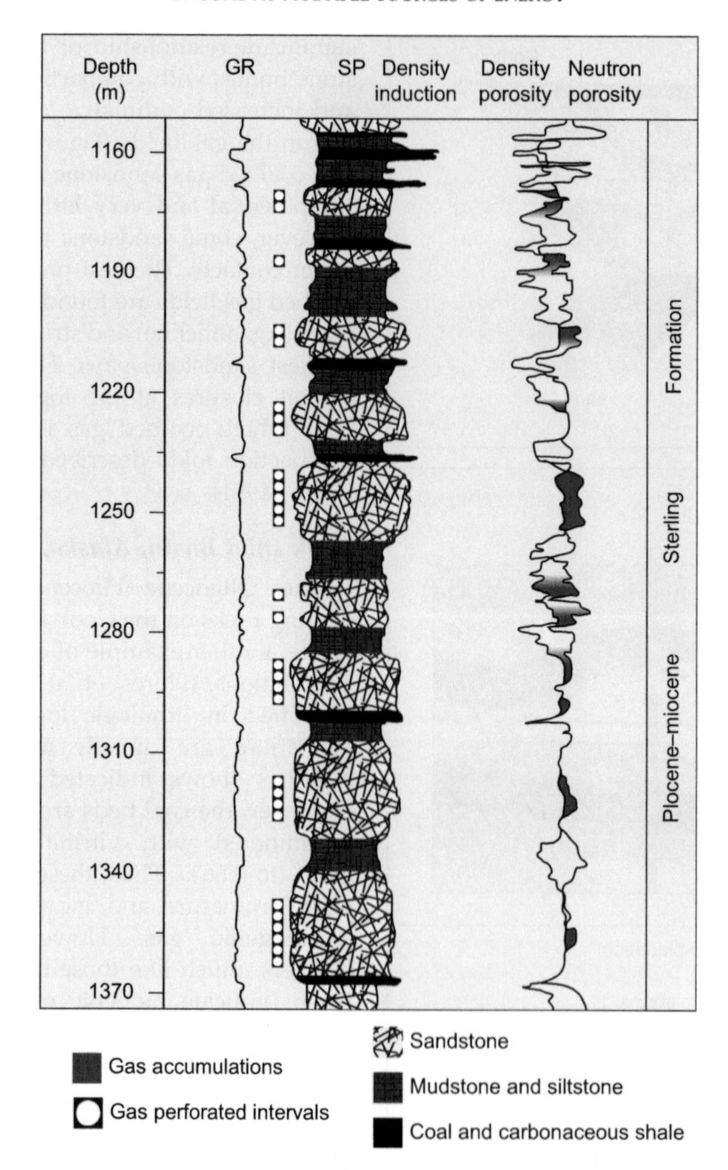

FIGURE 2.33 A stratigraphic section and related geophysical and lithological logs of the Pliocene–Miocene Sterling Formation in the Cook Inlet Basin, Alaska showing coalbed gas accumulations in fluvial channel sandstones, which are sourced from adjoining coal beds. GR, gamma ray; SP, spontaneous potential; m, meter. *Source: Modified from Flores et al. (2004).*

groundwater levels in the coal beds. But more importantly, the fluvial channel sandstones interbedded with the coal beds are gas-charged reservoirs, which migrated from the coal sources (Figure 2.33). The vertically stacked gas-charged sandstone reservoirs have provided multiple targets of natural gas development in the basin.

The Neogene gas play in the Cook Inlet Basin includes 18 gas fields mainly along the northern and eastern part of the basin with discovered reserves of 0.17 Tcm (Magoon, Molenaar, Bruns, Fisher, & Valin, 1995). The first field was discovered in 1957, and exploration programs continuing into the early 1970s. The three largest gas fields range in reserves from 0.07 to 0.02 Tcm and many of the other gas fields are small (e.g. 0.17 Bcm), which are very similar in size to those in the Powder River Basin. The U.S. Geological Survey (2011) reestimated the total undiscovered gas resources of the Tertiary sandstones to be a mean 336 Bcm. Most of the gas is produced from the Miocene–Pliocene Sterling Formation, followed by the Miocene Beluga Formation and Oligocene Tyonek Formation. The traps are mostly structural (e.g. anticlines and faulted anticlines) and some combined structural and stratigraphic (e.g. sandstones) traps (Magoon et al., 1995). The small sizes of the gas fields may be because the traps are mainly in compaction folds much like those in the Powder River Basin. The Neogene stratigraphic section is thermally immature and unable to generate methane; thus, the biogenic gas was generated locally from coals and has migrated to adjacent structures and stratigraphic traps. According to Sherwood et al. (2006) most of the gas reserves have been depleted with about 18% of the original reserves still remaining as of late 2003.

The discovery and development small-size gas fields sourced from low-rank coals in the Powder River and Cook Inlet Basins demonstrate the potential of other gas-charged sandstone reservoirs in coal-bearing sequences around the world. The prospects of coal-derived, gas-charged sandstone reservoirs are best developed in shallow areas where opportunities of gas desorption from coals have permitted expulsion by changes in groundwater hydrostatic head, like in the Powder River Basin. The shallow depths of these gas-charged sandstone plays and the ability to delineate these reservoirs from well control, particularly from coalbed-producing wells, make them attractive economic secondary or primary targets for exploration and development. However, gas-charged sandstones from coal sources in deeper areas as demonstrated in the Cook Inlet Basin also may be attractive targets for exploitation with delineation of large structural traps. However, stratigraphic traps with delineation of the depositional dip of fluvial sandstone belts deposited by the same channel systems may also provide similar or better size target for development. In the Powder River Basin, Flores (2004) described numerous fluvial sandstone belts as much as 15 km wide and more than 100 km long. Often times, the developmental strategy of this gas resource in coal-bearing rocks, is recovery and completions from multiple sandstone and coal beds. In the shallow eastern margin of Powder River Basin, Flores (2004) estimated the size of undiscovered gas fields to range from a minimum of 0.08 Bcm to a maximum of 5.6 Bcm.

SUMMARY

As energy mix, coal is best known as a solid fuel and associated coalbed gas. Coal is less known for its hydrocarbon byproducts such as coal-derived oil and condensate fuels. Also, coal liquefaction through the CTL technology is not as much considered even though it is a proven process used in South Africa and Europe as alternative energy source. Major coal-producing and consuming countries such as China, which has the largest coal reserves in the world, is investing in CTL technology to compliment hydrocarbon production. Coal is a significant resource for natural gas or coalbed gas with estimated GIP resource from 67 to 252 Tcm. Coalbed gas is a major part of the unconventional gas, which includes shale gas and tight sand gas, making up 38–46% of the natural gas production in the United States. Coalbed gas comprises 8–10% of the total natural gas production in the United States. Most of the recoverable coalbed gas resources are in Russia, United States, China, Australia (includes New Zealand), Canada, and Indonesia. These six countries have a total of about 24 Tcm of recoverable coalbed gas resources with Russia comprising the most at about 24% of total and Indonesia comprising the least at 6% of total.

Coalbed gas is known for the past 150 years as a coal mine hazard with potential explosions or outbursts occurring worldwide. More importantly, in a carbon-constrained world, coal as solid, liquid, and gaseous fuels has become an environmental challenge as emitter of greenhouses gases. For example, as a solid fuel, which is combusted to generate electric power, it has become a lightning rod to environmental activists. Coal is produced in more than 50 countries with 40% of production from surface coal mines and 60% from underground coal mines. China is the largest coal producer, which is 2.95 trillion metric tons, mainly from underground coal mines in 2010 compared to the nearest producer of less than 1 trillion metric tons by the United States in the same year but mainly produced from surface coal mines.

Large-scale worldwide coal production from underground mines has caused deadly mine outbursts by instantaneous violent ejection of coalbed gas, coal, and rock during mining operations. More than 30,000 coal mine explosions occurred worldwide since 1996 but with more than half in China. However, as of 2009, there are 14,673 coal mines in China of which 3412 or 23% of the total are considered to be high-gas content and gas-outburst prone mines. Fugitive coalbed gas emissions of CMM from surface and underground mines and AMM from closed or shut-in mines are commonly released to the atmosphere. Coal-producing countries continue to exploit and recover CMM and AMM for industrial and commercial use.

As petroleum system, coal-derived hydrocarbons are explored and developed as conventional energy worldwide. Coal-derived oil and condensate are geologically distributed in Carboniferous, Permian, Jurassic, and Cretaceous coal-bearing rocks but are commonly exploited in Cenozoic coal-bearing sequences, which are more oil prone. CTL or coal liquefaction technology can convert coal to synthetic fuels, which are refined into gasoline and diesel fuel. Conventional coalbed gas, which is expelled, trapped, and sealed commonly occur in sandstone reservoirs in coal-bearing sequences. Gas-charged sandstones are best developed in shallow areas of coal basins where gas migration from coals are controlled by changes in groundwater levels.

CHAPTER

3

Origin of Coal as Gas Source and Reservoir Rocks

OUTLINE

http://dx.doi.org/10.1016/B978-0-12-396972-9.00003-3

KEY ITEMS

- Origin of coal as a source and reservoir of coalbed gas requires understanding biological, physical, and chemical properties of peat as well as processes and environments of peat accumulation.
- Study of genesis of peat to coal has led to understanding by coal geologists of their depositional environments and by biologists/botanists their ecological environments.
- Peat-forming environments or peatlands cover 3 million km^2 or about 2–3% of the world's total landmass. Peatlands make up 60% of global wetlands, which are huge carbon sink and not well known as source of greenhouse gas.
- Peatlands are known as mires, bogs, swamps, marshes, moors, fens, muskegs, and pocosins. Peatlands are grouped into bogs, swamps, and marshes (former two are mires) and their deposits accumulated in subsiding coastal, deltaic, and alluvial plains. Not all peat is precursor of coal being eroded and degraded before burial.
- Surface hydrology, climate, nutrient supply, and external aquatic environments control accumulation and classification of peatlands. These factors are used to identify mires from rheotrophic vs ombrotrophic, minerotrophic vs oligotrophic, and topogenous vs ombrogenous.
- In addition, these factors control the evolution of peatlands from marsh to swamps and bogs as well as the processes of peatification.
- Peat is organic soil classified as fibric, hemic, and sapric depending on the degree of decomposition. Groundwater level and fluctuations control aerobic and anaerobic microbial zones of decomposition and extent of decay of plant litter to form peats.
- During peatification, biogenic gas is generated through methanogenesis by anaerobic microorganisms below groundwater table. Biogenic methane,

carbon dioxide, and nitrous oxide are produced and emitted to the atmosphere amounting to about 100 Tg annually.

- Preservation of early stage biogenic peat gas as coalbed gas is not known. The dynamics indicate formation of gas bubbles in pore waters, which are held on walls of organic matter by atmospheric and hydrostatic pressures. Gas is trapped in overpressure zones and is released when hydrostatic pressure is changed.
- Laboratory studies of peats inoculated with anaerobic microorganisms generated carbon dioxide and methane; however, only carbon dioxide is retained in the peat. Coals were heated to simulate early stage coalification and large volume of methane formed at 35 °C but decreased at higher temperatures. The gas may be preserved by rapid burial and seal conditions.
- Depositional peat-forming systems such as alluvial, coastal, and delta plains and lacustrine system are prone to rapid burial and sedimentation amenable to preservation of peat and associated gas.
- Peat deposition is in equilibrium with erosion, oxidation, degradation, and subsidence providing accommodation space from sea level transgression, and peat compaction and autocompaction.
- Tropical peat bogs in coastal and alluvial plain systems in Southeast Asia are presented as modern analogs of coal-forming environments. These modern analogs are used for the formation of economic coals and reservoir rocks for coalbed gas.
- Modern peat deposits to form mineable beds and coalbed gas reservoirs have to accumulate from >87,000 to >175,000 years and undergo peat:coal compaction ratios of 4:1 to 13:1 from lignite to anthracite coals.

COAL AS SOURCE AND RESERVOIR OF COALBED GAS

Organic origin of coal is the key to serving both as source and reservoir rocks of early- to late-stage coalbed gas generated through geologic time. Early stage gas formed during the deposition of the original organic matter in mires and late-stage gas formed after deformation of the basin of deposition. It is now believed that even younger late-stage gas is formed well after uplift and erosion of the basin of deposition particularly when the groundwater systems were established in the coal aquifers/reservoirs. In the interim, however, gas is lost and regenerated at various stages between the times when early and very late gas stages were formed. In addition, gas generated in the coal depending on the hydrostatic pressure either remains in the reservoir as unconventional gas or expelled from the coal as pressure changed into adjoining porous reservoir rocks (e.g. sandstones) as conventional gas. Unconventional and conventional gas are trapped in stratigraphic and structural loci and sealed much like traditional hydrocarbon accumulations.

The organic composition and rank of coal dictate the type of coalbed (peat) gas (e.g. thermogenic and biogenic) that is generated before (e.g. peatification), during, and after coalification. Coalbed gas is primarily composed of methane (CH_4) and secondarily of carbon dioxide (CO_2), nitrogen (N_2), and trace amounts of higher hydrocarbon gases in variable proportions of mainly

C_2–C_4 (e.g. ethane, propane, butane). The proportions of CH_4 relative to CO_2, N_2, and higher hydrocarbon gases depend partly to the elemental composition of the macerals or organic matter contained in the coal.

Figure 3.1 shows the quantities of coalbed gases generated by sapropelic (Types I and II) and humic (Type III) organic matters based on changes in elemental composition (e.g. carbon, hydrogen, oxygen) during coalification (Clayton, 1998). The thermogenic coalbed gas is shown in the lower part of the diagrams and biogenic gas, as post coalification product, is shown in the upper part of the diagrams. Thermogenesis of coalbed gases generally occurs at temperatures greater than 110 °C and methanogenesis at temperatures less than 100 °C (Flores, 2008; Gorody, 1999). Thermogenic coalbed gas is considered to be the major nonassociated gas sourced mainly from thermally mature coal (e.g. bituminous). Other natural gases derived from other source rocks (Types II and III occur in more variable thermal maturation stages (Figure 3.2). Although biogenic or microbial gas is depicted in Figure 3.2 as forming in a limited thermally immature stage, additional study of coalbed gases worldwide indicates that biogenic gas occurs in all ranks of coal (Rice, 1993). Specifically, the composition of the coal plays a major role in the conversion of organic matter molecules to methane by anaerobes (Zinder, 1993).

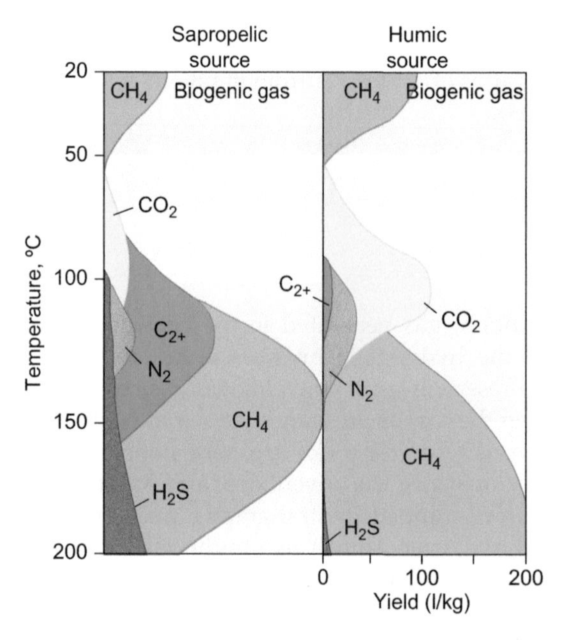

FIGURE 3.1 Diagram showing calculated volumes of gases generated from sapropelic (Types I and II kerogens) and humic (Type III kerogen) organic matter. Biogenic or microbial gas is shown to develop at low temperature with highly variable volumes. CH_4 = methane, CO_2 = carbon dioxide, C = carbon, N = nitrogen, H_2S = hydrogen sulfide; l/kg = liter per kilogram. *Source: Adopted from Clayton, 1998, which was modified from Hunt (1979) and Rice (1993).*

More importantly, the character of coal as a gas reservoir is influenced by the maceral composition (Clarkson & Bustin, 1997; Close, 1993). The ability of the reservoir to transmit or flow gas (e.g. permeability) to the borehole is controlled by the occurrence, distribution, and amount of fracture systems in the coal. For example, the frequency, spacing, and height of fractures vary with the maceral composition, thickness, and mode of stacking or lithotype (Dawson & Esterle, 2010). The development of pore spaces in coal, which in turn influence gas adsorption, is dependent in part on maceral composition (Crosdale, Beamish, & Valix, 1998). That is, vitrinite-rich or woody coals have greater adsorption capacity suggesting more pore spaces. Moreover, the pore structures (e.g. size and shape) are related to maceral composition and depositional environments (Zhang, Tang, Tang, Pan, & Yang, 2010). For example, the size and shape of pores are mainly reflected by the nature of preservation of residual cell structures of precursor plants and/or the size and packing distributions of maceral detritus. Also, the relative proportions of woody and nonwoody macerals may be connected by rate of peat accumulation and variation in the accommodation space during the natural life of the peatland system but more importantly related ecology.

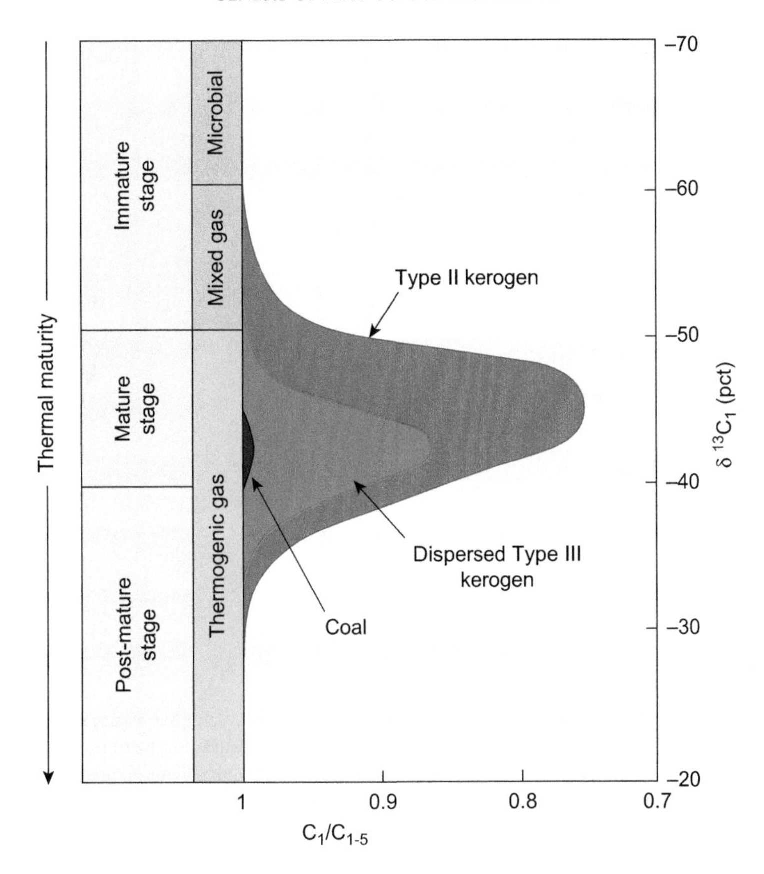

FIGURE 3.2 Diagram showing different source rocks (e.g. coal, Types II and III) in relation to changes in thermal maturation and chemical and methane carbon isotopic composition (δ^{13}C). C_1 = methane; C_2 = ethane, C_3 = propane, C_4 = butane, C_5 = pentane. *Source: Adopted from Rice, Clayton, and Pawlewicz (1989).*

GENESIS OF PEAT TO COAL: CONCEPTS

Fundamental to understanding coal genesis as a source and reservoir of coalbed gas is linking to the forerunner peat deposit. There are two concepts of investigating the genesis of coal and peat, which in the past do not or seldom interact: depositional and ecological systems. The disconnect between the study of depositional and ecological systems of coal and peat deposits may have led to the misunderstanding and erroneous interpretations of their origin. As Moore (1989) alarmingly put, "there has been an unfortunate lack of information exchange between coal geologists and ecologists concerned with modern peatlands to the detriment of both areas of study." Regrettably after almost 25 years, this observation still rings through and more so with coalbed methane development in which petroleum geologists and engineers should enlist more cooperation from coal geologists, petrologists, sedimentologists, and stratigraphers. The situation is exacerbated by lack of or sparse collaboration within (e.g. coal petrologist, paleobotanist, paleoecologist, sedimentologist, stratigrapher) and between (biologist/botanist, chemist, sedimentologist, hydrologist, petroleum developers) the areas of study as well as lack of common platforms of publication in which interdisciplinary flow of ideas prevails. This is not an indictment of these areas of study but rather a call for more interdisciplinary research and cooperation to provide a rounded and better-founded scientific knowledge of the origin of peat as applied to coal and hydrocarbon energies. Multiauthored publications of coal depositional environments with symbolic papers on ecological origin of peat as applied to coal abound throughout the literature (Dapples & Hopkins, 1969; Lyons & Alpern, 1989; McCabe & Parrish, 1992; Morgan, 1970; Rahmani & Flores, 1984).

Depositional Concept

Depositional system is a large-scale and all-encompassing study by geologists, mainly by sedimentologists and stratigraphers, of the environments of deposition of multilayered sedimentary rocks that include coal beds. In this sense, the concept focuses on the "megafacies" of sedimentary rocks. Geologists interpret the depositional environment of sedimentary rocks by focusing on criteria such as fossil type and abundance, rock type, sedimentary structures, etc. (Figure 3.3). Because of this emphasis the origin of coal as interpreted in relations to the environments of deposition of the interbedded sedimentary rocks often is superficial. Coal sedimentology and stratigraphy or coal facies as well the coal paleoecology, petrology, and chemistry are not considered in detail due to the lack of backgrounds of sedimentologists and stratigraphers. Nevertheless, this concept of uncovering the origin of coal often led to broad-brushed application of depositional environments such as "riverine and coastal peatlands". This broad focus on physical processes and geologic events provided only a superficial understanding of the origin of coal. The direct and best paths to reconstruction of past coal depositional environments are connections to the modern peat environments (e.g. "present is key to the past"), which require the understanding of ecological systems of peatlands (Figure 3.4). Coal deposition has been poorly related to modern peatlands due to coal geologists' unfamiliarity with biological, botanical, chemical, and hydrological processes.

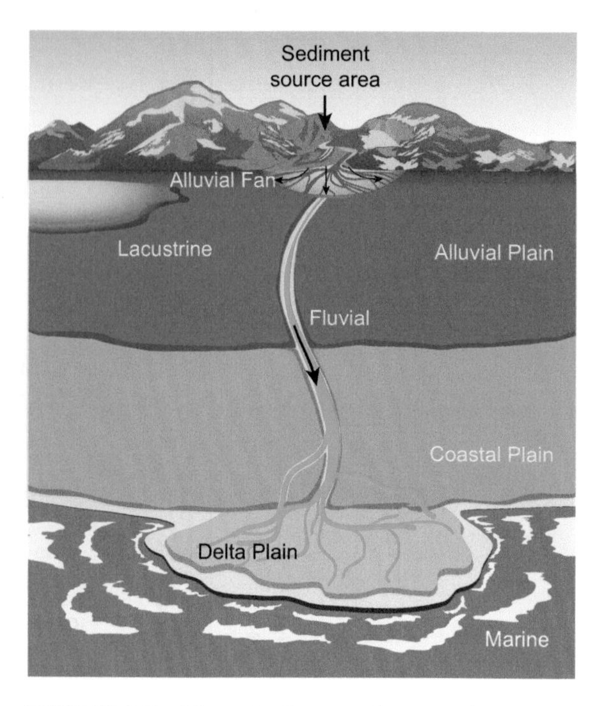

FIGURE 3.3 Schematic diagram of a generalized depositional system showing from mountain-front alluvial fans to coastal–delta plains. This depositional system is meant only to show general settings of potential areas (green) of peat accumulations. Specific areas are discussed in the succeeding sections. *Source: Modified from Schlumberger (2012).*

Ecological Concept

Ecologists, on the other hand, view peat-forming environments as wide-ranging ecosystems (Quinty and Rochefort, 2003) where the dynamic of production of biomass (e.g. living or biological materials) is the result of combined and interactive biological, chemical, physical, and hydrological processes. According to Moore

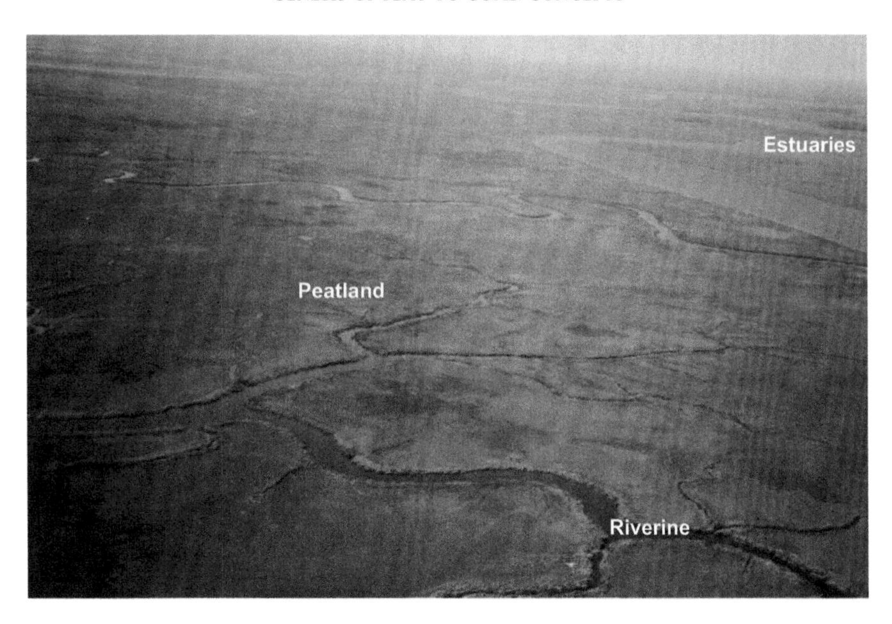

FIGURE 3.4 Coastal peatlands drained by rivers, creeks and estuarines toward the sea. *Source: From http://www.photolib.noaa.gov/htmls/line0095.htm.*

(1989) the sun's energy is converted by photosynthesis to chemical energy in plant organic molecules partly used in their metabolism, which in turn, becomes dissipated during transpiration (Figure 3.5). Furthermore, the residual energy in plants finds its way to the leaves, stems, flowers, fruits, and roots. The total energy budget of the ecosystem is complemented by other ecosystems brought in by water as rainfall and surface and subsurface flows (e.g. rivers, groundwater), and by air as pollens and spores as well as by transported microorganisms and higher forms of life. The biological and chemical energy of the ecosystem is balanced by physical energy as manifested by the inanimate substrate, which mainly provides mineral nutrients, anchor for roots, inorganic catalysts, and medium for surface water and groundwater circulation for plants. However, energy inputs to ecosystems are far from balanced because they are variable between and within environments (Moore, 1989). That is, an extensive ecosystem (e.g. coastal wetland) provides larger source of energy compared to a restricted ecosystem (e.g. riverine peatland) (Figure 3.4). Moreover, interactions and activities of various life forms within the same ecosystem may vary such that a plant leaf may be foraged upon before it can be decomposed by other life forms (e.g. microorganisms). The rate of decomposition of different plant remains (e.g. leaves, stems, bark, and roots) in a peat-forming system varies depending on the (1) make up of the litter, (2) additional supply of plant materials from living biomass , (3) kinds of decomposer acting on the organic remains above and below the surface, and (4) position of the organic remains with respect to the groundwater level, which controls the mode of degradation (Moore, 1989). The vertical and lateral changes of organic matter deposits produced by the dynamic microcosm of peat-forming ecosystems during a period of time represent a "macrofacies", which is a small entity of the depositional system. Nevertheless, variations of energy inputs to the peat-forming ecosystems or peatlands provide a better insight

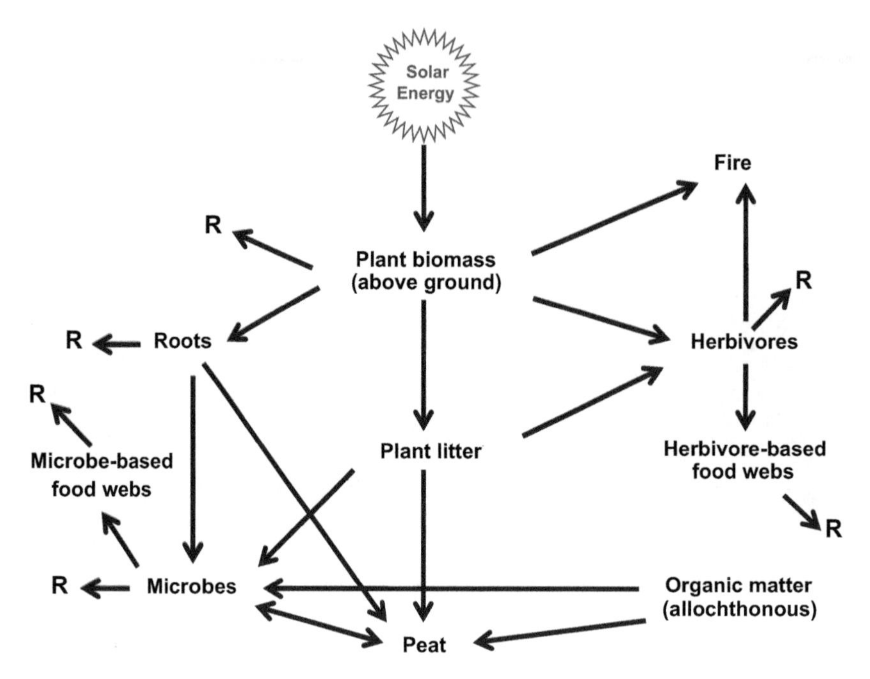

FIGURE 3.5 Generalized chart of energy flow from solar energy to the peatland ecosystem. R = respiration. *Source: Modified from Moore (1989).*

into the nature of peat formed as influenced by topography, hydrology, biology/botany, and chemistry.

In order to understand the exploitation and development of coalbed gas, it is critical to unite investigations of both depositional and ecological systems because much like human's characteristics are a function of genome (e.g. peat organic materials as influenced by ecosystems) and environments (e.g. shape, thickness, and extent of coal beds as influenced by depositional systems) (Moore, 2012). That is, the makeup of the organic matter (e.g. wood, leaves, roots) determines reservoir properties such as porosity and permeability and the coal volume estimated from thickness and extent determines the amount of coalbed gas resource. The "megafacies" of depositional systems and "macrofacies" of peat-forming ecosystem are each part of the whole but combined yield a more complete analyses and interpretation of the geological, physical, biological, chemical, and hydrological processes of the gas source and reservoir rocks. For example, the "megafacies" approach is important to the characterization of conventional coalbed gas reservoirs (e.g. sandstones) in contrast to the "macrofacies", which is important to the characterization of unconventional coalbed gas source and reservoir rocks. The depositional system controls the thickness, geometry, aerial extent, occurrence, and distribution of the detrital rocks and coal. The ecosystems significantly contribute to the chemistry and biology of the peat organic matter, which in turn, translate to how effective coal is as source and reservoir of gas and other hydrocarbons. Thus, the origin of organic matter in coal is directly related to the plant composition as affected by various agents of decomposition in the peat-forming ecosystems. Consequently, the critical test to understanding the origin of coal depends on the accurate characterization of the ecosystems of modern peat analogs as applied to ancestral depositional systems. Countless books, special publications, and individual and review papers have shed light on the facets of peat-forming and depositional environments of coal for many decades (Diessel, 1992; Ferm & Horne, 1979; Greb, DiMichele, & Gastaldo, 2006; Lyons & Alpern, 1989; Moore, 1989; Moore & Bellamy, 1973; Spackman, 1958; Teichmuller, 1989; Wanless & Weller, 1932; Weller, 1930). However, as indicated by the literature cross-expertise collaboration to shed better light on the origin of peat to coal remains limited.

PEAT-FORMING ENVIRONMENTS

Concepts and terminologies of peat-forming environments or peatlands vary from user to user with different backgrounds, from literature to literature of various specializations, and from one continent to the other. For example, a sedimentologist may use and propagate concepts of peat-forming ecosystems in relation to contemporaneous detrital depositional systems such that in this context an "allochthonous" (derived from outside the ecosystem) peat or coal deposit is interpreted as degraded organic matter washed in or transported by water from the adjoining depositional environment (McCabe, 1984). In contrast, an ecologist looking at an ecosystem rich in "allochthonous" inorganic matter as brought in by groundwater flow typical of rheotrophic (permanent high water table) mires (Moore, 1989). These two situations used the same "allochthonous" term each as different in sense and process as the other. Even in the same publication with scientists of the same background (e.g. ecologists, biologists) the term "swamp" is applied in the European sense as dominated by herbaceous plants in permanent water table above the peat surface in contrast to the American sense as dominated by trees in a shallow water wetland (Cameron, Esterle, & Palmer, 1989; Moore, 1989). Thus, applications of many terms to describe and classify peat-forming ecosystems are inconsistent and confusing, consequently the reader inquiring further into peat-forming ecosystems is directed to Moore and Bellamy (1973), Clymo (1978, 1983, 1984), Gore (1983), and Greb et al. (2006).

BOX 3.1

DEFINITIONS OF WETLANDS—CONCEPT AND MISCONCEPTION

The definition of wetlands is nonspecific including hydrology, vegetation, and soil as well as nonuniversal with different countries having their own definitions (Shepard, 2006). Wetlands are between terrestrial and aquatic systems where groundwater table is at or near the surface or the land is covered by shallow water. Wetlands are not always peat-forming environments and should not be used synonymous to peatlands (Schwerdtfeger, 1980). Wetlands are all-encompassing term to include both nonpeat- and peat-forming environments.

Wetland attributes (1) periodically support hydrophytes, (2) are undrained hydric soil for a substrate, and (3) are nonsoil substrate, which is saturated with water or covered by shallow water at some time during the growing season of each year (Cowardin, Carter, Golet, & LaRoe, 1979). The major confusion about wetlands is regarding nomenclature, which can vary regionally or nationally, and the same term can describe different conditions, depending on where that term is used as follows (Shepard, 2006):

(Continued)

BOX 3.1 *cont'd*

Bog—A peat-accumulating wetland that has no significant inflows or outflows and supports acidophilic mosses, particularly sphagnum as well as herbaceous vegetation (temperate climate). In contrast, bog in tropical climate supports forest vegetation.
Fen—A peat-accumulating wetland that receives some drainage from surrounding mineral soil and usually supports marsh-like vegetation.
Marsh—A frequently or continually inundated wetland characterized by emergent herbaceous vegetation adapted to saturated soil conditions. In European terminology a marsh has a mineral soil substrate and does not accumulate peat.
Mire—Synonymous with any peat-accumulating wetlands such as swamps and bogs (European definition; Moore, 1989).
Moor—Synonymous with peatland (European definition). A high moor is a raised bog, whereas a low moor is a peatland in a basin or depression that is not elevated above its surrounding area.
Muskeg—Large expanses of peatlands or bogs; particularly used in Canada and Alaska.
Peatland—A generic term of any wetland that accumulates partially decomposed plant matter or peat.
Swamp—Wetland dominated by trees or shrubs (U.S. definition). In Europe a forested fen or reed grass-dominated wetland is often called a swamp, for example, reed swamp. Depending on where this term is used in North America and Europe, swamp may or may not accumulate peat.

TYPES OF PEATLANDS

Peat-forming environments or peatlands make up more than half of the world wetlands saturated with water (e.g. surface and groundwater), which control development of peat soil and the types of plant and animal ecosystems. Types of peatlands are known by various terms such as mires, bogs, mires, swamps, moors, fens, muskegs, and pocosins along with other similar settings (Cameron et al., 1989; Clymo, 1983; Gore, 1983; Moore, 1987, 1989; Moore & Bellamy, 1973; Masing, 1975). The types of peatlands according to ecologists, biologists, and botanists (Cameron et al., 1989; Moore, 1987, 1989) differ in the variety of vegetation, water availability, nutrient supply, natural processes of surrounding environments, topography, geography, and climate (Figure 3.6).

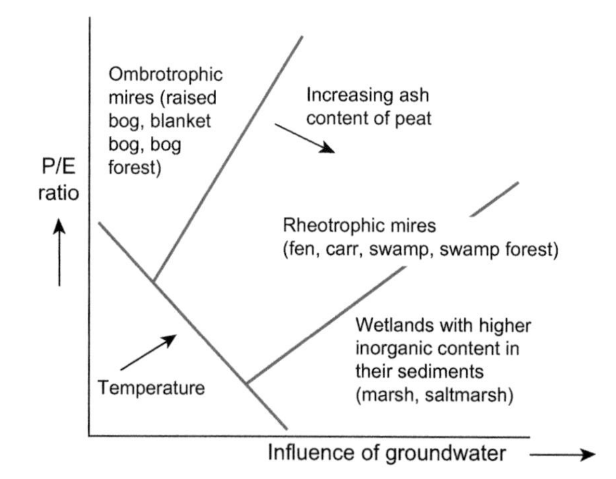

FIGURE 3.6 Relationship between the influence of groundwater, precipitation (P), evaporation (E), and temperature in the development of types of peatlands. Arrow designates direction of increase. *Source: Modified from Moore (1987).*

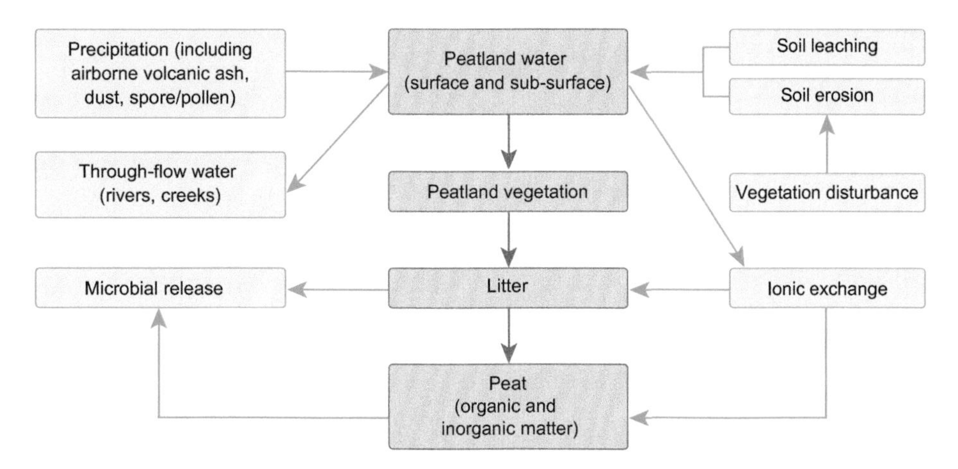

FIGURE 3.7 Generalized flow patterns of nutrients in the peatland ecosystem. *Source: Modified from Moore (1987).*

Varieties of vegetation (e.g. biodiversity), such as floating vs. nonfloating, trees vs. herbaceous, and forest vs. nonforest, differentiate various forms of peatlands. Water availability as surface and subsurface waters controls the types of vegetal growth, decomposition of vegetal matter, and mode of nutrient supply. The flow patterns of nutrients in peatlands are shown in Figure 3.7. Nutrient supply, in turn, influences the types and rates of biomass production. The nature of external environments such as rivers, lakes, and seas whose sediment-laden waters inundate peatlands govern whether peat turns into "muck" or "gyttja" (e.g. peaty clay, clayey peat) or "pure" organic deposit, and if it is oxygenated. Topography such as depressions may accumulate water that can sustain and support peatlands, which in turn, derive their nutrients from surrounding highlands. Geographic location (e.g. latitudinal) controls the types and distribution of vegetation, productivity, rate of accumulation, and morphology of peatlands. Climate changes such as seasonal weather patterns, temperature, evaporation, precipitation, hydrology, and sea level rise will affect peatlands.

Based on the above factors, Moore (1987) has organized and simplified classifications of the various peatlands to fit the following combined elements: (1) water availability particularly groundwater, (2) climate affected by temperature, precipitation, and evapotranspiration, (3) nutrient supply from the substrate or from the atmosphere (e.g. precipitation), and (4) nature of external environments. What are the important roles of each of these factors toward controls to formation of peatlands?

CONTROLS ON DEVELOPMENT OF PEATLANDS

Groundwater

In general, groundwater, mainly derived from meteoric water, is important to the distribution of vegetation, biomass productivity, decomposition of plant parts, and preservation of the organic matter in the types of peatlands (Figure 3.6). Peat hydrologists subdivide the peat layers into an upper aerobic layer (acrotelm) and a lower waterlogged and compacted layer (catotelm) based on seasonal and/or annual fluctuations of the groundwater table (Figure 3.8) (Ivanov, 1948; Ingram, 1978; Peatland Ecology Research Group, 2009). The acrotelm is affected by fluctuating

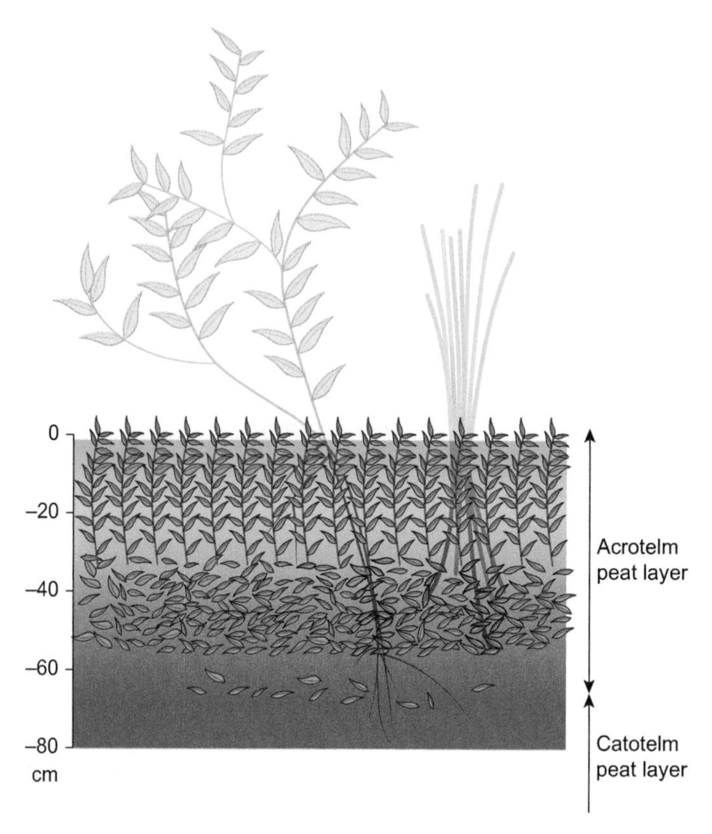

FIGURE 3.8 Acrotelm and catotelm peat soil layers controlled by hydrological regimes. cm = centimeter. *Source: Adopted and modified from Peatland Ecology Research Group (2009).*

groundwater table, has high hydraulic conductivity, and abundant peat-forming aerobic microorganisms. The catotelm is permanently below the groundwater table, has low hydraulic conductivity, is devoid of aerobic microorganisms, and is dominated by anaerobic microorganisms. The dichotomy of peat layers implies that nutrient transfer is best developed in the acrotelm layer (concentrated exchange of water with atmosphere and surrounding areas). This model suggests that the acrotelm layer receives plant materials from biomass productivity and the catotelm layer receives partly decayed materials from the acrotelm. Based on the peat accumulation model of Clymo (1984) decay of plant materials in the acrotelm is rapid due to the large populations of aerobic microorganisms. Loss of materials produced on the surface is high when it enters the catotelm where decay is slow due to anaerobic microorganisms and becomes significant over time assuming that biomass productivity on the surface is constant. In addition, peat hydrologists have demonstrated that groundwater flow and drawdown through the acrotelm influence evapotranspiration sufficient to reduce hydraulic potential in the local peatland flow system and cause discharge of regional groundwater through the mineralized substrate to the peatland (Fraser, Roulet, & Lafeur, 2001). This flow reversal continues seasonally and annually, which can alter peatland biogeochemistry with episodic emissions of CH_4 into the atmosphere and changes in pore-water chemistry (Romanowicz, Siegel, & Glaser, 1993; Siegel, Reeve, Glaser, & Romanowicz, 1995).

Climate: Temperature, Precipitation, and Evaporation

Broadly put, climate is the distribution and variation of weather patterns in an area over long periods of time as influenced by temperature, precipitation, and evaporation (Figure 3.6). Climate changes control almost all attributes of peat-forming ecosystems such as energy flux, plant dynamics, and soil and nutrient cycling (Bridgham et al., 1999). Climatic variables such as temperature and precipitation are the most important factors determining distribution, biodiversity, and growth of vegetation as well transportation of nutrients. The rate and depth of decomposition of organic matter of peat are controlled by temperature. For example, types of peat differ between temperate climate regions (e.g. partly decayed herbaceous plant remains producing fibric to hemic peats) and tropical climate regions (e.g. highly decayed woody plant remains producing hemic to sapric peats). Peat-forming ecosystems are susceptible to availability and retention of water such that the ratio of precipitation and evaporation directly control plant growth. For example, in tropical peatlands, the rate of precipitation is more than the rate of evaporation (except during drought periods) such that the perennial water in the peatlands promotes biomass productivity. Increase in evaporation caused by high temperature will boost biomass productivity in peatlands.

Development of peatlands is controlled by water surplus such that more precipitation (e.g. rain, snow) is needed than evaporation via evapotranspiration, which is loss of water to the atmosphere through combined evaporation and plant transpiration. Ideal condition for optimum development of peatlands is precipitation of about 30–50% more than evapotranspiration. Peatlands have the need of a perennial humid climate with a constant allotment of precipitation combined with relatively low intensity sunlight for longer growing season resulting in maximum biomass productivity and peat accumulation.

Nutrient Supply

Nutrients are chemical elements (e.g. nitrogen, potassium, phosphorus, etc.) and compounds (e.g. ammonium, nitrate, phosphate) as well as metals derived from the substrate, atmosphere, and adjoining environments necessary for growth metabolism for plants (Rydin & Jeglum, 2006). The type and growth rate of plants in the peatlands vary from where they are found in the spectrum of high to low nutrient supply (Figure 3.7). High nutrient supply represents sediment influx from river floods and low nutrient supply represents rainfall, dust, and aerosols from the atmosphere. This spectrum of nutrient supply affects plant growths either enhancing or reducing biodiversity. Perhaps the most important role of nutrients is the effects on productivity and rates of decay in peatlands. Aerts, Van Logtestlin, Van Staalduinen, and Toet (1995) reported the effects of increased nitrogen supply on productivity and potential litter decay rates of Carex species (e.g. sedges), which are the dominant vascular plant species in peatlands in the Netherlands. Under conditions of nitrogen-limited plant growth, increased nitrogen supply leads to increased productivity but does not affect carbon:nitrogen ratios of plant litter and potential decay rates of that litter. In addition, under conditions of phosphorus-limited plant growth, increased nitrogen-supply does not affect productivity but leads to lower carbon:nitrogen ratios in plant litter and to a higher potential decay rate of that litter.

External Aquatic Environments

Contemporaneous aquatic environments such as rivers, creeks, lakes, and seas are important to accumulation and preservation of peat deposits (Figure 3.4). Peatland-draining rivers and creeks are important suppliers of nutrients and oxygenated water. An earlier discussion in nutrient supply indicated that sediment-laden floodwaters from rivers and creeks contain mineral nutrients to enhance vegetal growth in the peatlands. Floodwater influxes from fluvial networks, which are controlled by microtopography of the peatlands, affect the acidity of the water and the dynamics between the nutrients and water types. For example, plant communities may be more tolerant to higher acidity than others. More importantly, oxygenated water from these fluvial networks facilitates decomposition by aerobic microorganisms and fungi (Cameron et al., 1989). Groundwater table is controlled by base level or sea level fluctuation, which is dictated by transgression and regression of the sea. That is, transgression of the sea causes the groundwater table to raise gradually promoting the accumulation of thicker peat deposits. The commonly held view is that the high groundwater table is the primary cause for low peat decay rates leading to the preservation of deep peat deposits. However, according to Morris and Waddington (2011) groundwater residence time in peatlands may be detrimental to slow peat decay and accumulation because of slow pore water turnover leading to high concentration of dissolved inorganic carbon and phenols. Morris and Waddington (2011) documented that enough concentrations of these chemicals may slow and halt microbial respiration, which in turn, arrest peat decay and accumulation. Finally, an important aspect of the fluctuation of the sea level is the preservation of the peat deposits as a result of rapid transgression causing burial by sediments (e.g. high accommodation space).

The types of peatlands based on the above parameters are summarized in Figure 3.6, which is modified from Moore (1987) to include only three broad categories of bogs (e.g. raised, blanket, forest), swamps (e.g. fen, carr, forest), and marshes (e.g. salt marshes). Moore (1989) defined the term mire, which was a part of his 1987 classification, to include bogs and denotes any nonsaline water or freshwater peat-forming environments irrespective of vegetation, chemistry, and hydrology (Gore, 1983; Moore & Bellamy, 1973). Thus, mires in this book include mainly bogs and swamps and nonmires include only marshes to follow the definitions of Moore (1987, 1989). In general, marshes are highly influenced by surface water and groundwater compared to bogs, which is highly influenced by the ratio of precipitation and evaporation and temperature with swamps falling between these peatland types (Figure 3.6). The most important factors that separate these types of peatlands according to Moore (1987) are that bogs mainly derived nutrients from precipitation (e.g. rainfall) or sometimes called ombrotrophic although not all bogs are rain prone and that the swamps and marshes derived nutrients from groundwater or sometimes called rheotrophic.

EVOLUTION OF PEATLANDS

Balanced Groundwater and Rainfall

The difference between peat mires (e.g. bogs, swamps) formed in the temperate and tropical regions is related vegetation in balance of freshwater supplied by groundwater and rainfall. Water transports nutrients that are critical to plant growths, which in turn, supply plant litter on the peatland surface and decays in the acrotelm layer and accumulate in the catotelm layer (Figure 3.8). The water or hydrological balance toward biomass productivity, decomposition, and peat accumulation was expressed by Moore and Bellamy (1973) in the following basic equation:

$$\text{Inflow} = \text{Outflow} + \text{Retention}$$

But when modified by climatic template is expressed in the following multifaceted equation:

$$\text{Inflow} + \text{Precipitation} = \text{Outflow} + \text{Evapotranspiration} + \text{Retention}$$

Simply put, the basic hydrological equation suggests that plant growth and peat accumulation start with the water inflow some of which is retained after which the rest of the water flows out of the peat mire. Thus, the peat holds a volume of water from retention and it behaves like an active aquifer, which displaces its own volume of groundwater. Peat accumulation begins within the retained volume of water, which continues to grow up to the level in which the groundwater drains above the regional water table forming as a perched water table fed only by precipitation (e.g. rainfall, snow). The hydrological equilibrium may become unbalanced when precipitation declines leading to lowering of the groundwater table and exposure of the peat above the water table to oxidation and degradation. On the other hand when precipitation increases, vegetation and peat are drowned by rising groundwater table, which in turn, drowns the roots that need oxygen to function properly and move nutrients to supply the plants. An important part of the hydrological cycle is the release of water to the atmosphere from the plant leaves or transpiration. Plants grow roots into the peat soil to tap water and nutrients to support the stems and leaves. Transpiration rates depend on temperature, humidity, sunlight, precipitation, peat soil type and saturation, wind, and surface topography. During dry seasons, transpiration can contribute to the loss of moisture in the upper layer (e.g. acrotelm) of the peat soil, which can have an effect on peatland vegetation. Water lost from the peat ground surface to the atmosphere is defined as evaporation. Combined lost of water from the plants and ground is called evapotranspiration.

Hydrological Perspectives

Moore and Bellamy (1973) developed a classic model using successive hydrological conditions to describe the evolution of types of peatlands. The model (Figure 3.9) is still most relevant, appropriate, and applicable in characterizing successions of combined hydrological, biological, and physical processes giving rise to types of peatlands. Figure 3.9 shows the types of peatlands that cover both ends of the spectrum from the marshes to the bogs. The succession of peatland types can be interrupted anytime by erosion and preservation by sediment cover; however, the peat thickness will vary from thin to thick depending on the completion of the entire succession. That is, if the peat accumulation during the early stages is interrupted by erosion, oxidation, and/or sedimentation the resulting peat deposit is thin. If peat accumulation is uninterrupted and has the benefit of sustained growth for the full extent of time the resulting peat deposit is thick.

The first stage (Figure 3.9) is the initial peat deposition in a topographic low in which surface water flows either bringing in large or small amounts of sediments from outside such as rivers into a floodbasin in coastal and alluvial plains. The first scenario combined with a slow rate of peat formation caused by the strong oxygenation of the peat ecosystem through the large influx of water concentrated near the surface. The influx of sediments results in supplying nutrients to luxuriant vegetation such as trees and production of a heavy sinking peat. In contrast, the second scenario of small water flow and less material that is added from outside results in a faster rate of peat growth by a light-floating mat below which the water flows. This stage is interpreted to develop the marsh type of peatland (e.g. freshwater and salt-water marshes). In the second stage (Figure 3.9) peat accumulation serves to divert and concentrate water flow in adjoining river channels. Peatlands either continue to be flooded by sediment-laden waters from the channels or remain high but not inundated while peat continues to grow. The peatland is also subject to the effects of moving groundwater during inundation. This stage is interpreted to develop swamp type of peatland (e.g. fens, carrs, swamp forest). The third stage (Figure 3.9) represents sustained vertical and horizontal growth of peat, which elevates the

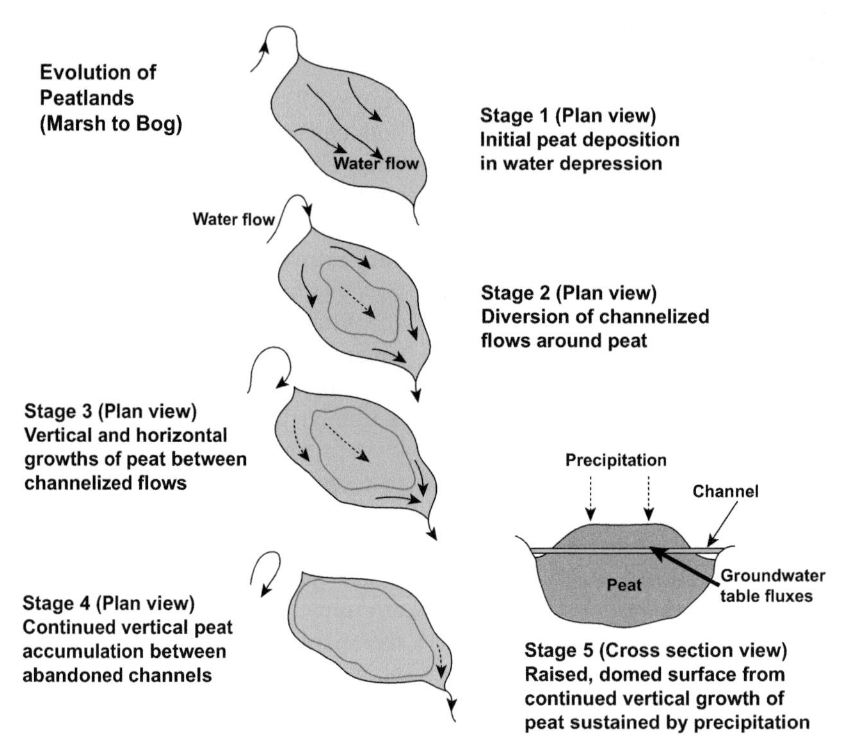

FIGURE 3.9 Diagrams (not to scale) of plan views or maps (stages 1–4) and cross section (stage 5) of the evolution of marsh to bog applicable to both temperate and tropical peatlands. *Source: Modified from Moore and Bellamy (1973).*

central part of the peatland beyond the reach of water inflow from river channels. Water supply is mainly from rainfall on the peatland surface combined with leakage from surrounding peat areas. Also peatland areas immediately lying along the adjoining river channels probably display a sluggish continuous flow of water, which brings in supply of oxygenated water and nutrients. The water and nutrient supplies permit more luxuriant growth of plants along the margins than in the central part of the peatland where rainfall is the main source of nutrients. This stage is interpreted to develop early phase bog type of peatland. In the fourth stage (Figure 3.9) vertical peat growth continues with most of the peatland unaffected by water from adjoining channels and inundation occurring only during groundwater table rise from heavy rainfall. The fifth stage (Figure 3.9) represents the development of a raised or domed peat surface from sustained vertical growth such that it is no longer affected by seasonal fluctuations of the groundwater and surface water. The dome-shaped peat surface possesses a perched groundwater table fed by rainwater. Stages 4 and 5 are interpreted to develop a late-phase bog type of peatland (e.g. blanket, raised, and forest bogs) such as in tropical regions.

The evolution of the types of peatlands as described above and modified from Moore and Bellamy (1973) is an all-inclusive view of how peat accumulation may gain more stored energy resources with increased biomass productivity. It also explains the different types of peatlands worldwide (e.g. Europe vs North America). For example, the succession of peat accumulation reflects the kinds of vegetation that grew at different stages such as from herbaceous plants to forest peatland types results in more woody composition, which is in turn more fuel-grade deposit. This succession may be interrupted and reversed such that the forest peatland type may be succeeded by moss-dominated plant producing less fuel-grade deposit. The interruption of the normal succession of peat accumulation from stages one to five (Figure 3.9) may be the result of external processes such as rapid rise of groundwater table as a result of rapid subsidence of the substrate or transgression of the sea causing burial by sediments. On the other hand, when groundwater table falls as a result of uplift of the substrate or regression of the sea causing erosion, oxidation, and degradation of the peat as well as erosion of the peat. Climatic changes such as increased temperature causes drought resulting in peat fires.

MARSH TO BOG PROGRESSION

The youthful nature of Holocene peat deposits hardly buried and barely accumulating since 10,000 years before present is not perfect modern analogs for coal. Under these circumstances, diagenesis in anaerobic conditions and coalification has not begun. Whether peat transforms to coal or not, the basic processes of peat accumulation remain the same. Thus, the study of modern analogs of mires especially from marsh to swamps to bogs can be working analogs for producing economic deposits as well as source and reservoir of coalbed gas.

Moore and Bellamy (1973) suggest that peat accumulation can occur from the head of valleys to the coastal plain where water can collect in depressions such as ponds, lakes, ponds, floodbasins, abandoned channels, estuarines, lagoons, and bays (see Figure 3.9 different stages as examples). However, based on the geological records from Carboniferous to Tertiary Periods, coals are mainly associated with lacustrine, alluvial, deltaic, and coastal plain depositional systems. Thus, the preservation potential of peat in water-filled depressions is best with peatlands associated with lakes, alluvial, deltaic, and coastal environments (Figure 3.9) exposed to various geological processes (e.g. erosion vs deposition) in depositional systems found in selected temperate and tropical regions.

From Marsh to Swamp

The successional development of peatlands from marshes to bogs generally occurs in temperate and tropical regions with the main difference being the types of vegetation, hydrology, and nutrient supply. In both regions, the initial peat-forming environment is a marsh developed in salt water or freshwater, which is vegetated by aquatic herbaceous plants, rushes, grasses, and floating mats of rhizomes (e.g. cattails and sedges in temperate and tropical climates in Europe, United States, Canada, and Africa) (Cameron et al., 1989; Moore, 1989). Marshes may contain dotted thickets of trees (e.g. cypress, long-leaf pine) particularly occurring as hummocks as it evolved into a swamp (Figure 3.10) (Spackman, Dolsen, & Riegel, 1966). Floating mats vegetation, which is buoyed

FIGURE 3.10 North Carolina coastal marshes and swamps dotted by thickets of cypress, long-leaf pine, scrubs, and grasses drained by rivers and estuaries. There is a marked change in vegetation from marsh to swamp. *Source: Adopted from Frankenberg (1999).*

above the substrate, is normally not rooted unless found in shallower water and principally extract nutrients from water flowing below and through the mat. Thus, a marsh may be composed of a complex of grass fields, together with floating mats, aquatic plants, and shrubs. The marsh vegetation thrived from nutrients brought in by groundwater and surface water washing in from adjoining aquatic areas, which is typically rheotrophic or minerotrophic. Marsh peat is characterized by rapid rates of decomposition, but little of the initial organic matter is preserved although the deposits may be rich in transported organic materials and siliceous materials from plants deposited as limnitic sediments resulting in high ash content. Cameron et al. (1989) describes the high ash (>50%) marsh peats associated with sediments at the bottom of depressions where sediments settled and mixed with transported organic materials as "mucks".

Swamp peats commonly overlie marsh peats, which are the normal sequence of succession (Figure 3.11). Swamps are dominated by trees

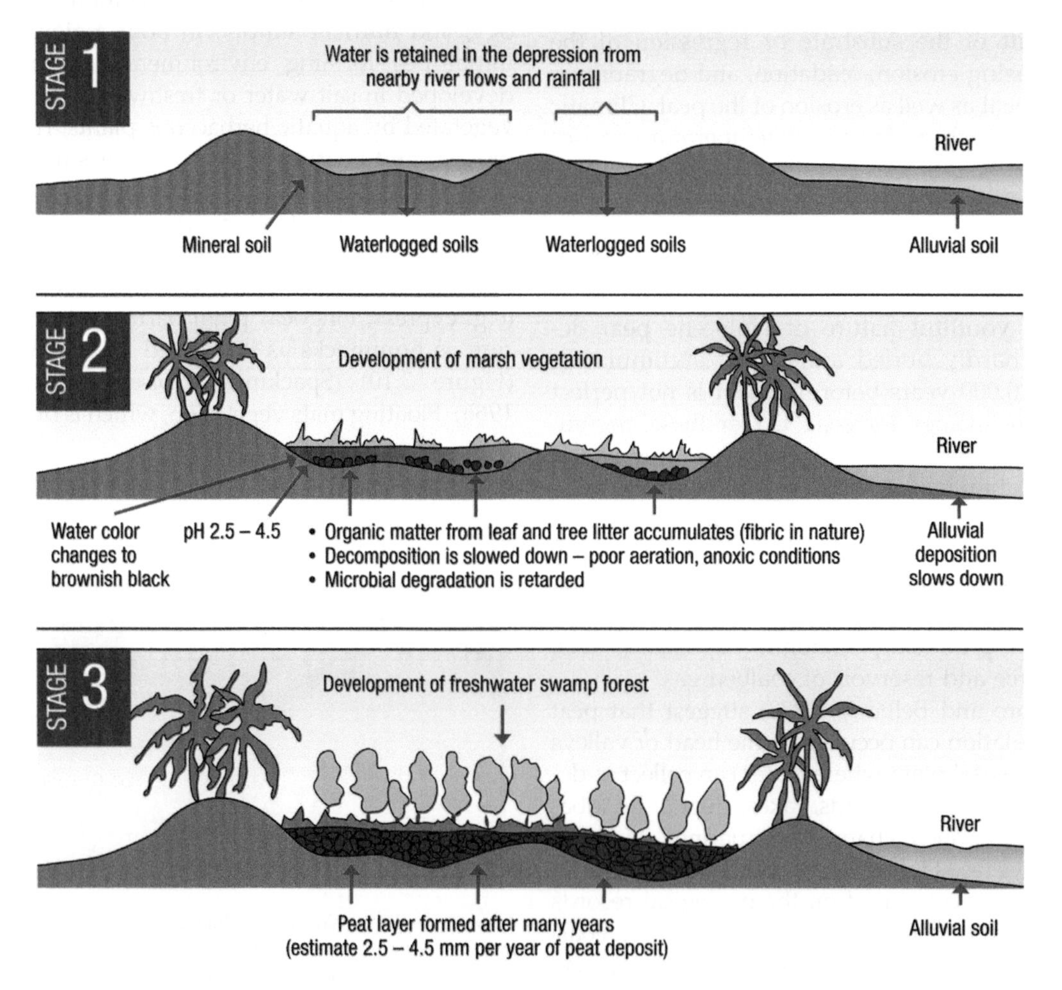

FIGURE 3.11 Tropical peat swamp that evolved from water depressions nearby rivers to marsh vegetation and freshwater swamp forest. *Source: Adopted from UNDP (2006).*

and maintain high water table; hence the name "swamp forests" (Moore, 1989; UNDP, 2006; Wetlands International, 2012). Unlike the marshes, swamps are mainly developed and supported by freshwater but rheotrophic or minerotrophic in nature as the vegetation thrives on nutrients from water runoff and groundwater. Thus, if the marsh was developed in a salt-water depression (e.g. near the sea), the succession of peat deposits will be salt-water marsh peat succeeded by freshwater marsh and swamp forest peats (Figure 3.11). On the other hand if the marsh was developed in a freshwater depression (e.g. lake, alluvial floodbasins, abandoned channels), the succession will be freshwater marsh and swamp forest peats. However, the marsh to swamp successions may be preceded by peaty clay or clayey peat or organic mud or "muck" (gyttja or gyttjae).

The different types of vegetation in the marshes and swamps produced different types and amounts of peat. Thus, the succession of vegetation from marsh to swamp is from generally aquatic herbaceous plants, grasses and rushes overcome by dominantly trees. Moore (1989) argued that the successions do not directly translate to biomass productivity (e.g. vegetation) in the swamp forests, which is much greater than in herbaceous- and aquatic-plants marshes, and nonforested swamps. The amount of above-ground productivity tend to be lower in the swamp forests because a large proportion of tree biomass is nonproductive support tissues (e.g. woody trunks, branches) in contrast to the marsh herbaceous plants, sedges, rushes, and grasses in which all growth stems and leaf tissues promote more plant productivity. However, the peat accumulations in swamp forests consist or mainly woody tissues, which may consist of microbial protection (e.g. resins) and less prone to anaerobic decay resulting in slow but efficient decomposition. The biomass productivity suggests that in terms of economic accumulation of peat transforming to fuel-grade coal, that the swamp forests peat (coal) is better for power combustion and coalbed gas reservoir because of their higher wood content. This viewpoint is supported by the studies of peat swamps in temperate and tropical regions in North America, Europe, and Southeast Asia by Andriesse (1988), Cameron et al. (1989) and Cohen, Raymond, Ramirez, Morales, & Ponce (1989).

From Swamp to Bog

Bogs are formed mainly in both temperate and tropical regions but consist of different types of peatlands, vegetation, rate of vertical peat accumulation, and topographic expression (Table 3.1). Bogs are wet, spongy, organic matter residue beyond which are a dichotomy of descriptions based on how nutrients are derived and elevation from the adjoining surface and/or water level; hence, topographic expression. The European use of bogs is restricted to peat-forming ecosystems that depend on rainfall for water and nutrient supplies otherwise called "ombrotrophic bogs" (Gore, 1983; Moore 1987, 1989; Moore & Bellamy, 1973). Under this usage, bogs are above-ground biomass (e.g. raised and blanket-like) that succeeded low-lying swamps and in the course of biomass production displays plant successions of dominantly small stature or herbaceous plants (e.g. mosses, sedges, shrubs). However, other ombrotrophic bogs such as in the boreal region (e.g. Canada, Scandinavia, Russia) are vegetated by forests (e.g. coniferous, cypress trees). In these bogs, the net biomass productivity is low but the rate of vertical peat accumulation or accretion is high (Moore, 1989). However, Gorham, Janssens, and Glaser (2003) reported that "bog islands" vary in vertical peat accretion from 0.87 to 1.12 mm/yr in the interior part compared to 0.56–0.93 along the margins. In addition, the rate of vertical peat accumulations varies from 0.18 to 0.38 mm/yr in Alaska to 0.50–0.79 mm/yr in Minnesota, USA. In cool temperate climate with consistent rainfall,

TABLE 3.1 Comparison of Temperate and Tropical Peatlands, Vegetation Types, and Rates of Vertical Peat Accumulation Formed During the Holocene Compiled from a Review of the Literature

Climatic Region	Geographic Location	Type of Peatland	Types of Vegetation	Time (Years BP)	Rate of Vertical Peat Accumulation (mm/yr)	References
Cool temperate	West central Alberta, Canada	Fen	Mosses to trees to moss-dominated	8000–0	2	Yu, Vitt, Campbell, and Apps (2003)
	Northwestern Canada	Fen-bog	mosses, low shrubs, even trees, sedges	2000–0	0.282 ± 0.052 0.563 ± 0.070	Robinson and Moore (1999)
Warm temperate	Northern New Zealand	Bog	Forest, loss of tall trees and expansion of subcanopy species and seral vegetation in forests	11,700–0	0.9–1.7	Newnham, de Lange, and Lowe (1995)
	North Carolina coast, USA	Swamps	Aquatic plants to cypress forest	5000–2000	0.3–1	Cameron et al. (1989)
Tropical	Kalimantan, Indonesia	Bog	Forest to stunted trees	13,000–8000 9530–8590 8000–500	0.9 2 0.15–0.38	Page et al. (2004)
	Sarawak, Malaysia	Bog	Forest to stunted trees	4300–3900 3900–2300 2300–0	4.6 3.0 2.2	Anderson (1964)
	Sumatra, Indonesia	Dendang bog	Forest to stunted trees		1.6–2	Cameron et al. (1989)
	Sumatra, Indonesia	Puding bog	Forest to stunted trees	3790–4490	0.17	Flores and Hadiyanto (2006)

mm, millimeter; yr, year.

some bogs form a blanket shaped in contrast in warm temperate climates raised bogs with convex upper surfaces are more common. In North America, the use of the term bogs is not limited to the raised topographic expression ("raised bogs") and they may or may not succeed low-lying swamps (Cameron et al., 1989).

Most bogs are distributed in temperate and cold climates, mainly in North America, Europe, Russia, North Asia, and South America. Bogs are mainly classified according to the basal topography and thickness of peat. A blanket bog is described as consisting of peat that is moderately deep accumulation that averages 2.6 m thick deposited on gentle to steep slopes (Hammond, 1981). In contrast, raised bog is a deep peat accumulation that averages 9 m thick deposited on flat central plains (Hammond, 1981). In temperate regions such as in Germany bogs are classified based on surface elevation: low, transitional, and high moors (Andriesse, 1988). Low moors are developed in low elevations in contrast to high moors, which are formed at higher elevations; a classification similar to the raised bogs of Ireland. High moors rise above the groundwater level relying on precipitation for nutrient supply. Low moor develops with groundwater and sediment influxes for nutrients. Thus, there is a close connection with the genetic origin of peat bogs and topography; the latter either based on basal topography or topographic elevation upon which the peat develops.

Truly raised bogs are used to describe forest bogs in the tropical and subtropical regions where high rates of rainfall exceed evapotranspiration. The tropical and subtropical bogs, which will be all-inclusively classified as tropical bogs in this book, lie within latitudes 35° North and South including those at high altitudes (Andriesse, 1988). Unlike bogs found in cold and warm temperate regions there is more rainfall to sustain the forest vegetation. Recent information on tropical and subtropical peat resources indicates that they may cover about 8.2% or 35.8 million hectares of the total global peat soil of 436.2 million hectares (Andriesse, 1988); however, the peat soil may extend from 420 to 500 million hectares worldwide (Kivinen & Pakarinen, 1980). Table 3.2 shows updated statistics of tropical peatlands worldwide and Southeast Asia based on Immirzi and Maltby (1992), Rieley, Ahmad-Shah, and Brady (1996), and Rieley (2007). The largest concentrations of tropical peat bogs are around the South China Sea in Southeast Asia, South America (Amazon), Mainland Asia, and Central Africa (Andriesse, 1988; Sielfermann et al., 1992; Rieley, 2007). Southeast Asia (e.g. Indonesia, Malaysia, Papua New Guinea) covers about 26 million hectares or 69% of all the tropical peatlands (Figure 3.12; Table 3.2).

Tropical peat bogs in Southeast Asia particularly in Sarawak, Malaysia, Borneo, are produced mainly from forest-type vegetation represented in the Baram River tide-dominated, cuspate-shape delta plain to the inland coastal and alluvial plains (Figure 3.13). The study of the peat bogs in the Baram River drainage basin by Anderson (1964) shows in the coastal lowlands areal distribution of mangroves at sea level followed landward by brackish water and freshwater plant communities, which are replaced by forest-type vegetation (e.g. *Shorea*-dominated) on the ombrogenous (see next section for definition) raised peat domes. Figure 3.14 shows the areal vegetation zones of the forest communities of the ombrogenous peat bogs in the drainage basin of Baram River (Anderson, 1961; Cameron et al., 1989). The central part of the peat bogs is characterized by open-bog plain of stunted, thin trees (3–4 m high and <50 cm girth), shrubs (Figure 3.15(A)), and pandan trees broken up by open lakes covered by algae, moss, and pitcher plant. The vegetation zonation is caused by lower pH and nutrient supply toward the center of the bogs. Figure 3.15(B) shows tall trees mainly distributed along the margins of the bogs in which floods from juxtaposed river

TABLE 3.2 Area Coverages of Peatlands Worldwide

Region (Subregion)	Area Coverage (Range in Hectares)	Subarea Coverage (Range in Hectares)
Africa	2,995,000	
Central America	2,276,000–2,599,000	
Mainland Asia	1,351,000–3,351,000	
Pacific	36,000–45,000	
South America	4,037,000	
Southeast Asia	36,627,000–45,965,000	
Indonesia		17,853,000–20,073,000
Malaysia		2,730,000
Papua New Guinea		500,000–2,890,000
Thailand		64,000
Brunei		110,000
Vietnam		24,000
Philippines		10,700

Source: Modified from Immirzi and Maltby (1992), Rieley et al. (1996), and Rieley (2007).

channels served as nutrient supply. The peat bog deposits (Figure 3.15(C)) consist of dense woody peat derived from decomposition of remains of large trees in the lower part and fibrous peat derived from remains of stunted trees in the upper part, which mimics the areal zonation of the vegetation. Anderson (1964) described that the peat accumulation in the peat bogs is a result

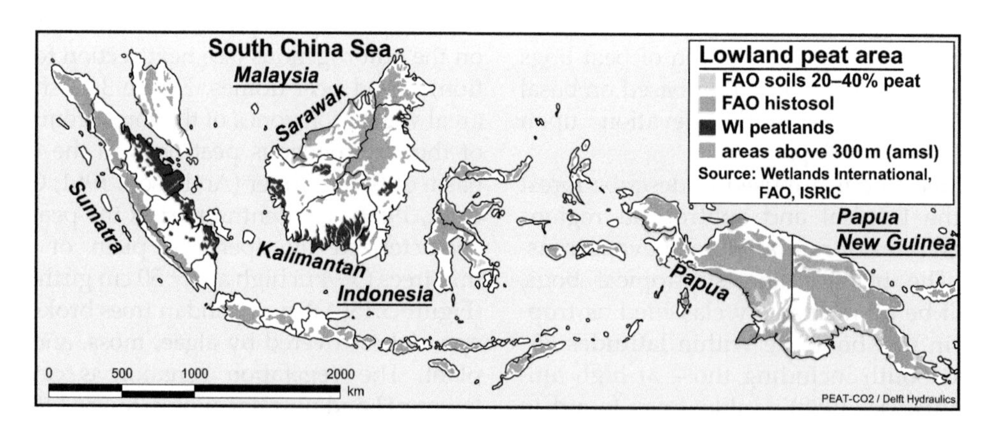

FIGURE 3.12 Distribution of Southeast Asia tropical peat bogs around South China Sea in Malaysia, Indonesia, and Papua New Guinea. FAO = Food and Agriculture Organization; WI = Wetlands International; amsl = above mean sea level. *Source: Adopted from Rieley (2007).*

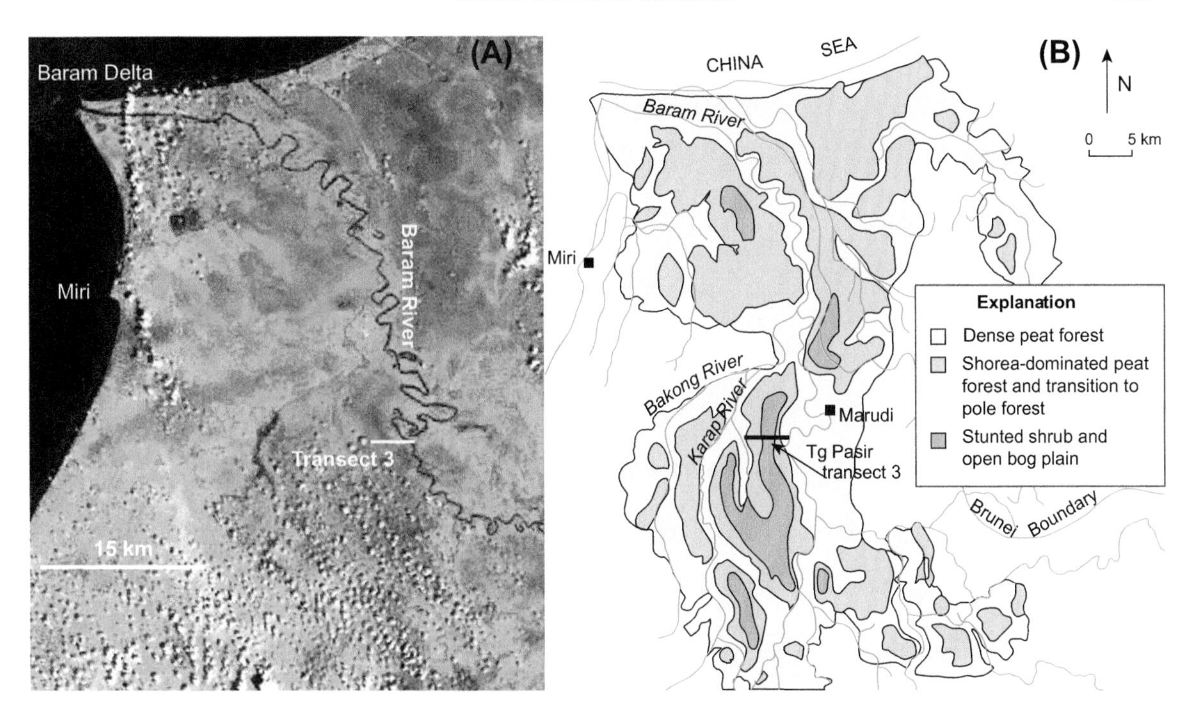

FIGURE 3.13 Satellite photograph of the cuspate-shaped, tide-influenced delta, coastal, and alluvial plains of the Baram River, near the city of Miri in Sarawak, Borneo (A). Map showing zoned vegetation on peat in the Baram River alluvial, coastal, and deltaic plains, Sarawak, Borneo (B) also shown in (A) by contrasting dark to light green representing variations in vegetation. Dense peat forest is represented by tall trees (60 m high) at the margins of peat domes, changing to intermediate high trees (Shorea; 40–60 m high) toward the center of the peatland where stunted trees (<20 m high) grow (see Figure 3.14). See Figure 3.16 for transect cross section. *Source: Adopted from Hart and Coleman (2004) and Cameron et al. (1989) as modified from Anderson (1961, 1983).*

of the Baram River progradation in which the wave-dominated delta is fringed by mangroves and successively replaced by brackish water and freshwater plant communities. Staub and Esterle (1994) interpreted the Baram River delta to include a coastal plain system from microtidal, mesotidal to macrotidal, which includes a tide-dominated macrotidal estuarine embayment along a 450 km stretch of coastline. Continued river aggradation is accompanied by the construction of levees that were colonized by freshwater vegetation, which in turn formed interfluves where peat accumulated and supplied with nutrients by intermittent floodwaters.

Figure 3.16 is a cross section of the peat bog which shows the lower peaty layers with 2% to >10% clay and ash contents, which represent the beginning of the interfluve peat accumulation. The interfluve peat was isolated from the river channels by upward buildup of organic matter and led to elevation and doming of the peat surface from the drainage level (see Figure 3.11). The termination of floodwater incursions is represented by the upper peat layers with <1% to 2% ash content (Figure 3.16). Sustained isolation from fluvial incursions and decay of the peat resulted in increased acidity, which decreased microbial activity, at the central part of the bog where only stunted vegetation survived. The sulfur content, which decreases upward, reflects the initial brackish influence and evolution into freshwater peat bog. The western tide-influenced coastal and deltaic plains of Sarawak, Borneo, is dotted with these peat bogs developed from the Lupar

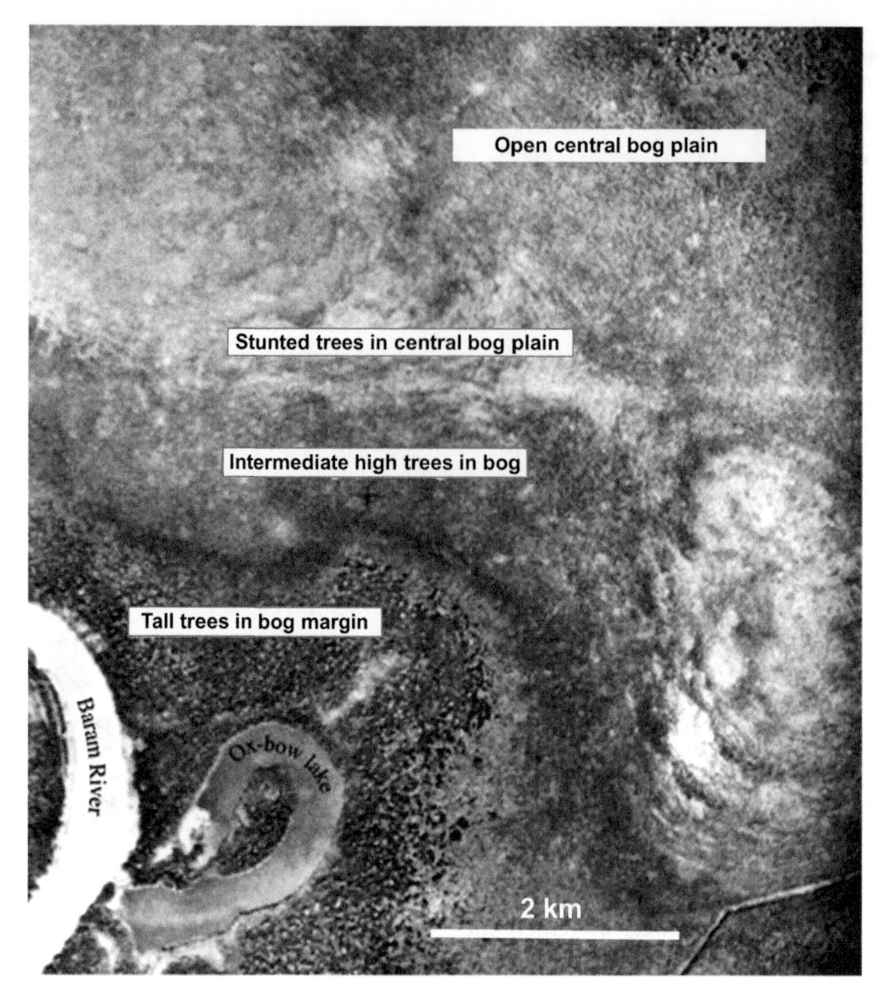

FIGURE 3.14 Areal photograph of the domed peat bogs in the alluvial plain of the Baram River and oxbow lake (abandoned channel), Sarawak, Borneo, showing the zonation of vegetation from the bog's margin to the central part of the bog plain (see Figure 3.15). *Source: Modified from Anderson (1983).*

embayment, Rajang River, to the Baram River (Figure 3.17) (Staub & Esterle, 1994; Staub, Among, & Gastaldo, 2000).

METAPHORS FOR PEATLANDS

Various terms such as rheotrophic vs ombrotrophic (Moore, 1989), minerotrophic vs oligotropic or eutrophic (Cameron et al., 1989), and topogenous vs ombrogenous (Diessel, 1992) peatlands have been propagated to reclassify peatlands in relation to nutrients, hydrologic conditions, topography, and types of vegetations. Much like the metaphors that proliferated and that were discussed in Chapter 1, these peatland terms have been used interchangeably. Hydrology, which is the major criteria of the Moore and Bellamy (1973) peatland model, in association

FIGURE 3.15 Photographs of Indonesian peat-bog vegetation types. (A) Peat bog vegetation of short, small-trunk trees and (B) peat bog vegetation of tall and large, buttress-trunk trees. (C) Photograph by the author of an exhumed peat bog deforested and drained for agriculture in Jambi, Sumatra (see Fig. 3.33 for location and Fig. 3.20 for peat types at this site). *Source: Photographs A and B courtesy of Esterle (1990) and Tim Moore.*

with nutrient supply and topography play an important role in understanding the definitions of these terms.

Rheotrophic vs Ombrotrophic

Types of peatlands such as bogs and swamps (or mires) and marshes exist mainly dependent on the source of water runoff, groundwater, and precipitation, which bring in nutrients for biomass productivity for peat deposition. These processes refine the types of peatlands into rheotrophic and ombrotrophic. Rheotrophic is defined by Moore (1989) in the European sense as a peatland type, especially marsh, which has relatively high or low water table (e.g. surface runoff, groundwater) irrespective of the presence or absence of types of vegetation. Thus, surface water and groundwater are the most important attribute and sediments and inorganic matter brought in by these media to the peatland from external aquatic environments serves as important source of nutrients for biomass productivity (Figure 3.7). That is, the water table is always above the peat surface even during the dry season, which assumes peat accumulation on a flat surface or low-lying topography.

Ombrotrophic type of peatland is dependent entirely on precipitation (e.g. rainfall, snowfall) for water and nutrient supply (Gore, 1983; Moore, 1989; Moore & Bellamy, 1973). This indicates a nutrient-poor ecosystem. The biomass of ombrotrophic mires such as bogs is formed "above ground" (Moore, 1989) because it remains above the groundwater table. Consequently, the ombrotrophic bog is "raised" or elevated above adjoining drainage level (e.g. river, lake, sea). The rheotrophic type peatland may be succeeded by and in juxtaposition with the ombrotrophic type peatland. Thus, in the successional development of types of peatlands the ombrotrophic bog may follow the rheotrophic swamp and marsh.

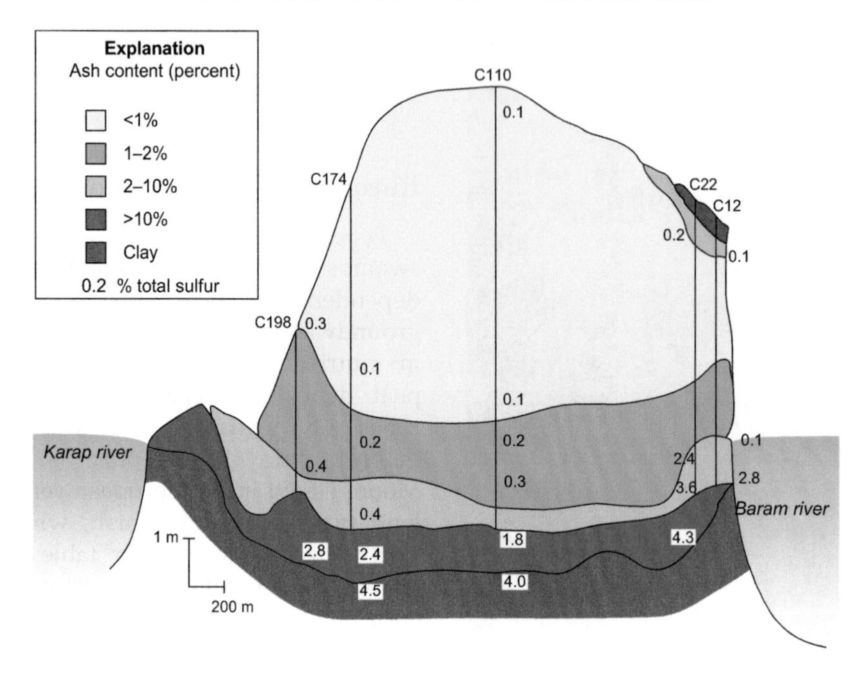

FIGURE 3.16 Cross section of the Tg Pasir peat bog in the interfluve of the Baram and Karap Rivers. See location of cross section in Figure 3.14. *Source: Adopted from Cameron et al. (1989).*

Minerotrophic vs Oligotrophic

According to Cameron et al. (1989) minerotrophic peatlands derive water for their vegetation from rock or soil, which serves as the substrate (Figure 3.18). This suggests that the nutrients and minerals are derived from the underlying rock or soil substrate, which in turn, is dependent on the bedrock geology, hydrology, and water quality. Thus, the peatlands are primarily fed by groundwater recharge but secondarily may be fed by runoff or flood-rich nutrients from adjoining river channels (Figure 3.18). Verhoeven (1986) suggests that water of minerotrophic peatlands is a mixture of rainwater and groundwater both of which supply nutrients from the peat soil. The availability of nutrients to the vegetation is determined by the mineralization of organic matter, which is the result of bacterial breakdown.

Oligotrophic peatlands are synonymously identified as bogs and raised bogs, which is nutrient-poor in comparison to mesotrophic and eutrophic with increase in nutrient status. In comparison, ombrotrophic bogs, which is rain fed, is similarly nutrient poor because precipitation contains very small amount of nutrients and minerals from the atmosphere. Thus, both oligotrophic and ombrotrophic bogs are both nutrient poor and are among the least biomass-productive (e.g. low-biodiverse and low-stature plants) of all the types of peatlands. These types of peatlands are identical in terms of nutrient poor ecosystems, which is why they are used interchangeably in the literature.

Topogenous vs Ombrogenous

Diessel (1992) categorized peatlands into topogenous and ombrogenous types based on their origin either from surface and groundwater or rainfall inputs, respectively. Lode (1999) redefined ombrogenous as surface waterlogging by

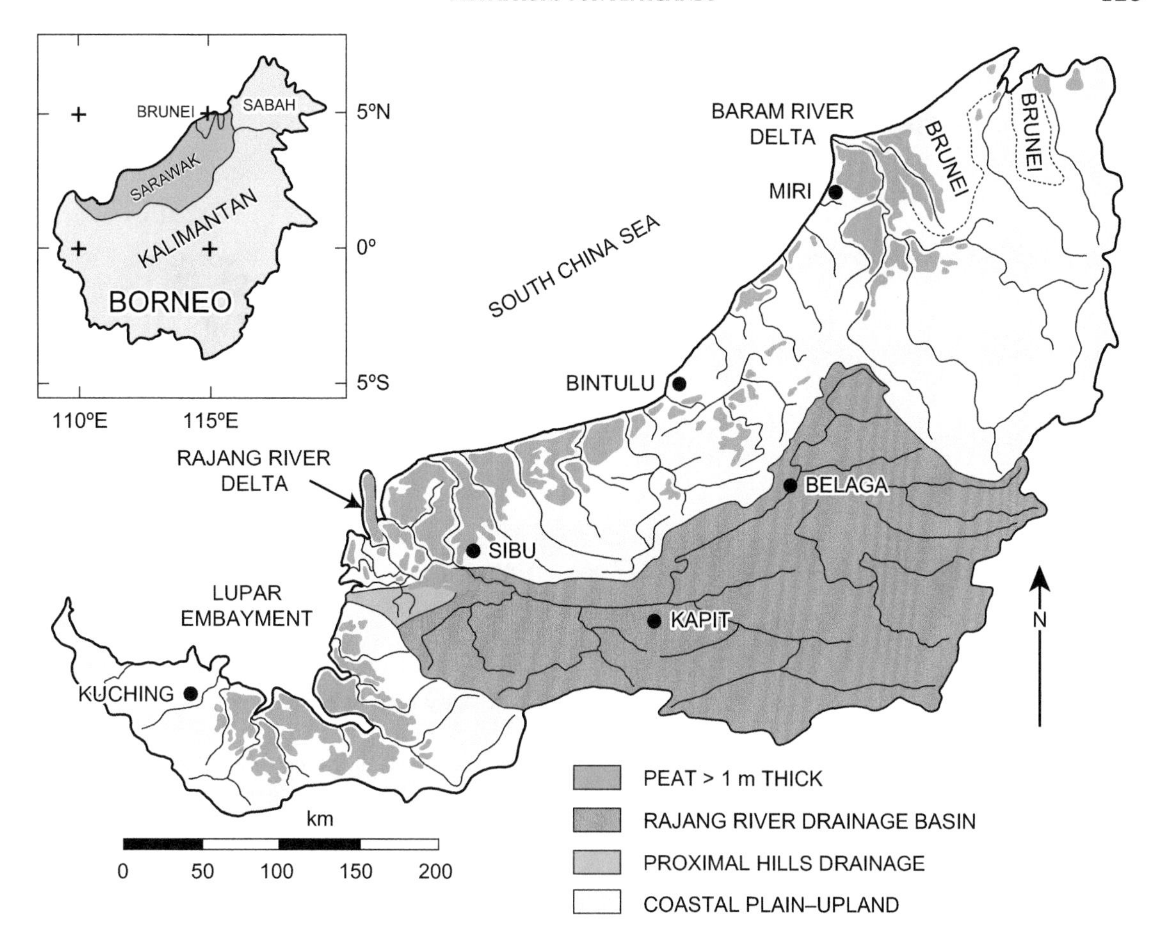

FIGURE 3.17 Map of peatlands (>1 m thick peat) along the western coastal region of Sarawak, East Malaysia and Brunei showing the locations of the tide-influenced Rajang and Baram deltaic plains, and adjoining coastal plains. *Source: Adopted from Staub et al. (2000).*

precipitation and topogenous as water accumulated in topographic depressions, mainly by mineral soil and surface water applied in temperate peatlands. These definitions differ from Driessen (1978) classification of tropical topogenous peat as similar to low-moor peat in a temperate region, which begins with aquatic vegetation and ends with buried organic debris in a submerged depression. Furthermore, Driessen (1978) defined tropical topogenous peat with closed canopied vegetation, year-round rainfall, and very high humidity, which gradually expands into adjoining areas and rises into dome shape eventually ending as nutrient-poor ecosystem.

A review of the literature exposes the conflicted use of topogenous and ombrogenous mires (Figure 3.18). For example, Wheeler (1999) referred the use of "ombrogenous wetlands" to be restricted "flattish surface". Yerima and Van Ranst (2005) referred to the tropical ombrogenous peat as lowland deposit. However, a large part of the literature agrees with acceptable use of the difference in topographic expressions between tropical topogenous and ombrogenous mires (Figure 3.18).

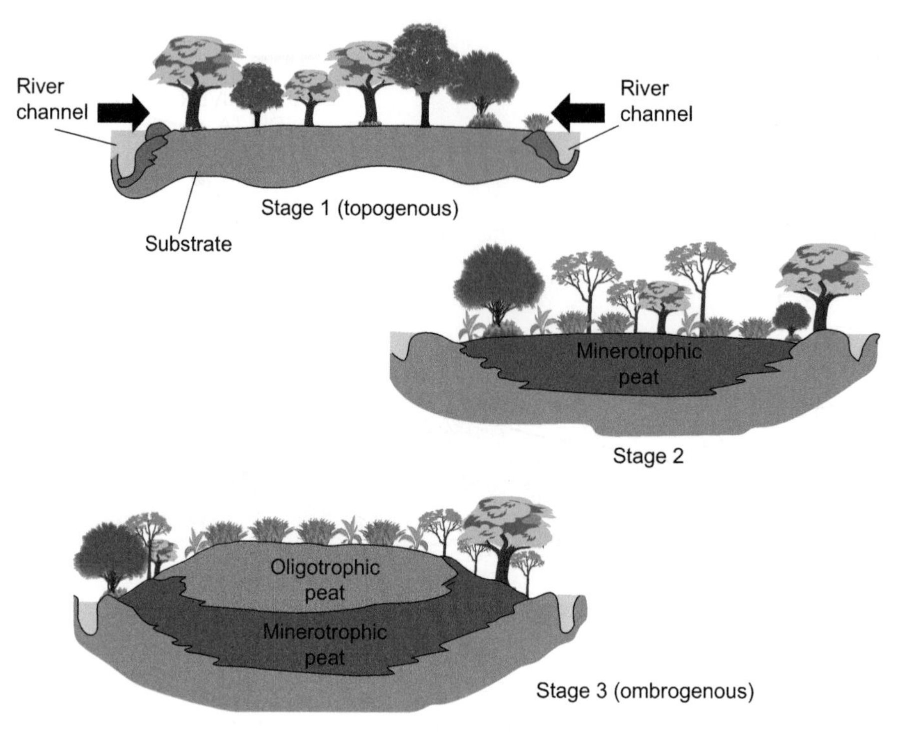

FIGURE 3.18 Diagrammatic cross sections (stages 1–3) showing evolution of minerotrophic to oligotrophic and topogenous to ombrogenous peat mires. Stage 1 is plant colonization of an interfluve nourished by floodwaters (arrows). Stage 2 is peat accumulation simultaneous to levee aggradation. Stage 3 is peat buildup resulting in perched groundwater table and domed surface. *Source: Modified from Anderson (1961), Cameron et al. (1989), and Esterle (1990).*

For example, topogenous peat mire is "planar" confined to topographic depressions with horizontal upper surface and the ombrogenous peat mire is "domed" with outward surface drainage (Neuzil & Cecil, 1985). Staub and Esterle (1992) used the terms topogenous swamps in a tidal-flat setting coalescing with ombrogenous mires. It is assumed from the depositional environment point of view that both mires include low-lying swamps succeeded by raised ombrogeneous mires. Of the various terminologies the terms topogenous and ombrogenous have the widest range of definitions, which lead to confusion.

BOX 3.2

PEATLANDS AS RESOURCES AND CARBON SINKS—CONCEPTS AND MISCONCEPTIONS

Peatland resources are huge covering a total area of about 3–4 million km^2 or about 2–3% of the world's landmass along bodies of freshwater and saltwater (NIEA, 2010; World Energy Council, 2007; Pravettoni, 2009; Holden, 2005). Peatlands make up about 60% of the world's wetlands (Figure 3.19) but only about 7% of all world's peatlands is exploited or commercialized for agriculture, forestry, and energy resources (Andriesse, 1988; Ekono, 1981; Hook, 2007). Peat is cut, stacked, and dried for energy in lieu of wood for cooking and heating particularly in the European boreal region (e.g. England, Ireland, Netherland, Finland, Sweden, and Russia) for at least 2000 years. Also, peat has been cultivated as agricultural peat soil for as long as 6000 years especially in the Southeast Asian tropical region (e.g. Indonesia, Malaysia, Brunei, and Singapore). Thirty percent of global peatlands is in the tropics and 60% of this is found in Indonesia where peatland trees are logged. Peat was used for electricity generation in small units in the range of 20–1000 kJ/s (Andriesse, 1988).

Table 3.3 shows combustion properties of peat in comparison to wood and lignite as determined from their moisture content, calorific value, bulk density, and volatile matter.

Peat has received considerable attention for its diverse form of natural resources; however, the decayed organic deposit is not well known as a carbon sink (12% of the global carbon pool, IPCC, 1996; IPCC, 1997) of mainly methane (CH_4), carbon dioxide (CO_2) and dissolved organic carbon (DOC). For example, the Southeast Asian peats hold 42,000 megatonnes of carbon (Hooijer, Silvius, Wösten, & Page, 2006; Pearce, 2002). As peatlands develop they store atmospheric CO_2 and as the organic matter deposit continues to decompose, it emits CH_4 ("marsh gas") to the atmosphere. The global terrestrial carbon resources are as much as 528,000 megatonnes (Gorham, 1991; Immirzi and Maltby, 1992). According to Hooijer et al. (2006) these carbon resources are equivalent to about 33% of the world's soil carbon and to 70 times as of the 2006 annual global CO_2 emissions from burning fossil fuel. In addition, the production and consumption of DOC in peatlands play a major role in the carbon budget of adjoining aquatic ecosystems where DOC is transported and exported by surface and groundwater into adjoining aquatic ecosystems (e.g. rivers, lakes), which in turn contributes to global warming (Pastor et al., 2003). Fluxes of DOC, CH_4, and CO_2 in peatlands resulting from variations in time, space, temperature, precipitation, and hydrology play a major role to the global greenhouse gas budget (Lohila et al., 2011). Impacts of anthropogenic greenhouse gas emissions from peatlands are exacerbated by peat fires resulting from drainage and deforestation in preparation for transmigration and resettlement (e.g. 5% of 20 million hectares converted to agricultural in Indonesia), and droughts due to microclimate change.

PEAT TYPES: FIBRIC, HEMIC, AND SAPRIC

Peat, which is considered as organic soil and classified as histosol or histos (Greek) for tissue (Andressie, 1988), is characterized according to its degree of decomposition or degradation. According to Moore and Shearer (2003) changes in groundwater level mainly control the types of peat that are produced from the decomposition. Furthermore, depositional environment, tectonic setting, and climate have no correlation to peat types. However, change in climate (e.g. precipitation, temperature) will certainly influence the shifts of the groundwater level and, thus, the position of aerobic and anaerobic zones of decay of the organic matter. The consensus is that peat types formed in marshes, swamps, and bogs are functions of types of vegetation and extent of decomposition in which the plant litter have undergone decomposition (Cameron et al., 1989; Farnham & Finney, 1965; Henderson & Doiron, 1981). The types of vegetation and extent of decomposition may be measured by the texture (e.g. coarse vs fine grained) of the organic matter of the peat. Moreover, based on the classification recognized by soil taxonomy, peat soil may be based on wet bulk density, fiber content, saturated water content, and color (Table 3.4) (Andriesse, 1988).

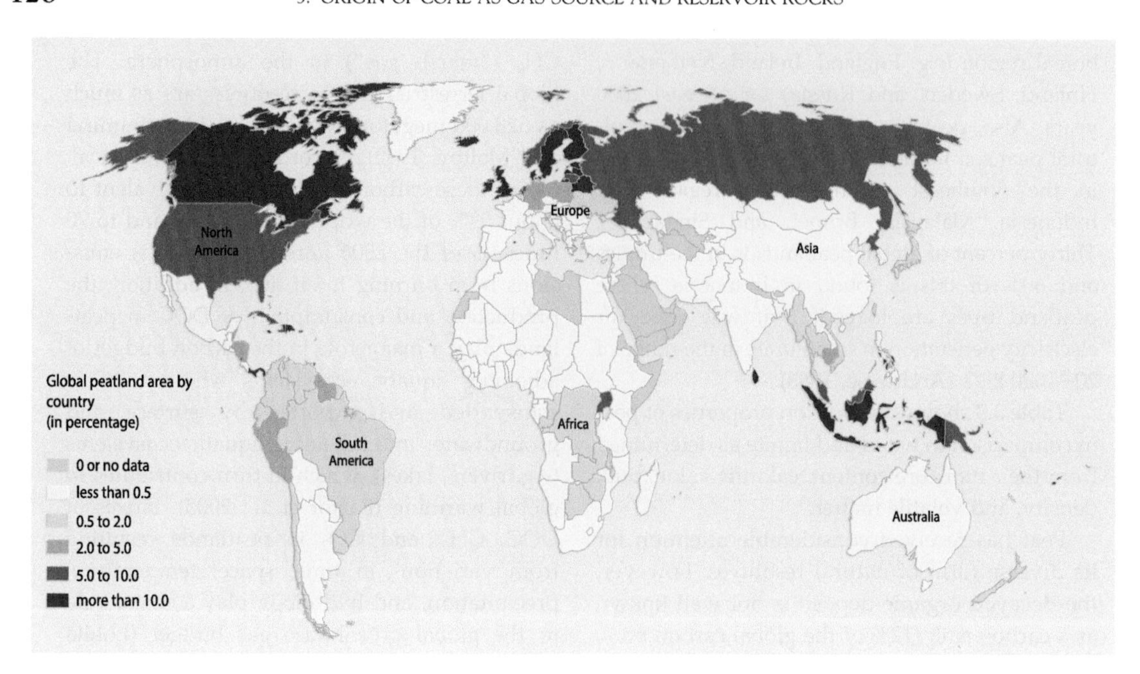

FIGURE 3.19 Global distribution and percentage of peatlands by specific country locations. *Source: Adopted from Pravettoni (2009).*

TABLE 3.3 Combustion Properties of Peat in Comparison to Wood Biomass and Lignite Coal

		Peat			
Combustion Properties	**Wood (Biomass)**	**Mill**	**Sod**	**Briquette**	**Lignite**
Effective calorific value of dry matter (mean MJ/kg)	18–19	18–22	18–22	18–22	20–24
Effective calorific value at operating moisture content (mean MJ/kg)	12–13	7–12	11–14	17–18	11–14
Volatile matter (mean % dry matter)	75–85	65–70	65–70	65–70	50–60
Bulk density at operating moisture content (kg/m^3)	320–420	300–400	300–400	700–800	650–800
Operating moisture content (%)	30–35	40–55	30–40	10–20	40–60

MJ/kg = megajoule/kilogram; kg/m^3 = kilogram/cubic meter; % = percent.
Source: Modified from Ekono (1981).

There are three basic types of peat soil recognized by soil taxonomy (Soil Survey Staff, 1975): (1) fibric, (2) hemic, and (3) sapric based mainly on the degree of decomposition of the original plant material, which is indicated by their fiber content (Andriesse, 1988). Fibers are fragments (retained on a 100-mesh sieve) of plant tissue, minus live roots, with recognizable cellular

TABLE 3.4 Criteria of Classification of Types of Peat Soils

Criteria of Classification	Fibric (L. *fibra*; Fiber)	Hemic (Gk. *hemi*; Half)	Sapric (Gk. *sapros*; Rotten)
Wet bulk density	<0.1	0.07–0.18	>0.2
Fiber content	2/3 vol. before rubbing, 3/4% vol. after rubbing	1/3–2/3% vol. before rubbing	<1/3 vol. before rubbing
Saturated water content as percent of oven-dry material	850 to >3000	450 to >850	<450
Color	Light yellowish brown or reddish brown	Dark grayish brown to dark reddish brown	Very dark gray to black
Degree of decomposition	Lowly decomposed	Moderately decomposed	Highly decomposed

L = Latin; Gk = Greek; % = percent, vol. = volume
Source: Modified from Soil Taxonomy (Soil Survey Staff, 1975) and Andriesse (1988).

structure decomposed enough to be crushed and shredded with the fingers. Large wood fragments 2 cm or greater in diameter are not classified as fibers. Highly decomposed plant materials do not contain fibers and moderately decomposed materials have high fiber content. Slightly decomposed materials have more than 50% by volume with wood comprising the remaining materials. Peat types may be best recognized in the field by the difference of their color, which is caused by their degree of decomposition (Cameron, 1973, 1975).

Fibric Peat

Fibric peat is the least decomposed with more than two-thirds recognizable plant fibers, coarse-grained texture, and when compressed the peat discharges tea-colored water with little or no organic matter. In temperate climate fibric peat is mainly derived from sphagnum mosses and fine root system of shrubs. In the tropical climate, woody fibric peat is mainly derived from slender, extensive root systems and horizontal stems of trees (Anderson, 1964). In the field, after rubbing the peat, less than one-third of the peat ooze between the fingers (Institut Pertanian Bogor Team, 1980). Fibric peat is characterized by light yellowish brown, dark brown or reddish brown colors resulting from exposure to the air above the water level.

Hemic Peat

Hemic peat is moderately decomposed with one-third to two-thirds recognizable plant fibers predominantly composed of fragments of roots, stems, leaves, wood, bark, and seeds and, thus, medium-grained texture. According to Cameron et al. (1989), freshwater and salt-water marshes produce hemic peat mainly from sedges, rushes, and grasses as well as from shrubs and arboreal forests in swamps as woody hemic peat. In the field, after rubbing the peat, less than one-third to two-thirds of the peat ooze between the fingers (Institut Pertanian Bogor Team, 1980). Hemic peat is characterized by commonly dark grayish brown to dark reddish brown colors (Figure 3.20(A)) caused by partial exposure to air below the water level.

Sapric Peat

Sapric peat is the most decomposed, fine-grained texture, less than one-third recognizable plant fibers, and when compressed retains moisture and behaves like a paste. In the field, after rubbing the peat, less than two-thirds of the peat oozes between the fingers (Institut

FIGURE 3.20 (A) Moderately decomposed woody fragments and grayish brown hemic peat from a Sumatran (Jambi) peat bog. (B) Highly decomposed, paste-like, dark gray to black sapric peat from a Sumatran (Jambi) peat bog.

Pertanian Bogor Team, 1980). Sapric peat is very dark gray to black colors (Figure 3.20(B)) due to nonexposure to air. According to Cameron et al. (1989) sapric peat is mainly derived from small and easily decomposed algal and herbaceous aquatic plants of ponds and marshes. However some sapric peat is derived from woody plants of forest swamps.

PROCESSES OF PEATIFICATION, GASIFICATION, AND DIAGENESIS

The accumulation of plant litter in peatlands is a small but significant part in the transformation process of peat to coal. Peatification is a process of transformation of plant materials through chemical and microbial changes into peat at the surface and near surface. The alteration process begins on the surface with physical breakdown of plant parts (e.g. leaves, trunks, stems, barks, roots) into small organic particles or debris. Vertebrate and invertebrate animals (e.g. mites, springtails, insects, arthropods) contribute to the mechanical breakdown by consuming, mixing, and transporting materials. In addition, worms (e.g. termites) and insects (e.g. wood ants) play an important role in mechanical disintegration of the plant litter. Burrowing animals mix and aerate peat materials. Mechanical breakdown may be performed by surface water flow on the peatlands particularly during increased precipitation in tropical regions eroding and breaking up plant litter. In temperate regions, ice may form in cracks and wedge into peat fissures and heave plant materials. In both regions, expansion and contraction from seasonal heating and cooling break up plant material. Microbial activity during peatification generates early stage biogenic gas or gasification. Upon burial, peat undergoes the first stage of thermal maturation (Ro $<0.5\%$) of diagenesis (Horsfield & Rullkotter, 1994).

Aerobic vs Anaerobic Decay of Peat

A large part of peat breakdown by microbial and chemical processes occurs below the surface. The most intensive transformation of the plant debris by decomposers or aerobic and anaerobic microorganisms occurs in the "peatigenic layer" from below the surface to about 0.5 m at depth (Stach et al., 1982). However, anaerobic microbes cease to exist at greater depths at about 10 m (Stach et al., 1982) below which chemical changes predominate. Microorganisms and fungi are active within the peatigenic layer, which includes the acrotelm layer immediately below the surface and the underlying catotelm layer (Figure 3.8). Aerobic microbes and fungi are active above the groundwater table in the unsaturated acrotelm accompanied by humification process. Anaerobic microbes are active below the groundwater table in the saturated catotelm accompanied by gelification process.

Humification vs Gelification

Humification is a process of formation of humic substances (organic matter that has reached maturity) decomposed from plant remains. Humification results from progressive oxidation by addition of oxygen from oxygenated water and lowering of the groundwater table, pH values of the water in the peat, and heat exposure or peat temperature. In the presence of oxygen, microbes and fungi attack lignin or organic substance that binds the cells, fibers, and vessels of wood and converted to humic substances. The oxygen-rich materials in the peatigenic layer such as cellulose and hemicellulose are decomposed readily by microbes resulting in the enrichment of carbon-rich lignin and humic acids (Zeng, 2005). During early peatification, depolymerization of polysaccharides or complex of carbohydrate polymers by decomposers occurs (Stout, Boon, & Spackman, 1988). In addition, hemicellulose or constituents of the cell walls of plants and are simpler structure than cellulose, are rapidly removed followed by conversion into glucose (Stout et al. 1988). Lignin is resistant but a proportion is degraded mostly under aerobic conditions yielding large amounts of aromatic, phenolic and carboxylic acid (COOH) (Killops & Killops, 2005). In addition to these residues microbial metabolites and the decayed remains of fungi and bacteria contribute to the organic matter (e.g. fungal carbohydrates). Fundamentally, peats formed in different temperature, climate, and types of vegetation result in various degrees of decomposition of the organic substances (Table 3.5) (Kurbatov, 1968).

Gelification succeeds humification at depth in which the humic substances pass through a soft, plastic stage from peat to brown coal (lignite). The original plant material, water and ion supplies, amount of alkalinity, and nature of oxidation influence gelification. Carbon content increases rapidly with depth and increases as much as 60%, which stabilize at greater depths as well as moisture content decreasing rapidly. As indicator of diagenesis, free cellulose can be found in the deeper part of the peatigenic layer, which distinguishes peats from soft brown coal. Also, higher degree of gelification distinguishes brown coal or lignite from peat.

ORIGIN OF PEAT GAS: A BIOGENIC GENERATION

During biochemical changes by microorganisms in the peat, gases such as CH_4, CO_2, NH_3, N_2O, N_2, and H_2S together with H_2O are the main expelled products. Methane and CO_2 are the predominant products from the peat (Zeng, 2005; Stout et al. 1988). A large portion of the CO_2 is derived from breakdown of carbohydrates. Cellulose is readily decomposed by microorganisms to CO_2, ethanol, and various acids. Demethylation of the methoxyl groups of lignin is probably a significant contributor of CH_4. Lignin is an aromatic polymer that anchors

TABLE 3.5 Degree of Decomposition of Organic Substances in Various Types of Vegetation in Peat Swamps

	Peat Swamps	
Organic Substance	**Sedge or *Carex* and Reed or Grass-Like Vegetation (30–40% Decomposed)**	**Birchwood or Medium-Size Tree or Shrub Vegetation (55% Decomposed)**
Bitumen	1.1–3.3	8.8
Humic acids	32.2–33.6	52.2
Hemicellulose	8.6–15.0	1.0
Cellulose	3.5–3.7	0.0
Lignins	12.9–18.6	0.0
Cutin	5.2–11.9	16.0
Not determined	21.2–29.2	22.0

% = percent
Source: Modified from Kurbatov (1968).

cellulose fibers and strengthens plant walls; however, decomposition of lignin by microorganisms is not well known. Much of the lignin in peatified wood may be only slightly altered and becomes the primary precursor of vitrinite (Stout et al., 1988). However, the groundmass (matrix material) in low rank coal generally contains some macerated wood, which exhibits greater alteration, involving depolymerization and defunctionalization (Killops & Killops, 2005).

The large volume of gas (e.g. CH_4, CO_2, N_2O) generated, stored, and released in peatlands has produced interest worldwide in relation to understanding the total greenhouse gas budget, in particular and long term effects on global climate change, in general. Also, peat gas is an integral part of coalbed gas that is generated in the early stage of peatification and potentially retained during coalification. The origin and dynamics of generation, storage, and desorption or release of peat gas have been largely ignored as an energy resource. Peat gas is generated during decomposition of plant biomass in the acrotelm layer (Fig. 3.8; Fig. 3.21) by aerobic microbes mainly as carbon dioxide and in the catotelm layer by anaerobic microbes mainly as methane. Methane, which forms as gas bubbles, occludes interstitial pores of the peat reducing the hydraulic conductivity and water saturation that further intensify degradation of biomass due to the lack of nutrients (Brown, 1998; Brown & Overend, 1993). However, the quality of organic matter affects microbial decomposition in peatlands. Reiche, Gleixner, and Kusel (2010) observed that highly decomposed organic matter is directly related with lower CO_2 generation and reduction of methanogenesis. Results of the work of Reiche et al. (2010) suggest that undecomposed organic matter is a requirement for generation of biogenic CH_4 and CO_2 indicating that gas is generated at shallow depths in the peat.

Peat Methanogenesis

Methanogenesis is the process of generation of methane by methanogens, which are strictly anaerobic microorganisms or archaeons (Prescott, Harley, & Klein, 2002). Methanogens derive energy by converting carbon dioxide, hydrogen, formate, acetate, and other compounds from organic matter of peat to methane

BOX 3.3

MARSH GAS: LOST RESOURCE TO THE ATMOSPHERE–FACTS OR FICTION

Marsh or swamp gas, which is composed mainly of methane but also include carbon dioxide, nitrogen oxide, and hydrogen sulfide, is a vernacular for natural gas that forms when plant materials decompose in peatlands. "Marsh gas" is known by a variety of names in English folklore as "will-o'-the-wisp", "ignis fatuus", "jack-o'-lantern", "hinkypunk", "hobby lantern", or "ghostly light", the latter seen at night over bogs, swamps, and marshes (Trevleyan, 1909). Although methane does not ignite spontaneously, the chemical lore has it that the presence of phosphine or diphosphine in the gas is spontaneously flammable in air. Phosphine is the common name for phosphorus trihydride (PH_3), also known as phosphane (also phosphamine), which is a colorless, flammable gas with a boiling/condensation point of −88 °C at standard pressure and auto-ignition temperature of 99.85 °C (Airgas, 2012).

Peatland gases are liberated to the atmosphere since more than 360 mya through: (1) natural desorption, (2) agricultural activities, and (3) peat fires. The main peatland greenhouse gases in the order of decreasing importance are: methane, carbon dioxide, and nitrous oxide. Peatlands are the largest natural source of methane desorption to the atmosphere with elusive estimates of total global emission. Early estimated emission was from 100–500 Tg (Teragrams or 10^{12} g) annually (Augenbraun, Matthews, & Sarma, 1997; Augenbraun and Smith, 1998) out of the total global natural methane emission of as much as 380 Tg methane per year (Chappellaz, Fung, & Thompson, 1993). However, more recent estimates indicated that methane emission is 503–610 Tg per year, and more than 70% originates from activity of methanogens (IPCC, 2007). Regardless of the different estimates, the huge estimates of peat methane releases are a compelling reason for potential energy resource assessment. Finally, the growing land use of peatlands for agriculture or for horticulture has led to draining for deforestation, farming, and extraction as fuel (Fig. 3.15(C)). Draining of peatlands leads to release of carbon dioxide. For example, in 2006 peatland fires caused by droughts in Indonesia released up to about 900 million metric tons of carbon dioxide (ScienceDaily, 2009) with hydrogen sulfide more common in brackish and salt-water peatlands. Global CO_2 and N_2O emisions of more than 2 Gt (gigatonnes) CO_2-eq./yr result in drainage of peatlands; however, the CH_4 emission from the peatlands worldwide is much larger (Couwenberg & Fritz, 2012; Joosten, 2011).

or methane and carbon dioxide. Methanogens achieved biomethanation in association with fermentative and hydrogen-producing, acetate-forming microoganisms (Miyamoto, 1997; Page et al., 2004). These microbial consortia form a pathway of reactions (Miyamoto, 1997; Page et al., 2004) in which the fermentative microorganisms (fermenters) secrete enzymes, which hydrolyze polymeric materials (e.g. cellulose, hemicellulose, starch, proteins) to monomers (e.g. glucose and amino acids) and in turn convert to higher volatile fatty acids, hydrogen, and acetic acid. This reaction is followed by the hydrogen-producing, acetate-forming microorganisms (acetogens), which convert the fatty acids and produce hydrogen, carbon dioxide, and acetic acid. Finally, the methanogens convert the hydrogen, carbon dioxide, and

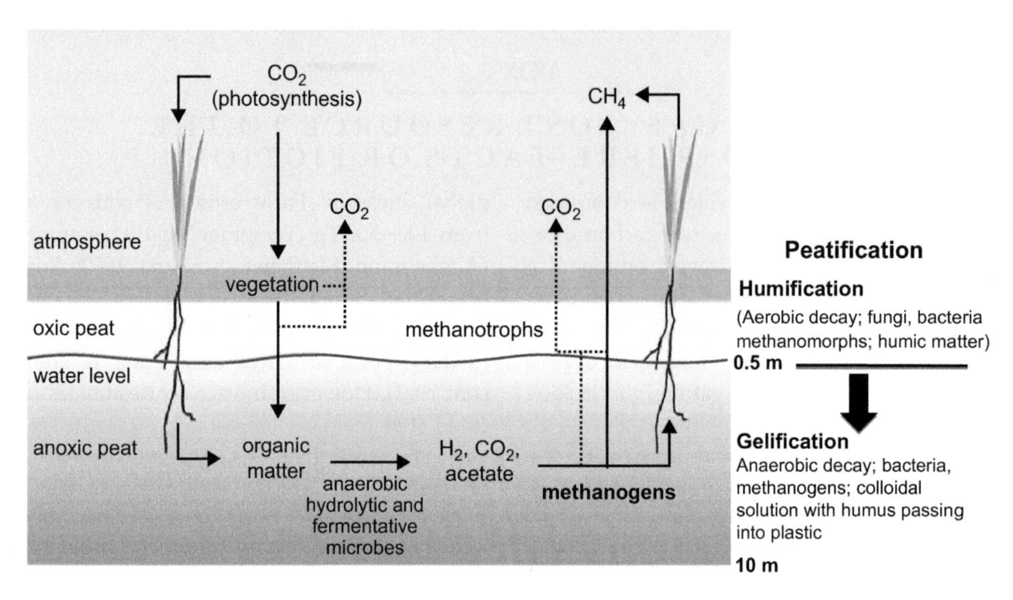

FIGURE 3.21 Schematic diagram showing the carbon dioxide cycling and generation of methane by anaerobic microbes through methanogenesis in peatlands. CH_4 = methane; CO_2 = carbon dioxide; H = hydrogen. *Source: Modified from Page et al. (2004).*

acetate, to methane and carbon dioxide (Figure 3.21). This methanogenic pathway is methyl-type fermentation or aceticlastic methanogenesis. Another pathway of methanogenesis is by carbon dioxide reducing microorganisms.

Generation of biogenic peat gas is developed in boreal and tropical peatlands (Galand, Fritze, Conrad, & Yrjala, 2005; Metje & Frenzel, 2007). In the subarctic and arctic peats, methanogenesis is controlled by temperature. The maximum temperature for methanogenesis was between 26 and 28 °C (4.3 μmol methane) but the activity was even higher at 4 °C (0.75 μmol methane) (Metje & Frenzel, 2007). Aceticlastic methanogenesis accounted for about 70% of the total methanogenesis. In comparison, the vegetation types control methanogenesis in the boreal or northern region. Galand et al. (2005) reported that the contributions of aceticlastic methanogenesis to total methane production differed among the three-peatland ecosystems with various vegetation types. The roots of vascular plants (e.g. sedges) are shallower penetrating into the peat than nonaerenchymatous (no air channel in the roots) peatland plants. The root penetrations into anoxic peat layers allow access of acetate into the deeper anaerobic peat layers, which in turn, favors aceticlastic methanogenesis. Results suggest that methanogenic conditions in peat soils rely on a constant supply of easily decomposable metabolic substrates.

Strict anaerobic bacteria (methanogens) produce methane in tropical peat such as the Indonesian (Sumatra and Kalimantan, Borneo) peat bogs (Sabiham, 2010; van der Gon & Neue, 1995). The rate at which anaerobic decomposition of tropical peats proceeds as well as the resulted ratio of carbon dioxide to methane depends on the pH of the peat soil. The methanogenic pathway of anaerobes generates methane through the reduction of carbon dioxide with hydrogen or organic molecules as the hydrogen donor if soil pH is very low (Sabiham, 2010). The activity of methanogens is usually optimum around neutrality or under slightly

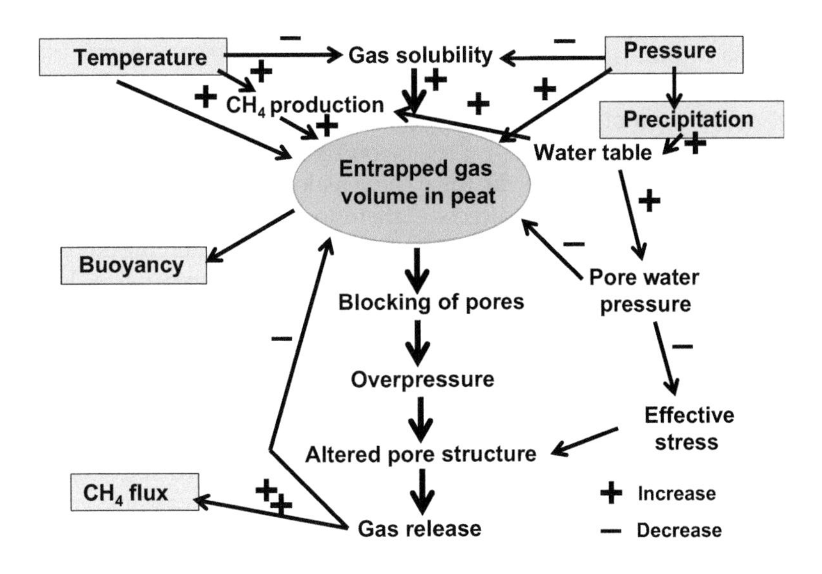

FIGURE 3.22 Dynamics of entrapped gas in peat. CH_4 = methane. *Source: Adopted from Strack et al. (2005).*

alkaline conditions and is very sensitive to variations in soil pH (Garcia, Patel, & Ollivier, 2000; Wang, Delaune, Masscheleyn, & Patrick, 1994). The carbon-reduction methanogenic pathway may be supported by the study of carbon fluxes in peat bogs in the interior Kalimantan, Borneo (Vasander, 2005). High carbon dioxide and low methane fluxes led Vasander (2005) to speculate that the lack of sedges like those in the temporal and boreal peatlands, which contributed to aceticlastic methanogenesis (Galand et al., 2005) may be one of the causes of low methane flux.

Dynamics of Peat Gas

A review of the literature indicates that the gas, mainly CH_4, is supersaturated in peat water and its entrapment is related to hydrostatic and barometric pressures (Strack, Kellner, & Waddington, 2005). This mechanism of entrapment and desorption of CH_4 is suggestive of hydrostatic pressure as a major cause of CH_4 being held and stored in the surface areas (e.g. pores, fractures) of coal. Strack et al. (2005) proposed a conceptual model to describe the factors that affect behavior and volume of peat gas in peatlands, biogeochemistry of the gas, and mechanism of gas release (Figure 3.22). The conceptual model, based on previous studies (Beckwith & Baird, 2001; Brown, 1998; Buttler, Dinel, Lé vesque, & Mathur, 1991), shows that as a result of peat decomposition gases (e.g. CH_4, CO_2, H_2S) reach supersaturated concentrations in the peat pore water. In order for dissolved gas in water to form gaseous bubbles the partial pressure of the dissolved gas must be above the hydrostatic pressure of the peat (Rothfuss & Conrad, 1998). According to Strack et al. (2005), the partial pressure is governed by Henry's Law, which states that the solubility of a gas in water at a particular temperature is proportional to the pressure of that gas above the water. Thus, gas solubility decreases resulting in net transfer of gas from aqueous to gaseous phase as the peat temperature increases. Changes in atmospheric (e.g. barometric) and hydrostatic pressures as well as gas generation and consumption will affect the gas pressure in the peat (Fechner-Levy & Hemond, 1996). Also, a decrease of atmospheric pressure and a lowering of the groundwater table will reduce pressure at depth promoting gas expulsion (Baird & Gaffney, 1995).

Gas concentration in the peat results in the formation of gas bubbles, which in turn, affects pore-water pressure creating vertical and horizontal hydraulic pressure gradients (Glaser et al., 2004; Kellner, Price, & Waddington, 2004). The hydraulic gradients promote gas flow from high to low pressure areas, which according to Strack et al. (2005) do not necessarily promote immediate flow of gas upward through the peat to be released at the surface. The gas flow may be dependent on the physical properties of the peat matrix, groundwater table level, and CH_4 concentration (Romanowicz, Siegel, Chanton, & Glaser, 1995). The flow of gas upward may be prevented by a confining layer containing pores in the peat matrix that are smaller in diameter than the gas molecules as well as bubbles trapped within the pores, which in turn, increases the CH_4 volume. Thus, gas flow is best developed where a confining layer is not formed where the peat matrix contains pores large enough for the CH_4 can flow through.

A conceptual model of the dynamics of formation and trapping of gas bubbles in peat is shown in Figure 3.23 proposed by Kellner, Waddington, and Price (2005). The gas bubbles adhere to or are suspended between pore walls of premacerals or organic components (e.g. wood fragments, roots, spores, pollens, fibers, etc.) as long as their buoyant force do not cause them to rise (Figure 3.23(A)). According to Kellner et al. (2005) when the gas bubbles grow and/or coalesce to diameter sizes larger than the pore sizes, the gas bubbles are trapped (Figure 3.23(B)). Thus, trapped gas bubbles clog pores and develop confining layers, which locally trap gas creating overpressure zones (Figure 3.23(C)). Gas breaks out from these overpressure zones with sudden changes in the hydrostatic pressure causing gas to be released to the peat surface (Strack et al., 2005). When pressure decreases the gas volume increases due to increased gas expulsion and bubble expansion. The low-density CH_4 developed as gas bubbles play a major role in the buoyancy of "floating" and "quaking" peatlands. The size of the bubbles increase until their buoyancy exceeds the hydrostatic and barometric pressures that hold them in place, at which time methane bubbles move upward and are released at the peat surface through violent outbursts (Tretkoff, 2011). This type of event or ebullition accounts for a large proportion of methane lost to the atmosphere as greenhouse gas from global peatlands.

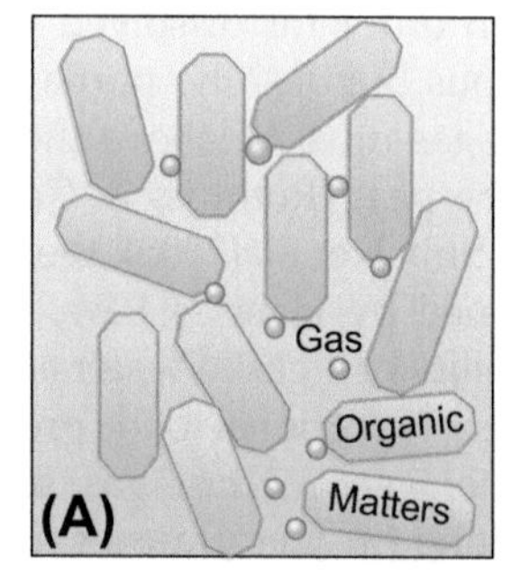

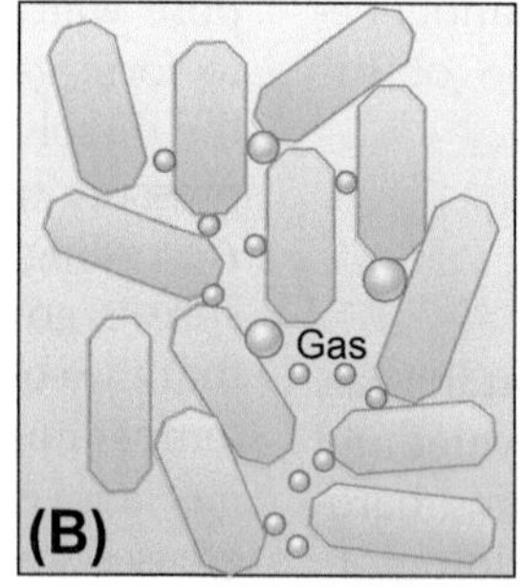

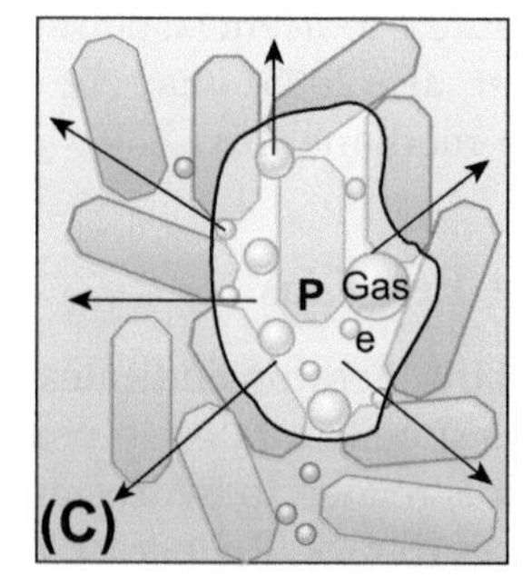

FIGURE 3.23 Entrapment model of gas bubbles in peat. (A) Gas bubbles are trapped in voids between organic matters and adhered to the pore walls. (B) Gas bubbles accumulate and grow leading to restriction of both water and gas flows. (C) Results in development of closed zones or confining layers, which may further trap locally generated gas creating overpressure zones in the peat (Pe). *Source: Adopted from Kellner et al. (2005).*

The changes in the atmospheric and hydrostatic pressures, which affect the dynamics of peat gas, are directly related to climatic regimes such as precipitation, evapotranspiration, and temperature. Precipitation minus evapotranspiration, which supplies water to the peatlands, controls the rising and falling of groundwater table. For example, during summer rainfall increases surface and subsurface water supplies, which in turn raise groundwater table and hydrostatic pressure. During the same season, increased temperature, which promotes water evaporation from the peat, results in lowering of the groundwater table as well as of the hydrostatic pressure. Thus, within the summer season if both factors are acting independently, each will influence retention or desorption of peat gas, which demonstrates microclimatic and hydrologic controls to the dynamics of peat gas. These factors control the seasonal development of gas bubbles in the peatlands.

Peat Gas: Precursor of Early Stage Coalbed Gas?

The volume of biogenic peat gas preserved as coalbed gas is not known and according to Rice (1993) the peat gas is largely lost prior to burial. Rice and Claypool (1981) indicate that most of the initially formed biogenic gas is lost by aerobic microbial oxidation or escape to the atmosphere due to low hydrostatic pressure to hold the gas. Thus, one of the requirements of gas preservation is a holding mechanism to retain the gas in solution in the pore waters until the layers of peats and sediments are sealed, compacted, and trapped. According to Rice (1993) biogenic gas can be generated during coalification after peatification (e.g. gelification). The most likely period of preservation of biogenic peat gas is at the early stage of burial from peat to subbituminous coal (low rank with Ro of <0.5%) during diagenesis (Rice, 1992; Rice & Claypool, 1981). However, it has been demonstrated in laboratory experiments that burial of water saturated, anoxic peat bog leads to inactivation of respiration of anaerobic microorganisms and methanogenesis (Blodau, Diems, & Beer, 2011).

Gas from Peat to Coal Reservoir: Generation During Diagenesis

Diagenesis follows the biochemical stage of peatification and begins with inundation of the peat by deposition of detrital sediments (Kim, 1978). As burial by the sediments and subsequent subsidence continue microbial activity is terminated by increased temperature and pressure as well as by accumulation of toxic substances. The effects of diagenesis toward coalification of peat organic matter cannot be accessed during early burial conditions but can be tested in the laboratory. Because the microbial degradation of cellulose and lignin start off coalification, Kim (1978) used sterile samples of these organic matter and wood inoculated with microorganism to study gas formation analyzed by gas chromatography in the laboratory, which resemble the biogenetic phase of peatification. The result of the study is summarized in Figure 3.24, which shows CH_4 and CO_2 are the dominant gas products from all the samples. Carbon dioxide and H_2 were detected first, followed by CH_4, and over a period of several months the gas composition average 95% of CH_4 and 5% for CO_2. However, additional analysis by Kim (1978) indicates that the gas retained in peat is predominantly CH_4 is retained in peat with some CO_2 and heavy hydrocarbon gases.

In order to compare the biogenic gas formation during peatification to diagenetic phase of gas formation during early coalification, Kim (1978) used coal samples of various rank (e.g. lignite, subbituminous, bituminous) from various U.S. coal mines to analyze gas generation with increased temperature. The volume of hydrocarbon gases (e.g. commonly methane, isobutane, and pentane) increased as temperature increased but correspondingly CH_4 decreased at higher

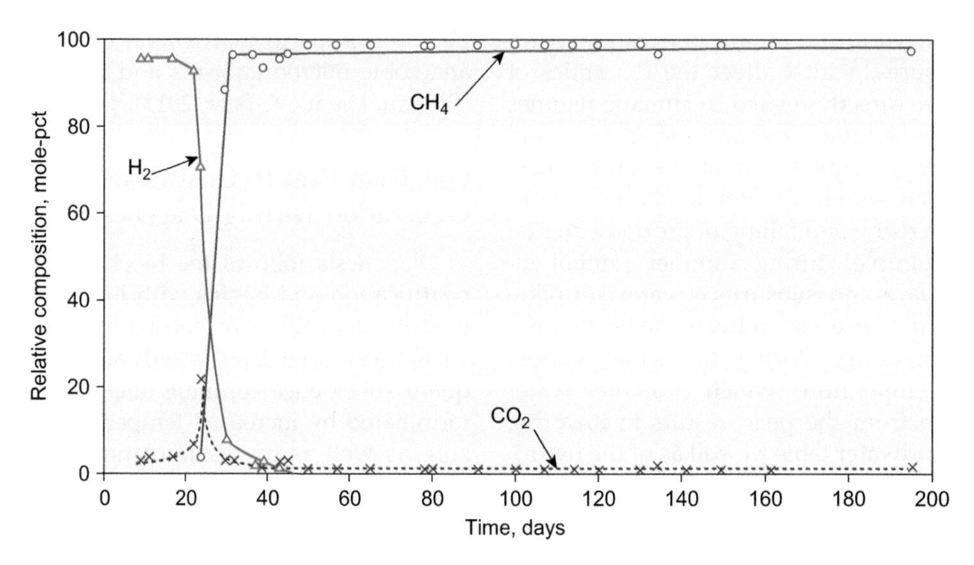

FIGURE 3.24 Biogenic gases (e.g. methane and carbon dioxide) produced from wood, cellulose, and lignin samples inoculated by microorganisms. CH_4 = methane; CO_2 = carbon dioxide; H = hydrogen; pct = percent. *Source: Adopted from Kim (1978).*

temperature. The generated hydrocarbon gas during the experiments contained average of more than 95% CH_4 at 35 °C, 50% at 125 °C, and 30% at 150 °C (Figure 3.25). Thus, the demethylation during diagenesis occurs at a very low temperature (35 °C) and the reaction may involve more complex reactions with heavy hydrocarbons.

Although the laboratory experiments of Kim (1978) proved that CH_4 is generated from cellulose and lignin during biogenic phase of peatification and diagenetic phases of early coalification, sustaining gas generation and accumulation in the peat coal reservoir may not be a *fait accompli.* With increasing diagenesis the lignin and

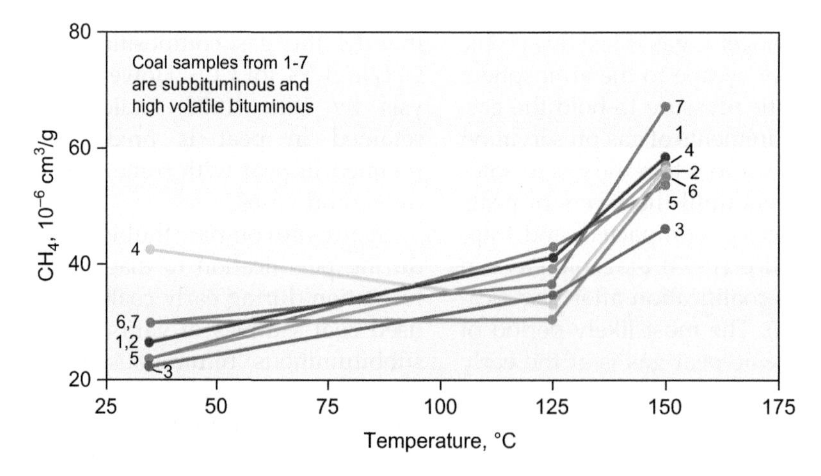

FIGURE 3.25 Methane emitted from coal with increased temperatures (from 35 to 150 °C) under laboratory conditions. CH_4 = methane. *Source: Adopted from Kim (1978).*

polysaccharide content of peat decreases but the content of humic substances increases. Cellulose is still found in peat but is absent from brown coal (e.g. lignite), which is formed at a more advance phase of diagenesis. At the end of diagenesis the resulting brown coal contains no carbohydrates and the amount of relatively unaltered lignin has decreased to <10% (Killops & Killops, 2005). Thus, the cellulose and lignin, which are the common source of CH_4 according to Kim (1978) may be depleted toward the advance phase of diagenesis.

Gas preservation from peat to coal may rely entirely on rapid deposition, burial, and entrapment by impermeable seal rocks (e.g. mudstone, shale). These factors are controlled by the depositional environments, and tectonic and eustatic settings associated to the peat accumulation. For example, peat accumulation in alluvial floodplains can be inundated and buried gradually by sediments during intermittent flooding and/or rapidly during flash flooding. In addition, rapid sediment burial of peat accumulation may develop in river deltas coupled with subsidence and transgression by the sea.

DEPOSITIONAL SYSTEMS OF PEAT (COAL)

Coal originates as peat that forms in bogs, swamps, and marshes with the former two types called mires, which are more likely to accumulate economic deposits. The peat bogs, swamps, and marshes are subsequently buried, compacted, and lithified by deposition of other sediments. The sedimentary layers with the peat are deposited by a contemporaneous depositional system, which comprises various environments (Figure 3.3). A depositional environment is a geographic complex with a distinctive set of physical, chemical, and biological conditions and is characterized by sediment accumulation rather than by erosion processes.

Deposition occurs due to progradation, lateral accretion, or aggradation in which eroded sediments are delivered (e.g. rivers) below base level or sea level (Ethridge, 2011; Schumm, 1993) and subsidence or drop of the area beneath the environment of deposition. However, landward from the shoreline sediments may be delivered below temporary base levels such as lake, river, and floodplain levels. Low-lying areas in coastal, delta, and alluvial plains such as rivers floodplains, lakes, bays, estuarines, and lagoons provide local sediment accommodation spaces, which are measured by the distance from the surface of deposition (e.g. bottoms of river floodplains, lakes, bays) to the local water or base level (Jervey, 1988). Abandoned low-lying depressions in the alluvial and coastal plains, which are far removed from sediment influxes, provide peat accommodation spaces.

Coastal Plain Depositional System

A review of the literature indicates that the most important and common coal-bearing rocks (Cobb & Cecil, 1993; Dapples & Hopkins, 1969; Ferm & Horne, 1979; McCabe & Parrish, 1992; Rahmani & Flores, 1984;) and peat-forming environments (Cameron et al., 1989; Catuneanu, 2006; Cohen et al., 1989; Staub & Esterle, 1993, 1994; Staub et al., 2000;) are deposited primarily in coastal, deltaic, and alluvial plains and secondarily in lacustrine depositional systems. Coastal and deltaic plains peat (coal) depositional systems (Figure 3.26) are developed in low-lying areas behind shorelines, which are prone to modification by processes related to fluctuations of sea level. Specifically, the coastal plain depositional system consists of delta plains, which are flanked by strand plains consisting of barrier and back-barrier environments, and tidal environments (Figure 3.27). These peat-coal-forming environments, which are in a flat area behind the shoreline, overlap each other depending on the influence of sea level rise and fall. Thus, the organic deposits (Figure 3.27),

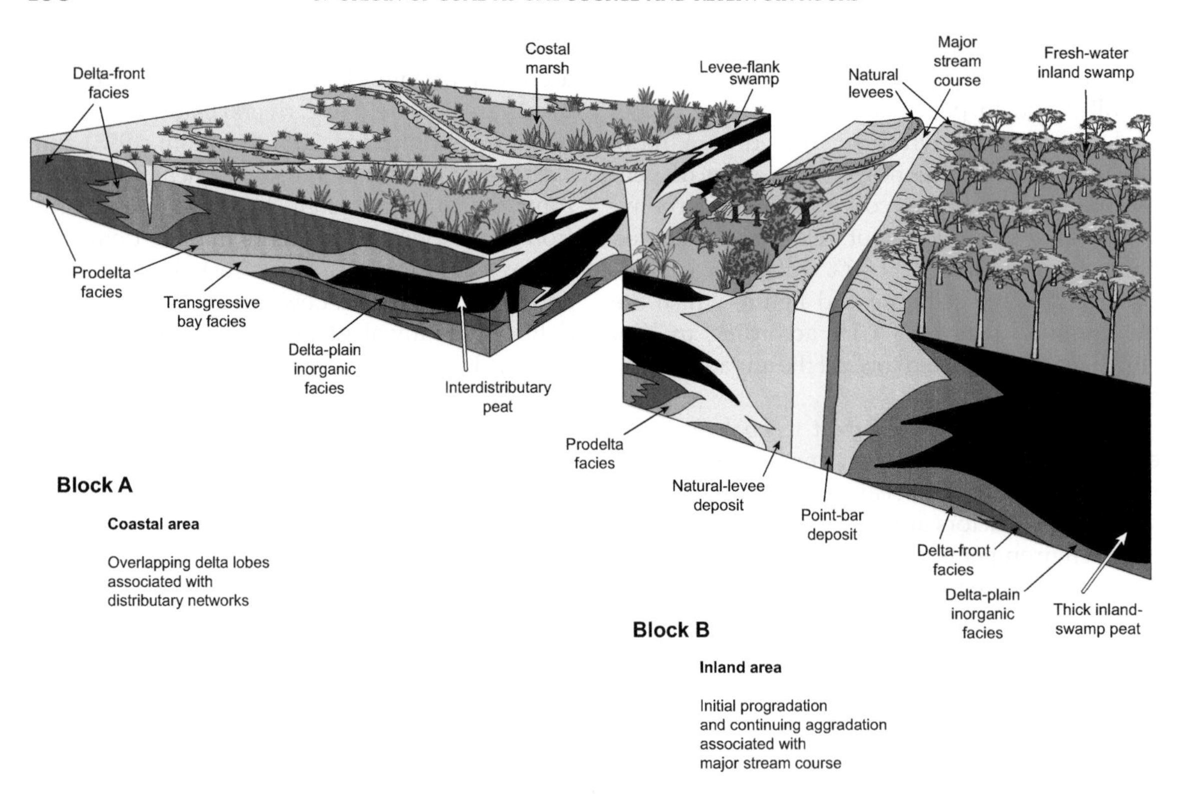

FIGURE 3.26 Block diagrams of (A) deltaic plain and (B) alluvial plain peatlands. *Source: Modified from Frazier (1967).*

which are in close proximity to the base level or sea level, are prone to rapid burial and preservation.

The changes in base level at the shoreline as a result of eustatic fluctuations and tectonic conditions have triggered creation of accommodation spaces amenable to coal accumulation (Bohacs & Suter, 1997; Catuneanu, 2006; Flores & Cross, 1991; Shanley & McCabe, 1998;). The concept of accommodation space is described as the area (length, width, and depth) in which sediments can infill (Jervey, 1988). For more information on accommodation and excellent summaries of related stratigraphic sequence concepts, the reader is directed to Catuneanu (2006) and references therein. The significance of accommodation space is providing a depositional area in which significant peat accumulation and sediments infill is associated with maximum flooding surfaces (Figures 3.28 and 3.29) (Catuneanu, 2006; Hamilton & Tadros, 1994). According to Catuneanu (2006), the optimum conditions for peat accumulation is the time at the end of a shoreline transgression and at the lower part of the highstand systems tract during alluvial aggradational sedimentation (Figure 3.29). Accommodation space in paralic environments is governed by base level (or water level) and subsidence (e.g. tectonic, sediment-peat compaction and autocompaction). Critical to the accumulation of peat is groundwater table, which is controlled by the rising and falling of the base level or sea level coupled with subsidence of the area beneath peatlands. That is, as long as the rates of biomass and organic productivities keep pace with upward and downward changes in base level and subsidence, peat accumulates over the space available below groundwater level. Thus, the most ideal conditions for accumulation of thick

FIGURE 3.27 Tidal channel at low tide exposing low-lying peat deposit in the Indonesian coast plain. *Source: Photograph courtesy of Tim Moore.*

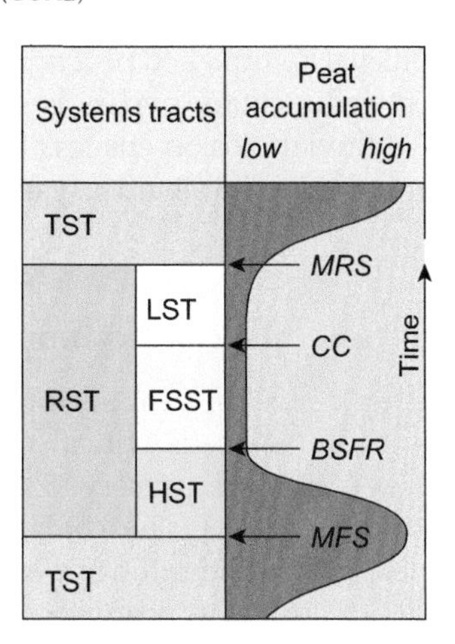

FIGURE 3.29 Chart showing the trend of peat accumulation during different stages of base level changes. Not to scale. TST = transgressive systems tract; RST = regressive systems tract; HST = highstand systems tract; FSST = falling-stage systems tract, LST = lowstand systems tract; MFS = maximum flooding surface; BSFR = basal surface of forced regression; CC = correlative conformity; MRS = maximum regressive surface. *Source: Adopted from Catuneanu (2006).*

peat are at high accommodation rates coincident with rising groundwater table. These conditions are best represented during the transgressive systems tract (Catuneanu, 2006). However, when the limit of accommodation space with lowering

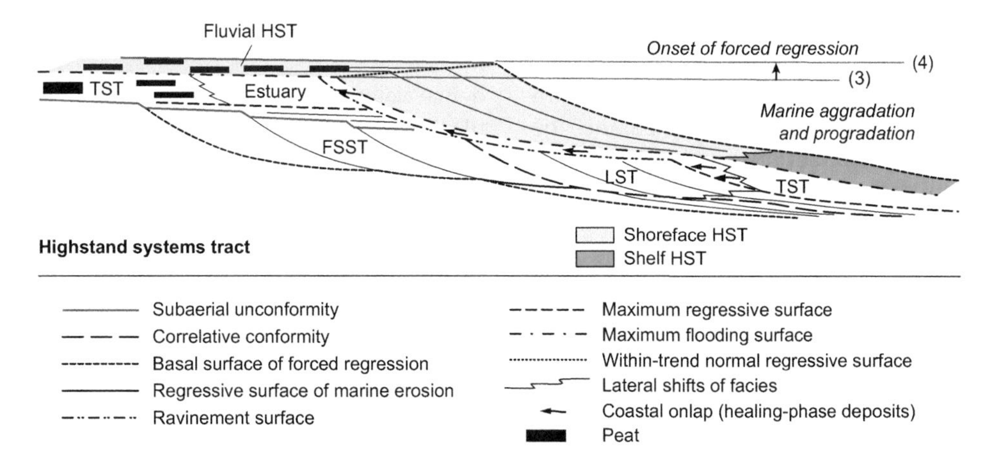

FIGURE 3.28 Schematic diagram of highstand system tract or prism (yellow) consists of fluvial, coastal, and shoreface deposits. Peat accumulation occurs in the fluvial and coastal environments. HST = highstand systems tract; LST = lowstand systems tract; TST = transgressive systems tract. *Source: Modified from Catuneanu (2006).*

groundwater level is surpassed peat accumulation is terminated, sedimentary bypass is developed, and extensive fluvial erosion ensues. This condition is typical in lowstand systems tract (Catuneanu, 2006).

"Balancing Act" of Peat Accumulation

The "balancing act" of peat accumulation in an accommodation space is critical to whether the organic deposits are preserved as thin or thick beds (Esterle, 2011). The "balancing act" comes about between the rates of peat erosion, oxidation, degradation, accumulation, and subsidence, which in turn are overprinted by transgression/highstand or regression/lowstand systems tracts. The transgression/highstand systems tract is characterized by a high accommodation to sediment supply ratio and represents the optimum condition for thick peat accumulation (Catuneanu, 2006). The decreasing sediment supply from denuded terranes presents less likely condition of choking peatlands by sediment-laden floodwaters. The regression/lowstand systems tract is characterized by high sediment supply in low accommodation space with base-level fall and represents minimum condition for thin peat accumulation. During this time floodplain and related peatlands are prone to erosion because of laterally moving rivers.

In addition, peat accumulation in the abandoned low-lying depressions is a balance between their subaerial oxidation and degradation due to lowering of groundwater level and fluvial erosion during lowstand systems tract vs subaqueous drowning and sedimentation during highstand systems tract. These environmental conditions are particularly sensitive in the coastal plain, which is juxtaposed and receptive to the shoreline shifts, experience intermittent flooding by marine water as a result of regional or basin subsidence. Therefore, for thick peat to accumulate and be preserved there must be equilibrium of the rates of peat accumulation and erosion, oxidation, and degradation, which are dependent on the rates of subsidence and sedimentation. If subsidence exceeds the rate of peat accumulation the resulting deposit is thin due to rapid burial by sediments brought in by for instance by transgression of the sea. If the rate of peat accumulation exceeds the rate of subsidence, the result is also thin deposits due to erosion, oxidation, and degradation because of lack of sediment cover from regression. Thus, thick peat accumulates when growth keeps up with subsidence. Large-scale regional subsidence results in vertically stacked peat-coal beds landward to an equally thick storage and buildup of other sediments. Two processes that are underestimated in assisting the "balancing act" of peat accumulation in relation to accommodation space, subsidence, transgression, burial, and preservation are peat compaction and autocompaction.

Peat Compaction

Fundamentally, subsidence below local peat accommodation spaces such as in abandoned fluvial belts and floodplains in coastal, delta, and alluvial plains may overprint the effects of regional or basin subsidence. Differential compaction of fine- and coarse-grained sediments in which mud compacts about three times more than sand during burial causes local subsidence below peat-forming environments particularly on floodplain sediments. Peat is highly compressible due to high water content compared to mud, silt, and sand; thus, peat compaction may contribute considerably to the total subsidence in coastal, delta, and alluvial plains. A model that quantifies spatial and temporal trends in peat compaction within Holocene fluvial-dominated deltaic floodbasin sequences by van Asselen, Karssenberg, and Stoouthamer (2011) estimated subsidence due to peat compaction to be highly variable in time and space, with local rates of up to 15 mm/yr, depending on the sedimentary sequence (e.g. sand dominated vs mud dominated). Additionally, lowering of groundwater level may cause subsidence. Subsidence due to peat compaction may exceed estimates of relative sea level rise, which may cause drowning of delta plains (van Asselen, et al., 2011).

Alluvial floodplains are also susceptible to compaction-induced subsidence due to peat compaction (van Asselen, Stouthamer, & Smith, 2010). Results of work by these authors in the Cumberland Marshes, east-central Saskatchewan, Canada, show peat compaction of as much as 43% within a few centuries and compaction rates of as much as 6.08 mm/yr. This rate of compaction is two-times less than the previously estimated peat compaction in deltaic floodbasins by van Asselen et al. (2011). Peat compaction in the alluvial floodplain is controlled by: (1) organic matter composition, (2) stress imposed on the peat, and, (3) varieties of plants. In the Cumberland alluvial plain, crevasse splays and natural levees are determined most receptive to compaction of underlying peats at short timescales (100–102 years). Peat compaction varies from uniform loading by sheet-like crevasses splay sediments to differential compaction of variable thickness of natural levee sediments. According to van Asselen et al. (2010) local subsidence from peat compaction results in additional accommodation space; thus, locally increasing rates of floodplain sedimentation. Similar compaction (also defined as autocompaction) was observed in marginal marine and estuarine environments in which Holocene peaty sediments compacted creating subsidence of 56 cm (e.g. vertical dimension of accommodation space) and vertical accretion rates from 0.99 to 6.84 mm/year of sediments (Bird, Fifield, Chua, & Goh, 2004).

Peat Autocompaction

Autocompaction is the consolidation of a column of sediment caused by its own weight and is a significant process in many coastal marsh environments (Allen, 1999; Pizzuto & Schwendt, 1997; Williams, 2003). Autocompaction is also defined as the tendency of sediments to increase in bulk density in relation to particle-size distribution and organic matter content through time caused by compaction as additional sediments are deposited (Allen, 2000; Bird et al., 2004). Williams (2003) indicated that autocompaction is a significant part of elevation change that results in the decrease or lowering of relative marsh surface elevation. Peat autocompaction is compression of the deposit under its own weight because of high organic content (Kaye & Barghoorn, 1964). The dewatering and compressibility of peats, which contain from 200% to 2000% water that is expelled during loading, enhance autocompaction. Peat autocompaction may be measured with the reduction of porosity at depth (Drexler, de Fontaine, & Brown, 2009). Compressibility is a function of the organic matter composition, degree of decomposition, and chemical composition (Ajlouni, 2000; Huat, Kazemian, Prasad, & Barghchi, 2011). For example, fibrous peat has very high water content because of its coarse-grained or texture and organic particles are hollow. This is exemplified by the West Malaysia tropical peat, which contains 200–700% water content (Huat et al., 2011).

In contrast, temperate fibrous peat was found not related to any measure of compressibility (Price, Cagampan, & Kellner, 2005). The increase in effective stress or loading cause water to be expelled or squeezed from pore spaces (Price et al. 2005). Artificially and naturally drained peatlands show that fluctuations in groundwater level cause: (1) effective stress changes large enough to alter the peat volume significantly (e.g. causes peat subsidence or autocompaction), (2) decreases in hydraulic conductivity of the peat, (3) water retention in the peat, and (4) variability in peat moisture (Chow, Rees, Ghanem, & Cormier, 1992; Kennedy & Price, 2005; Oleszczuk, Szatylowicz, Brandyk, & Gnatowski, 2000; Price, 2003; Price et al., 2005; Roulet, 1991). Moreover, it is known that deeper, more decomposed peat is more consolidated (Clymo, 1983; Ingram, 1983). Thus, the high water content of peat results to more compressibility than the detrital sediments, which contributes to autocompaction during peatification, subsequent burial, and local subsidence. Autocompaction may be aided by reduction of peat volume resulting in elevation collapse from loss of swelling by roots when vegetation dies and oxidation of root organic matter (Day et al., 2011).

Alluvial Plain Depositional Systems

The literature is replete of coals interpreted to have formed in alluvial plain peat-forming environments such as floodplains, lakes, and abandoned alluvial belts (Ferm & Cavaroc, 1968; Fielding & Webb, 1996; Flores, 1981; Flores & Sykes, 1993; Roberts, Stanton, & Flores, 1994; Staub & Esterle, 1994; Staub & Richards, 1993). Alluvial plain peat-coal forming environments occur landward of the

BOX 3.4

DROWNING THE WORLD'S COASTS: CAUSES AND EFFECTS–CONCEPT AND MISCONCEPTION

Many of the coasts and deltas worldwide are drowning. Overeem, Syvitski, Kettner, Hutton, and Brakenridge (2010) reported that 85% of 33 world deltas suffered severe flooding during the last decade. A total of 260,000 km^2 is submerged temporarily, however about 10% of this flooded area is below sea level. As a result of drowning the coasts, wetlands worldwide are predicted to disappear by the end of twenty-first century if sea level rise is faster than the amount of sediments available for deposition (Kirwan & Guntenspergen, 2010). In the Mississippi River delta, Blum and Roberts (2009) attributed the loss of wetlands to reduction of river sediment load by 50% from dam construction and global sea level rise. In addition to loss of wetland ecosystems, drowning of the global coasts will affect cities, communities, farmlands, and industrial and commercial developments.

Although global sea level rise is an overprinting cause (e.g. melting ice sheets due to global warming) of coastal drowning additional underlying causes are tectonic subsidence of the earth lithosphere, and sediment and peat compactions of Holocene strata (Allen, 1999; Long, Waller, & Stupples, 2006; Overeem et al., 2010; Tornqvist et al., 2008;). Analysis of compaction rates of Holocene strata in the Mississippi River delta ranges from 5 mm/yr on a millennial scale primarily with associated peat deposits to 10 mm/yr or more on decades and centuries scale (Tonrqvist et al., 2008). Rapid compaction of Holocene coastal wetlands in southeast England, United Kingdom lowered the peat surface by at least 3 m (Long et al., 2006). The strong influence of compaction on the evolution of late Holocene landscape in southeast England may extend into other coastal lowlands in northwest Europe. The effects of sediment loading and peat compaction have been shown to cause rapid rates of subsidence (Allen, 1999). Overeem et al. (2010) suggested that sediment compaction from water, oil and gas developments, sequestration of sediment in upstream reservoirs, and floodplain engineering are contributing factors. Thus, wetlands in general and peatlands in particular in coastal and deltaic plains are in uncertain grounds as they are prone to accelerated sea level rise, the magnitude of which exceeds longer-term eustatic and/or crustal effects.

seaward sloping coast. It merges with coastal and delta plains as a part of the "coastal progradation prism" (Figures 3.4 and 3.28), which may shift landward when progradation is accompanied by aggradation (Catuneanu, 2006; Posamentier, Allen, James, & Tesson, 1992). In this setting, the "sedimentary prism" is a continuum of alluvial and coastal/delta plain landward. Thus, much like the coastal/delta plain sensitivity to base level or sea level changes so is the landward alluvial plain and its ability to develop accommodation spaces for peat accumulation during progradation and aggradation.

Alluvial plains are flat surfaces aggraded by meandering, anastomosing, and/or braided river channels, which are bordered by flat-lying areas consisting of floodplains, lakes, and peatlands (Figure 3.30). During river floods, these flat areas received suspended sediments either through overbank (above levees) and/or crevasse channels breached through the levees. The peat-forming environments occupy depressions with stagnant pools or ponds of water fed by groundwater such as swales in old river channel deposits and floodplain lakes. These depressional wetlands may have started with floating and submerged vegetation, evolving into peatlands with rooted and herbaceous plants encroached by shrubs and trees sustained by nutrients from surface and groundwater flows. Here, the groundwater level fluctuation effect distinct zonations of vegetation with varieties organized consistent with their tolerance and adaptability to sediment-laden floodwaters. As nearby rivers abandoned channel courses forming alluvial ridges accompanied by subsidence due to regional tectonism and/or compaction/autocompaction of the sediments and peats accommodation space permitted expansion of mires (e.g. swamps and bogs) over abandoned parts of the alluvial plains. The mires were transformed into forested peatlands resulting in increased biomass productivity and peat accumulation periodically saturated by surface and groundwater increased water flow followed by decreased water movement or stagnation favorable to organic decomposition.

FIGURE 3.30 Photograph of peat-forming alluvial plain drained by meandering river and abandoned channel. *Source: Photograph courtesy of Esterle (1990) and Tim Moore.*

The mechanics of preservation of peat accumulation in alluvial plains is much like those in coastal and deltaic plains. Besides regional tectonic subsidence, compaction and autocompaction of peat plays an important role in the evolution of alluvial plains. These mechanisms provide extra accommodation space, which affects temporal and spatial fluvial sedimentation patterns (van Asselen, 2009). It has been suggested in ancient alluvial coal-forming deposits that differential sediment and peat compaction affect floodplain gradients and permitted the process of fluvial avulsion (Ferm & Cavaroc, 1968; Flores, 1981). In a similar process peat autocompaction contributed to sediment and peat compaction and affected the framework architecture of both modern and ancient alluvial sequences. van Asselen (2009) reported that on about 100-year timescales peat compaction occur at high magnitude as a result of sediment loading by crevasse splay and levee deposits. This creates local accommodation space resulting either in river avulsion into low-lying area or rapid infilling by fluvial sediments and peat accumulation. Peat compaction by sand deposits of river channels creates vertical aggradation and fixation of fluvial channels. Also, well-developed peat deposits lateral to river channels prohibit lateral migration, particularly in raised ombrotrophic bogs. Finally, avulsion of long stretches of river channels and large alluvial belts may occur after longer period of time (Flores, 1981).

For some time, the merit of ancestral alluvial plains far removed from marine influences as important peat (coal) forming depositional environments has been questioned. The argument against interior alluvial plains (e.g. intermontane and foredeep of foreland basins) forming peat (coal) is that they are more prone to high rates of erosion and aggradation. Alluvial plain successions or "fluvial accommodation-based sequence" with abundant, thick coal beds is described as high accommodation systems tract (Catuneanu, 2006; Leckie & Boyd, 2003). The alluvial successions are tectonically and climatically controlled and independent of marine base level or sea level changes. As an example, the northern U.S. Rocky Mountains and Great Plains contain successively northeast younging interior basins (Flores, 2003). These Late Laramide Paleogene basins (Figure 3.31(A)) are separated by arches of crustal wedge pop ups and flanked by back-thrust and forward thrust faults, which are connected by low-angle master thrust faults in the lower crust (Figure 3.31(B); Erslev, 1993). These interior basins were drained by Paleogene alluvial depositional system continuum with associated peat-forming and lacustrine environments (Figure 3.31(A)), which produced abundant, economically, thick coal beds that are mined for power generation and developed for coalbed methane.

Lacustrine Depositional System

Lacustrine depositional systems are defined as small to large open bodies of water with wetlands formed in topographic depressions or dammed river channels. Upland of the lacustrine system consists of wetlands, which include shrubs, trees, and persistent emergent vegetation. In addition, floating and sunken vegetation are common aquatic plants. The lacustrine system is classified and distinguished primarily by trophic or state of nourishment, alkalinity, annual cycles of thermal stratification, circulation, morphometry (e.g. size and shape of the lake basin and drainage area; water permanence), and water chemistry (e.g. including salinity) (Cole, 1994). Based on nutrient state, lacustrine system is classified as oligotrophic (or nutrient poor), eutrophic (or nutrient rich), and mesotrophic (or intermediate nutrient supply). These lakes may be dimictic (or twice a year turnover of water), monomictic (or once a year turnover of water), and meromictic (or no annual turnover of water).

Lacustrine system occurs in tidal and nontidal coastal plain, alluvial plain, and continental

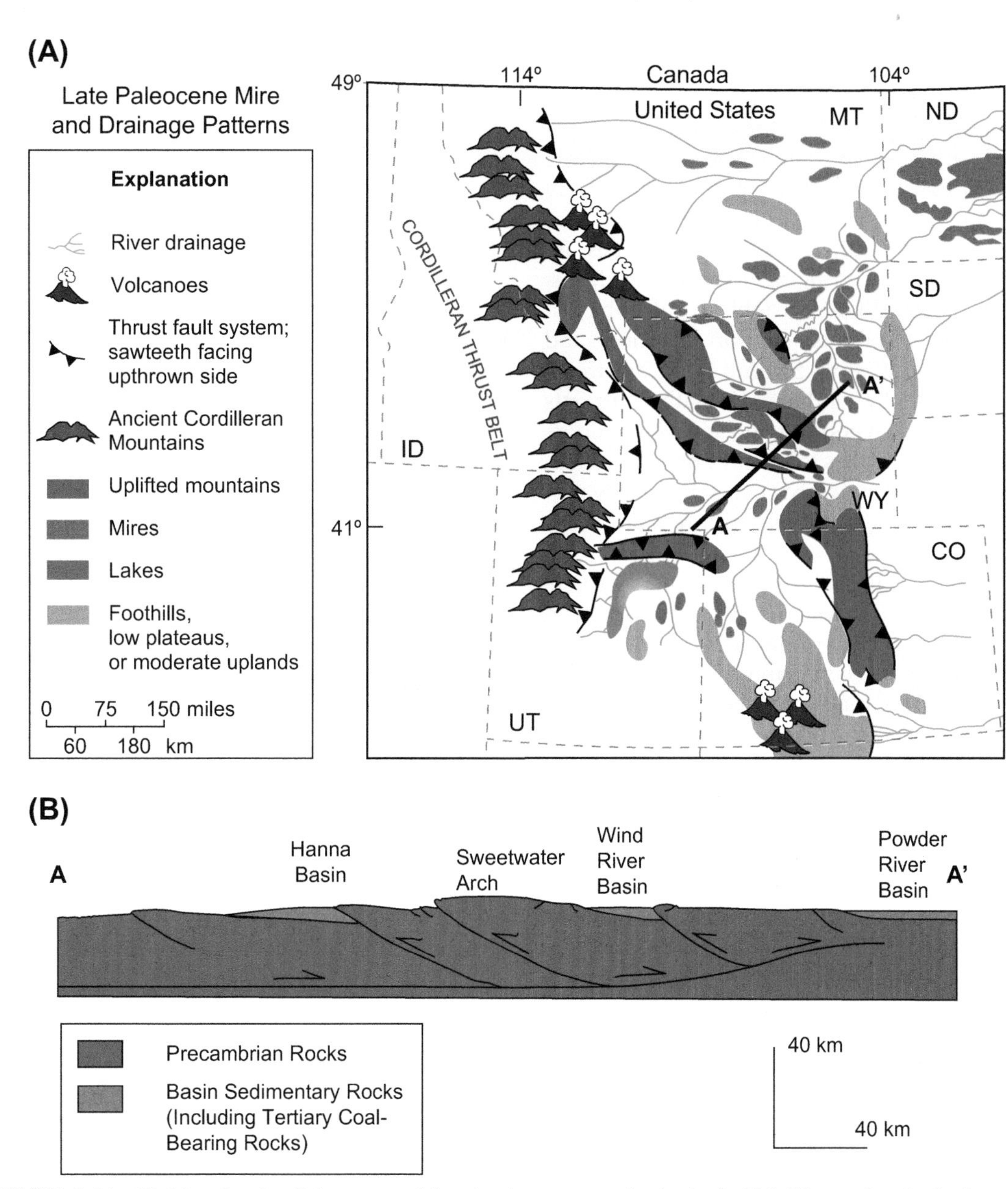

FIGURE 3.31 (A) Map showing Paleogene coal-forming intermontane basins in the U.S. Western Interior basins, which represent fluvial accommodation-based sequences tectonically and climatically controlled independent from marine influence. (B) Cross section of Precambrian surface through Tertiary coal-bearing basins. See location in A. *Source: Adopted from Flores (2003) and modified from Erslev (1993).*

areas separated from marine influence. Coastal plain lakes are often hydrologically connected by groundwater or surface water flow in a small stream. Water is usually acidic, dark (coffee-like) stained, and low transparency and the substrate is composed of sand and/or muck (or peaty mud or gyttja). Examples of coastal plain lakes are marsh and swamps lakes. Marsh lakes are tidally influenced bringing in nutrients transforming the lake to mesotrophic to eutrophic environments. Nutrients support floating and submerged vegetation. Swamp lakes are freshwater, nontidally influenced with nutrients exported from rivers and forest swamp floors transforming the lakes to mainly eutrophic lakes. Rooted-plant organic sediments fill in swamp lakes, which is a traditional progression into a peatland.

Coastal and alluvial plains drained by rivers produce lacustrine systems of fluvial origin in abandoned river meander or oxbow lakes. These lakes from when the meander bend is cut off from the main river course by deposition of a levee or buildup of marsh/swamp at the headward (or upper) end of the meander. Additionally, floodplain lakes in alluvial plains are formed with buildup of levees that are overtopped or breached by crevasses of the associated river, which periodically replenish hydrological and nourishment of wetland vegetation. In alluvial plains shrubby swamps occur along sluggish streams, floodplains in temporary ponds related with depressions, and old oxbows, which are initial areas of peatland formation as they as succeeded by forested swamps.

Related to lacustrine system is palustrine, which include all freshwater, nontidal wetlands dominated by trees, shrubs, and persistent emergents. Palustrine wetlands mainly occur shoreward of lakes, rivers, river floodplains and lakes, catchments, and as islands in lakes or rivers. Palustrine wetlands are not all peat-forming environments. Rivers usually empty into a lacustrine system with the extension forming riverine-lacustrine-deltaic coastline often forming palustrine wetlands. Lacustrine delta system may derive organic debris from highly vegetated floodplain immediately upstream. The transported organic matter is deposited in lakeward-prograding delta lobe that exhibits features typical of allochthonous peat (Treese & Wilkinson, 1982).

Ancestral lacustrine coal-forming systems are not as common as coastal and alluvial plains. Most ancestral lacustrine coal-forming systems are associated with tectonically, rapidly subsiding active basin margins (Cabrera & Saez, 1987), "pull-apart" basin (Whately & Jordan, 1989), "pull-apart" basins, which developed alluvial fans along fault lines (Heward, 1978; Li, Li, Yang, Huang, & Li, 1984); extensional subbasin (Flores & McGonigle, 1991); and intermontane grabens affected by strike-slip fault (Demirel & Karatigit, 1999). The alluvial basin-fill successions in these basins probably reflect syndepositional tectonics and climate fluctuations typical of high accommodation fluvial sequences (Catuneanu, 2006). Climate plays a crucial role in formation of coal deposits in lacustrine basins. Tectonics and changes in accommodation and sedimentation also play an important role in a lacustrine basin.

TRANSFORMATION OF PEAT TO COAL: DIFFERENCES IN CONCEPT AND TIME

The most obvious difference between how ecologists and geologists perceive the origin of peat and coal formation is the amount of time involved in the process. The "macrofacies" and "megafacies" concepts for viewing peat and coal formation, respectively, highlight this difference in the perspective of time. Modern ecologists observe biological, chemical, physical, and hydrological processes, which produce peat within just a short period of time (e.g. several thousands of years) (Table 3.1). In contrast, coal geologists have the luxury of observing the final product, which represents a deposit of a longer period of time (e.g. tens of thousands to millions of years). This concept is best demonstrated by the rate of peat accumulation within a period of time. In a review of a number of studies of peat accumulations in boreal, temperate climate, Lucas (1982) summarized that it requires from 600 to 2400 years for 1 m of peat to accumulate with an average of 1500 years. Perhaps the short period of time during which peat is deposited in modern times led Cameron et al. (1989) to express eloquently that "most modern peat deposits in the world are not precursors of economic coal beds because they are too thin, too local in extent, and/or too far from sea level to be buried by marine sediments before destruction by erosion and decomposition."

Cameron et al. (1989) studied the geology, botany, and chemistry of an impressive selection of modern peat-forming environments from temperate and tropical latitudes. These modern peatlands began to form and accumulate peat following the melting of continental glaciers some 13,000 years ago in the temperate region and following the final sea regression or Wurm glaciation some 11,000 years ago after which the sea slowly began to rise in the tropical region. In the cold temperate latitudes (e.g. Maine and Minnesota, USA) within about 8000 to 12,000 years, maximum peat accumulation ranges from 4.5 to 8.5 m thick or a rate of about 1 m per 1412 to 1778 years, which is within the time periods estimated by Lucas (1982). In contrast, in the tropical latitudes (e.g. Sarawak, Borneo) within about 5400 years, maximum peat accumulation is about 17 m thick or a rate of about 1 m per 450 years. More specifically, Anderson (1964, 1983) studies in the same Sarawak peatlands have shown that the peat deposited within: (1) 4300 years accumulated at a rate of 1 m in 214 years; (2) 3900 years accumulated at a rate of 1 m in 333 years; (3) 2300 years accumulated at a rate of 1 m in 455 years. These estimates indicate that peat in tropical latitudes accumulates at least three times as fast as in temperate latitudes (Table 3.1).

Perhaps one of the most underestimated factors about the rate of peat accumulation is related to the stage of development of peatlands. This can be demonstrated with increasing impoverishment of the peatland ecosystem, biomass accumulation will be slowed down and as a consequence also peat accumulation. Winston (1994) proposed scenarios in which the development of tropical peat bogs ceased to accumulate peat particularly in the central bog plain as a result of the hydrology and morphology. Peat accumulation can proceed only as long as the groundwater table continues to rise to a maximum height depending on water influx and internal conductivity of the peat aquifer. As the peat bog grows vertically and laterally, peat accumulation slows down because the groundwater table ceases to rise and cannot keep up to preserve available plant liter. Thus, the rate of peat addition to the top of the peat body and net peat accumulation ceases.

ANALOGS OF ECONOMIC COAL AND COALBED GAS RESERVOIR

Not all the modern peat-forming environments produce coal based on their biological, chemical, hydrological, and physical attributes. However, in terms of the potential of peat transforming into economic coal deposits, Cameron et al. (1989) and Moore (1989) suggested that peat bogs have better prospects than peat marshes to accumulate thick coal beds because the conditions of preservation of plant material in the peat bogs are extremely good and composed of fuel-grade, high-quality organic matter. Cameron et al. (1989) and Cohen et al. (1989) assessed the fuel potential of various peat deposits in bogs, swamps, and marshes in coastal and alluvial plains in temperate and tropical regions, which are frequently reported in the literature (Dapples & Hopkins, 1969; Diessel, 1992; Lyons & Alpern,

TABLE 3.6 Comparison of the Quality and Types of Peats and Peatlands in Temperate and Tropical Climates

Peat Quality	Cold Temperate	Warm Temperate	Tropical
Moisture (AR)	92	84	93
Ash yield (%) (dry basis)	<5–8	4–7	1–14
Volatile matter (%)	60–75	58–60	58–64
Fixed carbon (%)	20–30	35–40	24–41
Sulfur (%)	0.4–0.6	0.2–0.7	0.1–7.7
Hydrogen (%)	5–6	4–6	5–6
Nitrogen (%)	0.6–2	1–1.5	0.9–1.2
Oxygen (%)	28–41	27–28	30–32
Heating value (kJ/kg)	20,000–24,400	22,000–23,000	22,000–26,000
Peat types	Fibric and hemic	Hemic and sapric	Hemic and sapric
Peatland types	Marsh	Marsh and swamp	Swamp and bog

AR, as received; %, percent; kJ/kg, kilojoules per kilogram.
Source: Modified from Cameron et al. (1989) and Cohen et al. (1989).

1989; McCabe & Parrish, 1992; Rahmani & Flores, 1984) as major coal-forming environments in the geological record. Table 3.6 compares the peat quality and chemical composition (e.g. moisture, ash, volatile matter, fixed carbon, sulfur, hydrogen, nitrogen, oxygen, heating value) summarized from Cameron et al. (1989) and Cohen et al. (1989), which are very similar to the proximate and ultimate analyses to determine the energy-potential for coal. These parameters are compared against the climatic regions, types of peatlands, and types of vegetation that dominate the peat-forming environments. More specifically, the woody peat bogs and swamps (mires to use the definition of Moore, 1989) in the warm temperate and tropical regions possess lower moisture, ash, and volatile matter, and higher heating value and fixed carbon due to woody composition. However, when the thickness of peats formed within the same period of time (Holocene Period) are compared between the various regions, the tropical peats (as much as 17 m) have "twice" the potential for economic deposits than the temperate peats (as much as 8 m). A major consideration in understanding the fundamentals of peat transformation to coal is the potential for coal to become coalbed gas reservoir. The reservoir prospect lies in the inherent inorganic and organic matter composition of the peat, which in turn, is controlled by the types of mires (e.g. bogs, swamps), peat types, and relationship with processes of external aquatic environments.

ATTRIBUTES OF PEAT BOG RELEVANT TO COAL

The use of the Southeast Asian Holocene bogs as analogs of economic coals have been spurred on by the views of Cameron et al. (1989) that most world peat deposits are not very good candidates for economic coal beds, which in turn have a poor potential as coalbed gas reservoirs. As a caveat, it is by no means that the peat deposits of these bog analogs represent the entire population of peat-forming environments, which are transformable into economic coals. However, these tropical peat bogs have been used prominently as analogs of Carboniferous, Cretaceous, and Tertiary coals (Cecil, Delong, Cobb, & Supardi, 1993; Esterle, 1990; Esterle & Ferm, 1994; Flores, 1986; Holdgate, 2005; Staub & Esterle, 1992). Whether or not these bogs are ideal analogs, the basic processes of peat accumulation and contemporaneous depositional system may provide an understanding of how coal serves as a source and reservoir rock for coalbed gas. In addition, the contemporaneous depositional system, which includes alluvial, coastal, and deltaic plains, represents a time slice of landward to seaward continuum of peat-forming environments not often duplicated in records of ancient coal-bearing rocks.

Characterization as Potential Coalbed Gas Reservoirs

Peat bogs in the tropical region are not as well investigated as are temperate bogs. Tropical peat bogs and accompanying depositional systems have become important analogs for economic coals, coalbed gas, and petroleum reservoirs with the former deposits potentially developing thick, low ash yield, and fuel-grade deposits (Cameron, et al., 1989; Flores, 1981; Mazumder, Sosrowidjojo, & Ficarra, 2010; McCabe, 1984; Shoup, Lambiase, Cullen, & Caughery, 2010; Staub & Esterle, 1992; Whitmore, 1975). The physical and chemical properties of the peat bogs contribute significantly to the combustion of clean coal but more importantly potential reservoirs of coalbed gas. Alluvial, coastal, and deltaic plains in tropical regions exhibit facies relationships that may be favorable for the development of petroleum reservoirs. Thus, for these reasons the geological, morphological, depositional, and ecological attributes and rates of peat accumulation of Southeast Asia, in general, and tropical peat bogs in Indonesia, in particular, must be adequately described.

Geological Attributes

The tectonic setting prior to the Holocene, when the Southeast Asian peatlands formed, was inherited from the Neogene when the Sundaland (Sumatra–Java) underwent regional subsidence followed by uplift related either to the subduction of the Indian Oceanic plate beneath Sumatra and Java or possibly to microcontinental collision in southeast Kalimantan (Cole & Crittenden, 1997; Friederich, Langford, & Moore, 1999; Koesoemadinata, 2000; Sudarmono and Eza, 1997). Backarc and fore-arc basins were formed, and rifting opened up the Makassar St between Borneo and Celebes, resulting in a passive margin basin in East Kalimantan, Borneo (Figure 3.32). Rifting also occurred in continental Southeast Asia where northwest–southeast and northeast–southwest trending strike–slip faults formed accommodation spaces between extensional basins that remained separated (Friederich, Guanghua, Langford, Nas, & Benjavun, 2000). Accommodation spaces and active subsidence of the Southeast Asia basins, overprinted by Holocene glacial/interglacial periods, provided favorable conditions for the development of domed ombrogeneous peat deposits.

The Southeast Asia bogs, which accumulated peat in the Holocene, developed on fluvial, deltaic, tidal, and marine sediments related to the last regression during the Würm glaciation about 11,000 years before the present (Anderson, 1983). Lowering of sea level exposed an extensive landmass comprising Sundaland, of which the islands of Sumatra, Java, and Borneo, and the Malaysian peninsula are parts. Subsequently, the sea level rose until it reached a peak level about 5500 years ago. Depositional settings during the Würm glaciation, for example, consisted of rivers with their watersheds in the Barisan Mountain Range in eastern Sumatra and the central massif of Borneo (Figure 3.32). These fluvial systems created

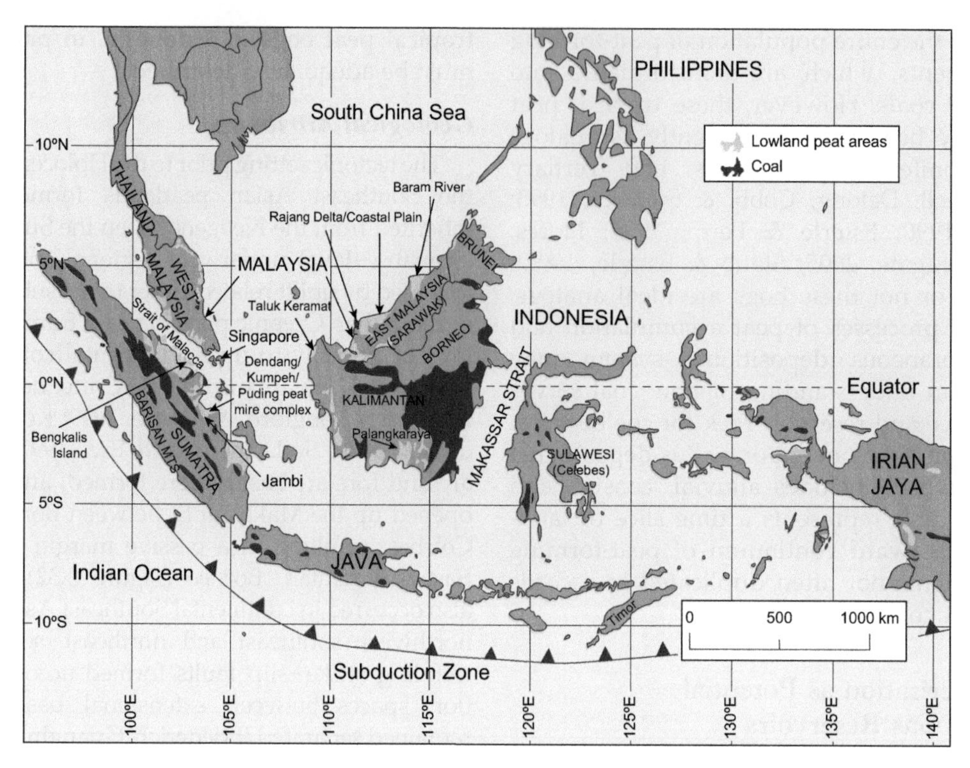

FIGURE 3.32 Geological setting and location of peatlands and coals in Indonesia, Borneo, and Malaysia. *Source: Modified from Andriesse (1974), Anderson (1983), van de Meene (1984), and Hardjono and Supardi (1989).*

coastal plains as much as 100-km-wide areas (e.g. west Sumatra) that were partly constructed by prograding deltas. Reworking of the fluvio-deltaic sediments by diurnal sea level fluctuations and waves resulted in deposition of tidal and barrier sediments that fringed the coastal plains, and carved out bays and estuaries.

Depositional and Ecological Attributes

Fundamental coalbed gas reservoir properties such as thickness, size, shape, distribution, permeability, and porosity are directly linked to the depositional and ecological settings of peat deposits. Important features of the Southeast Asia peat bog deposits include their thickness, morphology, and rate of accumulation. Andriessie (1974) reported a maximum peat thickness of 4 m in coastal mires. In comparison, peat thickness in the inland part of the mires is as much as 12 m. However, peat thickness in the inland parts of peatlands in Sumatra is as much as 16 m and in Sarawak as much as 17 m (Anderson, 1983; Polak, 1950; van de Meene, 1984). Anderson (1961, 1964), Andriesse (1974), Tejoyuwono (1979), Cameron et al. (1989), and Diemont and Supardi (1987) have described the domed structures particularly of the inland or basin peat mires in Sumatra and Borneo.

The domed morphology, and vertical and lateral variations of peat deposits are best demonstrated by profiles for a distance of about 110 km from the coastal to alluvial plains of the Rajang River, in Sarawak, Borneo, Malaysia (Figure 3.33(A)–(D)). The basal surfaces of the peat bogs, dip 2–4 m below the base level from the margins to the center of the peat bogs. The

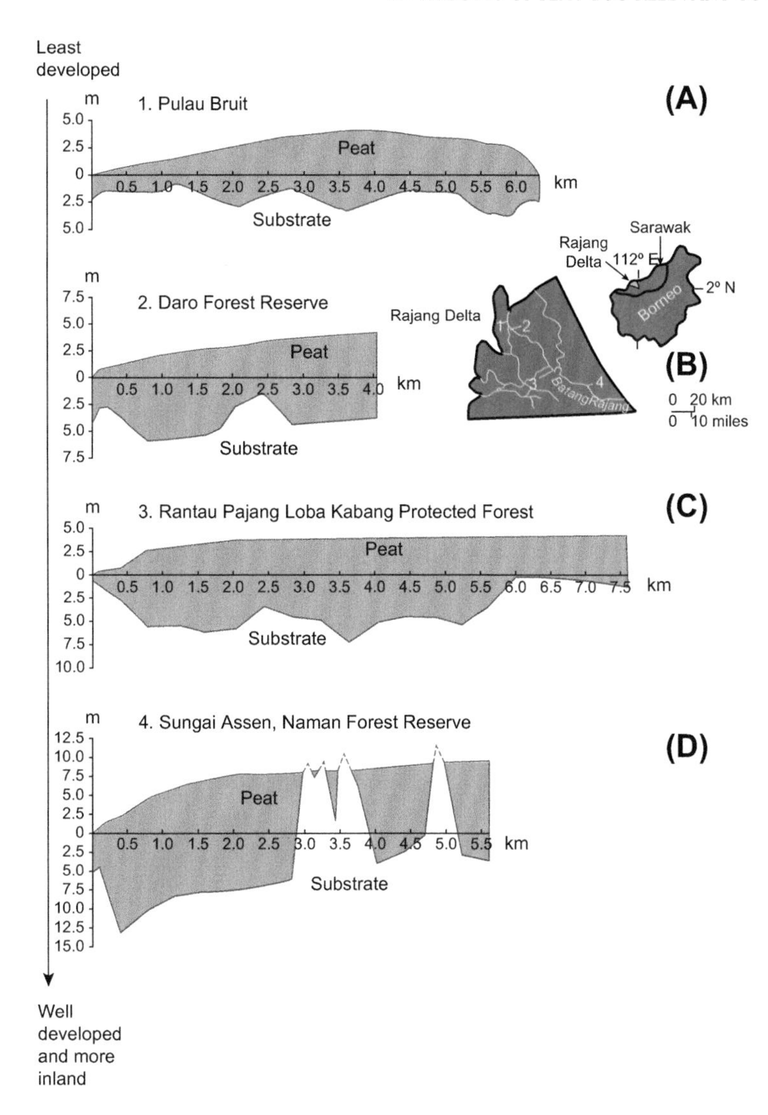

FIGURE 3.33 (A–D) Cross sections of domed peat bogs showing variations of their convex upper surfaces indicating from least to highly developed deposits along the Rajang River delta and coastal plain in Sarawak. See Figures 3.17 and 3.32 for additional locations of Rajang River delta. *Source: Modified from Anderson (1983).*

upper surfaces of these peats rise 3.5–7.5 m above the base level from the margin to the center of the deposits. Thus, the cross sections of the peat bogs are shaped like a "pillow" or domed with either a sharp crest or a flat plain. They are lenticular to lens shape, more than 7.5 km wide and 27 km long. The high elevation of the peat bogs above river drainage suggests that when accumulation of organic matter exceeded influx of detrital sediments along the levees and bog margins, the peat bogs developed convex upward surfaces. The peats became exceedingly acidic toward the center without detrital influx, which reduced microbial activity and decomposition, resulting in the central flat plain. In general, the doming effect is best developed in older inland peat bogs in the upper deltaic and alluvial plains of the Rajang River (Figure 3.33(C) and (D)).

Sumatran Peat (Coal) Analogs

The most commonly cited analogs of tropical peat bogs are the deposits in Borneo (e.g. Sarawak and Kalimantan). However, the Sumatran peatlands, which make up as much as 70,000 km^2 of the eastern coast at the central and southern parts of the island (Figure 3.34(A) and (B)), have been estimated to contain up to 20 m thick peat and 19 Gt of carbon, which are important ecological resources (Wahyunto & Subagjo, 2003). Studies of the Jambi peatlands in Sumatra, Indonesia by Cameron et al. (1989) and Flores and Hadiyanto (2006) observed similar features to those in the Borneo peatlands (Figure 3.34(A) and (B)). The most important feature of the Jambi peatlands is that they extend as a continuum depositional and ecological systems from the alluvial plain to the tidal-influenced coastal plain shoreward to the Strait of Malacca. The alluvial plain peatlands include the Kumpeh and Dendang peat bogs and are drained by the meandering and anastomosed Batang Hari and Sungai Kumpeh rivers (Figure 3.34(B)). The Kumpeh peat bog is similar to that of the Rasau Jaya peat bog of West Kalimantan, Borneo with well-developed levee and floodplain sediments flanking the domed peat deposit, which is juxtaposed with the Batang Hari and Sungai Kumpeh rivers.

Depositional Controls of Peat (Coal) Reservoirs

It is apparent from the profiles of peat mires from the Jambi peatlands in Sumatra that the rivers and associated levees play a major role in the domed shape of the peat deposits. A contrast of several domed peat bogs reveals that fluvial processes along the rivers in the alluvial plain, from upstream reaches to the coastal plain to the tidal-influenced coast, control the areal geometry, distribution, and dimension of the deposits. The Kumpeh domed peat bog is about 7 m thick and 240 km^2 area, which is

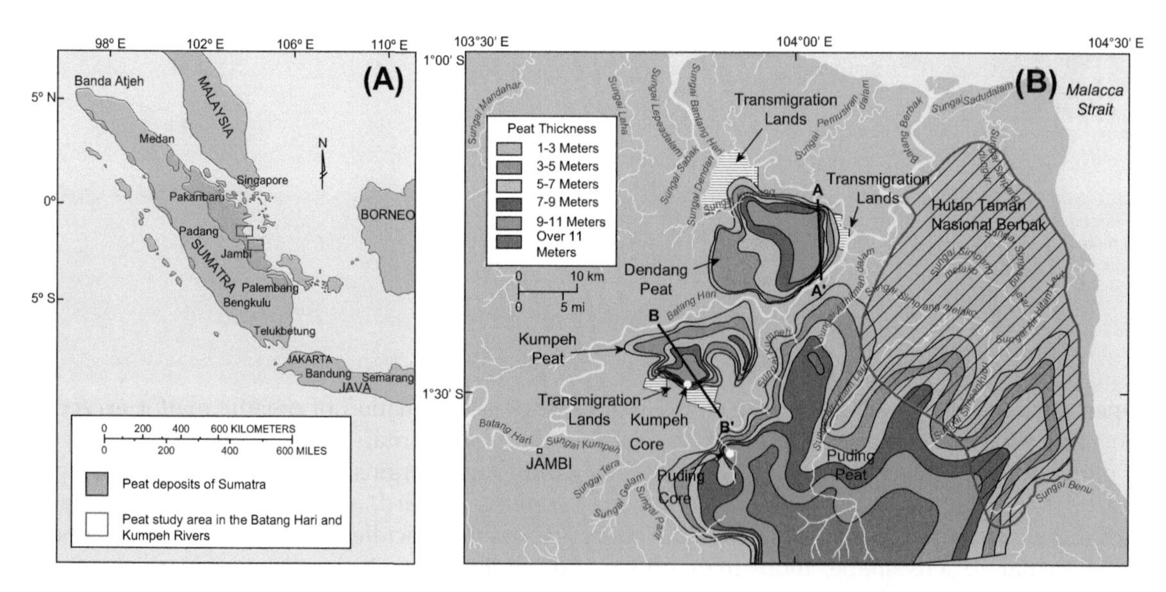

FIGURE 3.34 (A) Geographic distribution of Sumatran peatlands. (B) Isopach map of peatlands in the Jambi area. Lines of cross sections A–A' and B–B' are shown in Figures 3.36 and 3.35, respectively. *Source: Modified from Flores and Hadiyanto (2006).*

partly being drained for agricultural transmigration (Figures 3.35 and 3.34(B)). The peat bog occupies an interfluve between the anastomosing Batang Hari and Sungai Kumpeh rivers. The Dendang domed peat bog, downstream of the Kumpeh peat bog, is about 12 m thick and 540 km^2 area, which is also partly being drained for agricultural transmigration (Figures 3.36 and 3.34(B)). The upper surface of the Dendang peat rises about 5 m above the river level from the margin to the center of the deposit, which is scoured and infilled by an abandoned channel. Both the Kumpeh and Dendang bog margins are low lying, which are covered by floodplain sediments about 2 km beyond the river edge at one end and at the other end the elevated thick peat constrains the river channels.

Stream patterns superimposed on isopach maps of Sumatra peat deposits (Figure 3.34(B)), reported by Hardjono and Supardi (1989), show that the meandering, anastomosing, and erosional characteristics of the fluvial and distributary channel systems determine the areal lenticularity, distribution, and dimension of the peat. Figure 3.34(B) depicts the variations in the areal geometry, distribution, and dimension of domed peat mires in the upstream part of the fluvial system, where the anastomosed pattern of the meandering Batang Hari and Sungai Kumpeh river channels define an elongate areal geometry of the Kumpeh peat deposit. In addition, the isopach map of the peat is recessed northward at the south-central part of the mire mimicking meander loops on the modern Sungai Kumpeh River. The interfluvial pattern of the Batang Hari River and tributaries influenced the ovoid-like geometry of the Dendang peat deposit. Branching of the tributary at the watershed along the mire margin and subsequent headward erosion may have caused a southward recession of the isopach map on the northeast part of the deposit. Fluvial erosion of these domed peat deposits is best shown in the Dendang peat bog (Figure 3.34(B)) where the southeastern part of the peat is eroded by the Batang Hari River. In addition, the Dendang peat was also eroded, as exhibited by an abandoned river channel (Figure 3.36), the fluvial erosion and aggradation of which is another example that contradicts the belief that rivers do not erode or aggrade on a peat deposit, much less a domed peat. The lateral progradation and aggradation by the Batang Hari River and its tributaries control the general

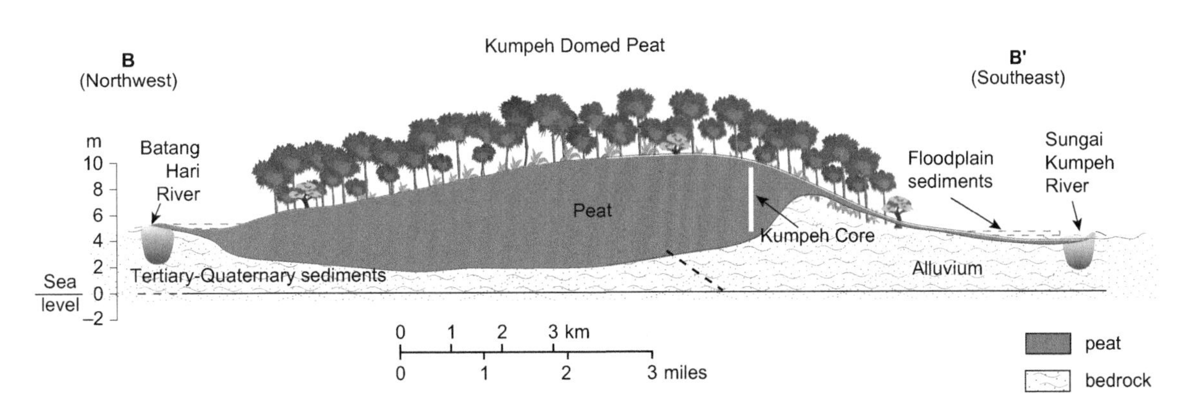

FIGURE 3.35 Cross section showing the convex upper and concave lower surfaces of the Kumpeh domed peat bog in the Jambi area. Also note floodplain sedimentation along the margins of the Sungai Kumpeh River. A part of Kumpeh core is shown in Figure 3.20. Location of cross section in Figure 3.34(B). *Source: Modified from Cameron et al. (1989), Hardjono and Supardi (1989), and Flores and Hadiyanto (2006).*

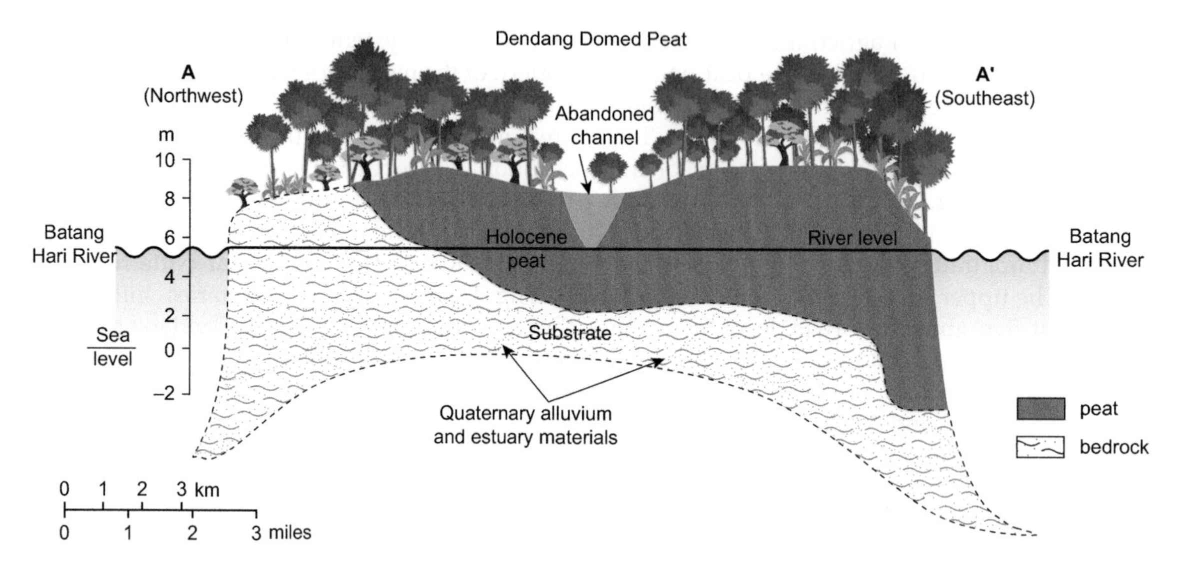

FIGURE 3.36 Cross section showing the convex upper and concave lower surfaces of the Dendang domed peat bog in the Jambi area. Note erosion of the Dendang domed peat by the Batang Hari River at the southeast part of the deposit. Location of cross section in Figure 3.34(B) *Source: Modified from Cameron et al. (1989), Hardjono and Supardi (1989), Esterle (1990), and Flores and Hadiyanto (2006).*

north-south areal distribution and dimension (240−540 km^2) of the Dendang and Kumpeh peat deposits.

The areal dimension of the Kumpeh and Dendang peat bogs is relatively small in contrast to the Endapan peat bog, which is more than 12 m thick and covers an area of 4000 km^2 (Figures 3.37 and 3.38). The Endapan peat bog, which is found east and south of the Batang Hari-Sungai Kumpeh fluvial system, represents a peat deposit

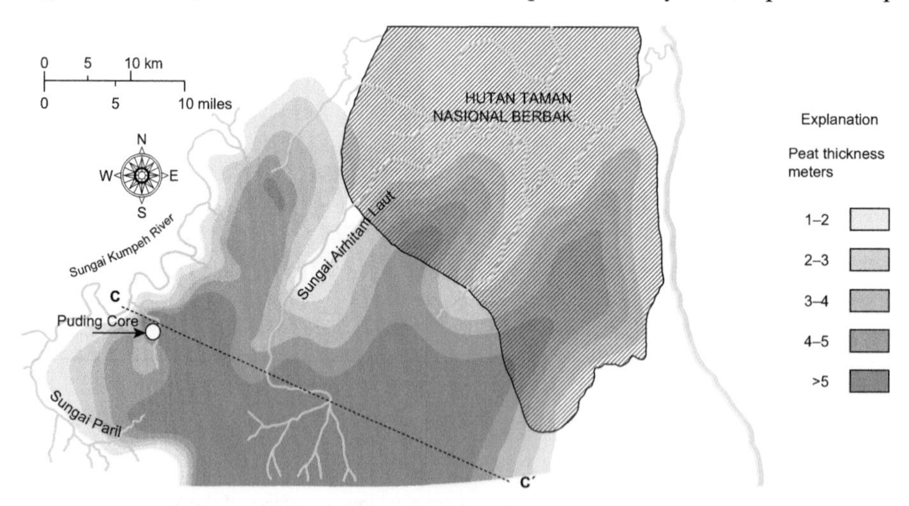

FIGURE 3.37 Isopach map of Endapan peat bog west of the Batang Hari River in the Jambi area (see Figure 3.34(B)). Line C−C′ is location of cross section in Figure 3.38. *Source: Modified from Hardjono and Supardi (1989) and Flores and Hadiyanto (2006).*

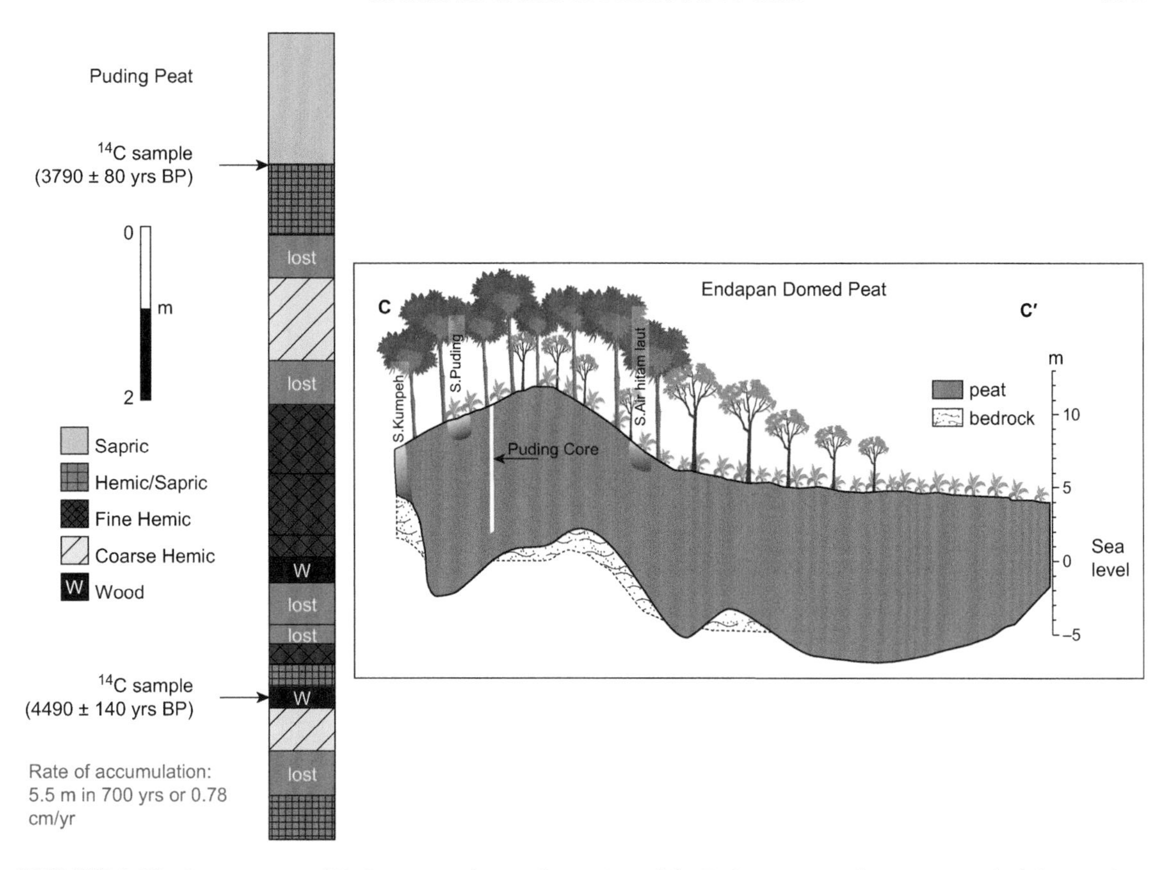

FIGURE 3.38 Cross section of Endapan peat bog and a section of the Puding core on the western end of the peat bog. Carbon isotope (^{14}C) age dates of two samples from the upper and lower parts show rate of peat accumulation of 0.78 cm/yr. Location of cross section in Figure 3.37. BP = before present. *Source: Modified from Hardjono and Supardi (1989) and Flores and Hadiyanto (2006).*

unaffected by river aggradation. Thus, the Endapan peat bog developed and grew unchanged by active fluvial processes as well as serve to confine the Batang Hari and Sungai Kumpeh rivers to the west of the coastal area (Figure 3.38). Here the Sungai Kumpeh River is restricted by the high elevation of the west margin of the Endapan domed peat bog, which gradually declines to the coastline to the east and south.

Toward the tidal-influenced coastal plain drained by the Sungai Kuantan and Sungai Reteh Rivers northwest of the Jambi peatlands (Figure 3.34(A)), the branching, meandering, and headward aggradation of estuaries and tidal incursions also controlled the areal geometry and distribution of domed peat deposits. Figure 3.39 shows isopach maps of the Enoki I, Enoki II, Reteh, and Kualacinaku peat deposits (Hardjono & Supardi, 1989) that lie between the Sungai Kuantan and Sungai Reteh Rivers. The areal geometry of these interconnected peat deposits displays an overall elongate shape that widens and exhibits a dendritic pattern eastward or on its seaward margin (Figure 3.39). Tidal incursion and erosion along estuaries probably cause the dendritic pattern. Fingerlike embayments into the peat mires are caused by headward aggradation and erosion by estuaries and tidal

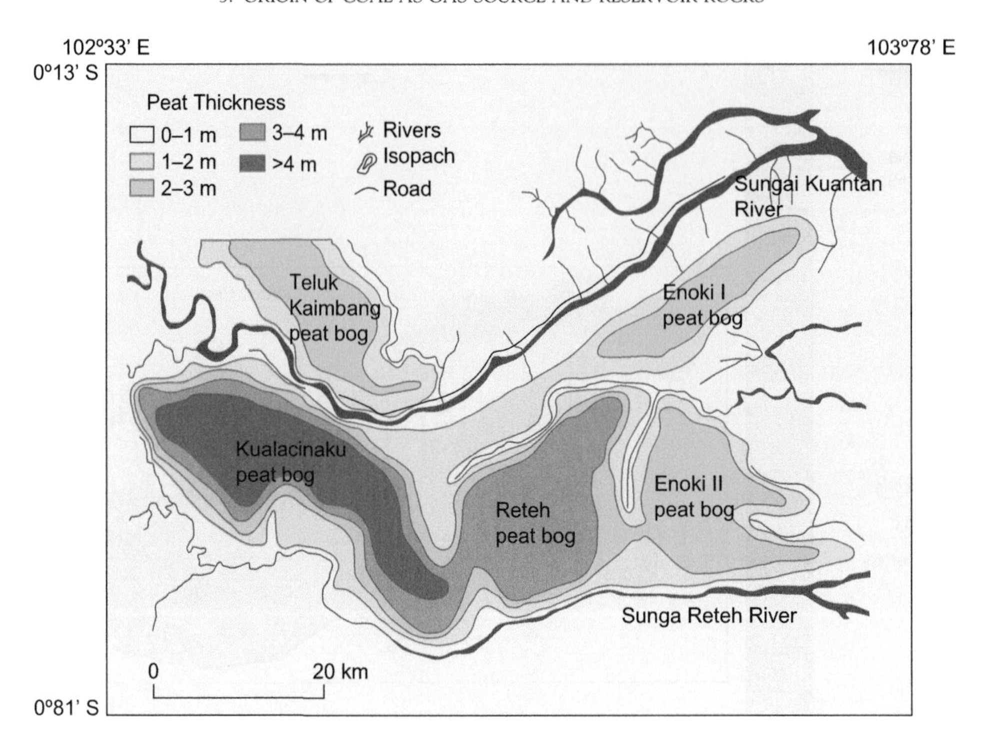

FIGURE 3.39 Isopach maps of peat bogs in the tidal-influenced coastal plain drained by the Sungai Kuantan and Sungai Reteh Rivers northwest of the Jambi peatlands. Location of map in Figure 3.34(A). *Source: Modified from Hardjono and Supardi (1989) and Flores and Hadiyanto (2006).*

fluctuations. The landward thickening of the Kualacinaku peat (Figure 3.39) is similar to peat bogs in the Rajang River deltaic and coastal plain in Sarawak, Borneo (Figure 3.33(A)–(D)), which show better developed domed profiles in the inland rather than the seaward part of the coastal plain (Anderson, 1983).

Ecological and Climate Controls of Peat (Coal) Reservoirs

The thick peat deposits of the Southeast Asia are highly influenced by the equatorial climate characterized by mean rainfall of more than 100 mm/mo, seasonally uniform temperature with annual range of <25 °C, low wind velocities, and cooler nights than days. Along the coastal areas mean annual rainfall is from 2340 to 2883 mm (Samingan, 1980). The rainfall varies locally with the wettest months of the year occurring from November to January and the driest period from June to July (Oldeman, 1978). The daily evapotranspiration rate on land (lower than 4.34 mm) can be slightly more than that on open waters (4.12–4.2 mm). However, the minimum rainfall (<100 mm/mo) exceeds evapotranspiration in the domed peat-bog forests. Morley (1981) indicated that peat–mire forests receiving rainfall of >60 mm/mo for more than 2 consecutive months do not form domed surfaces. Thus, domed peat–mire forests only form in an "ever-wet" climate (Anderson, 1964).

These climatic conditions promote accumulation of peat from vegetation in anaerobic, ever-wet, humid environments. Vegetation consists of more than a dozen species, most of which are tall (40–60 m high) trees mixed with ferns, herbs, and shrubs that make up the understory of the peat bog forest (Anderson, 1983). Tree vegetation adapts to the ever-wet, damp, and highly humid mire environments by developing kneed, stilt, and spreading buttress root types. Shrubs, herbs, and ferns form impenetrable thickets and mosses heavily encrust stems of trees and shrubs. Freshwater stands over much of the mires year-round; however, mangrove and nipah palm trees generally grow along the coast and along rivers where affected by tidal incursions. Near the coast mangroves dominate and along inland waterways nipah palms dominate.

The forested peat bogs in Sumatra are very similar ecologically and morphologically to the bogs of the Sarawak, and Kalimantan, Borneo, which display the most pronounced concentric zonation of vegetation from the margins to the centers (Figure 3.14). Anderson (1961, 1964, 1983) recognized six ecological types or phasic communities of forest vegetation varying from mixed forest vegetation along the margin of the domed peat bog, to well established forest vegetation dominated by very large trees of *Shorea albida* (up to 60 m high) away from the margin, and to dense small trees (up to 12 m high) and open savannah woodland with stunted trees (up to 3 m high) at the center (Figure 3.14). Initial studies on Sumatran peat bog forests noted decreased in stature from the periphery toward the center of the peat dome (Sewandono, 1937) with large (25–60 m in height) emergent trees dominated by diptocarps decreasing and undergrowth shrubs, sedges, palms, and grasses increasing as typified in the Jambi peatlands (Figure 3.40(A) and (B)).

The concentric zonation of vegetation types in the Borneo and Sumatra forested domed peat bog produced unique petrographic and organic materials as recognized by Cameron et al. (1989), Esterle, Ferm, and Yiu-Liong (1989), Moore and Hilbert (1992), Esterle (1990), and Esterle, Ferm, Durig, and Supardi (1987). Cameron et al. (1989), Esterle and Ferm (1994), and Esterle et al. (1989) described the petrology

FIGURE 3.40 Vegetation changes in the Endapan peat bog from (A) dense to (B) open vegetation in the central bog plain.

of the Dendang domed peat deposit (Figure 3.41) in Sumatra, which consists of woody hemic and sapric peat at the periphery of the deposit and mainly woody hemic in the center.

The lateral and vertical variations of plant material in the peat deposits and the associated lateral changes in modern floral assemblages were also identified by palynological studies (Anderson & Muller, 1975; Morley, 2000). Anderson and Muller (1975) used pollen types to document the sequence of evolution of peat deposits in Sarawak, Borneo. Mangrove pollen collected from the clay below the peat decreases into the peat and contains genera entirely confined to the modern peat bog. The only wind-dispersed pollen found in the peat is that of conifers. Demchuk and Moore (1993) reported that, based on palynomorph study of Miocene lignite from Kalimantan, Borneo, the floral

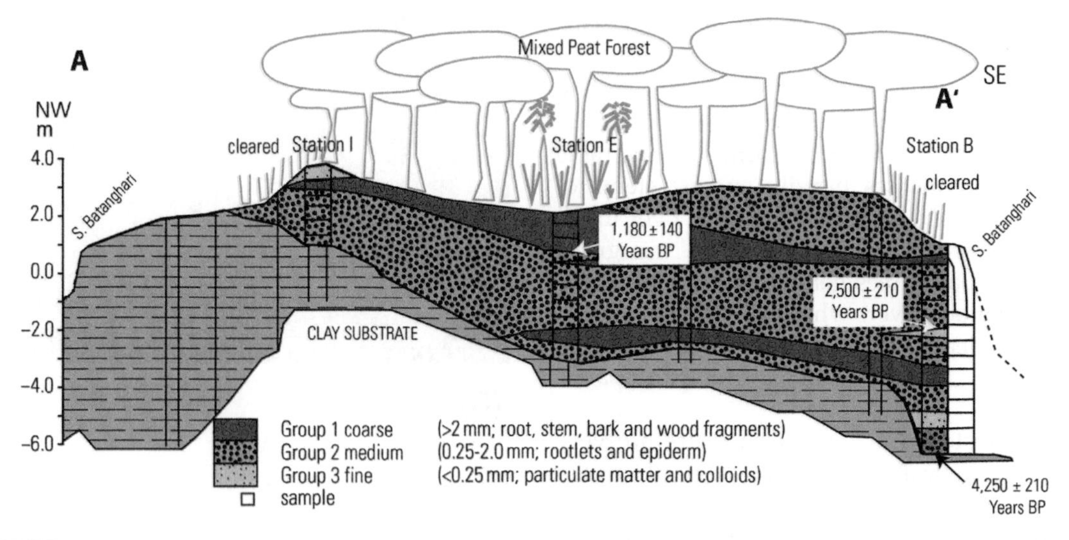

FIGURE 3.41 Cross section of the Dendang peat bog showing the peat facies based on types and textures of the peat. Carbon isotope age dates are shown indicating rate of peat accumulation of about 0.82 cm/yr. Location of cross section in Figure 3.34(B). BP, before present. *Source: Modified from Cameron et al. (1989) and Esterle (1990).*

composition is identical to the present-day peat-forming vegetation. However, it is difficult to clearly correlate the plant successions seen in modern domed peat due to the highly variable tropical vegetation, variable pollen production, mixing of pollen between the phasic communities, and coarse nature of the organic material.

RATES OF VERTICAL PEAT (COAL) ACCUMULATION

Radiocarbon (^{14}C) dates of the peat deposits in Sarawak, Borneo, show an overall rate of vertical accumulation of 2.81 mm/yr (Anderson, 1964; Wilford, 1960). The accumulation rate ranges from 4.76 mm/yr during the early period, which gradually decreased to 3.14 mm/yr in the intermediate period, to 2.22 mm/yr in the late period of peat development. The rapid peat accumulation rate in the early period is related to dominantly forest-type vegetation that formed a shallow convex surface, followed by a slower rate of peat accumulation in a flat-plain center of the bog initiated by stunted trees and finally by advanced forest bogs that covered a greater part of the flat plain. The average rate of accumulation of Sumatra peat deposits (Dendang peat bog) observed by Cameron et al., (1989) is 1 m in 500–600 years or 2–1.6 mm/yr (Table 3.1).

^{14}C age determinations of two samples from the Puding core in the Endapan peat bog (Figure 3.38) collected, at depths of 1.5 and 7.0 m by the author and personnel of the Directorate Mineral Resources of Indonesia from the Puding peat mire, immediately southwest of the Dendang and Kumpeh peat mires, Sumatra, and analyzed by Geochron Laboratories, yielded ages 3790 ± 80 years BP and 4490 ± 140 years BP, respectively. Based on these data, the accumulation for the Puding peat −5.5 m in 700 years or a rate of 0.17 mm/yr—is about 10 times slower rate than the Dendang peat mire. It therefore appears that, within the same alluvial plain, a major difference in accumulation rates of peat deposits exists probably due to local microclimate conditions, variations in autocompaction of the peat, differential compaction of underlying sediment, and differential peat degradation. Most importantly, the Puding peat bog was

FIGURE 3.42 Remnant of the 1997 peat forest fires resulting from drought in the Endapan peat bog.

unaffected by fluvial aggradation and progradation. These variations in accumulation rate of peat may explain the variations in the thickness of coal deposits in a temporal fluvial setting. When the accumulation rate of the Puding peat deposits is plotted on the graph of latitude (both hemispheres) vs maximum peat accumulation rate by Diessel, Boyd, Wadsworth, Leckie, and Chalmers (2000), the position of our data is off the scale indicating an unusually rapid rate of accumulation. This rapid rate of accumulation is in contrast to the slower rate of accumulation of the upper sapric peat deposits between the present and 3790 ± 80 years (Figure 3.38). The slow rate of accumulation for the uppermost part of the Puding peat deposit may be explained by degradation probably due to a reduced and/or fluctuating water table either from natural or man-made causes (e.g. deforestation). In addition, during our trip to the mire evidence of forest fires from the 1998 drought was observed (Figure 3.42). These fires lowered the surface of the peat deposit, which in turn was drowned by the water table, contributing to the most recent degradation of the peat.

TRANSFORMATION OF PEAT BOGS TO COALBED GAS RESERVOIRS

Tropical forest bogs produce thick peat deposits as much as 17 m; have extensive areal distribution; and low ash or detrital inorganic matter content (<10% moisture-free basis) (Cameron et al., 1989; Esterle & Ferm, 1994; Flores & Hadiyanto, 2006). These characteristics combined are critical to the preservation and accumulation of ancestral coal beds as well as their potential as source and reservoir for coalbed gas. Accumulation of thick and extensive peat precursor leads to coal reservoir with possible large gas holding capacity and gas resources. The buffering of sediment influx of the raised peat topography yields low ash content of the coal reservoir.

This principle is best depicted in Figure 3.6 in which the peat marshes with higher inorganic content due to sediment influx from floodwaters will yield more ash content, which makes the resultant coals poor quality for solid combustion and lower coalbed gas content. McCabe (1984) reported that "most active clastic environments (e.g. back-barrier salt marshes, deltaic swamps) are not sites where coal-forming peat accumulate, even though they contain extensive swamps." This despite the fact that countless papers have been published about coal forming in various depositional systems such as coastal plain (e.g. delta, back-barrier, tidal environments), alluvial plain (e.g. fluvial, floodplain, floodplain lake), and lacustrine systems (Dapples & Hopkins, 1969; Ferm & Horne, 1979; Lyons & Rice, 1986; McCabe & Parrish, 1994; Rahmani & Flores, 1984).

For peat-forming environments to be formed coal, the right mix of peat thickness, quality of organic composition, burial of peat prior to destruction by erosion, oxidation, and degradation, and contemporaneous clastic environments are important. The first two conditions are present in peat bogs and the last two conditions are imparted in rapidly subsiding basins providing accommodation space influenced by tectonism and/or sea level fluctuations. In this scenario, peat bog ecological and contemporaneous depositional systems have to be explored.

Compaction and Timing of Peat to Form Coalbed Gas Reservoir

The most direct comparison from peat to coal thickness is compaction ratio, which is largely dependent on the organic matter and moisture contents as well as the methodology used in measurement of compaction (e.g. petrographic, density, stratigraphic, inclusions) (Elliott, 1984; Ryer & Langer, 1980). For example, the peat comprising mainly, moss, reed-sedge, stunted plants/trees accumulated peatlands in the cool temperate latitudes will compact more than the peat comprising dominantly of forest or woody trees accumulated peatlands in the tropical latitudes. The peat:-coal compaction ratio varies widely from 1.4:1 (Ashley, 1907) to 30:1 (Zaritsky, 1975). The compaction ratio for peat to bituminous coal varies between 4.1:1 and 30:1 (Ryer & Langer, 1980). A peat:coal compaction ratio of 10:1 may be assumed for the subbituminous coals (Petersen, Nielsen, Koppelhus, & Sorensen, 2003). Precise measurements of 54 deformed tree trunks within the subbituminous Wyodak coal seam in northeastern Wyoming were made at eight locations in surface mines along the outcrop north and south of Gillette (White, 1986). Theory of elasticity applied to the individual measurements gives ratios of peat to coal thickness, which ranges from 1.7:1 to 31:1 and average 7.1:1.

Given the range of peat to coal compaction ratios average about 4:1 for lignite, 7:1 for subbituminous, 10:1 for bituminous, and 13:1 for anthracite (Table 3.7), the original thickness of the precursor peat can be estimated. Thus, a 33 m thick subbituminous coal (e.g. Wyodak coal in the Powder River Basin, Wyoming, USA) would have originally been about 231 m thick of peat and a 7 m thick bituminous coal (e.g. San Juan Basin, New Mexico, USA) would have been originally about 70 m thick of peat. A 120 m thick peat in the Baram River may have formed from 4000 to 5000 years (Cameron et al., 1989). Thus, using an average 4500 years the 231 m thick peat of the subbituminous coal would have taken 86,625 years to accumulate in contrast to the 70 m thick peat of the bituminous coal would have taken about 26,250 years to accumulate.

Peat Types vs Coalbed Gas Reservoir

The key to understanding internal characteristics of coalbed gas reservoirs, which are important to reservoir engineers and petroleum geologists, is the peat lithofacies development in terms of types and textures of the peat bog. The initial peat formation of a coal bed is represented by the nutrient-rich ecosystem or environment, which mainly resulted in arborescent plant growth followed by a high degree of organic preservation. Toward preservation of plant remains, decomposition yielded varying types of peat (e.g. fibric, hemic, sapric) and the texture of resulting organic matter (e.g. coarse vs fine grained). The vertical and lateral variations of peat types and their texture result in peat lithofacies, which reflects the internal properties (e.g. maceral composition) of the coal, which in turn, control the reservoir characteristics (e.g. permeability, porosity, gas affinity).

The development of peat facies is demonstrated with the macroscopic investigation of peat types and textures of the Dendang peat bog (Cameron et al. 1989). The peat lithofacies variations are based on the presence and correlation of coarse

TABLE 3.7 Compaction Ratios of Peat to Lignite, Subbituminous, Bituminous, and Anthracite Coals Based on Methodologies and Review of 20 Publications

Coal Rank	Methodology	Peat: Coal Thickness Ratio		References
		Range	Average	
Anthracite	Density (compressibility)	13:1; 12.5:1	13:1	Lewis (1934), Mott (1943)
Bituminous	Coal balls (concretions of carbonate minerals and pyrite)	5:1–40:1		Thiessen (1920),
	Macerals (microspcopically recognizable constituents of coal, e.g. tissue, cell, spore, pollen, wood)	3.5; 20:1; 30:1 7:1–19:1; 7:1–20:1; 1.2:1–2.2:1 3.5:1; 11.3:1	10:1	Zaritsky (1975), Stach et al. (1982), Winston (1986); Nadon (1998) USGS (1982)
	Density (compressibility)			Ashley (1907) Lewis (1934), Falini (1965)
		10:1		
	Stratigraphic			Greb et al. (2003)
Subbituminous	Macerals Tree trunks	10:1; 7:1 1.7:1–31:1	7:1	Petersen et al. (2003), Warwick and Stanton (1988) White (1986)
	Stratigraphic	10:1		Roehler (1987)
Lignite	Density Wood	2.5:1; 10:1 4:1; 3:1	4:1	Volkov (1965); Falini (1965); Ting (1977); Stout and Spackman (1988)

(e.g. >2 mm root, stem, bark, and wood fragments), medium (0.25–2 mm rootlets and epiderm), and fine (<0.25 mm particulate matter and colloids) hemic, sapric, and fibric peat types across the Dendang peat bog (Figure 3.41). The coarse (group 1) is the most abundant and thick peat lithofacies, which will result in mainly woody coalbed gas reservoir in contrast the mainly common and thin medium (group 2) and least common fine (group 3) peat lithofacies.

SCALING PEAT FACIES TO RESERVOIR LEVEL

Peat is succeeded by coal facies as a result of coalification during burial and exposure to temperature at depth. For example, woody fragments of the forest peat bogs, which are decay resistant, give rise to woody-textured coal facies after coalification (see discussion in the next chapter). In addition, the different grain sizes (e.g. coarse, medium, fine) of the peat facies result in the nature of lamination of the coal such that the coarse wood-dominated fragments results in coarsely laminated facies in contrast to the fine fragments or attritus results in finely laminated facies (Cameron et al., 1989; Figure 3.41). The characterization of the composition and the texture of the peat-coal facies determine the reservoir properties of the coal in terms of variation in gas content, permeability, and regional gas flow.

However, there is a scale difference between peat and coal facies as shown in Figure 3.43 as

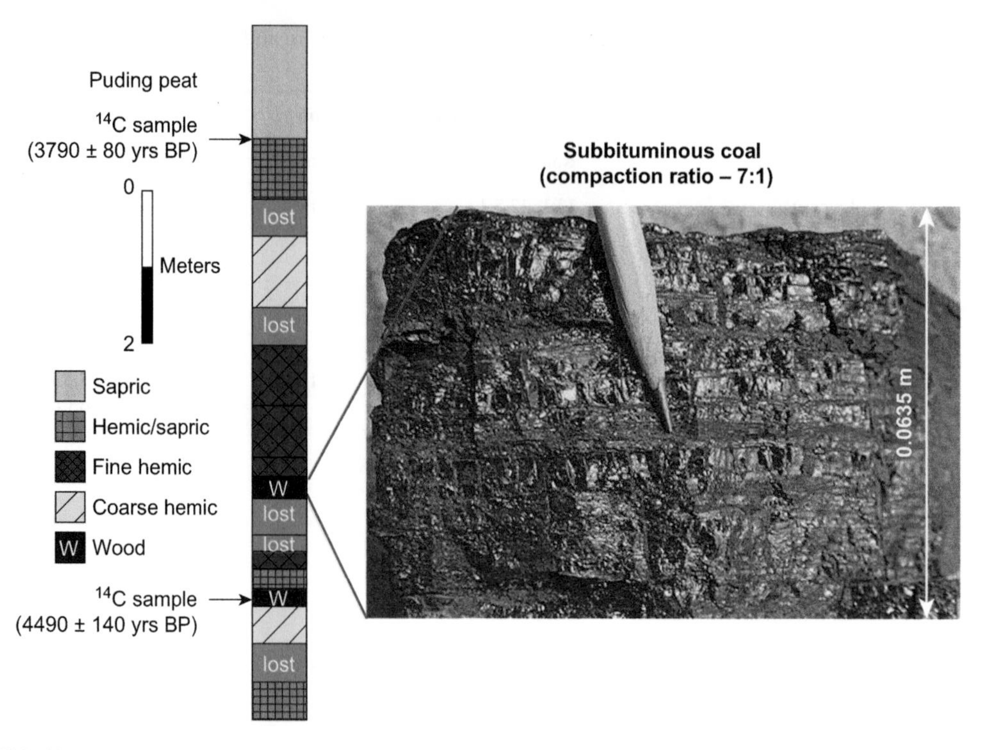

FIGURE 3.43 Comparison of Puding peat core to subbituminous coal after burial, compaction, and coalification. Location of Puding core in Figure 3.38. ^{14}C = carbon isotope; BP = before present. *Source: Modified from Flores and Hadiyanto (2006).*

influenced by compaction and compressibility of the peat during the process of coalification. This process is illustrated in Figure 3.43 of original peat deposit typified by the Puding peat in the Endapan peat bog, which consists of thin to thick woody, sapric, hemic, and fibric peat layers transformed to a subbituminous coal with 7:1 peat:coal compaction ratio (Table 3.7). The Puding peat, which is about 11 m thick, when compacted to subbituminous coal becomes about a 1.6 m thick bed. Thus, the 0.0625 m thick subbituminous coal is about 1/26 order of magnitude of the compacted peat (1.6 m thick), which is correlated to one woody peat facies. The subbituminous coal consists of woody (huminite or vitrinite macerals) vitrain (black, shiny) bands.

An idealized model (Figure 3.44) was developed by Stanton, Flores, Warwick, and Gluskoter (2001) for peat deposits of temperate and tropical climates. The model has implications for coalbed gas reservoirs. In this model the peat bog shows vertical and lateral variations of the peat facies influenced by vegetation types and nutrient supply. The peat bog starts as a low-lying swamp supplied by water influx laden with sediments for nutrient supply of largely arborescent vegetation (e.g. cypress) produces a minerotrophic peat. The resulting peat facies is mainly composed of high ash (from the detritus) and coarse-grained woody peat. As vertical peat accumulation continuous sediment-laden floodwaters are restricted to the bog margins, which supported arborescent vegetation resulting in ashy, woody peat facies. However, nutrient supply is limited mainly from precipitation, toward the central pat of the peat bog, which form acidic open-bog plain. These conditions caused relatively better preservation of vascular plant tissues, which in

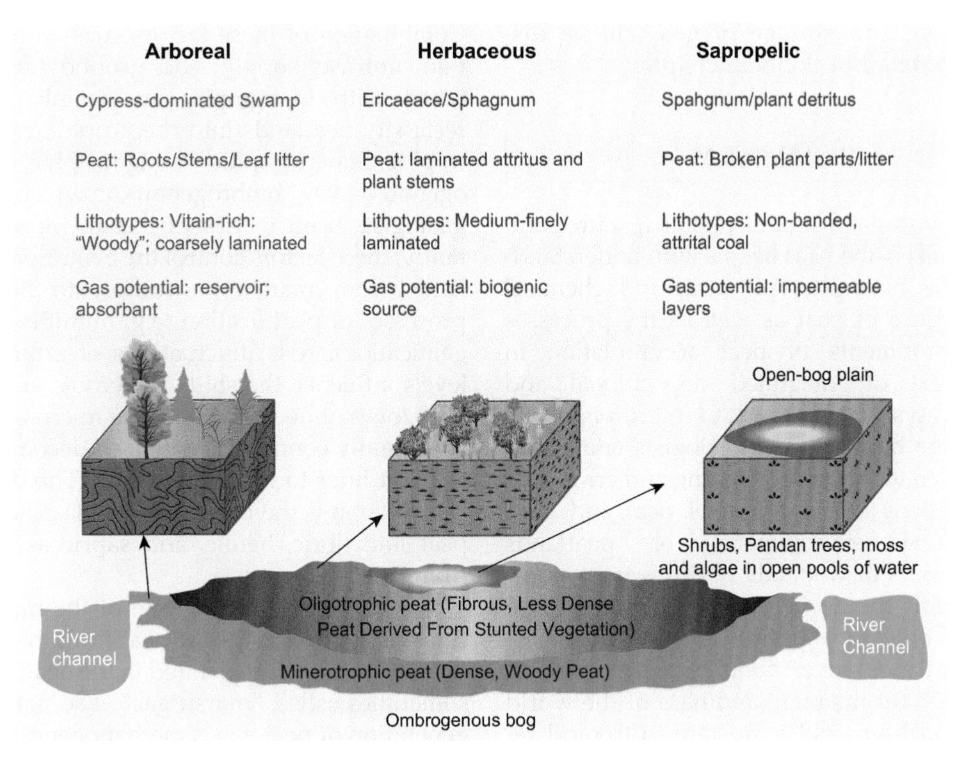

FIGURE 3.44 Conceptual model of temperate and tropical bogs and peat/coal lithotype, vegetation, and gas content. *Source: Modified from Stanton et al. (2001).*

turn increase the degree of doming effect of peat bogs (Polak, 1950). The increasing acidity toward the central bog plain supported vegetation decreasing in herbaceous plants (e.g. Ericacea, sphagnum) to stunted shrubs (associated with sphagnum) in comparison to the bog margins, which continued to be supplied by nutrients from sediment laden floodwaters and supported arborescent vegetation. This resulted in oligotrophic peat, which produced a very variable peat facies controlled by vegetation types. After coalification the arboreal or woody and coarsely laminated part of the peat bog becomes a potential reservoir gas absorbant. The herbaceous, attrital-rich, medium to finely laminated part of the peat bog becomes a biogenic source of gas in the reservoir. The sapropelic or sphagnum/plant detritus rich, nonlaminated (nonbanded) part of the bog becomes impermeable layers, which serve as seal rocks.

Thus, the significance of the "bog model" to coalbed gas reservoir engineers and developmental geologists is the heterogeneity of coal as a source and reservoir originates from the precursor peat. The lateral or aerial variability and uneven vertical distributions of gas content and flow in the coal reservoir lie in the peat-coal facies controlled by the types of vegetations and the degree of decomposition of the organic matter. The inherent difference in the reservoir characteristics of coal from a sandstone or limestone reservoir is the organic composition, degree of decomposition, mulching, and stage of coalification all play an important role in the effectiveness and efficiency of the reservoir. Coalification, gasification, and storage of gas will be discussed in detail in the next chapter.

SUMMARY

The key to the origin of coal as a source and reservoir of coalbed gas begins with understanding of the biological, physical, and chemical characteristics of peat as well as the processes and environments of peat accumulation in the context of integrated depositional and ecological systems. The lack of interdisciplinary cooperation between coal geologists and ecologists has led to misunderstanding and erroneous interpretations of the origin of peat and coal. Peat-forming environments or peatlands comprise 60% of wetlands where water saturation controls the development of peat soil and growth of types of vegetation and animal ecosystems. Peatlands cover about 3 million km^2 or about 2–3% of the total landmass of the world and extend from cold temperate to tropical regions. The peat deposits serve as huge carbon sink and source of greenhouses gasses of primarily methane secondarily carbon dioxide and nitrous oxide.

Types of peatlands are variously recognized as mires, bogs, swamps, marshes, moors, fens, muskegs, and pocosins. Peatland types are simplified and grouped into bogs, swamps, and marshes, the former two called mires following the ecological classification. These peatlands tend to be preserved in geologically subsiding coastal, deltaic, and alluvial plains and lacustrine depositional systems. Cameron et al. (1989) suggested that all peat deposits are not precursors of coal because they are too far from sea level to be buried by marine sediments before destruction by erosion and decomposition.

The formation and classification of peatlands are greatly controlled by hydrology (e.g. surface and groundwater), climate (e.g. precipitation, temperature, and evapotranspiration), nutrient supply, and external aquatic environments. Combination of these factors such as precipitation, nutrient supply, and groundwater, combined with topography are mainly used to reclassify peatlands into rheotrophic vs ombrotrophic, minerotrophic vs oligotrophic, and topogenous vs ombrogenous, which many geologists used to describe coal. More importantly, these factors control the evolution of peatlands from marshes, swamps, to bogs and processes of peatification (e.g. humification and gelification). Also, fluctuations of groundwater levels influence the shifts of aerobic and anaerobic zones of decomposition by microorganisms that mainly control the extent of decomposition of plant litter to form peat types. The degree of degradation is the main basis of classification of peat into fibric, hemic, and sapric as adopted from Soil Taxonomy.

Gasification of peat is one of the major processes that develop during peatification. Biogenic or microbial gas is generated in the peat, which is sometimes called "marsh gas". The pathway to generation of peat gas is methanogenesis or biomethanation in which methane and carbon dioxide are the byproducts of anaerobic microorganisms or methanogens. Methanogens derive energy by converting carbon dioxide, hydrogen, formate, acetate, and other compounds from the peat organic matter to either methane or methane and carbon dioxide. Peat gas generation occurs below the groundwater table. Collectively, methane, carbon dioxide, and nitrous oxide are produced in peat, which are mainly released as greenhouse gases. However, methane is the most abundant gas, which is largely desorbed to the atmosphere with estimated emission of about 100–500 Tg annually.

Whether this early stage biogenic peat gas is preserved and how much as coalbed gas is not well known. The dynamics of peat gas have been observed to form as gas bubbles in pore waters of the groundwater and is held on walls of organic matter by atmospheric and hydrostatic pressures. The gas is trapped between confining layers creating overpressure zones, which release the gas when there is sudden changes in the hydrostatic pressure. Laboratory studies of peat materials inoculated with anaerobic microorganisms generated first carbon dioxide followed by large amounts of methane but further analysis indicates that only carbon dioxide is retained in the peat. In comparison, coal materials of various ranks were heated to resemble catagenesis (e.g. early stage coalification) and produced large volume, as much as 95%, of methane forming at 35 °C but drastically decreased (e.g. 50–30%) at higher temperatures (e.g. 125–150 °C).

Early stage peat gas may be preserved with rapid burial and sealed by sediments in major depositional peat-forming systems such as alluvial, coastal, and delta plains as well as lacustrine system. In these depositional systems peat accumulation is in equilibrium with subsidence, sedimentation, and erosion controlled by creation of accommodation space, which is assisted by peat compaction and autocompaction. Case studies of tropical peat bogs in the coastal and alluvial plain systems in Southeast Asia are presented as potential modern analogs of coal-forming environments. These modern analogs have been widely used as examples of formation of economic coals, which also served as reservoir rocks for coalbed gas worldwide. In order for these modern peat deposits to form coalbed gas reservoirs and mineable beds, they have to accumulate much longer period of time (e.g. >87,000 to >175,000 years) and undergo peat:coal compaction ratios of 4:1 to 13:1 from lignite to anthracite coals.

Finally, the variations of peat types and facies with specific properties reflecting unique process and environment of formation are important to coalbed gas reservoir engineers and developmental geologists. Facies distribution in the peat bog controls the heterogeneity of coal as a source and reservoir rock. The lateral or aerial variability and uneven vertical distributions of gas content and flow in the coal reservoir are dependent on the peat–coal facies, which is in turn controlled by the types of vegetations and the degree of decomposition of the organic matter.

CHAPTER

4

Coalification, Gasification, and Gas Storage

OUTLINE

http://dx.doi.org/10.1016/B978-0-12-396972-9.00004-5

KEY ITEMS

- Peat transformation occurs on the Earth's surface from tens to hundreds of thousands of years involving biochemical and microbial processes but coal transformation occurs in the subsurface over millions of years involving chemical, physical, and structural processes.
- "Metamorphic" change of organic matter in coal during burial through geologic time is coalification, which is interchangeably used with maturation through chemical and physical changes producing low- to high-rank coal.
- Principal causes of coalification are temperature and pressure inherent to burial at increasing depths through geologic time accompanied by tectonic deformation.
- Effects of temperature on coalification are straightforward. Pressure is confused with tectonic stress, which affects physical structures but promotes changes in chemical composition in coal.
- Coalification includes dehydration, bituminization, debituminization, and graphitization. Peatification is not a part of coalification, being a surficial process during deposition.
- Processes of coalification are chemical and physical involving changes in chemistry and physical properties of organic matter toward denser, drier, more carbon-rich, and harder coal. Chemical and physical changes include color, luster, hardness, water, bulk density, aromaticity, and composition in coal ranks.
- Coal rank (e.g. lignite, sub-bituminous, bituminous, and anthracite) depends on the degree of coalification expressed by levels of changes in fixed carbon, moisture (water), volatile matter, and calorific value.
- Despite efforts to internationalize coal rank classification designed for producers, sellers, and buyers of coal and users based on industrial and technical purposes, individualized systems continue to be developed based on coal properties for each country.
- Chemical and physical changes of organic matter are expressed in maceral composition of the coal. Microscopic properties of macerals permit grouping into vitrinite, liptinite, and inertinite.
- Depending on the macerals and stages of coalification, hydrocarbon by-products are generated in coal such as coalbed gas formed during bituminization.
- Coalbed gas is thermogenic gas generated at high temperatures during late-stage coalification and biogenic gas is generated at low temperatures by microbes (methanogens) when coal beds are exposed to meteoric groundwater recharge after basin deformation.
- Thermogenic and biogenic gases are differentiated according to their carbon isotopic composition of methane.

Thermogenic gas contains heavier carbon isotopic composition of methane than biogenic gas.

- Coalbed gas is stored in pore and fracture systems. Pores in coal matrix, which hold 90% of the gas, contain six to seven times more than a conventional natural gas reservoir of equal rock volume can hold.
- Coalbed gas is stored by physical and chemical adsorptions on solid surfaces in coal by dispersion–repulsion forces (e.g. van der Waals and Pauli repulsion) and electrostatic, covalent, and ionic bonds.
- Gas is desorbed from internal surface areas in coal by depressurization (e.g. reduced hydrostatic pressure). Gas flows from micropores to mesopores and macropores by bidisperse diffusion.
- Fick's law of gas diffusion governs gas flow in coal matrix and Darcy's law governs gas flow in coal fractures.

TRANSFORMATION OF PEAT TO COAL

The transformation of peat to coal is a complex process that involves various degrees of chemical, physical, and structural changes of plant organic matter exposed to progressive effects of temperature and pressure through geologic time. The transformation begins at the surface as plant litter of peat and continues during burial of peat transforming into coal though millions of years. This process is in contrast to peatification, which is mainly microbial and biochemical in nature and occurs on the surface and within a few meters below the surface, through thousands of years. The transformational processes of peat to various grades or ranks of coal or stages of maturation are the concepts of coalification (Figure 4.1). The general sequence of coalification from low- to high-rank coals is lignite, subbituminous, bituminous, semianthracite, and anthracite. To the extent that the transformation affects chemical and physical changes suggesting that coalification involves the process of diagenesis. In the same context, because alterations of the plant organic matter are intensive, coalification also involves the process of metamorphism. The alterations of plant organic matter occur at depth concurrent with increasing burial, thermal, and tectonic conditions. Thus, the chemical, physical, and structural changes, although not observed directly during the process, are interpreted to occur at various stages of coalification. The causes and effects of these processes will be discussed in succeeding sections.

The chemical, physical, and structural changes during peat to coal transformation are accompanied by increase in carbon content and aromaticity as well as decrease in oxygen and loss of porosity, which are collectively important in terms of coal as source and reservoir of hydrocarbons (Levine, 1993). One of the more important process simultaneous with coalification is the generation of methane as well as other byproducts enriched in hydrogen and oxygen. The nature and amount of the expelled hydrocarbons and whether they are mainly composed of gases or liquids is a matter of considerable debate (Boreham, Blevin, Radlinski, & Trigg, 2003; Wilkins & George, 2002). Nonetheless, the huge amount of organic matter in coal makes it a very attractive potential source and reservoir of hydrocarbons produced during coalification. More importantly, coal possesses an affinity to adsorb methane, which may be generated as thermogenic and biogenic gases.

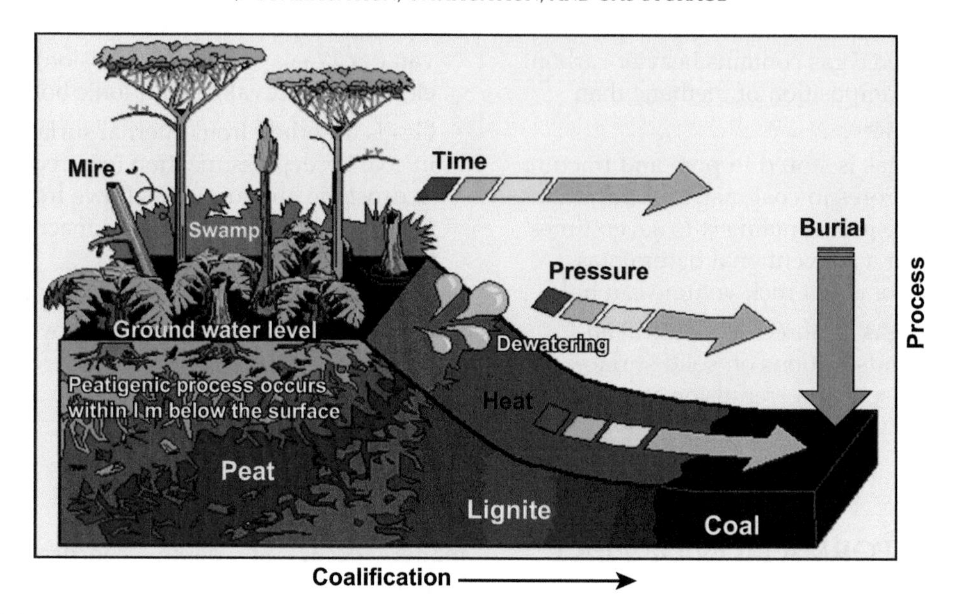

FIGURE 4.1 The process of peatification, which occurs on the surface is followed by coalification of the peat upon burial in the subsurface where it its dewatered and exposed to heat and pressure. *Source: Modified from Greb, et al. (2006).*

Coalification by-products, which are not expelled such as methane, are stored in the coal and potentially entrapped as coalbed gas. Methane is stored in open pores, which Levine (1993) estimated to comprise 5–20% of the coal volume depending on the coal rank. Coal porosity is typically very low (1–5%) and normally occupied by water. Most of the essential coal porosity resides in the open molecular structure, which can provide ease of movement for small methane molecules in and out of the structure under the influence of reservoir pressure (e.g. hydrostatic) (Levine, 1993). Fractures in coal are additional storage space where "free" and dissolved gases can be held by confining stress and/or hydrostatic pressure. The ability of methane to diffuse and/or flow from the coal matrix where the gas is mainly stored is directly influenced by the permeability and pressure of the reservoir. The role of coalification in gasification and creation of storage spaces in coal will be discussed in the succeeding sections.

Thus, coalification is viewed differently from various standpoints of specialties such as coal geologists, coal petrologists, geochemists, and petroleum geologists and engineers. A coal geologist mainly views coalification in terms of coal ranks particularly in relation to assessments of coal resources and reserves. A coal petrologist views coalification in terms of the alteration and classification of organic matter composition or macerals (e.g. analogous to mineralogy in "hard" and "soft" rocks). A geochemist (e.g. coal or organic chemist) views coalification in terms of chemical composition of the coal and their macerals in relation to fuel grade and hydrocarbon potential. The petroleum geologist and engineer view coalification in terms of maturation in relation to expulsion and sorption of hydrocarbons and reservoir properties. Thus, coalification is a rather involved and complex process borne out by the heterogeneity in the composition of coal, which is altered in various ways through geologic time (Taylor et al., 1998).

BITUMINIZATION, DEBITUMINIZATION, AND GRAPHITIZATION OF ORGANIC MATTER (METAMORPHISM)

Postpeatification processes (e.g. humification and gelification), which are followed by coalification of peat, include three successive and overlapping stages: bituminization, debituminization, and graphitization of organic matter. Each stage is a product of combined chemical, physical, and structural reactions of organic matter to causes of coalification.

Causes of Coalification

The transformational processes, mainly biochemical to physicochemical, that are responsible for the conversion of peat to coal are attained during burial at depth concurrent with increasing thermal, pressure, and tectonic conditions (Figure 4.2). These conditions are governed mainly by temperature and pressure enhanced through the geologic time (e.g. thousands to millions of years). According to Teichmuller and Teichmuller (1966), during coalification, temperature is most influential to chemical reactions in terms of duration of exposure and attaining optimum heat. That is, maturation of the same rank coal may be attained either by exposures to prolonged low temperature or to fast-acting high temperature. The phenomenon of more rapid rates of heating of peat (or coal) by, for example, tectonism, thermal fluids, and igneous intrusion, was termed as contact metamorphism (Stach et al., 1982). Pressure (e.g. burial-confining stress) causes alterations in the physical structure as well as slows down the chemical reaction of the coal (Cao, Li, & Deng, 2009; Cao, Li, & Zhang, 2007; Faiz et al., 2007a). In contrast, pressure from geological tectonism (e.g. Earth's crust deformation) may cause infrequent and limited increase of coal rank. Given these conflicting observations, the debate on which of the causes of coalification truly contribute significantly to the physicochemical reactions began with the acceptance of concept of Teichmuller and Teichmuller in the 1960s. The debates in Europe and North America centered between temperature and tectonism. Because of the complexity and heterogeneity of the composition of coal, this debate may never be resolved satisfactorily. The readers are referred to the far-reaching papers by Hower and Gayer (2002), Lyons and Alpern (1989), and Ruppert et al. (2010) on the case studies and historical perspectives of the mechanisms of coalification.

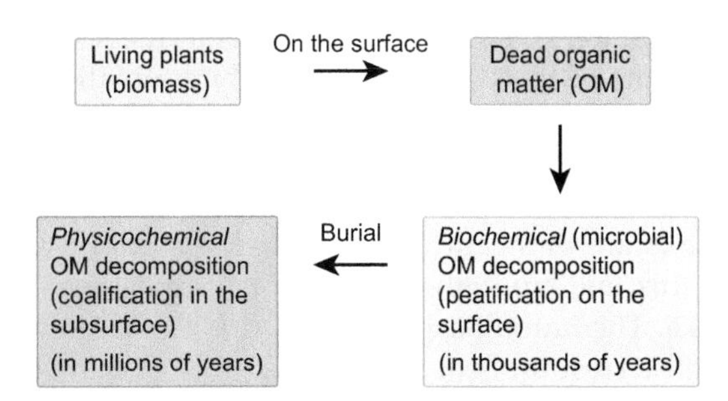

FIGURE 4.2 A simplified flowchart of the transformational processes of biomass of living plants through peatification on the surface in thousands of years and coalification in the subsurface in millions of years. OM, organic matter.

Temperature

Increase in temperature that is directly or indirectly applied to the peat (or coal) during burial is the most commonly accepted mechanism of coalification. The geothermal gradient is the rate of increase in temperature per unit depth in the Earth from the core heat flow, which has not changed from the paleogeothermal gradient (Kanana & Matveyev, 1989). Heat can be applied indirectly by an increase in temperature related to the Earth's paleogeothermal gradient and paleothermal conductivity of rocks with increasing depth of burial. Also, direct application of heat may be from thermal fluids and igneous intrusive bodies (e.g. laccoliths, dikes, and sills). Rising temperature with depth caused by paleogeothermal gradient, which promotes increased chemicophysical changes and rank of coal, is called Hilt's law. For example, Stach et al. (1982) demonstrated the influence of paleogeothermal gradient in Tertiary rocks of Upper Rhine graben where "warm" areas (7°–8 °C/100 m) have reached bituminous coal at 1500 m depth compared to "cold" areas (4 °C/100 m) at 2600 m depth.

A review of the literature indicates that temperatures during coalification ranges from 50 °C to >300 °C from brown or lignite to anthracite coal (Diessel, Brothers, Black, 1978; Kanana & Matveyev, 1989; Shimoyama & Iijima, 1974; Stach et al., 1982; Taylor et al., 1998). Bituminous coal thermally forms from 85 °C to 150 °C and anthracite from 190 °C to >300 °C. In order to attain a certain rank coal, higher temperatures with more rapid rates of heating (e.g. contact metamorphism) rather than with slower heating rates (e.g. gradual subsidence) are necessary. The influence of temperature on coalification during burial is not always a straightforward phenomenon because of the influence of tectonism, thermal fluids, and igneous intrusive bodies, which are either applied solely or in combination with coal-bearing basins (Bustin & England, 1989; Cao, Davis, Liu, Liu, & Zhang, 2003; Flores & Bader, 1999; Golitsyn, Courel, & Debriette, 1997; Harrison, Marshak, & Onasch, 2004; Hower & Gayer, 2002; Middleton & Hunt, 1989; Ruppert et al., 2010; Sachsenhofer, Jelenb, Hasenhuttl, Dunkl, & Rainer, 2001; Snyman & Barclay, 1989). However, faults served as avenues of migrating thermal fluids as short-term high-temperature sources of heat for coalification. For example, in the Blanzy-Montceau Basin at Massif Central, France, convective and conductive heat flow transfer through fault tectonics played an important role in the coalification. In the Appalachian Basin, in the eastern United States, heat carried by the migrating fluids through the Pennsylvanian coal-bearing strata elevated the temperature to anthracite-grade conditions (~250 °C), that is, assuming the thermal-fluid event lasted about 1 year (Harrison et al., 2004). Here, hot fluids migrated from the Alleghanian orogenic belt toward the foreland, carried heat into the coal, and caused anthracitization.

Igneous intrusive bodies play a major role in regional and local coalification particularly in volcanically active areas (Cooper, Crelling, Rimmer, & Whittington, 2007; Flores & Bader, 1999; Herudiyanto, 2006; Sachsenhofer et al., 2001; Snyman & Barclay, 1989). Emplacement of plutonic rocks increased temperatures in the surrounding areas, while dikes and sills intruded into coal forms contact metamorphism (Flores & Bader, 1999; Sachsenhofer et al., 2001). In general, local igneous intrusions through coal beds have a minimal overall basinal effect on coals as it has been shown in the Raton Basin in Colorado and New Mexico (Figure 4.3) (Flores & Bader, 1999). Tertiary dike and sill intrusions (circled areas in Figure 4.3) have transformed some coals into bituminous rank and natural cokes in local areas in the basin. However, the effects of the emplacement of larger Tertiary intrusive stocks (e.g. Spanish Peaks) and laccoliths (e.g. Vermejo Park and Morley domes) affected the rank of the Tertiary Raton Formation and Upper Cretaceous Vermejo Formation coals on a large scale that rank increased from high-volatile bituminous C to A (on a dry basis). However, there is a gradational change in coal rank from a high-volatile A bituminous in the southern part of the Raton Basin to a high-volatile C bituminous in the northern part that averages about 30,238 kJ/kg (on a dry basis). For example, the coals north of a line just south of Spanish Peaks are of steam quality and the coals south of this line are high-quality metallurgical-grade coke. The Raton Basin is an area of anomalously high terrestrial heat flow related to the Spanish Peaks and associated underlying magmatic intrusions. The presence of high heat flow and low-volatile coal in the Raton Basin indicates the potential generation of substantial amounts of dry methane from both the Tertiary and

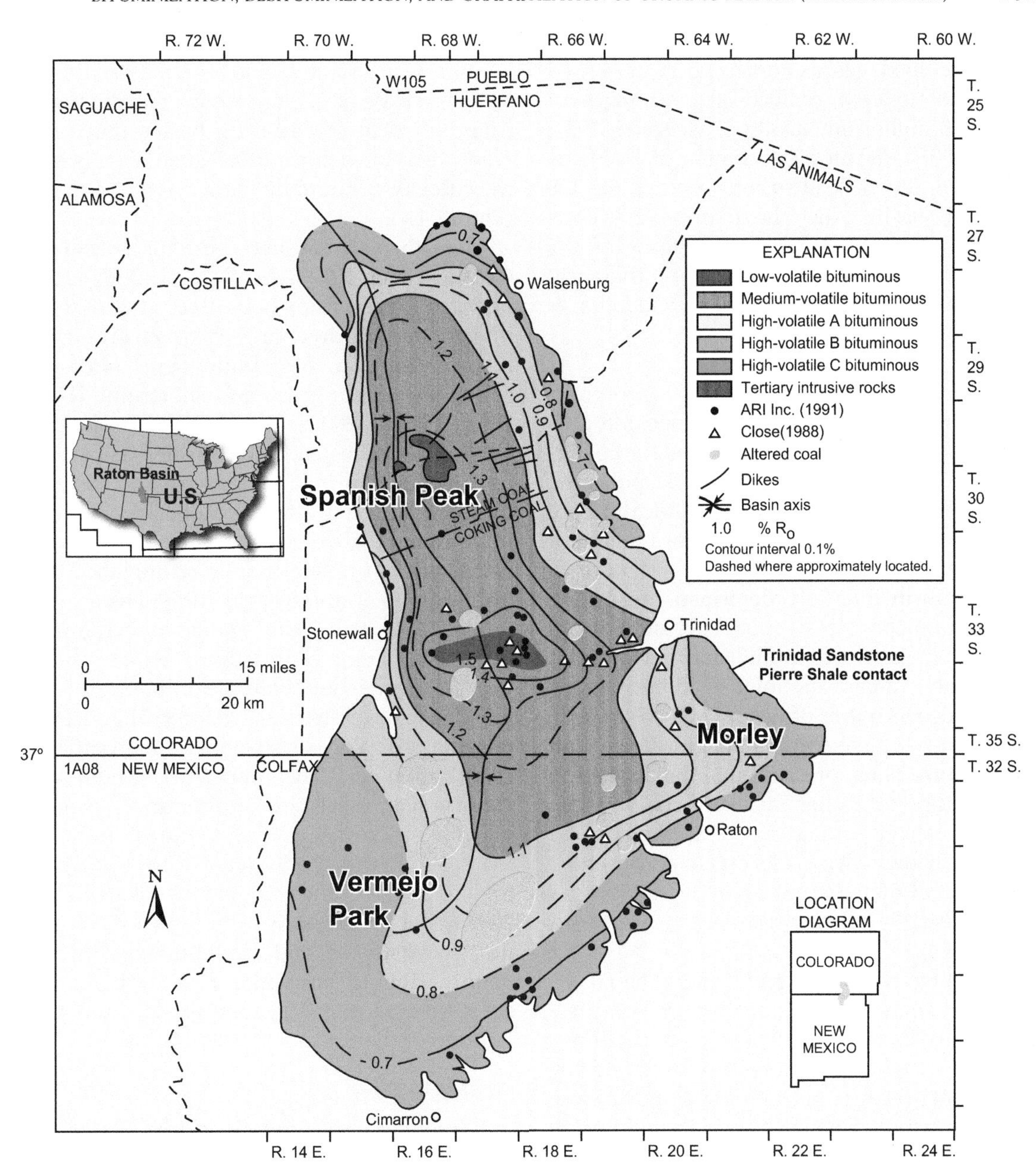

FIGURE 4.3 Map of coal rank based on ASTM International (2011) classification of the upper Cretaceous Vermejo Formation in the Raton Basin in southern Colorado and northern New Mexico in the United States (see inset map). Coal rank is locally altered by igneous intrusions in the gray areas. Coal rank in the southwest part of the basin is affected by laccoliths. *Source: Modified from Flores and Bader (1999).*

Upper Cretaceous coals, which are estimated to contain coalbed gas as much as 35 cc/g and mean undiscovered coalbed gas resources of about 45.3 million m^3 (Dolly & Meissner, 1977; Higley, 2007). Elevated gas content of coal beds heated by igneous intrusions occurs in the Gunnedah Basin, Australia (Gurba & Weber, 2001), although the coal-rank patterns of the Permian basins are a combination of burial, heat flow from hot fluids, and tectonics (Middleton & Hunt, 1989).

Pressure

Pressure is defined as the force per unit area applied in a direction perpendicular to the surface of an object. Damberger (1991) indicated that the effects of pressure on coalification are difficult to separate from the effects of temperature. It is generally accepted that coals are of higher rank in intensely deformed areas than in undeformed areas; thus, pressure from folding contributes to coalification. However, geologically, areas of intense deformations occur in deeper parts of basins, which are also affected by paleogeothermal gradients and burial depth. Thus, pressure may be applied in combination with other causes of coalification. In geological terms, force may be applied during (1) burial by overlying rocks due to compaction, (2) folding of the adjacent rocks, (3) faulting of rocks causing local pressure, and (4) igneous intrusion.

According to Cao, Li, and Deng (2009), Cao, Li, and Zhang (2007), the controversy on pressure causing coalification is confusing with tectonic stress and other pressures exerted by mechanical energy (e.g. confining and overlying strata). Pressure benefits physical coalification but retards chemical coalification, whereas tectonic stress affects both the physical structure and promotes chemical composition changes of the coal. Compressive or shear stress affects the aromatic-ring layers of the coal, influences their parallel arrangement, and enhances directional growth. Thus, directional pressure or shear stress is a necessary condition for graphitization (Cao, Li, & Deng, 2009). The effects of tectonic stress (e.g. compressive stress) may be exemplified in coal-bearing basins that underwent crustal deformation such as those in the Rocky Mountain intermontane basins (Figure 4.4).

Here, the Late Laramide structural basins (e.g. Greater Green River, Wind River, Bighorn, and Powder River Basin) bounded by a system of interconnected, diverging and converging, and anastomosing arches underwent northeast–southwest compression and shortening (Erslev, 1993; Flores, 2003; Flores, Roberts & Perry, 1994). The arches and intervening basins collectively were deformed by forward and backward thrust faults, often stacking of backward thrusts in the manner of deformation in the southeastern Cordillera of Canada (Bustin & England, 1989). The Late Laramide structural basins were infilled with Paleocene coal-bearing sediments (e.g. Fort Union Formation), which showed varying degrees of coalification based on vitrinite (e.g. huminite; see Chapter 5) reflectance (Figure 4.4). The succession of maturation or coalification of Paleocene Fort Union coals expressed in mean, maximum, and percentage vitrinite reflectance ($\%R_{omax}$) ranges from >0.6 to <0.3 from the interior (e.g. Green River basin) to the exterior (e.g. Williston) basins (Flores, 2003). The progression from high to low coalification eastward and northeastward may be mainly due to disparate deformation-related successive partitioning, compression, and thrust faulting of the basins.

Geologic Time

Geologic time expressed in millions of years provides a prolonged duration to ensue coalification. Upper Miocene peat deposited 17 mya and buried to depth of 5440 m at 140 °C in the Gulf of Mexico produced high-volatile bituminous coal (35–40% volatile matter (Stach et al., 1982)). Pennsylvanian peat deposited 270 mya at the same depth and temperature in the Ruhr

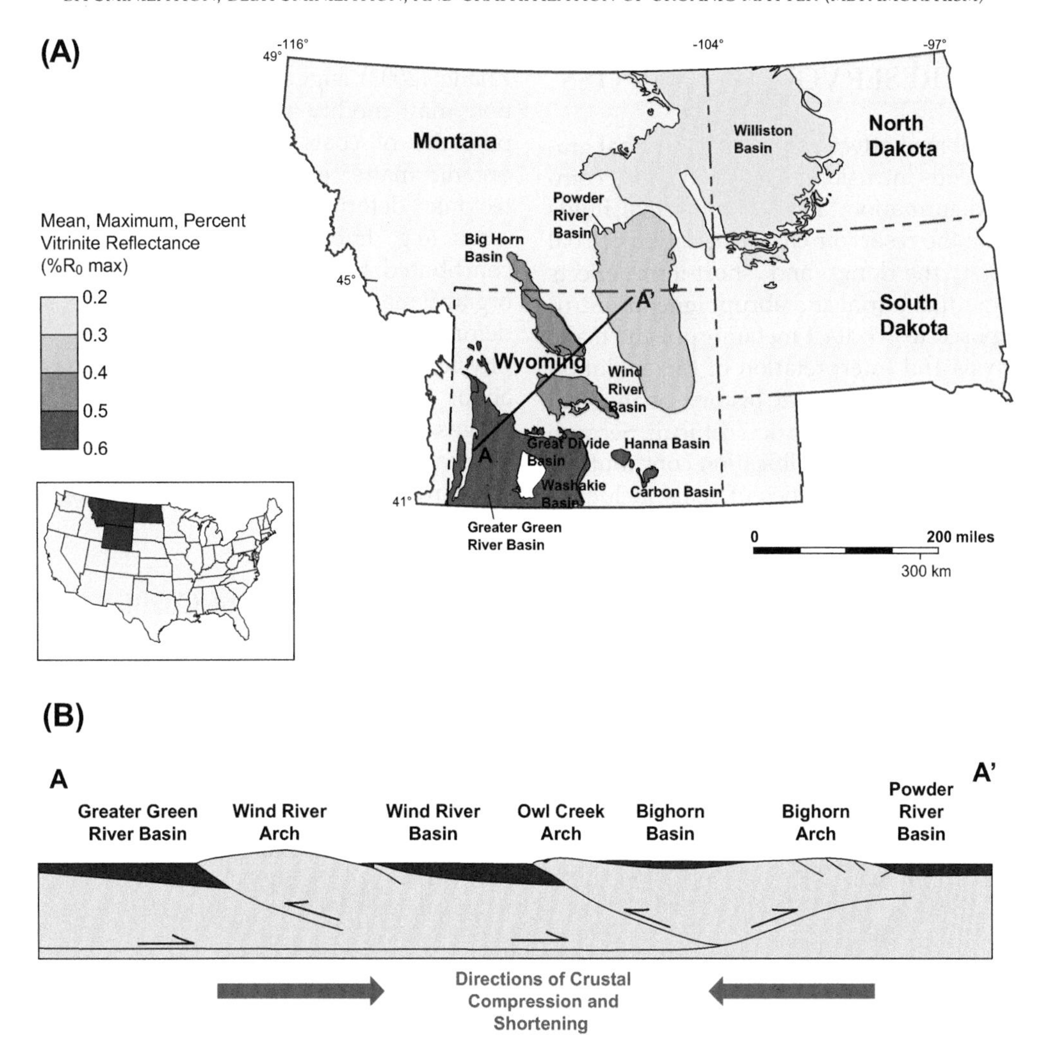

FIGURE 4.4 (A) Map of coal basins in the northern Rocky Mountains and Great Plains in the United States (see inset map) showing decreasing vitrinite reflectance (%R_0) of coals in the Paleocene Fort Union Formation from southwest to northeast. (B) Cross-section *A*–*A'* of Precambrian through Tertiary Greater Green River, Wind River, Bighorn, and Powder River Basin. Coal rank was enhanced by crustal compression and shortening. *Source: Modified from Erslev (1993), Flores (2003), and Flores et al. (1994).*

Basin produced low-volatile bituminous coal (14.16% volatile matter). If the rate of burial is too high, sediments will not have enough "cooking time" to produce high-rank coal from peat. During the long process of geologic evolution, all types of tectonic–thermal activities resulted in the diversification of coal in terms of physics, chemistry, textures, and structures. The organic and inorganic components in coal are affected by various geologic agents, which sequentially influenced the coal seam distribution, rank, and quality.

INFLUENCE OF COALIFICATION ON GAS RESERVOIR PROPERTIES

The combined effects of paleogeothermal gradients, igneous intrusions, tectonism and accompanying deformation, and hydrothermal fluids complicate the reservoir characterization of coal beds. First, the long- and short-term effects such as gradual burial and abrupt igneous intrusions exposed to "contact metamorphism" make the analysis and interpretation of the evolution of maturation and thermal history of the coal and accompanying gasification difficult. Second, the multiple causes of coalification contribute to complex reservoir properties of coals including porosity and permeability. Third, the heterogeneity of the coal composition reacting to a combination of coalification mechanisms may play a role in the origin of coalbed gas. Creedy (1988) suggested that the gas content of coal may not exceed the methane sorption capacity of the reservoir, which is determined at a temperature equal to the maximum burial depth during coalification. However, permeability of coal reservoirs decreases with burial depth such that it levels off about 1800 m below the surface (Figure 4.5) (McKee, Bumb, Way, Koenig, & Brandenburg, 1986). Cao, Davis, Liu, Liu, and Zhang (2003) suggested that tectonic deformation may modify the hydrocarbon-generating potential of coals. The adsorption of soluble organic matter expelled during and following tectonic deformation onto enhanced surface areas (e.g. fractures, pore spaces) may have contributed to higher extract yields of soluble organic matter of the deformed coal. Tectonic deformation not only may affect the original physical structure of a coal but also may bring about changes in chemical properties critical to gas generation. Finally, thermal evolution of coal during coalification can affect pore structures (e.g. size and morphology) and connectivity.

Processes and Stages of Coalification and Gas Generation

The processes of coalification are mainly chemical and physical as they involve changes in the chemistry and physical properties of organic matter from peat to denser, drier, more carbon-rich, and harder coal. Although biological processes are recognized as part of coalification in some publications (Levine, 1993;

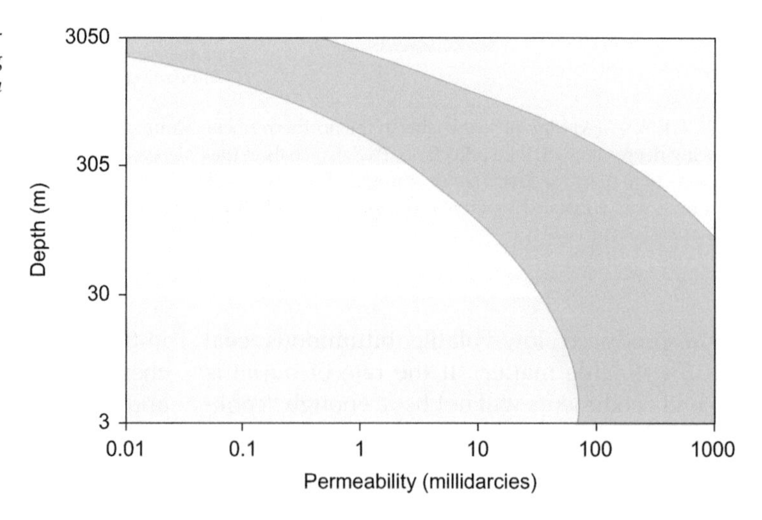

FIGURE 4.5 Diagram of the relationship of permeability and depth indicating decrease at depth. *Source: Modified from Flores (1998) and McKee et al. (1986).*

Thomas, 2002), these processes are considered in this book as a significant part of peatification on the surface in contrast to chemical and physical processes of coalification in the subsurface. A summary of the chemical and physical changes during coalification of a tropical peat to lignite is given in Table 4.1. The chemical and physical changes are mainly associated with color, luster, hardness, water, bulk density, aromaticity (e.g. huminite and liptinite macerals), and composition (e.g. carbon, hydrogen, oxygen, and volatile matter) (Andriesse, 1988; Levine, 1993). The volatile matter and proportions of the atoms of hydrogen, carbon, and oxygen are accepted measures of various stages of coalification. However, it is noteworthy to indicate that according to Black (1989), the chemical changes in soft brown coal or lignite in terms of the atomic ratios of hydrogen:carbon and oxygen:carbon are somewhat simplistic and may not be valid because the humic material is very heterogeneous. That is, chemical reactions through gelification and depolymerization of mixed plant tissue and humic materials may yield opposing trends of the hydrogen/carbon and oxygen/carbon ratios (Hatcher, Berger, Szeverenyi, & Maciel, 1982; Russell & Baron, 1984).

Chemical Process of Coalification

In this book, the chemical process of coalification is principally keyed to its relation and application to coal as a source of hydrocarbons. Thus, to this end, the traditional and universally accepted concept of chemical process of coalification involves increased maturation of coal with accompanying increase in fixed carbon, reduction of volatile matter, and changes in proportions of hydrogen, carbon, and oxygen. Hydrogen, carbon, and oxygen represent the elemental composition of organic matter or macerals (Table 4.2) (e.g. vitrinite, liptinite, and inertinite), which compose humic and sapropelic coal types (see Chapter 5 for definition of maceral types). Carbon content increases and oxygen

TABLE 4.1 Changes in Physical Properties and Chemical Composition of Peat to Lignite

	Changes in Properties and Composition							
	Physical				Chemical			
Organic Deposits	Color	Luster	Hardness	H_2O Vol.	Bulk Density	Aromaticity	VM	Composition
Peat	Lighter	Earthen (soil-like)	Soft (plasticlike)	Higher	300–400 kg/m^3	Lower (hum. and lip.)	60–70 wt%	C = 50–60 wt% H = 5–7 wt% O = 30–40 wt% Cal. Val. = 20–23 mJ/kg (db)
Lignite	Darker	Mainly dull and partly vitreous	Hard	Lower	650–780 kg/m^3	Higher (hum. and lip.)	50–60 wt%	C = 65–75 wt% H = 4.5–5.5 wt% O = 20–30 wt% Cal. Val. = 20–24 mJ/kg (db)

mJ/kg, megajoule/kilogram; kg/m^3, kilogram per cubic meter; hum., huminite; lip., liptinite; C, carbon; H, hydrogen; O, Oxygen; Cal. Val., calorific value; VM, volatiler matter; db, dry basis; wt%, weight percent.
Source: Modified from Andriesse (1988) and Levine (1993).

TABLE 4.2 Petrographic Nomenclature and Genetic Groupings of Coal Macerals

Maceral Group	Identification Criteria	Origin	Maceral
Vitrinite (huminite)	Fluorescence; response to etching	Cell walls, gelified	Telinite
		Cell structure, poorly defined	Collotelinite
		Vitrinite fragments	Vitrodetrinite
		Vitrinite groundmass, mottled	Collodetrinite
		Vitrinitic colloidal	Gelovitrinite
		Infills	
		Cell infills	Corpogelinite
		Infills of cracks and voids	Gelinite
Liptinite	Characteristic fluorescence in blue light	Spores and pollens	Sporinite
		Leaf cuticle, roots, and stems	Cutinite
		Algae	Alginite
		Corkified cell walls	Suberinite
		Plant resins	Resinite
		Liptinite fragments	Liptodetrinite
		Cavity-filling resins, cell lumens, and originating after oil generation	Exsudatinite
		Plant oils	Fluorinite
		Amorphous bitumen	Bituminite
Inertinite	High reflectance in white light	Highly oxidized wood	Fusinite
		Less-oxidized wood	Semifusinite
		Oxidized humic gel	Macrinite
		Oxidized fragments of inertinites	Inertodetrinite
		Fungal hyphae, spores, and other fungal remains	Funginite
		Oxidized resin	Secretinite

Source: Compiled from ICCP (1971), ICCP (1998), ICCP (2001), Stanton et al. (1989), and Sykorova et al. (2005).

content decreases with increasing maturation or coalification. Hydrogen content decreases with increasing maturation in the same coal type. In order to determine the variability of coal rank and maceral composition, the proportions or ratios of the atoms of hydrogen/carbon and oxygen/carbon were originally plotted in a diagram by Van Krevelen (1981) (Figure 4.6). During coal maturation, it initially loses oxygen and lastly hydrogen in relation to carbon (Box 4.1).

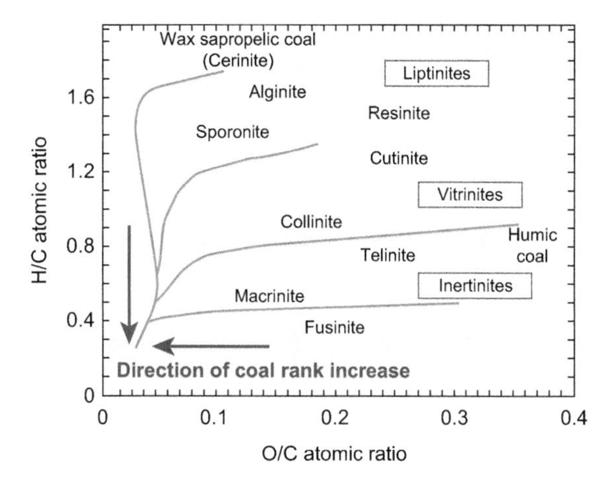

FIGURE 4.6 Diagram of coal macerals in relation to hydrogen/carbon and oxygen/carbon atomic ratios, coal types, and direction of coal rank increase. H, hydrogen; O, oxygen; C, carbon. *Source: Modified from Hunt (1991), Mukhopadhyay and Hatcher (1993), and Van Krevelen (1981).*

The atomic ratios of hydrogen/carbon and oxygen/carbon as shown by the van Krevelen diagram vary considerably between different coal maceral groups (e.g. liptinites, vitrinites, and inertinites) (Figure 4.6) and coal types (e.g. humic and sapropelic). Mukhopadhyay and Hatcher (1993) described the differences as the following: (1) The vitrinite group (e.g. collinite and tellinite macerals) is oxygen rich and hydrogen poor. (2) However, the vitrinite group is richer in hydrogen compared to the inertinite group (e.g. macrinite and fusinite macerals). (3) The liptinite group of macerals is enriched in hydrogen compared to the vitrinite and inertinite groups. (4) Nevertheless, alginite macerals are more hydrogen-rich than the resinite and sporinite macerals within the liptinite group (Figure 4.6).

The differences in the proportions of hydrogen, carbon, and oxygen using the van Krevelen diagram are described by Levine (1993) as the following. (1) In the humic and sapropelic coal types, carbon increases as hydrogen and oxygen decrease. (2) For both coal types, carbon initially amounts to 70 wt% and then increases to 100 wt% at the highest rank coal. (3) Various coal types start with different carbon, oxygen, and hydrogen wt%. (4) Humic coals start as oxygen rich and hydrogen poor. (5) In contrast, the sapropelic coals are initially hydrogen rich and oxygen poor. (6) The coal types become indistinctive toward the highest rank coals despite their initial differences in proportions of hydrogen, carbon, and oxygen.

Fundamental to the observations on carbon enrichment with increase of rank is the exclusion of organic matter that is hydrogen rich and oxygen rich during the entire processes of coalification. According to Levine (1993), the changes in elemental composition can be followed through a pathway of chemical reactions during the coalification processes. Associated with these reactions is devolatilization, which is the progressive loss of by organic matter undergoing coalification process. The pathway of reactions are an oversimplification of the general chemical evolution of coal and for details, which are beyond the limit of this book, the reader is referred to the classical seminal works of Juntgen and Karwell (1966), Kopp, Bennett, and Clark (2000), Levine (1993), Lewan (1992), Tissot and Welte (1984), and Van Krevelen (1981).

The pathway of coalification processes follows a sequential order of reactions of polymerization, depolymerization, cracking, and expulsion (Levine 1993). Polymerization is a process in which small molecules called monomers combine to produce a large network of molecules called a polymer. Depolymerization is a process that converts the polymers (macromolecules) into component monomers (smaller molecules). Cracking is a process that breaks up heavy molecules into lighter molecules. Expulsion is a process in which molecular products formed during coalification are expelled from the coal. Central to these chemical reactions are changes in molecular structures throughout the coalification process (Figure 4.8; Mukhopadhyay & Hatcher, 1993; Stach et al., 1982; Taylor et al., 1998).

BOX 4.1

COAL MACERALS ANALOGOUS TO MINERALS—CONCEPTS AND MISCONCEPTIONS

Contrary to popular beliefs in noncoal disciplines, "black coal" is not a homogeneous rock. For example, much like granite or sandstone, which is composed of quartz, feldspar, and mica minerals, coal is composed of analogous organic "macerals". However, a mineral is homogeneous crystalline (or grain for sandstone) characterized by a chemical composition in comparison to a maceral, which varies widely in chemical composition and physical properties. The concept of macerals was first introduced by Stopes (1935) to represent coal constituents studied by reflected light microscopy (e.g. reflectance, color, shape, and relief or polishing hardness) and isolated by maceration (to soften in Latin verb *macerare*).

Macerals are classified into three groups, vitrinite, liptinite (or exinite), and inertinite, based on their origin, morphology, structure, light reflectance, and chemical composition (Figure 4.7) (ICCP, 1998; ICCP, 2001; Mukhopadhyay & Hatcher, 1993; Stach et al., 1982; Sykorova et al., 2005; Taylor, Liu, & Shibaoka, 1983; Taylor et al., 1998; ASTM, 2011). The vitrinite group of macerals consists of vascular land plant remains such as cell walls, cell fillings, tissues, detritus, and amorphous gel (Figure 4.7; Table 4.2). Liptinite (exinite) group of macerals consists of diverse plant and animal remains derived from phytoplankton (algae), bacteria, cuticles, spore, pollen, resin, waxes, chlorophyll, and detritus (Figure 4.7; Table 4.2). The inertinite group of macerals consists mainly of charred remains formed from peat fires and bacterial surface oxidation such as cell walls, fungal cell walls, fish, and other animal relics (Figure 4.8; Table 4.2). ICCP (1998) defined and differentiated the vitrinite maceral group as gray color and reflectance that is generally between the darker liptinites and lighter inertinites over the rank range in which the three respective maceral groups can be readily recognized. Generally, inertinites are not potential sources of energy for combustion and hydrocarbon generation. The maceral groups consist of many individual macerals classified according to their vegetal matter origin (Table 4.2).

The ICCP (1998) document also presented a term huminite maceral group to designate a group of medium gray macerals with reflectances generally between those of the darker liptinites and the lighter inertinites. The huminite maceral was first introduced by Szadecky-Kardoss (1946) for brown coals commonly know as lignites. The introduction of the term huminite by ICCP (1998) provided a classification covering both low and high coals, which is useable as huminite and vitrinite for low-rank coals (Sykorova et al., 2005).

According to Levine (1993), the process of coalification in low- to medium-rank coal is characterized by the depolymerization reaction predominating over the polymerization reaction. Furthermore, the cracking reaction dominates during bituminization, whereas the expulsion process dominates during debituminization (see Figure 4.12). Bituminization is defined as a process of generating mobile by-products (e.g. thermogenic gas and oil) and producing aromatization and condensation of the solid by-products (e.g. coal and kerogen) (see Figure 4.12). However, the continuous decrease of the hydrogen/carbon and oxygen/carbon

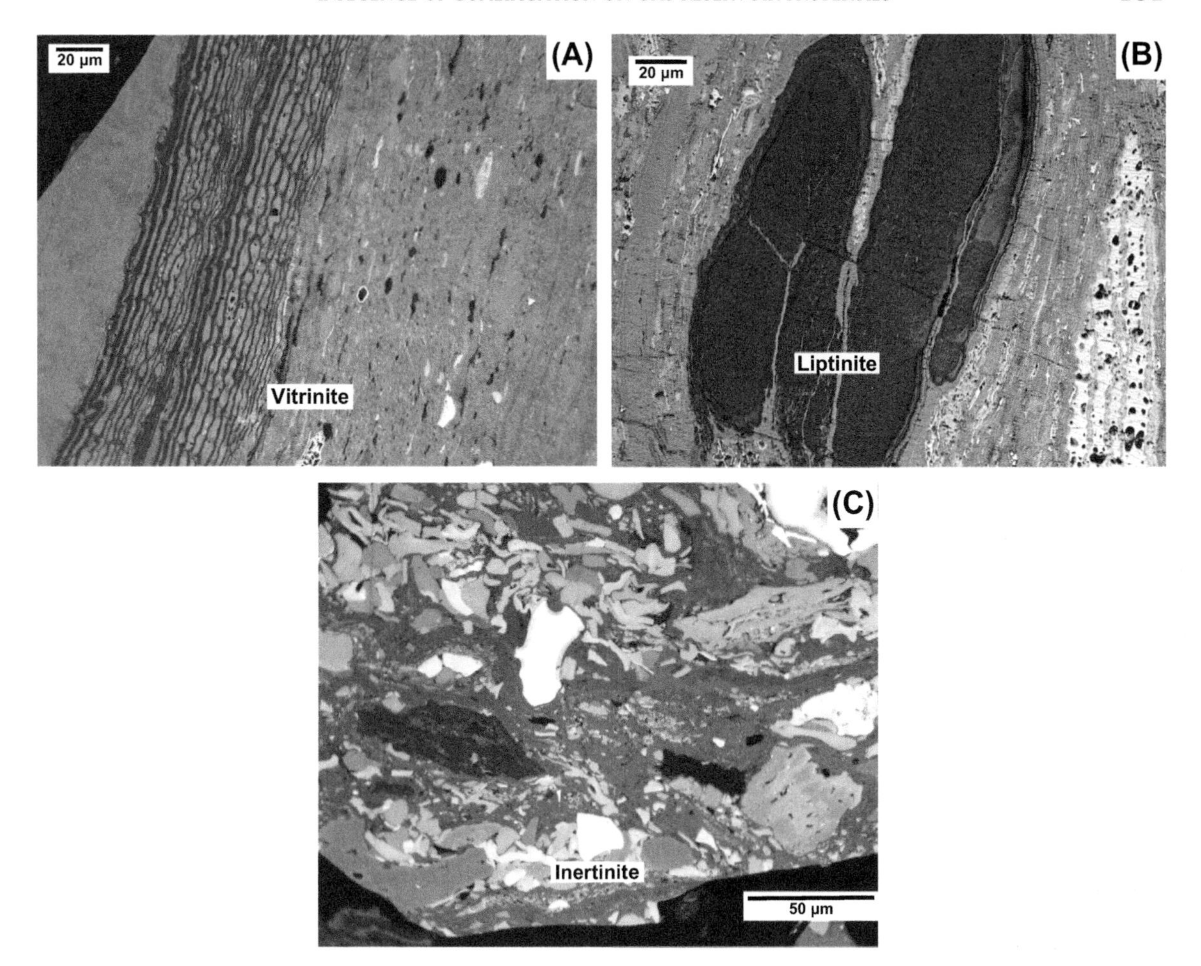

FIGURE 4.7 Photomicrographs of bituminous coal macerals: (A) vitrinite, (B) liptinite, and (C) inertinite. *Source: Photomicrographs by Tim Moore.*

ratios in residual coal suggests some expulsion of hydrocarbons throughout the coalification process. It is noteworthy that devolatilization reaction in which coals become enriched in carbon with increased coalification is not considered by Levine (1993) as representative of the general coalification processes because it falls short of accounting for the evolution, entrapment, and change in molecular structure of the carbon-enriched residue. However, devolatilization is an important factor in low- to medium-rank coal in which hydrogen-rich organic matter (e.g. liptinites) has the capacity to generate larger volumes of hydrogen-rich by-products.

Physical Process of Coalification

As in the chemical process of coalification, the physical processes are key to understanding its relationship to coal as a reservoir of hydrocarbons, mainly gas. During coalification, peat is altered physically and structurally, which occurs at the early stages caused by pressure preceding the chemical processes brought on by increase in temperature (Stach et al., 1982). In an attempt to

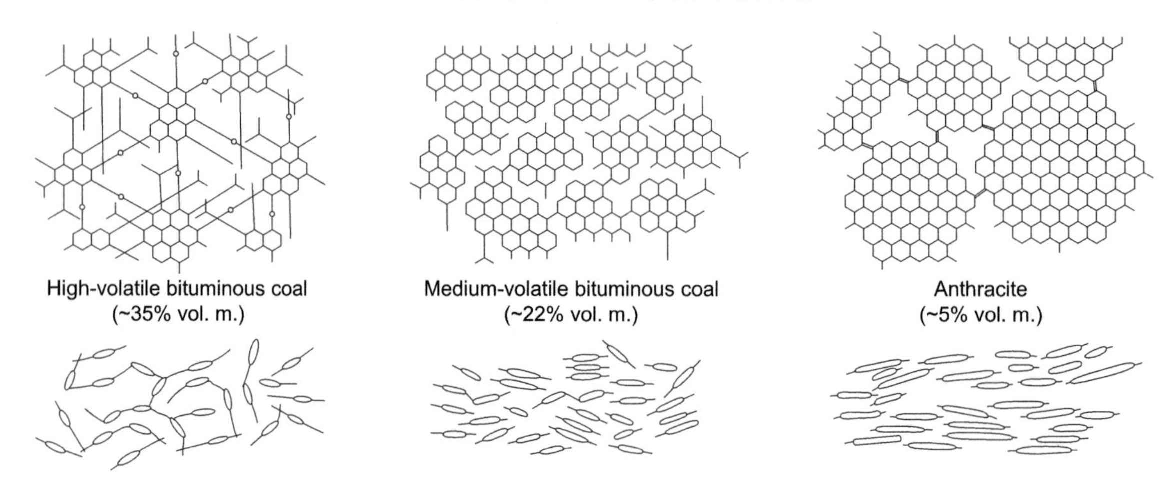

FIGURE 4.8 Simplified diagrams showing progressive changes in molecular (upper level) and physical (lower level) changes of vitrinite group macerals during various stages of coalification from bituminous to anthracite. Vol. m., volatile matter. *Source: Adopted from Stach et al. (1982).*

simulate changes that can occur during the early stages of coalification, Rollins, Cohen, Bailey, and Durig (1991) subjected a series of peat samples from different depositional and vegetational settings to increasing pressure and temperature. Compaction, microbands, color, destruction of cell walls in certain tissues, and formation of new macerals manifested purely physical changes during coalification (Shearer and Moore, 1996; Shearer et al., 1995). In addition, chemical changes were evident with destruction of the cellulose components in the absence of microbial activity. Similar compression experiments of peat by Cohen and Bailey (1997) duplicated creation of microbands and associated microcracks due to dewatering perpendicular to the microbands, which is relevant to the formation of permeability in coal.

Hatcher, Wenzel, and Faulon (1993) in a study of lignin (e.g. wood or precursor of vitrinite) during coalification determined modification of the chemistry and three-dimensional helical molecular structure. The random, large-scale molecular structural reorientations of lignin during coalification will likely destroy delicate physical attributes such as bordered pits and cellular walls. These physical attributes of wood persist through subbituminous coal, which suggests that the early coalification reactions are likely to be those which cause minimum distortion of the three-dimensional molecular structure. The most readily apparent reaction observed during the vitrinization process is the demethylation (or removal of methyl group from a molecule) of lignin. The vitrinitization of humic organic matter results in coals becoming black and lustrous. The microscopic study by Ting (1977) of transformation (e.g. diagenesis) from peat to lignite revealed detailed changes in anatomical structures of woody tissues and cells. A compaction of 4:1 from peat to lignite is accompanied by loss of volume due to loss of cellular cavities and intergranular spaces and the consequent loss or redistribution of moisture.

COAL RANK CLASSIFICATION SYSTEM

The classification system of coals by rank depends on the degree of coalification or progressive changes from the original peat precursor. However, coal rank classification is not a universal system because of the different causes of coalification, criteria on coalification changes, and uses or applications of coal as required by various countries. Classification of coal is keyed toward (1) the geological assessment of coal resources and coal mining and development; (2) producers, sellers, and buyers of coal and users; as well as (3) for industrial and technical purposes. Each has a different vision of coal but the classification system provides guidelines with regard to the coal quality as required and applied to respective applications.

Historical Perspectives

Attempts were made to establish coal classification system earlier but it was not until the nineteenth century that a formal system was introduced with the advent of using coal for steam boilers, coke ovens, and blast furnaces. The applications of coal as a fuel during the Industrial Revolution provided the need to codify coal according to its "heat value" instead of just simply classifying bright, brown, and black coals. This led to the introduction of terms such as soft and hard coal, lean and fat coal, and bituminous coal. In addition, the classification of coal was introduced based on the composition of the coal residue after combustion with terms such as caking and sand coals. Before the end of the nineteenth century, C.A. Seyler a noted chemist of Great Britain introduced a coal classification revised in 1924 (Seyler, 1924) based on the basic composition of coal but was not readily accepted in commercial and technical groups. This was superseded by a classification system designed by the Coal Board of Great Britain based on volatile matter and caking power of coal. In the early twentieth century, coal classification in both Europe and United States focused on the carbon residue, volatile matter, fixed carbon, and calorific value properties of coal.

It was obvious that every country used its own coal classification system during the early twentieth century. However, after World War II, with renewed international trade for fuel coal, a demand for a widely subscribed international system was established through the United Nations (United Nations, 1988a,b). The first international coal classification system of hard coals was published in 1956 by the Economic Commission for Europe of the United Nations (Geneva), which was revised to include brown coals in 1957. These international classification systems were designed for commercial and technological applications based on chemical parameters and/or physicochemical tests. In addition, both systems are expressed by numerical codes considered as codifications. Both systems are also, at present, totally out of date and inapplicable due to developments in coal technology and world commerce in solid fuels. Furthermore, both systems cited are based mainly on Northern Hemisphere coals and therefore do not correspond to all coals presently known and utilized worldwide, particularly regarding hard coals. For example, the vitrinite maceral group is the most abundant group, which makes as much as 90% in North American coals. In contrast, Australian, South American, African, and Indian Gondwana coals are generally vitrinite poor.

International Coal Classification

The United Nations introduced an international codification system for medium- and high-rank coals in 1987 recognized by the Economic Commission for Europe (United Nations, 1988a,b). Low-rank coal is defined as consisting of a gross caloric value (e.g. heating value) of less than 24 MJ/kg (moist, ash-free basis) and mean vitrinite reflectance as less than 0.6%. Higher rank coals (medium and high rank) are defined as consisting of gross calorific value equal to or more than 24 MJ/kg (moist, ash-free basis) and vitrinite reflectance equal to or more than 0.6%. The 1987 international coal classification, also expressed in numerical codes, replaced the 1956 system in order to meet renewed worldwide interest and increased trade of thermal and metallurgical coals. The original intent of the international classification system is to make it applicable to all coals of different genesis, rank, geological age, deposits, and from single to multiple beds of run-of-mine and washed coals. Furthermore, it is based on reflectance and reflectogram of vitrinite, maceral composition, volatile matter, ash and total sulfur contents, gross calorific or heating value, and free swelling index (suitability for production of coke or residue from distillation of low-ash, low-sulfur bituminous coal). The general hard and brown coal categories of the International Coal Classification (Figure 4.9) of the Economic Commission for Europe (United Nations, 1988a,b, 1998, 2004) have been recognized by the International Energy Agency (IEA, 2009).

Despite the United Nations' effort to internationalize coal rank classification, individualized systems by France, China, Australia, and New Zealand continued to be developed and keyed toward the coal properties of each country (Alpern, Lemos de Sousa, & Flores, 1989; Peng, 2000; da Silva, 1989; Suggate, 2006). For example, Alpern et al. (1989) developed an international

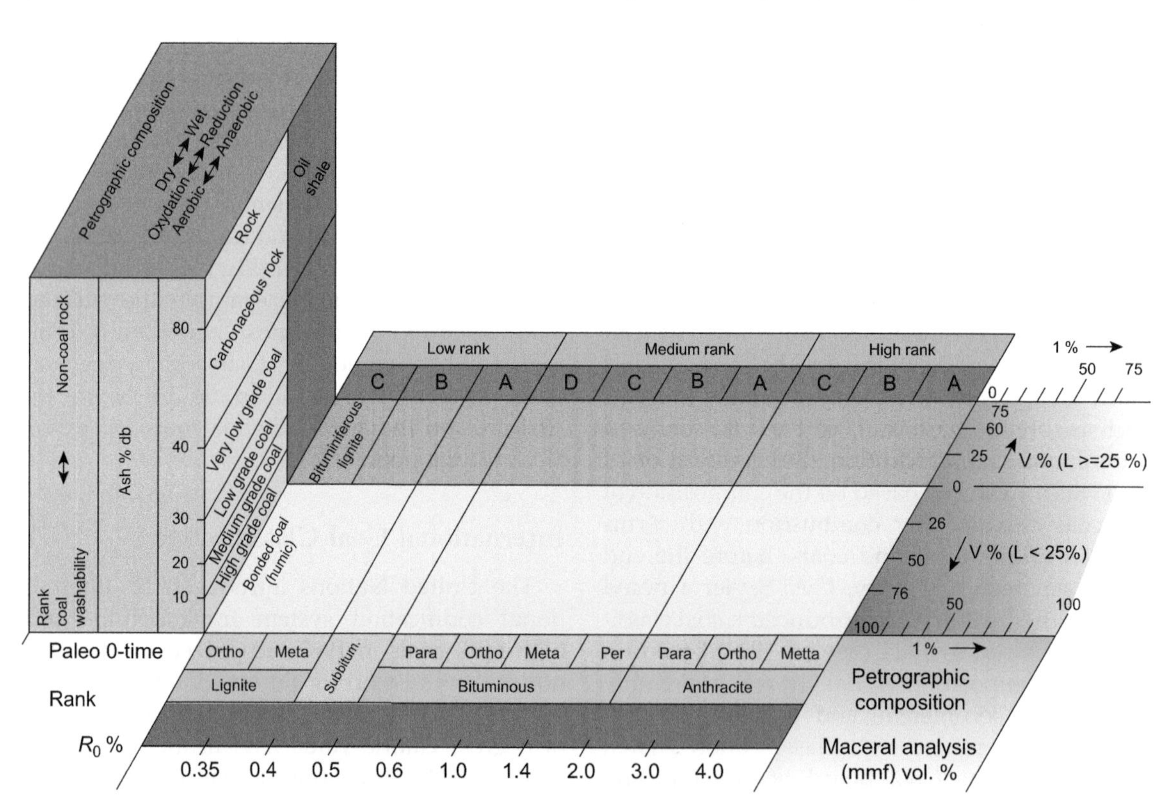

FIGURE 4.9 General version of the United Nations Economic Commission for Europe in-seam coal rank classification. Classification is based on rank (horizontal), petrographic composition (breadth), and grade (vertical). Rank is related to percentage vitrinite reflectance (%R_0) and gross calorific value. Composition is related to percentage vitrinite (V), liptinite (L), inertinite (I), and mineral matter free (mmf). Grade is related to ash content on dry basis (db) and washability test. For more detailed discussions, the reader is referred to United Nations (1998a,b) from which this figure is adopted. % = percent.

petrographic classification for France based on light reflectance of coal (e.g. vitrinite), maceral composition, and ash content as the rank parameters independent of trade and technical use. This system relied mainly on coal petrographic criteria and, in general, the International Committee for Coal and Organic Petrology (ICCP) adopted coal classification of four organic components (e.g. graphite, semi-graphite, natural coke, natural char) (Kwiecinska and Petersen, 2004), Economic Commission for Europe (UN/ECE), and Standards Association of Australia. As additional new data and computerization of data are collected, coal classification system continues to evolve.

The Alpern international coal classification was offered in contrast to the coal classification of American Society for Testing Materials (ASTM), which emphasizes volatile matter. Alpern et al. (1989) suggested that volatile matter is not a good rank parameter because of variability in liptinite and inertinite maceral groups. However, da Silva (1989) argued that the chemical and physical properties of Brazilian Permian Gondwana coals are different from those of Northern Hemisphere coals such that rank determination can be more meaningful to use different standard specifications from ASTM, Deustche Industrie Normen, and Alpern's universal classification. These specific standards are measurements of reflectance of vitrinites along with measurements of fluorescence of selected liptinites and coal extracts. Lessons learned from these differences are that worldwide coal rank classifications are widely variable depending on the coal characteristics and that these variabilities will reflect on the estimations of coalbed gas resources and reservoir properties.

American Society for Testing Materials (ASTM) Classification

The ASTM provides standards and guidelines to test and analyze coal to determine coal rank classification based on properties progressively altered depending on the degree of metamorphism (ASTM, 2002; 2011). The United States and Canada have followed the guidelines for coal rank classification developed by the American Standards Association (ASA) and ASTM since 1936. Through time, ASTM continues to improve and develop guides, practices, and test methods, which do not conform to the international coal rank codification system (Alpern et al., 1989; IEA, 2009; United Nations, 1988a,b, 2004).

The ASTM coal rank classification is based on a number of parameters obtained by various prescribed tests, which include calorific or heating value, volatile matter, moisture, ash, and fixed carbon (Figure 4.10). Calorific or heating value of coal is defined as heat energy released as it undergoes complete combustion with oxygen. Volatile matter is released gas when coal is heated in the absence of air at set conditions. The gases include carbon dioxide, volatile organic and inorganic gases containing sulfur, and nitrogen. The moisture represents water intrinsically contained in the coal, which is measured as the amount of water released when a coal sample is heated at set conditions. The moisture does not include free water on the surface of the coal, which is removed by air-drying the coal. Ash is an inorganic residue left over after a coal is completely combusted and is largely composed of silica, aluminum, iron, calcium, magnesium and other compounds. The ash yield may be different from the original clay, quartz, pyrites, and gypsum contents in the coal prior to combustion. Fixed carbon is a solid residue of organic matter after the release of volatile matter and moisture after complete combustion. Fixed carbon is composed mainly of carbon and minor amounts of hydrogen, nitrogen, and sulfur. It is calculated by subtracting from 100 the percentages of volatile matter, moisture, and ash. The ASTM coal rank classification system is presented in Table 4.3.

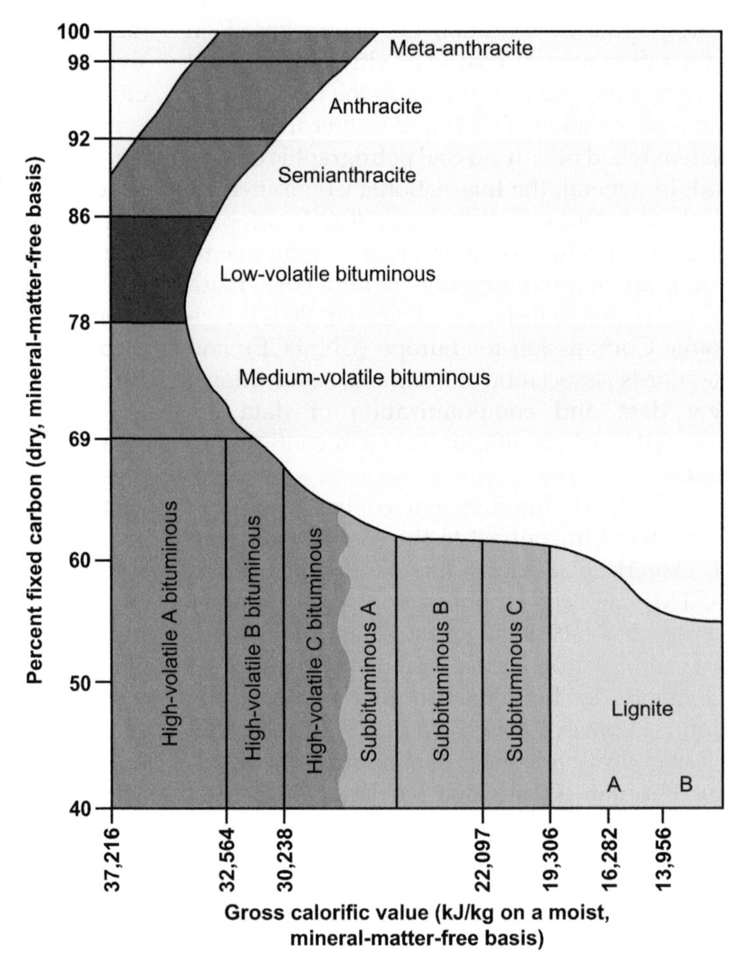

FIGURE 4.10 Diagram showing coal rank classification based on calorific value and percentage fixed carbon following ASTM International (2002) and recognized by United States and Canada. Values in parenthesis are the calorific values in kilojoules per kilogram (kJ/kg). *Source: Adopted from Schweinfurth (2002).*

EFFECTS OF MATURATION ON COAL PROPERTIES

The degree of changes undergone by a coal during various stages of maturation from peat to anthracite has important effects on physical and chemical properties of maceral groups and coal ranks. The most obvious changes in low-rank coals, such as lignite and subbituminous coals, are physical manifestation such as being softer, and friable with a dull and earthy appearance. In addition, low-rank coals are characterized by high moisture and low carbon contents, which translate to low energy production and limited use. Higher rank coals are mainly harder, stronger, and have black vitreous luster. In addition, the high-rank coals contain more carbon and lower moisture contents, thus producing more energy and having broad applications (Figure 4.11). The ultimate coal rank is anthracite, which has the highest carbon and energy contents and lowest moisture content but the most calorific value. In the United States, coals are classified according to the percentages of fixed carbon, moisture (water), volatile matter, and calorific value in British thermal units (or

TABLE 4.3 Coal Classification by the American Society for Testing Materials

rank and group	fixed carbon percentage (dry, mineral-matter-free basis) equal to or greater than	fixed carbon percentage less than	volatile matter percentage (dry, mineral-matter-free basis) greater than	volatile matter percentage equal to or less than	caloric value (noist, mineral-matter-free basis)* British thermal units per pound equal to or greater than	British thermal units per pound less than	megajoules per kilogram equal to or greater than	megajoules per kilogram less than	agglomerating character
Anthracitic									
meta-anthracite	98	...	...	2	...	...	...	...	
anthracite	92	98	2	8	...	...	...	...	nonagglomerating
semianthracite†	86	92	8	14	...	...	...	...	
Bituminous									
low-volatile bituminous	78	86	14	22	...	...	...	...	
medium-volatile bituminous	69	78	22	31	...	...	...	...	commonly
high-volatile A bituminous	...	69	31	...	14,000 ‡	...	32.6	...	agglomerating §
high-volatile B bituminous	...	...	...	...	13,000 ‡	14,000	30.2	32.6	
high-volatile C bituminous	...	...	...	...	11,500	13,000	26.7	30.2	
					10,500	11,500	24.4	26.7	agglomerating
Subbituminous									
subbituminous A	...	...	...	...	10,500	11,500	24.4	26.7	
subbituminous B	...	...	...	...	9,500	10,500	22.1	24.4	
subbituminous C	...	...	...	...	8,300	9,500	19.3	22.1	nonagglomerating
Lignitic									
lignite A	...	...	...	...	6,300	8,300	14.7	19.3	
lignite B	...	...	...	...	...	6,300	...	14.7	

* Moist coal contains natural inherent moisture but does not include visible water on the surface. † if agglomerating, classify in low-volatile group of the bituminous rank. ‡Coals having 69 percent or more fixed carbon on the dry, mineral-matter-free basis are classified by fixed carbon, regardless of calorific value. § There may be nonagglormerating varieties in these groups of the bituminous rank; there are also notable exceptions in the high-volatile C bituminous group.

Source: *2000 Annual Book o ASTM Standards*, section 5, volume 5.06.

Source: Adopted from ASTM (2011) and ASTM International (2002).

joules) after the sulfur and mineral matter contents are subtracted. Coals are classified into lignite, subbituminous, bituminous, and anthracite, each subdivided as shown in Figure 4.11 (Box 4.2).

Stages of Maturation: Coal vs Kerogen

The traditional concept of describing the progression of maturation or coalification is to include five stages from peatification to graphitization (Levine, 1993; Moore. 2012). However, in this book, peatification is described in detailed in Chapter 3 in association with the origin of peat formed on the surface. Maturation is considered in this book to begin upon burial by sediments in the subsurface. Thus, the stages of maturation processes (e.g. chemical and physical) of peat to coal include progression from (1) dehydration, (2) bituminization, (3) debituminization, to (4) graphitization (Figure 4.12).

Stages in Coal Maturation

Dehydration is the initial stage of maturation or coalification of peat upon burial, which is characterized by the following chemical and physical properties summarized from Kim, 1978; Levine (1993); Moore (2012), Stach et al. (1982), and Taylor et al. (1998). (1) Significant expulsion of water, carboxyl acids (e.g. COOH), and carbon dioxide. (2) Physically, the organic

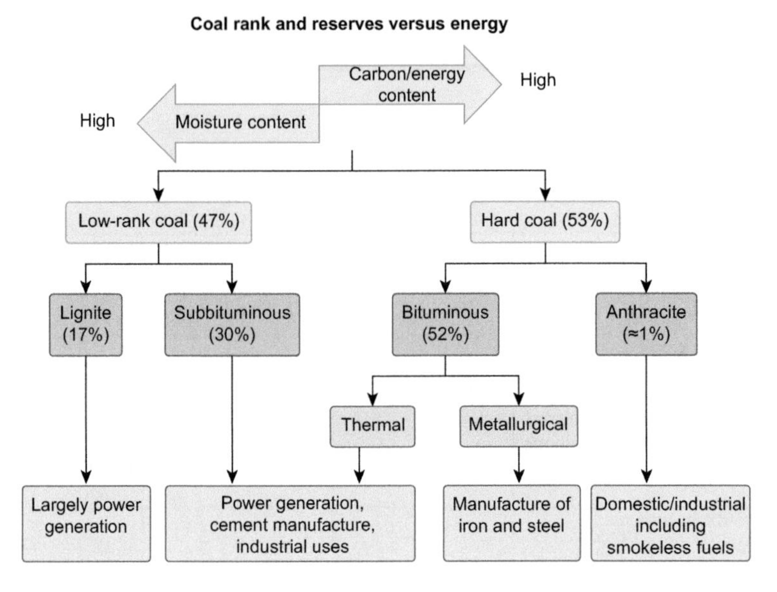

FIGURE 4.11 A flowchart showing relative moisture and carbon content with respect to coal rank, percentage of reserves, and energy use. *Source: Modified from World Coal Association (2012).*

BOX 4.2

METAPHORS IN LOW-RANK COAL CLASSIFICATION: FACTS VS FICTION

The first attempt to systematize coal classification internationally was for "hard coal" or high-rank coal in 1956 by the United Nations Economic Commission for Europe. This was followed by systematization of "brown" coal or low-rank coal by the same international organization. Both classifications were later adopted. In Europe, "brown" coal is usually recognized as opposite of "hard" coal. In many countries, the term "brown" coal is used in varied descriptions. Thus, terms such as low-rank coal, brown coal, and lignite are used interchangeably worldwide. These terms are basically identical, with overlapping definitions in different coal classification systems. Lignite is the US ASTM rank terminology, while Australian, International Organization for Standardization, and most European systems use 'brown" coal. Germany often correlates "braunkohle" to lignite.

The fiction is the implied "brown" color, which is a misnomer because the coal can vary from yellow to brown to black as coalification increases, the last of them particularly applied to subbituminous coal. In many countries, subbituminous coal is considered "brown" coal. The overlapping colors are resolved by using differences in hardness (e.g. soft and hard brown coal) and sheen (e.g. dull and lustrous brown coal). Thus, "soft" coal is at times applied to "brown coal". More importantly, the color and hardness properties are related to the moisture content, which varies from 50% to 60% and is lost readily on exposure to the atmosphere (e.g. during surface mining).

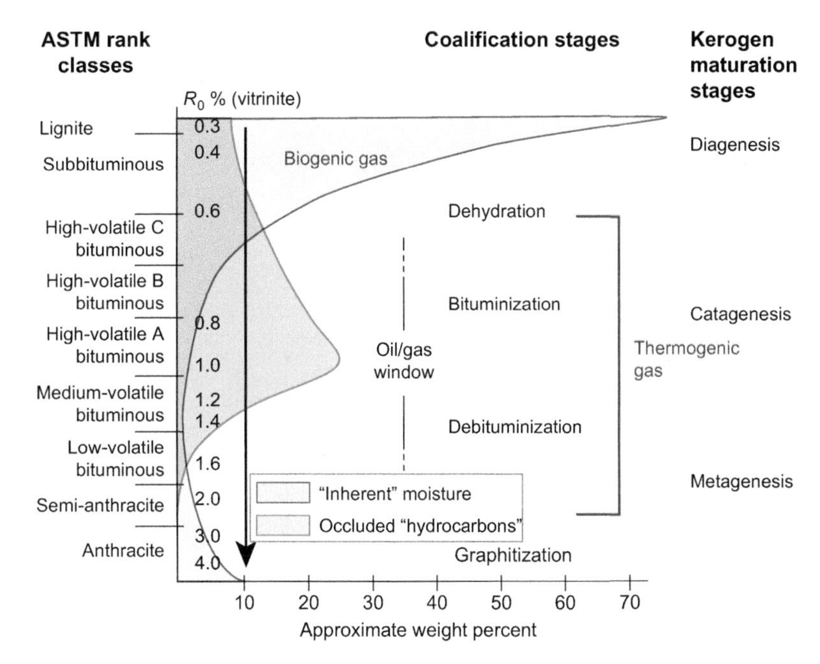

FIGURE 4.12 Chart showing moisture and hydrocarbon trends in relation to coal rank, coalification stages, kerogen maturation stages, vitrinite reflectance, and gasification. Hydrocarbons especially thermogenic gas develop mainly in bituminous coals. Subbituminous and lignite coals are high in moisture content and are prone to develop late-stage biogenic gas. *Source: Modified from Levine (1993).*

matter undergoes compaction and increase in density and heating value. (3) Expulsion of water results in: (a) orders of magnitude loss of porosity (b) organic matter entering into lignite through subbituminous, and (c) holding capacity of methane increasing exponentially. (4) Bed moisture in the low-rank coals varies from 35 to 75% for lignite and from 8 to 30% for subbituminous. (5) Bed moisture in high-rank coals varies from less than 8–22% at the lower end of the bituminous rank. (6) Inherent bed moisture may be critical to: (a) demethylation process, (b) generation of biogenic gas, and (c) adsorption and storage of methane especially in low-rank coals. (7) "Kim recipe" (see Chapter 3) indicates generation of gas in microorganism-inoculated "peat wood" from 95% to 30% methane at 35 °C to 150 °C, respectively. (8) Potential generation of methane during dehydration is made possible by the presence of bed moisture possibly inherited from the peat vegetal matter precursor that acts as hydrostatic pressure to hold the gas in the coal matrix. How much of this early-stage gas is retained through maturation is cause for more research.

Bituminization is the next stage of maturation as burial of peat (coal) continues and is characterized by the following chemical and physical properties summarized from Hunt (1991), Levine (1993), Moore (2012), and Wilkins and George (2002): (1) generation and entrapment of hydrocarbons; (2) increased hydrogen bonding, strength, vitrinite reflectance (R_0), extract yield, and fluorescence; (3) mobilization of the hydrocarbons, which enter the "oil window"; (4) generation of thermogenic gas (e.g. methane), which occurs any time and throughout the subbituminous A to high-volatile A bituminous rank; and (5) significant increases in maximum gas holding capacity.

Debituminization is the subsequent stage of maturation, which is distinguished by the following chemical and physical properties summarized from Levine (1993), Moore (2012), and Taylor et al., (1998): (1) expulsion of

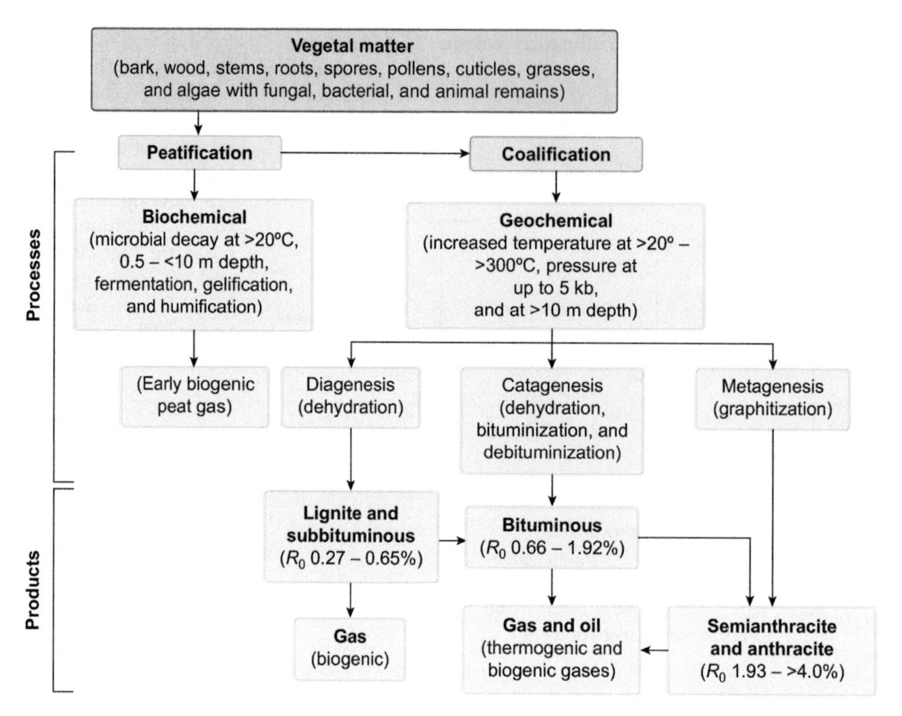

FIGURE 4.13 Flowchart showing processes and products during peatification and coalification. R_o, vitrinite reflectance; kb, kilobar (pressure).

low-molecular-weight hydrocarbons, (2) maximum generation of thermogenic gas (e.g. methane) and other oil substances, (3) decreased strength, (4) growth of cleats or fractures, (5) exiting of the oil window, (6) moisture values at their lowest point, and (7) reappearance of water.

Graphitization is the final stage of maturation and is characterized by the following chemical and physical properties summarized from Bustin, Ross, and Rouzaud (1995), Levine (1993), Moore (2012), and Taylor et al. (1998): (1) organic matter composed of almost all carbon with loss of hydrogen and nitrogen, (2) capacity to adsorb and hold gas continues to increase, (3) gas (e.g. methane) generation is virtually at its nadir, (4) remnants of gas are probably expelled, (5) moisture increases slightly, (6) aromatic lamellae (e.g. condensed, ordered carbon in flat plates) develop from the semianthracite to meta-anthracite rank, (7) increased strength, and (8) cleat (or fracture) healing.

The flowchart showing the processes and products of coalification of peat derived from plant remains is shown in Figure 4.13. The process of dehydration is important to the formation of lignite and subbituminous coals in which late-stage biogenic gas is the ultimate product. The processes of bituminization and debituminization are valuable to the formation of bituminous coals in which gas (e.g. late-stage thermogenic and biogenic gas) and oil are the final products. Finally, the process of graphitization is prominent in the formation of anthracite coals in which some thermogenic gas is potentially the last product.

Stages in Kerogen Maturation

The stages of coal maturation, in general, are parallel to the thermal maturation of petroleum deposits from kerogen precursors (Tissot & Welte, 1984) (Figures 4.12, and 4.14). The pathways of maturation and evolution of kerogen include (1) diagenesis, (2) catagenesis, and (3) metagenesis. Microscopic examination by Correia (1969) supported the progressive degradation and alteration through diagenesis, catagenesis, and metagenesis much like in the stages of coal maturation. For example, kerogen (insoluble mixture of organic chemical compounds and petroleum precursor) shows gradational alterations of spore and pollen from yellow to orange, red, and finally brown at increasing depths (Tissot & Welte, 1984). In addition, ultraviolet fluorescence of organic matter decreases at depth, which is the result of condensation (Levine, 1993).

Diagenesis is the first stage of kerogen degradation with the decrease of oxygen content and correlative increase of carbon content at increasing depth (Tissot & Welte, 1984). In terms of petroleum exploration, diagenesis corresponds to immature kerogen and only small amount of hydrocarbons are generated from the source rock. Catagenesis is the second stage of kerogen degradation with marked decrease of hydrogen content due to the generation and release of hydrocarbons with increasing depth (Tissot & Welte, 1984). In petroleum exploration, catagenesis corresponds to the zone of oil generation or "oil window". Also, catagenesis marks the beginning of cracking, which generates "wet gas" with increasing proportion of methane. Metagenesis is the third and final stage of kerogen alteration with the slow elimination of the residual kerogen consisting of two carbons or more out of three atoms (Tissot & Welte, 1984). In terms of petroleum exploration, metagenesis occurs mainly in the dry gas zone.

Kerogen maturation, sometimes referred to as thermal maturation, is measured by vitrinite reflectance, which increases gradually from 0.5% to 2%. Horsefield and Rullkotter (1994) determined that the values of vitrinite reflectance for the stages of kerogen maturation are diagenesis

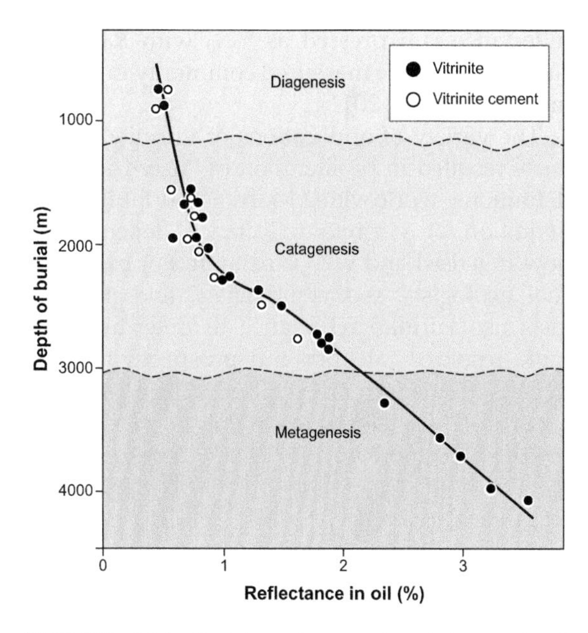

FIGURE 4.14 Vitrinite reflectance in relation to depth during the progress of maturation of Type III kerogen of the Upper Cretaceous Douala Basin in offshore Cameroon. m, meters; %, percent. *Source: Adopted from Tissot and Welte (1984).*

($R_0 < 0.5\%$), catagenesis ($0.5\% < R_0 < 2.0\%$), and metagenesis ($2.0\% < R_0 < 4.0\%$). This is demonstrated in Figure 4.14 showing vitrinite reflectance as a function of depth (Tissot & Welte, 1984).

ROLE OF VITRINITE REFLECTANCE

The maturation of coal and kerogen shares a common change in the physical and chemical properties of organic matter that is universally used as a "thermal maturity or rank" indicator to assist in coal, coalbed gas, and petroleum explorations. This universally accepted "thermal maturity" indicator is vitrinite reflectance and as the term implies, is a microscopic determination of the percentage of vertically incident reflected light on the maceral from a polished surface of coal pellets or from disseminated organic matter of sedimentary rocks. Thus, the vitrinite reflectance is expressed as $\%R_0$ with R_0 referring to reflectance measured commonly in oil immersion (ASTM, 2011).

The variety of applications of vitrinite reflectance resulted in proliferation of "user-friendly" definitions worldwide. More importantly, the definition of vitrinite reflectance depends on how it is used and who is using it. For example, coal geologists, sedimentologists, and petrologists use vitrinite reflectance to determine the rank, resource category and prospectivity, and depositional environment (e.g. nature of plant material) of particular coals (DeVanney, 2001; Diessel & Gammidge, 1998; Hower, Greb, Kuehn, & Eble, 2009; Quick & Tabet, 2003). Petroleum geologists and geochemists use vitrinite reflectance to determine the expulsion and kinetics of hydrocarbons as well as source rock evaluation of play appraisals of hydrocarbon deposits (Borrego et al., 2006; Dembicki, 2009; Jasper, Krooss, Flajs, Hartkopf-Fröder, & Littke, 2009). A sedimentary basin modeler uses vitrinite reflectance to determine the thermal maturity and history of sedimentary basins (Carr, 2003; Dewing & Obermajer, 2009; Higley, 2007; Mukhopadhyay & Dow, 1994; Naeser & McCulloh, 1989; Pahari, Singh, Prasad, & Singh, 2008).

Why use the reflectance of vitrinite and not other macerals of the liptinite and inertinite groups? The reflectance of the three macerals groups depends on chemical composition such as carbon content and "aromaticity" (e.g. class of compounds, such as benzene and toluene, with distinctive pleasant smells). According to Levine (1993), the carbon content and aromaticity of liptinite macerals are originally lower than those of vitrinite; however, this composition increases through debituminization along with a corresponding increase in reflectance. Furthermore, the reflectances of liptinite and vitrinite macerals converge at high ranks ($R_0 > 1.0\%$) and therefore it is very difficult to differentiate microscopically at higher ranks. The carbon content and "aromaticity" of inertinite macerals are originally high; however, the composition changes slightly during maturation (Levine, 1993). Thus, the reflectance of inertinite macerals is not reliable at higher thermal maturity indicator.

Despite significant differences in the carbon content and aromaticity of the three groups of macerals, it is possible to use their reflectance in low-rank coals (e.g. subbituminous) because these compositions are high. The chemical and petrographic properties of the subbituminous coals in the Powder River Basin, Wyoming and Montana, show that based on purely petrographic and relative reflectance categories, liptinite < huminite (vitrinite) < inertinite (Rich, 1980; Stanton, Moore, Warwick, Crowley, & Flores, 1989). The relative reflectance reflects different plant materials for huminite (e.g. cell walls), liptinite (e.g. waxes and resins), and inertinites (e.g. carbonized remains). The International Committee for Coal Petrology (ICCP, 1971, 1998; Stanton et al., 1989) suggested that reflectance of huminite maceral instead of vitrinite maceral be used in low-rank coals. Humic macerals are dominated by low-reflectance humic materials, which according to ICCP (1971) should be referred to as huminite macerals. The huminite maceral group becomes vitrinite maceral group at higher rank levels at a reflectance of 0.5%. However, the Australian Standard has combined the two materials within the vitrinite group.

Applications of Vitrinite Reflectance to Coalbed Gas

The importance of vitrinite (huminite) reflectance is its application particularly in coalbed gas exploration and development. Application of vitrinite reflectance relies on the concept that, in general, it increases gradually in lower rank coals and rapidly in higher rank coals (Levine, 1993). The increase of vitrinite reflectance is a result of a succession of physical and chemical changes. In lignite and subbituminous coals, the gradual increase of vitrinite reflectance is caused by compaction of molecular structure

TABLE 4.4 Conversion of Organic Matter at Depth Related to Increasing Vitrinite Reflectance (% R_0)

Occurrence (Depth, km)	Vitrinite Reflectance (% R_0)	Conversion of Organic Matter (%)
0.10	0.28	0.0
0.15	0.31	3.6
0.49	0.47	11.8
1.48	0.59	21.7
2.90	1.00	33.9
4.25	1.89	42.0
4.60	2.73	45.0

Source: Adopted from Barsky et al. (2011).

and increase in density. An experimental coalification of "brown" coal by Barsky, Vlasov, and Rudnitsky (2011) shows the mass loss corresponding to vitrinite reflectance at depth (Table 4.4). In bituminous coals, rapid increase of vitrinite reflectance is caused by replacement of water by aromatic hydrocarbons. In higher rank coals, increase of vitrinite reflectance is caused by condensation of the coal structure resulting in increased adsorption index (Levine, 1993).

Local and Regional Vitrinite Reflectance Variations

The most common application of the increase of vitrinite reflectance is mapping isoreflectance contours to depict local and regional variation of maturation. Mapping vitrinite isoreflectances delineates rank where coal classification is not available and may highlight occurrences of potential coalbed gas prospects. Also, it may draw attention to local anomalies, which may or may not favor coalbed gas accumulations at the same time determining the cause of coalification. Vitrinite reflectance may be used to predict potential coalbed gas content and to calculate gas volume. An example by Kalkreuth and Holz (2000) in San Terezinha coalfield, Brazil, estimated gas content based on vitrinite reflectance of 0.70% at 400 m and 0.89% at 950 m depths with coalification gradient of 0.04%/100 m, moisture content of 5 wt%, and ash yield of 50 wt%. Estimated coalbed gas content at 400 m depth is 3.19 m^3/t and at 950 m depth is 5.53 m^3/t.

An invaluable application of vitrinite reflectance is determining the burial history of a basin in relation to maturation of coals and petroleum source rocks. An example selected for this purpose is the Raton Basin where as previously discussed from work of Flores and Bader (1999), thermal maturation of the Cretaceous Vermejo Formation and Paleocene Raton Formation was locally and regionally elevated by igneous intrusions. Higley (2007) used petroleum system-based assessment to investigate the coalbed gas and related petroleum resources in the Raton Basin in Colorado and New Mexico, USA. However, because the vitrinite reflectance is applied to coalbed gas, the results from Higley (2007) focused on the Cretaceous and Paleocene coal beds. The vitrinite reflectance data of the Vermejo and Raton Formations are summarized in Figure 4.3, which vary from 0.57% along the eastern, northern, and southern parts to 1.58% in the central part of the Raton Basin (Higley, 2007; Johnson & Finn, 2001). The high vitrinite reflectance in the south-central part of the basin may be due to laccoliths in the Vermejo Park dome (Flores & Bader, 1999). In general, the areal distribution of the vitrinite isoreflectance is in agreement with the structural configuration of the basin.

Modeling of the burial history (Figure 4.15) based on vitrinite reflectance indicates that the Vermejo and Raton coals are thermally mature, with the highest maturity in the south-central and deepest part of the basin. The burial history was adjusted to the R_0 data using (1) 1500 m of erosion (exhumation after the Laramide orogeny) from 40 mya to present and (2) a heat

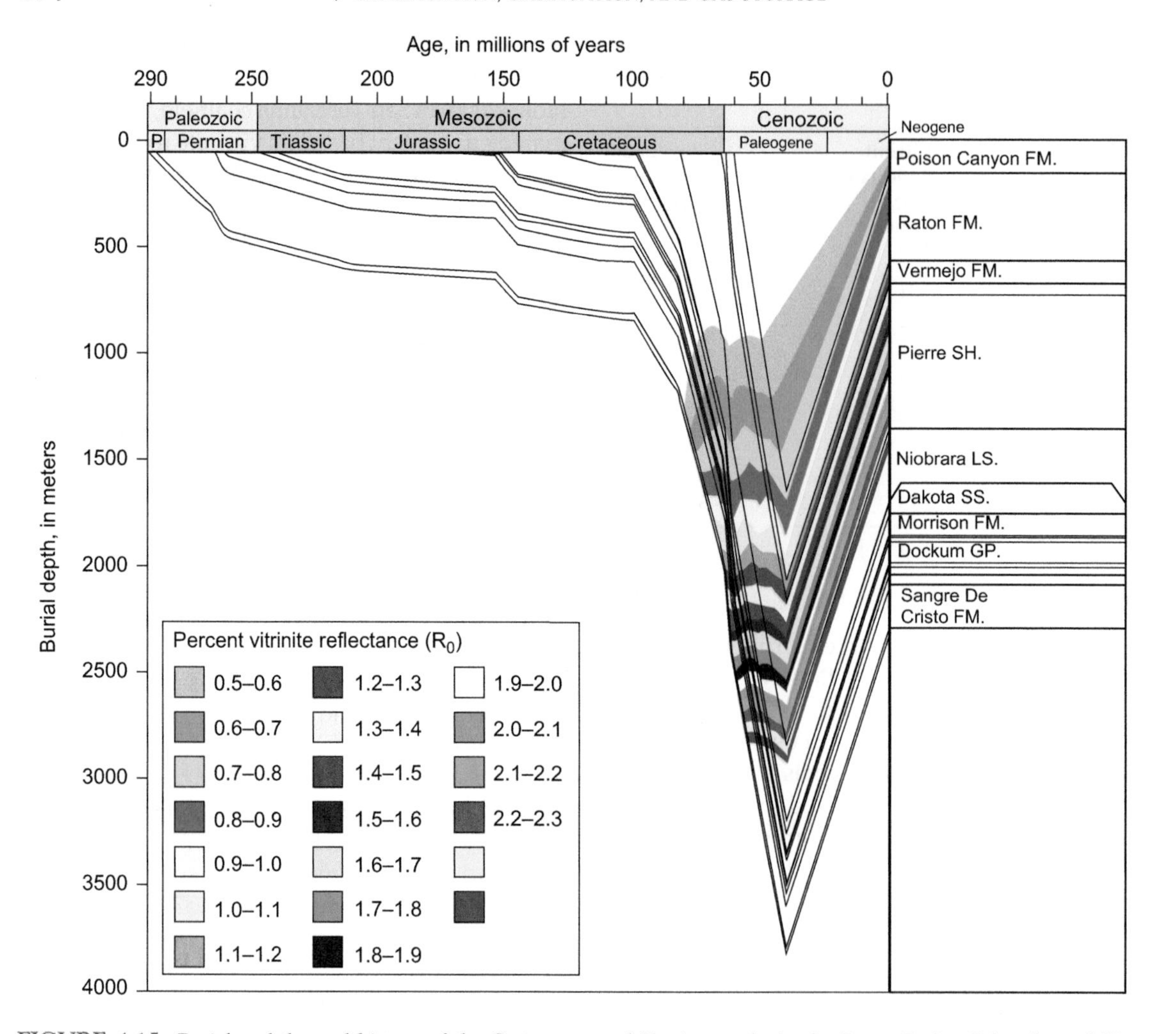

FIGURE 4.15 Burial and thermal history of the Cretaceous and Tertiary rocks in the Raton Basin, Colorado and New Mexico, United States. Vitrinite reflectance of the Upper Cretaceous Vermejo and Paleocene Raton coals ranges from 0.6% to 1.2%. Thermogenic coalbed gas generation window was interpreted by Higley (2007) at about 55 mya. FM = Formation, SS = Sandstone, SH = Shale, LS = Limestone, GP = Group. *Source: Adopted from Higley (2007).*

flow of 55 mW/m^2 from 290 mya to 65 mya, at which time it changed to 121 mW/m^2. The coalbed gas is thermogenic in origin with average $\delta^{13}C$ values of −44.8% (Carlton, 2006). Higley (2007) indicated that based on the thermal burial history, the onset of gas generation from coals of the Raton and Vermejo Formations started about 50 mya and kerogen from the coals is 45–75% transformed to gas with the potential of generating more thermogenic gas in case of deeper burial.

The influence of sills and dikes on vitrinite reflectance is very local but drops off to background at approximately one intrusion width away (Cooper, 2006; Cooper, Crelling, Rimmer, & Whittington, 2007). The mean vitrinite reflectance of coals associated with dikes varies from 0.95% to 7.1% and those associated with sills from 0.74% to 6.93%. These workers noted that the effect of the localized intrusions to methane generation within the contact metamorphic zone is significant as indicated by isotopic (e.g. $\delta^{13}C$) shifts (Figure 4.16). Volatiles generated during contact metamorphism with coal include methane and carbon dioxide. According to Cooper (2006) and Cooper et al. (2007), most of the carbon dioxide may have been generated during burial maturation before emplacements of igneous intrusions, while the methane was generated during the sill/dike intrusions. Sills are more important on a regional scale for generating methane than dikes. In addition, vitrinite reflectance may be locally elevated in structurally deformed areas such as faults and folds.

Limitations of Vitrinite Reflectance

Determination of thermal maturity of coal beds and other source rocks may be compromised by anomalously low vitrinite reflectance or

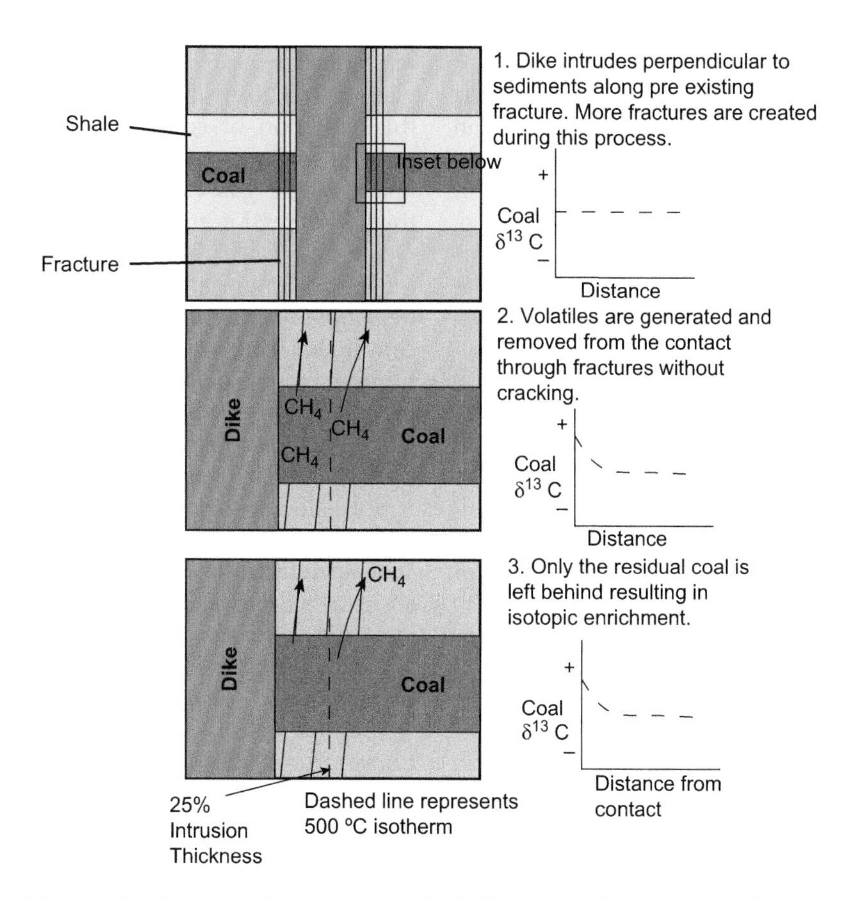

FIGURE 4.16 Diagram showing generation and escape of volatiles (e.g. methane or CH_4 and pyrolytic carbon) during contact metamorphism with dike in the Raton Basin in Colorado and New Mexico, United States. Steps 1–3 show progressive enrichment of carbon isotope of methane ($\delta^{13}C$) in residual coal. That is, during contact metamorphism, volatiles will have higher (heavier) $\delta^{13}C$ at the contact zone and lower (lighter) $\delta^{13}C$ farther away from the zone. *Source: Adopted from Cooper (2006).*

suppressed compared to a "typical" vitrinite that has undergone similar coalification processes (e.g. temperature, pressure). Suppression of vitrinite reflectance may be from 0.1% to 0.4% depending on the rank and maturation conditions (Levine, 1993). However, vitrinite reflectance values can also be suppressed and caused by prevalence of perhydrous vitrinites and lack of orthohydrous vitrinites. The most serious impact of vitrinite reflectance suppression is the underestimation of heat flow, thus the maturity of source rocks lead to errors in predictions of transformations and quantity of hydrocarbons (Samuelsson & Middleton, 1998). Suppressed vitrinite reflectance is frequently caused by organic petrographers having problems in finding and measuring appropriate vitrinite materials in hydrogen-rich kerogens, extracting organic matter for analysis and analytical procedure (Barker, Lewan, & Pawlewicz, 2007; Bostick, 2011; Lo, 1993). Vitrinite reflectance suppression is commonly developed in many sedimentary basins and may happen more frequently than recognized.

Carr (2000) defined vitrinite reflectance suppression as the reduction in reflectance resulting mainly from the maceral composition of the source rock. Thus, suppression of vitrinite reflectance is due to (1) high liptinites associated with vitrinite, (2) occurrence of litptinite-derived aliphatic lipids and bitumens, (3) production of perhydrous (hydrogen/aliphatic-rich) vitrinite in anaerobic and/or alkaline depositional environments, and (4) presence of hydrogen- and aliphatic-rich vitrinite originating from a particular flora. Vitrinite reflectance suppression occurs on deposition of sediments and reflectance is reduced through burial until the hydrogen/bitumen is removed from the structure. In contrast, Carr (2000) defined vitrinite reflectance retardation, by contrast, as thermochemical reduction resulting from the effects of overpressure in a sedimentary basin. Suppression occurs at depth in a basin due to the development of a pressure seal in contrast to retardation, which continues to occur as long as the overpressure remains. The presence of overpressured sediments in many sedimentary basins means that vitrinite reflectance retardation may occur more frequently than is commonly recognized. How does guesswork on vitrinite reflectance suppression affect coalbed gas exploration? For example, vitrinite reflectance of the Upper Cretaceous Ferron coals in Utah was found to be higher in the north than in south of the coalfield (Quick & Tabet, 2003). These vitrinite reflectance variations were determined inconsistent to the coalfield-wide coal rank. In the Ferron coalbed, gas fairway indicates that coals in the north are of higher rank than coals in the south. Measured vitrinite reflectance does not accurately show this variation of coal rank. In addition, suppressed vitrinite reflectance is observed in the north where gas contents are relatively high; thus, suppression may affect coalbed gas yield. Quick and Tabet (2003) related the suppressed reflectance to burial history where overpressure developed during the coalification stage, which continued until recent uplift and cooling. This finding may provide a model for developing prospective coalbed methane targets elsewhere.

What does it take to get the conjecture out of the analysis of vitrinite reflectance to solve the suppression problem? Newman (1997) and Newman, Eckersley, Francis, and Moore (2000) suggested the following techniques: (1) diagnose and correct vitrinite reflectance suppression using bulk chemistry for coals and (2) petrologic approach for dispersed organic matter in sedimentary rocks. These techniques are developed to combine reflectance with quantitative vitrinite fluorescence, whose plot and graphical relationships allow objective differentiation of Types III and IV organic matter into suppressed vitrinite, normal vitrinite, recycled vitrinite, and inertinite. Newman (1997) suggested that measurements should be made in a nitrogen atmosphere to avoid fluorescence alteration. Other detection and replacement methods for vitrinite reflectance were proposed by Fedor and Vido (2003), Lis,

Mastalerz, Schimmelmann, Lewan, and Stankiewicz (2005), Petersen et al. (2009), Wilkins, Wilmshurst, Hladky, Ellacott, and Buckingham (1995), and Zhong, Sherwood, and Wilkins (2000).

TYPES OF GAS GENERATION DURING AND POST COALIFICATION (MATURATION)

Large amounts of coalbed gas (e.g. methane) are generated from coals during the latter stages of coalification (e.g. bituminization and debituminization). Early laboratory and numerical studies of predicted gas yields of high-rank coal estimated coalbed gas (e.g. methane) varying from 100 to 300 cc/g of coal (Hunt, 1979; Juntgen & Karwell, 1966; Juntgen & Klein, 1975; Karweil, 1969). These values were revised to 150–200 cc/g of coal depending on the elemental and organic matter composition and rank (Rice, 1993). For example, the liptinite maceral group is mainly hydrogen rich and is capable of generating oil. In contrast, the vitrinite maceral group is mainly oxygen rich (e.g. Type III kerogen) with some hydrogen-rich macerals capable of generating wet gas. Hydrocarbons in coalbed gas are derived from thermal breakdown of kerogens (e.g. thermogenesis) and bacterial generation through fermentation and carbon dioxide reduction (e.g. methanogenesis) in coal. Thus, based on the origin, coalbed gas is primarily classified into thermogenic and biogenic gases. In a ground-breaking paper on coalbed gas in the San Juan Basin, Scott, Kaiser, and Ayers (1994) subdivided these gases to include: (1) early and migrated thermogenic gas, (2) primary and secondary biogenic gas, and (3) mixed thermogenic and biogenic gases.

Thermogenic Gas

Thermogenic coalbed gas may be defined according to the coal rank, maturation (e.g. vitrinite reflectance), and temperature of generation. These parameters, however, vary from coal to coal, stratigraphic section to section, basin to basin, and country to country (Alsaab et al., 2008; Lo, Wilkins, Ellacott, & Buckingham, 1997). Therefore, it is best to define thermogenic coalbed gas on the basis of one parameter at a time. Also, defining thermogenic gas based on a new parameter in comparison with another using a different parameter is not advisable. For example, a geochemist described thermogenic gas and onset of thermogenesis at coal rank approximately equivalent to vitrinite reflectance ($\%R_0$) of about 0.6 (high-volatile bituminous C and at about 110 °C) (Clayton, 1998). To a coal geologist using ASTM, coal rank classification of high-volatile bituminous C is defined purely based on high carbonaceous matter (e.g. >60% fixed carbon on dry, mineral matter-free basis), more than 14% volatile matter on dry, mineral matter-free basis, and calorific value of more than 24,423 kJ/kg (10,500 Btu/lb) on a moist, mineral matter-free basis (Figure 4.10). Regardless of the rigid premise of the onset of thermogenic gas, it is believed that generation will have variable timing, in relation to coal rank, depending on organic composition and specific basin and burial history (Moore, 2012).

Origin

The key to the origin of thermogenic gas is temperature in combination with geologic time according to Levine (1993). A review of the literature on early experimental studies to simulate gas (e.g. methane) yields in coal complimented by numerical modeling and the timing of gasification during coalification, which are the cruxes to the origin of thermogenic gas, point to a continuing debate (Behar, Lewan, Lorant, & Vandenbroucke, 2003; Clayton, 1998; Juntgen & Karwell, 1966; Kotarba, 2004; Kotarba & Lewan, 2004; Levine, 1993; Lewan, 2010; Su, Shen, Chang, & Huang, 2006). The debate centers around the quantification of hydrocarbons especially gas from coals of varying maceral compositions and ages during maturation or coalification. The assumption that most coals singularly generate gas has been proved to be incorrect by the assumption that liptinite-rich Tertiary coals are more oil prone than gas prone (see Chapter 2). Microscopic studies indicate that oil generated by the coal permeates from the liptinite source to pore systems where it is stored (Alsaab et al., 2008; Erdmann & Horsefield, 2006; Levine, 1993). The trapped oil is transformed into gas (e.g. methane) as coalification proceeds and coal matures under increasing thermal conditions. It is suggested that gasification begins relatively late, where the trapped oil in the pores of the coal matrix is subsequently transformed by continued cracking generating lighter hydrocarbons (Ritter & Grover, 2005). The thermal cracking of the entrapped oil in the coal matrix at high temperatures is defined as secondary thermogenic coalbed gas in contrast to primary thermogenic gas, which results from thermal cracking from sedimentary organic matter into liquid and gaseous hydrocarbons. Thus, the debate about the timing of gas generation and coalification stage(s) as well as were this process formed during bituminization, debituminization, or before continues. Figure 4.12 shows that much of the thermogenic gas is generated during the bituminization and debituminization stages. Moore (2012) suggested that gasification continues as maturation exits the bituminization stage.

Additional debate persists regarding mass balance of expulsion of hydrocarbon byproducts during maturation (Behar, et al., 2003; Clayton, 1998; Kotarba & Lewan, 2004; Levine, 1993; Su et al., 2006; Levine, 1987; 1993). These authors postulated that in determining the methane-generation curves (e.g. Juntgen & Karwell, 1966) it was assumed that the entire coal portion present as volatile matter in low-rank coal was finally expelled as volatiles in high-rank coals reflecting a negative balance for water. Levine (1993) recalculated coalbed gas yields using mass balance approach and the model of Juntgen and Karwell (1966) from lignite to anthracite ranging from 116 to 280 l/kg, respectively. These estimates differed from the calculations of Juntgen and Karwell (1966) of actual coalbed gas yield in nature of 120–150 l/kg.

Clayton (1998) reviewed the literature and compiled data on theoretical coalbed gas yields using laboratory-heating experiment to simulate generation and expulsion of coalbed gas. Figure 4.17 summarizes the results of the laboratory-generated gases and for some exceptions the methane yields are significantly less than quantifications from early well-established models (Juntgen & Klein, 1975; Karweil, 1969). For example, at an experimental temperature of 400 °C, which represents vitrinite reflectance of about 1.8–2.0% (equivalent to low-volatile bituminous and semianthracite rank), about 122 l/kg of methane would be generated from most quantification models. However, no experiment, except a few, generated >150 l/kg at about 600 °C and majority of the experiments yielded about 50 l/kg. According to Clayton (1998), the variations in the experimental gas yields, in addition to the mass balance problem (Levine, 1993), are the results of (1) differences in types of organic matter (Types I, II, and III); (2) retention of gas in the pore spaces instead of expulsion as free gas; (3) high total organic carbon and gas-to-oil ratio; (4) complications of coal generating wetter and liquid hydrocarbons; (5) complications from sample collections and laboratory procedures; (6) differences in maceral composition, organic matter particle size, and rank; (7) laboratory experiments running at higher temperature than coal reservoir desorption; and (8) conditions (higher water solubility versus hydrocarbons and natural versus closed systems) in the experiments.

More recent studies, which weighed in on the controversy, proposed additional innovative ideas to assist in the theoretical laboratory experiments to estimate gas yields from coal.

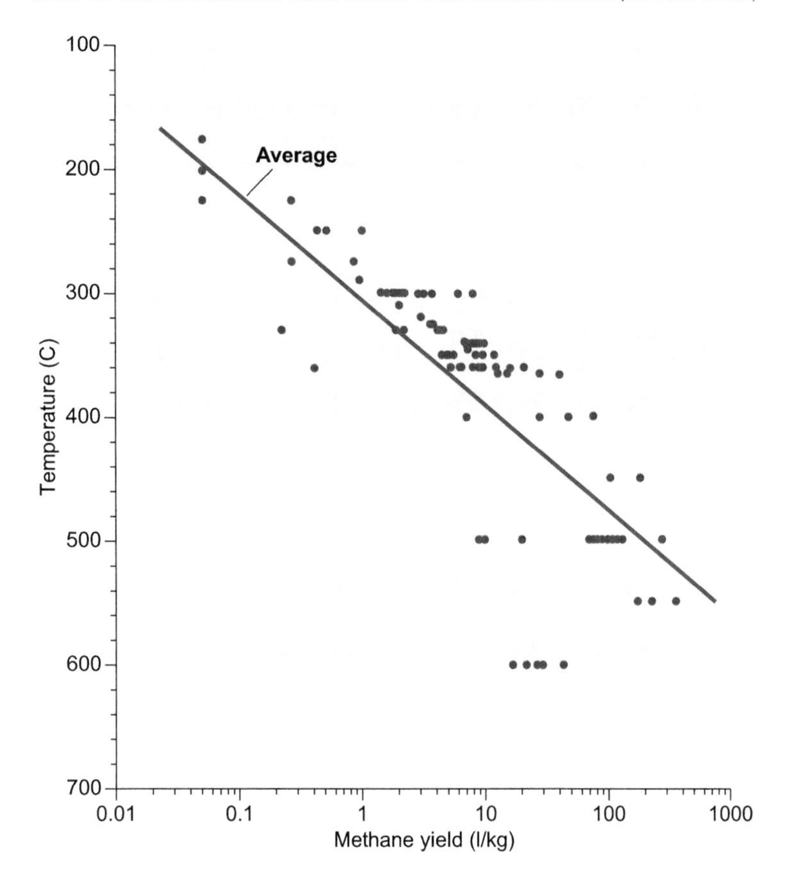

FIGURE 4.17 Diagram showing relationship of methane yield (l/kg) with temperature for laboratory pyrolysis experiments on lignites and other coals under anhydrous and hydrous conditions at changeable times. l/kg, liter per kilogram. *Source: Modified from Clayton (1998).*

According to Butala, Medina, Taylor, Bartholomew, and Lee (2000), the prediction of volumes of gas through laboratory-heating experiments underestimate the gas yields derived from organic matter because kinetic models are taken in mathematical isolation. Zhang, Hill, Katz, and Tang (2008) proposed that kinetic and/or laboratory predictive models must be coal-basin specific. Their study in the Piceance Basin in Colorado, United States, shows that with increased vitrinite reflectance, carbon dioxide generation decreased substantially corresponding with increased methane generation. Perhaps, the most interesting concept was proposed by Tang, Jenden, Nigrini, and Teerman (1996), who used Paleocene lignite from North Dakota in their pyrolysis experiments to determine methane yields in low-rank coals. Their studies were conducted because most published studies have provided insufficiently detailed estimates of gas contents in lignite, subbituminous coal, and high-volatile B bituminous coal. Their study shows that in most cases methane concentration in these lower rank coals is attributed mainly to migrated gas. The idea of relating the theoretical coalbed gas yields to lower rank coals was further reinforced by Duan et al. (2011), who suggested that in order to understand the isotopic fractionation of coalbed methane during staged accumulation, it is necessary to use peat in pyrolysis experiments. Their study demonstrated that the higher the initial thermal maturity of the starting peat, the heavier the carbon and hydrogen isotopic compositions of methane (C_1) and ethane (C_2). Also, the carbon isotope composition of methane differed considerably depending on the original thermal maturity of the peat in contrast to hydrogen isotope composition, which did not differ as much. These observations compliment results of a previous study of Duan, Wu, and Zheng (2005) in which peat and coal were used and compared for pyrolysis experiments. Results of their study demonstrate that coals generated mainly hydrocarbons gases and peat generated hydrocarbon gases, which are lower than the nonhydrocarbon components. However, in some cases, peat has higher regenerative potential of coalbed gases than coals depending on organic matter content. In the hydrocarbon gases, methane is predominant and heavy hydrocarbons (C_2–C_4) are present in small amounts.

Process

Kotarba and Lewan (2004) and Su et al. (2006) found that amounts of thermogenic gases produced by nonhydrous pyrolysis experiments of Polish lignites and Miocene Chinese coals are 1.29 and 1.6 times greater than gases produced by hydrous pyrolysis, respectively. Kotarba (2004) provided a detailed characterization of origin and volume of thermogenic gas generation from Polish coals. The process of thermogenesis (at 360 °C) is simulated by hydrous pyrolysis experiments on coals from lignite (0.3% R_0) to semianthracite (2.0% R_0) (Kotarba, 2004). Under thermogenesis, the lignite reached a medium-volatile bituminous rank (1.5% R_0). A high-volatile bituminous coal achieved a low-volatile bituminous rank (1.7% R_0). A semianthracite reached an anthracite rank (4.0% R_0). Estimated thermogenic methane yields vary from 20 dm^3/kg for lignite to 0.35 dm^3/kg for the semianthracite. At a vitrinite reflectance of 1.7% R_0, about 75% of the maximum potential for the coal to generate thermogenic methane has been exhausted. At a vitrinite reflectance of 1.7% R_0, more than 90% of the maximum potential for a coal to generate carbon dioxide has been exhausted. Lewan (2010) conducted a series of experiments that involved hydrous pyrolysis of lignite to anthracite coals from Poland, Germany, Ukraine, and the United States. The experiments were conducted at 360 °C for 72 h to determine their gas-generation potential with results indicating that about 80% of methane generation is realized by a vitrinite reflectance of 1.5% R_0. Lewan (2010) concluded that the gas potential remaining after thermal maturities at 2.0% R_0 is much lower than proposed and Type III kerogen in coals is no longer a significant source of gas in shallow parts of a sedimentary basin.

Potentially, the carbon dioxide generated during thermogenesis is retained in the coal source rock. As the sedimentary rocks associated with the coal beds are exhumed through uplift and erosion conditions, such as establishment of a groundwater system, they are favorable to methanogenesis by microbes in the coals. The resulting methanogenic pathway for generation of biogenic gas in the coal reservoirs is carbon dioxide reduction. According to Kotarba (2004), the volume of biogenic gas (e.g. methane) generated by carbon dioxide reduction can exceed the quantities of thermogenic methane generated from the same coal bed by a factor of two to five. The large amounts of generated carbon dioxide can be used quantitatively to interpret mixing of indigenous thermogenic gas with microbial methane or exogenous thermogenic gas from other source rocks. The biogenic gas generated during this process is interpreted here as equivalent to the primary biogenic gas of Scott et al. (1994) because the carbon dioxide is generated during coalification, even though methanogenesis occurs after exhumation. Also, this is in contrast to the secondary biogenic gas generated after exhumation and through methanogenesis by fermentation, which formed along basin margins in contrast to the biogenic gas from carbon dioxide reduction in basin centers (Flores, Rice, Stricker, Warden, & Ellis, 2008) (see section below for details). Hydrous pyrolysis of a coal, regardless of the rank, provides a diagnostic $\delta^{13}C$ value for its thermogenic hydrocarbon gases and is discussed in the following section.

Isotopic Composition

The concept of isotopic composition is important for the following reasons: (1) determination of the origin of the coalbed gas, (2) employment of strategies in exploration and development of coalbed gas, and (3) deployment of appropriate management plan for coproduced water with production of coalbed gas. Coalbed gases can be differentiated according to the source of their carbon and hydrogen isotopic composition (Figure 4.18). Coal bed thermogenic gas, which is either a dry (almost pure methane) or wet (components of ethane, propane, butane, and condensate) gas, contains isotopically heavier carbon compared to gas generated at a lower thermal maturity and by microbial activity. The isotopic composition of coalbed gases differs extensively based on (1) thermal maturation or temperature of coal formation, (2) composition of the source organic matter, (3) Eh and pH conditions determining microbiological activity, and (4) secondary processes (e.g. diffusion, migration, adsorption, desorption, and mixing of gases of different origin) (Kotarba, 2001).

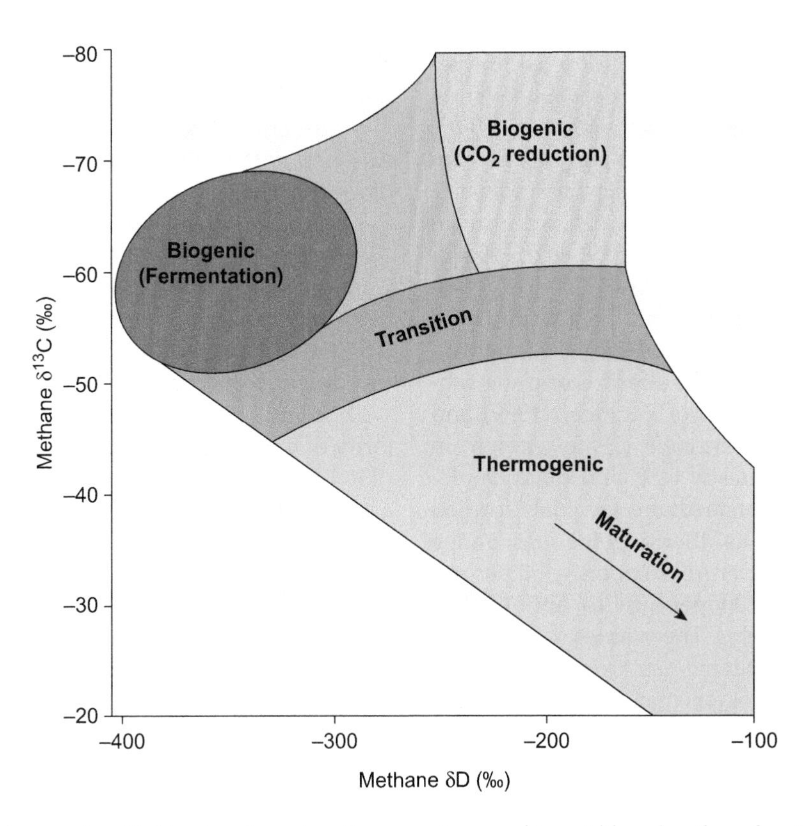

FIGURE 4.18 Compositional fields of gas types (e.g. thermogenic versus biogenic) based on the carbon isotope of methane ($\delta^{13}C$) and deuterium isotope of methane (δD). *Source: Modified from Whiticar et al. (1986).*

The hydrocarbons of coalbed gas are defined by variable proportions of C_1 to C_5 hydrocarbons (C_1 = methane, C_2 = ethane, C_3 = propane, C_4 = butane, C_5 = pentane) (Clayton, 1998; Rice, 1993). In addition, carbon dioxide, nitrogen, oxygen, and hydrogen make up smaller portions of the gas. The $\delta^{13}C$ (delta isotope of carbon) for each hydrocarbon permits discrimination of thermogenic gas. Rooney, Claypool, and Chung (1995) proposed that $\delta^{13}C$ values of methane, ethane, and propane are a function of fractional conversion of kerogen and oil to gas. The fractional conversion is in turn related to temperature of gas generation in conjunction with the compositional kinetic model. In coalbed gas, the carbon isotopic composition of methane is related to the thermal maturation level and maceral composition of the coal (Clayton, 1998). The $\delta^{13}C$ values of coal macerals vary about 3.5 per mil (Whitecar, 1999) but much more in isotopic carbon composition between different modes of formation and thermal maturation (e.g. thermogenic versus biogenic pathways). The large difference may result from (1) greater molecular or isotopic heterogeneity and (2) differences in generation and/or accumulation of gases from the differing types of organic matter (Rooney et al., 1995). Also, Clayton (1998) proposed that enrichment of ^{13}C (as much as 25 per mil) might be a result of increased maturation due to kinetic isotope effect. Using the universally accepted model of Whiticar (1999) and Whiticar, Faber, & Schoell (1986) shown in Figure 4.18, the values of $\delta^{13}C$ and delta Deuterium isotope (δD) of methane for coals formed by thermogenesis as thermogenic gas range from less than −20 per mil to about −50 per mil and more than −100 per mil to about −325 per mil, respectively. The values of $\delta^{13}C$ of methane for thermogenic gas are much heavier than those of the biogenic gas (see next section) and the heavier the values, the more mature the gas becomes.

The most interesting observation about one of the potential causes of the variability of the isotopic composition of coalbed gas is maceral composition of the coal. This factor may be broader ranging than "meets the eye" because as pointed out in Chapter 3, coal macerals depend on the vegetation types (Table 4.2), which in turn depend on the depositional environments of the peat precursor. Moreover, as discussed in Chapters 1 and 3, vegetation types changed through geologic time whereby vascular plants evolved and diversified through Cretaceous Period and by Tertiary time flowering plants became dominant. Thus, geologic ages of the coals matter in the final analysis of the isotopic composition of coalbed gas, a fact that is undervalued. This is not a criticism but a plea for more interdisciplinary cooperation between geologists, geochemists, paleoecologists, and paleobotanists.

A notable attempt to relate isotopic composition using the $\delta^{13}C$ of methane and ethane to the geology was performed by Rice (1993). The data used were mainly collected from Canada and United States. Figure 4.19 is modified with updated data from Rice (1993) to show the plots of all the Tertiary, Cretaceous, Jurassic, Triassic, Permian, and Pennsylvanian coals worldwide. According to Rice (1993), coalbed gases in Canada and United States are chemically dry over the range of methane $\delta^{13}C$ values from −61 to −29 per mil and generally wettest over the range of methane $\delta^{13}C$ values from −48 to −34 per mil. It is interesting to note that the majority of the Cretaceous (68%) and Pennsylvanian (25%) coals plot within the wettest area, which are sapropelic and thermogenic in origin. Only a few Cretaceous coals with dry gas plot in the humic composition. The Tertiary coals plot in the driest, more methane, and mixed biogenic and thermogenic area. However, updated data on Tertiary coalbed gas have shown that microbial activity has played a major role in generation of Tertiary coals in North America (Flores et al., 2008; Warwick & Flores, 2008). Carboniferous coals in Europe (e.g. Germany, Poland) range in wetness (C2+) from 0 to 70.5 and methane $\delta^{13}C$ values range from −16.8 to −70.4 per mil

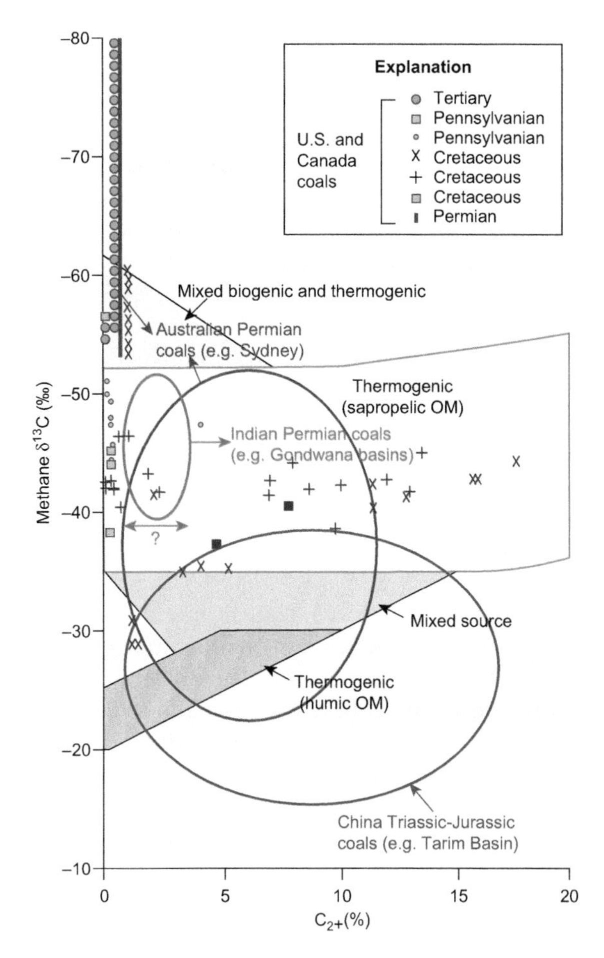

FIGURE 4.19 Methane $\delta^{13}C$ versus wetness (C_{2+}) worldwide modified from Rice (1993). Note that in the Indian coal, the C_{2+} (%) is subjective as indicated by the limited data source (Raina et al, 2009). OM, organic matter. *Source: Modified from Faiz and Hendry (2006), Raina et al., (2009), Rice (1993), Flores et al. (2008), and Zhang et al. (2007).*

(Rice, 1993). Permian coals in Australia are <10 in wetness (C2+) and have methane $\delta^{13}C$ values from −24 to −80 per mil (Rice, 1993). Faiz and Hendry (2006) indicated that thermogenic gases in Australia contain as much as 12% ethane and higher hydrocarbons and that methane $\delta^{13}C$ values range from −23 to −78 per mil. However, a more recent report by Faiz, Murphy, Lidstrom, and Ashby (2012) suggests that ethane and higher hydrocarbon content of Bowen Basin coals can be up to 30%. The Australian coals would probably plot in Figure 4.19 within the thermogenic gas in both the sapropelic and humic organic matter. The Triassic and Jurassic coals of China (e.g. Tarim Basin) contain <1−18% ethane and higher hydrocarbons and methane $\delta^{13}C$ values range from about −16 to −37 per mil (Zhang, Liu, Dai, & Tang, 2007), which plot in the thermogenic, humic organic matter in Figure 4.19. Some Permian coals (e.g. Gondwana) in India according to Raina, Pande, Sharma, Singh, and Singh (2009) contain small amount of heavy hydrocarbons (C_{2+}) and their methane $\delta^{13}C$ values range from −41.5 to −52.2 per mil, which plots within the thermogenic, sapropelic organic matter in Figure 4.19. Also, these authors state that some evidence of biogenic gas is indicated by methane $\delta^{13}C$ values from −22.7 to −69 per mil.

Biogenic Gas

The Earth is about 4.6 billion years old and the most well-known oldest sedimentary rock formed by ancient microorganisms are the stromatolites, which are composed of trapped sediments and secreted calcium carbonate by cyanobacteria and algae. This fossilized remnant of microbial- and algal-formed sedimentary rock represents the oldest in the geologic record, about 3.5 billion years, of microbial activity on the planet surface. Microorganisms are ubiquitous minuscule life forms of our planet present in water (e.g. deep-oceanic hydrothermal vents at 400 °C), rocks (e.g. in a fracture fluid with gold-bearing mines at 3100-m-deep South African mine), animals (e.g. in stomachs of ruminants such as cattle, deer, and elk, in which microbes degrade thick cellulose walls of grass), soils (e.g. from ever-wet tropical peat to dry, hot desert soils), and other varied microenvironments (Brahic, 2008; Kotelnikova, 2002; Peters & Conrad, 1995; Petersen et al., 2011; Prescott, Harley, & Klein, 2002). During the early geological past, the ancient microorganisms were divided into two major groups, which in turn, were subdivided into three branches or domains (e.g. *bacteria, archaea,* and *eukarya*) depicted in a universal phylogenetic tree of life shown in Figure 4.20 (Olsen & Woese, 1993; Prescott et al., 2002).

Archaea was not recognized as a major domain until the late 1970s when Dr Carl Woese and colleagues at the University of Illinois in a pioneering research discovered this new group based on a significant difference in molecular sequences (e.g. known as 16s rRNA), which consist of codes for part of the cell's ribosome (complex structure that makes up protein). They found that microbes that survive in high temperatures or produced methane genetically cluster together and distinctly different from the usual bacteria and the eukaryotes. Since the discovery of this "third form of life" archaea have been found to thrive in diverse microenvironments from extreme aquatic and terrestrial to cold habitats such that in the surface waters of coastal Antarctic it makes up to 34% of the microbial biomass (Prescott et al., 2002). Many scientific books and journals still refer to these "bacteria" as "archaebacteria" but the use of the term has been dropped in favor of archaea. The microorganisms capable of generating methane as a part of their metabolism in various sedimentary rocks have been recognized to come from a subgroup of the domain archaea (Prescott et al., 2002).

The discovery of archaea had a profound impact on exploration and development of fossil energy in general and hydrocarbons in particular. Microbes are injected into petroleum reservoirs to enhance oil recovery as well as to biodegrade hydrocarbon-impacted ecosystems with the latter advancing to use genetically engineered microbes for bioremediation (Leahy & Colwell, 1990; Ollivier & Magot, 2002; Van Hamme, Singh, & Ward, 2003). More appropriately, natural gas in certain sedimentary rocks (e.g. coals and shales) in shallow margins of basins have been analyzed and interpreted as unconventional shallow biogenic gas accumulations. According to Schurr and Ridgley (2002), these accumulations, which are dominantly

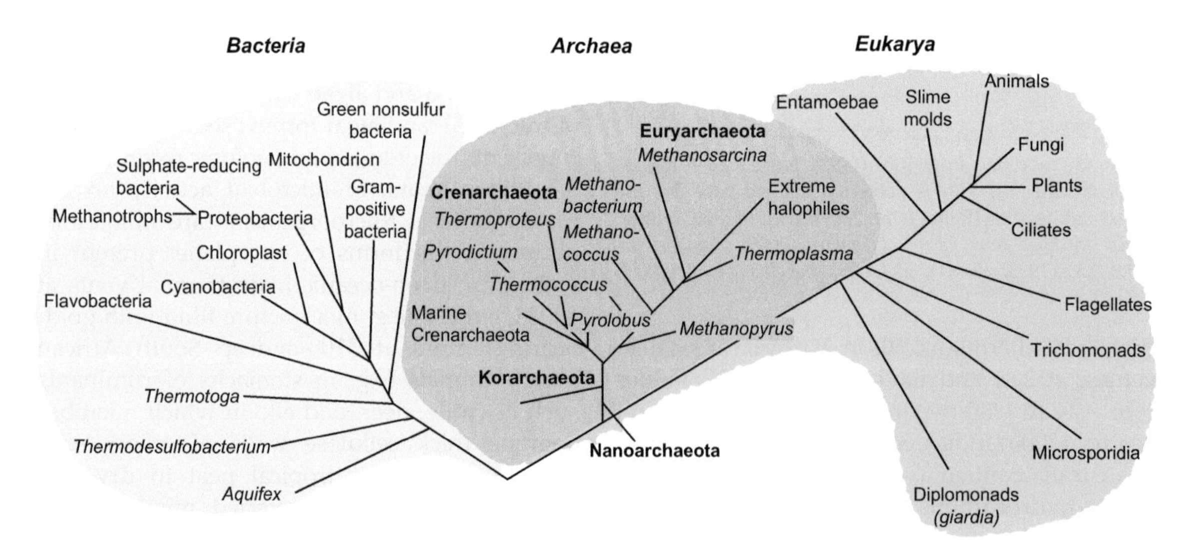

FIGURE 4.20 Phylogenetic tree of the domains of bacteria, archaea, and eukarya. The methanogens are in the archaea. *Source: Adopted from Prescott et al. (2002).*

methane, are not associated with thermally mature source rocks. The shallow biogenic gas accumulations have become viable targets for exploration and development using unconventional technology but less constrained by recoverability and economically compared to deeper conventional gas accumulations. This impetus has led to understanding the exploitation of unconventional coal bed and shale gases for the past decade as well as in enhanced recovery of coalbed gas (Golding, Rudolph, & Flores, 2010; Jenkins & Boyer, 2008; Flores et al., 2008b; Law & Rice, 1993; Martini et al., 1998; Scott et al., 1994).

It is well known that biogenic gas is mainly found in shallow coal beds from 50 to 850 m, possibly even as deep as 1100 m below the surface, which is the threshold of the zone of alteration from thermogenic gas (Faiz & Hendry, 2006; Flores, 2008b; Golding et al., 2010; Kotarba, 1988; McVay, Ayers, & Jesen, 2007; Pashin, 2007, 2010; Rice, 1993; Tao et al., 2005; Zhang & Chen, 1985; Hosgormez et al., 2002; Penner et al., 2010). According to Kotarba (1988), the zone of alteration must be controlled by seals, faults, and groundwater movement. It is widely accepted that the critical geologic event when biogenic gas is generated is after uplift and exposure of coal-bearing rocks along basin margins to meteoric water and subsequent establishment of groundwater systems (Faiz & Hendry, 2006; Flores et al., 2008; Rice, 1993; Rice, Flores, Stricker, & Ellis, 2008). Thus, the shallow occurrence of biogenic gas may be directly related to its formation during posttectonic deformation, although this may have been superseded by expulsion of thermogenic gas that was microbially altered during deformation (Faiz & Hendry, 2006; Faiz et al., 2003).

Although biogenic gas is believed to occur in younger Tertiary and Cretaceous coals, recent studies have demonstrated their occurrence in older Permian and Pennsylvanian coals (Faiz & Hendry, 2006; Faiz et al., 2007a; Flores, 2008a; Flores et al., 2008; Formolo, Martini, & Petsch, 2008; Rice, 1993; Strapoc et al., 2008; Warwick & Flores, 2008; Li et al., 2008; Midgley et al., 2010). Like thermogenic gas, which is mainly generated in high-rank coals (e.g. bituminous and anthracites), biogenic gas is primarily generated in lower rank coals at low temperature (e.g. subbituminous) (see Figure 4.12). Strapoc et al. (2011) suggest that, in general, the higher the coal rank, the lower the rate of conversion to methane. These workers indicate that the fraction of coal that is converted to methane from high-volatile bituminous coal (e.g. 1%R_0) to lignite (e.g. 0.2 %R_0) is from about 0.05 to 0.4 per year. In addition, they suggest that not only coal rank but also coal macerals directly related to precursor vegetation may influence the rate of biodegradation.

The paradigm of methane generation in low-rank coals is contradicted by studies demonstrating the accumulations of secondary biogenic gas in Cretaceous and Permian high-volatile to low-volatile bituminous coals, which contain mixtures of biogenic and thermogenic gas (Boreham, Draper, & Hope, 2004; Faiz & Hendry, 2006; Faiz et al., 2003; Scott et al., 1994; Smith & Pallasser, 1996). Generally, major amounts of heavy hydrocarbons (C_2–C_5) in wet gases associated with biogenic methane indicates that it is overprinted by late-stage gas generation in the higher rank coals. This is demonstrated in the Permian high-volatile and low-volatile bituminous coals in the Sydney Basin, Australia. Here, Faiz et al. (2003) established that a large part of the initial thermogenic wet gases were expelled and microbially altered from the shallow coals during post-Cretaceous uplift (Figure 4.21). These Permian coals were subsequently refilled with secondary biogenic gas presumably by introduction of meteoric water recharged from outcrops along the basin margins. Furthermore, Faiz and Hendry (2006) demonstrated that gas composition and isotope data indicate that biogenic gas accumulation decreases toward the basin center due to reduced accessibility to groundwater recharge.

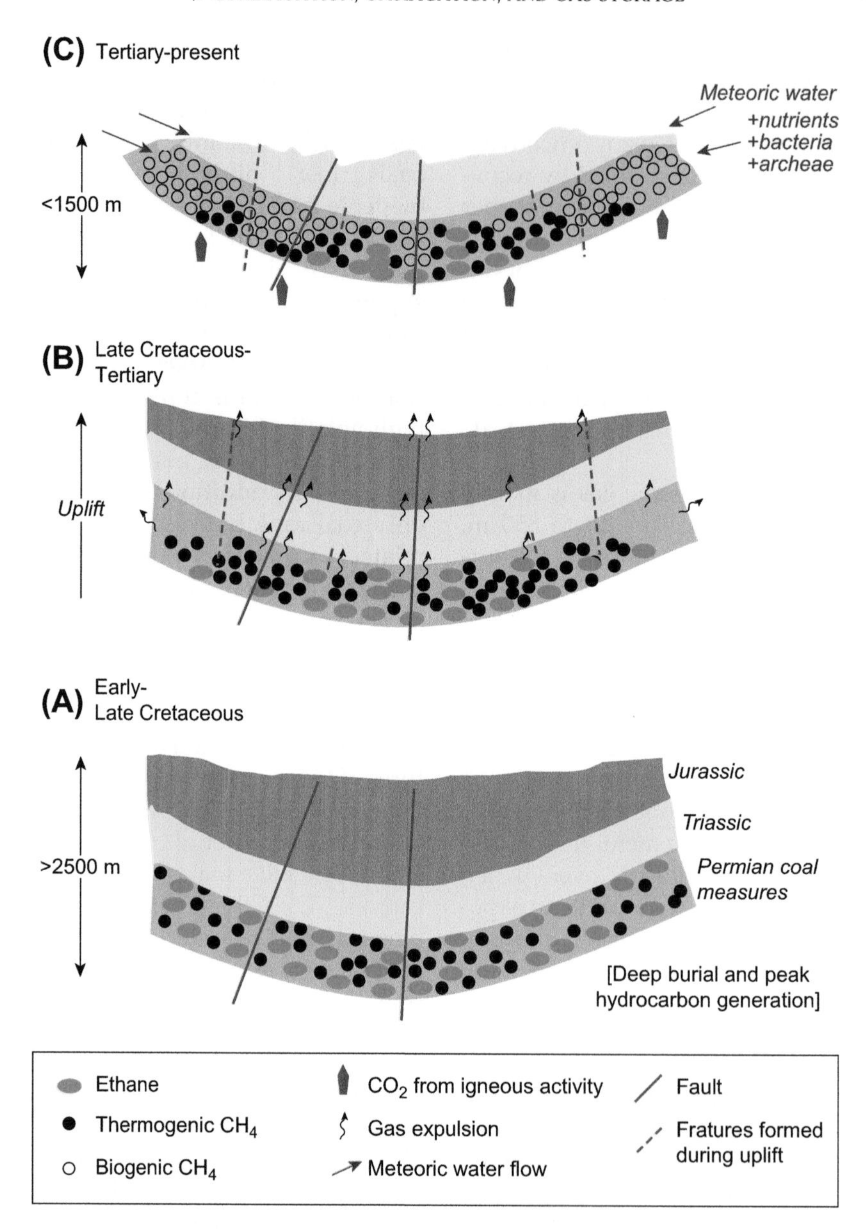

FIGURE 4.21 Model for generation of biogenic methane for post-Cretaceous uplift in the Sydney Basin, Australia. The model shows expulsion of gas during the Late Cretaceous and Tertiary uplift followed by biogenic gas generation in the Tertiary with influx of meteoric water along shallow basin margins. *Source: Adopted from Faiz and Hendry (2006).*

Similar conditions existed in the San Juan Basin in the United States where Cretaceous medium-volatile bituminous coals contain mixed thermogenic and biogenic gas (Scott et al., 1994). The Black Warrior and Arkoma Basins in the United States also contain mixed thermogenic and biogenic gases in Pennsylvanian medium- to low-volatile bituminous coals (Pashin, 2007,

2010; Rice, 1993). Therefore, if biogenic gas can accumulate in higher bituminous rank coals, can it accumulate in much higher rank anthracite coals? According to Ryan (2012), Cretaceous–Jurassic anthracite coals in the northern Bowser Basin in British Columbia, Canada, can readsorb expelled thermogenic gas or generate biogenic gas presumably during or after basin deformation.

Origin

The basic concept that much of the coalbed gas is thermogenic in origin has prevailed because of its close relationship to and analysis of gas from gassy coal mines, which developed high-rank coals (Chapter 2). However, the advent of chemical analysis of gas from deep wells has proved the biogenic origin of coalbed gas (Rice, 1993; Smith & Pallasser, 1996). Biogenic gas is typically high in methane and a dry gas compared to thermogenic gas, which is lower in methane and a wet gas with the presence of heavy hydrocarbons. In addition, compared to thermogenic gas, biogenic gas uniquely contains light carbon isotope composition (see following section). These characteristics and recent geochemical studies of new coalbed gas from the burgeoning industry have indicated that a large part of this unconventional gas is a result of microbial activity or is biogenic in origin (Ahmed, Smith, & George, 1999; Boreham et al., 2004; Clayton, 1998; Faiz et al., 2003; Flores, 2000; Flores et al., 2008; Rice, 1993; Scott et al., 1994; Smith & Pallasser, 1996; Strapoc et al., 2011; Faiz et al., 2012; Klein et al., 2008).

It was discovered through either fieldwork and/or laboratory experiments that early-stage biogenic gas is produced in peat and lignite (Chapter 3) and late-stage biogenic gas is produced after coalification. Late-stage biogenic gas has been proposed to occur during the past tens of thousands to millions years as a result of the incursion of nutrient-carrying meteoric water in coal beds (Faiz & Hendry, 2006; Faiz et al., 2003; Flores et al., 2008; Kotarba & Rice, 1995; Rice, 1993). The inflow of meteoric water into basin margins leading to generation of biogenic gas is exemplified in the Sydney Basin in Australia and Powder River Basin in Wyoming, United States (Figures 4.22, 4.23). In these two basins, the biogenic gas is identified by their light carbon isotopic composition of

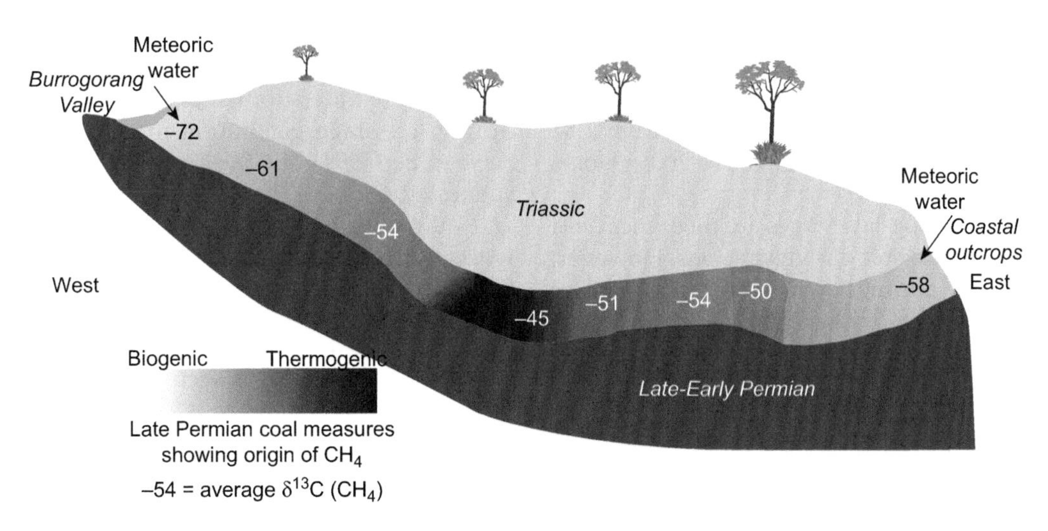

FIGURE 4.22 Cross-section of the coals of the Illawarra Coal Measures showing distribution of methane $\delta^{13}C$ across the Sydney Basin, Australia. CH_4, methane. *Source: Adopted from Faiz and Hendry (2006).*

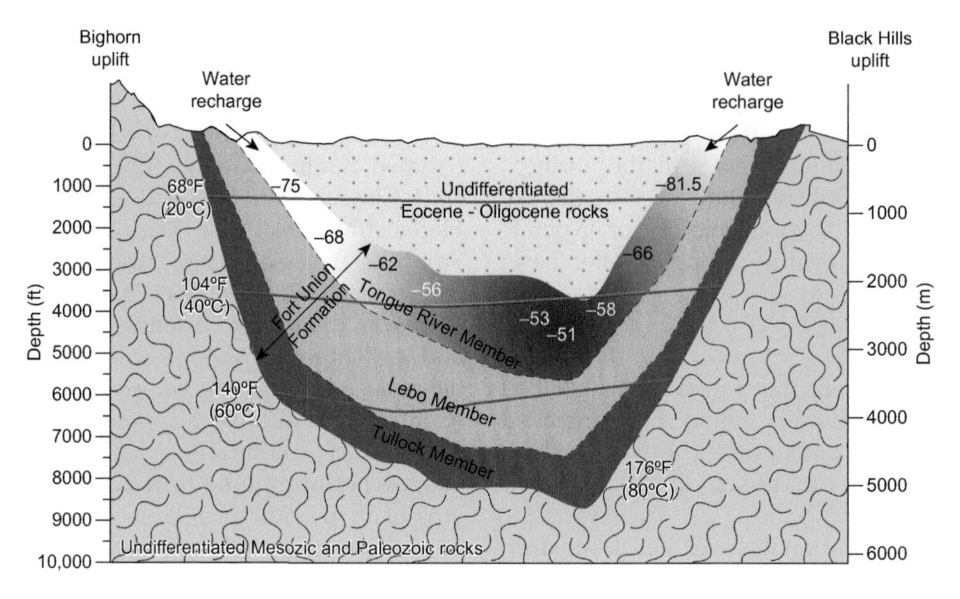

FIGURE 4.23 Cross-section of the coal-bearing Paleocene Fort Union Formation showing distribution of methane $\delta^{13}C$ in the coaly Tongue River Member across the Powder River Basin in Wyoming, United States. *Source: Modified from Flores (2004) and Flores et al. (2008).*

methane (see section on biogenic isotopes). However, the carbon isotopic composition in the center of the Powder River Basin is much lighter than in the Sydney Basin. The Sydney Basin center gas is interpreted as thermogenic with heavier $\delta^{13}C_{CH}$ (e.g. from −45 to −51 per mil) in contrast to the Powder River Basin, which is produced by carbon dioxide reduction (biogenic) in origin and has lighter $\delta^{13}C_{CH}$ (e.g. from −51 to −58 per mil) (Faiz & Hendry, 2006; Flores et al., 2008).

Thus, early and late stages are theoretical end members of biogenic gas generations. However, the long span of time between the initial stage of coalification (e.g. lignite rank) and late-stage coalification (e.g. bituminous and anthracite rank) in which biogenic gas can be generated and potentially mixed with thermogenic gas is still not well known. For example, Scott et al. (1994) suggested that in the San Juan Basin in New Mexico and Colorado, which is well known for thermogenic gas generation, bacteria were transported basinward in groundwater flowing from the elevated northern basin margins. These bacteria metabolized wet gas components, *n*-alkanes (or long-chain hydrocarbons cuticle wax of plants), and organic compounds in the coal and generated secondary biogenic methane and carbon dioxide following coalification, uplift, erosion, and cooling. Thus, biogenic gas may supplement thermogenic gas during late-stage generation in the manner of Scott et al. (1994) but Bustin and Bustin (2008) suggested that this process is not pronounced and may only be expressed in high-rank coals with low Langmuir pressure (see following section for definition). The timing of late-stage gas generation may be complicated by tectonic and burial history of the basin. This may be demonstrated in the Black Warrior Basin (Pashin 2007, 2010) where thermogenic gas is dominant in high-rank coals (e.g. bituminous). Burial history of the basin indicates an influx of freshwater, which sustained methanogenic activity, that may have started early during the Cretaceous when the Pennsylvanian coal-bearing rocks were exposed on the surface. However, according to Pashin (2010), burial was renewed in Late Cretaceous, which covered the coal-bearing strata from groundwater recharge. Succeeding exhumation exposed these coal-bearing strata to groundwater recharge accompanied by methanogenic activity.

Methanogens against Harsh Microenvironments

The debate about the two theoretical end members of gas generation, especially the role of methanogens during the early and late stages, is not well known and may be clarified by the ability of these obligate anaerobic microbes or microbes that live and grow in the absence of oxygen to survive harsh conditions. The two major conditions critical to coalification based on pyrolysis experiments are temperature and pressure. Another condition is water or moisture. In a landmark, long-term pyrolysis experiments of peat to coal by Van Heek, Juntgen, Luft, and Teichmuller (1971) indicated the following conditions. Based on extrapolation from degassing curves from these experiments, Van Heek et al. (1971) estimated conditions of early coalification at temperatures about 20 °C–50 °C; water is first lost. This is followed by generation of some methane at about 30 °C–70 °C and of carbon dioxide at about 70 °C–100 °C. Last, at about 160 °C–200 °C, large amount of methane is generated. Stach et al. (1982) suggested that the temperature for anthracitization is about 300 °C. Based on pressure and temperature (P/T) conditions for carbon metamorphism from abundant carbonaceous material of paraschists (or sedimentary schists), Diessel et al., (1978) indicated progressive coalification to anthracite up to 3 kb/225 °C where graphite first appears. Furthermore, continued coalification and graphitization occurred at P/T conditions from 3 kb/225 °C to 5 kb/335 °C. Bed moisture at the beginning of high-volatile bituminization according to Levine (1993) is 22%.

The debate about methanogens, which produced peat and coalbed gases, is that they cannot survive conditions greater the 50 °C and increased heat during coalification (maturation) is common in the literature (Aravena, Harrison, Barker, Abercrombie, & Rudolph, 2003; Kim, 1978; Moore, 2012; Scott et al., 1994; Strapoc et al., 2008). However, Faiz and Hendry (2006) suggest that methanogenic activity can exist over a wide range of temperatures. The most common temperature range is ~4 °C to ~100 °C (Jones, Leigh, Mayer, Woese, & Wolfe, 1983). Thus, based mainly on temperatures at which methanogens could have survived through thermal conditions up to 100 °C, Van Heek et al. (1971) suggested generation of early-stage methane and carbon dioxide during coalification. Methanogens that may have survived at these temperatures and harsh environments are classified as extremophiles (or microbes adapted to extreme temperature, pressure, or chemistry) (Figure 4.24) (NASA & Stsci, 2006).

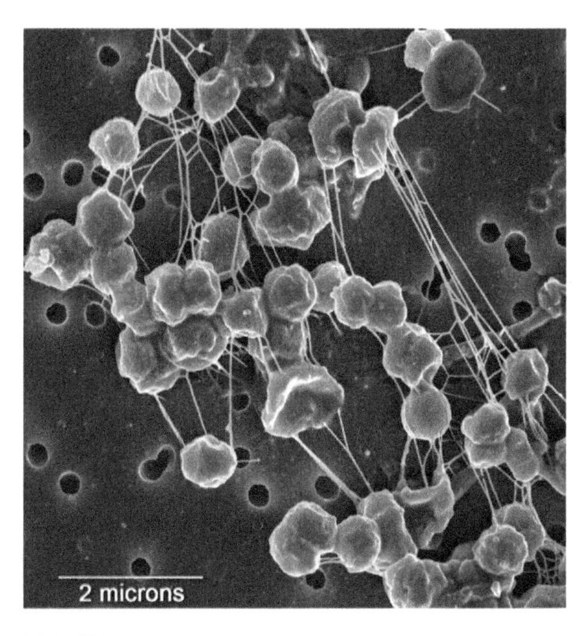

FIGURE 4.24 Photograph taken from electron microscope of the methanogen that can survive below freezing temperature. *Source: Adopted from NASA and STSci. (2006).*

TABLE 4.5 Cell Counts of Methanogenic Archaea and Rates of Methane Production Pre- and Postexposure to Martian Conditions

Cultures	Cell Counts 10^7	Survival Rates (%)	CH_4 Production (nmol/h ml)
Methanosarcina sp. SMA-21, control	6.1 ± 0.6	100	48.61 ± 6.57
Methanosarcina sp. SMA-21	5.5 ± 0.8	90.4	44.11 ± 5.08
SMA-16, control	6.2 ± 1.1	100	52.77 ± 6.18
SMA-16	4.2 ± 0.9	67.3	45.37 ± 0.03
SMA-23, control	7.8 ± 1.4	100	22.13 ± 1.94
SMA-23	4.7 ± 1.2	60.6	13.92 ± 3.87
Methanobacterium sp. MC-20, control	8.1 ± 1.3	100	27.38 ± 3.09
Methanobacterium sp. MC-20	0.09 ± 0.01	1.1	0.03 ± 0.001
Methanogenium frigidum, (DSM 16458) control	2.3 ± 0.8	100	2.76 ± 0.07
M. frigidum (DSM 16458)	0.1 ± 0.04	5.8	0.003 ± 0.005
Methanosarcina barkeri (DSM 8687), control	3.7 ± 0.5	100	20.43 ± 2.38
M. barkeri (DSM 8687)	0.01 ± 0.00	0.3	0.01 ± 0.01

nmol, nanomole (10^{-9}), Mean ± standard error, n = 3; SMA = Cells of strain; MC = Non-permafrost; DSM 8687 = Peat bog from Germany; DSM 16458 = Antarctica, Ace Lake water column; % = percent.
Source: Adopted from Morozova et al. (2006).

A ground-breaking study by Morozova, Mohlmann, and Wagner (2006) may shed some light on the adaptability and capability of methanogenic archaea to survive in very harsh microenvironments. These researchers collected methanogenic archaea from peaty soils (e.g. fibrous, coarse silty organic matter) from the floodplain of the Lena Delta in the Siberian permafrost. The permafrost methanogens and other well-studied methanogens from nonpermafrost microenvironments were exposed to conditions simulating Martian conditions, which mimicked conditions on the Earth 3.8 billion years ago (Morozova et al., 2006). After 22 days of exposure to thermophysical conditions at Martian low- and midlatitudes, up to 90% of the Siberian methanogenic archaea survived in pure cultures and in environmental samples. In contrast, up to 99.7% of the methanogenic archaea that originated from nonpermafrost microenvironments did not survive the same conditions. Table 4.5 shows the percentage survival rate and the methane produced by the methanogenic archaea under experimental Martian conditions. The arctic permafrost environment is characterized by low annual air temperature varying from −48 °C to 18 °C and mean annual precipitation of 190 mm. The experiments indicated that methanogenic archaea from the permafrost are more resistant to temperature shifts between −75 °C and 20 °C than well-studied methanogens from other microenvironments. This is a total temperature shift of 95 °C compared to the presumed survival thermal condition of 50 °C (Scott et al., 1994). These experiments and observations do not completely simulate physicochemical and thermal conditions of coalification but they demonstrate the capability to survive by certain subgroups of methanogenic archaea and other nonmethanogenic bacteria. Therefore, an interesting point to bring out is that cold and

BOX 4.3

DIVERSE LIFE STYLES OF METHANOGENS—CONCEPTS AND MISCONCEPTIONS

The marginalization of methanogenic archaea or methanogens to anoxic environments during the buildup of oxygen on the Earth's surface 2.4 billion years ago would have been enough to limit their microenvironments (Liu, Beer, & Whitman, 2012). However, advances in discovery of these obligate anaerobic microbes have found varied life styles based on tolerance to temperatures from mesophiles to thermophiles and hyperthermophiles or microbes with growth optimum of 20°−45 °C, 55 °C and higher, and 80°−113 °C, respectively (Prescott et al., 2002). A 2-year laboratory study of NASA and STsci (2006) shows that some microbes such as methanogens (Figure 4.25) adapted to cold temperatures survived and reproduced at −2 °C. In addition, the microbes developed defense mechanism for protection from cold temperatures. The methanogens were considered potentially adaptable to life on cold, dry Mars and could survive on a planet without oxygen. It is proposed by some scientists that methanogens could have produced the methane detected in Mars' atmosphere.

Other bacteria were exposed to higher and more variable temperatures (−80 °C−80 °C) or a total of 160 °C temperature shift and survived (Olsson-Francis, de la Torre, Towner, & Cockell, 2009). These microorganisms that thrive in extreme thermal conditions at relatively high temperatures (45 °C−80 °C) are thermophiles and those that thrive in extreme temperatures (>80 °C) are hyperthermophiles (Rampelotto, 2010). They are found from volcanic soils permeated by hot vapors and deep-sea hydrothermal vents in which extremely thermophylic methanogens (Figure 4.25) can survive at maximum of 98 °C and will grow up to 110 °C (Prescott et al., 2002). Thus, methanogenic archaea may be more resilient to extreme conditions than originally thought and are able to survive harsh microenvironments.

FIGURE 4.25 Close-up photograph of the methanogen *Methanopyrus kandleri* that lives near extreme temperatures of thermal vents on the ocean floor. *Source: Adopted from Yale News (2009).*

temperate peats are more prone to generate early-stage biogenic gas during coalification (Box 4.3).

In addition, methanogenic activity can occur in very diverse aqueous settings despite the two end-member model that was traditionally accepted where CH_4 is produced by acetate fermentation in freshwater settings and by carbon dioxide reduction in marine sediments (Whiticar et al., 1986). In fact, methanogenesis by CO_2 reduction occurs in many freshwater settings (Lansdown, Quay, & King, 1992). In a petroleum system, biogenic gas forms in aqueous environments with carbon dioxide reduction forming methane in marine environments (Rice & Claypool, 1981; Whiticar et al., 1986). Also, biogenic gas can form in hypersaline lake environments (Dang et al., 2008). However, coalbed gas forms in brackish water with carbon dioxide reduction and fermentation in freshwater (Flores et al., 2008; Rice, 1993; Warwick, Breland, & Hackley, 2008).

Methanogenic microorganisms thrive mainly in aquifers found in various geologic structural, rock, and hydrocarbon settings worldwide (Table 4.6) (Boivin-Jahns, Ruimy, Bianchi, Daumas, & Christin, 1996; Glossner, Schmidt, Flores, & Mandernack, 2008, 2009; Kleikemper et al., 2005; Olson, Dockins, McFeters, & Iverson, 1981; Schmidt, 2007; Schultse-Makuch, Goodell, Kretzchmar, & Kennedy, 2003; Shimizu et al., 2006; Kotarba and Pluta, 2009). The aquifers found in faulted and intermontane areas include sandstone, clay, and coal aquifers varying from 100 to 1800 m depths. The microorganism biomass ranges from a minimum of 1.0×10^6 cells/l to a maximum of 1×10^9 cells/l found in a shallow (e.g. 224 m) clay aquifer in Belgium. In the Powder River Basin in Wyoming, the microorganism biomass of both archaeal methanogens and bacteria is in deeper (e.g. 400 m) coal aquifer and is as much as 8.74×10^7 cells/l (Figure 4.26). (Mandernack et al. (2007), Schmidt (2007), and Glossner (2008; 2009)).

Process

The biogenic process of generating coalbed gas involves groups of strict anaerobic microbes with a variety of metabolic life styles. Coalbed gas is produced by consortia of methanogenic archaea (e.g. methanogens), which acquire energy by converting carbon dioxide, hydrogen, formate, methanol, acetate, and other compounds to methane or methane and carbon dioxide (Prescott et al., 2002). These organic materials consumed by methanogens are collectively called "substrates". Methanogens flourish in anaerobic environments high in organic matter (e.g. coal, shale) such that degradation by consortia of microbes of 1 kg of organic matter can yield up to 600 l of methane (Prescott et al.,

TABLE 4.6 Microorganism Biomass (e.g. Bacteria and Methanogens) in Coal, Sandstone, and Clay Aquifers in Various Geological Settings Worldwide

Location (Aquifer)	Depth (m)	Microorganism Biomass (cells/l, min.)	Microorganism Biomass (cells/l, max.)	References
Montana, US (deep sandstone aquifer)	1200–1800	Not available	1×10^6	Olson, et al., (1981)
Studen, Switzerland (oil-contaminated aquifer)	Not available	6.6×10^7	1.5×10^8	Kleikemper et al. (2005)
New Mexico, US (intermontane basin aquifer)	>100	5.0×10^5	3.3×10^6	Schultze-Makuch et al. (2003)
Wyoming, US (Powder River Basin, coalbed gas aquifer)	134–611	1.05×10^6	8.74×10^7	Glossner et al. (2008; 2009); Schmidt (2007)
Mol, Belgium (Clay aquifer)	224	1.0×10^4	1×10^9	Boivin-Jahns et al. (1996)
Japan, north (fault-bordered, coalbed gas aquifer)	900	3.08×10^5	9.88×10^5	Shimizu et al. (2006)

m, meter; l, liter; min., minimum; max., maximum.

Source: Compiled from Schultse-Makuch et al. (2003), Boivin-Jahns et al. (1996), Glossner (2008, 2009), Kleikemper et al. (2005), Olson et al. (1981), Schmidt (2007), and Shimizu et al. (2006). Modified from Glossner et al. (2008, 2009).

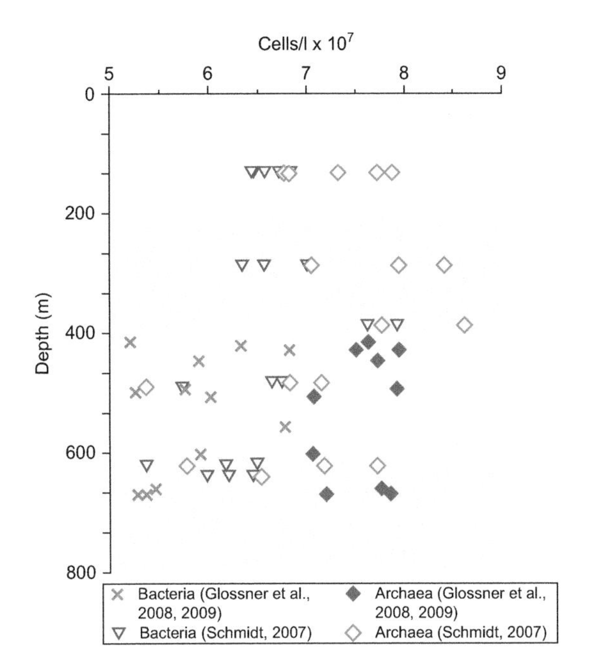

FIGURE 4.26 Archaeal and bacterial biomasses in coalbed gas producing coals in the Paleocene Fort Union Formation in the Powder River Basin, Wyoming, United States. l, liter, m, meters. *Source: Adopted from Glossner et al. (2008, 2009), Mandernack et al. (2007), and Schmidt (2007).*

2002). For example, and simply put by Prescott et al. (2002), anaerobic microbial digesters will degrade particulate waste in a sewage treatment plant to hydrogen, carbon dioxide, and acetate. This biochemical process is followed by carbon dioxide-reducing methanogens forming methane from the carbon dioxide and hydrogen, while aceticlastic methanogens cleave acetate to form carbon dioxide and methane. Two-thirds of the methane produced by microbial digesters comes from acetate. The conversion of acetate to methane and carbon dioxide is expressed by the following equation (Abbanat, Aceti, Baro, Terlesky, & Ferry, 1989):

$$CH_3COOH \rightarrow CO_2 + CH_4 \quad \text{(Aceticlastic Reaction)}$$

According to Ferry (1993), the aceticlastic (or fermentation) reactions of carbon dioxide (CO_2) and methane (CH_4) are mainly formed from acetate and methyl groups and conversion of acetate to methane is activated by acetyl-coenzyme A. For aceticlastic reaction (e.g. methyl-type fermentation), 75% of the hydrogen in methane is derived from the original organic matter or macerals and 25% from formation water, producing a fractionation of the isotopes between methane and water (Whiticar, 1999).

In coal, the organic matter or macerals (e.g. plant remains from the peat precursor), which are composed of cellulose and lignin (e.g. wood), are degraded by anaerobic microbial digesters into chemical compounds and molecules such as acetate, fatty acids, carbon dioxide, ammonia, hydrogen, and hydrogen sulfide (e.g. monomers and oligomers) (Figure 4.27) (Zinder, 1993). The long-chain fatty acids are in turn converted by hydrogen-producing acetogens into formate or hydrogen and carbon dioxide and acetate. These compounds and molecules are transformed by hydrogen-consuming methanogens of the domain archaea into methane through energy generation, which include enzymes and coenzymes (Deppenmeier, 2002). Faiz and Hendry (2006) described very well the processes involved in methanogenesis in Australian coal substrates and the roles of groups of associated microorganisms toward generation of methane.

Thus, this example of stepwise pathway of methane production or methanogenesis is accomplished by successions of biochemical breakdown of monomers and oligomers (Zinder, 1993). This author proposed a methanogenic pathway of step-by-step and collective interactions of a hierarchy of metabolic groups of microorganisms toward generation of methane and carbon dioxide, which include fermentative anaerobes (e.g. fermenters), acetogens (e.g. hydrogen producers), and methanogens (e.g. hydrogen users and acetate converters to methane and carbon dioxide or acetotrophic) (Figure 4.27). The reader is referred to Demirel and Scherer (2008) for an excellent review of the roles of acetotrophic and

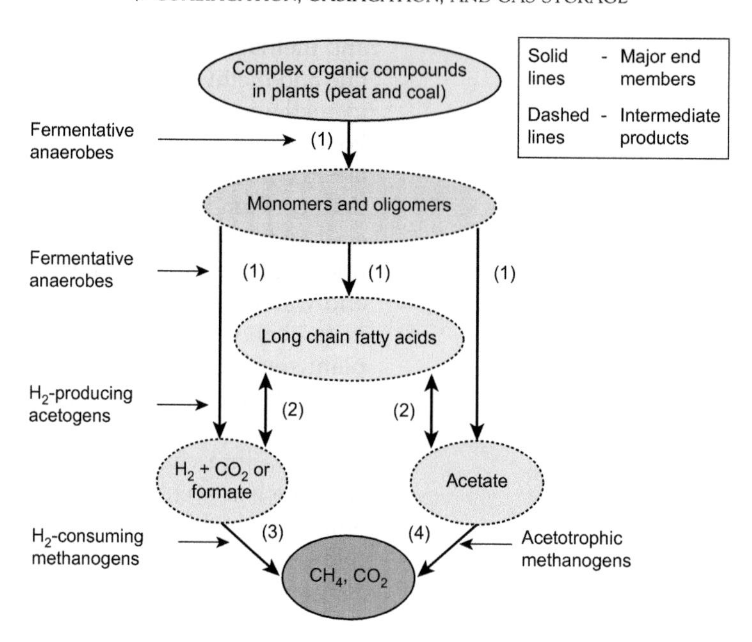

FIGURE 4.27 Flowchart of stepwise (e.g. (1), (2), (3)) methanogenic pathways for conversion of organic matter to methane in freshwater environment. CH_4, methane; CO_2, carbon dioxide; H, hydrogen. *Source: Modified from Zinder (1993).*

other methanogens during anaerobic conversions of various biomass substrates to methane. Each of these anaerobes plays an important role in establishing a stable microenvironment at various levels of the methane fermentation or aceticlastic methanogenic pathway. Table 4.7 shows the various species of methanogens that thrive on different "substrates" (e.g. reacting molecules) toward breakdown of organic matter to produce methane (Prescott et al., 2002). The majority of methanogens use carbon dioxide and hydrogen "substrates" for methanogenesis compared to limited number of methanogens, which use acetate "substrate" (Faiz & Hendry, 2006; Zinder, 1993).

Another microbial methanogenic pathway toward generation of methane is through carbon dioxide reduction with hydrogen or formate as the electron donor. Majority of the anaerobes are able to reduce carbon dioxide at a rate faster than acetate conversion to methane (Blaut, 1994; Keltjens, 1984). This methanogenic pathway is

TABLE 4.7 Selected Methanogens and Their Morphology and Methanogenic Substrates (e.g. Reacting Molecules) Used

Methanogenic Genera	Morphology	Methanogenic Substrates Used
Methanobacterium	Long rods or filaments	$H_2 + CO_2$, formate
Methanothermus	Straight to slightly curved rods	$H_2 + CO_2$
Methanococcus	Irregular cocci or spherical	$H_2 + CO_2$, formate
Methanomicrobium	Short curved rods	$H_2 + CO_2$, formate
Methanospirillum	Curved rod or spirilla	$H_2 + CO_2$, formate
Methanosarcina	Irregular cocci, packets	$H_2 + CO_2$, methanol, methylamines, acetate

Source: Adopted from Prescott et al. (2002).

expressed by the following equation (Abbanat et al., 1989):

$$4H_2 + CO_2 \rightarrow CH_4 + 2H_2O$$
(Carbon Dioxide Reduction)

The hydrogen in the methane in carbon dioxide reduction is derived solely from the formation water, which produces methane fractionated with respect to formation water (Whiticar, 1999).

According to Faiz and Hendry (2006), the amount of methane produced by the aceticlastic and carbon dioxide reduction pathways is controlled by availability of nutrients, temperatures, nature of "substrates", and coal rank. Strapoc et al. (2011) suggest that lower rank coals may generate more methane compared to higher rank coals because of increased rate of bioconversion. However, in the Powder River Basin, the subbituminous A to B coals in the basin center generate more gas through carbon reduction pathway than the subbituminous C coal in the basin margin, which is generated by combined aceticlastic and carbon reduction pathways (Flores, 2004; Flores et al., 2008) (Box 4.4).

BOX 4.4

METHANOGENS: EQUAL-OPPORTUNITY ENERGY PROVIDERS—FACT VS FICTION

Ever since the discovery of methanogenic archaea, methanogens were touted as producers of clean energy source of methane (Prescott et al., 2002). Methanogens play varied roles as the world's methane providers from waste digester to biodegrading oil to gas. Anaerobic digester microbes (e.g. bioreactors) degrade sewage sludge to produce biogas (e.g. carbon dioxide and methane). Since 2000–2001, anaerobic bioreactors have been applied in 1400 treatment plants in 65 countries for bioconversion of varied types of industrial effluents to biogas (Mosquera-Corral, Gonzalez, Campos, & Mendez, 2005). Although this biotechnology has not been perfected, it appears to be a promising source of methane as long as more organic carbon is reduced during the bioconversion process. Sowers and Ferry (2002) suggested that because many industrial wastes contain high concentrations of salts that inhibit freshwater inoculum, perhaps marine methanogens, which have the ability to adapt to varied solute conditions, may be more appropriate to use as bioreactors in treatment plants.

Jones et al. (2008) suggest that methane recovery that is difficult from oilfields might be economically achieved by methanogenic biodegradation enhancement. Their study indicates a methanogenic biodegradation in subsurface degraded oil reservoirs worldwide, which resulted in hydrocarbon alteration and association of dry gas with severely degraded oils. Furthermore, they observed that biodegradation of crude oil in subsurface petroleum reservoirs have adversely affected the majority of the world's oil, making recovery and refining of that oil more costly. With the advent of the shale gas industry, the significance of methanogens in methane generation has become part of the development and enhanced recovery strategies. Worldwide, organic-rich shale has become a successful target for exploration for natural gas and methanogenesis has played a dominant role in the generation of methane in low thermally mature and highly fractured reservoirs (Martini et al., 1998; Selley, 2005). Koide and Yamazaki (2001) proposed using methanogens to sequester carbon dioxide and enhanced methane recovery in coal beds in Japan. They argue that this methanogenic process potentially alleviates the growth of greenhouse gases in the atmosphere.

Isotopic Composition

Methane is composed of two stable carbon isotopes ^{13}C and ^{12}C and their ratios in hydrocarbon gases are established at the time of formation and remain unchanged under most conditions. The biogenic gas is generally isotopically lighter than thermogenic gas because of the enrichment of isotope ^{12}C, which is about 8.3% lighter than isotope ^{13}C, by microorganisms (Rice, 1993). Biogenic methane generated by aceticlastic reaction (or fermentation) and carbon dioxide reduction pathways can be differentiated based on the isotope of carbon of methane ($\delta^{13}C_{CH_4}$) and isotope of deuterium of methane (δD_{CH_4}) using Whiticar et al. (1986) compositional fields (see Figure 4.18). The biogenic methane generated by the aceticlastic reaction or fermentation ranges from about -50 to -70 per mil in $\delta^{13}C_{CH_4}$ and is depleted in δD_{CH_4} from -250 to -400 per mil. Biogenic methane generated by carbon dioxide reduction ranges from -60 to -80 per mil and is enriched in δD_{CH_4} from -150 to -250 per mil. However, the $\delta^{13}C_{CH_4}$ values can be even heavier for carbon dioxide reduction depending on the extent of biodegradation. According to Schoell (1980), deuterium concentrations (δD) of biogenic methane from worldwide occurrences range from -180 to -280 per mil and are found to be depleted in deuterium at about -160 per mil. The values of the isotope of deuterium of methane depend on the formation water and the biochemical pathway.

For example, in the Powder River Basin, the values of $\delta^{13}C_{CH_4}$ range from -50 to -83 per mil and -283 to -327 per mil indicating generation by aceticlastic or fermentation reaction (Flores et al., 2008). However, the fractionation of deuterium isotopes of water and methane that indicate about 67% of the gas sample was generated by carbon dioxide reduction pathway. This indicates that the dominant metabolic pathway in the Powder River Basin is carbon dioxide reduction and not aceticlastic (fermentation) reaction as previously interpreted by Rice (1993). Thus, the isotopes of formation waters play an important role in the biochemical reactions with methanogenesis, which uses the water, as a part of the reactions.

Role of Groundwater

Isotope composition of formation waters in coal-bearing rocks has been used to understand the hydrogeology and biogeochemistry of the coalbed gas reservoirs (e.g. coal and sandstones) (Ayers et al., 1991; Rice et al., 2002). This is a particularly important study because coals and sandstones serve as aquifers, which are saturated with methanogens and their nutrient supply. In this aqueous setting, according to Rice et al. (2008), a series of microbe-mediated redox reactions develop in reducing environments, progressing from reduction of nitrate, to manganese and iron oxides, then to sulfate, and finally to carbon dioxide (e.g. substrate for methanogenesis). Thus, the overall effect of water composition of reduction reactions on the methanogenic pathway is depletion of sulfate, increase of bicarbonate, and establishment of a reducing environment, all representative of coalbed gas-coproduced water.

The important isotope compositions of formation waters that are closely related to methanogenesis are isotope of deuterium of water (δD_{H_2O}) and isotope of oxygen of water ($\delta^{18}O_{H_2O}$). Isotopic studies by Rice et al. (2008) on the formation of coproduced waters in the Powder River Basin indicate that isotopic variations may be directly related to the origin of the water along the Global Meteoric Water Line (GMWL; Craig, 1961). GMWL is the average relationship of hydrogen and oxygen isotope ratios of terrestrial water worldwide. Studies by Rice et al. (2008) show that values of δD_{H_2O} and $\delta^{18}O_{H_2O}$ fall on the lower portion of the GMWL (lighter δD_{H_2O} and $\delta^{18}O_{H_2O}$), which indicate that some of the water in Powder River Basin may be older than modern times, with an isotopic signature more representative of a cooler climate. Cross-plots of these isotopes show that within coalbed gas-producing reservoirs (e.g.

aquifers), the values vary along the GMWL over fairly large ranges. The δD_{H_2O} and $\delta^{18}O_{H_2O}$ within a coal bed trends from heavier (enriched in deuterium) to lighter (depleted in deuterium) values in the direction away from groundwater recharge areas similar to that observed by Gorody (1999). Rice et al. (2008) interpret this trend as representing a mixture of more recent water with older basinal water along recharge flow paths. The trend of the water with these values reflects the oldest in the Powder River Basin and represents the isotopic signature of the original formation water after methanogenesis and/or thermogenesis have altered the isotopic signature (Rice et al., 2008).

The Powder River Basin paradigm is indeed very similar to that modeled by Faiz and Hendry (2006) and Faiz et al. (2003) in the Permian coals of the Sydney Basin in Australia. The deuterium isotope (δD) for methane, in conjunction with carbon isotope, was used to distinguish between carbon dioxide reduction and aceticlastic reaction pathways. These workers determined that the δD values of biogenic methane are mainly related to that of the interstitial water and the proportions of methane derived from carbon dioxide and acetate. Deuterium isotope analyses of methane generated from aceticlastic reactions indicate that the hydrogen is mainly derived from methyl groups of the organic matter and only 25% is derived from formation water (Faiz & Hendry 2006; Valentine, Chidthaisong, Rice, Reeburgh, & Tyler, 2004; Whiticar 1999). In contrast, most of the hydrogen is derived from sources associated with formation water for carbon dioxide reduction reactions (Daniels, Fulton, Spencer, & Orme-Johnson, 1980; Faiz & Hendry, 2006; Whiticar, 1999).

GAS SORPTION, STORAGE, AND DIFFUSION

The big attraction toward coal as gas reservoirs is its potential to store large gas-in-place resources. This view has been propagated by the idea that coal consists of a large internal surface area in which it can store huge volumes of methane. This is reinforced by comparison of the coal surface area to that of conventional reservoirs. Nuccio (2000) suggested that an unconventional coal reservoir has from six to seven times more surface area than a conventional natural gas reservoir of equal rock volume can hold. Another axiom used is 1 g of coal can have as much surface areas as several football fields. Moore (2012) calculated that 5 million m^3 of coal contain internal pore surface area of more than 1500 km^2, which is equivalent to the size of the state of Rhode Island of the United States.

In order to make these assumptions true, coal must be at full gas saturation and the mechanism to hold the gas in place is present in the coal reservoir. Coal reservoirs are not always fully gas saturated, that is, the in situ storage capacity is equal to the gas content (Mavor & Nelson, 1997). However, coal reservoirs are commonly more undersaturated in field conditions than saturated and this state of the reservoir may be mainly influenced by temperature and pressure (e.g. stress confining and hydrostatic), which holds the gas in the coal similar to the mechanism that holds gas in the peat precursor (see Chapter 3). More importantly, these characteristics of the coal reservoir may control the recoverability of the gas; thus, characterization of the physical properties of the coal reservoir is important to the reservoir engineers and modelers (Chapter 5). The concept of the relationship between gas generation and storage as related to coal rank was best described by Rice (1993) and summarized in Figure 4.28. Rates of gas generation increase but the capacity of gas storage decreases with coal rank. Gas storage in the molecular structures is slightly higher in anthracite coals than in the high-volatile bituminous coals. However, the total storage capacity is exponentially higher (e.g. from >25 to >100 cm^3/g or about four times greater) in the high-volatile bituminous coal than in the anthracite coals.

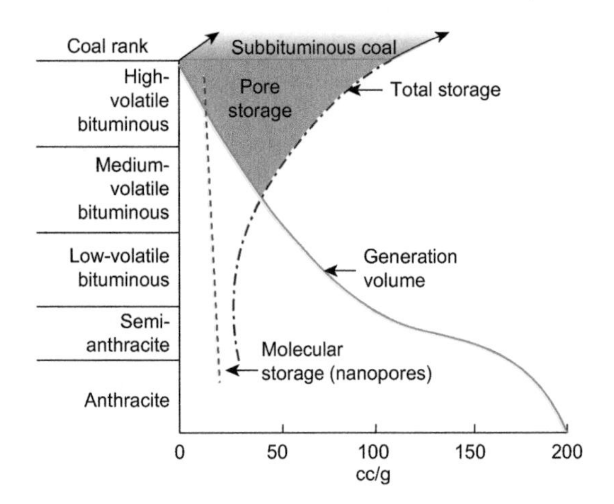

FIGURE 4.28 Diagram showing amount of methane generated and stored in the coal pore system in relation to coal rank. cc/g, cubic centimeter per gram. *Source: Modified from Rice, 1993.*

The storage capacity of the pore systems in coal matrix increases from medium-volatile to high-volatile bituminous coals. The increase of pore system storage capacity is extended into the low-rank coals (e.g. subbituminous coals). The generation volume of gas is higher in the anthracite coals (e.g. about 100–200 cm^3/g) than in the high-volatile bituminous coal (<25 cm^3/g). Rice (1993) suggested that these observations are given with the caveat that the (1) potential gas-in-place resource can be estimated, (2) the amount of gas desorbed out of the coal can be estimated, and (3) the migrated gas is either trapped in adjoining reservoirs (e.g. sandstones) or released to the atmosphere.

Dynamics of Gas Sorption and Storage

The advent of coalbed gas development and potential enhanced coalbed gas related to carbon dioxide sequestration has brought about renewed and landmark studies of sorption or adsorption of gas. In general, gas sorption is studied for the past 240 years but it was not until the past two decades that it underwent concentrated experimental and numeral modeling studies in relation to coal-absorbent surfaces (Florentin, Aziz, Black, & Neighm, 2009; Harpalani & Chen, 1995; Karacan, 2003; Larsen, 2004; Mosher, 2011). Excellent reviews of how adsorbed gas played an important role to coalbed gas and carbon dioxide-related sorption processes and gas content calculations are provided by Busch and Gensterblum (2011) and Moore (2012).

Coalbed gas, which is generated either by thermogenesis or by methanogenesis, is stored in coal reservoirs by absorption and adsorption on or within macromolecular nanopores, pores (e.g. micropores, mesopores, and macropores), cleats, and/or fractures (Clarkson & Bustin, 1999; Juntgen & Karwell, 1966; Laubach, Marrett, Olson, & Scott, 1998; Levine, 1993; Levine, Schlosberg, & Silbernagel, 1982, Marks, 2006; Mavor & Nelson, 1997; Medek, Weishauptova, & Kovar, 2006; Rice, 1993; Van Krevelen, 1981; Yi, Akkutlu, & Deutsch, 2008). Gas absorption is a process of molecular assimilation within a liquid phase or in the latticework of a solid (Mavor & Nelson, 1997). In contrast, gas adsorption is a process of attachment to a surface, of fluid phase molecules, by chemical and/or physical bonds (Mavor & Nelson, 1997). Gas sorption as used here encompassed both absorption and adsorption processes.

More often, the absorption and adsorption terms are used interchangeably but adsorption is the more common process given that the majority of the gas molecules are attached on the internal surface area of the pore system in the coal matrix and that fractures (or cleats) where gas molecules are absorbed make up only 2–3%. How much gas is sorbed on or within macromolecular nanopores is not well known. Thus, there are three states or mechanisms of gas storage in coal (Karacan, 2003; Larsen, 2004; Medek et al., 2006; Romanov, Goodman, & Larsen, 2006; Yi, Akkutlu, Karacan, & Clarkson, 2009): (1) adsorbed state in the fracture (or cleats) system, (2) adsorbed state in the pore system, and (3) absorbed/dissolved state in intramolecular nanopores within the macromolecular structures of the coal material.

The absorption and adsorption processes are controlled by changes in temperature and/or pressure (e.g. confining overburden stress and hydrostatic) conditions. That is, increase in temperature results in decrease storage capacity and increase in pressure results in increase storage capacity (Rice, 1993). In addition, coal maturation influences storage capacity, which is more in low-rank coal than in high-rank coal. Thus, gas sorption and storage are more complex than in conventional reservoir such as sandstone. The coal reservoir has multifaceted and hierarchical characteristics influenced by heterogeneity of the coal structures (Figure 4.29). That is, the coal is a network of pore and fracture systems in which as the confining pressure is reduced, the first gas to flow is "free gas" held by compression in fractures. The gas molecules in the "nanopore" and/or microporosity systems in the coal matrix are diffused and flow to the mesopores and macropores, which in turn migrate to the fractures (or cleats).

Molecular Structures vs Gas Sorption

The intramolecular and/or macromolecular structures of the coal matrix play a role in the gas sorption–storage system of the coal. How significant is this role is not well understood. However, some advances are being made, which are keyed to carbon dioxide sequestration related to enhanced production of coalbed gas.

The relationships of molecular structures to gas sorption and storage in coal have been investigated for a long time but their practical applications to coalbed gas development are not well known. Studies have been experimental using high-transmission electron microscopy and wide-angle X-ray scattering and numerical modeling applications. With the use of these instrumentations, it has been determined that the coal matrix consists of layered structures of covalently linked polycyclic

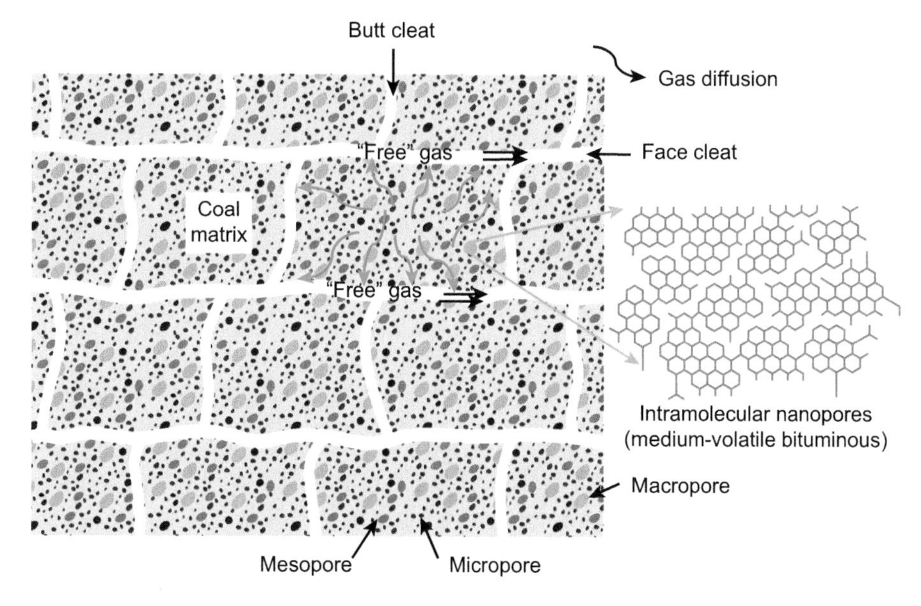

FIGURE 4.29 Diagram showing the intramolecular nanopores, pore and cleat (fracture) systems, and gas diffusion in the coal matrix to "free" and dissolved gas in the face and butt cleats. *Source: Modified from Kolesar (1986) and Stach et al. (1982).*

aromatic macromolecular units and that the distance (e.g. 3.5–4 Å) between these layers accounts for the "nanopores" and/or microporosity of the coal matrix (Figure 4.29) (Cartz, Diamond, & Hirsch, 1956; Marks, 2006; Medek et al., 2006; Ozdemir, 2004; Sharma, Kyotani, & Tomita, 2000). Thus, the term "molecular sieve" is aptly used for the microporosity of coal matrix.

Understanding the complex organic composition in terms of their intermolecular and macromolecular structures and bonding of coals has been demonstrated in the laboratory relating to absorption and adsorption of gas (Marks, 2006; Medek et al., 2006). Marks (2006) studied the small hydrocarbon molecules trapped in the macromolecular network in order to understand the complex composition and structure of bituminous coals. These bituminous coals are composed partly of layered network of polycyclic aromatic units with differences in average interlayer and ordering of layers, which may be correlated to absorption and swelling of coal. Marks (2006) suggests that the degree of coalification as well as maceral and paleobotanical contents of the coals have a significant influence on the structure.

Medek et al. (2006) using numerical modeling to combine adsorption and absorption isotherms on coal suggested occupation of intramolecular cavities in the macromolecular coal matrix. They designated these cavities as a nanoporous system. An isotherm or sorption isotherm of a gas (e.g. methane and carbon dioxide) attached on a surface of solid (absorbent), which describes sorbed gas storage capacity, is a function of pressure at constant temperature (Mavor & Nelson, 1997). Medek et al. (2006) demonstrated from the difference of spatial parameters calculated from the adsorption and absorption isotherms that the effective size of intramolecular cavities maybe estimated. Research by Marks (2006) and Medek et al. (2006) have demonstrated the presence of intramolecular voids but their relationship to microporosity volume is not well understood.

Pores vs Gas Sorption

A pore is an opening in the coal matrix (absorbent) that is more deep than wide and varies in width and shape along its length. Pores may be isolated or interconnected with a network (Figure 4.30) expressed by relative size distribution described ably by Clarkson and Bustin 1999, Gan, Nandi, & Watler, 1972, Levine et al., 1982, Radlinski et al., 2004, Sing, 1982, Yi et al., 2009, and Zhang, Tang, Tang, Pan, & Yang, 2010. The pore size spectrum of the pore system is shown in Table 4.8 compiled from Gan et al., (1972), Senel, Guruz, and Yucel (2001), and Sing (1982). In coal, micropores make up an average of about 77%, mesopores an average of about 5%, and macropores an average of about 15% with the remaining average 3% consisting of cleats/fractures (Mardon, Takacs, Eble, Hower, & Mastalerz, 2007). However, average porosity differs with coal rank. Measurements of coal porosity and their internal surface area are subjective and do not precisely depend on the kind of fluid injection used (e.g. mercury, helium) and compressibility of the coal at high temperature and pressure, which control fluid penetration (Zhang et al., 2010). Also, all pores are not open and certain amount of pores may be closed by mineralization.

The origin of coal matrix pores is not too well understood except that they may be inherited from the original peat structures or formed during coalification. The study of Zhang et al. 2010 may have shed some understanding of the origins of macropores, mesopores, and micropores. Macropores (Figure 4.31) are probably voids or spaces of residual cell structures of precursor plants or in the maceral detritus, which may be termed primary pores (Wildman & Derbyshire, 1991; Zhang et al., 2010; Lin, 2010). The origin of mesopores is very similar to the macropores,

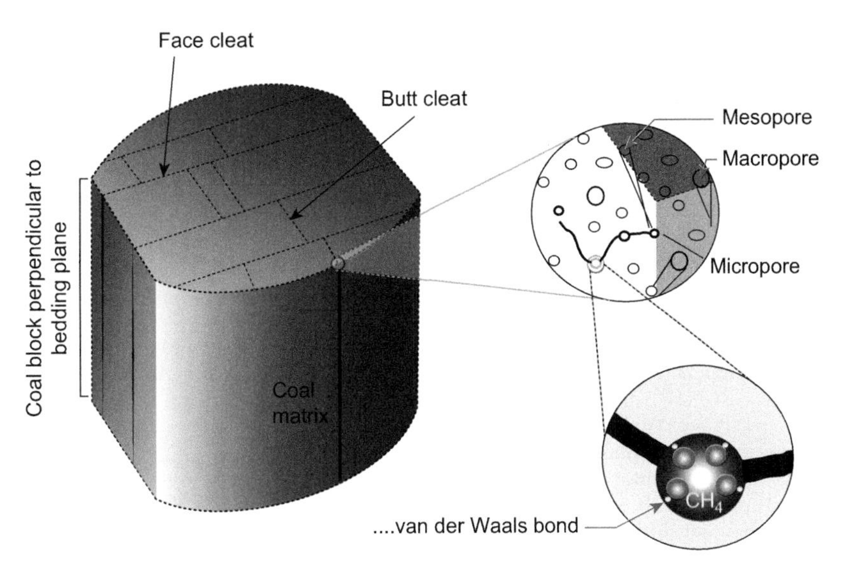

FIGURE 4.30 Diagram showing pore distribution and connectivity in the coal matrix, and adsorbed methane in a micropore.

some of which may be smaller compressed macropores (Zhang, 2001; Zhang et al., 2010). Therefore, macropores and mesopores are related to the original cellular structures of the plant remains (e.g. woody versus nonwoody) or primary pores, which are inherited from the precursor peat and its depositional environment. The origin of micropores is probably related to the intermolecular spaces in the coal macromolecular structures (e.g. nanopores) developed by gelification of the organic matter during coalification and compression of some primary pores to smaller pores (Clarkson & Bustin, 1996; Marsh, 1987; Zhang, 2001; Zhang et al., 2010). Thus, the origin of micropores is controlled by coalification or is coal rank sensitive and may be secondary in origin. These observations may directly relate to the original concept that the total volume of micropores increases with rank (e.g. lignite to anthracite) from about 10% to 75% (Levine 1993; Moore, 2012). However, the micropore volume decreases from subbituminous to bituminous from 0.08 cc/g to 0.02 cc/g (Figure 4.32) (Levine, 1993). This suggests that although more micropores are formed during coalification, the pore volume decreases toward the higher rank coals. Finally, the micropores, which are the primary sites of gas adsorption and diffusion, are directly related to the gas storage capacity.

TABLE 4.8 Coal Pore Types and Sizes (Width) in Angstrom Unit and Nanometers

	Size (Width)	
Pore Types	**Angstrom**	**Nanometers (nm)**
Micropore	2–20	<2
Mesopore	20–500	2–50
Macropore	>500	>50

One angstrom is one-ten billion of a meter; one nanometer (nm) is 1/1,000,000,000 of a meter.
Source: Modified from Clarkson and Bustin (1996), Flores (2008a), Gamson et al. (1993), Gan et al. (1972), Sing (1982).

In general, the percentage total volume of micropores sharply increases and that of macropores sharply decreases from lignite to anthracite coals (Levine, 1993; Moore, 2012). The question

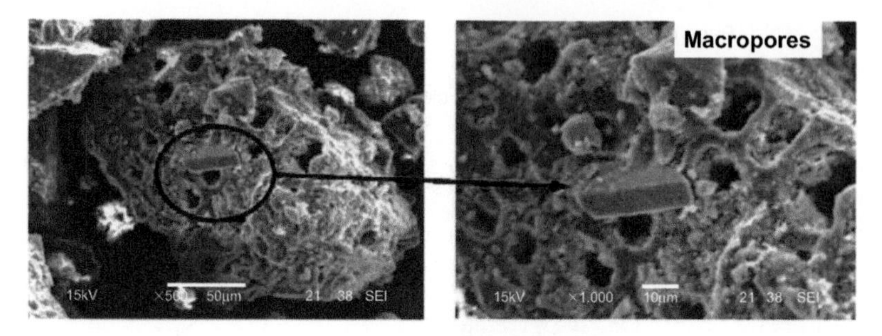

FIGURE 4.31 Scanning electron microscope images showing very large pores (>10 μm) of coal; magnification: 500× for left image and 1000× for right image. μm = micrometer or 1/1000000 of a meter. *Source: Adopted from Lin (2010).*

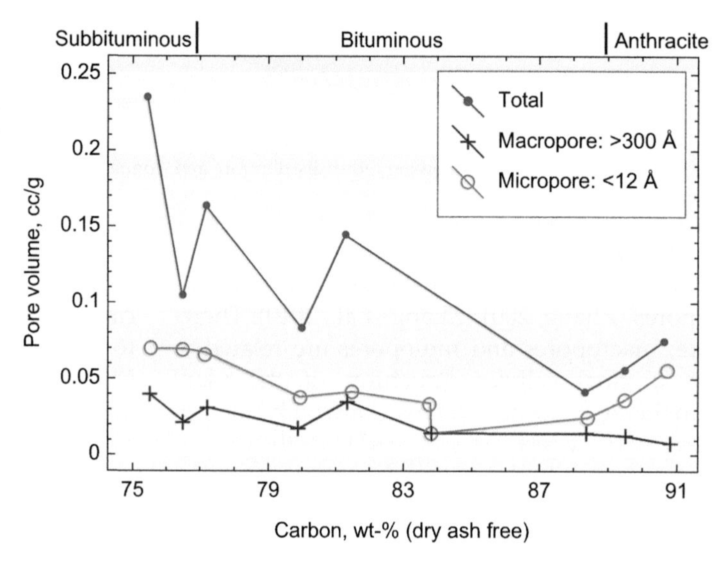

FIGURE 4.32 Diagram showing variations in micropore and macropore volumes in relation to carbon content and coal rank. cc/g, cubic centimeter per gram; Å, angstrom; wt-%, weight percent. *Source: Modified from Levine (1993).*

arises as to the origin of these pores, which Levine (1993) suggested to be inherited either from the precursor peat (e.g. primary porosity) or during coalification (e.g. secondary). However, the pore volume of micropores and macropores and total pore volume slightly decrease from subbituminous to anthracite coals, and the total pore volume increases from the low-volatile bituminous to anthracite (Figure 4.32). According to Levine (1993), the increased total and micropore volumes may be explained by partial occlusion of the pores by oil, which are opened by dissolution at higher temperature.

Although it is generally recognized that the coal porosity system includes both large cleats or fractures (macroporosity) and small pores (microporosity) (Clarkson & Bustin, 1999; Cui, Bustin, & Dipple, 2004; Harpalani & Chen, 1997; Rodrigues & Lemos de Sousa, 2002), in this book, the two are classified as separate systems with the cleat (fracture) system mainly related to permeability. Also, this dual system should avoid confusion of terminologies between "macropore" describing the cleats/fractures and "macropore" relating to the largest pore of the micropores. The large percentage of the micropores (<2 Å in diameter/width) does not give account of the potential inaccessibility of clogged and/or smaller diameter (width) pores to gas adsorption and consequently gas flow.

The "molecular sieve" and pore character of the coal matrix make the total sorption–storage system a suspect, as argued by previous studies (Thomas & Damberger, 1976; Levine, 1993; Rice, 1993). Levine (1993) argued about the suppression of pores from oil occlusion in high-rank coals, which are difficult to dislodge. Thus, suppression of both coal nanopore and pore systems contribute to the uncertainty about their role in the gas sorption and storage capacity of the coal matrix. This coal matrix characterization may partly account for the undersaturated gaseous nature of the reservoir and cast some doubt to the accuracy of the gas estimates from enhanced recovery and sequestration.

Levine (1991) suggested that gas storage capacity of coal is unfavorably affected by the presence of liptinite and hydrogen-rich vitrinite macerals. These coal macerals are prone to produce hydrogen-rich kerogen or high-molecular-weight oil during increased maturation or coalification. These liquid hydrocarbons tend to clog pore systems of high-rank coals such as high-volatile bituminous coals in which part of the surface area is lost leading to reduction of storage capacity (Thomas & Damberger, 1976; Rice, 1993). However, increased storage capacity may result in thermal cracking of the liquid hydrocarbons in higher rank coals from low-volatile bituminous to anthracite (Thomas & Damberger, 1976). However, according to Hunt (1991) and Littke and Leythaeuser (1993), heavy hydrocarbons (e.g. C_{2-5}) have stronger adsorption similarity relative to methane, which suggests that they may be more difficult to dislodge from occluded pores.

A new experimental approach to determine the actual accessibility of pores in the coal matrix to methane and carbon dioxide was unveiled in a breakthrough study by Melnichenko et al. (2012). The study concentrated on comparison of low-volatile bituminous (79–84% carbon, dry ash-free basis) coals with differing porosity (7–16% Hg porosity) from the Illinois Basin, United States, and Bowen Basin, Australia. The experiments demonstrated the existence of closed pores in coals, which are not accessible to methane and carbon dioxide. Accessibility for each coal is a function of pore size and maceral composition. The part of the pore system accessible to carbon dioxide and methane in each coal is high in more porous inertinite-rich coal and low in less-porous vitrinite-rich coal. Melnichenko et al. (2012) concluded that the amount of accessible pores in coal probably is related to the total porosity and formation of the interconnected network of pores accessible to fluids, which is a function of pore size and made possible in more porous coals.

Thus, sorption of gas especially in high-rank coals is suppressed particularly near the stage of bituminization (Figure 4.12) (Levine, 1991). This results in competition of molecular sites for sorption between the heavy hydrocarbons, methane, and carbon dioxide. Also, the storage capacity for methane decreases with increased coal rank contrary to the long held concept that storage capacity increases with coal maturation. Faiz et al. (2007a) reported that the methane sorption capacity decreases with rank for some Sydney Basin coals. Thus, molecular structures have partly negative impact on gas storage capacity in coal reservoirs particularly in high-rank coals.

Gas sorption and storage depend on the molecular size and weight generated during gasification and coalification (Levine, 1991, 1993; Rice, 1993; Van Krevelen, 1981). Van Krevelen (1981) originally described the microstructures of coal as a molecular "sieve" within which methane is stored. The molecular structure of coal is such that it can only accommodate small molecules of gases like methane and carbon dioxide (Rice, 1993). This depends on the molecular size as well as the pore size of the microstructures. The pore sizes and pore throat or

TABLE 4.9 Sizes of Gas Molecules in Nanometer

Gas Molecules	Size (Nanometer)
Methane (CH_4)	0.38
Carbon dioxide (CO_2)	0.33
Nitrogen (N_2)	0.36

Source: Modified from Flores (2008a) and Kurniawan, Bhatia, and Rudolph (2006).

opening may control the amount of stored adsorbed and desorbed methane, carbon dioxide, and nitrogen with molecular diameters shown in Table 4.9 (Flores, 2008a). The equivalent sizes of these molecules and others are in angstrom. The adsorbed layers of gas molecules, in turn, depend on the size and internal surface area of the pore system, which may be interconnected.

Cleats (Fractures) vs Gas Sorption

Cleats are two sets of open and/or closed (e.g. mineralized) fractures or joints perpendicular to bedding planes of coal beds (Close, 1993; Laubach et al., 1998). The two sets of cleats include the face cleat that is more continuous than the butt cleat that is orthogonal to and terminates against the face cleat (Figure 4.33). The long, high, and wide cleats or cracks are often identified as "macropores" by noncoal geologists, which in this book should be reserved and belong only to the pore system with pores having ovoid-like apertures. The length, which is straight or curvilinear, of the cleat is parallel to the surface of the bedding plane of the coal bed. The height is perpendicular to the bedding plane of the coal bed. The width of the cleat is the opening or aperture along the length and is parallel to the surface of the bedding plane of the coal bed. The distance between two parallel cleats (e.g. face and butt) is the spacing. These properties define the three-dimensional network of cleats and potential connectivity or

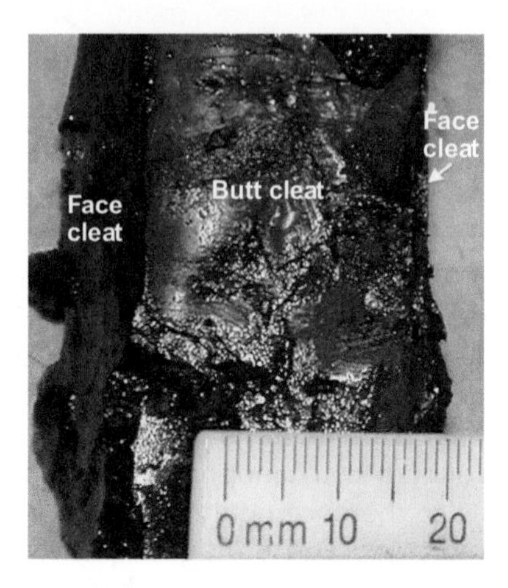

FIGURE 4.33 Close-up view of a cleat (fracture) system including orthogonal face and butt cleats (about 90°) from a core in the Powder River Basin.

permeability, which permits the coal reservoir to flow water and gas to the borehole (more discussions on reservoir characterization in Chapter 5).

In the context of reservoir characterization, cleat or fracture system partitions blocks of coal matrix, which contain the pore system as well as the intramolecular nanopores. It is suggested that gas sorbed in the nanopores and pores of the partitioned coal matrices are the primary storage system and sorbed gas in the cleats or fractures are the secondary storage system (Mavor & Gunter 2004).

Gas sorption in cleats or fractures is in two states: (1) "free" or vapor stored by compression and (2) solution dissolved in water (Mavor & Nelson, 1997). Thus, the cleat system contains water, the saturation of which will control the amount of gas adsorbed and/or in solution. Water is considered to compete with gas molecules for storage sites in the cleat system. According to Mavor and Nelson (1997), the volume of sorbed gas in both states is small compared to the volume of gas sorbed in the nanopore and pore systems in the coal matrix. However, these free and dissolved gases are the first to be desorbed and produced in the well.

Principle of Adsorption

The principle of gas adsorption is based on theoretical kinetics and statistical mechanical concepts, which will not be discussed in depth in this book but the reader is referred to the classic works by Aranovich and Donahue (1995), Brunauer, Emmett, and Teller (1938), Dubinin (1966), and Langmuir (1918). In addition, modifications and clarifications of this early concept are found in Busch and Gensterblum (2011), Clarkson, Bustin, and Levy (1997), Mosher, (2011), Ozdemir, Morsi, and Schroeder (2004), and Yao, Liu, Tang, Tang, and Huang (2008). The importance of the adsorption concept is its application to carbon dioxide sequestration and enhanced recovery of methane (Lin, 2010; Mosher, 2011; Ozdemir, 2004).

Gas adsorption on the internal surface area of coal includes three models: monolayer (Type I), multilayer (Type II), and pore filling (Type III) (Figure 4.34). The monolayer adsorption by Langmuir (1918) is the simplest model, which assumes that gas molecules are adsorbed as a set number on a solid surface. The Langmuir isotherm (see next section) model was modified by Brunauer et al. (1938) by proposing a multilayer adsorption model adding a second layer of gas molecules on the monolayer of Langmuir (1918) as well as a third and overlying layers. Both adsorption models are applicable to a flat surface or a pore surface (e.g. curve) when the radius of the pore is big (Ozdemir, 2004). However, these models are not applicable if the pore size is a few gas molecules wide, in which case the pore-filling model by Dubinin (1966) is applied.

When the gas molecules come in contact to a solid surface they are attached or bonded by chemical or physical reactions (Ozdemir, 2004; Ozdemir et al., 2004). Chemical adsorption (e.g. chemisorption) is when the gas molecules are bound to the surface by chemical bond. Physical adsorption (e.g. physisorption) is when the gas molecules are bound by the sum of attractive or repulsive forces between molecules and electrostatic forces. Castellan (1983) demonstrated the difference between physical and chemical adsorption by the reaction of nitrogen on an

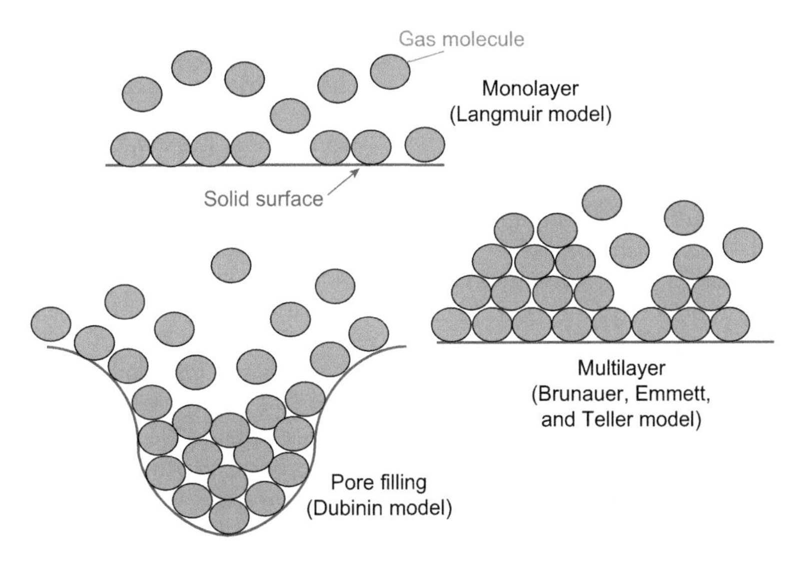

FIGURE 4.34 Simplified diagrams of adsorption mechanisms as proposed by the Langmuir model; Brunauer, Emmett, and Teller model; and Dubinin model. *Source: Compiled from Brunauer et al. (1938), Dubinin (1966), Langmuir (1918), and Ozdemir (2004).*

iron surface. Nitrogen is physically adsorbed as nitrogen molecules on the iron surface at −190 °C. However, the amount of adsorbed nitrogen molecules decreases with increased temperature such that at about 500 °C, only nitrogen atoms are chemically adsorbed on the iron surface (Castellan, 1983).

According to Mosher (2011), for the methane–carbon interaction, the most dominant intermolecular forces are the dispersion–repulsion forces or van der Waals and Pauli repulsion, respectively. It is believed that the repulsion forces are stronger than the dispersion forces, but combined, they are both dominant forces. These forces are weaker than chemical bond but are a reversible mechanism. It is assumed that the monolayer adsorption occurs at low pressure and the multilayer adsorption occurs at high pressure with increasing bonding force. Normally, decreasing the pressure weakens the bond between the solid surface and the gas molecules. For example, if hydrostatic pressure (e.g. hydrostatic head) is the principal force of bonding, the water is usually removed or dewatered to flow the gas molecules. In addition, it is believed by Bustin and Bustin (2008) and Crosdale, Moore, and Mares (2008) that adsorption may be sensitive to temperature, which can be specific to highly variable.

Langmuir Isotherms

In 1916, Irving Langmuir developed an equation to describe the relationship between the interdependence of gas molecules adsorbed as monolayer on a solid surface (plane) area held by pressure at a set temperature. Equilibrium is attained when molecules in the gas phase ("free") and the corresponding gas molecules or atoms are adsorbed on the solid surface (e.g. adsorbed as monolayer) and it is simply expressed by the following relation (Adeboye, 2011; Langmuir, 1918):

$$V = V_L P / P_L + P$$

where:

V = volume of gas adsorbed, standard cubic feet/ton or cubic meter/ton.
P = Pressure, pounds per square inch or pascal.
V_L = Langmuir volume, standard cubic feet/ton or cubic meter/ton; the monolayer adsorption capacity of the coal under investigation.
P_L = Langmuir pressure, the pressure at half the Langmuir volume, pounds per square inch or pascal (Langmuir, 1918).

Additional proposed adsorption equations to describe the behavior of gas molecules on a solid surface are recommended to the reader specifically for multilayer sorption (Brunauer et al., 1938) and for pore-filling volume (Dubinin, 1966). The multilayer and pore-filling volume equations calculate the surface and micropore volume areas, respectively, on which the gas molecules are adsorbed. For additional excellent explanations of adsorption isotherms, the reader is referred to Adeboye (2011), Byrne and Marsh, 1995, Gregg and Sing (1982), Marsh (1987), Mosher (2011), Ozdemir (2004), Parkyns and Quinn (1995), and Stoeckli (1995).

The commonly used isotherm model to measure the storage capacity of the porous coal matrix (e.g. sample no less than 100 g) is the Langmuir isotherm. Langmuir storage capacity is the maximum volume per sample weight of adsorbed phase gas corresponding to monolayer surface coverage. The storage capacity is the ideal holding capacity of a coal sample measured in the laboratory at a constant temperature (e.g. isotherm) equal to the reservoir temperature at depth and subjected to different pressures outside the coal matrix. Thus, Langmuir isotherm relationship provides an insight into the pressure dependence of the degree of surface gas adsorption and therefore, the storage gas capacity in the coal reservoir. The Langmuir pressure is equivalent to the pressure at which the storage gas capacity is equal to half the Langmuir capacity (Mavor & Nelson, 1997). That is, the coalbed gas storage capacity is only an ideal holding gas capacity and the measured gas content is the actual gas, which is the real gas charge according to Moore (2012) in the reservoir (see Chapter 9). This relationship of the Langmuir isotherm is important to reservoir engineers and modelers because it predicts (1) the pressure of gas desorption, (2) the volume of gas as pressure is reduced or gas recoverability, and (3) the residual gas in the reservoir on abandonment (Mavor & Nelson, 1997).

Many laboratories used various equipment setups according to their specifications to measure adsorption isotherms, particularly Langmuir isotherms. However, one principle they have in common is the use of an apparatus called a gas pycnometer or helium pycnometer very similar to the schematic diagram shown in Figure 4.35 (Adeboye, 2011; Cui, Bustin, & Bustin, 2009). The laboratory apparatus measures the volume of a granular, porous (or nonporous) solid sample based on gas displacement governed by Boyle's law of inversely proportional relationship between volume and pressure of gas (Cui et al., 2009; Webb, 2001). The Boyle law is expressed as the follows:

$$\rho V = k$$

where:

ρ = pressure of the system, pounds per square inch or pascal.
V = volume of the gas, standard cubic feet/ton or cubic meter/ton.
k = constant value representative of the pressure and volume of the system.

The apparatus may consist of two chambers with a valve to admit a gas under pressure

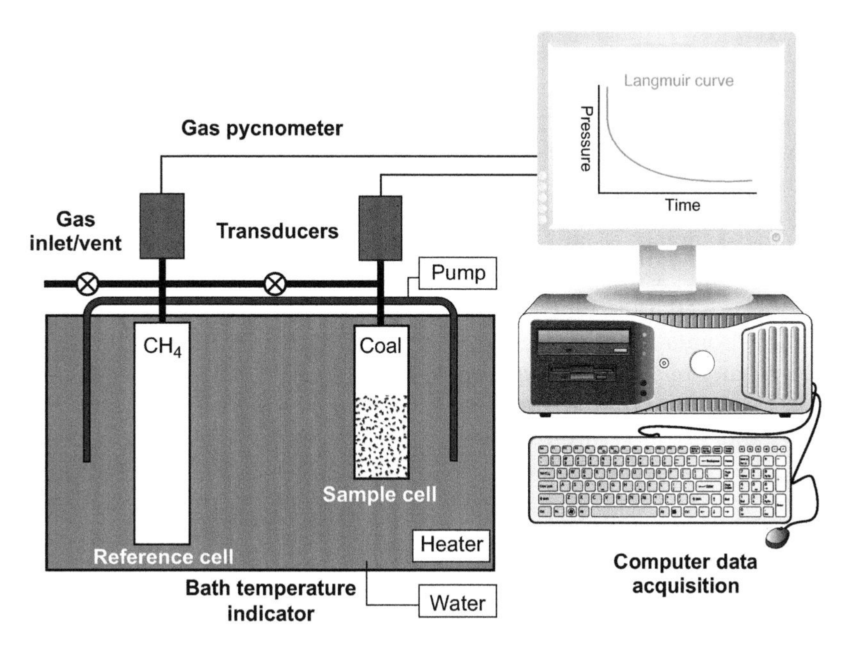

FIGURE 4.35 Simplified diagram depicting a generic adsorption isotherm setup. *Source: Compiled from Adeboye (2011), Cui et al. (2007), and from Gary Stricker (2010), (Personal Commun.) with communications with Marc Bustin.*

in one chamber and a transducer pressure-measuring device connected to the second chamber. A valve pathway connects the two chambers with a valve vent from the second chamber. That is, one chamber of the apparatus contains a known or reference volume of gas, which is used to douse a sample cell (e.g. coal sample) in the second chamber of the apparatus. The amount of gas adsorbed in the sample cell is determined using Boyle's law based on the change in pressure in the sample cell (RMB Earth's Science Consultants, 2005). The temperatures of the reference and sample cells are maintained at prescribed reservoir temperature from well data. The transducers, which are integrated to a computer equipped with specialized software, measure the pressures in the reference and sample cells. The computer keeps track of the transducers and switches for dousing and purging the reference cells (RMB Earth's Science Consultants, 2005). Also, the computer checks the state of equilibrium of gas adsorption of the sample and pressure such that no more gas is adsorbed by the sample at a specified pressure.

The Langmuir isotherm measurement is performed under high-pressure methane (or carbon dioxide) volumetric adsorption method. Several and separate pressure points are collected each after equilibrium is attained. Figure 4.36 shows an example of the Langmuir adsorption isotherm plots where 10 separate pressure points were collected of four sets of subbituminous coal samples on: (1) as-received, (2) dry, ash-free, (3) ash-free moisture-included, and (4) moisture-free ash-included bases. The coal sample on as-received basis is interpreted to have lowest adsorbed methane compared to the coal sample on dry, ash-free basis, which has the highest adsorbed methane. For example, at pressures of about 4138 kPa, the as-received basis sample has about 1.9 cc/g of adsorbed methane compared to the dry, ash-free sample, which has about 2.8 cc/g of adsorbed methane.

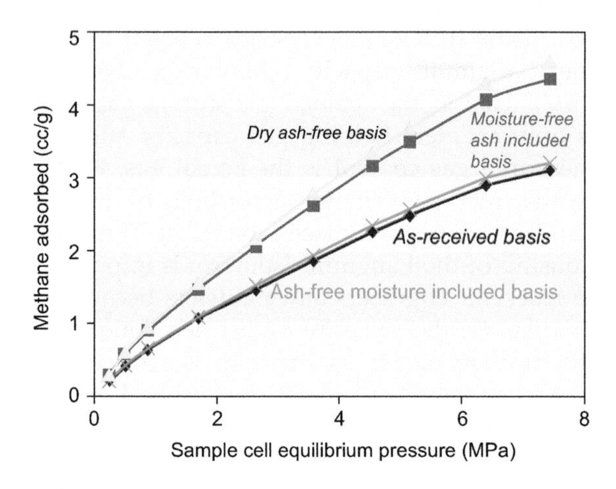

FIGURE 4.36 High-pressure methane adsorption plot of subbituminous coal samples on as-received, dry ash-free, ash-free moisture-included, and moisture-free ash-included bases. cc/g, cubic centimeter per gram; MPa, megapascal.

Although the dry ash-free coal sample contains more adsorbed methane, it is interpreted that the as-received coal sample is a more realistic measurement because it best represents and resembles the reservoir conditions.

Gas Diffusion and Flow

Gas diffusion and flow in coal beds may be described as a two-phase or binary transfer of gas diffused from the pore system (intramolecular nanopores and micropores–macropores) in the coal matrix to the cleat (fracture) system followed by flow of the gas through the reservoir to boreholes. This gas diffusion-flow system is known as "dual porosity system" and "primary and secondary porosity system" (Mavor & Gunter, 2004; Shi & Durucan, 2005). These workers detailed representation of the order of diffusion within the multiscale pore system in the coal matrix is called the bidisperse diffusion model. The model is a two-step gas diffusion process in the coal matrix occurring as (1) surface gas diffusion in the micropores (e.g. <2 nm) and (2) pore diffusion in the mesopores (2–50 nm) and macropores (>50 nm).

Simply put, gas diffusion in coal occurs in the matrix following Fick's law in contrast to the cleat (fracture) system in which gas is transported in laminar flow obeying Darcy's law. These two laws explain how the sorbed gas is transported from the pores in the coal matrix to the cleat (fracture) system and eventually to the open borehole. The Fick's diffusion law postulates that the diffusive flux goes from a high-concentration area to a low-concentration area proportional to the concentration gradient. The Darcy's law hypothesizes that the apparent velocity of a flowing fluid in a permeable medium is directly proportional to the applied pressure gradient. Thus, the concentration gradient drives the gas diffusion across the porous coal matrix and once the gas enters the cleat system, gas flow is driven by pressure gradient. This two-phase mechanism of gas flow is what separates the unconventional coal from the conventional reservoir (e.g. sandstone).

Although Fick's law is applicable to fluids, in this book, the principle focuses on its application to gas diffusion in coal. Several publications have described Fick's law as applied to coal; however, the reader is specially referred to a recent article by Moore (2012) for an additional excellent explanation. Moore (2012), and Zarrouk (2008) explained Fick's law as

$$F = -D\,\mathrm{d}C/\mathrm{d}x$$

where:

F = diffusion flux (kg/[m^2 s]),
D = effective diffusivity (m^2 of coal control surface area]/[s]),
$\mathrm{d}C/\mathrm{d}x$ = concentration gradient ([m^3 of gas]/[m^3 of coal]/[m length along gradient]),
C is in [m^3 of gas]/[m^3 of coal], and
x is in meters

The diffusion coefficient is the most critical part of the equation because it is in part a function of the properties of the gas. For example, a gas mixture of light and heavy molecules when passed through a porous–permeable medium is assumed to transport the light gas faster than the heavy gas at the end of the porous–permeable medium. This phenomenon is comparable to the coal bed reservoir, which consists of a binary gas system consisting of methane and carbon dioxide with proportional concentrations of about 9:1. The gases, however, are mainly transported in water-saturated coal reservoirs that behave as aquifers.

The study of Cui et al. (2004) on the differential transport of methane and carbon dioxide in the coal bed aquifers demonstrates some revealing results. Their work indicates that the differential transport of methane and carbon dioxide through the coal aquifer is a function of adsorption equilibrium and water solubility. They argue that although carbon dioxide is 20 times more soluble at lower temperature (50 °C), it is only several times more efficiently transported than methane. They account for the difference with the stronger adsorption of carbon dioxide than methane. This leads to enrichment of carbon dioxide such as seen in the San Juan and Powder River Basins. However, in the Powder River Basin, the enrichment may be complicated by nitrogen concentration probably derived from meteoric water as a result of groundwater recharge (Stricker et al. 2006).

Coal reservoirs are considered aquifers, which contain and transmit quantities of water under normal field condition. Therefore Darcy's law, which governs gas flow in the cleat (fracture) system of water-saturated coal reservoirs, has been unanimously suggested in the literature as an appropriate model to apply in the gas flow transport beyond the porous coal matrix (see Figure 4.29). However, the most recommended model of Darcy's law to apply in the coalbed gas flow is the one-dimensional, single-phase or linear dimensional model (Adeboye, 2011; Pakham, Cinar, & Moreby, 2009). Thus, once the diffused gas from the porous coal matrix enters the cleat (fracture) system, the gas flow or transport is governed by the following Darcy's law equation (Adeboye, 2011):

$$\nu = -\kappa/\mu_f(\nabla P - \rho_f g)$$

where:

ν = velocity flux (volume per time, e.g. ft^3/s or m^3/s),
κ = function of coal bed permeability (darcy/millidarcy),
μ_f = fluid viscosity (kg/(m s),
∇P = pressure gradient,
ρ_f = fluid density, and
g = acceleration due to gravity.

Darcy laminar flow is valid if cleat or fracture walls are smooth and the openings are unimpeded. However, in natural conditions and more often than not, they are closed by mineralization and/or pinched. Gamson, Beamish, and Johnson (1993) observed that cleats are partly blocked by diagenetic minerals (see Chapter 5). These workers suggested that combined diffusion and flow probably transport the gas under this condition depending on the size of the unfilled channels.

In general, the mechanism of gas flow in the porous and permeable coal is complex in that about 10% of methane in the cleat systems is as occluded gas and about 90% of the methane in the matrix pore system is as adsorbed gas (Basu & Singh, 1994). Presumably the large volume of gas from the microporous coal matrix must be diffused and flow out through the cleat system. However, there is still some debate about the mechanics of gas transport and the idea that gas flow through the cleat systems is the last stage. It has been demonstrated through computer and numerical modeling and field application that when a borehole is drilled through the coal reservoir, the pressure (e.g. confining stress, hydrostatic head) on the hole sidewall drops with respect to the in situ pressure causing radial or linear flow to the borehole (Basu & Singh, 1994). The pressure distribution is a function of time. Flow through the porous coal reservoir is treated at steady state if conditions are unchanged with time. However, depletion of reservoir pressure (e.g. dewatering) causing gas desorption is an unsteady state. The initial gas desorption produces the "free" gas from the cleat (fracture) system, which is at the early stage and not late stage as proposed. When flow becomes stabilized, reservoir conditions return to a steady state. Thus, Darcy's law applies when the initial gas pressure is high; then once the pressure decreases and stabilizes, Fick's law of diffusion applies.

During this later stage, radial gas flow or diffusion following Fick's law is applied within the blocks of coal matrices toward the cleat system (Figure 4.29). Because the face cleats are more continuous than the butt cleats, the ratio of permeability in face cleats over butt cleats is about 1:1 to 1:17.

ROLE OF HYDROSTATIC PRESSURE IN GAS DESORPTION

More often than not, the most common cause of coalbed gas reservoirs' unexpected low gas content can be traced to depletion of hydrostatic pressure in the reservoir. In this section, the role of hydrostatic pressure is discussed in relationship to gas diffusion and flow or transport within the coal reservoir. The ability of the coal reservoir to move gas is controlled by the porosity and permeability properties as well the maceral composition of the coal. However, the most important external factor that influences the gas movement is hydrostatic pressure. The hydrostatic pressure is important to gas production and additional discussion of this process is given in Chapter 7.

The concept of hydrostatic pressure, which is demonstrated in Figure 4.37, is applied by a

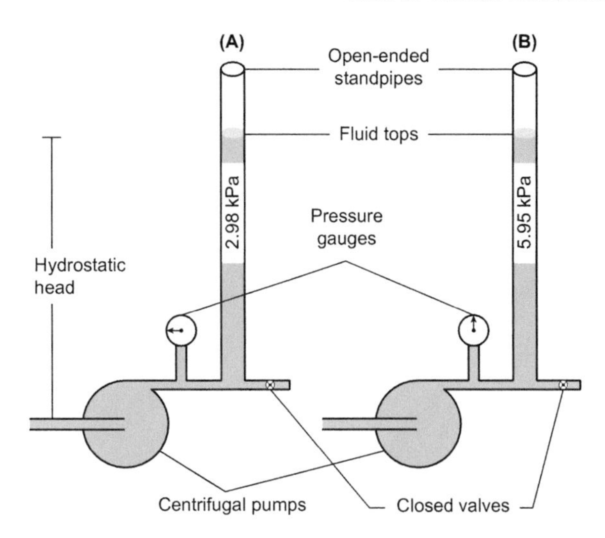

FIGURE 4.37 Diagram of hydrostatic head and pressure. The tops of the fluid columns (A) and (B) are the same but the pressure gauge indicators are dissimilar due to different fluid densities. The hydrostatic head is the vertical column and hydrostatic pressure is the force exerted by the fluid. *Source: Modified from Schlumberger (2012).*

static column of water at a given point within the fluid caused by the force of gravity (Schlumberger, 2012). The pressure is measured from the elevation of the top to the bottom of the water column known as hydrostatic head. Thus, hydrostatic pressure increases in proportion to depth because of the increasing weight of water column exerted as a downward force from above. In addition, the weight of the water column is controlled by the density of the fluids (e.g. freshwater versus brackish water) as shown in Figure 4.37 (Schlumberger, 2012). Normally, the coal reservoir is overlain by the water column whose weight is applied on a unit area of the reservoir at depth in addition to pressure from weight acting on the surface of the water. For example, the hydrostatic pressure is lower along the shallow basin margins in contrast to the higher hydrostatic pressure in the deeper basin centers.

Gas stored in the coal reservoir is held in the pore and cleat (fracture) systems by the hydrostatic pressure acting as a confining force (Busch & Gensterblum 2011). Once the hydrostatic pressure is reduced, for example, by dewatering, the gas is released or desorbed from the reservoir. This process of releasing the gas from the coal is termed desorption. However, the use of the term "gas desorption" pertains mainly to the release of the gas from the cleat and pore systems. Initially, the "free" gas from the fracture system flows to a low-pressure area (e.g. dewatering well), which is followed by diffusion of gas in the mesopores and macropores to the cleat systems. Last, the gas from the nanopores and/or micropores diffuses to the larger pores, described as bidisperse diffusion (Figure 4.29).

Reducing hydrostatic pressure of the coal reservoir is a standard procedure worldwide to desorb gas. Thus, the relationships of adsorption isotherms, in general, and high-pressure Langmuir isotherms, in particular, are important concepts to coalbed gas reservoir engineers. Mavor and Nelson (1997) suggested that the Langmuir isotherm relationship is used to determine (1) initial pressure of gas desorption, (2) amount of desorbed gas or recoverable gas on pressure reduction, and (3) amount of gas left or residual gas in the reservoir once abandoned. The most important use of the Langmuir isotherm relationship is prediction of the initial pressure at which the gas desorbs from the coal reservoir. Figure 4.38 is an example of a Langmuir isotherm relationship of a subbituminous coal in the Powder River Basin, Wyoming, United States. The estimated total gas content of the coal after desorption is about 2 cc/g with corresponding critical desorption pressure of about 3275 kPa and depth of about 67 m. Thus, the understanding of the in situ gas storage capacity of the coal eventually leads to estimation of gas-in-place and recoverable reserves. This topic is discussed in Chapter 8 on resource assessment methodologies.

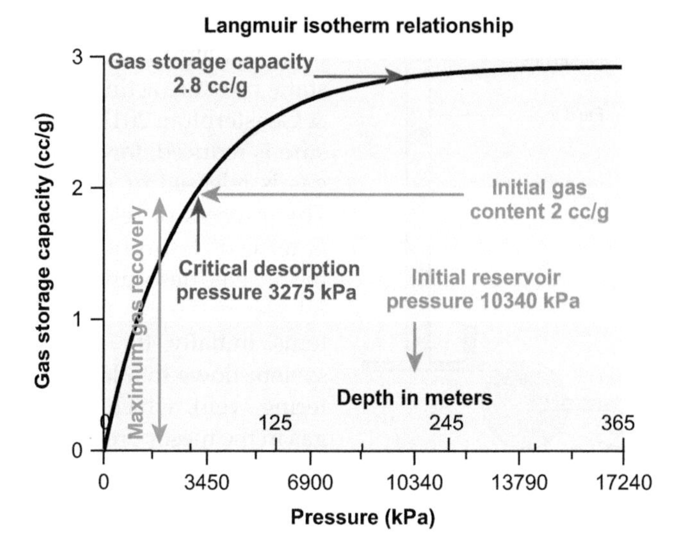

FIGURE 4.38 An example of Langmuir adsorption isotherm determining the desorption gas pressure, recovery, and residual of a subbituminous coal. cc/g, cubic centimeter per gram; kPa, kilopascal.

SUMMARY

The logical step after peat deposition and peatification (see Chapter 3) is their transformation to coal. Peatification occurs on the Earth's surface, from tens to hundreds of thousands of years, and involves mainly biochemical and microbial processes. Coal transformation occurs below the surface of the Earth, in millions of years, and involves chemical, physical, and structural processes. The succession of "metamorphic" changes of organic matter in coal during burial is termed coalification. Coal maturation through chemical and physical changes of the organic matter produces low- to high-rank coals. The causes of coalification are temperature and pressure, which are inherent to burial of the organic deposits at increasing depths, and modification by geologic time, tectonic deformation, and other external factors (e.g. igneous intrusions).

Although the influence of temperature on coalification is straightforward, the influence of pressure is confusing. Pressure is often confused with tectonic stress and other pressures exerted by mechanical energy (e.g. confining stress). Also, the influence of pressure is contradictory because it is believed to benefit the physical process of coalification but retards the chemical process. Furthermore, tectonic stress both affects the physical structure and promotes chemical compositional changes in the coal. However, the combined effects of temperature and pressure control the stages of coalification. The stages of coalification include dehydration, bituminization, debituminization, and graphitization with increased temperature and pressure. The processes of coalification in general involve changes in the chemical and physical properties of organic matter toward denser, drier, more carbon-rich, and harder coal. In addition, the chemical and physical changes are associated with color, luster, hardness, water, bulk density, aromaticity, and carbon, hydrogen, oxygen, and volatile matter composition of the coal. The degree of chemical and physical changes is expressed in various ranks of coal.

Classification system of coals by rank depends on the degree of coalification or progressive changes from the original precursor peat. The level of changes in percentages of fixed carbon, moisture (water), volatile matter, and calorific value are the criteria for recognition of lignite, subbituminous, bituminous, and anthracite coals as adopted by the ASTM and used mainly in North America. Despite the efforts of the United Nations Economic Commission for Europe to internationalize and codify coal rank classification directed toward producers, sellers, buyers and users of coal based on industrial and technical purposes, individualized systems continue to be developed according to the coal properties of each country. An internationalized classification that is based on analytical data from the Northern Hemisphere coals is not amenable to Gondwana coals (e.g. Australia, China, India, South America, and South Africa).

Major chemical and physical changes of the organic matter are expressed in the maceral (e.g. "organic mineral") composition of the coal. Based on the origin and microscopic properties, these macerals are grouped into (1) vitrinites, (2) liptinites, and (3) inertinites. Hydrocarbon by-products (e.g. oil and gas) are generated from the coal depending on the maceral types during various stages of coalification. Bituminization is the most important stage of coalification in which coal bed oil and gas are generated in the coal. The origin of coalbed gas is thermogenic and biogenic. Thermogenic gas is generated at high temperatures during the later stages of coalification. Biogenic gas is generated at low temperature by microbial activity mainly after deformation and uplift when coal beds are charged with meteoric water from groundwater systems. Thermogenic and biogenic gases are differentiated according to their carbon isotopic composition of methane. Thermogenic gas mainly contains the heavier carbon isotopic composition of methane compared with biogenic gas. Biogenic methane is generated by methanogens of the domain archaea, which are obligate anaerobic microbes that can survive in diverse microenvironments from extreme cold to hot temperatures.

Thermogenic and biogenic gases are stored in the pore and fracture systems of coal. The internal surface area of pores in the coal matrix is about six to seven times more than a conventional natural gas reservoir of equal rock volume can hold. About 90% of the gas is held in the coal matrix pore system (e.g. micropore, mesopore, and macropore). This coal matrix pore system does not account for the intramolecular nanopores in the macromolecular layers of the organic matter composition. The intramolecular nanopores of coal are not well known but theoretically may store both oil and gas molecules. The coalbed gas is stored by absorption within and adsorption on the void spaces in the coal. Sorption is by physical reaction (e.g. physisorption) and chemical reaction (e.g. chemisorption). Gas molecules are attached (or bonded) on the solid surface by dispersion–repulsion forces (e.g. van der Waal's and Pauli's repulsion) and electrostatic, covalent, and ionic bonds.

Gas is desorbed from the internal surface areas of the coal by depressurization (e.g. lower confining stress and dewatering to reduce hydrostatic pressure). Fick's law of gas diffusion governs gas flow or transport in the coal matrix. The gas flow is described as bidisperse diffusion in which gas is diffused from the micropores and migrates to the mesopores and macropores. The diffused gas in the coal matrix is moved to the permeable fracture system where Darcy's law governs flow. The gas stored in the fracture system is often called "free" gas (includes dissolved gas), which is the initial gas transported to the borehole and eventually first produced. About 10% of the coalbed gas is stored in the fracture system.

CHAPTER

5

Coal Composition and Reservoir Characterization

OUTLINE

Coal and Coalbed Gas
http://dx.doi.org/10.1016/B978-0-12-396972-9.00005-7

KEY ITEMS

- Coal is a multicomponent combustible and reservoir rock in terms of related elements and compounds of organic and inorganic or mineral matter constituents.
- Organic matter is a major constituent, which determines the combustible energy, source of hydrocarbons, and internal surface area in which gas is adsorbed and stored.
- Mineral matter or "inorganic minerals" is a minor component important in combustion and reservoir characterization.
- Mineral matter consists of elements, compounds, and minerals of environmental concerns that are hazardous air pollutants during combustion in coal-fired power plants.
- Mineral matter controls (1) internal architecture and compartmentalization, (2) pore availability to gas adsorption, (3) ability of gas to flow, (4) shrinking and swelling, (5) gas content, (6) gas recoverability, and (7) potential for enhanced gas recovery and carbon dioxide sequestration in coals.
- Mineral matter affects coal performance as a gas reservoir infilling pores and cleats or fractures preventing gas flow to production wells.
- Mineral matter consists of silicate, carbonate, sulfide, sulfate, phosphate, and other groups of minerals, which form from low to high temperatures, and forms fly and bottom ashes, gas effluents, and trace elements of environmental concerns.
- Organic composition of coal can be grouped as macerals, and their association as microlithotypes and macrolithotypes or lithotypes.
- Coal composition controls reservoir properties such as gas adsorption capacity, gas content, porosity, permeability, and gas transport.
- In general, not considering rank, vitrinite or woody coals adsorb more gas due to microporous characteristics. Low-rank coals are mainly composed of macropores and high-rank coals are dominated by micropores.
- Coal permeability is a function of width, length, height, aperture, spacing, frequency or density, and connectivity of cleats or fractures. They are endogenetic and exogenetic in origin based on formation during burial, compaction, and coalification and during deformation, uplift, and erosion of the basin of deposition.
- Coal permeability affects drilling, completion, stimulation, and production essential to successful coalbed gas development. The origin and properties of coal permeability are different from conventional reservoirs.
- Coal composition and physical properties impact reservoir characterization such as permeability and porosity, which in turn affect reservoir performance and sustainability of gas production.

- Coal shrinkage and swelling are caused by gas desorption and adsorption, impacting permeability and porosity. Coal shrinks as gas desorbs from pores and fractures lowering confining and pore pressures.
- Coal swelling is induced by gas adsorption, increasing pore pressure and expanding coal matrix. Coal swelling affects enhanced gas recovery with injection of carbon dioxide for sequestration.
- Reservoir characterization in relation to macrolithotypes, gas content, gas adsorption, porosity, and permeability is exemplified by the Powder River Basin coals.

INTRODUCTION

The core of the characterization of coal reservoirs is evaluation of the composition (e.g. organic and mineral matters) and internal structures (e.g. porosity and cleat or fracture systems). Knowledge of these two major coal properties will impact the success or failure of gas flow and deliverability to production wells. However, depending on one's perspectives, coalbed gas producibility may rely on other factors such as reservoir pressure, gas content, coal rank, and coal geology as well as drilling, completion, and production techniques (Logan, 1993). Nevertheless, these factors and techniques will be more efficiently applicable with better comprehension of the coal composition and internal structures of the reservoir.

The organic matter composition of coal is not uniform vertically and laterally throughout the reservoir. Consequently, the amount of gas is not uniform across the entire coal reservoir. For example, the ash content, which occurs as a dispersed component or discrete partings, usually reduces the gas content of the coal (Stricker et al., 2006). Furthermore, the composition of the organic matter may influence the nature and pattern of the porosity of the coal. A thought-provoking observation by Yao, Liu, Tang, Tang, & Huang (2008) is that the pore patterns resemble those of the original pores of plant materials. Whether this observation is related to intermolecular nanopores of organic matter, open cell walls of plants, or otherwise is not well understood. Another interesting note from personal experience in coring in the Powder River Basin was that certain coal intervals, which were difficult to core, were called "tight and dense" by well site geologists, but upon our megascopic examination of the cores, they were seen to be composed of "rubblelike" coal fragments that are shiny and have a high amount of vitrain.

Discussions in Chapter 4 have demonstrated that the vast majority or about 90% of the gas is stored in the coal matrix, which has low permeability, in which gas flow is primarily through diffusion from micropores to the larger pores. There is a debate about whether these micropores are really nanopores within the macromolecular structure of the organic matter. The significance of the vast storage of gas in the coal matrix in nanosize pores, which according to Radlinski et al. (2004), Mares et al. (2009), and Moore (2012) amount to an internal surface area of 3 m^2 for 1 cc of coal. Another important coal property in terms of reservoir characterization is the cleat and/or fracture system.

The underpinning of successful recoverability of gas is the shrinkage effect of the coal matrix, which is unique to a coal gas reservoir (Harpalani & Schraugnagel, 1990; Ozdemir & Schroeder, 2009). Shrinkage results from gas desorption and loss of moisture during dewatering and production. A countereffect of shrinkage in the coal matrix is swelling, which might be a problem for gas injection during enhanced recovery and carbon dioxide sequestration (Bustin, 2004). Another concern with swelling is injection of water to improve reservoir performance. Performing water injection–falloff to enhance permeability in the cleat (fracture) system, if not done properly, might fracture and ruin the integrity of the coal reservoir. A hindrance to this water-enhancement process is unexpected mineralization of cleats, which plug openings putting more stress to both the pore and the cleat (fracture) systems.

In comparison with Chapter 3 in which coal pores, permeability, cleats, and fractures were specifically related to gasification and gas storage, this chapter incorporates compositional and internal structural properties of coal applicable to its ability to produce coalbed gas. These factors encompass reservoir characterization with the analyses of the coal lithotypes in relation to coal composition, maceral groups, and megascopically recognizable features. The megascopic characterization of coal reservoir, which is linked to the analyses of coal facies of cores, is most pertinent to well site geologists, reservoir engineers and modelers, and drilling operators. Characterization of coal facies of cores is important for the determination of gas content, porosity, and permeability. In addition, reservoir integrity and performance are best understood with coal reservoir characterization.

COAL COMPOSITION

In a broader sense, coal is essentially composed of plant remains, which are coalified organic matter formed by exposure to high heat and pressure during burial. Organic matter comprises the major constituents of coal, which controls the heat of combustion (e.g. total calorific value) for steam generation and metallurgical use, source of hydrocarbons especially for gas generation or gasification, and the amount of internal surface area in which the gas is adsorbed and stored. As important as macerals are to hydrocarbon source and generation in coal, another constituent called organic matter or "minerals" are just as important if not more in terms of reservoir characterization (Figure 5.1). Mineral matter in coal is very important for utilization for combustion because if the elements, compounds, and minerals of concern are not cleaned up, they will make up much of the fly and bottom ashes affecting the efficiency as well as the release of hazardous air pollutants during combustion in the coal-fired power plants (Brownfield et al., 2004). Hower et al. (1998) suggested beneficiation or treatment of coal prior to coal utilization to reduce the release of trace elements. For additional in-depth and acknowledged reference of coal beneficiation, utilization, and environmental aspects, the reader is directed to a seminal publication edited by Suarez-Ruiz and Crelling (2008).

Mineral Matter

The role of mineral matter in coal reservoir is more far reaching than macerals to hydrocarbon generation and gasification. The mineral matter controls the (1) internal architecture and compartmentalization, (2) pore availability to gas adsorption, (3) ability of gas to flow, (4) tendency to shrink and swell, (5) gas content and in situ reserves, (6) gas recoverability, and (7) potential for enhanced gas recovery and carbon dioxide sequestration of the coal reservoir. Because of these impacts of mineral matter in the coal, their characterization should be a primary concern to reservoir engineers and modelers. If the development of minerals in coal is not well accounted for in reservoir characterization, unexpectedly poor well performance may result.

Minerals in coal range from a large component of clay minerals to trace elements. By definition, coal must contain less than 50% ash-forming mineral matter and is impure coal if it contains 30–50% mineral matter (Schweinfurth, 2002). Thus, the lowest threshold for mineral matter content of pure coal is less than 30%, but

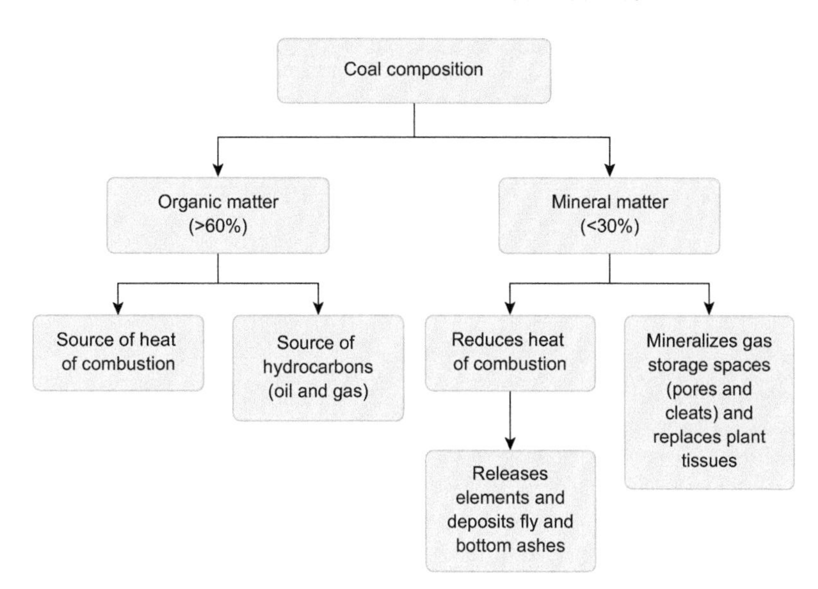

FIGURE 5.1 Flow chart showing the makeup of coal composition (e.g. organic and mineral matters) and their roles in coal utilization and as hydrocarbon source and reservoir.

perspectives on this may vary. Others may consider a much lower percentage of 20% (Stach et al., 1982). Regardless of this difference in observations, 20–30% is a significant amount to consider toward coal utilization (Taylor et al., 1998; Brownfield et al., 2004) and coalbed gas reservoir production (Gurba et al., 2001). According to Mukhopadhyay and Hatcher (1993), coal contains as much as 30% by volume of mineral matter. For additional comprehensive information on organic petrology and organic matter, the reader is directed to the outstanding references herein (Finkelman, 1994, 1995; Finkelman & Stanton, 1978; Hower & Wagner, 2012; Taylor et al., 1998; Ward, 2002).

There are more than 120 inorganic compounds including 76 elements in coal, but most of the elements are present in trace amounts (Schweinfurth, 2002). The trace elements are measured in concentrations on the order of parts per million. Although not common, some trace elements such as silver, zinc, or germanium are concentrated in some coal beds, which make them valuable resources (Finkelman & Brown, 1991). However, certain trace elements (e.g. arsenic, mercury, lead, chromium, selenium, cobalt, cadmium, etc.) of environmental concern are potentially hazardous, particularly if they are concentrated in more than trace amounts and cannot be cleaned or removed from combustible coal. In addition to these compounds and elements, there are about 120 minerals in coal (Schweinfurth, 2002). The mineral matter is grouped into silicates, carbonates, sulfides, sulfates, phosphates, and others (Table 5.1). The most common of these minerals in coal are the silicates and carbonates, and the least common are the sulfides, sulfates, and phosphates. By far, the most abundant group of mineral matter in coal is clay minerals or aluminosilicates (Speight, 2005).

Origin of Mineral Matter

The traditional concept of grouping minerals in coal follows that of Stach et al., 1982 and has been improved by Mukhopadhyay and Hatcher (1993), Taylor et al. (1998), Ward (2002), and Schweinfurth (2002). This involves grouping according to their origin as follows: (1) deposited as detritus with the precursor peat, (2) inherited

TABLE 5.1 Mineral Groups of Mineral Matter, Associated Common Minerals and Occurrence. Selected Minerals are Applied to the Coal Reservoir as Mainly Pore and Cleat Infillings

Mineral Groups	Minerals (Common)	Occurrence	Reservoir Application
Silicates	Quartz, chalcedony Clays (kaolinite, illite, smectite, chlorite) Feldspars (orthoclase, plagioclase)	Abundant to common	Pore, cell cavity, and cleat infillings
Carbonates	Calcite, dolomite, siderite	Common	Mainly cleat infilling
Sulfides	Pyrite, marcasite; sphalerite	Rare	Cleat/fracture infilling
Sulfates	Anhydrite, barite; gypsum	Rare to common	Mainly cleat infilling
Phosphates	Apatite, crandallite, goyazite, monazite	Rare	Cell and pore infillings
Others	Heavy minerals (anatase, rutile, zircon)	Rare	Not determined

Source: Modified from Demchuk (1992), Kortenski and Kostova (1996), Rao and Walsh (1999), and Ward (2002).

from the original plants, and (3) precipitated during and after coalification and postdeformation. The first two modes are primary origin and the last mode is secondary origin. However, a far-reaching paper on mineral matter in coal by Ward (2002) covered a more detailed description of their origin and description.

Mineral matter in coal is best defined by Ward (2002) as encompassing dissolved salts in pore water, inorganic elements and compounds, as well as discrete, crystalline, and noncrystalline minerals. The discrete mineral grains or particles include epiclastic and pyroclastic particles introduced into the peat during deposition. Epiclastic particles are detrital sediments and grains transported by water into the mires during flood events from nearby rivers or creeks (Ruppert, Stanton, Cecil, Eble, & Dulong, 1991). These sediments ranged from clay to sand-size particles. The clay particles may be the result of flocculation from acidic or saline waters. The sand and silt particles may be part of suspended load of floodwaters that infiltrate the mire vegetation. Floodwaters have the tendency to invade low-lying mires compared to the raised bogs (Chapter 3) (Flores, 1986; McCabe, 1984). The pyroclastic particles are airborne, derived mainly from volcanic ash falls, which are termed tonsteins. Tonsteins being airfall in origin and rich in kaolinite and smectite are more widely distributed than the epiclastic mineral particles; thus, they make better markers for correlation of coal beds (Bohor & Triplehorn, 1993; Knight, Burger, & Bieg, 2000).

The other group of inorganic minerals are formed as authigenic syngenetic and epigenetic minerals. The syngenetic minerals are formed during peatification or early diagenesis (Scott, Mattey, & Howard, 1996), whereas the epigenetic minerals are formed during coalification through postdeformation (Ward, 2002). The epigenetic minerals, in turn, are formed by fluids of various compositions and temperatures, which moved through the coal beds, remobilization of syngenetic minerals, igneous intrusions or magmatic influx, postdepositional fluid movement, hydrothermal processes, and groundwater flow (Baker, Bai, Hamilton, Golding, & Keene, 1995; Crowley, Ruppert, Stanton, & Belki, 1994; Demchuk, 1992; Finkelman, Bostick, Dulong, Senftle, & Thorpe, 1998; Hower et al., 2000; Hower, Williams, Eble, Sakulpitakphon, & Moecher, 2001; Karayigit, Spears, & Booth, 2000; Querol, Chinenon, & Lopez-Soler, 1989). These inorganic minerals play a major role in reservoir characterization and determine the poor or high capacity of the reservoir to adsorb and flow gas Box 5.1.

BOX 5.1

MINERALS IN COAL: THE MATTER BEHIND AIR POLLUTANTS—FACTS VS FICTION

Coal is a multicomponent combustible, composed predominantly of organic and subordinately of inorganic minerals (mineral matter), which in turn, contain many elements and compounds. During combustion at high temperatures, the elements and compounds of the minerals are transformed into new gaseous and solid matter. Coal combustion products in power plants consist of fly ash, bottom ash, boiler slag, and flue gas desulfurization material (Brownfield et al., 2005). Fly ash is the most important product that is released along with hazardous trace elements. Flue gas desulfurization is a process by which chemical (e.g. calcium carbonate) is injected into the flue gas to trap the sulfur. However, sulfur in the form of sulfur dioxide has created acid rain and its emission impacts global climate change because of its cooling effect (Kaufmann et al. 2011; Lu et al. 2010). The sulfur is mainly derived from unprocessed organic and inorganic mineral called pyrite. Fly ash is a residue produced during coal combustion consisting of fine particles derived from minerals (Brownfield et al., 2005). The composition of the fly ash reflects the original elements (e.g. calcium, iron, magnesium and phosphorus) and minerals (e.g. quartz, kaolinite, clays, pyrite, calcite, and others) of the feed coal. Minerals in feed coal and fly ash by-product are important to coal utilization because the more variability in the coal, the more change in the fly ash product. Elements in coal and the forms in which they occur in the coal (e.g. organic bound, associated with minerals) determine the ease of release during ash utilization and disposal. Trace elements (e.g. arsenic, barium, cadmium, cobalt, chromium, mercury, nickel, lead, strontium, and uranium) measured in part per million are classified as elements of environmental concern Schweinfurth (2002). Minerals in coal have been the focus of intense study for decades as the major source of pollution from coal-fired power plants.

Inorganic Minerals in Reservoir Characterization

The impact of inorganic minerals on reservoir properties is not well understood because there is not much interdisciplinary cooperation between coal petrologists and reservoir engineers and modelers. This is not for the lack of opportunity but perhaps more due to unawareness of the importance of mineralization as it applies to coal reservoir characterization. The scale of the inorganic minerals may have influenced this. At a larger scale, interbedded mineral matter serves to compartmentalize horizontally the coal reservoir and these "partings" can be used as markers for correlation of coal beds, which can be traced to the river channel point sources (Flores, Spear, Kinney, Purchase, & Gallagher, 2010). At a smaller scale of characterization, mineral matter serves to mineralize pores and cleats, reducing the reservoir performance, as well as decreases the internal surface area of the coal. As shown in Table 5.1, various inorganic mineral groups infill pores, cellular cavities, and cleats (fractures) in coal. Thus, mineralization of internal surfaces in the coal plays a major role in coalbed gas generation, storage, and flow.

Pore and Cellular Infillings and Plant Tissue Replacements

Pore and cellular infillings by inorganic minerals start as early as peatification. Authigenic (e.g. formed in place) inorganic minerals, which formed with the peat, are called syngenetic minerals. These minerals are crystalline and noncrystalline intergrown with the organic matter or macerals. For example, as a result of strong leaching conditions during peatification, water and airborne minerals (e.g. feldspar and volcanic minerals) may be dissolved or altered to form syngenetic minerals. Remobilization of dissolved or altered minerals infiltrate, precipitate, and infill open pores and cell cavities prior to compaction during burial. Common syngenetic minerals infilling cellular structures and pores are kaolinite, quartz, pyrite, and phosphatic minerals (Table 5.1) (Ward, 2002). Quartz, derived from dissolution of basement rocks, is a common infill of cell cavities of plant tissues during early-stage peatification (Lindquist & Isaac, 1991; Sykes & Lindquist, 1993; Ward, 2002). The soluble silica from the groundwater in the peat is absorbed by some plants and then deposited in the cellular structures. Casts of authigenic quartz, which infilled and replaced the plant cells as well as bipyramidal beta-form quartz from volcanic ash fall, were isolated from the Wyodak feed coal for the Powder River Basin (Figure 5.2(A)) (Brownfield et al., 2005). According to Ward, Crouch, and Cohen (1997) quartz-infilled pores developed prior to peat compaction suggesting that the original pores are still open and the pore fills survived through coalification with the final coal product.

Syngenetic clay minerals in the form of kaolinite may also infill pores and cellular structures as well replace plant tissues (see Table 5.1) (Bohor & Triplehorn, 1993; Spears, 1987; Ward, 2002). Kaolinite, which is a common pore filling in coal, may have been derived from volcanic ash (Bohor & Triplehorn, 1993 and Ward, 2002). Syngenetic phosphates commonly fill pore and cell cavities and replace plant tissue (Ward et al., 1997). Like the syngenetic kaolinite, phosphates may have been derived from volcanic ash. Syngenetic sulfides are commonly represented in coal by pyrite, which occurs as framboidal spherical shape (Figure 5.2(B)) and euhedral–anhedral minerals. Pyrite commonly fills pore and cell cavities and replaces plant tissues (Ward, 2002). Pyrite is formed under reducing conditions in the peat mires and bogs, which produced interactions of iron with hydrogen sulfides in the groundwater that percolated and precipitated the syngenetic mineral (Querol et al., 1989; Ward 2002). A consensus idea is that pyrite formed in peat is influenced by marine waters (Rao & Gluskoter, 1973).

The role of syngenetic inorganic minerals in infilling pores and replacement of plant tissues in coalbed gas reservoir characterization has been underestimated. Reservoir modeling of coal beds has assumed the idealized storage capacity with maximum pore availability for gas adsorption. Although coal petrologists have understood for a long time that inorganic mineralization influences combustion, the influence on gas adsorption, storage capacity, content, and coalbed gas reserve is not well understood. Also, modeling coal beds for enhanced gas recovery and carbon dioxide sequestration has focused on the potentially large amount of internal surfaces of pores for gas storage. The impact of pore infillings by inorganic minerals is the destruction and elimination of the available open spaces for gas adsorption. The influence of syngenetic minerals is probably significant because infillings occur during peatification and prior to burial when the pore cavities are still open. Thus, mineralization of pores reduces the amount of internal surface area in the coal matrix, which accounts for about 90% of the gas volume in an ideal open-pore condition storage capacity. The reduction of adsorbed gas leads to lower gas content and potential reserves. Pore infillings decrease storage capacity of the coal for carbon dioxide sequestration.

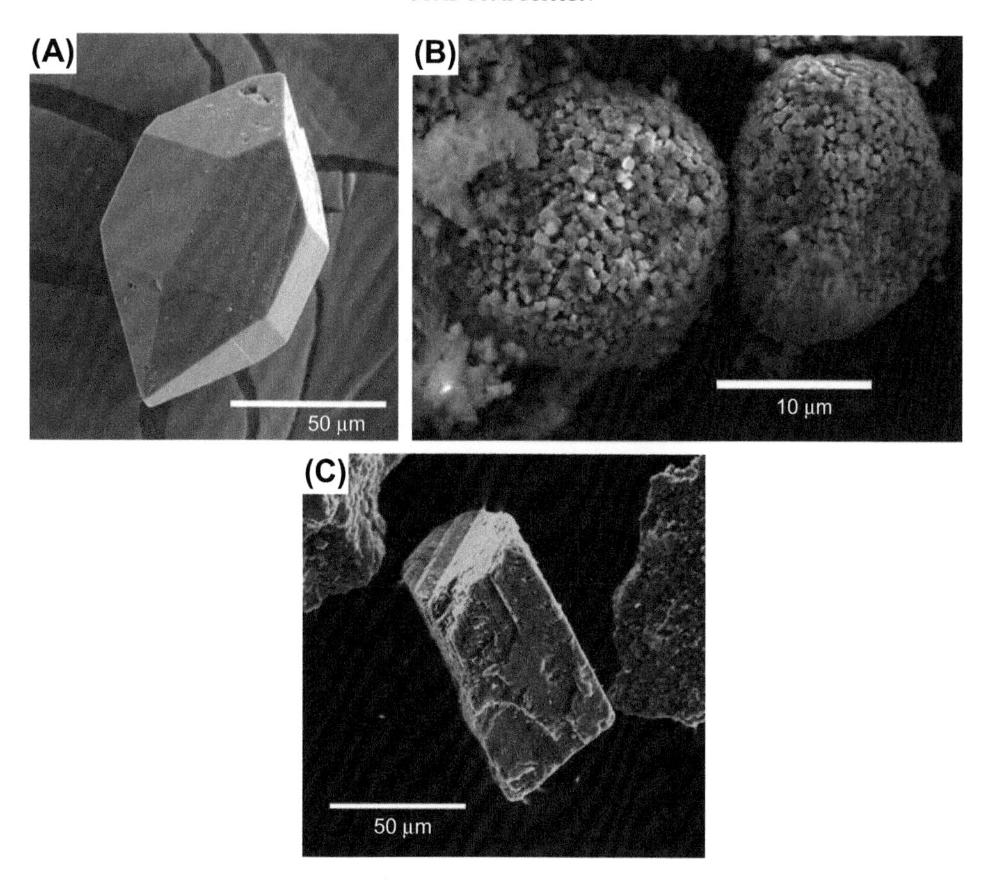

FIGURE 5.2 Scanning electron microscopic images of mineral matter. (A) Euhedral beta-form quartz of the silica group, (B) framboidal pyrite of the sulfide group, and (C) calcite of the carbonate group. *Source: Adopted from Brownfield et al., (2005).*

Cleat or Fracture Infillings

Epigenetic inorganic minerals, which formed after the coal formation, are the most common infillings of cleats or fractures; thus they are commonly perpendicular to the coal bed. Carbonates, quartz, clays, and sulfides are common cleat or fracture infillings (Ward, 2002). The consensus idea has been that epigenetic minerals infilled cleats or fractures after compaction of the coal (Ward, 2002). However, because cleats are formed by desiccation, the epigenetic minerals infilling cleats or fracture more than likely formed after dehydration of the coal. This may be supported by the appearance of "fractures" in vitrinite maceral, which were interpreted by Ward (2002) as resulting from brittle failure under stress. These "fractures" are much narrower and less persistent compared to cleats (Figure 5.3). The epigenetic minerals have a wide range of sources such as magmatic influx, igneous intrusion, hydrothermal processes, expulsion of inorganic compound from the organic matter, and groundwater flow (Brownfield et al., 2005; Crowley et al., 1994; Faraj, Fielding, & Mackinnon, 1996; Ward, 2002).

Calcite, dolomite, and siderite (Figure 5.2(C)) are common epigenetic minerals infilling cleats or fractures in coal (Kolker & Chou, 1994;

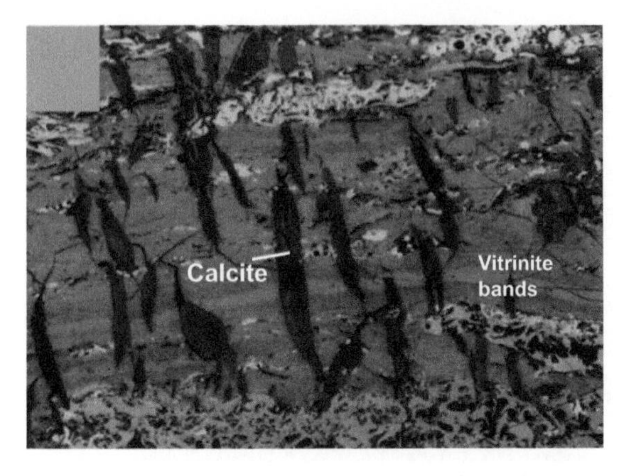

FIGURE 5.3 Optical microscopic image calcite en-echelon and sigmoidal veins in vitrinite bands (field width 1 mm); all images in polished sections. *Source: Adopted from Ward (2002).*

Ward, 2002). It is not uncommon to find these epigenetic minerals, especially carbonates, to be mineralized in the cleat in as many as five stages resulting in mineral zonation. According to Kolker and Chou (1994), carbonate mineralization in cleats in the Illinois Basin coals occurred at low temperatures (e.g. 15–70 °C). However, it is possible that mineralization by fluid injections from magmatic and hydrothermal sources may occur at higher temperature. For example, calcite and dolomite are abundant at the coal contacts where hydrothermal fluids were introduced by igneous intrusions (Finkelman et al., 1998; Querol et al., 1997). Thus, carbonate minerals are formed at lower temperatures compared to silicates, which formed at temperatures from 70 to 170 °C (Ward, 2002), which may have occurred at a higher temperature of 550–650 °C (Kweicinska, Hamburg, & Vleeskens, 1992). The high temperature of formation exemplified by silicates such as illite and kaolinite is commonly found in semianthracite and anthracite coals.

The anisotropy of cleat infillings is not only controlled by thermal and external influences but also more importantly by the geological and tectonic histories of the coal basins. As demonstrated by Faraj et al. (1996) in Surat Basin in Australia, the second cycle of burial of sediments promoted widespread carbonate mineralization in butt cleats and joints. Calcite was precipitated from meteoric water at about 80 °C. However, the extensive carbonate mineralization in butt cleats and the near lack of carbonate mineralization in the face cleats is interpreted to be caused by the change in the direction of lateral compressive stress during Jurassic–Cretaceous times relative to that during an earlier Triassic orogeny. A more recent work by Solano-Acosta et al. (2007) has reinforced the concept that thermal and tectonic history play an important role in mineralization of coal cleats or fractures in coal as demonstrated in the Illinois Basin in the United States. Here, a first generation of mineralization by kaolinite in coal cleats occurred during maximum burial at about 78 °C and a second generation of calcite mineralization in coal cleats occurred after uplift at about 43 °C. The sensitivity of cleats or fractures to mineralization by inorganic minerals should be given the importance that it deserves as a tool in characterizing coal reservoirs.

The major effect of infillings of cleats or fractures in coal by inorganic minerals is mainly on the gas flow. Reservoir engineers, geologists, and modelers pay much attention to the cleats or fractures as permeability avenues for flow of gas and water to the production well. Reservoir modeling of gas flow traditionally has assumed a concept of very slow gas diffusion from the coal matrix into "regular" cleat or fracture system (Gilman & Bekle, 2000). Thus, modeling does not account for the importance of the mineralization of cleat or fractures, which lowers the permeability and more notably impedes gas and water flow to the production well. Permeability anisotropy as measured by variable mineralization of coal cleats or fractures, if ignored, can impact considerably the optimum designs of drilling, completion, stimulation, and production operations. For example, the suitable hydraulic fracturing techniques adopted for drilling and configuration of production wells depend on the permeability anisotropy of the coal reservoir. For horizontal coal bed wells, orientation can generally be chosen to take advantage of anisotropy by drilling perpendicular to the major horizontal permeability component (Wold & Jeffrey, 1999).

Clastic Mineral Matter in Reservoir Characterization

The discussions in Chapter 3 on peat formation emphasized the role of detrital sediments comprising mineral grains being brought into the mires and bogs as water and airborne particles. These mineral grains, in contrast to the syngenetic and epigenetic crystal or noncrystalline minerals, are deposited as discrete continuous layers or dispersed constituents. According to Mukhopadhyay and Hatcher (1993), they are generally fine grained and as much as 5 μm in diameter. However, sand to clay grains (>2–>1/256 mm in diameter) make up those transported by floods into the peat mires. Water-laid minerals grains or epiclastic minerals are the most common mode of deposition of mineral matter in the peat mires. When the sediments are laid down as distinct layers in the peat, they show lateral continuity as coal partings (thinner than intervening coal beds) in the final coal product partings, which are interbedded with the coal, are often responsible for compartmentalizing the coal reservoir, and serve as impermeable layers that impede vertical flow of gas and water. The impermeable partings are often composed of carbonaceous clays, volcanic ash, and silty clay (Figure 5.4). Dispersed mineral grains may have been brought to mires and bogs by surface water or wind. Volcanic ash fall (pyroclastic mineral grains) and other fine-grained particles are carried by wind and dust storms, which are generally the mode of transportation of inorganic minerals in the central part of bogs. Although the mineral grains are originally deposited across the peat surface

FIGURE 5.4 Photograph showing a Tertiary coal bed with interbedded detrital or epiclastic partings and volcanic ash (e.g. pyroclastic) or tonstein partings in the State of Washington, United States. *Source: Photograph courtesy of Michael Brownfield.*

as very thin layers, they may have been reworked by animals, groundwater, and rainfall, which percolated and dispersed downward into the peat deposit. When the mineral grains (considered as a part of coal ash content) are dispersed throughout the peat and become part of the composition of coal, these inorganic minerals lower the gas content of the coal because methane is not adsorbed onto nonorganic matter.

Coal is essentially considered a continuous and homogenous unconventional gas reservoir; however, the epiclastic and pyroclastic layers trap gas as well as confine water flow horizontally. Figure 5.4 demonstrates the vertical architecture of coal with interbedded detrital mineral layers. The vertical partitioning imparts heterogeneous, layered, and multiple coal reservoirs each containing different sets of porosity and permeability properties. The vertical partitioning transforms the coal reservoirs into separate completion zones. Spatial partitioning of the coal reservoirs by the pyroclastic inorganic minerals complicates the assessment of gas reserves. Thus, the heterogeneity of the coal reservoir makes reservoir quality delineation more difficult. As a general rule, the partings comprising clastic inorganic mineral grains (e.g. epiclastic, pyroclastic), which accumulate with the peat, are more laterally continuous than the syngenetic minerals, which have closely intergrown with the organic matter in the peat. The partings consisting of pyroclastic minerals grains or volcanic ash are more laterally continuous than the epiclastic mineral grains. The pyroclastic partings are very reliable markers for correlation of coal beds. The brief duration (e.g. days or weeks) of volcanic ash falls make isochronous time horizons ideal for correlation of coal beds (Lamarre, 2003).

The lateral extent of the vertical reservoir compartmentalization by partings is not well understood in the subsurface coalbed gas development for lack of reservoir data even with the aid of closely spaced, high-resolution geophysical logs for correlation and drilling conditions. However, Karacan and Mitchell (2003) numerically modeled partings in a mineable coal and illustrated the effects on borehole gas production and mining emissions. Figure 5.5 shows the grid distribution representing parting layers as cut away from the coal randomly distributed in the middle part of the coal bed around a borehole location (Karacan & Mitchell, 2003). The blocks within the modeled bed were assigned values of 0.05 millidarcy (mD) fracture permeability and 2% porosity. Also, the properties of partings were allocated "null" values of effectively no porosity, no permeability, no gas storage capacity, and no fluid flow. Figure 5.6 (Karacan & Mitchell, 2003) shows curves of gas productions from horizontal and vertical boreholes in the presence of partings in comparison with those of the base coal bed. The gas production curves show that the area in the coal bed with partings resulted in a decreased production in both vertical and horizontal boreholes. Both borehole production rates showed decrease of almost half that of a continuous coal bed, which may be due to the presence of the partings and changes in the permeability around the boreholes. Since both boreholes were drilled through the low-permeability partings, the effective permeability of the coal bed that the boreholes intersected was lower than that in the base coal bed case (Karacan & Mitchell, 2003).

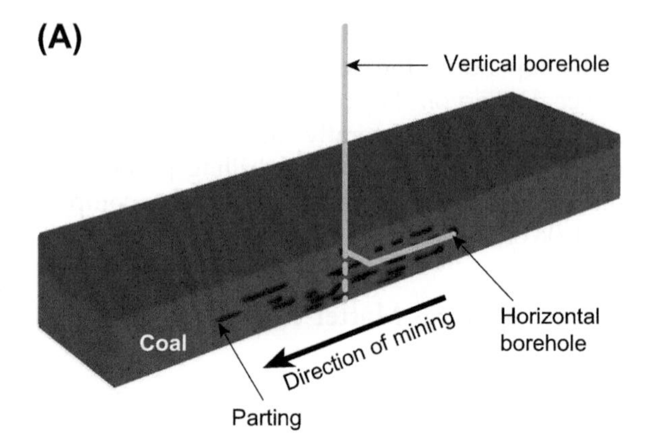

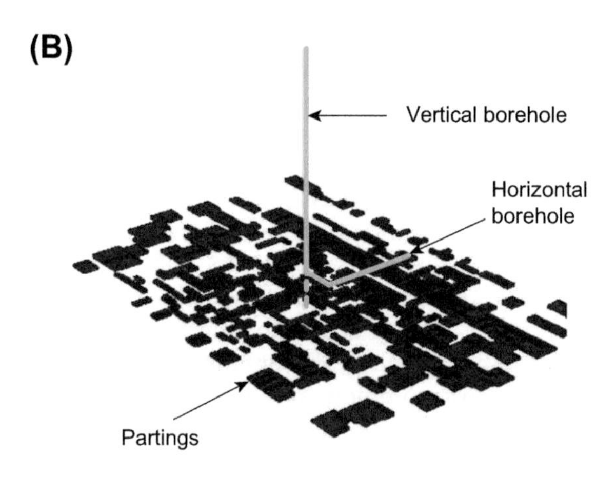

FIGURE 5.5 Numerical models of coal partings in a US coal mine around vertical and horizontal boreholes (A) showing the distribution of grids representing the properties of parting layers cut away of the coal bed (B). *Source: Adopted from Karacan and Mitchell (2003).*

For the period of degasification of the coal in front of mining, boreholes drilled in an area of extensive coal partings encountered irregular flows of gas (Karacan & Mitchell, 2003). A horizontal borehole drilled between extensive partings, which effectively divide a coal bed into separate reservoirs, drained methane only from that portion of the coal reservoir. The coal beds

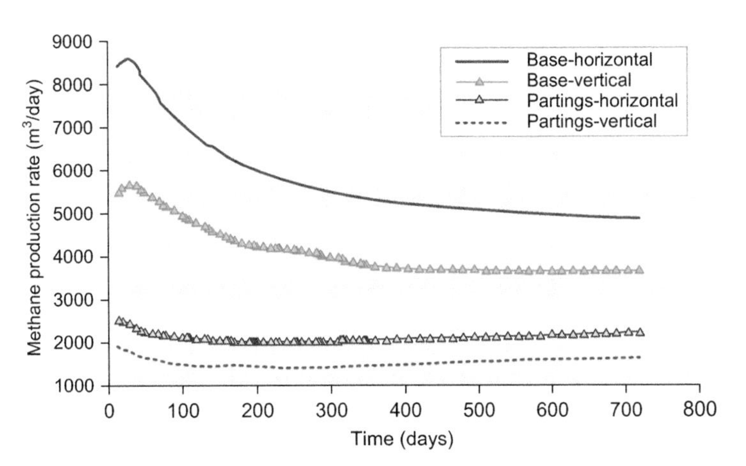

FIGURE 5.6 Curves of gas production from the horizontal and vertical boreholes in the presence of partings in comparison with those of the base of coal bed. The gas production curves show that the area in the coal bed with partings resulted in a decrease in both vertical and horizontal boreholes. *Source: Adopted from Karacan and Mitchell (2003). m^3/day, cubic meter per day.*

above and below the partings remained undrained. The remaining gas represents a potential hazard for outburst during future mining or would be unavailable for commercial production (Diamond, 1982; Karacan & Mitchell, 2003). According to Karacan and Mitchell (2003), in the case of vertical boreholes, which are commonly utilized for completion wells in a typical coalbed gas field, coal partings can also be problematic to the flow of gas and water. Vertical borehole is much shorter than a horizontal borehole, which penetrates more gassy zones and more partings rendering the well ineffective. In addition, if the fracturing treatment does not penetrate effectively above and below the parting, the flow of both gas and water is reduced. Partings that compartmentalize the coal bed hydraulically are also important from a mine ventilation standpoint because they can isolate large volume of gas that can suddenly be liberated when penetrated by a mine entry. Partings create variations in the coal density as a function of mineral matter or ash in the coal (Beaton, Langenberg, & Pana, in press; Marchioni, 2003). Most importantly, partings create strong discontinuities in the petrophysical fabric (e.g. porosity and permeability) within the coal bed reservoir. These may also be related to gas and water concentrations within the coal bed.

As demonstrated by Stricker and Flores (2009), the volcanic ash partings, greater than 0.3 m thick, show distinct signatures in gamma ray logs of the Wyodak coal in the Powder River Basin, Wyoming, United States (Figure 5.7). Analysis of the ash yield, which is as much as 8%, shows good correlation with the presence of trace elements phosphorus, strontium, and uranium, which are diagnostic minerals such as crandallite group related to volcanic ash (Brownfield et al., 2005). Using similar parameters, Stricker and Flores (2009) analyzed the ash yield and trace elements for the Wyodak coal that is mined in the eastern margin of the Powder River Basin in Wyoming (Figure 5.8). The coal, which produced gas, was studied in correlative wells (1–5 km apart) and contains a succession of five volcanic ash partings and corresponding trace elements. Flores et al. (2010) has used these volcanic ash partings to correlate the Wyodak coal across the Powder River Basin for as far as 80 km. Within this extent, the Wyodak coal is very variable in gas content and productivity. However, the number of compartmentalized multiple gassy zones in the coal, which vary from 25 to 40 m in thickness, is difficult to ascertain (Chapter 7).

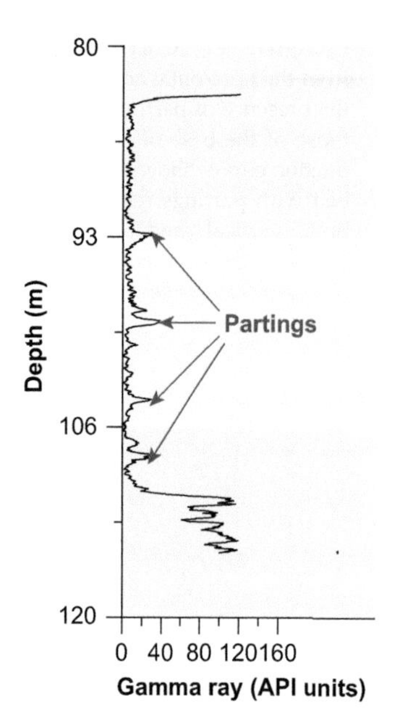

FIGURE 5.7 Partings indicated by Gamma ray "kicks" in the thick Tertiary Wyodak coal bed in the Powder River Basin, Wyoming, United States. *Source: Adopted from Stricker et al., (2006).*

Organic Matter

Coal macerals are defined according to their microscopic characteristics under transmitted and reflected lights and fluorescence (Table 5.2). The differences in reflectance between macerals are indicative of their physical and chemical properties as controlled by coalification. However, the underlying difference between coal macerals is recognition of the original plant parts that produced groups of macerals. For coal petrologists, identification of compressed, degraded, and altered plant parts under the microscope are not as difficult as for a noncoal petrologists. Thus, coal macerals aredivided into microlithotypes, which are related to their megascopic lithotypes or macrolithotypes. The megascopic constituents in coal are more readily identifiable, practical, and usable by noncoal geologists and engineers for reservoir characterization in the field and/or laboratory Box 5.2.

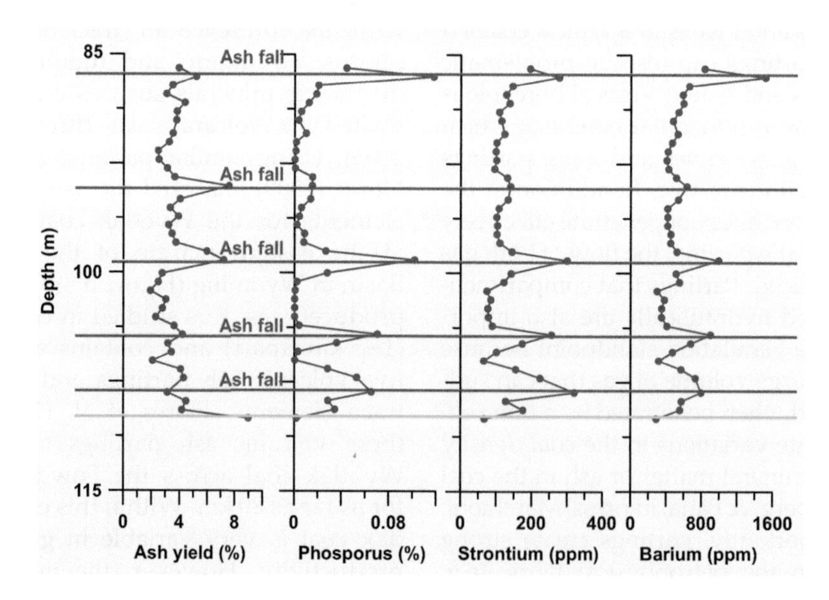

FIGURE 5.8 Chemistry of the volcanic ash fall (e.g. pyroclastics) partings in the Tertiary Wyodak coal bed in the Powder River Basin, Wyoming. The ash falls, which can be correlated, contain relatively high phosphorus, strontium, and barium. ppm = parts per million; % = percent. *Source: Adopted from Stricker and Flores (2009).*

TABLE 5.2 Maceral Groups and Modes of Optical Identification

Maceral Group	Transmitted Light	Reflected Light	Flourescence
Vitrinite	Red brown	Intermediate gray	None of yellow (lignites) brown (bituminous)
Liptinite	Yellow	Dark gray	High intensities; very variable with rank
Inertintite	Opaque	White, yellowish, light gray	None

Modified from Bustin et al. (1983), Styan and Bustin (1983), Stach et al., 1982, and Taylor et al., (1998).

BOX 5.2

LITHOTYPES IN COAL: ORIGIN AND NATURE—CONCEPTS AND MISCONCEPTIONS

Lithotype in coal is a concept that distinguishes various bands in coal originally designed for "macroscopic" identification (Stach et al. 1982). The macroscopic identification of lithotypes was designed to be preliminary to petrographic examination whereby coal bands can be visually examined with the naked eye, especially in coal mines. The original criteria for recognition of lithotypes are the thickness of the bands, color, and luster. The similarities of the macroscopic descriptions of the coal bands or layers (e.g. bright vitrain, semibright clarain, dull durain, and charcoal-like fusain) served as markers to correlate coal sections in mine areas. The methodology is very similar to description and correlation of similar physical properties of stratigraphic intervals in other sedimentary rocks.

Tasch (1960) related the origin of the coal bands to their level of deposition in a subsiding swamp, which was termed "seam formation curve". The curve started from "wet" condition at the bottom to "dry" condition at the top. The wet swamp bottom contains "floor rock" deposits of inorganic sediments overlain by carbonaceous shale. This succession is overlain by dull and dull-banded coal overlain by banded coal. Toward the top, these successions are overlain by bright and bright-banded coal, which are capped by fusain representing the dry top of the swamp. The origin and nature of lithotypes were debated ever since, the debate especially revolving around depositional environments and relationship to vegetation especially for Australian, Indonesian, and New Zealand peat and coal (Bolger, 1991; Cameron, Esterle, & Palmer, 1989; Esterle, Ferm, & Yiu-Lion, 1989; Holdgate, 2005; Luly, Sluiter, & Kershaw, 1980; Moore, Esterle, & Shearer, 2000; Moore, Shearer, & Esterle, 1993; Sluiter, Kershaw, Holdgate, & Bulman, 1995). Thus, the concept of lithofacies is not uniformly applied worldwide.

Microlithotypes

An earlier attempt to relate petrographic characteristics of organic and inorganic constituents of coal and their lateral and vertical variations within a single bed or different coal beds of a particular coalfield was made by Chao, Minkin, and Thompson (1982). The purpose of this attempt is to relate definitive descriptions of coal characteristics to coal facies in order to interpret their depositional environments, diagenetic changes, burial history, and coalification or metamorphism. Descriptions were performed from coal cores, which were studied in sufficient detail to describe and define coal microlithotypes, macrolithotypes, and coal facies and their variations. Chao et al's. (1982) automated scheme for megascopic analysis is based on distinctive bands 2 mm or more in thickness, which are first demarcated by visual inspection. These bands consist of either nearly pure microlithotypes/macrolithotypes such as vitrite/vitrain or fusite/fusain or assemblages of microlithotypes. Megascopic analysis is next performed to determine volume percentages of vitrite, inertite, minerals, and microlithotype mixtures in bands 0.5–2 mm thick. Other computer-based schemes of megascopic identification of the components of coal from cores were devised by Ferm, Smith, Weisenfluh, and DuBois (1985) to assist and facilitate descriptions by geologists and mining engineers. Both schemes of Chao et al. (1982) and Ferm et al. (1985) were codified for computer use. These megascopic or macroscopic studies and schemes are more relevant now than ever with coalbed gas reservoir characterization particularly for noncoal trained workers.

The term microlithotype is based on maceral and/or maceral–mineral association and percentage as observed under incident light microscope within the bandwidth of 50 μm (0.05 mm) in which each maceral or maceral group comprises more than 5% by area (Hunt, 1989; Hower, Esterle, Wild, & Pollock, 1990; Hower & Wagner, 2012; ICCP, 1963; Mukhopadhay & Hatcher, 1993; Stach et al., 1982; Taylor et al., 1998). Hunt (1989) described microlithotypes as the "degree of microscopic mixing" of various macerals. Microlithotype is comparable to a lamination in clastic sedimentary rock particularly in shale, which ranges from 0.05 to 1 mm in thickness. However, most clastic laminae range from 0.1 to 0.4 mm in thickness and are based on grain size alternation of sand, silt, and clay particles. Microlithotype was devised to use much like in clastic sedimentary rock the designation of "lithic" or rock types in coal, which is defined by the percentage of maceral composition and association at a microscopic scale. As shown in Figure 5.9 and in Chapter 4, the suffixes "inite" and "ite" are attached at the end of the words for the maceral groups and macerals and for microlithotypes, respectively.

The concept of microlithotype is based on petrographic observations that macerals are mainly associated with one or two macerals from the same maceral group or from the other two maceral groups. That is, a vitrinite maceral may be associated with another vitrinite maceral, a liptinite or an inertinite maceral, and other types of permutation. Stach et al. (1982) originally defined these associations as "monomaceral," "bimaceral," and "trimaceral," respectively. Depending on the maceral associations, such as bimaceral of vitrinite and inertinite, the microlithotype is termed "vitrinertite" by combining both maceral names (Figure 5.9). Another example of trimaceral association of liptinite, vitrinite, and inertinite gives rise to the term "vitrinertinite". The hierarchical succession of the names is related to the percentage of the macerals. This concept of maceral associations is very similar to "lithotype associations" in clastic sedimentary rocks, which are grouped giving rise to "lithofacies associations" as determined from their depositional environments.

Following Hower (1990), Hower and Wagner (2012); Mukhopadhay and Hatcher (1993); Stach et al. (1982), and Taylor et al., 1998 bands greater than 0.05 mm in thickness and consisting of greater than 95% vitrinite are termed vitrinite (ICCP, 2011) or adopting Stach et al's. (1982) original terminology, is called "vitrite". This one-maceral-dominated association is in contrast to the two-maceral association or bimaceral (e.g. clarite, durite, and vitrinertite) and three-maceral association or trimaceral (e.g. duroclarite, clarodurite, and vitrinertoliptite) (Figure 5.10). The monomaceral association

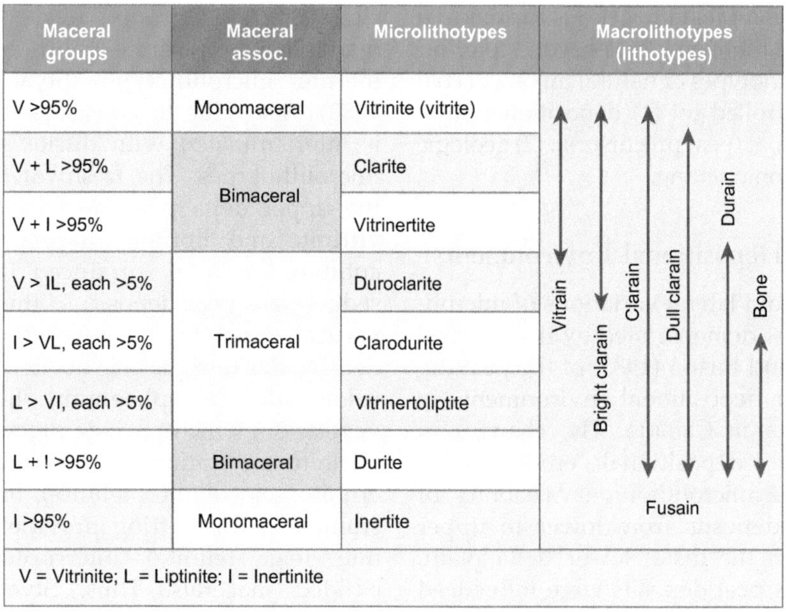

Maceral groups	Maceral assoc.	Microlithotypes	Macrolithotypes (lithotypes)
V >95%	Monomaceral	Vitrinite (vitrite)	
V + L >95%	Bimaceral	Clarite	Durain
V + I >95%		Vitrinertite	
V > IL, each >5%	Trimaceral	Duroclarite	Vitrain, Clarain, Bone
I > VL, each >5%		Clarodurite	Dull clarain
L > VI, each >5%		Vitrinertoliptite	Bright clarain
L + ! >95%	Bimaceral	Durite	
I >95%	Monomaceral	Inertite	Fusain
V = Vitrinite; L = Liptinite; I = Inertinite			

FIGURE 5.9 Relationship of maceral groups and associations, microlithotypes, and macrolithotypes. %, percent; assoc., association. *Source: Modified from Hower et al. (1990) and Mukhopadhyay and Hatcher (1993).*

contains not less than 95% vitrinite and no more than 5% liptinite and/or inertinite to be a vitrinite or vitrite microlithotype. The bimaceral association is composed of vitrinite and liptinite and vitrinite and inertinite, both maceral combinations accounting for more than 95% and each maceral including more than 5% of the whole. The trimaceral association consists of all three groups of macerals, each of the macerals comprising more than 5%. The importance of these maceral associations to reservoir characterization will be more apparent in the macrolithotypes section below especially in relation to gas content and development of cleat or fracture system.

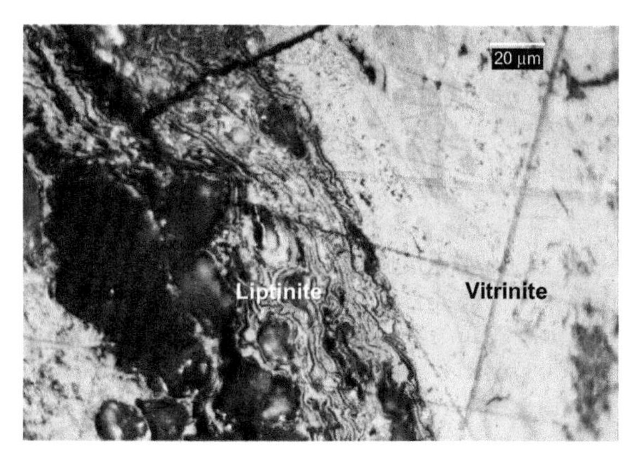

FIGURE 5.10 Maceral group consisting of liptinite and vitrinite (bicameral association) or clarite microlithotype. *Source: Photograph by Tim Moore.*

VARIATIONS AND VALUES OF MICROLITHOTYPES IN COALBED GAS RESERVOIRS

The value of microlithotypes (e.g. maceral association) is better characterization of properties (e.g. mechanical, rank) of coal important in mining and utilization (Hower, Graese, & Klapheke, 1987; Stach et al., 1982). For example, monomaceral microlithotypes increased the grindability of coal in contrast to trimaceral microlithotypes, which decreases grindability (Hower et al.,

1987). That is, the more homogenous the maceral composition, the more energy is needed to grind the coal. Microlithotypes or natural mixing of coal macerals is controlled by (1) depositional environment of the coal (peat precursors), (2) geologic age, and (3) tectonic setting.

Variations in Depositional Environments

The vertical and lateral variations of microlithotypes are best demonstrated by the classical work of Styan and Bustin (1983) of the peat deposits and their depositional environments in the Fraser Delta in Canada. The changes in vegetation and depositional environments demonstrate the microlithotype variability of temporal peat deposits from lower to upper delta plains. On the distal lower delta plain, thin sedge-grass peat deposits were influenced near the base by marine conditions and higher in the deposit by freshwater conditions. High cellulose to lignin ratio and limited exposure to degradation of these tissues in marsh plants produce mainly vitrinite-rich, with minor liptinite, macerals (e.g. cutinite, alginite) from algal and sedge lipids. On transition to freshwater peat, fusinite (e.g. oxyfusinite, pyrofusinite) partially replaces liptinite- (exinite) and vitrinite group macerals. Subsequent marine transgression has promoted degradation of the peat. Styan and Bustin (1983) proposed that the resulting coal bed from this thin and discontinuous deposit would develop bands of vitrite microlithotypes near the base and grade up section into interlaminated durite and vitrinertite.

The transitional peat-forming areas of the upper and lower delta plain contain maceral composition similar to those in distal delta plain deposits. However, with the presence of crevasse splays, the peat fabric and vertical gradation are disrupted by floods bringing in oxygenated waters followed by extended desiccation and exposure to peat fires (Stracher et al., 2010) resulting in inertinite-rich maceral group (e.g. inertodetrinite, macrinite, sclerotinite, and oxyfusinite) or inertite microlithotype (Styan and Bustin, 1983). Also, due to crevasses floods, the peat is interlaminated with durite and vitrinertite microlithotypes. The freshwater conditions in the upper delta plain also resulted in mixed vitrinite and liptinite macerals (e.g. telinite, cutinite, cerinite) produced by freshwater sedge-grass peat deposit. In this setting, rapid vertical variability in microlithotypes is typified by clarite replacing durite as well as local colonization by sphagnum, shrubs and pine vegetation, which provide precursors of mixed vitrinite and liptinite maceral groups (e.g. suberinite, telocollinite, telinite). In addition, tree stumps in this setting provide massive vitrinite (e.g. telinite) interruptions of these banded macerals. Thus, Styan and Bustin (1983) proposed that the coal bed originating from such variable peat would be thick, laterally extensive and characterized by vitrite with laminae of liptite and lenses of clarite in the upper part. However, when the resulting coals formed across the delta plain are correlated, the deposits will consist of vitrite with interlaminated vitrinertite and durite microlithotypes in the lower delta plain and vitrite–inertite microlithotypes mixed with liptite, clarite, and durite microlithotypes in the upper delta plain. Thus the upper delta shows more vertical and lateral variability in microlithotypes than that in the lower delta plain. The vertical and lateral variations in vegetation, depositional environments, and resulting peat/coal microlithotypes and macrolithotypes are shown in Figure 5.11. The microstratigraphic variations of the microlithotypes and macrolithotypes are similar to those described by Markic and Sachsenhofer (2010) of the Pliocene Velenje lignite in the Velenje Basin, Slovenia. These authors hypothesized that the coal fabric is the result of aggradational patterns of wood-rich trees and less-resistant cellulose-rich (e.g. herbaceous) dwarf

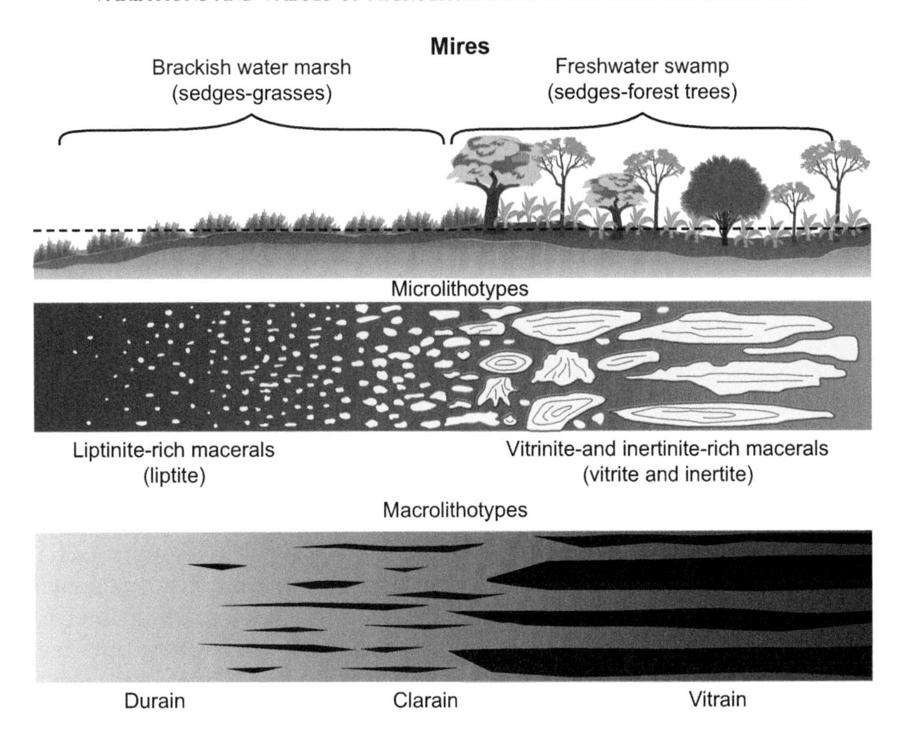

FIGURE 5.11 Relationship of microlithotypes and macrolithotypes with depositional environments in a Pliocene lignite in Slovenia. Lateral distribution of microlithotypes and macrolithotypes are controlled by vegetation types. *Source: Adopted from Markic and Sachsenhofer (2010).*

plants. The resulting microlithotype products are vitrite and liptite.

Variations in Geologic Age

There is a fundamental difference in maceral composition of Carboniferous coals in the Northern Hemisphere (e.g. Europe and North America) and the Permian Gondwana coals in Antarctica, Australia, India, South Africa, and South America (Chapter 4). In general, the vitrinite group is in contrast to the Permian coals, which are dominated by the inertinite group, dominates the Carboniferous coals. The basic difference may be explained by the cooler and drier paleoclimate in which the Permian coals developed. Thus, the different maceral assemblages reflect the variation in microlithotypes or maceral association in these coals. Also, it reflects the influence of paleoclimatic conditions and geological age.

The variations in microlithotypes in coal according to geologic age are controlled by the changes in vegetation through geologic time (Chapter 1). The evolution of woody, vascular plants in the Carboniferous to the rise of flowering plants in the Cretaceous into the Tertiary periods have introduced more diversity in the vegetation and resulting microlithotypes through geologic time. For example, a wide range of late Paleozoic coal microlithotypes in the form of vitrite, clarite, and durite and inertite are not as variable in composition when compared with those of the Carboniferous and Permian Periods (ICCP, 2011). A major difference in microlithotypes compared with the late

Paleozoic coals is the general presence of inertinite as durite in the Carboniferous and as inertite in the Permian Gondwana coals (ICCP, 2011). The microlithotypes of Australian Permian coals was redefined by Hunt (1989) to consist mainly of vitrinite (vitrite), inertite, and vitrinertite. The amount of liptinite in Australian Permian coal is small (5%) such that the most common maceral association is derived from the vitrinite and inertinite maceral groups. The maceral association gives rise to mainly vitrinite (vitrite + clarite), inertinite (inertite + durite), and combinations of vitrinite and inertinite (vitrinertinite + duroclarite + clarodurite) (Figures 5.9–5.11). The underlying cause of the microlithotype variation of the Australian Permian coal according to Marchioni (1980) is caused by the significant difference in the arborescent vegetation from forest moor to reed moor vegetation. When the Australian Permian coal microlithotypes are compared with the Permian Indian coals described by Navale and Saxena (1989), they consist of (1) fusic- or inertite-rich, (2) vitro-fusic- or vitrinite-rich, and (3) fuso-vitric- or inertinite-rich most common microlithotypes. Despite the varying terminologies, both Permian coals are dominated by microlithotypes with vitrinite and inertinite maceral association.

Tertiary coals in contrast to older coals such as Eocene coals in the sub-Himalaya of India show richness of vitrinite with low concentrations of inertinite and rare occurrences of liptinite; thus, they are vitrite-rich microlithotypes (Singh & Singh, 1995). These Tertiary Indian coals are very similar to Tertiary coals described by Singh, Singh, Singh, and Arora (2010) in Indonesia in which the microlithotypes are mainly vitrite (huminite rich) and have low concentrations of liptite and inertite (liptinite and inertinite macerals). The maceral group association of the Tertiary coals in the Powder River Basin, United States, is on the whole compositionally similar to the Indian and Indonesian coals in terms of richness in vitrinite or huminite composition. However, based on the work of Stanton, Moore, Warwick, Crowley, and Flores (1989), the Paleocene and Eocene coals in the basin are composed mainly of vitrinite (huminite) and liptinite maceral associations. It is interesting to note that the most gas-prone coal bed (e.g. Wyodak coal) has greater variability in the inertinite maceral composition. Also, the least gas-prone coal bed (e.g. Eocene Felix coal) contains the most amounts of liptinite and inertinite maceral group association. Based on the vertical and lateral variations in maceral group association, Stanton et al. (1989) interpreted the thin to thick coal beds to have formed in various stages of peat mire to bog, respectively. The thin coal beds are represented by the initially low-lying topogenous mires with either low-statured herbaceous plants or arborescent plants, which were flooded by detrital minerals that terminated peat accumulation. The thick coal was interpreted to begin much like the thin coals but developed as a domed peat with thickened margin resulting from detrital influx derived from adjoining river channels and a convex center characterized by nutrient-poor oligotrophic bog.

The microlithotypes of Cretaceous coals are exemplified by the San Juan Basin, in the United States (Table 5.3; Buillit, Lallier-Verges, Pradier, & Nicolas (2002)). The microlithotypes are mainly represented in the order of high to low percent by vitrite, clarite, duroclarite, and vitrinertite. Buillit et al. (2002) explained the microlithotype variations due to the variable depositional environment in a delta plain setting, which is exposed to base level changes caused by progradational, aggradational, and retrogradational events during regression and transgressive cycles. This maceral association was originally interpreted by Pawlewicz (1985) to indicate a terrestrial forest containing mainly woody vegetation in a slightly fluctuating groundwater level. The less-common microlithotypes suggest deposition in an open moor or deep water usually inhabited mainly by herbaceous plants. Macerals from both environments are weathered, suggesting infrequent dry periods or periods of lower

TABLE 5.3 Maceral Association and Corresponding Microlithopyes as well as Minimum, Maximum, and Mean Occurrence in Cretaceous Coals in the San Juan Basin, U.S.

Maceral Association	Microlithotypes	Min–Max (%)	Mean (%)
Monomaceral	Vitrite	3.5–60	30
	Liptite	0–0.5	0
	Inertite	0–5	5
Bimaceral	Vitrinertite	0.5–39	14
	Durite	0–4	0
	Clarite	0–51	23
	Sporoclarite	0–9	2
Trimaceral	Duroclarite	0–62	23
	Clarodurite	0–8	1
	Vitrinertoliptite	0–3	0.5

% = percent; Min = minimum; Max = maximum.
Source: Modified after Buillit et al. (2002).

water table levels where the peat was exposed to subaerial oxidation.

The approach toward using maceral or microlithotype indices for interpreting depositional environments of Australian Gondwana coals originated by Diessel (1986) has generated spirited debate in attempts to prove or disprove the concept. The common method used to determine the depositional environments of microlithotypes or maceral association is based on gelification and tissue preservation indices (TPIs) by Dehmer, 1989; 1995, Diessel (1986), and Kalkreuth, Kotis, Papanicolaou, and Kokknakis (1991). These studies are applicable to high-rank and low-rank coals. The gelification index (GI) is a proportion of gelified and nongelified maceral composition (vitrinite + macrinte/semifusinite + fusinite = inertodetrinite), which measures the wetness of the peat (Diessel, 1986; Kalkreuth et al., 1991). The TPI is the proportion of preserved to degraded plant tissues (telinite + telo collinite + semifusinite + fusinite/desmocollinite + macrinite + inertodetrinite), which measures the density and suitability for tissue preservation of trees (Dehmer, 1989, 1995). However, Smyth (1989) suggested that no particular coal microlithotype is significantly related with specific paleodepositional environment for Australian Permian and Jurassic coals and dispersed organic matter (DOM). More specifically, microlithotype relationships do not hold from one geologic age coal or DOM to another age. Crosdale (1993) applied the method to New Zealand Miocene subbituminous coals and found that the low inertinite and ubiquitous bituminite composition present difficulties of interpreting maceral ratios. The problem is further exacerbated by the facts that (1) absence of good woody tissue structures indicate absence of arborescent vegetation, (2) inertinite is unrelated to vegetation, (3) gelification is used as geochemical sense in contrast to biochemical sense and does not indicate groundwater level, and (4) detrital macerals and microlithotypes are the result of in situ fragmentation and not transportation. Although the controversy continues, success of application of the methodology was reported with modification of the original formulas to adopt to specific coal types for Tertiary lignite and subbituminous coals in Greece and Indonesia by Antoniadis, Mavridou, Papazisimou, Christanis, and Gentzis (2006) and Suwarna and Hermanto (2007); Permian–Triassic coals in India by Singh, Singh, and Singh (2011); and Pennsylvanian and Permian coals in

China by Zhang, Tang, Tang, Pan, and Yang (2010). Zhang et al. (2010) modified the Diessel (1986), Kalkreuth et al. (1991), and Dehmer (1989; 1995) formulas in order to relate with various coal facies expressed by the following relations:

Eqns:

$$GI = (\text{Vitrinite} + \text{macrinite})/(\text{semifusinite} + \text{fusinite} + \text{inertodetrinite})$$

$$TPI = (\text{Telinite} + \text{collotelinite} + \text{semifusinite} + \text{fusinite})/(\text{collodetrinite} + \text{macrinite} + \text{inertodetrinite})$$

$$GWI = (\text{Gelinite} + \text{corpogelinite} + \text{minerals} + \text{vitrodetrinite})/(\text{telinite} + \text{collotelinite} + \text{collodetrinite})$$

$$WI = (\text{Telinite} + \text{collotelinite})/(\text{collodetrinite} + \text{vitrodetrinite})$$

where:

GI = gelification index
TPI = tissue preservation index
GWI = groundwater index
WI = wood index

Microlithotypes in Reservoir Characterization

It is commonly reported in the literature that vitrinite-rich coals have high gas adsorption capacity and thus high gas content (Beamish & Gamson, 1993a,b; Chalmers & Bustin, 2007; Lamberson & Bustin, 1993). This is attributed to the high micropores in vitrinites. On the other hand, this composition leads to generally slow diffusion of gas. Thus, it stands to reason that microlithotypes with high vitrinite maceral association will have high microporosity and gas content and slow diffusion capacity. In contrast, Harris and Yust (1976) found that in bituminous coals, the inertinites comprise mainly mesopores and the liptinites are composed of macropores, indicating that the inertinites are the most porous of the maceral groups and the liptinites are the least porous. Thus, these observations suggest that microlithotype association of vitrinite- and inertinite-rich coals (e.g. vitrinertite) would be the most favorable combination of reservoir to maximize porosity and gas adsorption. The least ideal microlithotype association would be all-liptinite maceral group such as durite and clarite to minimize the effects of porosity and gas adsorption. In addition, as will be discussed in the following section, microlithotype and macrolithotype control permeability properties in coal.

The mixing and vertical successions as well as lateral variations of microlithotypes play a major role in the coal reservoir properties. Also, knowledge of the microlithotype variations allows better reservoir production modeling of the coal. Smyth and Buckley (1993) indicated that the analyses of the combination of sequential microlithotypes may be used to predict permeability in relation to gas drainage in the coal bed reservoir. Microlithotype analyses provide the amount, vertical position, and thickness and spacing of thin to thick bands of vitrite, vitrinertite, inertite, and mineral-rich zones in the coal. Hierarchical ordering of microlithotypes provides strategic modeling of gas flow and recovery plan. According to Smyth and Buckley (1993), the permeability of vitrite-rich Permian coal in the Sydney Basin, Australia, is 30 mD compared to 2.3 mD for inertite-rich coal. Thus, this data on the variation of microlithotypes can be used to predict gas and water flow in the coal reservoir.

The variations of microlithotypes in coal control the behavior of gas adsorption. Karacan and Mitchell (2003) reported that areas in the Pittsburgh coal rich in inertite and clays stored more carbon dioxide gas in simulating sequestration. Pore diffusion is higher in inertite and clays than in other organic microlithotypes. Inertite microlithotype or inertinite-rich maceral contains more pore spaces and in some cases adsorbs more gas than other maceral/microlithotype (Mastalerz, Gluskoter, & Rupp, 2004; Stanton, Flores, Warwick, & Gluskoter, 2001). According to Beamish and Crosdale (1998), mine outbursts may be controlled by microlithotype association such that coals rich in vitrinite and inertodetrinite increase outburst probability. Variations in coal microlithotypes create different gas desorption rates and gas content gradient. Presumably, vitrinite- and inertodetrinite-rich microlithotypes do not desorb gas rapidly in contrast to fusinite- and semifusitnite-rich microlithotypes, which desorb gas rapidly.

TABLE 5.4 Classification of Macrolithotypes Based on Proportion of Bright and Dull Bands in Bituminous Coal

Macrolithotype	Proportion of Bright and Dull Bands
Vitrain	<10% dull bands
Clarain (bright)	65–90% bright (vitrain) bands
Clarain	35–65% bright bands
Clarian (dull)	10–35% bright bands
Durain (dull)	<10% bright bands
Fusain	Fibrous; silky luster
Bone	High ash

% = percent.
Source: Modified from Hower et al. (1990).

Macrolithotypes

Macrolithotype is defined here as megascopically (e.g. visible to the naked eye) recognizable constituents of coal, which were originally described and identified as lithotype by ICCP (1963; 1971). Stach et al's. (1982) original description of lithotype is limited to humic coals to distinguish from sapropelic coals (e.g. cannel, boghead), which are rare. In addition, lithotypes of humic coals are further distinguished from sapropelic coals by Stach et al. (1982) as differently recognizable coal "bands" in contrast to the sapropelic coal, which is generally "homogeneous" and "unstratified." Thus, in order to differentiate between the "microscopically" identifiable maceral components of coal and the recognizable "bands" of coal, it is appropriate to apply the term macrolithotype for that lithotype in this book. Furthermore, coal macrolithotypes are more suitable for use in coal bed reservoir characterization in the field and laboratory by geologists and reservoir and production engineers.

Macrolithotypes like microlithotypes are the product of the vegetation types that make up the precursor peat. The composition of macrolithotypes like microlithotypes depends on the macerals and maceral association and percentage (Table 5.4).

Historical Perspectives

The concept of macrolithotypes has evolved from the classic work of Stopes (1919) on the British Paleozoic bituminous coals (hard coals) to include four "ingredients", namely, vitrain, clarain, durain, and fusain in hand specimens. Davis (1978) indicated that these coal ingredients were originally described by Stopes (1919) under the microscope such that the "ingredients" clarain and durain contain high proportions of exinite or liptinite maceral group. Thus, it was clear that in early inception of macrolithotype, composition of the coal was determined by microscopic identification and that it was a determinant factor. In a subsequent publication, Stopes (1935) related the earlier macrolithotype classification with additional maceral composition such that cutinite and resinite of the liptinite maceral group are included in the durain. A subsequent publication by Wandless and Macrae (1954) added two more "ingredients" of dull bituminous coal consisting of black and gray durain described under the microscope. An earlier publication by Cady (1942) introduced the terms clarodurain and duroclarain toward better megascopic description. Thus, the terms for the "ingredients" of banded, bituminous coal coined by Stopes (1919) have expanded and become more confused when the ICCP (1963), in an attempt to standardize coal petrology nomenclature, adopted these terms and used vitrite, clarite, and durite as microlithotype terms. Although these macrolithotype terminologies by Stopes (1919, 1935), Wandless and Macrae (1954), and later by Stach et al. (1982) were based on descriptions from Carboniferous bituminous coals, subsequent applications to other coals that have similar properties further introduced uncertainty into the classification (Davis, 1978). For example, the use of the Stopes (1919, 1935) and Wandless and Macrae (1954) terminologies is untenable for Gondwana Permian coals in Antarctica, Australia, India, South Africa, and South America, which do not contain some of the "ingredients" (e.g. durain or durite) (Davis, 1978).

In order to avoid the confusion and make the macrolithotype description more useable in the field, a U.S. Geological Survey classification (Table 5.5) was introduced by Schopf (1960). The US Geological Survey classification was simplified to include only three "ingredients" of constituents of banded bituminous coal, namely, vitrain, fusain, and attrital coal. The attrital coal matrix or "attritus" in contrast to the associated vitrain and fusain, is defined by distinct thickness and percent. Attritus is derived from a variety of vegetal matter and occurs as a fine-grained, tightly compacted mass (Theissen, 1920). Thickness of the bands is classified as thin to thick with modifiers such as moderately thin bands, which are 0.5–2 mm thick. These "bands" may be represented and estimated as percentage within the coal such as a range of 15–30%. Nomenclature as described by Davis (1978) may be cumbersome such as moderately thin-banded vitrain in moderately dull attrital with sparse fusain bands.

TABLE 5.5 Classification of Macrolithotypes with Emphasis on Attrital or Attritus in Bituminous Coal in the United States

Macrolithotype	Luster
Vitrain	Bright
Fusain	Silky
Attrital	Bright Moderately bright Midlustrous Moderately dull Dull

Modified from Schopf (1960).

Other macrolithotype classifications have emerged such as the ASTM (1991), which classified macrolithotype into banded, nonbanded, and impure coal (Figure 5.12). Hower et al. (1990) modified the classification by deleting the terms clarodurain and duroclarain of Cady (1942) and replaced it with bright clarain and dull durain based on the degree of preservation from peat to bituminous coal of the wood and matrix (Table 5.4). Moore et al. (1993) proposed that texture such as grain size keyed into plant materials in various stages of decay plays a major role in understanding macrolithotypes.

Macrolithotypes in coal are important for the characterization of high-rank coals and less so in low-rank coals in which coalbed gas is economically accumulated only in subbituminous coal. However, the key to understanding the macerals, microlithotypes, and macrolithotypes of high-rank coals is connecting these coal characteristics to those of the low-rank coals. The standard practice in characterizing coal rerservoirs has generally been to study separately the properties of the high-rank coals from low rank coals a very good reason that a large volume of the gas (e.g. thermogenic) is generated in bituminous and anthracite coals. Although only a small volume of gas (e.g. biogenic) is economically generated in subbituminous coal and none to date in lignite, it has become apparent that the key to understanding reservoir properties of high-rank coals is directly connecting them to those of the

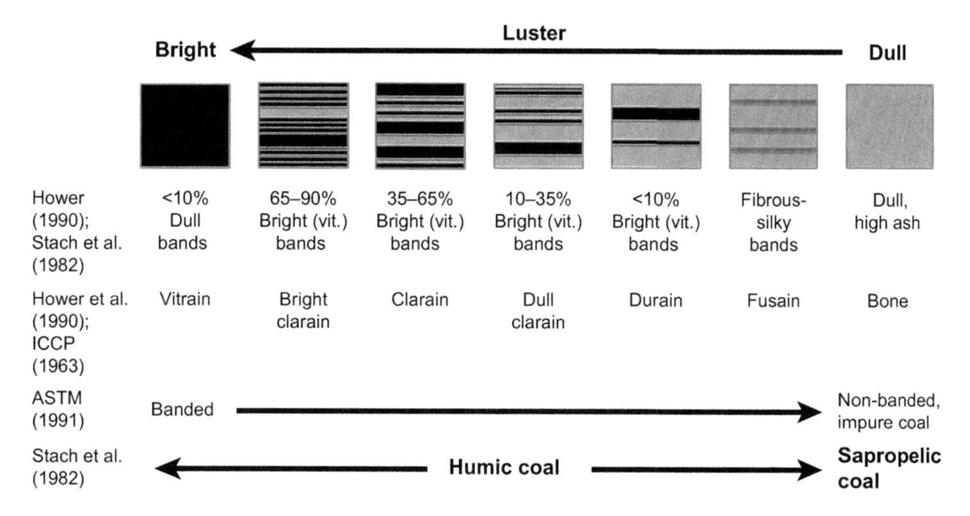

FIGURE 5.12 Classification of macrolithotypes ranging from vitrain to bone coal. Vit., vitrain; %, percent. *Source: Modified after ICCP (1963), Stach et al. (1982), Hower et al. (1990), ASTM (1991), ASTM (1994), and Esterle (2011).*

low-rank coals, including especially the lignites. There is a lack of "connecting the dots" in terms of reservoir properties between these two coal ranks. This idea is borne out from that of Cameron et al. (1989) discussed in Chapter 3 in which these workers argued admirably that for economic coals to form, these deposits generally have to be originally thick peat, preferably peat bogs. That is, if the initial peat is very thick, the resulting deposit will survive burial, compaction, and coalification yielding thinner but economic coal. Although this is an arguable idea, what is not is that through all these processes, the biological, chemical, and physical characteristic of the coal, although transformed at various degrees, are inherited from peat to low-rank and high-rank coals. Thus, there is a better need to connect the original to transformed properties in order to understand better coalbed gas reservoir characterization.

Macrolithotypes of High-Rank Coals

The fundamental macroscopic components of humic medium- to high-rank coals (e.g. bituminous and anthracite) are vitrain, clarain, durain, and fusain. The alternating characteristics of these macrolithotypes create "layers" or "bands," which are the building blocks of megascopic descriptions of coal and should be used for reservoir characterization. It is widely known that coal "bands" are directly related to the precursor plant assemblages; state and nature of preservation under biological, chemical, and physical conditions; and depositional settings. In this respect, Hower et al. (1990) interpreted that well-preserved precursor woody and related tissues yield vitrain bands and least-preserved mixed plant fragments yield durain bands. However, how these properties are connected in the transitional lower rank coals is not clear. Additionally, the alternations of macrolithotype bands produce textures, which have distinctive color, luster, fracture, and thickness. Thus, vitrain is described to be black, have brilliant to bright luster, be shinny and glassy to massive texture, have conchoidal fracture, and be brittle and occurs in at least several millimeter-thick bands. Clarain is semibright, has silky luster, is duller or less bright than vitrain, has smooth surface, is fractured at right angle to bedding, and is finely banded. Durain is black or gray, has greasy luster, is always

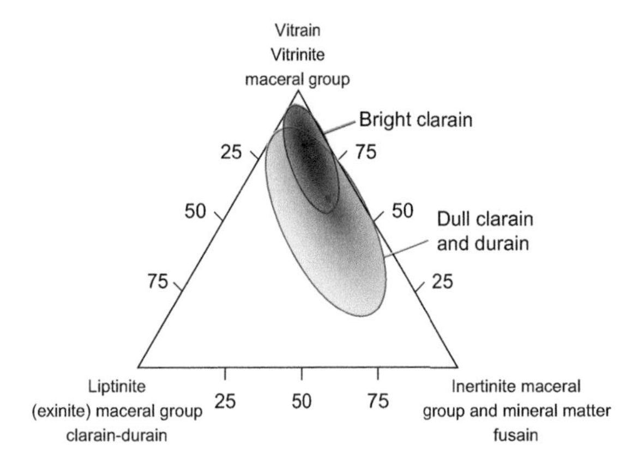

FIGURE 5.13 Ternary diagram showing the compositional overlap of bright clarain, dull clarain, and dull durain suggesting potential area of misidentification of both groups of macrolithotypes. *Source: Modified from Cameron (1978).*

dull, is hard but breaks into big lumps or blocks, and has rough fractured surface and relatively thicker bands. Fusain is black, has silky luster, is fibrous looking, resembles wood charcoal, is soft and friable, and has sooty to powdery texture and thick bands.

The categories of bright clarain and dull durain are best demonstrated in a classical study of the Illinois Basin coals by Cameron (1978), which are placed according to their maceral composition (e.g. vitrinite, liptinite, and inertinite maceral groups) in a ternary diagram (Figure 5.13). The diagram shows compositional overlap, which occur in the area of the triangle above 75% vitrinite where bright clarain clusters (darker shades) and a large number of dull clarains (darker shades) also occur. The dull clarains have a more broad compositional range than the bright clarians. Thus, Cameron (1978) questions the validity of using these macrolithotype terminologies given the degree of overlap in maceral group composition. Cameron (1978) further suggests that the used of the terms clarain and durain may also be doubtful, although he proposed that their used must be tempered by the objective of the research work. Cameron (1978) cautioned, however, that placing the clarain and durain layers in a strict sense of compositional difference may not be the correct way to use these terms, but if used as Stopes (1919) originally defined based on luster and texture, may be the approach to take. Perhaps, the US Geological Survey method of simplifying the macrolithotype constituents to vitrain, fusain, and attrital coal may resolve the clarain–durain problem.

A composite chart of the macrolithotypes in bituminous coal inclusive of the original definition and modifications of Schopf (1960), Stopes (1919), Wandless and Macrae (1954), ASTM (1991), ASTM, 1994, Hower et al. (1990), ICCP (1963; 1971), and Stach et al. (1982), are shown in Figure 5.12. The vitrain and clarain macrolithotypes are mainly bright compared to the durain and fusain macrolithotypes. The macrolithotypes from vitrain to fusain are banded and bone coal is nonbanded and impure (ASTM, 1991; ASTM, 1994). The transitional macrolithotypes such as bright and dull clarains are based on the amount of coalification and percentage (10–65%) vitrain bands (Hower et al., 1990). The bright (vitrain) and dull (nonvitrain) bands reflect the vertical and lateral heterogeneity of the coal. More importantly, in terms of coalbed gas reservoir characterization, each macrolithotype comes with its own set of physical properties, which can enhance or impede production of coalbed methane. It is important to emphasize that the macrolithotype classification was designed only for bituminous and anthracite coals, although attempts were made to use it in low-rank coals.

Macrolithotypes of Low-Rank Coals

Macrolithotype classification for low-rank coals (e.g. lignite and subbituminous) is more complex than for higher rank coals because of the presence of the original peat organic matter structures with little modification from degradation and coalification. Lignite is the first product of transformation during initial coalification from peat, and subbituminous coal is the subsequent product. Although lignite is too low rank for coalbed gas accumulation compared to subbituminous coal, it is still useful toward understanding macrolithotype development in particular and coalbed gas reservoir characterization in general.

The complexity of lignite macrolithotypes was pointed early on by Hagemann and Hollerbach (1980), who classified macrolithotypes based on the presence or absence of xylites (e.g. plant structures) or minerals, banding types, color and hue, gelification, inclusion, and surface texture. This classification later was modified by Mukhhopadyay (1986) to include pure and impure detrital or groundmass. Hagemann and Hollerbach (1980) recognized that lignite consists of bands of different color, structure, texture, and gelification, which are products of the original plant materials and different levels of biogeochemical degradation. These workers used the macroscopic and petrographic features to classify lignites into several macrolithotypes. Various macrolithotypes of lignite were analyzed for organic geochemical characteristics and compared with the petrographic and chemical properties. The chemical and soluble matter composition shows a correlation with different macrolithotypes. For example, Hagemann and Hollerbach (1980) found that variations in aliphatic hydrocarbon content are related to the intensity of biochemical degradation of the parent plant material as exemplified in the Lower Rhine area. Succeeding attempts to describe macrolithotypes of lignite in the United States took a different direction using modified classification by the US Geological Survey (Schopf, 1960). For example, macrolithotype analysis of Paleocene lignites of the northern Great Plains of the United States recognized vitrain, fusain, and attrital coal (Kleesattel, 1984; Schobert, Karner, Kleesattel, & Olso, 1985). The attrital coal is composed of mixed huminite, liptinite, and inertinite. Despite the difference in macrolithotype classifications between the European and US Northern Great Plains lignites, the constant and important criterion is directly related to the increasing loss of cellulose and lignin (wood).

Attempts were made to simplify macrolithotype descriptions of lignites (Kolcon & Sachsenhofer, 1999; Kwiecinska & Wagner, 1997; Markic, Zavšek, Pezdič, Skaberne, & Kočevar, 2001; Ticleanu et al., 1999). More importantly, attempts were made to codify descriptions only for lignites of Europe (e.g. Poland) based on lithofacies nomenclature devised by Miall (1977) based on textural and structural features. Widera (2012) proposed that the term lignite macrolithotypes should only refer to macroscopically identifiable and homogeneous stratum with a thickness of more than 5 cm, which is characterized by recognizable and percent wood fragments, detrital matrix, and fossilized charcoal (e.g. fusitic lignite) in relation to texture (e.g. grain size) and sedimentary structure (e.g. stratification). Thus, following lithofacies characterization of Miall (1977), lignite should be codified for grain size (e.g. gravel, sand, fines such as silt and clay, diamict, organic deposits, etc.) and structure (e.g. massive, horizontally stratified, planar cross-stratified, and deformed) (Kolcon & Sachsenhofer, 1999; Kruger & Kjaer, 1999; Markic & Sachsenhofer, 1997; Mastalerz, 1995; Rhee & Chough, 1993). Thus, much like the previous macrolithotype classifications proposed by Hagemann and Hollerbach (1980), Kleesattel (1984), Mukhhopadyay (1986), Schobert et al., (1985), and Schopf (1960), the recognizable plant materials remain the most important criterion.

An important aspect of proposed lignite macrolithotype classification is following a purported "10% rule", which is used by European coal scientists and is recommended by ICCP (1993) (Widera, 2012). The rule recommends that if the total volume of constituents contains 10% of any single constituent, then it is included in the macrolithotype nomenclature. Lignite includes (1) xylitic (e.g. plant structures) fragments, (2) fine detritus matrix, and (3) fusitic particles (Kolcon & Sachsenhofer, 1998, 1999; Kwiecińska & Wagner, 1997; Markic & Sachsenhofer, 1997; Taylor et al., 1998; Widera, 2012; Wolf, 1988). These compositions may be used to name a macrolithotype by a combination of constituents in association with textures and structures governed by the "10% rule". For example, a "xylitic" lignite macrolithotype association consists of at least 90% xylites (e.g. stumps, trunks, and branches) >1 cm. The rest of the macrolithotype association is up to 10%, which may include fine detritus (<1 cm) (e.g. organic matter) (Widera, 2012). Structures of this macrolithotype may be massive, horizontally stratified, gelified, and deformed (Table 5.6). Another example is detritic lignite macrolithotype association consisting of more than 90% of fine-grained (<1 cm) homogeneous organic matter with the rest, up to 10%, being xylitic fragments (Widera, 2012). The structures of this lignite macrolithotype association includes horizontal to subhorizontal stratification and gelified structure.

An excellent example of field descriptions of an open-cut mine highwall of Miocene coals in central Poland using the modified Miall (1977) textural and structural clastic sedimentary rock lithofacies classification is shown in Figure 5.14, described by Widera (2012). The purpose of applying Miall (1977) lithofacies classification to lignite macrolithotype association is to relate the textures and structures to depositional facies and environments much like its original use in clastic sedimentary rocks. Widera (2012) interpreted the lignite macrolithotypes as deposited in a mire system that replaced an anastomosed fluvial system. The xylitic-dominated macrolithotype with different structures was formed in a forested dry mire characterized by groundwater table deep below the peat surface. It is presumed that the fusitic macrolithotype is associated mainly with the xylitic macrolithotype. Both the xylodetritic and detroxylitic macrolithotype associations were deposited in bush to forested dry to wet mires characterized by fluctuating low to high groundwater table. The detritic macrolithotype formed in open water vegetated by reeds and sedges as well aquatic plants. Thus, based on

TABLE 5.6 Classification of Macrolithotypes based on Composition, Grain Size, and Sedimentary Structure for European Lignite Coal following Miall (1977) Lithofacies Classification

Macrolithotype	Composition	Grain Size	Structure
Xylitic	Wood fragments (Trunks, branches twigs); >90% Matrix (detrital matter)	>1 cm; coarse grained <1 cm; fine grained	Horizontal, massive deformed (e.g. folded, fractured/ faulted), gelified
Detritic	Homogeneous detrital organic matter; >90% Woody fragments	<1 cm; fine grained >1 cm; coarse grained	Massive, deformed (folded/ fractured) subhorizontal, gelified
Xylodetritic	Relatively more xylites (wood fragments) than detrital organic matter	Mixture of more coarse than fine grained	Mixture of the above structures
Detroxylitic	Relatively more detrital organic matter than xylites (wood fragments)	Mixture of more fine than coarse grained	Mixture of the above structures
Fusitic	Fossil charcoal	Needle-shaped particles	Massive, lenticular

Source: Modified from Miall (1977), Wolf (1988), Ticleanu et al. (1999), Kolcon and Sachsenhofer (1999), Markic and Sachsenhofer (2010), and Widera (2012).

these environmental characterizations, Figure 5.14 is described by Widera (2012) as vertical successions of detrital macrolithotypes of open water to mixed xylitic–detrital macrolithotypes of wet forest mires with increasing grain size of plant fragments upward. Based on the color, the detrital layers are dark compared to the mixed xylitic–detrital layers, which are light except for the uppermost layer of dominantly xylitic layer, which is dark. The succession of color is interesting to note when compared to the Australian lignite or brown coal in the Gippsland Basin where potential biogenic gas has been described by Midgley et al. (2010).

The European criteria of lignite macrolithotype classification are closely similar to those of the Australian brown coal. According to Perry, Allardice, and Kiss (1984), macrolithotypes of the Latrobe Valley coals in Gippsland Basin, southeast Australia, result from differences in the botanical assemblages, depth, and nature of the mire groundwater level and degradation of plant material. This genetic approach of macrolithotype classification is demonstrated by works

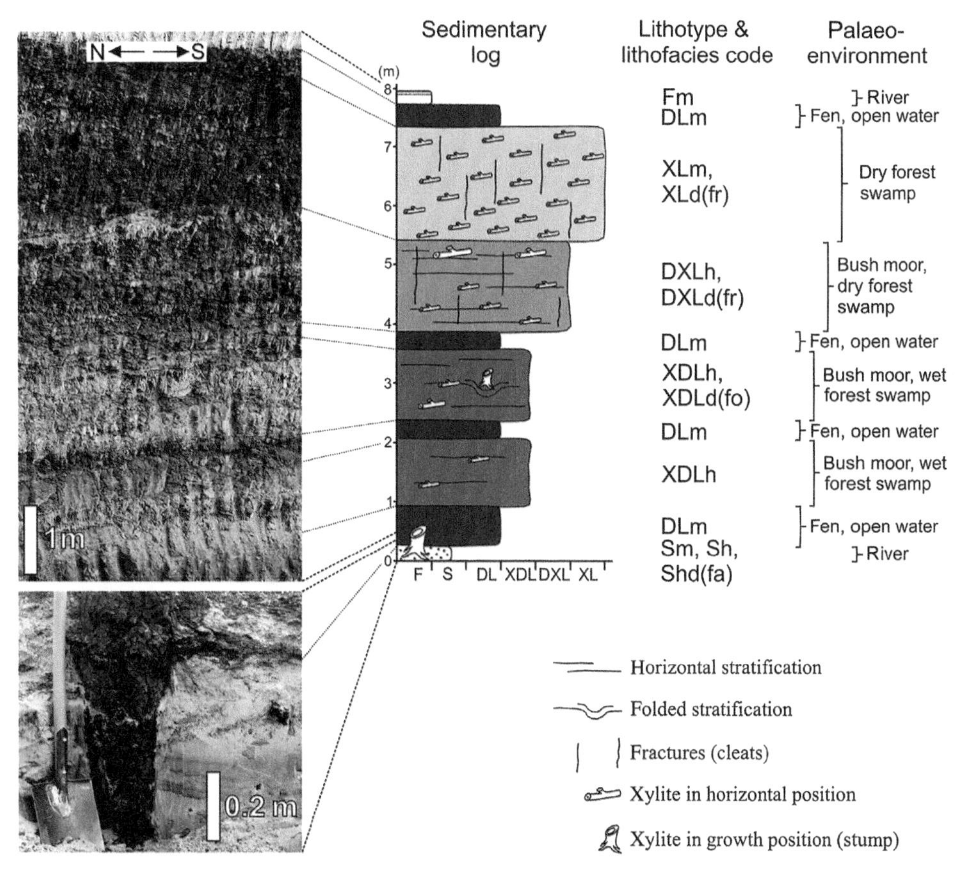

FIGURE 5.14 Macrolithotype log description for Tertiary lignites in central Poland using coal composition, grain, size, and structure adopted from Widera (2012). Log description adopted similar descriptions of clastic sedimentary as emphasized by textural differentiation of the rock types proposed by Miall (1977). X, xylitic; D, detritic; XD, xylodetritic; DX, detroxylitic; F, fusitic; L, lignite; m, massive; h, horizontally stratified; d, deformed; fr, fractured; fa, faulted; fo, folded; S, sand.

of George and Mackay (1991), Holdgate (1984, 2003, 2005), and Holdgate, Wallace, Gallagher, and Taylor (2000). Detailed sequence stratigraphic and megascopic descriptions of the Tertiary (Oligocene–Miocene) Latrobe Valley brown (lignite) coals, more than 100 m thick, reveal distinct macrolithotype coal layers (5 cm–5 m thick) mainly on partially dried and weathered open-cut mine highwall faces. The layers are characterized by different color, texture, gelification, and weathered shrinkage patterns (Holdgate, 2003). The Latrobe Valley brown coal layers are interpreted as resulting from variations in depositional environments controlled by groundwater level, paleobotanical assemblage, and climatic and eustatic sea-level changes (George, 1975; Holdgate, Kershaw, & Sluiter, 1995). George (1975) originally described five macrolithotypes based on color of the coal bands, which include dark, medium dark, medium light, light, and pale. The gradation from dark to light color reflects a loss of moisture of as much as 5%. In addition, each macrolithotype is recognized based on texture and jellification as well as wood and charcoal contents (Holdgate et al. (2000). For example, the medium light and medium dark macrolithotypes contain wood fragments and tree trunks in contrast to dark macrolithotypes, which are gelified and contain charcoal (Figure 5.15). However, Holdgate (2005) observed that the medium dark to dark, high moisture content macrolithotypes increase seaward or toward marine influence in contrast to the medium light to pale, low moisture content macrolithotypes increase landward. Thus, the darker macrolithotypes are more waterlogged and the light or pale macrolithotypes are more weathered (Figure 5.15) (Holdgate, 2003). The vertical successions of the variegated colored macrolithotypes displayed in

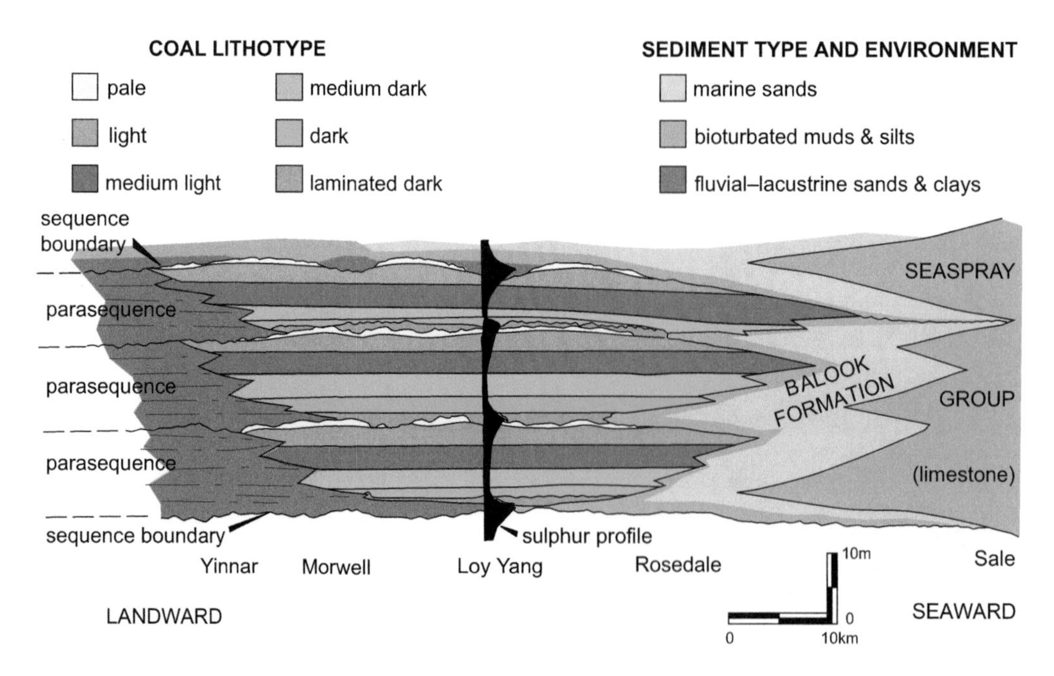

FIGURE 5.15 A model of the lignite sequences across Latrobe Valley, Gippsland Basin, Australia, showing upward lightening macrolithotype cycles and associated parasequences at the seaward wedge of the model. Increasing upward cycles of sulfur content reflect relationship with seaward movement. *Source: Adopted from Holdgate (2003).*

three open mine cut faces in the Latrobe Valley are shown in Figure 5.15. The generally medium light to medium dark Australian brown coal macrolithotypes are similar to the mixed xylitic–detrital macrolithotypes of the central Poland lignites. Also, the dark-colored Australian brown coal macrolithotypes are much like the detrital-dominated macrolithotype of the central Poland lignites, which are formed in open water mires, which are also "waterlogged."

The depositional concept proposed for the variegated layers of Latrobe Valley brown coal is initially traditional topogenous mires (e.g. swamps) stretching from the coastline to inland, each with differing paleobotanical communities controlled by the fluctuations of the ground water table (Figure 5.16(A) (Holdgate, 2003, 2005; Holdgate et al., 1995; Holdgate & Sluiter, 1991). The most interesting aspect of the proposed depositional model is the evolution of the coastal topogeneous mires to raised bogs very similar to the tropical Indonesian bogs (Figure 5.16(B)). These two models presumably explain the changes in the colors of the macrolithotypes in

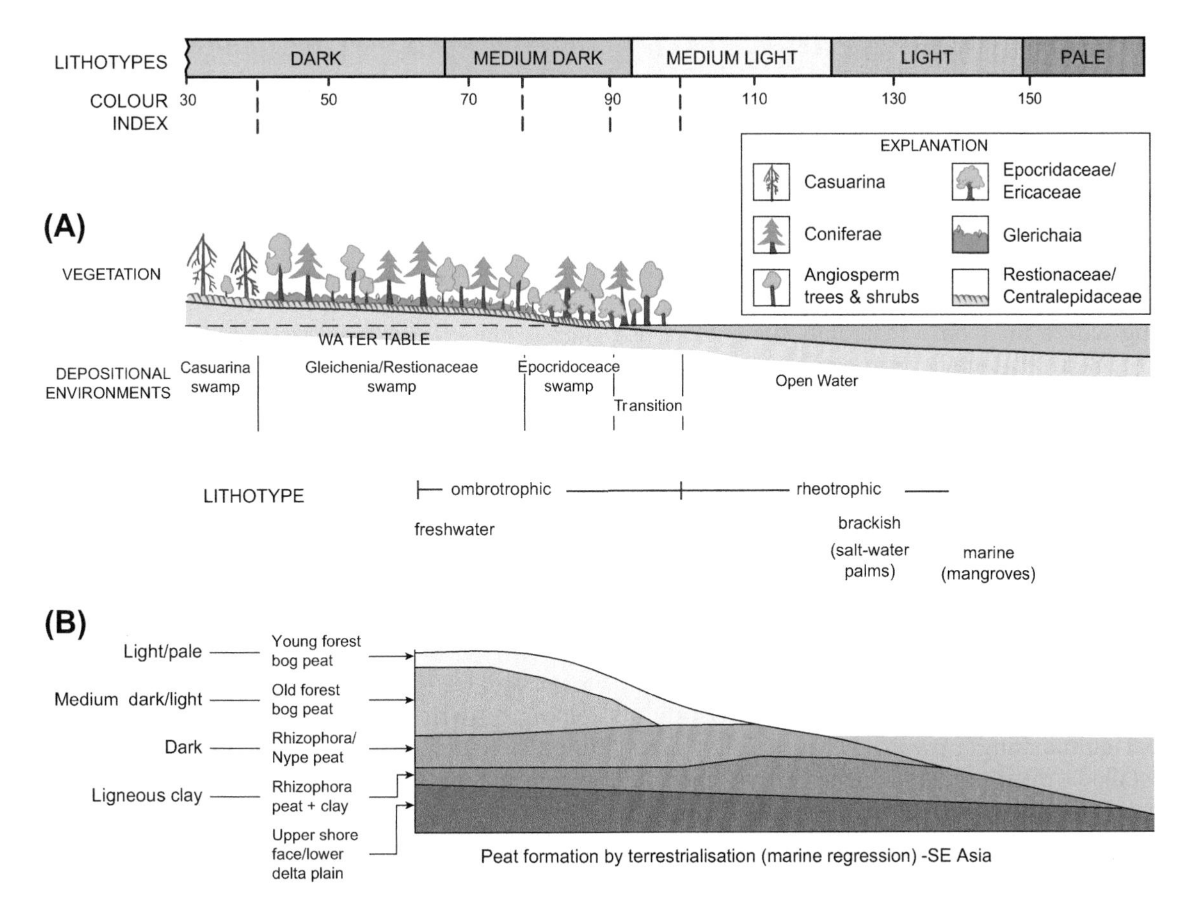

FIGURE 5.16 (A) Relationship of depositional environments and vegetation analyzed from pollen groups related to color changes (e.g. dark to light colors) of the lignite macrolithotypes in the Latrobe Valley, Gippsland Basin, Australia. (B) Peat dome depositional model patterned from Southeast Asia based on the relationship of the macrolithotypes and water depths for the Latrobe Valley lignite in the Gippsland Basin, Australia. *Source: Adopted from Holdgate (2003).*

which the dark macrolithotypes are formed in the "waterlogged", anaerobic part of the topogenous peat and the light (pale) macrolithotypes are formed in the "open-canopied", anaerobic part of the peat bog. These peat successions are, in turn, controlled by groundwater and sea-level fluctuations as well as basin subsidence. If these models hold true, we come partway in "connecting the dots" from peat to lignite, the formation of "economic coal" according to Cameron et al. (1989), and the process of macrolithotype development.

Brown coals are known for their unique change in color of macrolithotypes from generally dark to light from bottom to top of the coal bed. This phenomenon has been called "upward lightening" (Holdgate, 2003) or "dulling up" by Crosdale (1995). The factors controlling the dulling or lightening up of brown coal macrolithotypes have stimulated debate (Esterle et al., 1989). Figure 5.15 shows this "upward lightening" succession of macrolithotypes in the Victoria brown coal. Holdgate (2003) described each succession beginning with a thick dark macrolithotype, which is correlated regionally to equivalent marine-derived interbedded sedimentary facies. The medium dark to medium light macrolithotypes make up most of the successions, which lightened upward and terminated by erosion and weathering surfaces resulting in a pale macrolithotype (Figures 5.15 and 5.16). According to Holdgate (2003), each succession represents a parasequence.

Crosdale (1995) described the Miocene Maryville coals in New Zealand and showed well-developed dulling up successions or maceral associations, each defined by a decrease of bright coal culminating upward in sapropelic coals. Detailed microscopic investigations showed that during upward dulling, the maceral composition remains uniform at 80–90% huminite (vitrinite), 10–15% liptinite, and 1–4% inertinite, except for the liptinite-dominated sapropelic coals. The dull coal bands are characterized by maceral fragments in contrast to the bright bands, which are composed of intact plant fragments. Crosdale (1995) interpreted the dulling up maceral associations as related to the development of ombrotrophic peat bogs where growth of large trees and whole plant fragments are preserved during early stages of peat formation and around the bog margins. However, Esterle et al., 1989 show that peat bogs in Malaysia show more well-preserved large plant fragments in the central part than in the base and margins of the bog. Thus, these workers recommend that tropical peat bogs are poor analogs for dulling up coal macrolithotype sequences.

Subbituminous Coal Macrolithotypes

The macrolithotypes in subbituminous coal can be viewed as being between lignite in which the original plant structures are still preserved and bituminous coal in which the plant structures are degraded and coalified. Due to the transitional nature of the subbituminous macrolithotypes, workers have resorted to either using and/or modifying both high rank and lignite classifications in association with chemical and physical properties (Moore et al. 2001; Puttmann et al. 1991; Stricker, Flores, & Trippi, 2010; Stanton et al., 2001; Trippi et al., 2010). Moore and Shearer (1999) reported that this situation is overshadowed by previous four major macrolithotype classifications, three of which are devised for high-rank coals (Schopf, 1960; Stopes, 1919; Davis, 1978) and one devised for brown coal (George, 1975). These classifications emphasized the "brightness versus dullness" and "texture" attributed to the coals as underpinning criteria. However, there is an underlying confusion of what "texture" means. In brown coal, "texture" means the fragments of plant materials and/or wood (George, 1975; Widera, 2012). In high-rank coals (e.g. bituminous), "texture" means bands, which invariably represent wood remains (Moore & Shearer, 1999). Furthermore, comparison of high-rank coals of Paleozoic and Mesozoic ages and subbituminous coals of Tertiary age shows that the high-rank coals are always banded and low-rank coals contain unbanded units on account of degradation (Moore & Shearer, 1999).

Arguably, the key to characterization of subbituminous coal macrolithotypes is the presence and/or absence of the unbanded and banded coal units. The work of Mares and Moore (2008) on New Zealand Tertiary subbituminous coals show bright unbanded and banded units, which vary in thickness up to 30 mm. Vitrain bands are as much as 70% by volume. In comparison, the work of Flores, Moore, Stanton, Stricker (2001) and Moore, Flores, Stanton, and Stricker (2001) on Tertiary coals in the Powder River Basin, United States, show similar characteristics of band and unbanded units. However, the vitrain bands are more than 160 mm in thickness and as much as 63% by volume. Moreover, in the Powder River Basin, major macrolithotypes were identified ranging from attrital- to vitrain- dominated composition and from finely laminated to very coarsely laminated with the latter lamination consisting of more than 50% vitrain by volume. These are very similar to the New Zealand subbituminous coals (Mares & Moore, 2008). The term attrital was modified from the classification of Schopf (1960).

For lack of a standard macrolithotype classification for subbituminous coals, it is recommended that characterization should follow the approach used for the New Zealand and Powder River Basin coals (Flores et al., 2001; Mares & Moore, 2008; Trippi et al., 2010). It is interesting to note that both coals are gas prone with the Powder River Basin subbituminous coals being a prolific coalbed gas producer. Detailed characterization of macrolithotypes of the Powder River Basin subbituminous coals are provided in Flores et al. (2001); Moore et al., 2001; Stricker and Flores, 2009; Stricker et al., 2010; and Trippi et al. (2010). The coal macrolithotype description involved the use of three major components: vitrain, fusain, and attritus or attrital coal. Vitrain consists of shiny, bright bands representing coalified wood remains (Figure 5.17). Fusain is silky,

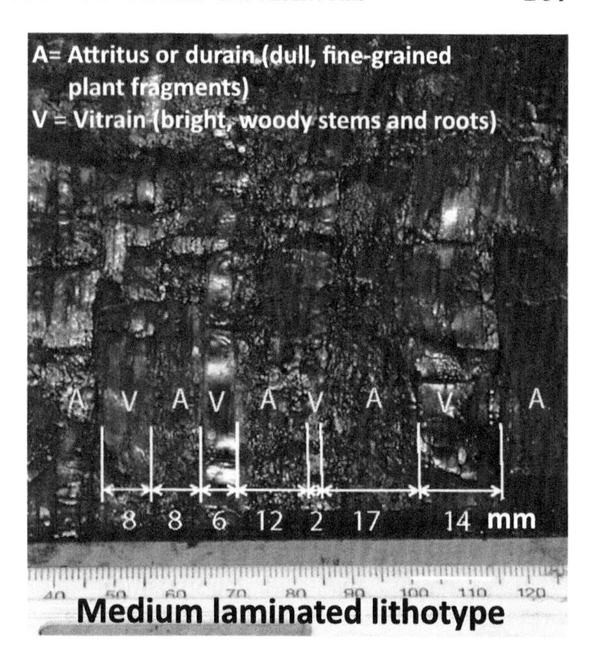

FIGURE 5.17 Photograph of interbedded vitrain and attrital macrolithotypes in the Tertiary coals cored in the Powder River Basin, Wyoming, United States. Coal bedding plane is parallel to bands of macrolithotypes.

dull, fibrous, and a soft to friable layer, which represents burnt charcoal (Taylor et al., 1998). Both vitrain and fusain layers or bands are embedded in the attritus, which is the coal matrix or groundmass. Attritus is a fine-grained mixture of vegetal matter (e.g. liptinites) (Figure 5.17). Attritus is duller than vitrain and brighter than fusain. The classification by Schopf (1960) of attritus ranged from dull to bright with the bright attritus equivalent to "clarain" of Stopes (1919) and the dull equivalent to the "durain" of Stopes (1919).

The most important feature of the subbituminous macrolithotypes is the bands or laminations (laminae) separated by attritus, which display varying thickness and are parallel or subparallel to the bedding plane. For example, most subbituminous coals in the Powder River Basin contain mainly alternating vitrain and attrital laminations intermingled with sparse fusain lamination (Figure 5.17). The thickness

TABLE 5.7 Classification Matrices of Macrolithotypes based on Thickness of Vitrain and Attrital Bands and Laminations Devised for Subbituminous Coals in the Powder River Basin in Wyoming, United States

COAL Macrolithotype Matrices		Average Attrital Band Thickness				
		0 mm	**<5 mm**	**5–20 mm**	**20–40 mm**	**>40 mm**
Average vitrain band thickness	0 mm	Fusain	Attritus	Attritus	Attritus	Attritus
	<5 mm	Finely laminated	Finely laminated	Finely laminated	Attritus	Attritus
	5–20 mm	Medium laminated	Medium laminated	Medium laminated	Medium laminated	Medium laminated
	20–40 mm	Woody vitrain	Coarsely laminated	Coarsely laminated	Coarsely laminated	Coarsely laminated
	>40 mm	Woody vitrain	Woody vitrain	Very coarsely laminated	Very coarsely laminated	Very coarsely laminated

mm = millimeter.
Source: Ron Stanton and Mike Trippi based on analysis of more than 900 core coal samples.

of these bands and their composition (e.g. vitrain, fusain, and attritus) form the basis of the macrolithotype classification. Table 5.7 is a subbituminous coal macrolithotype matrix for a proposed classification of banding thickness (vitrain versus attrital). Each of these bands is divided into various average thicknesses ranging from 0 to >40 mm. The compositional matrices and cells, in turn, are classified as finely laminated with an average thickness of <5 mm, medium laminated with an average thickness of 5–20 mm, coarsely laminated with an average thickness of 20–40 mm, and very coarsely laminated with an average thickness of >40 mm. The category of woody vitrain contains little or no attritus in contrast to the attritus category, which contains little or no vitrain. Fusain category is 100% fusain or fusinite. For example, to use the table, a vitrain band that has an average of 5–20 mm thickness is categorized as medium laminated vitrain. In addition, an attritus band that has an average of 20–40 mm thickness is categorized as coarsely laminated attritus. Figure 5.18 exemplifies the woody material, which contains woody texture of chevron-like striations and massive appearance.

The methodology of the macrolithotype classification system for the subbituminous coal is described by Trippi et al. (2010). The reader is referred to this work for more details of the system. The system was initially devised for purely macroscopic description and recognition of "lithotypes" toward modeling coal facies (Figure 5.19). The systems evolved from macroscopic descriptions in the field of entire sections of mine highwalls of as much as 30-m-thick coals (Stanton, Moore, Warwick, Crowley, & Flores, 1989) to coal cores of coalbed gas wells in the Powder River Basin in collaboration with the author (Stanton et al., 2001). Stanton's pioneering work on visual description of subbituminous coal cores from gas wells resulted in the development of the

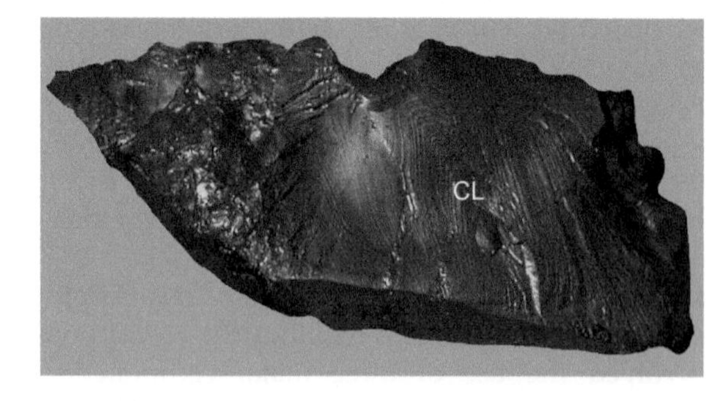

FIGURE 5.18 Photograph of a woody textured (e.g. convoluted laminae, CL) macrolithotype in the Tertiary coals in the Powder River Basin, Wyoming, United States.

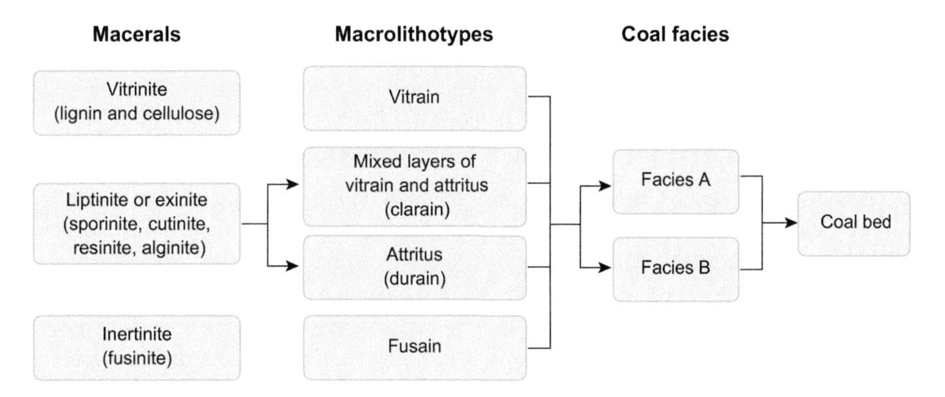

FIGURE 5.19 A flow chart showing relationship of macerals, macrolithotypes, and coal facies toward lithofacies analysis of a coal bed. Descriptions start from microscopic macerals to macroscopic lithotypes and facies association of a coal bed. *Source: Modified from Stanton et al. (2001).*

coal macrolithotype matrix (Table 5.7). The benefit of the macrolithotype system was a practical, functional, and fast methodology solely adopted for coal cores. However, the methodology was initially conceived to measure and describe vitrain, attritus, and fusain bands and laminations but not to account for their percentages (Trippi et al., 2010). Moore et al., (2001) used point-count method to estimate the compositional and textural components of the coal cores. Measurements were made of the largest dimension of the smallest axis of the point-counted bands in millimeters but were transformed to a phi (ø) scale for statistical analysis (Moore et al., 1993; Shearer & Moore, 1994).

The benefit of describing the thickness of macrolithotype bands and laminations and relating these attributes to their organic components is its applicability to reservoir characterization and use as predictive tool for gas content.

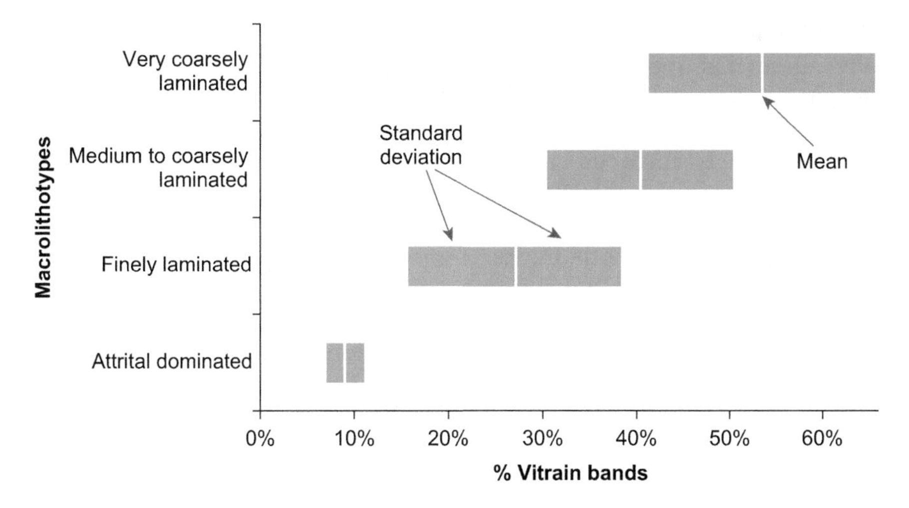

FIGURE 5.20 Chart showing relationship of macrolithotypes and percentage vitrain in the Tertiary coals in the Powder River Basin, Wyoming, United States. Standard deviation is the statistical measure by which the values differ from the mean. The laminated macrolithotypes have more dispersion than the attrital macrolithotype. *Source: Modified from Moore et al. (2001).*

BOX 5.3

CONNECTING ORIGINAL PLANTS TO GAS CONTENT—CONCEPTS OR MISCONCEPTIONS

It was originally believed by most workers that coal rank is the major contributor to gas content until Bustin and Clarkson (1998) revealed, "globally there is no or little correlation between coal rank and adsorption capacity". Also, these authors indicated that there is a good correlation between microporosity, vitrinite, and high-pressure methane adsorption capacity. Despite these observations, many researches find no relationship between coal composition and gas properties (Mares et al., 2009; Moore, 2012). Nevertheless, many studies have correlated coal composition to microporosity (Busch & Gensterblum, 2011; Ceglarsk-Stefańska & Zarebska, 2002; Chalmers & Bustin, 2007; Zhang et al., 2010). The key to connecting the dots from coal composition to gas content is partly tied to vitrinite, which originated from woody plants in specific mire/bog depositional settings. However, there is nonwoody coal composition (e.g. inertinite and mixed macerals) that shows positive correlation with gas content (Mastalerz, Goodman, & Chirdon, 2012; Stricker, Flores, & Trippi, 2010). The similarity of gas adsorption capacity of woody and nonwoody composition suggests that both have similar pore architecture (Chalmers & Bustin, 2007). Gas content or volume is controlled by the internal surface area (e.g. pores and fractures) of the coal (Lamberson & Bustin, 1993). Thus, there are fundamental differences in adsorption capacity worldwide, which cannot be accounted for by coal rank or composition.

For example, Figure 5.20 demonstrates the direct relationship of vitrain with band thickness in that there is a higher percentage of vitrain in thicker bands and laminations. Many investigators generally agree that methane adsorption, and thus gas content, is controlled by macerals, microlithotypes, and macrolithotypes. That is, more gas is adsorbed mainly due to higher microporosity in vitrinite- or vitrain-rich rank-comparable coals. The applicability of this concept to coals with thicker bands and laminations of vitrain is the ability to predict higher methane adsorption capacities for these subbituminous coals Box 5.3.

Coal Composition and Quality vs Gas Analysis

Why is coal composition and quality important to gas analysis? To begin with, the chemical composition of coal is analyzed by proximate and ultimate analyses (ASTM, 1993). Proximate analysis determines by using set procedures regarding moisture, volatile matter, fixed carbon, and ash content (Figure 5.21). Proximate analysis is reported by percent on as-received, moisture-free, and moist- and ash-free bases. Ultimate analysis is determined also using set procedures for ash, carbon, hydrogen, nitrogen, oxygen, and sulfur contents, which are reported as percent on as-received and moisture- and ash-free bases. Ultimate analysis is important for classification of coal by rank. Both analyses are important for utilization, commercial, and industrial purposes. However, for the purpose of coalbed gas, the proximate analysis, particularly moisture, ash, and volatile matter, is the most important for gas content and adsorption.

The primary reason for analyzing coal chemistry is to determine specific application for reservoir characterization especially for gas

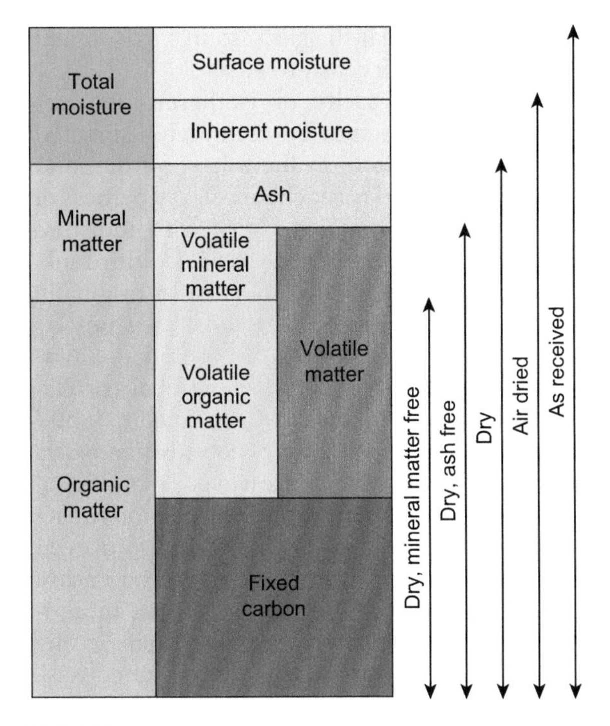

FIGURE 5.21 Diagram showing different proximate chemistry analytical bases for coal. *Source: Modified from Ward (1984).*

content and adsorption analyses. For this purpose, coal samples are received for laboratory analyses in different conditions (Schweinfurth, 2002). Consequently, results of the analyses are reported in various forms conditional on the state of the coal samples when received. Coal samples may be delivered to the laboratory in fresh (e.g. coal cores) conditions immediately after collection. As a result, the samples only need minimal preparation (e.g. grinding) and processing (e.g. splitting). The results of the analyses of these coal samples are reported on "as-received basis" (Figure 5.21). Other coal samples are delivered dry from field conditions such as not properly bagged and wrapped to preserve moisture, prolonged storage, long transport, and poor handling of the samples. The results of the analyses for these coal samples are reported as "dry basis". When the coal samples are delivered with excessive moisture of wet condition from the original coal bed, the results of the analyses of these samples are on "moist basis." Analyses also may be reported on a "mineral matter-free" or "dry, mineral matter-free" bases for use in calculating coal rank, which is controlled by maturity of the organic matter. "Mineral matter-free basis" is when the amount of mineral matter in the coal sample is deducted from the total analytical results to provide only the amount of organic matter. "Dry, mineral matter-free basis" is when the sample was delivered dry, nearly dry, or dried out before analysis. The "mineral matter-free basis" analytical results are used for comparison with results of "as-received basis" analysis because fresh coal samples may contain mineral matter or ash from partings in the coal, roof and floor rocks of the coal maybe be included in the sampling, and mineral matter may fill the pore and fracture systems as well as be dispersed in the coal matrix. Mineral matter in the coal may increase the ash content of the coal, which in turn affects the gas content and adsorption capacity analyses.

In order to avoid the different conditions of analyses and determine an accurate relationship between coal quality and gas content and adsorption capacity, mistake-free handling of coal samples is paramount. Handling of coal samples for coalbed gas analyses from the field is best described by Stricker et al., (2006); Stricker et al., (2010); and Carroll and Pashin, (2003). Fresh samples for coal quality and adsorption capacity (e.g. Langmuir isotherm) analyses are best collected separately from the core barrel before putting the coal in the canisters (Moore, 2012; Stricker et al., 2006). The amount of samples (e.g. 500 g for coal quality and 100 g for adsorption isotherm) for analyses varies between laboratories. Avoid storing samples for prolonged period (e.g. days) before delivering and/or mailing the samples to the laboratories. The key is to deliver samples fresh with moisture intact as much as possible. It is critical to record the location, depth, and

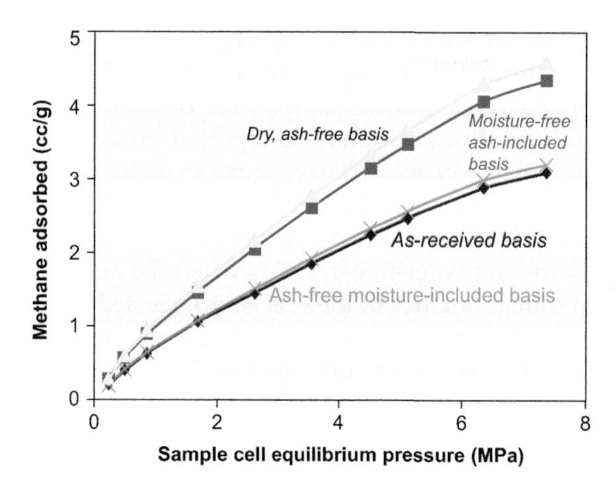

FIGURE 5.22 High-pressure methane adsorption isotherms for various proximate analytical bases (e.g. as-received, dry ash free, moisture free ash included, ash free moisture included). Subbituminous coal sample from the Powder River Basin. cc/g, cubic centimeter per gram; Mpa, megapascal.

thickness of all sampled coal on the plastic bag and/or desorption record book.

The most relevant aspect of condition of analysis of coal to coalbed gas is in reporting results of adsorption capacity or isotherms. Analytical results of adsorption isotherms are reported on "as-received; dry, ash-free; ash-free moisture-included; and moisture-free ash-included bases. As shown in Figure 5.22, the "dry, ash-free basis" in all the analytical results indicates a higher adsorbed gas (e.g. methane) or ideal storage capacity than "as-received basis" for any given pressure. The high moisture content of the coal analyzed on "as-received basis" is interpreted to block methane for adsorption. On the other hand, when moisture is removed as in the "dry, ash-free basis", more surface area is available for adsorption of methane. The other adsorption isotherm analyses lie in between the results of "as-received" and "dry, ash-free" bases. However, the "moisture-free ash-included" basis always has higher adsorbed gas than the "ash-free moisture-included" basis. This observation suggests that moisture content plays a bigger role than ash yield in storage of gas or methane. Thus, it is recommended that the "as-received basis" analytical results be used to achieve a more realistic gas resource assessment than using the "dry, ash-free basis".

Adsorption capacity or isotherms and gas content of coal cannot be accurately estimated without information on the ash, moisture, and volatile matter. These chemical properties of coal are interrelated especially with moisture and volatile matter, which decrease with rank. Although gas adsorption capacity was originally thought to be correlated with rank, the study by Clarkson and Bustin (1996) of Australian, Canadian and US coals indicated no global correlations of Carboniferous, Permian, and Cretaceous deposits, although correlation exists in specific coal basins. However, the role of coal quality to gas content and adsorption capacity is not always a clear-cut relationship even within a basin. For example, Carroll and Pashin (2003) reported for the Pennsylvanian bituminous coals of the Black Warrior Basin in the United States that there is a significant correlation (e.g. correlation coefficient of 0.87 for methane and 0.68 for carbon dioxide) between volatile matter (e.g. rank) and adsorbed gas in the coal (Figure 5.23(A)). Nevertheless, in the same study, there is markedly insignificant correlation (e.g. 0.24 correlation coefficient for methane and 0.00 for carbon dioxide) between ash yield (e.g. mineral matter) and amount of adsorbed gas (Figure 5.23(B)). Mineral matter probably in the form of infilling pores and fractures such as clay minerals, quartz, pyrite, and calcite affected the adsorption capacity of the Black Warrior Basin coals (Carroll & Pashin, 2003). Clarkson and Bustin (1996) reported that moisture might block methane adsorption in relation to carbon dioxide due to difference in chemical and physical properties and availability of microporosity. Adsorption capacity of coal presumably decreases proportionally with increasing ash and moisture content. During high-pressure methane adsorption analyses, equilibrium moisture and ash content are required to simulate the reservoir conditions. Thus, based on these observations, the most important coal quality analysis that is most

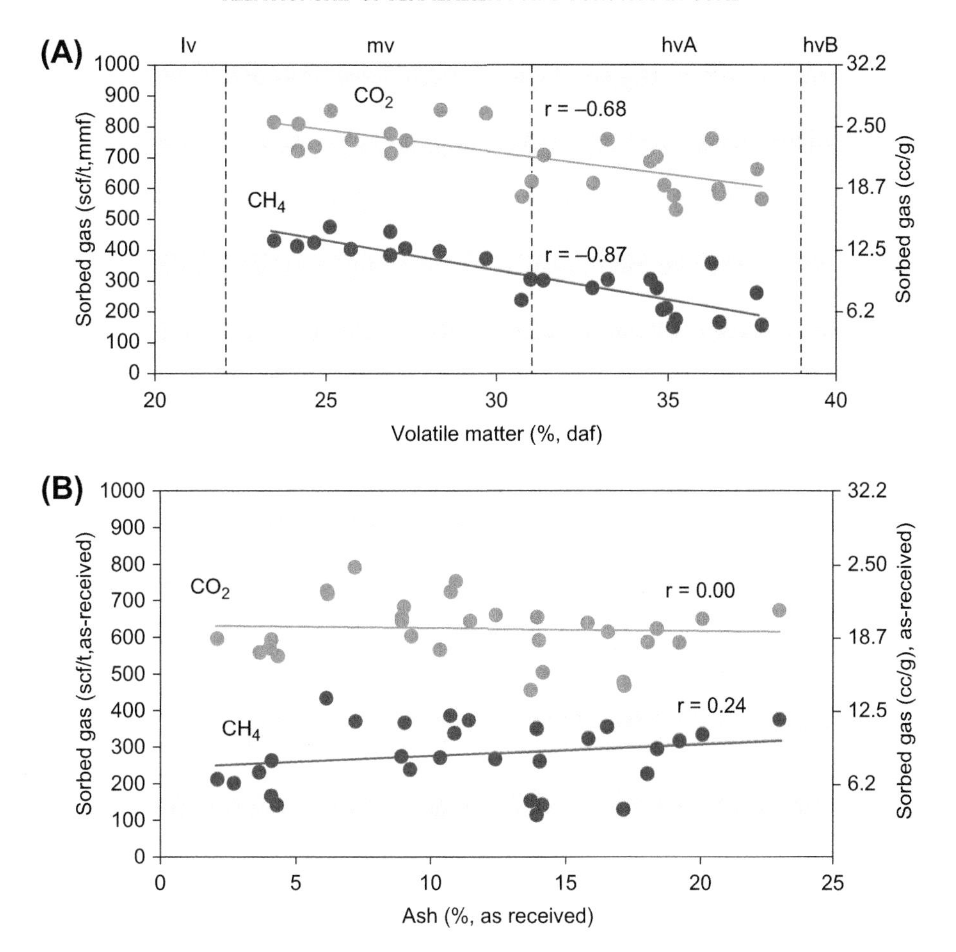

FIGURE 5.23 Diagrams showing (A) adsorbed methane and carbon dioxide on mineral-matter free (mmf) basis in relationship to percentage volatile matter on dry, ash-free (daf) basis and (B) adsorbed methane and carbon dioxide on as-received basis versus percentage ash on as-received basis for Pennsylvanian bituminous coals in the Black Warrior Basin, Alabama, United States. Lv, low-volatile bituminous; mv, medium-volatile bituminous; hvA, high-volatile A bituminous; hvB, high-volatile B bituminous; r, regression coefficient. cc/g, centimeter/gram; scf/t, standard cubic feet per ton; %, percent. *Source: Modified from Carroll and Pashin (2003).*

appropriate in the interpretation of coalbed gas assessment is proximate analysis.

RELATIONSHIP OF PERMEABILITY AND POROSITY IN COAL

Successful production of coalbed gas is extremely affected by cleat or fracture permeability and matrix porosity in the coal. In this chapter, permeability and porosity are discussed in terms of their function in the quality of performance of the coal reservoir. The rudimentary test of performance of a coal reservoir is its ability to flow gas that is economically recoverable from production wells. Coal reservoir performance is controlled by pores (e.g. the pore system) and their ability to diffuse gas into the adjoining cleat or fracture system where Darcy flow of gas and water governed their transport to production wells. The ability of the cleat or fracture system to transport the diffused gas from the coal matrix is controlled by the openness, density, connectivity, and length of cleats or fractures. The capacity of pores to diffuse gas is highly influenced by the density of pores (e.g. micropores) in the coal matrix, which in turn controls the internal surface area that adsorbs and concentrates the gas and the distance of transport from the concentration area to the producing well. More importantly, the quality of the coal reservoir may be affected by the relationship between the cleat or fracture and pore systems during gas desorption and production.

Porosity in Coal

It is mostly accepted by workers in the field (Clarkson and Bustin, 1996; Cui, Bustin & Dipple, 2004; Moore, 2012; Rodrigues & Lemos de Sousa, 2002) that the term "porosity" in coal is a dual system of pores and cleats or fractures. The magnitude of the amount of pores in coal as described by Moore (2012) is the single most important property that makes coal a unique reservoir among the unconventional and conventional reservoirs. Also, what is more unique of the porosity in coal is that the majority of the gas is adsorbed in the porosity system in the coal matrix. What is more interesting is that a large part of the volume adsorbed in the matrix is held in micropores in contrast to mesopores and macropores. The sizes of the pores are determined by either the width or the diameter with the latter preferred because pores are "holes" with either long or short dimensions. On the basis of the degree of sizes, pore system is classified in increasing size as micropores of <2 nm (nanometer), mesopores (2–50 nm), and macropores (>50 nm) (Gan, Nandi, & Walker, 1972; Senel, Gürüz, & Yücel, 2001). The term macropore system has often been used in the literature as the cleat or fracture network in the dual porosity classification (Van Krevelen, 1993). Walker (1981) reported that a majority of the micropores in coal are voids about 0.5–0.6 nm in size, which was classified as ultra micropores, and this led to the characterization of coals displaying "molecular sieve" behavior to gases.

Porosity is mainly related to the coal composition and rank. Vitrinite maceral (e.g. vitrain macrolithotype) predominantly contains micropores and inertinite (e.g. fusain macrolithotype) mostly contains mesopores and macropores (Adeboye, 2011; Gan et al., 1972; Harris & Yust, 1976; Lamberson & Bustin, 1993; Mastalerz, Drobniak, & Rupp, 2008; Unsworth, Fowler, & Jones, 1989; Zhang et al., 2010). Vitrinite content is the single most important maceral composition because of high coal total pore area due to the abundant microporosity (Adeboye, 2011). Harris and Yust (1976) related the microscopic composition of bituminous coals to pore size and porosity and established that vitrinite is mainly composed of micropores and mesopores, inertinite of mesopores, and liptinite of macropores. These researchers concluded that inertinite is the most porous maceral group and liptinite is the least porous. Mastalerz et al. (2008) studied the mesopores and micropores of Pennsylvanian bituminous coals in the Illinois Basin and found coal lithotypes (e.g. fusain, clarain, and vitrain) differ in pore volume and surface area. For example, micropore surface area for fusain is the smallest (72.8–98.2 m^2/g) and for vitrain is the largest (110.5–124.4 m^2/g). Thus, vitrain has higher adsorption capacity than fusain and clarain. However, maceral composition is not the main controlling factor of microporosity. Interestingly, the work of Faiz, Aziz, Hutton, and Jones (1992) demonstrated the contribution of microporosity in coal related to mineral matter. These workers found in the Australian Permian coals that the volume of macropores increases with the increase in mineral content. Nevertheless, the pore sizes and porosity vary significantly in the mineral matter-rich coals for reasons that are not clear.

Gan et al. (1972) described coals related to porosity in lignite to anthracite coals in which pore volume and frequency are controlled by rank. The average pore size of low-rank coals is larger than that of high-rank coals (Yao et al. 2011). An earlier report by Rodrigues and Lemos de Sousa (2002) indicated that high-volatile bituminous A and higher rank coals are dominated by micropores. The high-volatile bituminous C and B coals are dominated by mesopores. Lignite and subbituminous coals are dominated by macropores. However, Parkash and Chakrabartty (1986) argued that micropores instead of macropores are significant in low rank coals, which continues to be debated. Gurdal and Yalcin (2001) indicate that the volume of micropores in Turkish bituminous coal decreases with the increase in coal rank up to 1.0% vitrinite reflectance (R_0) and then increases with the increase in rank. This micropore and coal rank relationship is also duplicated with the proportion of macropore porosity decreasing with increased coal rank (Rodrigues and Lemos de Sousa, 2002). Thus, there are contradictory data on the relationship of porosity and rank of coal, which are unique to coal basins and may not necessarily correlate worldwide. The correlation of the porosity volume with coal rank expressed by carbon content (air-dried basis) shown in Figure 5.24 (Levine, 1996) needs to be utilized with caution accompanied by awareness of local and regional conditions. Moore (2012) demonstrated inverse relationships of micropores and macropores (Figure 5.24). In general, macropores increase with decreasing coal rank and micropores increase with increasing rank. However, there are reversals in the pore volume relationships between subbituminous and bituminous coals at about 55–82% carbon content. Rodriques and Lemos de Sousa (2001) reported that the percentage of macropores decreases from 0.7 to >1.4% R_0 (e.g. high-volatile C to medium-volatile bituminous) and then increases as secondary porosity in higher rank coal (>3.0% R_0).

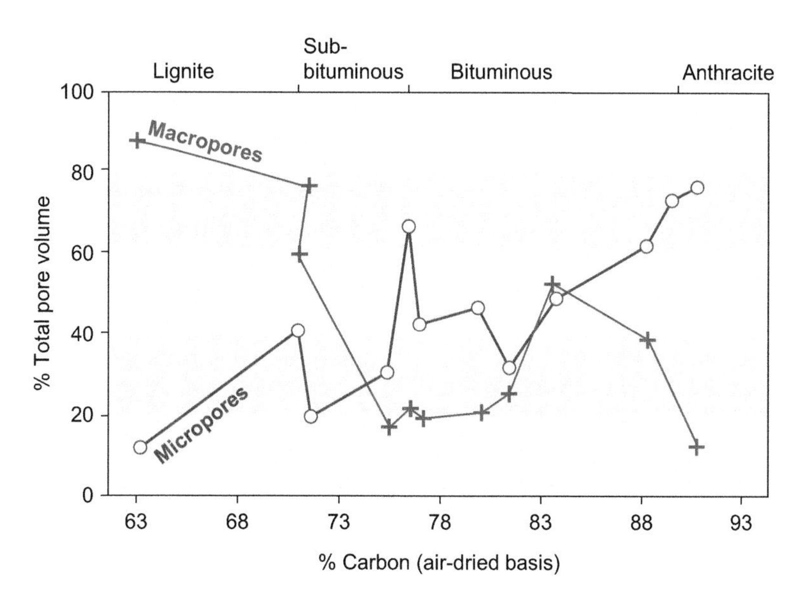

FIGURE 5.24 Diagram showing relationship of percentage total pore volume, percentage carbon on air-dried basis, and coal rank with micropores and macropores. *Source: Adopted from Levine (1996) and Moore (2012).*

Permeability in Coal

Perhaps, the most important property of the coal reservoir is permeability as expressed by cleat development and direction. Even though a coal may contain gas, the lack of permeability constrains gas recoverability and production. Additionally, permeability can have a significant effect on the drilling, completion, stimulation, and production strategies essential to successful development (Bilgesu & Ali, 2005; Close, 1993; Gentzis et al., 2009; Jenkins & Boyer, 2008; Laubach, Marrett, Olson, & Scott, 1998). Because procedures and strategies are important to the well performance and gas production, it is necessary to understand the origin and properties of coal permeability, which are vastly different from conventional reservoirs (e.g. sandstone, carbonate). On the other hand, coal permeability in mining, which is expressed in the cleat direction in coal, influences mine designs. Cleat directions of face and butt cleats represent zones of weakness in the coal. For example, room and pillar mining method may be designed with working faces parallel to the main cleat (e.g. face cleat) in order to smooth the progress of cutting the coal, as well as in longwall mining method, which changes and redistributes stress conditions in the surrounding coal causing failure (Durucan & Edwards, 1986).

Origin

Unlike the origin of porosity, which inherently originates within the organic matter and structures, the origin of permeability (e.g. cleat or fracture) is unrelated to the organic matter composition. However, permeability is controlled by coal bed moisture and volatile matter contents. Many workers have proposed dual origins of permeability: (1) inherent to progressive coalification and burial or endogenetic causes and (2) tectonic stresses during deformation or exogenetic causes (Ambrose & Ayers, 1991; Ammosov & Eremin, 1963; Close, 1993; Dawson & Esterle, 2010; Kendall & Briggs, 1933; Laubach et al. 1998; Law, 1993; McCulloch, Deul, & Jeran, 1974; McKee, Bumb, & Koenig, 1988; Solano-Acosta, Mastalerz, & Schimmelmann, 2007; Solano-Acosta, Schimmelmann, Mastalerz, & Arango, 2007; Ting 1977; Wang, Massarotto, & Rudolph, 2009). Other workers suggest a combination of both causes (Ammosov & Eremin, 1963; Close & Mavor, 1991; Laubach et al., 1998; Tremain, Laubach, & Whitehead, 1991; Wang et al., 2009). According to Kendall and Briggs (1933), cleats have been studied for nearly 180 years. It is only in the past 100 years that workers have focused on the endogenetic and exogenetic origins of coal cleats (Kaiser, 1908; Hoefer, 1915).

Endogenetic Cleats

Endogenetic cleats are early features formed during increased coalification with progressive loss of moisture, which caused desiccation of coal (Hucka & Bodily, 1989; Ting, 1977; Tremain et al., 1991; Law, 1991; 1993). Another proposed cause of endogenetic cleats is differential compaction causing folding around channel sandstones (Figure 5.25) (Ambrose & Ayers, 1991; Ayers & Zellers, 1991; Laubach et al., 1998).

Shrinkage from Loss of Moisture

Loss of moisture during progressive coalification is demonstrated by about 65% bed moisture loss from peat to the boundary of subbituminous A coal or to high-volatile bituminous C coal. Law (1991) proved shrinkage in coal caused by moisture loss with plots that are coincident between vitrinite reflectance versus cleat spacing and vitrinite reflectance versus bed moisture. In addition, Law (1993) described cleat spacing decrease with coal rank as a function of loss of moisture. The cleat spacing in coal versus fracture spacing in noncoal formations is believed to indicate the effects of desiccation (Close, 1993; Frodsham & Gayer, 1999; Huckla,

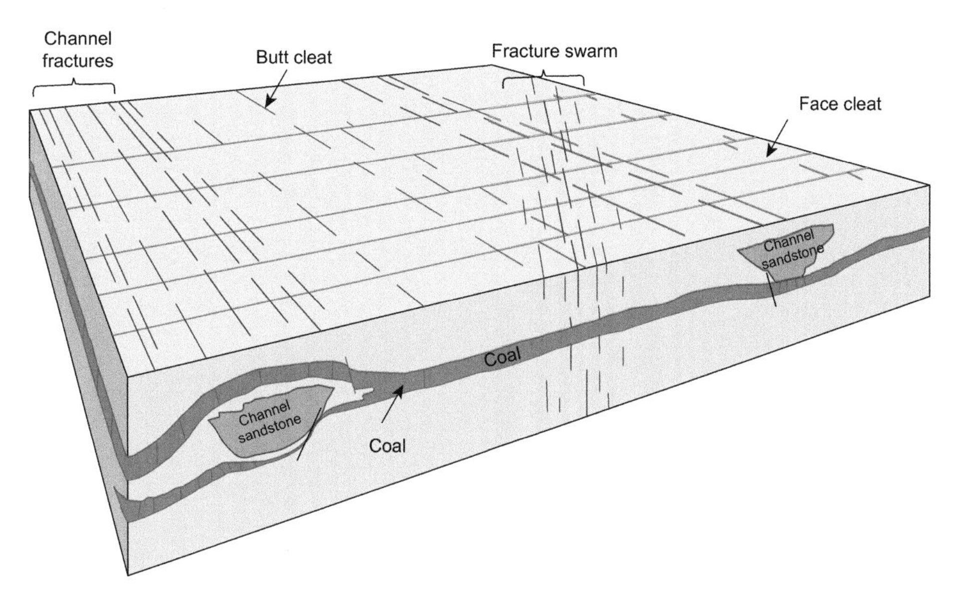

FIGURE 5.25 Face and butt cleat patterns and fractures associated with compactional folds around channel sandstones resulting from differential compaction and regional fractures. *Source: Adopted from Laubach et al. (1998).*

1989; Ting, 1977). Another factor, which is related to shrinkage, is elevated pore pressure in development of cleats in coal. Secor (1965) proposed that the development of cleats in coal is caused by high fluid pressure as applied by Hubbert and Rubey (1959) on overpressured fluids causing thrust faults. This phenomenon was proved by Burnham and Sweeney (1989), which reported significant increase of the rate of moisture loss from 0.3% to 0.5% R_o. Thus, pore pressure mechanism may occur as cleat formation from lignite B to subbituminous A or low rank coals.

A very good attempt to describe the formation of endogenetic cleats (e.g. face and butt) is found in a report by Ryan (2003), which suggested that cleats are related to shrinkage of the bulk coal caused by loss of both moisture and volatile matter during coalification. Figure 5.26 shows the decrease of percentage volume with rank increase by using average inherent water and as-received volatile matter contents for different ranks starting with lignite. Ryan (2003) estimated that when lignite is transformed to the rank with 0.7% R_o, the volume has decreased more than 50%. Recalculation from lignite to subbituminous coal from British Columbia, Canada, by Ryan (2003) demonstrates the incremental decrease of the coal mass with rank (Figure 5.27). The change of coal mass curve (green color) shows two peaks, which correspond to generation of carbon dioxide and water at about 50 °C. It is assumed that during these processes, gas adsorption as coal rank increases is less than the amount of generated gas. The first peak at 50 °C accounts for most of the volume decrease of the coal mass and according to Ryan (2003) is accompanied by development of the face cleat. Movement of water through these cleats enhances precipitation of mineral matter (e.g. clays). The second peak at 160 °C relates to the decrease of volume of coal mass with low water content and generation of methane. The low water content and more volume decrease of the coal may be accompanied by development of butt cleats (Ryan, 2003). This suggests the absence of butt

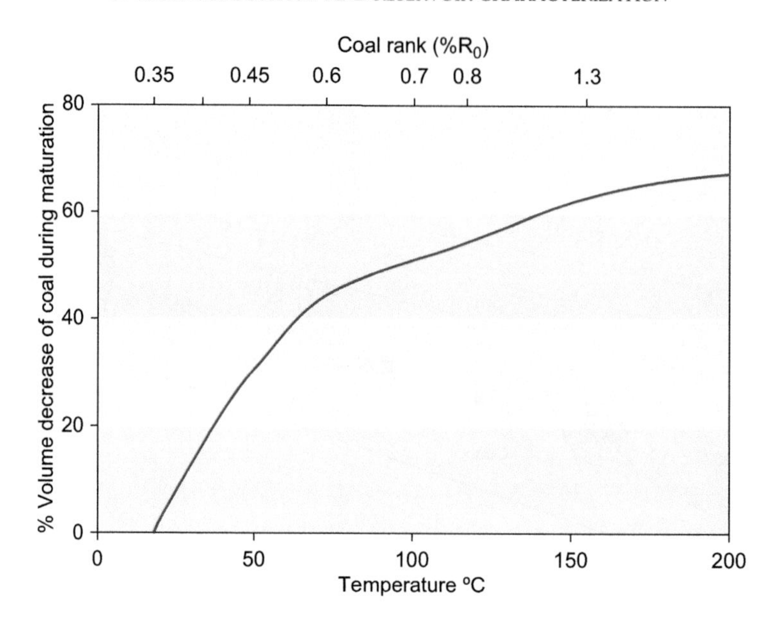

FIGURE 5.26 Diagram showing estimate of coal volume decrease related to rank expressed in vitrinite reflectance (%R_0). *Source: Modified from Ryan (2003).* R_0, *vitrinite reflectance.*

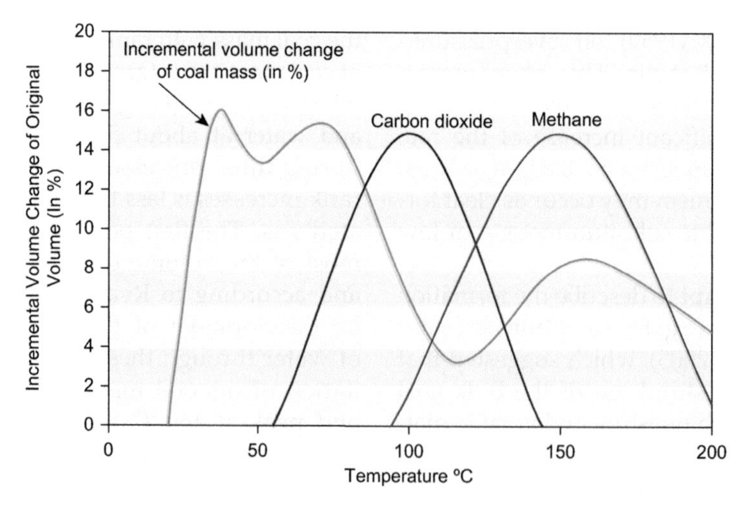

FIGURE 5.27 Diagram showing incremental decrease of coal mass volume during coalification and generation of carbon dioxide and methane with increase temperature. Peak generations of carbon dioxide and methane are reached at about 100 °C and 150 °C, respectively *Source: Modified from Ryan (2003).*

cleats in low-rank coals, which have not attained sufficient rank to generate methane. Also, the paucity of water movement through the cleats precludes precipitation of mineral matter, but, if present, they are mainly high-temperate carbonates.

Differential Compaction

Another mode of formation of endogenetic cleats is during burial. Differential compaction of interbedded channel sandstone, finer-grained siltstone and mudstone, and coal beds promotes development of coal cleats or fractures, master joints, and microfaults below and above sandstones (Figure 5.28(A)). The fine-grained beds compact more than the sandstone beds, thus creating "wrap-around" (e.g. compactional folds) sandstones and developing cleats in brittle coal beds (Ambrose & Ayers, 1991; Ayers & Zellers, 1991; Laubach et al., 1998). The brittle condition of the coal beds is presumed to be due to loss of moisture. In Cretaceous coal-bearing sequences, Ryer and Langer (1980) reported that the thickness of channel sandstones increases by an average of 2.1 m for 1 m decrease in the thickness of interbedded claystones or about 50% difference in compaction. Compactional folds are different from structural folds and must be confirmed with thinning and thickening of stratigraphic intervals. This stratigraphic relationship between channel sandstones and intervening fine-grained clastics and coal beds is well illustrated in the Paleocene Fort Union Formation in the Powder River Basin, Wyoming, United States (Figure 5.28(A); Flores, 1986). The differential compaction between the channel sandstone and interbedded siltstone, mudstone, and coal produced about 40% reduction of the interval along the margins of the channel sandstone. The coal beds below and above the thickest part of channel sandstones are densely cleated or fractured (Figure 5.28(B)). The fractures often show displacements forming microfaults into the sandstone. Usually, thick coal beds formed above the channel sandstones are more gassy than flanked coal beds due to "woody" composition or vitrain-rich macrolithotype. This resulted from preferential arboreal vegetal peat growth on elevated abandoned fluvial channel areas.

Exogenetic Cleats

The causes of exogenetic cleats or fractures are not straightforward because of multiple generations due to complex geology, tectonic deformations, and local structural conditions. Laubach et al. (1998) in reviewing the origin of exogenetic cleats proposed potential causes: (1) thermoelastic response of coal to regional tectonic extension and/of uplift of the basin and surrounding areas, (2) stress from disturbances due to local folding and faulting, and (3) high pore pressure causing expulsion of water causing stress. Many workers have attributed the development of cleats (e.g. shear fractures) to regional and local paleotectonic compressional stresses (Close, 1993; Dawson & Esterle, 2010; Laubach et al., 1998; McCulloch et al., 1974; McKee et al., 1988; Pashin. 1998; Solano-Acosta, Mastalerz, & Schimmelmann, 2007; Solano-Acosta, Schimmelmann, Mastalerz, & Arango, 2007; Wang et al., 2009). Shear fractures show striations and are mainly not perpendicular to bedding like tensional fractures. The coal cleats were assumed to be formed by external forces during deformation of the basin of deposition after coalification as a later-stage phenomenon compared with the endogenetic cleats (Solano-Acosta, Mastalerz, & Schimmelmann, 2007; Solano-Acosta, Schimmelmann, Mastalerz, & Arango, 2007).

Earlier, during studies of exogenetic causes of coal cleats, debate persisted about whether cleats or fractures are formed at shallow or deep burial (Laubach et al., 1998). The tensional stress applied to coal beds was compared to cooling, dehydration, and folding of igneous rocks near the surface (Engelder, 1987). The fact that cleat

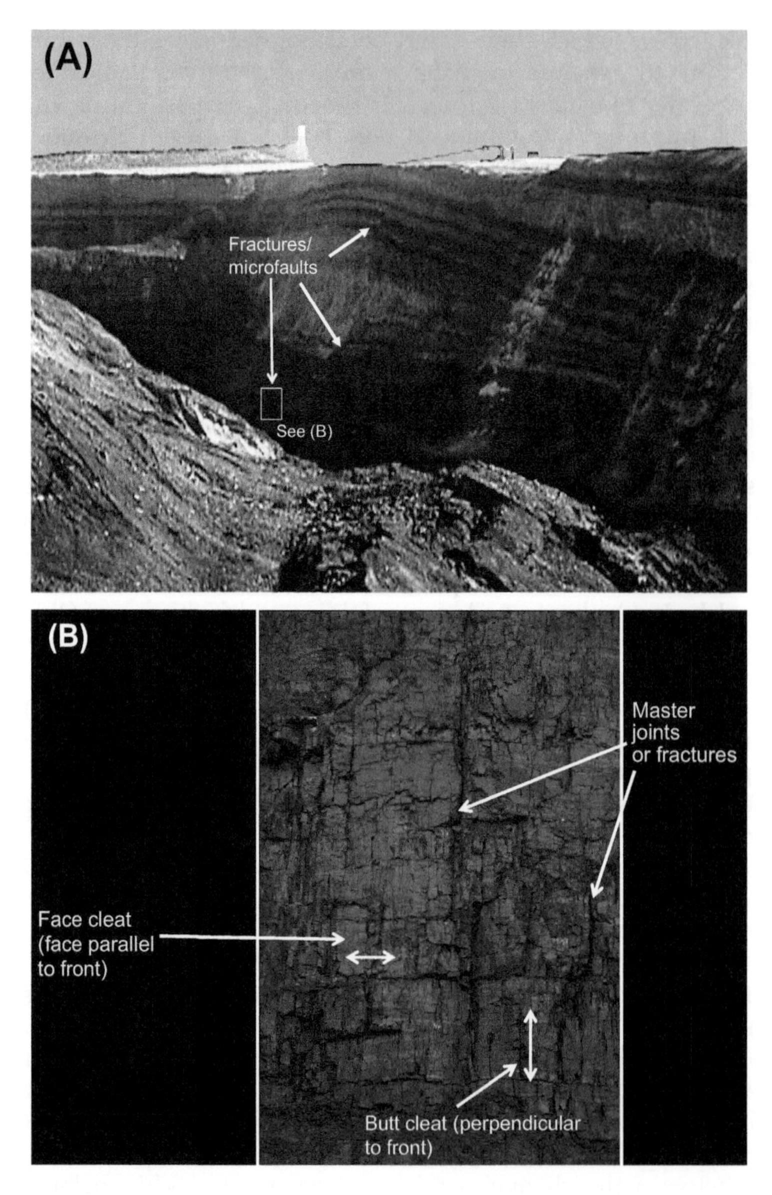

FIGURE 5.28 Photographs showing (A) fold around a fluvial channel sandstone due to differential compaction exposed in mine highwall and (B) coal cleats and master joints formed directly below the sandstone in the Powder River Basin, Wyoming, United States.

development in coals is a near-surface occurrence was supported by Olson and Pollard (1989), who argued that coal has increased strength to sustain differential stress in order to display planar shape of the cleats. In contrast, Hancock (1985) suggests that shear stress at depth may have played a role in cleat formation as "shear fractures".

Regional Tectonic Influence

Many workers have suggested that the regional uniform patterns of coal cleats reflect paleotectonic stress fields (Close, 1993; Kulander & Dean, 1993; Laubach & Tremain, 1991; McCulloch et al., 1974; Rippon, Ellison, & Gayer, 2002; Solano-Acosta, Mastalerz, & Schimmelmann, 2007; Solano-Acosta, Schimmelmann, Mastalerz, & Arango, 2007; Su, Feng, Chen, & Pan, 2001). For example, the east–northeast striking cleats or joints of the Appalachian Basin in the United States represent a rectilinear maximum stress field of more than 1500 km (Engelder & Whitaker, 2006). The stress field was formed during the final assembly of Pangaea as a result of plate–boundary interactions from convergence in which horizontal stress was parallel to closure between Laurentia and Gondwana (Engelder & Whitaker, 2006). Thus, the regional paleotectonic deformation of a coal basin (e.g. Appalachian), which caused cleat development, is a far-reaching mechanism applied from outside the basin of deposition. However, local structural deformation may have superimposed on the larger tectonism as indicated by the face cleat orientation of the Pittsburgh coal, which rotated from N 80° W in northwestern West Virginia to N 57° W in southwestern Pennsylvania (McCulloch et al. (1974).

Prediction of cleat orientation in coal basins, however, is not clear because of potential multistage and multidirectional nature of paleotectonic histories of coal basins. The stress and fracture history of the basin is critical for predicting subsurface coal cleat orientations, which can control permeability and coalbed gas production. For example, Rippon et al. (2002) demonstrated the Carboniferous coal cleat directions in the United Kingdom with the Variscan stress field across various tectonic conditions. The multistage and multidirectional cleat orientations are indicated by varied cleat orientations proximal to the Variscan Front in South Wales as evidence by a complex of stress deflection and block rotation. North of the Wales–Brabant High in central-northern England, the northwesterly optimum horizontal stress is parallel to the regional cleat direction. In the distal Midland Valley of Scotland, local structural deformation played a major role in cleat orientations. Rippon et al. (2002) observed all through Britain that major fault zones partitioned different sets of coal cleat orientations resulting from deep crustal compartmentalization during the Carboniferous Period. The structural evolution of England and Wales during the Carboniferous was primarily a consequence of an oblique (dextral) collision between Gondwana and Laurasia (Warr, 2002).

Influence of Local Structural Deformation

Local coal cleat orientations affected by folds and faults within coal-bearing rocks are just as complicated as in the regional cleat orientations (Laubach et al., 1998; Li, Lui, Yao, Cai, & Qui, 2011; Solano-Acosta, Mastalerz, & Schimmelmann, 2007; Solano-Acosta, Schimmelmann, Mastalerz, & Arango, 2007). Frodsham and Gayer (1999) reported that in the Carboniferous South Wales coalfield in the United Kingdom, large amounts of fractured coals are formed along coal-based thrust detachments. The distribution of the fractures in coal along the detached coal beds is heterogeneous. Shear stress along faults caused increase of fractures in Permian anthracite coals in the Qinshui Basin, China. However, the effect of permeability has been negatively impacted by mineralization of the fractures. Still, the permeability of the coals along the fault zones is higher than that of distant coals.

Solano-Acosta, Mastalerz, and Schimmelmann (2007) and Solano-Acosta, Schimmelmann, Mastalerz, and Arango (2007) reported that cleats in Pennsylvanian coals in the Illinois Basin in the United States were formed during early coalification and that cross-cutting cleats are at an angle to the present stress field (e.g. exogenetic) and parallel to lineaments. Both are endogenetic, however, because the early-formed cleats consist of more pronounced cleat mineralization (e.g. kaolinite) than the later-formed cleats (e.g. smaller apertures). Solano-Acosta et al. (2007) interpreted two stages of cleat formation resulting from tectonic deformations that were altered during coalification and basin history.

Influence of Basin Rebound

Coal basins rebound by uplift, which expends tremendous amounts of stress, which is subsequently relaxed and lessened by erosion. This process permitted development of cleats or fractures especially in shallow coal beds. The cleats formed in coal by basin rebound are prone to develop at shallower depths where stress is lesser than in deeper parts of the basin. Cleats formed by "exhumation" of coal basins are at right angle to earlier-generated face cleats (Engelder, 1987; Laubach et al., 1998). Laubach et al. (1998) reported that during "exhumation" of coal basins, stresses parallel to preexisting face cleats were not relaxed, which permitted development of secondary fractures perpendicular to these cleats. Open face cleats during "exhumation" behaved as surfaces against which butt cleats formed.

Coal cleats at shallow depths or along basin margins were prone to mineralization by low-temperature mineral matter (e.g. kaolinite and other clay minerals). Once coal beds were eroded and exposed at the surface or along basin margins, groundwater system was established by meteoric water. Also, it is at this time when biogenic gas is best formed by microbial activity because groundwater and nutrients to sustain the microbes are able to freely penetrate the coal beds from shallower and deeper zones (Flores, Stricker, Rice, Warden, & Ellis, 2008). Although cleats in coal are formed after uplift, evidence shows that cleats also formed prior to "exhumation" with mineralization in folded rocks (Tremain et al., 1991) (Box 5.4).

BOX 5.4

POROSITY, CLEAT/FRACTURE POROSITY, AND PERMEABILITY—CONCEPTS AND MISCONCEPTIONS

The literature is replete with the use of reservoir terms such as fracture porosity, cleat porosity, and fracture permeability, which are applied to describe coal (Cui & Bustin, 2006; Harpalani & Chen, 1995; Hucka and Bodily, 1989; Karimi, 2005; Li et al., 2011; Pan, Connell, & Camilleri, 2010). Engineers universally apply these terms to hydrocarbon reservoirs but the terms become confusing when applied to coal reservoir. The carryover of reservoir porosity terminologies from conventional to coal reservoirs has introduced inconsistency between pores and cleat/fractures to describe porosity. In general, fracture porosity in a generic reservoir is a function of fracture width and block size (Hensel, 1989). For example, wide, infrequent fractures in large blocks results in low porosities much like extensive fractures in small blocks. Hensel (1989) suggested that a formation must be fractured at high frequency to contain significant fracture porosity. It is commonly accepted in both coal and other hydrocarbon reservoirs that porosity is related to pores or voids per unit volume.

In coal, porosity is expressed in terms of the sizes (e.g. diameter) of the pores as micropores, mesopores, and macropores in the coal matrix (Clarkson & Bustin, 1996; Senel et al., 2001). Also, in coal, the pores occur in the coal matrix separated by the coal cleats or fractures. The coal matrix porosity and adjoining cleats or fractures are sometime referred to as dual porosity system. This dual porosity system was defined as coal fabric by Cui and Bustin (2006). The connectivity of the cleats or fractures controls the degree of permeability or ability of the coal to flow gas. Thus, mixing both coal matrix porosity and cleat or fracture porosity is a contradiction in terms.

Characteristics of Cleats

Cleat or fracture system in coal consists of two virtually orthogonal sets of face and butt cleats. The face cleat is a longer fracture and the butt cleat is a shorter fracture developed about 90° from the face cleat (Figure 5.29). The attributes of cleats in coal include aperture, length, spacing, height, frequency (density), and connectivity (Figure 5.29). These attributes control the permeability, which in turn, influence the producibility of gas in coal reservoirs.

Coal Cleat Aperture

Coal cleat aperture is the width of the crack and is microscopic to macroscopic in size (Figure 5.29) depending on whether the opening is under stressed (e.g. in situ confining pressure) or nonstressed conditions (MazumDer, Wolf, Elewaut, & Ephraim, 2006). For example, cleat apertures under confining pressure conditions in Carboniferous coal in Zonguldak Basin, Northwestern Turkey, vary from 0.05 to 0.0001 mm (Karacan & Okandodan, 2000).

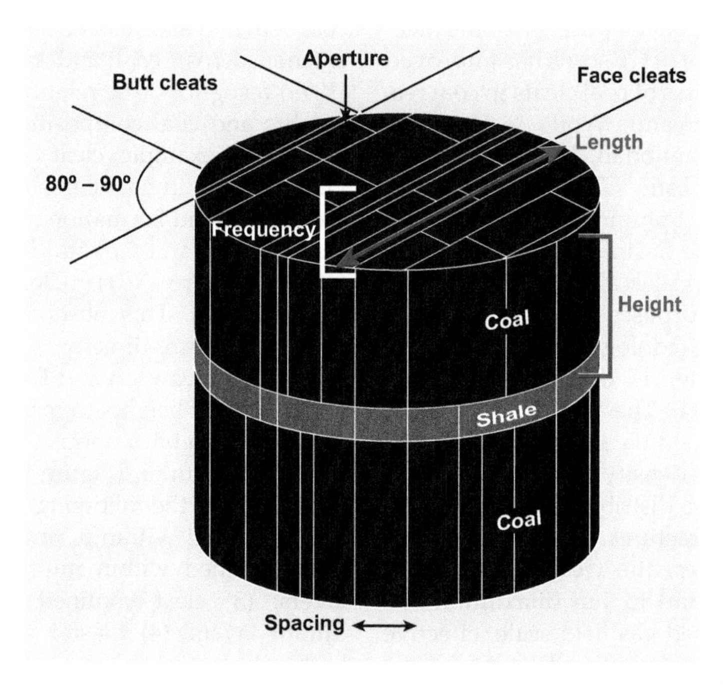

FIGURE 5.29 Diagram of a idealized coal core showing idealized face and butt cleat system consisting of aperture, length, height, spacing, and frequency.

Under nonstressed condition, cleat apertures vary from 0.01 to 0.30 mm for the Cretaceous Fruitland coal in the San Juan Basin, United States (Close 1993; Close & Mavor, 1991). Using computed tomographic (CT) scans MazumDer et al. (2006) studied Carboniferous coals in Poland, which vary from >0.04 to <0.5 mm in aperture width.

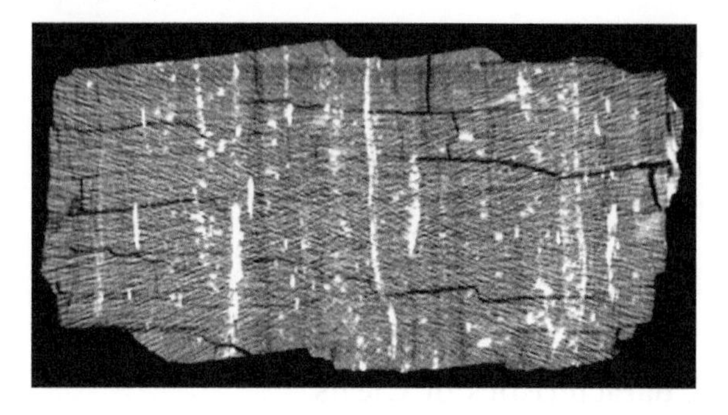

FIGURE 5.30 A computed tomography-scanned image of a coal sample (left) showing open closed macrocleats or fractures. *Source: Adopted from Bossie-Codreanu et al. (2004).*

Coal Cleat Length

Length is the extent or distance along bedding plane from end to end of cleats depending on the scale or magnitude of the coal "sample" being measured (e.g. from core to outcrop through field size). It is assumed that lengths of cleats are straight lines; however, studies of subbituminous coals in the Powder River Basin, Wyoming, United States, show curvilinear lengths across the 7.62 cm diameter of the core (see Figure 5.35(B)) (Stricker et al., 2006; Trippi et al., 2010). Lengths of coal cleats are either measured along the face cleat or the butt cleat depending on the purpose of the measurement. The advent of CT scans has improved the quality of attributes of coal cleats in coal reservoirs (Bossie-Codreanu, Wolf, & Ephraim, 2004; Karacan & Okandodan, 2000; MazumDer et al., 2006). A realistic CT scan by Bossie-Codreanu, Wolf, & Ephraim (2004) of a coal sample from Loraine Basin, France, from very fine scale to numerical scale (tens of meters) for simulation purposes are shown in Figure 5.30. For methodology and software applications, the reader is directed to Bossie-Codreanu et al. (2004). These analyses resulted in collection of detailed data on fractal number as well as coal cleat density, frequency, length distribution, aperture distribution, and average distance between fractures. For example, for two fracture families, the cleat lengths vary from 0.1 m (minimum) to 1 m (maximum). In comparison to coalbed gas field scale, effective fracture or cleat length induced by hydraulic fracture completion was not documented beyond 67 m (Olsen et al. (2003)).

Coal Cleat Height

A cleat height is the distance from the bottom to the top of a cleat perpendicular to bedding plane. Laubach et al. (1998) classified the height of coal cleats as primary when the fracture extends from bottom to top of a coal bed, as secondary when the fracture is confined to the coal bed, and as a master cleat when the fracture extends across coal and noncoal beds. Close (1993) recognized the relationship between cleat heights and coal composition (e.g. macrolithotypes). For example, cleat heights in individual vitrain layers in coal cores of the Upper Cretaceous Fruitland Formation in the San Juan Basin, New Mexico and Colorado vary from 1 mm to 60 mm (Figure 5.31) (Close, 1993; Close & Mavor, 1991). This observation indicates that cleat heights are directly related to macrolithotypes. Thus, Dawson and Esterle (2010) reclassified coal cleat heights to macrolithotype specific for Australian bituminous coals, which includes vitrain and durain (Figure 5.32). Cleats heights are related to the following macrolithotypes: (1) cleat confined within a single vitrain layer; (2) cleat confined within multiple stacked vitrain layers; (3) cleat confined exclusively within durain layers; (4) master cleat, which crosses both vitrain and durain layers; and (5) a large master cleat or joint that extends across many macrolithotype layers and partings.

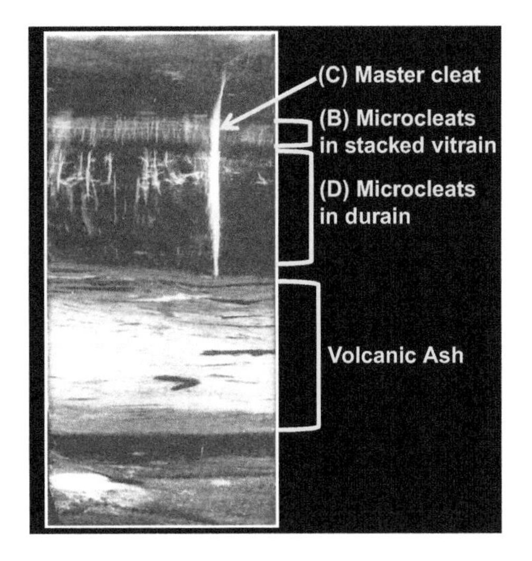

FIGURE 5.31 X-ray image of the Cretaceous Fruitland coal in the San Juan Basin showing macrolithotypes and associated cleats or fractures. *Source: Adopted from Close (1993).*

The Dawson and Esterle (2010) classification of cleat heights can also be applied to the macrolithotypes of the subbituminous coals in the Powder River Basin in Wyoming, United States. Figure 5.33 shows a core split of subbituminous B, which can be divided into two major layers: upper bright vitrain and lower dull durain (clarain or both) macrolithotypes. The lower dull durain macrolithotype can be divided using Table 5.7 into thin to medium laminated and coarsely laminated bands. The upper bright vitrain macrolithotype can be divided into thin to medium laminated and very coarsely laminated bands. All five classes of cleat heights of Dawson and Esterle (2010) can be recognized in this lower rank coal, which contains more than 50% vitrain. The master cleat or joint, which readily opens once the confining core barrel is removed, is common and more than 1 m in height (Figure 5.34). This joint cutting sinuously across numerous macrolithotypes when viewed on top of a core (perpendicular to bedding) displays curvilinear cleats.

Coal Cleat Spacing and Frequency

The spacing of coal cleats is the distance between two adjoining cleat surfaces (Figure 5.35(A)). Cleat spacing and aperture (Figure 5.35(B)) play a special role in determining

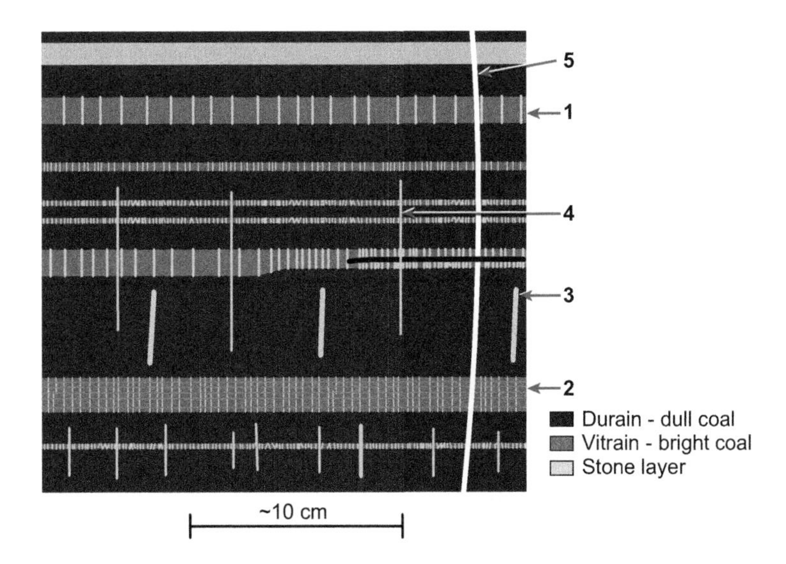

FIGURE 5.32 Cleat or fracture classification as related to macrolithotypes in Australian Permian coals. *Source: Adopted from Dawson and Esterle (2010).*

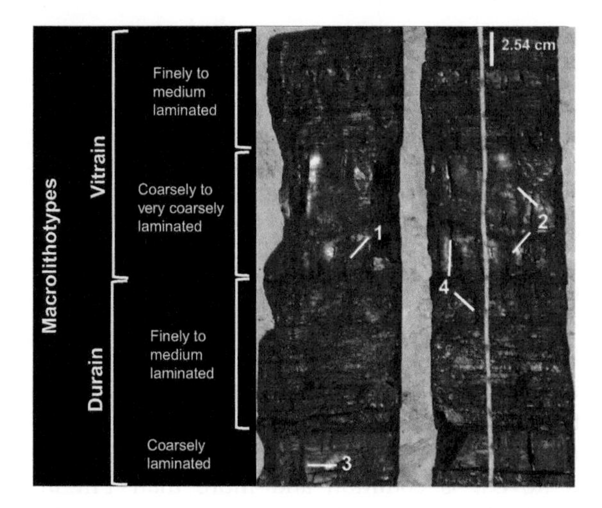

FIGURE 5.33 Photographs of a split core (8.25 cm diameter) of a Tertiary coal in the Powder River Basin, Wyoming, United States, showing macrolithotypes and associated cleats or fractures. (1) Cleat confined to a vitrain band, (2) cleat confined to stacked vitrain bands, (3) cleat confined to durain band, and (4) master cleat cutting across a few different bands of macrolithotypes.

FIGURE 5.34 A photograph of 3.3-m-long, 8.25 cm diameter Tertiary coal core in the Powder River Basin, Wyoming, United States, showing long master joints exposed after the core barrel was opened.

permeability, as reported by Laubach et al. (1998). The factors that control cleat spacing include coal rank, coal composition (e.g. macrolithotypes), bed thickness, and coal cleat height (Ammosov & Eremin, 1963; Dawson & Esterle, 2010; Jenkins, 1999; Laubach & Tremain, 1991; Law, 1993; Close, 1993; Ting, 1977; Yao et al., 2008). Cleat spacing decreases from lignite to bituminous coal as a result of loss of moisture and volatile matter (Ammosov & Eremin, 1963; Law, 1993; Scott, 1997; Ting 1977; Tremain et al., 1991). On the basis of vitrinite reflectance,

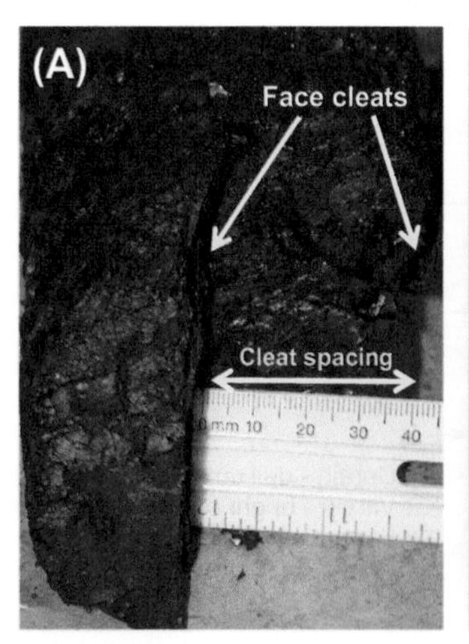

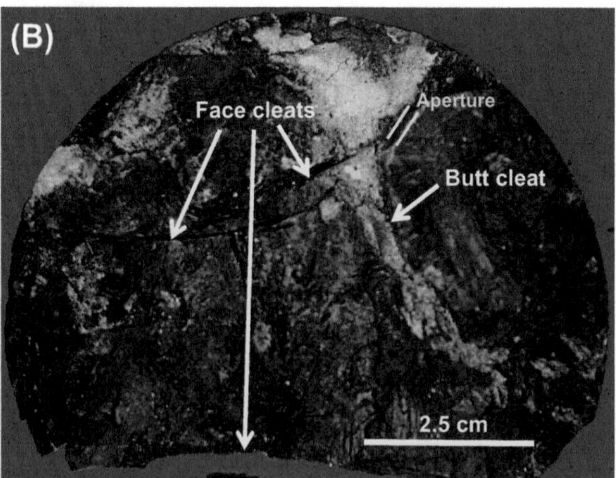

FIGURE 5.35 Photographs of Tertiary coal core (8.25 cm diameter) in the Powder River Basin, Wyoming, United States, showing cleat spacings (A) and top view of curvilinear cleats (B).

Law (1993) reported that coal cleat spacing is greater for lignite (0.25–0.38% R_0) than anthracite (2.6% R_0). The Cretaceous Fruitland subbituminous coals in the San Juan Basin in the United States possess cleat spacing of 2.5–6.3 cm compared to high-volatile C bituminous coal, which have a spacing of 0.3–0.6 cm. However, cleat spacing within the same coal rank such as the subbituminous coal A to C in the Powder River Basin, which range from 5 to 70 mm, do not differ significantly (Trippi et al., 2010). Furthermore, because subbituminous C coal is along the basin margins and subbituminous A is at the center of the basin, depth does not play a major role in cleat spacing. Laubach and Tremain, 1991 observed in the Fruitland coals that cleat spacing increases with increasing thickness of medium-bright bituminous coal. This concept was expanded by Jenkins (1999) observing that bright banded coal (e.g. 60–90% bright or vitrain) exhibits 3–5 mm cleat spacing, banded coal (e.g. 50% bright and dull or clarain/durain) consists of 2–10 mm cleat spacing, dull banded (e.g. 10–40% or durain) consists of cleats more than bright coals, and dull coal (e.g. <10% bright bands) consists of poor cleat spacing. Thus, in general, cleat spacing increases from bright to dull macrolithotypes.

Detailed core and mine studies of Permian bituminous coals in Australia by Dawson and Esterle (2010) show a significant relationship between cleat spacing and height. Using the height classification related with macrolithotypes shown in Figure 5.32, Dawson and Esterle (2010) show statistically significant difference of face and butt cleats with cleat height on various macrolithotypes (e.g. single vitrain and multiple stacked vitrain layers bounded by durain). For the "low-rank" bituminous coal (0.67% R_0), face and butt cleats show increasing cleat spacing with increasing thickness of vitrain layer. For "high-rank" bituminous coal (1.85% R_0), the cleat and height relationship is similar to that of the "low-rank" bituminous coal. However, the relationship in terms of the rate of change of the cleat spacing is much narrower for the "high-rank" than the "low-rank" bituminous coal for every change in coal height. An interesting note in the study of Dawson and Esterle (2010) is the development of the master cleats, which apparently occur once in the core sample suggesting cleat spacing to be much wider than the coal sampled, and their development is unrelated to the coal rank. Of the classes of cleat height, the master cleat or joint is the most important attribute in coalbed gas characterization because it represents the most continuous opening that crosses and connects many macrolithotypes of variable gas contents.

Coal cleat frequency or density is the rate of recurrence of cleat within a specified distance (e.g. per inch or centimeter) or square area. Much like cleat height, coal cleat frequency is controlled by coal composition (e.g. macrolithotypes), coal rank, and thickness. Spears and Caswell (1986) reported that cleat frequency is a function of macrolithotype (e.g. highest in vitrain) and is inversely related to bed thickness. Yao et al. (2008) reported significant difference in cleat frequency in relationship to the genetic origin of Pennsylvanian and Permian coals in the Webei coalfield in Ordos Basin, China. These workers reported that exogenetic cleats (e.g. microfractures formed by tectonic stress) have frequency of $>100/9\ cm^2$ with normally developed frequencies of $110–200/9\ cm^2$ and ultra developed greater frequencies. Endogenetic cleats (e.g. microfractures related to macrolithotypes and macrolithotypes) have frequencies of $<100/9\ cm^2$. Both genetically related cleats, except the ultra developed cleat frequencies, have significant relationship with permeability with higher frequencies having better permeability. Cleat frequency, which increases with coal rank, was demonstrated by Su et al. (2001) in Permian coals in western North China. These workers observed that when the vitrinite reflectance is less than 1.35% (e.g. medium low-volatile bituminous), cleat frequency increases. However, when the vitrinite reflectance is more than 1.35%, cleat frequency decreases, which Su et al. (2001) explained as due to cleats being destroyed or annealed by filling of secondary maceral and agglutination. Annealing also results from mineralization of the cleats mainly by calcite.

Connectivity

Coal cleat connectivity, which drives the permeability of coal, is lateral (one direction horizontal) continuity due to interconnection of the cleat network that includes face and butt cleats as well as other fractures formed by tectonic stress. Laubach et al. (1998) suggest that the network geometry of and connectivity of fractures enhance permeability. McCulloch et al. (1974) observed that permeability in the face cleat direction is 3–10 times greater than in other cleat directions. As suggested by Dawson and Esterle (2010), the master joints, which cut across many coal macrolithotypes, may be the key to connectivity as also suggested earlier by Chen and Harpalani (1995) and Tremain et al. (1991).

Regardless, the only way to test connectivity of coal cleats in the subsurface is from interwell communications from hydrostatic pressure responses during interference tests. In the San Juan and Black Warrior Basins in the United States, the pressure drawdown from the production to monitor wells along the face cleat direction was determined to be 328 m apart and along the butt cleat direction was determined to be 54 m apart (Close, 1993; Mavor, Close, & Pratt 1991). Thus, the connectivity is greater along face cleats than butt cleats in bituminous coals of these basins.

RESERVOIR CHARACTERIZATION

Coal composition and physical properties described in the previous sections impact gas content, gas transport, and well completion methods for coalbed gas reservoirs. More importantly, the behavior of the coal reservoirs from changes in permeability and porosity affects potential enhanced gas production from carbon dioxide sequestration. Thus, the coal properties and their changes applied to reservoir characterization are highlighted in the following sections.

Shrink and Swell Properties of Coal Reservoir

Fundamentally, the combined mechanisms of shrink and swell in coal affect both permeability and porosity. The shrink and swell behaviors in coal occur when gas is desorbed and adsorbed, respectively, from the coal matrix pores and cleats or fractures. Coal shrinkage is believed to increase permeability related to mechanical elasticity and/or ductility of coal (Durucan, Ahsan, & Shi, 2009; Gu & Chalaturnyk, 2006; Harpalani & Chen, 1995; Levine, 1993; Wang & Ward, 2009). It is hypothesized that once the gas is desorbed, accompanied by reduction of confining and pore pressures from cleats and coal matrix pores, the rock shrinks to the original mass (Pan et al., 2010; St. George and Barakat, 2001). The large volume of moisture lost during coalification especially in lower rank coals may have contributed to the potential shrinkage (Ozdemir & Schroeder, 2009). Thus, the shrinkage of coal is partly related to coal rank and composition (Laubach et al., 1998; Wang & Ward, 2009).

The combined effects of reduction of reservoir pressure and loss of moisture permit the coal mass to shrink and widen the cleats increasing the absolute permeability (Jenkins & Boyer, 2008). This mechanism may be counteracted by decrease in cleat aperture due to increased net stress from depletion of reservoir pressure. Absolute permeability is the ability of coal to transport only a single fluid when the reservoir is absolutely gas saturated (McGraw-Hill, 2003). However, coal bed reservoirs are not totally gas saturated and gas is held in the reservoir by hydrostatic pressure. Moore (2012) suggested that coal matrix shrinkage from reduction of pressure by water withdrawal from the reservoir might counterbalance cleat closure by the amount of gas saturation. Originally, Wang

(2006) reported that cleat closure is minimal from gas desorption in reservoir with high gas pressure, which is close to "absolute permeability" proposed by Jenkins and Boyer (2008). However, this reservoir condition is offset in undersaturated coals such as in the Powder River Basin in Wyoming, United States, where voluminous water is withdrawn in order to degas. Under this condition, the cleat or fracture apertures will collapse, which in turn lowers the permeability as well reduces the rate of recovery from gas diffusion in the pores of the coal matrix. Thus, the net effect of coal shrinkage is poor reservoir performance during gas production caused by "tight" permeability and porosity.

The mechanism of coal swelling is induced by gas adsorption, which increases pore pressure, permitting the coal matrix to expand. Many laboratory and modeling procedures are spent toward understanding coal swelling because it affects enhanced gas recovery with carbon dioxide sequestration (Cui, Bustin, & Chikatamarla, 2007, Gilman & Beckle, 2000; Gray, 1987; Massarotto, Golding, Bae, Iyer, & Rudolph, 2010; Palmer & Monsoori, 1998; Pan et al., 2010; Seidle & Huitt, 1995; Shi & Durucan 2005, Wang et al., 2009). Simply put, this process requires injection of carbon dioxide to replace residual methane molecules in the coal reservoir yielding enhanced gas production. However, gas adsorption was found to swell or expand the coal matrix mainly perpendicular to the bedding plane (Bustin, 2004; Day, Fry, & Sakurovs, 2012; Pan & Connell, 2011). This suggests that the cleat or fracture permeability, which is also perpendicular to bedding, might not be significantly affected by coal matrix swelling.

Coal swelling is controlled by coal rank and composition. Low-rank coal is relatively unaffected by swelling from carbon dioxide injection probably due to higher inherent bed moisture content than high-rank coals (Balan & Gumrah, 2009; Majewska, Majewski, & Zietek, 2010; Moore, 2012). The volumetric swelling difference between bituminous and subbituminous coals according to Day et al. (2012) may range from 1.9% to 5.5% in carbon dioxide depending on the rank. Karacan, 2003 reported that vitrinite- and liptinite-rich coals swell more than inertinite- and clay-rich coals. These compositional effects may be partly explained by changes in the pore properties during carbon dioxide injection. Massarotto et al. (2010) observed that injection of carbon dioxide in Permian bright and dull Australian coals led to increase in both micro- and mesoporosity. However, the macroporosity in the dull coal decreased, while it increased slightly in bright coal. Thus, current research indicates that coal swelling due to flooding of carbon dioxide reduces permeability, potentially affecting the feasibility of enhanced coalbed gas production and carbon dioxide sequestration.

Coal Macrolithotypes versus Gas Content

The relationship of coal composition to gas content is predicated based on the influence of gas adsorption capacity and desorption rate in various macrolithotypes (e.g. vitrain, clarain, and fusain). Although the coal composition and gas adsorption relationship is indirectly related to rank, it is also directly associated with porosity and permeability. Thus, based on this relationship, it is apparent that coal composition (e.g. macerals, microlithotypes, and macrolithotypes) impacts the final gas content of coal. This relationship has been confirmed by current research in both field and laboratory conditions with some success especially with vitrinite-rich coals that have higher methane adsorption than other maceral types (Crosdale, Beamish, & Valix, 1998; Mares & Moore, 2008; Mastalerz et al., 2012; Moore et al., 2001; Stricker & Flores 2009; Stricker et al., 2010; Trippi et al., 2010).

Ettinger, Eremin, Zimakov, and Yanovskaya (1966) reported that inertinite-rich coals adsorb more methane. Faiz, Aziz, Hutton, and Jones (1992) did not find significant relationship between coal composition and adsorption capacity. Most researchers, however, agree that vitrinite-rich (e.g. maceral and microlithotype) or vitrain-rich (e.g. macrolithotype) coals generally have high adsorption capacity, which directly relates to high gas content based on gas desorption in the field and laboratory by Mares & Moore, 2008, Moore et al. (2001), Stricker & flores (2009), Stricker et al., (2010), and Trippi et al. (2010)

According to Crosdale et al. (1998), gas (e.g. methane) adsorption capacity of Australian Permian bituminous coals in the Bowen Basin is directly related to micropores, which in turn depends on maceral composition and rank. Based on adsorption isotherms, bright coals (e.g. vitrain-rich macrolithotypes) have higher adsorption capacity than dull ones (e.g. inertinite-rich or fusain macrolithotypes). Investigations of these workers on methane desorption rate in the laboratory indicate that the bright vitrain-rich coals have the slowest desorption rates, which is directly related to high microporosity and cleat or fracture systems not filled by mineral matter. In contrast, the dull coals have faster desorption rate, which is related to development of macropores and micropores from mainly open cell structures. The relationship between macrolithotypes and porosity has been determined in an earlier study by Crosdale and Beamish (1995). That is, bright, vitrain-rich coal is more porous than dull coals.

Adsorption capacity of carbon dioxide was tested in the laboratory by Mastalerz et al. (2012) on Pennsylvanian bituminous coals in the Illinois Basin in the United States. Vitrain, clarain, and fusain macrolithotypes were analyzed for surface area using the Brunnauer, Emmett, and Teller multilayer adsorption model and for adsorption capacity using the monolayer Langmuir isotherm model. Two sets of coal samples from two different formations were studied, which showed surface area varying from 10.0 to 115.4 m^2/g in vitrain and from 6.0 to 15.9 m^2/g in fusain. Analysis of the carbon dioxide adsorption capacity on as-received basis varies from 42 to 51 cm^3/g in vitrain, from 37 to 42 cm^3/g in clarain, and from 24 to 34 cm^3/g in fusain. Thus, comparison of adsorption capacity indicates that carbon dioxide is adsorbed more in vitrain macrolithotype than in fusain macrolithotype (Mastalerz et al., 2012).

Coal Reservoir Properties Within and Between Basins

Although high-pressure methane and/or carbon dioxide adsorption isotherms are the normally accepted standard test of extrapolating gas volume in coal, it is an idealized maximum measurement. Another method to estimate gas volume or content from coal is gas desorption, which is less than estimated from adsorption capacity (Mares & Moore, 2008; Moore, 2012). Gas content is measured from coal cores collected in airtight desorption canisters and gas desorbed at reservoir pressure (see chapter for methodology). The gas content determined by this method starts in the field where coal cores are collected and ends in the laboratory at room temperature. The direct method of gas content analysis is a more realistic measurement than idealized volume of gas determined by adsorption analysis. The gas content data can be related to the coal composition of the cores using either microscopic (e.g. maceral and microlithotype) and/or macroscopic (e.g. macrolithotype) analyses. Because composition determined from microscopic analysis has been shown to strongly correlate with macroscopic analysis (Moore, 1995; Moore & Ferm, 1992; Moore et al., 2001), it is suggested that core description using macrolithotypes provides better data for correlation with gas content and/or adsorption isotherms of coal.

The investigation of Mares and Moore (2008) on the Tertiary subbituminous coals from the Huntly coalfield in New Zealand demonstrates the relationship of desorbed gas content to macrolithotypes. As previously described, the classification of macrolithotypes is based on the brightness/dullness properties and thickness or texture (e.g. 10–30 mm) of coal bands. Two coal beds were examined: one consisted of 57% bright, nonbanded coal, while the other consisted of 68% bright, banded coal. The bright, nonbanded coal contains the highest gas content (average 1.96 m^3/t) compared to the bright, banded coal (average1.51 m^3/t). In terms of adsorption capacity of the coals, the bright, nonbanded coal is approximately 15% (average) more than the banded coal and 20% (average) higher than the total average desorbed gas content. Mares and Moore (2008) explained the differences in adsorption capacity and gas content of the various coals by the variation in the original environmental settings (e.g. vegetation types and peat mire geochemistry) of the coals. The variability in composition and physical properties of coal between beds and within a formation and coal basin as demonstrated by the work of Mares and Moore (2008) seems to be more of a rule than not. More often, the variations of coal properties, whether local or regional in nature, play significant role in reservoir characterization, which will be demonstrated in the next section.

INSIGHTS OF RESERVOIR CHARACTERIZATION OF GAS PLAYS IN THE POWDER RIVER BASIN

Reservoir characterization can be a complex undertaking involving a broad range of techniques ranging from field mapping and hand sample description to complex geophysical analyses. Generally, it involves defining various properties of a reservoir rock that affect or control its ability to generate, store, and produce hydrocarbons. These properties can include gas (or oil) content, composition/mineralogy, porosity, coal lithotypes, permeability, thickness, lateral variability, structural controls, and other elements depending on the nature of a specific reservoir. Reservoir properties coupled with gas production data permit assessment of the variables that maximize production of coalbed gas as analyzed from the Tertiary subbituminous coals of the Fort Union Formation in the Powder River Basin. More than 900 cored coal samples (e.g. 0.67-m-long desorption canisters) were collected from 30 producing coalbed gas wells in the basin during a period of 5 years (2000–2005) in conjunction with the USGS-BLM Coalbed Methane project (Flores et al. 2006). All the cored coal samples were desorbed for gas content and 65 coal samples were analyzed for high-pressure methane and carbon dioxide adsorption capacities (Flores et al., 2001; Stricker et al., 2006; Trippi et al., 2010). The coal cores were described for macrolithotype composition and texture (e.g. brightness/dullness and band/lamination thickness) and permeability attributes (e.g. cleat spacing and frequency of density). These coal reservoir variables were related to gas content and coal rank.

Macrolithotypes vs Gas Content

The most basic relationship between macrolithotypes and gas content is best demonstrated locally and regionally. Moore et al. (2001) described a total of 55 m of a continuous coal core from a well in the Tertiary Fort Union Formation, Wyoming. The coal cores were collected from the Wyodak coal nearby a coal mine at the eastern margin of the Powder River Basin and were analyzed every 0.67-m intervals. Gas content was measured from desorption of the subbituminous coal initially in the field and in the laboratory. The cores were in turn described in the laboratory for macrolithotype attributes after gas desorption. Macrolithotypes were described and classified according to the thickness of coal bands (e.g. laminations) (Table 5.7). In addition, the abundance of vitrain macrolithotype was determined by macroscopic point-count method (Moore et al., 1993). The results of this study are partly summarized in Figure 5.36 showing vertical correlation of gas content with vitrain and fusain bands in the continuous coal core. In general, the percentage

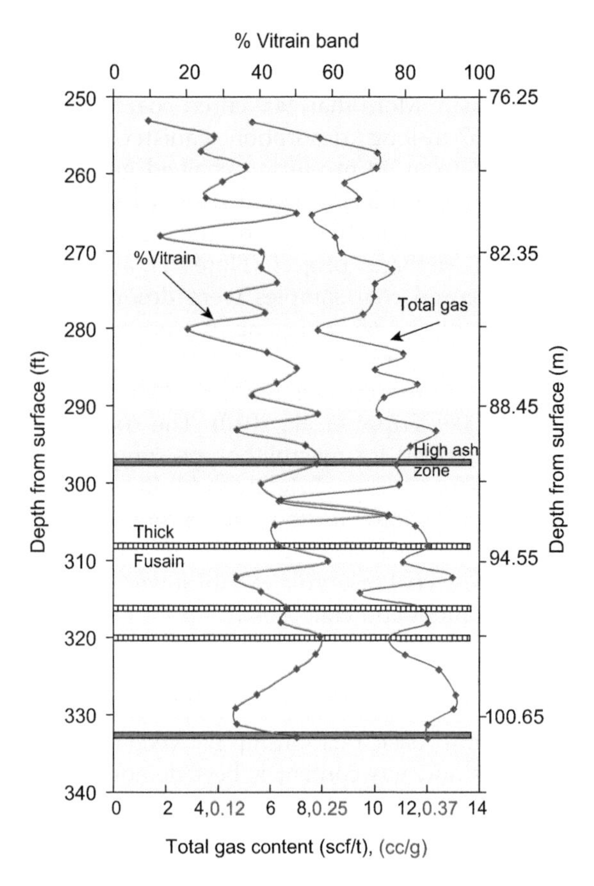

FIGURE 5.36 Diagram showing relationship of gas content with percentage vitrain band of a continuous 27-m-thick Wyodak coal in the Powder River Basin, Wyoming, United States. m, meter. ft, foot; cc/g. centimeter/gram; %, percent. *Source: Modified from Moore et al. (2001).*

of vitrain bands decreases from an average of about 50% in the lower part of the coal bed to an average of 30% in the upper part. Fusain is distributed in the lower part of the coal bed but the gas content does not appear to be low as previously observed by Crosdale et al. (1998) and Mastalerz et al. (2012). Correspondingly, the gas content averages 0.34 cc/g in the lower part and 0.28 cc/g in the upper part of the coal. The low gas content of the coal was due to previous desorption from coal mine dewatering process. Nevertheless, the study shows some evidence of reduction of gas content with decrease of vitrain bands from bottom to top of the coal bed. In addition, macrolithotype analysis indicates that the vitrain bands are common in the laminated macrolithotypes. The variable macrolithotype thickness indicates that the vitrain commonly occurs (40–50%) in the medium, coarse, and very coarse laminated coal. When this variable is applied to the coal bed, it was observed that the lower part is commonly composed of medium to coarse laminated macrolithotypes, which corresponds to the relatively high gas content. Thus, the vitrain-rich lower part is interpreted to have been formed in a topogeneous (low-lying) mire in which arborescent vegetation was sustained by detrital influx and the forested mire was intermittently subaerially exposed to form peat fires as represented by the fusain macrolithotype. The decrease of vitrain macrolithotype and increase of attritus macrolithotype in the upper part of the coal bed suggest transformation of the topogeneous mire to ombrogenous bog (raised bog).

The regional or basin-wide investigations by Flores et al. (2001), Stricker & Flores (2009), Stricker et al. (2010), and Trippi et al. (2010) of the same Wyodak coal that Moore et al. (2001) studied locally in the Powder River Basin show a similarity in relationship of woody and laminated macrolithotypes to high gas content (Figure 5.37). However, the basin-wide distribution of woody macrolithotype (Figure 5.38) shows variable but high percentage in the south-central part of the basin where gas content is locally as much as 3.89 cc/g (Stricker & Flores, 2009; Stricker et al., 2010). More importantly, total gas contribution from various macrolithotypes appears to be partly related to coal rank (Figure 5.39). In the laminated and woody macrolithotypes, the subbituminous C coal has the most variable high gas content compared with the subbituminous A and B, which have the least variable high gas content.

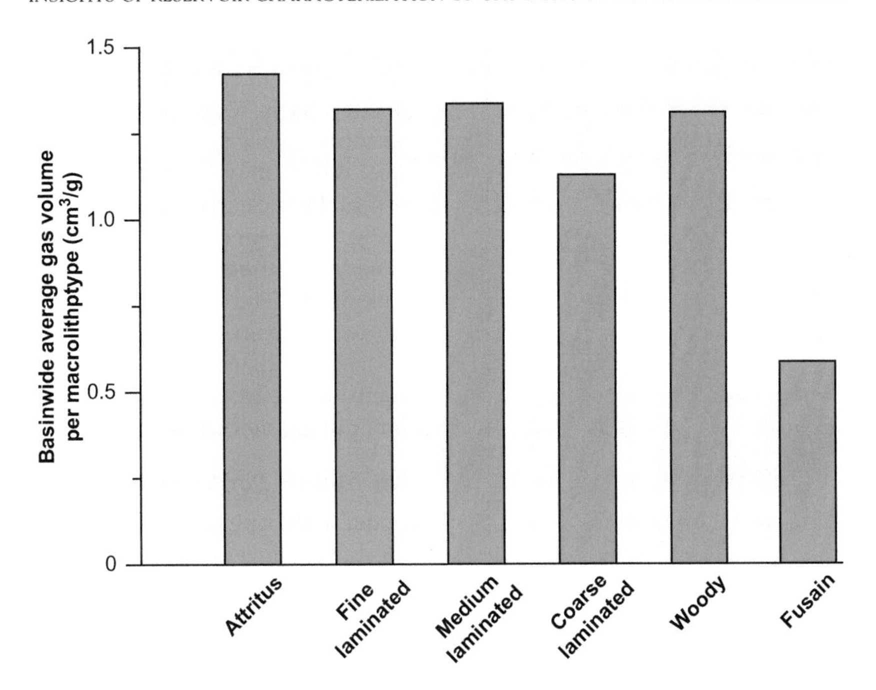

FIGURE 5.37 Histogram showing relationship of macrolithotypes and gas content of the Tertiary Wyodak coal based on 82 cored samples (0.67 m-long) in the Powder River Basin, Wyoming, United States. The attrital or attritus macrolithotype contains the most average gas content and fusain contains the least average gas content.

Coal Cleat (Fracture) Permeability

Thus far, the local and regional relationships of coal composition and gas content do not follow generally basic reservoir concepts. The addition of permeability attributes such as cleat spacing and frequency to the mix of coal properties can further test the validity of the traditional concepts. Cleat spacing or the width between two adjoining cleats or fractures is a good measure of permeability (Laubach et al., 1998; Solana-Acosta et al., 2007). As recommended by Dawson and Esterle (2010), the relationship of cleat spacing to macrolithotypes may be an association that directly affects the flow of gas. That is, traditionally vitrain-rich bands, especially in higher rank coals, normally develop closely spaced cleats providing more openings for gas adsorption and transport per square area (e.g. cleat frequency). Figure 5.40 shows the relationship of cleat spacing to macrolithotypes of the Wyodak coal in the Powder River Basin. The laminated and woody macrolithotypes have cleat spacing (average 25–34 mm) between fusain (average 20 mm) and attritus (average 47 mm). Regardless of the macrolithotypes, it is estimated that there is about <1 cleat per 25 mm (2.54 cm).

The relationships between cleat spacing, macrolithotype, and gas content in a continuous core of the 45-m thick Wyodak coal, which is the most productive reservoir in the Powder River Basin, are shown in Figure 5.41. The core is from a producing well at 383 m depth and is subbituminous A rank. The upper part of the coal bed shows 5–70 mm (average 30 mm) cleat spacing, common woody macrolithotype, and average 2.28 cc/g and as much as 2.81 cc/g gas content. Flores (2004) reported that cleat or fracture

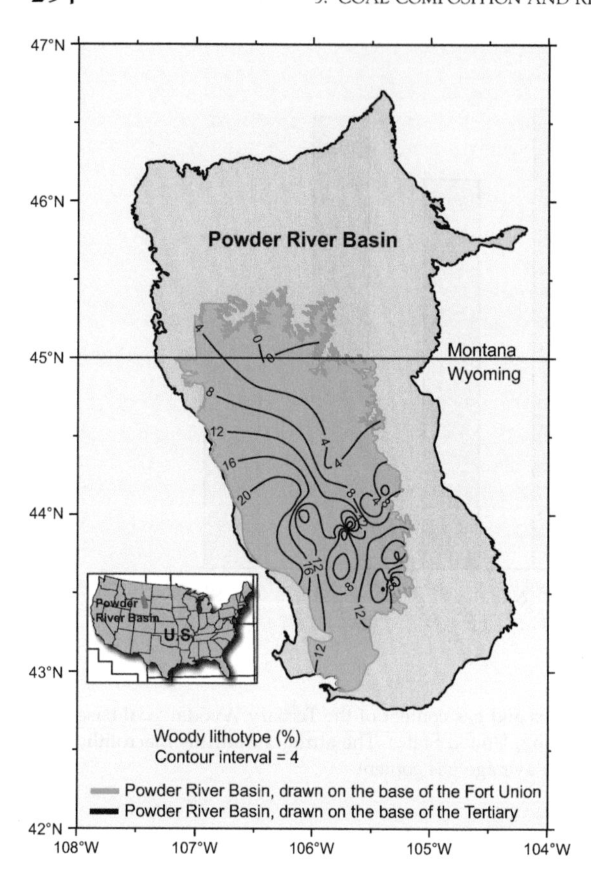

FIGURE 5.38 Map of the Powder River Basin in Wyoming and Montana, United States, showing percentage isolines of woody macrolithotype of the Tertiary Wyodak coal. *Source: Adopted from Stricker et al. (2010).*

permeability ranges from 1.5 D to 325 mD based on interference tests or injecting reservoir fluid into a well and monitoring fluid pressure from surrounding wells. The cleat spacing and permeability of the subbituminous coal in the Powder River Basin were compared to those in the bituminous coals in the San Juan and Black Warrior Basins (Laubach et al., 1998) to determine the aperture size, which is found to be 50 μ–>80 μm and 3 μm–40 μm, respectively (Figure 5.43).

The connectivity of the cleat or fracture system of the subbituminous coals in the Powder River Basin, Wyoming, based on the hydrostatic pressure drawdown from the production and monitor wells during interference tests is much greater than that of the Cretaceous coals in the San Juan Basin (Close, 1993; Mavor et al. 1991). The connectivity varies from 417 to 1450 m distance, which may be much longer than the extents of face cleat or fracture and interference among well pairs as far as 3.2 km apart. Thus, it is assumed that this long-distance connectivity is along master joints. Flores et al. (2010) reported a basin-wide network of southwest–northeast and southeast–northwest trending structural lineaments, which may have created joints during deformation of the basin.

Coal Matrix Permeability

Coal matrix permeability of the Wyodak coal zone based on effective stress tests of vertical and horizontal core specimens (52 mm in diameter) with introduction of methane, carbon dioxide, and nitrogen for each specimen ranges from 0.04 to 1.5 mD (Flores, 2004; Uniservices, 2002). Permeability varied depending on the type of gas with carbon dioxide displaying the smallest permeability and nitrogen displaying the largest permeability (Table 5.8). Also, coal matrix permeability varied between shallow Wyodak coals (98–196 m depths) and deep Wyodak coals (42–448 m depths). The matrix permeability (horizontal) of the shallow Wyodak coals for all three gases varies from 0.3 to 0.85 mD compared to that of the deep Wyodak coals, which ranges from 0.02 to 1.5 mD for both horizontal and vertical.

About 38% of the tested samples have coal matrix permeability of <0.1 mD, which is equivalent to tight gas sand permeability. This "tight" coal matrix permeability is reflected by >6 months of gas desorption in sealed canisters in the minority of the coal samples (Stricker et al., 2006). Thus, because sorption time is interrelated to diffusivity and shape of the microporous matrix, this indicates relatively very slow rate gas diffusion in the coal matrix for more than one-third of the Wyodak coal. Comparison of exponential reduction in permeability under

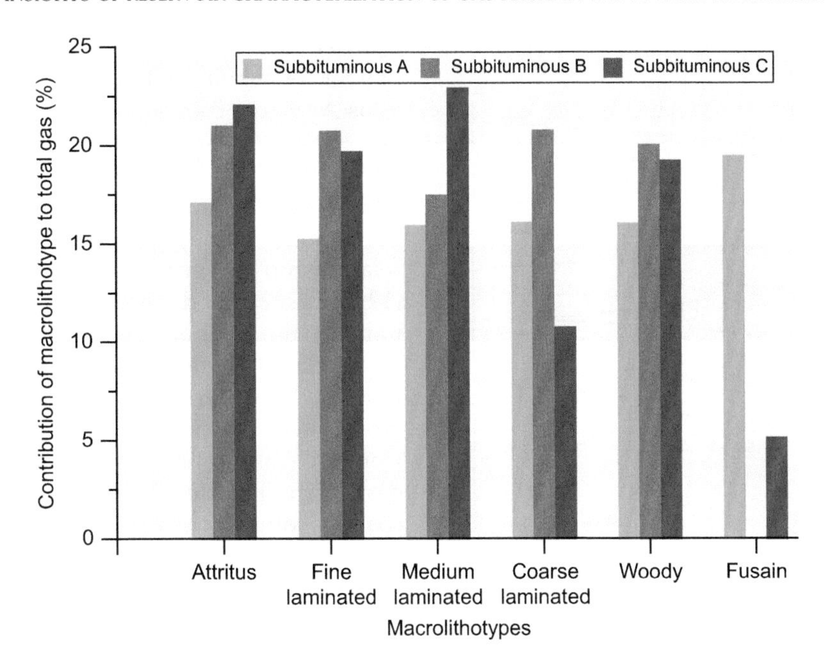

FIGURE 5.39 Histogram showing relationship of macrolithotypes, total gas content, and coal rank of the Tertiary Wyodak coal based on more than 900 core samples in the Powder River Basin in Wyoming, United States. %, percent. *Source: Data Modified from Trippi et al. (2010).*

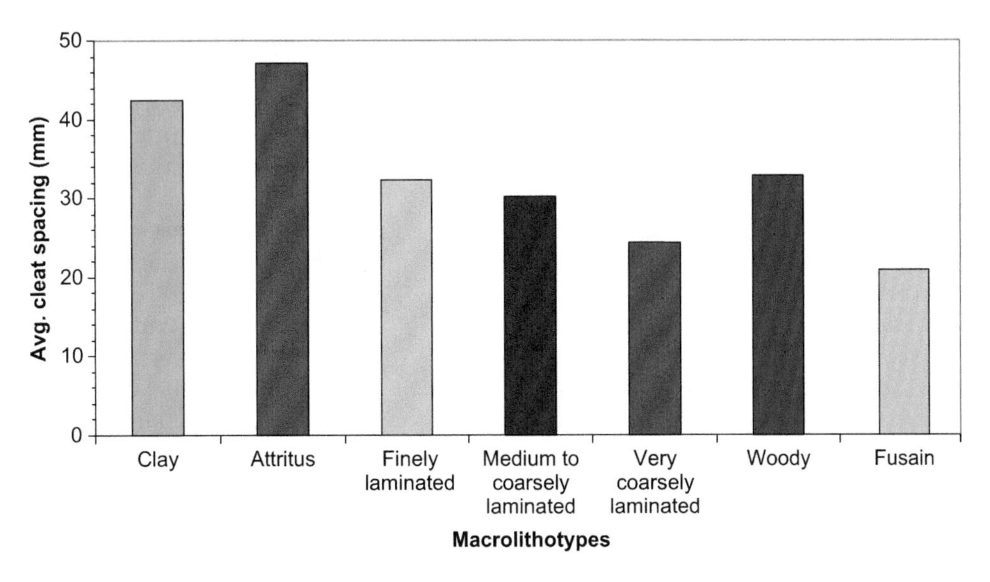

FIGURE 5.40 Histogram showing relationship of macrolithotypes and average cleat spacing of the Wyodak coal based on more than 900 core samples in the Powder River Basin in Wyoming, United States. The attrital or attritus macrolithotype contains the widest cleat spacing and the fusain macrolithotype contains the narrowest cleat spacing. mm, millimeter; avg, average. *Source: Modified from Trippi et al. (2010).*

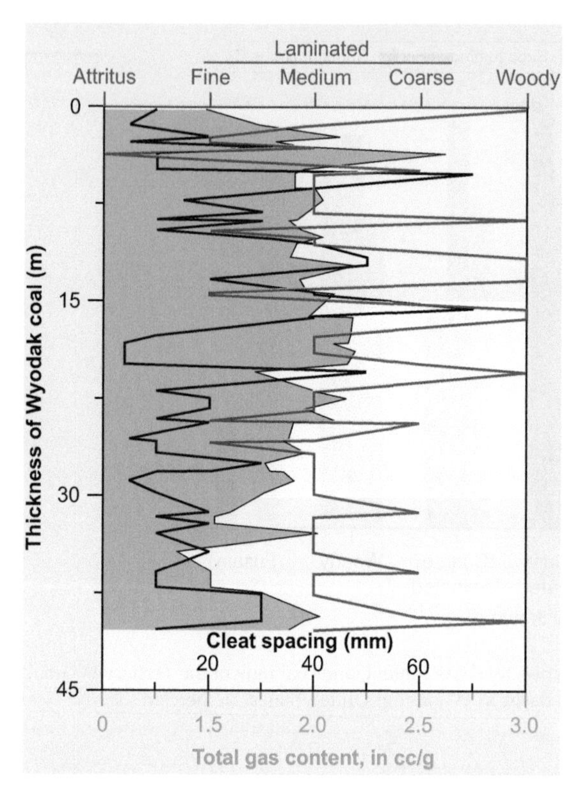

FIGURE 5.41 A chart showing relationships of cleat spacing, macrolithotypes, and total gas content of the Wyodak coal, 45 m in thickness (55 core samples), in the Powder River Basin in Wyoming, United States. *Source: Adopted from Stricker et al. (2010).*

increasing stress (or depth) shows that shallow coals have slower rate of increase than the deeper coal. The ratio of horizontal to vertical coal matrix permeability simulated under field conditions mainly showed more than 4:1 higher permeability in the horizontal direction (Flores, 2004). However, few samples showed vertical permeability greater than horizontal permeability by a factor of 2. Carbon dioxide (CO_2) adsorption showed the largest swelling strains for three gas adsorptions. Methane (CH_4) adsorption showed the largest swelling strains compared to nitrogen adsorption. The diffent adsorptions of these gases are probably influenced by the CO_2:CH_4 adsorption ratio of the subbituminous coals in the Powder River Basin, which varies from average of 9.5:1 for subbituminous C to 7:1 for sibbituminous A (Figure 5.43; Flores & Stricker, 2004).

Gas Content vs Coal Rank

The final validation of permeability attributes such as cleat spacing is its basin-wide relationship to coal and gas content. Law (1993) first studied cleat spacing of subbituminous coals in the Powder River Basin and then related them to higher rank coals elsewhere. This work has laid down the well-established concept that cleat spacing decreases from low-rank coals such as the subbituminous coal in the Powder River Basin to bituminous coals in other coal basins. The conclusion of Law (1993) on cleat spacing of subbituminous coals was based on mine areas in the eastern margin of the basin, which are mainly subbituminous C rank. The coals that Law (1993) studied along the eastern margin of the basin are some of the most prolific gas producers in the subsurface to the west. Well cores from these producing coals revealed that subbituminous B and A coals are more common basinward (Flores, 2004). The concept of cleat spacing decreasing from low- to high-rank coals should be applicable in the subbituminous C to A coals in the Powder River Basin. Study of the areal data in Figure 5.44 on cleat spacing indicates that their distribution basin-wide is not a clear-cut unidirectional change. Indeed, coal cleat spacing (15–25 mm) decreases toward the basin center and wide cleat spacing (>40 mm) in the northwest is in a locality with higher calorific value coals exposed to high heat flow in the subsurface (Flores, 2004). Thus, insights of the coal properties as applied to the reservoirs in the Powder River Basin, although unique, are lessons learned that single to multiple variables utilized in combination are not straightforward relationships. Local and regional factors also play a role in the overall reservoir characterization of coal properties.

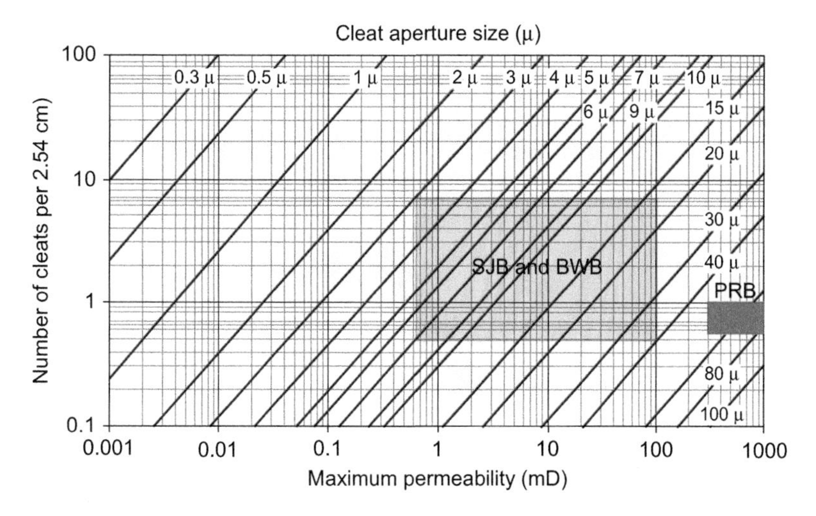

FIGURE 5.42 Relationship of cleat aperture, number of cleats (per 2.54 cm), and permeability, which shows in colored areas cleat apertures of the bituminous coals in the San Juan (SJB) and Black Warrior (BWB) Basins narrower than the sub-bituminous coals of the Powder River Basin (PRB). *Source: Modified from Laubach et al. (1998) and Trippi et al.(2010).*

TABLE 5.8 Coal Matrix Permeability of the Wyodak Coal in the Powder River Basin, Wyoming, United States. CH_4, methane; CO_2, carbon dioxide; N_2, nitrogen; MPa, megapascal; mD, millidarcy; NR, no results

Coal Samples (Direction of Stress)	Depths (m)	In Situ Stress (MPa) ⊥ to Flow	Coal Matrix Permeability (mD)		
			CH_4	CO_2	N_2
1. Horizontal	98	2.2	0.2	0.12	0.55
2. Horizontal	100	2.3	0.03–0.5	NR	NR
3. Horizontal	194	4.4	0.55	0.45	0.85
4. Horizontal	196	4.5	No flow	No flow	0.09
5. Horizontal Vertical	429	9.7 3–6	0.15 0.02–0.1	0.05 0.01–0.2	<0.01 0.01–0.2
6. Vertical	431	3–6	1.5	0.6–1.0	0.5–1.3
7. Horizontal Vertical	439	9.7 3–6	0.02–0.1 No flow	0.01–0.03 <0.01	0.03 0.1–0.4
8. Horizontal Vertical	448	9.7 3–6	No flow 0.04–0.1	No flow 0.01–0.03	No flow 0.15–0.5

Source: Modified from Flores (2004) and Uniservices (2002).

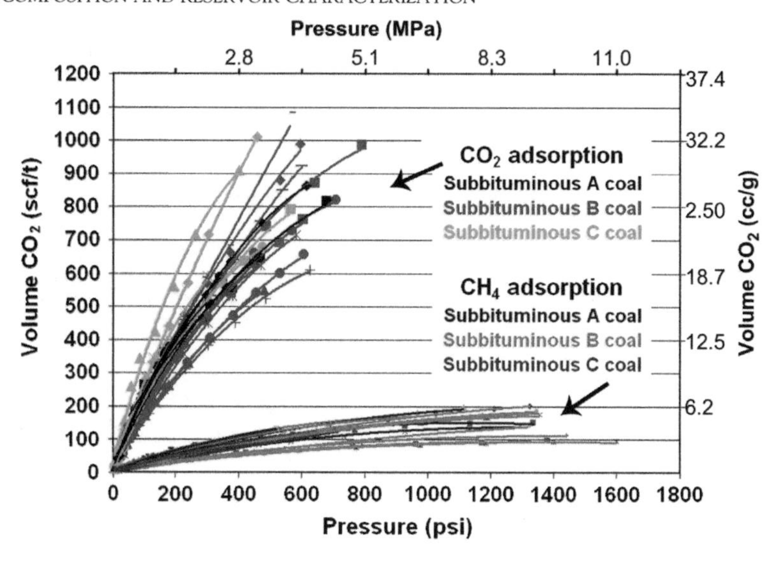

FIGURE 5.43 Adsorption isotherms of carbon dioxide and methane in subbituminous coals in the Powder River Basin, Wyoming. *Source: Flores and Stricker (2004).*

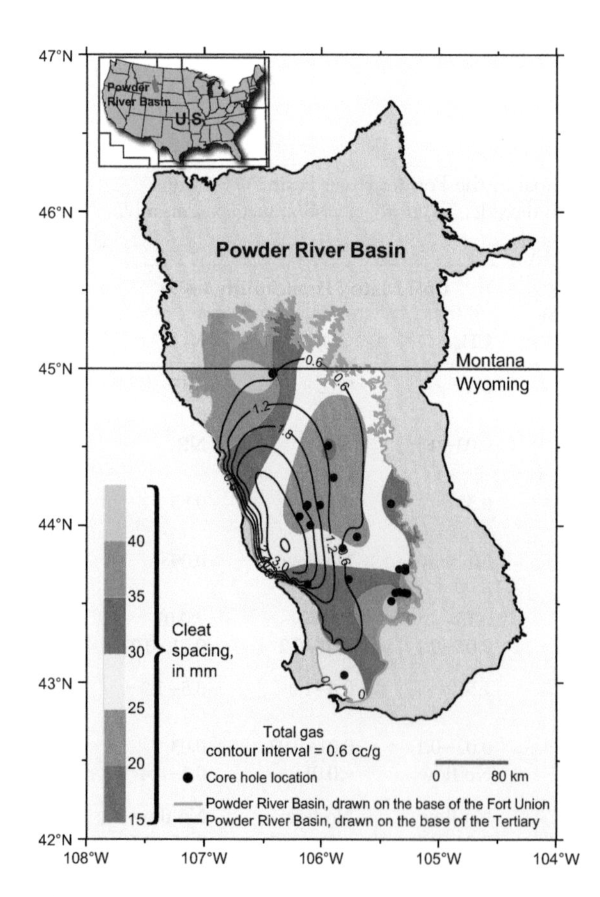

FIGURE 5.44 Map of the Powder River Basin in Wyoming and Montana, United States showing percentage isolines of total gas content superimposed on map of cleat spacing of the Tertiary Wyodak coal. *Source: Adopted from Stricker et al. (2010).*

SUMMARY

The multiconstituent composition of coal in terms of related elements and compounds can be condensed into organic and inorganic or mineral matter components for the purpose of reservoir characterization. Organic matter is the major constituent of coal, with fixed carbon the measure of energy during combustion (e.g. total calorific value) for steam generation and metallurgical use, source of hydrocarbons especially gasification, and the amount of internal surface area in which the gas is adsorbed and stored in the coal. Mineral matter or "inorganic minerals" are an equally or more important component of coal in terms of coal utilization for combustion and reservoir characterization. Mineral matter in coal is important for utilization because if the elements, compounds, and minerals are not cleaned up, their by-products will comprise the fly and bottom ashes, gas effluents, and chemical elements and compounds of environmental concern released during combustion in coal-fired power plants. The threshold of mineral matter content for coal to be considered impure is 30–50%. However, the threshold for ash content for combustible coal is about 10% depending on the clean coal technology of the power plant.

Mineral matter controls the (1) internal architecture and compartmentalization, (2) pore availability to gas adsorption, (3) ability of gas to flow, (4) tendency to shrink and swell, (5) gas content and in situ reserves, (6) gas recoverability, and (7) potential for enhanced gas recovery and carbon dioxide sequestration of the coal reservoir. The most important role of mineral matter affecting coal reservoir performance is infilling pores and cleats or fractures. Mineral infillings plug pores and cleats/fractures preventing gas flow. Mineral matter consists of silicate, carbonate, sulfide, sulfate, phosphate, and other groups of minerals. The most common mineral infillings are silicates, carbonates, and sulfides formed in low- to high-temperature conditions. Minerals from all groups serve as sources of fly ash, bottom ash, gas effluents, and chemical elements and as compounds of environmental concern.

The organic components of coal can be grouped as macerals (identified microscopically), microlithotypes (microscopic maceral associations), and macrolithotypes or lithotypes (identified megascopically). Macrolithotype description of coal cores for composition analysis is the most convenient method for reservoir characterization. Coal composition controls reservoir properties such as gas adsorption capacity, gas content, porosity, permeability, and gas transport. In general, not considering rank, vitrinite- or vitrain-rich (e.g. woody) coals adsorb more gas due to high micropore properties. Porosity in coal is also controlled by rank, with low-rank coals mainly comprising macropores and high-rank coals dominated by micropores.

Coal permeability is a function of the width, length, and height of the cleats or fractures as well as the spacing, frequency or density, and connectivity of the cleats or fractures. The origin of coal cleats or fractures is classified into endogenetic and exogenetic processes. Endogenetic cleats or fractures are formed during burial, compaction, and coalification of the coal. Exogenetic cleats or fractures are formed after coalification during deformation, uplift, and erosion of the basin of deposition of the coals. Permeability can have a significant effect on the drilling, completion, stimulation, and production strategies essential to successful coalbed gas development. Because procedures and strategies are important to the well performance and gas production, it is necessary to understand the origin and properties of coal permeability, which are vastly different from conventional reservoirs. Coal compositional and physical properties impact reservoir characterization of gas content and transport to well completion. Changes in permeability and porosity affect reservoir performance and sustainability of gas production.

Thus, coal reservoir characterization is highlighted. An important highlight of reservoir characterization is the mechanism of coal shrinkage and swelling caused by gas desorption and adsorption, respectively. Shrinkage of coal affects gas production because it impacts permeability and porosity. Gas desorption from the pores is accompanied by reduction of confining and pore pressures from cleats and coal matrix pores, permitting the rock to shrink to the original mass. The mechanism of coal swelling is induced by gas adsorption, which increases pore pressure, permitting the coal matrix to expand or swell. Coal swelling affects mainly enhanced gas recovery with injection of carbon dioxide for sequestration. Another important highlight of reservoir characterization is the relationship of macrolithotypes with gas content, which is related to gas adsorption capacity influenced by porosity and permeability. Finally, highlights on reservoir characterization of the coalbed gas in the producing Powder River Basin, United States, are offered as insightful examples.

CHAPTER

6

Resource Evaluation Methodologies

OUTLINE

Coal and Coalbed Gas

http://dx.doi.org/10.1016/B978-0-12-396972-9.00006-9

KEY POINTS

- Understanding coal resources and methods of assessments are important to planning coalbed gas projects.
- Coal resource and reserve estimates are based on geologic assurance determined by the distance between observation points where coal thickness is measured and samples are collected for analysis.
- Increasing geologic assurance with closer distance between observation points results in a progression of hypothetical–speculative resources (farthest spacing) to inferred, indicated, and measured reserves (least spacing), respectively.
- Geologic assurance-based methodology relies on the stratigraphic correlation of the same coal bed.
- Global assessments of coal resources and reserves universally conform to the geologic assurance method, but may be modified from country to country according to depth, distances between observation points, thickness, and complexity of the geology and structures of study areas.
- Coal resource–reserve methodology does not directly correlate to the petroleum industry's method of estimating coalbed gas resources–reserves.
- Methods of calculation of coalbed gas resources and reserves adopted by the petroleum industry include: volumetric, material balance, production data analysis, reservoir simulation, and analog methods. The volumetric and production data analysis are commonly used methods.
- Volumetric or gas-in-place (GIP) method is commonly used and calculated by gas drainage area, coal thickness, coal density, gas content, and other parameters such as coal quality, coal reservoir conditions, and adsorption isotherm of gas to modify and improve the calculation.
- An important parameter for GIP calculation is the gas content of coal, which can be collected from wireline cores and cuttings.
- However, gas content is measured from desorption of gas from coal cores or cuttings in sealed canisters by direct and indirect methods.
- The direct method is based on the analysis that gas released during desorption was proportional to the square root of time.
- The indirect method is based on analysis of methane adsorption isotherms, which measure the optimum adsorbed gas at varying pressures and at constant temperature. Use of wireline or geophysical logs (e.g. gamma ray, density, resistivity, sonic, neutron) is another indirect method of measuring gas content.
- Another specialized measurement of gas content of coal is the use of decline curve method.
- Gas production data analysis based on estimate of ultimate recovery (EUR) and spacing of coalbed gas wells has been adopted from the petroleum industry for estimating reserves.
- Case studies of Total Petroleum System (TPS) assessment of coalbed gas in the Powder River Basin, Wyoming and North Slope, Alaska, are reviewed for production data analysis and analog methods.
- The Petroleum Resources Management System guidelines have proposed standardization of coalbed gas resource–reserve categories to contingent resources and proved, possible, and probable reserves based on well spacings. Proved reserves are subcategorized into

undeveloped, developed, and proved reserves.

- Methodologies to estimate coalbed gas resources and reserves have room for improvement. A post mortem of methodologies with lessons learned from previous assessments is a good start for this process.

THE METHODOLOGY CONUNDRUM

The unification of the concepts of coal and coalbed gas resources and reserves with so many complicated attributes of coal as affecting minability and coal-to-gas resources is a difficult task. The conundrum is the intricate bridging of coal to a petroleum commodity. Assessment of coalbed gas resources is often simplified by utilizing the gas content of coal beds multiplied by the coal resources. Simplification of assessment methodology underscores the complexity and role of reservoir parameters, which are important to the accurate estimation the volume of coalbed gas. Nevertheless, the simplified method highlights the significant function of coal resources and reservoir properties toward the calculation of coalbed gas resources and reserves. The basic information on coal resources and reserves is found in Chapter 1 as well as in many technical publications referred therein. An important dynamic in the viability of economic development of coalbed gas is the vast coal resource endowment worldwide. The key to assessment of coalbed gas reserves is the amount of coal resources available for development. Estimates of coalbed gas resources and reserves vary widely depending on the methodology and parameters used for calculations and the purpose of the assessment. For example, the conservative estimate of coalbed gas-in-place (GIP) resources in the conterminous United States by the U.S. Geological Survey (USGS) is more than 19.8 trillion cubic meters (Tcm) (Rice, 1997). The GIP estimate is the amount of gas in the coal beds based on the total measureable coal resources in the conterminous United States coal basins, which include both technically recoverable and nonrecoverable gas. Technically recoverable gas is reproducible resource using recent advances in technology irrespective of economic conditions. Nuccio (2002) reported about 2.8 Tcm of the GIP is recoverable in the conterminous United States coal basins. In addition, the GIP estimate is based on the total gas content of the coal beds, which is controlled by the amount of adsorbed gas based on rank, composition, porosity, and other reservoir properties (Chapter 5).

Fundamental to the calculations of the coal resources is knowledge of the rank, depths, overburden, thickness, lateral extent, vertical variability, and areal distribution of the coal beds. These parameters are traditionally determined from investigations of the coal geology, stratigraphy, and sedimentology of coalfields or basins. Simply put, coal resource estimates significantly control estimates of the volume of gas contained in the coal beds. That is, calculations of GIP and other methods of gas volumetric and material balance approaches rely on the accurate estimation of the amount of coal, which in turn, is controlled by precise areal delineation of the "same" coal bed(s). The areal extent of the coal bed(s) is determined by several methods to trace continuity of a bed(s) avoiding the trappings of the old principle of "layer cake" coal geology. This modern approach generally uses the basic principles of coal stratigraphy and sedimentology, and is related to the depositional environments of the coal and related sedimentary rocks. The conventional tradition of tracing continuity and areal occurrence of coal bed(s) from maps, mine workings, drill holes, and through widely spaced points are insufficient to determine accurate coal and coalbed gas volumetrics. Because coalbed reservoirs are internally heterogeneous and beds are highly variable in lateral and vertical extent their thickness and distribution can change rapidly over short distances. This fact prevents capture of the exact representation of the geometry of the coal body and the precise establishment of the coal volume and proper application of coal and coalbed gas volumetric calculations. Thus, coalbed gas volumetric estimates (e.g. total gas content in standard cubic meter) must account for rapid variations laterally and vertically within a coal reservoir in a well, between wells, within a gas drainage area and field, and within a basin (Stricker et al., 2006).

The volumetric estimate of coalbed gas not only depends on the assessment methodology but also on the quality, quantity, and recency of the data. For example, estimates of the volume of coalbed gas in the Powder River Basin in Wyoming and Montana, United States, have changed several times from 1984 during predevelopment to the 2004 during peak development. Choate, Johnson, and McCord (1984) estimated the volume of coalbed gas from the Fort Union coals in the Powder River Basin based on 1 trillion metric tons of coal resources calculated from oil and gas, geophysical logs, and average gas content of 0.70 m^3 per ton of coal as 0.85 Tcm GIP. ARI (2002) indicated that estimates of coalbed gas resource in the Powder River Basin ranged by a factor of five between assessments of the U.S. Department of Energy, U.S. Bureau of Land Management, and the USGS. These federal agencies estimated undiscovered recoverable coalbed gas from 0.22 to 1.10 Tcm, which is explained by different methodologies, geologic models, and resource assumptions (ARI, 2002). The USGS estimated technically recoverable coalbed gas using petroleum system methodology and estimate of ultimate recovery from production history of more than 1000 coalbed gas wells in 1999–2000 as about 0.35 Tcm for the Fort Union coals in the Powder River Basin (Flores, 2004). The differences in estimates of undiscovered recoverable coalbed gas resources are anticipated as a result of addition of more data (e.g. coalbed gas wells, gas content, gas production, stratigraphic control). Thus, as more information becomes available and applied to assessment methodologies, it is expected that the estimates of undiscovered recoverable coalbed gas will continue to vary through time.

Overprinted on the local and regional variations of coalbed gas calculations is the employment of different assessment methodologies utilized within a country and worldwide between countries. Thus, the magnitude of variability of coalbed GIP estimates by a factor of three from various coal basins worldwide differs due to a variety of assessment methodologies utilized from country to country. Boyer and Qingzhao (1998) reported that the estimates of coalbed GIP or volumetric for major coal-bearing basins in the world ranges from 89 to 269 Tcm based on methods developed in the United States. However, there are various methods of assessments developed in the United States, which vary from State to State, industry to Federal agencies, and Federal agency to agency. Thus, understanding the concepts behind specific assessment methodologies is critical to understanding coalbed gas resources. According to Boyer and Qingzhao (1998) assessment methodologies frequently exclude application of resource estimation methods for conventional hydrocarbons. The suggestion of applying assessment methodology for conventional gas by Boyer and Qingzhao (1998) to coalbed gas has been supplanted by reclassification of coalbed gas, shale gas, tight gas sand, and gas hydrates as unconventional hydrocarbons (Chapter 2), which deserve a separate methodology of resource assessment (Schmoker, 2002, 2004; Schmoker & Dyman, 1996).

The assessment concepts advocated by Boyer and Qingzhao (1998) significantly differ from that proposed by Schmoker and Dyman (1996) and Schmoker (2002). Boyer and Qingzhao (1998) supported the used of volumetric and material or mass balance methods for assessments. The volumetric method or "macroassessment of the total GIP" employs the simple formula combining the volume of stored gas per unit volume of coal reservoir with total reservoir volume (e.g. amount of coal resources). The material balance method utilizes the law of conservation of matter formulas combining reservoir pressure conditions, gas adsorption/desorption analyses, and porosity/permeability properties, which requires the knowledge and understanding of the characteristics coalbed reservoirs (King, 1990; Penuela, Ordonez, & Bejarno, 1998; Walsh, 1995).

In contrast, the methodology proposed by Schmoker and Dyman (1996) and Schmoker (2002) is a knowledge-based assessment of Total Petroleum System (TPS), which was adopted by the USGS for world oil and gas resources and reserves. The modification of the concept toward assessment of coalbed gas as unconventional gas system required recognition of coal beds as a continuous accumulation, which occurs independent of the water column and/or owe their accumulation to buoyancy of gas in water (Schmoker, 2002). In addition, as typified by conventional hydrocarbon accumulation, unconventional gas accumulation does not correspond to a field or pool defined by a down-dip water contact (Schmoker, 2002). As such, the customary resource-assessment methods for conventional gas accumulation cannot be used and a specialized method is required for the unconventional, continuous gas accumulations such as coalbed gas, shale gas, tight gas, and gas hydrates. This chapter will focus on methods specific to coalbed gas. The approach suggested by Schmoker (2002) requires an estimate of total GIP connected with a recovery factor to focus on the assessment from total volumetric to prediction of additional undiscovered gas reserves. The GIP method relies on the measurement of the in situ gas content and the amount of coal within an area.

Another approach proposed by Schmoker (2002) is based on the production history of continuous gas reservoirs, in general, and coalbed gas reservoir, in particular, as displayed empirically by production wells and reservoir-simulation models. In this method based on production characteristics in contrast to GIP or volumetric, is the basis for forecasts of potential additions to reserves. This method is to some extent similar to the production-data-analysis method described by Clarkson and Barker (2011) that uses production type curves and flowing material balance. Although Clarkson and Barker (2011) considered the geologic assurance-based approach to estimate resources/reserves by the coal mining industry, the authors utilized the petroleum industry well spacing approach with the premise of continuous coal thickness within the gas drainage area. Rushing, Perego, and Blasingame (2008) used reservoir simulation of constant flowing pressure to determine the impact of coal reservoir properties on decline behavior of production wells.

Integrated with the various approaches is the probabilistic reserves assessment that uses for example a Monte Carlo method, which incorporates estimates of well decline and recovery, volumetric, material balance, well pressure analysis caused by production, and other reservoir parameters. The Monte Carlo simulation method for the reliability assessment of undiscovered energy resources and reserves is commonly used worldwide (Charpentier & Klett, 2000; PRMS, 2011; Sarmiento & Steingrimsson, 2007; Stoltz & Jones, 1998; Swinkles, 2010; Yongguo, Yuhua, Yong, & Qiuming, 2008). The probabilistic method employs reservoir parameters to calculate the statistical uncertainty of GIP volumetric and recoverable resource volumes (Swinkles, 2010). Stochastic method such as the Monte Carlo equations is applied to generate probabilities, which lead to various quantitative risk analysis and decision-making methods (Swinkles, 2010). Probability levels of the total recoverable gas volume can then be related to 1P, 2P, and 3P resource/reserve categories.

Barker (2008) applied the 1P, 2P, and 3P categories for coalbed gas assessment. A case study by Yongguo et al. (2008) shows that results of resource/reserve estimates using the Monte Carlo method have a narrower error range compared to the volumetric method. The uncertainty analysis is a key to lessening the high potential risks of economic assessment of coalbed gas. According to Moore (2012) the reserve categories play an economic viability and commerciality of the coalbed gas prospects.

The following methods have been adopted for unconventional or continuous accumulations of coalbed gas: (1) GIP or volumetric, (2) material or mass balance, (3) production performance or history, and (4) reservoir simulation. According to Clarkson and Barker (2011) these methods are applied during various stages of coalbed gas development. The application of these methods depends on the availability of data, which in turn, controls the accuracy of the assessment methodology. Generally, the GIP or volumetric method is applied during the early stage although it can be used throughout the development period. The material balance, production history with decline curves, and reservoir simulation can be applied when data on production, permeability and porosity properties, and flowing and shut-in pressures are available.

COAL RESOURCES VS GAS RESOURCES

Estimates of coalbed gas resources and reserves are rooted in the understanding of methods of assessment of coal resources and their classification. More specifically calculations of the GIP or volumetrics of coalbed gas are directly related to the mass of the coal. Coal resources are the total amount or tonnage (e.g. short or metric tons) of coal in the ground within the limits of thickness, lateral continuity, and overburden thickness or depth. The amount of coal is usually reported as original resources that are in the ground prior to mining production. The overburden thickness or rock interval from the top of the coal to the surface is an important factor in the economic extraction or mining of coal. However, coalbed gas extraction is not encumbered by depth depending on drilling/well costs as long as permeability exists in the coal reservoirs.

Coal Resource Classification System

A comprehensive coal resource classification includes identification of the areal coverage, spacing of data points, thickness of coal and overburden, density, rank and quality, and estimate of the quantity of coal. All these parameters are important to coal development. However, the most important parameters for coalbed gas development are the areal coverage, thickness, density, and rank of coal. The coal resource classification system is intended to quantify total amounts of coal prior to and after mining. In addition, the classification system is intended to identify coal quantities that are (1) known, (2) undiscovered, (3) economically recoverable, (4) potentially recoverable, and (5) subeconomic resources at varying distances from data points of measurements of coal thickness.

The key to the classification of coal resources and reserves, which includes measured, indicated, inferred, and hypothetical, is the distance of data points from which coal thicknesses are measured varying from close to far apart, respectively. The distance of data or control points demonstrates the perceptible continuity of a coal bed between measurement points assuming correlation of similar physical characteristics and stratigraphic position of the bed. Another factor that contributes to the reliability of resource and reserve classifications is the distribution of points. Often times coal resource and reserve estimates are performed without regard of stratigraphic correlation of the same coal and enclosing beds. The degree of confidence in correlation increases with the knowledge and abundance of stratigraphic control points (e.g. oil and gas wells, coal drill holes, outcrops). This is termed geologic assurance or the condition of certainty and confidence of the occurrence of a quantity of resources based on the data and spacing of control points from where the coal thickness is measured, described, and sampled. However, the degree of geologic assurance is controlled by the quantity and quality of the geologic data; that is, the more available data the closer the spacing of data points and more reliable the coal resource estimates. Once the data points indicate that the stratigraphic correlation and continuity of a coal bed and enclosing rocks is possible between data points within an area, then an estimate of the coal resources and reserves of that coal bed can be calculated within that area. Thus, a total coal resource and reserve estimates of an area, coalfield, or a coal basin is the compiled data and calculations of mapped and correlated individual coal beds. In the absence of well-established correlations of coal beds, coal resources calculations are often performed in the form of bulk volume of coal, which is not an accurate representation of the quantity of coal in an area.

Coal Resource Categories vs. Reliability Estimates

Coal resources are classified according to the abundance and reliability of data used to perform the resources estimates. The USGS provides the methods that are universally used for reliability estimates of coal resources, which often are used as a guide (Averitt, 1974; Wood, Kehn, Carter, & Culbertson, 1983). Figure 6.1 demonstrates the areas of reliability using coal thickness measured from points in outcrops and well data.

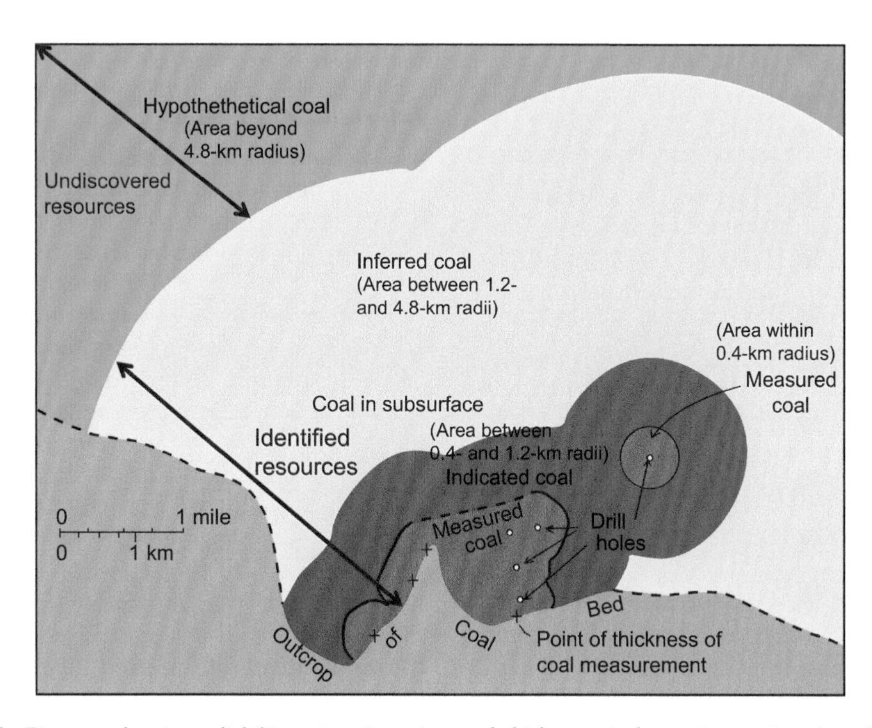

FIGURE 6.1 Diagram showing reliability categories using coal thickness at observations points (e.g. data) along the outcroppings and subsurface. Radius originated from observations points. *Source: Modified from Wood et al. (1983).*

Measured Coal Resources

Estimation of measured coal resources is on the basis of mappable and correlation of individual coal beds whose thickness is measured from outcrops, coal drill holes, and oil and gas wells. Measured coal resource estimates are based on closely spaced data control points or high degree of geologic assurance (Figure 6.1). The measured coal resources of a bed are calculated within 0.4 km radius (0.8 km diameter) around a control point. Thus, the resources are calculated by projection of the thickness, rank, and quality of the coal beyond the point of measurement. The thickness of coal beds measured from the control points is determined by rank, which includes anthracite and bituminous coals 35 cm or more thick and lignite and subbituminous coals 75 cm or more thick. These coal resources extend up to 1800 m in depths.

The close spacing of control points demonstrates better capability to trace the lateral continuity of the coal beds. Thus, geologic assurance is at a high degree. According to Averitt (1974), the close spacing of control points permits close delineation of the coal body and the estimation of tonnage for measured coal resources to be accurate within 20%of the true tonnage. This estimated tonnage, however, does not represent the true tonnage of the coal bed because where the spacing of control points are farther apart on the same bed, resource calculations are included as indicated and inferred with decreasing abundance of data.

Indicated Coal Resources

Estimation of indicated coal resources is much the same method as that used in measured coal resources, which is based on mappable and correlation of individual coal bed(s). Data points where coal thickness is measured also are from outcrops and subsurface observations. However, unlike the measured coal resources, the control points for the indicated coal resources are farther apart or moderate degree of geologic assurance. As shown in Figure 6.1 the indicated coal resources of a bed are calculated from 0.4 to 1.2 km radius (0.8–2.4 km diameter) around a control point. Thus, the thickness and lateral extent of the coal bed is projected farther from the control point than that for the measured coal resources. The distance between control points of thickness measurements for indicated coal resource is from 1.6 to 2.4 km apart to assume coal bed continuity.

The zone of indicated coal resources as shown in Figure 6.1 is about 2.4 km wide beyond the zone of the measured coal resources. Indicated coal resources include anthracite and bituminous coals 35 cm or more thick and lignite and subbituminous coals 75 cm or more thick. Much like the measured coal resources, the depth limit is up to 1800 m. The coal bed calculated for indicated resources may be extended beyond the indicated resource boundary for estimates for inferred resources.

Inferred Coal Resources

Calculations of inferred coal resources are based on presumed lateral extension or continuity of coal beds beyond the zones of measured and indicated coal resources. Usually the inferred coal resources are in inaccessible areas with limited control points and a low degree of geologic assurance. Thus, the lateral extent of coal beds is based on the geologic character of the beds and enclosing rocks. The zone of inferred coal resources lies between 1.2 km and 4.8 km radius from the control point of coal thickness measurement. Estimates of the inferred coal resources are calculated by projection of the coal thickness beyond the indicated coal resources area. Because the control points used in the assessment of inferred resources are far apart, the degree of geologic assurance is low.

Undiscovered Coal Resources (Hypothetical and Speculative)

The undiscovered coal resources are postulated coal deposits either separate from or extension of the measured, indicated, and inferred coal resources. Based on the degrees of reliability, undiscovered resource is categorized as hypothetical and speculative coal resources. Estimates of hypothetical coal resources are based on a very low degree of geologic assurance and usually occur in unmapped and unexplored areas. Calculations of tonnage of hypothetical coal resources are based on: (1) extrapolated coal beds beyond the inferred coal resources beyond the radius of 4.8 km from the control point and (2) geological character of the coal. Much like the measured, indicated, and inferred coal resources, hypothetical coal resources consist of anthracite and bituminous coals 35 cm or more thick and lignite and subbituminous coals 75 cm or more thick to a depth of up to 1800 m. Speculative coal resources have the lowest geologic assurance and include possible areas outside coal basins and coalfields deeper than 2000 m (Averitt, 1974). Usually, data of presumed occurrence of coal below 2000 m are not collected because they are uneconomic and beyond the technical capability of being mined at the present time.

Other Countries

Other countries such as India consider the depth of occurrence in the economic extraction, scope for exploration, and coal resources categorized in the depth ranges of: 0–300 m, 30–600 m, and 600–1200 m. Germany, Poland and United Kingdom like India categorized coal resources at maximum depths of 1500, 1000, and 1200 m for hard coal, respectively, at specific thickness (0.35–1.5 m), maximum sulfur content (2%), a minimum calorific value (15–24 MJ/kg) (United Nations, 2008) (Ersoy, 2005; Schmidt, Gerling, Thielemann, & Littke, 2007; Smakowski & Paszcza, 2010).

UNIVERSAL GUIDELINES TO COAL RESOURCES ASSESSMENT

Guides to coal resource assessment are very similar through out the world. The guidelines proposed the same resources categories with minor variations in implementation given favorable geological conditions. The broad guidelines summarized from the United Nations (UN), USGS, Joint Ore Reserves Committee (JORC) (Coalfields Geology Council of New South Wales and the Queensland Mining Council, 2001; UNFC, 2010; Wood et al., 1983) all have common recommendation such as:

1. Maps for each coal bed or zone that contain coal resources,
2. Size of unit area or areal extent of coal resources,
3. The resource category(s) on which the coal resources estimate is based,
4. The coal quality and rank,
5. Thickness of the coal as measured from stratigraphic control points or points of observation,
6. The thickness of overburden to determine minability of coal beds overburden, and
7. Economic considerations are emphasized such as (a) proposed mining methods, (b) criteria used to limit, the reserves, and (c) mining recovery.

Of the three guidelines (USGS, JORC, UN), the USGS recommended a strategy for the estimation of resources in vicinities of where coal bed bifurcates into tongues (Wood et al., 1983). This distinction is important because the main requirement for coal resource assessment is the preparation of coal bed maps for each bed or seam (see number 1 in the above list). The USGS guideline recognizes the lateral variability and common occurrence of coal splitting, and does not treat coal beds as a proverbial "layer cake" rock unit. The splitting of coal causes difficulty in resource estimation, which requires the delineation of the boundary between the main coal bed and the tongues. Thus, the areas of the main coal body and coal bed tongues are mapped separately with the tongues being individually estimated for resources. Unfortunately this guideline has not been followed religiously. Moreover, failure to recognize this concept has serious repercussions in the precise estimation of coalbed gas reserves.

According to Wood et al. (1983), when estimating coal resources in such a geologic condition it is necessary to delineate the "coal split line" between the areas where the coal resources are estimated separately. The "geologic condition," however, requires the sophistication ability to recognize the split and interpret the depositional environment of the interbedded rock units (see following sections). The "split line" is defined where the coal bed tongues become thinner than the intervening parting of the main coal bed. The parting is a non-coal layer in a coal bed, which does not exceed the thickness of either the underlying or overlying parts of the coal bed (Wood et al., 1983). All these points where the thickness of the main coal bed and the tongues are measured are located in the coal bed map.

Wood et al. (1983) proposed the following approach of extrapolated coal bed map method to estimate coal resources controlled by depositional environments.

1. Collect all geologic data in the area adjacent to the coal being estimated for resources.
2. Collect all data on coal and overburden thicknesses, coal quality, and coal rank.
3. Evaluate all data from (1) and (2). Construct coal isopach map noting thickening and thinning directions. Create coal quality (e.g. calorific value, ash yield, trace elements) isopleths maps; the latter two parameters may indicate direction and transport of sediments.
4. Distinguish depositional and erosional trends and features adjacent to and in the coal being estimated for resources. Emphasis is placed, using these trends and features, on the nature of ancestral rivers, deltas, and coastal environments and related peat-forming swamps and/or bogs. In addition, directions and distances of point source of sediments of these environments are considered.
5. Expand stratigraphic work from known to unknown areas using surface (e.g. outcrops) and subsurface (geophysical logs of oil, gas, and water wells) data points. Establish stratigraphic correlations and depositional models of coal beds from these data points.
6. Construct structural sections and contour maps to use in estimating overburden thickness and rank of coal beds in both known and unknown areas.
7. Project all data (e.g. isopach and isopleths maps, depositional, stratigraphic, and structural trends/features) of the above steps from known to unknown areas.
8. Determine the average coal thickness per unit area in each coal category (e.g. measure, indicated, inferred, and hypothetical).
9. Calculate and sum estimated tonnages for each category, which represent the best information for coal resources in the known area (e.g. identified resources) to unknown area (e.g. undiscovered resources).

Geologic Assurance

Fundamental to the resource assessment of coal and coalbed gas resources and reserves is the concept of geologic assurance (Wood et al., 1983). Geologic assurance is controlled by spacing, quantity, and quality of control points as affected by correlation and depositional environments of coal beds (Figure 6.2). That is, the more control points the closer the spacing between data points and increase the certainty of correlation of coal beds. Although this is an ideal situation, the standard operating procedure is that initially resource assessments are normally

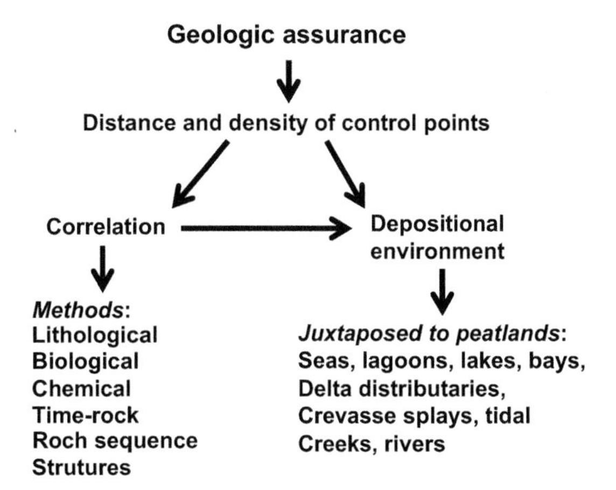

FIGURE 6.2 Diagram showing a flow chart of the hierarchy of variables that determine geologic assurance.

performed in virgin areas with minimal or no stratigraphic data are available. Thus, correlation of coal beds is governed by availability of stratigraphic data points, which are far apart with the potential for miscorrelation of coal bed being very high. As exploration and development proceed, additional data points are collected and the reliability of the stratigraphic correlations and mappability of the coal beds are better established.

Spacing Guidelines

Spacing or distance of data control points to determine and ensure geologic assurance is a universally accepted concept, which is dictated by the geology of the coal, coalfield, and coal basins. These factors are different worldwide, which are reflected by variable guidelines promulgated by different countries. Although spacing of control points for geologic assurance are different worldwide, most countries create and modify their guidelines in conformity with the UNFC (2010), USGS resources classification (Wood et al., 1983), and Joint Ore Reserves Committee code (JORC Code) and Guidelines (2004), Geological Survey of India (2012), Indonesia Ministry of Energy & Mineral Resources (2012). For example, Table 6.1 shows that the Indonesian standard for geologic

TABLE 6.1 The Geologic Assurance on Coal Resources Classification for Other Countries Such as Indonesia is the Same Concept as Stated by Wood et al. (1983). However, it Also Depends on Sedimentary and Tectonic Parameters, Which Lead to Classification of Geologic Conditions into Simple, Moderate, and Complex

	Geologic Conditions		
Parameters	Simple	Moderate	Complex
1. Sedimentary			
Coal thickness	Little variation	Some variation	Large variation
Coal continuity	Thousands of meter	Hundreds of meter	Tens of meter
Coal splitting	Almost none	Some	Many
2. Tectonic			
Faulted	Almost none	Moderate	Highly
Folded	Almost none	Moderate	Highly
Dipping	Gently	Moderate	Steeply
Intrusion (igneous)	No influence	Moderate influence	High influence
3. Quality variation	Little variation	Some variation	Large variation

TABLE 6.2 Indonesian Geologic Assurance Ranging from Simple to Complex Geologic Conditions Determines the Distance between Observation Points for Measured, Indicated, Inferred, and Hypothetical

	Coal Resources			
Geologic Conditions	**Measured**	**Indicated**	**Inferred**	**Hypothetical**
Simple	500 m or less	>500–1000 m	>1000–1500 m	No limit
Moderate	250 m or less	>250–500 m	>500–1000 m	No limit
Complex	100 m or less	>100–200 m	>200–400 m	No limit

assurance on coal resources classification is mainly controlled by the sedimentary and tectonic parameters, which dictate categorization of geologic conditions into simple, moderate, and complex. Sedimentary parameters are distinguished according to the thickness variability, continuity, and splitting of coal beds to measure the complexity of the geologic conditions. In addition, tectonic parameters include faults, folds, dipping beds, and igneous intrusions to characterize the various geologic conditions. Based on the range of simplicity and complexity of the geologic conditions of the coal geology and tectonism, the distance between control points is determined and the degree of geologic assurance is assigned to each coal resource (e.g. measured, indicated, inferred, and hypothetical) (Table 6.2).

The simple geologic condition with little variability in coal continuity and thickness, no coal splitting, limited or no faulting, folding, intrusions, and gently dipping beds can be assessed with control points that are far apart. In contrast, complex geologic conditions with highly variable coal continuity and thickness, coal splits, faulting, folding, intrusions, and steeply dipping beds can be assessed with closely spaced control points. The concept of geologic assurance, which reflects estimates of uncertainty, dictated by the degree of complexity of the geology and deformation is exemplified by India where spacing from control points is 200 m for proved (e.g. measured) resources, 1000 m for indicated resources, and 2000 m for inferred resources. The Australian JORC Code of coal resource classification based on geologic assurance (Table 6.3) does not vary much in distance between points of observation to that of the United States, Indonesia, and India (Stoker, 2009).

The Stratigraphic Factor

The stratigraphic factor is the foundation of geologic assurance toward estimation of coal and coalbed gas resources. Catuneanu (2006) defined stratigraphy as the science of rock strata, which encompasses their properties and attributes important to interpretation of the geologic origin and history. Although the science of stratigraphy covers a broad concept, it is the lateral and vertical characteristics of the rock strata that are most applicable to coal resource assessments. More specifically stratigraphic correlation and depositional environments of the rock strata greatly impact the geologic assurance. The classic definition of stratigraphic correlation, which is still applicable today, is the demonstration of equivalency of rock units (Krumbein & Sloss, 1963). However, stratigraphic equivalency of rock units in the more modern context of sequence stratigraphy may be demonstrated by time-rock continuity despite changes in lithofacies beyond the scale of individual depositional systems as suggested by Catuneanu (2006). Nevertheless, in coal resource assessment in which data collection of thickness of correlated individual coal beds is a requirement in the preparation of coal bed maps, conventional stratigraphy based on lithological or lithostratigraphy,

TABLE 6.3 Evolution of the Concept of Geologic Assurance Based on Distance between Points of Observation, Which Changed from Longer to Shorter Distances for Measured, Indicated and Inferred Coal Resources

	Coal Resources		
JORC Code/Guide	**Measured**	**Indicated**	**Inferred**
1986	<1 km between PO; extrapolate < 500 km	<2 km between PO; extrapolate <1 km from PO	Presence of coal unambiguously determined
2001	<500 m to may be 1 km between PO; *extrapolate <500 m	<1 km to may be 2 km between PO; *extrapolate <1 km from PO	<4 km between PO; *extrapolate <2 km from PO
2003	<500 km between PO; should not be extrapolated more than half the distance between PO	<1 km between PO; should not be extrapolated more than half the distance between PO	<4 km between PO; should not be unreasonably extrapolated beyond the last line of PO

* *Trends in coal thickness and quality should not be extrapolated more than recommended minimum kilometers from points of observation.*
PO, points of observation.
Source: Modified from Joint Ore Reserves Committee code (JORC Code) and Guidelines (2004).

lithofacies association, and biological or biostratigraphy related to individual depositional systems is still the time-tested approach (Figure 6.2). Rock units may be traditionally recognized by their physical (e.g. color, texture, thickness, internal structures), biological (e.g. plant and animal fossils), and chemical (e.g. chemical composition) properties. However, in coal physical and biological attributes (e.g. black and plant-rich) are more a detriment than a benefit following the axiom of they all look alike.

Stratigraphic correlations of coal beds and associated rock strata revolve around their depositional environments and related sedimentologic processes by which they form. Conventional stratigraphic correlation of rock strata requires knowledge and interpretation of depositional environments. Simply put correlation of coal beds and associated rock strata form a base for studies of depositional environments (Ferm & Horne, 1979; Fielding, 1984; Flores, 1986; Flores, Spear, Kinney, Purchase, & Gallagher, 2010; Flores, Spear, Purchase, & Gallagher, 2010; McCabe, 1984; Ryer, 1984). However, according to Wood et al. (1983) it is a means to an end to establish maps of coal bed or coal zone, which are known to contain coal resources, as a part of documenting the estimates of resources. Correlation is the underpinning of coal bed or coal zone maps, which record the following information presumably of the same coal bed: (1) trace of the outcrop of the coal bed, (2) control points where thickness is measured from the same bed at surface exposures, and (3) control wells, drill holes, and subsurface points where coal thickness of the same coal was measured. The problem is as difficult where the coal beds are steeply dipping and/or highly deformed (e.g. faulted).

In coal-bearing rocks, correlation is a demonstration of perceptible continuity of a coal bed between control points by showing correspondence in character and stratigraphic position (Wood et al., 1983). The traditional method of correlation is to trace and map a coal bed with distinctive lithologic, biologic, and chemical features along surface exposures from one locality to another locality (Figure 6.2). In addition to these features, recognizing the contacts between the beds and adjoining units may aid in the mapping of a coal bed. Also, mapping a coal bed is practiced in the subsurface using geophysical logs and/or lithological samples (e.g. rock cuttings and cores) from well to well (Flores, Spear, Kinney, et al., 2010; Flores, Spear, Purchase, et al., 2010; Wood et al., 1983). This may be possible in areas where closely spaced wells permit mapping, however, this method may be restricted in areas where lithology changes occur rapidly laterally. In both surface and subsurface conditions, the density of control points (e.g. outcrops, wells) and rate of lithological variation influence the accuracy of coal bed correlations. Thus, other methods may be applied in support of coal bed correlation such as biological (e.g. maceral composition, palynology) and chemical (e.g. inorganic and trace elements) attributes of the rock unit (Ayers, 1986; Crowley, Ruppert, Stanton, & Belki, 1994; Flores, Spear, Kinney, et al., 2010; Flores, Spear, Purchase, et al., 2010; Gochioco, 1992; Nichols, 1999; Scholes & Johnston, 1993; Schweinfurth, 2002; Stanton, Moore, Warwick, Crowley, & Flores, 1989). Other methods such as sequential position of the rock unit with respect to underlying and overlying rock units and unconformities as well as structural features (folds, faults, igneous intrusions, etc.) may be of value in establishing correlations (Figure 6.2) (see Catuneanu, 2006 and references therein).

Thus, coal bed correlation is based on understanding the stratigraphy, however, it demands more sophistication in interpreting the stratigraphic data of the coal and enclosing rock units to develop depositional correlation-based models. This deposition correlation approach is complicated by the areal extent of coal beds limited by erosion and/or nondeposition. This problem is exacerbated by lateral changes of the coal beds into other lithological units with differing properties sometimes called lithofacies change (Reading, 1996). Figure 6.3 demonstrates the lateral and vertical traceability of coal bed successions, which are restricted by lithofacies change in the Cretaceous Fruitland Formation in the San Juan Basin (Fassett, 2010). The Fruitland coal beds are separated above and below as well as laterally from other rock units by their depositional continuity and extent, which are controlled by surrounding depositional environments.

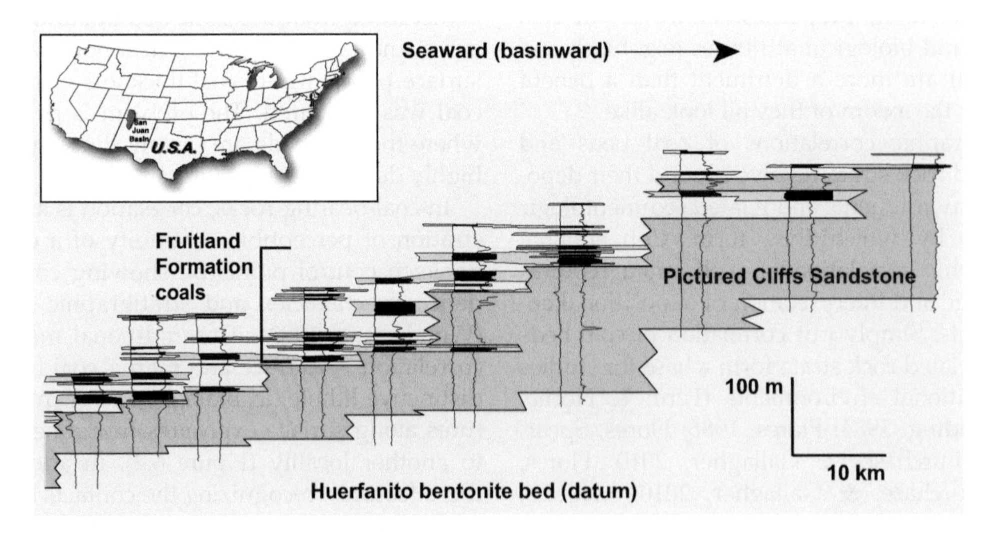

FIGURE 6.3 The lateral and vertical variabilities of coal beds in the Cretaceous Fruitland Formation in the San Juan Basin, New Mexico, United States. *Source: Adopted from Fassett (2010).*

The Depositional Connection

Geologic assurance, stratigraphy, and distance of control points to estimate coal resource categories are all directly connected to depositional environments. However, the depositional factor has played minor and secondary roles in coal and coalbed gas resources assessment for the simple reason that resource specialists have not expended their time identifying the depositional component of coal deposits. Although the principle of coal deposition should be the underpinning test of geologic assurance, resource analysts generally lack the experience or time for this type of investigation. All too often resource assessment is on a short fuse and based on sparse to readily available data, which are not amenable to exhaustive depositional study.

A depositional environment is defined as a site where sediments (e.g. detrital, chemical) accumulated, governed by physical, biological, and chemical processes related to modern and applied to ancient environments, and lithified into sedimentary rock units. The areal extent and peat thickness of peatlands are controlled, in part, by the areal distributions and processes of surrounding depositional environments. In addition, because peatlands are genetically linked and grade into these contemporaneous, physically active environments, the peatlands respond to the changing depositional trends. For example, a shift in the paleoshoreline in the basin of deposition controls the seaward or landward position of the peatlands and resulting thickness of coal deposits.

Figure 6.3 demonstrates this phenomenon where base level lowering produced seaward stratal stacking of marine sandstones and juxtaposed coal beds formed on back-barrier and delta plain peatlands (Flores & Erpenbeck, 1978; Fassett, 1989). Along this setting Cretaceous Fruitland coals are continuous but thin and thicken as well as split with a 20-km distance perpendicular to depositional strike. More importantly the stratigraphic architecture of the coal-sandstone is overlapping one another in an en echelon pattern and vertically stacking was controlled by relative sea-level fluctuations. Continuity of coal beds was determined by Fassett (1989) based on a time-rock unit (Huerfanito bentonite bed), which is volcanic in origin. The conventional approach would have used the Pictured Cliffs Sandstone as a time-rock unit, which would have correlated the coal bed immediately on top of the sandstone as a single, very continuous bed (e.g. layer cake) beyond its limited depositional site controlled by seaward regression.

Coal bed Variability

Definition of the areal extent and thickness of coal beds is the primary objective of resource assessment, which can only be delineated with the use of stratigraphic control points and by proper correlation as a result of their depositional environments. The rate of lateral and vertical variability of coal beds depends on their depositional environments and changes in depositional trends with respect to related peat-forming environments. Thus, the degree of certainty of geologic assurance of coal resource assessment is as variable as the area in which peat/coal accumulated. The consensus is that size or extent of the area (e.g. swamp, mire, bog) of peat/coal accumulation is controlled by the types of depositional environments (e.g. delta plain vs alluvial) (Ferm & Cavaroc, 1968; Ferm & Staub, 1984; Ferm et al., 1979; Flint, Aitken, & Hampson, 1995; Flores, 1986; Stanton et al., 1989; McCabe & Parrish, 1992; Moore 2012). However, Fielding (1987) suggested that depositional control is overshadowed by subsidence and sediment supply.

Research by Ayers (2002), Ferm et al. (1979), Greb, Eble, Hower (1999a,b), Horne, Ferm, Carrucio, and Baganz (1978), McCabe (1984), Ryer (1981), and Warwick and Stanton (1988) demonstrated small- and large-scale variability of the areal extent, thickness, and shape of coal beds. Many of these workers have advocated the need for more stratigraphic data and additional closely spaced control points in order to demonstrate apparent continuity of coal beds. They suggest that the most informative cross sections of coal-bearing rocks are those in which control points are closely spaced in which there can be no doubt about lateral continuity of coal beds. The next best thing in which no interpretation is required between control points to trace coal bed continuity along a continuous road cut or other large-scale excavation. Demonstration of the scale of lateral continuity, thickness variability, and areal extent of coal beds from ancient deltaic to fluvial coal-bearing deposits in relationship to stratigraphic control during the past 30 years is represented by the works of Ferm et al. (1979), Flores, Spear, Kinney, et al. (2010), Flores, Spear, Purchase, et al. (2010), and Greb et al. (1999a, b).

Small-Scale Variability

Milici and Houseknecht (2009) demonstrated this concept in Figures 6.4 and 6.5 in which the detailed stratigraphy of a Pennsylvanian (Hartshorne) coal beds within a 100,699 km^2 area in the Arkoma Basin reveals the shape, anatomy, and pattern of development of the deposit in a fluvial setting. The isopach map (Figure 6.5) shows a coal body about 31−62 km wide and 187 km long with two linear coal subbodies as much as 1.8 m thick trending east to west (Figure 6.5). The linear coal subbodies are flanked on the southwest and east by no coal. Between these linear coal subbodies is an area of about 0.6−1.8 m thick coal separated by a east-west trending, ribbon-like area of no coal or sandstone (yellow color).

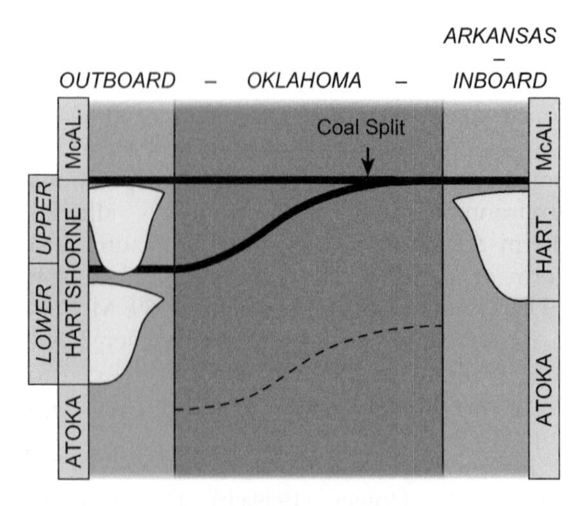

FIGURE 6.4 Stratigraphic cross section of the Lower and Upper Hartshorne coal beds in the Pennsylvanian Hartshorne Formation and associated rock units in the Arkoma Basin in Arkansas and Oklahoma United States. HART. = Hartshorne, McAL. = McAlester. *Source: Adopted from Milici and Houseknecht (2009).*

A stratigraphic cross section (Figure 6.4) showing detailed lithofacies descriptions and variations provides a clue to the depositional settings of the coal body and laterally equivalent sedimentary rock units. The 1.8-m thick coal subbodies along the margins results from the presence of an upper coal bench, which splits from the lower (main) coal bench and thins toward the center or toward the fluvial channel (Figures 6.4 and 6.5). The narrow east-west belt of fluvial channel sandstone (yellow color) with no coal represents temporal deposition with the merged lower and upper Hartshorne coal beds formed in the adjoining mires (Figure 6.5). The ribbon-like area of no coal in the center of the coal body represents the split of the lower and Upper Hartshorne coal beds with the deposition of sands in the fluvial channel on the lower bed and encroachment of the mire of the upper bed upon abandonment of the channel. From the stratigraphic cross section (Figure 6.4) and coal isopach map (Figure 6.5) it would appear that the depositional environments were two linear, flat peat islands flanked on the north and south by a through-flowing fluvial channel.

Although the areal linear cross-sectional shape is a diagnostic feature for fluvial setting coal bodies, other studies by Ferm et al. (1979)

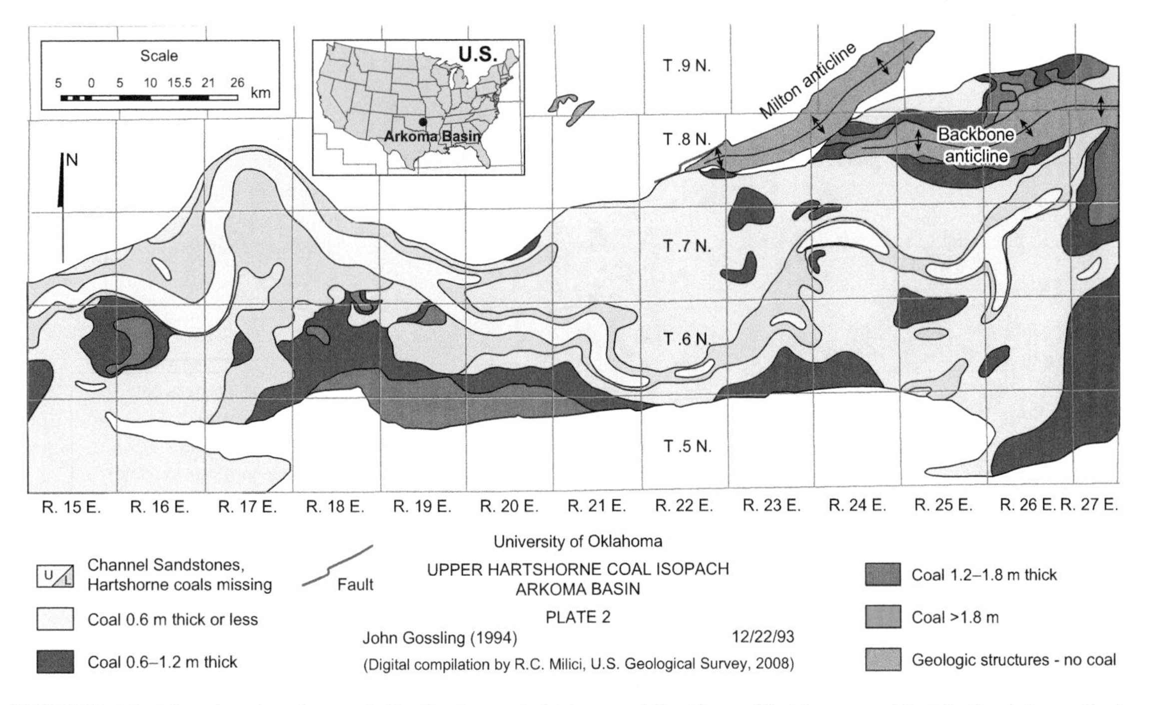

FIGURE 6.5 Map showing the areal distribution and thickness of the Upper Hartshorne coal bed in the Arkoma Basin, Arkansas and Oklahoma, United States. The areal distribution of the Upper Hartshorne coal bed depicts the split by a fluvial channel sandstone (see Fig. 8.4). U = upper; L = lower. *Source: Adopted by Milici and Houseknecht (2009).*

and Pedlow (1977) in delta plain depositional environment show more variable shapes. The Allegheny coal beds in the Appalachian region in western Pennsylvania in a marginal delta plain occur in five elongate subbodies ranging from 3.2 to 6.4 km wide and 9.6–16.1 km long within 434-km^2 area. The cross-sectional shape of the coal subbodies is tabular decreasing in thickness rapidly toward the mire edges from 0.6 m to less than 15 cm and no coal. Ferm et al. (1979) interpreted the coal subbodies formed in interfluve areas and the sites of no coal as tidal creeks infilled by siltstones and shales. In this setting the lateral continuity of coal beds may be extrapolated up to 3.2 km between control points. In these case studies by Ferm et al. (1979) and Pedlow (1977), the general characteristics of the rocks surrounding the coal bodies are important in mapping the shape and margins of the coal beds. In addition, mapping of split lines of the coal along channel margins provide a better and more precise coal bed map of merged single versus split beds for separate resource assessment. Finally, larger data sets averaging 1.6 km apart made for better delineation of the small, elongate coal subbodies.

Greb et al. (1999a, b) demonstrated the value of large data sets (3800 locations within 1860 km^2 area) in mapping Middle Pennsylvanian (Duckmantian) coal bed in the central Appalachian region (Figure 6.6). Like in the western Pennsylvania Alleghenian coal bed, the Duckmantian coal bed shows similar stratigraphic vertical and lateral variations (Figure 6.7). That is benches of the coal bed

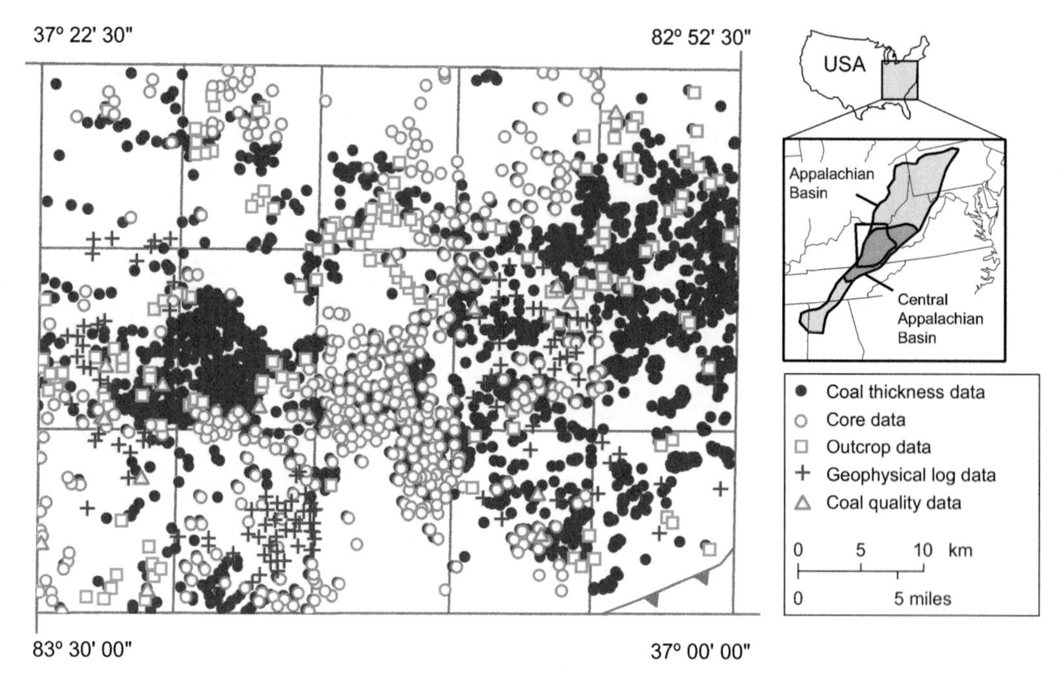

FIGURE 6.6 A map showing the density of observations points of a Duckmantian coal bed in the central Appalachian Basin in eastern Kentucky, United States. *Source: Adopted from Greb et al. (1999a, b).*

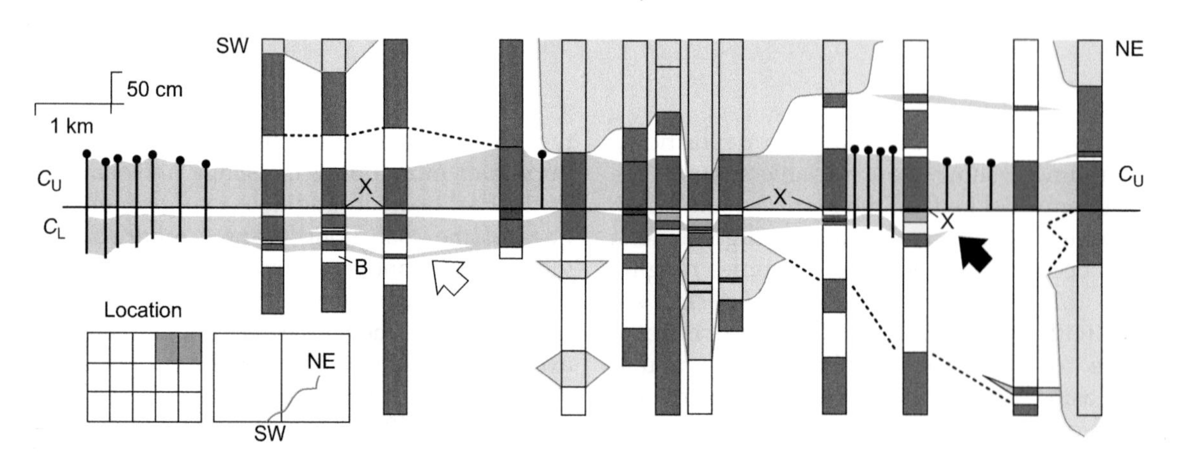

FIGURE 6.7 A cross section of the Pennsylvanian fire clay coal beds showing the upper (C_U) and lower (C_L) beds separated by clay parting in eastern Kentucky, United States. *Source: Adopted from Greb et al. (1999a, b).*

pinch out into adjoining detrital rocks. However, the closely spaced data points permitted isopaching of the coal benches into elongate to lenticular shape much more complex than the coal subbodies of the Alleghenian coal. Figure 6.8 shows the isopach map of the lower bench as an east–west oriented lenticular coal sub-body (38.6 km long and 23 km wide)

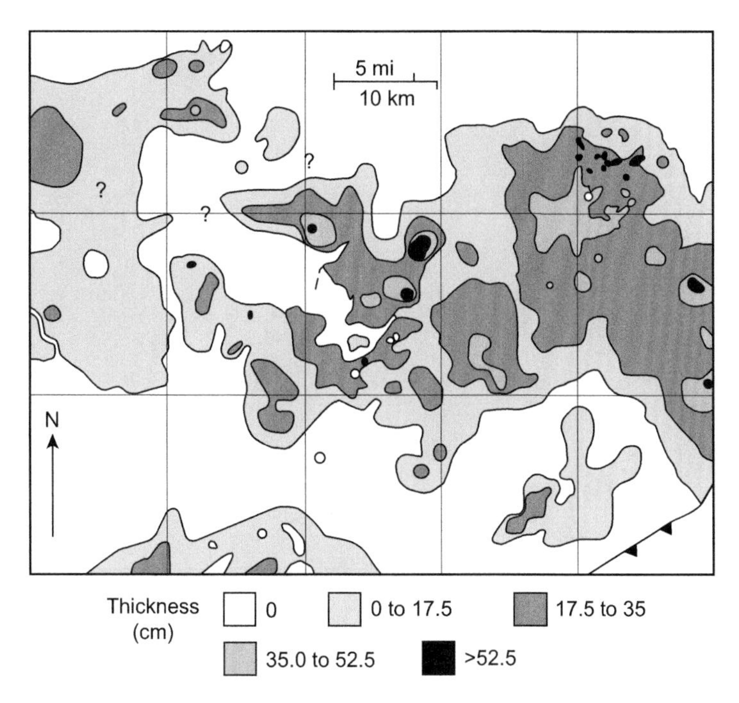

FIGURE 6.8 An isopach map of the upper part of the fire clay coal indicating a prominent elongate, east–west trend. *Source: Adopted from Greb et al. (1999a, b).*

surrounded by small elongate subbodies (412 km long and 4 km wide). The larger lenticular subbody has a very serpentine boundary suggesting more detail interpretation resulting from closely spaced control points. Figure 6.9 shows the isopach map of the upper bench as a southeast–northeast lenticular body along the thickest part (from 35 to >105 cm) with branching smaller lenticular coal subbodies to the northwest; both subbodies delineated by also very serpentine boundaries.

Greb et al. (1999a, b) interpreted the vegetation during accumulation of the upper bench of the coal bed as shifted from planar to domed mires. Furthermore, Greb et al. (1999a, b) suggested that faulting also controlled depositional environments resulting in channel sandstones with angular changes in orientation. In addition, channel sandstones and rider coal benches with high ash yields, which occasionally merged with the upper coal bench, draped lateral lithofacies. These depositional and tectonic controls may explain the right angle reorientation of the directions of elongation between the lower and upper benches. The fine-tuning of the shape and changes in orientation of coal subbodies are critical to certainty of geologic assurance in coal resource assessment, which can only be documented by more data sets and close spacing of control points as demonstrated by the works of Ferm et al. (1979) and Greb et al. (1999a, b).

Large-Scale Variability

Ferm et al. (1979) and Greb et al. (1999a, b) have demonstrated the variability of a single coal bed within peat/coal depositional

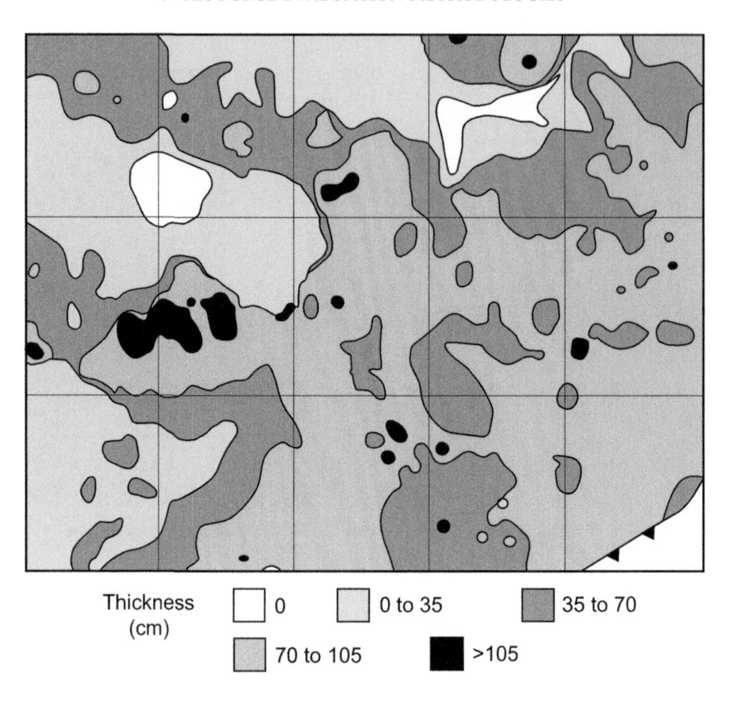

FIGURE 6.9 An isopach map of the lower part of the fire clay coal indicating a prominent lenticular, southwest–northeast trend. *Source: Adopted from Greb et al. (1999a, b).*

environments with sizes or areal extents from 450 to 1860 km^2 areas with 250 to 3800 locations where coal thicknesses were measured, respectively. The density of control points in these areas varies from 0.5 to 2.0 per km^2 area. Control point densities are within spacings required for estimation of measured and indicated coal resources categories. A larger scale peat–coal depositional environment of a single coal is described by Flores, Spear, Kinney, et al. (2010) in the southern part of the Powder River Basin in Wyoming. This area is about 7770 km^2 underlain by the Wyodak–Anderson coal bed, which is 6–46 m thick, and overlain by 61–366 m thick overburden (rock interval above the coal to the surface). There are 14,000 coalbed gas, oil and gas wells, and coal drill holes with density of control points of about 1.8 per km^2 area but only 10,000 geophysical logs of these wells were used to measure the coal thickness (Figure 6.10). The stratigraphic data sets are about four times greater but cover four times more area than that of Greb et al. (1999a, b).

Figures 6.11 and 6.12 show a merged Wyodak–Anderson coal bed, which either splits or pinch out along its western margin. The Wyodak–Anderson coal bed splits into as many as five coal beds, which in turn split, merge, and pinch out (Flores, Spear, Kinney, et al., 2010). Splitting caused the merged Wyodak–Anderson coal bed to expand into an interval of as much as 305 m thick, which contains thin to thick coal beds interbedded with thick mudstone, siltstone, and sandstone beds (Flores, Spear, Kinney, et al., 2010). This single Wyodak–Anderson coal bed is one of the largest coal deposit accumulated in a peat-forming depositional environment interpreted as a domed or raised bog (Flores, Spear, Kinney, et al., 2010). However, the coal deposit

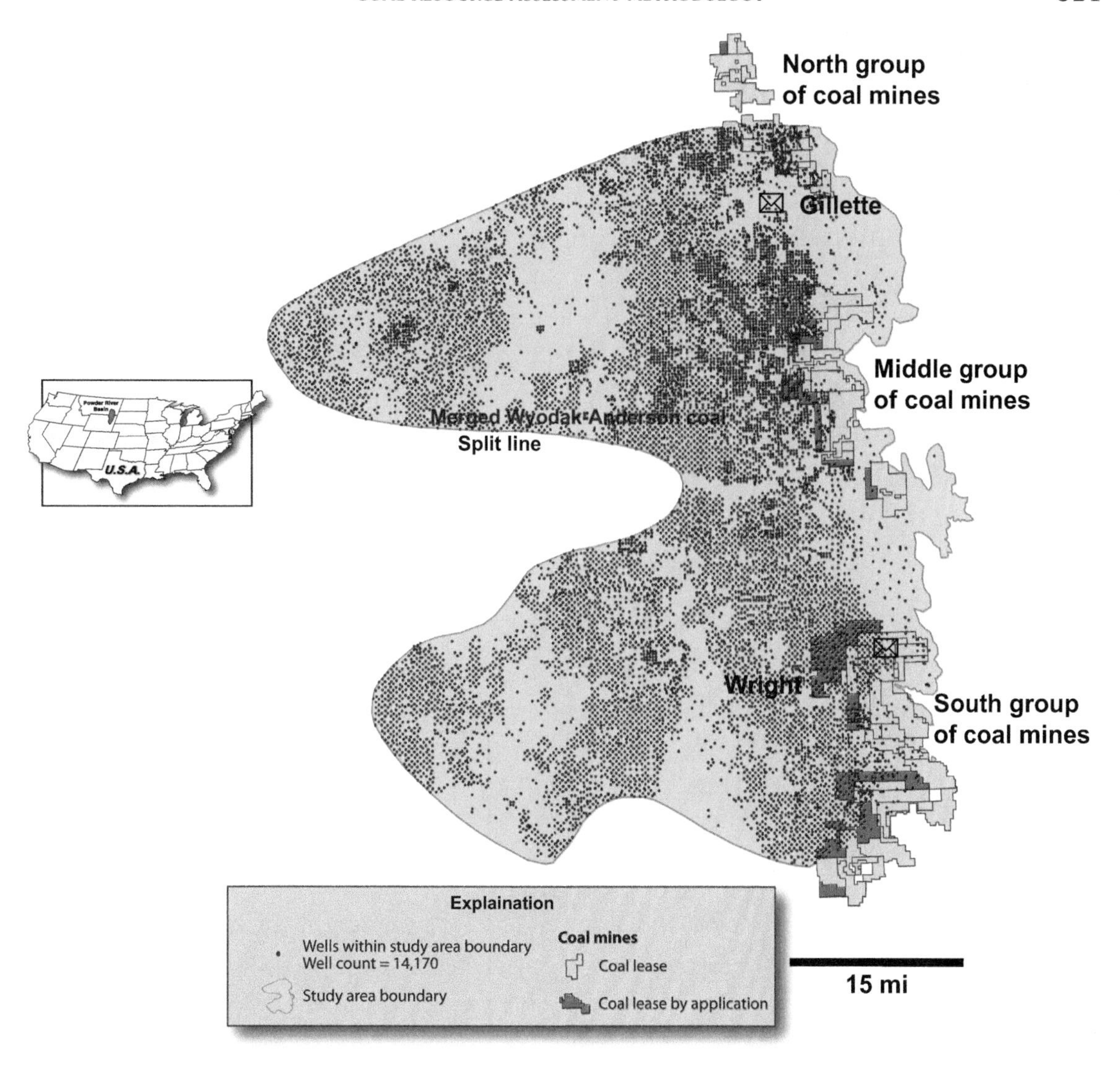

FIGURE 6.10 A map of the merged Wyodak-Anderson coal zone of the Paleocene Fort Union Formation in the south and east of the Powder River Basin, Wyoming. See index map for location in the United States. *Source: Modified from Flores, Spear, Kinney, et al. (2010), Flores, Spear, Purchase, et al. (2010).*

is locally interrupted by deposits of minor, confined fluvial channels (see Figure 6.11). The coal thickens along its western edge where the coal is interbedded with numerous partings corresponding to the split line marginal to major fluvial channel sandstones (see Figure 6.11).

COAL RESOURCE ASSESSMENT METHODOLOGY

Coal resource assessment methodology is an integral component of estimation of coalbed gas resources. Kuuskraa and Boyer (1993) suggested that the necessary requirement for

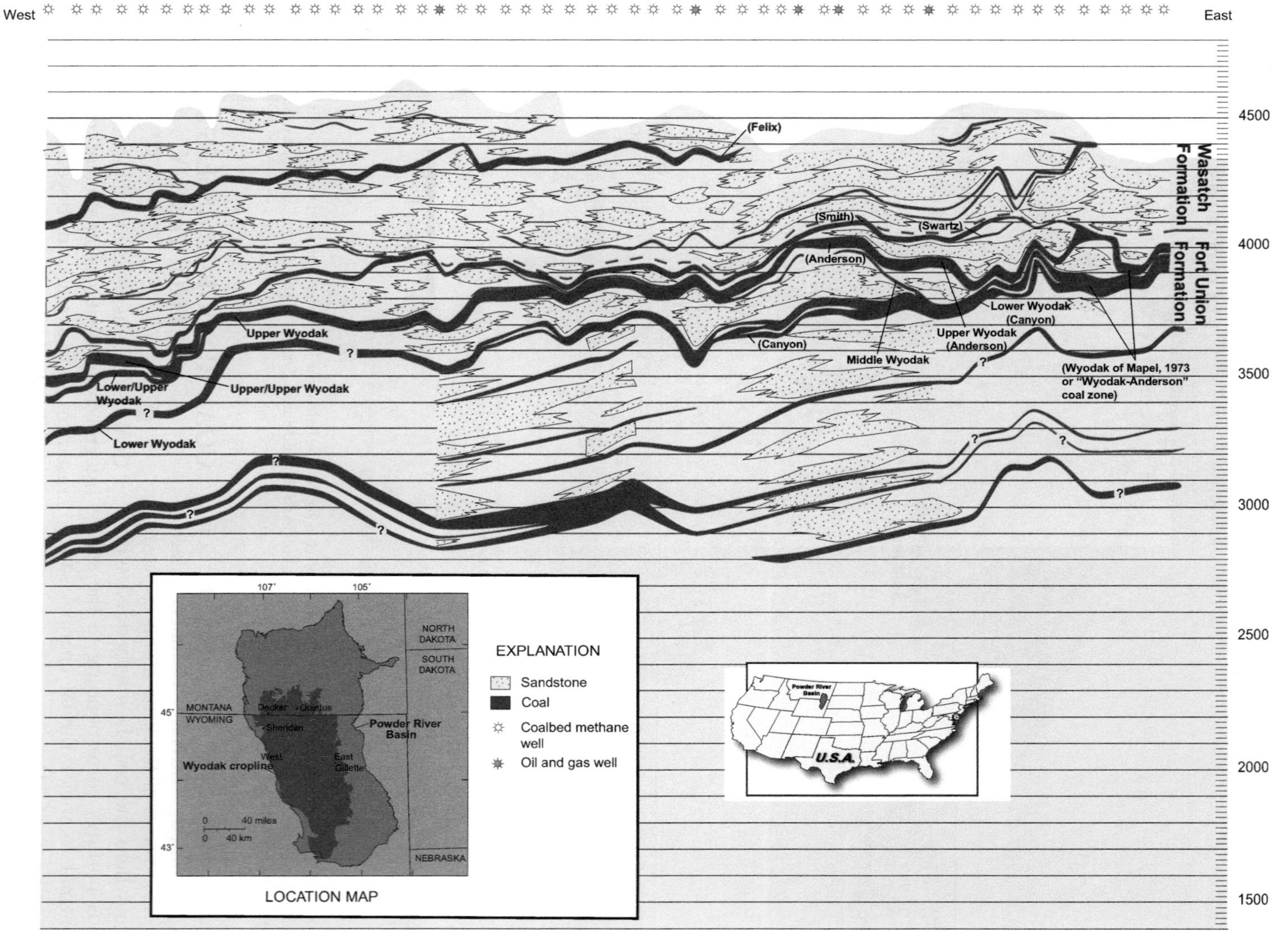

FIGURE 6.11 An east to west cross section across the central part of the Powder River Basin in Wyoming, United States (see index map for location). The merged Wyodak coal zone splitting on the easternmost part of the cross section. The reader is referred to Flores et al. (2010) for the references cited in the cross-section. *Source: Modified from Flores, Spear, Kinney, et al. (2010), Flores, Spear, Purchase, et al. (2010).*

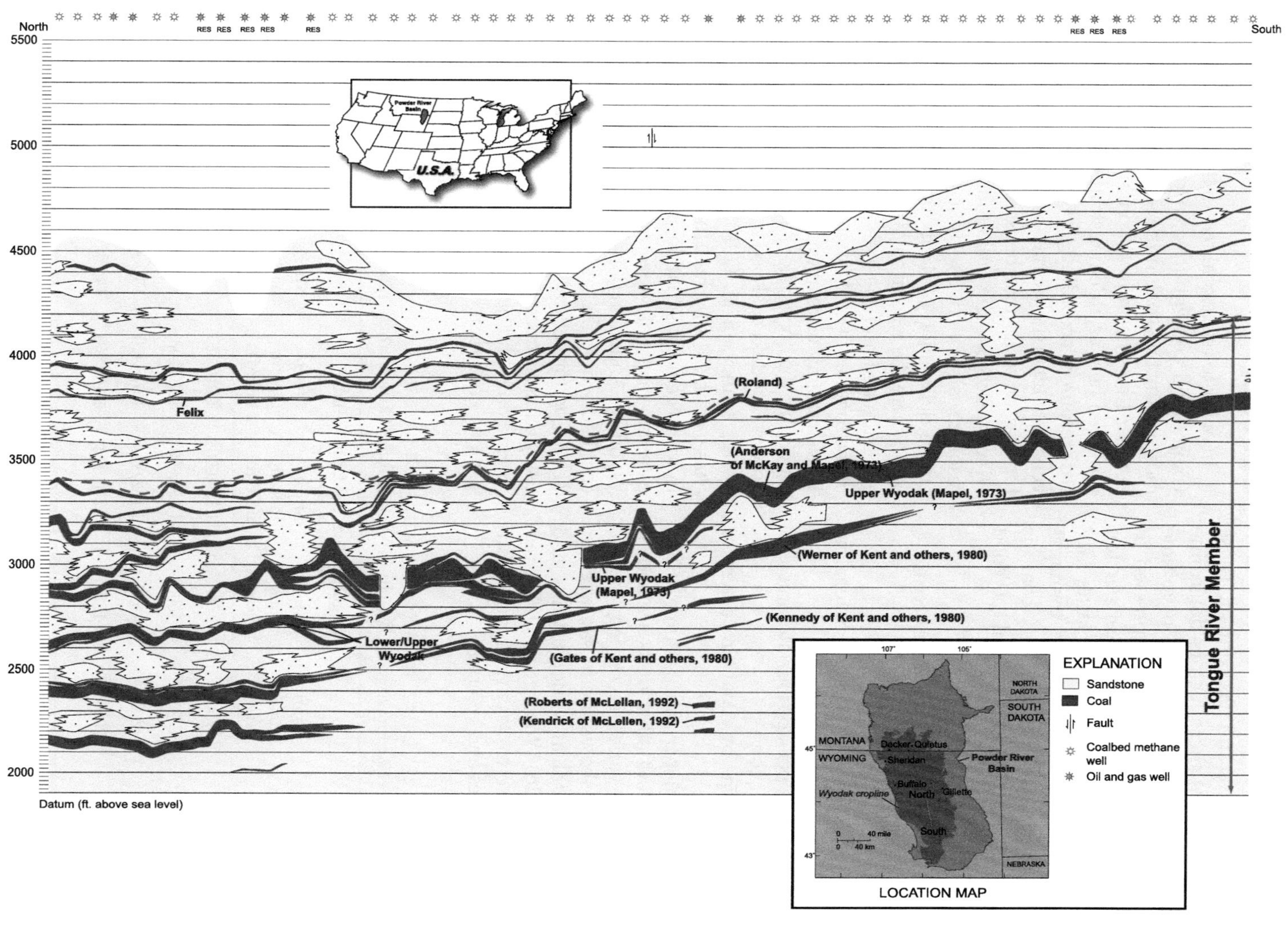

FIGURE 6.12 A north to south cross section across the central part of the Powder River Basin in Wyoming, United States (see index map for location). The merged Wyodak coal zone splitting northward. The reader is referred to Flores et al. (2010) for the references cited in the cross-section. *Source: Modified from Flores, Spear, Kinney, et al. (2010), Flores, Spear, Purchase, et al. (2010).*

sustained gas flow is resource concentration based on coal thickness and gas content. These authors indicated that these factors determine the GIP for a given area expressed in volume of gas per unit area. GIP is also expressed in volume of gas per standard cubic feet per ton or cubic meter per metric ton. The latter GIP estimate requires calculation of the tonnages of coal. Tonnage or volume of coal is estimated by the following formula (Wood et al., 1983):

$$A \times B \times C = \text{tonnage or volume of coal}$$

where:

A = Weighted average thickness of coal in inches, feet, centimeters, or meters.

B = Weight of coal per appropriate unit volume in short or metric tons.

C = Area underlain by coal in acres or hectares.

The weight of coal per unit volume (B) in short or metric tons and other units are shown in Table 6.4, which varies according to coal rank. Also, the table shows the *in situ* average specific gravity of coal at various ranks and differences in ash yield. Wood et al. (1983) recommended using these values in preparing estimates of coal resources and reserves.

Case Study

In order to relate coal resources to coalbed gas volume and concentration, a case study of the most productive coal in the southern part of the Powder River Basin in Wyoming, United States, is offered here as a demonstration. Estimation of coal resources according to Wood et al. (1983) is very difficult in the vicinity of where a coal bed bifurcates or splits into beds or tongues, each of which exceeds the minimum thickness for resource calculation. The difficulty arises from accurate delineation of the boundary between the area where the resources of the merged coal bed are estimated and the areas where the resources of the tongues of coal are individually estimated. This stratigraphic relationship resembles the condition of the Wyodak–Anderson (Upper and Lower Wyodak) coal bed as described by Flores, Spear, Kinney, et al. (2010), Flores, Spear, Purchase, et al. (2010). Thus, in estimating resources in such a geologic situation, it is necessary to delineate a map of the area of the merged Wyodak–Anderson (Upper and Lower Wyodak) coal bed where the resources will be calculated. This is performed by correlation of the coal bed using interconnected network of east–west (Figure 6.11) and north–south (Figure 6.12) cross sections constructed with coalbed methane, oil,

TABLE 6.4 Average Specific Gravity and Weight of Various Rank Coals in the United States

		Weight of Coal per Unit Volume					
Rank	**Specific Gravity (ave)**	**Short tons/ acre-ft**	**Metric tons/ acre-ft**	**Short tons/sq mile-ft**	**Metric tons/sq mile-ft**	**Metric tons/sq hectometer-meter**	**Metric tons/sq hm m**
Lignite	1.29	1750	1588	1,120,000	1,016,320	12,900	1,290,000
Subbituminous	1.30	1770	1605	1,132,800	1027,200	13,000	1,300,000
Bituminous	1.32	1800	1633	1,152,000	1,045,120	13,200	1,320,000
Anthracite and semianthracite	1.47	2000	1814	1,280,000	1,160,960	14,700	1,470,000

ave, average; ft, foot; sq, square; hm, hectometer; m, meter
Source: Modified from Wood et al. (1983).

gas, and coal drill holes averaging 0.8 km apart (Flores, Spear, Kinney, et al., 2010). The bottom and top of the coal bed was mapped by creating surfaces in ArcMap – ArcInfo (Flores, Spear, Purchase, et al., 2010). Similar coalbed surfaces were mapped for wells between the cross sections and were fitted within the coalbed surfaces mapped from the cross sections. All the correlated wells were selected to construct an isopach map of the Wyodak–Anderson (Upper and Lower Wyodak) coal bed from which the measured, indicated, inferred, and hypothetical resources were estimated on a 0.25-acre (or hectare) grid cell (Flores, Spear, Purchase, et al., 2010).

The merged Wyodak–Anderson (Upper and Lower Wyodak) coal bed in the southern Powder River Basin makes it a desirable assessment unit because it can be mapped as a single tooth-shaped body and is an extension of the coal bed mined in the Gillette coalfield (Flores, Spear, Kinney, et al., 2010) (Figure 6.12). Resource estimation is difficult in those areas where the Wyodak–Anderson (Upper and Lower Wyodak) coal bed splits or bifurcates into thinner coal beds, which exceed the minimum thickness compared to the interburden (rocks between coal beds) thickness for resource calculation. About two thirds of this coal bed, 6–46 m thick, is overlain by as much as 388 m thick overburden composed of interbedded mudstone, siltstone, sandstone, and thin coal beds of the Fort Union and Wasatch Formations. The remaining one third of the coal bed, 15–46 m thick, is overlain by as much as 457 m of overburden at the southwest margin of the body and as much as 671 m at the northwest margin, both areas underlain by as much as 46 m thick coal. Thus, the ratio of overburden to the merged Wyodak–Anderson coal bed west of the Gillette coalfield ranges from <10:1 to >200:1 (Figure 6.12) (Flores, Spear, Kinney, et al., 2010).

The southern Powder River Basin includes as many as 14,000 coalbed methane, oil, gas, and coal drill holes with accompanying geophysical logs (e.g. gamma ray and resistivity), which were drilled on the average of 0.8 km apart. These closely spaced wells permitted accurate lateral correlations using the well logs of the Wyodak–Anderson coal bed along several interconnected regional cross sections to overview the three-dimensional areal distribution of the coal body (Flores, Spear, Purchase, et al., 2010). In order to extend the stratigraphy and lateral distribution of the Wyodak–Anderson coal bed from the known cross sections into and across the unknown areas (between cross sections) correlations used all well records with reported top/bottom intervals from the Wyoming Oil and Gas Conservation Commission (WOGCC) using ArcMap – Arc Info (ESRI, 2009; Flores, Spear, Purchase, et al., 2010). That is, the surfaces of the top and bottom of the Wyodak–Anderson coal bed in the cross sections were areally mapped across the southern Powder River Basin. These surfaces were then used to construct a latticework to extrapolate by latticework the top and bottom of the coal beds in the WOGCC database to extend correlations between cross sections (Flores, Spear, Purchase, et al., 2010).

The first step was to filter the reported top and bottom of the coal beds identified as gas producing zones in the WOGCC database to fit within the mapped interval of the Wyodak–Anderson coal top and bottom of the cross sections, which are then selected for inclusion in the data to isopach the bed. Another important step is converting the depths to elevations to eliminate the effects of surface topography.

The second step was to incrementally filter the top and bottom surfaces of gas producing zones in wells from the WOGCC database. With the coal beds identified as above, the reported producing intervals were assigned to specific coal beds by selecting the producing intervals that matched within ±1.5 m, ±3.0 m, ±4.5 m, ±6.0 m, up to ±9.1 m of the top and bottom of the Wyodak–Anderson coal. These wells were then added to the database for construction of the isopach map.

The 9.1 m threshold was selected because the nearest coal bed above and below the Wyodak–Anderson coal bed is about 10.7–18.3 m. Filtering for wells that are duplicated and for Wyodak–Anderson coal bed greater than 49 m due to splits netted 6630 wells from the original 14,000 wells. Additional filtering was performed on this data by buffering wells used along the cross sections reducing the total to 2989 wells, which were used to isopach by simple Kriging method allowing weighted averaging of the thickness of the Wyodak–Anderson coal bed. The isopach map (Figure 6.13) was used to calculate the coal resources of the merged Wyodak–Anderson coal bed in the southern Powder River Basin.

Cell-Based Coal Resource Calculations

The calculation methodology used for this assessment was a cell-based modeling adaptation of Wood and others (1983). Initially, a hierarchy of data points was established to create a coal thickness isopach. The order of importance, or confidence, was (1) cross-section wells, (2) drill holes outside a 1.2-km buffer zone of cross-section wells, (3) coalbed gas wells (e.g. drilled holes, cored drill holes reported in the WOGCC database) outside a 1.2-km buffer zone of drill holes and the previously established cross-section buffer zone, and (4) additional wells outside a 1.2-km buffer zone of State wells and the two previously established buffer zones. The wells were filtered for greater than 45 m thick of the merged Wyodak–Anderson coal bed along the split boundary because of over thickening cause by introduction of interburden (for example, sandstone, siltstone, and mudstone). This hierarchy resulted in a decrease of approximately 9000 wells used for resource calculations and created a method to assess confidence in the data.

The wells point data set was modeled in ESRI's GeoStatistical Analyst (ESRI, 2009) extension using simple Kriging to interpolate and extrapolate weighted-average thickness values. The coal bed thickness model was converted to a fixed raster data set, or GRID file format, with a 1/4-acre (0.10 ha) cell size. The thickness GRID was used as input in ESRI's Spatial Analyst (ESRI, 2009) extension to create a raster tonnage data set using Wood and others' (1983) and Mercier and others' (2010) formula in Raster Calculator (Scott Kinney and Tracey Mercier, Personal Commun., 2010):

Thickness GRID (in ft or m) × 1770 (weight of coal per acre ft or per hectare) × 0.25 (area of each cell, in acres or hectares)

or

Thickness GRID (in ft or m) × 1605 (weight of coal per acre ft or per hectare) × 0.25 (area of each cell, in acres or hectare)

The resultant GRID contained tonnage values per cell can be converted to international units used to summarize resources by any given zone. By using a cell-based instead of a vector-based methodology, coal thickness and tonnage estimates can be generated for each 1/4-acre (0.10 ha) cell in the GRIDs as opposed to larger polygonal areas. This can provide a more efficient mechanism to report summary statistics as any polygonal area can be overlain on the thickness or tonnage GRIDs to generate summary statistics summarized. Examples of other polygonal zones that can be summarized include (1) counties, (2) 7.5-min quadrangles or topographic maps, (3) measured, indicated, and inferred buffer zones, and (4) overburden categories (Figure 6.14) (Wood et al., 1983).

Coal resources were calculated for the entire merged Wyodak-Anderson coal bed delineated by split boundary after removing mined-out and clinker areas. After resources were calculated, a cell-based Euclidean allocation method and a vector-based Voronoi method (De Berg, van Kreveld, Overmars, & Schwarzkopf, 2000) were used to validate

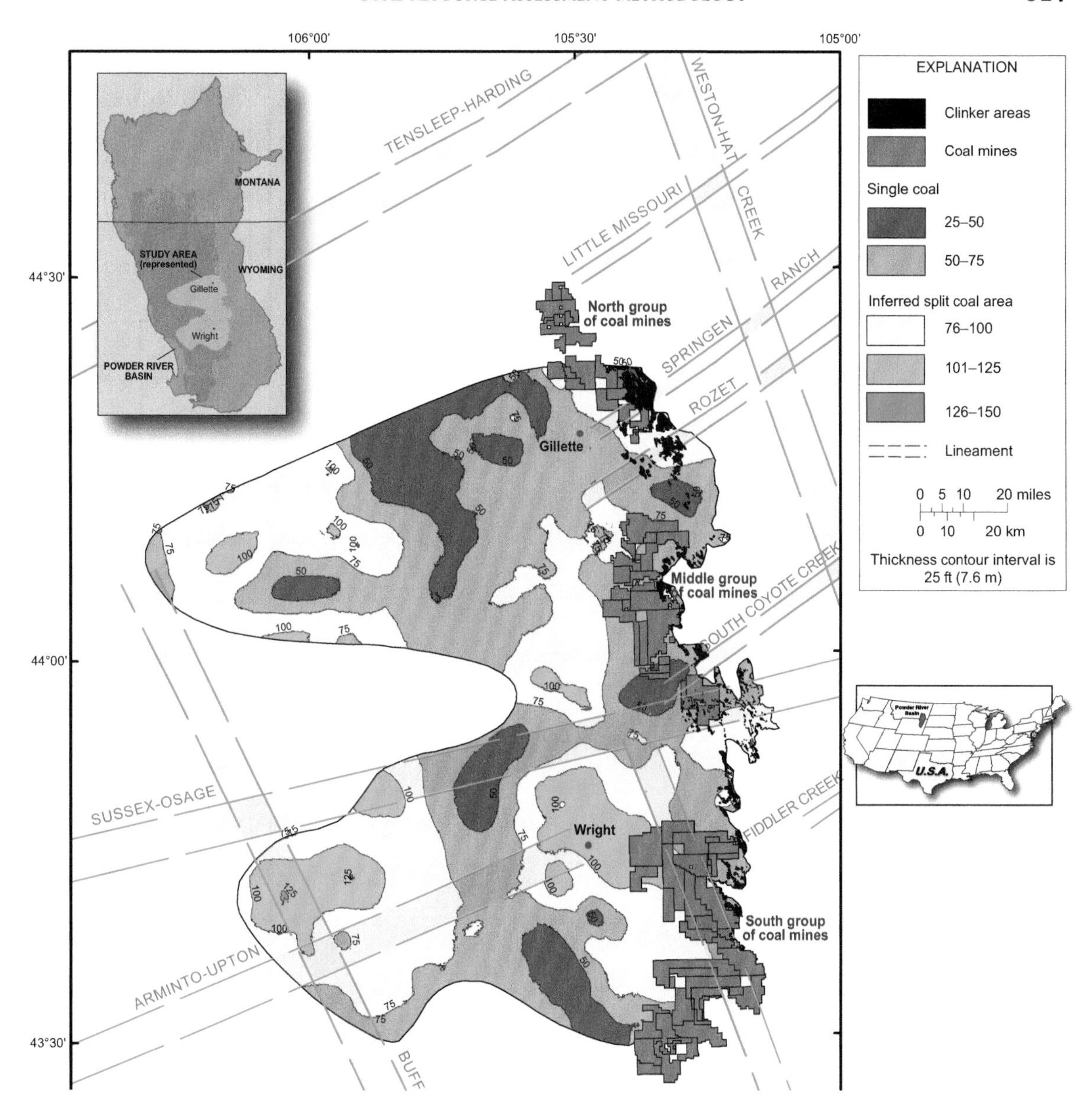

FIGURE 6.13 An isopach map of the merged Wyodak-Anderson coal zone of the Fort Union Formation in the southern part of the Powder River Basin in Wyoming. The thickness of the merged Wyodak-Anderson coal is unaffected by northwest-southeast and northeast-southwest basement lineaments. *Source: From Flores, Spear, Kinney, et al. (2010), Flores, Spear, Purchase, et al. (2010).*

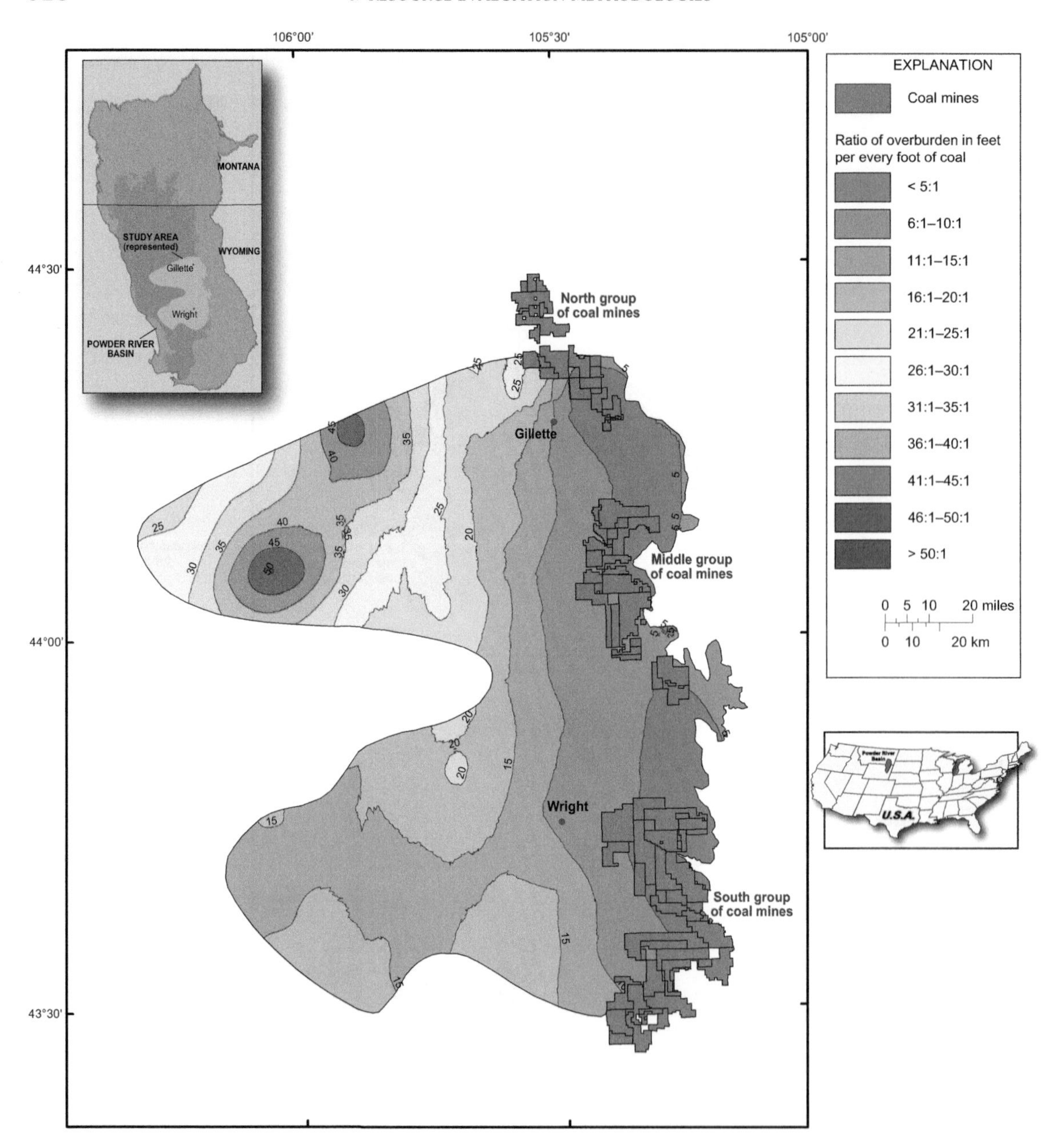

FIGURE 6.14 A ratio map of the overburden rock per thickness of the merged Wyodak-Anderson coal in the southern part of the Powder River Basin in Wyoming. The overburden-coal thickness ratio increases toward the northwest.

the calculated totals. Both methods generated a less than one percent difference in resource and reserve totals than the cell-based Kriging method. Wood et al., (1983) prepared guidelines for standardized reporting categories in order to have rational and uniformity of procedures to follow as well as for accurate inventory of resources. The purpose of these guidelines is to create coal resource estimates prepared by various workers, which are meaningful comparison world-wide, and for a nation, state, county, region, basin, township and range and quadrangle. Federal coal and surface ownership are categories also included because the Powder River Basin contains split-state ownership and large amount of subsurface mineral ownership. The guidelines also require reporting according to overburden (e.g. from 0–100 ft or 0–30 m to >500 ft or >150 m), and net coal thickness (e.g. from 2–5 ft or 0.6–1.5 m to >40 ft or 12 m). Thus, the merged Wyodak–Anderson coal body, excluding the "tongues," in the southern Powder River Basin was estimated to contain measured resources of 53.5 billion metric tons, indicated of 95.2 billion metric tons, inferred resources of 35.4 billion metric tons, and hypothetical resources of 748.4 million metric tons or a total resources of 184.9 billion metric tons.

DATA COLLECTION

Precision of data collection dictates the degree of accuracy of resource assessment. Coalbed gas resource assessment of GIP or gas concentration involves complicated data collection and analyses. This process mainly includes collection of data to estimate gas content and reservoir properties to determine gas saturation. These estimates rely on data from acquisition of coal samples from coalbed gas exploration wells for gas desorption and isotherm analysis. The procedure involves the collection of samples of fresh coal and desorbing gas over a period of time is necessary to measure and estimate the volume of gas charge in a coal bed reservoir. Coal samples for gas desorption can be from drill cuttings and core (Montgomery & Barker, 2003). However, accurate measurements of gas content are best obtained from coal cores. Because of the cost of drilling and coring, time constraints from prolonged coal desorption and laboratory analyses, loss of gas during transport from the field to the laboratory, and demand by companies for speedy initial appraisal, the conventional procedure of gas desorption from coal cores is being replaced by new technology of in situ gas measurement in the wellbore (Lamarre & Pope, 2007; Carlson, Hartman, Obluda, & Miller, 2008). Lamarre and Pope (2007) described this new technology as using a downhole chemical-sensing tool called Raman spectrometer, which is lowered by wireline into a 12.7 cm-diameter wellbore, developed specifically for coalbed gas measurement. The Raman spectrometer provides the critical desorption pressure of gas dissolved in coal bed reservoir fluid (e.g. water) in the wellbore.

Despite the new technology for in situ gas measurement conventional coring of coal for gas desorption at the well site still provides the most reliable gas content data. However, reliable collection of data depends on the methods followed at the well site, which include use of proper equipment (e.g. drill rig, core barrels, laboratory trailer, canisters), techniques (e.g. preparation and handling of cores samples), procedure of measurement of gas and management of data, and deployment of appropriate number of field personnel. When a coalbed gas drilling and coring project is properly implemented, it is a major undertaking as typified by the cooperative program between the USGS, Wyoming Bureau of Land Management (BLM-WY), and 29 gas operators in the Powder River Basin from 1999 to 2009 (Flores et al., 2006). The cooperative field-laboratory project, with the author as project chief, was created to minimize cost and provide real time data from coal cores donated by the coalbed gas companies and operators collected, analyzed, managed, and interpreted by the USGS for policy decision makers in the Wyoming Reservoir Management Group of the BLM-WY.

Coal Sampling and Handling Strategies

Critical to coal sampling from the wellbore is the collection of fresh and continuous core samples for gas desorption. More importantly, in order to minimize lost gas the core retrieval must be performed with the least amount of time of travel to the surface. These sampling issues depend on the core-recovery systems, core barrel equipment, experience of operator on drilling and coring through coal beds, and physical characteristics of the coal bed (Figure 6.15). The experience of the author in the Powder River Basin indicates that operators with experience in drilling for coal mines have a better "feel of the coring rpms" of the rock than others and highly fractured coals are difficult to retrieve continuous samples. In the Powder River Basin because of the shallow depths (e.g. <1000 m) of coal beds the rigs for drilling coalbed methane in most cases are usually moderate size very similar to that used to drill conventional water wells. Elsewhere for deeper coal beds drill rigs used are much like those in conventional oil and gas development.

Core-Recovery Systems

Core-recovery systems for collecting coal samples from wellbores include conventional, wireline, sidewall, and pressure cores as well as drill cuttings. Each of these systems has advantages and disadvantages in terms of cost, quality, quantity, reliability, and time constraints (Table 6.5). Waechter, Hampton, and Shipps (2004) suggested that each of these core recovery systems may be utilized under different sets of geological conditions. Conventional coring method, which requires numerous trip times out of the wellbore, are amenable for use in shallow wells with less than two coal beds. In order to retrieve the core from a conventional core drill, the core barrel is detached from the hole, which is time consuming process because each rod is removed one at a time (Figure 6.16). Wireline coring method is amenable for use in deeper wellbores with more numerous coal beds. The mechanism of wireline coring is such that after coring the

FIGURE 6.15 Truck-mounted rig used for drilling and coring, coal beds in coalbed gas wells in the Powder River Basin, Wyoming, United States. Coring the Wyodak-Anderson coal (see bottom of core barrel) for the USGS-BLM coalbed gas project in the Powder River Basin.

TABLE 6.5 Advantages and Disadvantages of Various Drill Core and Cutting Recovery Systems as to Cost, Quality, Quantity, Reliability, and Time Constraints

Core Recovery System	Advantages	Disadvantages
Conventional core	Less expensive; potentially large and continuous samples; possible evaluation of discrete coal benches for cleats, fractures, and petrology; minimal contamination	Potential lost of gas; numerous trip times
Wireline core	Better percent recovery; continuous samples; coring is fast; minor trip times; better sample collection from deep wellbores	Expensive; minimal trip times
Sidewall core	Reduce cost; minimal trip time; minimize lost gas; sampling units can be accurately picked from wireline logs; minimal lost gas; amenable to sampling deep wellbores	Small size samples; need more samples for accurate representation; limited potential for additional analyses
Pressure core	Minimize lost gas; direct measurement of gas; minimal contamination of sample; provide large sample for other analyses; coal geology evaluation of samples such as cleats and petrology	Very expensive
Drill cuttings	Inexpensive; collection of samples during regular drilling operation; rapid but unreliable analysis	Very small, rubblelized samples; indescribable samples; maximum lost of gas; sampling units not accurately picked; highly contaminated samples

Source: Modified from Waechter et al. (2004).

core barrel is left at the bottom of the hole while the inner tube with the rocks is detached and pulled up rapidly to the surface. This process permits repeated coring through thick intervals of coal and rock units and core retrieval at a very fast rate (e.g. in minutes depending on the depth of the wellbore). Rotary cores collected from the side of the wellbore after completion of wireline or geophysical logging, as the name implies, cuts sidewall core. This method is more appropriate for in deeper wellbore but is constrained in washout areas of the sidewall.

Pressure coring is a self-contained system, which captures and maintains the core at or near bottomhole pressure, preventing gas expansion and fluid loss (Eaton, Redford, Segrest, & Christensen, 1991). A gel displaces drilling mud from the core barrel to keep in situ pressure constant and inner barrel frozen for transport to the laboratory. Owen and Sharer (1992) and Diamond and Schatzel (1998) suggested that pressure coring is an alternative to direct measurement method for obtaining coal sample is reservoir pressure, eliminating lost gas. Drill cuttings are small granular particles of crushed coal produced by surface rotary-drilled wellbores. Cuttings have shorter distance to the center of the coal fragment, which promotes rapid rate of gas diffusion. Xue et al. (2011) devised new

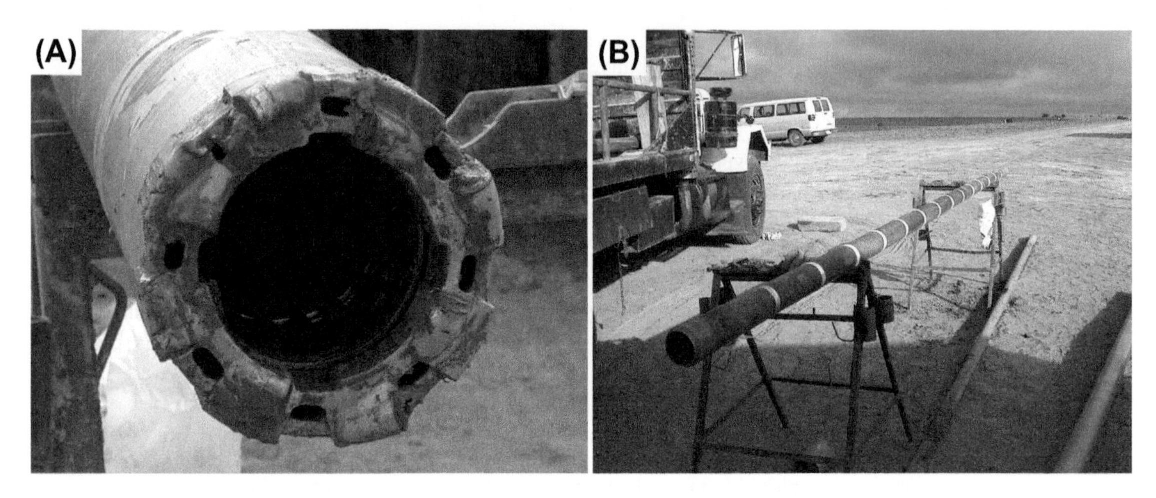

FIGURE 6.16 Conventional core barrel assembly (A) containing a split inner tube (B) for collecting continuous coal core in the Powder River Basin, Wyoming.

drill cuttings sampling-while-drilling system based on double tubing drill holes and reversed circulation of pressurized air. The drill cuttings sampling system is recommended for soft coal, which may be used in direct measurement of gas content. Coal samples are difficult to obtain in soft coal due to wellbore deformation and collapse.

Prior to collection of conventional cores, each borehole will be drilled near the top of the coal bed and will run a typically 15.2 cm steel pipe casing. The coring will attempt recovery of a continuous core, from bottom to top, of the coal bed by using a 7.6 cm diameter, 4.5 m long split core barrel. Coring runs will be performed depending of the thickness of the coal. That is, if the thickness of the coal is 4.5 m, one run is only attempted. If the thickness of the coal is 30.5 m, seven runs will be attempted and so on. As part of the calculation of the total gas content, during drilling the depths, time when coring started, time when the core barrel started up the hole after coring, and time when the core barrel arrived on the surface are recorded for use in calculation of lost gas. Once on the surface, the split core barrel is opened and the coal is divided into 0.6 m sections (Stricker et al., 2006). Long canisters were used specifically for the USGS-BLM cooperative project in the Powder River Basin, because of the thick coal beds. Using the long canisters permit more desorption of the thick coals. Standard procedure suggests 0.3 m long canisters to be used for desorption and description of the cores but this author suggests that the sooner the coal core is inserted in the canister and sealed the least gas is lost and coal oxidized. Each coal section is inserted into a airtight canister especially designed and constructed by the USGS for gas desorption (Figure 6.17) (Stricker et al. 2006). The canister is sealed and purged with helium (inert gas) to make sure that there is no gas leak from the canister. Gas leakage is tested by submerging the bottom and top of canisters into the water bath. The canister is immersed in 76.2×91.4 cm-diameter plastic tanks filled with water (water baths) heated by thermostatically controlled by water heaters and maintained to constant temperature to simulate reservoir temperature of the coal bed in the ground. The water baths conduct more even heat to the plastic canisters than the aluminum canisters.

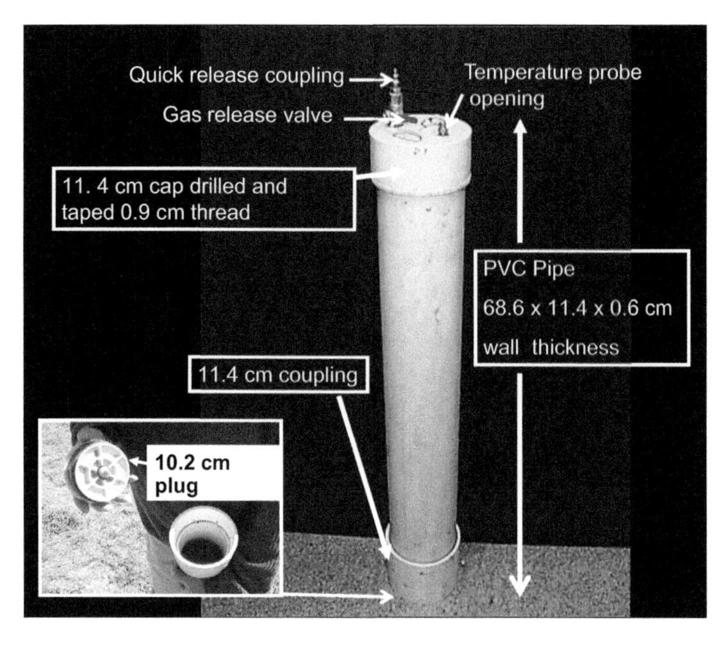

FIGURE 6.17 Design and dimensions of PVC pipe gas canister utilized by the USGS (Stricker et al. 2006) and Bureau of Land Management coalbed gas cooperative project in the Powder River Basin, Wyoming, United States. *Source: Flores et al., 2006.*

BOX 6.1

ALUMINUM VS PVC CANISTERS: CONCEPTS AND MISCONCEPTION

The old U.S. Bureau of Mines (USBM) designed the aluminum canisters to collect gas samples and monitor outburst potential from coal mines. However, the USBM direct method of gas measurement eliminated the multiple canisters and replaced it with one canister. The advent of polyvinyl chloride (PVC) plastic pipes provided an alternative to aluminum canisters (Barker et al., 2002; Moore et al., 2004). The basic difference between the aluminum and PVC plastic pipe canisters is the cost with the aluminum more costly to manufacture. PVC plastic canisters are easy to make with readily available materials. The USGS-BLM project in the Powder River Basin used (see Figure 6.17) plastic containers to desorbed gas, which was constructed as follows: (1) cut canister from 11.4 cm outside diameter body schedule 40 PVC using 30.5 cm chop saw with fine tooth blade; (2) PVC cap is drilled and tapped with a 0.6 cm NPT pipe thread; (3) cap is glued onto the cut pipe using PVC cement; (4) after the cement cures, a self-closing quick connect and thermocouple well are threaded into the tapped holes in the cap using Teflon tape as a sealant on the thread surfaces; (5) plastic ring with larger diameter to fit the bottom of the canister as a lip is cup and glued at the base; and (6) an appropriate plug with pressurized (5 psia) rubber ring is used as a base.

Plastic pipes are widely available even in remote locations and can be fabricated on-site. However, care should be taken to makeup the canister ahead of time so the solvents used in construction have evaporated to reduce or eliminate interference with measurements of gas composition (wetness).

Aluminum canisters are relatively heavy and bulky. The aluminum canisters are not easily fabricated in the field and most designs require welding. A significant advantage of aluminum is its high heat conduction leading to faster temperature equilibration in the water tank. A major disadvantage is that aluminum is reactive metal, easily corroded especially by the low ph water associated with coal. The low pH water may react and produce hydrogen gas and alter gas composition and isotopic results (Barker et al., 2002).

Gas Desorption Systems

Two major preparations must be met for gas desorption in the field and laboratory. First, a powered, temperature-controlled, wet-laboratory trailer is necessary for well site work (Figure 6.18(A)). The trailer must be equipped with water baths and heating systems as well as gas desorption measurement system or manometer (Figure 6.18(B)). Barker et al. (2002) designed a manometer to work with zero-head measurements of desorbed gas volume at ambient temperature and atmospheric pressure. The measurement system of Barker et al. (2002) uses a hand-held graduated cylinder, sliding reservoir tank, and hand-held reservoir modified from Diamond and Levine (1981), Diamond and Schatzel (1998), and Ryan and Dawson (1993). Diamond and Schatzel (1998) described examples of manometer apparatuses modified by Prada, Rodriquez, Flores, Fuentes, & Guzman, 2010 for multiple laboratory measurements of gas volume of Colombian coals. According to Barker et al. (2002) the quickest method to measuring gas volume at ambient atmospheric pressure is the zero head measurement. This is performed by manually lifting the hand-held graduated cylinder until the water levels in the reservoir tank and in the graduated cylinder are at the same level or height thus, at zero head. The canisters are placed in temperature-controlled water baths at reservoir or

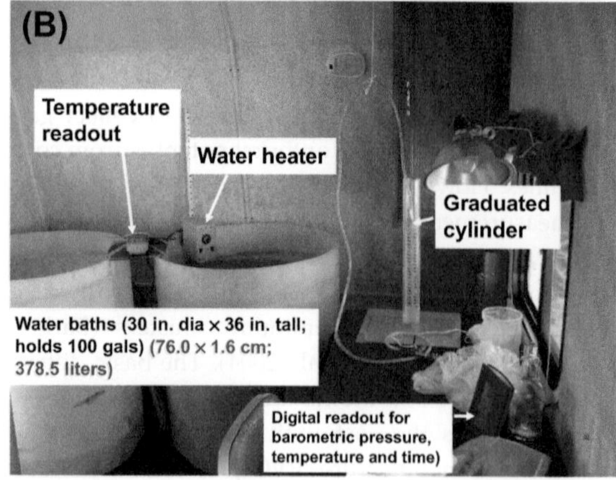

FIGURE 6.18 Field laboratory modified from a trailer (A), which include gas desorption equipment used by the USGS–BLM-WY in the Powder River Basin. *Source: Flores et al., 2006.*

near-reservoir conditions measured from water and oil/gas wells or extrapolated from thermal gradients (Figure 6.18(B)). However, Barker et al. (2002) suggested that there is a close correspondence between the temperature of the drilling mud and coal sample (e.g. cuttings, cores). These authors suggested that there must be a heat transfer from the drilling mud to the coal cuttings or cores. Thus, the temperature of the drilling mud must be the preferred temperature to use for gas desorption rather than water temperature in the coal reservoir measured or extrapolated by thermal gradient or from water and oil/gas wells.

Analytical Procedures

Desorbed gas measurements are made at timed intervals for 12 to 24 hours after being collected. The procedures used for desorption of coal cores in the canisters include (1) desorb and measure every 15 min during a 9-h period and (2) record the following parameters at each measurement: (a) time, (b) ambient temperature (atmospheric), (c) barometric pressure (atmospheric), (d) internal temperature of the canister, (e) temperature of water is occasionally measured, and (f) volume of gas. These measurements are recorded in a spreadsheet (Figure 6.19(A)) used to manage the gas measurements, data, and determine the volume of gas per canister in standard cubic feet (scf) or standard cubic meter (scm) (Figure 6.19(B)) (Mavor & Nelson, 1997). After the 9-h period, the coals in the canisters are continuously degassed progressively every 30 min, 1 h, 3 h, and 6 h depending on the gassiness of the coal. This desorption process takes place on-site in a CBM wet laboratory for a few days until such time that the canisters are ready to transport to the USGS laboratory facilities in Denver, Colorado. In Denver, additional measurements of gas (degassing) are progressively performed every day, few days, and week until all the coals in the canisters are bled free of their gas. It takes as long as 3–4 months to completely degas all the canisters per well. When all the coalbed gas in the canisters is desorbed each canister is weighed and filled with distilled water and re-weighed.

The corrections of the raw gas desorption volume for effects of gas expansion and contraction in the void volume in the canister as well as corrections of gas volume from ambient temperatures and pressure to standard conditions are calculated using equations reported by Mavor and Nelson (1997). The reader is referred to this report for exceptional descriptions of the methodology and calculations because the mathematical equations are not reproduced in this book. The results of using the equations are best demonstrated in a simplified spreadsheet for reporting collected raw gas volume data and calculations of gas content (see Figure 6.19).

Canister Headspace Measurement Procedure

Headspace is the open space between the coal core and the roof of the cap of the canister as well as spaces around and spaces of internal spaces crevices. Fully filling the canister with distilled water and comparing it with weight of the canister without water plus weight of the coal measures the headspace. However, as part of the procedure by some laboratories and field operators use water (e.g. tap, distilled, coalbed gas produced) on the well site before gas desorption. There are advantages and disadvantages of this procedure and this author do not recommend this procedure. The benefit of using distilled water to fill the headspace is that it keeps the coal moist, purges gas and air in the headspace and internal spaces; thus, minimizes any reaction and degradation of the coal. However, it optimizes the volume of water-soluble gases, such as CO_2. Barker et al. (2002) suggested using helium to purge the headspace of air. However, this procedure also has the following disadvantages: (1) compressed helium tanks are heavy

(A)

	Time		*Uncorrected Data*		*Measurement Conditions*		
Date & Time	Desorption Time	Square Root of Desorption Time	Incremental Desorbed Volume	Desorbed Volume	Canister Temperature	Ambient Temperature	Ambient Pressure
mm/dd/yy hh:mm:ss	hours	hours*0.5	cc	cc	Deg. F	Deg. F	Inches Hg
8/13/99 9:15	0.999	0.999	0	0	71.1	72.5	26.02
8/13/99 9:30	1.249	1.118	72	72	70.7	72.5	26.02
8/13/99 9:45	1.499	1.224	75	147	72.3	74.7	26.02
8/13/99 10:00	1.749	1.322	71	218	74.1	77.5	26.03
8/13/99 10:15	1.999	1.414	68	286	75.0	78.6	26.03
8/13/99 10:30	2.249	1.500	68	354	76.6	80.1	26.03
8/13/99 10:45	2.499	1.581	57	411	77.2	76.1	26.03
8/13/99 11:00	2.749	1.658	56	467	77.7	76.3	26.03
8/13/99 11:15	2.999	1.732	53	520	79.3	80.2	26.03
8/13/99 11:30	3.249	1.802	49	569	81.0	84.0	26.02
8/13/99 11:45	3.499	1.871	47	616	81.7	84.2	26.02
8/13/99 12:00	3.749	1.936	44	660	81.0	79.9	26.01
8/13/99 12:15	3.999	2.000	38	698	81.1	81.5	26.01
8/13/99 12:30	4.249	2.061	32	730	81.5	82.6	26.01
8/13/99 12:45	4.499	2.121	31	761	82.6	84.7	26.00
8/13/99 13:00	4.749	2.179	34	795	82.2	81.7	26.00
8/13/99 13:15	4.999	2.236	34	829	81.9	81.7	26.00
8/13/99 13:30	5.249	2.291	33	862	82.0	80.2	25.99
8/13/99 13:45	5.499	2.345	32	894	82.9	84.9	25.98
8/13/99 14:00	5.749	2.398	30	924	82.9	84.4	25.99
8/13/99 14:15	5.999	2.449	28	952	82.4	80.8	25.98

(B)

Air-Dry basis results for DDH-16: canister 11		
Parameter	units	Value
Intercept at Time Zero	scm/mt	-0.04
Slope	scm/mt-hour*.5	0.04
Lost Gas Content	scm/mt	0.040
Measured Gas Content	scm/mt	0.543
Residual Gas Content (Estimate)	scm/mt	0.364
Total Gas Content	***scm/mt***	***0.947***
	scf/t	***30.333***
Diffusivity	1.077E-07 per second	
Spherical Sorption Time		171.90 hours
Sorption Time (63.2% of total gas has desorbed)		406.50 hours

FIGURE 6.19 Spreadsheet used for data to be recorded, collected, and managed, which are summarized in a table for the lost, measured or desorbed, and residual gas as well as other information during desorption of a subbituminous coal in the Powder River Basin (Stricker et al., 2006).

and cannot be transported on passenger planes as baggage by US law regulation, (2) not readily available, (3) only low pressure helium can be used with safety in plastic-plugged PVC canisters, getting a vigorous flow of gas the valves of the canisters.

A disadvantage to water-filled headspace is that water promotes biological growth and chemically reactive. For example, it was known early in the testing for gas in the Powder River Basin that operators add water in the canisters to minimize oxidation. Our experiments tested this process and found that during measurements of gas for a period of time there was a spike in gas volume (Stricker & Flores, 2002). This was interpreted to be a methanogenic spike, which was determined as a significant part of methanogenic pathway of generation of biogenic gas in the Powder River Basin (Flores, McGarry, Stricker, 2005; Chapter 4). Thus, as water in the canisters promotes methanogenesis in low rank coal, carefully filling the headspace with water replacing empty space in the coal and canister is still a beneficial procedure to measure water mass and headspace as long as the procedure is done at the end and after the gas measurements when water is withdrawn as rapidly as it is filled. No lasting effects to the coal quality are manifested by this procedure. With just using distilled water for headspace measurement, it is possible to calculate easily the apparent sample density, once the headspace volume in the canister is measured.

Filling the headspace with water and measuring the mass of the water added is a common procedure. However, according to Barker et al. (2002), the headspace volume is assumed equal to the weight of the water added to the canister converted to volume by assuming that 1 gram (g) of water = 1 cubic centimeter (cc) of headspace volume. These workers suggested purging with helium reduces microbial activity and that adding water is only required if the headspace is not purged by helium. Measurement of coal density of a low-ash bituminous coal using the water-filled headspace method is exemplified according to Barker et al. (2002) of the following:

Measured by water addition: canister headspace = 1948g

Total canister volume = Empty canister filled with water − canister weight = 2762 g

Therefore, within this canister, coal volume is equal to (2762 g − 1948 g) × (1 cc/g) = 814 cc

$$\text{Canister} + \text{coal mass} = 2122\ \text{g};$$

$$\text{Canister mass} = 1028\ \text{g}$$

$$\text{Coal mass} = 2122\ \text{g} - 1028\ \text{g} = 1094\ \text{g};$$

$$\text{Coal density} = 1094\ \text{g}/814\ \text{cc} = 1.34\ \text{g/cc}$$

The above measurements plus the weight of the empty canisters are used to calculate the headspace volume and density of the coal in each canister. These measurements plus the desorption data collected on-site and in the Denver laboratory, in turn, will be used in the final calculations of the gas content in cc/g or scf per ton of coal.

The calculations will utilize the Direct Method determination of gas content, which was developed originally to estimate the gas content of coals to be mined (Kissell, McCulloch, & Elder, 1973; Diamond, Schatzel, Garcia, & Ulery, 2001; Australian Standard, 1999). The USBM Direct Method determines the most accurate estimate of lost gas volume (gas lost from the coal sample during the elapsed time from the start of desorption to the time that the canister is sealed) compared to other methods. Thus, the total gas content for each coal in the canister will be estimated using the Direct Method lost gas procedure plus the measured gas content (gas volume released from the coal sample during desorption measurements divided by the coal mass) and residual gas content (volume of gas remaining in the coal sample at the end of canister desorption measurements divided by the coal mass). The USGS has determined that the residual gas content for all the Powder River Basin coals is approximately 2.5% of the total gas content, although other analyses contain more residual gas depending on the various laboratories employed. Using coal bed reservoir GIP analysis for a specific purpose of gas resource assessment follows these calculations.

Types of Gas Measurements

The commonly used analytical methods for measurement of total gas content include determination of lost, measured or desorbed, and residual gases. Each of these measurement types is measured separately and summed up estimate the total gas content of the coal. The works of Mavor and Pratt (1996), Mavor, Pratt, Britton (1994), and McLennan, Schafer, and Pratt (1995) present more comprehensive analyses and methodologies of these types of gas measurements.

Lost Gas

The lost gas is volume of gas that cannot be directly measured because it is desorbed during drilling, coring, travel to the surface, and preparation of the coal sample before sealing the canister. Lost gas makes a part of the total gas content that escaped from the cores or cuttings (e.g. small particles) caused by the gas diffusion, hydrostatic pressure variability, the time of retrieval, and depth of the borehole. Waechter et al. (2004) suggested that sample preparation and drilling/coring methodology also are important factors to lost gas. Lost gas is extrapolated from the rate of gas desorption of the coal in the pressurized canister during the early stages of desorption (e.g. first 9 h). Extrapolation is made from the beginning of

BOX 6.2

SIMILARITY BETWEEN COALBED AND SHALE GAS: CONCEPT AND MISCONCEPTION

Worldwide the resource potential of shale gas is about 56% more than coalbed gas (Kawata & Fujita 2001). These gas resources have characteristics more in common than not. The concepts of coalbed and shale gas both as source and reservoir of the gas are similar. Both source rocks generate thermogenic and biogenic gas and the conversion process of the gas from the organic content is similar. However, the composition differ in that the coal source rock contains more than 66% total organic matter and the shale source contains less than 33% total organic matter. The gas is adsorbed with the coal source adsorbing more gas in the matrix (e.g. 90%) than in the shale source in which more free gas in the fractures is adsorbed. The depositional origin of shale in deep marine environments is in contrast to the subaqueous freshwater and/or brackish water settings of coal. The depositional setting of the shale is more dispersed over a larger area than coal.

The methodologies used to estimate gas resources and reserves are generally related for both coalbed and shale gas. GIP estimates may be determined from total gas content determined from gas desorption. Also, storage capacity of shale gas is determined from adsorption isotherms. However, Hartman et al. (2008) suggested that due to the difference in the quantity and quality of inorganic and organic composition, the mineralogy of the shale gas impacts on the gas adsorption beyond the total organic content of the rock. As a result methods used for both reservoirs may not satisfactorily translate due the more complex mineralogy and lithology.

gas desorption during coal sample recovery to time zero or beginning of desorption (Figure 6.20). Perhaps, the most important factor that controls lost gas estimate is the technique of extrapolation to time zero, which is interpretive. It varies from operator to operator and it varies where the operator places that line of extrapolation or linear fit on the concave portion of the gas curve. Barker et al. (2002) suggested that time zero is the most critical factor in determining lost gas, which they interpret as related to the exchange of heat of the drilling mud to the coal samples.

Kissell et al. (1973) proposed two procedures to follow in estimating time zero and cumulative lost gas time, which may account for the drilling medium. These workers suggested that if the drill hole is cored with water or drilling mud, desorption is assumed to start when the coal core was halfway to the surface. The cumulative lost gas time (tlg) would then be expressed as (Diamond et al., 2001):

$$\text{tlg} = (t_4 - t_3) + (t_3 - t_2)/2$$

where:

t_2 = time core retrieval starts,
t_3 = time core reached surface, and
t_4 = time core sealed in desorption canister.

If the hole was cored by air or mist, pressure release and gas desorption were assumed to begin at the first penetration of the coal bed by the core barrel. In this case, tlg would be expressed as (Diamond et al., 2001):

$$\text{tlg} = t_4 - t_1,$$

where:

t_1 = time coal bed first penetrated and
t_4 = time core sealed in desorption canister.

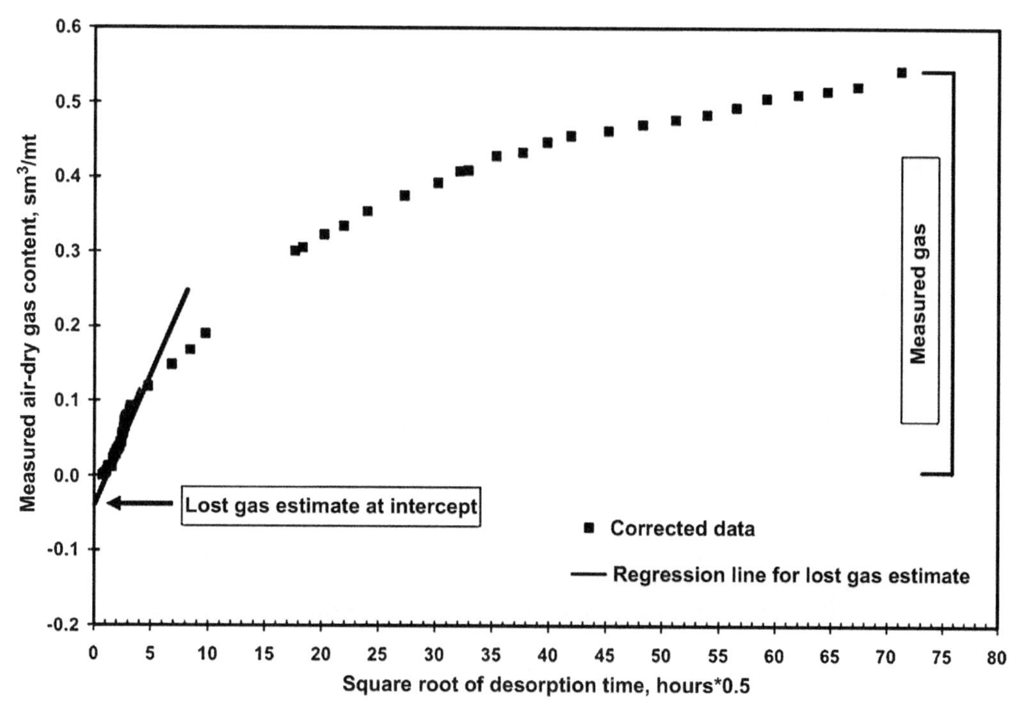

FIGURE 6.20 Direct method graph showing cumulative gas measurements during desorption through time of a subbituminous coal in the Powder River Basin (Stricker et al., 2006). Straight line is fitted to the gas desorption data collected during the first 9 h for lost gas. Gas continuous to be desorbed until it is depleted as measured gas. sm^3/mt, standard cubic meter per metric ton.

Desorbed or Measured Gas

Bertard, Bruyet, and Gunther (1970) originally described desorbed gas or desorption of gas accumulations in the sealed canister as directly measured (first introduction of the concept of Direct Method) from variations of water displacement. Also, Bertard et al. (1970) observed the rate of gas desorption is relative to the square root of time for the first 20% of total gas volume. Kissell et al. (1973) later applied and modified the Direct Method to deep unmineable coals in boreholes. The purpose of this study is to predict coalbed gas emission as applied in front of underground coal mining. Schakel, Hyman, Sainato, and LaScola (1987) further refined the Direct Method by measuring pressure differentials of canisters as gas is desorbed by utilizing ideal gas law (Boyle's Law: $P_1V_1 = P_2V_2$; where, P_1 is the initial pressure, V_1 is the initial volume, P_2 is the final pressure, V_2 is the final volume) to determine the desorbed gas volume, which has evolved as a more accurate U.S. Bureau of Mines measurement of low volume gas desorbed from oil shale.

Residual Gas

Gas volume and desorption from the coal in the canister slowly declines with time. Extended gas desorption measurements as gas decline are terminated at an arbitrary low rate of desorption. This rate varies from coal sample to sample depending on the size of the samples. Small coal samples (e.g. drill cuttings), which diffuse faster reach a low rate of desorption in a matter of days in contrast to a blocky and continuous coal samples (e.g. coal cores), which diffuse slowly reach low rate of desorption in several months (e.g. 6 months in the Powder River Basin coals). Some volume of gas or residual gas remains in the coal when the direct measurement of gas is terminated. Residual gas has been thought of trapped as gas in the coal molecular structure; however, Bertard et al. (1970) and Levine (1992) suggested residual gas is that which remains in the coal in equilibrium under one atmospheric pressure of methane in the canister. A graphic method of calculating residual gas called the Decline Curve Method was introduced by McCulloch, Levine, Kissell, and Deul (1975). Diamond and Levine (1981), and Diamond, LaScola, and Hyman (1986) suggested a method of measurement of residual gas by crushing coal into small particles by using steel balls while measuring the desorbed gas much like direct method (Figure 6.21). Analysis of U.S. coals by Diamond et al. (2001) shows that residual gas comprises 40–50% of the total gas content in high volatile-A bituminous, blocky coal beds. In comparison medium to low volatile bituminous coal beds typically contain less than 10% residual gas.

Direct vs Indirect Measurement Methods of Gas Content Determination

The total gas content may be measured by direct and indirect methods. Other specialized methods that may fall under both methods as decline curve, Australian, Amoco, and CSIRO methods are devised for use by the industry to speed up the procedure (Moore, 2012; Australian Standard, 1999).

Direct Methods

Bertard et al. (1970) developed the original direct method to determine gas contents from drill cuttings from horizontal wells in underground coal mines. The methodology was based on the assumption that gas released during desorption was proportional to the square root of time. The analysis resulted in the determination of the volume of gas lost from the moment when the coal cuttings were drilled from the coal bed to when it was moved into a container. The lost gas was estimated using a formula and the gas measured by desorption meter or U-tube manometer, which was connected to a tube into a sealed glass flask containing the coal samples. The emphasis in this methodology is rapid

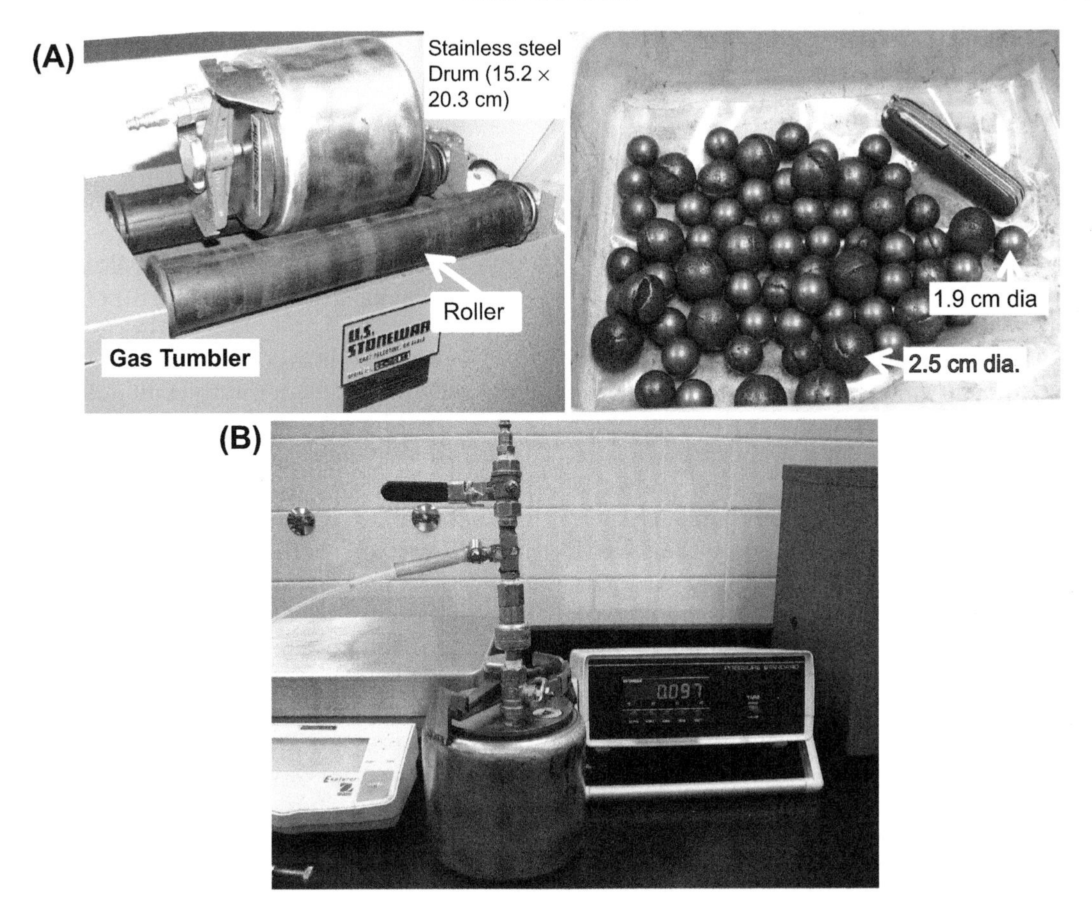

FIGURE 6.21 Equipment used to measure residual gas from crushed coal varying in size from particles to fragments. Coal is crushed by steel balls in a stainless steel drum on a tumbler (A) from which desorbed gas is measured by digital pressure equipment (B). *Source: Photographs courtesy of Gary Stricker.*

recovery and sealing of the coal samples in the container to minimize lost gas, which Bertard et al. (1970) considered suitable as long as it is less than 20% of the total gas content. After determination of the lost and measured gas, the coal sample was crushed to measure the remaining gas volume using a graduated cylinder suspended over a pan of water into which gas released by crushing was collected. The lost gas plus measured gas plus residual gas is the total gas content of the coal expressed in cc/g or scf/t.

The direct method was developed by the U.S. Bureau of Mines (USBM) in the early 1970s to measure gas content in coal mines for general safety applications (Diamond et al., 2001). The USBM direct method was modified from Bertard et al. (1970) methodology but adopted for core coal samples. In addition, the coal sample was collected in standard aluminum canister (0.3-m long and 10.2 cm diameter) for gas desorption. Gas volume was measured by periodically releasing the pressure resulting from accumulated gas in the canister into a water-filled graduated cylinder. The lost gas, measured or desorbed gas, and remaining or residual gas were added to make up the total gas content (Diamond et al. 1986). According to Diamond et al. (1986) the lost gas volume (Q_l) was estimated from several direct measurements of the early gas desorption immediately after the coal was sealed in the canister, which were made every 15 min for up to 9 h. The gas content data are plotted on a graph of cumulative desorbed gas against the square root of time of desorption. The estimated lost gas volume is then extrapolated from the initial linear portion of the desorption curve through the point on the x-axis (see Figure 6.20). Ulery and Hyman (1991) suggested improvement of the USBM direct method by making notes on gas temperature and ambient pressure to correct gas measurements to standard temperature and pressure (STP).

The measured or desorbed gas (Q_m) is calculated periodically using the water displacement procedure with the water-filled graduated cylinder. After an extended period of time gas desorption reaches very slow rate with longer periods of time of measurement. Recommendation for termination of gas desorption varies from (1) daily rate of less than 0.05 cm^3/g by McCulloch et al. (1975), (2) average rate of 10 cm^3/g daily for one week period by Diamond and Levine (1981) to 0.05–10 cm^3/g per day for one week period by Ulery and Hyman (1991). After calculations of measured or desorbed gas are discontinued, the residual gas (Q_r) is analyzed by crushing as much of the desorbed coal sample to a powder (−200 mesh) to release the remaining in situ gas (Diamond & Levine, 1981). The volume of gas released is measured by the water displacement method after the sample cools to ambient temperature. The gas volume released by crushing is attributed only to the portion of the sample crushed to a powder. The total gas content Q_t of the sample is then estimated according to Diamond and Schatzel (1998) expressed by the following equation:

$$Q_t = (Q_l + Q_m)/M_t + Q_r/M_c$$

where:

Q_l = lost gas volume.

Q_m = measured or desorbed gas volume.

Q_r = remaining or residual gas volume.

M_t = total air-dried mass (weight) of the coal sample.

M_c = air-dried mass (weight) of the coal sample crushed to a powder in the ball mill.

The USBM modified the direct method in the early 1980s as a result of assessment of low volumes of gas adsorbed within the organic matter associated with oil shale. The original direct method was not a sufficiently precise procedure to measure of the low gas content of oil shale such that a new technology was developed to more accurately measure the gas. Also, the new direct method increased the accuracy for estimation of gas content of coal with measurement of differential pressure buildup between readings and by the ideal gas law to calculate the volume of gas released into the desorption canister (Diamond et al., 1986). The new methodology abandoned the water displacement procedure and relies on an apparatus with pressure transducers and digital readouts. This eliminates problems of water-soluble gas and headspace discrepancies.

The versatility of the direct method is that gas content of coal samples can be measured using whole coal samples from wireline conventional cores and cuttings or crushed samples. Whole or core coal samples are used for prolonged desorption technique and crushed samples are used for quick desorption technique. The prolonged desorption technique is mostly utilized in the United States to determine coalmine methane to predict outburst, coalbed gas resource assessment (e.g. GIP), and basic research on reservoir properties (Diamond & Schatzel, 1998). The whole or core coal samples are usually obtained from boreholes drilled on the surface or through the coal mine face. The prolonged desorption method allows the release of gas to a very low rate of desorption, which often takes from weeks to months (Stricker et al., 2006). Thus, results of gas content measurements are often long in coming, which is not amenable to the quick results demanded by the industry. However, Stricker et al. (2006) devised a technique by which in two week preliminary results are provided as soon as the gas content (lost plus part of measured) curve is plotted and extended by a predictive formula. Stricker et al. (2006) recommended that the cutoff rate of desorption is 0.003 cc/g (0.09 scf/t). After prolonged gas desorption, the coal sample is then crushed into small granules in order to release rapidly residual gas.

Instead of waiting for an extended period of time to obtain results of gas content measurements, the coal samples (e.g. core, cuttings) are crushed for rapid analysis, which is commonly used in Australia and elsewhere (Saghafi & Williams, 1998; Saghafi, Williams & Battino, 1998; Moore, 2012). The measured gas content and estimated total gas content is available within days. Thus, the quick-crushing method is suitable for coal mine applications and sometimes coalbed gas resource assessment by the industry. Sawyer et al. (1987) indicated that only 63% of the total gas content is desorbed with the quick-crush method because the lost, measured, and residual gas are difficult to determine completely. Saghafi and Williams (1998) and Saghafi et al. (1998) indicated the accuracy of measurement of gas content of coal using rapid crushing technique. The rapid crushing technique accelerates the rate of gas desorption from coal, which allows measurement of gas content in hours instead of days or weeks as necessary for the traditional method of desorption (Saghafi et al., 1998). The quick crushing technique is used when gas content estimates are needed for either prevention of mine outburst or immediate calculation of coalbed gas reserves.

Smith and Williams (1981) devised a direct method procedure to resolve the problem of accurately measuring the total gas content of coal cuttings from rotary-drilled boreholes. The lost gas component of the total gas content was central to the water displacement measurement procedure. Smith and Williams (1981) assumed that no gas remains in the coal cuttings at the end of desorption period; thus, residual gas is not determined. Due to the speed of gas desorption from coal cuttings, final results are generally obtainable in one week.

Indirect Methods

Kim (1977) indirectly estimated gas contents from adsorption isotherm data. Adsorption isotherms measure the optimum adsorbed methane at varying pressures and at constant temperature (Chapter 4). Thus, an adsorption isotherm is a measure of the maximum methane sorption or idealized storage capacity of the coal. Mavor and Nelson (1997) suggested if the in situ storage capacity is equal to the gas content if the coal is saturated with gas. However, most coal reservoirs are undersaturated with gas; thus, the gas content will be overestimated in relationship to the adsorption isotherm. This assumption is particularly correct in coal beds at shallow depths. Nevertheless, Mavor and Nelson (1997) and Mavor, Pratt & Nelson (1995) suggested that the adsorption isotherms should serve as useful checks for gas contents, which should be much lower than the adsorbed gas.

Another indirect method of estimating gas contents includes using wireline logs (Diamond & Schatzel, 1998; Mullen, 1988, 1991; Hawkins et al., 1992). Scholes and Johnston (1993) defined wireline logs as measurements of one or more physical properties of rocks as a function of depths. The metal cable or wireline contains electrical conductors in the core that powers the downhole tool and communicates the toll signature back to the surface (Scholes & Johnston, 1993). The signatures are displayed as geophysical logs such as resistivity, gamma ray, acoustic or sonic, neutron, and density (Scholes & Johnston (1993). The tool does not measure the gas content of the coal beds in the wellbore but rather the logs are utilized as empirical estimation curves. Diamond, Murrie, and McCulloch (1976) and McFall et al. (1986) determined gas contents from empirical estimation curves of measured gas content results plotted against depth and apparent rank of coal beds. These workers used algorithms relating gas content to coal rank derived from actual total gas content and to proximate analysis of coal cores collected in the study area. The log-derived coal property data are then used to estimate gas content in relation to coal rank and depth.

Recent workers have proposed an indirect method of estimating gas content using geophysical logs, which was reviewed by Saghafi, Hatherly, and Pinetown (2011). The premise of this concept is that coal consists of organic and inorganic composition and that the gas content is related to the abundance of these components. Saghafi et al. (2011) suggested that the vertical and lateral changes of the inorganic component of the coal might respond to geophysical logs. Thus, the changes of the log responses might provide an estimate of the gas content of the coal. The log responses might be typified by the density log but Saghafi et al. (2011) indicated that using density log alone not in conjunction with other logs (e.g. gamma ray, sonic or induction, photoelectric absorption), which also respond to variations in the inorganic content of the coal, is an ineffective method. Saghafi et al. (2011) suggested that more experimental and theoretical work is required before the relationships between the various geophysical logs and gas content can be established (Chapter 7).

Bowler and Tedesco (2004) proposed that bituminous coals have a low bulk density and are easily identified. Based on high volatile B and A bituminous coal cores in the U.S. Illinois Basin, Illinois, and Forest City Basin, Kansas gas desorption and adsorption measurements in the laboratory can be correlated to low bulk density permitting an estimate of both adsorbed and desorbed gas volumes in coal. The relationship of maceral composition (e.g. vitrain-rich or woody, fusain, attritus) versus clay partings in a subbituminous coal in the Powder River Basin to a density log may yield more sensitive responses than originally thought (Figure 6.22). A frequency diagram of measured coal densities of 590 coal cores of the subbituminous coal in the Power River Basin is shown in Figure 6.23. This shows that the coal density mostly occurs within a relatively narrow range between 1.25 and 1.3 g/cc.

Fu, Qin, Wang, and Rudolph (2009) use multivariate regression to establish The relationships between measured gas content and depth, as well as density, natural gamma, resistivity and sonic logs were established by Hou, Hong and Hao (2002) and Fu et al. (2009) conducted in various coalfields in China. Results were mixed from high correlation coefficients ($R^2 = 0.9$) to negative correlation. Bhanja and Srivastava (2008) estimated gas content for Jharia coalfield in India from sonic, density, photoelectric absorption index (Pe), and gamma ray logs. These authors asserted that the estimates of gas content of bituminous and lignite coals determined from the logs were calibrated with gas data from the studied coalfield. Bhanja and Srivastava, 2008 had proposed the following empirical equation to estimate the methane gas content of coal using sonic, density, gamma ray and Pe logs for Jharia coalfield:

$$V = 0.767\ \{\Delta t/(\rho c \times \mathrm{GR} \times \mathrm{Pe})\} + 10.67$$

where:

V = Volume of gas (cc/gm).

Δt = Sonic log response for coal = 120 μs/ft (for Jharia coalfield).

ρc = Density log value for coal seam (g/cc).

GR = Gamma ray response for coal (cps).

Pe = Photo-electric absorption index value for coal (1.8 b/electron for Jharia coalfield).

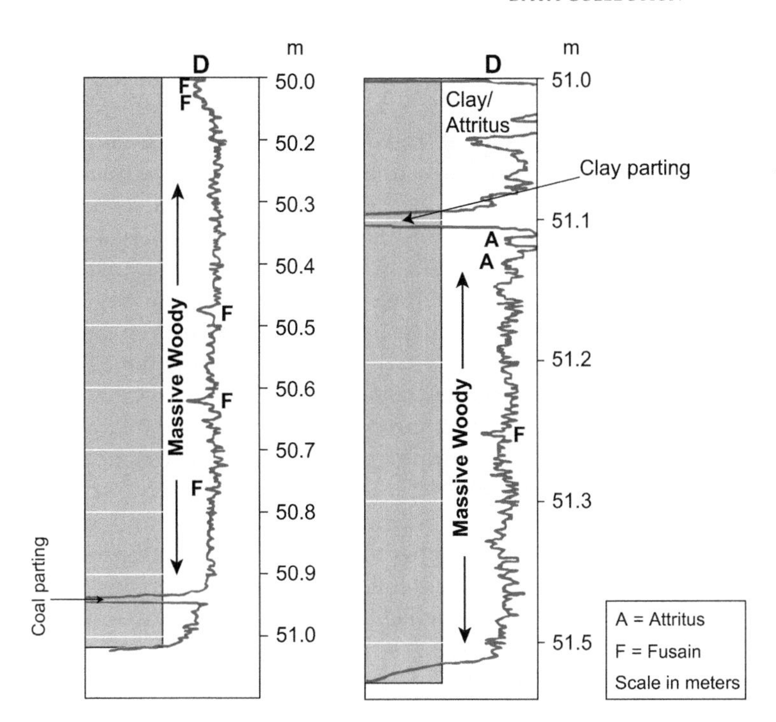

FIGURE 6.22 Continuous core section of the Wyodak-Anderson subbituminous coal in the Power River Basin, Wyoming with a correlative density log. Coal maceral and inorganic partings control the behavior of the log. D, density.

Specialized Methods

The specialized methods such as the decline curve, Amoco, CSIRO and Australian methodologies of measuring gas are proposed for the use of the gas industry. Chase (1979) proposed the decline curve method to estimate gas volumes, which are desorbed from exploration coal cores. The method is unique in that it eliminated the measurement of residual gas content because this part of the total gas content does not impact production from degasification or coal mining. Consequently, the lost and measured of desorbed gas parts of the total gas content were estimated. According to Diamond and Schatzel (1998) the decline curve method requires plotting on a semi-log graph paper the volume of gas release rate in cc/g per week against cumulative volume of desorbed gas in cm^3 in order to generate a linear trend with least-squares regression analysis (Figure 6.24). The lost gas

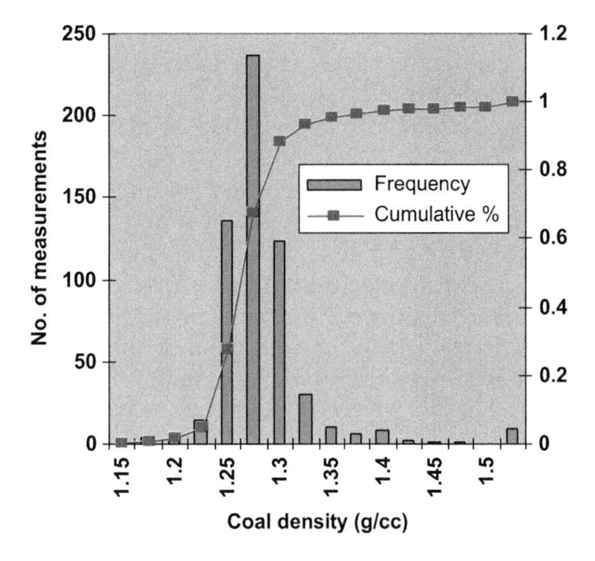

FIGURE 6.23 Frequency diagram of the density of subbituminous coals from 590 cores collected in the Powder River Basin, Wyoming constructed by Dwain McGarry of BLM-Wyoming. g/cc, gram per cubic centimeter.

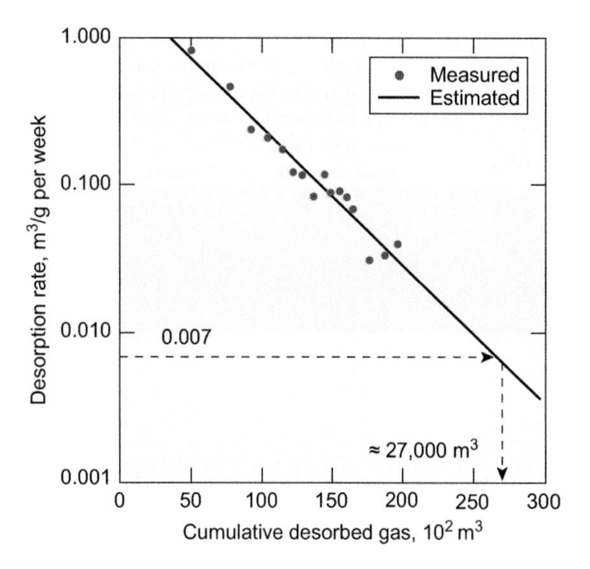

FIGURE 6.24 A graph showing a decline curve to estimate the volume of desorbed gas. *Source: Adopted from Diamond and Schatzel (1998). cm^3/g = cubic centimeter per gram.*

part of the total gas content is determined by water displacement much like the USBM direct method. The volume of measured or desorbed gas part of the total gas content is determined from the decline curve (Figure 6.24) by using a selected desorption rate (0.001 cc/g per day or average of 0.007 cc/g per week) to estimate complete desorption (Diamond & Schatzel, 1998). In order to determine the predicted volume of desorbed gas in cm^3 on the x-axis, the 0.007 cc/g per week desorption rate is projected to the regression line constructed from the measured desorption values (Figure 6.24). According to Diamond and Schatzel (1998) the total volume of desorbable gas is expressed by the equation:

$$Q_d = Q_l + Q_p$$

where:

Q_d = desorbable gas volume.
Q_l = lost gas volume.
Q_p = predicted desorbed gas volume.

The graphical procedure for estimating residual gas was introduced by McCulloch et al. (1975) as a decline curve method. Although this method eliminates the measurements of lost and measured or desorbed gas part of the total gas content McCulloch et al. (1975) still recommended that all the parts of the total gas contents (e.g. lost. measured or desorbed, and residual gas) need to be measured at standard temperature. In contrast the Amoco method is a non-linear curve-fit method to estimate lost gas from all the measured or desorbed gas data and is similar to the direct method (Yee, Seidle, & Hanson, 1993). The Amoco method fits a direct method equation to non-linear regression to all the measured or desorbed gas data and requires a computer program to apply the methodology (Mavor & Nelson, 1997). In contrast to the Amoco method Shtepani, Noll, Elrod, and Jacobs (2010) recommended a new method based on non-linear regression of measured or desorbed gas content, which offers a more accurate estimation of lost gas in tandem with adsorption isotherms.

The USBM direct method has been modified to fit specific applications by organizations in the United States, Australia, and other countries (Diamond & Schatzel, 1998). In the United States the USBM direct method was modified by the Gas Research Institute as a part of their coalbed gas research investigations and the USGS to assess the National and International coalbed gas resources (Barker et al., 2002; Flores, 2004; Flores et al., 2006; Mavor & Pratt, 1996; Mavor et al., 1994; 1995; Prada et al., 2010; Stricker et al., 2006). In Australia, the USBM direct method was adopted and modified by the Standards Association of Australia and the Commonwealth of Scientific and Industrial Research Organization for determining coalbed gas emissions from coal mines.

The Gas Research Institute modification to the USBM methodology is with the estimation of lost gas resulting in testing at ambient temperature (Mavor & Pratt, 1996). Mavor et al. (1994) during work for the Gas Research Institute found that the Amoco method estimate of total gas content averaged 21% more than the direct method. These workers determined that the error is solved and the Amoco method accurate when only the first half of desorption data was utilized. The USGS methodology significantly departed from the USBM method by using PVC plastic tubes canisters for replacement of the aluminum canisters. The PVC plastic canister was fitted with an internal headspace temperature measurement system and was constructed from off-the-shelf components at less cost and less effort to build in the laboratory and field conditions (Barker et al., 2002). Stricker et al. (2006) modified the PVC plastic canisters into 0.61 m long to accommodate desorption of thick coal beds in the Powder River Basin. The USGS canister design in turn was adopted and modified by the CRL Energy of New Zealand (Moore, 2012).

The Australian modification of the USBM direct method includes a unique change in the gas volume measurement using an acidified brine solution to prevent carbon dioxide going into solution in the water displacement system (Standards Association of Australia, 1991; Diamond and Schatzel, 1998). Lost gas (Q_1) is determined in the field much the same way as the USBMdirect method with time zero estimated when the core is halfway to the surface. The measured or desorbed gas (Q_2) is determined in the laboratory using two interconnected gas volume desorption apparatuses; the second is used when the desorption rate is low whereby the desorbed gas is allowed to free flow into the second apparatus. Gas desorption is terminated when the there is no observed change in the rate in a week or when the gas desorption curve (e.g. gas volume versus the square root of time) reaches asymptotic (Diamond & Schatzel, 1998). The residual gas (Q_3) may be measured from crushed coal samples. In addition to the Standards Association of Australia methodology the Commonwealth of Scientific and Industrial Research Organization recommended a quick-crushing gas content analysis applied to cores collected from underground mines (Saghafi et al., 1998). These workers indicated that this method eliminates or minimizes the problem of carbon dioxide going into solution in the water of the measurement system.

The modifications of the USBM direct method by academic, governmental, and industry users have led to improvement to the assessment of the total gas content, which includes the lost, measured or desorbed, and residual gas. However, the efficiency and accuracy of this assessment rely on the preparation and handling of the coal samples summarized by Testa and Pratt (2003). Most of all, accuracy of the assessment of total gas content is controlled by coal properties, which vary from coal to coal (e.g. rank, composition, permeability/porosity, etc.), coal geology which vary from coalfield to coalfield or coal basin to coal basin (e.g. depths, structure, distribution, etc.), and conditions during collections of the samples in the field (e.g. summer, winter, drilling equipment, drilling operators, desorption personnel, etc.).

COALBED GAS RESOURCE AND RESERVE ASSESSMENTS

The size of the gas supply in a given area depends on whether the area is in a coalfield, coal region, and coal province. Wood et al. (1983) defined the differences of these hierarchies of coal-bearing areas based on the number of coal beds and coal areas. A coalfield is a discrete area underlain by rocks with one or more coal beds. A coal region is an area containing one or more coalfields. A coal province is an area containing two or more coal regions. A coal basin, which is defined in a depositional and structural context, may contain either a coalfield or coal region. A coal province may contain one or more coal basins. The coalfield area is probably equivalent to the prospect-size area (10^5–10^7 m^2) and the coal province-basin area is comparable to the basinal or regional-size area (10^8–10^{10} m^2)

of Boyer and Qingzhao (1998). These workers suggested that the prospect size area is important to gas producers and the basinal or regional area is significant to national and international institutions.

Although the estimates of coal resources and reserves are keyed to the coal areas as defined by Wood et al. (1983), coalbed gas resources and reserves are often standardized using conventional oil and gas, and unconventional gas assessment guidelines in terms of Petroleum Systems. For example, assessment of coalbed gas resources and reserves in the Powder River Basin in Wyoming and Montana, United States, which is identified as a geologic-based province, assessed various Tertiary and Cretaceous coal-bearing and associated sandstone units as unconventional and conventional assessment units (Flores, 2004). In contrast, the gas industry in the Powder River Basin emphasized coal zones such as the Wyodak and Big George coals as resource and reserve units and recognized the coal zones as "plays." Magoon and Dow (1994) defined a play is one or more prospects and a prospect is a possible trap to be investigated for commercial accumulations of hydrocarbons in the context of Petroleum Systems. Elsewhere, in the Raton Basin, Colorado and New Mexico, United States the USGS assessed the Petroleum System in the province as combined conventional and unconventional assessment units (Higley, 2007). Thus, despite the sincere attempt to standardize methodology and terminology for resource and reserve assessments of coalbed gas in relationship to conventional oil and gas, the process remains fluid. Boyer and Qingzhao (1998) correctly noted that the distinct characteristics of coalbed gas resources and reserves prevent direct application of established assessment guidelines for conventional oil and gas.

BOX 6.3

UNDERESTIMATING AND OVERESTIMATING RESOURCES/RESERVES – FACT VS FICTION

It is a fact that evaluation of coal and coalbed gas resources can be underestimated or overestimated as a result of improper characterization of the source and reservoir rock. Past experiences of gas operators in the U.S. coal basins indicate that gas reserve estimates have not been reliable, which significantly affects the economic valuation and certification (Moore, 2012). That is, coalbed gas production does not match the GIP estimates. For example, produced coalbed gas volumes are either smaller or larger than the GIP.

The inaccuracy in estimation of GIP is controlled by heterogeneity of reservoir properties (e.g. permeability, porosity, coal rank and composition, coal geology and stratigraphy, etc.) and by gas saturation and pressure to flow the gas in the reservoir. Diamond and Schatzel (1998) indicated that an adsorption isotherm, which measures the storage capacity of gas, overestimates the actual gas content; thus, the reservoir is undersaturated with gas. According to Rushing, Perego, and Blasingame (2008) the largest errors of reserve estimates take place during the period of the first few years after peak production is attained and initial production declines. Using the Arps decline curve methodology, Rushing, Perego, and Blasingame (2008) determined that for coalbed gas wells with long-term hyperbolic behavior, the initial errors of reserve estimates are underestimated by 20–30%.

The accuracy of assessment of coalbed gas resources and reserves is controlled by various coal properties (e.g. rank, composition, etc.) that are very different from conventional oil and gas accumulations. The variability of coal properties as well as inherent lateral and vertical unpredictability of coal beds have overestimated or underestimated coal resources and reserves. For example the ash content and density of coal may have been inconsistently and incorrectly applied to the equation for resource calculations. Furthermore, the stratigraphic variation and distribution of coal beds and coal zones may have been misinterpreted due to splitting and merging behaviors. Most importantly, miscalculations of coalbed gas resources and reserves may have been caused by the misuse and lack of data. The availability and quality of data ranging from none in virgin areas to limited and good in producing areas significantly influenced the methodology used and quality of results in estimating coalbed gas resources and reserves. Thus, the assessment methodologies of coalbed gas resource and reserves include using (1) analogs in virgin areas, (2) GIP in areas of limited data, and (3) petroleum system utilizing gas production history in producing areas.

Gas-in-place Resources Using Coal Resources

The generally accepted definition of coalbed gas resource is volume of gas in the coal bed, which is adsorbed in the matrix and fractures, free state in the fractures, and dissolved in the reservoir water (Boyer & Qingzhao, 1998; Kuuskraa & Boyer, 1993; Meneley et al., 2003; Moore & Friederich, 2010; Potential Gas Committee, 1995; Scott et al., 1995). Zuber (1996) referred to this all-encompassing volume of coalbed gas in the reservoir as GIP. Mavor and Nelson (1997) redefined GIP as volume of gas stored within a specific volume of coal reservoir, which includes carbonaceous partings. Boyer and Qingzhao (1998), King (1990), and Zuber (1996) suggested two methods of estimating the volume of gas in the coal: material balance and volumetric.

Material Balance

According to Boyer and Qinzhao (1998) the volumetric assessment method is commonly and easily utilized and the material balance is least utilized as it involves knowledge of the coal reservoir properties such as permeability, composition, etc. Morad (2008) reported that material balance is application of mass balance to a producing reservoir with accompanying pressure reduction. The author suggested that the remaining GIP and OGIP can be determined by monitoring the cumulative gas production, average reservoir pressure, and using the PVT properties of gas. A number of material-balance equations have been developed that include adsorbed gas storage (Ahmed, Centilmen, & Roux, 2006; Clarkson & McGovern 2001; Jensen & Smith, 1997; King 1990, 1993; Seidle 1999; Walsh, 1994; Walsh, Ansah, & Raghavan, 1994), but the degree of complexity of the equations increases as free-gas (or compressed-gas) storage, water and pore volume compressibility, and water production and encroachment are accounted for. The method developed by King (1990) remains the most rigorous, although the equations may be difficult to apply in practice because of the need for iterative calculations. Since 1997, starting with Jensen and Smith (1997) work, approximations have been developed that ease the use of material balance for coalbed gas reservoirs, without necessarily sacrificing significant accuracy. Since then material balance methodology has advanced to demonstrating production characteristics and estimating gas reserves for a coalbed gas well (Hsieh, 2003). . More recent advances have analyzed producing coalbed gas reservoirs, which exhibited flow characteristics including single-phase flow of gas in dry reservoirs, single-phase flow of water in undersaturated reservoirs, and two-phase (gas and water) flow in saturated reservoirs (Clarkson. Jordan, Gierhart, & Seidle, 2008; Gerami, Pooladi-Darvish, Morad, & Mattar, 2007).

Volumetric

By definition, estimation of GIP theoretically has to include the following: (1) volume of gas stored in the pore system of the coal matrix, (2) volume of gas stored in the fracture or cleat system of the coal, and (3) volume of dissolved gas in the water that fills the fracture or cleat and pore systems, and (4) the volume of coal per unit area. Boyer and Qingzhao (1998) recommended that all these properties of the reservoirs must be measured before the estimation of the coalbed gas resources. The equation to determine the volume of gas or GIP is expressed according to Boyer and Qingzhao (1998) as follows:

$$G_i = Ah\left[\varnothing_f(1 - S_{wfi})/B_{gi} + C_{gi}\, p_c\right]$$

where:

G_i = gas resource volume or GIP (standard cubic feet per tons or standard cubic meter per metric ton).

A = study area (hectares, acres, square kilometer or square meter).

h = coal thickness (feet, inches, centimeters, meters).

$\varnothing_f$ = interconnected fracture porosity (fraction).

S_{wfi} = interconnected fracture water saturation (fraction).

B_{gi} = gas formation volume factor at initial pressure (cubic meter, gas per cubic meter of reservoir).

C_{gi} = initial in situ sorbed gas concentration.

p_c = pure coal density.

In order to standardize the equation in which the gas content and coal density are reported as dry, ash-free basis Boyer and Qingzhao (1998) modified the equation as follows:

$$G_i = Ah\left[\varnothing_f(1 - S_{wfi})/B_{gi} + C_{gi}\, p_c(1 - f_a - f_m)\right]$$

where:

f_a = average weight fraction of ash (fraction).

f_m = average weight fraction of moisture (fraction).

The GIP equation is further simplified to remove the free gas portion because it is negligible as follows:

$$G_i = Ah(C_{gi}\, p_c)$$

Some of the elements of the equation of Boyer and Qingzhao (1998) such as the interconnected fracture porosity and water saturation are very much dependent on acquisition of the properties from coal cores, specialized wireline fracture identification logs (e.g. combination of resistivity and laterolog, etc.). However, a note of caution is in order regarding fracture porosity system, which varies significantly laterally and vertically within a borehole, coal mine, and coal bed. Nelson (1999) described other heterogeneities that would contribute to variability of coal bed reservoirs. Moreover, the effects of shrinking of coal resulting from dewatering have to be accounted as it controls interconnectedness of fracture system (Chapter 5). Stricker, Flores and Trippi (2010)and Trippi et al. (2010) studied fractures of lignite to bituminous coals in the Powder River, Green River, and Williston Basins in the U.S. Rocky Mountain region from 37 coreholes totaling about 330 m thick collected from 37 gas operators. These workers observed that fracture systems in terms of cleat spacing and frequency are related to rank and maceral composition, which changes within and between coal beds. Thus, application of a single value of these parameters does not capture the entire variability of properties of the reservoir. Because of the high variability of the fracture system within a coal bed and in order to achieve high accuracy, this methodology requires measuring the reservoir properties in each. Thus, application of this formula for estimation of the GIP may be more rigid and labor intensive than other simplified methodologies.

Nevertheless, the value of the Boyer and Qingzhao (1998) equation is in maintaining the in situ condition of the reservoir; thus, coal quality analysis is performed on as-received basis. However, problems arise when the coal quality is reported as dry, ash-free basis. In an ideal situation when all the elements of the equation are available and applied the model will account for most of the volume of gas per unit weight or mass of the coal. The caveat is that there is no one-equation model can guarantee estimation of 100% of the in situ GIP of coal beds. Given that each of the methodologies has inherent inaccuracies from collection and handling of coal samples to desorption of gas and to estimation of GIP, the cumulative errors yield an incomplete assessment of coalbed gas resources and reserves. The best protection from methodological inaccuracies is to minimize the errors.

A variety of assessment equation models have been presented at about the same time and after the Boyer and Qingzhao (1998) model in order to simplify, modify and improve the GIP model (Chen, Yang, & Luo, 2012; Crockett et al., 2001; Dallegge & Barker, 2000; Kong, Irawan, Sum, & Tunio, 2011; Mavor & Nelson, 1997; Simpson, 2008). Simplification of the GIP equation model compared to the Boyer and Qingzhao (1998) equation is typified by the model proposed by Mavor and Nelson (1997) expressed by the equation:

$$\mathrm{GIP} = 1359.7\, Ah\overline{\rho}\overline{G}_c$$

where:

GIP = gas-in-place.

A = study or drainage area.

h = coal; reservoir thickness (meters).

$\overline{\rho}$ = average in situ coal density at the average in situ coal composition (g/m^3).

$\overline{G}_c$ = average total gas content at the average in situ coal composition (scm/mt).

According to Mavor and Nelson (1997) this GIP equation model is applied to groups of coal reservoirs in close proximity stratigraphically with similar rank. The equation is simplified compared to Boyer and Qingzhao (1998) to exclude interconnected fracture porosity saturation, ash and moisture contents, and initial sorbed gas concentration and initial pressure.

Other modifications and improvement to the GIP equation model are using total gas content in tandem with adsorption isotherms and burial history (Chen et al., 2011; Dallegge & Barker, 2000). The adsorption isotherm measures the ideal gas storage capacity of the coal at constant temperature (Chapter 4). Thus, given the total gas content measured by direct method at present reservoir temperature and reservoir pressure (e.g. hydrostatic) the maximum gas content can be estimated. Thus, the adsorption isotherm can be used to estimate the gas content, which in turn can be applied to the coal mass estimate (see equation below) of the coal zone. According to Dallegge and Barker (2000), the gas estimate can be improved by using the burial history of the coal reservoir. According to Dallegge and Barker (2000) the GIP is corrected for moisture and ash content of the coals as accounted by their experiments, which calculate the total gas content adjusted for the carbon content of the coal zone. The gas content determined from these analyses and methods can then be calculated from the coal mass to determine the GIP as follows (Dallegge & Barker, 2000):

$$\mathrm{GIP} = \mathrm{CM} \times \mathrm{GC}$$

where:

CM = coal mass.

GC = gas content.

$$\mathrm{CM} = Z \times A \times D$$

where:

Z = coal zone thickness.

A = drainage or study area.

D = coal density.

In order to determine the volume of gas generated in the study area, Dallege and Barker (2000) used burial history and kinetic models of gas generation (using BasinMod software). These workers assumed total gas adsorption of the source and reservoir coal. The GIP was estimated by multiplying the computed total gas content of the coal zone by the total coal mass.

Adsorption isotherms, which define the adsorbed gas, were also used by Simpson (2008) in estimating the original gas-in-place (OGIP) expressed as:

$$\mathrm{OGIP} = 0.031214\, AhV_{\mathrm{m}}y\rho(bP_{\mathrm{i}}/1 + bP_{\mathrm{i}})$$

OGIP = Original gas-in-place (scm/mt).

A = Drainage or study area (hectares, acres, km^2, mi^2).

h = thickness of coal (meter, feet).

V_{m} = total gas content (scm/mt, scf/t).

y = mineral-matter free mass fraction of total coal (fraction).

ρ = coal density (g/cc).

b = Langmuir shape factor (psi^{-1}).

P_{i} = initial reservoir pressure (psia).

According to Simpson (2008), the parameters of the equation are acquired during drilling, coring, and logging or from analogs. A unique value of the equation is potential estimation of residual gas by replacing P_i with present reservoir pressure. The other parameters do not change significantly according to Simpson (2008).

Most of the GIP or OGIP equation models can be further simplified by elimination of the coal area, density, and thickness parameters. More often in study or drainage area (e.g. coalfield, coal region or basin) in the United States and other countries, the coal resources already have been estimated by government agencies. For example, assessment of GIP in the Powder River Basin in Wyoming, United States, was made possible by previous estimates of coal resources (volume of coal) in conjunction with the 1999 USGS National Coal Resources Assessment Project (Ellis, et al., 1999). The USGS coal resource assessment methodology (see previous sections in this chapter) included calculations of coal density, rank, and quality as well as depths (e.g. overburden), quadrangle areas (e.g. 15 min-quadrangle, and land subsurface mineral ownerships (e.g. Federal, State, Private). Normally, coal resources have been previously assessed from coalfield to basin size in the United States and in other countries. Thus, the volume of coal in these coal areas in conjunction with total gas content desorbed from the coal can be used to estimate the GIP using the Powder River Basin equation model. The USGS and U.S. BLM cooperative work (Crockett et al., 2001; Flores, Stricker & Kinney, 2004) on estimations of GIP and recoverable resources in the Powder River Basin, Wyoming, yielded the following simplified equation.

$$\mathrm{GIP} = A \times B \times C$$

where:

GIP = gas-in-place (Scm/mt or Scf/t).

A = volume of coal (short tons, metric tons).

B = amount of gas per unit volume (Scm, Scf).

C = area underlain by coal (mi^2, km^2).

ASSESSMENT OF COALBED GAS AS A PETROLEUM SYSTEM

In recent years the petroleum industry has focused on guidelines for estimation methodologies and procedures of resources and reserves of conventional reservoirs, in general and unconventional reservoirs, in particular. The guidelines were published by the Petroleum Resources Management System (PRMS, 2011) and sponsored by the industry (e.g. Society of Petroleum Engineers, American Association of

Petroleum Geologists, World Petroleum Council, Society of Petroleum Evaluation Engineers, and Society of Exploration Geophysicists). Guides to deterministic procedures for estimating recoverable quantities of oil and gas using volumetric, production, and analogous performance analysis as well as probabilistic reserves and resources methodologies are discussed in detailed by Yongguo et al. (2008), Senturk (2011), and Swinkels (2011).

In the PRMS guidelines, estimation methodologies for unconventional resources, which require different approaches from conventional resources, were discussed by Chan (2011). The relationship between the conventional and unconventional resources is demonstrated in Figure 6.25 as the resource triangle by Chan (2011) and modified from Holdtich (2001). The tight gas and heavy oil formations sit astride the boundary between the conventional and unconventional resources. Clarkson and Barker (2011) discussed the methodologies of estimation of coalbed methane or coalbed gas part of the resource triangle, which is the primary interest in this book. These workers mainly applied methods developed for conventional reservoirs as follows: (1) volumetric, (2) material balance, (3) production data analysis, and (4) reservoir simulation.

The volumetric or GIP and material balance methods have been discussed in the previous sections. The production data analysis and reservoir simulation methods require acquisition mainly of production and reservoir data (e.g. gas/water production, flowing and shut-in pressures, etc.). The assessment method for coalbed gas in terms of a petroleum system as modified from conventional resources and utilized by the USGS and other assessors probably comes close to the production data analysis and reservoir simulation methods.

The concept of petroleum system is not a new concept but rather an old one that has been used all along by industry practitioners and resurrected for petroleum resource assessment for economic and commercial considerations. The TPS defined by Magoon and Dow (1994) as a natural system includes a source-rock pod and related oil and gas, which incorporate all geologic elements and processes necessary for the accumulation of the hydrocarbons. TPS was mainly designed for resource and reserve assessments of conventional oil and gas but was adopted and modified for unconventional

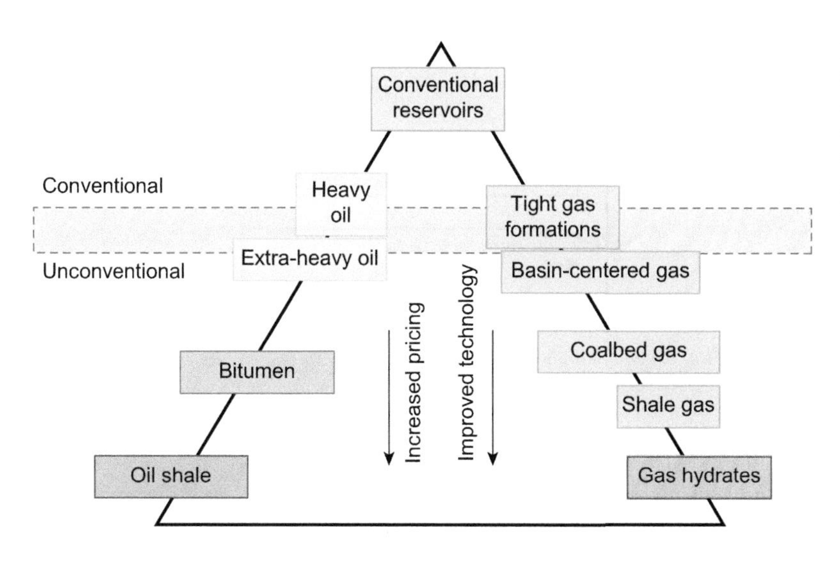

FIGURE 6.25 Diagram showing the triangular relationship between conventional and unconventional resources with emphasis on the unconventional resources. *Source: Adopted from Chan (2011) and Holditch (2009).*

gas particularly estimation of economically and technically recoverable quantity of coalbed gas (Attanasi & Rice, 1995; Rice, Young, & Paul, 1995; Schmoker & Dyman, 1996; Schmoker & Klett, 2002, 2007). According to Attanasi and Rice (1995) the coalbed gas assessment is based on the geologic elements of the TPS assessment distinct in the province, which include the geologic and stratigraphic frameworks, source and reservoir rock types, hydrocarbon generation, migration, trap and seal formations, and timing. The assessment is prepared to assess coalbed gas resources within an assessment unit (AU) consisting of a volume of rocks with similar characteristics and including discovered and undiscovered coalbed gas resources. The AU is organized to estimate continuous accumulation (e.g. coalbed gas) and/or conventional accumulation (e.g. sandstones as structural and stratigraphic traps) in which EUR, and gas and water production profiles of wells vary significantly across the assessment province (Attanasi & Rice, 1995). The AU was evaluated by the (tested) cells, which are determined by the spacing of coalbed gas wells permitted by the State. The TPS was analyzed in the context of probabilistic methods by Schmoker and Klett (2002), which is best applied in risk analysis for petroleum exploration ventures by Rose (2001) (Figure 6.26).

Case Study: TPS and Coalbed Gas Assessment

The USGS has assessed coalbed gas TPS in conjunction with the 2000 National Oil and Gas Assessment program in various coal basins in the conterminous United States and Alaska (Flores, 2004; Higley, 2007; Roberts, 2008). For example, the coalbed gas TPS in the Powder River Basin in Wyoming and Montana was assessed in 2000 consisting of the Tertiary-Upper Cretaceous

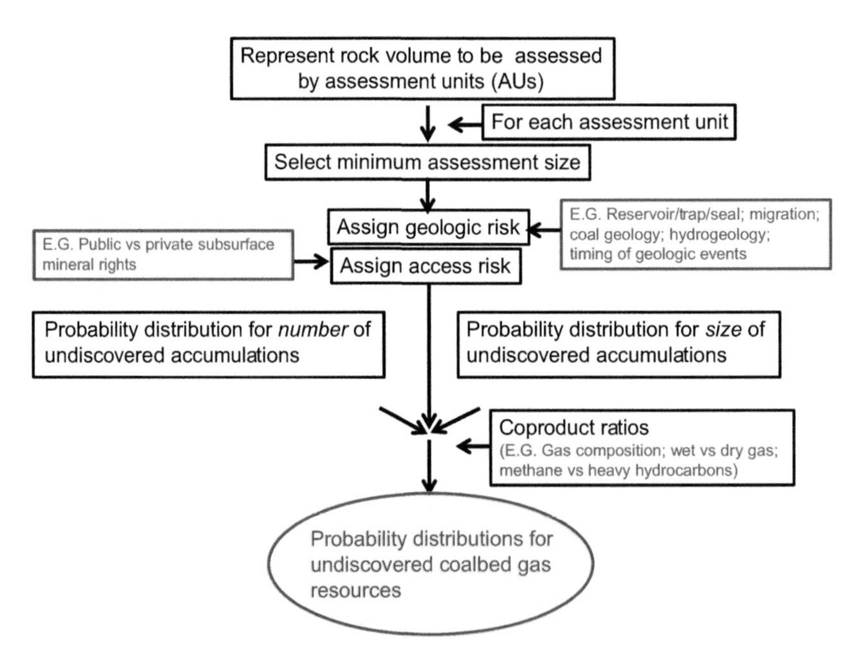

FIGURE 6.26 Flow chart showing the sequence of investigation using the TPS toward assessment of coalbed gas resources and reserves. *Source: Modified from Schmoker and Klett (2002).*

coalbed gas TPS consisting of four AUs (Flores, 2004). The AUs are based on the thickness and distribution of coal beds and the characteristics of associated fluvial channel sandstones. In three of the AUs the subbituminous coal beds served as source and reservoir rocks and in the fourth AU the coal bed served as source for gas, which migrated to the sandstones.

The coalbed gas development, which started in the 1980s (Hobbs, 1978), was just maturing when the USGS assessed the Powder River Basin in the year 2000 (Flores, 2004). Presently most of the coalbed gas production (e.g. 70%) is mainly from two coal zones or plays (e.g. Wyodak and Big George plays) in the upper part of the Paleocene Fort Union Formation (Flores, Spear, Kinney, et al., 2010, Flores, Spear, Purchase, et al., 2010)). The Wyodak coal zone (Wyodak-Anderson) is shallow coalbed gas play along the eastern margin (see Figures 6.10, 6.11 and 6.12) and the Big George (Wyodak-Anderson equivalent) is a deep coalbed gas play in the central part of the Powder River Basin. The number of evaluated (tested) cells as of April 2000 in the upper part of Fort Union Formation AU was 3090 of which 2449 have EUR (Figure 6.27). The EURs from these cells are from 1179 producing coalbed gas wells, which were completed at the end of 1999; however, only 638 wells at the time had sufficient production history to calculate EURs (Cook, 2004; Flores, 2004). The 638 coalbed gas wells were completed mainly from the Wyodak coal zone play along the shallow eastern margin of the basin (Figure 6.27). The 638 wells were organized into groups of thirds with the first group of one-third comprising the oldest and longest producing (1980–99) coalbed gas wells. The second group of one-third coalbed gas-producing wells was developed from 1998 to January 1999. The third group of one-third comprised the newest and shortest producing (01/1999–11/1999) coalbed gas wells. The EURs of all the coalbed gas wells greater than the minimum size (0.56 million m^3) analyzed in the groups of thirds are shown in Figure 6.28 (Cook, 2004). Based on this analysis the median total recovery per cell is 5.4 million m^3 for the first one-third, 6.8 million m^3 for the second one-third, and 5.9 million m^3 for the third one-third (Flores, 2004). The increase in EURs is mainly explained by extensive draining of gas from neighboring undrilled leases and subordinately by the use improved technology for well completions. The primary life span of coalbed gas wells in the eastern part of the basin from 2000 to 2004 is about 7–8 years after which the well life was extended by booster pumps (Figure 6.29).

The total area of the Upper Fort Union AU in the Powder River Basin varies from a minimum of 8,447,000 acres (3,418,379 ha), a median of 8,892,000 acres (3,598,464 ha), and a maximum of 9,337,000 acres (3,778,549 ha). Estimated coalbed gas drainage area per cell having potential for additions to reserves in the next 30 years is a minimum of 40 acres or 16 ha (16 wells per square mile), median of 80 acres or 32 ha (8 wells per square mile), and a maximum of 640 acres or 258 ha (one well per square mile) as adopted from permitted well spacing by the Wyoming Oil and Gas Conservation Commission. However, because of increased issues of management of co-produced water the Wyoming Oil and Gas Conservation Commission later changed the well spacing from 40 to 80 acres.

Analog Method

Many coalfields and coal regions of the world contain coal resources data but rarely coalbed gas data. Moore et al. (2009) and Moore and Friedrich (2010) described the uncertainty that comes with estimation of coal and coal bed resources with limited or no data. Traditionally and historically, in a given coalfield or coal region in the world, coal resources data are more available having been investigated longer than coalbed gas data. Thus, estimation of coalbed

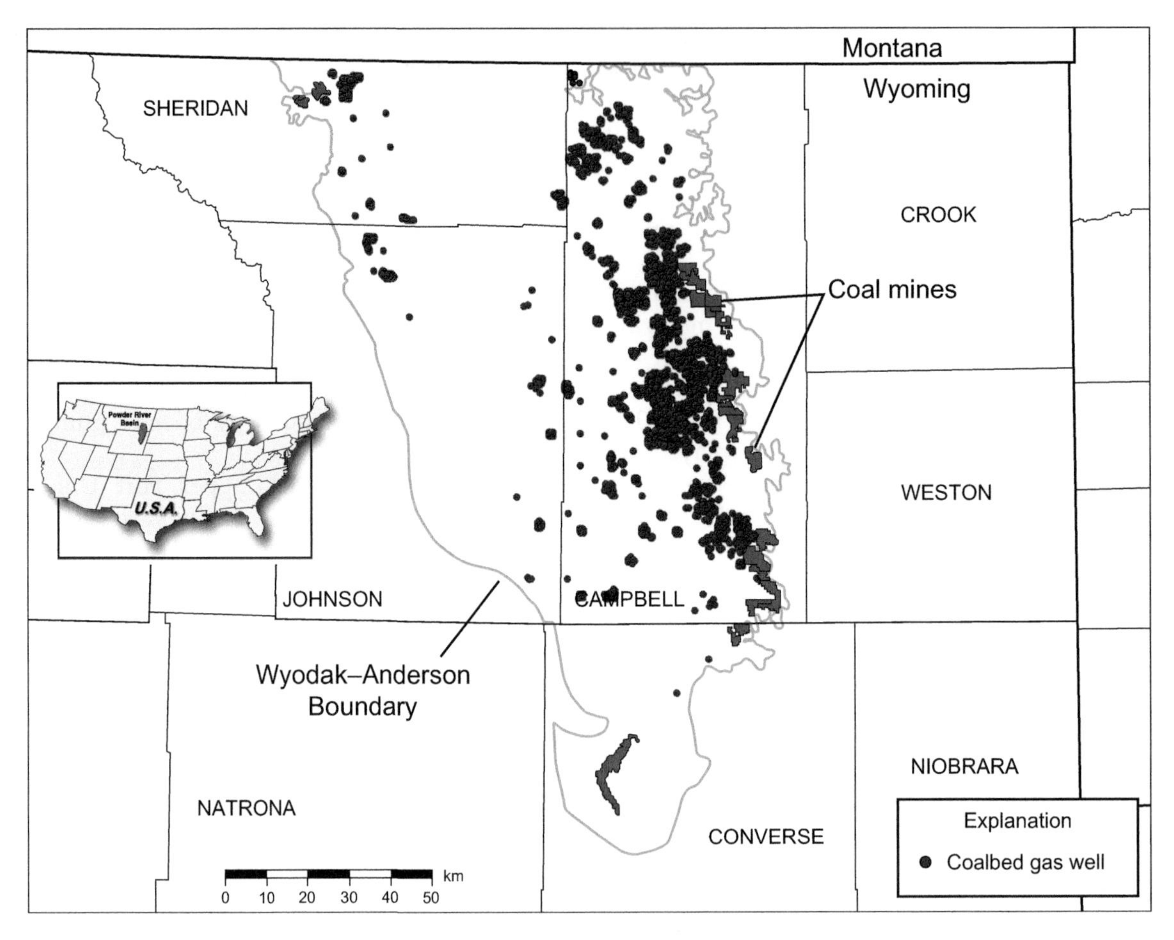

FIGURE 6.27 Map showing coalbed gas wells in the eastern part or shallow margin of the Powder River Basin, Wyoming at the end of 1999. *Source: Data source from Wyoming Oil and Gas Conservation Commission website. Blue dots are coalbed gas wells. Red areas are coal mine sites.*

gas resources contains more uncertainty than estimation of coal resources. Although coalbed gas exploration and development have been instituted for the past 20–30 years, data collection have been concentrated in coalfields in the United States, Australia, and Canada. During the last decade additional coalbed gas data have been collected in China, Colombia, India, Indonesia, Japan, and some European countries. Coalbed gas data (e.g. adsorption isotherms, gas content, gas production, etc.) were collected from lignite to anthracite rank coals in different geologic and tectonic settings from Pennsylvanian to Miocene ages.

The analog method is used when there is no or limited coalbed gas data (e.g. production, adsorption isotherm, gas content, gas composition, etc.), which typifies virgin, unexplored, and undeveloped areas. These coalbed gas data may be used in virgin areas with the same coal properties such as thickness, rank, composition, quality, thermal maturity, and

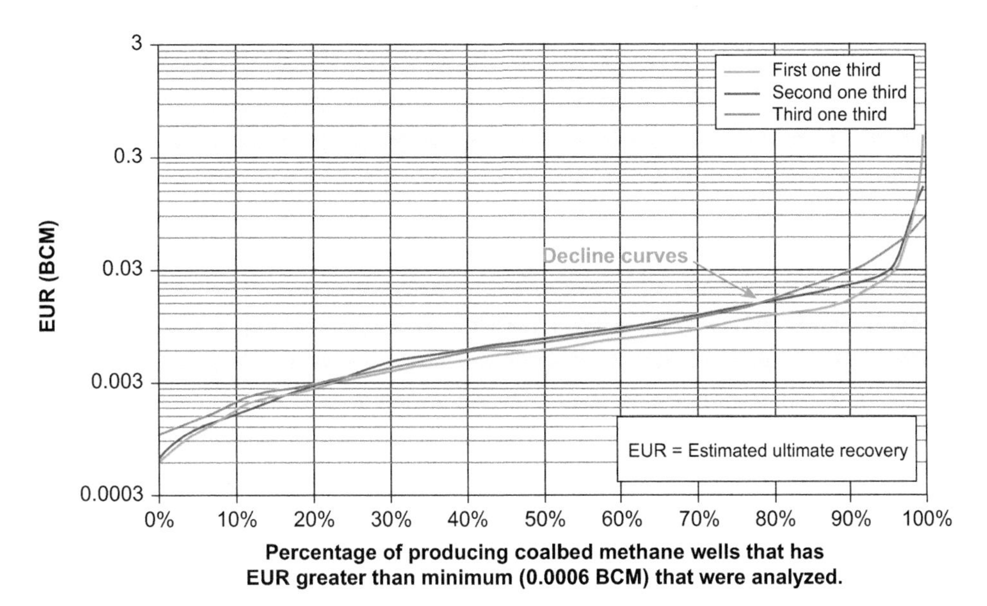

FIGURE 6.28 Distribution of estimate ultimate recovery for coalbed gas wells in the first, second, and third on-thirds analyzed in the upper part of the Fort Union Formation in the Powder River Basin, Wyoming. EUR, estimate ultimate recovery; MMCF, million cubic feet. *Source: Adopted from Flores (2004).*

similar geologic and tectonic settings. For example, Flores et al. (2004) assessed the potential of coalbed GIP resources of Alaska with emphasis on the Cook Inlet Basin using the Powder River Basin coal bed analogs. The two basins are similar in coal rank, composition, and quality but slightly different in geologic and tectonic settings. What is known of the Cook Inlet Basin is that the sandstones contain biogenic gas, which migrated from the coal beds much like the Powder River Basin (Magoon & Anders, 1990; Flores et al. 2004; Warwick & Flores, 2008; Dawson et al., 2012). In addition, limited information of gas content is available (Smith, 1995), which can be used in the GIP assessment. This information was complimented by gas content and adsorption isotherms of comparable Powder River Basin coal beds in terms of rank and vitrinite reflectance.

Even in the Powder River Basin where coalbed gas data abound in the Fort Union Formation there is a general lack of data for the overlying Eocene Wasatch Formation AU, which was assessed by Flores (2004) as a hypothetical AU. The majority of the test coalbed gas wells in the Fort Union Formation drill though the overlying Wasatch Formation. That is, a test well is one that has penetrated and tested the Upper Fort Union AU. Because there is only very limited or no coalbed gas wells that tested the Wasatch Formation AU at the end of 1999, the assessment used coalbed gas production data as an analog from the tested coal beds in the Upper Fort Union Formation AU. The coal beds included the Anderson coal bed of the Wyodak coal zone, which is stratigraphically closest bed, similar rank, coal quality, and thickness to the Wasatch coal beds. The Wasatch Formation AU had to have a minimum EUR per cell of 0.56 million m^3 for assessment; however, 164 cells were estimated to have EURs greater than 0.56 million m^3. The percentage of total assessment unit area in the Wasatch Formation AU that has potential for

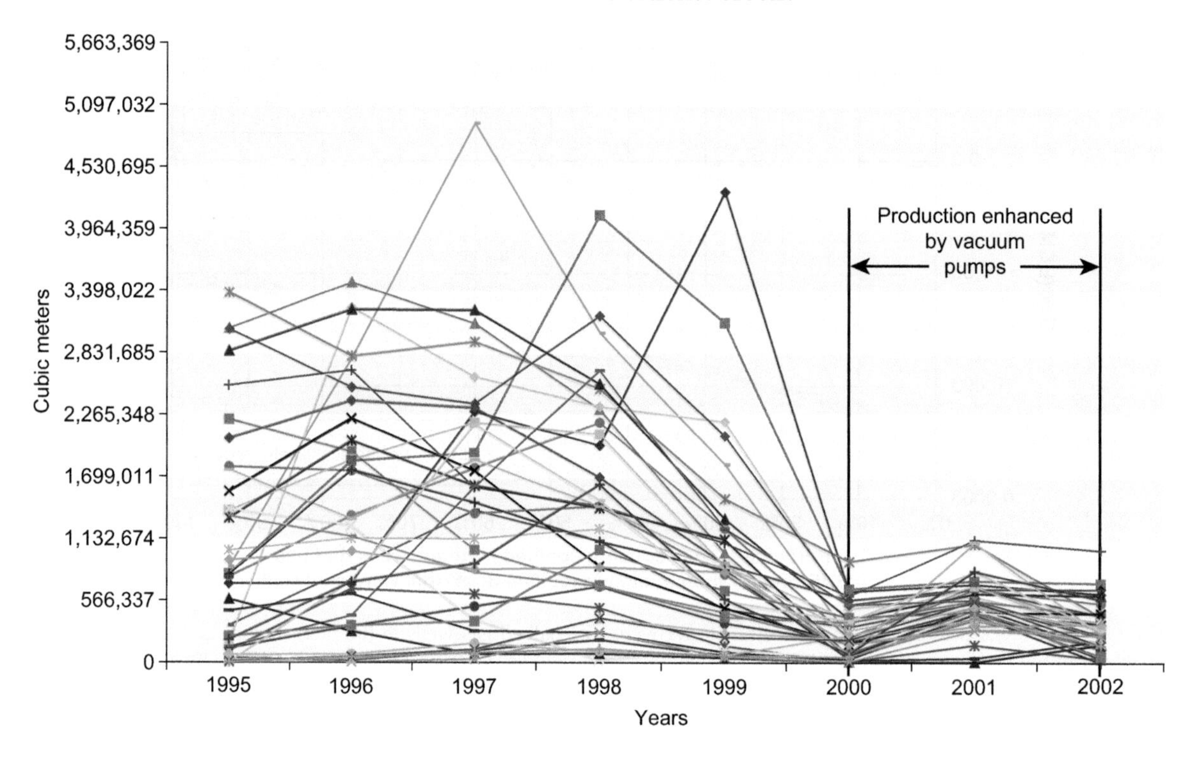

FIGURE 6.29 Distribution of annual production of 30 coalbed gas wells in the eastern part of the Powder River Basin from 1995 to 2002. During the 7 year period the coalbed gas wells peaked and declined production in 2000 after which the wells were connected to booster pumps to prolong their life. *Source: Data source from Wyoming Oil and Gas Conservation Commission website.*

additions to reserves in the next 30 years was estimated at a minimum of 1%, a median of 13%, and a maximum of 24%. This is based on the occurrence of net coal thickness of 15 m in the west-central part of the Powder River Basin in Wyoming. The total recovery per cell for untested cells having potential for addition to reserves was estimated a minimum of 0.56 million m^3, median of 5.1 million m^3, and a maximum of 85 million m^3.

Roberts (2008) assessed the undiscovered, technically recoverable coalbed gas resources in the Cretaceous and Tertiary rocks of the North Slope and adjacent State waters where limited or no coalbed gas production is occurring. The lack of coalbed gas production from reservoir accumulation necessitated cell drainage areas to be derived from producing wells from known continuous accumulations with analogous geologic characteristics, which Roberts (2008) described as analog areas. The method of TPS assessment is similar to that described in the Powder River Basin by Flores (2004) in which each continuous accumulation is assigned to four coal-bearing formations as AUs. Roberts (2008) indicated that data necessary for the assessment of coalbed gas consist of: (1) area per cell of untested cells, (2) percentage of the untested AU area, and (3) total recovery of petroleum per cell for untested cells having the potential for additions to reserves in 30 years. Comparison of vitrinite reflectance (R_0) of Tertiary lignite and subbituminous coal beds in the Sagavanirktok Formation ranges 0.33–0.52% R_0 and is analogous to the Tertiary Fort Union coal beds in the Powder River Basin, which ranges from 0.31 to 0.47% R_0 (Flores, 2004). Thus, Roberts (2008) selected the producing coalbed gas wells from Fort Union Formation as analogs for estimation of drainage area and total recovery per cell of untested cells. The Powder River Basin coal bed analogs were limited to the Anderson or Canyon coal beds, which belong to the Wyodak coal zone (Flores, 2004). The estimates of total recovery per cell in the Sagavanirktok Formation coalbed gas AU were based on Powder River Basin analog wells with the following: (1) a minimum EUR of 0.56 million m^3 and (2) a median EUR of 5.1 million m^3 was applied to untested cells in the AU. Roberts (2008) assumed that much like in the Powder River Basin, gas was generated from coal beds by microbial activity (Flores et al., 2005). The Powder River Basin coal bed analogs were also used in the Cretaceous coal-bearing AUs in which the gas volumes per unit of acreage and ultimate recovery for individual wells might be slightly comparable. Roberts (2008) maximum estimate of 42.5 million m^3 is consistent with maximum EURs from coalbed gas well analogs in the Powder River Basin (Figure 6.30).

Petroleum Industry PRMS

The concepts of petroleum resources classification were discussed and defined by Ross (2011) and for additional information the reader is referred to the petroleum industry's guidelines for application of the PRMS (PRMS, 2011). Ross (2011) defined petroleum-related projects such as a coalbed gas according to its chances of commerciality by using: (1) reserves, (2) contingent resources, and (3) prospective resources (Figure 6.31). Thus, for a coalbed gas project that satisfies the requirements for commerciality as set out in Ross (2011) and PRMS (2011), reserves may be assigned by three estimates according to recoverable sales quantities and adopted by Barker (2008) and Carlson and Barker (2011) as follows: 1P (proved) and 2P (proved plus probable) reserves, and 3P (proved plus probable plus possible) resources (Figure 6.32). Barker (2008) and Clarkson and Barker (2011) defined reserves in the context of PRMS (2011) as: (1) demonstrated commercial production potential; (2) marketable and commercial gas composition, gas content, and coal thickness; (3) depth within accepted economic limits of a coalbed gas fairway; (4) feasible development plan that is economically viable under existing market; and (5) development within a realistic time period.

Barker (2008) and Clarkson and Barker (2011) have discussed the current practices of classification and estimation of coalbed gas resources and reserves with adoption of the incremental approach to delineation and development used by the coal mining industry and well spacing used by the petroleum industry. The coal mining industry approach is the same as the Wood et al. (1983) approach of coal resources and reserves with increasing uncertainty as the geologic assurance decreases or as the distance of control points (e.g. data) increases consequently the progressive classification of proved, probable, and possible reserves. According to Clarkson and Barker (2011) the coal geologic assurance-based method applied to the petroleum well spacing method provides no precise assessment of the range of uncertainty in recovery efficiency, which resulted in large reserves and different reserve growths. Thus, in order to comply with the PRMS (2011) principles, Barker (2008) and Clarkson and Barker (2011) offered the recent practices to define resource/reserve areas.

According to Barker (2008) and Clarkson and Barker (2011) this definition presumes that the group of wells is established in the coalbed gas fairway that is composed of a laterally continuous coal thickness with satisfactory gas content and permeability. Whether the lateral continuity of coal thickness is within the same correlated coal bed(s) cannot be ascertained but discussed in the following section.

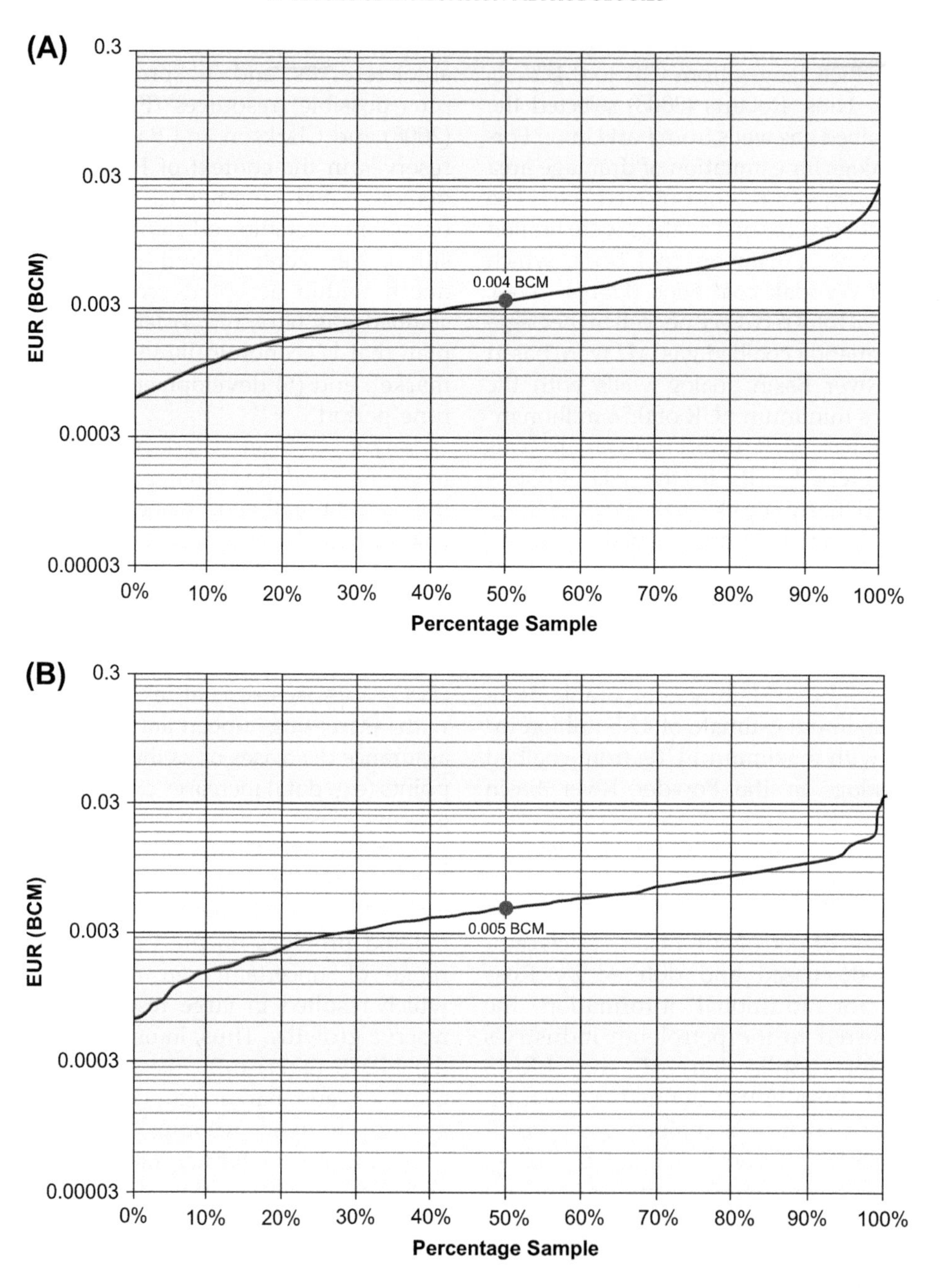

FIGURE 6.30 Distribution of estimate ultimate recovery (EUR) for the coalbed gas wells in the Anderson (A) and Canyon (B) coal beds in the upper part of the Fort Union Formation, Powder River Basin, Wyoming. The EURs were used as analogs for the Tertiary and Cretaceous coals in the North Slope, Alaska. Median EUR in each graph represented by the red dot. 0.13 BCFG = 3.7 million m^3. *Source: Adopted from Roberts (2008).*

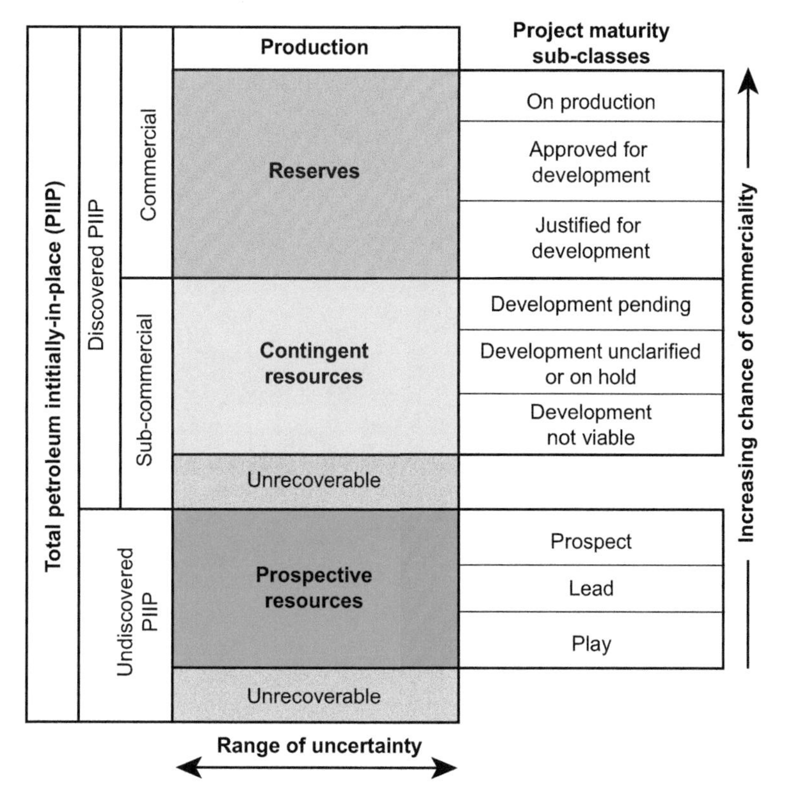

FIGURE 6.31 The Petroleum Resource Management System classification is based on the range of uncertainty and increasing chance of commerciality. *Source: Adopted from Ross (2011).*

These resources/reserve areas are well demonstrated in Figure 6.32 which starts at the largest area or 3P with contingent resources determined from drilling, logging, testing, and sampling coalbed gas content and thickness to establish enough presence of potential gas and gas rate, which may be undemonstrated and uneconomic (Barker, 2008). The 3P area contains mainly possible resources, wherein according to well spacing rule the distance from probable locations is two-well spacings. However, these can be extended to farther distances as supported by coal and quality. The next smaller reserve area (Figure 6.32) contains probable and proved reserves in which probable reserves according to well spacing rules are two-well spacings between proved locations. These also may be extended to farther distances as supported by coal geology and quality. The smallest reserve area or 1P (Figure 6.32) contains proved reserves, which include undeveloped, developed nonproducing, and developed producing. Proved undeveloped is a small drainage area for producing and nonproducing wells proven to contain commercial and recoverable coalbed gas (Barker, 2008; Clarkson & Barker 2011). Well spacing varies from 80 to 320 acres (e.g. 1 acre = 0.40 ha) in the United States and as much as 550 acres in Australia (Jenkins and Boyer, 2008; King, 2008). Moore (2012) discussed the certification of reserves of a coalbed gas project based on the PRMS (2011) methodology. In order to achieve certification at the reserve level, commercial flow rates must be

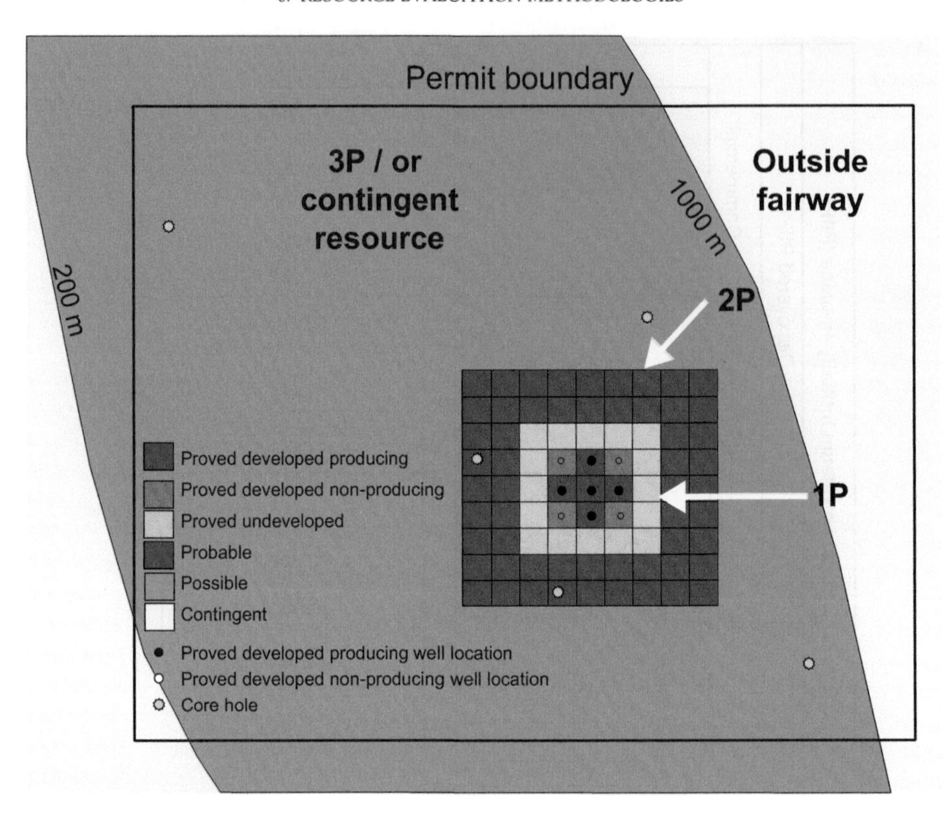

FIGURE 6.32 Classification of resource and reserve areas (1P, 2P, and 3P) for coalbed gas with the depth limits of commercial production ranging from 200 to 1000 m. 1P, proven reserves; 2P, proven plus probable reserves; 3P, proven plus probable plus possible. *Source: Adopted from Barker (2008; 2011).*

demonstrated and be supported by the wells containing gas data throughout the permit area, measured permeability tests, and demonstrable areal extent of the reservoir (Moore, 2012). Moore (2012) noted that the value of most coalbed gas projects is based on the 2P reserves in contrast to 1P for conventional gas projects.

METHODOLOGIES: ROOM FOR IMPROVEMENT

Improvement of methodologies for estimation of coalbed gas resources and reserves has been suggested by many workers more recently by Jenkins (2008), Moore et al., (2009) and Moore and Friedrich (2010). These workers and others have emphasized more often than not that the sources of inaccuracies of resource and reserve estimates begin from the inherent heterogeneity of the coal beds and the variability of associated noncoal rocks. Jenkins (2008) suggested the parameters (e.g. areal extent, thickness, gas, content, coal density, coal rank, adsorption isotherms, etc.), which serve as input to determine gas resources and reserves using probabilistic techniques are very variable physically and compositionally. Coal bed heterogeneity, in turn, depends on depositional environments; hence the importance of understanding non-coal lithologies associated with the coal beds. Moore et al. (2009) indicated that the heterogeneities of these parameters and other input data introduce several levels of uncertainties, which affects the accuracy of estimations of coalbed gas resources such as GIP. According to Moore et al. (2009) the sources of parameter uncertainty include the following: (1) various measurement errors, (2) simplification of a complex geological model, (3) insufficient data to distinguish the complex geological model, and (4) uncertainty of the geological model itself. The ideal situation is to provide a measure of value such as "worst," "bad," and "best" case scenario to express the uncertainty of each parameter (Moore et al., 2009). For example, when the heterogeneity of a parameter such as coal thickness is low (e.g. nonsplitting, nonmerging, nonpinching laterally), the range of value from "best" to "worst" is low. Thus, assigning a measure of value of uncertainty of parameters is a good start of improvement of methodology.

The diversity of GIP (see previous section) is indicated that the methodology continues to be improved to achieve better estimate of the coalbed gas resources. However, recent improvements toward the estimation of coalbed GIP are focused on the internal properties of the coal reservoir such as adsorption isotherms, chemistry, reservoir pressure, burial, etc. (Chen et al., 2011; Dallegge & Barker, 2000; Simpson, 2008). Although these parameters are important to the GIP equations, one parameter on the other side of the equation, which has been scrutinized toward its application to GIP estimation, is the A or study area or drainage area. This parameter is defined as the areal coverage and extent of a coal bed to be drained for gas. This definition assumes a high degree of lateral continuity and uniformity of thickness (t in the equation) of coal bed in a three-dimensional framework.

In order to avoid introducing more error than necessary to the A and t parameters of the GIP equation, use of coal resources values is suggested for these two parameters. The A and t parameters of the GIP equations have been measured for coal resources in many coalfields and coal regions throughout the world. During this process, the degree of uncertainty of the lateral continuity and thickness of coal beds relies on the distance between observed points of the same coal bed (Wood et al. 1983). That is the less the distance (<0.5 m) between observation points for the thickness and occurrence of the same coal bed the more likely ("best" of Moore et al., 2009) is the coal bed to be identified as coal reserves. The more the distance ($>0.5-1.25$ m) between observation points of the same coal bed the less certain ("bad to worst" of Moore et al., 2009) the coal bed occurrence; thus the coal should only calculated as probable reserves ("bad") or hypothetical ("worst") resources. Thus, the degree of uncertainty of the estimates is expected to decrease with closer observation points (e.g. data). Therefore, the geologic assurance-based coal resource and reserve estimates should be adopted for calculations of the coalbed GIP resources and reserves where no well production data is available.

Post Mortem with Hindsight

More than three decades have passed in the evaluation and extraction of coalbed gas and we seem to be no closer to understanding a multidiscipline assessment of coal and coalbed gas resources and reserves. The concept that coalbed gas is an integral part of the natural gas resource is valid; thus, the commodity must be assessed using the methodologies applied to petroleum system, however, the methodologies may be misplaced without knowledge of the coal system (Warwick, 2005). The last decades of coalbed gas assessment should pave the way to improve the methodology in terms of lessons learned from coalbed gas producing basins. The mature coalbed-gas producing basins in the United States, Canada, and Australia, provide lessons in exploration and development (Chapters 7–9). For example, the Powder River Basin has been producing coalbed gas since early 1980s or about 30 years. Coalbed gas resource assessments of the basin have been performed twice during the past 18 years (Flores, 2004; Rice & Finn, 1996). The difference between Rice and Finn (1996) and Flores (2004) coalbed gas resources assessment is the amount of EURs of producing wells and more advanced analysis of the coal stratigraphy and resources and coal reservoir data, which was performed by the National Coal Resources Project, and the USGS–Bureau of Land Management Cooperation Project (Fort Union Coal Assessment Team, 1999; Flores et al., 2006).

The coalbed gas assessment by Flores (2004) and succeeding development in the Powder River Basin has provided insights for improvement of methodologies. The most critical parameter to understand at the outset of any assessment, exploration, and development is the coal geology and stratigraphy. Even though, the basin contains abundant coal resource and stratigraphic data, the information was not fully examined before use in terms of coal reservoir continuity and characterization. The failure to define controls and limits of coal beds led to inconsistent reservoir correlation and nomenclature, which in turn, led to misidentification and confused reporting of gas production. Furthermore, the problem created difficulties in gas drainage and protection of correlative rights, which are interest of different ownerships to develop the same reservoirs in adjoining leases. The issue of coal bed correlation or miscorrelation also leads to miscalculation of GIP resources and reserves, which is exacerbated by the merging, splitting, and pinching out of coal beds to form thick to thin beds, respectively. Coal bed correlation influenced by availability of data is a single most important factor in the variability of estimates of coal and coalbed GIP resources and reserves. The accuracy of gas content applied to rapidly thickening and thinning coal beds will cause overestimation or underestimation of GIP resources and successive calculations of recoverable resources as well as proved and unproved reserves, which in turn, are provided for certification (Moore, 2012). The stratigraphic distribution of coal beds controls the gas drainage area and gas volume, which plays a major role in calculation of GIP and production. It is suggested that for greater accuracy to resource estimations to separate coal bodies of split beds from merge coal bodies and to make separate calculations after correlations are properly performed using depositional environments as a guide.

Based on production history the coalbed gas model in the Powder River Basin is expected to add reserves into the next 30 years (Flores, 2004). This expectation is deemed an insufficient assessment without additional information on reservoir properties from the work of Flores et al. (2006). For example, the Flores (2004) assessment of the TPS Upper Fort Union AU, which drew production data for EURs of wells concentrated in the eastern margin of the basin has not taken into consideration of rapid development, which saw the increased infill drilling around producing wells in the area from 2000 to 2009 adding more than 19,000 wells in 9 years. The work of Stricker et al. (2006) showed that Wyodak coal zone targeted for production the shallow eastern margin of the basin consisting of subbituminous C coal beds in contrast to the deeper central part of the basin, which contains subbituminous B and A coal beds. Coalbed gas content values have been positively verified by workers (Stricker et al., 2006; Higley, 2007) showing that higher rank coals contain more adsorbed gas. Thus TPS assessment unit based mainly on the shallow, lower rank coals to estimate addition to reserves for the next

30 years with the higher rank and deeper coals (e.g. Big George coal zone) that still remain to be developed may have pitfall. In addition according to Flores (2004) the subbituminous coal in the Powder River Basin is typically undersaturated. The subbituminous C and B are 2–3 times more undersaturated at 66% and 57%, respectively, than subbituminous A coal at 23%. Thus, the saturation of coal beds plays an important role in determining a more realistic coalbed gas resources/reserves either in the form of gas-in-place or production history models.

The use of the coalbed gas assessment in the Powder River Basin as an analog to similar rank coal beds in the United States and worldwide should be tempered with moderation. Recent studies of Flores, Spear, Kinney, et al. (2010), Flores, Spear, Purchase, et al. (2010) and Flores and Stricker (2012) comparing the high-pressure methane adsorption isotherms of the same age and rank coals of the Powder River Basin and those in Colombia, Indonesia, and the Philippines show significant differences in the gas storage capacity. Figures 6.33, 6.34 and 6.35 show the differences in adsorption isotherms of the subbituminous C, B, and A coals in the Powder River Basin and Tertiary coal basins in Colombia, Indonesia, and the Philippines. The gas storage capacity of subbituminous coals based on the isotherms in Colombia, Indonesia, and the Philippines is about twice as much as the same coal rank in the Powder River Basin. The major controlling factor of the difference in adsorption isotherms is the vitrinite composition, which average about 10–15% more in Colombia, Indonesia, and the Philippines (Flores & Stricker, 2012). Finally, based on adsorption isotherms of all coal ranks from these countries, the coals in

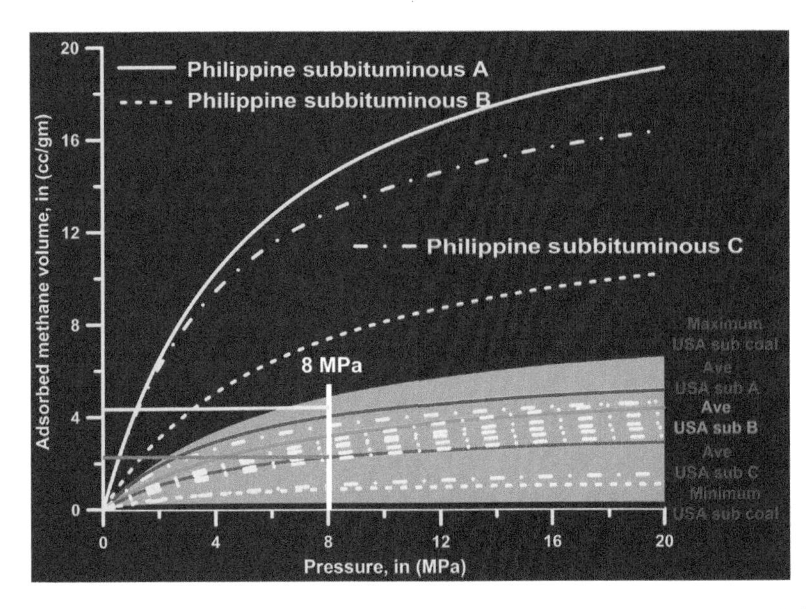

FIGURE 6.33 High-pressure methane adsorption isotherms of Tertiary subbituminous C coals of the Philippines and Powder River Basin in the United States. Arbitrary comparison of the adsorbed gas at 8 MPa shows the Philippine coal store about twice more gas than the Powder River Basin coals. MPa, megapascal; cc/gm, cubic centimeter per gram; sub, subbituminous. *Source: Adopted from Flores and Stricker (2012).*

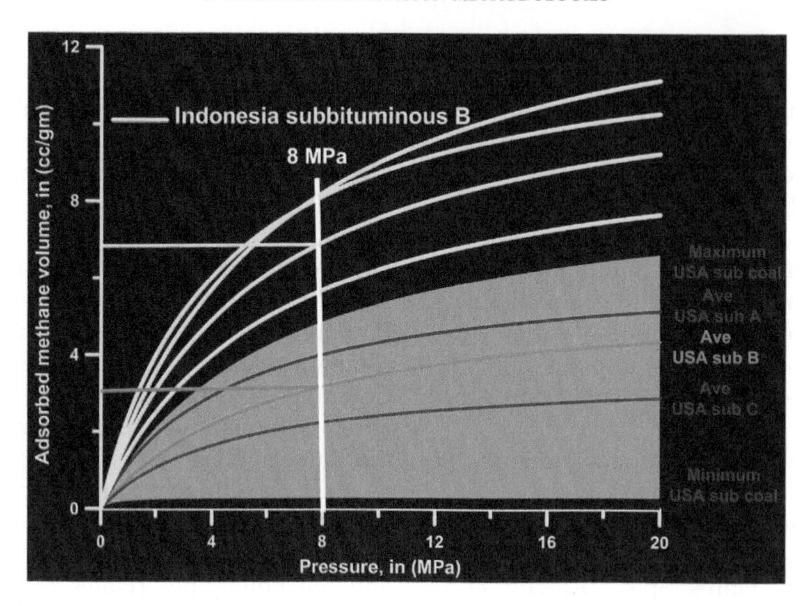

FIGURE 6.34 High-pressure methane adsorption isotherms of Tertiary subbituminous B coals of Indonesia and Powder River Basin in the United States. Arbitrary comparison of the adsorbed gas at 8 MPa shows the Indonesian coal store about twice more gas than the Powder River Basin coals. MPa, megapascal; cc/gm, cubic centimeter per gram; sub, subbituminous. *Source: Adopted from Flores and Stricker (2012).*

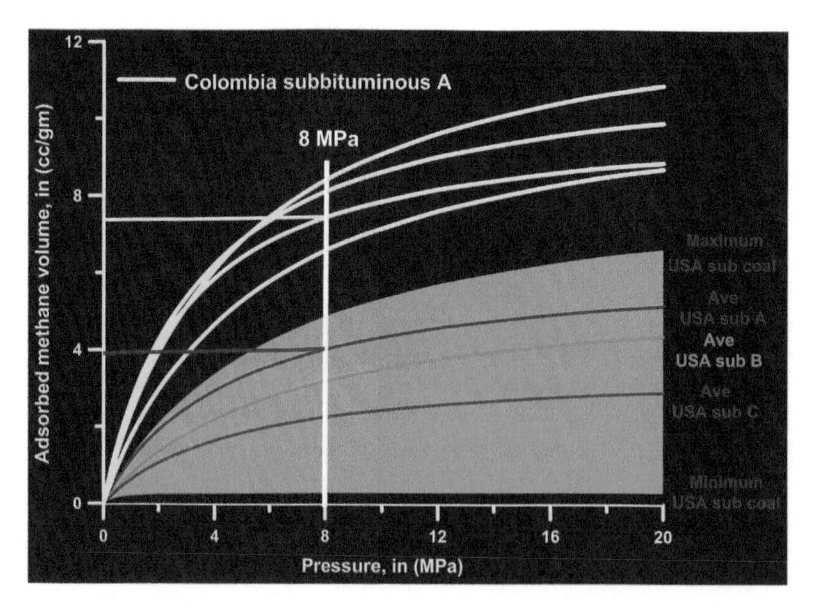

FIGURE 6.35 High-pressure methane adsorption isotherms of Tertiary subbituminous A coals of Colombia and Powder River Basin in the United States. Arbitrary comparison of the adsorbed gas at 8 MPa (megapascal) shows the Colombia coal store about twice more gas than the Powder River Basin coals. MPa, megapascal; cc/gm, cubic centimeter per gram; sub, subbituminous. *Source: Adopted from Flores and Stricker (2012).*

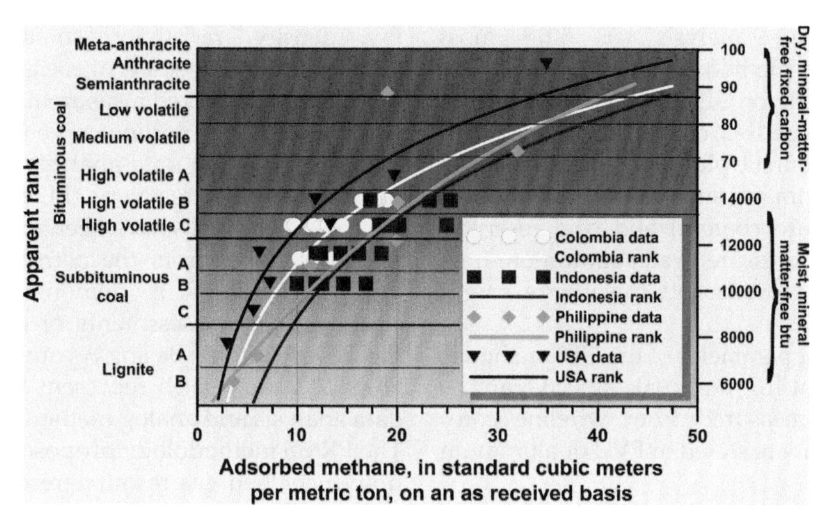

FIGURE 6.36 Relationships of adsorbed methane and apparent rank from lignite to anthracite of Pennsylvanian to Miocene coals in the United States, Colombia, Indonesia, and the Philippines. Adsorbed methane in Indonesian and Philippine coals at the boundary of subbituminous A to high volatile C bituminous rank is about 7.5 Scm/Mt more than the United States coals. *Source: Adopted from Flores and Stricker (2012).*

Colombia, Indonesia, and the Philippines hold more gas than coals in the United States (Figure 6.36).

SUMMARY

The key to implementing plans for coalbed gas exploration and development projects is the knowledge of coal resources and methods of assessments. Understanding coal classification and categories will assist in the estimation of coalbed gas resources. The critical feature of coal resource and reserve estimates is the use of geologic assurance approach to impart certainty. Geologic assurance reflects the spacing between observation points where coal thickness is measured and samples are collected for coal quality analysis. The degree of assurance increases with more geologic and stratigraphic data, closer data points, and better quality data. The farther the distance between data points the lower the certainty of existence of coal resources (e.g. hypothetical–speculative resources). The closer the distance between data points the higher the certainty of existence of coal resources (e.g. inferred, indicated, and measured reserves). Geologic assurance methodology is based on accurate stratigraphic correlation for measurement of coal continuity, thickness, and coal quality of the same coal bed. Assessments of coal resources and reserves worldwide conform to the geologic assurance principle, which is modified from country to country according to depth, distance between observation points, thickness, and complexity of the coal geology and structures of the study areas.

The coal resource–reserve methodology does not directly translate to the petroleum industry's method of estimating coalbed gas resources–reserves because of lack of a range of uncertainty of gas recovery. The methods of estimation of coalbed gas resources and reserves adopted by the petroleum industry include: volumetric, material balance, production data analysis, reservoir simulation, and analog methods. The volumetric and production data analysis are the most commonly used methods.

The volumetric or GIP method is calculated using gas drainage area, coal thickness, coal density, and gas content plus other parameters to improve the estimates such as coal quality (e.g. ash and moisture content) and coal reservoir conditions (e.g. pressure, water saturation, fracture porosity), and adsorption isotherm or gas storage capacity.

An important parameter of the GIP method is the gas content of the coal, which can be collected and measured from wireline cores and cuttings that are sealed in PVC or aluminum canisters.

The gas content is measured by desorption of gas from the coal cores or cuttings using the direct and indirect methods. The direct method is based on the desorbed gas from coal cores and cuttings and assumption that gas released during desorption was proportional to the square root of time. The indirect method is based on analysis of methane adsorption isotherms, which measure the optimum adsorbed gas at varying pressures and at constant temperature. Use of wireline or geophysical logs (e.g. gamma ray, density, resistivity, sonic, neutron) is another indirect method of measuring gas content. Other specialized measurements of gas content of coal utilize decline curve methods.

Gas production data analysis based on estimate of ultimate recovery (EUR) and spacing of coalbed gas wells have been adopted for estimating reserves from the petroleum industry. The methodology is demonstrated in case studies of TPS assessments of coalbed gas in the Powder River Basin, Wyoming and North Slope, Alaska, which represent the production data analysis and analog methods, respectively. The PRMS methodology proposed standardization of coalbed gas resource-reserve categories to contingent resources and proved, possible, and probable reserves based on well spacings. The proved reserves are, in turn, subcategorized into undeveloped, and developed and proved reserves based on well spacing. The various methodologies to estimate coalbed gas resources and reserves are not perfect and have room for improvement. The post mortem of the methodologies with hindsight of lessons learned from previous assessments is a good start for this process.

CHAPTER

7

Coalbed Gas Production

OUTLINE

Coal and Coalbed Gas
http://dx.doi.org/10.1016/B978-0-12-396972-9.00007-0

KEY ITEMS

- Drilling technology for coalmine gas drainage was inherited from Europe in the nineteenth century and was improved in the United States for coalbed gas development of unmineable coal beds.
- Coalbed gas production has extended into Australia, Canada, Europe, China, and India and exploration is being done in Indonesia, Russia, and other countries using advanced technologies in drilling, well completion, stimulation, logging, and gas gathering system.
- Drilling has advanced from vertical to horizontal wells increasing the efficiency of production. Vertical wells produce from single to multiple coal beds and horizontal wells penetrate coal beds parallel to bedding making more contact to cleat and fracture systems and extending gas drainage area.
- Horizontal wells are made more effective with the addition of lateral wells to the single horizontal well as multilateral wells, which can be extended as multibranched wells increasing recovery of coalbed gas reserves from a large drainage area.
- Surface directional drilling is a vertical well turned into a radius toward the coal bed penetrating parallel to bedding and united with a production vertical well. This technology reduces the footprint of drilling by being able to drill from one area many wells into the same coal bed or multiple coal beds.
- A critical aspect of drilling a coalbed gas well is creating minimal formation and reservoir damage by use or nonuse of drilling fluids. Drilling fluids run the risk of plugging coal cleats and fractures.
- Completion techniques are fundamental to coalbed gas wells because they influence gas flow and productivity of the coal reservoirs. The techniques may be performed by open- and cased-hole completions depending on the coal geology of the formation and reservoir.
- Stimulation of coal reservoirs enhances coalbed gas production by using fracture treatments such as hydraulic fracturing called "fraccing" or "fracking" often with water or combined with chemical additives and proppants.
- Reservoir characterization based on coal composition, rank, and geology plays a major role in completion and stimulation of wells. Anisotropism of the permeability and heterogeneity of coals are well known but it is still perceived as homogeneous reservoirs with simple orthogonal cleats and fractures.
- Wireline logging tools (e.g. resistivity, gamma ray, density, acoustic or sonic, and neutron) are used to identify coal beds in the subsurface due to low radioactivity, high resistance to electrical current, and low density.
- Advanced, high-tech logs have been devised to compliment wireline logs by calibration with coal quality, gas adsorption, and gas desorption data derived from coal cores. High-tech logs are used as predictive tools.
- Drilling, completion, stimulation, and logging are crucial to designing coalbed gas pilot project toward production. Essential to the pilot project is testing for gas content and reservoir producibility as dictated by the connectivity of cleats and fractures or permeability of the coal.
- Key to precommercial production of coalbed gas is testing the number of wells and their spacing to develop coalbed gas within a field and correlative rights protection. Groundwater and permeability of coal

reservoirs play important roles in determining well spacing.

- Important to development of coalbed gas is the design of gas gathering system, which includes gas measurement, pipeline, compressor, and treatment plant complexes. The gas gathering system feeds into downstream market delivery systems locally, regionally, nationally, and globally.
- Fueling the future with coalbed gas lies in the downstream delivery systems of liquefied natural gas (LNG) and compressed natural gas (CNG), both fast-growing domestic production from coal mines and unmineable coal reservoirs worldwide.

INTRODUCTION

The United States Bureau of Mines first applied the European coal mine drilling technology to desorb gas in order to prevent explosions or outbursts in coal mines in the United States (see Chapter 2). The initial tests to degas coal beds in US coalmines have proved successful in mitigation of coal mine outbursts as well as paved the way to potential recovery of coalbed gas from unmineable coal beds. The coal-mine-to-gas technology was adopted and improved by the petroleum industry as unconventional production technology. Since then the United States became a world leader in coalbed gas production and extended into Canada, China, Australia, Europe, and India. Development and production in these countries have emerged with the advance of drilling, logging, well completion, reservoir stimulation, and gas gathering technologies. In addition, the progress of coalbed gas production has resulted in the expansion and improvement of delivery-to-market systems from regional to national pipeline networks to LNG and CNG delivery systems worldwide. The Pacific Rim, which includes Australia, Brunei, Indonesia, Malaysia, and United States, leads the world in LNG production part of which includes coalbed gas comingled with conventional natural gas. In the United States, coalbed gas added as much as 8% of the gas reserves between 2000 and 2010, which is significant considering that coalbed gas production was first initiated in the 1980s.

The fundamentals of coalbed gas production are applied technologies from drilling to completion. In the early years of coalbed gas development, the United States invoked that the major constraints to coalbed gas production are completion technology and legal and institutional inconsistencies (Rightmire, 1984). Rightmire's insightful refrains in 1984 that are still relevant today are (1) coals are not conventional oil and gas reservoirs, (2) coals are thin, compressible, and chemically reactive, (3) coalbed gas production techniques are coal basin specific, and (4) multiple completions are necessary in some coal basins to maximize recovery from multiple coal beds. As a consequence, coals are treated as unconventional reservoirs, which differ in physical, chemical, and compositional properties from coal basin to basin. The physical continuity and distribution of the coals dictates completion methodology for coalbed gas production as adopted by operators in different basins.

The legal constraints to coalbed gas production foreseen by Rightmire (1984) have transpired and been resolved by the United States Supreme Court (1999), in general, and in the federal state governments and industry, in particular. The Court's decision was that coalbed gas does not belong to the coal and as such is separate from coal mining development and is regulated under the petroleum domain. However, as anticipated by Rightmire (1984), conflict between coal mine and gas companies occurred regarding desorption and loss of gas beyond mine properties during mining operations. Although this conflict was resolved amicably, similar conflicts have overflowed into other countries where coalbed gas emerged as a viable economic energy source. Even though countries (e.g. Australia, Canada, New Zealand, Indonesia, and Philippines) have used the US model, each country has developed its own laws and regulations from lessons learned of the conflict between coal mining and gas companies in the United States. Constraints of coalbed gas development were minimized between the two groups in these countries such that the best methods of producing gas are compatible and do not interfere with mining and development of coal.

DRILLING TECHNOLOGY

The technology of coalbed gas drilling has evolved since the US Bureau of Mines initially drilled to desorb gas in underground coal mines. According to Diamond (1993), coal companies are compelled in the United States under federal and state laws and regulations to control gas concentration in underground mines. An option to control gas concentration is to drain the coalbed gas through vertical wells giving way to horizontal wells. The precursor of vertical coalbed gas wells can be traced to coal mine drainage wells ideally in front of mining, which were used from the 1970s to 1980s in the Appalachian Basin in the eastern United States (Figure 7.1). Diamond (1993), described that these vertical wells with steel casings installed just above the coal beds for vertical gas drainage complete the holes. Installation of casings in

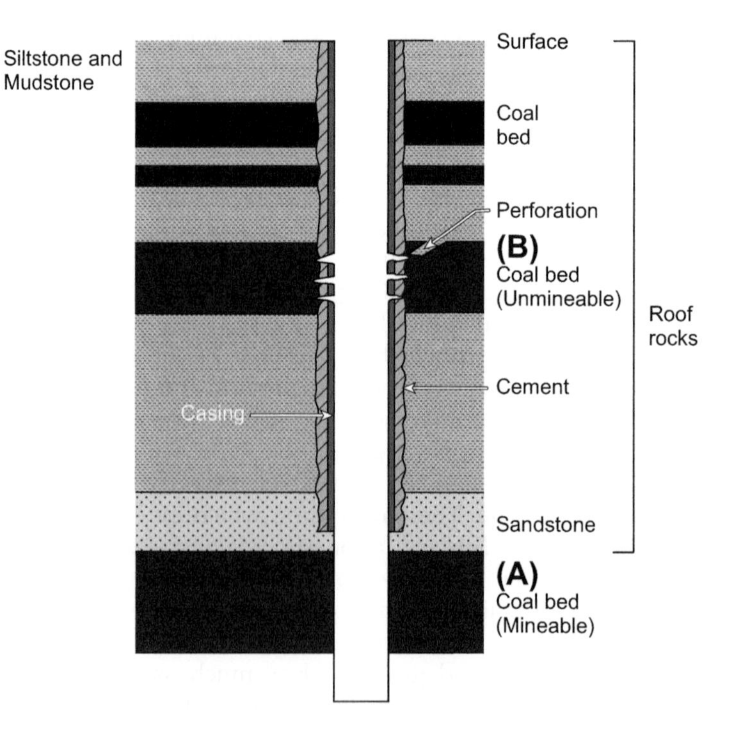

FIGURE 7.1 Schematic diagram showing vertical coalbed gas well used in the coal mining industry to extract coalmine methane in front of mining. (A) Open-hole completion in mineable coal bed. (B) Cased-hole completion in unmineable coal bed in the overlying roof rocks perforated for stimulation treatment. *Source: Modified from Diamond (1993).*

mineable coal beds is unacceptable for practical coal mine operations. However, Steidl (1991) reported that open-hole completions in coalbed gas wells with casing above the coal permit stimulations to be injected into the coal. When coal beds occur above the mineable coal bed, the casing is perforated to degas the coal. These techniques were eventually adopted with modifications by the gas industry.

Logan (1993) described various drilling techniques as adopted by the gas industry in shallow to deep coal basins (e.g. from <300 to >1000 m depth) in the United States. Logan (1993), noted that drilling coalbed gas in the United States posed the following problems: (1) large volumes of water; (2) although generally low gas pressured, some overpressured coal beds occur especially in deep coal basins; (3) stability of boreholes and caving of coal bed; and (4) formation damage. These problems required detailed understanding of the coal reservoir and geological characteristics of associated rocks in planning the strategy of drilling and completion of coalbed gas wells. An important part of the planning strategy is acquiring stratigraphic data on the coal and overlying strata during open-hole condition with the use of logging instruments. Another strategy used to collect data on the coal for reservoir characterization is core drilling.

According to Ramaswamy (2007), the major problems to be considered in planning for coalbed gas drilling program are (1) potential damage to the formation, (2) likely loss of circulation resulting from high permeability, (3) possible overpressurization of the reservoir, (4) probable conditions of water and gas flows, and (5) stability of the drillhole. Thus, the following geological and technological considerations are accounted for in designing coalbed gas wells (Boyer & Reeves, 1989; Ramaswamy, 2007): (1) lithology of the coal and associated rocks (e.g. sandstone, siltstone, and shale); (2) occurrence, distribution, and size of coal beds within the stratigraphic interval; (3) depth of the coal-bearing formation; (4) pressure and fluid saturation within the coal reservoirs; (5) casing program; (6) application of downhole logs; (7) drilling fluid treatments; and (8) stimulation procedures and completion techniques.

The lithological characteristics of the penetrated formation(s) are critical to determining the integrity and behavior of the open hole during drilling. For example, poorly case-hardened or cemented sandstones and mudstones or shales are prone to sloughing and cave-ins promoting loss of circulation as well as drilling and drill string recovery difficulties. In some coal basins, coal-bearing formations such as in the Paleocene Fort Union and Eocene Wasatch Formations in the Powder River Basin in Wyoming contain coals and sandstone, which serve as aquifers that are often connected with coal beds or other aquifers hindering drilling operations and potentially problematic stratigraphic intervals for completion and production (Cottingham, Gribb, Hays, & McLaughlin, 2012; Flores, 2004). The thickness, number, and vertical distribution of the coal beds control the casing, cement, and stimulation procedures. The depth of the well and nature of bounding strata determines the drilling techniques to be deployed. For example, according to Boyer and Reeves (1989), rotary percussion (air) drilling is normally used in stable, shallow coal-bearing strata in contrast to rotary mud system used in unstable and deeper strata.

Types of Drilling

Generally mobile, truck-mounted, single rotary rig outfitted with complementary vacuum and/or water trucks are used to drill and set casing for coalbed gas production wells, especially shallow wells with a cementing service employed to cement casing (Figure 7.2). However, drilling rigs vary from coal basin to basin in the United States. In the Powder River Basin in Wyoming and Montana, modest size rigs used to drill groundwater wells are modified and commonly employed in shallow wells (Flores, 2004) (e.g. depths from 75 m to more than

FIGURE 7.2 Photograph of a deployed mobile, truck-mounted drill rig in the central part of the Powder River Basin, Wyoming, United States. During peak drilling seasons from the late 1990s to the early 2000s, drill rigs many of which converted from groundwater drill rigs numbered more than 100 deployed basin-wide.

750 m). In eastern coal basins in the United States, most commonly used rigs are portable, self-propelled, hydraulically driven, and top-head drive drilling rigs (Logan, 1993). The shallow depths of coal reservoirs compared to conventional reservoirs permit well completion in a few days with single to three strings of surface casings cemented in position.

A standard drilling operation for a coalbed gas well in the western coal basins in the United States was described by Logan (1993), to include three stages (Figure 7.3) starting with a hole from the surface to about 91.5 m depth (or shallower) at about 31.1 cm diameter to set a 24.4-cm surface steel casing and set cement. This stage is followed by redrilling the hole using water-based mud medium (or air and other drilling fluids in various situations) to a 22.2-cm-diameter hole, which is bored to the top of the coal cased with a 17.8-cm-diameter steel tube and cemented to the top of the coal. Finally, the well is completed by drilling the hole with water through the coal and/or underlying coal beds to total depth. Logan (1993), described the drilling procedures and drill equipment specifications (e.g. size of drilling pipes and drill collars, weight of bits, etc.). For example, the size of drill collars, bits, and casing centralizers used in the Powder River Basin in Wyoming, United States, are shown in Figures 7.4 and 7.5. The depths of these stages vary from coal basin to basin and with topography and nature of rock formations.

In the eastern part of the United States such as in the Appalachian Basin and Black Warrior Basin, thin coal beds occurring in zones of multiple beds challenge drilling. Logan (1993), indicated that the thin coal beds at shallower depths than the western coal basins were drilled open hole to a single bed with mixed results of gas recovery. This drilling technique according to Logan (1993), was replaced by (1) multiple zone open hole, (2) single coal zone open hole, and (3) multizone cased hole. The drilling techniques for the coalbed gas wells in the eastern United States involved similar procedure and specifications as those in the western coal basins (Figure 7.6). However, setting casing and drill completions will vary as discussed in the following sections (Logan, 1993).

In western Canada, a standard drilling operation for a coalbed gas consists of boring a 20-cm-diameter hole down to the coal bed, which is partly lined by a steel or plastic casing (British Columbia Ministry of Energy and Mines, 2011). The annular space between the casing and rock formations are cemented. The steel or plastic casing and cement prevents exchange of fluids or gases between the well and the surrounding rocks or surficial materials and ensures that drinking water is protected.

Vertical vs Horizontal Drilling

Vertical wells are drilled from the surface straight downward into the overlying rocks above and into the target coal bed. The vertical borehole can be steered and drilled in a horizontal direction as it draws near the coal bed. When

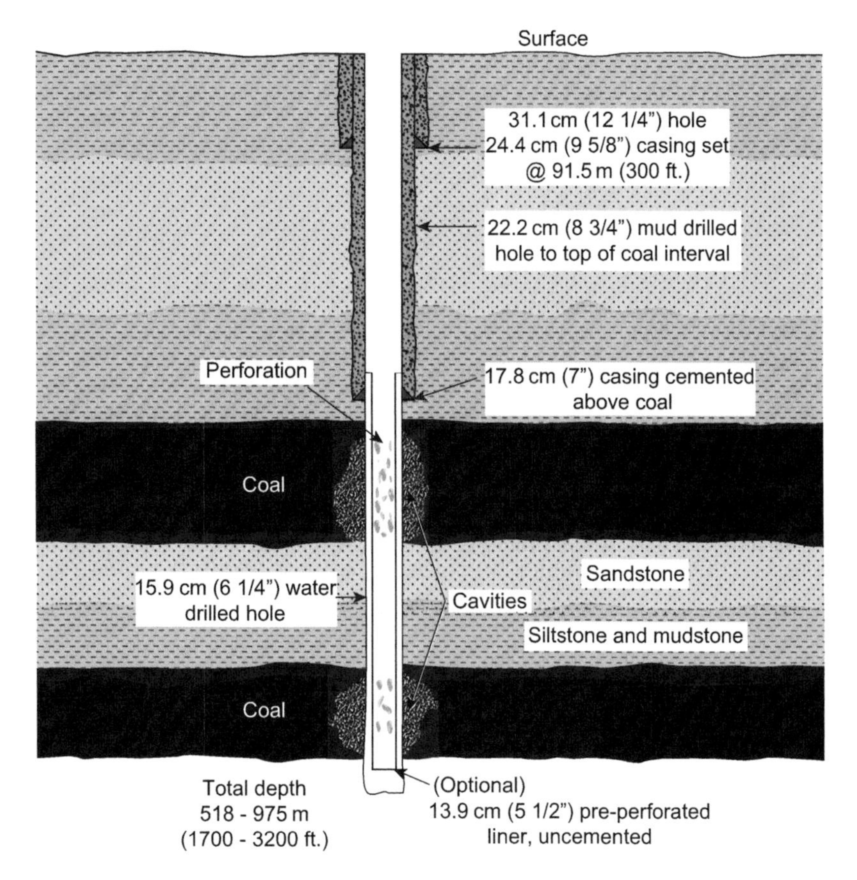

FIGURE 7.3 Schematic configuration of a cased-hole coalbed gas well showing three-sets of casings and perforation for stimulation treatment. *Source: Adopted from Logan (1993).*

the borehole is in the coal bed, it can be drilled laterally or horizontally for nearly 2000 m (Figure 7.7). Underground coal mines commonly employ horizontal drilling technique from the coal face and perpendicular to the face cleat to maximize gas drainage in front of mining. This technique is more efficient in draining gas compared to a vertical well, which has to be drilled for some vertical distance before reaching the coal bed.

The types of drilling techniques, vertical vs horizontal, utilized for coalbed gas production are controlled by (1) lithology of the penetrated formations, (2) thickness of the coal beds, (3) lateral continuity of coal beds, (4) reservoir properties (e.g. fractures or cleats), and (5) geological structures of the penetrated formations (e.g. folded and faulted versus flat bed attitudes). The strengths of the lithology of the formations dictate whether the vertical drilling of the coalbed gas well will utilize conventional rotary or rotary percussion drilling methods. Ramaswamy (2007) suggested that soft rocks require conventional rotary drilling and hard rocks need rotary percussion drilling for rapid penetration.

Ramaswamy (2007) suggested that the horizontal wells are more advantageous than the vertical wells based on the following: (1) longer distance of penetration of the coal; (2) consequently, a larger footprint in the reservoir; (3)

FIGURE 7.4 Representative drill bits used for drilling coalbed gas wells in the Powder River Basin in Wyoming, United States, showing various sizes. (A) 34.3-cm surface drill bit. (B) 22.2-cm longstring drill bit. (C) 15.9-cm drill-out bit. *Photographs courtesy of Tom Doll.*

flexibility of being oriented perpendicular to and intersecting the primary face cleats; (4) more effective in highly fractured coal reservoir; and (5) wells can be extended beyond horizontal into lateral and multilaterals wells (see following section).

Logan (1993) suggested that the utilization of drilling techniques depend on the coal rank, gas content, reservoir pressure, depth, and coal thickness, which differ from coal basin to basin in the United States (Table 7.1). For example, Logan (1993), indicated that bituminous coal contains gas that increases at depth from 8000 to 19,200 cc/g in the San Juan Basin, Mexico, United States. Here, the coal permeability in some overpressured areas (e.g. high-pressure gas kicks) in the coal basin ranges from about 5 to more than 50 mD, which controls both the methodologies to be applied in drilling and gas production. In contrast, Logan (1993), indicated that the subbituminous coal in the Powder River Basin in the United States contains gas from 800 to 2400 cc/g. Here, the coal permeability based on interference tests generally varies from

FIGURE 7.5 Representative centralizers used in the Powder River Basin, Wyoming, United States. The centralizers (17.8 cm or 7 inches) are placed strategically throughout the string to ensure casing centralization. *Photograph courtesy of Tom Doll.*

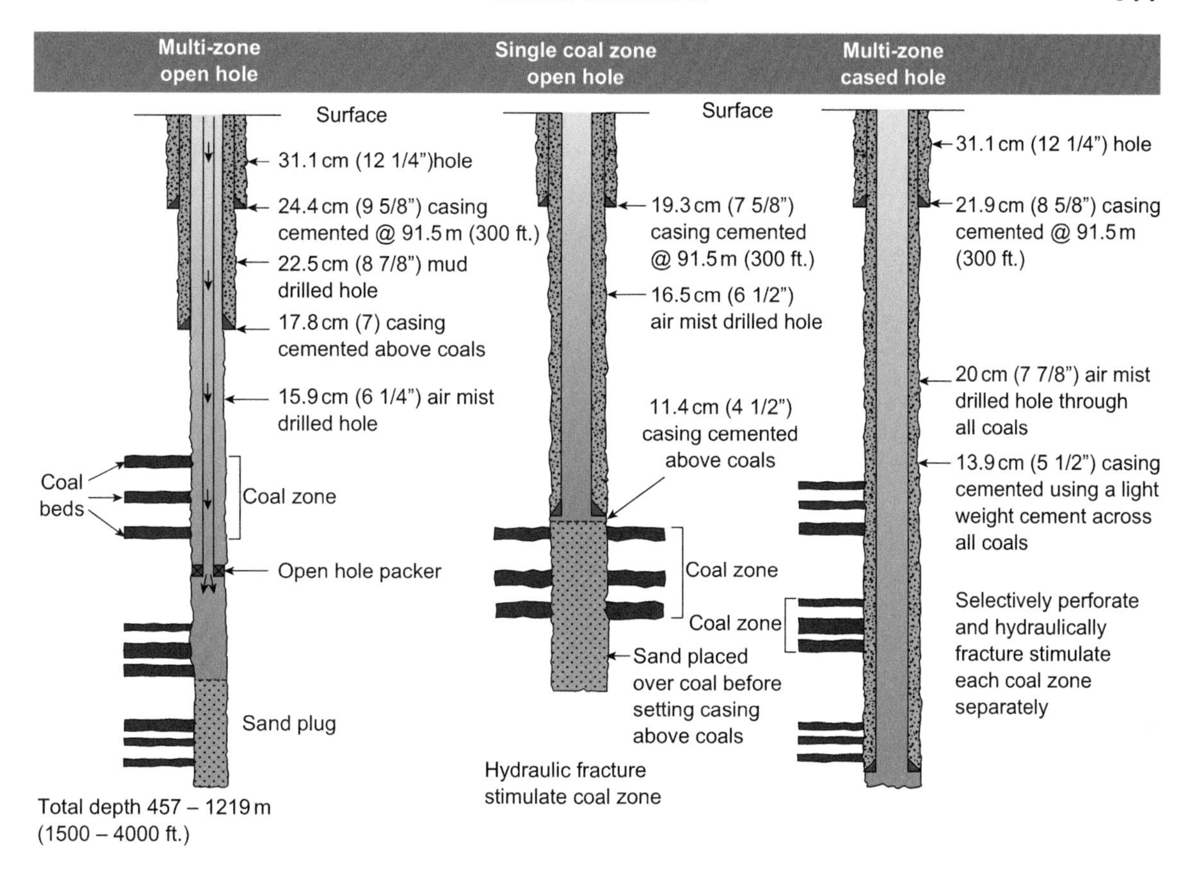

FIGURE 7.6 Schematic configurations of open-hole and cased-hole coalbed gas wells used for multiple coal beds of coal zones. *Source: Adopted from Logan (1993).*

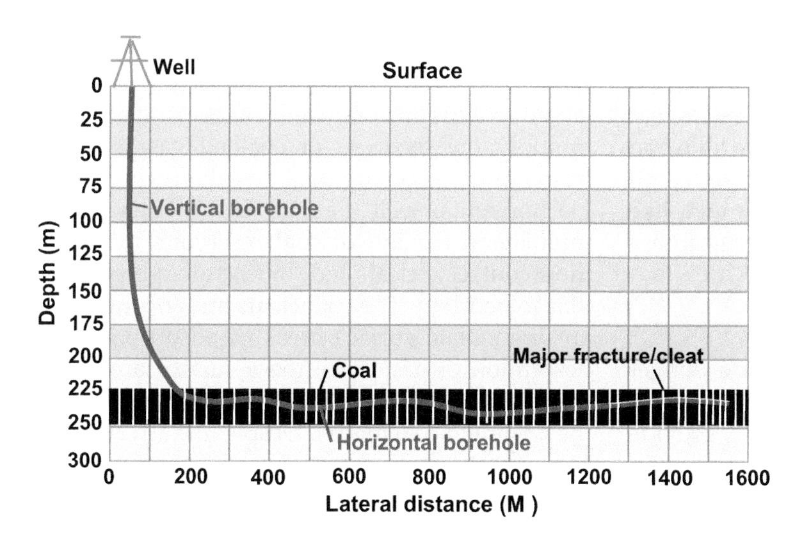

FIGURE 7.7 Schematic diagram of a vertical well steered and drilled horizontally for about 1600 m parallel to the bedding plane of the coal bed in order to make maximum contact to the coal cleat and fracture systems.

TABLE 7.1 Drilling Methods Utilized in United Sates Coal Basins Containing Different Age, Thickness and Depths of Coal Beds

US Coal Basins	Ages of Coal beds	Coal Thickness, Range (m)	Well Depth, Range (m)	Drilling Method
Central and Northern Appalachian Basins (Kentucky, Pennsylvania, Ohio, Virginia, West Virginia)	Carboniferous	0.5–3.0	200–585	Vertical, horizontal, directional
Arkoma Basin (Arkansas and Oklahoma)	Pennsylvanian	1–2	57–903	Vertical, horizontal
Black Warrior (Alabama)	Pennsylvanian	0.3–4	152–1375	Vertical
Illinois Basin (Illinois, Indiana, and Kentucky	Pennsylvanian	0.2–4.5	160–900	Vertical
Powder River Basin (Wyoming)	Paleocene	4–60	60–975	Vertical
Raton Basin (Colorado and New Mexico)	Cretaceous and Paleocene	0.3–4	137–1068	Vertical
San Juan Basin (Colorado and New Mexico)	Cretaceous	12–36	168–1220	Vertical, horizontal
Uinta Basin (Colorado and Utah)	Cretaceous	1–15	335–1200	Vertical

m, meter.

Source: Modified from Rieke (1980), Rightmire, Eddy, and Kirr (1984), Palmer et al. (1993), Pashin and Hinkle (1997), Pashin (2012), Lombardi and Lambert (2001), Huffman and Brister (2003), McCune (2003), Flores (2004), Milici (2004), Cardot (2006), Higley (2007), Ramaswamy, (2007).

325 mD to 1.5 D (Flores, 2004). Thus, the methods of drilling for coalbed gas in the Powder River Basin and San Juan Basin will be different, even though these basins contain thick coal beds. In the coal basins in the eastern part of the United States, Logan (1993), indicated that the most serious problem would be drilling in thin, multiple bituminous coal beds with increasing coal gas content at depth much like in the San Juan Basin but with permeability from 1 to more than 25 mD.

Lateral and Multilateral Drilling

Although horizontal wells provide longer distance and more effective means of coalbed gas recovery, the borehole does not provide maximum aerial coverage of the reservoir. Conceptually, the most efficient gas desorption and recovery methods employ more and longer horizontal boreholes in order to cover large areas in the coal beds. Thus, the borehole to be used in this situation is lateral and multilateral wells, which are forms of horizontal wells. Drilling lateral and multilateral wells off horizontal wells multiply the recovery of coalbed gas reserves from a much larger drainage area than in just a single horizontal well (Palmer, 2007). Lateral and multilateral drilling of coal reservoirs provides more contact to cleats and/or fractures perpendicular to bedding. The coal cleats and/or fractures are immediate surface areas of adsorption and desorption of gas. Also, these surface areas are parts of the coal where gas is diffused to from the adjoining coal matrix. Despite the advantages of horizontal wells, borehole stability analysis of stress conditions based on core plugs from a coal block for a horizontal coalbed gas well in southeast British Colombia, Canada by Gentzis (2009) predicted a few problems (e.g. stuck drill pipes and collapse of wellbore overtime).

Lateral boreholes can be drilled into various directions and patterns from an access borehole such as lateral to right and left of the original horizontal well (Figure 7.8(A)). This method of drilling is utilized in underground coal mines in the United States and China. Multilateral wells can be drilled from the single initial borehole from the coal face in the underground coal mine. Also, lateral wells provide greater area of degasification and methane capture near coal mines in China (Figure 7.8(B)) (Diamond, 1993; Meng & Tian, 2007). In contrast to the vertical coalbed gas wells, directional drilling of lateral boreholes eliminates intercepting overburden

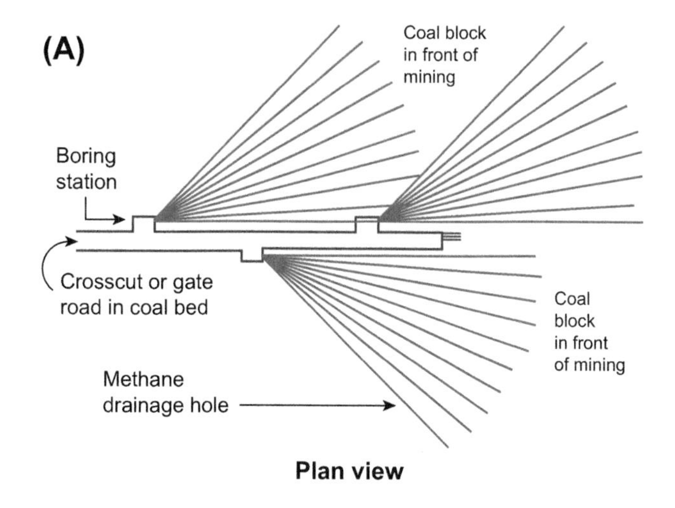

FIGURE 7.8 (A) Schematic plan view in an underground coalmine of coalbed gas drainage holes drilled from boring stations in a gate road (tunnel at the back of a longwall panel for degassing before mining starts) in front of mining along the bedding plane of the coal (Diamond, 1993). (B) Four horizontal wells producing from Permian coal beds off mine sites in China. *Source: Photograph courtesy of Lanyan CBM, Jincheng, China.*

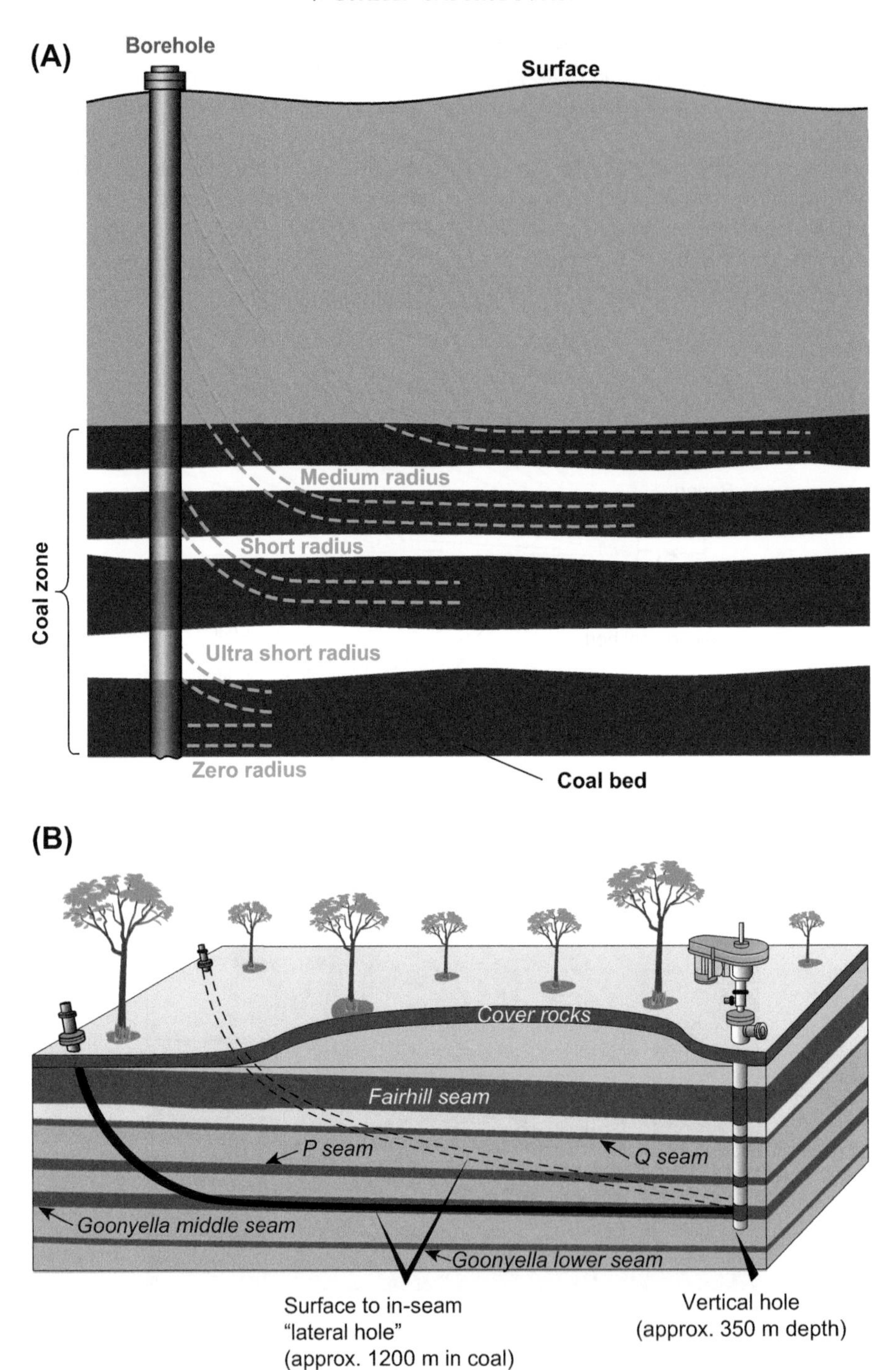

FIGURE 7.9 Diagrams showing (A) vertical and horizontal drillholes drilled at various radii or arcs into horizontal coal beds and (B) slant directional drillholes in front of coal mining in order to drain the coalbed gas from a vertical well in Australia. *Source: Diagram (B) courtesy of Dave Mathew.*

TABLE 7.2 Surface Directional Drilling Defined by Radius or Arc of Turn from the Vertical to Horizontal Wells

Radius Type	Radius of Turn (m)	Achievable Horizontal Distance (m)	Method
Zero	0	3	Telescopic probe with hydraulic jet
Ultrashort	0.3–0.6	60	Coiled tubing with hydraulic jet
Short	1–12	460	Curved drilling guide with flexible drill pipe; entire string rotated from the surface
Medium	60–300	46–1525	Steerable mud rotor used with compressive drill pipe; also conventional drilling can be used
Long	300–850+	600+–12,000+ (the latter is a record)	Conventional directional drilling equipment used; very long curve length of 850–1350 m needed to be drilled before achieving horizontal

m, meter; +, plus.
Source: Modified from USDOE (1993) and USEPA (2009).

rock of targeted coal beds. However, horizontal, lateral, and multilateral wells are more effective in flat lying than in folded, faulted, and thick coal beds. The disadvantage of drilling horizontal, lateral, and multilateral wells is that it increases the cost of coalbed gas development.

Directional Drilling

Surface directional drilling to drain coalbed gas from reservoirs is only recently used in the United States, Australia, and China, even though the technology, which was used for several decades by the oil and gas industry, can be traced to the nineteenth century. The coalbed gas and coalmine methane industries have increasingly relied on directional drilling to unite vertical and horizontal drilling of the same coal bed. For the coalbed gas industry, the technical concept of a long horizontal well theoretically draining coalbed gas from a single site in a 2.6 km^2 area can readily replace 16 vertical wells drilled at 0.16 km^2 spacing (USEPA, 2009). For the underground coalmine gas industry, surface-initiated directional drilling can be used in front of mining removed from mine operations. Environmentally, this drilling method reduces the footprint of gas production.

Surface directional drilling is classified according the radius of turn from the vertical well as shown in Figure 7.9(A) and described in Table 7.2 (USDOE, 1993; USEPA, 2009). Horizontal wells drilled from the surface using the curve configuration presented problems in pumping produced water. However, this problem was remedied by US and Australian drilling companies introducing a new technique, which includes directional drilling of several horizontal wells heading for a vertical well, which produces the gas and water (Figure 7.9(B)). This technique was referred to by USEPA (2009) as surface to in-seam (SIS) drilling. The surface wells drain the same thick coal reservoir as well as closely spaced multiple coal beds at various depths particularly used in Australia to drain gas in front of coal mining (Figure 7.9(B)). The SIS drilling is a common drilling technique in front of coal mining in Australia by slant rigs (Figure 7.10(A)). However, to minimize drilling time, increase coal reservoir permeability, as well as promote efficient and safe drilling operations, coiled tubing drilling is used in draining gas from coals in front of mining and from deeper unmineable coals in Australia (Figure 7.10(B)).

FIGURE 7.10 Photographs showing slant (A) coil tubing (B) drilling for the purpose of draining coalbed gas in front of coal mining in Australia. The coiled tubing, which is spooled on a large reel with downhole motor instead of jointed pipe, is also used in drilling for deeper unmineable coals for gas in Australia. *Source: Photographs courtesy of Dave Mathews and Mohinudeen Faiz.*

The ultimate method of directional drilling horizontal wells is the pinnate pattern of laterals or multilaterals. Pinnate wells show a treelike pattern where four major lateral wells radiate outward from a vertical well. Each of the major lateral wells, in turn, contains side laterals that can be drilled forming a 360° drainage pattern (CDX Gas, 2005; USEPA, 2009). The pinnate well pattern can drain as much as 5.180 m^2 and replace standard 80-acre locations (well spacing). Flourishing coalbed gas projects in the Appalachian and

Arkoma Basins advertise large initial production and quick dewatering programs resulting in draining as much as 90% of the reservoir in 2–3 years (Ramaswamy, 2007; USEPA, 2009). Multiple branched horizontal wells are most effective in highly fractured coals of Carboniferous and Permian ages such as in China (Qui, 2008). Maracic, Mhoaghegh, and Artun (2005) determined the maximum multilateral drainage pattern by taking into account the lengths of horizontal wells, spacing of laterals, and number of laterals.

In Australia, directional drilling is commonly used in front of coal mining in order not to hinder operations during gas and water extractions. Another advantage of drilling and gas extraction in front of coal mining is collecting geological data such as the character of faults and folds to assist in future mining development plans. Directional drilling or " SIS well" begins as a vertical or inclined (e.g. slant) borehole, which is steered through a medium-radius curve to go into the coal bed horizontally (Mitchell, 2005; New South Wales Parliament Legislative Council, 2012). The horizontal well is directionally drilled as much as 1200 m through the coal bed and the path of the borehole can be influenced to take account of the direction of main permeability of the reservoir. A single vertical well is drilled to intersect horizontal wells in multiple coal beds to produce gas and water, which results in infrastructure cost savings.

Drilling Fluids

Ideally, the most effective method of drilling coalbed gas wells with minimal formation damage and contamination is air drilling. Also, air mist, mud, chemical additives (e.g. foam), and coproduced formation water are used for drilling depending on the properties and pressure of the coal reservoir. Most importantly, drilling through overpressured areas, which presented problems in the San Juan Basin, can be performed by slightly overbalanced drilling mud (e.g. using water-based mud that is heavier than required to correspond to formation fluid pressure) to maintain well control while drilling (Logan, 1993). This is in contrast to underbalanced drilling where the pressure is kept lower that the formation fluid pressure by using air, air mist, foam, and formation water (Logan, 1993). The use of air and air mist not only minimized reservoir damage but also required less chemical additives and less environmental contaminations. This drilling method is probably important in areas such as in the Powder River Basin where large volumes of water from groundwater are coproduced with gas.

The use of drilling fluids risks injection into and plugging the fracture systems. Mud and additives used in drilling for coalbed gas wells can form mudcake in the cleat and fracture systems impacting permeability of the coal reservoirs (Gentzis, Deisman, & Chalatumyka, 2009). In general, plugging the fracture systems limits the gas delivery from the coal reservoir. The reliance on the fracture systems to flow gas through the coal reservoir favors drilling horizontal wells. The fracture systems developed normal to the coal-bedding (horizontal) plane are situated such that horizontal drilling parallel to the bed will penetrate the maximum number of fractures, leading to desorption of gas from huge surface areas in the reservoir. Thus, horizontal wells in coal beds provide optimum reservoir contact with the fracture systems to maximize gas flow, production, and recovery. Production from coal reservoirs depends on the fracture systems, which must not be impaired, for optimum gas recovery. However, based on laboratory studies of geomechanics (e.g. variations in applications of confining and axial stresses) of coal plugs and cores, horizontal well stability, changing permeability, and coal strength may be issues to consider (Deisman, Gentzis, & Chalatumyk, 2008; Gentzis, 2009).

BOX 7.1

IMPACT OF ADVANCED DRILLING TECHNOLOGY: FACT OR FICTION

Horizontal and directional drilling has been proved to be the most effective method of economically recovering and boosting coalbed gas production in unmineable and mineable coal beds in selected coal basins in the United States. USDOE (1993) reported that a horizontal well in the San Juan Basin had the advantage of producing seven times as much coalbed gas than a conventional well at the rate of 28,331–198,218 m^3/day. It is no coincidence that horizontal drilling technology has been applied worldwide by the oil and gas industries during the past two decades (Jiang et al., 2010; Luhowy, 1993; Saavedra & Joshi, 2002). USEPA (2009) indicated that directional drilling has the advantage of reaching farther into the coal reservoir from one access well site compared to a vertical well site. Thus, the advantage of directional over vertical drilling is the number of wells to drain gas from an area, which in turn, reduces the environmental "footprint" of coalbed gas.

In the Powder River Basin in Wyoming, United States, the BLM (2003) proposed to approve about 50,000 wells within about 55,500 km^2 area, which is about 1 well per km^2 area. Initially, the Wyoming Oil and Gas Conservation Commission (WOGCC) permitted well spacing of 16 wells per 2.6 km^2 area for dewatering to reduce the area pressure in the coal reservoir. However, due to the large volumes of coproduced water from these wells, which impacted ranchers and landowners, WCCOG ruled for 8 wells per 2.6 km^2 area. Arguably, directional drilling in the Powder River Basin was a potential technology to use in order to reduce the "footprint" of development. However, is the coal geology of the basin in which thick coal beds split and laterally vary within a short distance amenable to directional drilling (Flores, 2004). Is directional drilling effective with the structural geology overprinting on the laterally variable coal beds? More importantly, can improved technology balance the "footprint" with cost-effective gas development, water management, and risk of the local ecology (e.g. sage grouse) as reported by BLM (2003). The reality checks for these questions may be with the advances in modern drilling technology, which permits "laser-like" accuracy using computer modeling and visualization to steer and guide drilling through targeted reservoir rocks (Sanstrom & Longorio, 2002).

WELL COMPLETION

Drilling into coal reservoirs is the first step toward completion, stimulation, and production of the coalbed gas well. The technology required for well completion depends mainly on the characteristics of the coal reservoirs (e.g. thickness, number of beds, vertical separation of beds, lateral variability, permeability, gas content, and depth), stimulation techniques, and production strategies. Also, well completion must consider configuration for dewatering techniques in order to reduce the hydrostatic pressure and degas the coal reservoir. The best completion practices to optimize coalbed gas production were described by Manrique, Poe, and England (2001), Ramaswamy (2007), and Ramaswamy, Ayers, and Holditch (2008). Manrique et al. (2001) reported that in order to optimize coalbed gas production, all available technical, operational, and economic parameters must be integrated to complement coal reservoir characterization and geomechanical properties (e.g. pore pressures, saturations, coal quality, stress gradients and contrasts, Young's modulus, etc.). Manrique et al. (2001) used these parameters in a coalbed gas simulator in order to create optimal scenarios for reservoir-field development in association with increasing dewatering and production distinguishing completion and stimulation conditions. The integration of all these parameters is a step in the right direction but completion still needs a better coordination with coal geology.

Completion Methodology

Conventional completion methodology for coalbed gas wells includes open-hole cavity or under ream and cased hole for single or multi-beds coal reservoirs. Every completion requires preferred drilling techniques as dictated by specific set of problems. For example, the under-ream technique is best preferred for highly fractured, highly permeable, and overpressured coals such as high-rank coals (Logan, 1993). However, the under-ream technique is used in the Fort Union subbituminous coal, which is less-fractured cleats of bituminous and anthracite coal beds and mainly underpressured and very permeable reservoirs (Flores, 2004). Application of various completion techniques appears to vary with different situations such that in Australia, under-ream techniques are utilized in multiple coal bed situations, while in the United States, selective cased-hole perforations are recommended (Logan, 1993; Mitchell, 2005).

Open-Hole Cavity and Under Ream Completion

Completion of coalbed gas wells is usually performed by using open-hole techniques to under ream the coal reservoir. Normally, under reaming of the coal is achieved by drill bits with retractable flanges of various sizes and lengths (Figure 7.11). The under reamer is used to enlarge the hole and breaks the coal forming a cavity around the borehole; hence, this technique is sometimes called cavitation. Enlargement of the hole in the reservoir effectively increases the wellbore diameter and the surface area (e.g. fracture/cleat surfaces) in the coal for gas desorption. The procedure of cavitation is succeeded by a series of repetitive pressurization, depressurization, and clean up of the hole performed by a specially designed completion rig. Logan (1993), described the procedure as follows:

1. Drill string is pulled back into the 17.8-cm-diameter casing cemented to the top of coal and flow the well.
2. Trip back to bottom of hole and circulate out fill with formation water.
3. Pressurized by aerating water with compressor and increase air rate (e.g. 28.3–42.4 m^3/min) until coal returns are observed.
4. Flush drill pipe with water and pull back to 17.8-cm-diameter casing.
5. Shut in well and monitor surface pressure for 2 h.
6. Open well up through hydraulically operated relief valves on blooie lines and blow well down.
7. Repeat trip in to bottom of hole and clean out with formation water and air.
8. Blow hole with air at the rate of 71–85 m^3/min and with water at the rate of about 0.48 m^3/min. Continue to blow hole with air for 1–12 h and decrease water rates as condition in the hole allows.
9. Repeat pull back up into the 17.8-cm-diameter casing. Shut in well for about 2 h in order to permit buildup of reservoir pressure. Air may be injected to build up pressure.
10. Repeat open well up through hydraulically operated relief valves on blooie lines and blow well down.

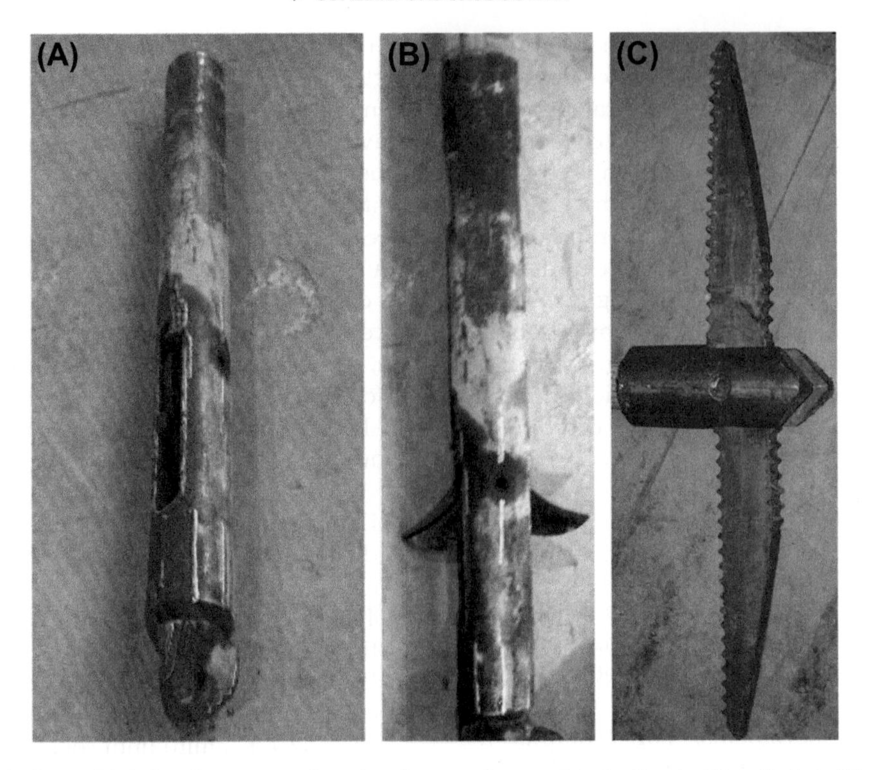

FIGURE 7.11 Scissor-bladed under reamers of varying diameter sizes used in the Powder River Basin in Wyoming, United States. (A) Retracted blade. (B) Deployed blade. (C) Double-serrated blade. *Photographs courtesy of Tom Doll.*

11. Continue operations for 10–20 days until hole stabilizes.
12. Complete as open hole or snub/strip a 13.9-cm perforated liner into place without cement setting cement.

Open-hole completion of wells drilled in the 1980s in the Black Warrior and San Juan Basins were the first type completed in single coal beds. This type of completion was selected because of simplicity and minimum risks. Halliburton (2007) indicated that open-hole completion provides unhindered contact to the coal surfaces from the wellbore. However, open-hole completion is frequently beset by the problem of hole collapse resulting in constant fill up of the cavity by fine to coarse-grained coal fragments. Although this problem can be remedied by packing the hole with gravel, the coal fines still accumulate, which in turn, fill up the fractures and water pumps. The open-hole completion is modified in the Powder River Basin by simply drilling to the top of the coal bed and then setting and cementing the casing. This can often be done in 1 day for shallow wells (Flores, 2004). This procedure is followed by drilling a normal hole through the coal beds and enlarged by drilling with an under reamer. The sharp flanges of the under reamer (Figure 7.11) are gradually extended during drilling such that the diameter of the hole is increased from top to bottom thereby increasing the surface area for gas desorption. The under ream technique significantly improved gas production over the conventional cased-hole completion in the Powder River Basin. However, the under ream technique is not always successful in other basins where coal reservoir permeability is lower than in the Powder River Basin.

Palmer, Lambert, and Spitler (1993) described successes of open-hole cavity in the San Juan and Black Warrior Basins where reservoir pressure and permeability are high. Cavitation in these basins has improved gas production to as much as 283,168 m^3/day. In order to prevent collapse and sloughing in the cavity, Palmer et al. (1993) described the same procedure as Logan (1993). Furthermore, knowledge of coal cavity geomechanics described in Palmer (1992) assist in understanding the phenomenon of borehole instability. The process of borehole instability as discussed by Palmer (1992) includes (1) coal fragmentation at the edge of the cavity or collapse zone, (2) shear failure in the shear zone beyond the collapse zone, (3) extrusion of coal toward the borehole, and (4) subsidence.

Cased-Hole Completion

Cased-hole completion is commonly used for single coal bed and is preferred for multiple coal beds within a coal zone or interval. Using cased-hole completion can improve problems of open-hole completion. According to Logan (1993), the advantages of cased-hole completion are (1) stabilization of the borehole; (2) selective completion of single coal beds; (3) use of fluid for drilling, which will control water flow and overpressured gas kicks; and (4) drilling a "rat hole" for a sump pump below the coal bed, which optimizes dewatering and gas production. The borehole is usually cased and cemented with either low-weight foam, or low-fluid-loss cements. Stage cementing across the coal bed or external casing packers are used to prevent the cement permeating into the coal in order to reduce damage of the reservoir (Boyer & Reeves, 1989; Logan, 1993). However, during the cement procedure formation damage can occur, which is a disadvantage of cased-hole completion.

The casing is perforated (Figure 7.12) through the coal reservoir to provide an avenue for stimulation treatments and gas production. Conventional jet-slotting tools (jet charged) using conventional wireline-conveyed perforating guns perform the perforation to access the coal beds (Figure 7.12). The coal bed thickness determines the number, size, and type of charge, which can vary extensively, but shot densities of 3 shots per 0.093 m^2 are common. For example, the number of perforations may be limited in multiple, thin coal beds. The effectiveness of the cased-hole completions may be reduced when the access points or perfs/slots are blocked by coal fragments behind the casing (Boyer & Reeves, 1989).

Cased-hole completion has been used for multiple coal beds. A major advantage of this completion is the ability to produce a number of closely spaced coal beds (e.g. 2–5) from one well as well as includes thin coal beds otherwise purposely excluded in the single-bed completion. Figure 7.13 shows that each coal in multibed completion is separately perforated. The problem with this technique is optimizing gas production from each coal reservoir resulting in different hydrostatic pressures (e.g. weight of groundwater applied to the coal beds or hydrostatic head). For example, if the groundwater surface for the entire coal zone is above the uppermost coal bed, the hydrostatic pressure will be higher in the lowermost than in the uppermost coal bed. This differential hydrostatic pressure imposes difficulty in uniformly degassing the underpressurized coal beds in the targeted coal zone. This coal-inclusive completion is commonly used in the United States, Canada, and Australia.

ROLE OF COAL GEOLOGY IN COMPLETION STRATEGY

In many developing coal basins in the United States and other countries, thin coal beds are bypassed because of insufficient gas content to be an operationally cost-effective and

FIGURE 7.12 Equipment used for peroration of coalbed gas well in the Powder River Basin in Wyoming, United States. (A) Perforation rig. (B) Perforation gun being lowered from the rig into the well. (C) Gun uses charge with shots penetrating casing with 9.65-mm-diameter perforation at 3 shots per 0.093 m^2 at 90° phasing (see arrows).

economical well. However, deployment of cased-hole completion of multiple coal beds technically makes production of coalbed gas from thin coal beds feasible. This concept makes more sense in coal basins where thick coal beds split into numerous thin coal beds and remerge (Flores, 1986, 2004). Conceptually, where the coal bed is thick, the strategy should be single-bed completion and where the same beds split, the strategy should be multiple-bed completion. However, this concept must be tempered by sedimentology and stratigraphy of the intervening intervals between the coal beds. Channel deposits consisting of sandstones and floodplain deposits consisting of siltstones and mudstones can split thick coal beds. Using the principle of differential compaction of the various sediment types, relatively noncompatible sandstones produce thicker or vertically more widely spaced coal-to-sandstone-to-coal interval. In contrast, the compactible mudstones and siltstones produce thinner or vertically closely spaced zones of coal-to-mudstone/siltstone-to-coal interval. Thus, the finer grained lithofacies-to-coal association may be best developed technically and economically using multibed completion. Based on this scenario, the most promising procedure for improvement of coalbed gas development with analysis of alternative completion technology to adopt is the use of coal geology based on sedimentology, stratigraphy, and depositional environments of the coal-bearing rocks.

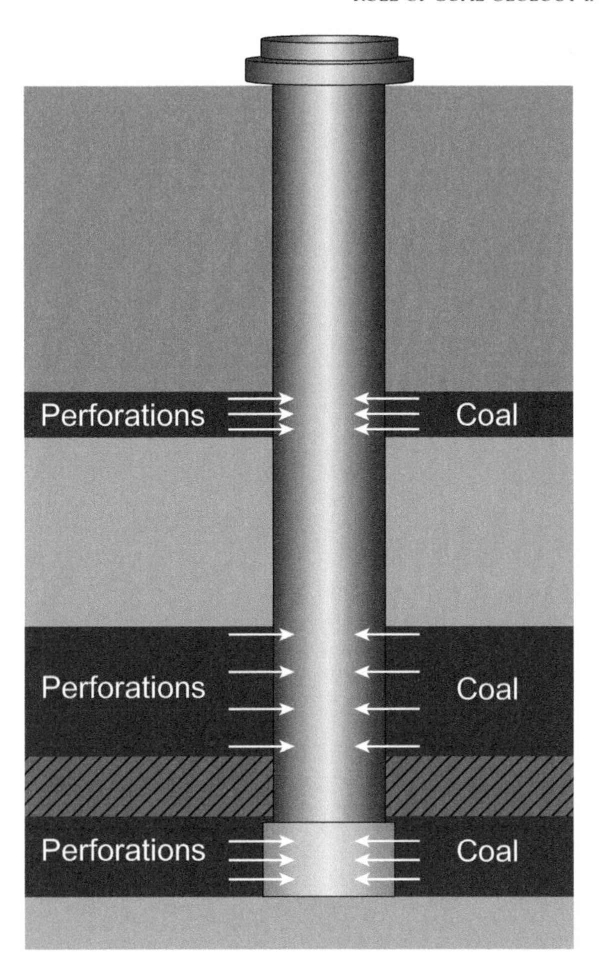

FIGURE 7.13 Perforation of multiple coal beds in the Powder River Basin in Wyoming, United States. *Source: Adopted from WOGCC (2001).*

FIGURE 7.14 Three coalbed gas wellheads each representing completion in a coal bed in the Powder River Basin, United States.

In the Powder River Basin, the initial coalbed gas plays were along the shallow eastern and northwestern margins of the basin. In these basin margin areas, low-cost drilling of shallow and thick (e.g. average >15 m) coal beds was developed using open-hole, single-bed completion. Single coal bed completions are commonly used at one site with as many as four completed wells each with its own set of surface facilities (Figure 7.14). Depletion of coalbed gas reserves in the shallow margins has caused migration of coalbed gas development to the deeper part of the basin where the coal geology is different. The majority of thick coal beds (e.g. Wyodak and Dietz) are split into thinner beds (>8 m) toward the central and deeper part of the basin. These coal beds in turn remerge to form an as thick or thicker coal bed (e.g. Big George) as those along the basin margins (Figure 7.15). Initially, the completion technology in the deeper part of the basin adopted was single-bed completion in the shallow margin. However, development bypassed thin coal beds (e.g. 6 m) arising from resplitting thicker beds due to low gas contents.

The USDOE (2003) conducted a feasibility study of multiple coal bed completion to include thin coal beds in the Powder River Basin. The study was to determine the application of multiple bed completion for the low-rank coals in the

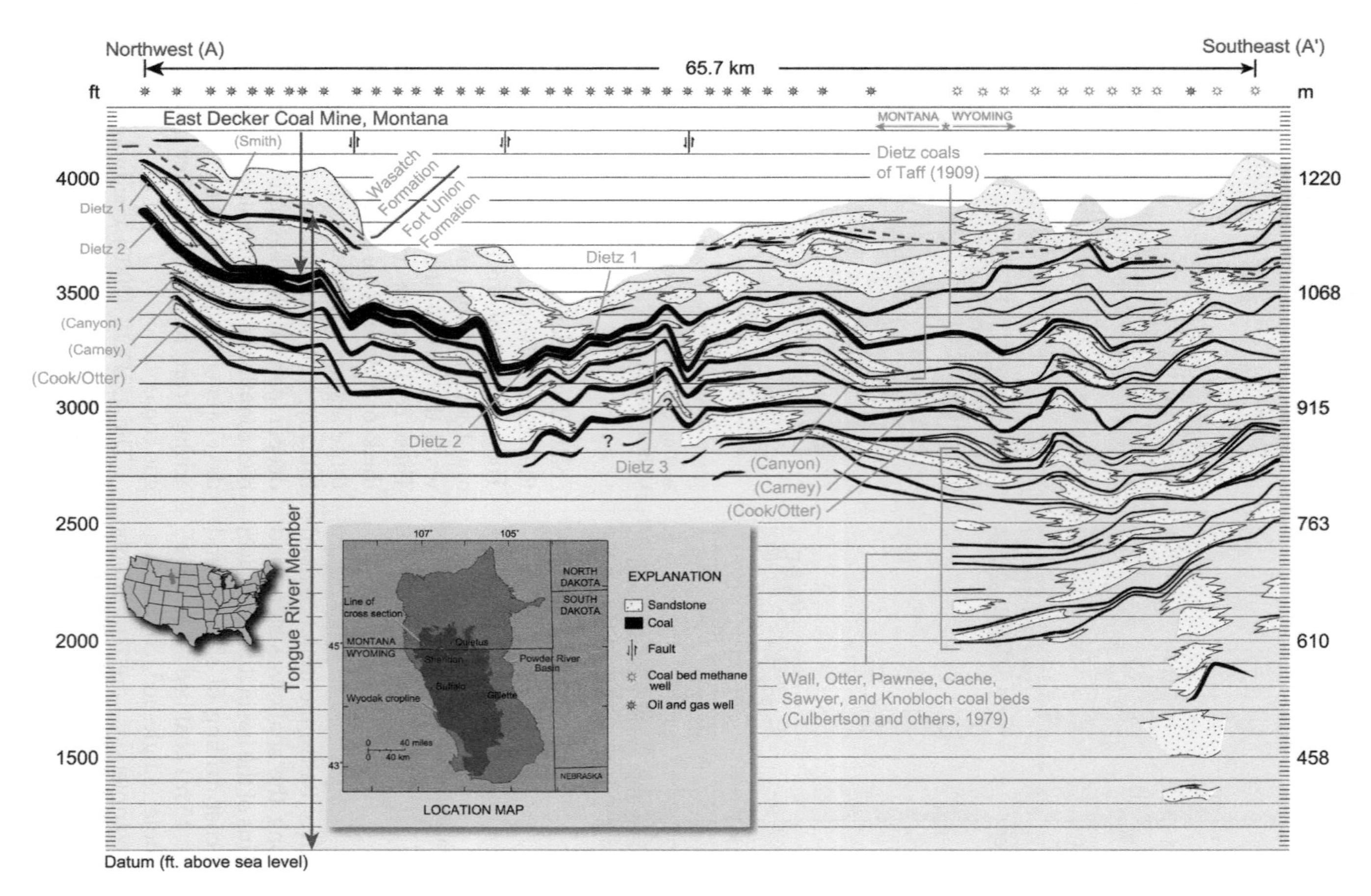

FIGURE 7.15 Stratigraphic cross-section A–A′ showing Dietz coal zone splits into multiple thin beds, which pinch out much like associated coal beds toward the south and east central part of the northern Powder River Basin. See Figure 7.16 for location of cross-section. The reader is referred to Flores et al. (2010) for the references cited in the cross-section. *Source: Modified from Flores et al. (2010).*

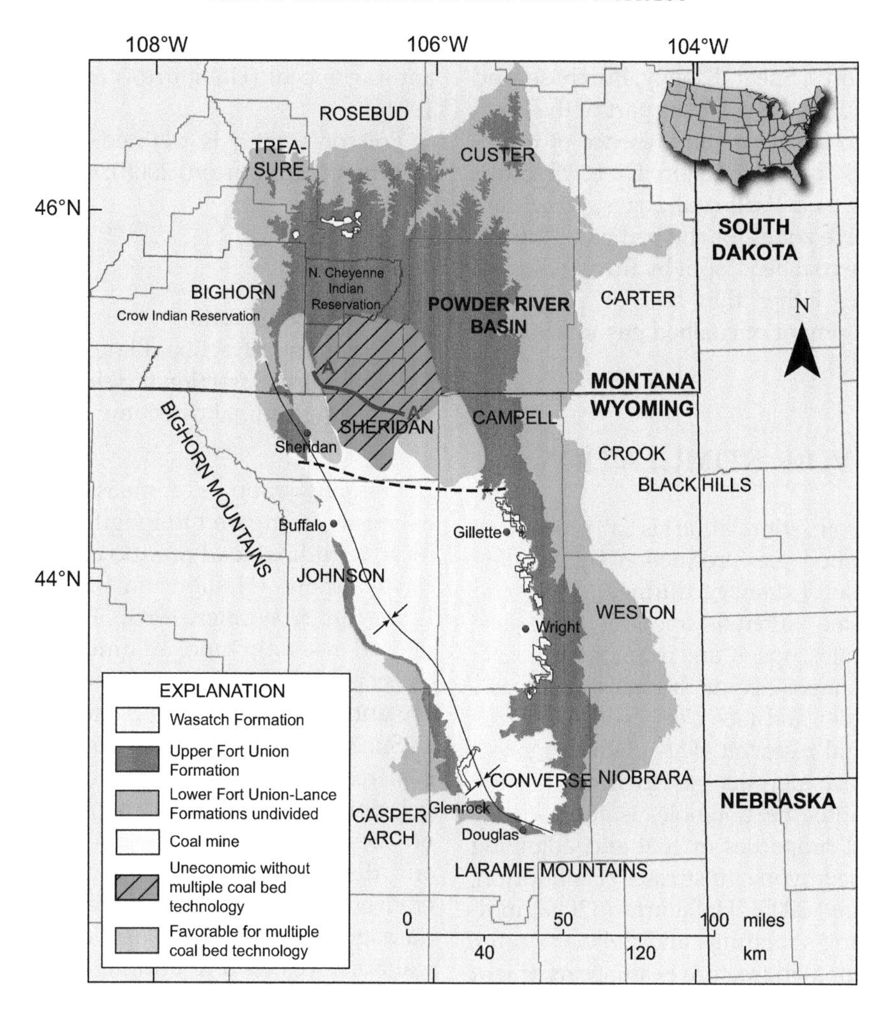

FIGURE 7.16 Map of the Powder River Basin in Wyoming, United States, showing areas where multiple coal bed well completion is uneconomic without and favorable with the technology. See Figure 7.15 for cross-section A–A′. *Source: Modified from USDOE (2003).*

basin with the purpose of increasing the technically recoverable reserves and economic benefits. The approach of the USDOE (2003) study in the Powder River Basin included the following: (1) obtain and analyze 270 new geophysical logs at the township level; (2) tabulate thin coals from 1.5 to 6.0 m by depth, stratigraphic interval, and basin area; (3) correlate these thin coal beds; (3) establish packages of multiple thin coal beds for multiple bed completion; (4) develop new gas content and adsorption isotherm analyses for comparative application to the coal beds; and (5) estimate ultimate recovery of the coalbed gas and water production.

The primary result of this study is recommendation of multiple completions of wells in the northern part of the Powder River Basin (Figure 7.16). One major benefit of the approach with the use of multiple bed completion is the reduction of development footprint in the basin. The work of Flores, Spear, Kinney, Purchase, and Gallagher (2010) in the northern part of the Powder River Basin supports the presence of multiple coal beds of the major Dietz-Wyodak, Cook, Pawnee, and Wall coal zones, which are major productive zones for coalbed gas. However, the multiple coal beds, split by fluvial channel sandstones, are either thin and/or pinch out making development of coalbed gas uneconomical (Figure 7.15).

RESERVOIR STIMULATION

Coal reservoir stimulation is a process of enhancing coalbed gas production by cleaning up the borehole of damage during drilling and completion using different fracture treatments. Stimulation dilates, opens, and extends the width, length, and connectivity of the natural fracture systems (e.g. butt and face cleats and megafractures) of the coal reservoir. This phenomenon of fracturing, which permits gas to flow through the fractures into the borehole, is founded by the mechanical properties of coal and adjoining rocks as imposed by in situ stresses (Halliburton, 2007; Olovyanny, 2005). Halliburton (2007) indicated that Young's modulus and Poisson's ratio, which are essential in fracture evaluations, might define the mechanical properties and related stresses producing fractures in coal. Young's modulus is an elastic property of rock expressed by Halliburton (2007) as follows:

$$E_x = \sigma_x/\varepsilon_x$$

where:

E_x = Young's modulus (psi or MPa),
σ_x = stress, x direction (psi or MPa), and
ε_x = strain, x direction.

The Young's modulus is a significant factor in forming the width of fractures in coal. The Young's modulus does not affect bituminous coal (e.g. from high-volatile B to low-volatile rank); however, the effect increases rapidly in anthracite coal (Halliburton, 2007; van Krevlen, 1981).

Poisson's ratio is defined by the following equation (Halliburton, 2008):

$$V = -\varepsilon_2/\varepsilon_1$$

where:

V = Poisson's ratio,
ε_2 = strain or fractional lateral expansion, and
ε_1 = strain or fractional deformation in longitudinal direction.

Poisson's ratio is a measure of the lateral expansion compared to longitudinal contraction for longitudinal load or ratio of strain to longitudinal strain (Halliburton, 2008; Warpinski, Branagan, & Wilmer, 1985). The Poisson's ratio of coal reservoir and adjoining rocks controls stress profile, which determines fracture boundary and orientation (Halliburton, 2008).

Successful fracture stimulation treatments in coal reservoirs commonly include plain water, gel, and foam fluids and sand proppant. Each of these treatments has its own advantages and disadvantage. For example, plain water is environmentally preferable and tends not to damage the skin or wall of fractures. The chemical additives tend to adhere along the surfaces of the fractures as cakes as well as have the tendency to plug the fractures. The additive fluids eventually rise from the reservoir to the surface due to differential internal pressures. In the United States, the recovered fluids or flowback need to be treated as regulated by the National Pollutant Discharge Elimination System prior to recycling into the surface water or underground injection (USEPA, 2012). Fracturing operations in coal reservoirs in the United States, Australia, and Canada and their results have been discussed best in numerous publications (Colmenares & Zoback, 2007; Davidson, 1995; Halliburton, 2008; Lambert, Niederhofer, & Reeves, 1990; Lehman, Blauch, & Robert,

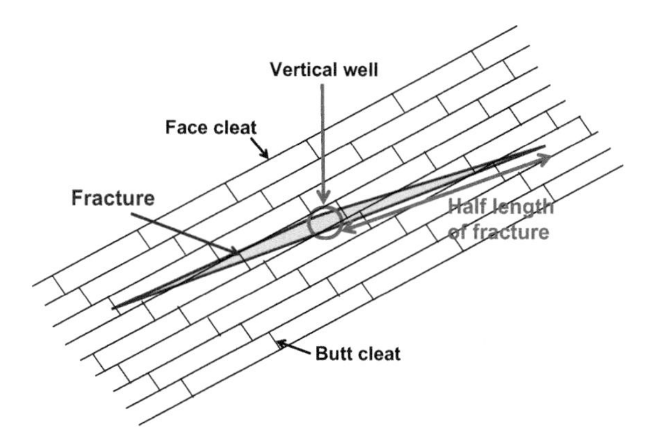

FIGURE 7.17 Generalized plan view showing location of a vertical well in a coal bed with orthogonal face and butt cleat orientations. The fracture created by stimulation shows a half-length from the well to the endpoint connecting face and butt cleats for gas to flow to the wellbore.

1998; Murray, 1996; Palmer et al., 1993; Ramaswamy, 2007; Wold, Davidson, Wu, Choi, & Koenig, 1995).

Hydraulic Fracturing

Hydraulic fracturing is the application of fluids consisting of plain water or water-based additives, which are pumped into the coal reservoir at high pressure. The fluid pressure normally exceeds the tensile strength of the coal reservoir such that the fluids open, lengthen, and enlarge fractures and cleats many of which extend tens of meters from the well, improving their connectivity (Figure 7.17). The opened fractures then permit the fluids to penetrate the network of fractures and cleats allowing avenues of water and gas to flow (e.g. Darcy flow) during dewatering and degasification.

The effects of fracture treatments on the Pittsburgh coal in Pennsylvania were demonstrated in a computer simulation of data for 5.5 years (Halliburton, 2008; Hunt & Steele, 1991). A comparison of gas production of the stimulated and unstimulated Pittsburgh coal reservoir in half-length fractures (45 m, 75 m) is shown in Figure 7.18. Half-length fracture is the length from the borehole (Holditch, 2009). The stimulated coal reservoir before the end of the first year outproduced the unstimulated coal reservoir by about six (45-m half-length) to eight (75-m half-length) times. In addition, the 75-m half-length fractures continued outproducing the 45-m half-length fractures for 3.3 years after which both continued to produce at the same rate. These treatments are usually pumped through perforated casing in the coal reservoir, although the same treatment may be performed in open cavity holes.

In the Black Warrior Basin, the relationship of cumulative coalbed gas production with half-length fractures and permeability is shown in Figure 7.19 (Halliburton, 2008; Spafford & Schraufnagel, 1992). The simulation, much like in the Pittsburgh coal, indicates that the longer half-length fractures (e.g. 121−182 m) outperformed the coalbed gas production of the shorter half-length fractures (e.g. 30−91 m). However, although the longer half-length fractures have improved coalbed gas production, permeability becomes less significant at the high and low ends as indicated from 0.5 to 6.0 mD (Halliburton, 2008). Thus, stimulation to lengthen fractures of coal reservoirs in the Black Warrior Basin with permeabilities below 0.5 and above 6 mD becomes trivial.

Hydraulic Fracture Fluids

The fluid system used with coal reservoir hydraulic fracturing varies from plain water, oil- and water-based treatments, to polymeric substances, which are mixed with proppants (USEPA, 2004). In order for the fluid system to function, each type of fluid requires unique properties and qualities as follows (Powell et al., 1999; USEPA, 2004): (1) viscosity for creating adequate width of the fracture, (2) optimal travel distance to lengthen the fracture, (3) carry enough proppants into the fracture, and (4) minimum gel for easier degradation to reduce cost.

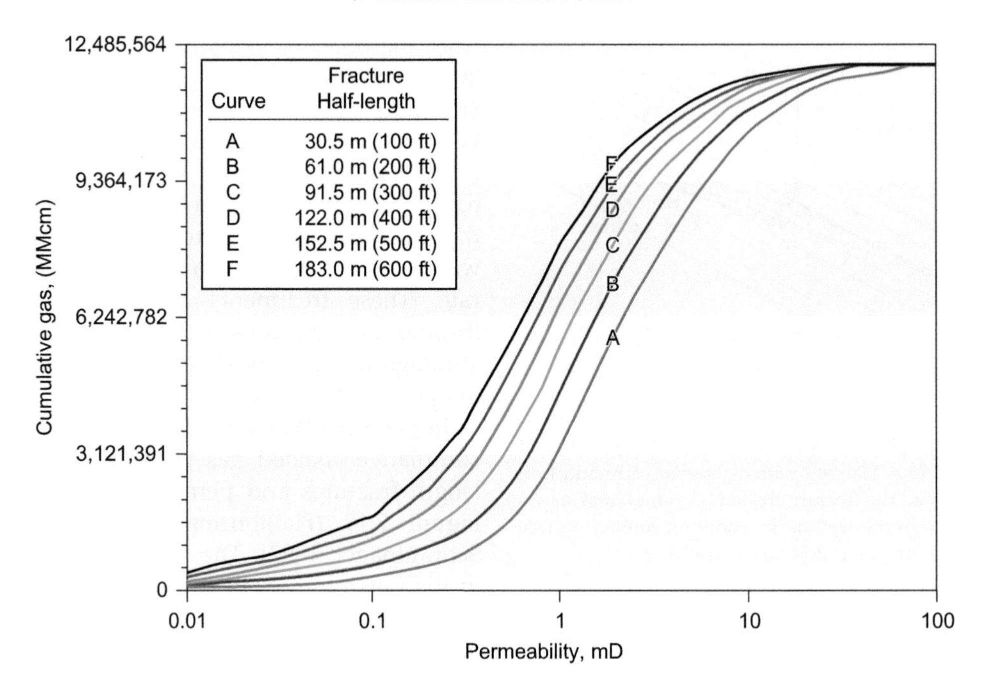

FIGURE 7.18 Chart showing computer simulation of the effects of various fracture half-lengths on the rate of coalbed gas production in relation to number of days. *Source: Modified from Hunt and Steele (1991) and Halliburton (2007).*

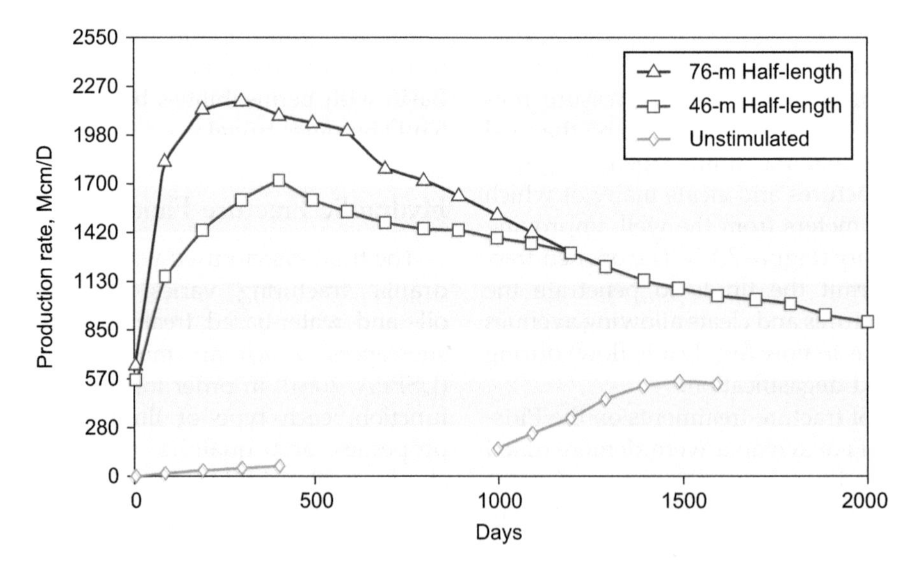

FIGURE 7.19 Chart showing the effects of coal permeability on coalbed gas production for various hydraulic fracture half-lengths in the Black Warrior Basin resulting from simulator estimates. Mcm/D = thousands cubic meter per day. *Source: Modified from Spafford and Schraufnagel (1992) and Halliburton (2007).*

BOX 7.2

HYDRAULIC FRACTURING: "FRAC" OR FICTION

Hydraulic fracturing is a technology used in conventional and unconventional oil and gas development worldwide. However, the United States and Canada represent majority of the hydraulic fracturing, which is popularly called "fraccing or fracking". Hydraulic fracturing is the practice of injecting fluids consisting of water, chemical additives, and proppants at high pressure into hydrocarbon reservoir rocks (Cooley & Donnelly, 2012). Hydraulic fracturing is designed to improve conventional oil and gas production but more so for unconventional gas such as coal bed, shale, and tight sand reservoirs. The unconventional gas reservoirs have especially low permeability to flow gas that they need enhancement by using hydraulic fracturing. However, hydraulic fracturing is not as widely practiced everywhere in unconventional reservoirs such as for the coal beds in the Powder River Basin (USEPA, 2004).

For the past decade, use of hydraulic fracturing technology has spawned intense controversy between advocacy groups and the energy industry and public concerns in the United States, Canada, Australia, Europe, and Great Britain. The controversy revolves around potential contamination of surface water and groundwater used for drinking by chemical additives in the fracture fluids, which either remain in the reservoir or leak into aquifers (Myers, 2012). Hydraulic fracture fluids that are recycled from the wells are stored temporarily at the well site and transported for treatment at disposal sites. However, the large volume of water used with these fluids for fracturing remains uncertain and reusing the treated water for fracturing would alleviate the problem of freshwater use (Nicot et al., 2011). According to Cooley and Donnelly (2012), the extent of groundwater contamination is indeterminate due to lack of baseline data. Although chemicals linked to hydraulic fracturing have been found in drinking water by USEPA (2011), groundwater contamination related to hydraulic fracturing has not been confirmed (API, 2010; Cohen, Grigsby, Bessinger, McAteer, & Baldassare, 2012; NYSDEC, 2011).

The main objectives of the hydraulic fracturing fluids are to initiate, dilate, enlarge, and lengthen fractures as well as convey the proppants to buttress or "prop" the fractures. The effectiveness of transfer and positioning of the proppants is controlled by the viscosity or resistance to flow of the fluid system. In coalbed reservoirs, cleats and fractures are formed in situ during coalification and deformation of the rocks. Thus, the role of hydraulic fracturing fluids is basically to expand, lengthen, and buttress the preformed cleats and fractures. Based on the coal reservoir characteristics, reservoir engineers may design formulas of fluids more effective for hydraulic fracturing. The mixtures of fluids designed for more effective hydraulic fracturing are provided in Table 7.3 by USEPA (2004). Detailed information on these fracturing fluid systems are found in Ely (1989), USEPA (2004), Ramaswamy (2007), and Haliburton (2007) who group them into the following categories: (1) plain water and potassium chloride water, (2) gelled fluids (e.g. linear and cross-linked gels), (3) foamed gels and gas, (4) acids (e.g. hydrochloric, formic), and (5) combination (two or more) or hybrid of the above fluids.

TABLE 7.3 Composition of Hydraulic Fracturing Fluids for Stimulation Treatment of Coalbed Gas Wells

Product	Chemical Composition	Hazard/Toxicology	Ecological
Linear gel			
Delivery system	30–60% by wt. guar gum derivative 60–100% by wt. diesel	Harmful when swallowed; combustible. Chronic effect/carcinogenicity.	Slowly biodegradable Not determined
Polymer	<2% by wt. fumaric acid; adipic acid	Flammable vapors	Partly biodegradable
Polymer slurry	30–60% by wt. diesel oil #2	Causes irritation when swallowed; flammable. Carcinogenicity	
Water gel	60–100% by wt. guar gum 5–10% by wt. water 0.5–1.5% by wt. fumaric acid	None Mildly eye irritant Chronic effect/carcinogenicity	Biodegradable
Cross-linker	10–30% by wt. boric acid; ethylene glycol; monoethanolamine; sodium tetraborate decahydrate	Harmful when swallowed; combustible Mildly irritant to eyes, skin;	Not determined Partly biodegradable Low fish toxicity
Foaming agent	10–30% by wt. isopropanol; salt of alkylamines; ethanol; 2–butoxyethanol; water 1–5% by wt. diethanolamine 25–55% by wt. ester salt 0.1–1% by wt. polyglycol ether	Harmful if swallowed or absorbed in skin; highly flammable Chronic effect/carcinogenicity Eye, skin, and respiratory irritation	Not determined Harmful to aquatic organisms
Acid treatment	30–60% by wt. hydrochloric acid 85% by wt. formic acid	Harmful if swallowed; Eye, skin, and respiratory burns Chronic effect/carcinogenicity	Not determined
Breaker fluid	60–100% by wt. diammonium peroxidisufhate	Harmful if swallowed; Eye, skin, and respiratory burns. Tissue damage	Not determined
Microbiocide/biocide	60–100% by wt. 2-bromo-2-nitrol, 3-propanedol; 2-dibromo-3-nitrilopropionamide 1–5% by wt. 2-Bromo-3-nitrilopropionamide	Harmful if swallowed Eye and skin irritation; causes severe burns and allergic reaction on repeated skin exposure. Discomfort to mouth, throat, and stomach Chronic effects/ carcinogenicity	Not determined
Acid corrosion	30–60% by wt. methanol; pyridinium, 1-(phenylmethyl)-, ethylmethyl derivatives, chlorides 5–10% by wt. propargyl alcohol, propan-2-ol 15% by wt. thiourea 1–5% by wt. Poly(oxy-1, 2-ethanediyl)-nonylphenyl-hydroxy 10–30% water	Fatal if swallowed; causes eye and skin irritation/burns, headache, dizziness, blindness, central nervous system effects, burns in respiratory tract; injury to lungs, throat, and mucous membrane; and tissue damage Chronic effects/ carcinogenicity	Partly biodegradable Toxic to aquatic organisms

wt., weight; % = percent.
Source: Modified from USEPA (2004).

Plain Water and Potassium Chloride Water

Plain water hydraulic fracturing is pumping coproduced or formation water and/or treated water to stimulate the coal reservoir. This treatment is preferred for highly fractured coal reservoirs where water pressure can be applied and still achieve enlarging and lengthening fracture, and cleat systems. The advantage of using plain water treatment is that it is an economical or low-cost application and is an environment-friendly process of reservoir stimulation. Proppants may or may not be used with plain water treatment. According to Ramaswamy (2007), the plain water transporting capacity or density is about 100–200 kg/m^3 due to low viscosity, which is adequate but not as efficient as the gelled fluids. This carrying inefficiency may result in lower rate of groundwater and coalbed gas flow and thus gas production.

A modification of plain water fracturing is water with potassium chloride with gelled fluids, polymers, and surfactants (Ely, 1989). Plain water and water-based fluids normally must be pumped at high rates or velocities and optimal pressure to be most effective. This process, in turn, results in friction damage to the coal face around the borehole and fracture and cleat surfaces found beyond the borehole. Thus, friction reducers (e.g. latex polymers and copolymers of acrylamides) are added to the water-based fracturing fluids to mitigate the friction damage. The various hydraulic fractured fluid systems used in coalbed gas-producing basins in the United States are shown in Table 7.3.

Gelled Fluids

The limitations of plain water treatment led to the introduction of water-based gelled fluids for more viscosity and longer conveyance of proppants. The used of water gelates, which are thickeners, is based on the permeability, porosity, pressure, temperature, and thickness of the coal reservoirs (USEPA, 2004). In order to increase the viscosity of plain water, the service industries introduced the gelled fluids composed of linear and cross-linked fluids.

A linear fluid is composed of gum and guar derivatives (e.g. hydroxypropyl guar and carboxymethyl hydroxypropyl guar) and cellulose derivatives (e.g. carboxymethyl guar and hydroxyethylcellulose) (USEPA, 2004). Ely (1989) indicated that guar is a polymeric substance derived from the guar gum plant. Guar gum or cluster bean is used as a vegetable in India and is a polysaccharide, a long chain made of sugars galactose and mannose, which is a food-grade product used in ice creams (Guar Gum, 2006). The chemical properties of guar gum, which are relevant for use in hydraulic fracturing fluids, are the capacity to (1) suspend solids, (2) bind by hydrogen bond with water, (3) control viscosity of aqueous solutions, and (4) form hard films (Guar Gum, 2006). These properties of guar gum are harnessed to formulate a gel to increase fluid viscosity for efficient conveyance of sand proppants and turbulent flow. Guar gum is biodegradable (Table 7.3). According to USEPA (2004), diesel, which is a petroleum distillate, is used instead of water to dissolve guar powder for more transportation capacity per unit volume.

Cross-linked fluid or gel is composed of metal-ion-cross-linked guar made possible by metal ions (e.g. aluminum, chromium, titanium), later complimented by low-residue cleaner such as cross-linked hydroxypropyl guar (Ely, 1989). Cross-linked gels are combined with linear gel fluid in order to permit more effective conveyance of the proppants and thus, increased coalbed gas production rates. The viscosity of cross-linked and linear fluids requires degradation by breaker fluids to improve recovery of the gelled fluids (e.g. flowback). Breaker fluids are either combined with the gelled fluids or applied separately as time- or temperature-dependent treatments (USEPA, 2004). Breaker fluids mainly consist of acids, oxidizers, and enzymes (USEPA, 2004). The introduction of enzymes and other organic-based fluids in the fracturing fluids provide a substance for bacterial growth, which in turn, reduces effectiveness of proppant transport and fracturing performance. Thus, biocides are introduced into mixing tanks with gelled fluids to destroy and/or inhibit bacterial growth. The deleterious enzymes may find their way growing into gas-gathering pipelines as discovered in a few instances such as in the Powder River Basin; thus, cleanup may be required. However, in the case of the Powder River Basin where the coalbed gas is generated by bacterial activity (Flores, 2008; Gorody, 1999; Rice, 1993), the proposed use of methanogens to convert organic matter to coalbed gas could be affected.

Perhaps the most important negative effect of the introduction of biocides into the coal reservoir is potential destruction of methanogens. This is especially effective in coal reservoirs such as in the Powder River Basin where biogenic coalbed gas is generated by methanogens. The biocides may greatly reduce or eliminate the potential use of the coal reservoirs for methane generation during the postcoalbed gas extraction. In addition, the negative effect of biocides would be more critical in higher, more-permeable virgin coals for the possibility of applying coal-to-coalbed gas technology to sustain and extend gas production in other basins.

Foam Gels and Gas

Hydraulic fracturing with foam, which is formed by trapped gas, is used to carry proppants into the fracture and cleat systems in coal reservoirs. Foams are much like beer bubbles from freshly opened bottle charged with gas as it pours out. Thin films of liquid substance separate the bubbles. The fracture fluids of foam utilize nitrogen and carbon dioxide as basic gas components (USEPA, 2004). In general, foam contains large amounts of gas bubbles of various sizes separated by liquid walls called films, which may become thinner or thicker with reduction or increase of the liquid, respectively. Thus, the state of foaminess is controlled by the gas/liquid, ratio forming either a gas-energized liquid or a true foam substance (Ely, 1989). The state of foaminess, in turn, reflects the ability of the medium to carry proppants and "frac" the coal reservoir. However, the amount of proppants used in the fracturing fluids reduces the liquid by about 75% in lieu of using linear or cross-linked gels (Ely, 1989). Foam and gelled fluids are frequently combined for more effective fracturing and transporting of additional volume of proppants (e.g. 798 kg/m^3) into the fracture and cleat system.

The combined foam and gelled fluids are valuable in coal reservoir with low permeability and hydrostatic pressure. A major advantage of using foam gels is less damage to the fracture permeability of the coal reservoir. However, foam mixtures may leak from the wellbore, which necessitates the use of fluid-loss additives such as fine-grained materials (e.g. 100 mesh sand and resin, silica flour, clay, and plaster materials) (Ely, 1989). The disadvantage of this is that the additives have a tendency to plug the fractures and cleats of the coal reservoir. Another disadvantage of the foam and gas mixture is carbon dioxide migrating into the pore systems (e.g. micropores, mesopores, and macropores), which in turn is adsorbed onto the pore surfaces, leading to plugging and swelling of the coal matrix and constricting passageways, thus ruining the permeability (Puri, King, & Palmer, 1991).

Acids

Acid stimulation or acidizing in association with hydraulic fracturing is a process of dissolving minerals (e.g. inorganic matter, see Chapter 5) in the coal reservoir. The acidizing process is applied only when there is pervasive mineralization of the fracture and cleat system. When applied, the acid dissolves the inorganic minerals that plug the fractures and cleats (Figure 7.20) resulting in opening and interconnecting passageways to flow water and gas through the reservoir. Acidic stimulation fluids include hydrochloric, acetic, and formic acids or combinations of these acids. The acids are

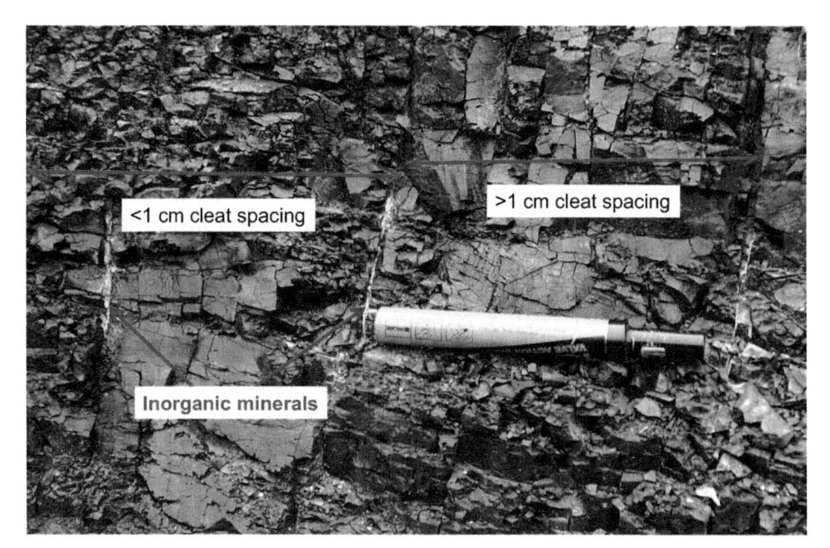

FIGURE 7.20 Inorganic mineral infilling of closely spaced face cleats (from <1 cm to >1 cm) in a subbituminous coal in the Powder River Basin. The permineralized cleat/fractures range 4–6 cm long. For scale, the color pen is about 12 cm long.

typically pumped into the reservoir at low concentration typically resulting from dilution with water- and gas-based fluids. Gelled and water–methanol fluids are combined with acidic fluids in order to increase the distance of transport to reach additional mineralized fractures and cleats (Ely, 1989). In addition to dissolution of inorganic minerals in the coal reservoirs, acid is used to break and degrade the viscosity of the fracturing fluids. This process, in turn, thins the fracturing fluids permitting easier flow to the surface or flowback for recovery and treatment. Because acids corrode steel tubes and casings, acid corrosion inhibitors such as acetone solvent are used as acid mixed fluids (USEPA, 2004).

Proppants

A significant solid additive to the hydraulic fracturing fluids is proppants, which are emplaced in the fracture and cleat systems for support. The proppants are in the form of silica sand, resin-coated sand, or ceramic sand. The properties of proppants in a pack are 20–40 mesh, 82 °C, and 27.6 MPa closure (Halliburton, 2007). The size of the proppants depends on the width of the fractures and cleats. Larger proppants allow for wider fractures for better Darcian gas and water flow and better fracture fluid flowback. Thus, the permeability of the coal reservoir secondarily depends on the size as well as the strength, roundness, and purity of the proppants (USEPA, 2004). The rate of transport and emplacement of proppants into a fracture was estimated to be 4000–11,000 kg of proppants for 17–57 m^3 fracturing fluids in 0.5 h (Powell, Hathcock, Mullen, Norman, & Baycroft, 1997).

One disadvantage of using proppants combined with hydraulic fracturing fluids of high viscosity is the production of fine-grained materials in the fracture and cleat systems and around the borehole. According to Halliburton (2007) fine-grained materials are created by the following processes: (1) adsorption of the fluids on the organic matter surface of the fractures and cleats, (2) etching by the fluids of the organic matter along the fractures and cleats, (3) damage to the skins of the fractures by high-pressure application, (4) abrasion by proppants mixed with fracturing fluids on the fracture and cleat surface, and (5) erosion of coal partings composed of mudstone and shale in powdered fine-grained materials.

The impact of the fine-grained materials is deposition and blockage of the fracture and cleat systems, which in turn, gradually deteriorates the permeability preventing water flow and gas production. The fine-grained materials are also recycled and mixed with the fluids, in turn, increasing viscosity, further damaging the fractures and cleats and eventually affecting reservoir pressures. The effects of fine-grained materials may be lessened by use of various grain sizes of the proppants in which the smaller size proppants are emplaced in smaller fracture and cleat systems leading to their blockage. This process of size redistribution reroutes the fracturing fluids into the larger fractures and cleats (Halliburton, 2008).

In order to demonstrate the role of hydraulic fracture treatments, Palmer (2010) wisely compared fraccing fluids and methods and proppant types applied to a number of coal beds at various depths of different US and Canadian coalbed gas-producing basins (Table 7.4). The fraccing methods are grouped into staging with three coal beds per stage and pinpoint with coiled tubing drilling. The pinpoint fraccing method targets a single coal bed for treatment. The fraccing fluids mainly call for treatments with gelled fluids and foams (e.g. linear and nitrogen) using 20–40 mesh proppants.

More specifically in the Raton Basin, Maccartney (2011) described hydraulic fracture stage using coiled tubing for pinpoint staging (Figure 7.21). The stage consisted of 17,886 l of foamed fluid (70% nitrogen), 30% water that is recycled coproduced water from coalbed gas wells, 27 kg of natural guar gel compound, 15 l of an organic enzyme to breakdown the gel, and 57 l of a mild detergent to create foam. Approximately 3628 kg of sand proppant is placed for every 0.3 m of coal stimulated. Based on volumetrics and computer modeling, hydraulic fraccing created fracture growth from 37 to 61 m in a half-length (Maccartney, 2011).

COAL RESERVOIR CHARACTERIZATION VS WELL COMPLETION AND STIMULATION

Understanding the characteristics of coal reservoirs and their application to well completion and stimulation is not an abstract pursuit.

TABLE 7.4 Comparative Hydraulic Fracture Treatments of Coalbed Gas Basins and Fields in the United States and Canada

Basin (Field)	Cherokee (1)	Central Appalachian	Raton	San Juan (Below Fairway)	Western Canada (Horseshoe Canyon)
Depth (m)	275–580	671	458–763	366–534	250–451
Number of coal beds	6–8 (As many as 12)	12–25	As many as 14	6	As many as 30
Frac method	Staged (3 coal beds/stage)	Pinpoint (coiled tubing)	Pinpoint (coiled tubing)	Staged (3 coal beds/stage)	Pinpoint (coiled tubing)
Frac fluid	Nitrogen foam plus small amount of gel	X-L borate 70%	Nitrogen foam plus linear gel	Hybrid (water pad then low-load X-L gel)	High-rate nitrogen injections
Proppant (mesh)	20–40	20–40	20–40	16–30 at 2981–4173 l/m min. (25–35 barrels/min)	None

Source: Modified from Palmer (2010).

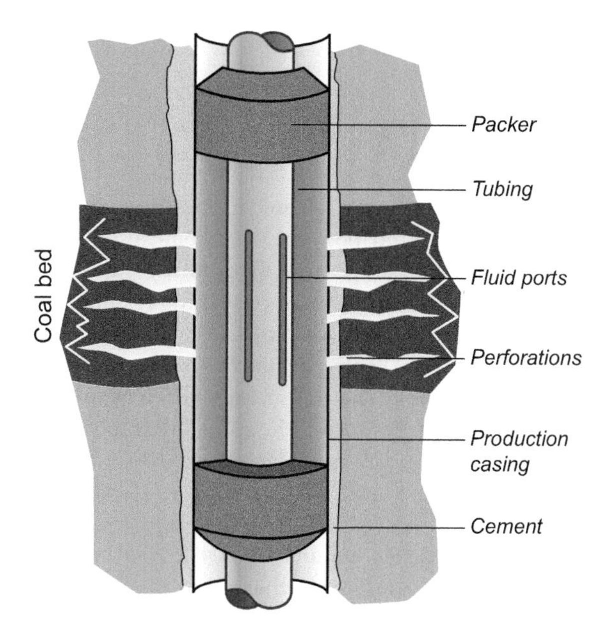

FIGURE 7.21 Diagram of coiled tubing used as a hydraulic fracture fraccing tool to treat coal reservoirs in the Raton Basin, Colorado. *Source: Adopted from Maccartney (2011).*

In order for both concepts to work, it needs not only knowledge of porosity and permeability but also understanding of the underlying controls of these parameters such as coalification, coal rank, composition, petrology, lithofacies, and geology. As discussed in the previous sections, the coal fracture and cleat systems are significant to the designing of well completion and stimulation. For example, for the horizontal drilling, the orientations of the fracture and cleat systems perpendicular to the coal-bedding plane have to be considered in order to penetrate the maximum surface of the reservoir for most effective gas production. In formulating hydraulic fracturing fluids, the conditions of the fractures and cleats such as inorganic mineralization, size, distribution, and extents play important role for stimulation. In both instances, permeability as a function of the fracture and cleat system is the primary concern for well completion and stimulation. Palmer (2010) discussed with excellent clarity the role of permeability or perm-based completion and stimulation as depicted in Figure 7.23. The perm-based strategy recommends under reaming and cavitation completion with very high permeability (>100 mD), cavitation and fraccing stimulation for medium-low volatile bituminous high permeability (20–100 mD), and pinpoint (single-bed treatment) fraccing stimulation for low-permeability (3–20 mD) coal reservoirs. In addition, the geomechanical properties (e.g. stress, strain, tensile strength, plasticity, and elasticity) of the coal reservoir are important factors to consider when planning well completion and stimulation (see Box 7.3).

Fundamental to consideration of fracture and cleat systems toward reservoir characterization is the knowledge that these properties are inherently variable vertically and laterally within the coal as dictated mainly by rank, organic composition, petrology, and lithofacies association. Frequently, reservoir-engineering characterization is performed from 8.9-cm cores and accompanying wireline logs that measure properties of the coal limited only to surrounding areas of the borehole. In addition, reservoir characterization is performed from interpretations of fracture treatments with measurements of fracture geometry and height. According to Palmer et al. (1993), if the fracture(s) is initiated in the coal around the borehole, when does the fracture(s) break out in the surrounding coal? This assumes that fractures are initiated and propagated beyond the borehole into bounding areas of the reservoir during stimulation.

Anisotropism of a Coal Reservoir

There is no doubt that the technology of completion and stimulation advanced the creation of coal reservoir permeability and enhanced coalbed gas production. However, as discussed in Chapters 3–5, permeability no doubt is controlled by the coal organic and inorganic composition, rank, and coalification as well as the sedimentology, stratigraphy, geology, and

BOX 7.3

COMPOSITION OF FRACTURE FLUIDS: CONCEPT AND MISCONCEPTION

With the onset of shale gas development and widespread public concern about using hydraulic fracture fluids, oil and gas service industries have voluntarily disclosed the composition of "fraccing" fluids (Cooley & Donnelly, 2012). The recipe of fracture fluids is conceptually designed according to reservoir characteristics, geology, depth, and extent of horizontal well completion. USDOE (2005) and NYDSEC (2011) reported that in general the fluids are composed mainly of water (90%), proppants (9%), and very minor chemical additives (1%) (Figure 7.22). Although the fluid mixture is designed for shale gas, the concentrations and volume, which may differ, probably apply to fracture fluids for coalbed gas wells. Much like the shale gas fracture fluids formula, the coalbed gas fracture fluids include fractions of mainly the gelled, acid, breaker, biocide, corrosion inhibitor, friction reducer, and other fluids (see previous discussions).

The fundamental debates between the media, industry, public, and environmental groups revolve around the 1% chemical additives. In order to understand how the 1% chemical additives are apportioned, distributed, and applied in an actual well completion, data submitted by Range Resources–Appalachia, LLC (2010) to the Pennsylvania Environmental Protection Agency. The well completion was in the Marcellus Shale, which used hydraulic fracturing fluids in the Marcellus Shale in Pennsylvania. The large volume of water stands out, which is almost 94% in comparison to three additives with a total volume of 0.09%. A study of USEPA (2004) indicated that some of these additives are known to be toxic and carcinogenic at very low amounts. Thus, the concerns about the chemical additives are not only their toxicity but also their hazardous components and environmental footprints. The reader is referred to the USEPA (2004; 2011), API (2010), NYDSEC (2011), Cohen et al. (2012), and Cooley and Donnelly (2012) for additional information of the pros and cons of the subject matter.

tectonic deformation of enclosing rocks. The best demonstration of the roles of these controlling factors is shown in a highwall of a coal mine where the entire coal reservoir and related sedimentary lithofacies are exposed and studied. The coal bed splits are exposed along the extent of the coalmine highwall, which is a 0.8 km long open pit or opencast mine (Figure 7.24). In order to put the coal reservoirs in stratigraphic and sedimentological contexts, the cross-section of the Dietz coal zone (merged coal beds separated by a parting) split fluvial channel sandstones and floodplain deposits down structural dip (10–15° southeast) into coal-beds is shown in Figure 7.15. The Dietz coal zone is about 15 m thick and splits into an upper bed (Dietz 1 or Anderson) 6–9 m thick and a lower bed (Dietz 2) 6–9 m thick (Flores, 1986; Moore, 1994). The lateral extents of Dietz 1 and 2 and interbedded fluvial deposits are shown in Figure 7.24. The coal beds wrap around the thinning and thickening fluvial channel sandstones as a result of differential compaction. During differential compaction, flexural bending of the coal bed below the thick sandstone resulted in formation of pronounced vertical fractures, which also

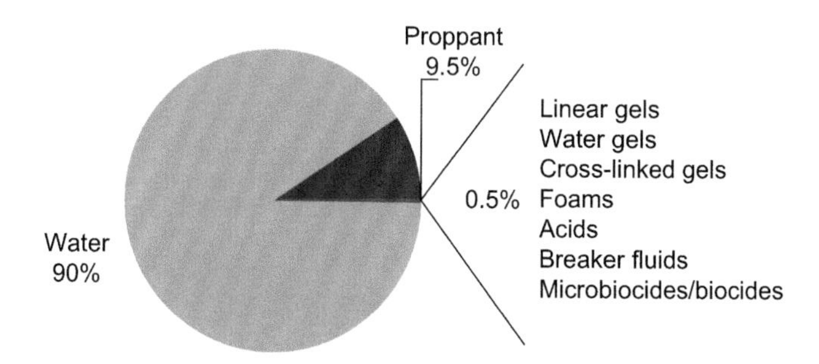

FIGURE 7.22 Pie chart showing the percent distribution of water, proppant, and chemical additives for coalbed gas hydraulic fracturing. *Source: Modified from USEPA (2004), USDOE (2005), and Cooley & Donnelly (2012).*

apply to similar deformation of the overlying coal bed. This locally enhanced permeability, which connects to the larger reservoir, is a prime target for coalbed gas production.

The anisotropism or heterogeneity of the fracture and cleat systems shown in Figure 7.25 is within a 10-m long and 7.5-m thick section of the Dietz 2 coal bed. The coal reservoir can be subdivided laterally into lower, middle, and upper layers displaying irregular contacts between the layers. The boundary or contact between the lower and middle layers is distinctive with relief as relict of differential compaction due to variable organic matter composition. This is a topographic

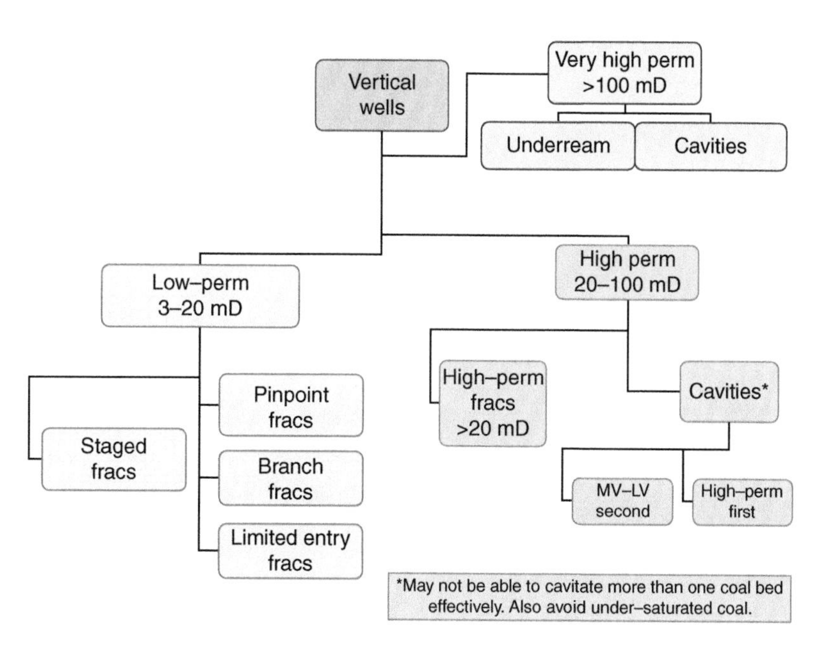

FIGURE 7.23 Flowchart showing strategy of perm-based completion and stimulation of vertical coalbed gas wells. mD = millidarcy; MV = medium volatile bituminous coal; LV = low volatile bituminous coal; perm = permeability; fracs = fraccings. *Source: Adopted from Palmer (2010).*

FIGURE 7.24 Surface coal mine highwall in the Powder River Basin, United States, showing differential compaction effects of coal beds and interbedded fluvial channel sandstones. The horizontal scale from foreground to the background of the mine highwall is about 0.8 km and the lowermost coal is about 6 m thick. *Source: Photograph courtesy of Decker Coal Mines.*

relief within the peat bog caused by deflation resulting from lowering of water table, aeration, and fungal oxidation (Moore, 1994). Normally, deflation of the bog surface results in a topographic low surface, which attracts flooding. However, in this case, the deflation of the peat surface is probably related to floral variations and/or degradational processes affecting peat-decay-resistant organic matter from different types of vegetation.

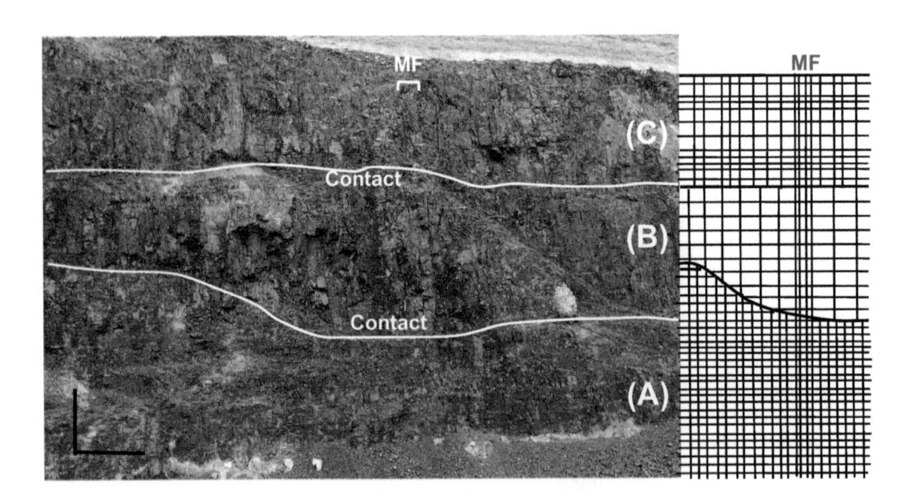

FIGURE 7.25 Coal face showing three layers (A, B, and C) of subbituminous rank with different patterns of cleat and fracture systems (see diagram on right). (A) Closely spaced cleat system in vitrain and durain coals. (B) Widely spaced cleat system in vitrain-rich coal. (C) Hybrid of closely and widely spaced cleat systems in vitrain and durain coal. MF is mega-fracture system (see Figure 7.27) connecting all three coal layers. Coal layers are separated by irregular contacts caused by differential compaction due to floral/vegetal variations.

A petrologic description of a core of the Dietz coal zone that included the Dietz 2 coal from the work of Moore (1994) shows organic maceral composition that indicates floral/vegetal remains of angiosperms (e.g. flowering plants) and gymnosperms (e.g. conifers and cycads). The remains of gymnosperms are resistant to decay compared to the remains of angiosperms, which are prone to yield more preserved plant tissues coalified as vitrain bands.

The proportions of decomposed fragments of cell walls, cell infillings, and oxidized cells walls in the coal layers of Dietz 2 represent the variation of vegetation types as vertical stacks of peat (Figure 7.26). Compositionally, these plant remains are shown in Figure 7.26(A) and (B) expressed as vertical profiles in a percent-based on study of Moore (1994). The vertical changes of floral remains indicate the ecological conditions and history of the peat bog through time as well as abundant plant tissues (Figure 7.26(C)) preserved as abundant vitrain bands in the coal face.

The three layers in the Dietz 2 coal contain variably spaced fractures and cleats in a mine face. The lower coal layer, 1.5–3 m thick, is characterized by closely spaced fractures and cleats (1.3–7.7 cm width) (Figure 7.20). Figure 7.20 shows highly variable fractures and cleats with spacing from <1 cm to >1 cm within 30-cm-long coal face. The middle coal layer, 1.5–3 m thick, contains mainly widely (7–23 cm width) spaced fractures and cleats (Figure 7.27). In addition, major or megafractures extend vertically from the lower to upper layers some of which show slickenside surfaces and mineralized fractures indicating structural deformation by differential compaction (Figure 7.28). The irregular boundaries of the partitioned fracture and cleat systems indicate lateral variability within the layers. The distinctive vertical partitions of the layers by characteristic fracture and cleat types suggest differences in the organic

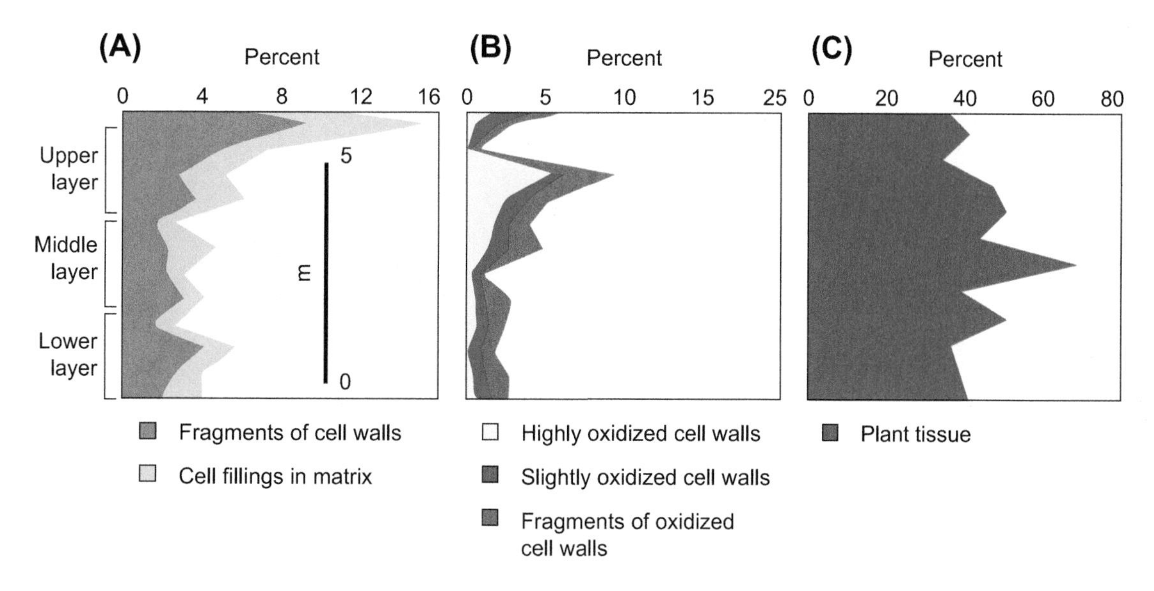

FIGURE 7.26 Vertical profiles showing organic compositional variations represented in the subbituminous coal (Dietz 2). (A) Cell wall fragments and filling in matrix. (B) Highly to slightly oxidized cell walls and fragments of oxidized cell walls. (C) Plant tissues. *Source: Modified from Moore (1994).*

FIGURE 7.27 Widely spaced "blocky" face cleats in the subbituminous coal of layer B in Figure 7.25. Layer B is dissected by megafractures. For scale, the color pen is about 12 cm long.

maceral composition as described in the previous paragraph. As shown in Figure 7.26(C), the middle layer appears to be composed of plant tissues or vitrain more than the lower and upper layers. However, the cleat spacing is wider in the middle layer. The lower and upper layers consist of 30–48% vitrain compared to the middle layer, which consists 40–65% plant tissues or vitrain (Figure 7.26(A)and (B)). The cleat spacing is narrower than that of the middle layer. The pattern of cleat spacing is contrary to previous observations by Close (1993) and Dawson and Esterle (2010). Correspondingly, the decomposed fragments of cell walls, cell infillings, and oxidized cells walls or durain are dominant (about 52–70%) in the lower and upper coal layers and subordinate (about 35–60%) in the middle coal layer. However, as shown in this case study, the fracture and cleat spacings are more variable than described by Law (1993) in the Powder River Basin. In addition, cleat and fracture permeability as described by Close (1993), Laubach, Marrett, Olson, and Scott (1998), and Dawson and Esterle (2010) is much more internally variable than the subbituminous Dietz coal. It is also different from the work of Trippi et al. (2010), which is based on coal cores in the margin and central part of the Powder River

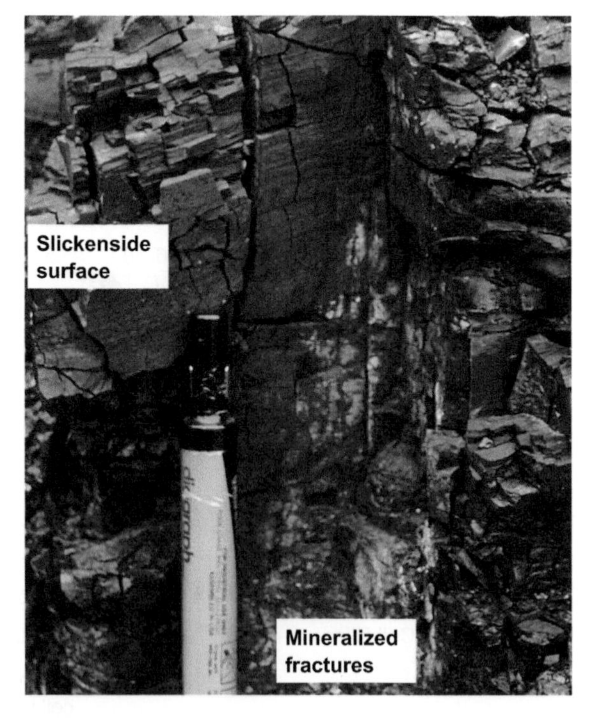

FIGURE 7.28 Megafractures with slickensides (striated surface) and inorganic mineralization (white area) in the Dietz 2 subbituminous coal. For scale, the color pen is about 8 cm long.

Basin. Lessons learned in this demonstration are that data applied across a basin or for that matter within a coal bed are not a "one-size fit all" proposition. Permeability reservoir characterization must include rank, composition, local area, field, and be basin specific and care must be exercised in applications of generalized cleat/fracture models.

The implications of the internal variability, vertical heterogeneity, and anisotropism of the coal fractures and cleats are directly linked to the design and application of well completion and stimulation in order to produce from a multilayer reservoir. Ali, Sarkar, Sagar, and Klimentos (2008) reported that the condition, nature of network, and relative connectivity of the cleat system vary from one coal to another and have an important bearing on gas production. Wicks, Schewerer, Militzer, and Zuber (1986) indicated that the anisotropism of the coal reservoir permeability controls the gas drainage pattern, that is, the basic idea that the orthogonal pattern of the coal cleat system should influence the gas drainage pattern. However, in the multilayer scenario, the state of the cleat and fracture systems is exacerbated by multivariability within a coal reservoir. Before gas production ensues, the complex nature of the reservoir properties must be studied to design an effective well completion. For example, during open-hole cavitation, the variegated layers behave differently with the closely spaced fractures and cleats of the lower layer more easily broken up as opposed to the more resistant, widely spaced fractures and cleats. The widely spaced fractures and cleats in the middle coal layer indicate a need for more stimulation to create more fractures and cleats to have effective water and gas flow conduits.

Production-wise, the closely spaced fractures and cleats in the lower coal layer have more disjointed connections or pathways through fractures and cleats, which lead to longer distance of transport of water and gas to the borehole. The megafractures provide surfaces where gas and water flow from the widely and closely spaced fractures and cleats, which in turn, serve as the most continuous avenues of water and gas transport for long distances to the borehole. The vertical variations in the fractures and cleats between coal layers or multilayer reservoir complicate the capability of horizontal and vertical wells to maximize gas production with such an intricate network of vertical and lateral permeability.

Internal anisotropism of the fracture and cleat systems of coal reservoir are well documented by McColluch, Deul, and Jeran (1974), Mavor, Dhir, McLennan, and Close (1991), Zuber and Boyer (2002a,b), Ali et al. (2008). Early in the history of coalbed gas development, anisotropism in the geometry and connectivity of face and butt cleat permeability of coal reservoirs was recognized by McCulloh et al. (1974) and Mavor et al. (1991). Reservoir anisotropism affects evaluation of production characteristics of a multilayer coal zone by presenting a wide range of reservoir pressure conditions, which in turn requires a versatile method of testing permeability (Zuber & Boyer, 2002a,b). As a result, the permeability of the fracture and cleat systems is a reservoir property of major significance to commercial levels of production, which cannot be attained without communication of these systems within the reservoir to the borehole (Mavor et al., 1991).

WIRELINE LOGGING TOOLS

Valuable subsurface data from wireline logging tools, which have been used for conventional reservoir for decades, are also acquired for coal reservoir characterization. Scholes and Johnston (1993) defined a wireline as a metal cable that holds electrical conductors, which power downhole instruments or tools and transmit signals from the tools to the surface to be recorded by logs as a function of lithology and depth. For detailed descriptions and discussions of wireline logging tools, the reader is referred to

Schlumberger (1991), Scholes and Johnston (1993), and Ellis and Singer (2010). Wood, Kehn, Carter, & Culbertson, 1983 specifically described logging tools useful and unique for coal beds in coal-bearing rocks.

BASIC LOGGING TOOLS

Wireline logs are most useful in identifying coal beds in the subsurface for resource assessment because of the beds' inherent low radioactivity, high resistance to electrical current, and low density (Figure 7.29). These distinctive properties are at variance with other lithologies, mainly shale, mudstone, siltstone, and sandstone in coal-bearing rocks. The basic or traditional tools required to measure properties of coal reservoirs and associated rocks as well as best suited for coalbed gas evaluation include the following (Asquith, 1982; Scholes & Johnston, 1993; Wood et al. (1983)): (1) resistivity

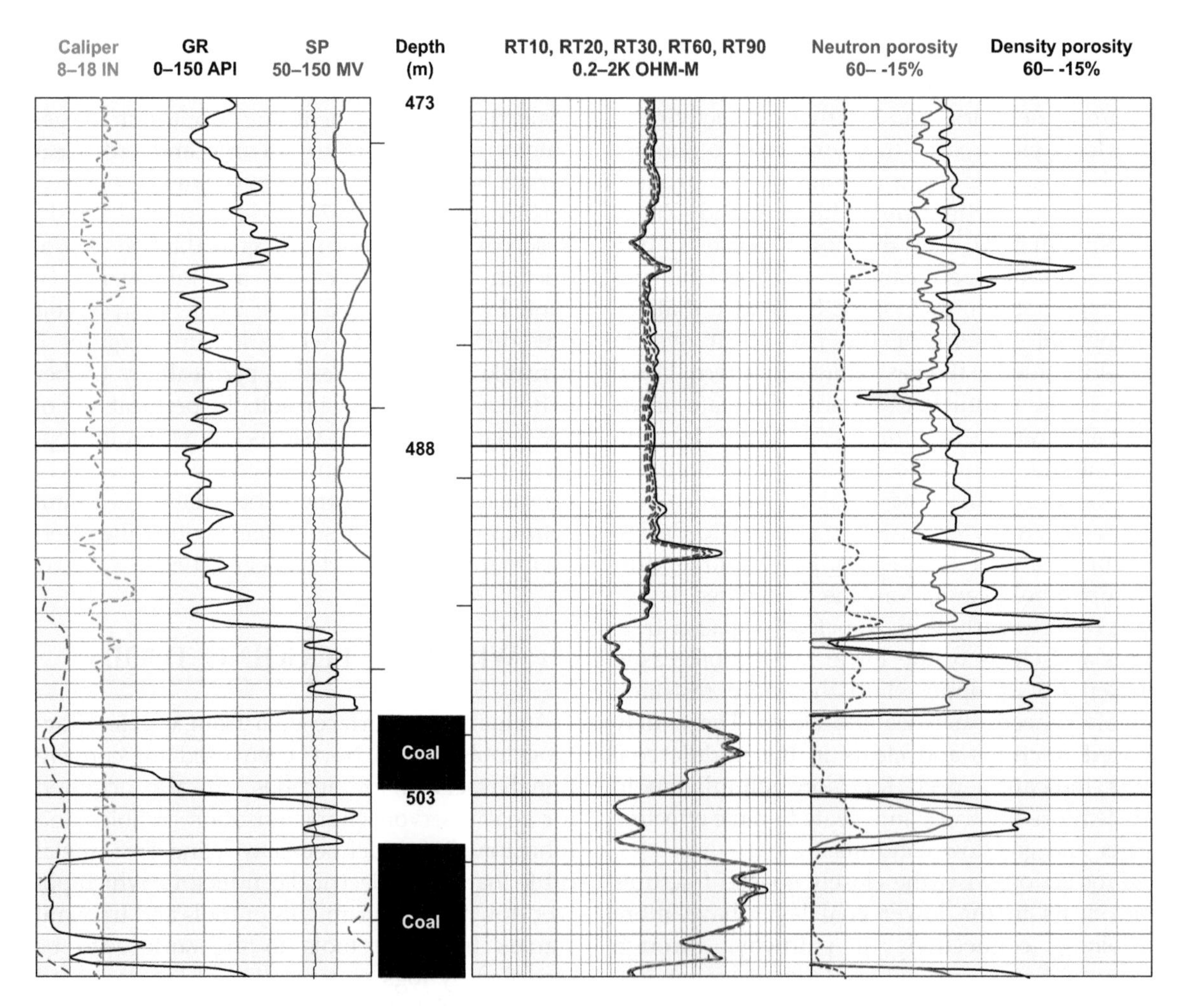

FIGURE 7.29 A suite of wireline logs (e.g. caliper, gamma ray (GR), spontaneous potential (SP), resistivity (R), neutron porosity, and density porosity) of a portion of the Fort Union coal-bearing rocks in the Powder River Basin in Wyoming, United States. *Source: Courtesy of Randy Caber, Anadarko CBM Group, Denver, Colorado.*

log, (2) gamma ray log, (3) density log, (4) acoustic (sonic) log, and (5) neutron log. The basic or traditional wireline logging tools can provide invaluable, rapid, economical, and detailed data on thickness, lithology, fluid content, correlation, depth, and structure of the beds penetrated by a coalbed gas well (Wood et al., 1983). Fundamentally, wireline logs are used to identify the thickness, continuity, and stratigraphic correlations of coal beds for coal resource assessment. The wireline log data may be available in coal basins from oil and gas logs but precaution should be made to insure that proper sets of logs are utilized in the assessment. Often, thin markers (e.g. partings) can be readily identified on wireline logs, even though it is difficult to recognize in coal outcrops and mine faces.

Resistivity Log

The resistivity log is acquired by measurements of the resistance of the coal and associated rocks to flow of electrical currents and by inducing currents into the reservoir rock (Figure 7.29) (Scholes & Johnston, 1993). Consequently, various combinations of electric and induction resistivity logs are deployed depending on the borehole diameter, drilling fluid, and salinity of the fluid. Thus, resistivity logs involve fluids to measure the resistance of the rocks surrounding the borehole. The resistivity is expressed in a "resistivity profile" or variation into the coal reservoir from the borehole. The high moisture of coal beds (e.g. lignites and subbituminous coal) account for the high-resistivity deflections on the resistivity profiles (Wood et al. (1983)). Limestones and sandstones such as in the Powder River Basin show high-resistance deflections and could be mistaken for coals. Limestones are relatively uncommon in coal-bearing rocks except where associated freshwater lacustrine and marine environments are juxtaposed with peat-forming environments (Flores, 1983). However, the sandstones, which are permeable and contain groundwater, can be distinguished from coal by the lithology that separates the two lithologies such as shale characterized by inherently low-resistivity deflections (Flores et al., 2010). Thus, the sandstones have nearly parallel resistivity profiles to the coal beds (Flores, Toth, & Moore, 1982).

Gamma Ray Log

The gamma ray log measures the natural gamma radiation emitted by the coal and associated rocks (Figure 7.29) (Scholes & Johnston, 1993). Gamma ray logs mainly differentiate the reservoir rock for shales, which have high natural gamma ray radiation. Characteristically, gamma ray logs are adaptable for use in coal-bearing rocks for the following reasons (Wood et al. (1983)): (1) they do not require fluids such that shales with relatively low or no permeability with no or minor fluids are amenable for gamma ray measurements, (2) they are not sensitive to variations of borehole diameter, (3) they can be used to measure through casing, and (4) they can be the least expensive logs to run. This is in contrast to resistivity logs, which require the presence of fluid for determining the resistivity of the rocks. Radiation from shales is from clay minerals containing radioactive elements such as potassium and thorium. In coal reservoirs, uranium may have been emplaced by deposition, which may be interpreted as shale in the gamma log (Flores et al., 1982). This complication may be resolved by using gamma spectrometry gamma ray logs (e.g. natural and induced gamma ray). Coal partings are especially distinguishable using gamma ray but oftentimes in the Powder River Basin, the thickness may be exaggerated to as much as seven times the true thickness (Flores et al., 1982; Wood et al., 1983). The radioactive property of partings (e.g. carbonaceous shale) in coal was used as tool to correlate coal beds in the Powder River Basin for long distances (Flores et al.,

2010). Also, the radioactive partings may provide information on the high-ash composition of the coal. Other rocks such as limestone and sandstone in coal-bearing rocks of some coal basins have as low radioactivity as coal.

Density Log

The density log measures the bulk density of the coal reservoir and associated rocks (Figure 7.29) (Scholes & Johnston, 1993). The density tool or gamma density tool utilizes a gamma ray source placed a distance from the gamma ray detector, which measures gamma ray count that is an inverse function of the density of the coal or associated rocks. The more gamma rays that are absorbed in denser rocks, the lesser is the ray count transmitted to the detector (Wood et al., 1983). The majority of coal ranks have low density (e.g. 0.7–1.8 g/cc) compared to adjoining lithologies, which make it a valuable tool to identify coal beds. The traditional use of density log is to measure the porosity of the reservoir (Scholes & Johnston, 1993). That is, if the matrix density is constant, the density of the coal and associated rocks is an inverse function of the porosity. In addition, the density log may provide information on the rank of the coal, that is, high-rank coal is denser than low-rank coal. The output of the gamma ray source and the gamma ray detector is sometimes confused with the natural gamma ray detector discussed in the previous section. The advantage of a density log is that it can detect coal partings much like the gamma ray log and thin coal beds. Unlike the gamma ray log, a density log must be run without the casing and varies with borehole diameter. Wood et al. (1983) suggested running a caliper log (see below in section on acoustic or sonic log) to complement interpretation of the density log. Because the density is sensitive to borehole rugosity, a correction curve should always be run. If the correction exceeds 0.20 g/cc, the density in that interval is considered invalid (Asquith, 1982).

Acoustic (Sonic) Log

Acoustic (sonic) logs can be use to measure travel time (amplitude of the wave) of a sound wave emitted from a logging tool and transmitted to a receiver tool (Scholes & Johnston, 1993). The velocity of the acoustic or sonic wave is controlled by the lithology and porosity of the coal and associated rocks (Wood et al., 1983). The main purpose of the acoustic or sonic log is to measure the porosity of the reservoir and associated rocks, which is calculated based on the linear relationship between effective (e.g. nonfracture or nonvug) porosity and travel time in clay-free reservoir (Scholes & Johnston, 1993). The decrease in velocity or increase in transit time is interpreted to be an increase in porosity (Wood et al., 1983). The short travel time of sonic waves in coal indicates high-rank coal. However, this technology is of less value than in oil and gas reservoir because in coal reservoir the gas is stored in the pore systems (e.g. micro-, meso-, and macropores). Wood et al. (1983) indicated that acoustic or sonic logs are of better value in detecting coals in deeper-than-unconsolidated shallower coal-bearing rocks. A few acoustic or sonic tools measure the diameter, condition, and stability of the borehole (Schlumberger, 2012), which may be used to complement density log.

Neutron logs measure the hydrogen ion concentration of the reservoir and associated rocks (Figure 7.29) (Asquith, 1982). As with the acoustic tool, the neutron tool uses a source and a detector. The rate of neutron counts in the detector, which is affected by borehole conditions, is the inverse function of the porosity (Scholes & Johnston, 1993). Like the acoustics, the neutron log is not directly applicable to coal reservoirs because gas is stored in the pore system. However, neutron log is of value in identifying coals because the tools measures high porosity due to the high hydrogen composition of the coal (Scholes & Johnston, 1993; Wood et al., 1983). The neutron profile or curve reflects high measurements adjoining fluid-filled pore and/or permeable fractures/cleats in coal and associated rocks due to high hydrogen content. However, the acoustic or sonic log records high readings adjoining coal beds due to high carbon content. According to Wood et al. (1983) clay filled with moisture adjacent to coal beds will record high readings obscuring the contact between both lithologies. Neutron log is affected by cave-ins and huge washouts in the borehole, which is why acoustic or sonic log must be a complementary log used for interpretation. Unlike other logging tools, neutron logs require log-specific charts for interpretation. One company's neutron log cannot be interpreted using another company's chart.

According to Zuber and Boyer (2002a,b), the most important basic wireline logging tools for evaluating coal reservoirs mainly includes a suite of resistivity, gamma ray, density, and caliper tools, although neutron logs are also incorporated on occasions. The caliper log is a mechanical device that measures the diameter of a borehole along its depth using acoustic or sonic devices (Schlumberger, 2012). The resistivity log is useful on occasion for estimating relative differences in reservoir permeability because of fluid invasion into the naturally fractured and cleated coal reservoir (Zuber & Boyer, 2002a,b). Gamma ray log can be used alone to identify coal in coal-bearing intervals, where no other rock types are known to have low natural radioactivity as coal (Wood et al., 1983). Conversely, in lieu of a gamma ray log, a combination of density and acoustic logs can be used in other coal-bearing rocks. Density logs, especially run at a high resolution, are most valuable for evaluating coal reservoir thickness because of the intrinsic differences in density between coal beds and surrounding sedimentary rocks. Similar results can be acquired with combined use of resistivity and neutron logs. The high-resolution density log can be used to determine the volume of inorganic or mineral matter) in coal (Zuber & Boyer, 2002a,b). Acoustic logs may provide valuable information for processing and interpreting seismic data.

VIRTUES OF HI-TECH LOGGING TOOLS

Logging tools especially for evaluating coalbed gas wells have advanced technologically during the past decade for the purpose of optimizing well production, in particular, and field development, in general (Anderson et al., 2003; Halliburton, 2008; Schlumberger, 2009). Zuber and Boyer (2002a,b) described newer state-of-the-art chemical logs, which are being used to identify basic chemical components in coal reservoirs. These advanced logs are re-created to provide a log-based coal compositional analysis of the coal reservoirs. More importantly, according to Zuber and Boyer (2002a,b), when the chemical logs are calibrated with core data, the tools can be used to evaluate the relative quality of coal reservoirs and assist in designing well completions. Thus, the new logging tools permit more detailed analysis of coal reservoirs behind the casings.

The wireline logging tools are sensitive to coal according to Scholes and Johnston (1993) due to their organic and inorganic (e.g. ash) composition, which induce positive log responses. Disregarding the multiplicity of indirect measurements of porosity, the traditional tools that may be most helpful in the exploration of coal reservoirs are the resistivity, gamma ray, and density. However, a combination of these logs may suffice for distinguishing distinct responses to coals compared to associated sedimentary lithofacies (Anderson et al., 2003; Schlachter, 2007; Schlumberger, 2009; Scholes & Johnston, 1993). Schlachter (2007) particularly indicates that reservoir permeability, pressure, and fluid samples among others have been successfully used to characterize coal reservoirs in Canada. Both Zuber and Boyer (2002a,b) and Anderson et al.

(2003) reported that the state-of-the art neutron-induced gamma ray spectroscopy log provides new capabilities for coalbed gas reservoir characterization (Figure 7.30). An indirect method is used by logging the chemical composition and ash mineralogy of the coal reservoir in lieu of direct measurements of gas content of coal (Schlumberger, 2009).

The greatest improvement in the advanced technology of wireline logging is the geochemical logging of coal. Wood et al. (1983) indicated that efforts were made to quantify ash

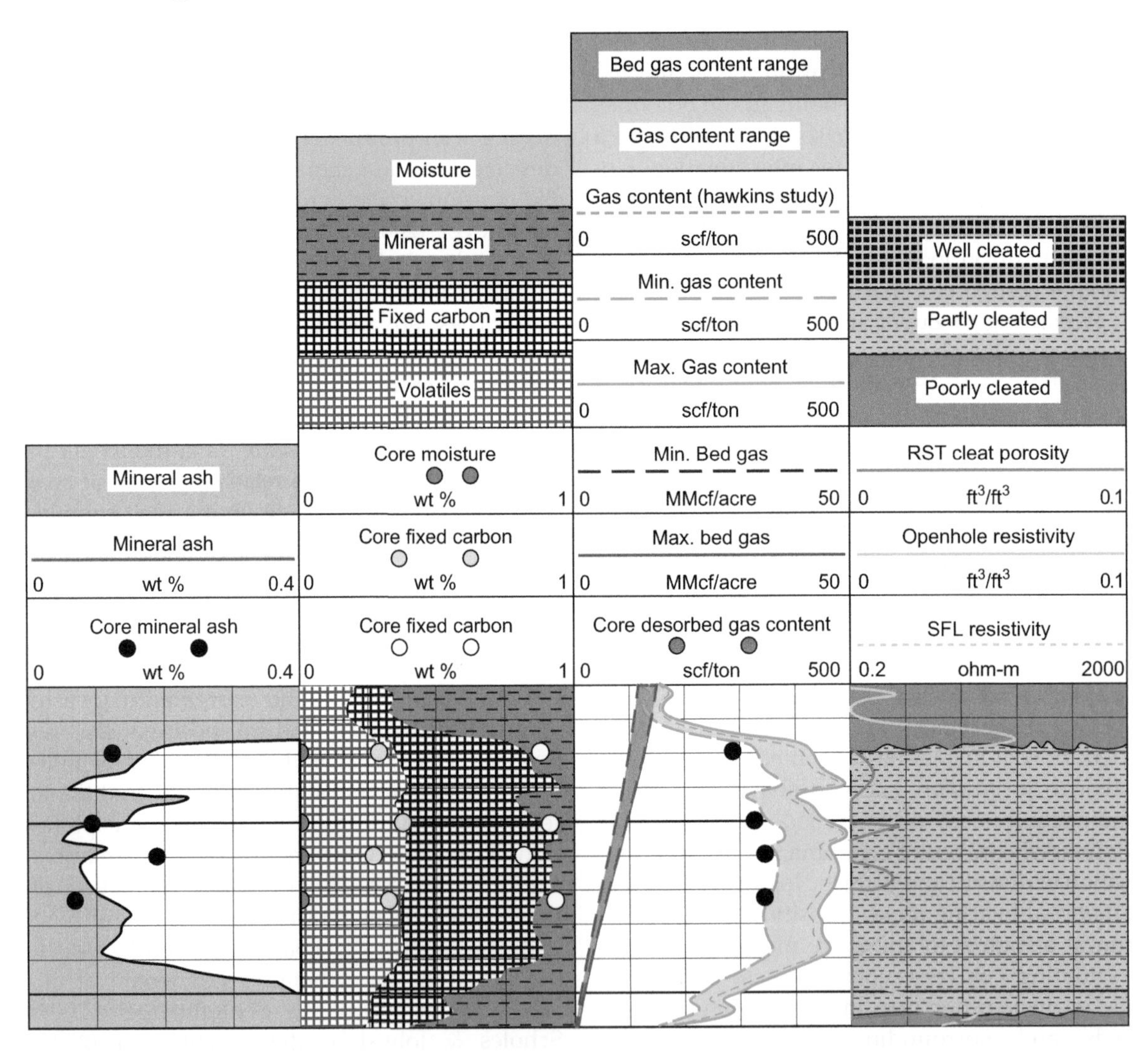

FIGURE 7.30 Coal evaluation logs using neutron-induced gamma ray spectroscopy from Reservoir Saturation Tool. From left to right: tracks 1 and 2 show coal proximate analysis from coal and logs, track 3 shows gas content and cumulative gas content from core and logs, and track 4 shows cleat intensity. SFL Resistivity in track 4 is induction resistivity measurement complimented by spherically focused laterolog (SFL) in the tool. SI–metric conversion factor: 1 cc/g = 32.07 scf/t (standard cubic feet per ton); MMcf = million cubic feet; 1 cubic foot (ft^3) = 0.03 cubic meter (m^3); wt % = weight percent. *Source: Adopted from Anderson et al. (2003).*

and moisture contents of coal in geophysical logs but the results were unreliable in the early 1980s. Since then, with the advent of the commercial evaluation of coalbed gas, a variety of reservoir characterizations tailored to coal have evolved that include the newest high-tech geochemical logs. Zuber and Boyer (2002a,b) and Anderson et al. (2003) first reported the high-tech geochemical logs that use Elemental Capture Spectroscopy (ECS) similar to that of Schlumberger technology (Schlumberger, 2009). The proximate analysis (e.g. moisture, volatiles, fixed carbon) and inorganic mineralogy (e.g. pyrite, clay or ash) of the coal calibrated from core data can be correlated to logs (Figure 7.31). Schlumberger (2009) described this new technology as open- and cased-hole geochemical logging using an empirical relationship determined from proximate and gas desorption/adsorption analyses calibrated from coal cores. The total gas content and gas adsorption isotherm are translated at separate depths as analyzed from metrics of the coal and ash yield. Correspondingly at the same depths, the extent of cleating and fracturing is indicated by the chemical log of the ash composition and mineralogy of the coal bed (Figure 7.31). The degree of cleating and fracturing, in turn, indirectly suggests the producibility of gas. High resolution of geochemical logs (e.g. ECS) enhances the identification of cleating and fracturing based on measurements of the ash yield, which may contain clay, carbonate, quartz, and pyrite compared to a standard log (Figure 7.32) (Schlumberger, 2009).

In addition to the hi-tech geochemical log, Zuber and Boyer (2002a,b) reported improved wireline logs such as high-resolution logs used for cased- and open-hole situations in coal reservoirs. Schlumberger (2009) indicated that basic wireline logs such as high-resolution density and gamma ray logs analyzed with innovative algorithms provide better delineation of the coal thickness because bulk density measurements are as high as 50 mm vertical resolution. Thus, with the calibration of the density and photoelectric absorption factor with core coal proximate analyses, these properties may be predicted (Schlumberger, 2009). The use of integrated logs such as resistivity, microresistivity, and neutron porosity compared to stand-alone density log provide better coal cleat and fracture porosity measurements (Schlumberger, 2009).

Application of High-Tech Logs

A study by the Oil and Natural Gas Corporation in the Permian coal-bearing Barakar Formation in India (see Chapter 9) has applied high-tech wireline logging tools for coalbed gas exploration from three pilot coalbed gas wells (Anderson et al., 2003). One of the wells was cored and the lithology was determined with high-resolution density, gamma ray, neutron, and resistivity logs integrated with high-tech tools (e.g. Fullbore Formation Micro-Imager, Dipole Shear Sonic Imager (DSSI), and ECS). Cores were collected from most of the 18 identified coal beds 1–20 m thick, which were analyzed for coal quality (e.g. proximate and rank analysis) and adsorbed gas content. These parameters along with cleat porosity were used to calibrate the analysis of the logs (Anderson et al., 2003). The results of the log analysis were mainly in agreement with the core data, although log measurement results were mixed. On the other hand, the ECS provided detailed data on the ash composition and improved the total ash yield for washout coal beds. Volatile matter analysis (e.g. organics, wax, carbon dioxide, and sulfur dioxide) was similar to analysis of the Barakar bituminous and anthracite coals. However, there are not enough independent measurements to estimate the volume of desorbed gas for each coal bed because the effect of adsorbed gas on the response parameters of coal is small (Anderson et al., 2003). Traditionally, a more reliable estimate of gas content is gas desorption from coal cores.

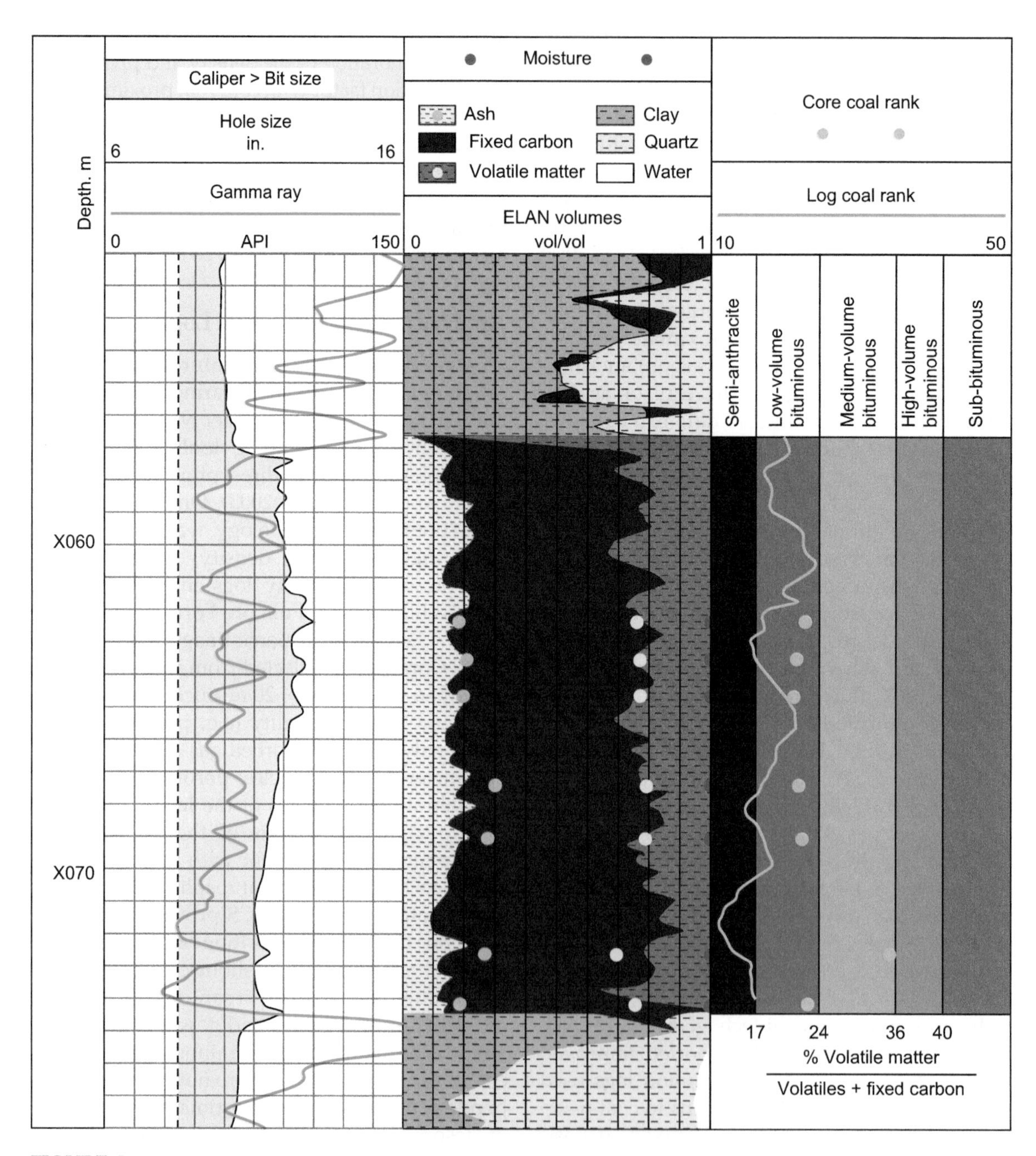

FIGURE 7.31 Determination of coal proximate and rank analyses. From left to right: track 1 shows cliper log indicating moderately washout borehole, track 2 shows good correlation of proximate analysis (see Table 7.5) and core data, and track 3 shows coal rank from logs, which are determined from proportions of volatile matter (dry, ash-free basis). % = percent. *Source: Adopted from Anderson et al. (2003).*

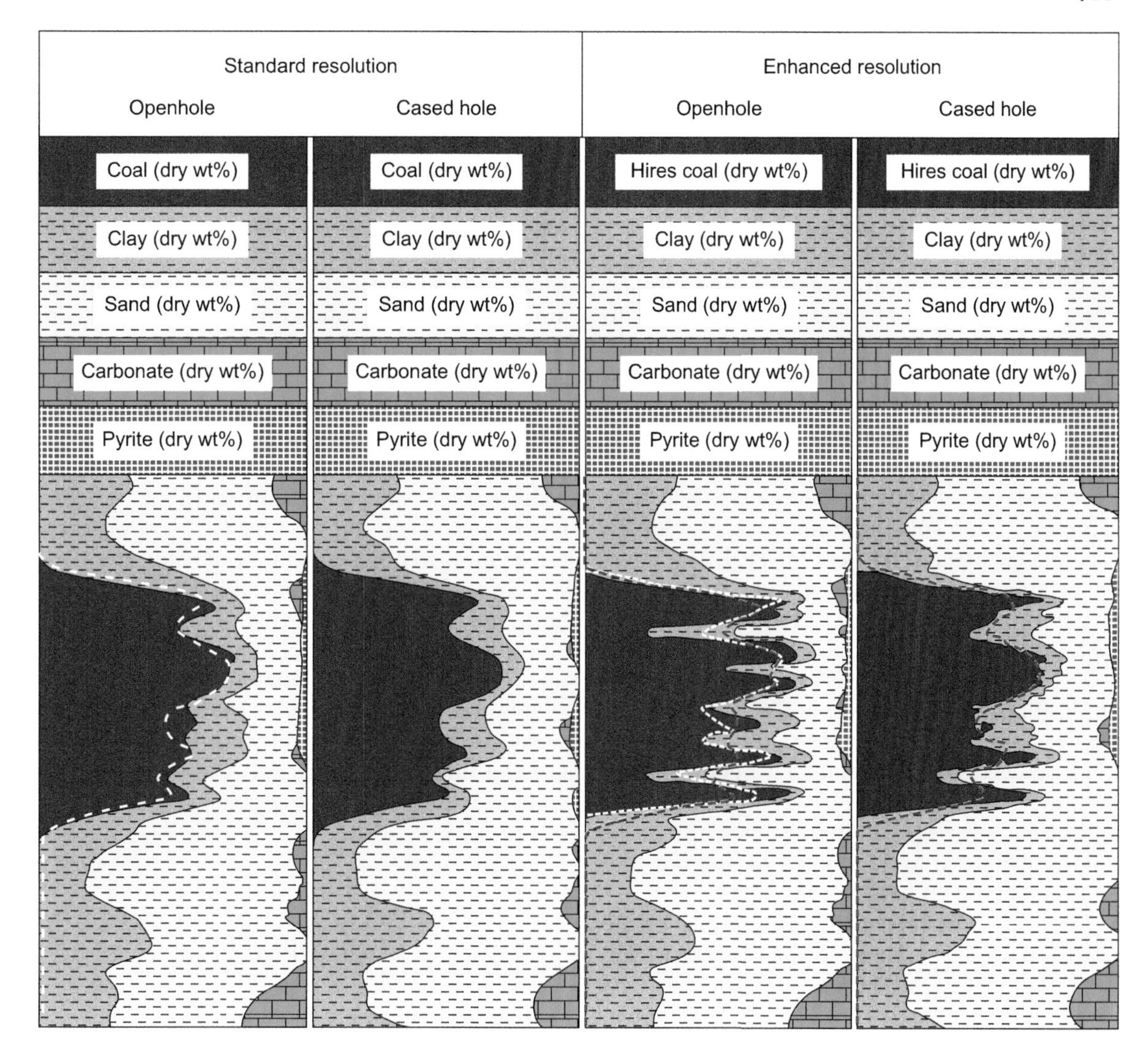

FIGURE 7.32 Comparison of standard and high-resolution open-hole and cased-hole elemental capture spectroscopy (ECS) of coal and associated rocks mineralogy. wt% = weight percent. *Source: Adopted from Schlumberger (2009).*

Estimating the coal rank from proximate analysis and gas content from coal rank, pressure, temperature, and adsorption isotherm may not be reliable enough to estimate gas volume. An indirect estimate of gas in place for coal is the use of rank and proximate analysis, which can be related to the suitable adsorption capacity of the coal. The cleat porosity was estimated by (Anderson et al., 2003) using a combination of (1) microresistivity log measurements, (2) separation of deep and shallow laterolog curves, (3) quantity and type of mineralization observed by the ESC tool, and (4) shear wave anisotropy measured by the DSSI data. Microresistivity is the most accurate gauge of cleat porosity, which can be used to calibrate the

ESC and DSSI data. Knowledge of cleat porosity provides data on flow capacity in addition to the estimate of gas volume.

Limitations of Logs

The balanced and objective discussions by Anderson et al. (2003) on the application of high-tech geochemical logs to the Indian Permian Barakar Coal Measures are subject to limitations. The key to this limitation is calibration with the coal quality, mineralogy, gas content, and adsorption isotherm from reference coal samples. The accuracy of replicating these properties is by quality and quantity of coal samples used for calibration. Oftentimes, coal core samples are too small in number to provide an accurate estimate of the coal interval. The internal heterogeneity within a coal reservoir in terms of these properties (Stricker et al., 2006) is such that closely spaced vertical samples have to be collected from continuous cores for accurate application. The lateral variability of coal beds also requires calibration of more coal cores of the same bed within a potential field. In addition, heterogeneity between coal beds, rank, and basins suggest that application of the technology will have to be tailored to specific coal and rank and on a basin-to-basin basis.

The condition and history of analyses of the coal samples as received in the laboratory are important controlling factors that affect analytical results (Schweinfurth, 2002). For example, fresh core coal samples arrived in the laboratory without additional processing, which are analyzed and results reported as "as-received" basis. Coal samples received dried out due to transport, storage, and mismanaged conditions are analyzed and results reported as "dry" basis. In addition, coal samples arrived "wet" compared to standard conditions are analyzed and results reported as "moist" basis. Other coal samples are analyzed as "dry, mineral-matter-free" basis for the purpose of calculating rank, which is controlled by the maturation of organic matter. That is, the mineral or inorganic matter is subtracted from the total analytical results to report only the amount of organic matter.

More importantly, determining the mineral and volatile matter composition of the coal requires destruction by burning the sample, which is not appropriate for retention of some volatiles. In order to alleviate this problem, coal samples are subjected to low-temperature ashing in which the organic matter is dissipated and inorganic matter preserved. However, some elements of inorganic matter may be lost during this process, thus requiring a different method of analysis (e.g. X-ray fluorescence of whole coal sample). The lesson learned in this process is understanding the history and purpose of the analysis to best suit analytical results for correlations with the geochemical logs. Despite the limitations, high-tech geochemical and mineralogical logging of coals still holds the key to optimizing coalbed gas production. The application of this innovative advanced technology worldwide can only have a significant role in characterizing and evaluating coal reservoirs toward enhanced development of coalbed gas in the future.

TOWARD COALBED GAS DEVELOPMENT

In general, there are three phases in achieving coalbed gas development: (1) early evaluation, (2) pilot project(s), and (3) commercial development (CAPP, 2006). The evaluation or exploration phase includes drilling wells and obtaining wireline logs in order to establish presence, depth, and thickness of coal beds especially in unexplored areas. Data collected from this phase is used to estimate coal and coalbed gas-in-place resources. Drilling, coring, completion, stimulation, logging, and testing for gas contents and production from initial wells that are spaced closer than developmental wells are performed during pilot project(s) work (Logan, 1993; Palmer et al., 1993; Schraufnagel, 1993). Essential to the pilot project study is testing for the producibility of coalbed gas as dictated by the connectivity of cleats and fractures or permeability of the coal reservoirs over a sustained period sometimes taking up to a year or two in unexplored areas. In high-permeability undersaturated, coals a larger number of wells may be needed to obtain valid results. Precommercial production of coalbed gas is the phase of testing how many wells, and at what spacing, are needed to develop coalbed gas within a potential field or pool. The pilot stage is the most practical aspect for evaluation and consideration in coalbed gas development critical to maximum recovery of gas reserves. Positive results from the pilot project lead to coalbed gas development in a larger than a pilot-size area such as from several sections to townships. Development involves many coalbed gas wells that are most favorably spaced to cost-effectively recover estimated gas reserves.

Coalbed Gas Well Spacing

The spacing of coalbed gas wells is the distance between boreholes and is measured by the quantity or density of wells per square area. Well spacing within a subsurface area accessed by a coalbed gas well varies from country to country as set by regulations of governmental agencies on conventional oil and gas development (BLM, 2003; British Columbia Ministry of Energy and Mines, 2011; CAPP, 2006; PRMS, 2011). In the United States, the well spacing for coalbed gas development inherited and modified from conventional oil and gas development is as follows: (1) 258-ha (640 acres) or 1 well per 2.6 km^2 (1 mi^2) area, (2) 129-ha (320 acres) or 2 wells per 2.6 km^2 (1 mi^2) area, (3) 64-ha (160 acres) or 4 wells per 2.6 km^2 (1 mi^2) area, (4) 32-ha (80 acres) or 8 wells per 2.6 km^2 (1 mi^2) area, and (5) 16-ha (40 acres) or 16 wells per 2.6 km^2 (1 mi^2) area. Different well spacing has been demonstrated by Wicks et al. (1986) to influence the percentage production of coalbed gas in the Black Warrior Basin, Alabama. Drilling one well in a 64-ha area will produce 25% of the gas in place compared to eight wells, which will produce 85% of the gas in place.

In Canada, coalbed gas well spacing varies from province to province governed by separate rules and regulations. For example, in British Columbia, coalbed gas well spacing is defined under the Petroleum and Natural Gas Act, which requires approximately 258 ha (British Columbia Ministry of Energy and Mines, 2011). However, in Alberta, coalbed gas wells due to low gas rates and pressures may necessitate location of wells at closer distance than established development of conventional oil and gas wells. Well spacing for coalbed gas wells varies from two to eight wells per section (e.g. 2.6 km^2 or 1 mi^2) per pool (or field) with four wells being the average (Alberta Energy, 2012; NEB, 2007). The typical gas well spacing for Alberta Province is one well per section per pool (or field); however, standard spacing regulations provide for increased well density through application to the Energy Resources Conservation Board of Alberta Province. The Energy Resources Conservation Board has amended Alberta's well density controls by removing the limitations for coalbed gas well, including coal seams with other interbedded lithologies and shale gas reservoirs throughout the province (ERCB, 2011). Thus, application of well spacing is no longer required for coalbed gas production, which is interpreted to promote development of coalbed bed in Alberta Province.

In other countries such as China and Australia, coal bed well spacing differs from that in the United States and Canada. In China, well spacing varies from 160 ha per well for vertical wells to 80 ha per well for horizontal wells (Hsi & Li, 2010; Meng & Tian, 2007). In the exploration phase in India, the cluster approach for test coalbed gas well drilling is at 24–32 ha spacing

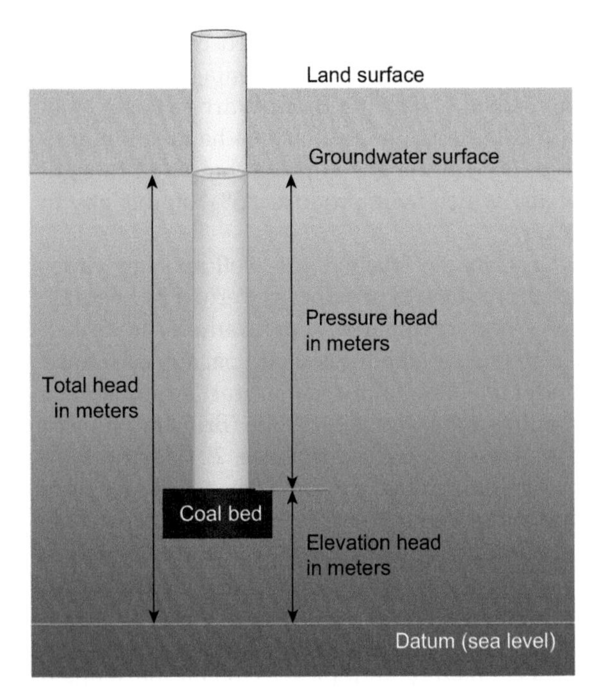

FIGURE 7.33 Schematic diagram showing groundwater pressure head, elevation head, and total head. *Source: Modified from Heath (1993), Alley et al. (1999), and NC Division of Water Resources (2012).*

(Sawhney, 2010). In Australia, well spacing varies within basins and between basins depending on the coal permeability (Mohinudeen Faiz, Origin Energy, Pers. Commun., 2012). In the Surat Basin (e.g. Jurassic Walloons sweet spots), well spacing is 750 × 750 m. In contrast, in the Bowen Basin (Permian Spring Gully sweet spots), the well spacing is 1.5 × 1.5 km. Producing areas in Permian Spring Gully field has highly permeable coals (10 mD to >1 D). Less-permeable areas will be drilled with reduced well spacing and infill drilling will be dictated in other areas depending on permeability. The government does not regulate well spacing in these Australian basins.

The Australian well spacing being dependent on the coal reservoir permeability is comparable to that in the United States (e.g. Powder River Basin) where well spacing was adjusted according to necessary water production. That is, more infill drilling was permitted in virgin exploration areas where the original hydrostatic pressure was preserved. In these areas in the Powder River Basin, as many as 16 wells per 2.6 km^2 well spacing were permitted for reduction of reservoir pressure.

Role of Groundwater

The well spacing within an exploration and developmental area (e.g. section, township) is influenced by the permeability, interwell connectivity, pressure, and aerial extent of the coal reservoirs. Inherent to the production of coalbed gas is the relatively low gas flow and pressure of the coal reservoir such that optimizing production requires wells to be drilled more closely than conventional oil and gas wells. The occurrence of groundwater within the coal is important in well spacing. More specifically, the hydraulic head (e.g. piezometric head), which is a measure of the water pressure (e.g. pressure head or hydrostatic pressure) above the coal reservoirs, is important to gas flow. The pressure head is the height (e.g. meters, feet) of a standing water column above the boundary of the coal bed(s) (Figure 7.33).

According to Wheaton and Metesh (2002), the amount of water that an aquifer (e.g. coal and sandstone) will yield to a coalbed gas well depends on hydrostatic pressure, hydraulic conductivity, saturated thickness, and storativity (Table 7.5). However, the hydrostatic pressure has the most significant impact on coalbed gas activities such as well spacing as well as gas flow and production. According to Wheaton and Metesh (2002), hydrostatic pressure, which is measured as groundwater level or head, imparts the energy to force the water through the aquifer (e.g. coal) to maintain flow to a coalbed gas well. Absence of hydrostatic pressure indicates that no water flows toward a coalbed gas well to replace water removed by pumping; thus following the principles of Darcian flow, gas production is either reduced or ceased.

TABLE 7.5 An Example of the Hydraulic Conductivity, Available Head and Storage Coefficient of Groundwater in the Coal-Bearing Fort Union Formation in the Powder River Basin, Montana, United States

Lithology	Horizontal Hydraulic Conductivity M/D (Ft/D)	Average Thickness m (Ft)	Available Head m (Ft)	Storage Coefficient
Sandstone and claystone	1.5 (5)	61 (200)	0.6–45.7 (2–150)	1E-01
Anderson coal	0.9 (3)	9.2 (30)	15.3–137.3 (50–450)	6E-05
Sandstone and claystone	0.03 (0.1)	76.3	45.8–137.3 (150–450)	1E-04
Canyon coal	0.4 (1.5)	6.1 (20)	61–137.3 (200–450)	6E-05
Sandstone and claystone	0.03 (0.1)	45.7 (150)	122–183 (400–600)	1E-04
Wall coal	0.6 (2)	3.1 (10)	106.7–183 (350–600)	6E-05

Source: Adopted from Wheaton and Metesh (2002).

The water flow to a coalbed gas well depends on the hydraulic gradient or the change in head per unit distance (e.g. m/m, ft/ft) determined from surrounding wells or area (Figure 7.34) (Alley, Reilly, & Franke, 1999; Heath, 1993). The basic groundwater principle states that the direction of groundwater movement is always in the direction of decreasing head, as shown in Figure 7.34 (NC Division of Water Resources, 2012). The direction of groundwater flow is determined by withdrawal of water by pumping from a coalbed gas well where the head is lowered in relation to the adjoining area. As the head in the coalbed gas well drops below the groundwater level of the surroundings wells water moves from the coal aquifer to the coalbed gas well. Groundwater level continues to be lowered increasing hydraulic gradient and rate of

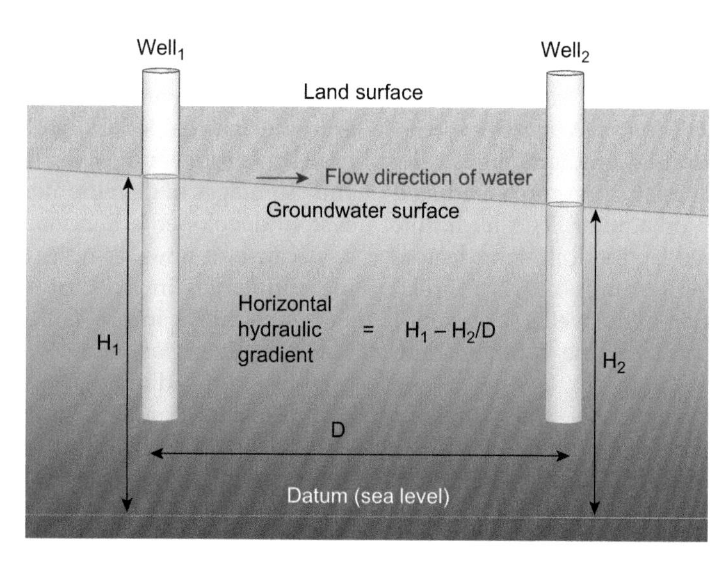

FIGURE 7.34 Schematic diagram showing the groundwater horizontal hydraulic gradient and direction of flow of water between two wells. *Source: Modified from Heath (1993), Alley et al. (1999), and NC Division of Water Resources (2012).*

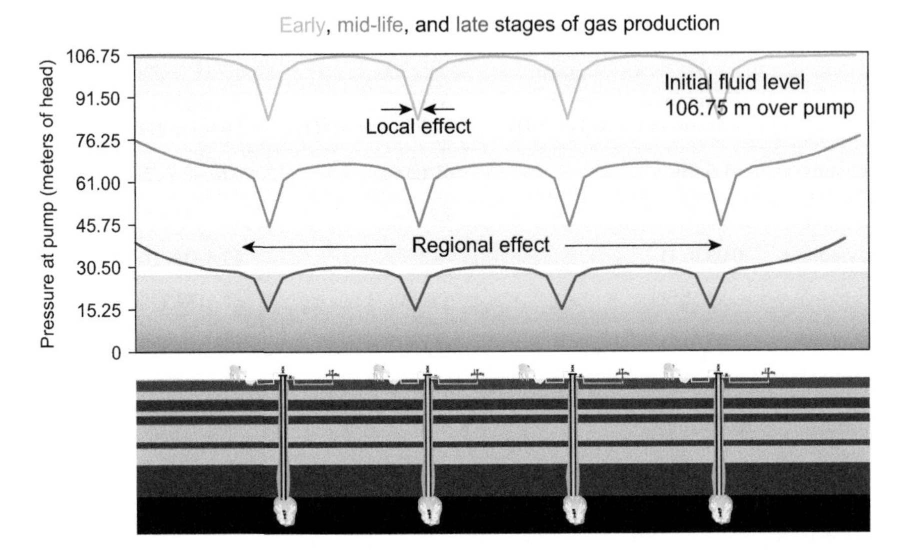

FIGURE 7.35 Schematic diagram of local and regional effects by withdrawing groundwater from coalbed gas wells. Lowering hydrostatic pressure profiles across the coal reservoirs in each well from pumping out water creates early to late stages of gas production, the latter with pronounced regional cone of depression. *Source: Modified from BLM, Casper District, Wyoming.*

flow of water until the inflow rate is equal to the withdrawal or pumping rate. Convergence flow of water from all directions toward the coalbed gas well creates a three-dimensional withdrawal surface with steeper hydraulic gradient toward the pumping well. This withdrawal surface is called the cone of depression and is a low-pressure area (NC Division of Water Resources, 2012). Thus, the hydraulic gradient moves the water and gas as hydrostatic pressure is reduced in the coal reservoir or aquifer. The hydraulic gradient is determined by the following factors (NC Division of Water Resources, 2012): (1) relative geographic position of the wells, (2) distance between the wells, and (3) the total head at each well.

Critical to determination of the hydraulic gradient is the distance of the wells, which in turn affects hydrostatic pressure and flow of gas associated with the water as dictated by Darcy flow. According to Alley et al. (1999), pumping a single well only locally affects the water flow system. However, pumping of many wells in a large area regionally affects the groundwater system as exemplified in Figure 7.35 in the coalbed gas pilot project in the Powder River Basin, Wyoming. Initial pumping in individual wells during the early stage of coalbed gas development shows a local effect of the groundwater system with drawdown around each coalbed gas well. As pumping continues from several wells through the mature and late stages of development, the groundwater system developed a regional drawdown effect.

Because of dewatering programs by the mining companies in front of surface coal mining since the 1960s in the Gillette coalfield of the Powder River Basin in Wyoming, hydrostatic pressures of adjoining coal reservoirs have been lowered resulting from regional groundwater drawdown. The main purpose of dewatering by the 13 surface coal mines along the eastern margin (e.g. Gillette coalfield) in the Powder River Basin is to deplete the groundwater-saturated sandstone aquifers adjacent to the mines, which prevented gravity drainage of water to the mine pits (Wheaton & Metesh, 2002). Consequently during the early 1980s when coalbed gas was explored adjacent to and basin-ward from the coal mines, well spacing was set mainly from four to eight wells per 2.6 km^2 area. In the basin center, the well spacing was set from 1 to 16 wells per 2.6 km^2 area. The 16.9 ha well spacing was permitted in unexplored areas in the basin center where the original hydrostatic pressure was unchanged by mining or nearby coalbed gas exploration and development. In a few cases in this area, it took as many as 16 months or longer to draw-down the groundwater level before gas desorption began. Thus, the well spacing in the Powder River Basin was a function of water withdrawal required from the coal reservoirs to start gas desorption. In addition, dewatering to degas the coal complicates these reservoir properties. In very permeable coal reservoirs, dewatering to lower below gas desorption hydrostatic pressures may require more closely spaced wells to be drilled. However, the withdrawal of the water and gas from the coal, in turn, is accompanied by shrinking of the coal matrix, which is counterproductive to gas desorption. As coal matrix, shrinks the voids and open fractures contract.

Gas Drainage and Correlative Rights

The role of groundwater flow in coalbed gas desorption and flow has shown that this process can establish large gas drainage areas across leased and/or unleased properties. As gas drainage is established by pumping coalbed gas wells, gas reserves from adjoining (leased or unleased) areas can be gradually or rapidly depleted depending on the hydraulic pressure, hydrostatic gradient, and well spacing. Correlative rights refer to the interests of different coalbed gas owners to develop the same coal reservoir in adjoining leases and unleased areas. Protection of correlative rights to capture gas within leased or unleased areas is of primary importance to coalbed gas development. However, in some countries, the rule of capture is recognized in situations where the offset mineral owner refuses to timely drill, or allow to be drilled, protective wells.

Figures 7.36(A), (B), and (C) demonstrate the principle of gas drainage and correlative rights through elapsed time of 90–270 days as it affected a hypothetical lease adjacent to unleased Federal acreage within a 15.5 km^2 area in the Powder River Basin (Larry Claypool, Personal commun., 2013). Figure 7.36(A) shows the leased tract with four coalbed gas wells in each of the four sections (a section is 2.6 km^2 or 1 mi^2 area) and the adjoining Federal tract located in the adjoining two sections. After an elapsed time of 90 days, the dewatering reached a hydrostatic pressure of 1.65–1.69 MPa creating a local cone of depression or drawdown around the wells and gas drainage to the four wells. The cone of depression and gas drainage area expanded around the four wells as dewatering continued at 176,022 l of water per day per well with the same hydrostatic pressure. After an elapsed time of 180 days of dewatering, the gas drainage expanded into the westernmost section of the unleased Federal acreage. With continued dewatering at the same hydrostatic pressure, the cone of depression and gas drainage expanded farther. After an elapsed time of 270 days of dewatering, the gas drainage expanded into most of the westernmost section and part of the easternmost section of the unleased Federal acreage. Without the protection of correlative rights, the coalbed gas in the unleased Federal acreage will continue to be drained and lead to loss of revenue (e.g. royalty) for the government.

Initially in the Powder River Basin, gas production was fraught with problems arising from rapid and chaotic development. This led to regional gas drainage and depletion of public and nonpublic coalbed gas reserves resulting in inequitable development of resources. Thus, the need arose for protection of

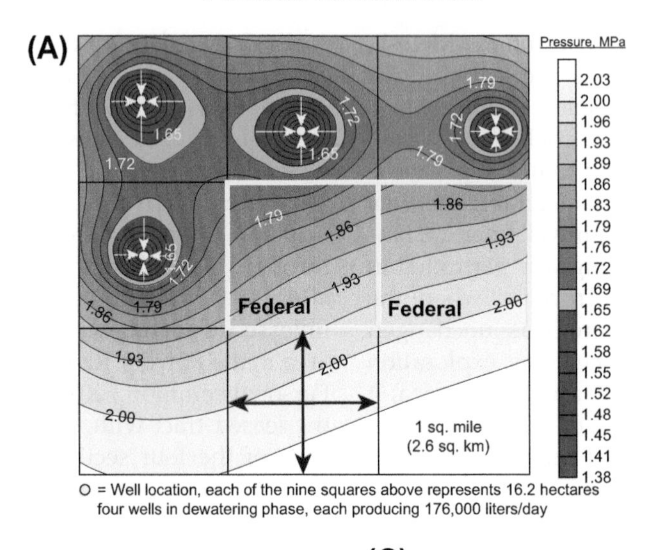

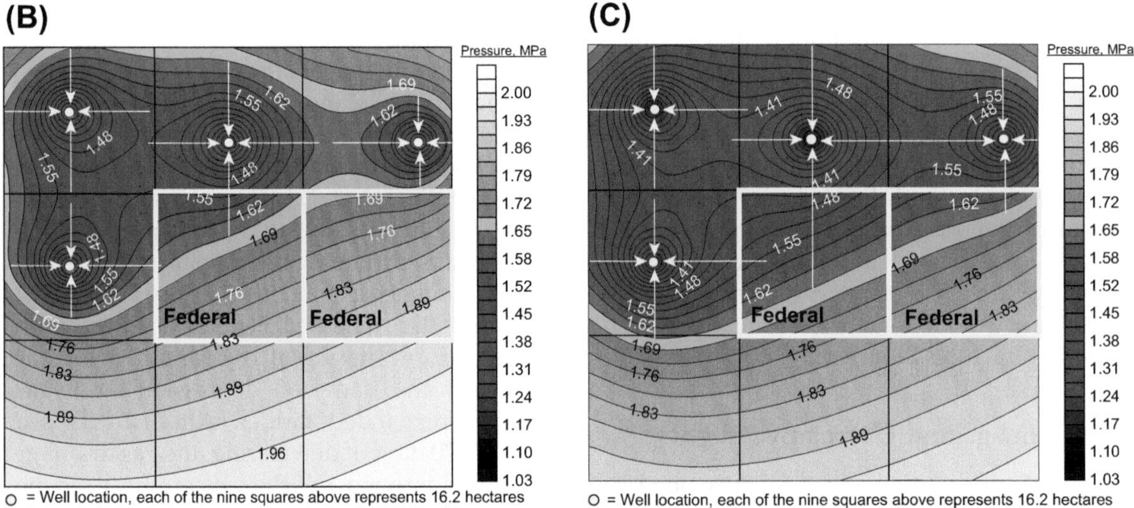

FIGURE 7.36 Aerial map showing hypothetical sections (e.g. square areas) and effects of dewatering from coalbed gas wells and on adjoining unleased Federal sections in the Powder River Basin. Graphic representations of local coalbed gas drainage (arrows) from groundwater withdrawal from each well after an elapsed time of 90 days at hydrostatic pressure of 1.65–1.69 MPa (240–245 psi) (A); elapsed time of 180 days with gas drainage expansion into one of the unleased Federal section (B); and elapsed time of 270 days with continued gas drainage expansion into both of the unleased Federal sections (C). *Source: Modified from Larry Claypool (Personal Commu., 2013), BLM, Wyoming, United States.*

correlative rights by means such as completion of offset wells in adjacent properties in the affected coal reservoir or entering into equitable agreement. In the Powder River Basin, to protect correlative rights of public coalbed gas reserves, the BLM-WY (Bureau of Land Management) formed coalbed gas units (e.g. unitization as large as 10,000 ha in which an operator is elected to develop each unit and participating lease owners proportionally share the gas production and development costs.

Coalbed gas wells are subject to the spacing regulations defined under the State and Federal Spacing Orders and Petroleum and Natural Gas Acts (British Columbia Ministry of Energy and Mines, 2011). State and/or Federal review boards (e.g. WOGCC, Colorado Oil and Gas Conservation Commission (COGCC), BLM, and British Columbia Ministry of Energy and Mines) issue gas field or pool spacing order which creates drilling units or a pattern of drilling units for a coalbed gas field or pool. The issuance of permits for unitization starts with the application of an operator or owner of coalbed gas for a permit indicating a plan to drill a coalbed gas well outside of such drilling units or pattern of drilling units and thereby potentially extend the coal bed pool or field. Diverse mineral interests are developed by the mineral owners, lessees, and operators joining together in a formal Unitization Agreement wherein drilling and production costs, revenues, royalty obligations, etc. are shared on acreage-based (or less commonly reservoir-based) proportions. The review boards for the purpose of prevention of waste of coalbed gas or to protect the correlative rights of the owners of coalbed gas have the power to establish or modify drilling units and field rules or unitize coalbed gas wells, pools, or fields. Drilling units are defined as uniform shape and size for an entire coalbed gas pool or field.

Due to the lateral variability of coalbed gas reservoirs, which may lack discrete geological limits across large areas, unitization has not always been sufficient to completely protect correlative rights. This has not been demonstrated anywhere more effectively than in the Powder River Basin where stratigraphic correlations of coal beds are complex and not consistent from one operator to another (Flores, McGarry, & Stricker, 2005). Often, this led to gas drainage of the same coal reservoir unknowingly by different gas operators in adjacent leases, which has been miscorrelated by misapplied coal bed nomenclature (Flores et al., 2005). Development occurred rapidly and extensively, with hundreds to thousands of wells being drilled over a relatively short time period. Extensive dewatering occurring from these wells over large areas resulted in widespread groundwater drawdown and accompanying reduction of the hydrostatic head on a regional scale. As a result, coalbed gas desorption was initiated over large areas where insufficient wells had been drilled to capture the gas resources. This caused gas to be drained from leases that lacked productive wells. This was evidenced by numerous later wells that required little to no dewatering to establish coalbed gas production. The primary means of protecting correlative rights (i.e. drainage protection) in the absence of unitization is drilling of protective offset wells on the affected leases. This requires either due diligence by the lessees or active enforcement by the lessors. In the case of Federal leases in the Powder River Basin, BLM-WY is the lessor. Federal oil and gas lease terms require protection from drainage. This is not necessarily the case for State or other leases.

GAS GATHERING SYSTEMS

Contemporaneous water withdrawal to reduce hydrostatic pressure in the coal reservoir in order to desorb and produce gas from the wellhead requires dual but separate gathering systems (Figure 7.37A). The common methods used to extract the water and gas from the coalbed gas wells include the use of beam pumps with gas separated by natural separation, packer separators, screening devices, "poor-boy" separators, etc. best described by Simpson, Lea, & Cox (2003). For example, the design of the "poor-boy" gas separator is shown in Figure 7.37(B), which was described by Simpson et al. (2003) as although commonly used in coalbed

BOX 7.4

CONFLICTS OF COALBED GAS OWNERSHIPS: CONCEPT AND MISCONCEPTION

Oil and gas ownership conflicts began with the Drake well drilled in Pennsylvania in 1859. Subsequently, the "rule of capture" has been an integral part of the law for development since (Brantly, 1971; Hardwicke, 1935). Hardwicke (1935) succinctly explained the "rule of capture" as an ownership producing wells on land with acquired oil and gas titles, which may be proved as part of migrated oil and gas from adjacent lands. Because of the nature of migration of oil and gas across property boundaries, the "rule of capture" gives the right to extract oil and gas that flow out of someone's land (USLegal, Inc., 2012).

Pertinent to coalbed gas are the legal disagreements over ownerships of coalmine and coalbed gas. Conflict between Amoco Production Company and the Southern Ute Indian Tribe of the various ownerships led the Supreme Court of the United States (1999) to rule on this dispute. In Amoco Production Company v Southern Ute Indian Tribe, 526 U.S. 865 (1999), the US Supreme Court overturned a lower court decision and ruled that coalbed gas is not a part of the coal. That is, coalbed gas is part of the gas estate and not the coal estate. The Court ruled that coal companies could degas during mining but the right to degas does not mean ownership of the gas. The Supreme Court ruling has been used by other countries (e.g. Australia, Canada, Indonesia, and Philippines) to formulate guidelines for the development of coalbed gas under the oil and natural gas estates.

Conflicts between coal mining and gas industries have delayed coalbed gas development in some countries. In the US, some coal companies have bought coalbed gas rights from gas operators at above market prices to continue unobstructed mining operations. Other coal mining companies (e.g. Powder River Basin) were embroiled in conflicts on depletion of coalbed gas reserves that have been deemed lost from the coal reservoir that was mined from dewatering in front of mining.

gas wells, it is rather an inefficient system. The method of water extraction mainly depends on the predicted rate of water production and permeability of the coal reservoir (Table 7.6). Submersible pumps (Figure 7.38) are utilized in the Powder River Basin where the volume of coproduced water is large and permeabilities are high. In comparison, conventional rod pumps are used in basins with lower volumes of coproduced water and permeabilities such as in the Raton, Uinta, Arkoma, and Appalachian Basins.

Gas separated at the wellhead enters the gas gathering system, which includes measurement equipment, pipeline networks, compressor stations, and treatment plants. In general, low-pressure coalbed gas from the wellhead is transported in plastic pipes to different compressor stations where the gas is pumped into an intermediate-pressure steel pipeline. The compressed gas eventually enters a main, high-pressure pipeline where it is carried to the market. Water is separately collected from the wellhead and is stored in tanks off well site, recycled for drilling, or moved by truck and pipeline to a disposal site for treatment (see Chapter 8). In some cases, water is reinjected to underlying reservoirs. Construction of related infrastructures for both gas and water gathering systems is a function of the

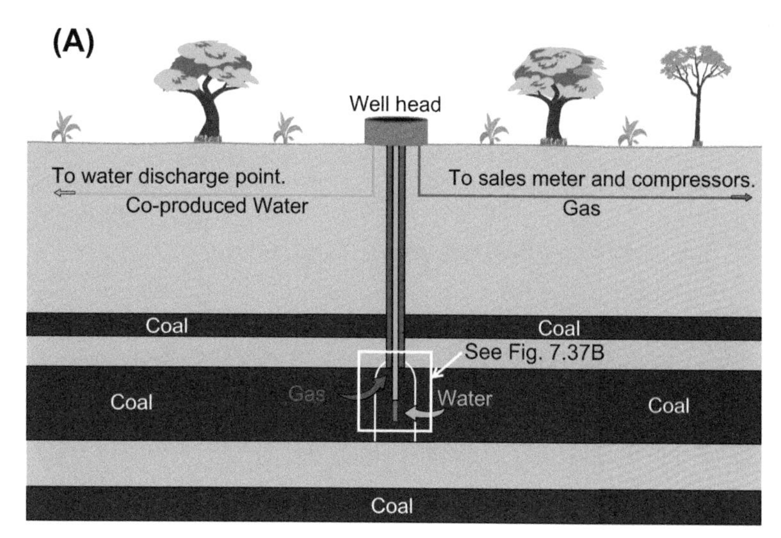

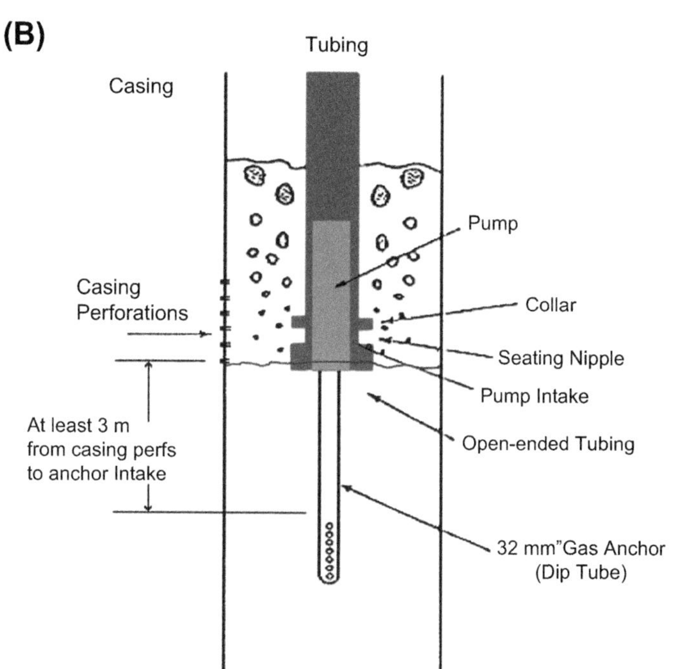

FIGURE 7.37 (A) Diagram of cross-sectional view of a completed coalbed gas well showing dual systems of water and gas drainage in the well production casing to the wellhead. (B) Design of "poor boy" gas separator, which relies on creating fluids to travel downward at >15 cm/s in order for bubbles rising to the annulus at high velocity and not entering the pump. *Source: Modified (A) from Dwain McGarry (Personal Commun., 2005) of BLM, Wyoming, United States. Adopted (B) from Simpson et al. (2003).*

landscape, land ownership, costs, and environmental constraints.

Building a gas gathering system begins with the knowledge of the gas flow rates, and plans on how to operate pressure limits and handle moisture in the gas (Schraufnagel, 1993; Simpson, 2008). Another consideration is the life span of coalbed gas wells, which is usually shorter

TABLE 7.6 Comparison of Coal Reservoir Properties (e.g. Permeability, Coproduced Water, Water Extraction) between Selected Coalbed Gas Producing Basins in the United States

Coal Reservoir Properties	Powder River Basin	Raton Basin	Uinta Basin	Arkoma Basin	Appalachian Basin
Water extraction	Submersible pump	Conventional rod pump	Conventional rod pump	Conventional rod pump	Conventional rod pump
Water disposal	Surface	Deep injection	Deep injection	Deep injection	Deep injection
Well spacing (acres)	80	160	160	80	60
Permeability (mD)	1–1000	ND	5–15	1–10	1–5
Rate of water production per day (m^3)	60	10–20	45	2	2–5

ND = no data; mD = millidarcies; m^3 = cubic meter.

periods, in comparison to conventional oil and gas wells. Life spans of coalbed gas wells vary from coal basin to basin and country to country. Coalbed gas wells in Australia may produce gas from 10 to 20 years; Canada, 10–40 years; and United States, 7–15 years (APPEA, 2012; Braun, Oedekoven, & Aldridge, 2002; British Columbia Ministry of Energy and Mines, 2011; New South Wales Parliament Legislative Council, 2012). In

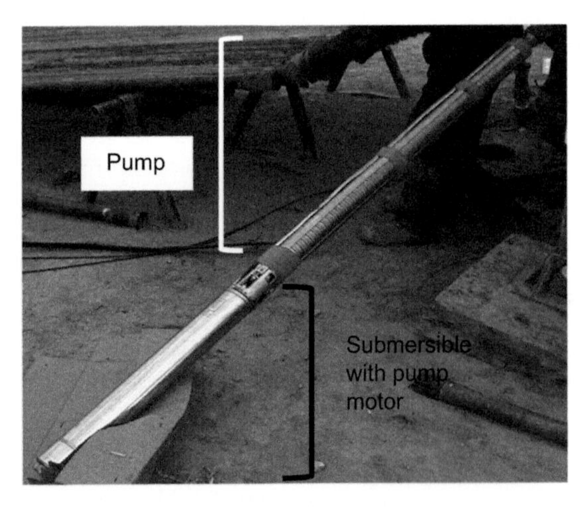

FIGURE 7.38 Submersible pump is one of several gas/water separators described by Simpson et al. (2003). Downhole electric submersible with pump motor in the lower part and pump in the upper part. This pump is one of many used in the Powder River Basin. Photograph courtesy of Tom Doll.

countries where major gas pipelines are not available to transport gas to the market, this must be considered in designing the intermediate gas gathering system. Thus, designing gas gathering system ranges from a simple to complex operation. The most simple gas gathering system is in coal mines where the coalmine gas is collected from the wellhead, treated for low water content, compressed, and directly used to run gas turbine engines for electricity. The most complex is collection of gas from the wellhead; metered for flow rates; transported to compressors that compressed the gas at various pressure stages; treated for water, sulfur, carbon dioxide, and nitrogen contents; and then finally transported to a high-pressure pipeline to market.

Gas Measurement Equipment

Gas measurement equipment and modifications for most efficient operations are discussed by Schrufnagel (1993) and Simpson (2003, 2008). Usually the gas flow rates are measured at or near the wellhead and at the various compressor stations. Coalbed gas is produced through the annulus between water production tubing and borehole casing (Figure 7.37). At the surface, the coalbed gas is piped through a positive-displacement meter where it is measured. Coalbed gas production is measured by mainly orifice, diaphragm, rotary, and turbine meters depending on volumes of gas produced (Lambert & Trevits, 1978; Schrufnagel, 1993). The diaphragm and rotary meters are normally used on nonstimulated and stimulated wells that produce less than 283 and up to 2379 m^3/day, respectively (Lambert & Trevits, 1978). The turbine meter is used to measure gas flows of more than 2379 m^3/day. According to Schrufnagel (1993), the orifice and turbine meters are mainly used for measurement of coalbed gas flow rates at the wellhead and the diaphragm and rotary meters for measurements of compressor fuel consumption. Comparison, advantages, and disadvantages of these measuring devices were discussed by Boyd (1991).

The orifice meter is a recorder inserted in the gas line to measure the difference in gas pressure between the upstream and downstream orifice plate (Schrufnagel, 1993). The difference in the gas pressure depends on the size of the plate, temperature, flow rate, and line pressure. An advantage of the orifice meter is that the differential pressure, line pressure, and temperature of the gas are continuously recorded and can be used for the interpretation of the production history of the well. Simpson (2003) has minimized measurement uncertainty and wasted horsepower by improving the size (e.g. the ratio of the orifice plate bore to the tube inside diameter) of the measurement equipment. According to Ward (2005), the coproduced water with the gas makes for a different set of measurement practices. That is, the orifice plate, which restricts the gas line, allows flow only through the center of the line and acts as a dam retaining water upstream of the meter. The water accumulation unfavorably affects the differential pressure measurement. This problem has been resolved by a meter in which the restriction in the center of the gas line permits gas flow around it and not through it (Ward, 2005).

The turbine meter measures the total flow and flow rate as the gas flows over rotor blades and magnets in the meter pickup signals linear to the gas velocity (Schrufnagel, 1993). The meter works with the principle that the total volume of gas is determined by the rotor blades between which are a predictable number of turns per unit volume. Ward (2005) indicated that as each rotor blade passes by a magnet, the excitation signal in the coil inside the magnet is amplified. The pulse is interpreted by an indicator or field controller as a unit of measure. Collection of pulses is used to determine flow rates and total accumulated volumes of gas. An advantage of the turbine meter is the optional use of a software to automate and calculate gas flow and total gas volume.

High-Tech Automation Measurement with Caution

According to Ward (2005), sensor technology used in gas measurement has advanced significantly during the past several years permitting users accuracies of 0.05%, which is a cost-effective investment of instrumentation. Manufacturers have invented high-tech multivariate transmitters, which provide processing of three signals as opposed to one signal, and users at reduced costs of installation have taken advantage of the Ethernet. Ward (2005) indicated that some manufacturers have integrated the functions of the multivariate transmitter with electronic flow computer into the same device minimizing cost of automation. This is all made possible by Ethernet communications with sensors including gas and water pressures ultimately affecting the onsite and off-site controls of the coalbed gas well.

The Wyoming Bureau of Land Management of the US Department of Interior has adopted requirements for producing oil and gas on Federal lands. One of the requirements (e.g. Onshore Order #5 amended by Notice to Lessee's dated March, 2004) includes Electronic Flow Computers, which are capable of generating, for example, gas volumes useable by accounting for royalty dispersion (Ward, 2005). This is particularly important in the Powder River Basin where about 55% of the subsurface mineral rights are owned by the federal government (Flores, 2004).

The intelligent functionality of well optimization (e.g. gas measurements) is located in a programmable logic controller, which can manage the required logic (Ward, 2005). Reduction of field visits on well sites is permitted by the combination of the programmable logic controller and electronic flow computer in one device or a hybrid controller. According to Ward (2005), the hybrid devices or field controllers can provide the ability to (1) remotely configure/program the controller, (2) remotely load new gas calculations, (3) share gas database, (4) display gas measurement and process the data locally, and (5) high-speed data logging of processed data permitting reservoir engineers to analyze and resolve well dynamics in real time.

Realities cannot be ignored when dealing with high-tech sensitive instrumentation as utilized in coalbed gas wells in conjunction with automation of field operations. Simply put, automation is necessary for optimizing production and efficiency of coalbed gas wells in a cost-effective manner. However, the conditions in coalbed gas wells are such that sensitive instruments (e.g. sensors) to conduct signals are installed in restricted narrow shallow to deep boreholes filled with groundwater and characterized by highly unstable walls of the coal reservoirs. Ward (2005) noted that the sensor technology is the weakest link in automation of coalbed gas wells. That is, the installation downhole of highly sensitive sensors in submersible transducers without physical damage affecting the accuracy to conduct signals is a common failure. Another challenge is interpretation of the pulsed signal, which may be modified under subaqueous condition in the coal reservoirs.

Gas Pipeline Network

Major gas producers and consumers such as Australia, Canada, Europe, and China have various levels of pipeline networks for gas transmission from gas in-field areas. The natural gas pipeline networks include the interstate, intrastate, and in-field pipelines as typified in the United States (Figure 7.39). The primary level of gas transmission is through the interstate pipelines across state boundaries connected to national markets and for exports. The secondary level of gas transmission is the intrastate pipelines within the state either linked to the interstate network or local markets and to gas producers within the state (Figure 7.39). These pipeline networks consist of large-diameter, high-pressure pipelines. For example, intrastate pipelines are operated in Texas, California, Wyoming, Colorado, and others with Texas and California being the top two natural gas consumers in the United States (USEIA, 2008; NPC, 2011). Texas has 93,380 km of natural gas pipeline compared to California, which has about 8636 km. USEIA (2008) reported that the intrastate pipeline network makes up about 29% of the total miles of natural gas pipelines in the United States.

Coalbed gas within a field is transmitted from the wellheads by small-diameter pipes, which are connected to downstream gas gathering facilities (e.g. processing and treatment facilities). A gas gathering system within a coalbed gas field extends over large areas in square kilometers or miles and connects hundreds or thousands of coalbed gas wells, each with their own production characteristics.

An in-field gas gathering system is exemplified in the Powder River Basin, Wyoming (Figure 7.40), in which a coalbed gas field has operated since 1999 south and west of Gillette, Wyoming (Flores et al., 2001). The gas gathering complex that serves this coalbed gas field moved a total of more than 1.3 million m^3/day in 2001 through 861 km of low-pressure lines from wells to pod stations, 56 km of intermediate gathering system from pod stations to the screw and reciprocating compressors and 30 km of high-pressure lines from the reciprocating compressors to the main pipelines. Within the complex, a gas gathering system consists of 10 pod stations with each station

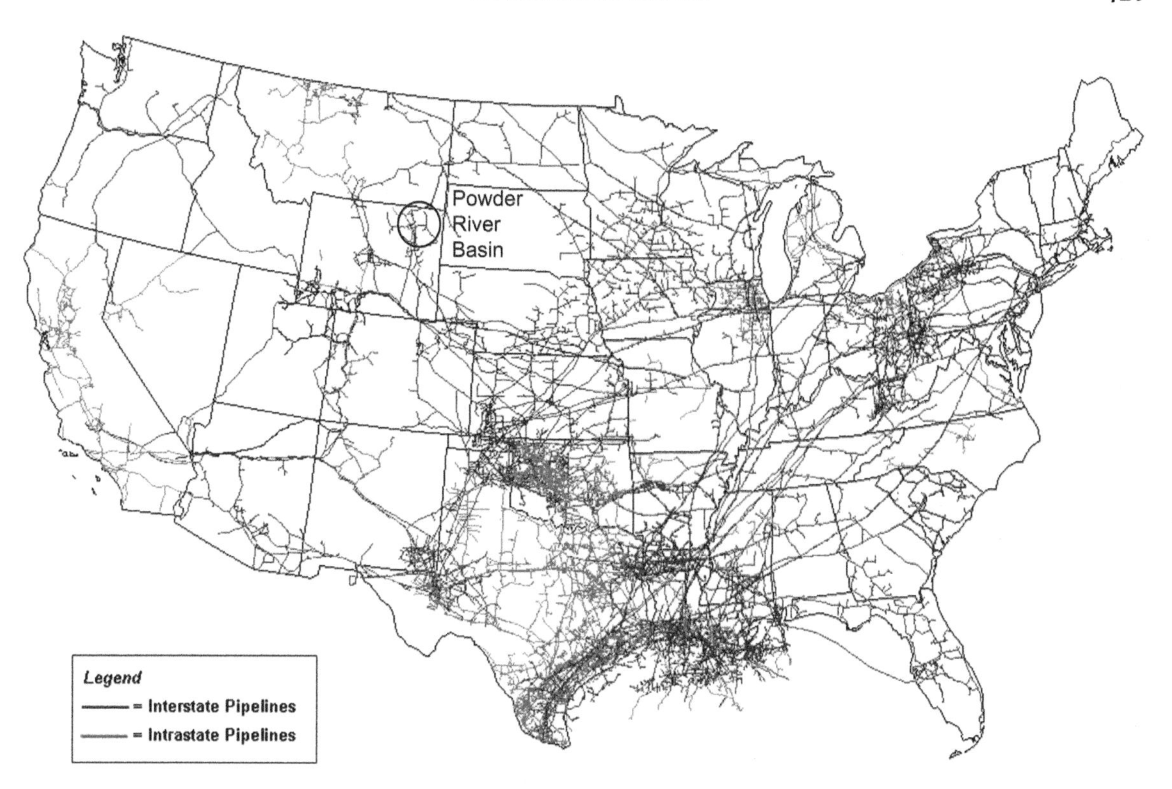

FIGURE 7.39 Network of intrastate and interstate natural gas pipelines in the United States. Circled area shows pipeline network in the Powder River Basin in Wyoming and Montana. *Source: Modified from USEIA (2008).*

serving 20–60 coalbed gas wells (Flores et al., 2001). The coalbed gas from each well is metered in the pod station and is commingled prior to being piped to the screw and reciprocating compressors (see next section). The gas is delivered into pipelines interconnecting to 61-cm pipelines (e.g. Thunder Creek) (Figure 7.39). The gas is either moved north to North Dakota through the Williston Basin Interstate pipeline or it is moved to a south pipeline (Nebraska, Kansas, the Front Range of Colorado, and California).

Gas Compressor Stations

The inherent low-pressure condition of coalbed gas from the wellhead necessitates compression to boost pressures at compressor stations built within the gas gathering system (Figure 7.41). The wellhead pressure of coalbed gas ranges from 0.01 to 0.03 MPa, which in turn is piped to rotary screw compressor station to increase the gas pressure to 0.48–0.55 MPa (70–80 psi). The compressed gas, in turn, is transmitted to a reciprocating compressor to further boost the gas to about 8.27 MPa. The compressed gas is transmitted through an output (tailgate) pipeline connected to either one or more of the major intrastate and interstate pipeline networks. The two levels of booster compressors have been used within the coalbed gas gathering system in comparison to single-level booster compressors (e.g. 6.89 MPa), which is used for pipeline-quality conventional natural gas directly produced at the wellhead.

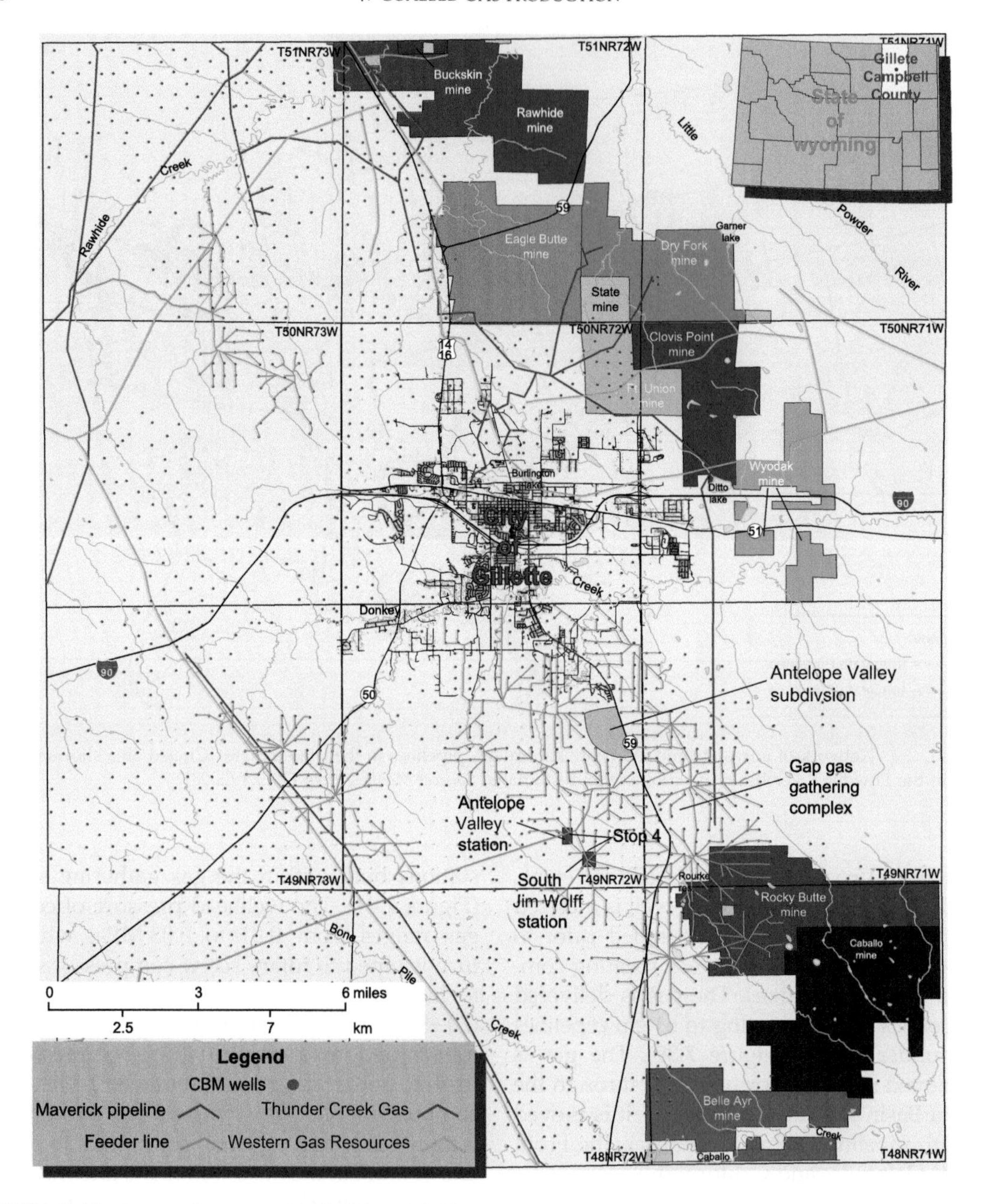

FIGURE 7.40 A map showing an infilled network of coalbed gas pipelines from the wellhead to gas compressors in stations south and west of the City of Gillette in the Powder River Basin in Wyoming, United States. There are several hundreds coalbed gas wells in this field. A section contains as many as 16 wells. *Source: Adopted from Flores et al. (2001).*

FIGURE 7.41 Coalbed gas gathering system in the Powder River Basin from the wellhead (A) where gas is piped to a metering pod (B) which in turn is piped to screw-type compressors to boost pressure to 0.55 MPa (C). The compressed gas is in turn piped to a reciprocating compressor station (D) to boost pressure up to 8.62 MPa for transport to trunk gas pipelines for intrastate and/or interstate market. Red arrows show a progression of gas gathering from upstream to downstream.

Simpson (2003) described a screw compressor as containing a pair of helical screws that interlock with each other as they rotate to compress gas. Thus, a screw compressor is sometimes called rotary screw compressor. The rotary screw compressor can function as oil-free or oil-flooded type. Oil-free screw compressor relies on the rotary screws to function without oil resulting in lower maximum discharge pressure capacity. The oil-flooded screw compressor requires oil to flood into the compression chamber to seal around the rotors, lubricate metal parts, and remove the heat of compression (Simpson, 2003). This process, in turn, results in a higher maximum discharge pressure capacity. The reciprocating (or recip) compressor uses a piston, which moves in a cylinder to pull gas into the compression chamber and raise the gas pressure to the required level when the piston reverses (Simpson, 2003). Reciprocating refers to repetitive back and forward motion or up and down motion of the piston. The function of the piston, which moves in a cylinder, is to draw a volume of gas through an intake valve during a downstroke motion and compress the gas during the upstroke motion. An outlet valve opens as the piston reaches the top of the upstroke motion allowing the gas to leave the compression chamber into the supply gas line. Repetition of the process results in the continuous production of compressed gas.

Maximizing the efficiency by minimizing compressor engine emissions, compression design focused for several years on designing lateral and transmission compression to operate in a narrow rate/pressure range (Simpson, 2003; Simpson & Lea, 2003). Single-stage reciprocating compressors have been used in gas gathering systems to avoid using three-stage machines at the transmission gas line to access high-pressure (6.9 MPa) transport system (Simpson, 2003). The use of oil-flooded screw compressors on well sites permits for very low compressor suction pressures at moderate to low discharge pressures and very good flow rates (e.g. 0.03–0.1 MPa).

Treatment Plants

The composition of coalbed gas varies within and between coal beds, fields, basins, and coalmine gas types. The variable composition of the coalbed gas requires upstream treatment in different degrees in plants within the gas gathering systems in order to meet the stringent standards of pipelines specifications for commercial transmission. As a significant part of the total commercial transmission downstream treatment of coalbed gas for liquefaction and compression mainly for global export was included, which has become a significant part of market and has expanded during the past decade. The composition of coalbed gas developed from virgin coal beds is higher grade, requiring little or no treatment compared to coalmine gas, which is contaminated with oxygen, carbon dioxide, and nitrogen.

Contamination may be treated by blending or spiking of low-quality gas with high-quality gas with an upgrading system. Upgrading low-quality gas to boost the methane concentration may assist but eventually it is necessary to use more advanced technology to remove the contaminants in order to bring the gas up to pipeline specifications (USEPA, 2008).

Upstream Treatment

In general, coalbed gas requires some degree of upgrading its quality such as extraction of water vapor, carbon dioxide, nitrogen, and treatment of other constituents along the gathering system in order to meet pipeline specifications (e.g. $<0.2\%$ oxygen, 2% carbon dioxide, 3% nitrogen, and 112 mg/m^3 water vapor; USEPA, 2008). Understanding the composition of biogenic and thermogenic coalbed gas from subbituminous and bituminous coals, respectively, can assist in a potential strategy to meet pipeline specifications. For example, thermogenic coalbed gas from bituminous and anthracite coals contain various concentrations of heavy hydrocarbons (e.g. ethane, butane, and propane), which may or may not require fractionation. Also, because of the large volume of water coproduced with biogenic coalbed gas especially with subbituminous coal (e.g. Powder River Basin), water vapor becomes an important component of the gas, which requires additional processing. A part of the operations in a coalbed gas gathering complex is in-line dewatering and cleaning through the process called "pigging" out the gas pipeline. That is, pig launchers and receivers are installed with the pipeline in which a device or "pig" is inserted and launched into the pipeline with compressed air pushing as the device internally cleans or purges the water vapor while traveling to the receiver. Another compositional parameter, which may arise as a problem, is hydrogen sulfide derived from the sulfur content of coal. Systems to remove oxygen, carbon dioxide, nitrogen, and water vapor can be stand-alone equipment, but according to USEPA (2008), an integrated enrichment facility can be installed to remove all four contaminants with a series of connected processes at one location.

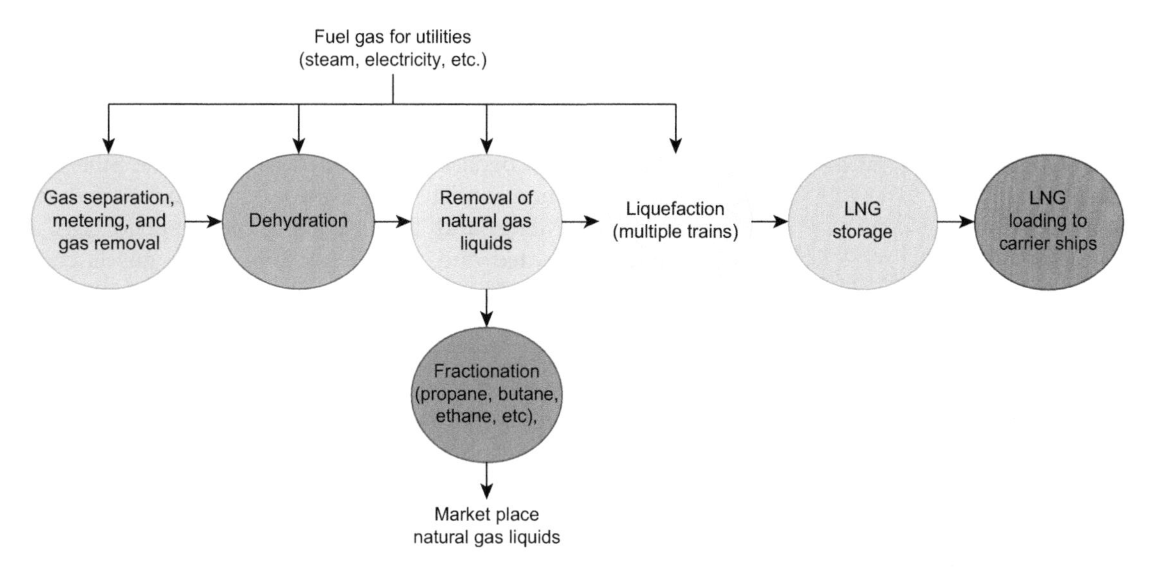

FIGURE 7.42 Flowchart showing processing of natural gas including coalbed gas into liquefied natural gas (LNG) for international markets. *Source: Modified from USDOE (2005).*

Downstream Treatment

Coalbed gas, which makes up about 8% of the total natural gas production of the United States is comingled with the conventional natural gas in the pipeline network. Thus, coalbed gas becomes part of the gas system transported to the market for domestic and global consumption. At the end of the downstream part of the market pathway is treatment of the gas for global markets. At the global end of the market, both the commingled coal bed and natural gas is liquefied as LNG) and/or compressed as CNG. The LNG is mainly methane converted to a liquid state for the purpose of storage and transport (USDOE, 2012). The process of liquefaction (Figure 7.42) requires extraction of oxygen, nitrogen, water vapor, carbon dioxide, acid gases, heavy hydrocarbons (e.g. ethane, propane, butane, pentane, and hexane), and other constituents much like the treatment that coalbed gas underwent in the upstream gas gathering system (see last section). The gas impurities are removed in order to prevent corrosion and blockage of equipment during the process of liquefaction. The natural gas is then condensed into liquid by cooling (refrigeration) to $-162\,^{\circ}C$ at atmospheric pressure using various processes including refrigeration with hydrocarbon refrigerants at a large scale. The liquefaction reduces the volume of the natural gas, which is about 47% as dense as water and heavier than CNG. The process is designed to upgrade to match pipelines specifications in gas gathering system of the receiving countries. LNG is stored in well-insulated and low-pressured tanks deployed onshore. LNG is stored as a "boiling cryogen" or as a very cold liquid at its boiling point for the pressure at which it is being stored. Storage of LNG utilizes a phenomenon called "autorefrigeration" where the LNG stays at near constant temperature if kept at constant pressure.

FUELING THE FUTURE WITH COALBED GAS

Fueling the future with coalbed gas has begun in Australia, which is the leading exporter of coalbed gas as a part of the LNG marketed worldwide and to Asia. The United States and Canada are developing LNG projects with the projection to export to Asian markets. Coalbed gas will be a part (8% of the total natural gas production in the United States) of the LNG to be exported by these countries. The traditional Asian LNG buyers, which include Japan and South Korea, will be joined by China and India in the coming decade.

China, which is producing coalbed gas from active coal mines and unmineable coal beds, has an LNG plant in Jincheng where the gas is locally marketed. Also, India is exploring and producing coalbed gas from active coal mines and unmineable coalbeds as a part of their natural gas resource (Ojha, Karmakar, Mandal, & Pathak, 2010). Indonesia, which is the world's largest LNG producer (USEIA, 2003), is currently exploring for coalbed gas to compliment the conventional natural gas reserves (Moore, Bowe, & Nas, 2012; Stevens & Hadiyanto, 2004). The Pacific Basin is the largest LNG producing region in the world supplying about 49% of the total global exports (USEIA, 2003). The Pacific Basin LNG exporters include Australia, Brunei, Indonesia, Malaysia, United States (e.g. Alaska), and Russia (Sakhalin Island). The other two exporters are the Middle East and Atlantic with LNG mainly composed of conventional natural gas (USEIA, 2003).

CNG involves compression of natural gas, which is mainly composed of methane, to less than 1% of the volume at standard atmospheric pressure. The natural gas is compressed and stored at high pressure of 24.8 MPa but can be stored at low pressure as adsorbed natural gas at 3.4 MPa (NSF, 2012). CNG is used for totally fueling light, medium, and heavy vehicles from sedans to trucks with novel storage tanks (Burchell & Rogers, 2002; NSF, 2012). However, natural gas vehicles can be bifuel in which vehicles are fueled by both compressed gas and gasoline.

The future of coalbed gas may lie with the marketing model developed by Shanxi Lanyan Coalbed Methane Group Co., Ltd (Lanyan CBM), in Jincheng, China, for CNG and LNG, which the author has personally witnessed while working with the company. Lanyan CBM produces coalbed gas from unmineable Permian coal beds in surrounding areas to the Sihe coal mine, which also produces coalmine methane (Meng and Tian, 2007). The coalbed gas is treated and processed: 25% of the gas is compressed, 25% is liquefied, and the remaining 50% is marketed for electrical power generation in the city, Shanxi Province, and adjoining areas. The company has constructed and owns CNG stations in Jincheng to fuel light and heavy vehicles. The company supplies coalbed gas to the Sihe coalbed methane power plant with an installed capacity of 120 MJ/s, which makes it the largest coalbed gas power plant in the world.

The rate of growth of CNG fueled vehicles is about 30% globally with India and Pakistan being the leading users. India is an emerging coalbed gas producer from potential Permian and Tertiary coal reservoirs. The United States and Canada have developed CNG transit buses and trucks but not as significantly as the leading countries. The Pickens Plan touted the enormous potential for use of CNG to fuel trucks and heavy vehicles as a means of reduction of foreign oil imports (Allen, 2008). Globally devices inserted for gas injection engine systems will soon come to your natural gas fueling stations.

SUMMARY

The basic technology of drilling to produce coalbed gas was inherited from Europe in the nineteenth century via coal mine drilling, which was used to drain gas for the prevention of outbursts or explosions. A century of technological improvements led to successful development and production of coalbed gas from mineable and unmineable coals in the United States, which extended into Australia, Canada, Europe, China, and India. Benefiting from the same improved technology, other countries (e.g. Indonesia, Russia) have begun to explore for coalbed gas. Thus, global production of coalbed gas emerged during the past two decades with advancement of drilling, well completion, stimulation, logging, and gas gathering technologies. Despite the improvements, successful drilling/development techniques still depend on rudimentary understanding of the lithologies of the coal-bearing formations, coal bed thickness, lateral and vertical variability of coal beds, coal reservoir properties, and geological structures. This requires acquired experience in coal geology in particular and geology in general.

Improvement in the efficiency of coalbed gas production is based mainly on the progress of drilling from vertical to horizontal wells and the advancement from single-bed to multiple-bed completions. Vertical wells were designed originally to produce from a single coal bed leading to production of more coal beds or multiple reservoirs. Well completion of multiple coal reservoirs increased the efficiency and volume of gas production at cost-effective operations. Horizontal wells extend gas drainage area by penetrating coal beds for longer distance and making more contact with the cleat and fracture surfaces, which is perpendicular to the bedding planes. Horizontal wells are made more effective for production with additional horizontal wells lateral to the single wells, thus extending aerial coverage to tap into the coal reservoir. Multilateral wells increase the recovery of coalbed gas reserves from a much larger drainage area compared to a single horizontal well. In addition, the multilateral wells can be expanded into multibranched or pinnate wells extended from the horizontal well for more effective gas drainage. Surface directional drilling is a conventional vertical well that is turned into a radius or arc toward the coal bed penetrating the reservoir parallel to bedding. Directional drilling unites the horizontal well with a vertical production well. This technology reduces drilling site footprints, which permits drilling several wells from one drill pad into the same coal bed or into multiple coal beds.

Critical to drilling a coalbed gas well is creating minimal formation and reservoir damage. Reservoir damage may be minimized by the use or nonuse of drilling fluids. Use of drilling fluids runs the risk of plugging the coal cleat and fracture systems. Key to the drilling of coalbed gas wells is completion techniques, which influence gas flow and productivity of the coal reservoir. The techniques may be performed by open- and cased-hole completions. Completion in coalbed gas relies in a large part on the coal geology of the formation and coal reservoir. Stimulation of the coal reservoir enhances coalbed gas production by clean up of damage formed during drilling and completion with the use of fracture treatments. These treatments are especially applied to "tight" coal reservoirs, which dilate, open, and extend the width and length and connectivity of cleat and fracture systems. Treatment by hydraulic fracturing called "fraccing" uses combined water, chemical additives, and proppants. Reservoir characterization based on the coal composition, rank, and geology plays a major role in completion and stimulation of coalbed gas wells. Although anisotropism of the permeability and heterogeneity of the coal reservoir are well known, it has not eliminated continued perception of coal beds simple and unilayered coals with homogenous orthogonal set of cleat/fracture system.

Basic wireline logs (e.g. resistivity, gamma, density, acoustic or sonic, and neutron) are useful universal tools for identifying coal beds in the subsurface due to inherent low radioactivity, high resistance to electrical current, and low density. High-tech logs have been newly devised to compliment basic logs by calibrating the logs with coal quality, gas adsorption, and gas content data derived from coal cores. However, the high-tech logs have both virtues and limitations as predictive tools for coal reservoir characterization. Drilling, completion, stimulation, and logging are important to establishing coalbed gas pilot project toward production. Essential to the pilot project is testing for gas content and its producibility as determined by the connectivity of cleats and fractures or permeability of the coal reservoirs. Crucial to precommercial development of coalbed gas is testing how many wells are required to produce gas within a field and optimum well spacing for production as well as correlative rights protection. Groundwater and permeability of coal reservoirs play valuable roles in determining well spacing. Finally, to develop coalbed gas, a gathering system must be designed to include gas measurement, pipeline, compressor, and treatment plant complexes. The upstream gas gathering system feeds into the downstream market delivery system locally, regionally, nationally, or globally. These markets are critical to fueling the future with coalbed gas particularly if it is to become a significant part of the liquefied and compressed gas systems.

CHAPTER

8

Co-Produced Water Management and Environmental Impacts

OUTLINE

Coal and Coalbed Gas

http://dx.doi.org/10.1016/B978-0-12-396972-9.00008-2

KEY ITEMS

- Co-produced water is extracted with natural gas during production involving initially large amount of gas replaced by more water as pressure is depleted for conventional natural gas in contrast to initially large amount of water replaced by more gas for unconventional gas (e.g. coalbed gas).
- Origin of coalbed gas co-produced water is endogenous groundwater from adherent, inherent, organically chemical-bound, and inorganically crystalline-bound moisture in the coal matrix in contrast to exogenous groundwater from meteoric water and water from other aquifers.
- Groundwater sandstone aquifers supply water into coal beds by leakage through fractures and contacts. Channel sandstones are not isolated hydrological aquifers, which are more extensive along their lengths or depositional dip than along their width and in direct contact with coal aquifers.
- Basinal groundwater flow through formational macro units can be mistaken for regional flow of groundwater along depositional dip of channel sandstone aquifers.
- Isotopes of oxygen [$\delta^{18}O$] and hydrogen (δ^2H [δD]) provide signatures critical to determination of origin of coalbed gas co-produced water. Isotope dating co-produced water determines renewability or nonrenewability of groundwater aquifers.
- A concern in coalbed gas development is management of co-produced water related to the amount and composition of water. Water management includes disposal, treatment, and beneficial use governed by state and federal regulations.
- Knowledge of the volume and composition of co-produced water can assist gas operators to estimate costs of management of co-produced water related to environmental, social, and economic impacts.
- Hydrostratigraphic units in coal-bearing rocks are continuous and discontinuous. Coal and sandstone aquifers are discontinuous as hydrologic units often interbedded with impermeable mudstones and siltstones making for compartmentalized aquitards and aquicludes.
- Composition of co-produced water may be viewed according to its environmental effects, beneficial uses, and geochemical processes.
- Co-produced water is sodium bicarbonate and sodium chloride types consisting of bicarbonate, calcium, chloride, magnesium, sodium, and sulfate unique to geochemical reactions and processes inherent to generation of methane in coals with a biological component.
- Geochemical reactions and processes are driven by freshwater and brackish and/or marine nature of the groundwater. Near-recharge areas meteoric water has elevated dissolved organics, high total organic

carbon, and low total dissolved solids (TDS) where residence is short. Away from recharge areas reflects groundwater composition of the original depositional waters, particularly in brackish and marine environments.
- A part of the geochemical reaction and process is reduction of sulfate downgradient, which is a biological component of methanogenesis or generation of methane.
- Composition of co-produced water affects management practices because water is discharged into surface waters, impoundments, and reinjected into subsurface aquifers. Water quality before disposal requires treatments that include aeration/chemical precipitation, reverse osmosis, ion exchange, and electrodialysis.
- Beneficial uses of treated and untreated co-produced water include augmentation of surface water; aquifer recharges, storage, and recovery for domestic/municipal, agricultural, environmental restoration/remediation, and industrial uses.
- Lessons learned of management practices of coalbed gas co-produced water in the Powder River Basin and other US Basins, in particular and worldwide, in general, indicate some similarities and no universal solution.

INTRODUCTION

Co-produced with coalbed gas is water, which is extracted and transported from the coal reservoir to the surface where it is disposed, treated, or reused. In many basins worldwide, coals serve as aquifers containing varied amounts of groundwater, which in turn apply hydrostatic pressure that holds the gas in the reservoir. Coal contains pore, cleat, and fracture surfaces where the gas is absorbed by the hydrostatic pressure as well as where groundwater resides. Pumping out volumes of groundwater over time reduces the hydrostatic pressure allowing the gas to desorb from the surfaces in the coal reservoir. Withdrawal of large volumes of groundwater to reduce the coal reservoir pressure is an important part of gas production, which makes coalbed gas wells unique compared to conventional natural gas wells. Co-produced water gradually increases over time and reaches peak volume during the later life of conventional natural gas wells. In contrast, co-produced water is at its peak during early life of coalbed gas wells. Thus, it takes withdrawal of groundwater to produce coalbed gas.

Production of large volumes of water has become an integral part of coalbed gas production, and poses management and environmental issues. Igunnu and Chen (2012) estimated that daily global co-produced water with oil and natural gas development is about 29.75 billion liters with more than 40% of this amount discharged into the environment. According to USEPA (2010), about 178 billion liters of water are co-produced with coalbed gas wells annually in the United States. About 40% of this co-produced water is discharged to waters in the United States. The issues are exacerbated by development of coalbed gas worldwide with varying climatic conditions from arid to tropical and from very warm to cold temperatures. For example, where rainfall and groundwater are in short supply such as in arid region, government laws from impacts of coalbed gas development especially protect water resources. Even in areas where rainfall and groundwater are plentiful such as in tropical regions, water resources are still guarded from impacts of development and controlled by government regulations. In cold climate where surficial water is frozen during certain parts of the year, disposal of co-produced water on the surface is of special concern. As coalbed gas development extended in these environmentally diversified regions, increased awareness and concerns by governmental land and resource managers, special interest groups, landowners, and land users have complicated management of co-produced water.

A significant part of planning coalbed gas development—management of co-produced water to protect the local and regional hydrology, ecology, landscapes, soil, and atmosphere—must be implemented during gas production. This developmental framework identifies co-produced water as another valuable resource to develop, treat, and not as a disposal waste byproduct. The value of the groundwater is controlled by the quality, which varies between freshwater, saline water, and seawater types. Consequently, coalbed gas development becomes more expensive with production cost added toward management of co-produced water and implementing beneficial use and/or recyclability. The diverse regions of development make management planning for co-produced water more complicated, as each region requires different plans of disposal. In addition, the characteristics of coal reservoirs in different development areas may or may not contain as much groundwater such that it requires variable details of management planning and cost. Cost-effective treatments of co-produced water are controlled by the ratio of coalbed gas to water production. Additional cost will be added with construction of infrastructure needed to manage co-produced water.

When planning for coalbed gas development, detailed understanding of the classification and standards of surface water and groundwater is necessary and they vary between local and national government domains. Classification of water is dependent of its characteristics, use, value, source, and locations (ALL Consulting, 2003). For example, the value of water is a function of its use for supplying the public (e.g. drinking and recreation), preservation of wildlife (e.g. aquatic, fowl, and plant), and agricultural and industrial utility. Standards or criteria are established to maintain the quality and beneficial use of surface water and groundwater. The water standards will significantly impact the management planning, treatment, disposal, and protection of both surface water and groundwater. Another important factor to consider during coalbed gas development is water rights declared by governmental constitutions and statutes. The water rights refer to the right of a user to beneficially, historically, or constitutionally use water from surface sources (e.g. rivers and lakes) or groundwater sources. Many states in the United States own the water rights and often State Engineers and Water Resources agencies are responsible for supervision, measurement, appropriation, distribution, and management of the state's surface water and groundwater. These agencies are influential in approving applications to dewater the coal reservoir, imposing restrictions and enforcing transportation of co-produced water through pipelines, storage in surface reservoirs, construction of reservoirs or dams for storage, and in-stream flow of co-produced water (ALL Consulting, 2003).

Worldwide, significant volumes of co-produced water from coalbed gas wells are generated during production but vary from country to country, basin to basin, field to field, and coal to coal. There are "wet coals" and "dry coals" with the former containing and generating more co-produced water than the "dry coals". This is illustrated by comparison of ratio of gas to water production between United Sates (e.g. Powder River Basin (PRB)) to those in the Qinshi Basin in China; Bowen, Surat, and Sydney Basins in Australia; and the Alberta Basin in Canada (Ingleson, McLean, & Gray, 2006; RPS Australia East Pty Ltd, 2011; Yang, Ju, Li, & Xu, 2011). During the life span of coalbed gas wells, which varies worldwide from 5 to 20 years, co-produced water from each well declines; however, it increases with migration and expansion of development areas. This very nature of exponential growth of coalbed gas co-produced water contributes the bulk of total onshore conventional and unconventional water production.

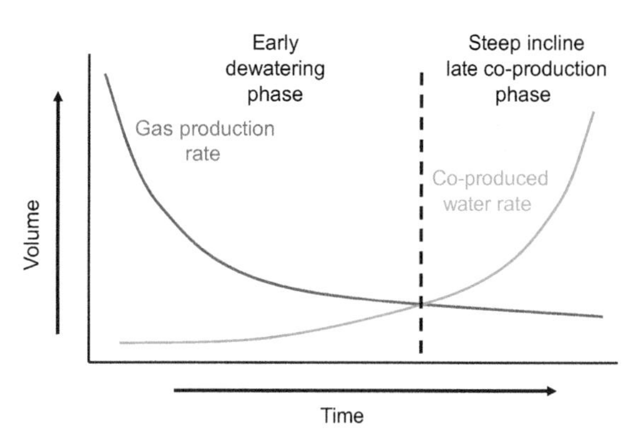

FIGURE 8.1 Gas decline and water incline production curves for conventional natural gas. Volume of co-produced water is low during the early phase and high during the late phase. *Source: Modified from Kuuskraa and Brandenburg (1989).*

CO-PRODUCED WATER

Co-produced water is defined as water or groundwater in the subsurface formations or reservoirs, which is extracted during production of oil and gas reserves. The term applies to production of both conventional and unconventional natural gas. However, the difference is the co-produced water from the conventional natural gas is from deeper formations and the coalbed gas is from shallower reservoirs. Co-produced water is otherwise known as produced formation water, produced water, and associated water (RPS Australia East Pty Ltd, 2011). In this book, co-produced water is used with coalbed gas production referring to groundwater extraction to reduce hydrostatic pressure in order to desorb gas from the coal reservoir. The groundwater by-product of coalbed gas production is transported, stored, treated, recycled, disposed on the surface, or reinjected in the subsurface.

In general, gas/co-produced water production and decline/incline curves of conventional natural gas wells and unconventional coalbed gas wells show inverse relationship (Figures 8.1 and 8.2). In conventional gas development, the

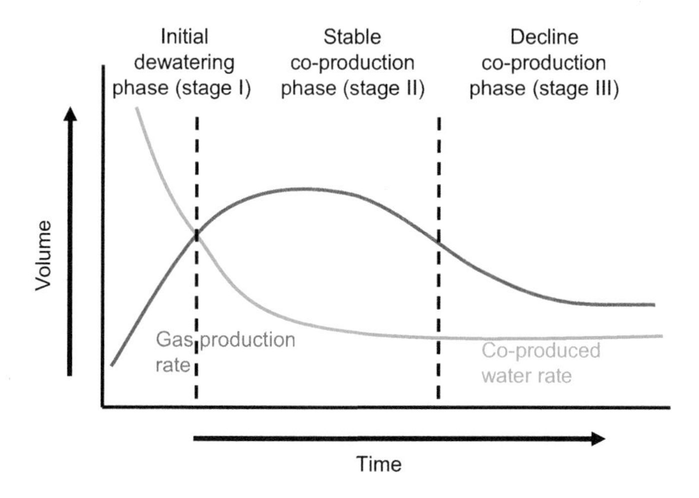

FIGURE 8.2 Gas incline and water decline production curves (stages I and II) for unconventional coalbed gas. Volume of co-produced water is high during the initial phase, and then stabilized and declined during the late phase. *Source: Modified from Kuuskraa and Brandenburg (1989).*

relationship varies mainly with variations in the geology, well distribution, and reservoir conditions. In coalbed gas, the coal geology, coal reservoir permeability, original reservoir hydrostatic pressure, reservoir gas and water saturation, well spacing, and production practices primarily control the relationship.

Conventional vs Unconventional

A pool of conventional oil and natural gas often is trapped (e.g. structural, stratigraphic, etc.) in deep reservoir rocks above the formation water due to their lighter density (RPS Australia East Pty Ltd, 2011). That is, the physical properties of these fluids are characterized by immiscibility of oil and gas in water and their lower density than water make them buoyant and rise above the formation water (Igunnu & Chen, 2012; Levorsen, 1967). In order to extract gas, it is necessary to drill into the reservoir rock containing a pool of natural gas, which includes gas and formation water. Usually the gas fills the pores and fractures in the reservoir rock immediately above the water contact. In contrast, an oil pool, which contains oil, gas, and water, includes a layer of gas filling pores and fractures in the reservoir rock above the oil layer, which in turn fill pores and fractures in the reservoir above the water contact (Levorsen, 1967). In a pool of natural gas drilling wells into the reservoir rocks, the virgin and water pressures drive the gas to the wellhead and onto the surface during production. Thus, the initial production involves a large volume of natural gas and as the pressure is depleted over time, water gradually replaces the volume of gas that is produced from the reservoir. The extracted formation water, which is separated from gas, is called co-produced water. During the early phase of production, natural gas comprises the bulk of the byproduct but over time co-produced water become a significant part of the production as shown by decline curves in Figure 8.1. The decline/incline curves have been a maxim for more than two decades and have been a tested contrast to decline/incline curves of co-produced water with coalbed gas wells (Kuuskraa & Brandenburg, 1989).

In the traditional concept of decline/incline curves of gas and water or groundwater production from coal reservoirs, the relationship is the inverse of that of the conventional natural gas (Figure 8.2). The concept is well accepted worldwide; however, comparisons to actual water production rates have mixed results (ARI, 2002; Cardott, 2002; Rice, Bartos, & Ellis, 2002; RPS Australia East Pty Ltd, 2011). The principle behind this concept is that coalbed gas is generated during coalification and postcoalification., which is absorbed on the surfaces of the pores, cleats, and fractures within the coal reservoir. The gas is held onto the coal surfaces by wellhead or hydrostatic pressure from the overlying groundwater with the coal serving as an aquifer. In order to desorb and release the gas from the surfaces of the coal reservoir, it is necessary to reduce the hydrostatic pressure, which is performed by producing water from the coalbed gas wells to lower the groundwater head (e.g. groundwater level to top of coal bed). Pumping out the groundwater in the reservoir breaks the physical–chemical bond between the gas molecules and the surfaces in the coal. The groundwater in the coal fractures is first to be produced from pumping and is accompanied by desorption of gas from the surfaces of the fractures. The initial production of gas from the coal fractures is called "free" gas. Continued dewatering of the coalbed gas wells permits desorption and diffusion of gas from pore surfaces in the coal matrix migrating into the coal fractures. The gas is in turn conveyed by groundwater through Darcy flow to the borehole, where it is separated from the water during production at the wellhead.

In principle, it takes time to reduce hydrostatic pressure to initiate gas desorption from the coal reservoirs, which normally involves withdrawal

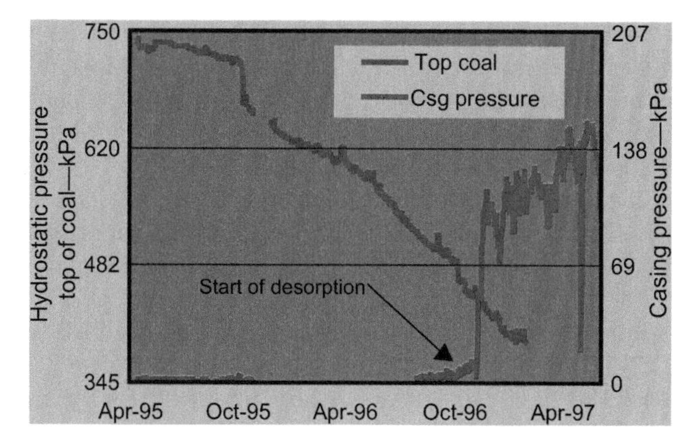

FIGURE 8.3 Hydrostatic and casing pressure curves from a monitor well in the Powder River Basin, Wyoming. The hydrostatic pressure is reduced 27% before coalbed gas began to desorb from the coal reservoir. Csg, casing; kPa, kilopascal.

of large volumes of groundwater during the first few weeks, months, or years (Figure 8.3). In some cases, it takes up to 2 years in the Powder River Basin (PRB). Dewatering the coal reservoir will take more time to reach critical point of gas desorption pressure in fields where the groundwater pressure head has not been previously reduced (e.g. by coal mining, oil and gas production, and groundwater wells). This reservoir condition involves withdrawal of a larger volume of groundwater. Dewatering the coal reservoir to reach gas desorption pressure will take less time in areas previously affected by depressurization (e.g. by coal mining, oil and gas production, and groundwater wells). Thus, this reservoir condition involves withdrawal of less volume of groundwater. In both scenarios, large amounts of groundwater are withdrawn from the coal reservoir during the initial phase of coalbed gas production, which theoretically decreases over time in contrast to the conventional gas production as originally depicted by Kuuskraa and Brandenburg (1989) and modified by Anderson et al. (2003) (Figure 8.2). Anderson et al. (2003) divided the co-produced water and coalbed gas production curves into the following: stage I, which is the initial production dominated by water; stage II or mature phase, in which water in the coal reservoir is produced as relative permeability to gas increases; and stage III with decline in both water and gas production.

Co-produced Water Rates with Coalbed Gas

A survey of actual water production rates with coalbed gas does not necessarily follow the trend as projected by Kuuskraa and Brandenburg (1989) and Anderson et al. (2003). The rates of water production vary considerably between and within wells, coal beds, fields, and basins. Examples from the Australian Bowen and Surat Basins and the US basins (e.g. Arkoma, Raton San Juan, and PRBs) demonstrate the variability of water production rates. According to RPS Australia East Pty Ltd (2011), water production in the Surat Basin in Queensland, Australia indicated initial amount of water extracted from a coalbed gas well range from 40,000 to 800,000 l/day before decreasing to about 100,000 l/day over a period of 6 months to a few years. At the Fairview coalbed gas field in the Bowen Basin, Queensland, initial daily water production rate averages 200,000 l/day/well and decreased to 20,000 l/day/well after 12 years (RPS Australia East Pty Ltd, 2011). Thus, in general, the water production rates from bituminous coals in these Australian basins behave much like that predicted by Kuuskraa and Brandenburg (1989) and modified by Anderson et al. (2003).

In comparison, the initial co-produced water production rates with coalbed gas in bituminous coals in the Arkoma Basin, Oklahoma vary from shallow (shelf) to deep (basin) parts. Cardott (2002) reported initial water production rates in the Arkoma Basin, Oklahoma range from 0 to 143,210 l/day (average 7154 l/day) from 492 wells in the shelf compared to 0–17,528 l/day (average 1311 l/day) from 256 wells in the basin from bituminous coal reservoirs. In the San Juan Basin, Colorado, the exponential decline of water in a single well is about 20,023 l/day during the first 10 months and between 7154 and 9539 l/day from 10 to 150 months from bituminous coal reservoirs (Ojeifo et al., 2010). In the Raton Basin, Colorado, the variations of water production rates in two coalbed gas wells are significant between the Vermejo bituminous coals (Papadopulus & Associates, Inc., 2008). For one well, water and gas production dominated the early period of the year 2001 after which both declined. In contrast, another nearby well water production rates declined from the initial production in late 2000 to the later part of 2006; however, gas production continued to rise stabilizing from middle of 2002 through late 2006. The variability in water production rates in these two wells mimics the overall water production in other wells in the basin, which continues without significant decline, although for some wells, water production declines over time. Water production data indicate that the average water production per well has not changed dramatically over the period 1999–2005 during which period, the total number of producing coalbed gas wells increased from 454 to 1665 (Papadopulus & Associates, Inc., 2008).

The PRB, which yields one of the highest water co-produced with coalbed gas in the United States (USDOE, 2002), demonstrates the variability of produced water within and between coal beds. More than 70% of gas production in the PRB in Wyoming is developed from the Wyodak coal bed, which splits into the Canyon (lower split) and Anderson (upper split) coal beds. Figure 8.4(A–C) are time zero plots showing the variations of the co-produced water, coalbed gas, and well count in these coal beds during the period of 1–4 years in the eastern part of the PRB west of the Gillette coal mines (ARI, 2002). The 13 coal mines, which produced about 42% of the annual US total coal production in 2011, develop the merged Wyodak coal along the Gillette coalfield on the eastern part of the basin. The coal mines dewatered in front of mining since 1970s and have created a regional groundwater drawdown on the western part of the coalfield, which lowered downdip the hydrostatic pressure of the Wyodak and split coal beds (Chapter 7).

In general, the co-produced water production curve in the Wyodak coal bed, which is based on 164 coalbed gas wells from 100 to 230 m depth and 21–24 m thickness, rises during the first 2 months and then declines gradually to the end of 4 years (Figure 8.4(A)). More specifically, according to ARI (2002), the initial rate of co-produced water is at 71,544 l/day, declines to about 47,696 l/day at the end of year 1 and 15,501 l/day at the end of year 4. The gas production curve rises during the first 3 months, stabilizes from 3 to 12 months, and declines gradually from 12 to 27 months to the end of 4 years. The initial rate of gas production increases gradually up to 8495 m^3 during year 1, plateaus for 12 months, and then declines to 4248 m^3 at the end of year 2 and 2265 m^3 at the end of year 4 (ARI, 2002). The rapid decline of co-produced water and gas production is probably due to the effect of dewatering of nearby coal mines, which created regional groundwater drawdown (Chapter 7).

The co-produced water curve in the Anderson coal bed, which is based on 89 wells from 143 to

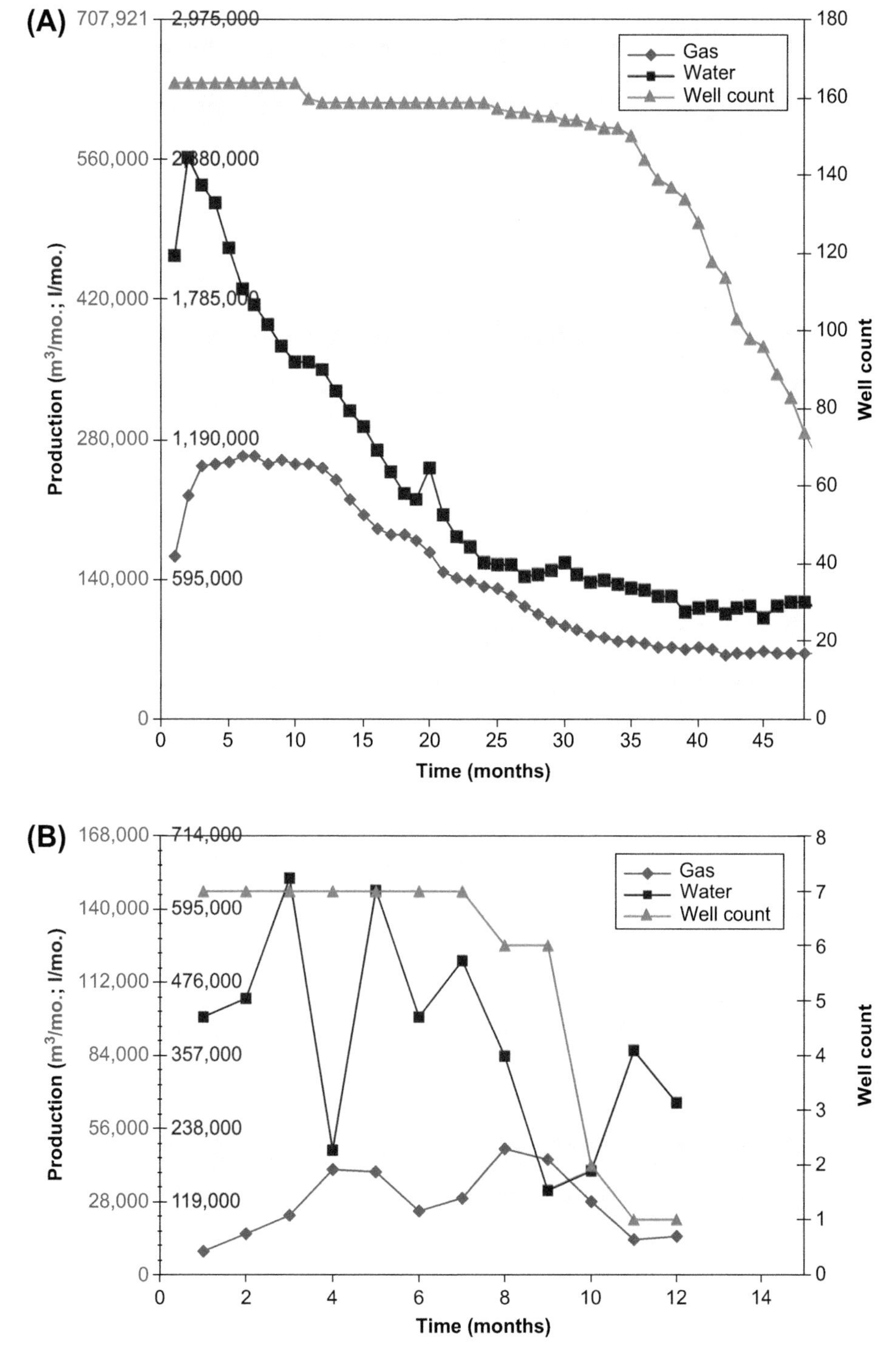

FIGURE 8.4 (A)–(C) are time-zero plots showing variations of the co-produced water, coalbed gas, and well count of the Wyodak (A), Anderson (B), and Canyon (C) coal beds during periods of 1–4 years in the eastern part of the Powder River Basin west of the Gillette coal mines. *Source: Adopted from ARI (2002).*

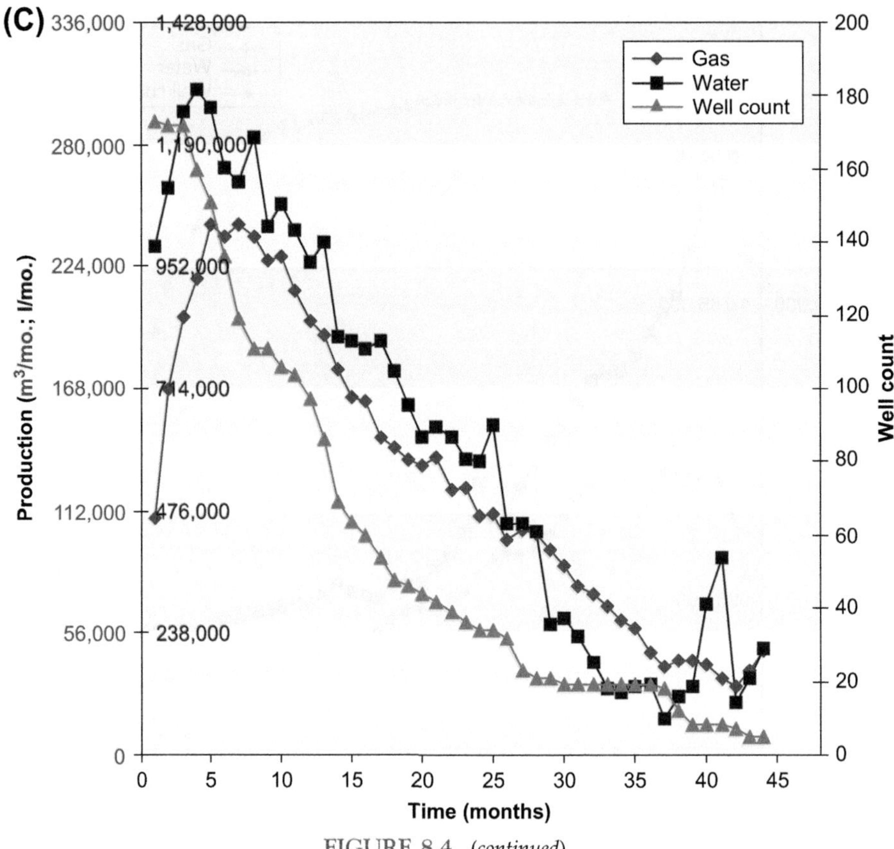

FIGURE 8.4 (*continued*).

213 m depth and 6–15 m thickness, rises during the first 3 months, declines and rises during the next 3–5 months, then declines to month 9, rises to month 11, and finally declines to the end of year 1 (Figure 8.4(B)). The initial rate of production of co-produced is 17,886 l/day and declines to 7154 l/day at the end of year 1 (ARI, 2002). The gas production curve rises gradually from 1 to 4 months, peaks and declines from 4 to 8 months, and finally declines to the end of the year. According to ARI (2002), gas production peaks early at 2832 m^3 indicating the presence of "free" gas. However, much like in the Canyon coal bed, the dewatering in front of coal mining in the Gillette coalfield probably influences the "free" gas.

The co-produced water curve in the Canyon coal bed, which is based on 173 wells from 168 to 235 m depth and 7–15 m thickness, rises steeply from 2 to 4 months and generally declines from 4 to 45 months (Figure 8.4(C)). More specifically, according to ARI (2002), the initial co-produced water rate is at 35,772 l/day during the first year and declines to 5962 l/day at the end of year 3. The gas production curve steeply rises from 2 to 6 months, stabilizes from 6 to 9 months, and declines from 9 to 45 months. In general, the amount of co-produced water from 29 to 45 months is lower than the gas production. The early peak of gas production in the Canyon coal at nearly

8495 m^3 during year 1 indicates the presence of "free" gas according to ARI (2002). However, dewatering in front of the coal mines in the Gillette coalfield could have influenced the gas peak production.

The co-produced water and gas production curves of the Canyon coal bed behave much like that predicted by the model of Kuuskraa and Brandenburg (1989) and Anderson, et al. (2003). This contrasts with the Wyodak and Anderson coal beds co-produced water and gas production curves, which generally decline after few months of development.

ORIGIN OF CO-PRODUCED WATER

A good start in understanding the origin of co-produced water is the clarification of metaphors on moisture and/or water contents in coal. Coal geologists, geochemists, and petrologists are familiar with the dilemma of the concept of moisture and water in coal. Is moisture mutually exclusive of water content in coal? As a part of the traditional analysis of the coal quality (e.g. proximate vs ultimate; Chapter 5), moisture plays an important part of the content of coal. Moisture is measured as weight percentage and a part of the total coal composition determined from proximate analysis. The most common method of measurement of moisture is by drying the coal sample to determine air-dried loss in relationship to the dry weight of coal. This procedure is an indirect measurement of the amount of water in the coal. Molecules of moisture or water are adsorbed on the internal surfaces of cavities as well as bound in the structure of the organic and inorganic composition of the coal. Filling moisture or water of these coal cavities plays an important role in coal being aquifers.

The origin of co-produced water of coalbed gas is more complicated than the co-produced water of conventional oil and natural gas. For conventional oil and gas, the traditional concept of co-produced water invokes meteoric, connate, and mixture of both as the origin (Levorsen, 1967). The meteoric water is rainfall that seeped into the pores and fractures of the reservoir rock. The connate water represents water (e.g. marine, brackish, and freshwater) in which the sediments were deposited and remained preserved in the pore and fracture systems of the reservoir rock. The origin of co-produced water in conventional oil and natural gas is generally similar to that of the coalbed gas, which has both internal and external components. However, the origin of water co-produced with coalbed gas is more complicated by the idea that coal serves as both source and reservoir rocks of gas and water. For the lack of existing terms to describe both internal and external components to the origin water in coal, terms are borrowed elsewhere in this book to characterize the genesis of co-produced water. The water derived internally or within the coal is endogenous in nature compared to water-introduced external from the coal as exogenous in nature.

Endogenous Origin

Knowledge of the origin of water co-produced with coalbed gas remains fragmented. It is not well understood despite many papers written about the presence of in situ water as a part of coal composition and how the component can be measured. Allardice and Evans (1978) reported that water or moisture is present in the coal matrix and is occurring in various forms that complicate measurement of moisture using proximate analysis (Chapter 5). Speight (2005) and Ryan (2006) discussed in detail the intricacies of the confusion on the methods and standards of measurements of coal moisture (Figure 8.5), which include free or surface, equilibrium (plus bed and capacity moistures grouped as inherent moisture), air-dried, and residual moistures. Total moisture or water in coal consists of the surface and residual moisture left over in the coal after determining the air-dried loss (Speight, 2005). Moisture loss is coal-maceral dependent and in

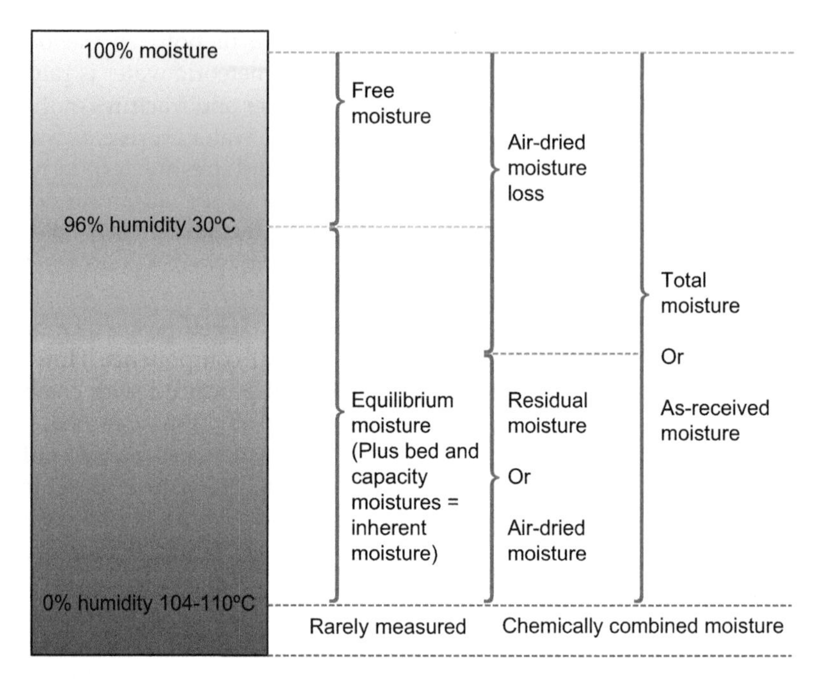

FIGURE 8.5 Components of coal moisture, which include free or surface, equilibrium (plus bed and capacity moistures grouped as inherent moisture), air dried, and residual moistures. *Source: Adopted from Ryan (2006).*

order of vapor pressures of moisture increases from vitrain (least loss) to clarain, durain, and fusain (most loss) (Speight, 2005). The underlying factor associated with these measurements of moisture is the status of coal sampling and handling prior to laboratory analysis (Chapter 7).

According to Allardice and Evans (1978), moisture resides in the fractures, pores, and capillaries of the coal matrix. Levine (1991) and Mukhopadhyay and Hatcher (1993) classified moisture in coal based in part on residency as well as organic and inorganic sources of water or moisture in the coal to include the following: (1) adherent of free moisture, which is retained as free state on surfaces, fractures, cracks, and cavities of the coal; (2) inherent moisture (bed, equilibrium, and capacity moisture), which is physically adsorbed on the surfaces of the pores and capillaries of the coal matrix; (3) chemically bound moisture or water of decomposition, which is organically associated in the molecular structure of the coal, which is released during coalification (this is sometimes known as thermal decomposition as temperature plays an important part of decomposition); and (4) water of hydration, which is associated with the inorganic minerals or mineral matter of coal such as clay minerals (water is included in the crystal structures of the inorganic minerals).

Classification of moisture in coal by Levine (1991) and Mukhopadhyay and Hatcher (1993) provides an important conceptual foundation of the origin of water sourced internally from the organic and inorganic composition of the coal. For example, the water of decomposition is a good example of endogenous water, which is released at 120 °C from oxygen-functional group (e.g. hydroxyl, carboxyl, and carbonyl) during coalification (Allardice & Evans, 1978). The water molecules bound to coal using oxygen in the coal and hydrogen bonds in the water molecules (Allardice & Evans, 1978). The released water of decomposition becomes part of the adherent and inherent moisture types for the bituminous coal, which undergone coalification from 100 to 150 °C (Stach et al., 1982; Taylor et al., 1998). How much does the organically and inorganically derived endogenous water make up the total co-produced water is not well understood. Attempts were made to relate the volume of water to the surface area because much like gas, water is absorbed on the surfaces of these cavities (Allardice & Evans, 1978; Ryan, 2006).

Another element of endogenous water that is difficult to account is the contribution from depositional water. That is, water derived from the depositional environments in which peat is deposited. The depositional water is indigenous to the original peat deposits, which varies from brackish water to freshwater in origin depending on the proximity of the accumulation of peat to the marine environment (Chapter 3). Peat accumulation in coastal and deltaic plains will contain marine-brackish water. Peat accumulation in alluvial plains and lacustrine coastal areas far removed from marine-brackish settings will contain freshwater.

The moisture content of peat varies from 75% to 90% and it is assumed that the water is a remnant from the depositional environment. Coalification reduces the moisture content of peat to about 30–40% in lignite, about 20–25% in subbituminous coal, and about 1–10% in bituminous coal; the latter decreasing from high to low volatile rank. Thus, the exogenous depositional water plays a significant role in the moisture content of lignite and subbituminous coal and minor role in the higher rank coals. Depths, however, complicate this relationship, which play a major role in reduction of moisture. For example, Germany's lignite is reduced from 65% to 48% at 400 m depth and in the Saar Region's bituminous, to semianthracite is reduced from 6% to 0.5% at 1500 m depth (Mukhopadhyay & Hatcher, 1993).

Exogenous Origin

Exogenous water is mainly derived from meteoric water and water from other aquifers. Meteoric water is derived from recent or modern atmospheric precipitation by mainly rain and snow, which provides surface runoff of rivers, streams, and creeks (Figure 8.6). Meltwater from snow and ice as well as lake water make up most of the remaining meteoric water. Topographic relief created by differential erosion of resistant strata and nonresistant coal-bearing rocks formed valleys where coal beds outcrop and subcrop. Coal beds often exposed to harsh climatic conditions permitting spontaneous combustion to form clinkers (e.g. porous and permeable red scoria-like, burned coal and roof rock) serving as groundwater recharge areas (Heffern & Coates, 1999). Meteoric water from precipitation, particularly in areas of concentration and recharge areas, infiltrates into shallow aquifers such as coal beds and interbedded sandstones (Figure 8.6). Depending on the continuity and/or discontinuity of these aquifers based on stratigraphy, sedimentology, structural dip, and directions of cleat and fracture systems, groundwater develops either local or regional flow.

Coal-bed aquifers form regional flow along laterally persistent coal zones in which individual coal beds merge and/or split. The regional flow of water is complicated by splitting of thick coal beds, which divides groundwater flow locally along thin individual beds. Moreover, the split coal beds often remerge and pinch out. In contrast, individual coal beds with low degree of lateral continuity form aquifers with locally limited groundwater flow. Regional groundwater flow is often overprinted by the orientations of face and butt cleats as well as fractures, which determine the preference of flow. Coal face cleat is more continuous than butt cleat for preferred direction of regional groundwater flow. However, the orientation of the fracture system is more preferred direction of water flow because it overprints the cleat system.

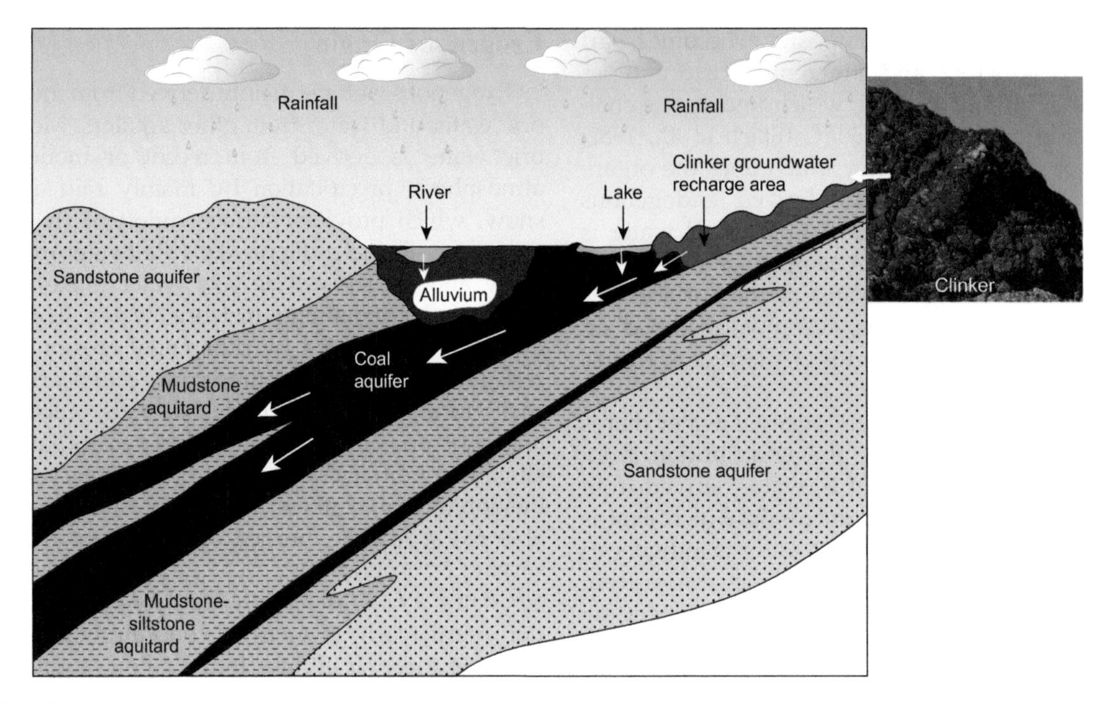

FIGURE 8.6 Coal and sandstone aquifers recharged with meteoric water (exogenous) from rainfall. Clinker consisting of red-colored, breccia-like rock burned from spontaneous combustion of coal and overlying sandstones, siltstones, and mudstones exposed on the surface. *Source: Modified from ALL Consulting (2003).*

Sandstones commonly interbedded with coal beds and impermeable aquitards/aquicludes of mudstones, shales, and siltstones develop as aquifers with local flow (Figure 8.6). However, when sandstones (e.g. channel sandstones) are traced along their depositional dip where they are more continuous, these aquifers form regional flow. When exposed on the surface, the sandstones are excellent recharge areas supplied by infiltration of meteoric water, which collects in pore and fracture spaces. The groundwater in turn vertically infiltrates into interbedded coal beds (Figure 8.6).

Another external source of co-produced water is from adjoining aquifers and recharge areas directly connected to the coal aquifers. Rainfall collects in groundwater recharge areas (Figure 8.6) such as burned coal or clinkers, which have high transmissivity, storativity, and permeability, are potential source of water being directly connected downdip to equivalent coal beds (Davis, 1976). Heffern and Coates (1999) suggested that clinkers in the PRB play a critical role in storage and recharge of water into coal aquifers adjoining downdip. Clinkers, which formed on the surface, are unconfined aquifers with the groundwater accumulation rising and falling with atmospheric pressure. High permeability of clinkers (Figure 8.6) provide a sponge-like property that absorbs rapidly infiltrating rainfall and snowmelt, which in turn directly infiltrates into coal or indirectly via streams, rivers, and lakes and eventually into coal and sandstone aquifers. As discussed in Chapter 4, clinker groundwater recharge areas also play an important role in supplying chemical compounds and dissolved solids becoming parts of the water composition as well as the nutrients for microbes that generate biogenic gas in downdip coal reservoirs (Faiz & Hendry, 2006;

Flores, Stricker, Rice, Warden, & Ellis, 2008; NRC, 2010; Rice, 1993).

Role of Sandstone Aquifers

An important source of co-produced water, which has been under appreciated, is vertical infiltration either from the surface or from adjoining aquifers such as permeable and porous sandstones. The traditional idea that sandstones are aquifers isolated by aquitards/aquicludes (e.g. impermeable shales, mudstones, and siltstone), although valid hydrologically, is not as compelling when viewed stratigraphically, sedimentologically, and structurally along depositional dip. However, sandstones juxtaposed against coal beds can be vertically interconnected aquifers and can serve as external source of co-produced water. Channel sandstones with basal scours formed by ancestral rivers, streams, and creeks often erode into the underlying fine-grained aquitards/aquicludes and onto the top of the coal bed providing connectivity between the two aquifers (Figure 8.7(A)). This sedimentological condition of two aquifers juxtaposed to each other often permits local vertical infiltration and flow of exogenous groundwater from the sandstone to the coal bed. Stratigraphically and architecturally, if properly corrected using closely spaced wells, the sandstone aquifers are continuous and connected along depositional dip but compartmentalized along depositional strike as demonstrated in the PRB (Flores, Spear, Kinney, Purchase, & Gallagher, 2010; BLM, 2012) (Figure 8.7(B)). Structurally, fractures connecting the sandstones, intervening aquitards/aquicludes, and coal beds also can serve as conduits for vertical infiltration of groundwater.

Figure 8.8 demonstrates the vertical infiltration, hydraulic conductivity, and connectivity of stacked aquifers of channel sandstones and the underlying thick Wyodak coal reservoir, which produced coalbed gas in the PRB, Wyoming. Pumping out the groundwater or co-produced water in the coal reservoir and overlying rocks resulted in the reduction of hydrostatic pressure (e.g. well-head gas pressure) and gas desorption (Figure 8.8(B)). Concurrently with extraction of the co-produced water, the groundwater level in the overlying uppermost channel sandstone was lowered (Figure 8.8(A)) indicating connection of water in the coal and overlying rocks. Vertical infiltration of water from the sandstone aquifer into the underlying rocks, which consist of stacked channel sandstones interbedded with siltstones, mudstones, and thin coal beds occurred into the immediately underlying coalbed gas reservoir (Figure 8.8(B)). The depth to the top of the groundwater level during the 7–9 years life of the well was lowered about 10 m for the uppermost channel sandstone and about 100 m for the coal reservoir (Figure 8.8(A,B)) indicating gradual vertical rate of infiltration. It took <2 years (e.g. 3/1/93–11/1/95) of pumping out co-produced water from the coal reservoir to lower the groundwater level 50 m in order to reach critical gas desorption point at well-head gas pressure. During the same 2-year period (2/18/93–2/18/95), the groundwater level in the channel sandstone was stable after which period the groundwater level gradually dropped through 2/18/02. It is interpreted that the connectivity of the stacked channel sandstone aquifers may be due to erosional contacts and through interbedded fine-grained sediments due to fractures permitting gradual leakage of water from the sandstone to the coal aquifer (Figure 8.7(A); section on right).

Vertical Hydraulic Connectivity of Sandstone–Coal Aquifers

Vertical stacking and hydraulic connectivity of aquifers may be more common in the PRB than as described stratigraphically and sedimentologically by Flores et al. (2010), which resulted in major problems of co-production of large volumes of water causing shut down of

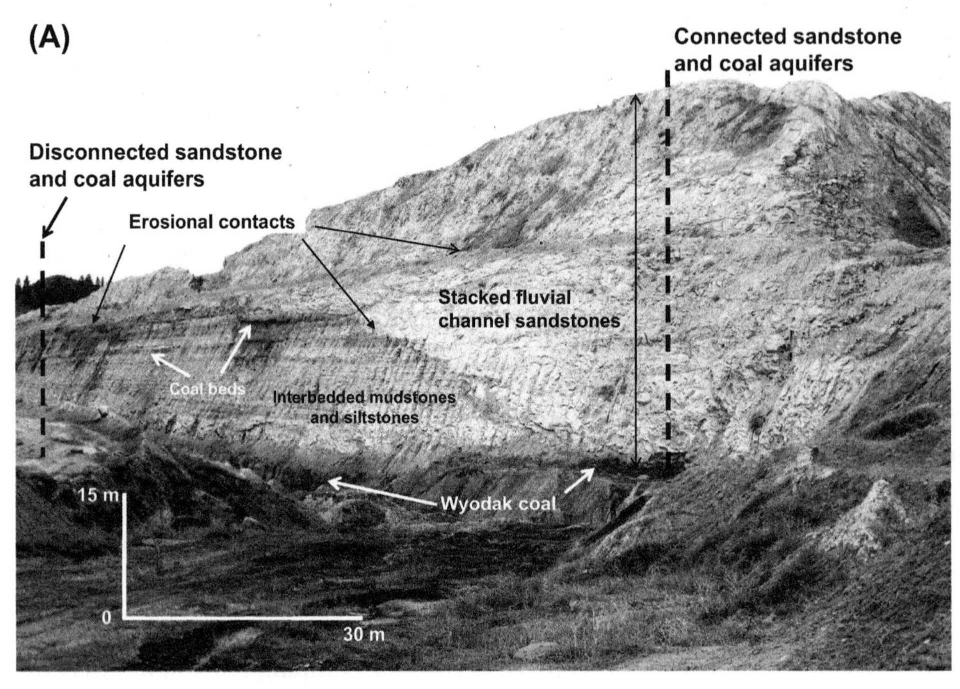
(A)
Connected sandstone and coal aquifers
Disconnected sandstone and coal aquifers
Erosional contacts
Stacked fluvial channel sandstones
Coal beds
Interbedded mudstones and siltstones
Wyodak coal
15 m
0
30 m

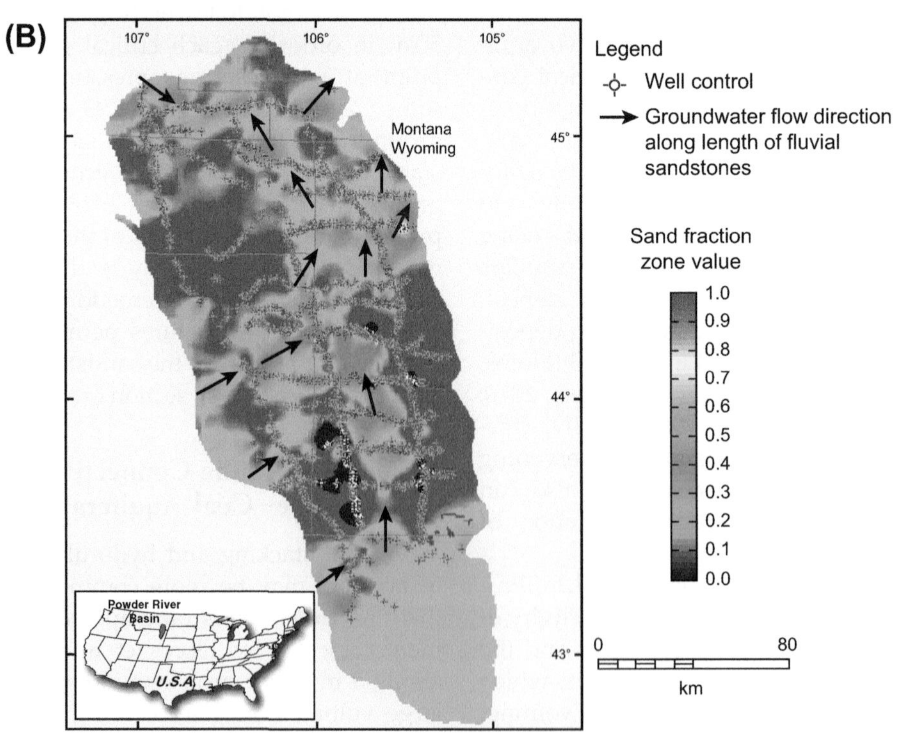
(B)
107°
106°
105°
45°
44°
43°
Montana
Wyoming
Powder River Basin
U.S.A.
Legend
Well control
Groundwater flow direction along length of fluvial sandstones
Sand fraction zone value
1.0
0.9
0.8
0.7
0.6
0.5
0.4
0.3
0.2
0.1
0.0
0
80
km

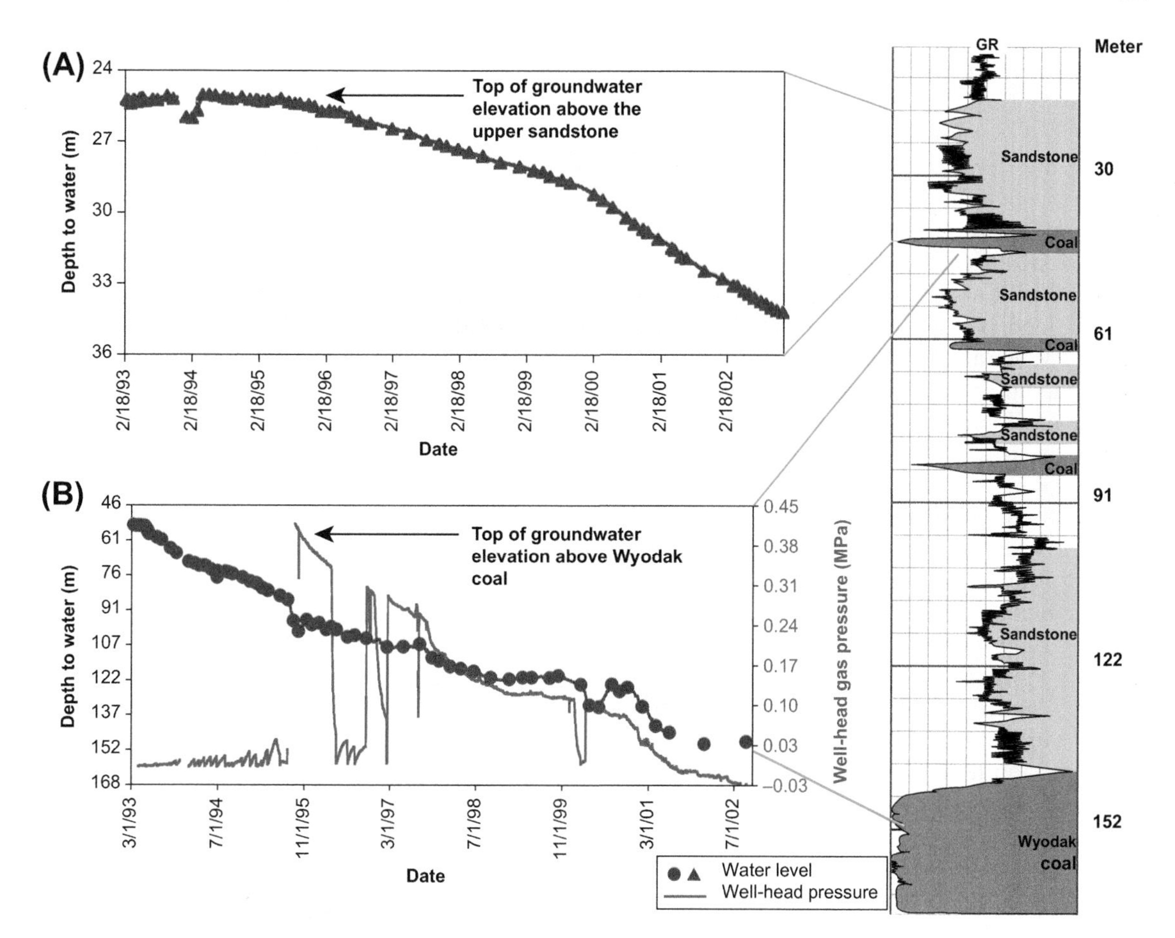

FIGURE 8.8 Stratigraphic section (right) showing the Wyodak coal bed (gas producer) and overlying fluvial channel sandstones interbedded with floodplain siltstones and sandstones and thin coal beds. Curves show changes in the elevation of groundwater level above the Wyodak coal bed (B) and overlying channel sandstones (A) and well head pressure, which demonstrate the vertical infiltration, hydraulic conductivity, and connectivity of stacked channel sandstone aquifer and the underlying thick Wyodak coal aquifer, which produced coalbed gas in the Powder River Basin in Wyoming, United States. GR, Gamma Ray; m, meter. *Source: Flores, 2004 and McGarry, and Flores, 2004.*

wells. Figure 8.7 illustrates the vertical and lateral variabilities of aquifers (e.g. fluvial channel sandstones and coal beds) and aquitards/aquicludes (e.g. mudstones and siltstones Figure 8.7(A)) very similar to the relationships described in Figure 8.8. One end (left of the photograph; Figure 8.7(A)) of a coal mine highwall in the PRB illustrates vertical isolation of

FIGURE 8.7 (A) Photograph of an abandoned coal mine highwall showing vertical and lateral variations a thick coal bed overlain by floodplain siltstones, mudstones, and thin coal bed scoured by a fluvial channel sandstones. On right of the photograph, the channel sandstone incised to the top of the Wyodak coal bed where the aquifers are connected. On left of the photograph, the impermeable fine-grained floodplain deposits separate Wyodak coal and sandstone aquifers. (B) Fluvial sandstone percent map (green–yellow areas) within the Wyodak–Anderson coal zone showing north-trending anastomosed and groundwater flow patterns (BLM 2012). Blue areas indicate mudstone-siltstone dominated floodplain areas.

fluvial channel sandstone and coal (Wyodak coal) aquifers separated by mudstone and siltstone aquitards/aquicludes interbedded with thin coal beds. However, within a short distance at the other end (right of the photograph; Figure 8.7(A)) of the coal mine highwall, the isolated aquifers are interconnected stratigraphically and sedimentologically. Here, the stacked fluvial channel sandstones overthickened from the other end of the mine and are directly on top of the Wyodak coal bed. Simply put, the incision by the fluvial channel indicated by the 15 m relief of erosion of the underlying mudstones and siltstones provided an accommodation space for deposition of the stacked sandstones directly on top of the Wyodak coal.

Aquifers are viewed in terms of their lateral continuity and hydraulic connectivity as a function of groundwater flow and seldom in terms of their vertical hydraulic connectivity with adjoining rocks and vertical infiltration of water (Figure 8.8). Disregarding the importance of vertical infiltration between sandstone and coal aquifers common in coal-bearing rocks minimizes the role of exogenous water from sandstones as a form of groundwater recharge of coal reservoirs. Mixing of vertical and horizontal flow paths away from recharge areas are recognized by Dockins et al. (1980), Hagmaier (1971), Lowry and Cummings (1966), and Lowry and Wilson (1986) in the PRB but this awareness is more of happenstance than a golden rule. In addition, the observation by Bartos and Ogle (2002) of cyclic geochemical alteration of groundwater as it migrates downward through interbedded deeper sandstone and coal aquifers in the Wasatch Formation in the PRB may be best explained by their hydrostratigraphic associations. As a result, information on the origin, volume, composition, and quality of co-produced water in coal reservoirs may have incomplete critical information and incorrect interpretations without the benefit of the stratigraphic and sedimentologic relationships of the reservoirs and interbedded rocks. For example, how much of the TDS and cation exchange, which are attributed to be sourced from contact with adjoining rocks, coal, and surface recharge areas, are from dissolution of minerals in the sandstones often interpreted as dense aquitards/aquicludes (Dahm, Guerra, Xu, & Drewes, 2011), which often make up more than 50% of coal-bearing rocks? This phenomenon is particularly significant where recharge areas and the source of fresh dissolved constituents are far removed.

ISOTOPIC SIGNATURES OF CO-PRODUCED WATER

Isotopic signatures of water co-produced with coalbed gas are critical to determination of its origin. Craig (1961) observed that the oxygen ($\delta^{18}O$) and hydrogen (δ^2H [δD]) isotopic values of meteoric water, which has not been evaporated, show linear relationship related by the following (USGS, 2004):

$$\delta D_{H_2O} = 8\delta^{18}O_{H_2O} + 10$$

where:

δD_{H_2O} = deuterium isotopic composition of hydrogen (δ^2H) of water;
$\delta^{18}O_{H_2O}$ = oxygen isotopic composition of water.

The equation is determined from data on precipitation around the world with very high correlation coefficient ($r^2 > 0.95$) and the linear relationship is called the global meteoric water line (GMWL) by Craig (1961). The high correlation coefficient indicates significant relationship of the oxygen and hydrogen isotopes in the water molecules. Thus, because precipitation is the basic source of groundwater, the oxygen and hydrogen compositions before and after recharge are useful tracers of water sources, processes, and histories (USGS, 2004).

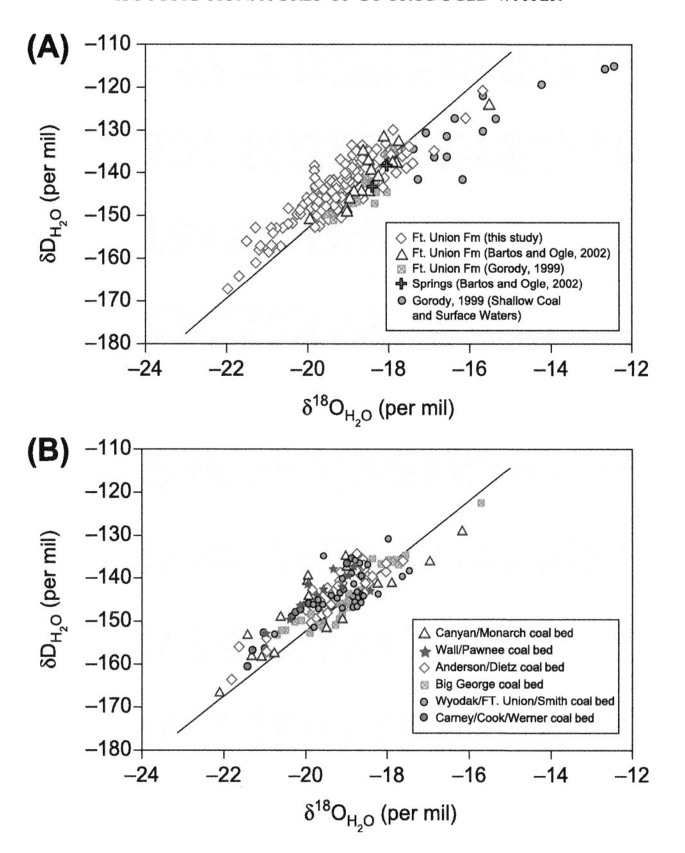

FIGURE 8.9 Isotopic composition of Deuterium (δ^{18}H) and oxygen (δ^{18}O) in coalbed gas reservoirs in the Paleocene Fort Union Formations, in the Powder River Basin in Wyoming, United States Ft., FT. = Fort. *Source: Adopted from Rice et al. (2008).*

For example, Rice et al. (2002), Rice, Flores, Stricker, and Ellis (2008) analysis of the $\delta^{18}O_{H_2O}$ versus δD_{H_2O} of co-produced water from coalbed gas wells in the PRB indicates clustering along the GMWL (Figure 8.9). The water samples are spread above and below the GMWL as defined by Craig (1961). Rice et al. (2002) interpreted the co-produced water samples above and below the GMWL as consistent with the isotopic signature of modern meteoric water. However, the co-produced water samples above the GMWL may have been influenced by methanogenesis and samples below the GMWL may have been mixed with near-surface water, which experienced some level of evaporation (Flores et al., 2008; Rice et al., 2008). In addition, the isotopic composition of the co-produced water samples from major producing coalbed gas reservoirs varies over a large range along the GMWL. That is, within a coalbed gas reservoir, the isotopic composition varies from heavier (enriched in deuterium) to lighter (depleted deuterium) values suggesting mixing of new recharge and old basinal waters (Rice et al., 2008). Flores et al. (2008) identified potential areas of groundwater recharge along the eastern and northwestern margins of the basin. According to Rice (1993), the groundwater system in the coal beds may have been established 10 mya after the uplift and erosion of the basin. Whether this was the last time groundwater was recharged or not is a source of debate.

BOX 8.1

IS GROUNDWATER RENEWABLE IN HUMAN LIFE TIME?—CONCEPT AND MISCONCEPTION

Groundwater is a critical element to water systems and supplies of small to large communities worldwide. Gleeson, Wada, Bierkens, and van Beek (2012) evaluated the balance of water moving out and coming into aquifers concluded that groundwater footprints, which are areas needed to sustain current demand, far exceed groundwater aquifer areas. Powell (2012) supported this conclusion and observed that groundwater levels are dropping globally in arid and semiarid regions. The withdrawal of co-produced water during coalbed gas development is an example of straining local groundwater supply, which makes up a small part of the groundwater footprint that is either renewable or nonrenewable.

Isotopic fingerprinting of co-produced water related to coalbed gas development as well as surface water has generated conflicting and confusing results in terms of whether groundwater is renewable or not. NRC (2010) reported that studies in the San Juan Basin using geochemical techniques dated co-produced water to be thousands to tens of millions years old. This observation suggests that groundwater of coal aquifers is old or "fossil" water, which accumulated slowly and when depleted, may not be replaced for millions of years, making the water fundamentally a "nonrenewable" resource (NRC, 2010). Similar studies by Oldaker and Fehn (2005) of groundwater of deep coal aquifers in the Raton Basin in Colorado and Susitna Basin in Alaska using Chlorine 36 isotope yielded more than 1.2–1.3-million-years-old "fossil" water. These workers dated near surface waters to be <60 years old. The combined observations suggest that deep "fossil" and shallow groundwater are not hydraulically connected and may not be renewable in human life time (e.g. 1.8 mya) and certainly not in hominid life time (3.2 mya) (NRC, 2010; Scientific American, 2012).

Sharma and Frost's (2008) work on carbon isotope ($\delta^{13}C_{DIC}$) and dissolved inorganic carbon (DIC) concentrations of co-produced water in the PRB can be used to evaluate hydraulic connectivity between coal-bed groundwater aquifers and surface water. These workers show that the carbon isotope values of the Wyodak coal zone groundwater is distinct from surface water due to methanogenesis during which ^{12}C is preferentially depleted by methanogens leaving DIC in the reservoir waters rich in ^{13}C. More importantly, carbon isotope and DIC demonstrate co-produced water is distinct from shallow groundwater and surface water in the PRB, which suggests no direct connectivity with the deep "fossil" water (Sharma & Frost (2008). Flores (2004) and Flores et al. (2008) identified "old" biogenic gas in the deeper part of PRB and "new" biogenic gas in the shallow part based on carbon isotope of methane. These workers interpreted that the "old" coalbed gas is generated by CO_2 reduction from long residency and the latter generated by fermentation methanogenesis resulting from recent groundwater recharge.

Rice et al. (2002) dated the coproduced water in the Wyodak coal zone and others using tritium concentrations from recharge after 1954 (mix modern and submodern). However, the work of Wheaton and Meredith (2009) suggests connectivity of surface water and shallow groundwater with recharge before 1954 of the 60-m-deep Dietz coal zone (e.g. Anderson, Dietz, and Canyon) in Montana part of the PRB. Near-recharge areas in the northwest basin margin, 75% recovery occurred within 5 years of the monitoring period when pumping was discontinued in the Anderson coal. In contrast, where pumping was most intensive, groundwater levels in the Anderson coal recovered 65% in 10 years (Wheaton & Meredith, 2009). These observations in the Dietz coal zone may be correct because it is not being as extensively dewatered for gas desorption as its counterpart coal reservoirs in Wyoming part of the PRB (Flores, 2004).

Dating Co-Produced Water with Isotopes

Dating of the introduction of the groundwater is critical to understanding the origin of the co-produced water and mixing/flushing of surface water with groundwater. The timing, especially of the mixing of waters in coal aquifers in the PRB, may have been much faster than originally postulated. As demonstrated by the juxtaposed fluvial sandstone and immediately underlying coal reservoir during pumping of water to reduce the wellhead pressure, mixing of groundwater from the shallower sandstone to the deeper coal ensued in a matter of few months during gas production (Figure 8.8). This brings up an important point about the concept of mixing waters, which was presumed to happen during the geologic evolution of the coal reservoir as opposed to during gas development. Hydrogeologically, Rice (1993) and Flores et al. (2008) suggested that groundwater in the Fort Union coals in the Powder River Basin (PRB) was established immediately after uplift and erosion of the basin (10 mya) and recharged during the Pleistocene Period (<2.5 mya), respectively. In both cases, early-formed coalbed gas was interpreted to be lost from the uplift and paleoclimatic changes (e.g. Pleistocene repetitive cooling and warming). Also, determining the timing is important in understanding the process of methanogenesis and the critical onset of gas generation in the Fort Union coals in the PRB.

Rice et al. (2002) used tritium to qualitatively estimate the time of groundwater recharge in the PRB. Tritium is a radioactive variation of hydrogen and is rare in the Earth's atmosphere although it is normally present in background levels in the environment due to atmospheric testing of nuclear weapon in the 1960s (Pennsylvania EP, 2012). Thus, the presence of tritium concentrations fingerprints its introduction in the atmosphere from the 1950s to 1960s. Groundwater samples from coal aquifers from Eocene and Paleocene coals in the basin were analyzed for concentrations of tritium to determine any "modern" imprint (Rice et al., 2002). Tritium was not detected at concentrations suggestive of modern water in any groundwater samples collected from the Paleocene and Eocene coal. However, tritium concentrations in the two coalbed gas wells from the Eocene coal aquifer suggest a mixture between submodern (recharged before 1952) and modern water, although the low concentrations suggest that groundwater in these two wells have very little modern water. The relative absence of modern water in all aquifers in the study area suggests that recharge processes to these aquifers are probably originated before the 1950s and 1960s. The method does not precisely indicate how much older the groundwater is than to the modern time period.

A major purpose of dating of co-produced water and groundwater using tritium, carbon-14 and chlorine-36 isotopes is to test the hydraulic connection between coal reservoir and surface and near-surface waters. Oldaker and Fehn (2005) reported that surface waters and shallow groundwater are <60 years old in the Raton and Susitna Basins, the latter in Alaska based on tritium and carbon-14 isotopes. However, dating co-produced water from coalbed gas wells at

563 m depth recorded at least 1.3 million years old based on chlorine-36 isotope. Using the same chlorine-36 isotope, the co-produced water of coalbed gas well at 793 m depth recorded at least 1.2 million years old. In both analyses, no "modern" waters were found indicating no hydraulic connection between surface water and coal reservoir water prior to middle-late Pleistocene time in the Raton and Susitna coal basins. Oldaker and Fehn (2005) interpreted the lack of hydraulic connection to mean that coal reservoirs contain a finite volume of water and can be depressurized. In addition, the lack of hydraulic connection between reservoirs to surface water may be environmentally significant.

The work of Snyder et al. (2003) on the water co-produced with coalbed gas of the Upper Cretaceous Fruitland coal beds in the San Juan Basin determined the age, origin, and history using ratios of iodine ($^{129}I/I$) and chlorine ($^{36}Cl/Cl$), and isotope values (δD and $\delta^{18}O$). Ratios of $^{129}I/I$ indicate minimum iodine ages of about 60 mya, which corresponds to the depositional ages of the Fruitland coals. The Fruitland coals were formed in peat bogs influenced by brackish water in back-barrier environments behind barrier deposits represented by the Pictured Cliffs Sandstone and in freshwater peat bogs landward and stratigraphically higher from the sandstone (Flores & Erpenbeck, 1981). The δD and $\delta^{18}O$ values, which plot along the GMWL (Figure 8.10), indicates a predominately meteoric source of the brine waters, probably derived from waters related with the back-barrier deposits. Snyder et al. (2003) suggested that the water was derived from brackish formation waters associated with backshore peat bogs, while the lightest values represent meteoric waters. The infiltration of

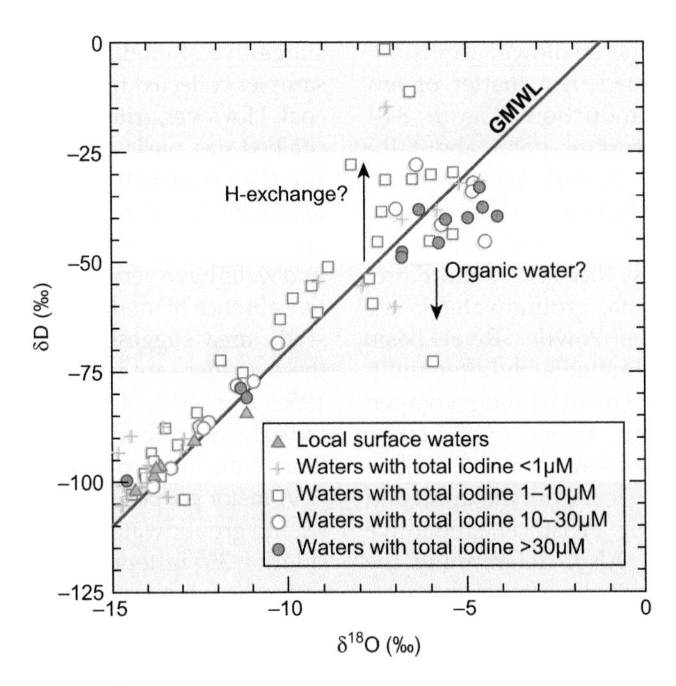

FIGURE 8.10 Values of δD and $\delta^{18}O$ for brines in the Upper Cretaceous Fruitland formation coals in the San Juan Basin in Colorado and New Mexico, United States. The samples with 1–10 μM total iodine plot above the global meteoric water line (GMWL) probably representing re-equilibration of the water in the coal aquifer. High-iodine samples plot below the GMWL probably representing addition of deuterium-depleted organic water. D, Deuterium; O, oxygen; μM, micrometer (10^{-6} m). *Source: Adopted from Snyder et al. (2003).*

recent waters is limited to the uplifted basin margins, which extend <5 km from outcrop (Snyder et al., 2003). The ratios of $^{29}I/I$ and $^{36}Cl/Cl$ indicate that the Fruitland coal has preserved its original isotopic signature. The ratios of iodine ($^{129}I/I$) and chlorine ($^{36}Cl/Cl$) and δD and $\delta^{18}O$ values indicate that the original brine water in the coal reservoirs derived from the brackish environment of deposition was mainly modified along the basin margins. The rest of the basin is characterized by groundwater, which has undergone varying stages of iodine enrichment through diagenesis and dilution (Snyder et al., 2003).

Previous isotopic analyses by Phillips et al. (1986, 1989) as well as more recent work of Riese, Pelzmann, and Snyder (2005) using iodine, chloride, carbon, oxygen, and hydrogen in groundwater of multiple geochemical systems documented that the residence time of coalbed gas water in the San Juan Basin ranges from thousands to tens of millions of years. Riese, Pelzmann, and Snyder, (2005) concluded that within the uncertainties of isotopic analysis, meaningful recharge of groundwater to all coal aquifers in the San Juan Basin, except for some of the basin margins in close proximity to outcrop areas, has not occurred within the scale of human time. Results of geological, geochemical, geophysical, biological, and ecological investigations have demonstrated that the last major recharge of water to the San Juan Basin coal systems occurred during Eocene time, approximately 35–40 million years ago (Riese et al., 2005).

The difference in methodologies in determining the origin, process, and history of the co-produced water between the PRB and San Juan Basin is the depositional environment of the coal beds. In the PRB, the Fort Union coals were deposited in freshwater peat bogs associated with intermontane alluvial plains (Flores, 1986). In the San Juan Basin, the Fruitland coals were deposited in brackish water back-barrier peat bogs along coastal plains (Flores & Erpenbeck, 1981). According to Moran, Fehn, and Hanor (1995) and Muramatsu and Wedepohl (1998), iodine is a biophilic element, which is associated with sedimentary organic-rich marine sediments and evolved brines. Thus, the Fruitland coals are different from Fort Union and other coals and the ^{129}I system is a more important tracer for determining ages and migration of brines associated with coalbed gas reservoirs in the San Juan Basin.

VOLUME AND COMPOSITION OF CO-PRODUCED WATER

The most important concerns in management of co-produced water of coalbed gas are the amount or volume and composition or constituents of the groundwater. Knowledge of the volume and composition of co-produced water is required to implement plans for management such as disposal, treatment, and beneficial use in compliance to governmental regulations; thus, helpful to land managers, regulators, reclaimers, resource assessors, and policy decision makers. Also, understanding the volume and composition of co-produced water can assist gas operators in estimating costs of managing co-produced water which includes pumping, electricity, pipeline infrastructure, permitting and implementing surface disposal, storage, and treatment systems, potential reinjection or transportation, and ecological and environmental abatements. The costs of management of co-produced water for oil, natural gas, and coalbed gas in general range from \$0.10 to several dollars per barrel of water (Veil, Pruder, Elcock, & Redweik, 2004). However, a survey for coalbed gas wells from different US basins indicates that the costs of handling co-produced water range from a few cents to more than \$1.00/119 l and can add significantly to the cost of gas production (ARI 2002; Rice & Nuccio, 2000; USEPA, 2001). The costs of handling co-produced water increases with the technology applied to store, dispose, and treat co-produced water with trucking and injection of residual concentrates.

TABLE 8.1 The Volume of Co-Produced Gas from Coalbed Gas Wells and Coal Rank in Selected Coal Basins in the United States as of 2005

U.S. Coal Basins	Coal Rank	Number of Wells	Average Volume Water Volume (l/Day/Well)
Appalachian	Bituminous	4000	3696
Arkoma	Bituminous	2180	1550
Black Warrior	Bituminous	4370	4889
Powder River	Subbituminous	17,000	14,279
Raton	Bituminous	1892	27,902
San Juan	Bituminous	5238	2504
Uinta (Ferron)	Bituminous	488	5723
Illinois	Bituminous	139	27

Source: Data for Black Warrior Basin from Alabama State Oil and Gas Board as of 2005; data for Powder River Basin from Wyoming Oil and Gas Conservation Commission as of 2005; data for Raton and San Juan Basins from Colorado and New Mexico Oil and Gas Conservation Commissions as of 2005; and data for Uinta Basin from Utah Division of Oil and Gas as of 2005. l, liter.

Volume of Water

The volume of co-produced water with coalbed gas varies widely within and between coal beds, fields, and basins (Table 8.1) depending on the hydrogeology and hydrostratigraphy. NRC (2010) estimated that at the height of coalbed gas production in all of 2008 prior to the economic slowdown, more than 5000 billions of liters of co-produced water were pumped out of coal beds in five western states (Colorado, New Mexico, Montana, Utah, and Wyoming) of the United States, which include the Piceance, Powder River, Raton, San Juan, and Uinta Basins. The quantity of produced water in 2008 depended on type of coal and the overall production history of the basin. Basins with a longer production history, such as the San Juan Basin, produced less total co-produced water and less water per well than the more recently developed basins, such as the PRB.

In general, the variations in the amount of water co-produced with coalbed gas wells are caused by differences on the coal rank, thickness, permeability, depth, and depositional environment as well as the duration of production and life span of coalbed gas wells. In addition, factors affecting the volume of co-produced water with coalbed gas production include those associated with the technology applied to well completion and forms of stimulation of the coal reservoir. A change in drilling technology from vertical to horizontal wells increases the gas drainage area and as a result increases the volume of water pumped out from the drainage area. The method of completion, cavitation versus perforation, will influence the amount of co-produced water with the perforated completion providing more control for multiple coal beds gas and water production. Stimulation technology utilized to create fractures and their connectivity make a difference in the volume of co-produced water.

The quality and hydraulic pressure of the coal reservoir play important roles in the volume of co-produced water with coalbed gas production. Physical properties of the coal enable it to shrink and swell, which affects pore throats and opening of fractures, and contribute to the capability of the reservoir to flow water. Withdrawal of water in front of coal mining, which produced from local- to regional-scale drawdown or cone of low-pressure areas, minimizes the potential for production of co-produced water because the water has already been removed. Pressure gradients and wellhead pressures, which determine the amount of time and volume of water to be extracted before gas desorption ensues are critical to predicting water production. In addition, the number of coalbed gas wells and well spacing to dewater and depressurize the coal reservoir also controls the volume of co-produced water. The type of pumps used to remove the water from the coal reservoir determines the rate of the reduction of hydraulic pressure in the wellbore.

Hydrostratigraphic (Micro) vs Hydrogeologic (Macro) Units

Volume and flow of groundwater in coal-bearing rocks are not completely defined by simple and classic concepts of basin-scale aquifer and confining layers (hydrogeologic or macro units). Within these macro units, discontinuous coal and sandstone beds interbedded with fine-grained mudstones and siltstones make up coal-bearing rocks that are a complex of hydrostratigraphic or micro units and compartmentalized aquifers. Detailed investigation and mapping of these hydrostratigraphic or micro units based on closely spaced control points of coalbed gas wells suggest better connectivity of fluvial channel sandstones than coal beds along their depositional dips (Flores, 1986; Flores et al., 2010).

Hydrostratigraphic or Micro Units

Typically, coal beds are permeable and hold and conduct groundwater, and can be described as aquifers. However, as a temporal depositional unit, coal aquifers are laterally discontinuous, pinched-out lenticular beds of local extent and vertical multidepositional continuum as well as a complex of hierarchically and laterally multisplit and multimerged coal beds within a coal zone of regional extent (Flores, 1986; Flores et al., 2010). In contrast, the interbedded mudstones and siltstones are less permeable, which retard vertical and lateral flow of groundwater, best described as characteristics aquitards/aquicludes or as classic confining layers with respect to the water saturated coal-bed aquifers. These fine-grained sedimentary rocks are equally locally discontinuous and regionally continuous as the coal beds. According to Bates and Jackson (1987), the fine-grained impermeable rocks sometimes known as aquicludes can absorb water but does not easily convey water to a spring or well. The coal and fine-grained sedimentary rocks only account for a small part of the volume of rocks in the coal-bearing sequences. These sequences also contain abundant interbedded sandstone beds depending on the depositional settings (e.g. fluvial or deltaic channels).

Sandstone beds are often porous and permeable rocks that store and transmit groundwater and are considered as aquifers (Ferris, Knowles, Brown, & Stallman, 1962). They are commonly bounded much like the coal beds by low permeability mudstones and siltstones (Figures 8.6 and 8.7). The common understanding is that sandstones are local aquifers independent of each other, have poor lateral connectivity, and are hydrologically isolated (Hinaman, 2005; Lowry, Daddow, & Rucker, 1993; Rice et al., 2002). The variability of the sandstone aquifers in relation to juxtaposed coal reservoirs in coal-bearing rocks could explain the unpredictability of coalbed gas well reservoir conditions (Figures 8.6 and 8.7). However, Figure 8.6 shows preferred lateral connectivity with the length of the fluvial channel sandstones or depositional dip and nonconnectivity along the width or depositional strike as demonstrated within the interval of the Wyodak–Anderson (Wyodak) coal zone and related hydrostratigraphic units in the PRB (BLM, 2012; Flores et al., 2010). Mapping of the sandstones benefitted by closely spaced coalbed gas wells in the basin show excellent evidence of compartmentalization across their widths (e.g. depositional strike) but when these porous and permeable rocks are saturated with groundwater they behave as aquifers along their lengths or deposition dip. The general concept of basin-scale groundwater flow (e.g. hydrogeology or macro units) in the PRB is toward the north and northeast directions. However, within the hydrostratigraphic or micro unit of the Wyodak–Anderson (Wyodak) coal zone aquifer, in particular, the groundwater flow is north and northwest as interpreted by Daddow (1986) and Lobmeyer (1985). This observation is supported by the presence of a northerly oriented, channel sandstones, which behaved like an aquifer system along their depositional dip (Figure 8.7(B)). This flow direction is contrary to that of the Wyodak–Anderson (Wyodak) coal aquifers, which split, merge, and pinch out to the north and west (Flores, 1986; Flores et al., 2010) (Chapter 6, Figures 6.11 and 6.12). In addition, this north–south channel-sandstone trend coincides with the occurrence of as many as six coal beds >6 m thick of the Wyodak–Anderson (Wyodak) coal zone (Rice et al., 2008). Thus, it begs the question is the groundwater flow more preferred along the length of the channel sandstones than along the trend of discontinuity of the coal beds? Also, the preferential groundwater flow may be controlled by the horizontal hydraulic conductivity (*k*) values of sandstones gravel (1.5 M/D) rather than coal (0.4–0.9 M/D) aquifers compared to impermeable mudstones–siltstones (0.3 M/D) (BLM, 2012; Ferris et al., 1962; Wheaton & Metesh, 2002).

As shown in Figure 8.8, the groundwater in the confined coal aquifer rises and links to an elevation (e.g. groundwater table) in the overlying fine-grained rocks above the top of the bed. This column of water applies pressure on the coal bed. A well drilled into the confined coal aquifer will have a water level rising to the same groundwater table elevation above the top of the bed (Figure 8.8). The artesian pressure is an indirect measure of the volume of water that is co-produced with gas desorption. Connectivity of the water column to an overlying sandstone aquifer complicates the ultimate volume of co-produced water in the well. Thus, it is important to determine the connectivity between coal and adjoining sandstone aquifers in order to understand the volumetrics of co-produced water. Consequently, the vertical and lateral hydrostratigraphic variations of confined aquifers and aquitards/aquicludes require detailed descriptions to understand their significant role in determining the volume and quality of water co-produced with coalbed gas wells. Furthermore, although the aquifers (e.g. coals and sandstones) appear to be vertically isolated and compartmentalized, laterally they may be interconnected to form stacked aquifers permitting vertical infiltration of water (Figure 8.8).

Hydrogeologic or Macro Units

Hydrogeologic units are defined as multistate in extent or macroscale, in which aquifers and confining layers are combined as regional aquifer and confining systems (Hotchkiss & Livings, 1986; Whitehead, 1996). Hinaman (2005) described hydrogeologic units to correspond with geologic formations but hydrogeologic units can include combined geologic formations and stratigraphic units can span multiple hydrogeologic units. For example, Hinaman (2005) described the Tertiary alluvial coal-bearing hydrologic units in the PRB in Wyoming as uniformly thick and nested bowl layers consisting from bottom to top of the Tullock aquifer, Lebo confining unit, and Wasatch–Tongue River aquifer (Figure 8.11). Stratigraphically in a descending order, the Tongue River, Lebo, and Tullock are members of the Fort Union Formation. According to Hinaman (2005), the basin-scale groundwater table does not intersect the land surface although intersected areas (e.g. A and B) may exist in the basin. The saturated thickness for the Tongue River–Wasatch (TRW) aquifer is designated *a*, which is the distance between the groundwater table and the base of the TRW aquifer. Where potentiometric surface of the Tullock aquifer (*b*) is above the

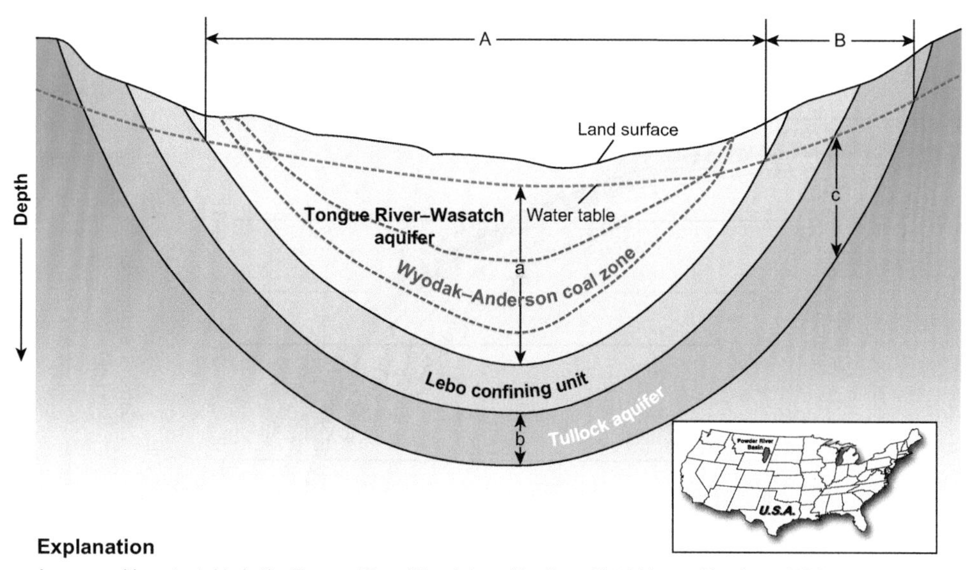

FIGURE 8.11 Diagram showing a general cross-section of the Tertiary hydrogeologic units, saturated units, Wyodak–Anderson (Wyodak) coal zone aquifer (red area), land surface, and groundwater table in the Powder River Basin in Wyoming (see inset map). *Source: Modified from Hinaman (2005).*

base of the TRW aquifer in an increasing amount toward basin center is an area where groundwater table is below the contact between the TRW aquifer and the Lebo (mudstone dominated) confining unit (B) (Hinaman, 2005). In this area (B), the TRW aquifer is completely unsaturated or absent. The amount of saturated thickness, *c*, is the difference between the groundwater table and the base of the Tullock aquifer. However, based on sedimentary facies analyses of Flores et al. (2010), the Tongue River Member is more complicated than stratigraphically envisioned by Hinaman (2005) because the member coal zones (e.g. Wyodak–Anderson) thicken 10 times from the basin eastern margin to the center where it is mud dominated.

Sequences of Upper Cretaceous marine-influenced sandstones and overlying coal-bearing rocks typical of coastal plain deposits such as those in the San Juan and Raton Basins across a state or states are classical examples of combined regional aquifers often confined by fine-grained sediments (e.g. marine and nonmarine shales; Figure 8.12). However, at a basin scale, rock formations are often combined and behave as traditional, regional aquifers being recharged with meteoric water in elevated areas along basin margins, which in turn flow downgradient in the deeper part of the basins. Water recharge occurs from within the human lifetime to millions of years. Riese et al. (2005) reported that this classic view of hydrological process is too simplistic based on geological, geochemical, geophysical, biological, and ecological investigation. Compartmentalized coal-bed aquifers and confining layers as defined within a basin or subbasin scale may be a more realistic model to apply for determining volume and

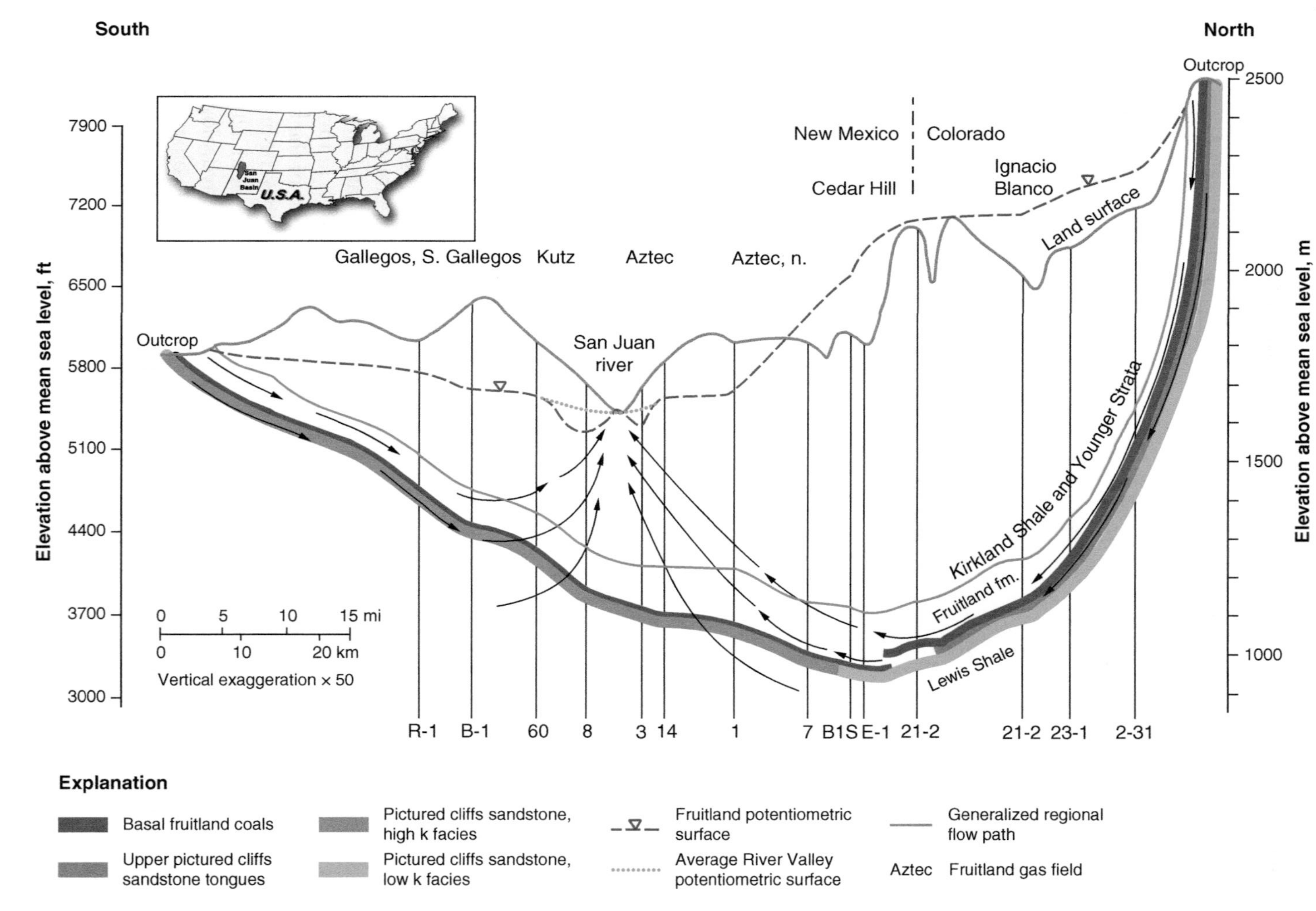

FIGURE 8.12 Groundwater pathways in the Upper Cretaceous Lewis Shale, Picture Cliffs Sandstone, and Fruitland formation in the San Juan Basin, Colorado and New Mexico, United States. *Source: Adopted from Kaiser and Swartz (1989).*

composition of water rather than at multistate scale level.

Measuring Groundwater Volume

The best attempt to estimate groundwater volume related to a coalbed gas producing basin was reported by Hinaman (2005). The detailed discussions of the methodologies and calculations used for water resource assessment are beyond the scope of this chapter, but the reader is referred to Hinaman (2005) for more comprehensive information. A synopsis of the assessment of Hinaman (2005) is presented here as a unique model because of the use of percentage of sandstones versus nonsandstones, in each of the hydrogeologic units referred to in Figure 8.11, which was advocated by the author. The author uses the term sandstone in contrast to "sand" by Hinaman (2005) because of cementation or case-hardening of the rock unit.

According to Hinaman (2005), the first method of estimating groundwater volume of the saturated rocks is to calculate by summation of the component of groundwater in "sands" and the component of groundwater in "nonsands". Hinaman (2005), assumed that a hydrogeologic unit is confined when it has more than 15 m of saturated thickness. The following equation was used to calculate the total amount of water in the "sands" for the unconfined aquifers in the TRW aquifers:

$$\text{Sandwater} = 1000 \times 1000 \times (\text{sattk} \times (\%\text{sand}/100) \times \text{sand} - \text{porosity})$$

where:

sandwater = total amount of water in the sands, in cubic meters;
1000 × 1000 = area of the grid cell, in square meters (see section on "Development of a Graphic Information System" of Hinaman (2005) for explanation of the creation of cell size of 304.8 m on a side);
sattk = saturated thickness (*a* in Figure 8.11), in meters;
%sand = percentage of sand; and
sand-porosity = porosity of the sand.

The equation used by Hinaman (2005) to calculate the total volume of groundwater in the "nonsand" of the cell in the TRW aquifers is the following:

$$\text{Non-sandwater} = 1000 \times 1000 \times (\text{sattk} \times ((100 - \%\text{sand})/100) \times \text{non-sand porosity})$$

where:

non-sand water = total amount of water in the non-sand component of the cell, in cubic meters;
1000 × 1000 = area of the grid cell, in square meters;
sattk = saturated thickness (*a* in Figure 8.4), in meters;
%sand = percentage of sand; and
non-sand porosity = the porosity of the non-sand. component (e.g. coal, mudstone, and siltstone).

The method of calculation by Hinaman (2005) is novel with respect to considering the total amount of groundwater in "sands" and "nonsands" but it could be more sensitive by taking into account the high variability of sandstones in the Wasatch and Upper Fort Union Formations and interbedded coal beds. The lack of detailed stratigraphic information on lateral and vertical variations of the thickness of sandstones and coal beds may have contributed to the constraints of the study. More importantly, accurate correlation of sandstones and coal beds, which is undervalued by nonstratigraphers and nonsedimentologists, certainly adds to the precision of coal thickness and volume predictions. Moreover, using the outcrop of the Fox Hills Sandstone to create a grid to apply to the Wasatch and Upper Fort Union Formations may not be similar. The hydrogeological modeling work of BLM (2012) of the PRB rectified this lack of data by using detailed stratigraphic and sedimentologic data from Flores et al. (2010). This is exemplified by Figure 8.7(B), which shows the detailed map of the channel sandstones in the Wyodak–Anderson (Wyodak) coal zone, one of the six hydrostratigraphic units defined by BLM (2012) basinwide. More importantly, the thinning–thickening of these hydrostratigraphic units was factored in based on splitting of coal zones and contemporaneous basin subsidence during their deposition, which control the calculation of sandstone percent within an interval (Figure 8.11).

COMPOSITION OF WATER

The composition of co-produced water of coalbed gas differs from that of the conventional oil and natural gas. In addition, the co-produced water of coalbed gas requires less treatment than oil and natural gas co-produced water. The difference in composition lies in the mixture of constituents in the reservoir, in which co-produced water has been in direct contact and developed chemical equilibrium with these hydrocarbon- and coal-bearing reservoirs over millions of years. Also, different chemistry between the conventional and unconventional co-produced water is largely dependent on the origin of the formation waters. A large part of oil and natural gas formation waters was derived from the original depositional waters, which were mainly from marine and brackish water environments. In contrast, the precursor coals were formed primarily in freshwater and secondarily in brackish water environments. Dahm et al. (2011) indicated that formation water involved during coalification includes chemical attributes of their original depositional waters. For example, coal deposits formed in continental environments contain higher concentrations of residual calcium and magnesium with lower concentrations of sodium and potassium than coal deposits accumulated in brackish water environments. In contrast, marine-influenced coal deposits contain higher concentrations of sodium and chloride with high boron concentration (Dahm et al., 2011). Moreover, Cheung, Sanei, Klassen, Mayer, & Goodarzi (2009) observed in coals of Alberta, Canada formed in brackish-marine water environments consist of 1.7 times more trace elements than coals formed in freshwater continental environments.

The composition of fluids used in drilling wells as well as seawater and groundwater used to maintain reservoir pressures in oil and gas wells is also a source of contaminants for co-produced water. According to NETL (2012a,b), the major constituents of co-produced water with oil and natural gas include (1) salt components such as salinity, TDS, and electrical conductivity (EC); (2) oil and grease as a measure of the organic chemical compounds; (3) inorganic and organic compound chemical additives used in drilling, completion, stimulation, and operation of the well, and; (4) naturally occurring radioactive materials (NORM) found in nature, formed by natural processes (e.g. uranium and thorium).

The most important constituent of co-produced water of oil and natural gas relevant to the debate on environmental impact and beneficial use is the chemical composition of the dissolved and dispersed organic compounds, chemical additives, and NORM (Veil et al., 2004). A comparison of the general composition of the conventional natural gas and unconventional coalbed gas co-produced waters is summarized by Sumi (2005) in Table 8.2. Based on pH value (4.9–9.9), the coalbed gas co-produced water values are more variable than co-produced water of oil and natural gas. However, the values of benzene, sodium (Na^+), barium, chlorine (Cl^-), and bicarbonate (HCO_3^-) for natural gas co-produced water are 4–120 times more than for the coalbed gas co-produced water. The salt content is mainly composed of chlorides and sulfides of calcium,

TABLE 8.2 Chemical Composition of Co-Produced Water from Conventional Natural Gas and Unconventional Coalbed Gas in the United States

Chemical Composition	Natural Gas Co-produced Water	Coalbed Gas Co-Produced Water
pH	6.5–8	4.9–9.9
TDS (mg/l)	20,000–100,000	160–170,000
Benzene (ppb)	1000–4000	<100
Na^+ (mg/l)	6000–35,000	500–2000
Barium (mg/l)	0.1–40	0.001–0.1
Cl^- (mg/l)	13,000–65,000	1000–2000
HCO_3^-	2000–10,000	150–2000

TDS, total dissolved solids; pH, logarithmic measure of hydrogen ion concentration; mg/l, milligram per liter; ppb, parts per billion.
Source: Adopted from Sumi (2005).

magnesium, and sodium with chlorides as much as 10 times that of seawater (Sumi, 2005). Thus, the co-produced water of coalbed gas generally exhibits better quality than unconventional oil and natural gas co-produced water.

Effects of Co-Produced Water Composition

The composition of co-produced water of coalbed gas may be viewed according to their environmental and beneficial effects as well as the geochemical processes that water undergo during its emplacement. The environmental and beneficial effects of co-produced water have been the focus of investigations during the period of 2000–2010 particularly with the background of tremendous amount of water generated from development of both conventional and unconventional gas (CWRRI, 2006; USEPA, 2010). The volume of co-produced water from conventional and unconventional oil, gas, and coalbed gas wells in the United States extracted annually is up to 2500 billion liters that is disposed, stored, treated, or recycled (Otton, 2005). The large volume of co-produced water from coalbed gas wells is best exemplified in the PRB where prior to the nationwide slowdown" of gas production, the basin yielded cumulative water production of about 96 BCM from about 29,026 coalbed gas wells as of January 2009 (Figure 8.13). The major question posed about co-produced water is that it can be economically and technologically treated and reclaimed into a water supply for beneficial use without environmental impact and liability for areas that need it the most.

Drinking Water Standard vs Co-Produced Water

ALL Consulting (2003) generally characterized co-produced water with coalbed gas as high salinity, sodicity, and trace elements. Environmentally, this is of concern as it determines the water's suitability for drinking water and its toxicity to plants and aquatic organisms. The USEPA (2012a,b) secondary drinking water standard is 500 mg/l TDS and seawater is approximately 35,000 mg/l TDS. The recommended TDS limit for potable water is 500 mg/l, and for beneficial use such as disposal into stock ponds

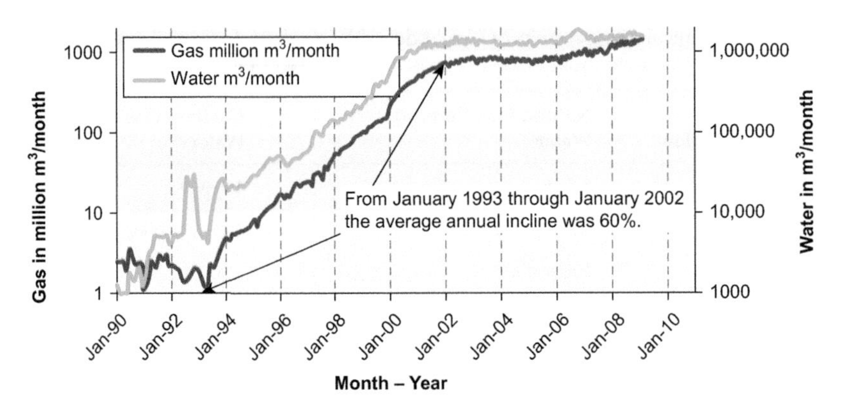

FIGURE 8.13 Chart showing production trends of coalbed gas and co-produced water in the Powder River Basin from January 1990 to January 2009. Steep incline of production of coalbed gas occurred from January 1994 to January 2002 and co-produced water from January 2001. Data are from WOGCC website.

or for irrigation, the limit is 1000–5000 mg/l (Jangbarwala, 2007). In addition, USEPA (2010) characterized the co-produced water of coalbed gas as containing trace pollutants of potassium, sulfate, bicarbonate, fluoride, ammonia, arsenic, and radionuclides, which affect the downstream disposal systems and potential drinking water quality standards. Salinity denotes the amount of the TDS and salts, which can be measured by EC because ions dissolved in water conduct electricity. TDS includes any dissolved minerals, salts (e.g. magnesium, calcium, sodium, and chloride), metals, cations, and anions in the water (Veil et al., 2004). On the basis of TDS, co-produced water of coalbed gas in the United States range from fresh to highly saline water.

Yang et al. (2011) studied TDS of co-produced water from coalbed gas wells in China (e.g. Qinshui Basin) in comparison to the PRB and reported values from 1046 to 5196 mg/l and a mean of 3121 mg/l. The TDS values, which are higher than the mean value of the PRB, are also more than the 1000 mg/l permissible according to the People's Republic of China Standards for Drinking Water Quality (GB 5749–2006) recommendation for potable water, Irrigation Water Quality (GB 5084–1992), and Environmental Quality Standards for Surface Water Standard (GB 3838–2002). The combination of coagulation–flocculation, sedimentation, filter, and reverse osmotic treatments can be used for recycling the co-produced water.

The US Environmental Protection Agency established national primary and secondary drinking water regulations. The national primary drinking regulations set mandatory and enforceable water quality standards for drinking water contaminants. However, the national secondary drinking water regulations are nonenforceable guidelines regulating contaminants. The co-produced water of coalbed gas contains most of the secondary contaminants. Table 8.3 shows the USEPA (2012a,b) levels for drinking water standards. The secondary contaminant levels are established as recommendations to water systems but not necessarily a requirement USEPA (2012a,b).

Sodium Adsorption Ratio

The salinity of co-produced water of coalbed gas relates to ionic species including the sodicity value, which pertains to sodium levels. Excessive concentration of sodium can occur in the form of sodium bicarbonate in co-produced water types (Table 8.4). The sodicity of co-produced water is expressed as the sodium

TABLE 8.3 Co-Produced Water of Coalbed Gas Contains Selected Secondary Contaminants with Concentrations Compared to Drinking Water Quality Standards

Contaminant	Secondary Standard (mg/l)
Aluminum	0.05–0.2 mg/l
Chloride	250 mg/l
Copper	1.0 mg/l
Fluoride	2.0 mg/l
Iron	0.3 mg/l
Manganese	0.05 mg/l
pH	6.5–8.5
Silver	0.10 mg/l
Sulfate	250 mg/l
Total dissolved solids	500 mg/l
Zinc	5 mg/l

mg/l, milligram per liter; pH, logarithmic measure of hydrogen ion concentration.
Source: Recommended by the USEPA (2012a).

adsorption ratio (SAR), which is the measure of the abundance of sodium relative to the concentration of calcium and magnesium in water. The formula for calculation of SAR is (Davis, Waskom, & Bauder, 2012; Jangbarwala, 2007; USEPA, 2010) as follows:

$$SAR = Na^{+}/\sqrt{1/2}(Ca^{2+} + Mg^{2+})$$

Where, SAR = sodium adsorption ratio; Na = sodium; Ca = calcium; and Mg = magnesium.

Quantifying SAR using the above equation results in unitless number, which can be directly related to soil conditions. The exchange of sodium for calcium and magnesium (SAR) indicated in the equation causes the phenomenon of swelling and dispersion in soils. Swelling of soil is caused by disproportionate distribution of pores, and results in decreasing soil structure and permeability, and soil infiltration problems (e.g. capacity) Gupta et al. (1984). Jangbarwala (2007) described dispersion as plugging of pores in the soil, which prevents and/or reduces the rate of water flow (infiltration) or hydraulic

TABLE 8.4 Co-Produced Water Types, Total Dissolved Solids (TDS), and pH in Various Coalbed Gas Producing Basins in the United States

Coal Basin	State	Co-Produced Water Type	TDS (mg/l)	pH	General Water Type*
Appalachian	Ohio, Pennsylvania, Virginia, West Virginia	No data	<10,000>	No data	Saline water
Black Warrior	Alabama	$Na-Cl-HCO_3$	160–31,000	5.34–9.9	Freshwater to saline water
Powder River	Wyoming	$Na-HCO_3$	270–2720	6.7–8.0	Freshwater to saline water
Raton	Colorado	$Na-HCO_3$	175–30,000	4.9–9.0	Freshwater to saline water
San Juan	Colorado	$Na-HCO_3-Cl$	410–100,000	5.2–9.2	Freshwater to seawater
Uinta (Ferron)	Utah	$Na-HCO_3-Cl$	6,350–42,700	7.0–8.2	Freshwater to seawater
Illinois	Illinois and Indiana	$Na-Cl-HCO_3$	2,532–83,920	7.0–8.1	Saline water to seawater
Piceance	Colorado	No data	1000–6000		Saline water

TDS (in mg/l) <1000 for freshwater, >1000 for saline water, and >35,000 for seawater (NRC, 2010).
TDS, total dissolved solids; pH, logarithmic measure of hydrogen ion concentration; mg/l, milligram per liter; <and>, less than and more than.
*Source: Modified from *NRC (2010) Ash and Gintautas (2009), Knutson et al. (2012), Otton (2005), Rice and Nuccio (2000), SeaCrest Group (2003), and USEPA (2010).*

conductivity through the soil. An approach to relate SAR and effects of salinity (EC) in terms of soil conditions is described by Hanson, Grattan, and Fulton (1999) and Jangbarwala (2007). Ayers and Westcot (1985) discussed the interaction of and relationship between SAR and EC on infiltration rates independent of texture properties of soils.

Formation of sodic or sodium-rich soil results in poorly drained, water-logged conditions, which affects aeration. These conditions reduce plant growth and organic decomposition causing infertile soils (Jangbarwala, 2007). High-pH soil may increase plant toxicity to certain elements and plant nutritional imbalances may develop (Davis et al., 2012). The salinity and SAR properties of co-produced water can present problems in beneficial use for agriculture. The toxic soil conditions can be treated and reclaimed by applying soil amendments such as gypsum (Jangbarwala, 2007; Davis et al., 2012). However, the removal of sodium is one of the primary purposes for treating co-produced water of coalbed gas, which can be determined following the flow chart shown in Figure 8.14 (Janbarwala, 2007).

Zeolite is another amendment for treatment of SAR proposed by a few studies (Huang & Natrajan, 2006; Zhao, Vance, Ganjegunte, & Urynowicz, 2008; Zhao, Vance, Urynowicz, & Gregory, 2009). Studies of Zhao, Vance, Ganjegunte, and Urynowicz (2008), Zhao, Vance, Urynowicz, and Gregory (2009) have shown that treatment of the co-produced water in the PRB with calcium-enriched zeolites reduced SAR and EC. The treatment involved using a naturally occurring sodium-rich zeolite to alternately soften the co-produced water. The exchange of sodium from the zeolite for calcium within the treatment water results in a calcium-rich zeolite. Sodium-rich co-produced water is then introduced into the calcium-rich zeolite, which results in reversed sodium removal reducing SAR.

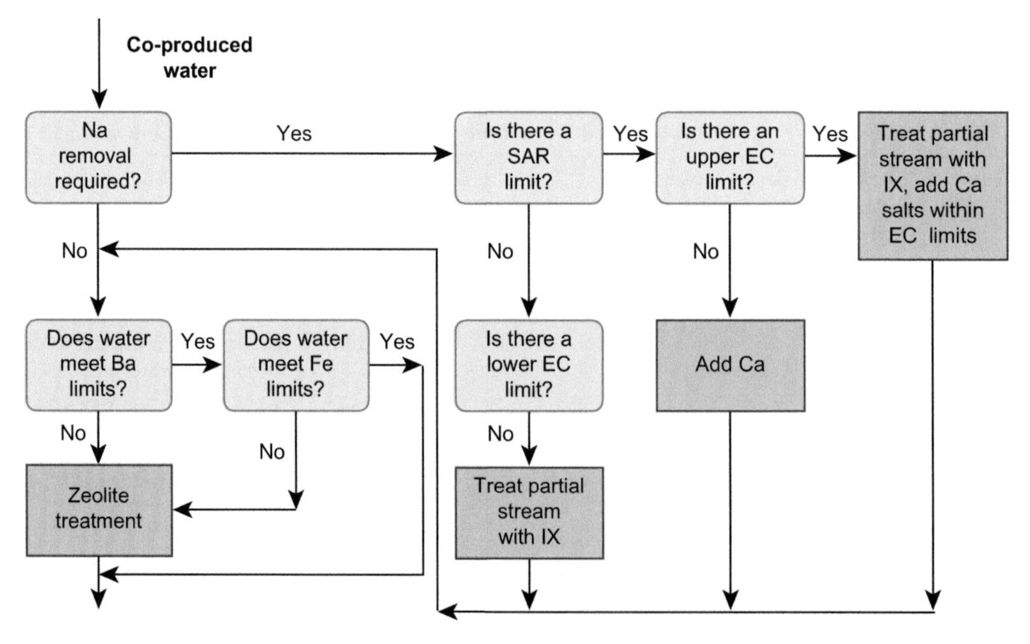

FIGURE 8.14 Flowchart showing various stages of treatment of co-produced water for removal of sodium. IX, ion exchange. *Source: Adopted from Jangbarwala (2007).*

Geochemical Process of Co-Produced Water

A contrasting view of co-produced water composition is a geochemically process-driven concept, which groups water into dominantly sodium bicarbonate (e.g. $NaHCO_3$) or sodium chloride (e.g. NaCl; Rice et al., 2002; Rice & Nuccio 2000; Otton, 2005; Table 8.4). The distinctive compositional groups consist mainly of bicarbonate, calcium, chloride, magnesium, sodium, and sulfate unique to geochemical reactions and processes intrinsic to generation of methane in coal reservoirs. According to Van Voast (2003), co-produced water associated with coalbed gas is geochemically modified by sulfate reduction, bicarbonate enrichment, and calcium and magnesium depletion. The modification results in water composition primarily of sodium and bicarbonate type influenced by freshwater settings and sodium and chloride type influenced by brackish-marine settings. Another component of the geochemical process is microbial reduction of sulfate toward generation of biogenic or microbial coalbed gas (Chapter 4).

Co-produced water of coalbed gas also contains trace amounts of aluminum, selenium, arsenic, iron, manganese, boron, barium, chromium, molybdenum, copper, and zinc as well as other chemical constituents (McBeth, Reddy, & Skinner, 2003). Dahm et al. (2011) presented a more detailed and comprehensive analyses of additional 27 inorganic constituents, which play a role in the evolution of water types. Trace element concentrations in co-produced water of coalbed gas water are commonly very low and are measured in micrograms per liter (μg/l) (McBeth et al., 2003). In general, the co-produced water composition may be looked at in terms of major dissolved ions including primarily sodium (Na), bicarbonate (HCO_3), chloride (Cl) and related calcium (Ca), and magnesium (Mg) as well as trace elements. Otton (2005) observed that most of the co-produced water of coalbed gas in US Western Interior coal basins is composed of predominantly sodium bicarbonate with less amount of chloride and wide-ranging TDS. The association with coalbed gas, which consists primarily of methane, minor amounts of carbon dioxide and nitrogen, and sparse heavy hydrocarbons, controls the composition of the co-produced water.

Rice et al. (2002) interpreted the bicarbonate component of co-produced water as being limited by the amount of calcium and magnesium precipitation of carbonate minerals. Coalbed gas co-produced water is relatively low in sulfate (SO_4) because the chemical conditions in coal beds favor the conversion of SO_4 to sulfide, which is removed as a gas or as a precipitate. According to NRC (2010), bicarbonate and sodium are more prevalent than magnesium, calcium, and sulfate in coalbed gas co-produced water because of the following geochemical processes: (1) sulfate, which is contributed by dissolution of mineral gypsum, is reduced by microbial activity or reduction at depth probably releasing bicarbonate; (2) calcium and magnesium are removed by ion exchange, which releases sodium, and by calcite ($CaCO_3$) precipitation; and (3) dissolution of sulfide minerals and organic matter oxidation in water recharge areas, both of which generate acid.

The geochemical processes described by NRC (2010) are reported by Brinck, Drever, and Frost (2008) to develop along groundwater flow paths within coal reservoirs in the PRB (Figure 8.15). The series of related biogeochemical reactions and processes developed along the flow path of the groundwater from surficial recharge areas to downdip of the coal aquifer. Brinck et al. (2008) described the evolution of the groundwater in the coal aquifer with the following stages: (1) salts such as gypsum, calcite, and epsomite, which accumulated in the unsaturated, semiarid soils of the basin are re-dissolved near the surface; (2) gypsum and salts precipitated along fractures are dissolved by water in the aquifer at shallow depth

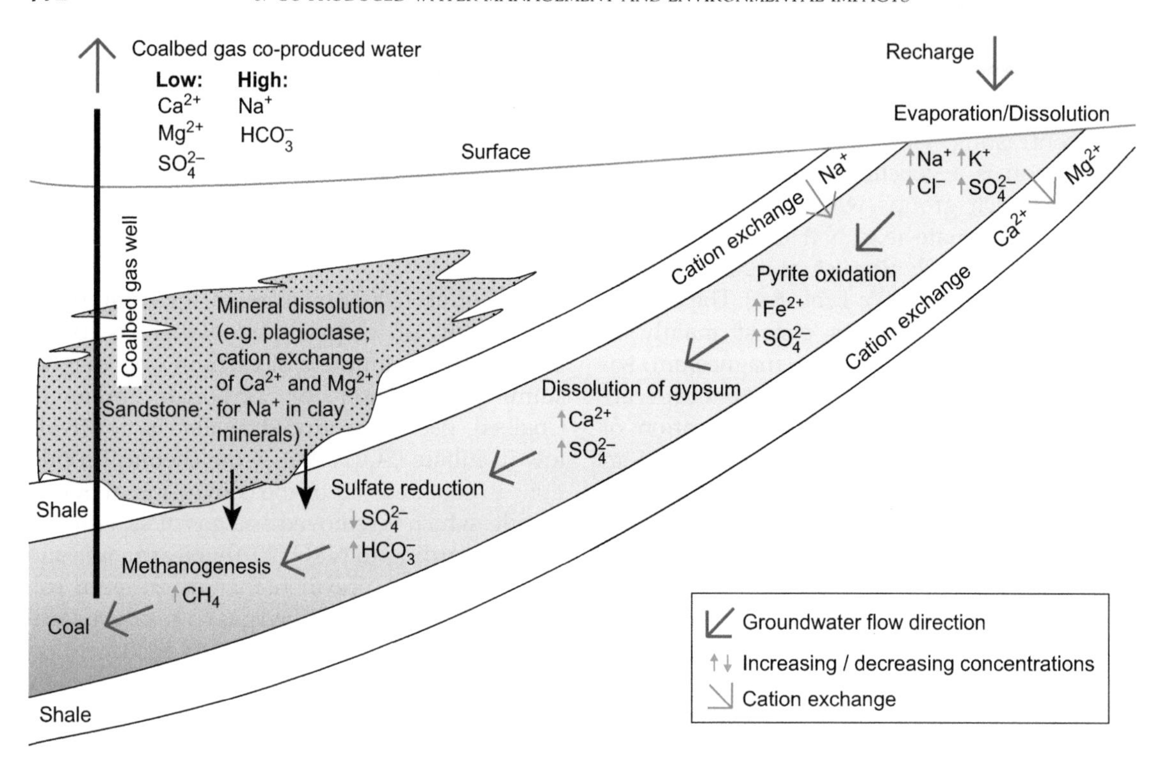

FIGURE 8.15 Series of biogeochemical reactions and processes developed along the flow path of the groundwater from surficial recharge areas to down dip of the coal aquifer. This biogeochemical model is modified to show potential contribution of exogenous source of water sourced from overlying and laterally juxtaposed fluvial channel sandstone aquifers. Ca, calcium; Cl, chloride; Mg, magnesium; Na, sodium; Fe, iron; K, potassium; SO_4, sulfate; HCO_3, bicarbonate; CH_4, methane. *Source: Modified from Brinck et al. (2008).*

releasing calcium and sulfate and incorporates products of pyrite oxidation, which in turn, releases iron and sulfate; (3) reduction of sulfate expends sulfate at greater depth producing bicarbonate, which in turn, increases bicarbonate causing precipitation of calcite and reducing calcium in solution; and (4) generation of methane by methanogenesis, which is desorbed, at great depth.

The model proposed by Brinck et al. (2008) for the evolution in composition of co-produced water is an excellent start for understanding biogeochemical reactions and processes of endogenous water internally within coal aquifers not only in the PRB but also in other coal basins worldwide, which generated biogenic coalbed gas. However, the biogeochemical model for the PRB by Brinck et al. (2008) leads to a more holistic view in the understanding of the role of exogenous water introduced from recharge areas along the basin margins and within the basin. An important point to make about the model is that biogeochemical reactions and processes are uniquely driven by freshwater in the groundwater system in the PRB, which is different from brackish and/or marine influenced water in other coal basins in the United States and in other countries.

Another geological feature that is unique to the PRB is the presence of about 4144 km^2 of

TABLE 8.5 Chemical Composition of Groundwater in the Clinkers in Eastern Margin of the Powder River Basin (PRB) in Wyoming and Northwestern Margin in Montana

	Composition of Groundwater in Clinker Recharge Areas (Average mg/l)					
Location in PRB	**TDS**	**Na**	**Ca**	**Mg**	**HCO_3**	**SO_4**
Eastern PRB, Wyoming	496–7268	25–574	48–499	20–548	84–802	70–4325
Northwestern PRB, Montana	320	14	52	27	285	31

TDS, total dissolved solids; Ca, calcium; Na, sodium; Mg, magnesium; HCO_3, bicarbonate; SO_4, sulfate; mg/l, milligram per liter.
Source: Adopted from Heffern and Coates (1999).

thick and permeable clinkers, which provides about 76 BCM of water storage from precipitation along the eastern and northern parts of the basin in Wyoming and Montana, respectively (Coates & Heffern, 1999).

These unconfined aquifers have high transmissivity and storativity of groundwater, which according to Heffern and Coates (1999) recharges the coals downdip. Table 8.5 shows the differences in the composition of groundwater in the clinkers in eastern margin of the basin in Wyoming and northwestern margin of the basin in Montana based on the work of Heffern and Coates (1999). The TDS concentrations of the groundwater generally are higher in eastern Wyoming compared to that in Montana. Also, the sodium, calcium, magnesium, bicarbonate, and sulfate concentration are higher in Wyoming than in Montana. These chemical parameters probably were derived by dissolution of soluble minerals from the clinkers and transmitted to surface drainages as well as into the adjoining subsurface coals via exogenous water.

A component of Brinck et al. (2008) biogeochemical model is the potential contribution of exogenous source of water sourced from overlying and laterally juxtaposed sandstone aquifers (Figure 8.15). The groundwater in sandstone aquifers has been in residence for a long time since their deposition, and has time to leach soluble minerals. For example, according to Rice et al. (2002), sodium enrichment in groundwater may occur by dissolution of plagioclases and cation exchange of calcium and magnesium for sodium in clay minerals. Alteration of other feldspars and volcanic ash can enrich sodium according to Lee (1981) as shown by the following (Dahm et al., 2011):

$$9H_2O + 2H_2CO_3 + \underset{\text{(albite)}}{2NaAlSi_3O_8} \Rightarrow 2Na^+ + 2HCO_3^- + 4H_4SiO_4 + \underset{\text{(kaolinite)}}{Al_2Si_2O_5(OH)_4}$$

Where H = hydrogen; O = oxygen; C = carbon; Na = sodium; Al = aluminum; and Si = silicon.

However, Brinck et al. (2008) reported that the low concentration of dissolved silica in the coal aquifer suggests that alteration of feldspars and other silicates is not an important factor. This conclusion does not negate the potential of dissolved solution containing enriched sodium from alteration of feldspars in overlying sandstones. The fluvial channel sandstones in the Fort Union and Wasatch Formation in the PRB contain abundant feldspars, which were unroofed from the Bighorn Mountains uplift (Whipkey, Cavaroc, & Flores, 1991). Another factor that may have been overlooked is the presence of volcanic ash partings in the coals as well as abundant montmorillonite–illite clay minerals derived from volcanic ash interbedded with mudstones and siltstones in the Fort Union Formation (Thorez, Flores, & Bossiroy, 1997), which are potential internal and external sources of dissolved sodium from their subsurface alteration. Also, Upper Fort Union sandstones commonly contain montmorillonite–illite clay minerals derived from volcanic ash. Calcite, dolomite, and hematite, which serve as chemically precipitated cement in sandstones, also undergo dissolution at depth by formation water containing carbonic acid (McBride, 1977). Thus, this geochemical modification of groundwater in sandstones at depth may explain the evolving alteration observed in the PRB by Bartos and Ogle (2002). It is a well-accepted concept that the water–rock interactions within the coal aquifers and between adjoining rocks contributed to dissolved solids in the co-produced water and exchanges of cations; however, cation exchanges from overlying sandstones are not fully understood due to lack of more detailed interdisciplinary investigations by coal petrologists, sedimentologists–petrologists, hydrogeologists, and geochemists.

Role of Recharge to Water Composition

Investigations of the role of recharge of surface and near-surface water to groundwater in coalbed gas producing basins have yielded mixed results. The results range from none during the human life time to as old as the time of the deposition of the coals (e.g. 60 mya) (Riese et al., 2005; Snyder et al., 2003). Various studies have documented that the composition of co-produced water is freshest near recharge areas compared to downgradient as indication of water recharge (Dahm et al., 2011; NRC, 2010; Van Voast, 2003). The water recharge from the surface dissolves salts and mixes with coal reservoir groundwater along its flow path, which becomes increasingly saline away from the recharge area. Dahm et al. (2011) reported that water near recharge areas reflects composition of meteoric water with elevated dissolved organics, high total organic carbon, and low TDS concentrations where residence is short. Far removed from the recharge areas, the composition of the groundwater reflects the influence of the original depositional waters particularly in brackish and marine environments (Van Voast, 2003). However, the composition can also reflect exchanges of mineral ions and precipitation, which reduce dissolved calcium, magnesium, barium, and sulfate concentrations (Dahm et al., 2011). Away from recharge areas, measured either geographically or depth, mixing with coal reservoir waters generally increases TDS, alkalinity, sodium, and chloride concentrations.

Coalbed gas waters of the five major basins listed in Table 8.4 are predominantly sodium bicarbonate waters with lesser amounts of chloride and TDS values are shown to range widely (Otton, 2005). For comparison, the EPA secondary drinking water standard is 500 mg/l TDS and seawater is approximately 35,000 mg/l TDS. The PRB waters exhibit modest TDS concentrations overall, which is one of the reasons why PRB water seems suited for beneficial use. In other basins, the quality of the water is not as good and beneficial use is more problematic. The San Juan Basin waters, for example, have highly varied TDS, and most operators simply reinject it. Likewise, most operators reinject Uinta Basin waters. In the Black Warrior Basin, the TDS concentrations are bimodal and operators are allowed to release the lower TDS waters to surface streams.

Van Voast (2003) differentiated the composition of the co-produced water between the Western Interior and Southern Appalachian coal basins (Figure 8.16). Co-produced water in all the coal basins (e.g. Black Warrior, San Juan, Raton, Powder River, Piceance, and Uinta basins) shows composition affected by original depositional settings of coals as well as mixing and flushing of groundwater from recharge areas to downgradient. According to Van Voast (2003), the major constituents of co-produced water of the Black Warrior, San Juan, Raton, Piceance, and Uinta basins are sodium and

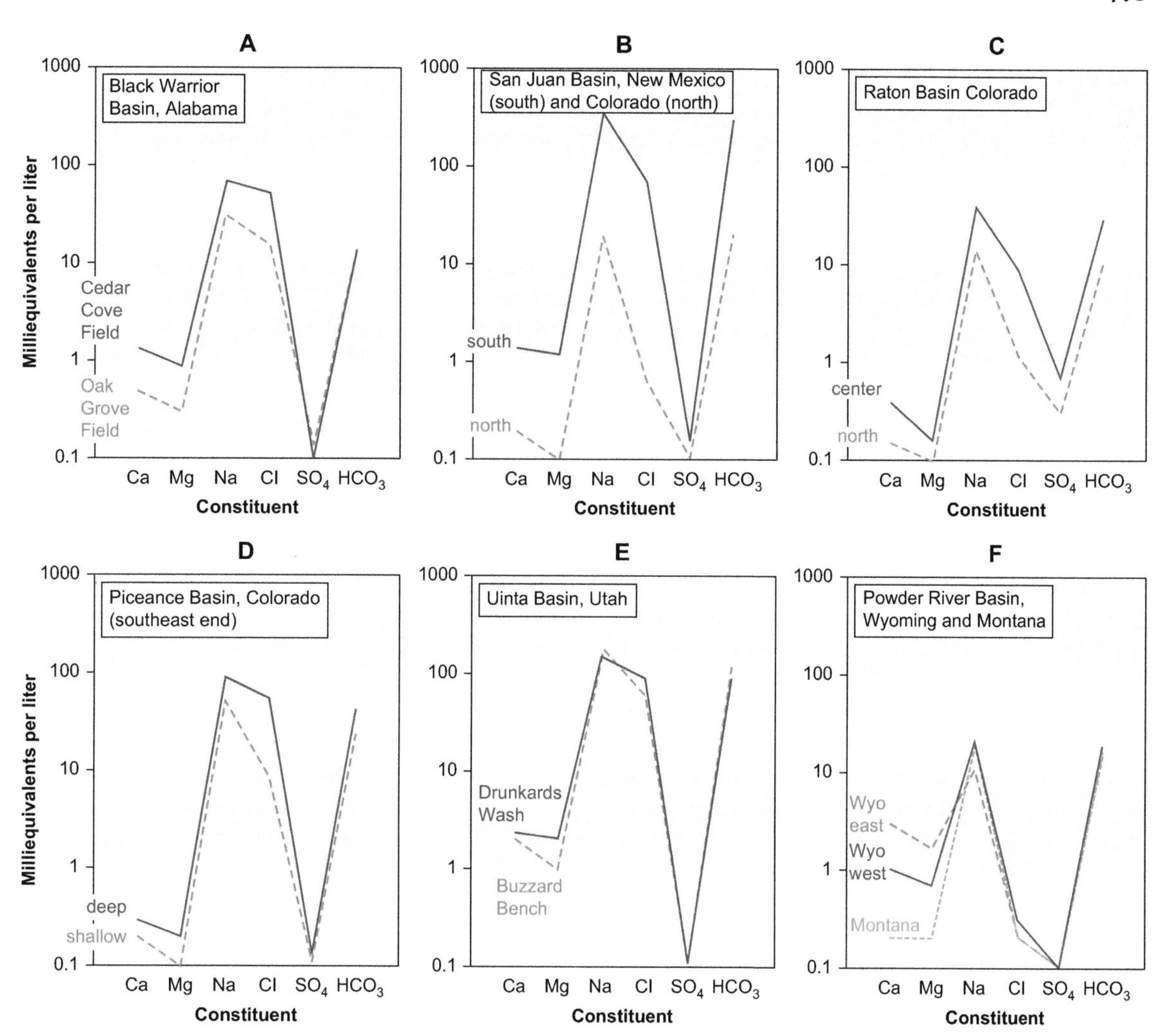

FIGURE 8.16 Diagrams A–F of the chemistry of co-produced waters from United States Western Interior and Southern Appalachian coal basins. Ca, calcium; Mg, magnesium; Na, sodium; Cl, chloride; SO_4, sulfate; HCO_3, bicarbonate. *Source: Adopted from Van Voast (2003).*

chloride, which reflect marine-transitional influence of the depositional settings of the coal beds (Figure 8.16(A–E)). However, the sodium and chloride concentrations of the co-produced water in the Raton Basin are considerably lower than the other Cretaceous coal-forming basins (Figure 8.16(E)). The difference in constituents may be explained by the coalbed gas producing coal beds, which were primarily formed in alluvial plain bogs (e.g. freshwater) landward to coastal plain mires (Flores, 1984). In contrast, the sodium and chloride of the co-produced water in the PRB include the lowest concentrations attributed to continental deposition of the coals (Van Voast, 2003), which is interpreted as deposited in intermontane alluvial plain bogs (Flores, 1986; Figure 8.16(F)).

Variations in composition of co-produced water in the coal basins can be attributed to mixing with and flushing by groundwater from recharge areas. For example, the calcium, magnesium, sodium, chloride, sulfate, and bicarbonate concentrations in the San Juan Basin are higher downgradient in the southern part compared to the upgradient in northern part where the groundwater recharge area is located (Kaiser, Swartz, & Hawkins, 1994; Van Voast, 2003; Figure 8.16(B)). Also, mixing and flushing of groundwater are demonstrated in the upgradient or near-surface parts of the Raton and Piceance Basins (Figure 8.16(C,D)). The low concentration of chloride supports freshwater origin of the co-produced water in the PRB (Figure 8.16(F)). Van Voast (2003) interpreted the low concentration or reduction of sulfate as (1) requirement reaction for the methanogenesis for generation of methane, (2) associated condition in thermogenesis, and (3) improved enrichment of dissolved bicarbonate resulting in depletion of calcium and magnesium.

The TDS of coalbed gas water range from fresh (160 mg/l) to saline (100,000 mg/l) and vary among and within basins (Table 8.4). The high variability and high content (e.g. 100,000 mg/l) of TDS in US coal basins may be due to inconsistent method of calculation (Van Voast, 2003). For example, according to Van Voast (2003) typical TDS in the San Juan Basin is 17,000–19,000 mg/l, Raton Basin about 2,200 mg/l, and the western PRB about 1,200 mg/l. For comparison, the recommended TDS limit for potable water is 500 mg/l, and for beneficial use such as stock ponds or irrigation, the limit is 1000–5000 mg/l. Average seawater has a TDS of about 35,000 mg/l. The TDS of the water is dependent upon the depth of the coal beds, the composition of the rocks surrounding the coal beds, the amount of time the rock and water react, and the origin of the water entering the coal beds. In general, most coalbed gas co-produced water is of better quality than waters produced from conventional oil and gas wells.

Major cations and anions are the most typical parameters measured to assess water quality and to determine water type. Of the four minor inorganic species listed, ammonium is significant because of nitrogen limits to surface and ground waters; the other minor species can impact plant growth. Several of the listed dissolved trace elements have primary drinking water standards established by the USEPA (2012a). Other impacts of the dissolved constituents include effects on water transmission equipment, cleanup technologies, pastureland or cropland being irrigated, and aquatic life; many aquatic life standards are lower than drinking water standards. If produced water is to be treated for drinking water, all these components of coalbed gas waters need to be evaluated, as are many of the same constituents in conventional produced waters.

In an ion exchange system (IX), wastewater passes through a system that contains a material (typically a resin) to extract and absorb specific constituents (USEPA, 2012a,b). In a typical setup, a feed stream passes through a column, which holds the resin. Pollutants absorb onto the resin as the water moves through the system. Eventually, the resin becomes saturated with the targeted pollutant requiring regeneration of the resin. A regenerant solution then passes through the column. For cation resins such as for sodium and metals, the regenerant is an acid, and the hydrogen ions in the acid remove the absorbed pollutant from the resin. The sodium and metals concentrations are much higher in the regenerant than in the feed stream. Therefore, the ion exchange process separates the sodium from the water and results in a concentrated brine stream and a treated produced water stream. Because the salt content of the produced water has been reduced, the treated stream can be discharged to surface waters or beneficially used (USEPA, 2010).

BOX 8.2

WHO OWNS THE WATER RIGHTS?—CONCEPTS AND MISCONCEPTIONS

Bushong (2006) raised an astute observation that in many Colorado coal basins, there are more water rights than water! The same observation can be made for many coal basins in the Western Interior of the United States in particular and worldwide especially in arid and semiarid regions. These areas and humid regions affected by droughts due to microclimate change have brought to the forefront the value of groundwater and the issue of its extraction with coalbed methane development.

NRC (2010) outlined the legal framework of fundamental water rights in the United States, which govern the management of co-produced water of conventional and unconventional hydrocarbons. The readers are referred to this reference and others for more informative details and nuances of who owns treated co-produced water rights (Bushong, 2006; Mudd, 2012). Because there are no National water rights, the states are left to administer them based on two systems: riparian for most of the eastern United States and prior appropriation water rights similar to rights to real property for the western United States (NRC, 2010). Each state interprets the prior appropriation of water rights based on culture, geography, custom, legislation, and law. Authorization for use of water rights is issued as permits by individual states. An important facet of appropriation water rights is the definition of beneficial uses such as for agricultural, irrigation, industrial applications (e.g. mining and petroleum), stock watering, wildlife and recreational activities, as well as domestic and municipal water and power supplies.

For example, for an industrial appropriator of water rights like a coalbed gas company, it is obligated to use only a certain amount of water for useful purpose and beneficial use. After the water is used for beneficial purposes, every co-produced wastewater is reverted to streams or reinjected. According to NRC (2010), the concept of "in-stream flow" is applied to the stream recycling process and is defined as the amount of water flow in a natural stream necessary to maintain in-stream values at satisfactory level. In-stream values are concerned with co-produced water pertaining to protection of fish and wildlife habitats; ecosystems, water quality resulting from waste integration; and recreation activities. Every state defines, practices, and implements in-stream values differently. The application of in-stream flow particularly to co-produced water of coalbed gas is management of discharge water to perennial and ephemeral streams. Thus, the treatment of co-produced water prior to release to these streams is important to their maintenance and preservation.

ENVIRONMENTAL IMPACTS AND CONCERNS ON CO-PRODUCED WATER DISPOSAL

During coalbed gas development, the co-produced water is directly discharged or disposed of into surface waters, applied to the land, stored in constructed impoundments and wetlands, or reinjected to the subsurface. Receiving surface waters range from dry washes to intermittently flowing ephemeral and continuously flowing perennial rivers, streams, and creeks as well as natural lakes and ponds (Figure 8.17). Land applications include agriculture, irrigation, and soil modification.

FIGURE 8.17 Photograph of ephemeral creek receiving co-produced water via an outfall in the Powder River Basin in Wyoming.

Constructed impoundments include artificial reservoirs and associated wetlands within and marginal to impoundments as well as in standing bodies of surface waters. Both surface application and unlined impoundments can be source for shallow groundwater recharge. Subsurface reinjection includes infiltration into sandstone and coal aquifers within, above, and/or below producing rock formations. Another disposal technique is enhanced evaporation with the use of sprinklers or "emitters" which spray the water up into the air. The use of these management strategies varies within and between basins as mainly influenced by the quality and volume of co-produced water as well as economics, technical considerations, and government regulations. The reader is referred to the Ingelson et al. (2006), RPS Australia East Pty Ltd. (2011), and USEPA (2010) reports, which conducted an excellent in-depth review of the environmental impacts and concerns of disposal of co-produced water of coalbed gas in the United States and Australia, respectively.

Surface Waters

The environmental impacts of direct discharge or disposal of co-produced water into surface waters concern aquatic organisms and vegetation, nonaquatic wildlife, ecosystems and habitats, water quality due to changes in salinity, sodicity, TDS, toxicity, temperature, and pH. The primary concerns include the alteration of composition of surface waters from co-produced water constituents in terms of salinity, sodicity, and toxicity, which directly affects aquatic plants and organisms and nonaquatic animals. Alteration in salinity of the surface waters can change communities from freshwater to saline water-tolerant plant species (Keith et al., 2003; Rawn-Schatzinger, Arthur, & Langhus, 2004). The effects of dilution by introduction of large volume of water and/or more or less saline water can cause alteration of habitats of aquatic species of the surface waters (Clearwater, Morris, & Meyer, 2002; Stanford & Hauer, 2003). Other constituents of co-produced water that are considered detrimental to aquatic plants and organisms include ammonia, bicarbonate, boron, cadmium, chloride, chromium hydrogen sulfide, selenium, and TDS concentrations (ALL Consulting, 2003; Davis, Zale, & Brhamlett, 2008; Farag & Harper, 2012; Fisher, 2001; Rice, Ellis, & Bullock, 2000; USDOI, 2005).

Exposure to high concentrations of sodium bicarbonate reduces survival of fish (e.g. fathead minnows and white suckers) and impacts freshwater streams and rivers by adversely affecting the ability of the fish to control their ion uptake (USGS, 2006). The increase of variations of water quality from co-produced water discharges to receiving streams, particularly with regard to pH, potentially can cause physiological stress to aquatic organisms (O'Neil et al., 1989). Streams receiving co-produced water with high concentrations of metals such as selenium, chromium, cadmium, aluminum, iron, manganese, lead, and copper can impact fish and birds (USDOI, 2005). High concentrations of selenium can accumulate in both fish and migratory aquatic birds causing low reproduction, high mortality rate, and deformity of embryos (Ramirez, 2005).

Environmental impacts and concerns also focused on the change of the morphology of the surface water due to change in flow, turbidity, sedimentation, and erosion of the rivers, streams, and/or creeks. Discharge of water volume to surface waters can cause modification of aquatic habitats. Water discharged into rivers, streams, and creeks increases flow and can increase channel-bank erosion, and generates significant amounts of suspended sediments as well as turbidity, which in turn can affect aquatic organisms (Arthur, 2001; Arthur et al., 2001; ALL Consulting, 2003). In addition, increased channelized flow with accompanying floods can destroy vegetation along riparian areas and stream-bed erosion impacts the availability of food and habitat for reproduction for bottom dwelling aquatic organisms (ALL Consulting, 2003; Davis et al., 2008; Rawn-Schatzinger et al., 2004; Regele & Stark, 2000). Finally, the changes in water flow cause erosion, provide sediment load, and increase turbidity in the surface water just favoring invasion of species that can outcompete native species exposed to the harsher environments (Bonner & Wilde, 2002; Davis et al., 2008; Gradall & Swenson, 1982).

Impoundments

Surface impoundments are constructed in on-channel and off-channel areas. Off-channel surface impoundments are excavated disposal pits constructed with sloping sides and bordered by elevated berms. Usually the impoundments are constructed nearby production areas to minimize costs of piping the water from the coalbed gas wells. On-channel impoundments (Figure 8.18) are situated within designated water features (e.g. perennial and

FIGURE 8.18 Photograph of on-channel impoundment in an alluvial plain in the Powder River Basin. Co-produced water is infiltrated and/or evaporated in the pit. *Source: Photograph courtesy of Cindy Rice.*

ephemeral streams, creeks, lakes, ponds, and dry washes) or within the alluvial floodplain of the water features as defined on a US Geological Survey 1:24,000 scale topographic map (WDEQ, 2006; WDEQ, 2007). Off-channel impoundments are not situated within the water features (e.g. perennial and ephemeral streams, dry washes, lakes, and ponds) and are constructed more than 152 m from the outermost alluvial floodplain deposits of any stream, river, or creek as defined by the US Geological Survey topographic map. Impoundment pits mainly serve as evaporating and/or infiltrating co-produced water depending on whether lined or unlined, respectively, as well as watering for livestock. In addition, impoundment pits are commonly used to minimize or eliminate co-produced water discharge or disposal into surface waters. The impoundment pits can also be used to manage co-produced water flow from the coalbed gas wells to an outfall (Figure 8.19).

In the PRB, Wyoming, more than 3000 impoundment pits receive co-produced water of which approximately 2500 are on-channel pits and about 200 are off-channel pits designed to store all co-produced water without discharge (WDEQ, 2006; WDEQ, 2007). Wyoming Department of Environmental Quality

FIGURE 8.19 Photograph of outfall, which is the discharge outlet at the end of a pipe, releasing co-produced water from several coalbed gas wells. Iron-stained pebbles and cobbles are caused by oxidation of co-produced water. That is, high concentrations of reduced species of iron and manganese in the co-produced water upon contact with oxygen and light on the surface oxidize and precipitate as iron and manganese oxyhydrates and oxides (Rice et al., 2002).

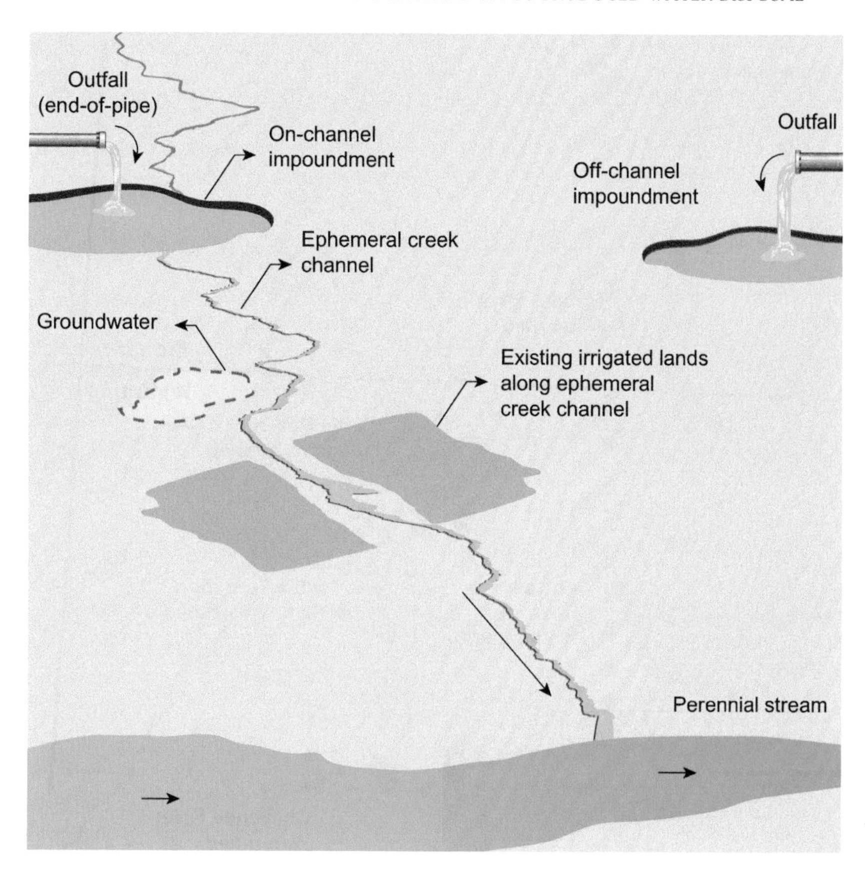

FIGURE 8.20 Diagram showing landscapes where impoundments are constructed in relation to drainages and surrounding areas. *Source: Adopted from WDEQ (2011).*

(WDEQ, 2011) permitting requirements for construction of impoundments apply to drainage areas (e.g. ephemeral creeks, perennial streams) and existing irrigation uses (Figure 8.20). Consequently, the WDEQ permitting guidelines cover all impoundments located in major hydrologic basins or watersheds of drainages basins (e.g. Powder River and Tongue River) in the PRB (Figure 8.21).

Potential environmental impacts result from leakage of contaminated co-produced water beneath the impoundments into alluvial valley sediments, soils, bedrock, and near-surface aquifers. The depth of groundwater levels in the PRB, Wyoming varies from 7 to 34 m from the surface (WDEQ, 2007). On-channel impoundments constructed on the alluvial deposits can directly communicate with streams through these deposits. Thus, any co-produced water leakage from the impoundment pit infiltrates back-and-forth into the stream, the water level of which is often influenced by changing climatic condition (Figure 8.22) (ALL Consulting, 2003). Potentially, water infiltrates the impoundment at a high rate when the stream is at high level (e.g. spring) and water infiltrates the stream at a low rate during low level (e.g. winter). The off-channel pits are also subject to leakage despite being located off of alluvial deposits

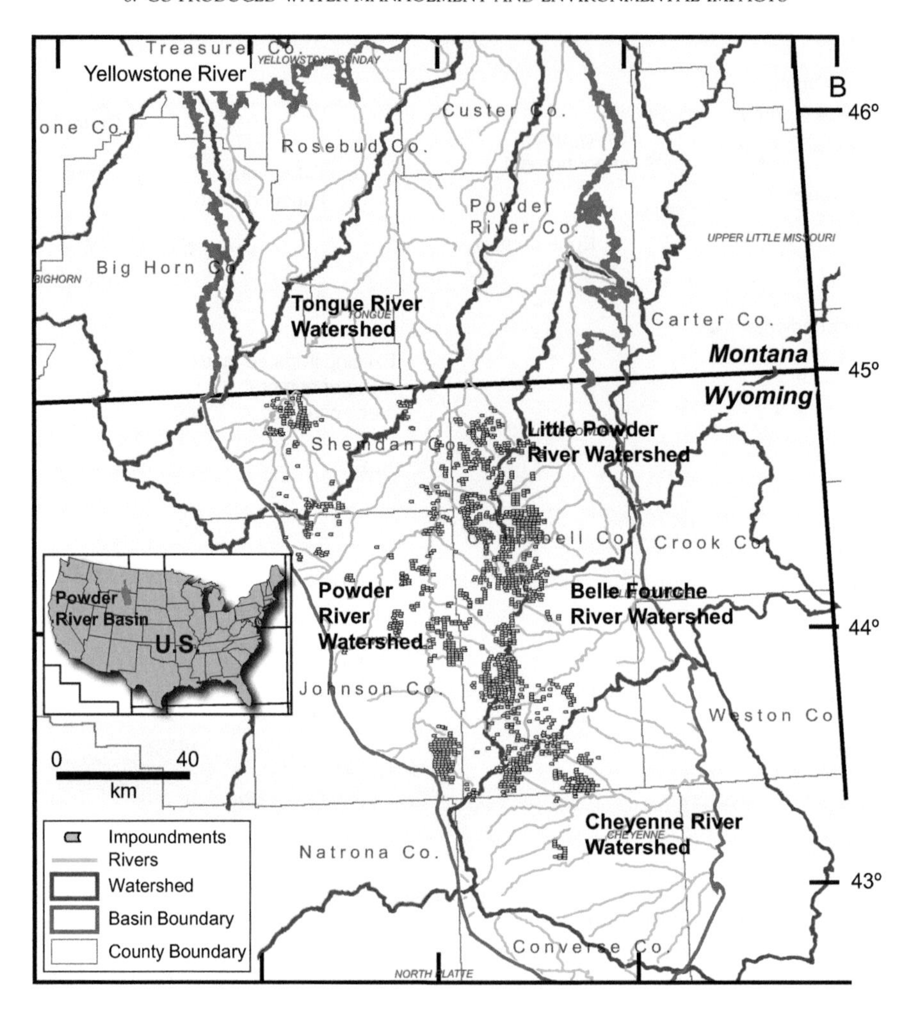

FIGURE 8.21 Location of existing impoundments in watershed areas of the Belle Fourche, Powder, Little Powder, and Tongue Rivers, which drain northward to the Yellowstone River and to the Powder River Basin, Wyoming and Montana. Cheyenne River drains east–northeast to the Missouri River. *Source: Modified from ALL Consulting (2008).*

often on bedrock. In most cases of leakage, the environmental impact involved leaching of salts to shallow aquifers. In both cases, the impoundment discharges can degrade water quality in receiving waters (ALL Consulting, 2003). Lipinski (2007) suggested that because of these problems, impoundments located in the alluvial valley should be as far away from rivers. Evapotranspiration by dense vegetation in alluvial point bars with shallow water table may result in major vadose zone salt accumulation, which upon leakage from impoundments co-produced water will dissolve, creating TDS plumes (Lipinski, 2007).

According to ALL Consulting (2003), extended water infiltration into soils and bedrocks results in the following potential reactions: (1) dissolution of minerals of host rock accompanied by changes in the quality of the infiltrating water; (2) mineralization of soils

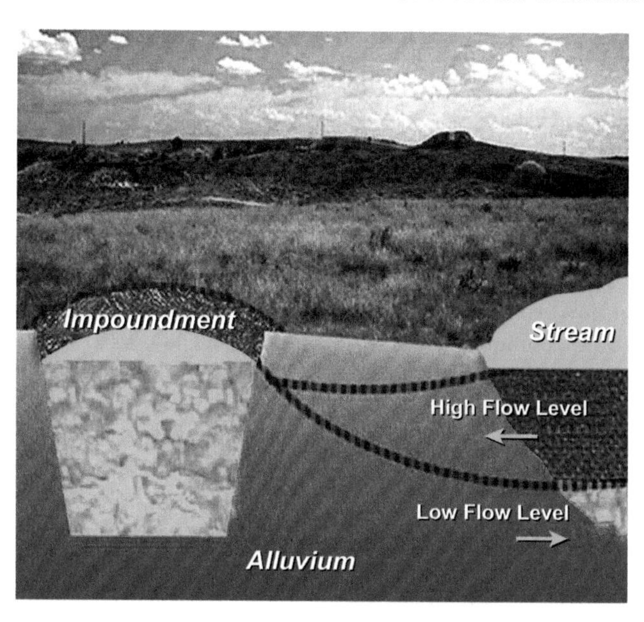

FIGURE 8.22 Diagram showing an on-channel impoundment interacting with season-changing stream level. Potential back-and-forth communications of water between the impoundment and stream with season changes produce two-way infiltration routes. *Source: Adopted from ALL Consulting (2008).*

with precipitation of salts, sulfates, and carbonates; (3) mixing of infiltrating water and shallow groundwater, which may improve, degrade, or have no substantial impact on the quality of the groundwater; (4) alteration in rate of infiltration with precipitation of minerals and dispersion of clay minerals in the soil, which in turn affect the operation of the impoundment; (5) redirection of flow of infiltrating water from vertical and horizontal to radial once a barrier is built; and (6) prospective extraneous issues (e.g. environmental) may affect the location of an infiltration system, which in turn may affect the impacts of the infiltration water. Understanding the chemical and mineral composition of the bedrock and mineralogy of the clays in the soil, chemistry of co-produced water in the infiltration system, and chemistry of the shallow groundwater and properly siting impoundments may mitigate most of the issues.

Chemical parameters of concern are dissolved sulfates from gypsum, selenium, and manganese as well as high concentrations of TDS beneath impoundments (WDEQ, 2007). In addition, evaporation of co-produced water especially in arid and semiarid climate exacerbates precipitation and concentration of pollutants (e.g. salts, heavy metals, or trace elements) decreasing the quality of the water. These concentrated pollutants may be released into the subsurface through infiltration or discharged into surface water during overflow (ALL Consulting, 2003). In addition to subsurface leaching, evaporation from impoundments can further concentrate pollutants in co-produced water, decreasing the quality of water released to the environment through infiltration or discharge (ALL Consulting, 2003). According to Doherty (2007), impoundments can also create new habitats for mosquitoes in coalbed gas production areas. These new habitats can introduce new species or invasive species and cause proliferation of the species already in the area (Davis et al., 2008; Doherty, 2007). The proliferation of species such as West Nile virus carrying mosquitoes in impoundments can cause human and wildlife health risks (Doherty, 2007).

BOX 8.3

IMPOUNDMENTS: COST BENEFIT OR BAD LEGACY—FACT VS FICTION

In Australia and in the Powder River Basin (PRB), the preferred method of water disposal is impoundments/evaporation ponds and surface discharge because of low cost. However, in Australia, the days of evaporation ponds are numbered because they cover large surface areas and leave behind a legacy of salts and other residue deposits (RPS Australia East Pty Ltd, 2011). These reasons prompted the Queensland government to adopt a policy promoting discontinuation of evaporation ponds in favor of other alternative disposal system. The promotion of more costly disposal systems will affect the feasibility of coalbed gas projects. Approximately 7600 GL of co-produced water will be extracted and treated from the Bowen and Surat Basins over the next 25 years.

In the PRB, the impact of impoundments is related to infiltration and its legacy to groundwater. The Wyoming Department of Environment Quality (WDEQ) as of January 2007 has issued more than 3000 impoundments. Evaluation of coring data by WDEQ as of summer 2007 from 836 impoundments out of 1263 exempted monitoring groundwater characterized by <5000 mg/l, no groundwater (<1.9 l/min yield), and less than 0.25 ha m capacity. Of 121 investigated impoundment sites, 8–11 impoundments exceeded their trigger values of TDS, sulfate, selenium, and arsenic (WDEQ, 2007).

Healy et al. (2008) studied one of the impoundments with co-produced water that infiltrated the groundwater prior to closure and reclamation. Large amounts of chloride (12,300 kg) and nitrate (13,500 kg) were released into shallow groundwater by infiltrating water, which accumulated in the underlying bedrocks (Eocene Wasatch Formation). Nitrate was more readily flushed from the bedrocks than chloride. Also accumulated are TDS concentrations, which exceeded 100,000 mg/l as a plume in the subsurface. TDS concentrations were greater (about 2300 mg/l) than those in the co-produced water from coalbed gas wells and in the shallow groundwater (about 8000 mg/l). Sulfate, sodium, and magnesium are the dominant ions in the plume. Thus, in both Australian and PRB, legacies are left behind on the surface and subsurface during the interim and long term after coalbed gas development.

Ecological Impacts: Wetlands vs Invasive Species

An unintended consequence of land application of co-produced water is the creation of wetlands with constructed impoundments as well as in natural lakes and ponds supplied by significant amount of water containing high TDS, sodium, and salts (Figure 8.23). A wetland is a sink in the landscape, which is susceptible to invasions of plants and organisms because it accumulates organic debris, sediments, water, and nutrients facilitating, creating, and accelerating growth and incursion of opportunistic species (Zedler & Kercher, 2004). Instability to wetlands, such as dispersal of materials that propagates organisms, salt influx, and hydrologic changes such as an influx of a large volume of chemically rich co-produced water, creates opportunities for

FIGURE 8.23 Photograph of a wetland-covered impoundment drowned by co-produced water in otherwise dry landscape in the Powder River Basin. Plant species (e.g. grasses and shrubs) along the wetland margin are adaptable to constant saturation of water.

wetland opportunists especially in arid and semiarid climates. Zedler and Kercher (2004) suggested that wetlands have the tendency to be dominated by invasive species, which is debatably an effect of the cumulative impacts associated with landscape sinks. Wetlands fed by co-produced water tend to have many invasive species (Galatowitsch, Anderso, & Ascher, 1999). Abundance of alien species may be related to road density, which is unique in coalbed gas development areas with construction of access roads and pipeline networks, indicating that landscape position interacts with dispersal routes and disturbances to facilitate establishment of new plants and organisms.

An example of infestations of invasive species via dispersal routes is the proliferation of noxious weeds in coalbed gas development areas. Seeds of invasive weeds are transported and spread by vehicles, persons, and construction and reclamation equipment. Native vegetation unable to compete with exotic weeds are either displaced or eliminated. Mitigation of exotic species include the following U.S. Bureau of Land Management protocol: (1) reseeding development sites promptly; (2) cleaning of equipment; (3) minimizing soil disturbances; (4) use of weed-free mulch and hay for reclamation; (5) use of livestock to control outbreaks of certain weeds; (6) use of approved herbicides; and (7) use of weed control instruction.

According to Gopal (1999), wetlands are best known as nature's kidney for being able to filter toxins and pollutants as well as absorbing nutrients. Opportunistic plant species thrive despite the harsh ecological conditions and tolerant species benefit the wetlands. Negri et al. (2003) examined the possible effectiveness of a plant-based system to optimize metal uptake and reduce the volume of water requiring treatment (e.g. co-produced water) by concentration of the salts. Settle et al. (1998) used halophytic species (e.g. salt marsh grass, *Spartina*) to reduce the volume of co-produced water to concentrate salts. Halophytes are salt- and sodium-tolerant plants, which grow under saline conditions, either by accumulating high-level salts in their tissues or by excluding salts from plant uptake. Settle, Mollock, Hinchman, and Negri (1998) used natural plant evapotranspiration and salt accumulation to reduce the volume of co-produced water requiring treatment or disposal while simultaneously reducing the total salt load by plant accumulation. The proposed process is a "bioreactor" which incorporates salt marsh processes with salt tolerant wetland plant species in a hydroponic growth environment.

Reinjection

A reinjection well is a borehole, which distributes fluids underground into porous rock formations or reservoirs (e.g. sandstones, limestones, and coals) below the shallow soil layer (USEPA, 2012b). The fluids are in the form of co-produced water, wastewater, brine or salt water, and water mixed with chemical additives, which involve treatment to suitable standards prior to reinjection. The purposes of reinjection wells include storing carbon dioxide for a long term (e.g. carbon sequestration), disposal of wastewater enhancing oil production, and preventing saltwater intrusion (USEPA 2012b). Co-produced water reinjection, in particular, refers to storage or disposal of underground by injection of water into a reservoir rock or formation (Figure 8.24). Contamination of groundwater aquifers is a major issue in any reinjection of contaminated co-produced water. Thus, the placement of the reinjection well above, within, or below the groundwater aquifer as well as the design of the well (e.g. multiple casings) plays a major role. Ideally, positioning the reinjection well below the groundwater aquifer provides the least risk of contamination.

The Safe Drinking Water Act guarantees that underground reinjection does not jeopardize sources of drinking water (USEPA, 2004). Underground source drinking water (USDW) is defined as groundwater aquifers which contain <10,000 mg/l of TDS and supply groundwater for public water system. Co-produced water usually is below 5000 mg/l of TDS, which is considered the threshold for safe drinking water. Under the Federal Underground Injection Control (UIC) Program, there are six well classifications for storage or disposal according to the type of fluid and location (USEPA, 2005; Table 8.6). According to USEPA (2010), coalbed gas operators can use the Class II wells to manage co-produced water, which improves oil and gas recovery and stores liquid hydrocarbons (Figure 8.25). The reinjection of co-produced water with high salt content or other impurities in Class II wells prevents the contamination of the soil and underground USDW. However, the

FIGURE 8.24 Photograph of coupled coalbed gas production and reinjection wells in the eastern Greater Green River Basin in Wyoming, United States.

TABLE 8.6 Well Classification and Description for Storage and Disposal According to the Types of Fluids Identified by the Federal Underground Injection Control Program

Well Class	Description of Reinjection Well	Inventory
I	Wells are used to reinject hazardous waste, industrial nonhazardous liquids, or municipal wastewater underneath lowermost formation containing underground source drinking water supply.	680 wells
II	Wells are used to reinject brines and other fluids associated with oil and natural gas production and liquid hydrocarbons for storage.	172,068 wells
III	Wells associated with solution mining (e.g. extraction of uranium, copper, and salts) reinjected beneath the lowermost underground source drinking water supply.	22,131 wells
IV	Wells are used to reinject hazardous or radioactive wastes into or above underground source drinking water supply. These wells are banned unless authorized under a federal or state groundwater remediation projects.	33 sites
V	Any reinjection well that is not included in classes I–IV. Also wells are used to reinject nonhazardous liquids into or above underground source drinking water supply and usually shallow, on-site disposal systems. There are some class V wells that reinject below underground source drinking water supply.	400,000–650,000 wells
VI	Wells are used to reinject carbon dioxide for long-term storage or sequestration in geologic formations.	6–10 commercial wells by 2016

Source: Adopted from USEPA (2005) and USEPA (2010).

normally low TDS concentrations of the co-produced water also reduce the risk of groundwater contamination. The ability of the rock formation to receive and store the co-produced water is mainly controlled by rock lithology, permeability, porosity, and geological structure as well as the presence of an impermeable cap rock to prevent leakage and location below saline aquifers to prevent contamination of USDW. A requirement by federal and state governments is reinjection of co-produced water into the original formation or into formations that are similar in properties to those from which it was extracted (Zimpfer, Harmon, & Boyce, 1988).

Class II wells can be newly drilled holes, existing coalbed gas wells that are marginally producing, plugged, shut-in, and abandoned wells, and dry holes depending on the economics and well conditions. The primary objective of using existing coalbed gas wells is the reinjection of co-produced water in the same coal and associated sandstone aquifers where the water was originally extracted. The quality and pretreatment of the co-produced water plays a major role in the reinjection relative to the coal aquifers. USEPA (2012b) suggested the following pretreatment for injection: (1) removal of iron and manganese by precipitation because these cations form oxides upon exposure to air, which may clog the well; (2) settling tanks with splash plates aerate the produced water, which oxidizes iron and manganese to insoluble forms that can precipitate in the tank; and (3) biocides may also be added to the produced water prior to injection to control biological fouling.

The reinjection of co-produced water of coalbed gas has played different roles compared to other modes of storage and disposal in various

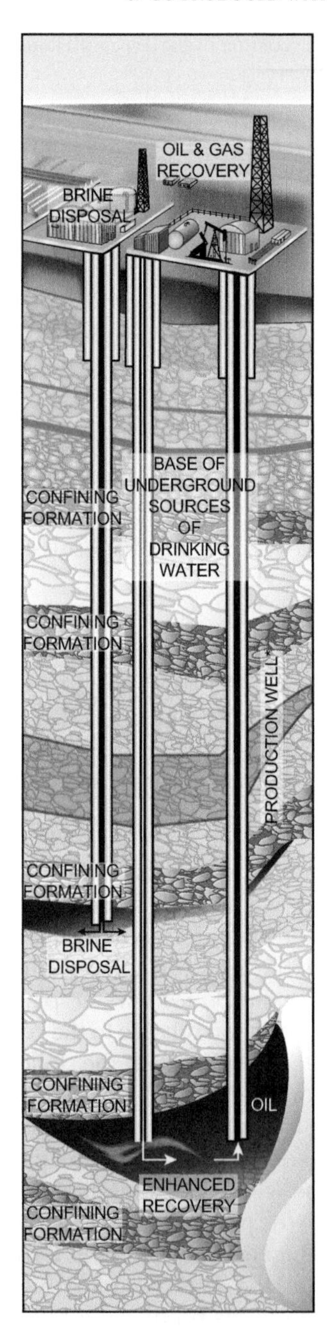

FIGURE 8.25 Class II well for reinjection of co-produced water to improve oil and gas recovery. *Source: Adopted from USEPA (2010).*

coal basins in the United States. For example, NRC (2010) reported reinjection of co-produced water in the San Juan Basin is 99.9%, Uinta Basin 97%, Raton Basin 28% in Colorado and 100% in New Mexico, Piceance Basin 100%, and PRB 3% in Wyoming. Where reinjection is not a major player of co-produced water storage and disposal such as in the PRB in Wyoming where majority of the coalbed gas is produced, impoundments comprise 64% and direct discharge to streams is 20% of the discharge (NRC, 2010). In the Montana part of the PRB with the least amount of coalbed gas production, direct discharge to streams is 61–65%, surface or subsurface irrigation is 26–30%, impoundment is 5%, and industrial dust control is 4–5%. In the Black Warrior Basin, stream surface disposal is mainly used because deep-well reinjection is not possible due to lack of permeable subsurface formations (Halliburton, 2007).

Treatment Management Practices of Co-Produced Water

Treatment management practices of co-produced water play a major role in the economics of coalbed gas development as storage and disposal costs are critical to project investment decisions as well as environmental and social impacts. The treatment management practices have to account for the varying degrees of water quality and volumes from potential coalbed gas development projects. The quality of the co-produced water varies from potable as in the PRB to saline as in the San Juan and Black Warrior Basins. The volume of water ranges from a few hundreds of liters per well per day in the Powder River and Raton Basins to much lesser amounts in San Juan Basin. If the coalbed gas project is located near coal mining areas, assume lesser volumes of co-produced water and desorbed gas resulting from the dewatering of the coal mines as compared to areas far removed from mining activities. In the past, lack of awareness of large volumes of water produced from coalbed gas wells resulted in inconsistent management of water flow and adapting to disposal regulations. Based on these variables, several factors need to be considered in establishing treatment management of co-produced water regarding the quality, quantity, decline rate, proximity of location to coal mines, and relevant regulations.

Coalbed gas operators in the United States coal basins use a variety of methods to manage, store, treat, and dispose co-produced water based mainly on the quality, availability, landscape compatibility, liability, and required regulations (Table 8.7). The number of projects, which employed co-produced water management, indicate preferential and multiple practices for different basins in the United States as surveyed by USEPA (2010) and summarized in Table 8.7. Co-produced water is normally produced and collected within different coalbed gas projects by operators in a basin. A coalbed gas project varies from a well to a group of wells, a lease to group of leases, and a coalbed gas unit leased by different operators but developed by a representative operator as required by unitized regulations. The co-produced water is gathered from each well through a pipeline system stored in a centralized system (e.g. water tanks), which is then treated and/or hauled to a commercialized disposal site, impoundment, or surface discharged. Each coalbed gas project managed the co-produced water from all the wells by using various storage, treatment, and disposal methods, which can vary from project to project.

Coalbed gas operators are required to treat co-produced water prior to disposal depending on the contaminants, use, and destination of the water. NRC (2010) and USEPA (2010) recognized potential technologies to treat coalbed gas co-produced water, which include aeration, chemical precipitation, reverse osmosis (RO), ion exchange, electrodialysis, and other

TABLE 8.7 Management Practices Utilized by Coalbed Gas Operators in Various Coal Basins in the United States

Basin	Direct Discharge	Land Application	Underground Reinjection	Evaporation–Infiltration Ponds	Haul to Commercial Disposal
Anadarko	0	0	14	0	6
Appalachian-Illinois	8	2	15	7	153
Arkoma	0	0	16	8	7
Black Warrior	13	0	0	1	2
Cherokee-Forest City	0	0	25	0	3
Green River	3	0	10	1	2
Powder River	149	29	31	145	4
Raton	3	0	0	3	2
San Juan	0	0	58	2	142
Uinta-Piceance	0	0	11	2	2

Source: Adopted from USEPA (2010).

technologies that are on pilot scale. Many of these treatment technologies are described by ALL Consulting (2003, 2005) in cooperation with the National Energy Technology Laboratory of the Department of Interior. The readers are referred to the ALL Consulting (2003, 2005), NRC (2010), and USEPA (2010) papers for more detailed discussion of various treatments for coalbed gas co-produced waters. The treatment technologies either reduce or remove the contaminants in the co-produced water, permitting beneficial use and discharge into surface waters. Figure 8.26 shows the potential management strategies and treatment pathway of co-produced water. Only the more commonly applied treatments are discussed below.

Aeration and Chemical Precipitation

Aeration of co-produced water is utilized to precipitate iron from the water, which reduces or removes the staining of the bed and coloring the receiving water of the stream (Figure 8.27). The process of aeration is demonstrated by a flow chart in Figure 8.28. The co-produced water is collected and passed through different systems of a series of treating aeration and sedimentation ponds, which in turn are either stored in a pond or directly disposed through a diffuser into surface water. The aeration process introduces oxygen into the co-produced water, which by nature lacks dissolved oxygen. This process is performed by mixing up air and co-produced

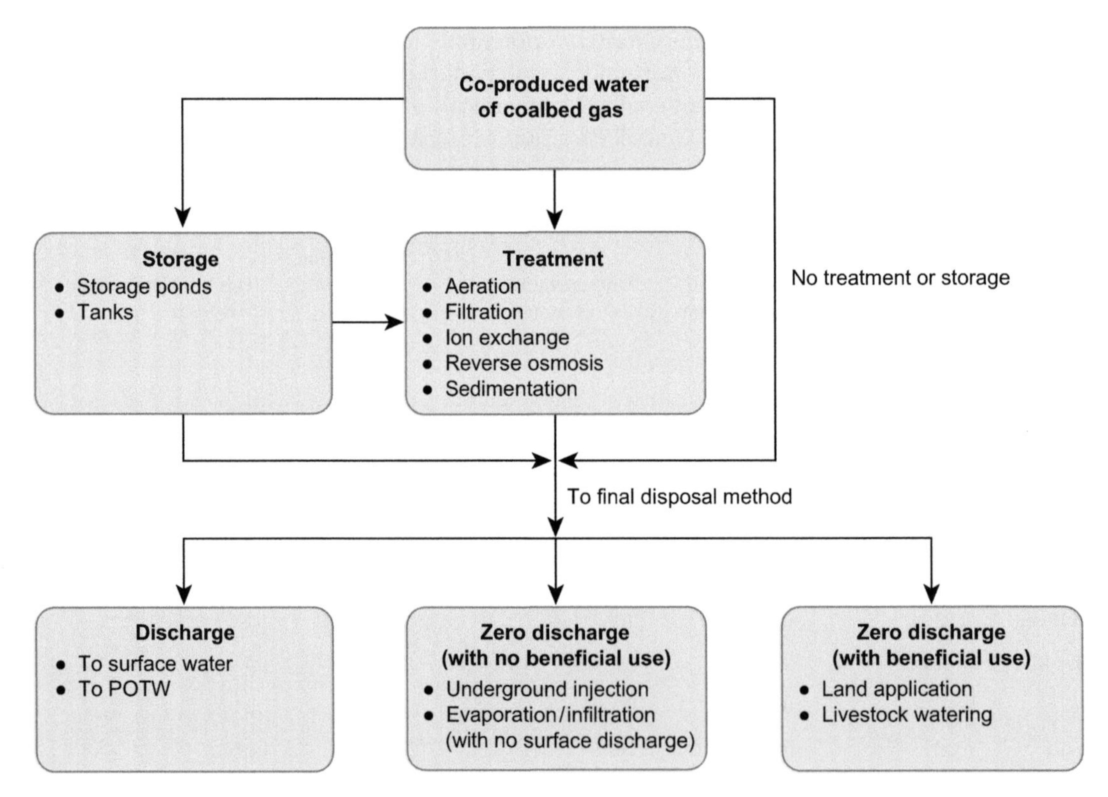

FIGURE 8.26 Flow chart showing potential management strategies and treatments of co-produced water. *Source: Adopted from USEPA (2010). POTW, publicly owned treatment works.*

FIGURE 8.27 Photograph of iron-stained water from oxidation of iron and manganese in the discharged co-produced water of an ephemeral creek in the Powder River Basin. *Source: Photograph courtesy of Cindy Rice.*

water by either injecting air into water, spraying water into the air, agitating the water mechanically, or allowing water to pass over an irregular water surface (USEPA, 2010). Halliburton (2007) described this treatment process as follows: (1) spraying the surface area of the water increases the absorption of oxygen from the air and (2) agitating provides repeated contact of the surface area of the water with the air permitting continual absorption of oxygen. The treatment technique uses spray nozzles, agitators, and bubble diffusers to aerate the water before discharge into the surface waters.

The series of treatments remove the contaminants from the co-produced water through oxidation, evaporation, or precipitation of suspended solids such as iron and manganese in the aeration and sedimentation ponds. The ferrous ion and ferric hydroxide precipitate after oxidation and flocculate gradually settling by virtue of specific gravity. Halliburton (2007), suggested that the following conditions be imposed in the treatment ponds to assist precipitation by gravity: (1) quiet settling waters, (2) >24 h detention time, (3) ponds of about 3 m depth that accumulate several years of precipitates before clean-up, and (4) baffled exit of water from the ponds. The discharge of the treated water from the ponds into the surface water flows over an effluent diffuser or riprap, which is composed of rocks arranged to further

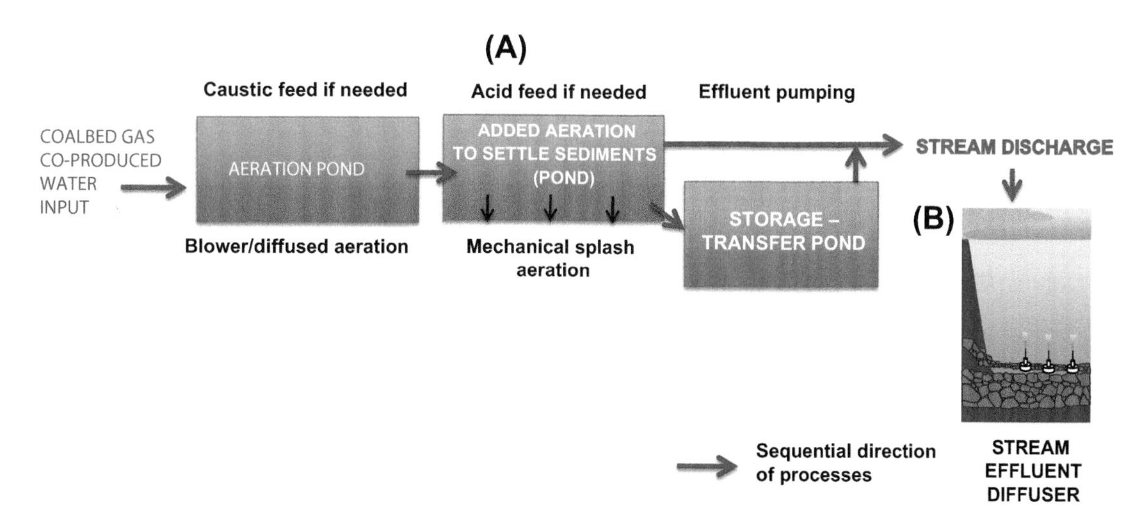

FIGURE 8.28 A generalized flow chart showing the sequence of processes for coalbed gas co-produced water from aeration, sediment settling, and storage ponds (A) to stream discharge (B) of treated water.

aerate the water before it enters the stream bed, as well as preventing or reducing erosion and further precipitation of pollutants from the water (USEPA, 2012a,b).

Reverse Osmosis

It has been more 200 years since a French Scientist introduced the concept of osmosis and since then the process of RO has advanced to a technology that provide purified water (e.g. desalination) and has common uses from kidney dialysis to battery water worldwide. Wastewater treatment with RO is the most significant treatment used to purify co-produced water from coalbed gas production. Thus, understanding reverse osmosis starts with the principle of osmosis. Normal osmosis functions through a semipermeable membrane, which allows low concentrated solution (e.g. lower contaminant concentration) to pass through to dilute a higher concentrated solution (Reid, 1966; WQA, 2012). The concentrations of both solutions are made more equal or in equilibrium when osmotic pressure is reached (see diagram as pressure is formed with the difference in water levels) (WQA, 2012). In reverse osmosis, the process is overturned or reversed in which the high-concentrate solution (e.g. greater concentration of contaminants) is "forced" through the semipermeable membrane into a low-concentrate solution (e.g. treated water; Figure 8.29; ALL Consulting, 2003; 2005; 2008). In order for reverse osmosis to work, the pressure causing "forcing" is applied on the high-concentrate solution to counteract the osmotic pressure and to filter through the membrane. Thus, reverse osmosis forms when pressure is applied to a highly concentrated solution causing the high concentrated solution to pass through a membrane to the lower concentrated solution. Thus, the process leaves a higher concentration of solute on one side, and only solvent on the other side.

The principle of reverse osmosis is applied to co-produced water as the high-concentrate solution by separating dissolved solids and other chemical constituents or large molecules and ions by passing through a semipermeable cellophane-like membrane (USEPA, 2010). Reverse osmosis is an established effective treatment for co-produced water removing TDS and other constituents such as arsenic and by

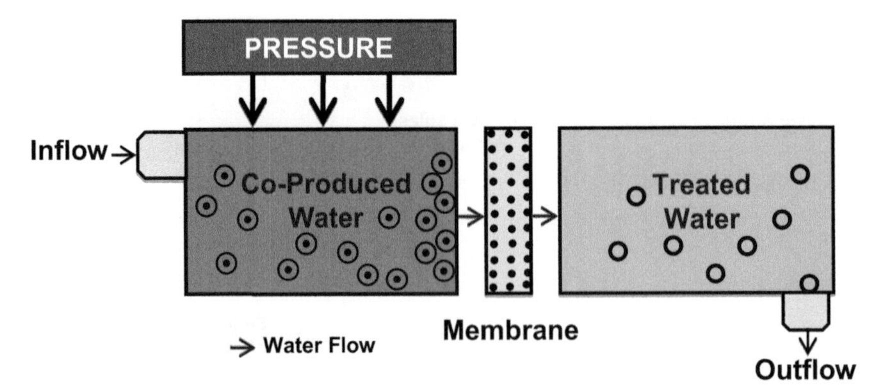

FIGURE 8.29 A generalized flow chart showing the sequence of processes of reverse osmosis for treatment of coalbed gas co-produced water.

converting brackish water/seawater (brine) to drinking water (e.g. desalination). Suspended solids of co-produced water need to be pre-treated for removal before use of reverse osmosis membranes to prevent blockage. Frequent flushing of the reverse osmosis system is required to make sure that suspended solids from the concentrate solution or scaling do not obstruct the membrane, which leads to higher pressure required to maintain a flow through the membrane. The membranes used for filtering co-produced water of coalbed gas are polymeric and zeolite membranes (Li, Ryan, Nenoff, Dong, & Lee, 2004). ALL Consulting (2003) described nanofiltration as a high-pressure desalting membrane process compared to microfiltration and ultrafiltration as low-pressure membrane filtration processes that are used to remove solid particles, which are often used in the pretreatment steps. In addition, ALL Consulting (2003) indicated that reverse osmosis is a primary desalting membrane process used in co-produced water treatment. The high-quality water resulting from the reverse osmosis process could be available for many beneficial uses (ALL Consulting, 2003).

Reverse osmosis for treating co-produced water of coalbed gas is used in the PRB (ARI, 2002). Welch (2009) reported on the evolution of two treatment plants, which incorporated reverse osmosis and recovery reverse osmosis. Together these treatment plants reduce waste by optimizing recovery using aeration to evaporate and concentrate the brine. In addition, Welch (2009) indicated that both plants get the most out of membrane performance by controlling filtration and scaling. The composition of co-produced water in the PRB is such that the treatment plants have to process wastewater high in sodium and bicarbonate, low in hardness, and may contain suspended solids, iron, silica, and barium. Consequently, sodium is closely checked and calcium carbonate solubility is predicted by bicarbonate carbonate and calcium concentrations to assess the potential for hardness scale formation using the Langelier saturation index (Welch, 2009). Although silica concentration in co-produced water is modest in the PRB, the high rate of recovery poses a scaling problem on the surface of the membrane. Moreover, the technique of controlling silica precipitation conflicts with that used for other ions. The original plant commissioned in 2006 was designed to allow discharge to have a blended sodium level and to meet Whole Effluent Toxicity standards (Welch, 2009). The proposed new plant is more advanced than the first plant by focusing on hardness and silica scale formation and acid feed. The main feature is ion exchange, which reduces polyvalent cations from the feedwater as well as calcium, magnesium, barium, and soluble iron to very low levels by exchanging them for sodium on the ion exchange resin (Welch, 2009).

Ion Exchange

Ion exchange otherwise called IX is a physicochemical reaction, in which ions are replaced or exchanged between solution and solid resin phases (USEPA, 2012a,b). In the context of co-produced water, the process of ion exchange generally reduces salts and inorganic constituents but more preferably reduces sodium. In an ion exchange complex, the co-produced water passes through a column that contains resin, which according to the USEPA (2012a,b) characteristically consists of elastic three-dimensional hydrocarbon network with large amount of ionizable groups electrostatically bound to the resin to absorb specific constituents. The resins are either natural inorganic zeolites or synthetic organic resins (ALL Consulting, 2008). The ionizable groups are swapped for similar charge ions in solution, which have a strong selectivity for the resin (USEPA, 2012a,b). Conceptually, co-produced water feed enters the column of resins, which are charged with sodium ions by passing a concentrated salt solution. As the co-produced water laden with iron and manganese goes through the column, the metals are attracted and bound to the resins. Consequently, the sodium is freed (e.g. exchanged) when the iron and manganese are attached to the resin. In addition to removal of metals, radium (NORM) from co-produced water can be removed using special ion exchange resins. NRC (2010) reported that ion exchange treatment effectively removed TDS in which data from one or more studies show some and up to 70% removal. The same report indicated that ion exchange very effectively removed (data from one or more study show 70% removal) sodium, bicarbonate, barium, fluoride, and ammonia.

Occasionally, the resin-ion exchange container or chamber is renewed by washing and refreshment of the salt solution (USEPA, 2010). Through time, the resins absorbed and become saturated with pollutants, which require rejuvenation by acid solution transmitted through the column. The acid solution used mainly for cations (e.g. sodium and metals) contains hydrogen ions that remove the absorbed pollutants from the resins. Because the concentrations of the pollutants are higher in the regenerating solution than in the feed stream, the ion exchange results in separation of sodium from water resulting in concentrated brine and treated co-produced water (USEPA, 2010). The reduction of salt of the co-produced water permits the treated water to be recycled for discharge to surface waters or for beneficial use (e.g. irrigation). According to NETL (2012a,b), the treated water contains more hydrogen ions resulting in low pH. The pH is normally increased using calcium carbonate, which in turn control the SAR detrimental to soil dispersion.

The most extensively used ion exchange treatment for co-produced water in the PRB and on a trial basis in other basins in the United States is the Higgins Loop (NRC, 2010; RPSEA, 2012). The Higgins Loop consists of four major chambers or operating zones separated by four major control valves or butterfly/loop valves as demonstrated in Figure 8.30 (Beagle & Dennis, 2007). Thus, the Higgins Loop uses a continuous countercurrent ion exchange contactor, which consists of a vertical cylindrical loop for liquid phase separations of ionic components. The co-produced water, resins, and process fluids are held in the chambers when the valves are closed. The resins are transmitted hydraulically (e.g. pulse) from chamber to chamber when the valves are opened. Pulsing of the resin bed through the loop is opposite to the direction of liquid flow. For example, valve A is closed in order to hold the co-produced water used to move the resin during the pulse cycle (NRC, 2010; RPSEA, 2012). The valve opens temporarily during the treatment and regeneration cycles; at the same time, the resins in the backwash chambers refill the pulse chamber. The contactor, which contains a packed bed of resin, is separated into four operating zones by butterfly or "loop" valves. These operating zones of adsorption, regeneration, backwashing, and pulsing function like four separate vessels.

The co-produced water for treatment is passed through the adsorbent resins such that sodium as well as potassium, calcium, and magnesium are adsorbed onto the resins. The Higgins Loop treatment can achieve 97–99% water recovery, lowers SAR up to 85%, and lowering TDS and bicarbonate concentrations; 1–3% of the volume ends up as concentrated brine that is deep-injected or transported offsite by truck for disposal by class II deep injection wells (Beagle, 2006; Matthews, 2007). The Higgins Loop continuous ion exchange process is usually limited to treatment of sodium-bicarbonate-type water and is not cost-effective in treatment

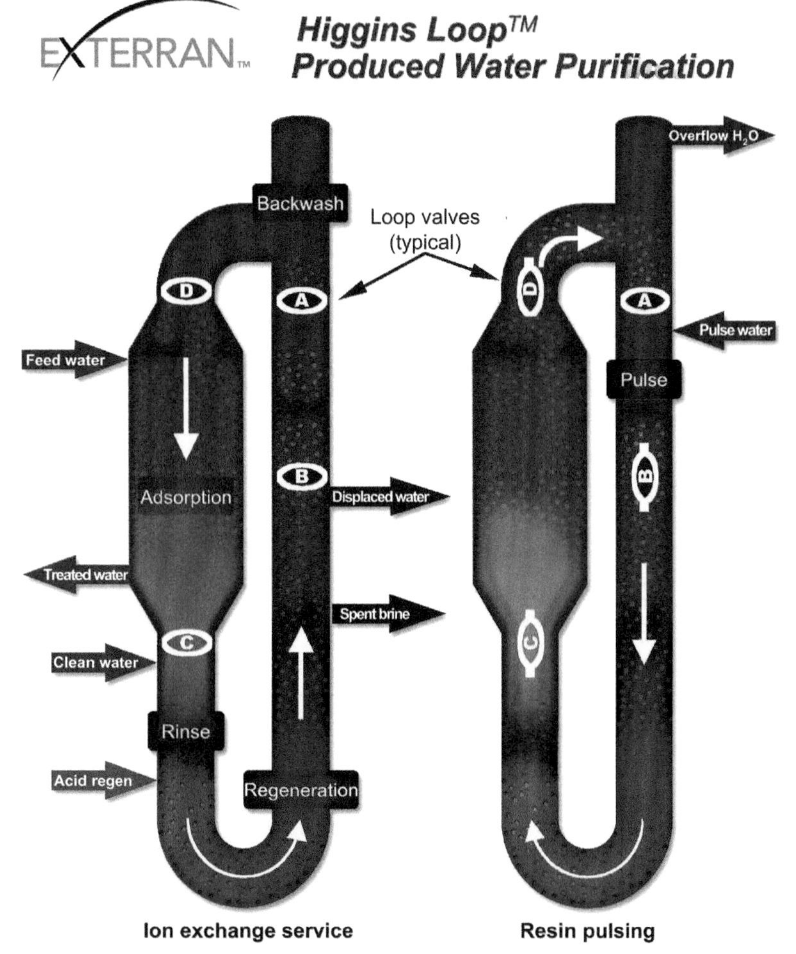

FIGURE 8.30 Diagrams showing the Higgins Loop process. The main control valves A, B, C, and D separate the four major chambers of the Higgins Loop. Resin, produced water, and process fluids are held in the chambers by the closed valves. The resin moves hydraulically or pulse, from chamber to chamber when the valves are opened. Both diagrams show valve A is closed. Valve A is closed during the pulse cycle to contain the water used to move the resin. It opens temporarily during the treatment/regeneration cycle while resin in the backwash chamber replenishes the supply in the pulse chamber. *Source: Adopted from NRC (2010), which used the diagrams with permission from Exterran Energy Solutions L.P.*

of other types of co-produced water, such as sodium chloride. Other configurations of ion exchange are designed to treat for fluoride, ammonia, barium, heavy metals or trace elements, radium, nitrates, arsenic, and uranium. Another process, continuous ion exchange process, treats for sodium bicarbonate (e.g. Drake Process). The Drake Process is a proprietary pilot-scale technology using an IX system that selectively removes sodium ions from coalbed gas co-produced water. For example, the PRB co-produced water is high in sodium and low in calcium and magnesium, which can yield high SAR values that limit beneficial use. Drake has four pending patents, which maximize the IX systems to treat PRB co-produced water (USEPA, 2010).

Electrodialysis

Brinck et al. (2008), Igunnu and Chen (2012), Rice et al. (2002), and Veil et al. (2004), have emphasized that the chemical composition of co-produced water, which includes mainly ions that are either positively charged (e.g. magnesium, calcium, sodium, etc.) or negatively charged (e.g. chlorine, sulfate, bicarbonate, etc.), plays a major role in their potential treatment. According to ALL Consulting (2003, 2005), the ions are attracted to electrodes with opposite electric charge (e.g. cathode and anode). Thus the concept of electrodialysis treatment is driven electrically, which takes advantage of the ionic characteristics (e.g. positive cation and negative anion electric charges) of the salts of the co-produced water and separates them with the electrodes (Figure 8.31). According to ALL Consulting (2005) and USEPA (2010), electrodialysis uses alternating cation and anion membranes placed between pair of oppositely charged electrodes. Thus, an electrodialysis unit is a stack of membranes, which includes several hundred cell pairs bound together with electrodes on the outside (Buros, 1990). According to UNEP (2012a), feedwater (e.g. co-produced water) passes concurrently through the cells providing continuous flow of treated desalted produced water and brine, which come out from the membrane stack.

Co-produced water is fed or pumped through spaces (e.g. feed or channel spacers) between the membranes forming parallel flow streams along

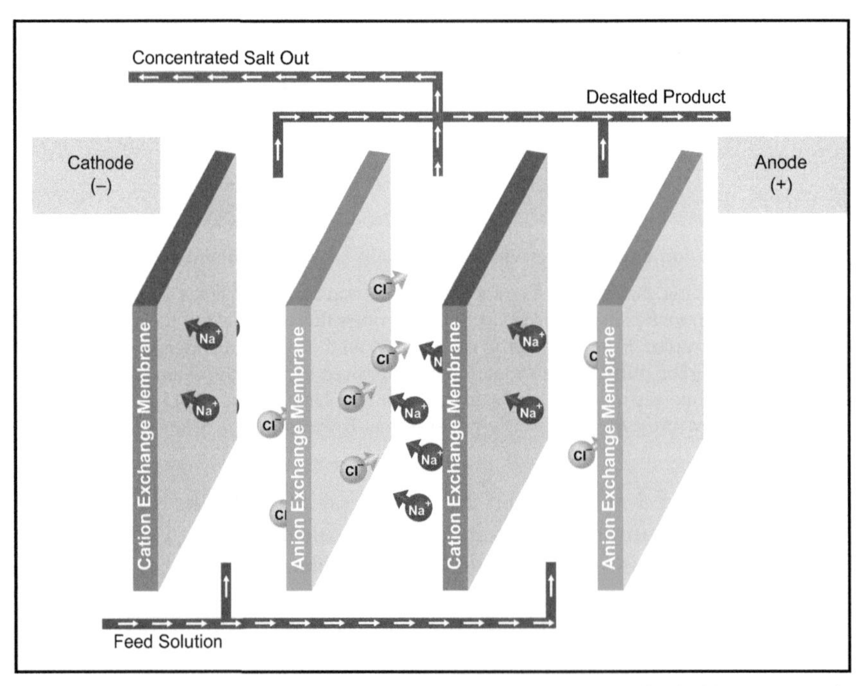

FIGURE 8.31 An electrodialysis assembly unit showing ion-exchange membranes flanked by opposite-charged electrodes and feed spacers. Adopted from Arthur et al., 2005.

the face of the membranes. Malmrose et al. (2004) indicated that application of electricity causes the positively charged ions of the co-produced water to migrate to cathode and negatively charged ions migrate to anode and are held in the polarized membranes. ALL Consulting (2003, 2005) suggested that during the migration, charged ions are rejected by similar charged ion exchange membranes (Figure 8.31). Consequently, co-produced water within the alternate feed or channel spacer becomes concentrated leaving desalinated water within the next spacer of the electrodialysis unit. The concentrate and desalinated water are continuously removed from the unit.

Other Treatment Technologies

USEPA (2010) reported that there are other proprietary and pilot-scale treatment technologies that treat co-produced water. One of these is the proprietary thermal distillation process, which concentrates TDS into a brine wastewater and low-concentrate TDS water is discharged. Pilot and patent-pending treatments include multiple technology applications from pretreatment to posttreatment technologies such as filtration, sedimentation, nanofiltration, reverse osmosis, ion exchange, and activated carbon (USEPA, 2010).

ALL Consulting (2005) reported new and innovative treatment technologies being developed for unconventional and conventional co-produced water. These include the capacitive deionization, electrochemical activation, electrodeionization, evaporation, rapid spray evaporation, freeze–thaw evaporation, pressure-driven membrane separation, high-efficiency reverse osmosis, oxidation reactor, and NORM treatment technologies. Other treatment technologies of co-produced water from oil and gas and coalbed methane are described by RPSEA (2012).

Igunnu and Chen (2012) reported on other innovative technologies that are applied to treatment of oil and natural gas, which may be applicable to coalbed gas co-produced water. The study reviewed current technologies used worldwide such as electrochemical techniques, which can be the future of management of co-produced water from conventional energy development. Igunnu and Chen (2012) argued that co-produced water could be treated based on electrochemistry because it behaves like potential electrolyte with relatively good conductivity. Furthermore, these researchers argued that application of photoelectrochemistry, co-produced water electrolysis, fuel cell and electrodeposition, and electrochemical engineering can produce clean co-produced water, and recover metals of with minimum co-produced water impact to the environment. The applications of these treatment technologies are often dependent on the disposal system and beneficial use that are employed in managing treated and concentrated waters.

BENEFICIAL USES OF CO-PRODUCED WATER

The Clean Water Act allows management of co-produced water for beneficial use as regulated and permitted by the National Pollutant Discharge Elimination System (NPDES). In general, the quality of co-produced water for beneficial use is low salts, low chlorides, no sulfates, low metals, and TDS from about 1500 to 3000 mg/l (Cline, 2006). The applications of beneficial uses of co-produced water of coalbed gas depend on the applicable regulations (e.g. county, state, and federal), quality/quantity of the treated, and economics. RPSEA (2012) reported the following matrices of beneficial uses of treated co-produced water: (1) augmentation of surface water uses; (2) aquifer recharge, storage, and recovery for domestic/municipal uses; (3) agricultural use; (4) environmental restoration use; and (5) industrial use.

BOX 8.4

FEUDING STATES ON SURFACE WATER DISCHARGES—FACT VS FICTION

Controversies are generated by widely divergent issues on the effects of the quality and quantity related to surface discharge of co-produced water from coalbed gas wells. The controversial issues are often litigated pitting states against each other. For example, the onset of the largest and most rapidly growing onshore coalbed gas play in the PRB has brought interstate water conflicts between Wyoming and neighboring states. Mudd (2012) described current United States Supreme Court litigation between Montana vs Wyoming, which claims that coalbed gas development in the PRB in Wyoming is depleting surface waters belonging to Montana under the 1950 Yellowstone River Compact. The Compact divided the interstate waters of the Yellowstone River system, which includes the Powder and Tongue Rivers as tributaries draining the basin, with the purpose of protecting the survival of irrigated agriculture in the arid region. The belief that the doctrine of prior appropriation governs the uses of water under the Compact led to litigation, which posed the following question: can coalbed gas development regulations be made consistent with traditional prior appropriation principles (Mudd, 2012)?

The US Supreme Court recently issued its first ruling addressing the separate question of whether the prior appropriation doctrine accommodates Wyoming irrigators who upgrade their irrigation efficiencies and reduce historic return surface water flows to Montana (Mudd 2012). It was concluded that groundwater withdrawals are within the scope of the Compact and that Wyoming is in violation of the Compact if it is allowing coal bed groundwater withdrawals to deplete surface waters belonging to Montana. The other issue is about Wyoming's coalbed groundwater pumping causing depletion of surface water supplies? Montana claims that Wyoming is permitting groundwater pumping, which resulted in the unlawful extraction of water belonging to Montana under the Compact. It was decided that Montana can sue Wyoming to enforce the terms of the compact and that groundwater pumping must be counted toward each state's water allocation affecting rivers covered by the compact (Mudd, 2012). However, the fact that groundwater is not even mentioned in the compact document is interesting because effects of groundwater pumping across the state line may be controlled by connectivity or non-connectivity of coal and sandstones aquifers addressed by Flores et al. (2010) and BLM (2012).

Thus, from the above controversy, the impact to natural resource law of surface waters is significant where interstate or intracountry groundwater withdrawal from coalbed gas development is involved. If the impact is significant, mitigating it may be difficult if not impossible by the time the fact becomes known (Mudd 2012).

Augmentation of Surface Waters

Discharges of treated co-produced water of coalbed gas to augment surface waters are permitted by NPDES program managed by the United States Environmental Protection Agency (USEPA, 2010). NPDES permitting requirements vary from state to state according to the quality of co-produced water and treated water and intended beneficial use. Also, each state has various standards for receiving waters and classification categories (e.g. downstream irrigation, wildlife, domestic and recreational). Some states have pollutant discharge elimination system programs (e.g. UPDES and WYPDES), which permit and regulate discharges of waste to surface waters. Thus, standards are designed to protect the beneficial uses of state waters, such as the support of aquatic life, public water supplies, recreation, and agriculture. More specifically, the potential effects of co-produced water with augmentation of inflow streams and/or lakes concern aquatic biota (e.g. fish, amphibians, reptiles, and macroinvertebrates.) as well as their aquatic and riparian habitats. For example, standards for salinity and suspended solids are developed in coordination with state agricultural agencies. States like Wyoming through WYPDES are controlling through site-specific permitting of coalbed gas co-produced water quality in proximity to downstream irrigation use and are planning a watershed-based permitting approach (Figure 8.21) (RPSEA, 2012).

Aquifer Recharge, Storage, and Recovery for Domestic/Municipal Uses

Aquifer recharge, storage, and recovery are used to replenish aquifers for future use such as storing water in the ground and recovering the water from the same reinjection well. Treated coalbed gas co-produced water can be used to recharge groundwater aquifers, which may impact drinking water standards. Consequently, authorization for disposal to deep/shallow well disposal systems is issued through the Underground Injection Control program so as not to impact the Underground Source of Drinking Water (USDW) (see section on reinjection from classes of wells). State/federal requirements make sure that the reinjected co-produced water do not degrade the quality of the groundwater within the injection zone, migrate into and degrade groundwater-bearing aquifers above and below the injection zone, or migrate to the surface via other wells within the area of influence of the injection well (RPSEA, 2012).

The constituents of co-produced water of coalbed gas that can contribute to water quality issues in terms of beneficial use include the following (Otton, 2005): (1) dissolved inorganic species—major ions—Na, K, Ca, Mg, HCO_3, Cl, SO_4; minor species—NH_4, B, Li, F; (2) dissolved trace elements—Fe, Ba, Mn, Se, Zn, Cu, Cd, Mo, Cr, As, Pb; (3) dissolved organic species—phenols and volatile aromatic compounds; (4) dissolved and dispersed hydrocarbons (far more common in conventional produced waters); (5) dissolved and suspended radionuclides; and (6) drilling and workover chemical additives.

The most beneficial use of treated or raw co-produced water of coalbed gas is augmentation of groundwater aquifers especially in US basins in arid and semiarid climates exacerbated by effects of long-term water droughts. If aquifer augmentation can be developed and depending on the volume of water intake, it could certainly alleviate not only short-term but also long-term problems of groundwater recharge and supply (Box 8.1). The best example of aquifer augmentation is in the PRB where most water is co-produced with coalbed gas and the drawdown of wells in the central part of the basin is more than 305 m (BLM, 2012). The drawdown from coalbed gas wells is also enhanced by drawdown from dewatering from coal mining along the basin margins. The City of Gillette according to Beeson (2006) planned artificial recharge using co-produced water discharge through reinjection into Fort Union aquifers (e.g. coals, sandstones) in order to meet its drinking water demands. Also, the State of Wyoming proposed to transport co-produced water to the City Casper for treatment for municipal use (Beeson, 2006).

Agricultural Use

Although agricultural water is commonly used in the United States and worldwide, the quality required is generally lower than for drinking and municipal uses. Agriculture is a major user of water worldwide estimated by UNESCO (2003) to be about 70%. Agricultural water is mostly affected by salinity, SAR, and toxic components of co-produced water. Co-produced water of coalbed gas has been used for watering livestocks (e.g. PRB). Guidelines for worldwide water quality for livestock and poultry uses are established by the United Nations Food Agricultural Organization (Ayers & Wescott, 1985). Water quality can have an adverse effect on soil and harvest yield. Agricultural irrigation using untreated co-produced water especially with high-sodium composition relative to the calcium and magnesium (SAR) may not be suitable use because of the deleterious effect on the soil. The problem is remediated by application of soil amendments (e.g. gypsum), which are effective in maintaining the low SAR values of the soil layer (Johnston, Vance, & Ganjegunte, 2008). The calcium in the gypsum counterbalances the sodium in the water and it is very similar to adding sulfur and gypsum in regular fertilizers. An acid-like sulfur neutralizes the co-produced water.

Another agricultural use of treated or untreated co-produced water of coalbed gas is drinking water for animals (e.g. livestock and wildlife) and as aquatic habitats for fish and waterfowl. In the PRB, ranchers use co-produced water for watering cattle. According to Bauder (2006), it takes water from two to three coalbed gas wells (function of water yield of wells) to supply 1000-head herd of cattle in the PRB. Ranching is one of the major agricultural operation in which co-produced water is put to beneficially use. In general, livestock can tolerate a wide range of drinking water constituents. Animals can often tolerate TDS from 750 and <5000 mg/l (Jangbarwala, 2007; Dahm et al., 2011). Some coalbed gas projects on ranchlands have built impoundments or watering sites to provide co-produced water as a drinking water for livestock.

Types of Irrigation

According to Veil et al. (2004), the irrigation of crops uses about 39% of the freshwater supply in the United States, which translates to 568 billion - liters/day (USGS 1998). Some basins (e.g. PRB and Raton Basin) contain potable or nearly potable water, which can be directly or passively treated prior to beneficial use such as irrigation of croplands (Figures 8.32 and 8.33). Paetz and Maloney (2002) described a useful treatment of coalbed gas water to mitigate salinity and sodicity problems so it can be used in a managed irrigation program. ALL Consulting (2003) summarizes the water quality requirements for crop irrigation as including critical parameters such as salinity and toxicity, which affect the plants and sodicity, which affects the soil. Other constituents of co-produced water, which

FIGURE 8.32 Photograph showing a cropland (background) irrigated by co-produced water discharged from an off-channel impoundment (foreground) fed by an outfall from coalbed gas wells in the Powder River Basin.

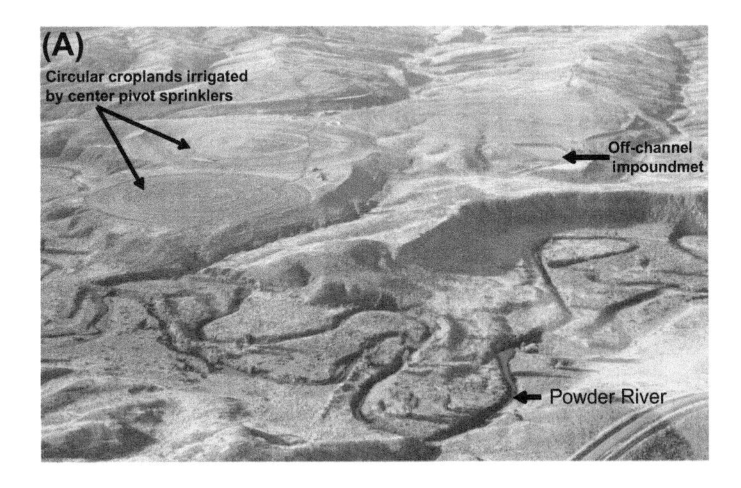

FIGURE 8.33 Photograph showing aerial view of circular-shaped cultivated croplands (A) irrigated by center pivot sprinklers drawing co-produced water (B) from off-channel impoundment above the Powder River alluvial valley in the Powder River Basin.

constrain agricultural use, include salinity, pH, alkalinity (carbonate and bicarbonate), concentrations of specific ions (e.g. chloride, sulfate, boron, and nitrate-nitrogen), heavy metals or trace elements, and microbial contaminants (RPSEA, 2012). Ayers and Wescott (1985) and USEPA (2005) provide recommended water criteria for irrigation and from treated water.

In the PRB, Wyoming irrigation with co-produced water has transformed the landscapes into productive agricultural areas and has become a popular alternative for management and reuse of water. Otherwise surface disposal requires an NPDES discharge permit (RPSEA, 2012). Examples of irrigation systems in the PRB include center pivot sprinklers, side-roll/wheel line sprinklers, hand-moved or fixed solid set sprinklers, big gun sprinklers, surface drip, subsurface drip, gated pipe flood and ditch flood (Figure 8.33; RPSEA, 2012). The center pivot and side-roll sprinklers are commonly used but subsurface drip irrigation (SDI) has emerged as an improved dispersal system for co-produced water. SDI is a network of polyethylene tubes laid down in the soil, which uniformly apply drips of water over a large surface. The co-produced water is filtered, treated, and pumped through the network such that the roots of vegetation derive the water in the subsoil and the topsoil remains dry (RPSEA, 2012). The sodium in the water is counteracted by the calcium and magnesium in the soil and the salts seep into lower depths.

SDI using co-produced water for irrigation makes up a small portion of the agricultural beneficiation of croplands in the PRB. Using the co-produced water with low salinity and high SAR creates soil problems as discussed earlier (see section on Sodium Adsorption Ratio). Studies have been conducted on the limited use of SDI in the PRB to determine the impacts of the use of treated co-produced water (Bern, Breit, & Healy, 2012; Engle, Bern, Healy, Sams, & Zuapancic, 2011). Engle et al. (2011) observed that much of the treated co-produced water infiltrated the vadose zone altering the soil flow dynamics and some of the water infiltrated the shallow aquifer system, which will magnify as more treated water is used in the future. In addition, these researchers reported that the composition of the treated co-produced water is more dilute than the shallow groundwater at the irrigation site, although significant mixing and geochemical reactions occur within the vadose system. High TDS concentrations and zone of high EC indicate that evaporation and dissolution of native salts control the composition in the soil column. Bern et al. (2012) studied an SDI irrigation, which mixes sulfuric acid with treated water applied via drip tubing buried 92 cm deep (Figure 8.34). SAR values vary from 0.5–1.2 for irrigated to 0.1–0.5 for nonirrigated soils at 0–30 cm depth after 6 years. Sodium accumulated only 8–15% more above the drip tubing. Sodicity has increased in soil

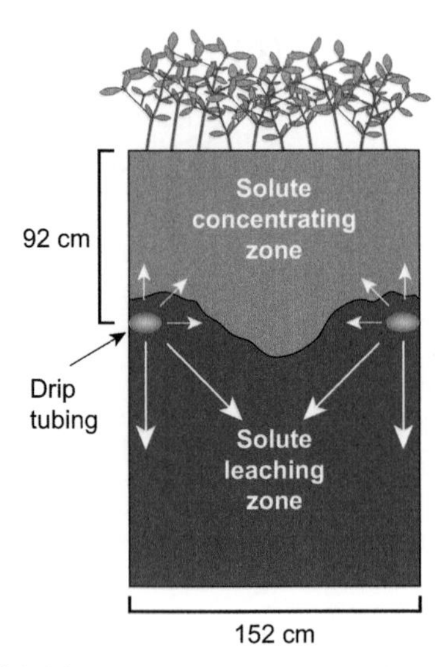

FIGURE 8.34 Diagram showing subsurface drip irrigation (SDI) in a column of soil zones governed by different treated co-produced water and solute movements. Arrows show relative movement of water from the drip tubing. *Source: Adopted from Bern et al. (2012).*

surrounding the drip tubing. In addition, sodium from the water accumulated as evapotranspiration concentrates solutes in the soil from 45 cm depth to the drip tubing. Sustained irrigation with downward flow of excess water, depleted gypsum, increased SAR to >14, and decreased EC to 3.2 mS/cm in soil water. Bern et al. (2012) concluded that increased sodicity in the subsurface and not in surface indicates that deep SDI is a viable use of irrigating with sodic waters.

ALL Consulting (2003) and RPSEA (2012) provide case studies from the PRB of irrigation projects conducted by gas operators to demonstrate the effects to different crops with the use of coalbed gas co-produced water. The first case study used livestock grass forage on irrigated plots sustained by pure co-produced water on one side and blended co-produced and surface waters on the other side. Although pure and blended co-produced water grew ample forage crops, the pure co-produced water had to be applied at a higher rate because the crops could not utilize it as efficiently as the surface water (RPSEA, 2012). The second case study was conducted on irrigated land, which grew drought-tolerant plants. The high SAR in the untreated co-produced water necessitated application of gypsum and other soil amendments. After application of co-produced water, the irrigated land grew hearty grass crops for livestock feed. The third case study is a test site, in which like the other two studies used untreated co-produced water for irrigation. After 2 years of application of soil amendments, the irrigated land was transformed from overgrazed rangeland to healthy grassland benefiting livestock and wildlife.

Environmental Restoration and Remediation Uses

The proliferation of thousands of impoundments especially in one basin such as the PRB, Wyoming has raised the issue of their environmental restoration and remediation. Impoundments receive 64% of the co-produced water in the PRB. These impoundments provide drinking water for livestock and wildlife, and are transformed into habitats serving as oasis for fish and waterfowl (NRC, 2010). ALL Consulting (2003) reported that untreated co-produced water of coalbed gas is used to restore privately owned fishponds in some states. For example, Wyoming discontinued fish stocking programs in certain ponds due to lack of water to sustain the aquatic habitat. Coalbed gas co-produced waters were used to benefit the ponds, allowing for continued fish stocking. The application and success of this beneficial use are governed by state guidelines, water quality and quantity, and proximity to coalbed gas producing basins (RPSEA, 2012).

The low salinity content of co-produced water in the PRB is environmentally tolerable; however, the high sodium carbonate content requires environmental remediation for the acute and chronic or long-term toxicity to aquatic life forms (e.g. fish and invertebrates) in impoundments (NRC, 2010). In addition, the exceedance of TDS, sulfate, and/or selenium groundwater standards established by the Wyoming Department of Environmental Quality may be adverse to fish and wildlife, which make impoundments their habitats (NRC, 2010). For example, coalbed gas co-produced water discharged in impoundments contains selenium in concentrations above 2 μg/l, which may cause bioaccumulation in sensitive aquatic species.

By design, surface impoundments store co-produced water for evaporation and infiltration. In arid and semiarid environments where evaporation is high such as in the PRB, the impoundments need to be replenished or restored with co-produced water, precipitation (e.g. rainfall and snowfall) or both. The construction of thousands of impoundments has created water reservoirs in which potentially large-scale artificial wetlands are formed, which in turn provide abundant, diverse, and reliable man-made wildlife and riparian habitats. The establishment of artificial habitats has created ecological systems that need to be maintained to continue to supporting exotic aquatic fauna and flora. These constructed wetlands require discharge of raw or treated co-produced water to safeguard both plant and animal life, which rely on sustenance from these artificial habitats. These impoundment wetlands may be in jeopardy with the reduction of gas production due to supply and demand as well as after cessation of coalbed gas production. NRC (2010) has highlighted cutting-edge research studies of management on co-produced water, water quality, beneficial uses, and effects of effluents to surface waters and groundwater, all relevant to present concerns. However, research on potential diminishing impoundment wetlands after coalbed gas development and their effects to resident and transitory aquatic and nonaquatic life forms lacks emphasis and should receive equal or more attention for future investigations. For example, are the new and/or exotic fauna and flora species being spawned in these man-made ecological habitats endangered with failure to maintain the impoundments?

According to RPSEA (2012), constructed wetlands associated with impoundments have several constraints on their beneficial usefulness: (1) large amount of land required by the wetlands; (2) sufficiently large volume of co-produced water is necessary to supply and support wetlands; (3) poor quality of co-produced water requires pretreatment or posttreatment; (4) in some agricultural and municipal uses, impaired co-produced water must be pretreated before entering a treatment wetland; and (5) contaminants may be released during high water flow events or periods when vegetation decomposition occurs. Potentially, if constructed wetlands are part of beneficial use of direct discharge, increased flows and erosion may change habitats. Additional impacts from the wetlands are from the accompanying pollutants of direct discharge.

Phytoremediation: Future Innovative Beneficiation

According to UNEP (2012a,b), there is increasing trend around the world to apply innovative remediation technology to contaminated watercourses using phytoremediation or plant remediation. RPS Australia East Pty Ltd (2011) suggests use of phytoremediation as part of the mix of treatments for co-produced water from coalbed gas in the next 25 years. *Phytoremediation* is defined by UNEP (2012a,b) as the use of living green plants for in situ removal, degradation, and containment of contaminants in soils, surface waters, and groundwater. Phytoremediation is an emerging biotechnology (McCutcheon & Schnoor, 2003) recommended for the following reasons (UNEP, 2012b): (1) low cost, solar-driven cleanup technology; (2) effective in shallow sites with low contaminant levels; (3) efficient in treatment of wide range of environmental pollutants; (4) useful in place of mechanical cleanup methods; and (5) environmentally "correct" treatment.

Plants have evolved with diverse genetic adaptations to control pollutant accumulations in the environment, in general, and particularly in wetlands. Phytoremediation entails growing plants in contaminated wetlands for a required period in order to remove contaminants or make possible containment and detoxification of the pollutants (Kirkpatrick, 2005; UNEP, 2012a,b). According to UNEP (2012a,b), plants break down, degrade organic contaminants and filter or trap metal pollutants. To remove pollutants from soil, sediment and/or water, plants can break down, or degrade organic pollutants or contain and stabilize metal contaminants by acting as filters or traps. The uptake of contaminants is through the root system, which absorbs and accumulates the moisture and nutrients as well as contaminants. The plant roots release root exudates (e.g. organic and inorganic compounds), which in turn increase or decrease the availability of contaminants in the root zone.

Kirkpatrick (2005) has explored remediation approaches using native plant species to deal with saline co-produced water from coalbed gas wells in constructed wetlands in the PRB. The most important issue with respect to co-produced water of coalbed gas wells deals with large volumes of moderately saline-sodic water. Kirkpatrick (2005) constructed closed-system wetland cells populated by native plant species established in hydrologically distinct communities in ephemeral channels with discharged co-produced water to simulate and assess similar conditions in the PRB. The selected native plant species found in areas of coalbed gas co-produced water discharge areas include the following three communities: (1) maritime bulrush, inland saltgrass, and creeping spikerush; (2) common cattail, prairie cordgrass, and Canada wild rye; and (3) American bulrush, Baltic rush, and streambank wheatgrass. Kirkpatrick's (2005) study indicates that constructed wetlands with salt-tolerant native plant species used large volumes of co-produced water while remaining are robust and viable. The native species potentially can accumulate high levels of sodium and other salts in their above-ground tissue as well as excrete the salts on leaf surfaces. Thus, phytoremediation using salt-tolerant native species can potentially provide benefits to lessen environmental degradation and to restoration of wetlands impacted by poor-quality co-produced water of coalbed gas wells.

Industrial Use

The industrial use of co-produced water of coalbed gas is voluntary but must be approved by the state regulatory agency that has the authority to dispose and use of treated or untreated water. Co-produced water is used in coal mines (e.g. PRB) for dust control. However, Wyoming Oil and Gas Conservation Commission following environmental standards related to surface and groundwater pollution prohibit water road spreading.

The practice of dust control with co-produced water is not applied beyond the roads, around streams, or near buildings. Co-produced water used for dust control must have TDS and sodium levels that meet regulatory standards. Environmental concerns associated with dust mitigation using co-produced water include (RPSEA, 2012) (1) salt build up along roadways, (2) migration of water and associated pollutants off roadways, (3) impacts to nearby vegetation, and (4) salt loading to nearby rivers.

Co-produced water may be used for drilling and hydraulic fracturing of coalbed gas wells. In coalbed gas drilling operations, co-produced water should not include hydrogen sulfide, which reacts with water to form a weak acid that is very corrosive to drilling equipment. Sulfides in the co-produced water may be derived from high sulfur contents of coals. The drilling water quality requirements may be different and are dependent on the drilling project. For example, the drilling water quality requirements include TDS $<20,000$ mg/l and calcium <400 mg/l for shallow drilling and calcium <100 mg/l for deep drilling (RPSEA, 2012). In addition to salt-related concerns, impacts of other pollutants in co-produced water include hydrocarbons, heavy metals or trace elements, and chemical additives used during drilling and stimulation. Coal reservoirs of coalbed gas wells can benefit by stimulation by hydraulic fracturing with the use of co-produced water. The purpose of hydraulic fracturing is to enhance production and the fraccing operation requires application of hundreds of thousands of liters of water. The fraccing flowback co-produced water can be treated for reuse for additional fraccing jobs. Recycling co-produced water for hydraulic fracturing should alleviate the stress of using local groundwater.

Other options for industrial use of co-produced water of coalbed gas are for coal washing and slurry piping in conjunction with in coal mine operations and coal transportation, cooling water for electric power plants, and enhanced oil recovery (EOR). According to Nummedal, Towler, Mason, and Allen (2003), EOR water-based technologies applicable to oil reservoirs include steam flooding, water flooding and spontaneous imbibition of brine, polymer-enhanced water flooding, and surfactant flooding.

MANAGING AUSTRALIAN CO-PRODUCED WATER

Lessons learned on management of co-produced water of coalbed gas from the PRB suggest there is no universal solution to its management and that it is a work in progress. However, there are management (e.g. beneficial, disposal, and treatment) similarities of co-produced water between the United States and Australia as best discussed in a report by

TABLE 8.8 Descriptions of Disposal Systems in Coal Basins in Australia

Disposal	Description	Australian Coal Basin
Evaporation ponds	For discharge into ponds for evaporation	Numerous
Re-injection	For injection into deep subsurface formations	Bowen (Queensland)
Surface water discharge	For direct discharge into a body of surface water	Bowen and Gunnedah (New South Wales)
Offsite disposal	For transport to offsite facility	Sydney (New South Wales)

Source: Adopted from RPS Australia East Pty Ltd. (2011).

RPS Australia East Pty Ltd (2011) prepared for the National Water Commission of the Australian government. This report was preceded by the CWIMI (2008) report for the Queensland government, which found considerable variability of co-produced water production in the Bowen and Surat coal basins within and between coal-bearing formations. The co-produced water of Australia in 2010 was 33 Gl per year in which 60% was from conventional oil and gas. To put into perspective, the estimated amount of co-produced water over the next 25–35 years will average 300 Gl/year with 3% to be contributed by conventional sources and the remainder from coalbed gas production for manufacture of liquid natural gas from the Queensland and New South Wales reserves.

RPS Australia East Pty Ltd (2011) reported potential management options for coalbed gas co-produced water, which include urban and industrial use, storage for future use such as managed aquifer reinjection, and agricultural use such for stock watering and replacing existing irrigation demand. Most of this co-produced water management will include treatment for quality (e.g. salinity, sodicity, etc.) previous to use or disposal. All the management options are very similar to practices used in the United States. Proposed beneficial uses such as for irrigation of forest crops for coal washing or coal preparation and coal-fired power plant cooling are innovative agricultural and industrial uses of co-produced water of coalbed gas. Table 8.8 is a list of disposal strategies for co-produced water from coalbed gas in Australia as of 2010. The most common disposal method of poor-quality co-produced water is via evaporation ponds. Much like in the PRB, evaporation ponds are shallow lined or unlined ponds using topographic features. Co-produced water from coalbed gas is not usually treated prior to discharge to evaporation ponds. Reinjection is not used in Australia as of 2010 but is being considered for future use. Issues such as geochemical incompatibility with receiving aquifers and locations of acceptable coal aquifers are prime concerns.

A significant part of proposed coalbed gas and liquids natural gas projects in Queensland coal basins is preparation of co-produced water management plans that include constraints analysis to identify the feasibility of potential strategies. These constraints include the following (RPS Australia East Pty Ltd (2011): (1) regulatory framework (restrictions imposed by legislation); (2) geography (distance to beneficial uses and disposal points from production sites); (3) water quality (required treatment, which presents economic, technological, and environmental challenges); (4) water quantity (amount of water for beneficial use, stability in demand, and level of uncertainty of projected estimates of amount of co-produced water that may affect ability of the producer to supply); (5) economic (cost related to management options and feasibility of particular management method); (6) environment (natural, social, and economic environments and short- and long-term effects of related management options used); and (7) technology (proven capability of co-produced water management and treatment technologies).

SUMMARY

Co-produced water is defined as extracted groundwater from aquifers during production of conventional (e.g. oil and natural gas) and unconventional (e.g. coalbed gas and shale gas) energy resources. The difference between conventional and unconventional resources is that in conventional production, large amount of natural gas is produced initially and replaced by more co-produced water over time as compared to unconventional production where large amount of co-produced water is produced during initial gas desorption and decreases over time. The origin of co-produced water from coalbed gas is either endogenous (derived internally) or exogenous (derived externally). Endogenous co-produced water consists of moisture, which resides in the fractures, pores, and capillaries of the coal matrix. Exogenous co-produced water is mainly derived from meteoric water and water from other aquifers. External aquifers such as sandstones play an important role in supplying co-produced water through vertical infiltration into interbedded coal aquifers. Vertical infiltration of groundwater is often underappreciated due to poor understanding of connectivity investigated from detailed stratigraphy and sedimentology of both sandstone and coal aquifers. Conventional belief that channel sandstones are compartmentalized and isolated hydrological aquifers is accurate across their widths by not their lengths or depositional dip, along which it is more continuous than associated coal beds and/or coal zones in the PRB. Basinal groundwater flow can be easily mistaken for the regional flow of groundwater along depositional dip of connected channel-sandstone aquifers as indicated in the PRB. Isotopes of oxygen [$\delta^{18}O$] and hydrogen ($\delta^{2}H$ [δD]) provide signatures of water co-produced with coalbed gas, which are critical to determination of its origin. Also, isotopic signatures and dating of co-produced water play a significant part in determining the renewability or nonrenewability of groundwater aquifers. A significant age gap exists between groundwater in deep "fossil or old (1.2–1.3 mya)" and shallow "submodern or young" (prior to 1952 and at least 60 years) coal aquifers indicating potential nonrenewability of co-produced water extracted from deeper parts of coal basins.

Important concerns in coalbed gas development are managing the volume and composition of co-produced water. Understanding the volume and composition of co-produced water is required in order to implement plans for management, disposal, treatment, and beneficial use in compliance with governmental regulations. This is valuable to land managers, regulators, reclamation specialists, resource assessors, and policy decision makers. Also, knowledge of the volume and composition of co-produced water can assist gas operators to estimate costs of management related to environmental, social, and economic impacts. The volume and flow of groundwater in coal-bearing rocks are not only influenced by regional aquifer and confining layers but also more importantly the discontinuity of coal and sandstones aquifers as localized hydrostratigraphic or micro units interbedded with impermeable fine-grained mudstones and siltstones aquitards/aquicludes making for a complicated compartmentalized lenticular aquifers and confining layers.

Composition of co-produced water of coalbed gas may be viewed according to its environmental effects and beneficial uses as well as the geochemical processes. Environmentally, the effects of composition of co-produced water directly relates to water quality standards and to reaction with soil upon discharge in the subsurface and/or surface. Co-produced water composition is a geochemically driven process and can be grouped into sodium bicarbonate and sodium chloride types. Distinct compositional groups consist mainly of bicarbonate, calcium, chloride, magnesium, sodium, and sulfate unique to geochemical reactions and processes inherent to generation of methane in coal reservoirs, with some biological component. Geochemical reactions and processes are uniquely driven by freshwater nature (e.g. sodium bicarbonate type) compared to brackish and/or marine nature (e.g. sodium chloride type) of the groundwater. The geochemical model involves surface water recharge, which dissolves salts and mixes with coal reservoir groundwater along its flow path becoming increasingly saline away from the recharge area. Water near recharge areas reflects composition of meteoric water with elevated dissolved organics, high total organic carbon, and low TDS concentrations where residence is short. Waters far removed from recharge areas reflects groundwater composition of the original depositional waters particularly in brackish and marine environments. A part of the geochemical model is reduction of sulfate downgradient, which is a biologically driven component of methanogenesis in generation of methane.

The composition of co-produced water significantly influences the management practices to be employed such as disposal, treatment, and beneficial use. Disposal of co-produced water includes primarily discharge into surface waters, impoundments, and reinjection into subsurface aquifers. The level of treatment of co-produced water prior to disposal depends on the water quality. Treatments include aeration/chemical precipitation, reverse osmosis, ion exchange, and electrodialysis. Beneficial uses of treated and untreated co-produced water include augmentation of surface water; aquifer recharge, storage, and recovery for domestic/municipal; agricultural; environmental restoration/remediation; and industrial uses. These management practices of co-produced water of coalbed gas have been partly adopted worldwide.

CHAPTER

9

Worldwide Coalbed Gas Development

OUTLINE

Coal and Coalbed Gas

http://dx.doi.org/10.1016/B978-0-12-396972-9.00009-4

KEY ITEMS

- The future of coalbed gas is linked to use of natural gas forecasted to increase more than 50% from the 2010 production and will account for more than 25% of global energy demand by 2035.
- Development and production of coalbed gas have been slowed by the 2008–2010 global recession with OECD countries (e.g. United States and Canada) reduced production. Australia plans for conversion of coalbed gas to LNG will supply expanding markets in Asia.
- NOECD countries (e.g. China, India, Indonesia, and Russia) collectively expanded exploration and development of coalbed gas during the global economic recession.
- Development of coalmine gas in the 1950s gave rise to exploitation of coalbed gas in the 1970s, which expanded in the 1990s and matured in the 2000s in the United States, Australia, and Canada.
- Exploration of coalbed gas in other countries (e.g. China and India) started in the 1900s with established development in the 2000s. Indonesia and Russia started exploration of coalbed gas in the 2000s with the first production after 2010.
- Coalbed and coalmine gas production in the conterminous United States is from Pennsylvanian to Tertiary age anthracite, bituminous, and subbituminous coals in 12 mature basins.
- Canada produces coalbed gas from Cretaceous age bituminous coals and potentially from Jurassic and Tertiary age bituminous–subbituminous coals in the Western Canada Sedimentary Basin (WCSB).
- Australia produces coalbed and coalmine gas from Permian age anthracite coal in the Sydney Basin, bituminous coal in the Bowen, Gunnedah, and Gloucester Basins, and Jurassic age bituminous coals in the Surat and Clarence-Moreton Basins.
- China produces coalbed and coalmine gas from Carboniferous and Permian age anthracite coals in the Qinshui Basin and potentially from the same age coals in the Ordos Basin.
- India produces coalbed gas from Permian age anthracite coals in four basins with 12–14 coalfields and potentially from Tertiary age subbituminous coals in six basins.
- Indonesia potentially will produce coalbed gas from Tertiary age semianthracite to subbituminous coals in two Sumatran basins and three Kalimantan basins in east Borneo.
- Russia produces coalbed and coalmine gas from Carboniferous and Permian age anthracite to bituminous coals in the Kuzbass Basin and potentially from the Pechora Basin.
- The OECD and NOECD coalbed gas-producing countries have invested in

coalmine gas or methane recovery and utilization projects.

- The impact of successful coalbed gas production in subbituminous coal has spurred search of another Powder River Basin model worldwide such as in Indonesia, India, Colombia, and the Philippines.
- Coalbed gas industry in OECD countries and emerging coalbed gas industry in NOCED countries benefitted from research and development on innovative drilling, completion, and stimulation of coalbed gas wells, which made production more efficient in the United States.
- Opportunities for new markets for coalbed gas such as Compressed Natural Gas (CNG) and Liquefied Natural Gas (LNG) have sustained development in Australia, China, and Indonesia.

INTRODUCTION

Worldwide development of coalbed gas has added to the overall understanding of coal as a source and reservoir for methane. This has also led to more effective drilling and completion techniques, and economical application of technology with environmentally sound management of co-produced water. Even with these challenges, the key to the future and the advantage of coalbed gas over conventional natural gas, tight gas, and shale gas development is the markedly low exploration, drilling, stimulation, and completion costs, which are attractive to developing countries with significant amount of shallow coal resources. The development of coalbed gas in the 1990s was at a fast-paced schedule breaking early forecasts of successes only to be slowed down during the 2008 economic downturn and drop of price of gas in the United States. It is difficult to predict how much and when the price of gas will rebound in the United States. However, the demand for a supply of clean energy and vast amount of coal resources in developing countries provide opportunities for a two-pronged development of coalbed gas in advance of and after coal mining as well as in deep unmineable virgin coals. The first development option provides extraction of coal and coalbed gas with the benefit of potential greenhouse gas credits making economically successful projects. More importantly, the second development option can be made more economically successful by lessons learned from innovative coalbed gas developmental techniques applied in the United States prior to the economic downturn. The future opportunities for coal and coalbed gas development will grow with the world's desire for more gas, clean energy supply, safe coal mines, and reduced greenhouse gas emissions.

Although the contribution of coalbed gas production to the global energy mix is presently only a small fraction, it is a vital complement to conventional natural gas especially when viewed as a clean energy source. As of 2010, unconventional gas comprised over 15% of the global gas production, which was 420 metric m^3 (Bcm), about 10 Bcm of which is from coalbed gas (IEA, 2013). Much of this global coalbed gas production comes from the United States, which makes about 8% of the national total gas production prior to the economic downturn. The United States, Russia, Canada, and Iran are the largest producers of natural gas (IEA, 2013). Of these countries, only the United States and Canada produce sizeable coalbed gas with Canadian production significantly lagging behind the United States. Australia is the most advanced major producer of coalbed gas outside the United States and will perhaps be a significant contributor to liquid natural gas (LNG) exports of gas from coalbed gas perhaps starting in 2014 or 2016 (IEA, 2013). Although not as advanced as the United States, Australia, and Canada, some countries such as China, India, Indonesia, and Russia are actively exploring and have budding production. In other areas, strong coalbed gas interest exists throughout Europe especially in coalmine gas recovery. However, development of deep coalbed gas is hampered by concern on additional environmental impacts of extraction of the methane. Still there are numerous other countries such as Bangladesh, Colombia, Philippines, Japan, United Kingdom, Germany, and New Zealand, which have some programs of exploration for coalbed gas.

IEA (2012) has predicted that China and India will surge in the use of coal, equally that of oil in 2017. However, both countries have rising coalbed gas development with China focused on exploiting and developing coalmine gas (e.g. coalmine methane (CMM) such as gob gas) to reduce mine explosions. Also, China uses coalbed gas recovered from coal mines to fuel power plants. The need to reduce coalbed gas content in advance of underground and strip coal mining to collect coalbed gas instead of venting to the atmosphere makes for a viable coal mine safety protocol and alternatively leveraging for reduction of greenhouse gases for carbon credits. Thus, the economic growth of China and India may benefit from a combination of both coalbed gas exploration as well as concomitant reduction in greenhouse gases.

The main component of coalbed gas is methane, which has about 25 times (Blasting, 2013) more global warming potential than carbon dioxide within a period of 100 years residence in the atmosphere (Chapter 2). That is, removing 1 ton of methane is equivalent to reducing 25 tons of carbon dioxide. Coalmine gas recovery and utilization are currently being conducted with great potential worldwide, including areas such as the United States, China, Poland, Australia, Ukraine, Germany, Russia, Mexico, United Kingdom, and India (Huang, 2010; Somers, 2012). Utilization includes electricity, boiler fuel, industrial fuel, town gas, LNG, and compressed gas (CNG). The United States Environmental Protection Agency (USEPA) coalbed methane outreach program includes 41 countries to recover and use coalmine gas in conjunction with global reduction of greenhouse gas and its impact on climate change (Somers, 2012).

GLOBAL GAS SUPPLY AND DEMAND

To a large degree, the supply and demand for natural gas consumption is driven by the need for cleaner and multi-applied energy. A report by the IEA (2011) has discussed the future role of natural gas in the global energy mix and projects that by 2035, global consumption of natural gas will rise by more than 50% from 2010 levels. Additionally, by 2035, natural gas will account for up to 25% of all energy consumed. However, the report (IEA, 2011) cautions that even with this shift to lower carbon producing fuel source, it might still not be enough to mitigate climate change.

The IEA (2013) estimated that worldwide gas demand in 2010 was approximately 3284 Bcm, which represents a 7.4% rise from 2009 levels. Gas demand has increased by about 800 Bcm over 2001–2010, which roughly translates into an annual rise of 2.7%. The global share of natural gas in the energy mix is about 21%. However, the USEIA (2011) reported that the global economic downturn from 2008 to 2009 resulted in about 4% decline for demand of natural gas followed by demand of 3.2 trillion m^3 (Tcm) in 2010 when economic growth resumed in 2010. The same USEIA report projected natural gas consumption from 2008 to 2035 to grow by about 4.7 Tcm partly from utilization by member countries of the Organization for Economic Cooperation and Development (OECD) (Figure 9.1(A)). It is predicted (USEIA, 2011) that the demand for natural gas will be strong

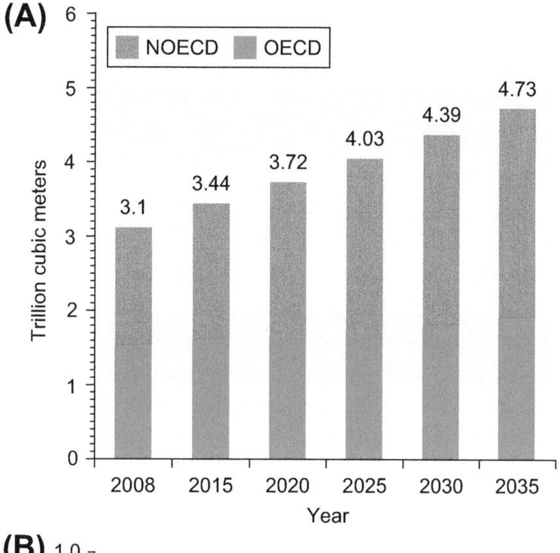

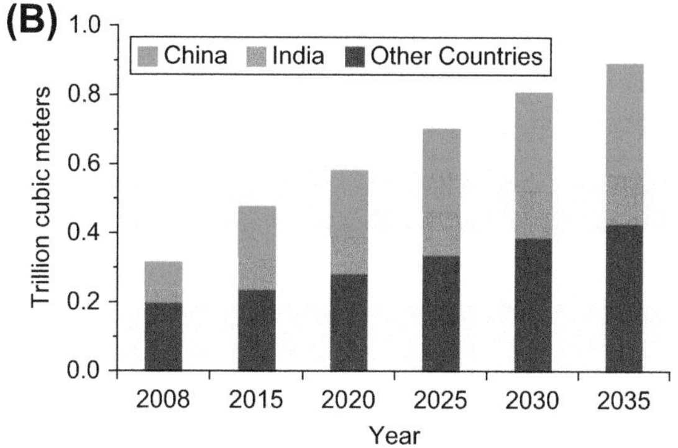

FIGURE 9.1 (A) Frequency diagram showing projected growth of consumption of natural gas between the OECD and NOCED countries from 2008 to 2035. Adopted from USEIA (2011). (B) Frequency diagram showing projected growth of consumption of natural gas in NOECD countries such as China, India, and other countries. *Source: Adopted from USEIA (2011).*

in economically growing non-Organization for Economic Cooperation and Development (NOECD) Asian countries such as China and India with average rate of consumption of 2.2% per year through 2035 accounting for about 35% of the total world consumption (Figure 9.1(A); USEIA, 2011). China and India will lead the growth of natural gas consumption from 2008 to 2035 with predicted increases of about 0.36 Tcm annually. In contrast, other NOECD Asian countries are predicted to increase by about 0.24 Tcm during the same period (Figure 9.1(B); USEIA, 2011).

USEIA (2011) indicated that in order to meet the consumption growth projected for 2035, additional gas supply of 1.7 Tcm or about 50% must be added mainly by gas producers from NOECD countries. It is thought that unconventional shale, tight, and coalbed gases will fill the gap in the predicted shortfall during this period. Production of unconventional gas will grow from 2008 to 2035 with production from OECD to average 3.2% per year from 0.39 to 0.92 Tcm, respectively. During the same projection period, NOECD unconventional gas production will grow from <0.03 to 0.37 Tcm. Much of the forecasts of unconventional gas production rely on the shale gas resources, which remains uncertain due to considerable variations and more so for insufficient data worldwide. However, coalbed gas could once again become a stable, economically viable source of natural gas especially for the NOECD Asian countries, which are projected to increase use of natural gas by an average of about 4% annually through 2035 (USEIA, 2011).

ROLE OF WORLD'S COAL TO COALBED GAS EXPLOITATION

The largest resources of coalbed gas are thought to be in Australia, Canada, China, Russia, and the United States. However, much of the world's coalbed gas recovery potential remains unexploited. Total global coalbed gas resources, as of 2006, were estimated to be 143 Tcm, but only 1 Tcm was actually recovered (WCA, 2013). The probable reason why so little coalbed gas had been extracted was because of lack of incentives in some countries to fully exploit the resource base, particularly in parts of the former Soviet Union where conventional natural gas is abundant. Although there are plentiful coal resources in the United States, Canada, China, Russia, the former states of federation of Russia, India, Africa,

BOX 9.1

NATURAL GAS RECESSION—CONCEPT AND MISCONCEPTION

The global gas market was depressed during the economic downturn ignited by the global financial crises of 2008. Gulik (2008) reported that the U.S. Economy entered into a recession at the end of 2007. The concept of economic recession according to Gulik (2008) is calculated from the National Activity Index, which measures the change in activity across 85 economic indicators in the United States. The start of the economic recession was indicated by the index falling during a 3-month moving average below −0.70 (Gulik, 2008). The misconception that energy is immune from this economic activity is shown in Figure 9.3, in which the price (e.g. Henry Hub) of natural gas slumped during 2008 and 2009, and remained almost flat through to 2012 (USEIA, 2013a). The economic downturn was preceded by a gradual price increase beginning in 2002 due to rapid growth of natural gas production associated with expanded interstate pipeline capacity. The expansion was spurred on by a plentiful supply of gas from coal and shale. The recession slowed industrial and electric power generation demands, which in turn curtailed natural gas drilling and production activities (Figure 9.2). Although the economic recession caused instability in the energy and gas markets worldwide resulting in decline of demand, new supply surged with the onset of LNG and gas from shale (Honore, 2011). The shift in the balance between supply and demand created new market paradigms.

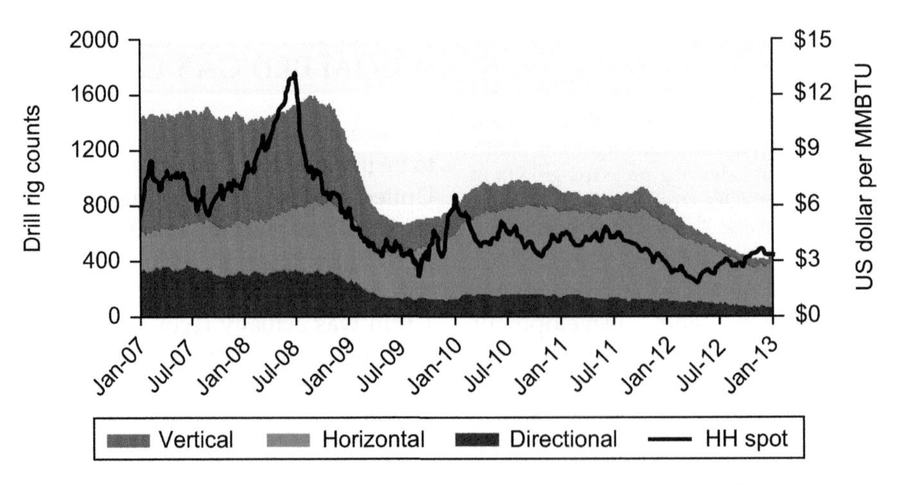

FIGURE 9.2 Chart showing the variations in Henry Hub (HH) spot price of natural gas from 2007 to 2013. Superimposed on the chart are the variations in the number of vertical, horizontal, and directional drilling rigs. MMBtu, million British Thermal Unit; 1 BTU, 1.055 kJ. *Source: Adopted from USEIA (2013a) with source from Baker Hughes.*

and even throughout Europe, the only countries with significant coalbed gas projects in progress are China, Australia, Indonesia, United States, and Russia. These countries possess the largest coal reserves and are major coal consumers (BP, 2012). However, despite the availability of coal for coalbed gas development as an inexpensive source of clean energy, investments in projects have been slowed by the global economic recession.

Global coalbed gas resources can come from coal beds pre- or postmining, as well as during mining (Chapter 2) actively and/or abandoned mines as well as from unmineable virgin coals. Gas recovery from unmineable coal beds refers either to production before mining (e.g. in front or in advance) or to technically uneconomic deposits. Production of coalbed gas from virgin coals refers to recovery from deposits too deep to be mined. Exploitation of CMM and abandoned mine methane (AMM) (see Chapter 2 for definitions) in advance of mining and coalbed gas from virgin coals are viable projects recommended in the global methane initiatives sponsored by USEPA (2010). Recovery of CMM and AMM is an important component in minimizing global coalbed gas emissions through the Methane to Market (M2M) programs (USEPA, 2010). Partnership in CMM and AMM recovery and utilization consist of 37 OECD and NOCED countries, which represent 70% of the world's emissions of anthropogenic methane (USEPA, 2010). According to USEPA (2010) as of 2005, CMM comprised 6% of the total global emissions and if that gas could be recovered, it would not only provide clean fuel but also improve mine safety and reduce greenhouse gas emissions.

GLOBAL COALBED GAS REGIONS: OECD VS NOCED

The coal resource-based development of coalbed gas is best summarized, reviewed, and analyzed according to its status and potential in the OECD and NOCED countries. The organization of this section of the chapter conforms, in general, to the structure of reporting of world energy resources by WEC (2007a,b), USEIA (2011; 2012; 2013a,b) and BP (2012). Selected member countries of the OECD and NOECD will be profiled according to the following topics: (1) coal-resource base, (2) rank of coal-resource base, (3) geologic age of resource base, (4) coalbed gas resource and exploitation activities, (5) resource distribution, (6) coalmine gas potential and emissions, and (7) opportunities beyond current status.

In addition, the level of research, technology, production, and consumption of unconventional gas, in general, and coalbed gas, in particular, has influenced the grouping of the OECD and NOECD countries. The NOECD countries such as China have more than tripled coal production compared to that of the United States. Also, China has more than doubled coal production compared to Australia and United States combined coal production in 2011 (BP, 2012). USEIA (2011) reported that the NOECD countries are projected to be the major consumer of coal and natural gas (e.g. coalbed gas) toward the year 2035 (Figure 9.1(A)).

The OECD includes 34 countries mainly from Europe as well as North and South America, and Australasia. The organization focuses on assisting member governments by sharing experiences to improve economic, social, and environmental policies. Additionally, the cooperation between OECD countries makes sharing and flows of technical expertise advantageous over NOECD. The NOECD includes the nonmember countries such as coal-rich China, India, Indonesia, Kazakhstan, Russia, South Africa, and Ukraine. As would be expected, coal-rich countries, whether OECD or NOCED, are the most prolific coal producers, consumers, and exporters of their resources.

OCED Countries

Out of 34 member countries of OECD, only 17 countries have some coalbed gas activities as summarized in Table 9.1 and demonstrated in

TABLE 9.1 Summary of Coalbed Gas Activities, Geologic Age, Coal Rank, Coalfields, and Coal Resource and Reserves of Various OECD Countries with Exploration and Development Potential. Compiled from Numerous References as Indicated in the Table

Country	Coalbed Gas Activities		Coal Age	Coal Rank	Basin/Region/ Coalfield	Coal Resource/ Reserve (Coalbed Gas/CMM Emissions)	References
	Coal mine	Virgin Coal					
Czech Republic	CMM AMM	CBG	Tert., Carb.	Lignite to bituminous	Bohemian, Silesian	Coal reserves: 1.1 Bt (CBG GIP resources: 100–500 Bcm; CMM emissions: 6.9 Bcm from 1990 to 2007; Recovered and used: 100 MMm^3/year)	Durica and Nemec (1997), USEPA (2010), BP (2012)
France	CMM AMM	CBG		Bituminous	Central Massif, Lorraine, Nord-Pas de Calais	CBG GIP resources: 28 Bcm; CMM emissions: 490 $MmtCO_2e$)	USEPA (2010)
Germany	CMM AMM	CBG	Tert., Carb.	Lignite to bituminous	Ruhr, Saar, Munsterland	Coal reserves: 40.7 Bt (CBG GIP resources: 3 Tcm; CMM emissions: 60.2 Bcm from 1990 to 2005)	Mosle et al., (2009), USEPA (2010), BP (2012), Somers (2012)
Hungary	CMM	CBG	Tert., Jur., Trias.	Lignite to bituminous	Bakony, Mecsek, Pannonian, Torony	Coal reserves: 1.7 Bt (CBG GIP resources: 152–159 Bcm; CMM emission: 167.3 MMm^3 from 1985 to 2008)	Landis et al. (2002), Lakatos et al., (2006), USEPA (2010), BP (2012)
Italy	CMM		Tert.,	Lignite to Subbitumious	Sulcis,	Coal reserves: 1000 Mt (CMM emissions: 10 MMm^3)	USEPA (2010), Somers (2012)
Japan	CMM AMM	CBG	Tert., Cret.	Bituminous	Ishikari, Kushiro-Oki, Joban-Oki, Nishisonogi-Oki, Northwest Kyusyu, Sanriku-Oki, Tenpoku, Tomamae	Coal Reserves: 350 Mt (CBG GIP resources: 27 Tcm: CMM emissions: 1.6 Bcm from 1990 to 2008)	Koide and Kuniyasu (2006), BP (2012), USEPA (2010), Somers (2012)

Mexico	CMM	CBG	Tert., Cret., Trias–Jur.	Lignite to bituminous	Sabinas-Saltillito-Monclova, Fuentes-Rio Escondido, Colombia-San Ignacio, Mixteca, Barranca, Cabullona, San Pedro, Corrallitos	Coal reserves: 1.2 Bt (CBG reserves: 119–212 Bcm; CMM emissions: 1.74 Bcm from 1990 to 2010)	UMEC (2009), USEPA (2010), Kelafant (2011), Somers (2012)
New Zealand		CBG	Tert., Cret.	Lignite to bituminous; minor anthracite	Northland, Waikato, Taranaki, West Coast South Island, Greymouth, Otago, Southland	Coal resources: 15.6 Bt; Coal reserves: 571 Mt (CBG reserves: 53 Bcm; CMM emissions: 158 MMm^3 from 1990 to 2008)	USGS (2004), Clark (2008), Mares and Moore (2008), BP (2012), USEPA (2010)
Poland	CMM	CBG	Tert., Carb.	Lignite to anthracite	Lublin, Silesian	Coal reserves: 5.7 Bt (CBG GIP resources: 1.3 Tcm)	BP (2012), USEPA (2010), Somers (2012)
Spain	CMM	CBG	Tert., Cret., Carb.	Lignite, bituminous to anthracite	Asturias, Castilla-Leon, Leon-Palencia, Ciudad Real, Cordoba, Teruel, Galicia	Coal reserves: 530 Mt (CBG GIP resources: 25.1 Bcm; CMM emissions: 837 MMm^3 from 1990 to 2008)	Fernandez-Rubio (1986), Martinez (2004), Cienfuegos and Loredo (2010), BP (2012), USEPA (2011)
South Korea	CMM		Tert.	Lignite and anthracite	Boeun, Bukpyong, Daechon, Dangyang, Kangmung, Jeonsun, Honam, Munkyeong, Pohang	Coal reserves: 126 Mt (CMM emissions 415,000 m^3 from 1985 to 2010)	USEPA (2010), BP (2012)
Turkey	CMM	CBG	Tert., Cret., Carb.	Lignite to bituminous	Amasra, Armutçuk, Elbistan, Zonguldak	Coal reserves: 2.3 Bt (CBG GIP resources: 3 Tcm; CMM emissions: 782 MMm^3 from 1990 to 2008)	Yalçin et al. (2002), Mustafa and Ayar (2004), USEPA (2010), BP (2012), (Somers, 2012)

CBG, coalbed gas; CMM, coalmine methane; AMM, abandoned mine methane; PPS, probable + possible + speculative; GIP, gas-in-place; Bt, billion tons; Mt, million tons; Tcm, trillion cubic meters; Bcm, billion cubic meters; m^3, cubic meters; Tert., Tertiary; Cret., Cretaceous; Jur., Jurassic; Perm., Permian; Penn, Pennsylvania; Carb., Carboniferous.

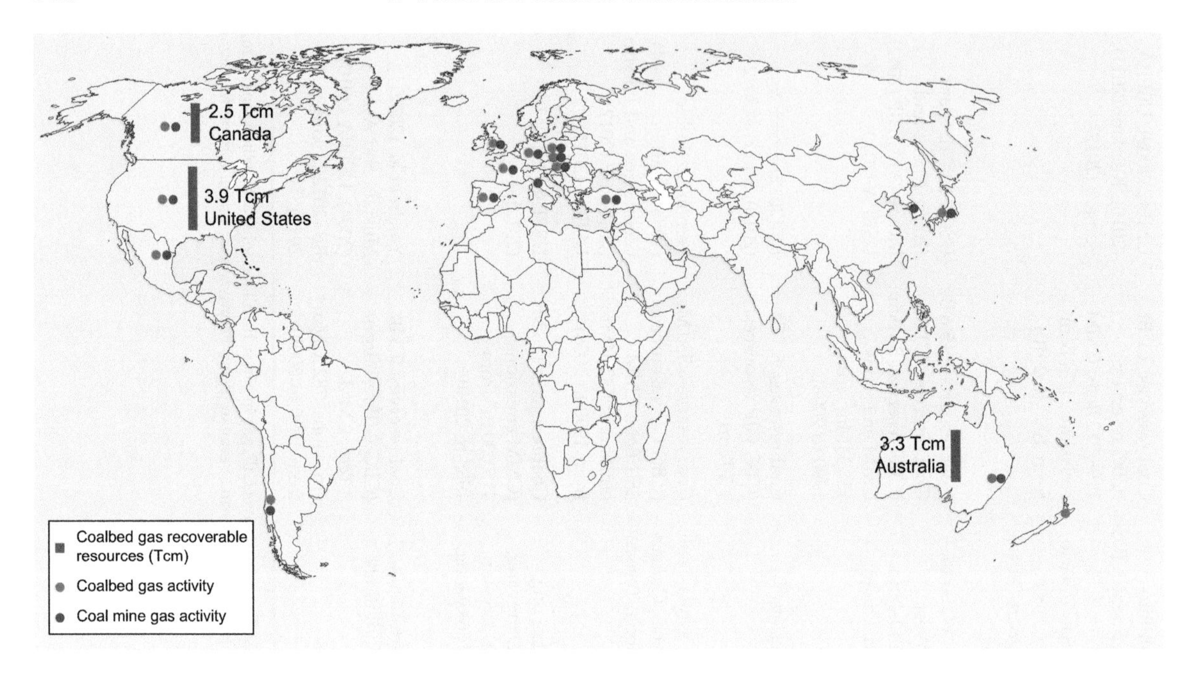

FIGURE 9.3 Map showing the worldwide distribution of coalbed gas and coalmine gas activities in OECD countries. The coalbed gas recoverable resources are of the United States, Australia, and Canada for the sake of comparison based on the estimates of Kuuskraa and Stevens (2009). Tcm, trillion cubic meter. *Source: The image of the world map is adopted from future.wikea.com.*

Figure 9.3. Coalbed gas activities are measured by the presence of exploration, development, and production of coalbed gas as well as documented coalmine gas emissions and potential recovery projects in relation to the global CMM initiative (Somers, 2012; USEPA, 2010). A comparison of the recoverable resources of coalbed gas worldwide as adopted from Kuuskraa and Stevens (2009) is shown in Figure 9.3. The OECD countries with recognized coalbed gas and/or potential coalmine gas projects (e.g. CMM and AMM) include Czech Republic, France, Germany, Hungary, Italy, Japan, Mexico, New Zealand, Poland, South Korea, Spain, and Turkey (Table 9.1). Other OECD countries not included in Table 9.1 which have coal resources but unrecorded coalbed gas activities include Austria, Belgium, Estonia, Greece, Ireland, Netherlands, Norway, Portugal, Slovak Republic, Slovenia, and Sweden. Many of these countries have lignite coal deposits with the exception of Belgium and Netherlands, which have previously extracted bituminous and anthracite grade coals (British Geological Survey, 2012). Not indicated in Table 9.1 and will be discussed in the following sections include the United States, Canada, Australia, and United Kingdom.

UNITED STATES

The United States has been commercially producing coalbed gas from coal mines since the 1950s and is presently the leading coalbed gas producer from unmineable or virgin coals worldwide since the 1970s. From early 1980s to present, 12 coal basins in the conterminous

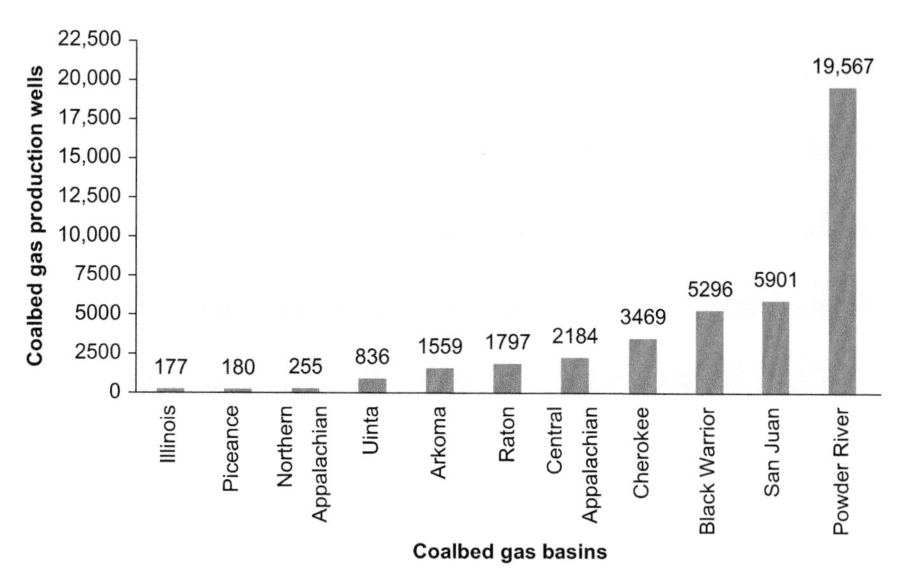

FIGURE 9.4 Frequency diagram showing coalbed gas production wells in 11 coal basins in the United States before 2008 global economic recession. *Source: Modified data from Cardott (2006), USEPA (2004), and database from COGCC, WOGCC, and USGS.*

United States have produced coalbed gas. No other countries have produced as much coalbed gas as the United States. However, since 2008, this production has slowed as a result of economic recession. Coalbed gas production peaked at about 55 Bcm in 2009 and declined the following year (USEIA, 2013b). As a result, operators of most of coalbed gas producing basins have reduced drilling activities and are taking this opportunity to evaluate the large volume of amassed reservoir data to assess reserves and maintain leases. The number of production coalbed gas wells for basins in the United States increased to 41,221 wells by the end of 2007 before the 2008 economic recession (Figure 9.4).

More than half of the 12 coal basins (Figure 9.5) are located in the Rocky Mountain region, which include the Powder River (Wyoming and Montana), Wind River (Wyoming), Greater Green River (Wyoming, Colorado, and Utah), Uinta (Utah), Piceance (Colorado), Raton (Colorado and New Mexico), and San Juan (Colorado and New Mexico) Basins. The other coal basins are located in the Central region of the United States, which include Arkoma (Arkansas and Oklahoma), Cherokee Platform (Oklahoma and Kansas), Illinois (Illinois and Indiana), and Gulf Coast (Texas, Louisiana, Mississippi, and Alabama) Basins. The Eastern region of the United States includes the northern Appalachian (Pennsylvania), central Appalachian (Virginia, West Virginia, and Ohio), and Black Warrior or southern most Appalachian (Alabama) Basins.

Geologic Ages and Rank of Gas Producing Coals

The geologic ages of the gas producing coal beds in the conterminous United States range from Pennsylvanian to Tertiary. The producing coal beds range from subbituminous to anthracite rank. Producing coal beds in the Appalachian

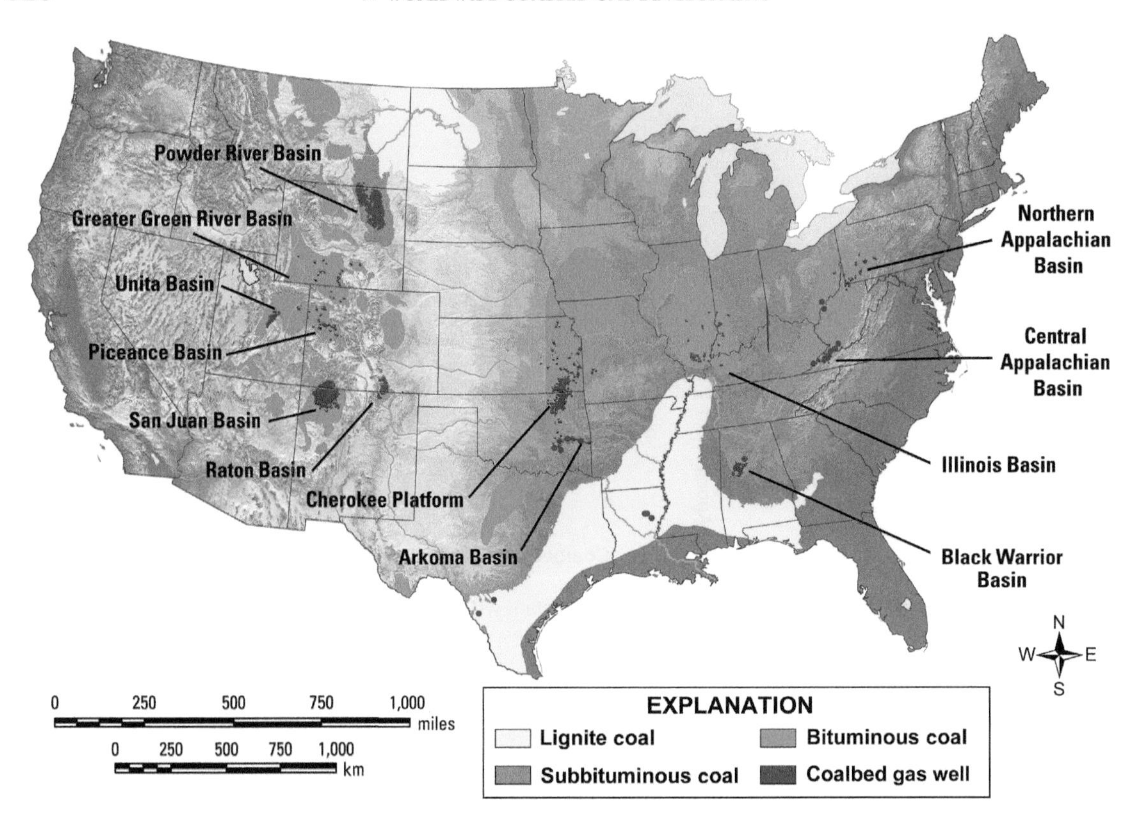

FIGURE 9.5 Location map of the 12 coal basins producing coalbed gas in the conterminous United States. Coal ranks are indicated in each basin. Map is modified from Tully (1996).

Basin (northern and central) and Black Warrior Basin are Pennsylvanian age and bituminous to semianthracite rank. The Illinois, Cherokee Platform, and Arkoma Basins produce gas from Pennsylvanian age bituminous coal although gas production from subbituminous coal occurs in the Illinois Basin. Late Cretaceous and Tertiary age subbituminous, bituminous, and anthracite coals produce gas in the Greater Green River and Raton Basins. The Piceance, San Juan, Uinta, and Wind River Basins produce from Cretaceous age bituminous rank coals. The Gulf Coast, Powder River, and Raton Basins produce from Tertiary age subbituminous rank coal although the Gulf Coast also has potential production from Cretaceous age bituminous coal.

Basin Hierarchy: Undiscovered Resources vs Production

The coalbed gas producing basins in the conterminous United States are best analyzed by arrangement into a series of hierarchy based on undiscovered, technically recoverable resources and production. Coal geology, coal reservoir properties, coal resources, and applied development technology, all discussed in Chapters 3–5 and Chapters 7–8, influence both criteria. Estimates of coalbed gas resources and reserves are variable based on methodologies applied in the assessment (Chapter 6). Thus, estimates are often not comparable. However, production estimates are much more consistent and more reliable for comparison.

Undiscovered Technically Recoverable Resources

The Potential Gas Committee (PGC, 2013) estimate of natural gas, in general, and coalbed gas, in particular, provides a realistic assessment of "technically recoverable gas resources potential" compared to gas-in-place resources (ICF Resources, 1990) of the United States. The PGC (2013) probable + possible + speculative resources for coalbed gas are 4.74 Tcm (statistically aggregated mean value). Since 2002, the U.S. Geological Survey (USGS) has an ongoing national assessment of geologic coal basin provinces, which uses a geology and petroleum system-based assessment methodology (Chapter 6). The results of the assessments of undiscovered, technically recoverable coalbed gas resources are presented below for comparison. The USGS assessment indicated that about 1.33 Tcm of coalbed gas resources (mean or F50 probability, that is 10 chances out of 20 that this volume will be recovered) exists in the Powder River, San Juan, Piceance, Greater Green River, Raton, and Wind River Basins (USGS, 2002). The Powder River Basin was estimated to contain more than 0.39 Tcm at the mean (F50 probability) of coalbed gas resources (Flores, 2004; USGS, 2002). The San Juan Basin, where coalbed gas has been produced for 30 years, was estimated to contain about 54% (0.68 Tcm at the mean or F50 probability) of coalbed gas resources (USGS, 2002). Minor basins such as the Greater Green, Piceance, Raton, and Wind River were estimated to be about 0.04, 0.06, and 0.01 Tcm at the mean (F50 probability) of coalbed gas resources, respectively (USGS, 2002). With the inclusion of recent USGS assessments of the Appalachian Basin with Black Warrior, Illinois, and Gulf Coast Basins and the reassessment of the Raton Basin (Milici, 2004; Milici & Hatch, 2004; Higley, 2007; USGS, 2007), a total of 1.87 Tcm is estimated for 10 out of 12 producing conterminous U.S. coal basins, which is more than one-third of the technically recoverable gas resources potential estimate by PGC (2013).

Higley (2007) estimated the Raton Basin as having 0.04 Tcm at the mean (F50 probability) of coalbed gas resources, which is 0.01 Tcm less than the 1995 technically recoverable resources (USGS, 2002). Warwick et al. (2007) reported the coalbed gas resources of the Gulf Coast Basin to be 0.11 Tcm at the mean (F50 probability). Overall about 0.43 Tcm of coalbed gas resources were assessed at the mean (F50 probability) in the Appalachian Basin and Black Warrior Basin Provinces (Lyons, 1996; Milici, 2004; Milici & Hatch, 2004). Of this total, the Black Warrior Basin was reported to contain about 0.04 Tcm at the mean (F50 probability) of coalbed gas resources. Thus, the conterminous U.S. coal basins can be classified into two resource-based groups: >0.1 and <0.1 Tcm based on the USGS estimates of undiscovered, technically recoverable coalbed gas resources. Basins with >0.1 Tcm coalbed gas resources include the San Juan, Powder River, Appalachian (combined central and northern), and the Black Warrior Basin in the southern part of the greater Appalachian Basin. Basins with <0.1 Tcm coalbed gas resources include Piceance, Greater Green River, Wind River, Raton, and Gulf Coast Basins.

Coalbed Gas Production

USEIA (2007) grouped the coalbed gas producing basins into four tiers based on the annual production from 1980 to 2006. Tier 1 basins, in descending order of production, include the San Juan, Powder River, and Black Warrior, which yielded 3.2–28 Bcm of coalbed gas per year. Tier 2 basins, also in descending order of production, include the Raton, central Appalachian, Uinta, Arkoma and Cherokee, which together yielded 0.8–3.2 Bcm of coalbed gas

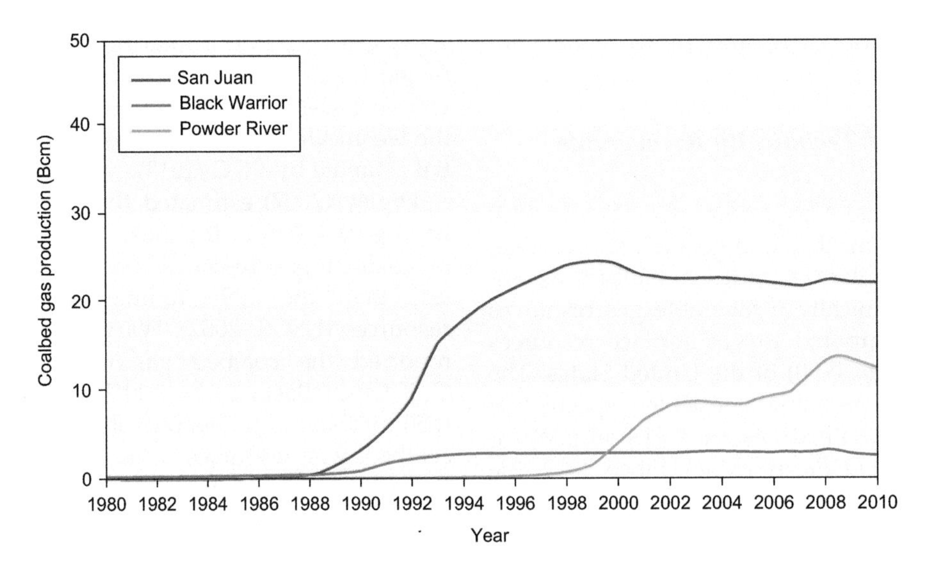

FIGURE 9.6 Coalbed gas production from 1980 to 2010 of Tier 1 coal basins of the United States. Bcm, billion cubic meter. *Source: Modified from USEIA (2007).*

per year. Tier 3 basins, in descending order of production, include Greater Green River, Piceance, northern Appalachian, and Gulf Coast yielded 0.014–0.140 Bcm of coalbed gas per year. Tier 4 basins (e.g. Illinois) yielded the smallest coalbed gas production of <280,000 m^3/year. The Tier system of basin hierarchical classification based on production coincides with the resource-based classification.

As shown in Figure 9.6, the 30 years (1980–2010) coalbed gas production of the Tier 1 basins started to increase for the Black Warrior and San Juan in 1988 and for the Powder River Basin in about 1997. The coalbed gas production in the Black Warrior Basin peaked in about 1995 and remained almost flat through 2010. The coalbed gas production in the San Juan Basin peaked in about 1999, declined and remained almost flat through 2010. The coalbed gas production in the Powder River Basin peaked in 2009 and declined through 2010. Since 2006, when USEIA (2007) last recorded production of Tier 1 basins, the coalbed gas production has not significantly increased through 2010 except for the Powder River Basin. The Powder River Basin remained the second most productive behind the San Juan Basin in the Tier 1 basins. By 2010, the total coalbed gas production from the Tier 1 basins (39 Bcm) made up about 78% of the total production in the conterminous United States, which declined from 2008 to 2010 as a result of the economic recession (Figure 9.6).

The coalbed gas production of the Tier 2 basins started in the early 1990s with the peaking of the Uinta Basin in 2002, Arkoma Basin in 2007, Cherokee Platform in 2008, and central Appalachian Basin in 2009 (Figure 9.7). In addition, production of Tier 2 basins remained below 3.2 Bcm except for the northern Appalachian Basin, which increased to about 4 Bcm within the limit for Tier 1 basins. Production for all the Tier 2 basins decreased in 2010. The total coalbed gas production of the Tier 2 basins in 2010 is about 10.25 Bcm, which was about 20.5% of the total production of all basins in the conterminous United States (Figure 9.6).

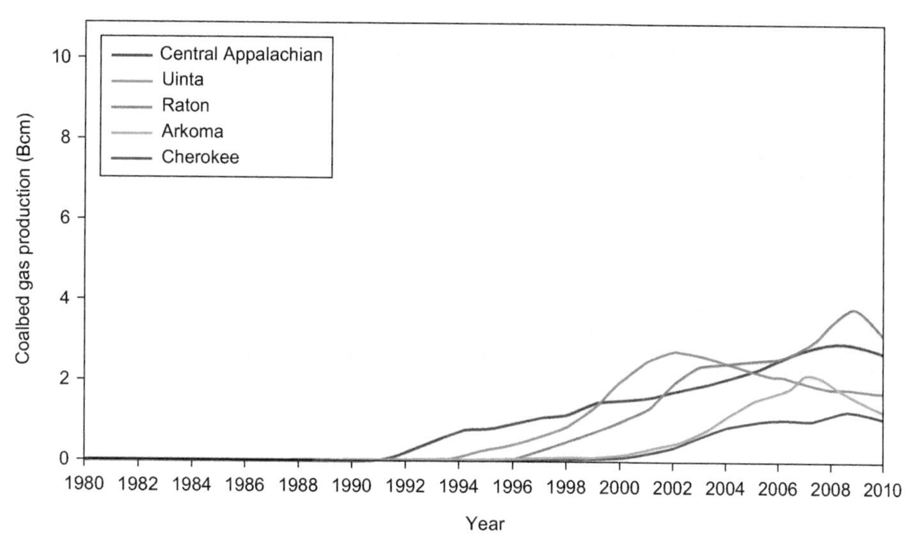

FIGURE 9.7 Coalbed gas production from 1990 to 2010 of Tier 2 coal basins of the United States. Bcm, billion cubic meter. *Source: Modified from USEIA (2007).*

The Tier 3 basins, which started coalbed gas production in late 1980s to early 1990s, have a binomial distribution of production (Figure 9.8). The production in the Piceance Basin peaked (0.10 Bcm) early in 1992 at the same time as the production from the other basins started to increase and subsequently remained flat to 2010. The Gulf Coast, Wind River, Greater Green River, and northern Appalachian all peaked in their production in 2009–2010. The Gulf Coast and Wind River Basins had a maximum production at about 0.025 and 0.075 Bcm, respectively. However, from 2006 to 2010, the coalbed gas production of the Greater Green River and northern Appalachian Basins rapidly increased and peaked at 0.52 and 0.43 Bcm, respectively. The total coalbed gas production of Tier 3 in 2010 is about 0.75 Bcm or about 1.5% of the total production of the coal basins.

Comparison of Mature vs Maturing Basin Characteristics

The comparison of the coal basins in the conterminous United States is complicated by the fact that some basins produced coalbed gas longer than others. Consequently, basins have matured while others have not. The common factor among coalbed gas producing basins is that each basin possesses unique set of characteristics (Table 9.2). Prospective operators place importance in the evaluation of the coal geology, coal rank, reservoir properties, and production potential before investments are considered in a coal basin. However, if advanced drilling, completion, and simulation technologies are applied, noneconomic basins can be made more profitable.

The characteristics of the Tiers 1–3 basins are compared in order to properly evaluate and understand the successful drivers for viable coalbed gas development. The success of coalbed gas development in the coal basins of the conterminous United States can serve to encourage the potential for commercial production worldwide. The successes of the Black Warrior and San Juan Basins initially developed in the 1970s and 1980s, respectively, are well known and the characteristics of their coal geology and reservoirs have served as proven

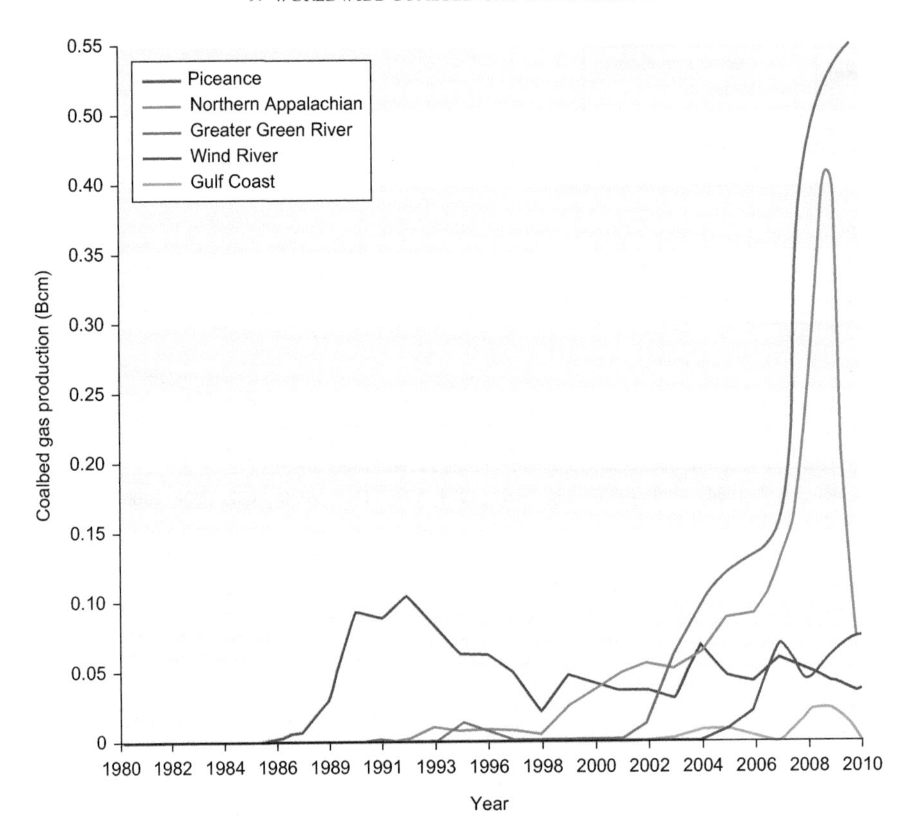

FIGURE 9.8 Coalbed gas production from 1985 to 2010 of Tier 3 coal basins of the United States. Bcm, billion cubic meter. *Source: Modified from USEIA (2007).*

models with the San Juan Basin highlighted as the most prolific producer worldwide (Ayers, 2003; Pashin, 2010). The proven technology and economics utilized in the Tier 1 coal basins paved the way for the sudden increase of coalbed gas development during the 1990s in the Tiers 2 and 3 basins. Thus, some of the basins developed later than these basins presented unique problems, which were overcome by improved drilling technology and became successful coalbed gas projects. For example, the Powder River Basin coals were originally deemed as uneconomical to develop in the 1980s due to their low-rank coals. However, the shallow thick coals were reevaluated for development in the 1990s with technologically improved completion (Flores, 2004).

The comparison of the characteristics of the coal geology, reservoir properties, applied technology, production, and coalbed mine gas recovery is summarized for six of the coal basins from Tiers 1–3 basins (Table 9.2). The coal geology includes the coal rank, thickness, and depths. Reservoir properties include permeability and gas content. The applied technology represents methods of drilling, stimulation, and water extraction. Production includes well spacing, types of water disposal, and daily well water and gas production. Active and abandoned mines coalbed gas recovery is a

TABLE 9.2 Summary of the Coal Geology and Reservoir Characteristics of Selected Tiers 1–3 Coal Basins Producing Coalbed Gas in the United States. Compiled from Numerous References as Indicated in the Table

Characteristics	Powder River Basin	Raton Basin	Uinta Basin (Ferron and Wasatch Plateau)	Arkoma Basin	Appalachian Basin (Central and Northern)	San Juan Basin
Coal Age	Paleocene–Eocene	Late Cretaceous–Paleocene	Late Cretaceous	Pennsylvanian	Pennsylvanian	Late Cretaceous
Coal thickness (range in meters)	1–75	1–15	1–10	1.2–2.4	1–6.7	1.5–12.5
Coal rank	Subbituminous C-A	High-volatile A–C bituminous	High-volatile B bituminou	Medium-volatile to low-volatile bituminous, and semianthracite	High-volatile A to low-volatile bituminous, semianthracite	Subbituminous A to low-volatile bituminous
Coal gas content (cc/g)	0.1–3.9	6.0–15.4	0.1–10.3	2.3–17.8	0.1–18.7	<3–25
Coal permeability (mD)	1–1000	1–5	5–20	1–10	0.5–1.5	<10–100
Well spacing, acre unit (hectare)	40–80 (16–32)	160 (64)	160 (64)	80 (32)	60 (24)	160 (64); 320 (130)
Well depth (range in meters)	55–975	137–1068	305–1345	87–1341	150–762	169–1220
Drilling	Vertical	Vertical	Vertical	Vertical, horizontal	Vertical, horizontal	Vertical, horizontal
Stimulation	Hydraulic fracturing not widely practiced	Hydraulic fracturing, gel, water, and sand proppant	Hydraulic fracturing, cross-linked borate gel, gel fracturing fluids, foams, and proppant	Hydraulic fracturing, linear gel, acid, nitrogen foam, and slick water	Hydraulic fracturing, nitrogen foam, water, scale inhibitor, microbicide additive, and sand proppant	Hydraulic fracturing, water, hydrochloric acid, slick water mixed with solvent, linear and cross-linked gels, and nitrogen- or carbon dioxide-based foams

(Continued)

TABLE 9.2 Summary of the Coal Geology and Reservoir Characteristics of Selected Tiers 1−3 Coal Basins Producing Coalbed Gas in the United States. Compiled from Numerous References as Indicated in the Table (*cont'd*)

Characteristics	Powder River Basin	Raton Basin	Uinta Basin (Ferron and Wasatch Plateau)	Arkoma Basin	Appalachian Basin (Central and Northern)	San Juan Basin
Co-produced water extraction	Submersible	Submersible	Conventional rod pump	Conventional rod pump	Conventional rod pump	Conventional rod pump
Co-produced water disposal	Surface discharge to streams, surface impoundments, irrigation, reinjection	Surface discharge to streams, surface impoundments, reinjection, evaporation pits	Reinjection, evaporation pit	Reinjection, commercial disposal	Reinjection, surface, discharge to streams, evaporation pit, commercial disposal	Reinjection
Gas production (m^3/D/Well); *Co-produced water (L/Day/Well)*	10,135−14,190 17,808 (ave); *3,577−57,235* *12,257 (Ave)*	3866 (ave); *27,846 (Ave)*	23,848−46,265; 25,637 (ave) *6.902 (Ave)*	4899 (ave.); *0−221,906*	12−1420 *596−5,962*	28,000−170,000 (Ave. 50,983); *4,770−59,620* *(4998 Ave)*
Coalmine methane (CMM and abandoned mine methane (AMM) recovery projects	Surface mines projects, estimated recovered emissions = 0.1 Bcm (2006−2009)	Abandoned mine methane (AMM) recovery project	Abandoned mine methane potential project	Coalmine methane, gassy mine potential recovery project	CMM and AMM recovery projects	Coalmine methane, gassy mine potential recovery project
Reference	BLM (2003), Flores (2004), USEPA (2004, 2005, 2008), Stricker et al. (2006), WOGCC (2013)	Close (1990), Tyler et al. (1995), Flores and Bader (1999), Bryner (2002), USEPA (2004, 2005, 2008), Ash and Gintautas (2009), Maccartney (2011)	Lamarre and Burns (1996) Gloyn and Sommers (1993), Dallege and Barker (2000), USEPA (2004, 2005, 2008)	Cardott (2001), Cardott (2006), Rieke and Kirr (1984), USEPA (2004, 2005, 2008)	Adams et al. (1984), Adams (1984), Lyons (1996), Conrad et al. (2006), Milici (2004), Milici and Hatch (2004), USEPA (2004, 2005, 2008)	Bryner (2002), Ayers (2003), USEPA (2004; 2005; 2009), McPherson (2006), Fassett (2000, 2010)

mD, millidarcy.

subset of development of the coalbed gas. Potential and ongoing CMM and AMM recovery projects are reported by USEPA (2005, 2008, 2010). The basins selected for comparative analysis include the Powder River and San Juan Basins for Tier 1 and Raton, Uinta, Arkoma, and Appalachian Basins for Tier 2. The Tier 2 central Appalachian Basin and Tier 3 northern Appalachian Basin were combined because of similarity in the coal geology. These basins were chosen to include as much variety in characteristics such as coal rank, gas content, permeability, well depths, drilling and completion technology, and production history.

Tier 1 vs Tier 2 Basins

The Tier 1 San Juan and Powder River Basins significantly differ in coal geology varying from bituminous to subbituminous, thin to thick beds, high to low gas content, low to high permeability, and deep to shallow wells (Table 9.2). Coal rank probably controls the difference in gas content and to a degree, the permeability between the Powder River and San Juan Basins. The well depths of the coalbed gas reservoirs between the two basins are about the same such that the coal rank difference is a result of the higher maturation of the San Juan Basin over that of the Powder River Basin. High permeability of the subbituminous coal of the Powder River Basin may have influenced the less aggressive stimulation applied compared the lower permeability of the bituminous coal of the San Juan Basin. However, the prolific gas production in the deepest part or fairway of the San Juan Basin may be all contributed by the high-rank coal reservoir. The thickness of the Powder River Basin coal reservoirs, which is as much as 6 times that of the San Juan Basin coal beds, contributed to the productivity.

The Tier 2 basins are very similar in coal bed thickness, rank, gas content, and permeability (Table 9.2). However, permeability is low in the Appalachian Basin and high in the Uinta Basin compared to the other Tier 2 basins. The permeability may have influenced the nature of stimulation applied to the reservoir with the Appalachian coal, which undergone more hydraulic fracturing treatment than the other coal reservoirs. The high coalbed gas production of the Cretaceous coals in the Uinta Basin is probably attributed to the high permeability. The gas productivity of the Tier 2 basins on the whole may be directly related to the high gas content directly related to the bituminous rank. The gassiness of bituminous coal and higher rank coal in general is typified in the central Appalachian Basin (Conrad et al., 2006; Figure 9.9(A),(B)).

Comparison of the characteristics of Tiers 1 and 2 basins yields no significant difference in the coal geology and reservoir properties (Table 9.2). The coal rank (high-volatile C-A bituminous and medium- to low-volatile bituminous to semi-anthracite), gas content (0.1–18.7 cc/g), permeability (1–20 mD), and well depths (87–1345 m) of Tier 2 basins are within the range of the same parameters of the San Juan Basin. The only distinctive difference is that the San Juan Basin is more permeable (<10–100 mD) and has high gas rates. However, the applied technology such as drilling, completion, and stimulation treatment between these basins influenced the production gas rates and volume of gas.

Technology: A Play Changer

Technological strategies to increase production of coalbed gas plays in the United States have focused in maximizing production using horizontal and infill vertical drilling as well as hydraulic fracturing. Horizontal drilling can increase gas recovery by penetrating the maximum coal cleat surfaces perpendicular to bedding planes (Chapters 5 and 6). In the San Juan Basin, adoption of horizontal access from vertical wells assists in maximizing gas production. However, infill vertical drilling also can increase gas recovery from an underpressured low-productivity coal reservoir by reconfiguring

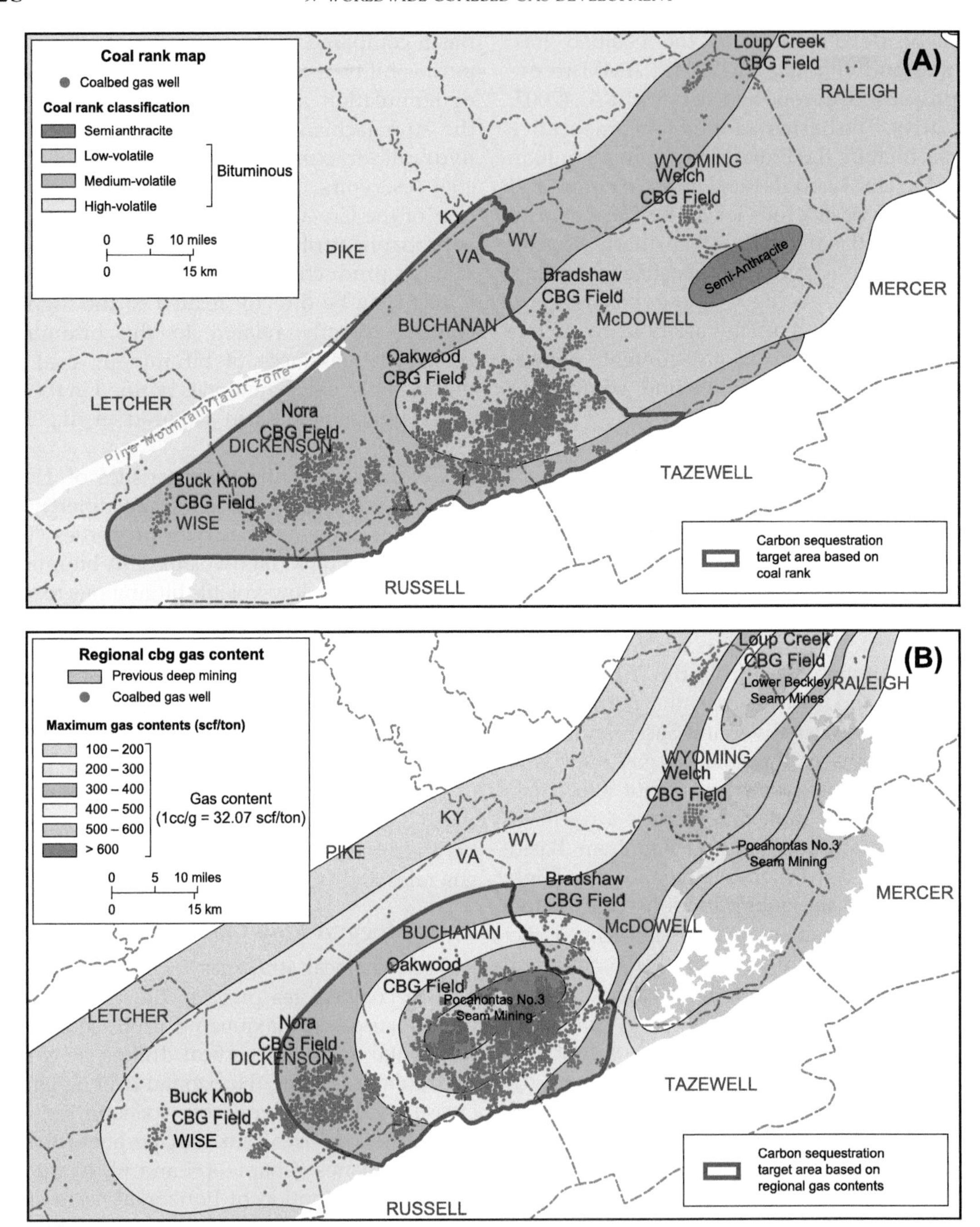

FIGURE 9.9 Coal rank (11A) and gas content (11B) maps of the Pennsylvanian coals in the central Appalachian Basin. See Figure 9.5 for location. cc/g, cubic centimeter per gram, scf/t, standard cubic feet per ton, cbg, coalbed gas. *Source: Adopted from Conrad et al. (2006).*

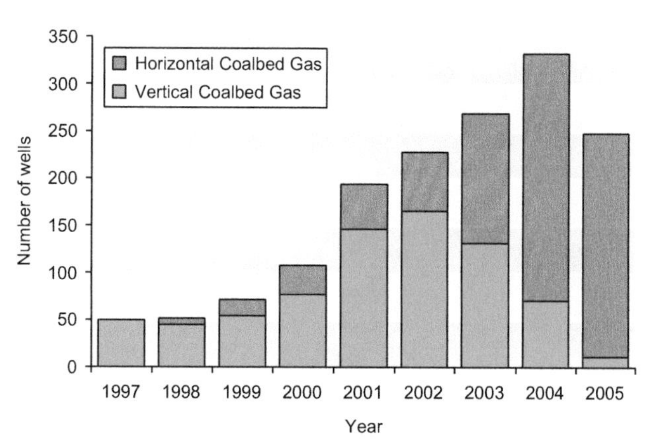

FIGURE 9.10 Frequency diagram showing the number of vertical and horizontal wells drilled in the Arkoma Basin from 1997 to 2005. Horizontal wells dominated (79–96%) in 2004–2005. *Source: Adopted from Cardott (2006).*

the permitted well-spacing unit. In the San Juan Basin, the change from the prevailing well spacing instituted in the 1980s of 320 (two wells per section) to 160 (four wells per section) acre units (1 acre = 0.40 ha) also improved insufficient to sufficient productive reservoirs. The use of horizontal drilling and multistage hydraulic fracturing with multistage stimulation treatment transformed inadequate to productive reservoirs (McDaniel, 1990). Thus, the effective application of available innovative technology can optimize production potential. Improved production through hydraulic fracturing has also been demonstrated in the Tiffany Area of the San Juan Basin (Ramurthy et al., 2003). Coalbed gas wells originally stimulated with a single-stage treatment were restimulated, which showed substantial improvements in production of 100–400%. Co-produced water through time decreased <5962 L/day in most coalbed gas wells. Although horizontal drilling and multistaged fracturing stimulation (Table 9.2) have been applied in the Raton, Piceance Arkoma, and the Central Appalachian Basins, coalbed gas production has not as yet reached the success of the San Juan Basin (Figure 9.10).

The Powder River Basin in comparison to Tiers 1 and 2 basins is unique because it contains the thickest (as much as 75 m) coal, is mostly low rank (subbituminous C-A), has low gas content, and is the highest permeability (>100s millidarcy to >1 darcy) of all the coalbed gas reservoirs (Table 9.2). The first coalbed gas wells drilled in the Powder River Basin were completed using conventional drilling, which produced little gas but huge volumes of water (Flores, 2004). In order to increase gas production, companies altered their drilling and completion technology adopting an underreaming method for the shallow coal beds (Chapter 7). Coalbed gas wells were drilled to the top of the coal, casing set, then a second phase of drilling went through the coal and then the coal was underreamed. Wells are mainly stimulated by pumping water to remove the damage by drilling. The high volume of extracted groundwater during production necessitated coalbed gas wells in the Powder River Basin to be spaced on 40- or 80-acre units (1 acre = 0.40 ha) compared to the 160–320-acre units in the other Tiers 1 and 2 basins. The 40-acre unit well spacing in the Powder River Basin required 16 wells to lift the water in the coal reservoirs often time leaking from overlying aquifers. The Powder River Basin is distinctive with regard to coalbed gas development in that the coals are overlain by large, thick

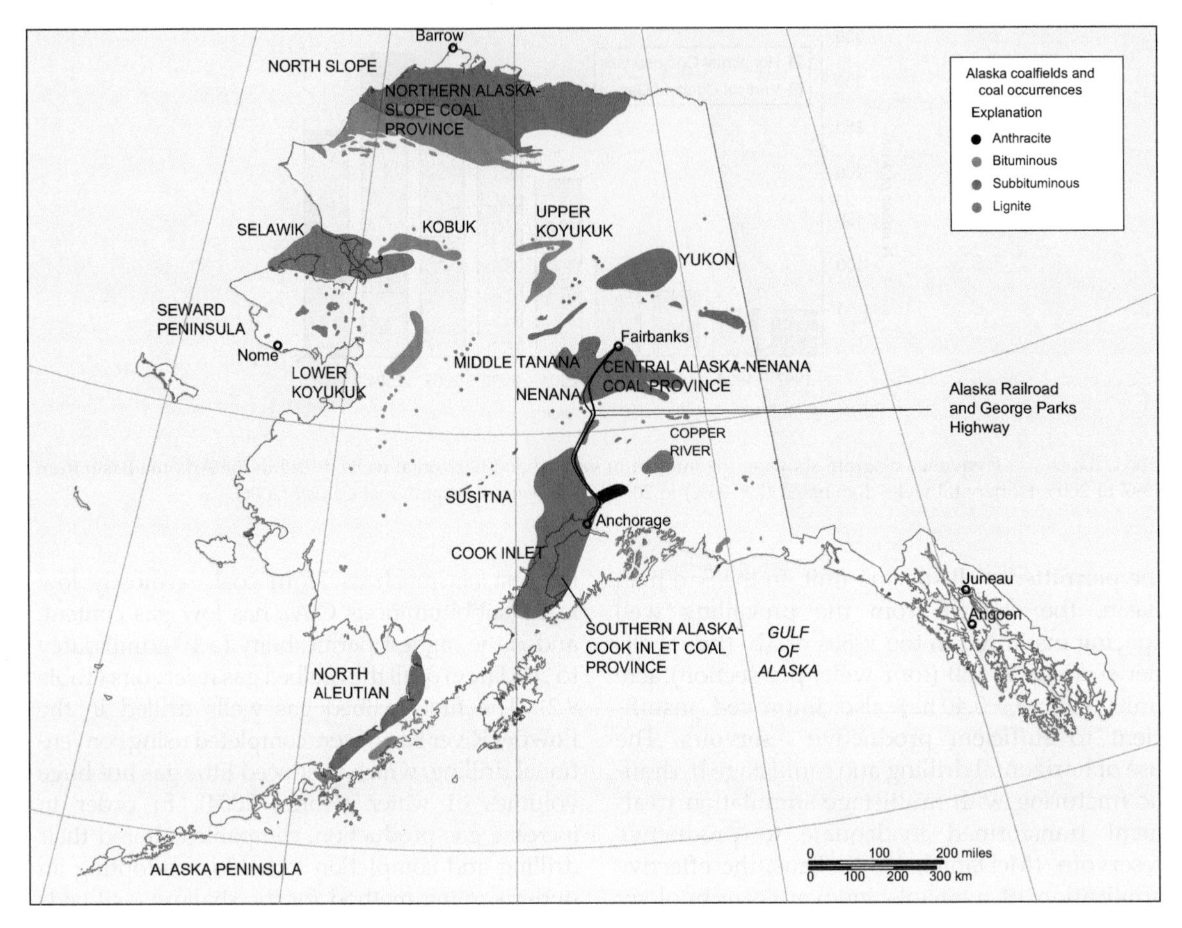

FIGURE 9.11 Map of Alaska showing distribution of coal basins and provinces as well as coal rank. Seventy eight percent of the total coal and potential coalbed gas resources of Alaska occur in the Northern Alaska-Slope and Southern Alaska-Cook Inlet coal provinces. *Source: Adopted from Flores et al. (2005).*

aquifers of the Wasatch Formation and underlain by aquifers of the Lower Fort Union Formation. The hydraulic connections between these large aquifers with intervening confining units and connections of coal reservoirs to overlying thick sandstone aquifers are complications to dewatering the coal reservoirs.

Emerging Coalbed Gas Basins in Alaska

Alaska's total coal resources are estimated to be 5203 billion metric tons, which surpass the total coal resources of the conterminous United States by 40% (Flores, Stricker, & Kinney, 2005). The coal resources are found in the Northern Alaska-Slope, Central Alaska-Nenana, and Southern Alaska-Cook Inlet coal provinces (Figure 9.11). The Northern Alaska-Slope coal province contains about 67% of the total coal resources of the State with about 50% each for subbituminous and bituminous coal. The Southern Alaska-Cook Inlet coal province contains about 28% of the total coal resources of the State represented by lignite and subbituminous rank with bituminous coal in the Matanuska coalfield. Approximately 81% of the coal resources in the Southern Alaska-Cook Inlet coal province is in the offshore and 19% is in the onshore margins of the Cook Inlet Basin.

The total coalbed gas-in-place resource for Alaska has been estimated to be 28 Tcm by Smith (1995); however, this estimate was later reduced to about 16 Tcm by Flores et al. (2005). The reduction was based largely on using other coalbed gas analogs (e.g. gas content, gas adsorption, and coal rank) from the conterminous United States (e.g. San Juan and Powder River Basins). The total coalbed gas-in-place resource of the conterminous United States has been estimated to be 21 Tcm (Rice, 1997). If the Alaska total coalbed gas-in-place estimate is added, the grand total coalbed gas-in-place of the United States is about 49 Tcm. Although, estimates of total coalbed gas-in-place resources are often reported as indication of gas total endowment, it is not necessarily a reliable measurement (Box 9.3). In addition, because 81% of the Cook Inlet Basin is in the deep offshore area, this considerably reduces the recoverable coalbed gas-in-place resources. The coalbed gas-in-place resources in the onshore Cook Inlet Basin best suited for development is estimated to be 0.23 Tcm, which is about 15% of the total gas-in-place resource of that area.

Further assessments of the coalbed gas resources for the Alaskan coal provinces have been reported by Roberts (2008) and Stanley et al. (2011). Both assessments estimated the undiscovered technically recoverable resources of gas from the Cretaceous and Tertiary coal-bearing rocks in the Northern Alaska-Slope and Southern Alaska-Cook Inlet coal provinces. Roberts (2008) estimated undiscovered recoverable resources of coalbed gas in the Cretaceous and Tertiary coal bearing rocks in the Northern Alaska-Slope coal province. Using the Total Petroleum System geology-based methodology, the total undiscovered recoverable resource has an estimated mean (F50 probability, that is there are 10 chances out of 20 that this volume will be recovered) of 505.6 Bcm. The lack of gas content data and producing coalbed gas wells to establish production history in the study region prompted Roberts (2008) to use the estimate of ultimate recovery for coalbed gas wells in the subbituminous coals of the Powder River Basin and bituminous coals of the Uinta (e.g. Drunkards Wash or Ferron Sandstone and Helper gas fields) as analogs for assessment. Although analog is the most logical method of assessment, characterization of gas content, adsorption isotherms, and coal petrology would have added value to the methodology because of the variability of the coal reservoirs.

Stanley et al. (2011) estimated a mean (F50 probability) of 130.8 Bcm of undiscovered recoverable coalbed gas resources of the Tertiary (e.g. Miocene to Pliocene age) coal-bearing strata in Cook Inlet. However, this estimates excluded coal-bearing units buried deeper than 1830 m, which probably contain coals with closed fractures prohibitive of gas migration (e.g. low permeability). Like the geology-based Total Petroleum System method used in the Northern Alaska-Slope coal province, analogs from the conterminous U.S. coal basins were used for assessment by Stanley et al. (2011). The Tertiary coalbed gas resources represent about 25% of the total conventional and unconventional oil and gas resources in the Cook Inlet. A large volume of natural gas is trapped in sandstones in the Tertiary strata as conventional gas natural gas resources (Stanley et al. 2011). However, it is thought that a large part of those conventional natural gas resources has migrated from the coal beds to the adjoining channel sandstones (Flores et al., 2005).

The reduction of coalbed gas-in-place resources to undiscovered, technically recoverable resources of Alaska mimics that of the

BOX 9.2

SEVERED MINERAL RIGHTS: COALBED GAS CONUNDRUM—CONCEPT AND MISCONCEPTION

In the United States, different individuals or companies own the surface and subsurface mineral rights and a split estate results when surface right ownership is different from subsurface right ownership (Kansas Geological Survey, 2002). Mineral rights are severed if the surface landowner does not possess all or part of the minerals below the surface of a tract of land. Federal, state, local governments or their representative agencies as well as private landowners can own mineral rights (BLM, 2012).

The concept of a split estate has been a source of controversy for coalbed gas development especially in the western United States where the federal government withheld the mineral rights under homestead lands. Also, previous landowners commonly sell the surface land and retain the subsurface mineral rights. The subsurface minerals include coal, oil, natural gas, and other minerals. The mineral rights ownership becomes complicated when the original landowner sells the land to one party, the oil and natural gas to an oil company, and retains the rights to the coal. The split ownerships by three parties of the surface and subsurface estates become a conundrum when the associated minerals are developed. It becomes more of a problem when the coal owner sells the rights to a coal company and who mines the coal but later a gas company leases and develops the coalbed gas, which by regulation is not a part of the coal.

Many "CBM Policies" devised by developing countries have hinged on resolving conflicts between coal owners and gas operators especially with coalbed gas development in adjoining concessions. The Courts and Legislature as exemplified by legal tiff often resolve conflict on coalbed gas trespass over coal rights over coalbed gas rights in the WCSB in Canada (Alberta Oil, 2011; Ingelson, 2010).

conterminous United States. That is, from the gas-in-place estimates of the Smith (1989) of 28 Tcm to Flores et al. (2005) estimate of 16 Tcm is a reduction of more than 55%. The assessments of the technically recoverable resources by Roberts (2008) in the Northern Alaska-Slope and Stanley et al. (2011) in the Southern Alaska-Cook Inlet have further reduced the resources to only 636.4 Bcm. The estimate is 3.9% of Flores et al. (2005) gas-in-place estimate and 2.2% of Smith (1989) gas-in-place estimate. The lesson learned in these assessments is that original coalbed gas resources are almost always overestimated and with new knowledge, technology, and economic assessments, will be significantly reduced over time.

CANADA

Coal basins in the conterminous United States and Alaska extend across their borders into Canada. There are 15 coal basins, which stretch from the Canadian west to east coasts and from the southern boundary with conterminous United States to the northwest boundary with Alaska (Figure 9.12). The largest coal basin is the Western Canada Sedimentary Basin (WCSB), which extends from west to east, in the Rocky Mountain Front Ranges and Foothills, and Interior Plains regions of Alberta, British Columbia, Saskatchewan, and Manitoba Provinces, respectively (Smith, 1989; Smith, Cameron, & Bustin, 1994, 2004; Vogler, 2006). Southward,

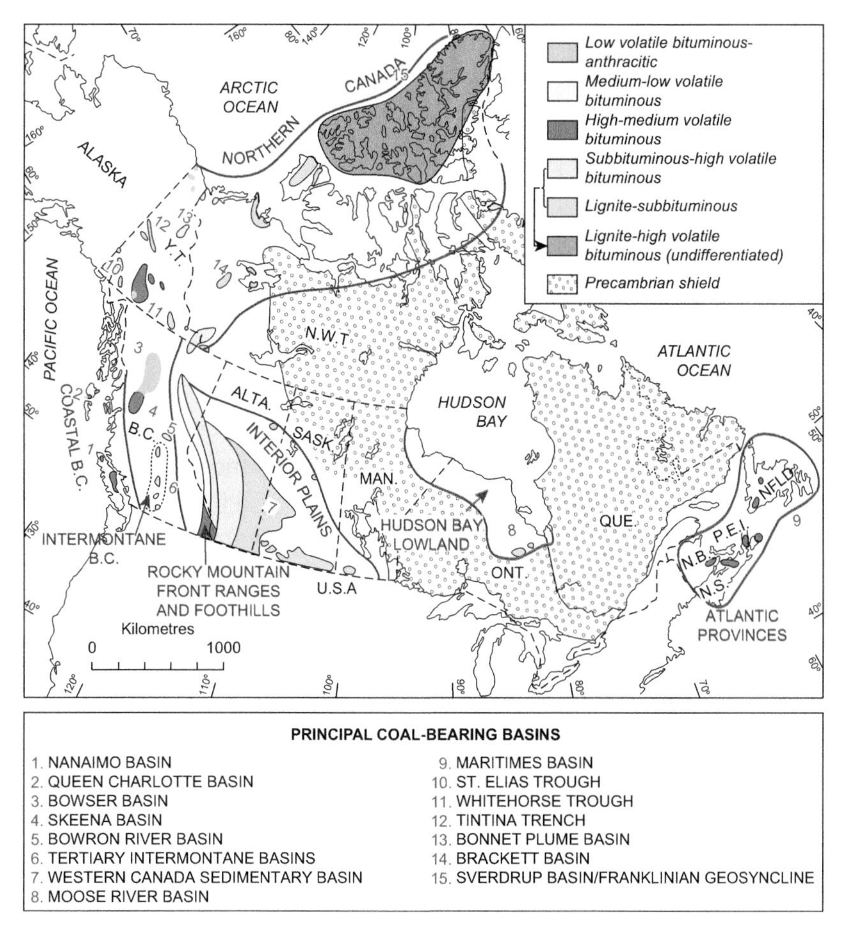

FIGURE 9.12 Map showing areal distribution and coal rank in coal basins in Canada. The Western Canada Sedimentary Basin (WCSB) is No. 7. *Source: Adopted from Smith, (1989).*

the WCSB continues into the north central coal region and Williston Basin in Montana and North Dakota of the conterminous United States (Figures 9.5, and 9.12). The tectonic setting of the WCSB is similar to that of the Rocky Mountains regions of the United States in that during the Columbian and Laramide Orogenies, a series of thrust sheets buried Cretaceous coal-bearing sedimentary fills in the deep foreland basin resulting in distinct coal characteristics.

There are 13 other smaller coal basins (Figure 9.12), which extend along the regions of coastal and intermontane British Columbia, northern Canada, and the Hudson Bay Lowland (Smith, 1989; Smith et al., 1994). The northern Canada region extends into the northern and central coal provinces of Alaska (Figures 9.11 and 9.12; Flores et al., 2005). The Atlantic Provinces of New Brunswick, Nova Scotia, and Newfoundland include the Maritimes Basin, which is correlative to the northern Appalachian Basin in the conterminous United States. These Canadian coal basins have inconsequential coalbed gas developments compared to WCSB

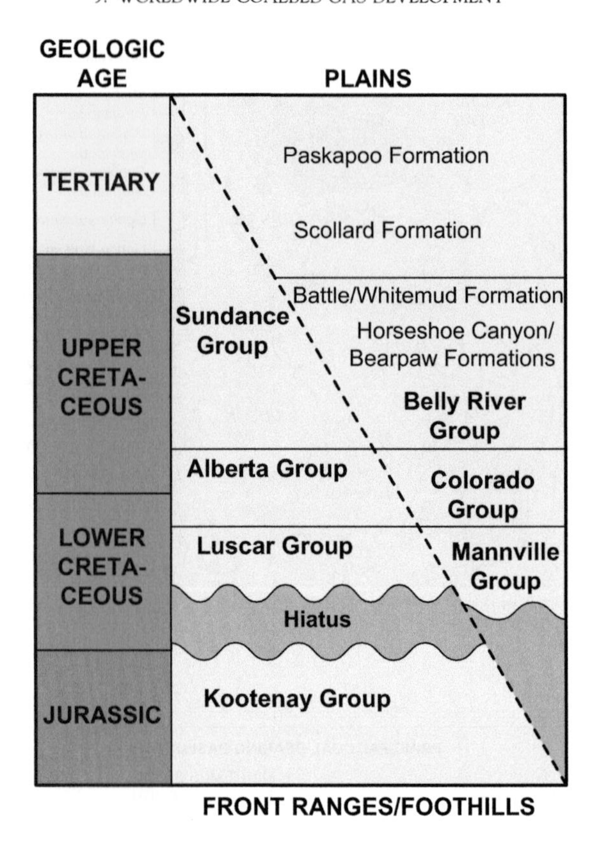

FIGURE 9.13 Generalized stratigraphic column showing geologic ages of rock units (e.g. groups and formations) from the Rocky Mountain ranges and foothills to the plains in Western Canada Sedimentary Basin. *Source: Modified from Beaton (2003a).*

where commercial coalbed gas development commenced in 2000 (Taylor, Bustin, Hancock, & Solinger, 2008).

Geologic Age and Coal Rank

The coal rank variations in the Cretaceous and Tertiary coal-bearing strata of the WCSB are mostly the reflection of the maximum depth of burial and deformation as related to (Smith et al., 2004) the following: (1) basin subsidence, (2) burial by thick Tertiary molasses and in part later eroded during the late stages of the Laramide Orogeny, (3) tectonic burial by stacked thrust sheets, and (4) proximity to maximum tectonic deformation areas. Thus, coal rank in the WCSB varies geologically and geographically. These variations in the patterns of subsidence, orogenesis, and stratigraphic/sedimentologic fill in the Cordilleran foreland basin have created distinct regions between which coal characteristics change and thus influence the mode of exploration and development of coalbed gas (Beaton, 2003a; Smith et al., 1994; Taylor et al., 2008).

The coal deposits in the 15 Canadian coal basins occur in strata ranging in age from Devonian to Tertiary (Smith, 1989; Smith et al., 1994; Taylor et al., 2008; Vogler, 2006). However, the major targets of exploration and development for coalbed gas are limited to the WCSB, which contains coal zones in the Jurassic to Tertiary (Paleocene) strata (Figure 9.13). The Jurassic coals in the southern Canadian Rocky Mountains and Foothills have ranks that are high- to low-volatile bituminous with sparse anthracites in the Kootenay Group (Figure 9.12 and Figure 9.13) (Smith et al., 1994). The Early Cretaceous coals in the Rocky Mountain Inner Foothills between the Front Ranges and Interior Plains range from medium- to low-volatile bituminous rank and from high-volatile bituminous to anthracite in the Gates Formation in the Luscar Group (Beaton, 2003; Smith et al., 1994; Figure 9.12 and Figure 9.13). Early Cretaceous coals in the Mannville Group, which underlies the Interior Plains range from subbituminous to high-volatile bituminous rank in the north and east, and high-volatile bituminous C-A in the central Plains (Beaton, 2003; Figure 9.12 and Figure 9.13). Coal ranks of Mannville coals increase with depth westward and these range from medium- to low-volatile bituminous coals. Thus, the regional vertical and lateral variations in coal rank demonstrate the influence of sedimentary and tectonic burial as well as proximity to deformation where folding and faulting might have increased rank.

Also, in the Interior Plains, the Late Cretaceous and Tertiary coals range from high-volatile bituminous to subbituminous and lignite rank (Figure 9.12). The coal rank increases at depth and toward the Foothills typified by the coal zone in the Upper Cretaceous Horseshoe Canyon Formation, which ranges from subbituminous to high-volatile bituminous (Beaton, 2003; Smith et al., 1994). The Drumheller coal zone in the Horseshoe Canyon Formation ranges from subbituminous B-A at shallow depths to high-volatile bituminous C in the deeper Central Plains region. The coal zone in the Upper Cretaceous-Tertiary Scollard Formation (e.g. Ardley) ranges in rank from subbituminous at the outcrop to high-volatile bituminous B at depth in the western Plains.

Coalbed Gas Plays in WCSB

According to the assessment report of the Canadian Gas Potential Committee (2005), the coalbed gas-in-place resource is about 14.8 Tcm and recoverable coalbed gas resource ranges from 0.3 to 1.3 Tcm. The coalbed gas-in-place resource estimates presume full gas saturation of the coal reservoir. The majority of the coalbed gas in-place resource of Canada (about 9.8 Tcm) is in the WCSB and specifically in the Cretaceous coals (e.g. Horseshoe Canyon, Mannville; Petrel Robertson Consulting, 2010; Taylor et al., 2008). Commercial development of coalbed gas in Canada started in 2000 and as of 2008, there are more than 6000 producing wells. Most of these coalbed gas wells were drilled during the years 2004–2007 and are in the Cretaceous age coalbed gas plays within the WCSB (Taylor et al., 2008). As of 2008, 28.1 million cubic meter (MMm^3) of the coalbed gas is produced from the WCSB coalbed gas plays. The coalbed gas production from 2000 to 2008 in WCSB in Alberta in the Cretaceous Horseshoe Canyon and Mannville is shown in Figure 9.14 (Forward Energy, 2009).

There are more than 40 potential coalbed gas plays found in coal zones in basins in Canada (Gatens, 2005). Coalbed gas plays are characterized according to specific parameters of basins and coal zones (Dawson & Sloan, 2001; Forward Energy, 2009). Basin parameters include depths of coal beds, rank, gas content, permeability, and degree of deformation (Dawson & Sloan, 2001). Based on overarching controlling factors of basin depth and deformation, Dawson and Sloan (2001) divided the WCSB, from west to east, into four basin plays: (1) restricted basins (British Columbia), (2) foothills and mountain regions (British Columbia and Alberta), (3) deep foreland basin (Alberta and Saskatchewan), and (4) shallow foreland basin (British Columbia, Alberta, and Saskatchewan). Coal zone parameters include depth, coal thickness, gas content, and gas-in-place resource (Forward Energy, 2009). The gas-in-place resource is estimated from the total gas content and total volume of coal (Chapter 6) per section.

A combination of the coal zone parameters is summarized in Table 9.3 to characterize selected coalbed gas plays in the WCSB. The Scollard, Belly River, and part Horseshoe plays are in what are

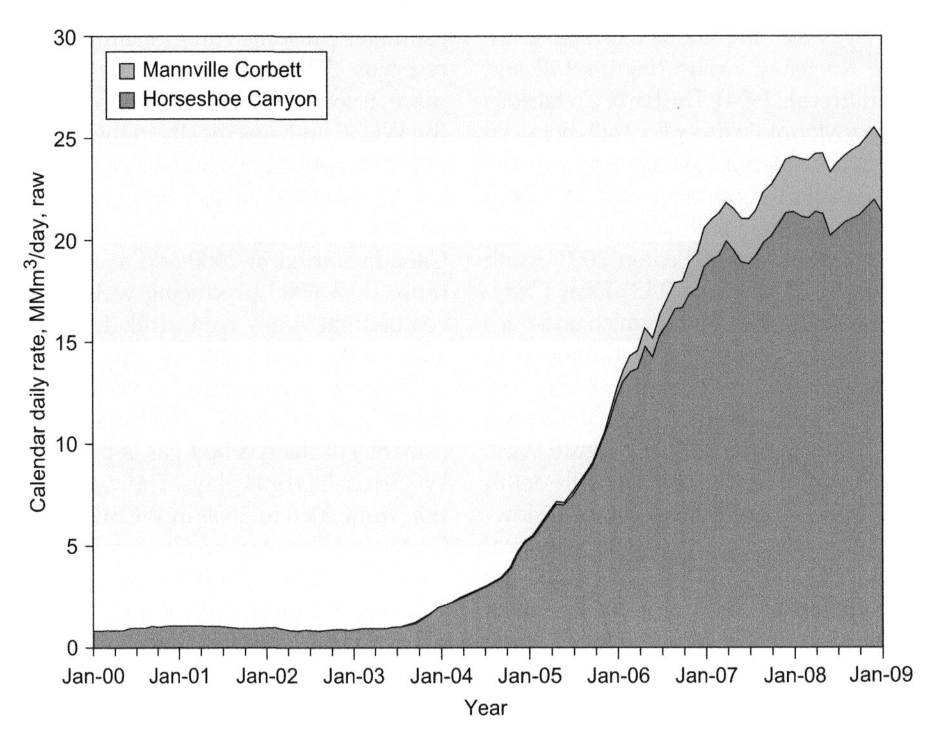

FIGURE 9.14 Chart showing 8 years coalbed gas production in the Cretaceous Horseshoe Canyon Formation and Mannville Group in WCSB, Alberta, Canada. *Source: Adopted from Forward Energy (2009).*

termed by Dawson and Sloan (2001) as shallow foreland basins. Part of the Horseshow and Mannville plays are thought to be within a deep foreland basin (Dawson and Sloan 2001). In all the selected coalbed gas plays discussed in this book, coal rank increases with depth (vertical trend) and geographic areas (lateral trend). Mostly coal rank increases toward the west or to the Foothills or toward the direction of deformation where the coal beds dip into the deeper part of the foreland basin. The increase of coal rank at depth and to the west is concomitant with increase of gas content. However, permeability, which conventionally decreases with depth due to overburden pressure, does not appear to be a major factor in the gas content (e.g. Mannville). The Horseshoe Canyon and Mannville coalbed gas plays have permeability up to 10 mD and the highest rate of gas production. However, Gentzis et al. (2008) reported that the Mannville has locally variable permeability ranging from 1 to 4 mD. According to Taylor et al. (2008), the daily production rate is 0.014 Bcm mainly from the Horseshoe Canyon (13.2 MMm^3) and Mannville (2.52 MMm^3).

AUSTRALIA

Australia has the most commercially successful development and advanced coalbed gas industry outside the United States and Canada. Although the initial exploration of Australian coalbed gas was in the mid-1970s, it was not until the 1980s and 1990s when substantial efforts were spent on exploration and commercial development paralleling that of the United States (Beamish and Gamson, 1992; Ham and Kantzas, 2008). However,

TABLE 9.3 Summary of Coal Reservoir Parameters in the WCSB, Which Includes the Coalbed Gas Plays, Depth, Coal Thickness and Permeability, Gas Content, Rate of Gas Production, and Drilling. Compiled from Numerous References

Coalbed Gas Plays (Formation; Group)	Coal Rank (Vertical and Lateral Trends)	Coal Bed Thickness (m)	Gas Content (cc/g)	Coal Permeability (mD)	Depth (m)	Rate of Gas Production (MMm^3)	Drilling	References
Scollard	Subbituminous to high-volatile bituminous C (*lateral trend*). Subbituminous to high-volatile bituminous B (*vertical trend*).	3–15	0.94–4.06	<0.1–7	<750	No data	No data. Used vertical wells for tests.	Smith et al. (1994), Dawson et al. (2000), Dawson and Sloan (2001), Beaton, 2003
Horseshoe Canyon	Subbituminous to high-volatile bituminous (*lateral trend*). Subituminous B-A to high-volatile bituminous C (*vertical trend*)	1–4	0.78–2.34	3–5	200–1000	13.2	Vertical; horizontal (multilateral)	Beaton, 2003, Taylor et al. (2008), Forward energy (2009)
Belly River	Subituminous C-B to high-volatile bituminous B (*lateral trend*). Subbituminuous B-A to high-volatile bituminous C (*vertical trend*).	1–3		1–10	<750	3.36	No data	Dawson and Sloan (2001), APF energy (2004), Taylor et al. (2008)
Manville	Subbituminous to high-volatile bituminous (*lateral trend*). High-volatile bituminous C–A to medium and low bituminous, anthracite (*vertical trend*)	<15	4.68–10.92 (8–15)	<0.1–10 (1–4)	750–2000	2.52	Vertical; horizontal	Dawson et al. (2000), Dawson and Sloan (2001), Beaton, 2003, Baltoiu et al. (2008) Taylor et al. (2008), Gentzis et al. (2008)

cc/g, cubic centimeter per gram; mD, millidarcy; MMm^3, million cubic meter.

coalbed gas production in Australia is not fully matured even with the first production in 1996 (Geoscience Australia, 2005). The country produces about 1.3% of the yearly coalbed gas output of the United States. The Geoscience Australia and BREE (2012) report that as of January 2012, the economically demonstrable resources (EDR) for coalbed gas in Australia were 0.99 Tcm. Additionally, Australia has exponentially larger estimates of subeconomically demonstrable and inferred resources than EDR. The potential in-ground or gas-in-place resources are even larger than the previous estimates with more than 7 Tcm. Coalbed gas makes up about 24% of the total natural gas EDR of Australia (Geoscience Australia and BREE, 2012).

Australia contains about 30 coal-bearing basins with coal deposits ranging in age from Permian to Tertiary. Six of these coal basins located in eastern Australia contain most of the coalbed gas EDR in the country (Baker & Slater, 2008; Geoscience Australia and BREE, 2012). The basins include Bowen, Clarence-Moreton, Gloucester, Gunnedah, Surat, and Sydney (Figure 9.15). The combined coalbed

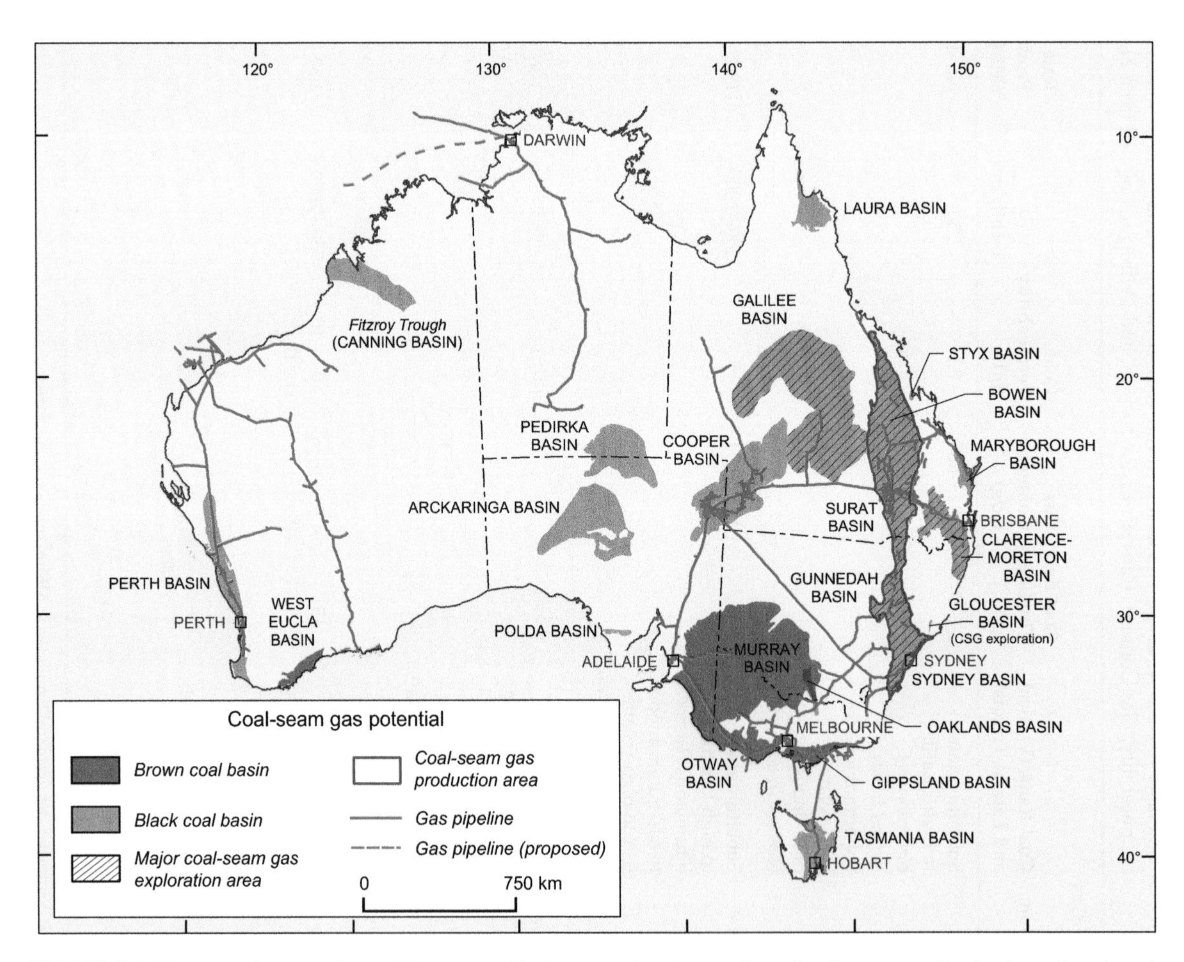

FIGURE 9.15 Map showing the coal basins, coalbed gas exploration and production areas, distribution of coal rank (brown and black), and pipeline network in Australia. *Source: Adopted from Geoscience Australia and BREE (2012).*

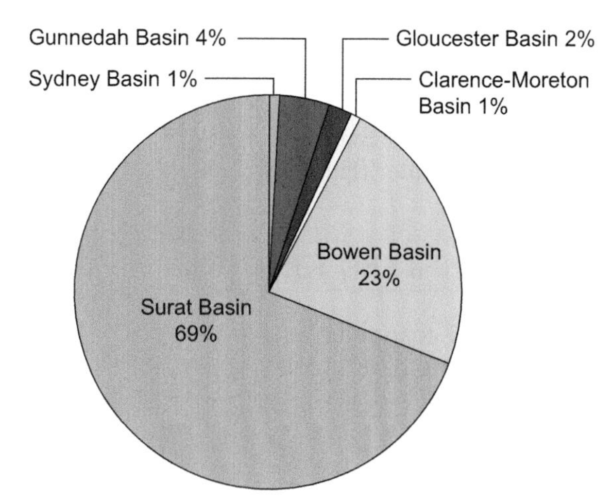

FIGURE 9.16 Pie chart showing percentage of coalbed gas reserves for each of the six coal basins in Australia. *Source: Adopted from DEEDI (2011; 2012).*

gas reserves of the Surat and Bowen Basins make up about 92% of the total for the six basins (DEEDI, 2011; DEEDI, 2012; Figure 9.16). Coalbed gas production from the Bowen and Surat Basins accounted for 8–9% of total natural gas production in Australia in 2008 (Australian Bureau of Agriculture and Resource Economics, 2010). The two basins provide the natural gas supplies for the region's demand for growth and to supply proposed LNG export projects. Baker and Slater (2008) reported about 86% of the natural gas used in the Queensland market comes from coalbed gas and accounts for 24% of the eastern Australian natural gas market.

Coal Geology, Age, and Rank

The most productive coal-bearing rocks for coalbed gas in Australia are mainly Permian age in the Bowen, Surat, and Sydney Basins with Bowen recording the highest production. Extensive exploration and testing activities occur in the Permian coals in the Gunnedah and Gloucester Basins and Jurassic coals (Walloon Coal Measures) in the Surat and Clarence-Moreton Basins. Over 2001–2010, coalbed gas exploration has increased substantially in Queensland and New South Wales as a result of the successful development in the Bowen and Surat Basins, which are the two major coalbed gas basins in Australia. Exploration has expanded beyond the Permian coals encouraged by the success in producing coalbed gas from low-rank coals (e.g. Powder River Basin) in the United States. The success has also promoted coalbed gas exploration in South Australia, Tasmania, Victoria, and Western Australia (Figure 9.15). Exploration has expanded in prospective basins such as in the Permian coal measures of the Galilee and Cooper Basins, Jurassic coal measures of the Maryborough Basin, Cretaceous coal measures of the Eromanga Basin, Tertiary coal measures of the Gippsland and Hillsborough Basins as well as in the Pedirka, Murray, Perth, Ipswich, Maryborough and Otway Basins (Geoscience Australia and BREE, 2012).

The Clareton-Moreton Basin, New South Wales contains high- to medium-volatile bituminous coals in the Middle to Upper Jurassic Walloon Coal Measures, Upper Triassic Evans Head and Red Cliff Coal Measures (equivalent to Ipswich Coal Measure in Queensland), and Middle to Lower Triassic Nymboida Coal Measures (Hutton, 2009; Stewart & Alder, 1995). The Gunnedah Basin in New South Wales contains medium-volatile bituminous coals in the Permian Black Jack Group and Maules Formation (Stewart & Alder, 1995). The most prospective for coalbed gas is the thick Hokissons Coal in the Black Jack Group. Also, channel sandstones (e.g. Clare Sandstone Member) directly above coal beds, which are charged with coalbed gas much like those in the Tertiary sandstones in the Powder River Basin in Wyoming and Cook Inlet Basin, Alaska provide potential gas production zones (Flores, 2004; Flores et al., 2005; Stewart & Alder, 1955).

The Sydney Basin is a part of the larger Sydney–Gunnedah–Surat–Bowen Basin, which extends from New South Wales to Queensland (Figure 9.15). In the Sydney Basin in New South Wales, the prime coalbed gas reservoirs are in the Late Permian Illawarra Coal Measures in the south and west, Tomago and Newcastle Coal Measure (Newcastle area) and Wittingham and Wollombi Coal Measures in the north (Hutton, 2009; Stammers, 2012; Stewart and Alder, 1995). Also, the Early Permian includes the Greta Coal Measures in the north and the Clyde River Coal Measure in the south of the basin (Hutton, 2009). The Gloucester Basin, 10-km-wide and 40-km-long infrabasin, in New South Wales contains Late Permian coals of more than 2000-m-thick Gloucester Coal Measure which contains targeted coalbed gas (Gloucester Gas Project of 110 wells in 2011; Gurba & Weber, 2000). The basin contains more than 1.3 Bcm (50 PJ) of coalbed gas resource in a 5 km^2 area (Gurba & Weber, 2000).

Gas Production and Reservoir Characterization

The coal reservoir characteristics of six eastern Australian basins are summarized in Table 9.4. The combined coalbed gas production from Australian coal basins from 1995 to 2010 is shown in Figure 9.17 (Geoscience Australia and BREE, 2012). Associated with gas production is co-produced water and as shown in Figure 9.18, about 99% comes from the Bowen and Surat Basins in Queensland (RPS Australia East Pty Ltd, 2011). The proposed increase of coalbed gas production with co-produced water for LNG projects in Queensland requires more improved knowledge of resource distributions, reservoir properties and compartmentalization (Underschlutz et al., 2010). These workers proposed accurate prediction of reservoir quality and gas content but more importantly using fracture stimulation to improve production from low-permeability coal reservoirs. Thus, much like in the United States and Canada, coalbed gas production will need effective hydraulic fracturing, completion, and stimulation of wells, which in turn will challenge management of the co-produced water (Ham and Kantzas, 2008).

UNITED KINGDOM

The fall of coal mining since 1990 is followed by the rise of recovery and use of coalmine gas (CMM and AMM) in the United Kingdom. United Kingdom has about 900 coal mines from which about 1.3 Bcm of coalmine gas was recovered and used from 1990 to 2008 (Coal Authority, 2008; USEPA, 2010). USEPA (2010) estimated that 31,000 tons CO_2e of methane have been recovered and used annually from abandoned mines. In 1990, the CMM production of United Kingdom is behind the United States, China, Germany, Russia, and South Africa.

There is a substantial volume of coalbed gas from virgin coals distant to the coal mines. On the basis of the distance from coal mines, depth, thickness, gas content, and rank, the British Geological Survey estimated a total gas-in-place resource of about 2900 Bcm (Jones et al., 2004; DECC, 2010). Thus, the evolution of consumption from coal to recovery and use of coalmine gas and potential exploitation of coalbed gas trade in the United Kingdom may be a model for fueling the future worldwide.

Coal Geology and Reservoir Characteristics

There are about 58 coalfields in the United Kingdom, which contain mainly Carboniferous coals (Figure 9.19; DECC, 2010). The source of the coalbed gas solely from Carboniferous coals is unique compared to Australia, Canada, and the United States, which contain younger and more diverse-aged coals serving as source rock of gas. Inevitably, the potential for coalbed gas

BOX 9.3

COALBED GAS LNG: FUELING THE FUTURE—FACT VS FICTION

Australia is poised to be the first OECD member nation to convert coalbed gas into LNG potentially supplying demand from East Asia (e.g. China, Japan, and Republic of Korea; Geoscience Australia and BREE (2012)). The proximity of Australia to the Asian market place is theoretically an advantage compared to LNG suppliers from distant parts of the world (e.g. Qatar, Nigeria, and Algeria). However, the emergence of Indonesia and Malaysia as LNG exporters in Asia (IGU, 2011) could pose a challenge for Australia's proposed LNG projects. Despite the competition, LNG has been touted as Australia's brightest economic prospects, which will lead the country to become a major LNG producer (Underschultz et al. 2010). The fact is that the global economic recession may curve the LNG prospects.

The downside of the Australian LNG boom is that more coalbed gas wells will be drilled accompanied by larger volume of co-produced water and expansion of disposal systems (RPS Australia East Pty Ltd, 2011). The RPS Australia East Pty Ltd (2011) report gives scenarios of potential total water production from Bowen (33.6%) and Surat (66.4%) Basins in southern Queensland for LNG over the life (>25 years) of their coalbed gas fields. Potential co-produced water for low-development scenario is 3150 Gl, probable development scenario is 4900 Gl, and likely development scenario is 7025 Gl. Underschultz et al. (2010) underscored the effects of increase of co-produced water particularly with coalbed gas reserves located in the Great Artesian Basin of Australia.

The boom on coalbed gas or coal seam gas as known in Australia and elsewhere has been slowed down by the global economic downturn and glut of natural gas. Also, the shale gas boom in the United States (IGU, 2011) and its emergence elsewhere poses a challenge for the Australian LNG projects. The expansion of coalbed gas LNG for export has given rise to the development in New South Wales and Queensland with potential market for greener and cleaner fuel of the future. The success of Australia in converting coalbed gas to LNG for global markets bodes well for the industry.

from the United Kingdom coalfields are compared to other countries especially with the coal geology and reservoir characteristics of the same age and coal rank as in the United States. For example, the Appalachian Basin is analogous to the coals in the United Kingdom. The Pennsylvania anthracite region in the northern Appalachian Basin occurs close to the Valley and Ridge fold and fault belt in contrast to the anthracite coal of South Wales, which occurs away from the Variscan fold belt (DECC 2010). Structural deformation in the Pennsylvania Anthracite region during the Allegany orogeny affected the coal rank. Prior to deformation, the Appalachian Basin was infilled by successions of peat-forming fluvio-deltaic clastics, marine-reworked sandstones, and shallow-shelf-platform-marine carbonates much like those in the Pennine Basin in northern United Kingdom (Ferm & Horne, 1979; Waters, 2009).

A number of important related factors regarding coal geology and reservoir characteristics such as coal rank or maturation, gas content, thickness, and depth in association with tectonic setting were used to assess the potential of coalbed gas from virgin Carboniferous coals. Although these coal reservoir parameters

TABLE 9.4 Summary of Coal Reservoir Properties of Currently Producing or Potentially Producing Australian Coal Basins. Compiled from Numerous References as Indicated in the Table

Coal Basin	State	Coal-bearing Strata	Coal Rank	Coal Thickness (m)	Depth (m)	Gas Content (m^3/t)	Permeability (mD)	Production Rate (By State, 1996–2005; MMm^3)	References
Bowen	Queensland	Coal groups 1–4	High- to low-volatile bituminous and semianthracite	<1–30		<1–11; 9–25	0.5–13.6	0.15–4.08	Mallet et al. (1995), Geoscience Australia (2005), Davidson (1992), Esterle et al. (2006), Hutton (2009)
Clarence-Moreton	Queensland, New South Wales	Walloon (shallow), Evans Head, Red Cliff, and Nymboida (deep) coal measures	Low- to high-volatile bituminous to medium-volatile bituminous	0.3–2.5	130–750; 1000–2000	ND	ND	ND	Stewart and Alder (1995), Ingram and Robinson (1996)
Gloucester	New South Wales	Gloucester coal measures	High-volatile A to low-volatile bituminous		<2000	ND	ND	ND	Gurba and Weber (2000)

Gunnedah	New South Wales	Black Jack Group and Maules Creek Formation	Medium-volatile bituminous	<1.0–17.5	Up to 900–<950	7–9	30–50	ND	Tadros (1993), Stewart and Alder (1995), Baker and Slater (2008), Santos (2010)
Surat	Queensland	Walloon coal measures	Subbituminous to high-volatile bituminous	3–4	100–400	1–8	0.07–>500	0.15–4.08	Stewart and Alder (1995), Geoscience Australia (2005), Pependick et al. (2011)
Sydney	New South Wales	Illawarra, Tomago, Newcastle, Wittingham, Wollombi, Greta, and Clyde River coal measures	High rank (bituminous)	2–12	650–850	Up to 18	5–30	0–0.84	Stewart and Alder (1995), Geoscience Australia (2005), Hutton (2009), Stammers (2012)

m, meter; m^3/t, cubic meter per ton; ND, not determined.

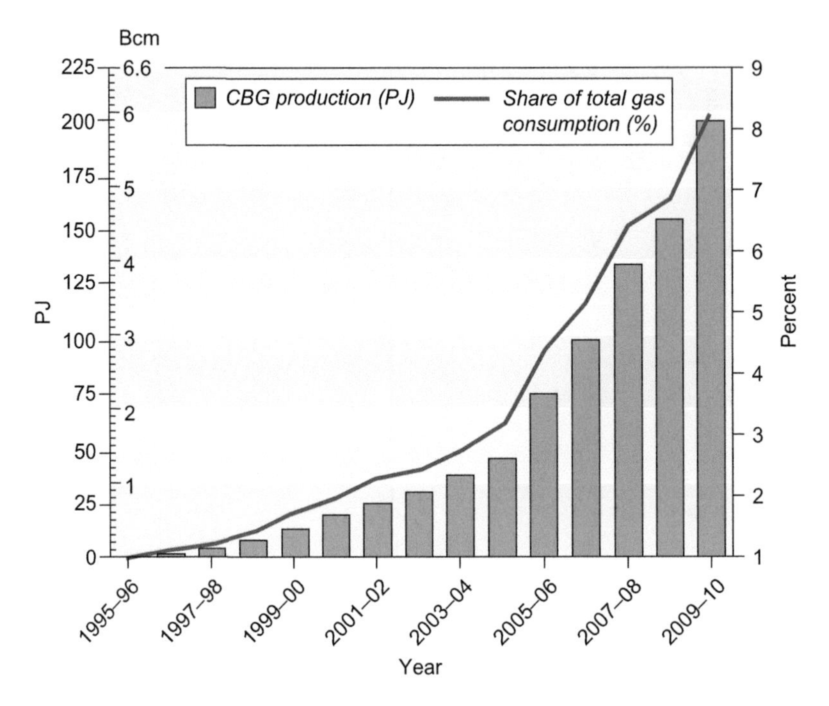

FIGURE 9.17 Frequency diagram showing the coalbed gas production history in Australia. CBG, coalbed gas; PJ, petajoule (1 PJ, 1015 J); %, percent; Bcm, billion cubic meter. *Source: Adopted from Geoscience Australia and BREE (2012).*

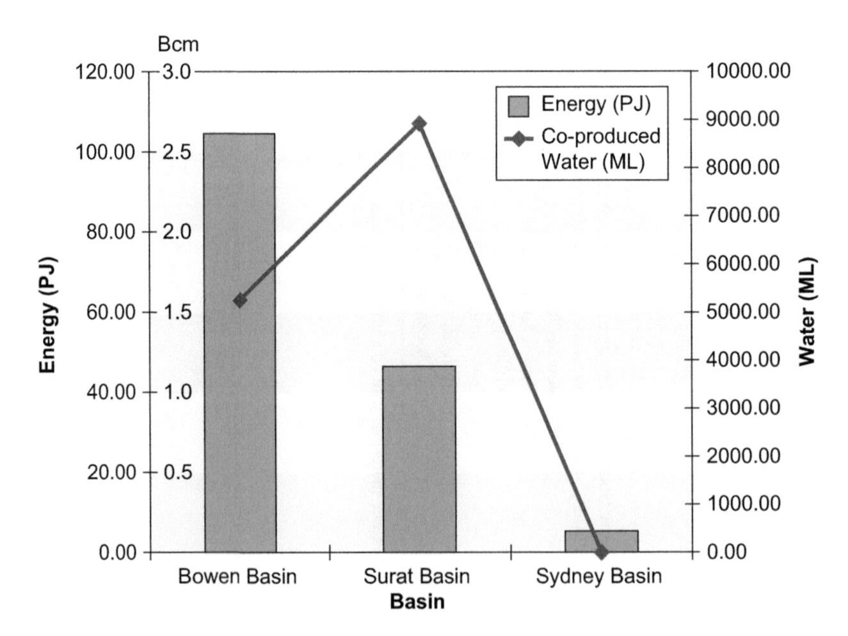

FIGURE 9.18 Frequency diagram showing coalbed gas production from the three major coal basins (e.g. Bowen, Surat, and Sydney) in Australia. The Surat Basin and the least volume by Sydney Basin generate the largest volume of co-produced water. Bcm, billion cubic meter; ML, megaliter (1 ML, 1 million liters; 1 PJ, 1015 J). *Source: Adopted from RPS (2011).*

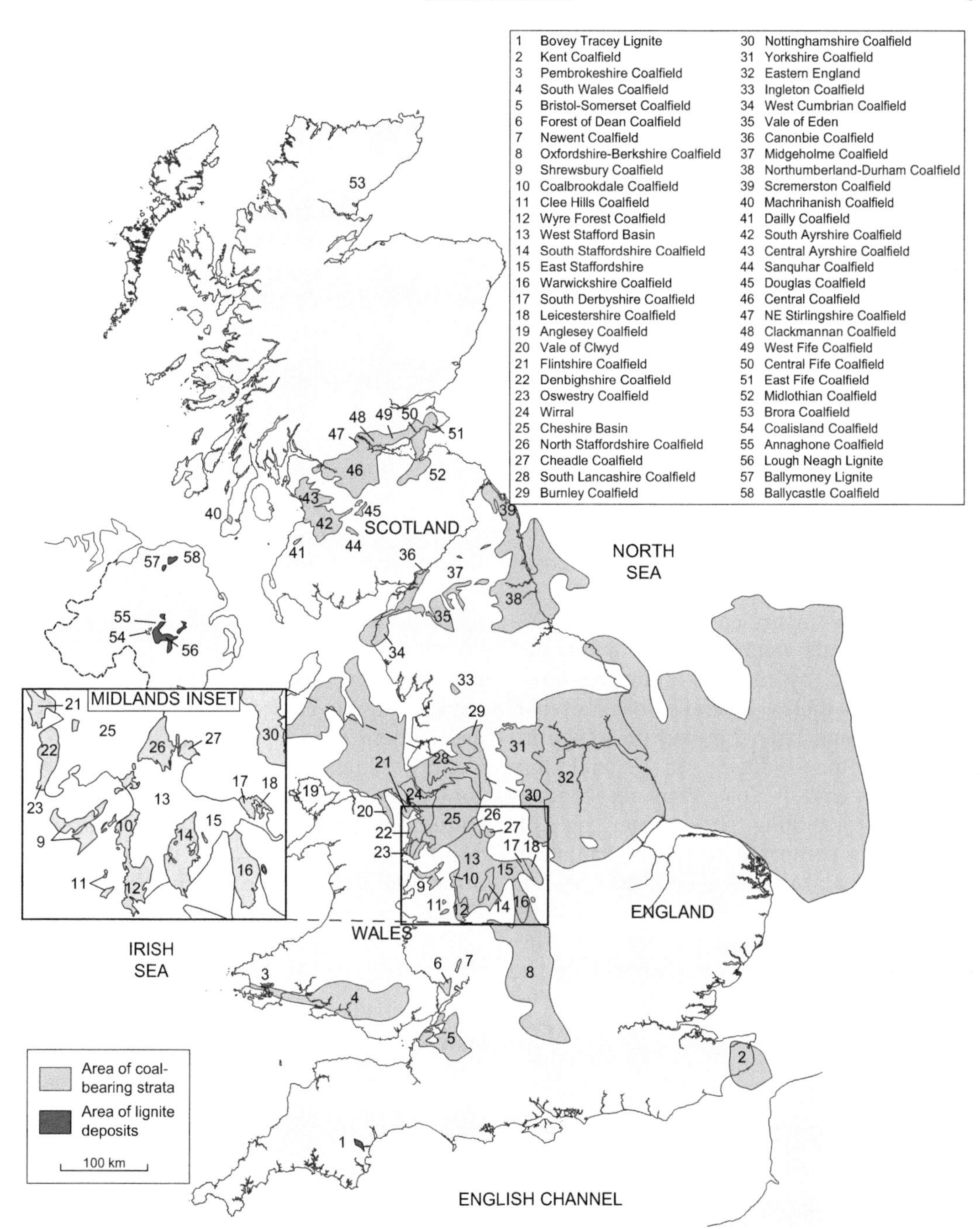

FIGURE 9.19 Map showing location of coalfields and areas (gray) containing Westphalian coals in the United Kingdom. Inset area is locations for Figures 9.20 and 9.22. *Source: Adopted from DECC (2010).*

are best described in ground-breaking investigations by Creedy (1983, 1988, 1994), Baily et al. (1995), and DECC (2010) utilized data from more than 400 coalbed gas boreholes and coal-face samples (National Coal Board database) characterize potential coal reservoirs in coalfields in the United Kingdom. The coal reservoir properties are compared to successful coalbed gas plays worldwide, particularly those of the United States.

Coal Reservoir Characterization

Coal Thickness vs Depth

The DECC (2010) report places the importance of the coal thickness, rank, gas content, and permeability in addition to the coal geology and tectonic setting when characterizing coalbed gas reservoirs found exclusively in the Carboniferous strata. The report emphasized that the Carboniferous coals were deposited in paralic basins subjected by marine transgressions, which led to coalfields generally forming coal beds mostly thinner than 2 m and up to 4 m in thickness in some coalfields. In general, the Westphalian A and B Coal Measures contain thicker coals resulting from merging of individual beds compared to the Westphalian C and D Coal Measures (DECC, 2010). Westphalian coal beds generally are thick to thin from the near-northern margin of the Wales–Brabant Massif where subsidence is slower than in the Pennine Basin depocenter to the north where subsidence is greater (Figures 9.19 and 9.20). The coals in the Westphalian Coal Measures occur between 200 and >1200 m in depth.

Gas Content vs Coal Rank

The variation of gas content of coals with rank in the United Kingdom Westphalian Coal Measures is exemplified by the western coalfields, which contain coals of higher maturity than those coals east of the Pennines, which were formerly part of the same Pennine Basin (DECC, 2010). The western coalfields contain gassy coals of the abandoned coal mines and represent the highest measured gas content of Westphalian coals in the National Coal Board boreholes. Figure 9.21 shows the successive increase in gas content northward from the Wales–Brabant Massif (Oxfordshire) to the Pennine Basin margin (Warwickshire and South Staffordshire) and depocenter (Lancashire) (DECC, 2010). Also, minor increase in gas content occurs southward from Oxfordshire into the foreland basin in Kent (DECC, 2010). The gas content gradients vary from $<1\ m^3/t$ (Oxfordshire) to $>1-<4\ m^3/t$ (Warwickshire and South Staffordshire) and $0-<16\ m^3/t$ (North Staffordshire) as well as $<7-<17\ m^3/t$ (Lancashire). Gas content in Kent varies from 0 to $<4\ m^3/t$. In addition, a Westphalian B coal bed (e.g. Nine Foot) in South Wales increases gas content to the northwest from high-volatile A bituminous (32% volatile matter) to anthracite (6% volatile matter). Creedy (1988) observed a corresponding increase in gas content from 5 to $20\ m^3/t$. The gas content gradients reflect the Westphalian Coal Measures depocenters, in which the gas content of coals is greater in North Staffordshire, Lancashire and NE Wales compared to South Staffordshire, and is higher in the anthracitic part of the South Wales Coalfield, which coincides with thicker Coal Measures (DECC, 2010).

Prospectivity of Coalbed Gas in United Kingdom Coalfields

The 900 coal mines in the United Kingdom coalfields are a major challenge for the prospectivity of coalbed gas more so than coal rank, thickness, permeability, and depth. Experience indicates that extensive degasification accompanies coal mining, which drain gas over large areas from the coal-mine face. In the Powder River Basin, surface mining drained coalbed gas for about 4–6 km from the coal-mine face since

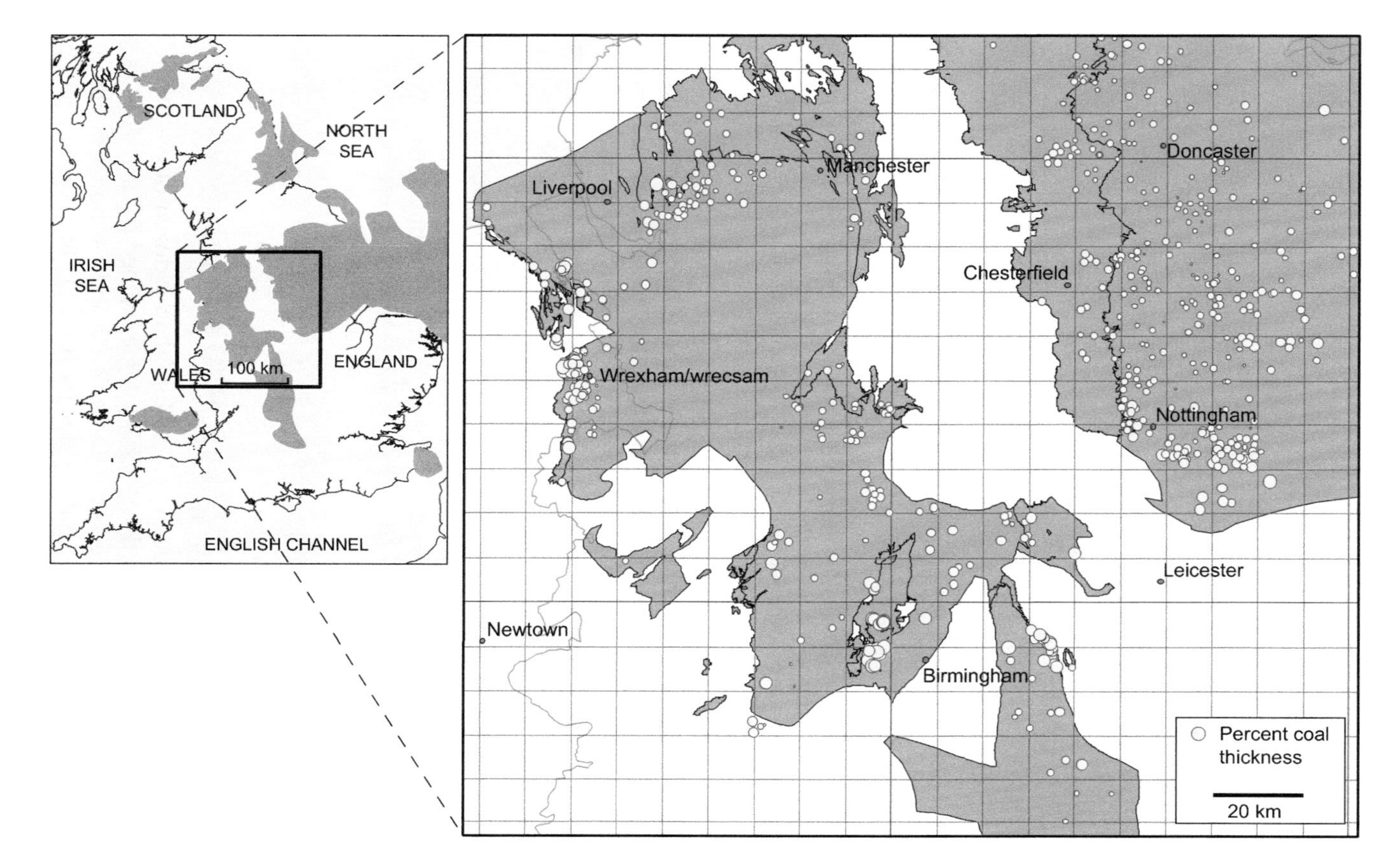

FIGURE 9.20 Map showing percentage of coal in the Westphalian Coal Measures. Diameter of circles indicates increasing percentage of coals (thicker). See inset map for location of the area. *Source: Adopted from DECC (2010).*

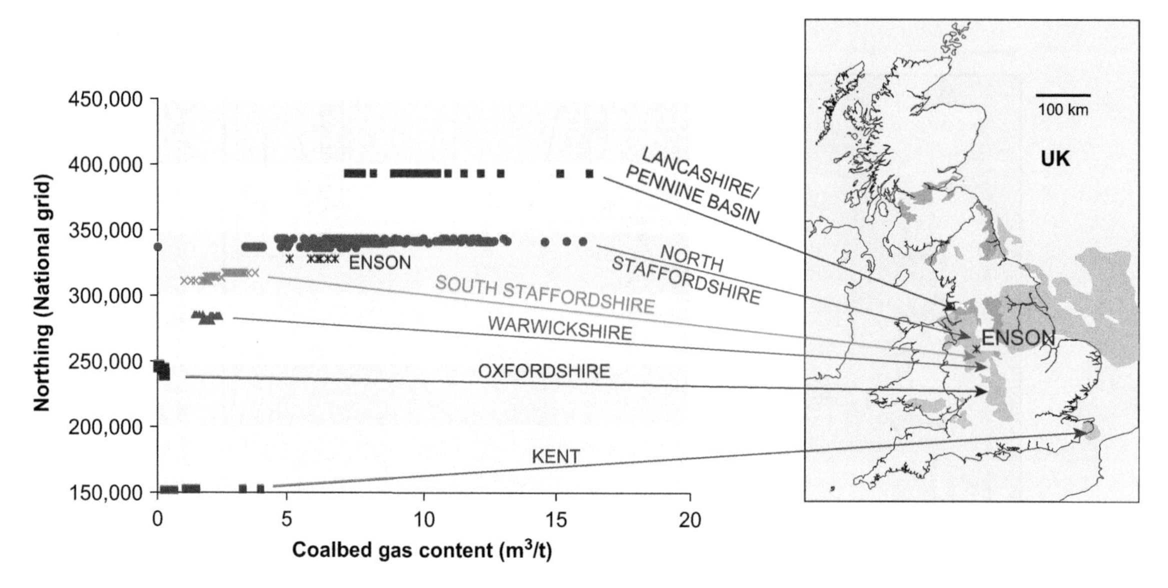

FIGURE 9.21 Combined map and graph showing northward increase of coalbed gas content for Westphalian coals by coalfields. Gas content increases northward from the Wales-Brabant Massif (Oxfordshire coalfield) to the Pennine Basin margin (Warwickshire and South Stafordshire coalfields), and basin depocenter (Lancashire coalfield). m^3/t = cubic meter per ton. *Source: Adopted from DECC (2010).*

mining started in the 1970s. However, in the United Kingdom, coal mining has gone on for much longer period of time for gas drainage in the form of CMM and AMM. Figure 9.22 shows an example of coalbed gas potential areas in the Cheshire and Stafford Basins in the Midlands and northwest England in the Westphalian Coal Measures (DECC, 2010). The prospect areas are surrounded by coal mines (collieries) and depths of the coal-bearing strata are divided into shallower and deeper than 1200 m.

Based on coalmine degasification caused by lowering of groundwater table from the Powder River Basin model, parts of the coalbed gas potential areas in the north Cheshire Basin and northeast Staffordshire Basin may be affected by dewatering of coal mines. However, propectivity of coalbed gas unaffected by coalmine gas drainage may be better in deeper areas (>1200 m) in both basins. Another challenge based on MacCarthy et al. (1996) report is the net coal thickness (about 40 m) thins to the northwest accompanied by increasing gas content toward the same direction. A few coal permeability reports in the area show another challenge to coalbed gas potential with the Great Row coal bed in the North Staffordshire measured to have permeability of 0.1–0.5 mD. In contrast, Durucan et al. (1995) recorded a permeability of 3 mD for a coal in a well in the Cheshire Basin. These challenges to prospectivity of coalbed gas are typical worldwide where coal mining and sparse reservoir data are a part of the history of the coal basins.

NOECD Countries

More than a dozen of the NOECD countries have some coalbed gas activities as summarized in Table 9.5 and as shown in Figure 9.23. Much like the OECD, the coalbed gas activities for NOECD countries are based on the presence of exploration, development, and production

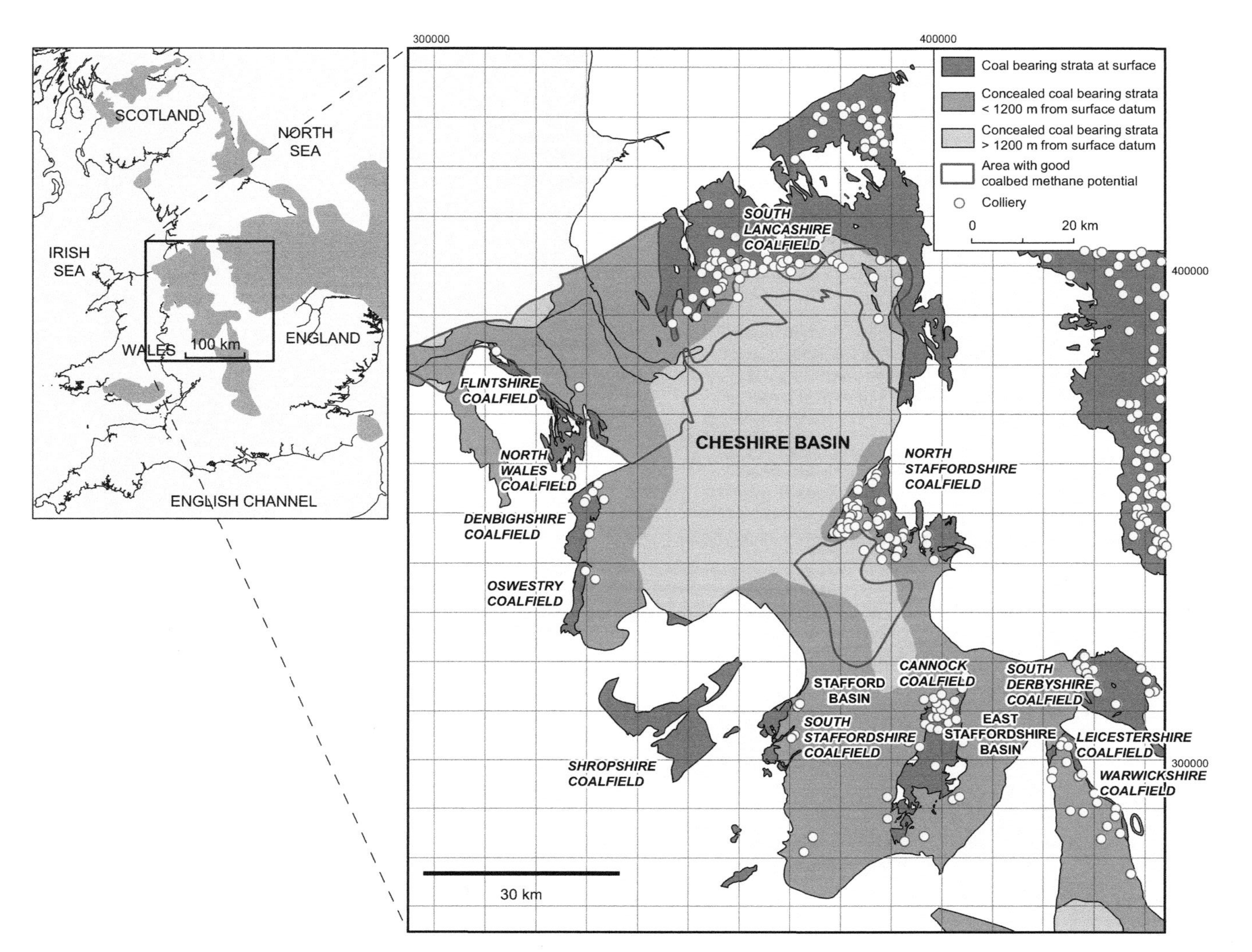

FIGURE 9.22 Map showing potential areas of coalbed gas in the Midlands and northwest England. Areas outlined in red represent good coalbed gas potential. See inset map for location of the area. *Source: Adopted from DECC (2010).*

TABLE 9.5 Summary of Coalbed Gas Activities, Geologic age, Coal Rank, Coalfields, and Coal Resource and Reserves of Various NOECD Countries with Exploration and Development Potential. Compiled from Numerous References Indicated in the Table

	Coalbed Gas Activities						
Country	Coal Mine	Virgin Coal	Coal Age	Coal Rank	Basin/Coalfield	Coal Resources/Reserves, (Coalbed Gas/CMM Emissions)	References
Botswana	CMM	CBG	Perm.	Bituminous to anthracite	Karoo (Morupule and Mamabula)	Coal resources: 200 Bt (CBG GIP resources: 5.6 Tcm; Recoverable reserves: 1.7 Tcm)	ARI, 2008b, USEPA (2010)
Brazil	CMM	CBG	Perm.	Lignite to bituminous	Parana	Coal reserves: 4.6 Bt (CBM GIP resources 19 Bcm)	Kalkreuth and Holz (2000), USEPA (2010), BP (2012), Somers (2012)
Bulgaria	CMM	CBG	Cret., Penn	Lignite to anthracite	Bobo Dol, Dobroudja, Katrishte, Maritsa Gotsedelchev, Oranovo, Svoge, Sofia, Pernik, Balkanbas	Coal reserves: 2.4 Bt (CBM GIP resources: 195.6 Bcm; CMM emissions: 1.8 Bcm from 1990 to 2010)	Marshall (2001), USEPA (2010), BP (2012), Somers (2012)
Czech Republic	CMM AMM	CBG		Lignite to bituminous	Bohemian, Silesian	Coal reserves: 1.1 Bt (CBG GIP resources: 100–500 Bcm; CMM emissions: 6.9 Bcm from 1990 to 2007; Recovered and used: 100 MMm^3/year)	Durica and Nemec (1997), USEPA (2010), BP (2012)
Georgia	CMM	CBG	Tert.	Lignite to bituminous	Akhaltsikhe, Tkibuli-Shaora, Tkvarcheli, Vale	Coal resources: 405 Mt (CBG GIP resources: 11.5 Bcm; CMM emissions: 18.8 MMm^3 from 1990 to 2006)	USEPA (2010), ARI (2009), Somers (2012)
Kazakhstan	CMM	CBG	Tert.	Lignite to anthracite	Ekibastuz, Karaganda, Maykuben, Shubarkol, Turgay	Coal reserves: 33.6 Bt (CBG GIP resources: 1.2–1.7 Tcm; CMM emissions: 13.8 Bcm from 1990 to 2009; CMM recovered and used: 121.7 MMm^3)	Alekseev et al., (2003); USEPA (2010); BP (2012); (Somers, 2012)
Mongolia	CMM AMM	CBG	Cret. Jur., Perm., Carb.	Lignite to bituminous	Choir-Nyalga, Orkhon-Selenge, Choybalsan, Sukhbaatar, Mongol Altay, Big Bogdyn, Douth Gobi	Coal reserves: 12.2 Bt (CBG GIP resources: 17–34 BCM; CMM emissions: 77.2 MMm^3 from 1990 to 2010)	USEPA (2010), Somers (2012)

Namibia		CBG	Perm.	Bituminous to anthracite	Karoo (Huab, Aranos, Owambo, Waterberg) greater Nama-Kalahari	No data	Piperi et al. (2011)
Nigeria	CMM		Tert., Cret.	Lignite to bituminous	Anambra, Makurdi	Coal resources: 2 Bt (CMM emissions: 617 MMm^3 from 1990 to 2005)	USEPA (2010), (Somers, 2012)
Pakistan	CMM		Tert.	Lignite to bituminous	Lakhra, Degari-Sor-Range, Duki, Shalrig-Khost-Harnai, Mach-Abegum, Thar, Salt Range, Hangu, Makarwai-Kurd-Sho	Coal resources: 185 Bt; Coal reserves: 2.07 Bt (CBG GIP resources: 1 Tcm; CMM emissions 69 MMm^3 from 1993 to 1994)	USEPA (2010), BP (2012), Somers, (2012)
South Africa	CMM	CBG	Perm.	Bituminous to anthracite	19 Karoo basins (70% of coal reserves in Highveld, Waterberg, and Witbank coalfields)	Coal reserves: 30.1 Bt (CMM emissions: 2.95 Bcm from 1990 to 2010)	USEPA (2010), Piperi et al. (2011), BP (2012)
Ukraine	CMM	CBG	Tert., Carb.	Lignite to anthracite	Donetsk, Dnieper	Coal reserves: 33.9 Bt (CBG GIP resources: 11−12 Tcm; CMM emissions from 1990 to 2010: 40.3 Bcm; recovered and flared: 1.7 Bcm)	IEA (2006), BP (2012), Somers (2012)
Vietnam	CMM	CBG	Tert., Trias.	Subbituminous, bituminous, anthracite	Da River, Ca River, Red River, Quang Ninh, Thai Nguyen, Backan, Na Duong, Nong Son	Coal reserves: 150 Mt (CBG GIP resources: 170−280 Bcm; CMM emissions: 396 MMm^3 from 1990 to 2010)	USEPA (2010), Somers (2012)
Zimbabwe		CBG	Perm	Bituminous to anthracite	Greater Nama-Kalahari	Coal reserves: 502 Mt	Piperi et al. (2011), BP (2012)

CBG, coalbed gas; CMM, coal mine methane; AMM, abandoned mine methane; PPS, probable + possible + speculative; GIP, gas-in-place; Bt, billion tons; Mt, million tons; Tcm, trillion cubic meters; Bcm, billion cubic meters; Tert., Tertiary; Cret., Cretaceous; Jur., Jurassic; Trias., Triassic; Perm., Permian; Penn, Pennsylvania; Carb., Carboniferous; Dev., Devonian; m^3, cubic meters.

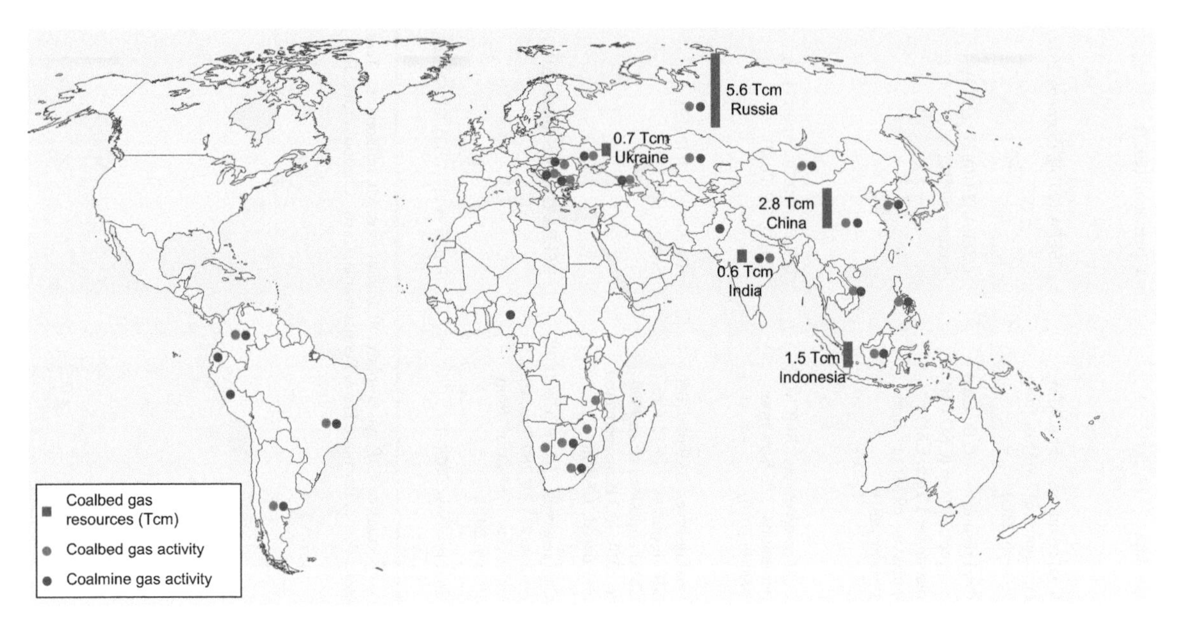

FIGURE 9.23 Map showing the coalbed gas and coalmine gas activities in NOECD countries. The coalbed gas recoverable resources are of the United States, Australia, and Canada for the sake of comparison based on the estimates of Kuuskraa and Stevens (2009). Tcm, trillion cubic meter. *Source: The image of the world map is adopted from future.wikea.com.*

of coalbed gas as well as documented coalmine gas emissions and potential recovery projects in relation to the global CMM initiative (Somers, 2012; USEPA, 2010). The major players of coalbed gas in the NOECD not indicated in Table 9.5 include China, Russia, Indonesia, and India are discussed below in order of advances in exploration and development, and demonstrated coalmine gas recovery projects.

CHINA

The success of China's coalbed gas exploration and development is related to coal mine safety and the Chinese central and local government subsidies for the extraction of coalbed gas much like the tax incentive provided by the United States government to gas companies in the 1980s (Flores, 1998). China is projected to produce coalbed gas of 30 Bcm by 2015, which includes production from CMM and virgin coals. China produced and utilized CMM about 2.5 Bcm in 2009 and about 3.5 Bcm in 2010 (IEA, 2013). Thus, in order to meet the projected production, China has to increase both CMM and coalbed gas from virgin coals about 10-fold in 5 years.

China's coalbed bed gas-in-place resources are estimated to be about 37 Tcm mainly found in the subsurface <2000 m deep (Qui, 2008). According to Liu (2008), there are 50 major coal-bearing basins in China in which total coal resources at <2000 m in depth are estimated to be 5.57 Tmt. The gas-in-place resources are contained in major coal basins and minor coalfields as shown in Figure 9.24 (Liu, 2007; USEPA, 2010). The coalbed gas-in-place is more than 10 times the country's proved reserves of conventional natural gas, which was estimated to be 2.4 Tcm in 2006 (Jia, 2009). Seven coal basins

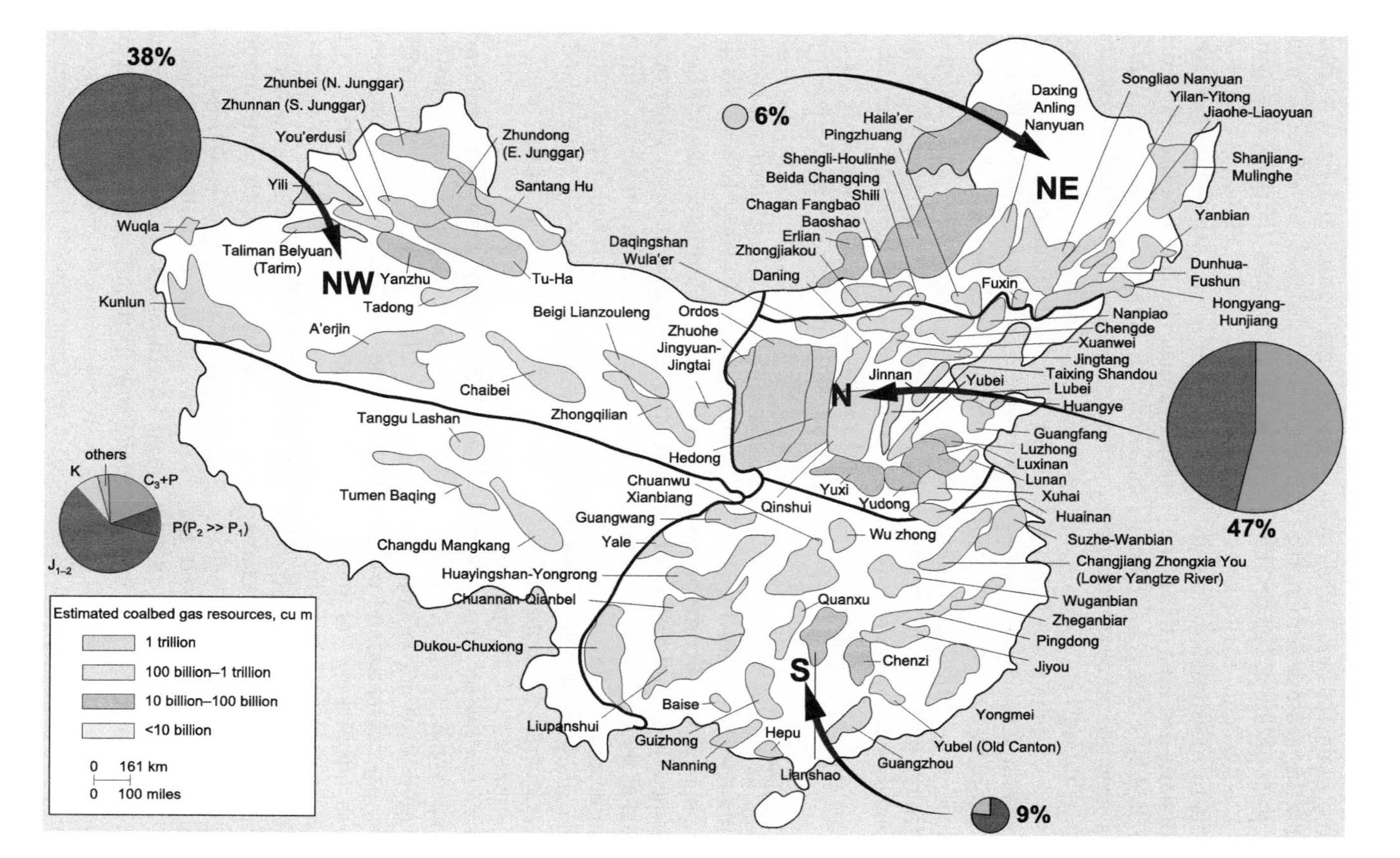

FIGURE 9.24 Map showing the distribution of coal basins and the estimated gas-in-place resources (colored areas) for each basin in China. Pie charts represent distribution of coal resources by geologic time: K, Cretaceous; J, Jurassic; P, Permian; C, Carboniferous. NE, northeast; N, north; NW, northwest; S, south. The percent (%) next to the pie charts represents the amount of coal resources in geologic time in each region with respect to the total coal resources of the country. *Source: Modified from Li et al. (1995); Liu (2007) and USEPA (2010).*

each contain about 1 Tcm coalbed gas resources in the northern and northwestern parts of China. Four coal basins each contain 100 Bcm to 1 Tcm coalbed gas resources in the northwestern, northeastern, and southern parts of China. Ten coal basins each contain 10–100 Bcm coalbed gas resources in the northern, northeastern, northwestern, and southern parts of China. Other minor coal basins/coalfields contain <10 Bcm coalbed gas resources stretched across the entire country (Figure 9.24). Although endowed with vast coalbed gas resources, China's gas-in-place still is behind Russia and Canada. However, if the coalbed gas-in-place of the United States is the combined conterminous United States and Alaska, China resources is behind that of the United States, which is more than 39 Tcm.

Coal Geology and Rank

The regional distribution of coalbed gas-in-place resources shown in Figure 9.24 corresponds to the coal districts defined by Li et al. (1995). The coal districts are related to the coal resources and periods of major coal accumulation during the Carboniferous, Permian, Early to Middle Jurassic, and Early Cretaceous. Generally, coal accumulations are Cretaceous-Jurassic to Carboniferous-Permian in northeast–northwest and south regions of China, respectively. Li et al. (1995) related the younging northward of coal accumulation to paleotectonic, paleoclimatic, and paleogeographic controls as affected by continental drift. The distribution of coal resources in the different Chinese regions associated with the geologic ages are depicted in the pie charts (Figure 9.24). The percent coal resources matching with geologic times are indicated next to the pie charts (e.g. 47 and 38%). The bulk (85%) of the coal resources is contained in the northwest and north regions of China. Wang (2011) reported that the Chinese coalbed gas-in-place might be apportioned according to geologic age as follows: Carboniferous–Permian about 49%, Jurassic about 50%, Cretaceous about 1%, and Tertiary about 0.04%.

The Carboniferous and Early Permian coals mainly accumulated in north China, which covers 12 provinces (Huabei depositional basin; Li et al., 1995). However, there are some coal resources in Middle-Lower Jurassic in north China. The coal resources in north China account for about 20% of the coal reserves compared to the mainly Late Permian coal resources in south China, which account for about 10% of the national coal reserves. The coal resources that are mainly found in north and northwest China and that make up about 60% of the coal reserves accumulated during the Carboniferous-Permian and Early-Middle Jurassic Periods. There are Middle Jurassic and Carboniferous-Permian coal basins in the northwest. The northwest region includes a few of the largest coal-bearing basins in China (e.g. Junggar, and Turpan-Hami Basins). Early Cretaceous coal-bearing basins, mostly fault-bounded, are distributed in northeast China and coal accumulated in this period accounts for about 7% of the coal reserve. Minor coal reserves are found in a few Triassic and Tertiary coal basins in northwest China (e.g. Tarim Basin) and northeast China (e.g. Fushun), respectively.

A wide range of ranks from anthracite, bituminous, subbbituminous, and lignite are represented in Chinese coal (Li et al., 1995). About half of the coal is low-rank bituminous coal and coking coal accounts for one-fifth of the total coal resources. The coking coal is a bituminous coal with a 17–22 caking index, which is converted into hard coke with high carbon content (75–80%) for steel production. Anthracite coal is mainly distributed in the north and south regions of China associated with the Carboniferous and Permian coals. However, anthracite is also associated with Jurassic and Cretaceous coals in the northwest and northeast regions of China. Coking bituminous coal is common in the south and north regions mainly associated

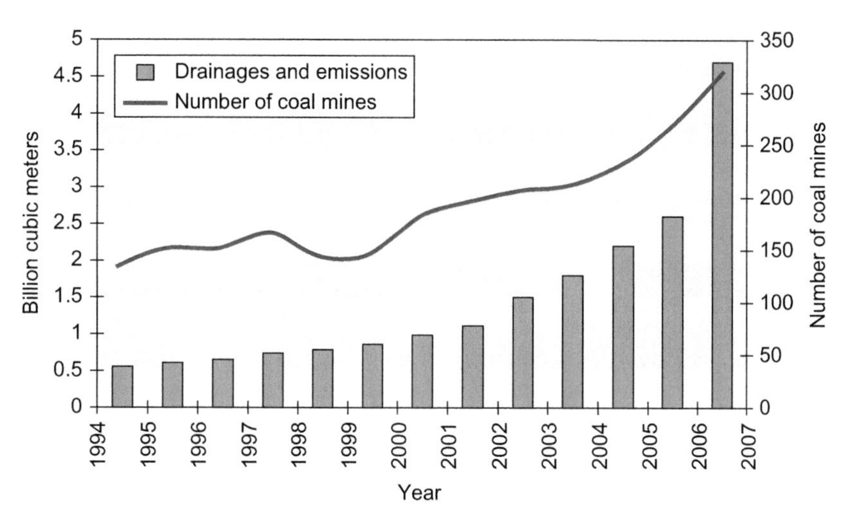

FIGURE 9.25 Frequency diagram showing gas drainage and emission as well as number of coal mines in China from 1994 to 2007. *Source: Adopted from IEA (2009).*

with Carboniferous and Permian coals. Noncoking bituminous coal is ubiquitous throughout the regions.

Coalmine Gas

A significant component of the coalbed gas resources of China is CMM, which is developed from active coalmines. China recovers coalmine gas from active mines (CMM), gob gas of abandoned mines (AMM), and predrainage of mineable coalbeds (CBM) (Hu et al., 2012). Mine gob gas recovery is in the early stages of development but active coalmine and predrainage gas recoveries are maturing stages of development in China. USEPA (2010) reported that in the 1990s, China had about 100,000 coal mines of various sizes mostly owned by villages and towns. The closure of under performing coal mines (e.g. <150,000 tons annual capacity) has reduced the number of operating mines to 12,122 at the end of 2009 (USEPA, 2010). Majority of the coal mines in China is underground mines, which made up >86% of CMM emissions (USEPA, 2010). Gas emissions from operating mines led to installation of increasing number of gas drainage systems. From 2006 to 2007, 300–350 coal mines installed gas drainage systems, which yielded combined drainage and emission gas of >4.5 Bcm of CMM (Figure 9.25; IEA, 2009). Hu et al. (2012) reported that 300 vertical and horizontal gob wells produced 70 MMm^3 of gas with extraction results from wellbores in the active mine area. Predrainage of coalbed gas before coal mining included 4576 vertical wells and 146 horizontal wells but only 2817 wells produced 1.57 Tcm (Hu et al., 2012).

According to Jiang (2007), the first CMM recovery and utilization project was inaugurated in 1991 at the Liuzhi coal mine in Guizhou Province, which used the coalmine gas for generating electric power. This was followed by Huainan coal mine in Anhui Province, where CMM was used for internal combustion engine and later for electric power generation (IEA, 2009). In 2006, two coal mines in Sichuan Province used CMM for power generation with expanded use in other coal mines in Liaoning and Shanxi Provinces in 2007. The largest CMM power plant (120 MW) in the world was built at the Sihe coal mine, owned by the Jincheng Anthracite

Coal Mining Group. CMM were collected at the mine and surrounding mines to fuel the power plant for distribution to residential, commercial, and industrial consumers in the Jincheng area (IEA, 2009). The CMM recovery and utilization expanded into recovery of coalbed gas from virgin coals by Lanyan CBM Inc., which compressed and liquefied the gas for domestic, residential and industrial use.

The aggressive implementation of the CMM recovery and utilization is driven by coal mine safety mainly focused on prevention of outbursts. The Chinese State Council formulated in 2006 coalmine gas predrainage and recovery policies to emphasize the mine safety-first approach using technology (Hu et al., 2012). According to the IEA (2009), the key points of the CMM policy are the following: (1) prior to mining, the coalbed gas (methane) must be predrained; (2) monitor and measurement systems for CMM must be installed; (3) CMM drainage system is required prior to coal production, which is suspended if CMM poses a significant problem; and (4) implementation of the policy is the legal responsibility of coal mine owners and operators.

The CMM policy was followed in 2008 by additional regulations of the Chinese Ministry of Environmental Protection, which include (IEA, 2009) (1) prohibition of emission of CMM by drainage systems, (2) mine drainage systems with 30% or higher concentrations of coalbed gas (methane) are prohibited, (3) to emit the gas, which is either used or flared, and (4) coalbed gas (methane) concentrations of <30% are permitted to be emitted.

Emerging Small and Large Coalbed Gas Basins/Coalfields

Several coal basins and coalfields in China have good potential for coalbed gas development but Qinshui and Ordos Basins have emerged as the most commercially productive basins (Huang et al., 2009; Figure 9.26). According to Wang (2011) at the end of 2010, coalbed gas production reached about 2.5 Bcm from about 5400

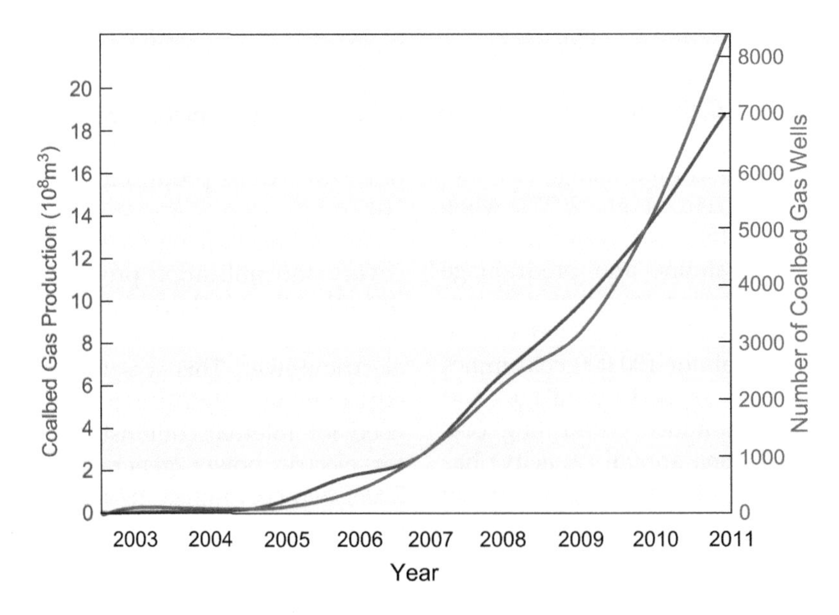

FIGURE 9.26 Diagram showing the coalbed gas production and number of wells from 2003 to 2011 for the Chinese coal basins (e.g. Qinshui, Ordos, Hubei, Fuxin). *Source: Data from Wang (2011) and Huang (2013).*

production wells in southern Qinshui and eastern Ordos Basins, and Huainan, Huabei, and Fuxin-Tiefa coalfields. Southern Qinshui Basin includes 3750 wells (e.g. vertical, multilateral, and horizontal), which produced 1.6 Bcm in 2010. Qinshui Basin covers about 23,000 km^2 in the southeast part of Shanxi Province and Ordos Basin covers about 370,000 km^2 in Shanxi, Shaanxi, and Gansu Provinces as well as vast regions of Inner Mongolia and Ningxia Hui (Figure 9.27). Ordos Basin has about 520 producing wells (e.g. vertical and multilateral horizontal), which reached production capacity of 0.5 Bcm at the end of 2010 (Wang, 2011). Although the areal size of Qinshui Basin is 12 times less than Ordos Basin, it contains coalbed gas-in-place of 5.5 Tcm compared to 9.8–10.7 Tcm for Ordos Basin (Feng, Ye, & Zhang, 2002; Gunter et al., 2005; Wang, 2011). More than one-third of the total coalbed gas-in-place resources of China are accounted by both basins. In addition, as of 2010 about 80% of coalbed gas production in China comes from these two basins (Wang, 2011). The coal geology, reservoir properties, and drilling techniques in the Qinshui and Ordos Basins are summarized in Table 9.6.

Qinshui Basin (Small)

The Qinshui Basin contains one of the world's highest rank coals (e.g. semianthracite and anthracite) that serve as coalbed gas reservoirs (Table 9.6). The first commercial production in the basin was in 2005 from 100 wells in the Qinshui County of Jincheng with daily production of 2000 m^3. The highest coalbed gas production is 16,000 m^3/day/well and the average production is 2000–4000 m^3/day/well (Wang, 2011; Yang Ju, Tong, & Xu, 2011; Huang, 2012). The production is from 1- to 6-m-thick coal beds of the Upper Carboniferous Benxi and Taiyuan

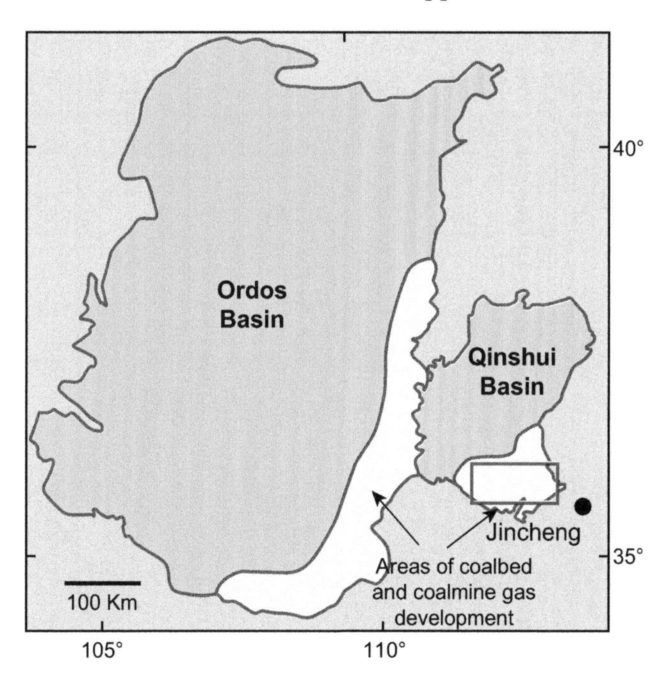

FIGURE 9.27 Map showing the Qinshui and Ordos Basins in north China. The major target areas for coalmine and coalbed gas development are in southeast and east Ordos Basin and south Qinshui Basin (see Figure 9.25 for location in north (N) China). The red-outlined area in the southernmost part of Qinshui Basin contains the most active gas fields in Chengzhuang, Fanzhuang, and Zhengzhuang.

TABLE 9.6 Summary of Coal Geology, Coalbed Gas Activities, Coal Reservoir Properties, and Drilling Techniques in the Qinshui and Ordos Basins, China. Compiled from Numerous References as Indicated in the Table as well as from Personal Work in China

Coal Basin	Coal Age	Coal Rank	Coal Thickness (m)	Reservoir Properties	Depth (m)	Drilling	References
Qinshui	Carboniferous, Permian	Semianthracite to anthracite	1–15	Permeability (range from 0.004–82.8 mD to 0.01–>3 mD) Gas content (15–25 m^3/t)	200–2000	Vertical and horizontal	Li et al. (1995), Ji and Yang (2007), Lin and Su (2007), Huang et al. (2009), Wang (2011); Yang et al. (2011), Hu et al. (2012)
Ordos	Carboniferous, Permian	Low-volatile bituminous	1–11	Permeability (range from 0.023–16 to 1–5 mD) Gas content (4–23 m^3/t)	100–1500	Vertical and horizontal	Li et al. (1995), Guo (2006), Yao et al. (2009), Wang (2011)

m, meters; mD, millidarcy; m^3/t, cubic meter per ton.

Formations and Lower Permian Shanxi Formation. The Bexin Formation contains two or more coal beds and Tiayuan Formation includes 5–10 coal beds (Ji & Yang, 2007; Lin & Su, 2007; Huang, 2012). The Shanxi Formation contains 2–7 coal beds with the #3 coal bed as a key gas reservoir. The Taiyuan and Shanxi Formations are the major coal-bearing units that contain the coalbed gas reservoirs. The coal reservoirs range from 200 to >1500 m deep with gas content increasing to more than 25 m^3/t with depth (Table 9.6). According to Ji and Yang (2007) and Lin and Su (2007), the coal beds generally thicken from 1 to 14 m from west to east and north to south in the basin.

The three existing coalbed gas fields in southern the Qinshui Basin (Figure 9.27) with active development include the Chengzhuang, Fanzhuang, and Zhengzhuang fields (Wang, 2011. The coalbed gas in the shallow (<600 m) Chengzhuang and Fanzhuang fields is developed from vertical and multilateral horizontal wells. The coalbed gas in the deeper (>1000 m) Zengzhuang field is developed mainly from vertical wells. The Zenghuang field located in the deeper part of the Qinshui Basin and toward the direction of coalbed thickening (Ji & Yang, 2007; Lin & Su 2007) fundamentally should contain gassy, thick coal reservoirs compared to the shallower Chengzhuang and Fanzhuang fields.

Ordos Basin (Large)

The Ordos Basin accounts about one-fourth to one-third of China's coalbed gas-in-place resources (Wang, 2011; Yao et al., 2009). Pilot coalbed gas wells were drilled in the southeast part of Ordos Basin (e.g. Weibei coalfield), which produced about 500–2500 m^3/t (Guo, 2006; Yao et al., 2009). However, according to Wang (2011), 520 coalbed gas wells with 11 multilateral horizontal wells were drilled in eastern Ordos Basin (e.g. Hancheng, Daning-jixian, and Sanjiao regions). Of the wells, 100 produced 100,000 m^3/day of coalbed gas or at a rate of 10,000 m^3/day/well (Wang, 2011). The information on Webei coalbed gas field was updated to contain 1051 wells with 771 wells producing by the end of 2012 (Yao et al., 2013).

Much like in the Qinshui Basin, the major coal-bearing rocks in the eastern Ordos Basin are the Upper Carboniferous Taiyuan and the Lower Permian Shanxi Formation (Yao et al., 2009). The Taiyuan Formation contains as many as three coal beds ranging from 1.5 to 10 m in thickness. The Shanxi Formation includes as many as three coal beds. The net coal thickness increases toward the north of the Weibei coalfield. The rank of the coals in the Weibei coalfield based on vitrinite reflectance (1.5–2.0%) is mainly low-volatile bituminous but projected to grade into semianthracite at depth (Yao et al., 2009). The adsorption capacity of the coals varies from 12.01–25.36 m^3/t to 13.91–29.54 m^3/t and gas content ranges from 8 to >15 m^3/t (Yao et al., 2009, 2013).

RUSSIA

Russia possesses the second largest coal reserves and is a leading coal producer in the World (BP, 2012). However, commercial coalbed gas development in Russia has lagged behind OCED countries. Geographically, Russia may be divided into regions east and west of the Ural Mountains, in which 22 coal basins are stretched across the country. Majority of the coal resources are located in the Asian part of Russia or east of the Ural with 80% in Siberia and 10% in the Far East region, and the remainder in East Europe part of Russia or west of the Ural (USEPA, 2010; Figure 9.28). The main coal-producing basins in Siberia are the Kuznetsky and Kansko-Achinsky Basins, along with the South Yakutsky Basins in the Far East region (IEA, 2009). Although Russia has a spectrum of anthracite to lignite coal deposits, more than 75% of coal production is from the anthracite and bituminous or hard coal. About 40% of the coal production is from underground mines and 60% from surface mines (USEPA, 2010). Coal mining

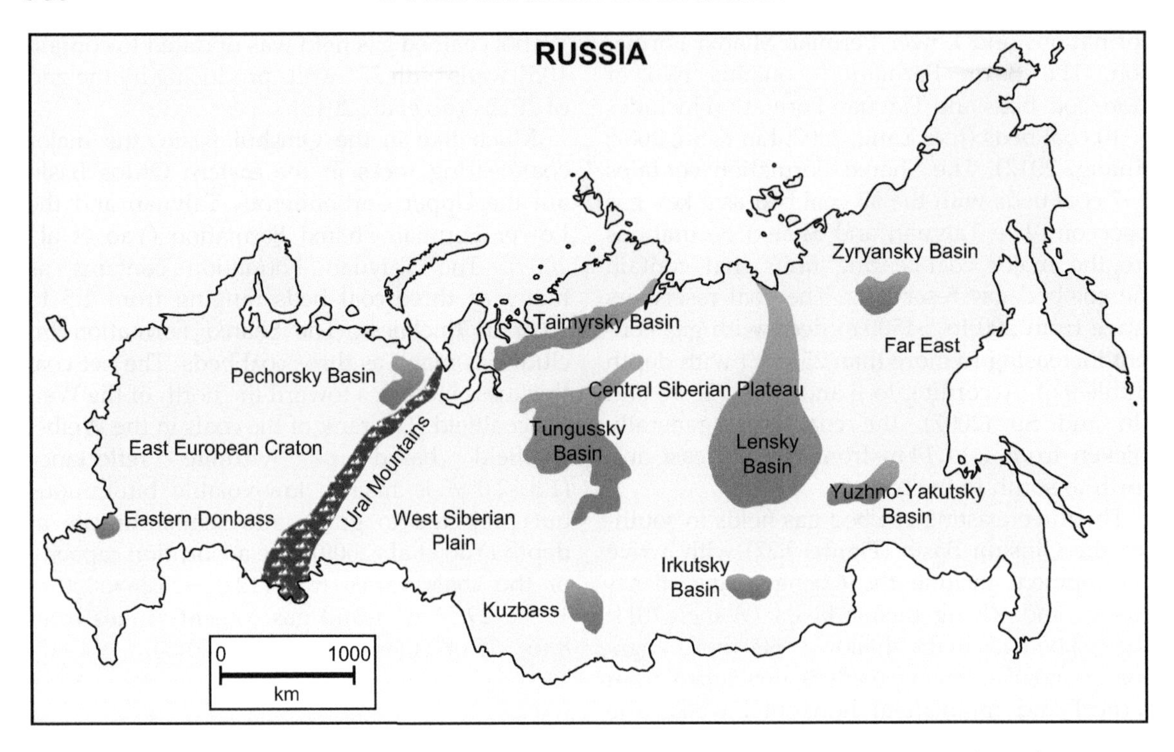

FIGURE 9.28 Map showing the nine coal basins with potential for coalbed gas in Russia. *Source: Adopted from Gazprom (2010) and USEPA (2010).*

especially in underground mines has employed degasification of the coal beds to prevent explosion. About 60% of potentially hazardous coal mines are considered to have abundant coalbed gas, which can lead to outbursts (Flores, 1998).

Coalbed Gas Basins and Resources

Seven coal basins (Donbass or Donetz, Pechorsky or Pechora, Kuzbass or Kuznetsk, Yuzhno-Yakutsky or South Yakutia, Tungussky or Tunguca, Lensk or Lena, and Taimyrsky or Taimyr) mostly in Siberia and East Europe regions have excellent potential for coalbed gas development from mineable and virgin coals (Figure 9.28; USEPA, 2010). The Donbass and Pechora coal basins are in the East European part of Russia. According to USEPA (2010), Russia's coal basins contain coalbed gas-in-place resources of about 83.7 Tcm, which accounts for one-third of the country's forecasted natural gas resources. More than half (47.6 Tcm) is contained in the Donbass, Pechora, Kuzbass, South Yakutia, Tunguska, Lensk, and Taimyr coal basins (Table 9.7). The Kuzbass Basin potentially contains one of the largest coalbed gas resources in Russia estimated to be 13.1 Tcm, which is accessible at 1800–2000 m depth (Table 9.7; Gazprom 2010). The Pechora Basin is another well-known basin for its coalbed gas-in-place estimated to be 1.9 Tcm. ARI (2008a) reported that the coalbed gas resources in the southwest Donbass region range from 90 to 107 Bcm/km^2. Two largely undeveloped coal basins (Tungusca and Lensk) untapped for coalbed gas have been estimated to contain

TABLE 9.7 Summary of Coal Geology, Coal Rank and Thickness, Depth, Gas Content, and Coalbed Gas Resources for Selected Russian Coal Basins. Compiled from Numerous Reference as Indicated in the Table

Basin	Geologic Age	Coal Rank	Coal Thickness (m)	Depth (m)	Gas Content (m^3/t)	Coalbed Gas Resources (Tcm)	References
Tungussky (Tungusca)	Carboniferous, Permian	Bituminous-anthracite	1–15	200–1500	16–35	20.000	Vorobyov et al. (2004), Gazprom (2010), USEPA (2010)
Kuzbass (Kuznetsk)	Carboniferous, Permian	High-volatile C bituminous to low-volatile bituminous	0.7–4.5; as much as 35	1800–2000	18–35	13.085	USEPA (2010), Vorobyov et al. (2004), Jones (2005), Gazprom (2010)
Lensk (Lena)	Jurassic-Cretaceous	Bituminous	1–20	300–2500	5–18	6.000	Ignatchenko (1960), Vorobyov et al. (2004), Gazprom (2010), USEPA (2010)
Taimyrsky (Taimyr)	Cretaceous	Bituminous-anthracite	Up to 10	800–1200	No data	5.000	Rovenskaya and Nemchenko, (1992), Gazprom (2010), USEPA (2010)
Pechorsky (Pechora)	Permian	Subbituminous; mainly bituminous–anthracite	3–20; 2.4 (ave)	500	18–38 18 (ave.)	1.942	Yakutseni et al. (1996), Vorobyov et al. (2004), Gazprom (2010), USEPA (2010)
Eastern Donbass (Russian Donetz)	Carboniferous	Subbituminous, bituminous–anthracite, metaanthracite	0.4–1.2	700–1200	12–22	0.097	Vorobyov et al. (2004), Gazprom (2010), USEPA (2010)
South Yakutia (Yuzhno-Yakutsky)	Jurassic–Cretaceous	Subbituminous; bituminous–anthracite	40–70	No data	10–19	0.920	Vorobyov et al. (2004), Gazprom (2010), USEPA (2010), Thomas (2012)
Ziryanka (Ziryansky)	No data	Bituminous	No data	No data	10–15	0.099	USEPA (1996), Vorobyov et al. (2004), Gazprom (2010), USEPA (2010)

m, meter; m^3/t, cubic meter per ton; Tcm, trillion cubic meter.

combined gas-in-place resources of 26 Tcm (USEPA, 2010).

The coalbed gas-bearing basins of Russia contain mainly Carboniferous and Permian coals of bituminous to anthracite rank (Table 9.7). Other coalbed gas-bearing basins are in Jurassic to Cretaceous age coals of mainly bituminous rank. The Carboniferous and Permian coals are established gassy coals as determined from Russian coal mines (Table 9.7). The mineable Russian coals are very gassy with gas content as much as 38 m^3/t. The depths of coal beds in the basins with potential gas development range from 200 to 2500 m. The coal beds vary in thickness from <1–35 m for hard coal and as much as 70 m for the soft or subbituminous coal (Table 9.7).

Coalmine Gas

Coalbed gas was extracted in Russia as a by-product from coal mines with the use of degassing systems including ventilation and wells drilled from the surface. The degassing systems recovered CMM about 1.5 Bcm from the Kuznetsky (Kuzbass) and Pechorsky (Pechora) Basins (IEA, 2009). The Kuzbass Basin accounted for about 78% of the CMM and the Pechora make up about 12% of the CMM. Despite the recovered CMM at an average of 25%, large volume of coalbed gas is vented or emitted. From 1990 to 2003, the underground mines in the Kuzbass Basin emitted more than 13 Bcm (USEPA, 2010). In 1998, the coal mines in the Pechora Basin vented 289.8 MMm^3 of coalbed gas, which was reduced to 42 MMm^3 in 2000 resulting to mine closures and abandonment, which in turn will emit AMM. Thus, the pathway to development of CMM and AMM in these basins is through predrainage or in front of mining drilling such as in China's Qinshui Basin.

The Kuzbass Basin includes the gassiest coal mines in Russia mainly from 73 active underground coal mines and 20 surface mines (USEPA, 1996). Jones (2005) reported as many as 36 working underground mines in the western periphery of the basin are potential for CMM recovery and utilization. In addition, there are 43 coal mines that are abandoned, which are potential for AMM recovery and utilization. Marshall et al. (1996) reported that there are more than 100 coal beds with an average thickness of 2.5 m, which have been mined at depths varying from 300 to 800 m in the Kuzbass Basin. Jones (2005) reported 124 coal beds in the Permian strata and 16 coal beds in the Carboniferous strata. The net coal thickness is 386 m in the Permian strata. The Permian coal beds have an average gas content ranging from 19 to 25 m^3/t with permeability of ranging from 20 to 150 mD (Uglemetan, 2002). According to Jones (2005), 6631 Bcm of coalbed gas-in-place resources is attributed to the Permian Upper Balakhonsky Stage and 6454 Bcm attributed to the Permian Kolchuginsky Stage. Gazprom in the Kuzbass Basin drilled exploration wells with production of 5.3 MMm^3 and in 2010, Gazprom opened the first coalbed gas production facility in the Taldynskoye field in the Kuzbass Basin (Gazprom, 2010). The Taldinskoye field forecasted resources are estimated at 95.3 Bcm.

The Pechora Basin is another coal mining region that has potential for coalmine gas development (Yakutseni et al., 1996). Much like the Kuzbass Basin, prospective coal beds are Permian in 1600-m-thick strata containing 150 coal beds as thick as 4 m in the Kungarian Stage. Yakutseni et al. (1996) reported that the coal beds range from subbituminous to anthracite with vitrinite reflectance from 0.5 to >2.5%. The high-rank hard coals indicate gas content in coal mines as much as 18–38 m^3/t and average 18 m^3/t (Table 9.7). The variable high gas content of coals in Pechora coal mines account for the CMM resources reported as 700 MMm^3 in 1996 and 1.5 Bcm in 2008 (IEA, 2009; USEPA, 2010; Yakutseni et al., 1996). Yakutseni et al. (1996) indicated that a problem in coalbed gas development in the Pechora

Basin is permafrost acting as a barrier to vertical movement of water during dewatering of coal bed reservoirs.

INDIA

The Ministry of Coal of India estimated the total coal resources of the country to be about 267 billion tons (MOC, 2010). The total coal resources estimate was updated by the Geological Survey of India to 293.5 billion tons with proved reserves of 118 billion tons (Bhattacharya, 2012). The estimated proved recoverable coal reserves are about 54.6 billion tons for the hard coal (anthracite and bituminous) and 4.6 billion tons of brown (lignite and subbituminous) coals (USEPA, 2010). This translates to 93% of the total proved recoverable reserves in 2010 are accounted by the hard coal and about 7% by the brown coal. The coal-bearing rocks in India are stratigraphically distributed in the Permian Gondwana Coal Measures comprising the hard coal and Tertiary (Eocene–Oligocene) deposits of brown coal (MOC, 2010).

The Permian Gondwana Coal Measures are mainly found in the Koel-Damodar, Son-Mahanadi, Pranhit Godavari, and Satpura Basins in northeast India (Figure 9.29). The coal resources in these four major Gondwana basins are contained in 17 major coalfields (USEPA, 2010; Verma, 2005). The Koel-Damodar Basin includes mainly Jharia, Ranijanj, Bokaro, Ramgarh, north and south Karanpura coalfields from east to west. The Son-Mahandi Basin includes mainly the Singrauli, Sohagpur, Korba, Ib-valley, and Talchir coalfields from northwest to southeast. The Pranhita–Godavari Basin includes mainly the Wardha and Godavari coalfields from northwest to southeast. The Satpura Basin includes mainly the Narmada coalfield.

In addition, 17 Tertiary coalfields containing lignite, subbituminous, and bituminous coals found in the northeast (Assam–Meghalaya coalfields), northwest (Cambay, Barmer-Sanchor, Bikaner, and Jammu–Kashmir coalfields), and southeast (Neyvell coalfield) India were offered for exploration and development (Figure 9.29; Verma, 2005).

Coalbed Gas from Virgin Coals

Areas beyond the gassy coal mines in Permian (Gondwana) and Tertiary coal measures contain favorable geological and reservoir characteristics for coalbed gas development (Table 9.8). According to Verma (2005) and DGH (2010), the Permian high-rank coals potentially store thermogenic gas in contrast to the Tertiary low-rank coal, which generates biogenic gas similar to the Powder River Basin. Gas content increases at depth and permeability (e.g. 0.3–3 mD) may be enhanced by network of faults in the Gondwana rift basins (Verma, 2005). The coal depths, which range from 300 to 1500 m, are projected to be gas-saturated deeper than 600 m (Verma, 2005). Twelve to fourteen Gondwana coalfields were targeted for coalbed gas exploration and development including four in the Koel-Damodar Basin, five in the Son-Mahanadi Basin, two in the Pranhita–Godavari Basin and one in the Satpura Basin (DGH, 2010; Dutta et al., 2011; Verma, 2005; Figure 9.34). A comparison of the coal stratigraphy, rank, thickness, depths, and gas content of the Permian Gondwana basins is summarized in Table 9.8.

Exploration of coalbed gas was initiated in 1990 mainly targeting the Permian Coal Measures (Peters, 2000). Encouraged by the successes in developing and producing gas from coal beds in OECD countries, the Indian government formulated a "CBM policy" in 1997 to utilize the coalbed gas potential of the country. The Ministry of Petroleum and Natural Gas administered and Directorate General of Hydrocarbons (DGH) implemented the "CBM policy". Initially, the DGH focused on the exploration in coalfields containing high-rank coals of the Permian Coal Measures. Implementation of the policy was

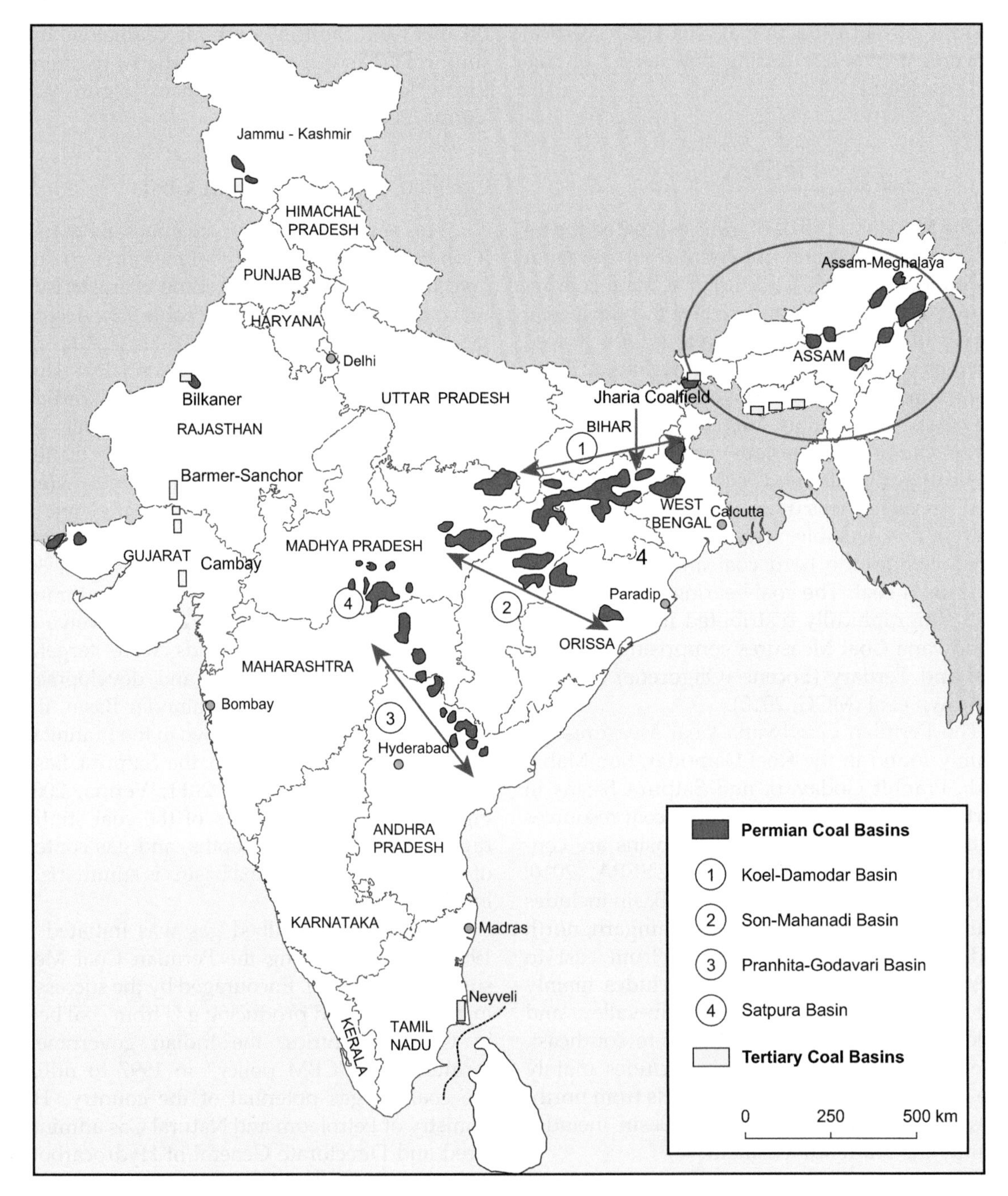

FIGURE 9.29 Map showing the Permian and Tertiary coal basins with potential for coalbed gas in India. *Source: Modified from Walker (2000) and USEPA (2010).*

TABLE 9.8 Summary of Coal Geology, Coal Rank and Thickness, Depth, and Gas Content in the Gondwana Coal Basins of India. Compiled from Numerous References as Indicated in the Table

Coal Basin	Geologic Age	Formation	Coal Rank	Coal Thickness (m)	Depth (m)	Gas Content (m^3/t)	References
Son-Mahanadi	Lower–Upper Permian	Raniganj and Barakar	High-volatile C–A bituminous	1–86	600–1200	3–10	Cassyhap and Tewari (1984), Verma (2005), Bhattacharya (2012)
Koel-Damodar	Lower Permian	Barakar	High-volatile A to low-volatile bituminous	1–64	300–1500	2–23.8	Cassyhap and Tewari (1984), Verma (2005), Ojha et al. (2011), Bhattacharya (2012)
Pranhita-Godavari	Lower and Upper Permian	Lower Kamithi and Barakar	High-volatile bituminous C	10–25	300–1500	4–6	Verma (2005), Mukhopadhyay et al. (2010), Sarate (2010), Bhattacharya (2012)
Satpura	Lower and Upper Permian	Bijuri and Barakar	Subbituminous C to high-volatile A bituminous	1–6; 3–11.5	300–1200	5–7	Singh and Rakesh (2004), Verma (2005), Mukhopadhyay et al. (2010)

m, meter; m^3/t, cubic meter per ton.

designed to attract global bidding and investment by following successful models (e.g. United States' "Crude Oil Windfall Profit Tax of 1980") by providing tax incentives such as 7-year tax holiday from the date of inauguration of commercial production (Flores, 1998; Verma, 2005).

Subsequently, the DGH offered a total of 26 coalbed gas or CBM blocks for global bidding in 2001, 2003, and 2005 to explore, develop (e.g. pilot assessment projects), and produce coalbed gas (Verma, 2005; USEPA, 2010). The CBM blocks (e.g. 70–1045 km^2 area) were estimated to contain 1.45 Tcm with projected total production of 39.7 MMm^3/day at peak production levels (DGH 2010; USEPA, 2010; Verma, 2005). In 2008, total coalbed gas production amounted to <30 MMm^3 (India Ministry of Petroleum and Natural Gas, 2009).

Figure 9.30 represents the CBM block, an 85 km^2 area in the Jharia coalfield, which contains 18 coal beds about 1–33 m thick, 3.01–5.93% porosity, 0.03–2.88 mD permeability, 6–16 m^3/t gas content at 300–1200 m deep in the Permian Barakar Formation, which are stratigraphically separated by the Barren Measure (Mandal et al., 2004; Verma, 2005). The stratigraphic and structural framework of Jharia coalfield indicate thickening of the Barakar Formation and Barren Measures from north to south, which is bounded by a major normal fault (Casshyap & Tewari, 1984). However, Ojha et al. (2011) suggested that the Jharia coals might contain 7.3–23.8 m^3/t within the depth range of 150–200 m. Additional gas analysis indicates that every 100 m increase in depth is related to a rate of 1.3 m^3/t increase of coalbed gas content. A test well showed gas production of 8000 m^3/day and water production of 2–5 m^3/day. There are a few of these small (70–95 km^2 area) coalfields (e.g. South Karanpura, Bokaro) but excellent storehouse of thermogenic coalbed gas in the Koel-Damodar rift basin.

Six CBM blocks from 790 to 1045 km^2 area in the Barmer Basin (e.g. Rajasthan and Gujarat coalfields) were offered in 2003 for exploration and development (Figure 9.29). The Eocene coal-bearing rocks contain 2–76 m net thickness with 4–4 m^3/t gas content at 600–1500 m depth (Verma, 2005). Other Tertiary coal basins such as in the Assam State contain Eocene-age coal beds up to 30 m thick, which range from subbituminous C to high-volatile bituminous C and Oligocene-age coal beds, which range from subbituminous C to high-volatile bituminous B (Mishra & Gosh, 1996).

Coalmine Gas

National coal production was increased from surface mines in shallow Tertiary coal resources compared early production from deep underground mines in Gondwana Coal Measures. As of 2008, there are 559 active coal mines with underground mines accounting for 337 and surface mines accounting 186, and 36 mixed underground and surface mines (USEPA, 2010). Eighty four percent of the coal production comes from the surface mines. About 61% of these coal mines are very gassy emitting at a rate of >0.01 to >10 m^3/t. Singh et al. (1999) reported that 40% of the coal mine disaster in India from 1908 to 1995 is caused by coalbed gas outbursts causing 839 fatalities. Coal Mining is reported to be contributing to about 9% of total coalbed gas emission of India (Pande, 1996), which have increased more than three times from 1995 to 2010.

Coalbed gas or CMM emissions from Indian coal mines increased from 763 MMm^3 in 1990 to 1.6 Bcm in 2010 (USEPA, 2010). Although the Indian government has demonstrated the commercial feasibility of recovery and utilization of CMM, no commercial production of coalmine gas is in place. Another potential source of coalbed gas is from AMM. The best potential for CMM recovery is before, during, and after mining with the later recovery from AMM. Five percent of abandoned coal mines in India are

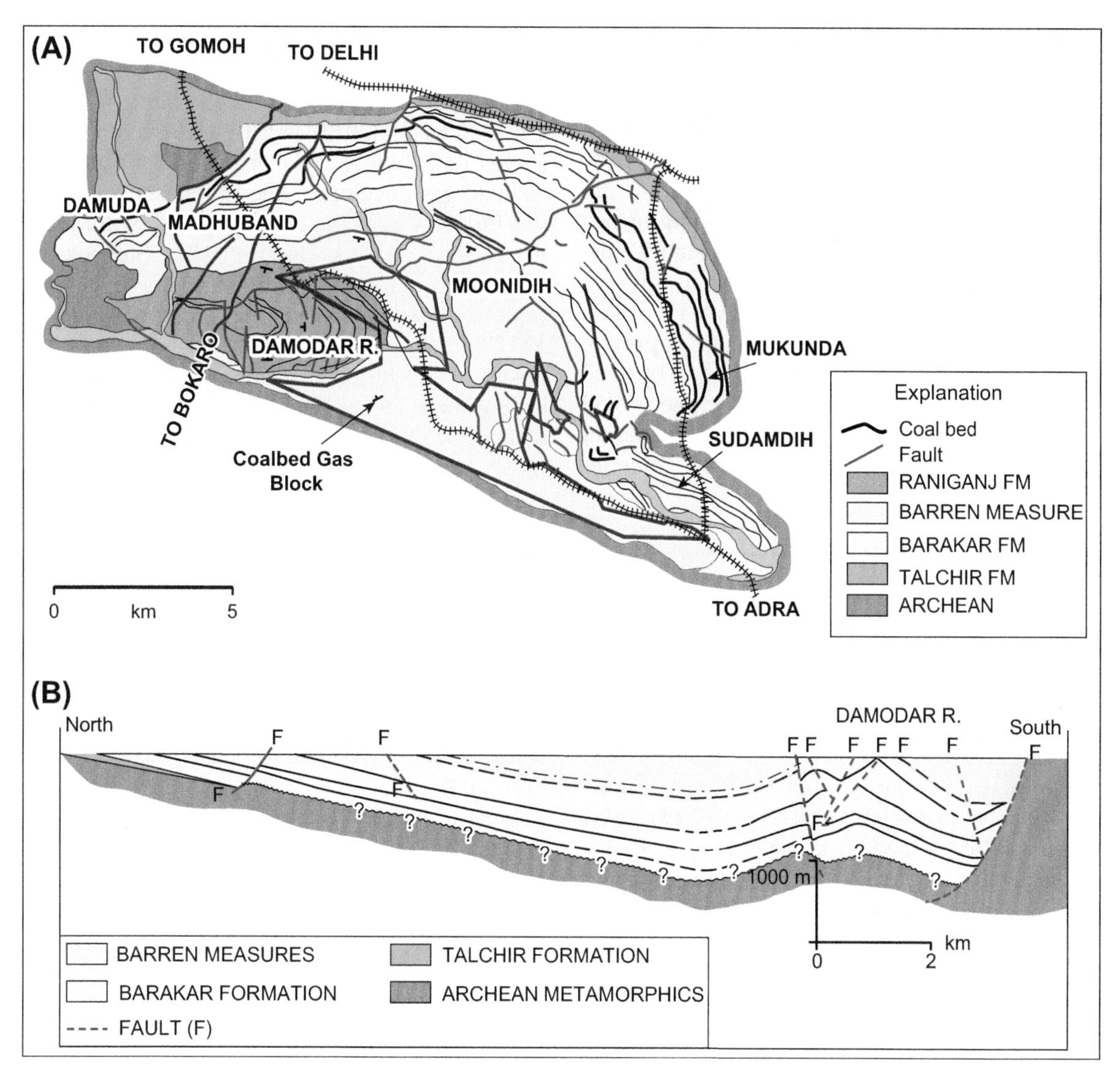

FIGURE 9.30 Map showing (A) the Jharia coalfield in the Koel-Damodar Basin in India and (B) the north–south cross-section across the coalfield showing the Gondwana Coal and Barren Measures. Blue-outlined area in (A) is the coalbed gas or CBM block area. FM, Formation. *Source: Adopted from Verma (2005) and Cassyhap and Tewari (1984).*

considered gassy (USEPA, 2010). Recovered CMM may be used to generate local power for electricity (with a maximum capacity of 500 kW), as well as being used in 50-ton mine dump trucks, powered by converted bifuel engines (UNDP, 2009).

USEPA (2010) reported that for the first time for India, prospective operators are being offered active coal mining blocks to extract and utilize CMM. India's Coal Mine Planning and Design Institute recently issued a notice inviting tenders for five CMM blocks held by Coal India Limited.

The blocks are located in the Jharia and East-Bokaro coalfields in northern India (Figure 9.29). The objective of the CMM projects is to drain gassy coal beds below the coal currently mined using predrainage technology of gas recovery. Original rules prohibited coal owners from extracting CMM released during mining operations. However, the Coal and Petroleum Ministries have agreed to allow coal owners (e.g. Coal India) to explore CMM. However, the Petroleum Ministry must approve commercialization of the coalbed gas by coal owners. The offering of the CMM projects will bring India's coalmine gas resources to market in an environmentally beneficial manner (USEPA, 2011).

INDONESIA

The large coal resources, which were assessed by BP (2012) as 13th in proved coal reserves worldwide, make Indonesia a potential coalbed gas developer. The Indonesian coalbed gas industry is at the early stage of exploration in 11 coal basins (Figure 9.31; Stevens & Hadiyanto, 2004). In addition to these basins, three sedimentary basins (e.g. Pembuang, Melawi, and Ketungau), which contain coal deposits, were added in Borneo from the study by Sorkhabi (2012). Unlike the coalbed gas resources in OECD and other NOECD countries, Indonesian coalbed gas resources are in archipelago basins, which contain about 85% of

BOX 9.4

BREAKOUT OF CMM CAPTURE: AN ASSET OR LIABILITY (FACT VS FICTION)

CMM capture for clean energy from coal mines is an asset in OECD countries (e.g. Australia, Germany, United States, United Kingdom, and South Africa, USEPA, 2008). CMM is a liability especially in the form of coal mine explosions and greenhouse gas emissions under the Kyoto Protocol (USEPA, 2006). However, breakout of CMM recovery in major coal producing NOECD countries (e.g. China and Russia) has used CMM to the next level of profitability. China has made a budding asset of CMM during 2001–2010 focused on safe operations of their more than 12,000 major operating coal mines (IEA, 2009). During this period of mitigating coal mine hazard, CMM was transformed to a valuable resource (Flores, 1998). Transformation was from fueling electric power plants on mine sites to potentially profitable domestic CNG and LNG for residential and transportation utilization. Russia harnessed CMM in coal mines in the Kuzbass Basin for beneficial commercial production and utilization and greenhouse reduction (Gazprom, 2010, 2013).

Karacan et al. (2011) extensively reviewed the state of CMM capture and utilization worldwide and found the countries with most CMM emissions are China, United States, Australia, Ukraine, India, and Russia. In these countries and other CMM emitting countries, implementation of CMM protocols depends on economic, social, and regulatory circumstances (Karacan et al., 2011; USEPA, 2010). These challenges may be overcome by partnerships with government entities (Karacan et al., 2011). However, a challenge that may have been overlooked is a legal issue as discussed by Bassett et al. (2011). These authors examined current United States laws of CMM and its capture and concluded that as of 2009, no federal law or regulation covers the release of CMM.

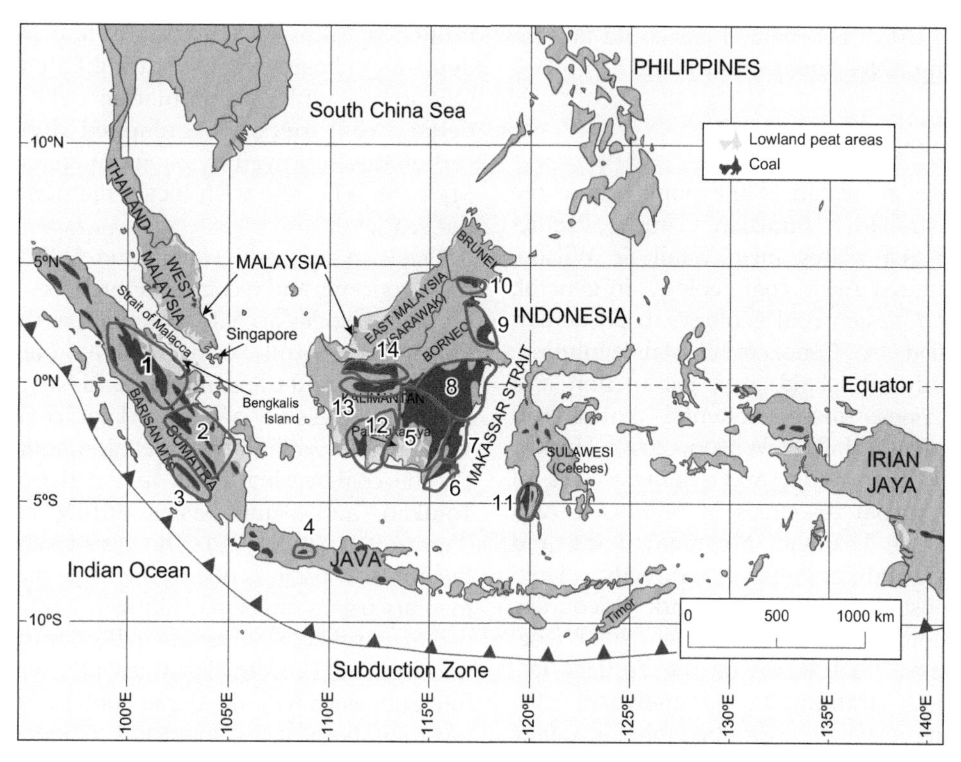

FIGURE 9.31 Map showing areas in Indonesia with Tertiary coals (black) in 11 basins identified by Stevens and Hadiyanto (2004) with potential coalbed gas. (1) Central Sumatra, (2) South Sumatra, (3) Bengkulu, (4) Jatibarang, (5) Barito, (6) Asem–Asem, (7) Pasir, (8) Kutei, (9) Berau, (10) Tarakan, and (11) Southwest Sulawesi Basins. Three sedimentary basins with coal deposits are added from Sorkhabi (2012), which include (12) Pembuang, (13) Melawi, and (14) Ketungau Basins. *Source: Map with peat and coal deposits is adopted from Flores and Hadiyanto (2006).*

low-rank coals (e.g. lignite and subbituminous) and 15% hard coal (e.g. mainly bituminous and a fraction of anthracite) proved recoverable reserves (USEPA, 2010). The coalbed gas potential of the low-rank coals of Indonesia is often related to the Powder River Basin of the United States in terms of similarities in coal thickness, rank, areal extent, and gas adsorption capacity (Flores & Stricker, 2010, 2012; Stevens & Hadiyanto, 2004).

Stevens and Hadiyanto (2004) estimated a vast amount of coalbed gas-in-place of about 13.6 Tcm, which is about 4 times the proven reserves (3.2 Tcm) of conventional natural gas of Indonesia. The proven coalbed gas reserves are estimated to be about 3.4 Tcm, which is still slightly more than the proven reserves of conventional gas. Cedigaz (2010, 2011) reported that Indonesia could also become the first country to produce LNG from a coalbed gas source, ahead of Australia's coalbed gas to LNG projects, which are scheduled to come on line around 2015. Indonesia has more than 20 active production-sharing contracts for ongoing coal bed drilling in pilot projects (Cedigaz, 2010). The Oil and Gas Journal (2011) reported a pilot well in the Sangatta West project in East Kalimantan started to flow gas on March 25, 2011 after the installation of a dewatering pump. According to

Cedigaz (2010, 2011) coalbed gas could flow to the Bontang LNG liquefaction plant.

Coal Geology

Indonesia is located at the point where the Indian-Australian, Eurasian, and Pacific-Philippine Sea plates meet (Hall & Wilson, 2000). Southeast Asian coal geology, in general, and the Indonesian coal geology, in particular, is controlled by Cenozoic crustal evolution, which produced island arcs, subduction and collision zones, microcontinental plates, and oceanic ridges (Hall & Wilson, 2000; Horkel, 1989). The distribution and characteristics of the Tertiary coal resources in Southeast Asia (e.g. Malaysia, Thailand, Myanmar, Laos, and Philippines) and Indonesia are directly related to a tectonostratigraphic framework (Friederich, Liu, Langford, Nas, & Ratanashien, 2000; Koesoemadinata, 2000). Based on the Tertiary tectonic and stratigraphic framework, the Indonesian coals may be grouped into Paleogene (e.g. Paleocene, Eocene, and Oligocene) deposits and Neogene (e.g. Miocene and Pliocene) deposits. The Paleogene is characterized by regional rifting resulting in sedimentation of syn-rift grabens and half-grabens from wrench faulting along the continental margins or rotation of the Sundaland (Sumatra–Java; Koesoemadinata, 2000). The Paleogene deposition includes forearc basin infilling by coal-bearing fluvio-deltaic sedimentary prisms prograding from magmatic arc, syn-rift infilling by coal-bearing fluvial and lacustrine deposits, and post-rift transgression with onlapping coastal, tidal, and deltaic coal-bearing deposits. Neogene was characterized by regional subsidence and uplift; the latter due to subduction of the Indian oceanic crust below the Sundaland, which in turn, resulted in creation of back arc or interarc and foreland (forearc or foredeep) basins (Hall & Wilson, 2000). Friederich et al. (2000) described similar evolution of coal basins from basement rifting during pre-Late Eocene infilled by alluvial fans, fluvial and lacustrine deposits to basin sagging during Late Eocene–Oligocene with accumulation of coastal–deltaic coal deposits onlapped by marine carbonates, followed by basin inversion during Miocene–Pliocene with local emergence above the sea level.

Davis, Noon, and Harrington (2007) refined these concepts and recognized three tectonostratigraphic episodes of coal accumulations during the Paleogene synrift transgressive, Paleogene–Neogene postrift transgression, and Neogene syntectonic regressive periods. Major coals basins were infilled with alluvial, deltaic, and lacustrine Eocene coal-bearing strata in the Barito, Kutei, Tarakan, and Berau basins during regression (Figure 9.31; Table 9.9). The Berau subbasin of the Tarakan Basin is one of the three major sedimentary basins developed along the eastern continental margin of Kalimantan in Borneo during the Eocene. The Tarakan Basin can be divided into four subbasins with the Berau subbasin being the most southern of the subbasins. Oligocene coal-bearing strata accumulated in the central and southern Sumatra. The Paleogene (Eocene) coals represent the oldest and first peat accumulations (Chapter 3) in the Indonesian major rift basins.

The Paleogene regression was culminated by subsidence and widespread Oligocene marine transgression (Davis et al., 2007). Early transgression reworked the Eocene continental deposits and redeposited along backstepping coastlines where coals formed. The landward or backstepping shifts caused onlapping of marine fine-grained sediments on the Eocene continental deposits. Tectonic compression and uplift followed the transgressive event, which led to basin redevelopment with infilling of Miocene alluvial and deltaic coal-bearing strata in the Barito, Kutei, Tarakan, Berua, and central and south Sumatra basins. Coal accumulation continued in the late Neogene (Pliocene) in a few of these coal basins. Whether the characteristics of the Indonesian Tertiary coal beds are influenced by climate, tectonic, and depositional

TABLE 9.9 Summary of Coal Geology, Coal Rank and Thickness, Depth, Gas Adsorption and Content, and Gas-in-Place in Selected Coal Basins of Indonesia. Compiled from Numerous References as Indicated in the Table

Basin	Geologic Age	Coal Rank	Coal Thickness (m)	Coal Depth (m)	Gas Content (m^3/t)	Gas Adsorption (m^3/t @ 5 MPa)	Gas-in-Place (Bcm)	References
South Sumatra	Upper Oligocene and Upper Miocene	Subbituminous, bituminous, and semianthracite	<40	150–1500	7.0	14–15	5178.9	Saghafi and Hadiyanto (2000), Stevens and Hadiyanto (2004), Davis et al. (2007)
Barito	Lower-Middle Eocene and Middle Miocene	Subbituminous to medium-volatile bituminous	10–60	Not determined	4.7	Not determined	2875.3	Stevens and Hadiyanto (2004), Davis et al. (2007)
Kutei	Upper Eocene and Middle-Upper Miocene	Subbituminous to bituminous	<0.5–4; <30	>1000	6.1; 0.91–13.52	2.6–10.5	2275.3	Stevens and Hadiyanto (2004), Davis et al. (2007), Moore et al. (2012)
Central Sumatra	Upper Oligocene to Lower Miocene	Subbituminous C to high-volatile bituminous	Not determined	<700–>1500	4.5	14–15	1485.7	Saghafi and Hadiyanto (2000), Stevens and Hadiyanto (2004), Davis et al. (2007)
North Tarakan	Middle Eocene, Middle Miocene, and Pliocene–Pliestocene	Lignite, subbituminous. and high-volatile bituminous	15–80	700	4.5	No data	495.2	Stevens and Hadiyanto (2004), Davis et al. (2007):
Berau	Eocene and Miocene	Subbituminous to high-volatile bituminous	0.5–>6	700	4.5		520.7	Stevens and Hadiyanto (2004), Davis et al. (2007)

m, meter; m^3/t, cubic meter per ton; MPa, megapascal; Bcm, billion cubic meter.

settings are not completely understood according to Davis et al. (2007). However, the influence of igneous activity has exerted marked changes on the coal ranks and probably gas content, adsorption capacity, and permeability of coal reservoirs.

Coal Reservoir Characterization

The characteristics of the coal reservoirs in six major coal basins out of 11 are summarized in Table 9.9. The coal reservoir characterization was limited to coal basins that have projected coalbed gas-in-place resources of more than 495 Bcm based on estimates of Stevens and Hadiyanto (2004) and where exploration and potential development are presently occurring. The North Tarakan and Berau basins can be lumped as one basin being Berau is a subbasin of Tarakan Basin (Figure 9.31). Based on the dichotomy of geologic age and coal rank, Indonesian coal reservoirs may be generally grouped as Paleogene bituminous and Neogene subbituminous with rank locally affected by igneous intrusions (Table 9.9). Sumatra (north and central) coal reservoirs are generally older and higher rank. The Kalimantan and Borneo (Barito, Kutei, and Tarakan-Berau) coal reservoirs are mainly younger and lower rank.

In general, the gas content is generally high from <1 to 13.5 m^3/t compared to the Powder River Basin (Flores, 2004). However, the overall coal rank of the Indonesian coals is mainly higher with vitrinite reflectance of 0.41–0.72% R_o of subbituminous C to high-volatile B bituminous (Saghafi & Hadiyanto, 2000). Saghafi and Hadiyanto (2000) studied the gas adsorption or storage capacity of Indonesian coal reservoirs and the results showed relatively low storage capacity (<10 m^3/t at 5 MPa). However, the same study showed gas storage capacity of the coal reservoirs in Sumatra coal basins (south and central) is higher at 14–15 m^3/t. In comparison, the adsorption capacity of the Indonesian subbituminous coals is much higher than the subbituminous coals (average 2.5 m^3/t at 5 MPa) of the Powder River Basin (Flores & Stricker, 2012). The difference in the adsorption capacity may be due to higher vitrinite macerals (83–93%) for the Indonesian coals (Moore, Bow, & Nas, 2012) compared to as much as 65% for the Powder River Basin coals (Flores & Stricker 2012). The thickness and stratigraphic variations of the coal reservoirs in the Tanjung Inim Formations are described in Figure 9.32.

Moore et al. (2012) reported the most detailed and informative reservoir characterization from three exploration closely spaced drillholes in the Sanggata coal field in north Kutei Basin, East Kalimantan, Borneo. The Miocene-age Balikpapan Formation contains up to 4-m-thick coal beds, which are the targeted coalbed gas reservoirs. Coal rank and organic composition as well as gas adsorption desorption, and gas saturation analyses were performed from coal core samples. The results of the coal reservoir characterization as related to depth in one of the cored wells, which is closest (4 km) to the Pinang Dome (igneous intrusive), are shown in Table 9.10 (Moore & Nas, 2013). The vitrinite reflectance and coal rank increase with depth, which corresponds to the increase of gas content from 233 to 730 m depth. Moore and Nas (2013) attributed the increase of coal rank and gas content to depth and heat flow from the nearby intrusive body.

Coalmine Gas

Indonesia is the fifth major coal producer of the world behind India, Australia, United Stats, and China, respectively (BP, 2012). Continued growth of coal production during the past few years is attributed to demand from neighboring countries such as China, Japan, South Korea, and the Philippines (USEPA, 2010). According to the Directorate of Energy and Mineral Resources of Indonesia, the country exports about 70% of its coal production in 2009 (Setiawan, 2009). The growth of coal production comes

FIGURE 9.32 Lithofacies association of Tertiary coal-bearing rocks with thin to thick coal beds that are gassy in the subsurface in the open pit mine of Tanjung Enim, South Sumatra Basin.

with the expanded use of surface mining method, which in turn, contributes to fugitive CMM emissions. USEPA (2010) reported that Indonesia's CMM emissions increased from 21 to 35 MMm^3 during the 1990–2005 period during which about 99% of coal production was from surface mines (Figure 9.32).

Indonesia signed the Kyoto Protocol in July 13, 1998 but has no national CMM capture projects. The minor use of underground mines (3 as of 2004) vs major use of surface mines where CMM is released to the atmosphere may lead to recovery and utilization projects. The best technique for CMM recovery as demonstrated in other mining operations of other countries is predrainage or gas desorption in front of surface mining. This gas recovery method may be permitted in the 2008 Indonesia Regulation of the Minister of Energy and Mineral Resources No. 36 enacted in 2008 to facilitate the development and commercially produce coalbed gas. The most important part of the government regulation relevant to CMM concerns the coal rights holders and holders of Petroleum Production Sharing Contract (PSC) of coalbed gas resources. At the mine site, the mine operators will have the first rights to assume gas exploration and development in their concessions. Where there are no existing concession holders, the area is opened for tenders, which will be submitted following current PSC tender rules. If a coal concession is astride with a PSC concession, the coal concession and PSC holders must agree to exploit the coalbed gas reserves in the common lease area. The agreement may be interpreted by cooperating holders as each have a 50:50 participation in any operations to develop the resource unless agreed on the contrary. The lack of agreement results in arbitration by the Indonesian Minister of Energy and Mineral Resources, which will determine the conflicts and set policies.

The implementation of Regulation No. 36 will permit coal operators to drain the CMM from a number of surface vertical wells into and in front

TABLE 9.10 Relationship between Coal Depth with Gas Content, Vitrinite Reflectance, and Coal Rank in the Sanggata Coalfield in North Kutei Basin, East Kalimantan, Borneo

Sample	Depth (m)	Gas Content (m^3/t)	Vitrinite Reflectance (%)	Coal Rank
1	233	0.91	0.53	Subbituminous A
2	327	5.56	↓	↓
3	355	2.95		
4	370	1.92		
5	402	5.11		
6	439	5.39		
7	493	6.55		
8	533	7.53		
9	585	7.22		
10	626	8.46		
11	648	9.61		
12	682	11.18		
13	714	11.63		
14	730	13.53		
15	798	10.25	0.69	High-volatile B bituminous

m, meter; m^3/t, cubic meter per ton.
Source: Adopted from Moore et al. (2012).

of the mining area, which is presently CBM present at the mine site already is being, or will be, released directly into the atmosphere during mining operations. The Project proposes to drain the CBM through surface drainage wells drilled into the mining areas. The recovery and utilization of coalbed gas may be for on-site power generation for electricity and vehicles. CMM extraction may be used for certified emissions reduction and abatement of global greenhouses gas.

OTHER POTENTIAL TERTIARY COAL BASINS

The potential coalbed gas production from Tertiary coals in Indonesia and India have encouraged the search worldwide for another Powder River Basin, which is the second most productive coal basin in the United States (see previous section). The conventional concept suggests that low-rank coals (e.g. subbituminous) common in Tertiary coal-bearing strata are not as productive coal reservoirs as high-rank coals. One of the criticisms against low-rank coals is the low gas content. However, this perception is offset by the accumulation of thick coals in the Tertiary, which provides more net coal thickness for multiple completions. The success in producing Tertiary subbituminous coals in the Powder River Basin has proved this point and demonstrated that coal permeability is 2–10 times better than the bituminous coals (Flores, 2004). However, the downside in coal production in the Powder River is the large volume of co-produced water.

Many countries possess low-rank coal resources, which accounts for about 50% of the global coal resources and can be converted into coalbed gas assets. The Powder River Basin coalbed gas model has set the standard for gas exploration in low-rank coals worldwide as demonstrated by the Colombia and Philippine examples.

Philippine Coal Basins

The Indonesian and Philippine archipelagoes are tectonically linked with their crustal evolution and timing of formation and distribution of coal-forming basins controlled by plate tectonic processes (Friederich et al., 2000; Hall & Wilson, 2000; Hall, 1997; Horkel, 1989). The Philippine archipelago is composed of modern and ancient island arcs and continental fragments (Hall & Wilson, 2000). The Tertiary coal basins in these countries are unique compared to other global coal basins in that associated plate tectonic and volcanic processes (e.g. igneous intrusions and volcanism) have affected the coal rank, gas content, and gas adsorption capacity of the coals (Flores & Stricker, 2012; Moore & Nas, 2013). Although igneous intrusions commonly elevate coal rank, their effect on coalbed gas desorption through contact metamorphism is not as common worldwide (e.g. Raton Basin in the U.S., Karoo Basins in the South African continent) (Chapter 4). During Paleocene–Eocene time, the Philippine arc system was located in the equatorial region about east of the Indonesian arc system (McCabe, Almasco, & Yumul, 1985). During the Early Neogene (e.g. Miocene), the Philippine arc tectonically translated northwestward. Collisions of the Sundaland-Eurasia plate on the west, Philippine Sea plate on the east, and Indo-Australian plate on the south created back arc and rift coal basins and subbasins (Friederich et al., 2000; Hall & Wilson, 2000; Horkel, 1989; Yumul, Dimalanta, Maglambayan, & Marquez, 2008). The areal extents of the coal basins in the Philippines are much more restricted than the Indonesian basins (Figure 9.32(A)). Although the main coal-forming ages are Eocene to Miocene, which is similar to Indonesia, only minor Eocene and Oligocene coals are found in the Philippines (Flores et al., 2006). There are 16 coalfields and more being discovered, which are scattered among eight sedimentary basins and subbasins and contain about 2.3 billion tons of coal resources (Flores et al., 2006). The coals range from lignite to semianthracite with only minor amount of the high-rank coals and most coal reserves are lignite and subbituminous coals. These coal beds were estimated to contain more than 16 Bcm of gas content based on gas storage capacity (100% saturation) (Flores et al., 2006).

Coal Geology

Horkel (1989) originally defined coal-forming sedimentary basins for both Indonesia and the Philippines as foreland, continental margins, inter arc, back arc, and rift, which are subduction- and collison-related basins (Figure 9.32(B)). The foreland or foredeep typified by the Indonesian basins (e.g. south Sumatra) accumulated laterally persistent coal deposits associated with fluvio-deltaic-shoreface and paralic peat complexes due to rapid subsidence adjacent to volcanic arcs (Davis et al., 2007; Horkel, 1989; Koesoemadinata, 2000). The sedimentary prism either laps onto the volcanic and basement rocks of the main arc or is dissected by a thrust fault. Continental margin basins infilled by similar fluvio-deltaic sedimentary prisms accumulated discontinuous coal deposits due to rapid subsidence adjacent to continental microcontinental plates. The inter arc (active) or back arc (inactive) basin is between an outer volcanic arc (e.g. subduction mélanges and ophiolites) and remanent arc and/or continental microplates infilled with alluvial, deltaic-shoreface, paralic, and shallow marine deposits. Continental rift basins (e.g. Cagayan Valley in northern Philippines) are infilled by alluvial, deltaic, and lacustrine deposits with paralic coals (Horkel, 1989). The Philippine coalfields are formed in foreland, rift, and back arc basins and subbasins bounded by subduction zones, active volcanic arcs, and on microcontinental plates (Figure 9.33(B)). Also, the basins and subbasins are dissected by major north–south trending thrust and strike-slip faults (Yumul et al., 2008). Unlike in Indonesia, the major coal-forming period in the Philippines basins and subbasins was Miocene age.

The tectonostratigraphic and sequence stratigraphic frameworks of Miocene coal-bearing rocks were characterized by marine transgressive–regressive stratigraphic sequences of delta front or shoreface and fluvial highstand systems tract (Figure 9.34). During the Early and Middle Miocene, these repetitive onlapping and offlapping stratigraphic sequences were followed by extensive transgression (lowstand systems tract) during the Late Miocene to Pliocene (e.g. Cebu coalfield) (Horkel, 1989). Early Miocene was characterized by coal accumulation in regressive systems of deltaic-shoreface and fluvial highstand systems tract, which infilled rift basins. The bituminous coals formed during this time period are gassy with estimated gas-in-place of up to 3.5 Bcm and outburst-prone in the Cebu coalfield (Flores et al., 2006). The Late Miocene coal accumulation was infilled by similar coal-forming systems but in back arc basins and subbasins. The Lower to Middle Miocene coal-bearing Semirara Formation in the Semirara coalfield represents accumulation in a backstepping coastline of a rapidly subsiding foreland subbasin. The Middle Miocene period of coal accumulation was culminated by extensive Pliocene marine transgression.

Semirara Coalmine Gas

Gassy subbituminous coals are found in the Semirara coalfield situated in the Semirara Island. The coalfield contains more than 3000 m thick coaly Semirara Formation, which is Lower-Middle Miocene in age and a testament to deposition in a rapidly subsiding subbasin, overlain by the Pliocene marine Buenavista Limestone (Figure 9.35(A),(B)). The Semirara Formation consists of three coal-bearing intervals interbedded with deltaic-shoreface sandstones (parasequences) with undetermined thickness of the lower coaly interval. The middle coaly interval is up to 1700 m thick and contains net coal thickness of about 337 m with as many as 32 coal beds and thickness up to 25 m, which are gassy with methane peak in mud and cuttings log ranging from 13 to 69.5% (Figure 9.35(B)). The upper interval is 500 m thick with as many as 18 coal beds and thickness up to 20 m. The coal beds of the upper interval of the Semirara Formation are mined in the Panian open pit (Figure 9.36). Based on the gas content and adsorption isotherms (gas storage capacity at 100% saturation) of the Upper Semirara coal beds and projected higher rank (0.36–0.61% R_o) of Middle Semirara coal beds at depth, Flores and Stricker (2012) estimated a gas-in-place for a 250 km^2 area in the Semirara Island and nearshore to be about 283 Bcm (Flores & Stricker, 2010). Flores et al. (2006) estimated the gas-in-place (100% saturation based on adsorption isotherms and gas content) of the upper coaly interval within the >6.5-km-wide and 15-km-long Semirara Island to be about 3.3 Bcm.

Gas adsorption and desorption analyses from coal beds in the upper interval of the Semirara Formation show results better than same-rank coals in the Powder River Basin. Comparison of adsorption isotherms for gas storage capacity of the subbituminous C coal for the Semirara coalfield and Powder River Basin is shown in Figure 9.37. The seven methane adsorption isotherms of subbituminous C coals in the Miocene Semirara Formation are above the average isotherm of the same-rank coal in the Paleocene Fort Union Formation in the Powder River Basin. The gas content is probably due to more virtinite macerals (as much 80%) in the Semirara coals

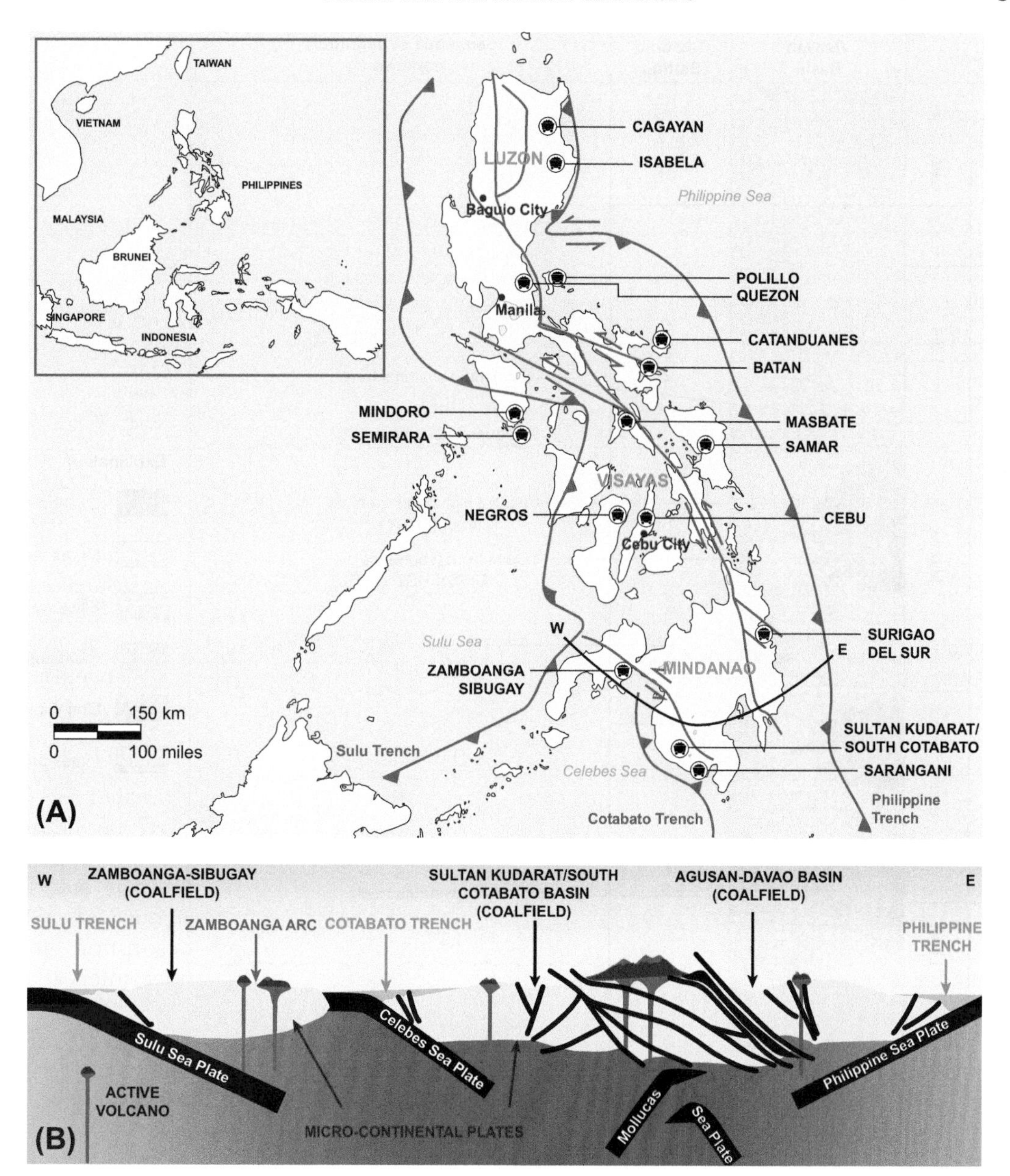

FIGURE 9.33 (A) Map showing the location of the 16 coalfields and tectonic framework of the Philippines. (B) Generalized W–E cross-section (see A for location) of the location shows the location of coalfields/basins with respect to active volcanic arcs, subduction zones, and microcontinental plates in Mindanao, southern part of the Philippines. *Source: Modified from Aurelio (2005), Flores et al. (2006), and Yumul et al. (2008).*

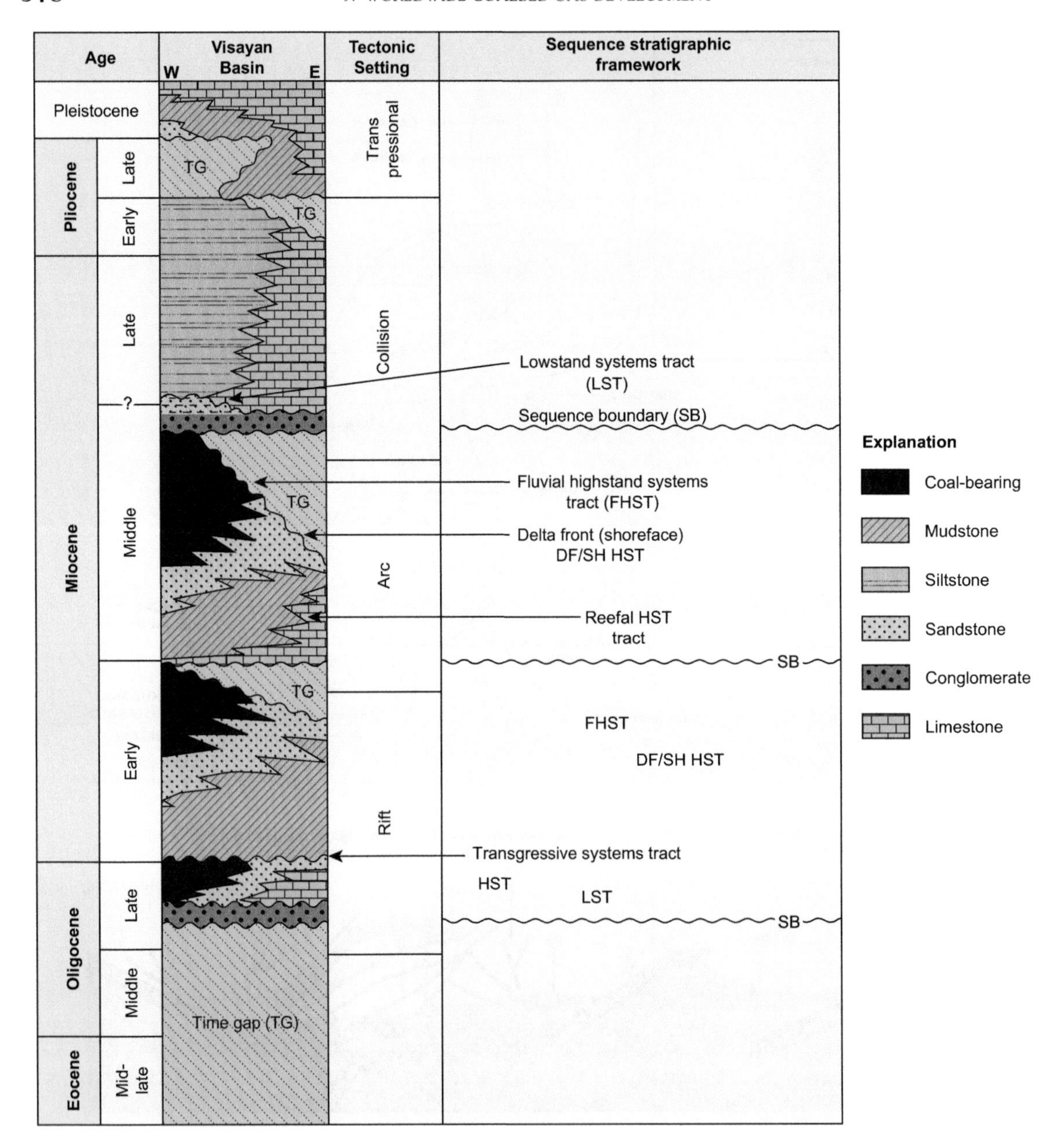

FIGURE 9.34 Tectonostratigraphic framework and stratigraphic sequences of Miocene coal-bearing rocks in Cebu coalfield in the central Philippines. See Figure 9.33(A) for location of Cebu coalfield. HST, highstand systems tract; SB, sequence boundary. *Source: Modified from Tanaka and Hamada (2005).*

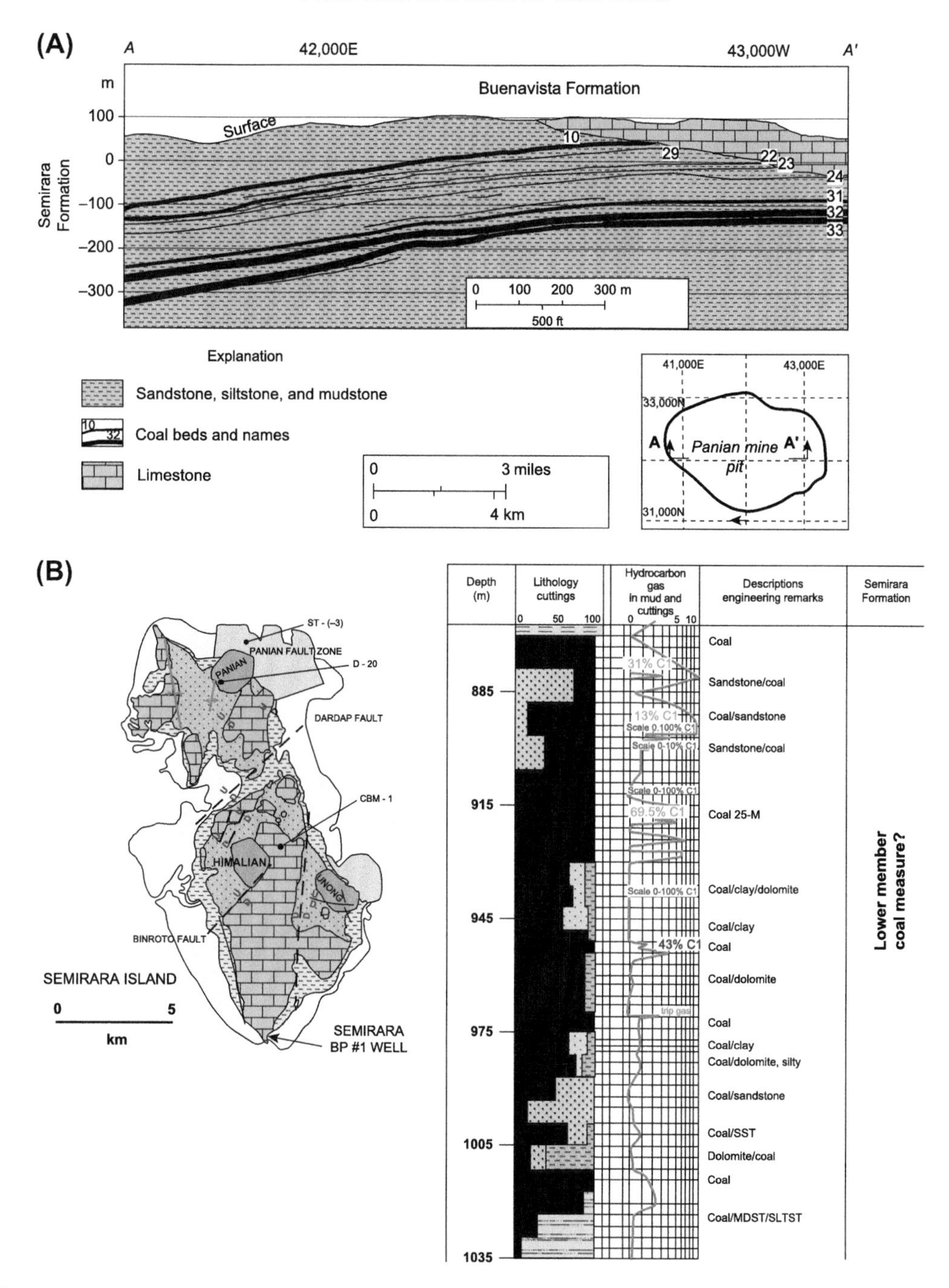

FIGURE 9.35 (A) Stratigraphic cross-section of the coal-bearing upper part of the Miocene Semirara Formation in Semirara coalfield (see Figure 9.33(A) and inset map for location). (B) Deep coals in the Semirara Formation, which shows gas kicks from mud and cuttings. *Source: Adopted from Flores et al. (2006).*

FIGURE 9.36 Panian surface coal mine in the Semirara Island, which exposed the thick coals of the upper Semirara Formation. See Figure 9.35(A) for the stratigraphic position of coal no. 32/33 in which the mine highwall is estimated to be about 250 m below the ground. See inset map in Figure 9.35(B) for location.

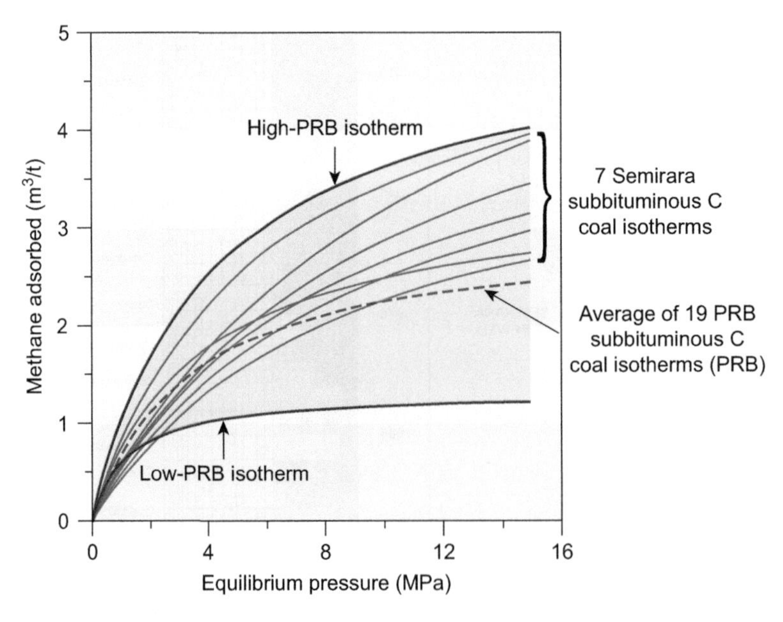

FIGURE 9.37 Adsorption isotherms of the Semirara subbituminous C coal in comparison to similar rank coal in the Powder River Basin (PRB). m^3/t, cubic meter per ton MPa, megapascal. *Source: Adopted from Flores et al. (2006).*

than in the Powder River coals (as much as 65%) (Flores & Stricker, 2012).

The Semirara surface mine is the largest coal producer in the Philippines accounting to about 92% of the national production. Small underground mines account for the remainder of the national domestic production. Many of the coal mines producing from semianthracite and bituminous coals are gassy and have been the cause of several mine outbursts. The CMM from the Semirara surface mine presents an opportunity to recover and utilize the gas for power generation in the mine site. The recovery of CMM may also count toward carbon trade credits under the Kyoto Protocol, which the Philippines is a signatory. More importantly, the predrainage recovery of CMM in underground mine sites would reduce the risk of mine explosion.

Colombian Coal Basins

The feasibility of coalbed gas from Tertiary low-rank coals in comparison to the Powder River Basin coals was investigated by Correa, Osorio, and Restrepo (2009) and Prada, Rodriquez, Flores, Fuentes, and Guzman (2010). Colombia is noted for producing and exporting hard coal (e.g. mainly bituminous rank), and is the largest producer in South America (USEIA, 2010a). The proven reserves of anthracite to lignite coals in Colombia are estimated to be 6 billion metric tons (Kelefant, 2011). The country's coalbed gas resources are estimated as low as 85–480 Bcm (Kelafant, 2011) and as high as 0.3–1.0 Tcm (Guzman, 2011).

Potential coalbed gas basins include the Cordillera Oriental Basin containing the coal-bearing Late-Cretaceous-Paleocene Guaduas Formation, Cesár-Rancheria Basin containing coal-bearing Paleocene Cerrejon Formation and Paleocene–Eocene Los Cuervos Formation, and Sinu-San Jacinto Basin containing coal-bearing Miocene–Pliocene Cerrito Formation (Garcia–Gonzalez, 2010; Prada et al., 2010; Figure 9.38). These coal basins along with other regions (e.g. Antioquia,

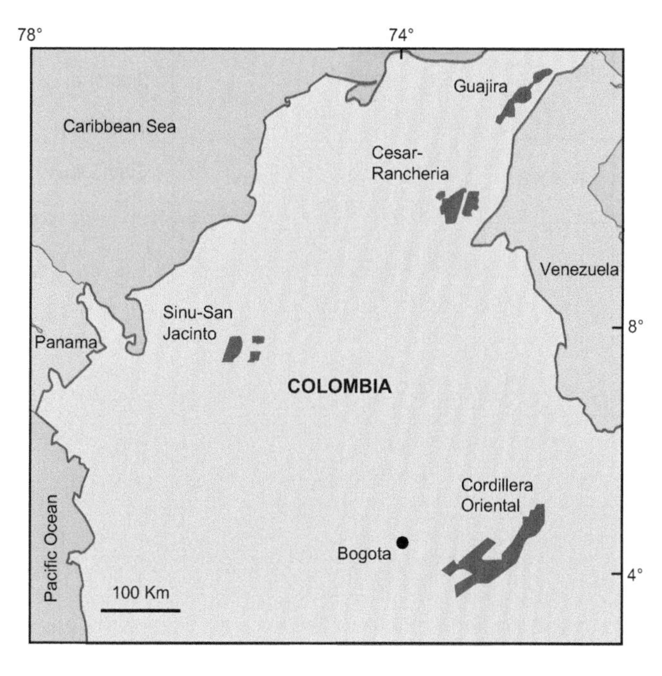

FIGURE 9.38 Map showing the locations of four coal basins (red colored areas) with potential coalbed gas in Colombia, South America. The coal basins are distributed in the eastern, central and western Cordillera Mountains. *Source: Modified from Prada et al. (2010).*

Boyaca, Cordova, Cundinamarca, Santander, Valle del Cauca, etc.) contain coalbed gas-in-place of about 85–480 Bcm (Correa et al., 2009; Guzman, 2011; Kelafant, 2011). The stratigraphy of the Paleocene–Eocene coal-bearing Los Cuervos Formation equivalent to the Tertiary coals in the Powder River Basin is shown in Figure 9.39. The formation includes as many as 15 coal beds ranging in thickness from 0.6 to 8 m. The coal beds range from subbituminous C to high-volatile B bituminous, which is higher rank than the subbituminous C-A of the Powder River Basin.

Prada et al. (2010) reported that the gas content increases to a maximum of 7.4 m^3/t

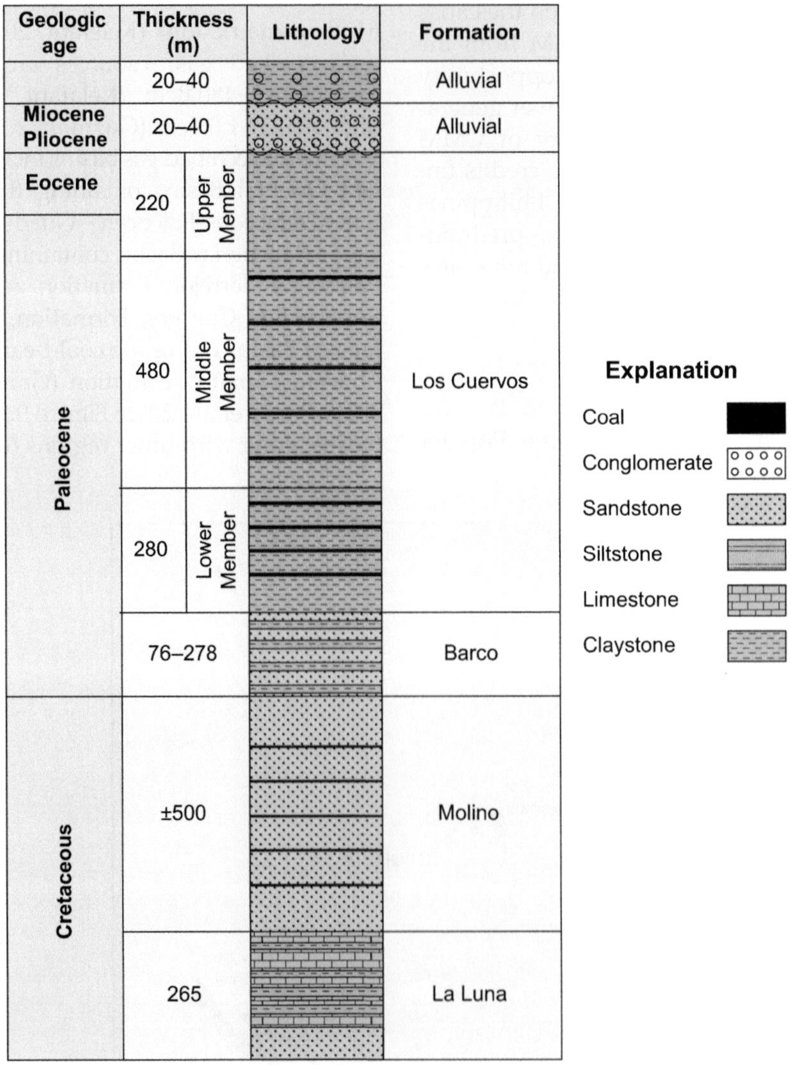
Geologic age
Thickness (m)
Lithology
Formation
20–40
Alluvial
Miocene
Pliocene
20–40
Alluvial
Eocene
220
Upper Member
480
Middle Member
Los Cuervos
Paleocene
280
Lower Member
76–278
Barco
±500
Molino
Cretaceous
265
La Luna
Explanation
Coal
Conglomerate
Sandstone
Siltstone
Limestone
Claystone

FIGURE 9.39 Generalized stratigraphic column of the Cretaceous and Tertiary rocks in Colombia. Coals of the Paleocene–Eocene Los Cuervos Formation in the Cesár-Rancheria Basin (see Figure 9.38) are prospects for coalbed gas. *Source: Modified from Prada et al. (2010).*

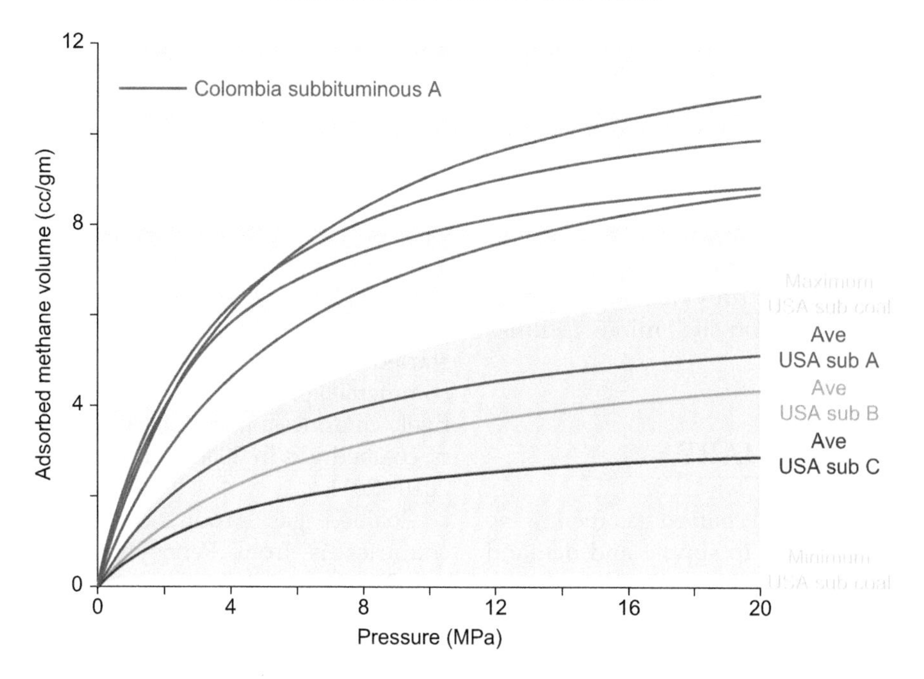

FIGURE 9.40 Chart showing the adsorption isotherm of subbituminous A coal of Colombia compared to the subbituminous A coal of the Powder River Basin. cc/gm = cubic centimeter per gram; MPa = megapascal. *Source: Flores and Stricker (2012).*

with depth, higher coal rank, and older geologic age. Generally, gas content varies with depth from 0.34 m^3/t at 149 m, 2.95 m^3/t at 271 m, and 4.6 m^3/t at 381 m. The lowest gas content is recorded by the subbituminous coal of the Cerrito Formation, which increases the high-volatile C bituminous coal of the Los Cuervos Formation and high-volatile C-B bituminous coals of the Cerrejon and Guadas Formation. In comparison, Schwochow (1997) reported that a well drilled for coalbed gas tests in the Cerrejon coalfield showed high gas content ranging from 5 m^3/t at 200 m to 12 m^3/t at 550 m depth.

The gas adsorption capacity of the low-rank coals of Colombia differs from subbituminous coals of the Powder River Basin (Flores & Stricker, 2012). The Colombian subbituminous coals are better than the coals in the subbituminous C-A coals of the Powder River Basin (Figure 9.40). More specifically, the subbituminous A coals in Colombia have a higher adsorption capacity than the same coal rank in the Powder River Basin. A similarity between the subbituminous coals of Colombia and Powder River Basin is the biogenic origin of the coalbed gas (Prada et al., 2010). The $\delta^{13}C$ of methane ranges from about −60 to −80 and δD of methane ranges from about −190 to −245‰. This isotopic values when plotted using Whiticar, Faber, and Schoell (1986) model, indicates bacterial CO_2 reduction process of generation of the coalbed gas.

Although 90% of coal production in Colombia is from surface mines, there are more than 3000 small underground mines that produce coal for the domestic market (USEPA 2010). USEPA (2010) reported that the underground mines have no methane drainage systems and with some mines >600 m depth, the coals may be gassy as indicated by gas content increasing at depth. Methane buildup in these mines may cause mine outburst. CMM emissions from the coal mines are estimated to be 0.28 Bcm/year (Kelafant, 2011). Thus, underground mines may be potential projects for CMM recovery and utilization. Drainage of coalbed gas in front of mining operations is another source of gas to provide fuel for generation of power for on-site mine facilities USEIA (2010a).

SUMMARY

The future of coal and coalbed gas fueling the future is directly linked to supply and demand for natural gas consumption and collective necessity to consume clean energy. It is projected the use of natural gas to increase by more than 50% from the 2010 production and will account for more than 25% of global energy demand by 2035. The report emphasized that the advantage of expansion of natural gas consumption may not be adequate to lower global carbon emissions enough to change the temperature rise of more than 2 °C to have an effect on climate change. A major component of the natural gas supply will be unconventional gas. Conversion of coalbed gas to LNG and CNG will play a major role in satisfying up-and-coming markets for OECD (e.g. Australia) and NOCED (e.g. China and Indonesia) countries to take advantage of projected gas consumption worldwide.

Development and production of coalbed gas have been dampened by the global recession in 2008–2010. OECD countries such as the United State and Canada have reduced production due to the low price of natural gas while Australia has initiated plans for LNG conversion of coalbed gas influenced by expanding natural gas markets in Asia. Countries such as China, India, Indonesia, and Russia have collectively expanded exploration and development of coalbed gas through the global economic recession. In addition, the NOED countries entered late and are still catching up with the exploitation of coalbed gas. The United States started development in the 1950s for coalmine gas and in the 1970s for coalbed gas from virgin coals, which matured in the 1990s–2000s. Coalbed gas exploitation in Australian and Canada emerged in the 1980s–1990s with commercial production in Australia and Canada in late 1980s and 2000, respectively. China and India started exploration of coalbed gas in 1990s with commercial production in 2005 and 2008, respectively. Indonesia and Russia started exploration of coalbed gas in 2000s with first production in 2010–2011.

Coalbed gas production from the OECD countries is from Pennsylvanian to Tertiary age. The conterminous United States coal basins produce from Pennsylvanian, Cretaceous and Tertiary age coals ranging from Pennsylvanian anthracite to Paleocene–Eocene subbituminous coals. A forgotten but important coalbed resource and endowed with more coal resources than the conterminous United States is Alaska. Large volumes of Tertiary age subbituminous to bituminous coals, which served as source of coalbed gas for conventional sandstone reservoirs (e.g. Cook Inlet Basin), are found in the three coal provinces of Alaska. Canada primarily produces from Cretaceous age bituminous coals with potential reservoirs in Jurassic and Tertiary age coals in the WCSB. Australia produces from Permian "Gondwana" anthracite coal of the Sydney Basin as well as bituminous coal in the Bowen, Gunnedah, and Gloucester Basins and Jurassic age bituminous coals in the Surat and Clarence-Moreton Basins. China mainly produces from Carboniferous and Permian age anthracite coals in the Qinshui Basin and potentially similar age and rank coals in the Ordos Basins. The Indian coal basins produce from Permian age anthracite coals and potentially from Tertiary age subbituminous coals. Indonesia produces mainly of Tertiary coals ranging from semianthracite to subbituminous rank with the first production from the Kutei Basin. Russia produces from Carboniferous and Permian age anthracite to bituminous coals in the Kuzbass Basin.

All the OECD and most of the NOECD coalbed gas producing countries have CMM recovery and utilization projects. The only OECD country with established CMM and AMM as well as good potential for coalbed gas from virgin coals is the United Kingdom. Since the reduction of the coal mining in the United Kingdom in 1990, CMM and AMM have been slowly replacing coal as an energy source. Potential coalbed gas development areas between coal mine areas are in Carboniferous age anthracite and bituminous coals. In addition, Tertiary age low-rank coals are an emerging source and reservoir of coalbed gas in other coal mining countries such as Colombia and the Philippines.

The established coalbed gas industry in major OECD countries (e.g. U.S., Canada, and Australia) and emerging coalbed gas industry in NOCED countries (e.g. China, Russia, India, and Indonesia) all benefitted from several factors. Previous research and development of recognized coal basins such as in the San Juan, Black Warrior, and Powder River Basins in the United States assisted in expanding exploitation of other basins worldwide. Development and production tax incentives in the U.S. designed for the coalbed gas industry and other unconventional energy have been adopted by other countries such as China, India, and Indonesia, thus spurring investments in the industry. The presence of infrastructure such as pipeline network and distribution systems to market as well as opportunities for new markets for coalbed gas such as CNG and LNG has projected expansion production in Australia, China, and Indonesia. The advanced drilling, completion, and stimulation technology have made production of coalbed gas more efficient worldwide.

CHAPTER

10

Coal and Coalbed Gas: Outlook

OUTLINE

KEY ITEMS

- Biogenic gas has become an important component of natural gas, which occurs in various settings on the surface, below the sea, and in the subsurface.
- Consortia of microorganisms generate biogenic gas by degrading organic matter deposited in various settings. Microbes play a major role in accumulations of natural gas in the subsurface of the planet.
- Biogenic or microbial gas endowment consists of primary, mixed, and secondary accumulations, which accounts for 40–50%

Coal and Coalbed Gas
http://dx.doi.org/10.1016/B978-0-12-396972-9.00010-0

of the global natural gas resource base in the subsurface stored in sedimentary basins.

- Microbial gas in mixed biogenic and thermogenic gas accumulation varies from 10% to 95%.
- Coalbed gas in bituminous and anthracite coals is mainly thermogenic in origin mixed with biogenic gas.
- Biogenic coalbed gas forms in subbituminous and high-rank coals and occurs in more than 30 basins worldwide.
- So far only the Powder River Basin in the United States is known to have almost pure (about 89%) biogenic gas in subbituminous coals.
- Biogenic gas in the Powder River Basin contributes about 15% of the total coalbed gas resource base of the United States. The proportion of biogenic gas in mixed coalbed gas is difficult to estimate but should contribute more to the resource base.
- The Powder River Basin is used as an analog to exploit biogenic gas in subbituminous coals, which along with lignite makes up more than 50% of the coal resources worldwide.
- Biogenic gas in the Powder River Basin served as a model to create coal to biogenic coalbed gas through methanogenesis leading to a transformational technology of sustaining gas production.
- Laboratory experiments and field tests by the industry have proved the feasibility of stimulating microbes in coal and groundwater using bioengineered nutrients to generate and sustain production of biogenic gas.
- Field tests and application of the coal-to-biogenic coalbed gas technology by the industry is performed in lignites of China and Indonesia with some measure of success.
- Although industrial research and development abounds on generation of biogenic gas in low-rank coals, there is a shortage of public research information by academia and government on practical application.
- Challenges in the environment and groundwater, changes in coal properties, gas storage potential, and gas generation–production sustainability need to be addressed for the future of gas bioconversion technology.
- Short and long-term outlooks of coal and coalbed gas science and technology are related to projected decline in coal consumption in the United States. Explosive growth of demand of coal and coalbed gas in Asia and western Pacific will ensure use of carbon-base energy and debate on greenhouse gases.

INTRODUCTION

Exploration and development of commercial petroleum accumulations have evolved from simple geologic structural and stratigraphic traps to complicated secondary–tertiary recoveries from reservoir rocks and technical-driven exploitation of source rocks. More specifically, the evolution from conventional to unconventional natural gas development has thrust coal to coalbed gas from the perceived “dirty” to “clean” energy. The past two decades have seen concerted efforts to develop natural gas from coal and shale beds known for their high organic content, which is greater than 66% and less than 33%, respectively. The high organic content of coal is transformed either by thermal cracking or low-temperature biodegradation into "clean" thermogenic or biogenic gas, respectively.

Natural gas has grown in demand globally as a "clean" alternative to power electric generation. In the past, natural gas was often a by-product of oil discoveries and "stranded" from recovery until infrastructure (e.g. pipeline network and liquefied natural gas (LNG)) had caught up with development (Katz, 2011). Development of natural gas remained "stranded" with the recent economic recession. However, the growing interest worldwide in unconventional natural gas from coal and shale source rocks is a shift toward expanding the gas supply particularly to meet demand from Asian markets (e.g. China, Japan, Taiwan, and South Korea). Until more than a decade ago, the world's coalbed gas was developed and produced from thermogenic gas of mature or high-rank coals (e.g. bituminous and anthracite). However, biogenic gas was exploited reluctantly from low-rank coals (e.g. subbituminous) due to presumed economic constraints. Where both types of gas occur mixed in the same coal reservoirs and basin, the biogenic gas is usually formed at shallow depths. However, "pure" biogenic gas can occur in a single basin such as in the Powder River Basin (Flores, 2004; Flores, Stricker, Rice, Warden, & Ellis, 2008) but this is a rarity. Nevertheless, the biogenic coalbed gas may turn out to be a transformational energy with potential renewability properties.

With the understanding of methanogenic processes by which biogenic gas is formed, the potential to transform coal to biogenic coalbed gas has gained worldwide interests from industry, governments, and academia. Thus, beyond biogenic coalbed gas is the possibility of gas generation in coal using stimulants (e.g. nutrients or amendments) to consortia of indigenous microbes or cultured microbes. The potential to renew coalbed gas production with this process in low-rank coals already producing biogenic gas and in other low-rank coals not producing coalbed gas, if successful, will be a transformational breakthrough in coal science and technology. Although methanogenesis has been known for more 100 years, application of the concept from the laboratory to the field is a step toward compressing geologic time to real-time generation of biogenic coalbed gas. This process might be possible with approved patents to stimulate and generate biogenic gas by bioconversion of coals (e.g. Clement, Ferry, & Underwood, 2012; Jin, Bland, & Price, 2010; Pfeiffer et al., 2008; Verkade, Oshel, & Downey, 2009; Finklestein et al., 2010).

Settings of Biogenic Gas

The traditional settings of biogenic gas range from on the surface to the subsurface as well as below sea and/or lake floors (Table 10.1). As discussed in Chapter 3, a large sink of biogenic gas is peatlands or peat soil rich in water-saturated organic matter and methane-generating microbes, a forerunner of coal. Peat-forming environments make up about 2–3% of the total surface of the global landmass, which extends from pole to pole and coastline to coastline. Another equally important setting of biogenic gas on the surface is soil (nonpeat), which is a mixture of weathered fragmented bedrocks, mineral grains, and decaying organic matter called humus. Soils vary geographically as controlled by climate and landscape. Soil is sometimes known as "Earth's living skin", which is the home to countless microorganisms and burrowing animals (e.g. earthworms, ants, termites) whose accomplishments vary from decomposition of organic matter to fixing atmospheric nitrogen (Planet Earth, 2013). Thus, peatlands and soils (nonpeat) overshadow other settings of biogenic gas such as man-made landfills and microbial digesters; rice and agricultural fields; animals, livestock and waste products; humans and waste products; and geothermal springs.

TABLE 10.1 Settings for Biogenic Gas in Natural Gas Worldwide

Surface (On Land)	Below Sea and Lake Floors	Subsurface (Shallow and Deep)
Peatlands	Marine sediments (gas hydrates)	Conventional reservoirs (oil and gas in sandstones, etc.)
Soils (other than soil from peatlands and agriculture)	Lake sediments (Connected to peatlands)	Unconventional sources and reservoirs (coal, shale, and sandstone beds)
Rice and agricultural fields		
Animals, livestock, and waste products		
Humans and waste products		
Man-made landfills		
Man-made digesters		
Geothermal springs		

Source: Modified from Katz (2011).

The largest sink of biogenic gas is beneath the sea floor where there are organic-rich marine sediments containing gas hydrates. Seawater covers about 70% of the global surface. Although not as aerially important as the sea floor sediments, lake floor sediments are high in organic matter derived from terrestrial plants and serve as a sink of biogenic gas. The lakes are often aerially connected to peatlands as well as isolated bodies of water (see Chapter 3). The shallow to deep subsurface settings of biogenic gas is common in sedimentary basins, which cover continental landmasses. These basins contain conventional oil and gas accumulations as well as unconventional oil and gas accumulations from coal and shale source. Consortia of microorganisms generated biogenic gas by degrading organic deposits in various settings on the surface and below the sea and lake waters as well as in the subsurface. The biogenic gas endowment, in general, and biogenic coalbed gas, in particular, in subsurface sedimentary basin settings is the focus of discussions in this chapter. Furthermore, the role and future of biogenic coal-to-coalbed gas technology will be addressed as well as the future of coal as a carbon-based energy.

HISTORICAL PERSPECTIVES ON NATURAL GAS AND ASSOCIATED BIOGENIC GAS

Rice and Claypool (1981) reported that biogenic gas was originally explored for their geological predictability and large aerial extent at shallow depths. Since then, evidence of microbial production of methane has been found in deep subterranean environments and subsequently commercial quantities of methane or paleohydrate was discovered in deep turbidite sandstone reservoirs occurring as deep as 3300 m (Chung, Yang, & Kim, 2011; Kotelnikova, 2002). Understanding the generation of biogenic gas in ancient sedimentary sequences was discussed by Claypool and Kaplan (1974), Rice and Claypool (1981), Zhang and Chen (1985), Rice (1992), and Clayton (1998). These workers found by themselves that the primary requirements for generation of large accumulation of biogenic gas after tens of thousands of years of burial in sedimentary basins were: (1) anoxic (oxygen depleted) environment with low redox potential (<300 mV), (2) low nitrate and sulfate concentrations, (3) mild temperatures, (4) abundant organic matter content and essential nutritional elements, (5) neutral pH values, (6) adequate pore space, (7) rapid sedimentation, (8) migration, (9) seal rocks, and (10) rapping mechanism.

Rice and Claypool (1981) explained that rapid deposition of marine organic-rich sediments leads to generation of biogenic gas in successive microbial ecosystems. Initially, oxygen is consumed by aerobic respiration, followed by nitrate and metal reductions, after which sulfate reduction dominates respiration. The depletion of sulfates, in turn, leads to generation and accumulation of methane, which is dissolved in pore water, resulting from aceticlastic pathway and carbon dioxide reduction by hydrogen produced by anaerobic oxidation of organic matter. Sulfate-depleted and anoxic environments control methane generation in marine sediments. The methane is held in the pore waters by hydrostatic pressure imposed by the overlying water column. Thus, biogenic gas is generated at low temperatures by decomposition of organic matter by anaerobic microorganisms.

The contribution of biogenic gas to the total global resource base of natural gas has evolved over time. Previous studies reported that microbial gas accumulations collectively account for more than 20% of the global natural gas resource base, which dominates individual sedimentary basins (Katz, 2011; Rice & Claypool, 1981). In addition, the contribution of biogenic gas is often not accounted for in mixed thermogenic and biogenic gas accumulations such that 10% more could be added to the total resource base (Granau, 1984). These estimates, which add up to more than 30% biogenic gas in the natural gas resource base does not include the contribution of secondary microbial gas, which is as much as 11% (Milkov, 2011). These calculations do not include biogenic gas endowment from source rocks such as coal and shale. Considering all these biogenic gas subsurface settings, it is clear that its contribution to the natural gas endowment is greatly underestimated, which probably minimally ranges from 40% to 50%.

BIOGENIC NATURAL GAS

Biogenic gas occurs worldwide as indicated by field accumulations with estimated reserves in individual gas fields varying from more than 6.6 Tcm in Russia (Urengoy field) to 28 Bcm in Indonesia (Terang–Sirasun field) (Katz, 2011) (Figure 10.1). The Russian (Urengoy field) biogenic gas field owes its origin partly to biodegradation and methanogenic process (see Chapter 4) of oil, which according to Milkov (2011) has affected half of the in-place oil and bitumen endowment of the world. However, the magnitude of methanogenesis in the world's petroleum accumulations, which results in generation of secondary biogenic methane toward the global natural gas endowment, is largely unknown (Katz, 2011; Milkov, 2011). The study of Milkov (2011) has shown that a total of 40 basins worldwide contain secondary biogenic gas, the majority of which are Cenozoic in age, varying in depths from 37 to 1834 m and temperatures of 12 °C–71 °C. Chung et al. (2011) reported that biogenic gas in Cenozoic basins could be found more than 3000 m deep. Using the methanogenesis of oil accumulations worldwide, Milkov (2011) estimated about 1855 Tcm of secondary microbial methane could have been generated throughout the geologic time. However, biogenic gas accumulations worldwide might be significantly underestimated owing to the highly variable proportions of mixed biogenic–thermogenic gas, which varies from 10% to 95% contribution by methanogenesis (Katz, 2011; Vandrè, Cramer, Gerling, & Winsemann, 2007). Thus, biogenic gas may contribute the majority

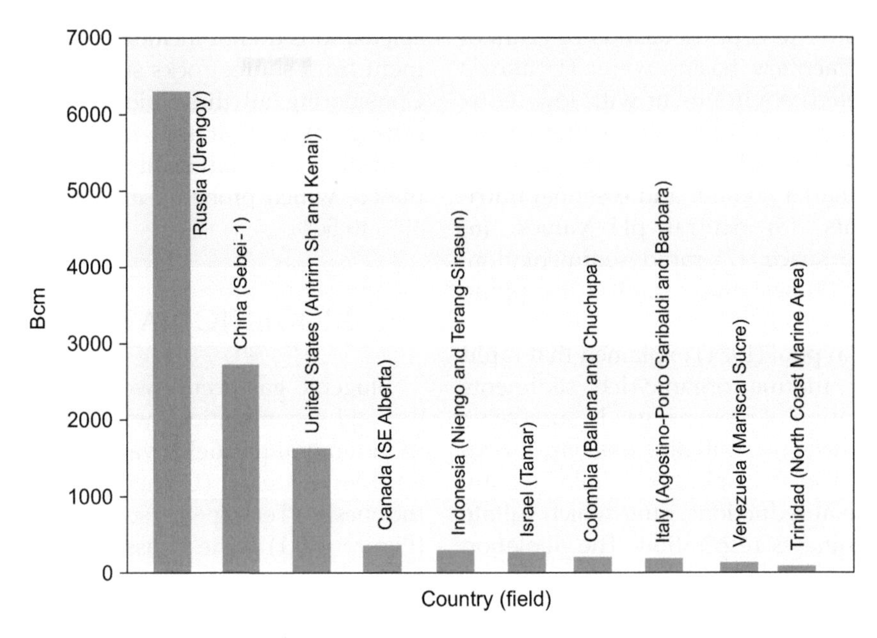

FIGURE 10.1 Histogram showing the biogenic gas reserves mixed with natural gas from selected countries as indicated by gas field accumulations. Bcm, billion cubic meters; SE, southeast; Sh, shale. *Source: Data adopted from Katz (2011).*

(95%) of the gas in an entire field or may contribute as low as 10% of the field.

BIOGENIC COALBED GAS

Although the understanding of unconventional biogenic gas resources in coal and shale has come to its own during the past 20 years, it is relatively recent compared to its counterpart conventional natural gas. A significant study by Formolo, Martini, and Petsch (2008) on biodegradation in coalbed gas reservoirs (e.g. Powder River and San Juan Basins) indicated that patterns of microbial biodegradation follow those of methanogenic, biodegraded black shale. However, these workers observed that loss of aromatic hydrocarbons in coal is accomplished before removal of saturated hydrocarbons in contrast to conventional observation in biodegraded petroleum systems or black shales. The principles of biodegradation of organic matter for biogenic gas generation in ancient sedimentary sequences are similar to that in coalbeds (Rice & Claypool, 1981; Rice 1992, 1993). The principles and pathways of methanogenesis in coalbeds are described in Chapter 4 and references therein. Unlike in other oil-bearing sedimentary sequences, biogenic gas in coalbeds can be generated in a few million years and in association with groundwater (Rice, 1993). The generation of biogenic gas in groundwater is a phenomenon known for a long time (Barker & Fritz, 1981; Carothers & Kharaka, 1980; Rice & Trelkeld, 1982), which plays a major role in current research on technology of real-time coal-to-biogenic coalbed gas production (Finklestein et al. 2010; Fallgren, Zeng, Ren, Lu, & Ren, 2013a; Fallgren, Jin, Zeng, Ren, Lu, & Colberg, 2013b; Pfeiffer et al., 2008).

Interest in biogenic natural gas production has grown substantially in recent years with the recognition that the operational life of depleted hydrocarbon reserves may be extended using technologies that enhance the activity of indigenous microbial communities. These technologies could also be applied to depleted coalbeds located throughout the Powder River Basin with the added economic benefit of an existing coalbed methane-related infrastructure. The responsible development of this technology requires the proper management of a wide range of complex scientific, environmental, economic and social issues.

Global Occurrence of Biogenic Gas

Historically, biogenic gas originally was reported by Rice (1993) to occur in various coal ranks, geologic ages, and coal basins worldwide. This benchmark study analyzed gas samples collected from coal core desorption, coal mine, pyrolysis, and production wells. Biogenic gas was recorded in subbituminous, bituminous, semianthracite, and anthracite coals from Carboniferous, Permian, and Cretaceous to Tertiary age. The Black Warrior, Powder River, Uinta, and Piceance Basins in the United States; the Bowen and Sydney Basins in Australia; western Germany; eastern China; and the Lower Silesian Basin in Poland included biogenic gas (Rice, 1993; Kotarba, 1998; Thielmenn et al., 2004). Results of the study indicated that biogenic gas from the US basins was mixed with thermogenic gas with methane $\delta^{13}C$ values ranging from -60.3 to $-29.1‰$ for the Piceance Basin and from -51 to $-41.9‰$ for the Black Warrior Basin. The Bowen and Sydney Basins had methane $\delta^{13}C$ values ranging from -80 to $-24‰$ and in eastern China, the values ranged from -66.9 to $-24.9‰$.

Biogenic coalbed gas has emerged as a viable target for exploration and development and has received considerable research interest worldwide. Research and development efforts have identified biogenic coalbed gas in more than 30 coal basins and about 35 sites worldwide (Figure 10.2). The global occurrence of biogenic gas updates the report of Strapoc et al. (2011), which recorded 33 sites. About half of the world's 30 coal basins with biogenic gas occurrence are producing coalbed gas. Biogenic gas is produced mostly mixed with thermogenic gas. Examples of production with mixed thermogenic and biogenic gas are found in the majority of coal basins in the United States (e.g. Black Warrior, San Juan, Appalachian, Illinois, Forest City, Piceance, and Uinta) (Faiz & Hendry, 2006; McIntosh, Martini, Petsch, Huang, & Nusslein, 2008; Milici, 2004; Pashin, 2007; Rice & Claypool, 1981; Rice, 1993; Scott, Kaiser, & Ayers, 1991; Stark & Cook, 2009; Strapoc et al., 2011; Strapoc, Mastalerrz, Schimmelmann, Drobniak, & Hedges, 2008; Volkwein, 1995). An exception is the Powder River Basin, which is mostly biogenic gas (Boreck & Weaver, 1984; Flores et al., 2008; Flores, 2000, 2004; Rice & Flores, 1991). The coal basins in Alaska such as the Cook Inlet have been tested to contain biogenic gas that is not produced directly from coalbeds but from interbedded fluvial channel sandstones that contain migrated biogenic coalbed gas (Dawson et al., 2012; Flores, 2004; Flores, Stricker, & Kinney, 2005; Rice & Claypool, 1981). Other mixed thermogenic and biogenic gas-producing basins are found in Australia (e.g. Sydney, Surat, and Bowen) and in Canada (e.g. Western Canadian Sedimentary Basin) (Aravena, Harrison, Barker, Abercrombie, & Rudolph, 2003; Bachu & Michael, 2003; Faiz & Hendry, 2006; Li, Hendry, & Faiz, 2008; Midgley et al., 2010; Papendick et al., 2011; Smith & Pallasser, 1996). Emerging coalbed gas-producing basins in China (e.g. Ordos) have been tested for biogenic gas (Guo et al., 2012). The other coal basins worldwide have been tested with varying degrees of occurrence of biogenic gas in nonproducing basins (e.g. Colombia, Indonesia, India, Japan, New Zealand, Russia, and Europe).

These basins contain thermogenic gas, which is mixed with biogenic gas. The dominant occurrence of mixed thermogenic and biogenic gas in producing coal basins makes it very difficult to determine the contribution of biogenic gas to the total coalbed gas production worldwide. Estimates of coalbed gas resources often do not

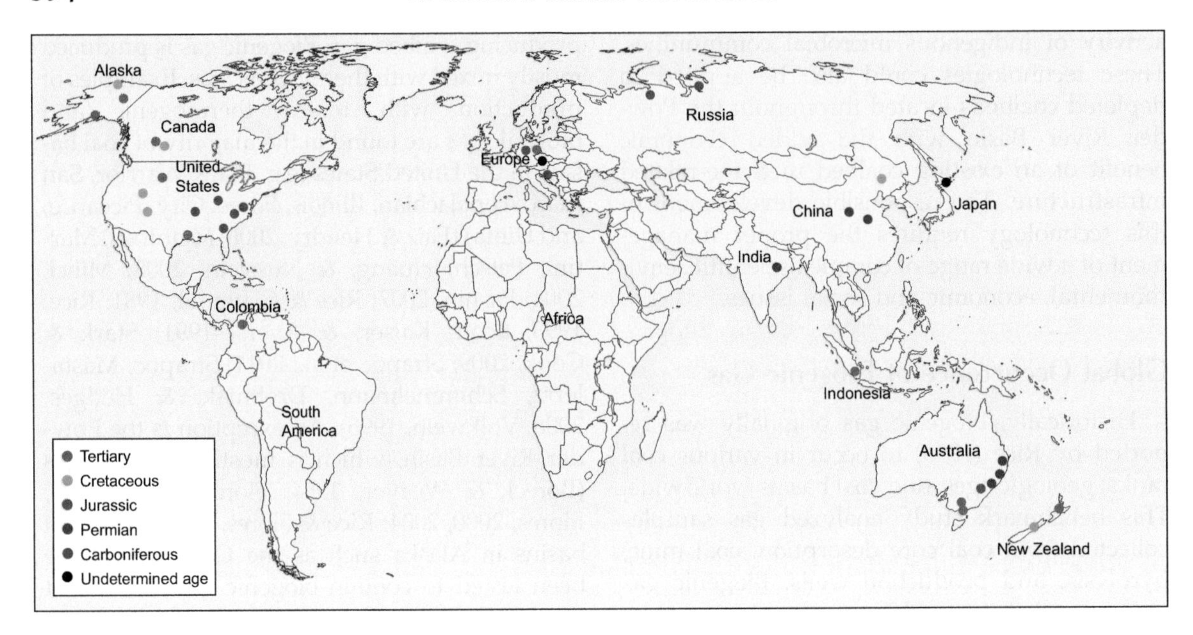

FIGURE 10.2 Worldwide distribution of biogenic gas in Tertiary, Cretaceous, Jurassic, Permian, and Carbonaceous coal basins. *Source: Modified from Yermakov, Lebedev, Nemchenko, Rovenskaya, and Grachev (1970), Smith and Pallasser (1996), Hosgormez et al. (2002), Veto et al. (2004), Zhang et al. (2005), Faiz and Hendry (2006), Shimizu et al. (2007), Butland and Moore (2008), McIntosh et al. (2008), Warwick and Flores (2008), Warwick et al., (2008), Prada, Rodriquez, Flores, Fuentes, and Guzman (2010), Midgley et al. (2010), Liu Song, Gao, Zheng, and Hong (2011), Papendick et al. (2011), Strapoc et al. (2011), Singh, Kumar, Sarbhai, and Tripathi (2012), and Susilawati, Papendick, Gilcrease, Esterle, and Golding (2012). See text for additional references.*

separate the volume of biogenic gas mixed with thermogenic gas to account for contribution from each. However, Strapoc et al. (2011) reported that about 40% of the coalbed gas production in the United States is contributed by biogenic gas due to the huge contribution from the Powder River Basin, which was described in Chapter 9 as the second most productive Tier 1 basin in the United States.

Powder River Basin Biogenic Gas Concept

The world's perception that the Powder River Basin is "purely" producing biogenic coalbed gas is a misconception. The biogenic gas concept was propagated by sparse molecular and isotope data collected initially from exploration and production wells. Boreck and Weaver (1984) initially analyzed the coalbed gas of the Powder River Basin and found that its methane $\delta^{13}C$ values ranged from −53.6 to −60.9‰ from desorbed samples and −60.1 to −62.4‰ from wellhead gas samples. These workers interpreted the light carbon isotopes as biogenic in origin based on the work of Rice and Claypool (1981). However, the gas samples were collected from core samples of the Anderson coal bed now known as the "Big George" coal from one well. Rice and Flores (1989, 1990, 1991) analyzed gas samples from the Wyodak coalbeds and interbedded sandstones collected from seven production wells along the eastern margin of the Powder River Basin. The gas samples had methane $\delta^{13}C$ values ranging from −53.8 to −56.6‰, which was confirmed to be gas of biogenic origin. Gorody (1999) collected additional eight samples from production wells along the eastern margin of the basin with methane $\delta^{13}C$ values ranging from −55 to −63‰ further establishing the biogenic origin of the gas. These initial gas samples collected from a total of 16 wells were the basis for the biogenic origin of the coalbed gas of the Powder River Basin, which dominated the concept until the work of Flores et al. (2008).

The onset of a rapidly growing number of production wells beginning in 2000 provided basin-wide collection of gas samples for compositional and isotopic analyses. Gas samples collected from 165 production wells were analyzed for composition (methane, carbon dioxide, nitrogen, and higher hydrocarbons) and stable carbon isotopes of methane, deuterium, carbon dioxide, deuterium isotope of methane, and coproduced water (Flores et al., 2008). In order to determine the potential contribution of thermogenic gas to the biogenic gas resource base, the following equation was applied (Flores et al., 2008):

$$C_1/C_2 + C_3$$

where:

C_1 = methane,
C_2 = ethane, and
C_3 = propane.

This formula is commonly used to differentiate between thermogenic gas (<1000) and biogenic gas (1000−4000). The $C_1/C_2 + C_3$ ratio ranged from 1000 to 4000 along the shallow margin of the Powder River Basin and from 98 to 1000 in the deeper part of the basin indicating biogenic and thermogenic gas distribution, respectively (Figure 10.3(A)) (Flores et al., 2008). Based on the formula, 24% of the total

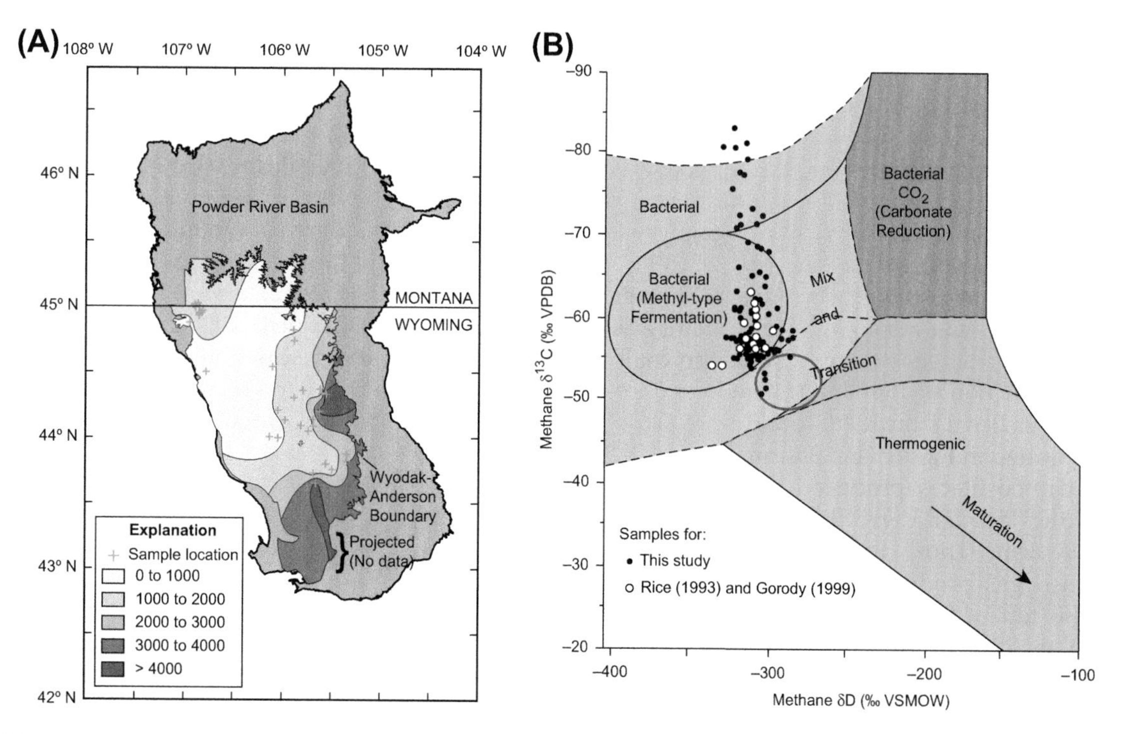

FIGURE 10.3 (A) Map of the Powder River Basin showing $C_1/C_2/C_3$ ratio ranging from 0 to >4000. (B) Diagram showing compositional classification of microbial and transitional thermogenic gases, which is circled in red. *Source: Adopted from Flores et al. (2008).*

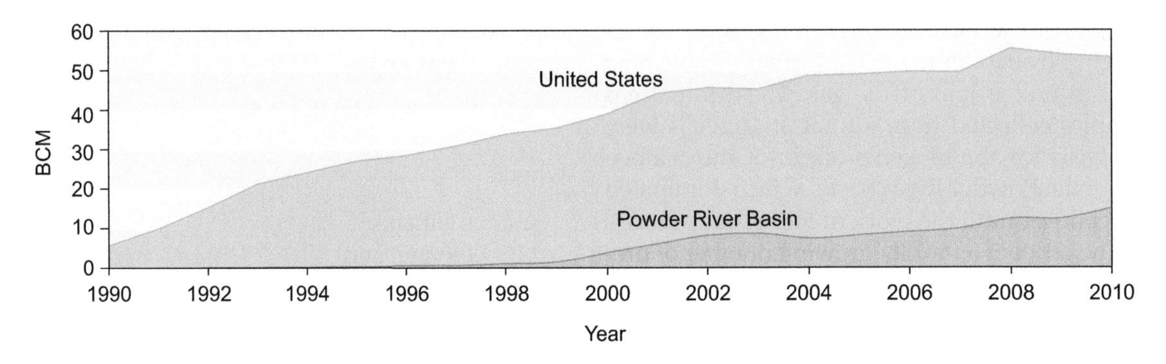

FIGURE 10.4 Diagram showing 20 years (1990–2010) of coalbed gas production in the United States in comparison with the production of mainly biogenic gas in the Powder River Basin, Wyoming. BCM = billion cubic meters. *Source: Data from USEIA (2012a) and WOGCC (2013).*

number of analyzed gas samples is thermogenic in origin and 76% is biogenic in origin. However, based on the methane $\delta^{13}C$ values ($<-55‰$) and methane δD values ($<300‰$), only 11% fall within the transitional or mixed biogenic–thermogenic gas range (Figure 10.3(B)) (Flores et al., 2008). The distribution of the small number of samples representing the 11% is shown within the $<1000\ C_1/C_2 + C_3$ ratio contour line, which includes majority of subbituminous A coal, in Figure 10.3(A). Additional data from the basin center needs to be analyzed to reaffirm the results.

In order to estimate the mixed biogenic–thermogenic gas resource base in comparison to the biogenic gas resource base, the coalbed gas production of the United States and Powder River Basin from 1990 to 2010 is estimated and plotted in Figure 10.4. During the 20 years, the United States produced about 757.8 Bcm and the Powder River Basin produced about 113.9 Bcm. Thus, the Powder River Basin coalbed gas production makes up about 15% of the total US coalbed gas production from 1990 to 2010. More specifically, of the total 113.9 Bcm coalbed gas production in the Powder River Basin, about 12.5 Bcm or 11% is thermogenic gas and 101.4 Bcm or 89% is biogenic gas indicating mixed but dominantly biogenic gas resource base. Flores (2004) assessed the technically undiscovered recoverable coalbed gas resources of the Wasatch and Fort Union Formation to be more than the mean (F50 probability; "mean" means there are 10 chances out of 20 that this volume will be recovered) of 396 Bcm, which is the largest single biogenic gas accumulation in the United States and the world. As of 2010 and after 20 years of production, almost 29% of the mean coal bed resource in the Powder River Basin base has been produced. Given favorable economic conditions, assuming that there is enough hydrostatic pressure left in the coal reservoirs based on the study by BLM (2012), and if the rate of production is maintained at an average of 5.7 Bcm/year, the production of the remaining 71% or about 282 Bcm of the mean coalbed gas resource base may continue in the near future.

Potential Powder River Basin Analogs Worldwide

The coalbed gas success in the Powder River Basin has spurred a worldwide effort to find other analogs from vast Tertiary low-rank coals. Since 53% of the world's total coal reserves are low-rank coals (e.g. subbituminous and lignite), there is a good possibility that another Powder River Basin will be found. The greatest potential for discovering a coalbed gas play similar to that of the Powder River Basin is in Tertiary coal basins in Australia, Canada, and Indonesia, which host large subbituminous and lignite resources. The Tertiary coals of southeastern Australia, especially in the states of New South Wales and Victoria, contain ideal shallow coalbed gas targets (Bunny, 2001). The Murray Basin in New South Wales and Gippsland Basin may contain biogenic gas (Midgley et al., 2010). Although a large volume of the Australian low-rank coals is lignite, deeper counterparts grade into subbituminous coals, which are excellent targets for exploration. In the Powder River Basin, a large part of the coal resources in the Eocene Wasatch Formation, which overlies the gas-rich Fort Union coals, is composed mainly of lignite but the same coals are transformed into subbituminous coal in the deeper part of the basin, which produces gas (Flores, 2004).

In the Canadian Western Sedimentary Basin in the Alberta Plains, Tertiary coals such as the Ardley coal zone in the Scollard Formation possess similar characteristics (e.g. thickness, gas content, depths, and coal rank) as in the Powder River Basin. The Ardley coal zone was compared to the subbituminous coals in the Powder River Basin, although short of being considered as an analog by Beaton (2003). The coal zone is subbituminous B–A with net coal thickness of as much as 20 m and gas content at 1.7–4.5 cc/g (Beaton, 2003). However, the permeability in the Ardley coal zone is much lower (<1–7 mD) than that of the Powder River Basin coals. Harrison, Gentzis, and Payne (2006) investigated the isotope geochemistry of the water and dissolved gas in the Ardley coal zone. These workers indicated the potential biogenic origin of gas with methane $\delta^{13}C$ ranging from depleted −46 to −51‰ but the $\delta^2H_{CH4}/\delta^2H_{H20}$ suggested CO_2 reduction pathway of potential methanogenesis. The $\delta^{18}O/\delta^2H$ indicates meteoric water much like that on the freshwater in the Powder River Basin.

Approximately 66% of the coal resources in Indonesia is composed of subbituminous and lignite ranks. Thus, 8.8 Tcm of the total 12.7 Tcm coalbed gas-in-place resources of Indonesia is mainly low-rank coals (Stevens & Hadiyanto, 2004). Although Tertiary coal deposits occur in two major ages (e.g. Eocene–Oligocene, Miocene), the majority of the low-rank coals are Miocene in age. The Miocene coals appear to be more akin to Powder River Basin coals with similarity in coal rank and 10–20% of the formations contain coals with more than 30 m net coal thickness. The Indonesian Tertiary coals have better gas content than the Powder River Basin coals (Stevens & Hadiyanto, 2004). Flores and Stricker (2012) reported that the adsorption capacity of the Indonesian coals is twice that of the Powder River Basin coals.

In order to understand the resource base, it is necessary to determine whether a gas accumulation is microbial, thermal, or mixed. This is accomplished through an analysis of both molecular and isotope composition. Primary microbial gas is dry with $C_1/C_2 + C_3$ values that are commonly greater than 1000. Methane generally accounts for 99% or less of the hydrocarbons present and is isotopically light ($\delta^{13}C < -50‰$) relative to the Pee Dee Belemnite (Vienna PDB) standard. Secondary microbial gas displays much broader ranges in molecular and isotopic character and will be discussed in more detail later.

The deluge of data generated during the past 20 years by teams of coal geologists and petrologists, biogeochemists, microbiologists, molecular engineers, and hydrologists (see Strapoc et al., 2011 for excellent review) have expanded the knowledge of biogenic coalbed gas beyond the coal basins investigated by Rice (1993). A survey from this study (Strapoc et al., 2011) expanded global occurrence of biogenic gas to 33 sites in 30 coal basins. Figure 10.2 shows the occurrence of biogenic gas throughout the world in Carboniferous,

Permian, Jurassic, Cretaceous, and Tertiary coals. The coal ranks in which biogenic gas occur ranges from anthracite to bituminous to subbituminous.

FUTURE SUSTAINABILITY OF COALBED GAS

Creating a more sustainable future for energy, in general, and coalbed gas, in particular, occurs in various forms depending on advantages and disadvantages. For example, sustainability for coalbed gas in China is to increase production from virgin coal and coalmine methane with financial support to the industry from the government in order to meet the country's demand for energy. In other producing countries such as Australia, Canada, and the United States, sustainability refers to production within environmental and regulatory framework that minimizes the effects on the landscape but still meets the demand for "clean" energy and "green" environment. Also, sustainability can be defined as finding a new source of gas by continuous generation and production of coalbed gas through biotechnology. This coal-to-biogenic coalbed gas technology, which stimulates coal to generate biogenic gas, is a transformational process toward renewability of energy. The advantage is generation of sustainable "clean" energy accompanied by a potential disadvantage of environmental concerns during its development.

In coalbed gas, sustainability comes with environmental concerns balanced with the need for "clean energy" and "green" environment supplying the increasing demands for natural gas in the domestic and international markets. The major environmental concerns to the production of coalbed gas are the associated extraction of large volumes of groundwater to desorb the gas. The quality of coproduced water, well stimulation such as fraccing to increase efficiency of production, and management of the coproduced water were discussed in Chapters 7 and 8. More importantly, extraction of groundwater and well stimulation with chemical additives can potentially degrade the water supply and aquifer. Discharge of large volumes of coproduced water disposed on the surface potentially has significant effects on the landscape.

Regulatory and legal framework governing coalbed gas consists of complex intergovernmental guidelines at the federal, state, and local levels. Often these guidelines are enacted with consultations from the industry, private, environmental, and public interest groups. The regulatory framework varies from water rights (see Chapter 8) to permitting rules and management of coproduced water. Establishing a regulatory framework for coalbed gas production and extraction of associated coproduced water is a dynamic process that is a continuous evolution throughout the development period. For example, in the Powder River Basin, well spacing regulation evolved from 16 to 8 wells per square mile/km unit, depending on the amount of water withdrawal and nature of its management. Thus, the complexity of the regulatory framework arises from different conditions within and between producing basins such that regulations are not a "one size fit all" process. Extensive research and lessons learned from coalbed gas developing countries (e.g. United States, Australia, and Canada) are valuable resources for emerging coalbed gas countries (Box 10.1).

Role of Powder River Basin in Sustainable Biogenic Coalbed Gas

The road to sustainable or renewable biogenic coalbed gas started about 30 years ago when biogenic gas was discovered from the Powder River Basin in the Fort Union coals by scientists in the US Geological Survey, Branches of Coal and Petroleum Geology. Research toward sustainability of this resource began in earnest during the turn of the century. The research investigations of Law, Rice, and Flores (1991) and Rice (1993) led to the fundamental principle

BOX 10.1

MICROBES DEVOURING COAL: FUTURE OF COALBED GAS—CONCEPT AND MISCONCEPTION

The industry has touted the concept of microbes "eating" coal to produce coalbed gas or methane as the future's answer in sustaining the generation and production of clean energy. The concept of methanogenesis (see Chapter 4) is a potentially transformational process of bioconversion of the organic matter of coal to gas as reported in the World Oil and MIT Technology Review (Bullis, 2012; Finklestein, DeBryuyn, Weber, & Dodson, 2005; LaMonica, 2010). Coal is thought to be a carbon-based friendly substrate in which indigenous microorganisms and associated groundwater make up an ideal microenvironment to generate new gas in abandoned coalbed gas wells or virgin coal beds, based on the history of biogenic origin of gas in the reservoirs (e.g. Powder River Basin coals). Stimulating with essential nutrients and amendments to generate coalbed gas might renew and enhance indigenous microbes, which originally generated biogenic gas in coal. Also, it has been found that critical volume of reservoir or formation groundwater is essential to this process of biogenic gas renewability (Jin, 2007). Methanogenesis can be duplicated in all coal ranks (Fallgren et al., 2013).

Although there is enormous potential to boost the natural gas reserves through methanogenesis, there are challenges involving technical and hydrogeological factors, reservoir properties, and environmental concerns (see preceding discussions). In addition, there are knowledge disparities regarding replicating the original microenvironment after gas production. Anaerobic microorganisms are the principal consortia to generate gas but during dewatering to degas, air may be introduced into part of the coal reservoir, which is deleterious to certain anaerobes when exposed to oxygen. Although such scenario has not shown any problems in the laboratory and field tests of biogenic gas production to date, it may be important to resume reduced and low redox potential conditions in certain "aerated" coal reservoirs and groundwater to restart or enhance methanogenesis for sustained generation and production after dewatering the reservoir. Also, with significant reduction of hydrostatic pressure, not much is known about if and how the "new" gas can be stored and produced from the depleted reservoir. More importantly, the prolonged and sustained geochemical effects of bioengineered nutrient and amendment injections to the coal and groundwater in small and large scales are not well understood. There is evidence that coalbed gas can be regenerated from abandoned wells and virgin coal beds in various geological settings; however, additional research is needed to address these potential issues as well as to optimize the coal-to-biogenic coalbed gas technology to sustain production in extended areas.

on the origin of the biogenic coalbed gas in the Powder River Basin. It was determined by Law et al. (1991) and Van Voast (1991) that the groundwater composition in the coal-bearing Paleocene Fort Union Formation in the Powder River Basin is a sodium carbonate type. The lack of sulfate in the groundwater suggests that sulfidogenic metabolism or inorganic pathways, which preceded methanogenic processes, could have depleted it. Another possibility is that sulfate depletion never existed because of the low sulfur content (about average 0.5%) of the coal reservoir (Flores & Bader, 1999). Rice (1993) suggested that uplift and erosion of the Powder

River Basin permitted the establishment of a regional groundwater system, which flowed in the thick coalbeds along the basin margin and resulted in microbial activities that generated late-stage biogenic gas. The biogenic gas was generated by CO_2 reduction as indicated by the isotopic composition of the gases and coproduced water. Thus, the key to unlocking the sustainability of biogenic coalbed gas is the coproduced water extracted during production.

Subsequent research investigations in the Powder River Basin by government, industry, and academia in the 2000s have proved that methanogens exist under anaerobic conditions in the coalbed gas-coproduced water and subbituminous coal (Figure 10.5) (Finklestein et al., 2005; Flores et al., 2008; Jin, 2007; Klein et al., 2008; Rice, Bartos, & Ellis, 2002). Thus, the coproduced water of the Powder River Basin served as the medium in the generation of biogenic gas. Ground-breaking research on treating coalbed gas-coproduced water with nutrients or "amendments" and substrate (e.g. coal matrix) enhanced the microbial consortia including methanogens in the laboratory microcosm studies (Finklestein et al., 2005; Green, Flanegan, & Gilcrease, 2008; Harris, Smith, & Barker, 2008; Jin, 2007; Ulrich & Bower, 2008; Pfiffer et al., 2008; Fnklestein et al., 2010). The treated coalbed gas-coproduced water can then be reinjected into gas-producing coal source/reservoir as an inoculum or "booster" to stimulate biogenic generation of methane, which forms the majority composition of coalbed gas.

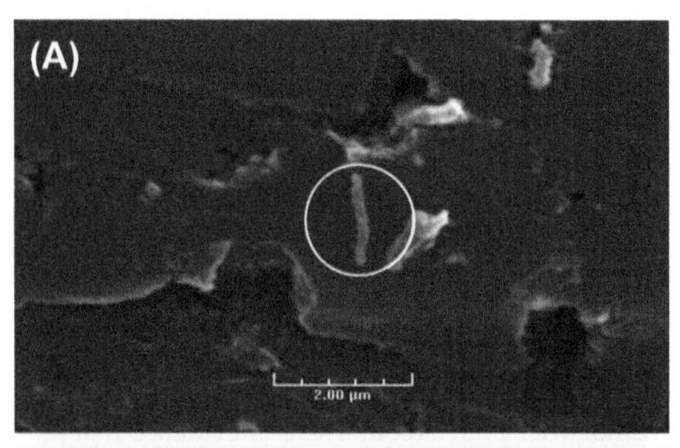

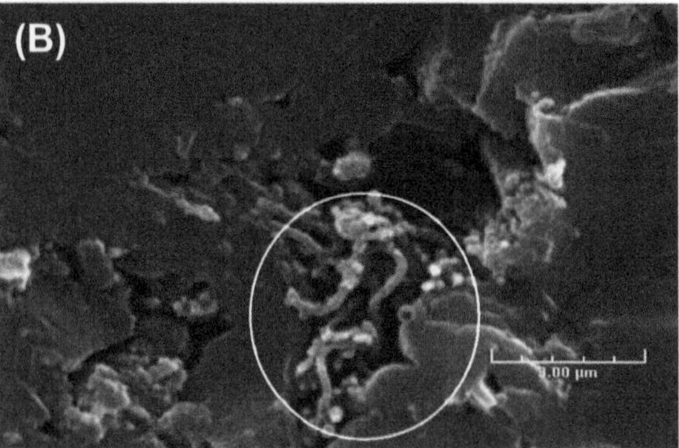

FIGURE 10.5 Scanning electron microscope photographs of methanogens in the Fort Union coal reservoirs sourced in the Powder River Basin, Wyoming and Montana, USA. (A) "Rodlike" and (B) "sinuous or wormlike" shaped microbes (see circled areas). *Source: Adopted from Moore (2012).*

Feasibility of Generation of Biogenic Coalbed Gas in Producing Coal Reservoirs

The feasibility of generating biogenic gas has been performed mainly on samples collected from subbituminous coals that have a history of production such as in the Powder River Basin. Finklestein et al. (2005) have demonstrated generation of biogenic coalbed gas using samples of the Dietz coal reservoir of the Paleocene Fort Union Formation in the northwest part of the Powder River Basin. Jin (2007) duplicated similar generation of biogenic coalbed gas using samples from the Wyodak coal reservoir collected in the North Antelope coal mine in the southeast part of the Powder River Basin. Both laboratory studies, which re-created field reservoir conditions, used collected formation waters (e.g. coproduced water), identified the consortia of methanogens, and tested the coal samples for potential generation of biogenic gas (e.g. carbon content and water content for chemistry) amenable to methanogenesis. The reader is referred to the work of Finklestein et al. (2005) and Jin (2007) for more detailed descriptions of the analyses, characterization, and discussion of the laboratory experiments. Also, the reader is referred to the Special Issue of the *International Journal of Coal* on "*Microbes,*

Methanogenesis, and Microbial Gas" (Flores, 2008) for additional laboratory studies on generation of biogenic coalbed gas.

The highly competitive nature of the technology of bioconversion of coal to biogenic coalbed gas has limited the public availability of the research results. However, the work white paper of Jin (2007) for the US Department of Energy provided examples of the methods, data, results, and discussions of such bioconversion technology. Although the study of Jin (2007) included analyses of oil shale, lignite, peat, and diesel-contaminated soil as substrates, the results of the subbituminous coal and coproduced water samples collected from the Powder River Basin will be summarized here. Microcosm was determined for the coal sample with each microcosm containing coalbed bed gas-coproduced water and groundwater collected outside the Powder River Basin. Also, a combination of enhancement and inhibitor was added to the microcosms containing coal samples (Table 10.2). Measurement of the headspace in the microcosms is expressed by the concentration of methane expressed in μmol/kg, which is plotted against the number of days of gas generation (Figure 10.6). The nutrient-amended, especially nutrient + substrate-amended, treatments produced the highest volume of methane (4.19–8.68 μmol/kg) from microcosms with coal and coalbed gas-coproduced water.

The treatments, which yielded high volume of methane include: milk + nutrient + vancomycin, nutrients, milk + nutrients, milk + nutrients + 2-bromoethanesulfonic acid, and inoculated + milk + nutrients. The nutrient treatment started generation of methane after 5 days in comparison to the other treatment, which started after 20 days. The rate of methane generation was considerably improved by nutrients after 60 days with microcosms that consist of coal and groundwater (Figure 10.7). Jin (2007) observed that the cumulative volume of methane generated from the treatments (7.40 μmol/kg) surpassed the live control (6.78 μmol/kg) after 102 days. The conclusions of the study of Jin (2007) indicate that nutrients can significantly stimulate generation of biogenic coalbed gas. Also, the results indicate that retreating coal may increase the availability of dissolved organic content (DOC) for microbial degradation, resulting in more gas generation. Jin (2007) suggested that pretreatment might be possible by adjusting the pH to either acidic or basic level to yield larger amounts of DOC, which provides more bioavailability substrates for proper microbial consortia to degrade and convert to methane. Perhaps, the most important finding of this study is the stress of baseline biogeochemical characterizations of coal, its coproduced and groundwater waters, and the adjustments to keep these parameters in favor of microbial metabolisms. These parameters are critical in assessing the potential of biogenic methane generation. The coal-to-biogenic gas technology continues to be improved. Laboratory research, field pilot tests, and full-scale trials are being conducted (Jin 2013; Personal Common.).

Feasibility of Generation of Biogenic Coalbed Gas in Lignite Coal

One of the interesting notes of Jin's (2007) study is the inconclusive result for potential gas generation from lignite. However, subsequent studies by Fallgren, Zeng, Ren, and Lu (2012) and Fallgren et al. (2013a) reported real-time biogenic gas generation from lignites with no history of gas production. This indicates that microbes including methanogenic microorganisms are abundant in this type of coalbeds and increase the range of this application. Fallgren et al. (2012, 2013a) conducted further laboratory studies by using indigenous lignites and formation water samples collected from Australia, China, and Indonesia, showing fresh methane production from the microcosms. These workers also conducted field pilot studies and generated methane from otherwise nonproducing lignites in China and Indonesia.

TABLE 10.2 List of Treatments or Nutrients and Inhibitors and Amount of Treatment Added to Microcosms in Coal to Generate Biogenic Gas

Microcosm ID	Coalbed Gas-Coproduced Water	Groundwater
1	Nothing	Nothing
2	0.0122 g 2-BESA	0.0122 g 2-BESA
3	0.0116 g vancomycin	0.0116 g vancomycin
4	0.0240 g $NaNO_2$	0.0240 g $NaNO_2$
5	0.0125 g NH_4Cl	0.0053 g NH_4Cl
	0.0043 g KH_2PO_4	0.0024 g KH_2PO_4
6	12.6 ml milk	12.6 ml milk
7	12.6 ml milk	12.6 ml milk
	0.3075 g NH_4Cl	0.3003 g NH_4Cl
	0.0794 g KH_2PO_4	0.0775 g KH_2PO_4
8	12.6 ml milk	12.6 ml milk
	0.3075 g NH_4Cl	0.3003 g NH_4Cl
	0.0794 g KH_2PO_4	0.0775 g KH_2PO_4
	0.0116 g vancomycin	0.0116 g vancomycin
9	12.6 ml milk	12.6 ml milk
	0.3075 g NH_4Cl	0.3003 g NH_4Cl
	0.0794 g KH_2PO_4	0.0775 g KH_2PO_4
	0.0122 g 2-BESA	0.0122 g 2-BESA
10	12.6 ml milk	12.6 ml milk
	0.3075 g NH_4Cl	0.3003 g NH_4Cl
	0.0794 g KH_2PO_4	0.0775 g KH_2PO_4
	0.0240 g $NaNO_2$	0.0240 g $NaNO_2$
11	0.0125 g NH_4Cl	0.0053 g NH_4Cl
	0.0043 g KH_2PO_4	0.0024 g KH_2PO_4
	0.0240 g $NaNO_2$	0.0240 g $NaNO_2$
12	Sterile solid	Sterile solid
	Sterile water	Sterile water
13	Sterile solid	Sterile solid
	Live water	Live water

TABLE 10.2 List of Treatments or Nutrients and Inhibitors and Amount of Treatment Added to Microcosms in Coal to Generate Biogenic Gas (*cont'd*)

Microcosm ID	Coalbed Gas-Coproduced Water	Groundwater
14	12.6 ml milk	12.6 ml milk
	0.3075 g NH_4Cl	0.3003 g NH_4Cl
	0.0794 g KH_2PO_4	0.0775 g KH_2PO_4
	Sterile solid	Sterile solid
	Live water	Live water

G, gram; ml, milliliter; PO_4, phosphate; NO_2, nitrogen dioxide; NH_4, ammonium; Cl, chloride; Na, sodium; K, potassium; H, hydrogen; BESA, Bromoethanesulfonic acid (methanogenic inhibitor); vancomycin is bacterial antibiotic.
Source: Adopted from Jin (2007).

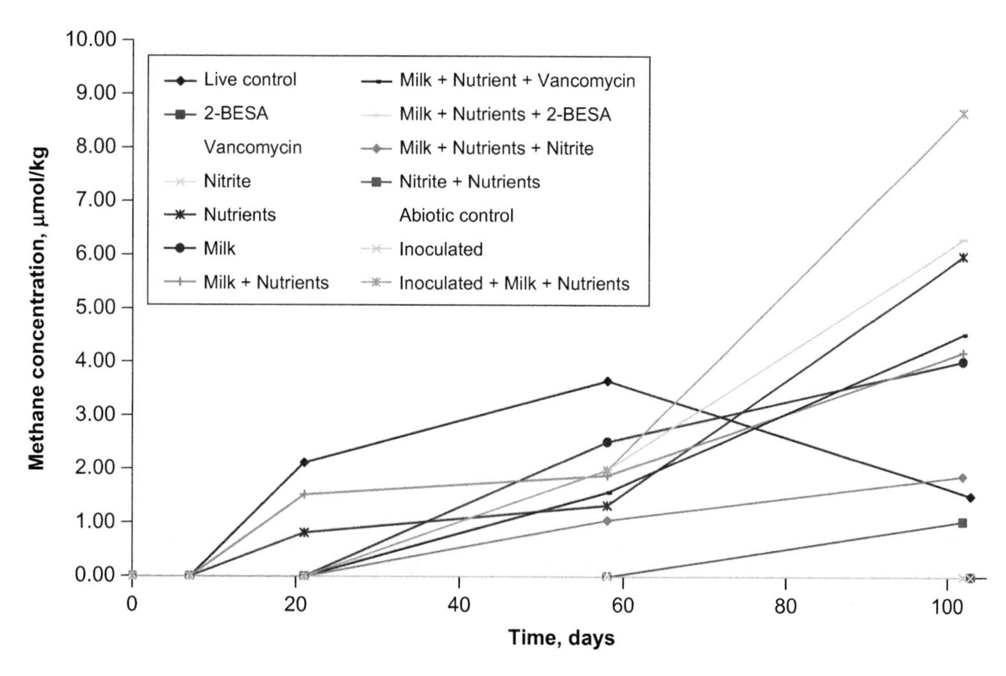

FIGURE 10.6 Chart showing cumulative methane produced by microcosms with coal and coproduced water from the Powder River Basin, Wyoming. Results show amendments such as nutrients, milk + nutrients, milk + nutrients + 2-BESA, milk + nutrients + vancomycin, and inoculated + milk + nutrients produced the highest amount of methane (4.19–8.68 µmol/kg) surpassing live control after 102 days. BESA, bromoethanesulfonic acid (methanogenic inhibitor); vancomycin is a bacterial antibiotic; µmol/kg, micromoles per kilogram. *Source: Adopted from Jin (2007).*

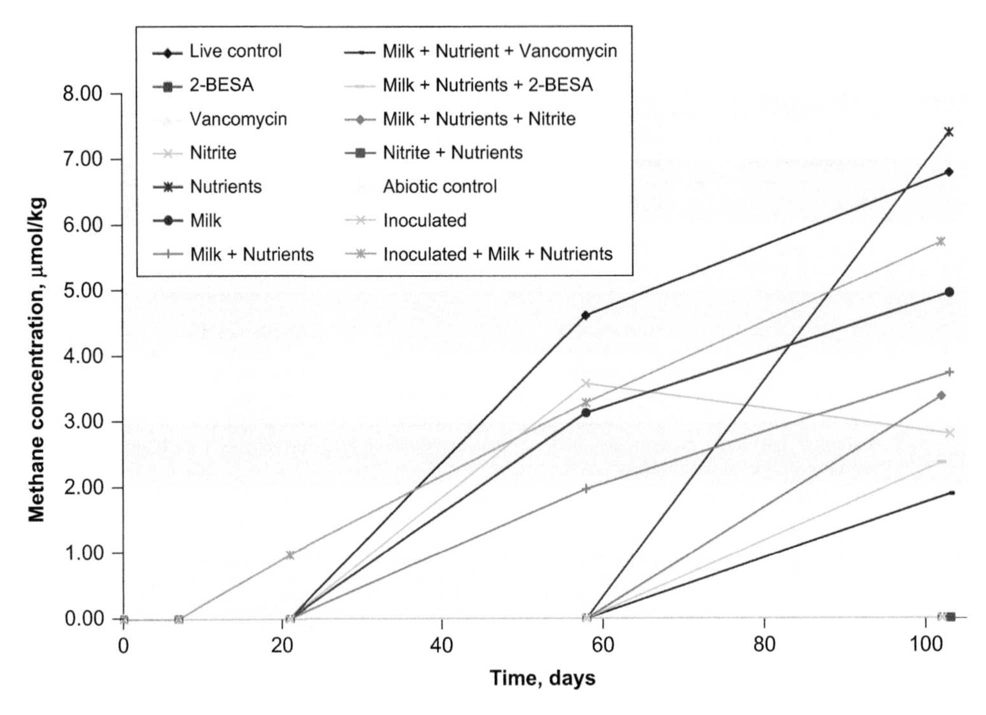

FIGURE 10.7 Chart showing cumulative methane produced by microcosms with coal from the Powder River Basin and groundwater collected outside the basin in Wyoming. Results show that amendments considerably increased the rate of methane production after 60 days compared to that in Figure 10.6. BESA, bromoethanesulfonic acid (methanogenic inhibitor); vancomycin is a bacterial antibiotic; μmol/kg, micromoles per kilogram. *Source: Adopted from Jin (2007).*

Their studies observed as much as a 336% increase in coalbed gas generation after the lignites were amended with nutrients, indicating presence of native methanogens and associated bacteria.

The results of both the laboratory and field tests are discussed in detail in Fallgren et al. (2013) and summarized in the following paragraphs. Figure 10.8(A)–(C) demonstrates the results of generation of methane during a period of about 80–100 days from the various lignites collected from the three countries. The treated Indonesian lignite yielded the highest rate of methane generation with as much as 0.34 mol/kg in about 75 days (Figure 10.8(A)). The treated Chinese lignite generated the lowest amount of methane at approximately 0.24 mol/kg in about 77 days (Figure 10.8(B)). The treated Australian lignite yielded as much as 0.33 mol/kg in about 100 days. Thus, it took longer to generate gas from the treated Australian lignite at about the same rate compared with the treated Chinese lignite for about 77 days. It is interesting to note that the untreated lignites (with just formation water) from all three countries generated slight amount of methane during the same period, presumably due to the presence of microbes and available nutrients during the sample processing during the microcosm setup.

Among the untreated lignites from the three countries, the Indonesian lignite contained the largest population of methanogens followed by the Australian lignite and then Chinese lignite, which had the least population. After nutrient treatment, the methanogen population of the Indonesian lignite increased to 8.76×10^8

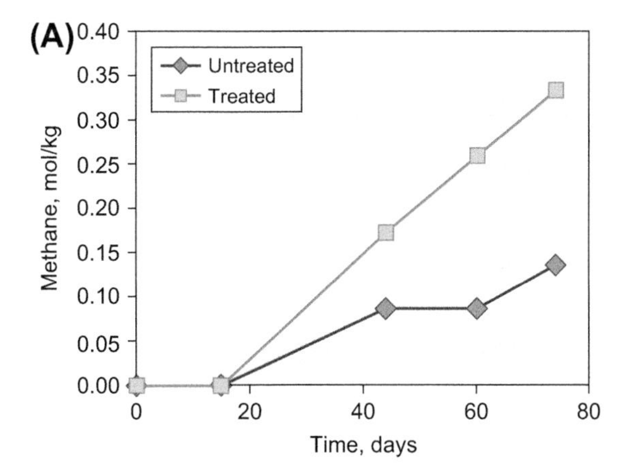

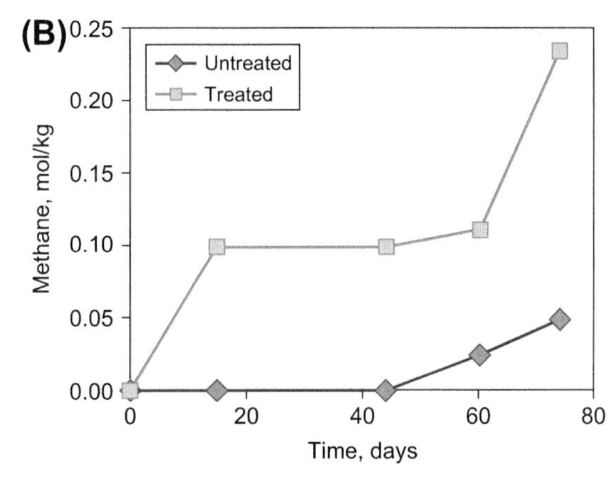

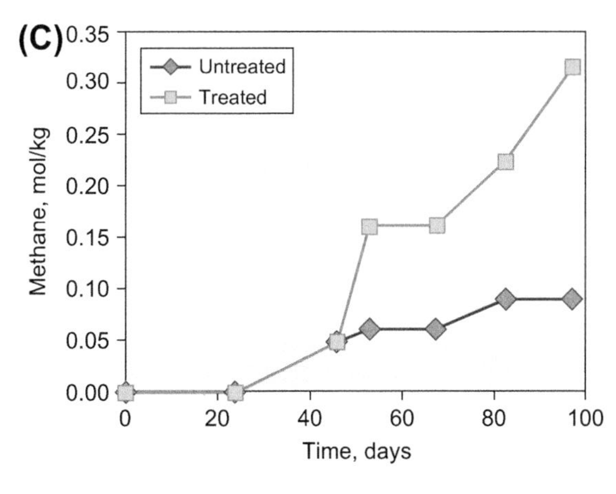

FIGURE 10.8 Chart showing cumulative methane produced by microcosms with treated and untreated lignites from Indonesia (A), China (B), and Australia (C). mol/kg, number of moles of solute per kilogram of solvent. *Source: Adopted from Fallgren et al. (2013a).*

cells/ml. In the nutrient-treated Australian lignite, the population of methanogens increased to 9.40×10^7 cells/ml. In contrast to the Indonesian and Australian lignites, the methanogen population in the treated Chinese lignite decreased, from 5.11×10^7 cells/ml in the untreated microcosms to 7.0×10^6 cells/ml in the treated (Figure 10.9). Thus, the study of lignites from Australia, China, and Indonesia by Fallgren et al. (2013a) proves that native methanogens in non-gas-producing lignites can be stimulated to generate biogenic gas. The Indonesian lignite was the first field-tested coal reservoir that produced biogenic coalbed gas (Fallgren et al., 2012, 2013a). The positive results in this singular study bode well for the future of biogenic coalbed gas in low-rank coals in general and in non-gas-producing lignites in particular.

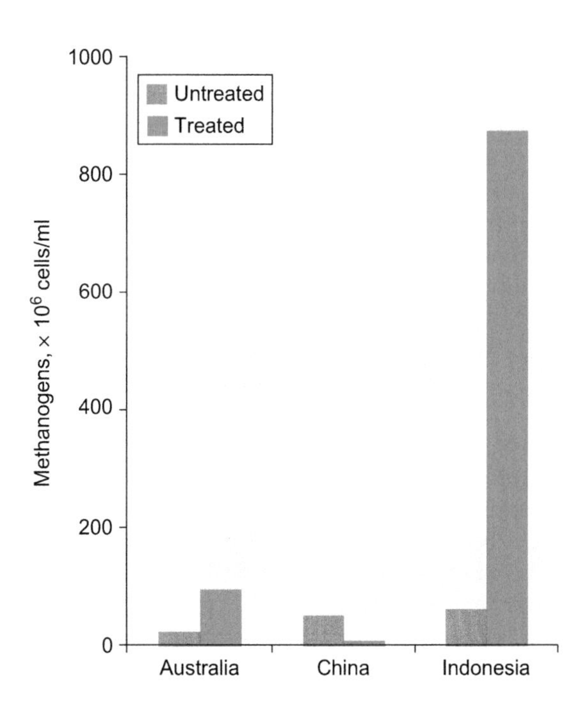

FIGURE 10.9 Histogram showing cell numbers of methanogens in microcosms contained in treated and untreated lignites from Indonesia, China, and Australia. *Source: Adopted from Fallgren et al. (2013a).*

FUTURE CHALLENGES OF COAL TO BIOGENIC GAS

Challenges to future large-scale application of the coal to biogenic coalbed gas may be due to technical and environmental factors, coal reservoir properties, and hydrogeological factors. Despite the ground-breaking successes to duplicate methanogenic processes in gas- and non-gas-producing coals to generate biogenic coalbed gas, it remains to be proved that the technology can be improved to sustain gas production. Fallgren et al. (2013a) appropriately pointed out that a major application of the technology is to extend the limited life span of coalbed gas wells. Does the 7–15 years life span of coalbed gas wells make production unsustainable? To prolong the life span of coalbed gas wells, the emerging coal-to-biogenic gas technology to generate new biogenic gas can potentially resume production for depleted wells and enhance production for existing wells. It was proposed by Finklestein et al. (2005) and Fallgren et al. (2013a) to "farm" biogenic gas by using the coal-to-biogenic gas technology in abandoned or low-producing wells and to rotate stimulation of different coalbeds to sustain production. Also, Fallgren et al. (2013b) recommended potential of stimulating microbial production in higher rank coals (e.g. bituminous).

How long biogenic coalbed gas production can be sustained is unpredictable. The sustainability of gas generation and production might be in part controlled by the groundwater supply (see Chapters 4 and 8; NRC, 2010). The very process of dewatering coalbed gas wells to desorb and produce gas from coal reservoirs has caused groundwater depletion (see Chapter 7) and might be the reason that will thwart production of fresh biogenic gas. That is, coalbed gas exhaustion in a well is caused by depletion of hydrostatic pressure as applied by the groundwater (e.g. water head) to the underlying coal reservoir. Depleted groundwater defeats the purpose of gas generation by water-based nutrient stimulation and subsequent production. More importantly, without hydrostatic pressure, potential newly generated biogenic coalbed gas cannot be held in the coal reservoir and becomes a challenge. Biogenic gas-producing basins such as the Powder River Basin have demonstrated groundwater draw down to as much as 305 m (BLM, 2012).

Reservoir properties such as permeability may be an important factor that controls biodegradation of organic matter in coal. Fallgren et al. (2013a, b) suggested that in order to permit large-scale microbial bioconversion of organic matter, more surface area of the coal matrix must be exposed. This can only mean increased permeability of the coal reservoir. It is not a coincidence that the Fort Union coals in the Powder River Basin have favorable biological and chemical settings for methanogenesis. The Powder River Basin coals have one of the highest permeability's (e.g. as much as 1 D or greater) of coalbed gas-producing reservoirs (Flores, 2004). However, gas producibility is locally variable within a well, between wells, and between and within gas fields, indicating that reservoir permeability is not uniform (see Chapter 5). Presented with a variable permeability, can well stimulation be adopted as a pretreatment method to increase coal surface availability to the methanogens? In addition, during stimulation, will coal fines lodged in the fractures serve as an obstruction or fodder to the methanogens? Jin (2007) and Green et al. (2008) suggested that pulverization of coal increases the amount of extractable DOC. Furthermore, the successes of generating biogenic coalbed gas in non-gas-producing lignites as reported by Fallgren et al. (2012, 2013) may be partly due to favorable reservoir permeability. In general, permeability is better in low-rank coals than in high-rank coals (see Chapter 5). The role of the physical properties of coal, in general, and reservoir permeability, in particular, as well as stimulation related to methanogenic favorability is not well understood.

A major challenge to bioconversion of organic matter in coal is chemical additives introduced in the coal reservoirs during stimulation (see

Chapter 7). Introduction of biocides as a component of hydraulic fracturing, which inhibits microbial growth, is potentially deleterious to microbial consortia use for coal-to-coalbed gas technology. The influence of biocides in the coal reservoirs should be determined in preparation for application of the technology. Coal reservoirs mainly stimulated by water hydraulic fracturing are least affected such as the Fort Union coals in the Powder River Basin.

It was pointed out by NRC (2010) that "fossilized" groundwater (see Chapter 8) may be withdrawn during gas desorption of coalbed gas wells in deeper part of coal basins (e.g. San Juan Basin). This might indicate groundwater recharge in geologic time proportions. Whether this condition is ubiquitous in all coalbed gas-producing basins or not is undetermined because of short supply of data and research investigation. In the Powder River Basin, groundwater drawdown and recovery from coalbed gas development in the northwest part of the basin in Montana (Wheaton & Metesh, 2002) may be much shorter compared to that in the intensely developed Wyoming part of the basin. Wheaton and Metesh (2002) indicated that in 30 years of coal mining (e.g. Decker mines), more than 3 m of drawdown at a distance of more than 8 km from the coal mines have been recovered from pre-mining levels after 3–4 years after reclamation.

In the Wyoming part of the basin, the groundwater drawdowns in highly productive coal zones of the upper part of the Fort Union Formation modeled by BLM (2003) are from much deeper. In the deep central and northwestern parts of the basin, maximum drawdown exceeds 244 m. The new BLM (2012) model analysis has shown that the same deep basin areas have reached drawdown of as much as 305 m. In the shallow southeast part of the basin, drawdowns were from 61 to 122 m over most of the active coalbed gas fields juxtaposed to active coal mining areas. Projected recharge from infiltration impoundments and surface discharge into perennial and intermittent creeks from precipitation only immediately relieves the shallow groundwater aquifers. How long the deep coal aquifers are recharged cannot be ascertained.

Other impediments to the use of coal-to-biogenic gas technology are the unknown potential impact of the treatment to the residual groundwater and the quality of the coal for future mineability. Short-term monitoring data from both laboratory and field tests have shown no observable influence on groundwater quality due to nonhazardous nature of the nutrients and their high biodegradability. This results in their consumption by and assimilation into microbes during multistep metabolisms of gas generation (Jin, 2013; Personal Commun.). However, the long-term effects of coal-to-biogenic gas technology on the groundwater are yet to be understood. In addition, pretreating the coal reservoir to release more DOC for biological conversion might also influence the quality of the groundwater. The long-term impact to the quality of the coal is also to be determined, although the industrial implementers of coal-to-biogenic gas technology suggest that only the minor components of coal, mostly volatile organic compounds, are converted to methane (Net Fuel website). Thus, the main structure of coal remains intact. This is particularly important to the future mineability and utilization of the coal in terms of potential change in calorific value and coal quality. The environmental concerns of the use of the coal-to-biogenic gas technology, if not addressed properly, may affect its large-scale commercialization. The current impediment is the lingering effects of the economic downturn, especially the price of natural gas.

COAL OUTLOOK

Coal remains the foundation of carbon-based energy both in developed and developing countries of the world. Coal fueling the future goes against the belief that its use be curtailed to reduce CO_2 emission. However, the reality is that coal utilization increased rapidly in developing countries (e.g. Brazil, India, China, Indonesia, Mexico, South Korea, and Turkey) from 1965 to 2011. Cattaneo (2013) reported that the percentage of increase of coal use during this period ranged from 699% in Brazil to 33,982% in Indonesia. Increase of coal use in India was 732% and the same in China with Hong Kong was up to 1543%. Thus, coal was the largest energy source for power generation of electricity during the past 46 years. Worldwide, the proposed 1199 new power plants will supply 1,401,278 MW (5,043,600 kJ/h) and the typical life span of coal-fired power plants is 40–60 years (Cattaneo, 2013).

Despite the pushback of burning coal for generation of electricity, advanced technologies to reduce carbon dioxide (CO_2), nitrogen oxide (NO_x), sulfur dioxide (SO_x), and particulate emissions have been employed to improve the efficiency of coal-fired power plants. Old small power plants and large inefficient plants are being replaced with installations that have new technologies at more expensive costs. IEA (2006) reported the potential reduction of CO_2 with utilization of new pulverized coal combustions systems (e.g. supercritical and ultrasupercritical) operating at higher temperatures and pressures to achieve higher efficiency (Figure 10.10).

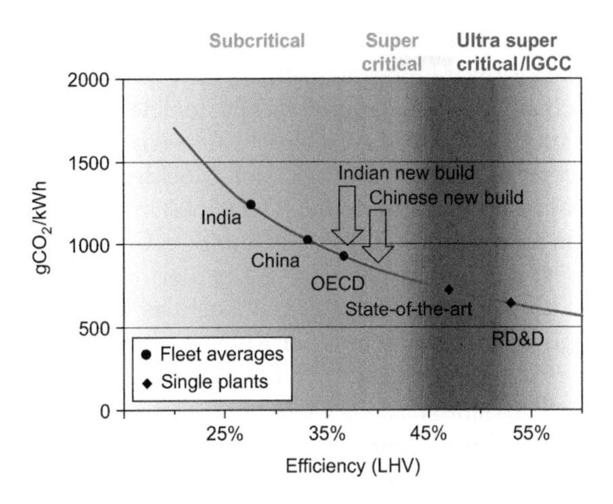

FIGURE 10.10 Carbon dioxide emissions from coal-fired power plants plotted against average efficiency of power generation in India and China compared to the supercritical and ultracritical power plants. LHV, lower heating value; gCO_2/kWh, carbon dioxide emissions per kW h of electricity generated. 1 kWh = 3600 kJ. *Source: Adopted from IEA (2006).*

For example, a new coal-fired power plant operating at 45%, reducing fuel use and CO_2 emissions by 22%, can replace an old coal-fired power plant with subcritical steam conditions operating at 35% efficiency. IEA (2006) calculated that as of 10 years ago coal-fired power generation in China was 33.2% and in India was 27% compared to an average in Organization for Economic Cooperation and Development (OECD) member countries of 36.7%. Since then, China has adopted constructions of the supercritical and ultrasupercritical power plants and has outpaced the United States in clean, advanced coal-fired power plants (Bradsher, 2009).

The future and sustainability of coal are constantly debated and was declared by MIT (2007) to be substantially reduced in use by 2050. MIT (2007) argued that the high cost of CO_2 emission from coal will constrain the use of coal. However, the CO_2 capture and sequestration (CCS) technology will permit coal to continue to be used beyond 2050 to meet the world's demand. This projection may be dashed by Rutledge (2013) who observed that coal production during the Kyoto Agreement (1997–2012) rose 66% and if this trend continues 90% of the coal reserves would be produced by 2067. This long-term coal use scenario may be compared to the short-, medium-, and long-term production and consumption outlook in the United States and the world by USEIA (2013) and IEA (2012), respectively (Box 10.2).

Short-Term Outlook in the United States

The short-term outlook for US coal production by USEIA (2013) is expected to increase to 934 million metric tons or by 1% in 2013 and to 947 million metric tons in 2014. The increased forecasts follow a downturn from 994 million

BOX 10.2

ROLE OF CO_2 SEQUESTRATION IN FUTURE COAL USE: FACT OR FICTION

There is a common belief that the success of CO_2 sequestration technology would extend the use of coal as electric power generator into the future (MIT, 2007; Moniz, 2007). The MIT multi-disciplinary report on "*Future of Coal: Options for a Carbon-Constrained World*" is an objective and insightful assessment of coal potentially remaining as an important energy source in the future in "any conceivable scenario". The report examines the socioeconomic and technical impacts of continued coal utilization. Also, the report scrutinizes the role of coal as an energy source, risks of global warming with CO_2 emissions, and availability of technology to achieve reduction of emissions such as CCS. The compelling argument that developed nations (e.g. United States, Europe) and economically emerging countries (e.g. China, India, East Asia) with large coal resource endowments will continue to rely on coal because it is abundant, cheap, and secure energy supply for power generation.

In order for coal to meet the global energy demand, the MIT report concluded that the CCS technology must be successfully deployed with coal-fired power plants. Spath and Mann (2001) reported that it is technically feasible to capture CO_2 from flue gas of coal-fired power plants but CCS consumes more coal and lowers the efficiency of the plant. Figueroa, Fout, Plasynski, McIlvried, and Srivastava (2008) reported that CO_2 from coal-fired power plants potentially could be captured postcombustion, precombustion, and during oxycombustion. In order to reduce costs, it is suggested that CO_2 captured from the plants must be transported and stored into geologic formations. Geologic storage options include depleted oil and gas reservoirs, enhanced oil recovery of reservoirs, deep unused saline water-saturated reservoirs, deep unmineable coals and enhanced coalbed gas recovery of coal reservoirs, and reaction with serpentine by CO_2 reacting with magnesium silicate minerals to form solid magnesium and carbonates that are ready for disposal (Chizmeshya, McKelvy, Marzke, Ito, & Wolf, 2007; Krevor & Lackner, 2011; Stasiulaitiene, Denafas, & Sliaupa, 2007; Stricker, Flores, Ellis, & Klein, 2008). Each of these geologic formations have there own advantages and disadvantages such as gas leakage (Bachu, 2008; Benson & Cole, 2008; Bustin, Cui, & Chikatamarla, 2008; Karacan & Mitchell, 2003; Karacan, 2007). CO_2 sequestration to enhance oil recovery so far has been a long-term solution successful for the past 4 decades adding 4–15% over primary and secondary recovery efforts (USDOE, 2010).

metric tons in 2011 to 925 million metric tons in 2012. During the same period, US total coal consumption was 909 million metric tons in 2011, 806 million metric tons in 2012 and has been forecasted to grow to 853 million metric tons in 2013 and 866 million metric tons in 2014. More than 90% of the total coal consumption is for electric power generation, which utilized 845 million metric tons in 2011, 748 million metric tons in 2012, and has been projected to increase by 795 million metric tons in 2013 and 803 million metric tons in 2014. In this short-term scenario, USEIA (2013) forecasts that coal production would grow by 1.3% in 2014.

Thus in general, total coal production and consumption decreased in 2012 but is expected to increase in 2013–2014 by about 9.8% and 9.3%, respectively. The short-term projected increase in US total coal production and consumption is attributed to the increase in price of both coal and natural gas, which was depressed during the economic downturn from 2008 to 2010. USEIA (2013) estimates that the delivered coal price averaged $2.40 per MMBtu (MM = million; Btu = 1.055 kJ) in 2012 and projected average delivered prices of $2.42 per MMBtu in 2013 and $2.45 per MMBtu in 2014. The coal price increase to be delivered in 2013–2014 to the electric industry is much lower compared to that a decade ago since 2011. The mine-mouth price for US coal is expected to increase by 1.5% per year, from $1.76 per MMBtu in 2010 to $2.56 in 2035, continuing the upward trend that began in 2000 (USEIA, 2012b).

US coal exports, which previously peaked at 102 million metric tons about three decades ago, rose to 114 million metric tons in 2012, surpassing the previous peak by nearly 12%. USEIA (2013) forecasts coal exports to decrease by about 14 million metric tons in 2013 and 2014. The ongoing economic downturn in Europe, which is the largest coal importer of the United States, which depressed international coal prices relative to natural gas price, and increased production in other coal-exporting countries (e.g. Australia, Indonesia) are the reasons for projected weakening in US coal exports in the short term (USEIA, 2013).

Long-Term Outlook in the United States

The volatile short-term outlook of US coal from 2012 to 2014 is followed by additional 4 years of instability as forecasted by USEIA (2013). As shown in Figure 10.11, the long-term outlook of US coal production declines for 4 years after 2014 resulting from projected low natural gas prices, rising coal prices, lack of growth in electricity demand, and increasing generation from renewables (USEIA, 2013). Also, demand for coal will be reduced due to new requirements to control emissions of NO_x, SO_2, and air toxics such as mercury and acid gases resulting in the retirement or stringent upgrades of some old coal-fired power plants.

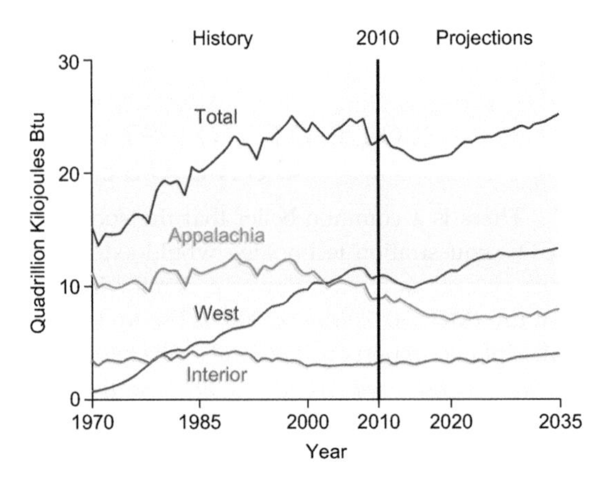

FIGURE 10.11 Historical and projected coal production by region in the United States from 1970 to 2035. *Source: Adopted from USEIA (2012b).*

USEIA (2013) forecasts that beyond 2015 coal production will grow at an average annual rate of 1.0% through 2035. However, the growth in terms of coal production and consumption will not be as robust as the previous three decades before 2010. Coal use for electricity generation after 2015 will increase as electricity demand grows and natural gas price rises. According to USEIA (2012b), coal production by US regions has declined and slowed down in the Appalachian and Interior. The substantial decline in the Appalachian region (e.g. central and northern) is mainly due to lower price of coal in other regions. The slowed down coal production in the Interior region (Illinois, Indiana, Kentucky, and Texas) is due to continued slow growth of demand. The coal production in the Western region, which included western states such as Wyoming, Colorado, Utah, Montana, New Mexico, and Alaska continues to grow at a slow rate. A large part (>40% of the total US coal production) of the production in the Western region comes from the Powder River Basin in Wyoming. The low-cost coal from the Western region provides much of the additional supply for coal-fired power plants east of the Mississippi River and supplies most of the coal used at new coal-fired power plants. US coal production is affected by plans to cut greenhouse gas emissions from existing power plants.

The long-term outlook is complicated by recent studies that predict that coal production in the United States will show a peak followed by a permanent decline during the middle of this century (see Chapter 1). Various prediction models have been used to forecast coal peak production in the United States to occur from 2030 to 2100 (Croft & Patzek, 2009; Hook & Aleklett, 2009; 2010; Milici, Flores, & Stricker, 2009; Milici, Flores, & Stricker, 2012; Patzek & Croft, 2009; Rutledge, 2011). As shown in Figure 10.11, the long-term coal production scenario projects flattening in the Appalachian and Interior regions, which puts pressure on production to meet supply in the western region of the United States. Milici et al. (2012) suggested that most of the production in the western region would be from the coal mines in the Powder River Basin. However, peak coal production in the basin is inevitable as indicated in Chapter 1 (Milici et al., 2009) and potential coal reserves are found in deeper part of the basin as indicated in Chapter 6. Additionally, estimates of coal reserves in the Powder River Basin, in particular, and the United States, in general, are unreliable due to lack of use of statistical measures to calculate the uncertainty of the recoverable reserves (Milici et al., 2012). Also, this uncertainty is generally exacerbated by assessments supported by old data, even though new data are available from recent petroleum-based wells and accompanying geophysical logs. The high-variability coal resource and reserve estimates fail to use coal bed correlations based on closely spaced data guided by its depositional settings as indicated in Chapter 6 as well as accurate economic-based statistical methodology.

Medium-Term Outlook Worldwide

Despite the appeal to reduce dependence on coal as a primary energy source in many developed and developing countries, the global demand continues to rise (IEA, 2012). By 2017, coal is expected to surpass oil as the world's top energy source (IEA, 2012). It is projected that during this period, coal demand will increase in every region of the world except in the United States, where coal use is generally slowing and supplanted by natural gas. The world will consume coal at about 1.2 billion metric tons more annually by 2017 compared to the present. This amount is equivalent to the current combined coal consumption of Russia and the United States (IEA, 2012).

China and India are projected to lead the expansion in coal consumption with demand for coal by China surpassing that for the rest of the world through 2017 and India will be the largest coal importer exceeding the United States as the second largest consumer in the world. The global coal consumption by 2017 is projected to be at 4.32 billion metric tons of oil equivalent (btoe) compared to 4.40 btoe for oil (IEA, 2012). The use of more "clean" gas fuel for generation of electricity in the United States presumably will reduce greenhouse gas emissions. However, the CO_2 or carbon capture systems (CCS) that reduce CO_2 in efficient gas-fired power plants may not be deployed during the medium-term outlook period or beyond. This means that CO_2 emissions will continue to grow for the lack of the CCS technologies. Without progress in CCS, and if other countries cannot replicate the United States experience and reduce coal demand with use of natural gas, coal will continue to be used worldwide.

As US coal supply and demand is projected to slow down, more US coal is exported to Europe, where low CO_2 and high gas prices have increased the use of coal for power generation (USEIA, 2012b). The report suggested that the coal demand will be close to peaking and demand by 2017 in Europe and is projected to drop above the 2011 level due to increase in renewable generation and decommissioning of old coal plants. The USEIA (2012b) concludes that even if the Chinese GDP growth were to slow to a 4.6% average through 2017, coal demand would still increase both globally and in China.

COALMINE GAS OUTLOOK

Coalbed gas with high methane content is emitted during active mining and when the mine is abandoned (see Chapter 2). A large part of the methane footprint in coal mining is from production; however, a small part comes from processing, storage, and transportation. Underground coal mines operate under 5% methane to maintain below explosion concentration, which is performed by large fans drawing air into the mine workings. The extracted mine air with methane is vented to the atmosphere through ventilation shafts or bleeders (see Chapter 2). In contrast to the ventilation systems, coalbed gas is also captured from surface and in-mine wells before and during mining. Abandoned underground coal mines also contribute to methane emission. The methane in abandoned coalmines is either residual gas or newly generated biogenic gas from mines flooded with groundwater. Surface coal mines emit as much as 14% methane according to the US Environmental Protection Agency, which is liberated to the atmosphere during dewatering operations.

It is impossible to account for all the vast amount of methane emitted by coal mines worldwide. However, the magnitude of global coalmine methane emissions may be gauged by that of the United States. According to the USEPA (2012), as much as 11% of the methane emission comes from coal mining compared to 37% from natural gas and petroleum system in the United States. Coalmine methane (CMM) is the fourth most dominant greenhouse gas emitted in the United States behind natural gas and petroleum, enteric (e.g. gastrointestinal processes by animals) fermentation, and landfill sources. Although the life span of methane in the atmosphere is shorter than that of carbon dioxide, methane traps radiation more efficiently than carbon dioxide. Thus, in terms of potency, the effect of methane on climate change is greater than that of carbon dioxide.

The outlook for coalmine methane is for more effective capture and utilization (see Chapter 2). As a result of the vast amount of coal production, China is the leading emitter of coalmine methane with more than 40% of the world's total emission (Figure 10.12) (IEA, 2009). According to the USEPA (2008) and IEA (2009), China leads the world on coalmine methane recovery and capture projects, with the purpose of safety first resulting from a history of very explosive coalmines. Recovery is emphasized in capturing coalmine methane in underground and abandoned mines. A wide variety of technologies are applied today in China to capture and utilize coalmine methane of variable concentrations (e.g. >80%, 30–80%, <30%, and 1% methane). The low concentration of methane mixed with oxygen, carbon dioxide, and nitrogen has been the main impediment in utilization of coalmine

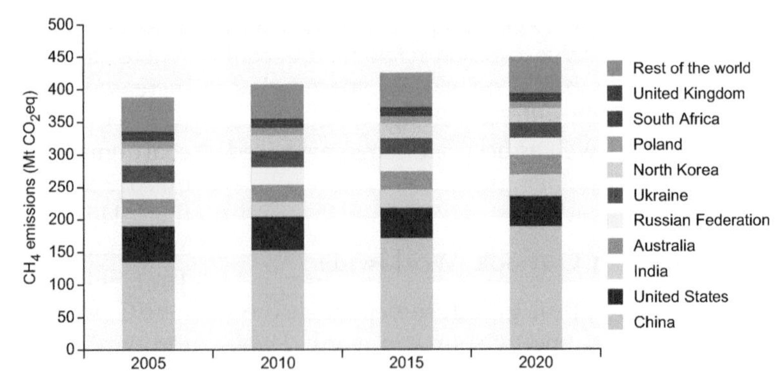

FIGURE 10.12 Methane emissions from coal mines worldwide increasing in China and decreasing in the United States to 2020. Mt = million tons; CO_2eq = equivalent carbon dioxide. *Source: Adopted from IEA (2009).*

methane. Thus, the Chinese model of recovery and use with technological advances can serve global knowledge resource. More importantly, the Chinese circumstances may serve as a model for global mitigation of non–carbon dioxide greenhouse gases.

CONCLUSIONS

Coal and coalbed gas, which are an energy mix in a one source–reservoir rock, gave rise to the old and new science and technology presented in this book. Coal is an old energy initially utilized during the Age of Industrial Revolution, which lasted to the present and apparently continues into the foreseeable future (see *Chapters 1–3*). However, the future of coal as an energy resource is different for OECD and Non-Organization for Economic Cooperation and Development (NOECD) countries (see *Chapter 7*). Consumption of coal is forecasted to slow down in the United States and in some OECD member countries that are attempting to reduce greenhouse gases. In contrast, consumption and demand for coal is on the rise in China and India as well as in other developing NOECD countries.

Coalbed gas is an energy born out of necessity for safe coal mine production during the Industrial Revolution but was rediscovered a few decades ago in search for cheap and clean energy from unmineable coalbeds (see *Chapters 2 and 9*). The development of both energy sources resulted in the evolution of the science and technology of coal and coalbed gas. New technologies are developed, which have resulted in huge progress in the exploitation of coal and coalbed gas (see *Chapters 7–9*). In addition, research and development have advanced the science of coal as a source rock and led to new concepts on coalbed gas in terms of its generation and production as well as sustainability or renewability in the future (see *Chapters 4–6*). It is clear that the science and technology of coal and coalbed gas have evolved into a truly multidiscipline endeavor in which no one discipline can exist without the other.

SCIENCE OF COAL AND COALBED GAS

Coal is an interdisciplinary science generally known as coal geology, which is an integration of stratigraphy, sedimentology, petrology, chemistry, and botany. The applications of stratigraphy and sedimentology in coal geology are basic for understanding the origin of peat in the depositional environments of coal at the local and basin-scale dimensions (see *Chapters 2–3*). Stratigraphy is important to correlations as sedimentology is to depositional systems with both significantly affecting the aerial distribution, thickness, and occurrence of coalbeds. The understanding of these principles is basic to coal resource–reserve assessments, where failure to recognize their role can mean an under- or overestimation of the total amount of coal in the ground (see *Chapter 6*). Oftentimes, the variability of coal resource estimates is caused by errors in correlations. This leads to miscalculations of coalbed gas resources and reserves as well as misunderstanding of the framework and characterization of coal reservoirs (see *Chapter 6*).

The petrology, chemistry, and botany of coal provide basic understanding of the composition of coal and its physical and chemical transformations from peat to coal during burial with exposure to heat and pressure at depths (see *Chapter 3*). The physicochemical transformations (e.g. coalification) lead to knowledge of evolution of coal from source to reservoir rock of oil and coalbed gas (see *Chapter 4*). More specifically, the concept of evolution of the organic composition of coal and its molecular changes during coalification are basic to coal properties as source and reservoir of coalbed gas (see *Chapter 5*). Coalification results in progressive loss of moisture and replacement by bitumens termed as bituminization, which may escape the coal matrix to form oil. Further coalification results in debituminization in which the coal molecular structure is dominated by "small molecules" such as methane, carbon dioxide, and nitrogen that are stored or escape the coal matrix. The process representing "cracking" is driven principally by temperature independent of pressure and the coalbed gas formed through this process is thermogenic in origin (see *Chapter 4*). This origin is in contrast to the biogenic origin of coalbed gas, which formed under low temperature driven by microbial activity and groundwater recharge after uplift and erosion of the coal basin.

TECHNOLOGY OF COAL AND COALBED GAS

The technology involving the production of coal and coalbed gas has evolved rapidly during the past two decades. The advent of technology provided a better understanding of petroleum engineering, hydrology, microbiology, molecular engineering, biogeochemistry, and environmental management in the production of coalbed gas. Emerging technologies applied to coalbed gas development such as horizontal well drilling, stimulation by fraccing, completion of multiple beds, and recovery have resulted in more efficient and maximized gas production (see *Chapter 7*). Advances in coalmine gas recovery and utilization technologies, which are driven by safe mine operations, have been successfully adopted to capture fugitive coalmine gas. However, fugitive mine gas remains a source of greenhouse gases, even though its collection has been proved to be an asset in China and the United States.

Perhaps, the most pivotal technological advances in coalbed gas are the generation of new biogenic coalbed gas and potential sustainability of such gas production. Stimulation of consortia of microbes in coal and in associated groundwater through engineered inocula or nutrients has been successful in the laboratory and field tests. Biogenic gas generation through microbial activity has been successful in biogenic gas-producing coals as well as non-gas-producing coals. Information abounds (see *Chapter 4* and *this chapter*) from laboratory experiments to re-create the processes of methanogenesis in coal. However, there is a shortage of information on the application of this technology either due to the unavailable proprietary data from the industry or lack of directed research in field application by academia and government. The innovations and breakthrough in coal-to-biogenic coalbed gas technology mainly originates from the industry, which is not wholly available to the public. Thus, the future of sustaining production of biogenic coalbed gas is controlled by the challenges discussed in this chapter.

Finally, it is early to declare the demise of coal as a basic global energy source. Although natural gas, of which coalbed gas is an integral part, has partially replaced coal for electric power generation in the United States, cleaner combustion of coal at very high temperatures using more efficient supercritical and ultrasupercritical power plants are emerging technologies in China and Europe. The projected growth in coal and natural gas demands in Asia and western Pacific (e.g. China, India, Japan, South Korea, and Taiwan) mainly supplied by exports of coal and coalbed gas as LNG from Australia and Indonesia assures a long-term future of the energy industry in that part of the world. The growth of CMM supply in China for the sake of mine safety ascertains a cleaner resource base. Thus, the Asia-Pacific region may become the "epicenter" of a reenergized Industrial Revolution in which advanced technology of electric power generation uses cleaner energy mix of coal and coalbed gas.

References

AAAS. (2011). *Turning coal to liquid fuel*. American Association for the Advancement of Science. <http://www.aaas.org/spp/cstc/briefs/coaltoliquid/> Accessed 05.12.11.

Abbanat, D. R., Aceti, D. J., Baro, S. F., Terlesky, K. C., & Ferry, J. G. (1989). Microbiology and biochemistry of the methanogenic archaeobacteria. *Advance Space Research, 9*, 101–105.

Abdullah, W. H., & Abolins, P. (2006). *Oil-prone paralic coals: A case study from the Balingian Province of Sarawak, Malaysia*. American Association of Petroleum Geologists International Conference and Exhibition, November 5–8, 2006, Perth, West Australia. <http://www.pesa.com.au/aapg/aapgconference/pdfs/abstracts/tues/OilfromCoalSources.pdf> Accessed 10.12.11.

Adams, M. A. (1984). Geologic overview, coal resources, and potential methane recovery from coalbeds of the central Appalachian Basin–Maryland, West Virginia, Virginia, Kentucky, and Tennessee. In C. T. Rightmire, G. E. Eddy, & J. N. Kirr (Eds.), *American Association of Petroleum Geologists Studies in Geology Series: Vol. 317. Coalbed methane resources of the United States* (pp. 45–71).

Adams, M. A., Eddy, G. E., Hewitt, J. L., Kirr, J. N., & Rightmire, C. T. (1984). Geologic overview, coal resources, and potential methane recovery from coalbeds of the northern Appalachian coal basin – Pennsylvania, Ohio, Maryland, West Virginia, and Ohio. In C. T. Rightmire, G. E. Eddy, & J. N. Kirr (Eds.), *American Association of Petroleum Geologists Studies in Geology Series: Vol. 317. Coalbed methane resources of the United States* (pp. 15–43).

Adeboye, O. O. (2011). Effects of coal composition and fabric on porosity, sorption capacity and gas flow properties in Western Canada sedimentary basin coals (M.S. Thesis), University of British Columbia.

Aerts, R., Van Logtestlin, R., Van Staalduinen, M., & Toet, S. (1995). Nitrogen supply effects on productivity and potential leaf litter decay of *Carex* species from peatlands differing limitation. *Oecologia, 104*, 447–453.

Agricola, A. (1556). *De ve metallica*. New York: Dover Publication, Incorporated. Translated by Hoover, H. C., and Hoover, L. H. and published 1912; republished 1950.

Ahmed, T., Centilmen, A., & Roux, B. (2006). *A generalized material balance equation for coalbed methane reservoirs*. Society of Petroleum Engineers, Annual Technical Conference and Exhibition, San Antonio, Texas, USA, Sept. 24–27, 2006. DOI:10.2118/102638–MS. SPE 102638, p. 11 <http://www.onepetro.org/mslib/servlet/onepetropreview?id=SPE-102638-MS&soc=SPE> Accessed 20.08.12.

Ahmed, M., Smith, J. W., & George, S. C. (1999). Effects of biodegradation on Australian Permian coals. *Organic Geochemistry, 30*, 1311–1322.

Ahmed, M., Volk, H., George, S. C., Faiz, M., & Stalker, L. (2009). Generation and expulsion of oils from Permian coals of the Sydney Basin, Australia. *Organic Geochemistry, 40*, 810–831.

Airgas Inc. (2012). *Material safety data sheet: Phosphine*. Radnor, Pennsylvania: Airgas Inc. <http://www.airgas.com/documents/pdf/001070.pdf> Accessed 08.02.12.

Ajdukiewicz, J. M., & Lander, R. H. (2010). Sandstone reservoir quality prediction: the state of art. *American Association of Petroleum Geologists Bulletin, 94*, 1083–1091.

Ajlouni, M. A. (2000). *Geotechnical properties of peat and related engineering problems*. Ph.D. dissertation, University of Illinois at Urbana-Champaign. 15–189.

Alberta Energy. (2012). *Coalbed methane development*. <http://www.energy.alberta.ca/NaturalGas/753.asp> Accessed 26.10.12.

Alberta Oil. (December 21, 2011). *First round settled in legal tiff over coalbed methane rights*. Alberta Oil Newsletter. <http://www.albertaoilmagazine.com/enews/> Accessed 08.03.13.

Alekseev, E. G., Mustafin, R. K., & Umarhajieva, N. S. (2003). *Coal methane: Potential energy prospects for Kazakhstan*. Methane Center, Kazakhstan. UNECE Ad Hoc Group of Experts on Coal in Sustainable Development 17 November 2003. <http://www.unece.org/fileadmin/DAM/ie/se/pp/coal/mustafin.pdf> Accessed 29.01.13.

Ali, M., Sarkar, A., Sagar, R., & Klimentos. (2008). *Cleat characterization in CBM wells for completion optimization*. Society of Petroleum Engineers, SPE Indian Oil and Gas Technical Conference and Exhibition, March 4–6, 2008, Mumbai, India, ID113600–MS, p. 10. <http://www.onepetro.org/mslib/servlet/onepetropreview?id=SPE-113600-MS> Accessed 16.10.12.

ALL Consulting. (2003). *Handbook on coal bed methane produced water: Management and beneficial use alternatives*.

Prepared for Grand Water Protection Research Foundation, U.S. Department of Energy, National Petroleum Technology Office, and Bureau of Land Management. Tulsa, Oklahoma: ALL Consulting. p. 322. <http://www.all-llc.com/publicdownloads/CBM_BU_Screen.pdf> Accessed 14.11.12.

ALL Consulting. (2005). *Summary of oil & gas produce water treatment*. Tulsa. Oklahoma: ALL Consulting. LLC. p. 53. <http://w.all-llc.com/publicdownloads/ALLConsulting-WaterTreatmentOptionsReport.pdf> Accessed 19.12.12.

ALL Consulting. (2008). *Anticipated and observed impacts to groundwater associated with construction use of infiltration impoundments in the Powder River Basin*. National Energy Technology Laboratory, Office of Fossil Energy, ALL Consulting. Tulsa, Oklahoma: U.S. Department of Energy. p. 185.

Allardice, D. J., & Evans, D. G. (1978). Moisture in coal. In C. Karr, Jr. (Ed.), *Analytical methods for coal and coal products* (Vol. 1; pp. 247–262). New York: Academic Press.

Allen, M. J. (1995). Exploration and exploitation of the east Pennine coalfield. In M. K. G. Whateley, & D. A. Spears (Eds.), *Geological Society, London, Special Publication: Vol. 82. European coal geology* (pp. 207–214).

Allen, J. R. L. (1999). Geological impacts on coastal wetland landscapes: some general effects of sediment autocompaction in the Holocene of northwest Europe. *The Holocene, 9*, 1–12.

Allen, J. R. L. (2000). Holocene coastal lowlands: autocompaction and the uncertain ground. In K. Pye, & J. R. L. Allen (Eds.), *Geological Society Special Publication: Vol. 175. Coastal and estuarine environments: Sedimentology, geomorphology and geoarcheology* (pp. 239–252).

Allen, M. (2008). *Pickens overlooks existing natural gas cars in energy plan: reality check*. Popular Mechanics. <http://www.popularmechanics.com/cars/alternative-fuel/4282954> Accessed 31.10.12.

Alley, W. M., Reilly, T. E., & Franke, O. L. (1999). *Sustainability of ground-water resource*. U.S. Geological Survey Circular 1186. <http://pubs.usgs.gov/circ/circ1186/html/title.html> Accessed 06.11.12.

Alpern, B. (1981). For a universal synthetic classification of solid fuels (in French). *Bulletin des Centres, de Recherches Exploration-Production, Elf-Aquitane, 5*(2), 271–290.

Alpern, B., Lemos de Sousa, M. J., & Flores, D. (1989). A progress report on the Alpern coal classification. In P. C. Lyons, & B. Alpern (Eds.), *Coal: Classification, coalification, mineralogy, trace-element chemistry, and oil and gas potential. International Journal of Coal Geology, 13*, 1–19.

Alsaab, D., Elie, M., Izart, A., Sachsenhofer, R. F., Privalov, V. A., Suarez-Ruiz, I., et al. (2008). Comparison of hydrocarbon gases (C_1–C_5) production from Carboniferous Donets (Ukraine) and Cretaceous Sabina (Mexico) coals. *International Journal of Coal Geology, 74*, 154–162.

Alvarez, L. A., Alvarez, W., Asaro, F., & Michel, H. V. (1980). Extraterrestrial cause for the cretaceous–tertiary extinction. *Science, 208*, 1095–1108.

Ambrose, W. A., & Ayers, W. B., Jr. (1991). Geologic controls on coalbed methane occurrence and producibility in the Fruitland formation, Cedar Hill field and coal site, San Juan Basin, Colorado and New Mexico. In S. D. Schwochow (Ed.), *Rocky Mountain Association of Geologists Guidebook Coalbed methane of western North America* (pp. 227–240).

Ammosov, I., & Ermin, I. V. (1963). *Fracturing of coal*. Moscow: IZDT Publishers. Office of Technical Services, Washington, D.E., p. 109.

Anderson, J. A. R. (1961). *The ecology and forest types of the peat swamp forests of Sarawak and Brunei in relation to their silviculture*. Unpublished Ph.D. thesis, University of Edinburgh.

Anderson, J. A. R. (1964). The structure and development of peat swamps of Sarawak and Brunei. *Journal of Tropical Geography, 18*, 7–16.

Anderson, J. A. R. (1983). The tropical peat swamps of western Malaysia. In A. J. P. Gore (Ed.), *Ecosystems of the world, Part B; mires, swamp. Bog, fen and moor* (pp. 181–199). Amsterdam: Elsevier.

Anderson, J. A. R., & Muller, J. (1975). Palynological study of a Holocene peat and Miocene coal deposit from N.W. Borneo. *Review of Palaeobotany and Palynology, 19*, 291–351.

Anderson, J., Simpson, M., Basinski, P., Beaton, A., Boyer, C., Bulat, D., et al. (2003). Producing natural gas from coal. *Oilfield Review*, 31. <http://www.slb.com/~/media/Files/resources/oilfield_review/ors03/aut03/p.8_31.pdf> Accessed 23.10.12.

Andriesse, J. P. (1974). *Tropical lowland peats in south-east Asia*. Royal Tropical Institute. Amsterdam, Department of Agricultural Research, Communication No. 63.

Andriesse, J. P. (1988). Nature and management of tropical peat soils. *FAO Soils Bulletin, 59*. <http://www.fao.org/docrep/x5872e/x5872e00.htm#Contents> Accessed 22.01.12.

Antoniadis, P., Mavridou, E., Papazisimou, S., Christanis, K., & Gentzis, T. (2006). Petrographic characteristics of the Karda lignites (core KT6A-3), Ptolemais Basin (Greece). *Energy Sources, Part A: Recovery, Utilization, and Environmental Effects, 28*, 373–388.

API. (2010). *Freeing up energy – Hydraulic fracturing: Unlocking America's natural gas resources*. American Petroleum Institute. <http://www.slb.com/~/media/Files/resources/oilfield_review/ors03/aut03/p.8_31.pdf>. <http://www.api.org/~/media/files/policy/exploration/hydraulic_fracturing_primer.ashx> Accessed 11.10.12.

APPEA. (2012). *Coal seam gas – Industry introduction.* <www.appea.com.au/csg/industry-facts/fact-sheels/969-csg-industry.html> Accessed 26.10.12.

Aranovich, G. L., & Donohue, M. D. (1995). Adsorption isotherms for microporous adsorbents. *Carbon, 33,* 1369–1375.

Aravena, R., Harrison, S. M., Barker, J. F., Abercrombie, H., & Rudolph, D. (2003). Origin of methane in the Elk Valley coalfield, southeastern British Columbia, Canada. *Chemical Geology, 195,* 219–227.

ARI. (2002). *Powder River Basin coalbed methane development and produced water management study.* Prepared by Advanced Resources International, Inc. for U.S. Department of Energy, Office of Fossil Energy and National Energy Technology Laboratory, DOE/NETL–2003/1184, p. 213.

ARI. (2008a). *Coal mine methane and coalbed methane development in the Donetsk region, Ukraine. Prepared for Donetsk Regional Administration and U.S.* Trade and Development Agency by Advanced Resources International, Inc. Arlington, Virginia, USA 2008.

ARI. (2008b). *Technical and economic feasibility study for coalbed methane development in eastern Botswana.* Prepared for Botswana Development Corporation and U.S. Trade and Development Agency by Advanced Resources International Inc. Arlington, Virginia, USA 2008.

ARI. (2009). *CMM and CBM development in the Tkibuli-Shaori Region, Georgia.* Advanced Resources International, Inc. Prepared for GIG/Saknakhshiri and U.S. Trade and Development Agency by Advanced Resources International, Inc. Arlington, Virginia, USA 2008.

Aristotle (340). "Meteorologica." English translation by H.D.P. Leew, 1952, Harvard University Press, Cambridge, Massachusetts, and William Heinemann Limited.

Arthur, D. (2001). Let's talk technical, Montana's CBM EIS status. In *Presentation from the August 29th Montana petroleum association's annual meeting.* EPA-HQ-OW-2008-0517, DCN 07229.

Arthur, J. D., Langhus, B., Epperly, D., Bohm, B., Richmond, T., & Halvorson, J. (September 22–26, 2001). Coal bed methane in the Powder River Basin. In *Proceedings of the 2001 ground water protection council annual Forum* (pp. 1–48). Reno, Nevada.

Ash, M. A., & Gintautas, P. (2009a). *Coalbed methane production, Raton Basin: Staff presentation.* Colorado Oil and Gas Conservation Hearing, August 2009, Trinidad, Colorado, p. 33. <http://cogcc.state.co.us/Library/Presentations/Trinidad_HearingAug_2009/Raton_Basin_CBM_Production_COGCC.pdf> Accessed 28.11.12.

Ash, M. A., & Gintautas, P. (August 18, 2009b). *Staff presentation coal bed methane production, Raton Basin.* Trinidad, Colorado: Colorado Oil and Gas Conservation Commission Hearing. < http://cogcc.state.co.us/Library/Presentations/Trinidad_HearingAug_2009/Raton_Basin_CBM_Production_COGCC.pdf> Accessed 10.02.13.

Ashley, G. H. (1907). The maximum rate of deposition of coal. *Economic Geology, 2,* 34–47.

Asquith, G. B. (1982). Basic well log analysis for geologists. In *Methods in exploration series* (p. 216). Tulsa, Oklahoma, USA: American Association of Petroleum Geologists.

van Asselen, S. (2009). *The effect of peat compaction on the evolution of alluvial plains,* 9th International conference on fluvial sedimentology, Tucuman, Argentina. IESGLO (Instituto de Estratigrafía y Geología Sedimentaria Global) - Universidad Nacional de Tucuman Fundación Miguel Lillo.

van Asselen, S., Karssenberg, D., & Stoouthamer, E. (2011). Contribution of peat compaction to relative sea-level rise within Holocene delta. *Geophysical Research Letters, 38,* 5. L24401. <http://dx.doi.org/10.1029/2011GL049835>.

van Asselen, S., Stouthamer, E., & Smith, N. D. (2010). Factors controlling peat compaction in alluvial floodplains: a case study in the cold-temperate Cumberland Marshes, Canada. *Journal of Sedimentary Research, 80,* 155–166.

ASTM. (1991). *Annual book of standards.* In *Gaseous fuels, coal and coke* (*Vol. 5.05*). Philadelphia, Pennsylvania: American Society for Testing and Materials. 19103–1187.

ASTM. (1993). *Annual book of ASTM standards. American Society for Testing and Materials, volume 05.05 Gaseous fuels; coal and coke.* Philadelphia: ASTM. p. 526.

ASTM. (1994). *Annual book of American Society for Testing and Material Standards, chapter D2796–94, standard terminology relating to megascopic description of coal and coal seams and microscopical description and analysis of coal: West Conshohocken.* Pennsylvania: American Society for Testing and Materials, 273–275.

ASTM. (2011). *Standard test method for microscopical determination of the reflectance of vitrinite dispersed in sedimentary rocks.* In *Annual book of ASTM standards: petroleum products, lubricants, and fossil fuels; gaseous fuels; coal and coke, sec. 5, v. 5.06, D7708-11.* West Conshohocken, PA: American Society for Testing and Materials (ASTM International). <http://dx.doi.org/10.1520/D7708-11>. 823–830. <http://www.astm.org/Standards/D7708.htm>. Accessed 20.03.11.

ASTM International. (2002). Coal and coke. In *Gaseous fuels; coal and coke Annual book of ASTM standards, section 5, petroleum products, lubricants, and fossil fuels* (pp. 163–626). Philadelphia, Pa.: ASTM International. v. 05.06.

Attanasi, E. D., & Rice, D. D. (1995). Economics and coalbed gas. In *1995 National Assessment of Oil and Gas Resources* (pp. 22). U.S. Geological Survey Open-File Report 95-75A.

Augenbraun, H., Matthews, E., & Sarma, D. (1997). Global methane cycle. In *Education: Global methane inventory*. National Aeronautics and Space Administration, Goddard Institute for Space Studies. <http://icp.giss.nasa.gov/education/methane/intro/cycle.html> Accessed 08.02.12.

Aurelio, M. A. (2005). *Tectonic complexities of Philippines volcanoes*. The Geological Society of the Philippines. Geocon 2005.

Australian Bureau of Agricultural and Resource Economics, Energy in Australia 2010, Canberra, Australia, April 2010.

Australian Standard. (1999). *Guide to the determination of gas content of coal – Direct desorption method*. AS 3980-1999 (2nd ed.). Standards Association of Australia. ISBN 0 7337 2906 1, p. 23.

Averitt, P. (1975). Coal resources of the United States, January 1, 1974. *U.S. Geological Survey Bulletin, 1412*, 131.

Averitt, P. (1981). Coal resource. In M. A. Elliott (Ed.), *Second supplementary volume, Chemistry of coal utilization* (pp. 58–59). New York: John Wiley and Sons.

Ayers, W. B., Jr. (1986). Lacustrine and fluvial–Deltaic depositional systems, Fort Union formation (Paleocene) Powder River Basin, Wyoming and Montana. *American Association of Petroleum Geologists Bulletin, 70*, 1151–1673.

Ayers, W. B., Jr. (2002). Coalbed gas systems, resources, and production and a review of contrasting cases from the San Juan and Powder River basins. *American Association of Petroleum Geologists Bulletin, 86*, 1853–1890.

Ayers, W. B., Jr. (2003). Coalbed methane in the Fruitland Formation, San Juan Basin, western United States: a giant unconventional gas play. In M. T. Halbouty (Ed.), *Giant fields of the decade 1990–1999* (pp. 159–188). American Association of Petroleum Geologists, Memoir 78, Chapter 10.

Ayers, W. B., Jr., Kaiser, W. R., Laubach, S. E., Ambrose, W. A., Baumgardner, R. W., Jr., Scott, A. R., et al. (1991). *Geologic and hydrologic controls on the occurrence and producibility of coalbed methane, Fruitland Formation, San Juan Basin*. Gas Research Institute Topical Rept. GRI-91/0072 (August 1987–July 1990), p. 314.

Ayers, R. S., & Westcot, D. W. (1985). Water quality for agriculture. In *Food and agriculture organization of the United Nations, irrigation and drainage paper, 29, Rev* (Vol. 1; p. 174).

Ayers, W. B., Jr., & Zellers, S. D. (1991). *Geologic controls on Fruitland coal occurrence, thickness, and continuity, Navajo Lake, San Juan Basin*. Gas Research Institute. Topical Report GRI–91/0072, 69–74.

Bachu, S. (2008). CO_2 storage in geological media: role, means, status and barriers to deployment. *Progress in Energy and Combustion Science, 34*, 254–273.

Bachu, S., & Michael, K. (2003). Possible controls of hydrogeological and stress regimes on the producibility of coalbed methane in Upper Cretaceous-Tertiary strata of the Alberta basin, Canada. *American Association of Petroleum Geologists Bulletin, 87*, 1729–1754.

Bagge, M. A., & Keeley, M. L. (1994). The oil potential of mid-Jurassic coals in northern Egypt. In A. C. Scott, & A. J. Fleet (Eds.), *The Geological Society Special Publication: Vol. 77. Coal and coal-bearing strata as oil-prone source rocks?* (pp. 183–200). London.

Baily, H., Glover, B. W., Holloway, S., & Young, S. R. (1995). Controls of coalbed methane prospectivity in Great Britain. In M. K. G. Whateley, & D. A. Spears (Eds.), *European coal geology* (Vol. 82; pp. 251–265). Geological Society Special Publication.

Baird, A. J., & Gaffney, S. W. (1995). A partial explanation of the dependency of hydraulic conductivity on positive pore water pressure in peat soils. *Earth Surface Processes and Landforms, 20*, 561–566.

Baker, G., & Slater, S. (2008). *The increasing significance of coal seam gas in Eastern Australia*. PESA Eastern Australasian Basins Symposium III, Sydney, September 14–17, 2008.

Baker, J. C., Bai, G. P., Hamilton, P. J., Golding, S. D., & Keene, J. B. (1995). Continental-scale magmatic carbon dioxide seepage recorded by dawsonite in the Bowen–Gunnedah–Sydney Basin system, eastern Australia. *Journal of Sedimentary Research, Section A, Sedimentary Petrology and Processes, 65*, 522–530.

Balan, H. O., & Gumrah, F. (2009). Assessment of shrinkage-swelling influences in coal seams using rank-dependent physical coal properties. *International Journal of Coal Geology, 77*, 203–213.

Baltoiu, L. V., Warren, B. K., & Natras, T. A. (2008). State-of-the-art in coalbed methane drilling fluids. *Society of Petroleum Engineers. Drilling and Completion, 23*, 250–257.

Barker, C. E., Dallegge, T. A., Clark, A. C., (2002). USGS coal desorption equipment and a spreadsheet for analysis of lost and total gas from canister desorption measurements. U.S. Geological Survey Open-File Report 02-496, Reston, Va. p. 13

Barker, G. J. (2008). *Application of the PRMS to coal seam gas*, 2008 Society of Petroleum Engineers Asia Pacific Oil & Gas Conference and Exhibition, SPE Paper 117124, Perth, Australia, p. 13.

Barker, G. J. (2011). Classification and reporting issues. In *Guidelines for application of the petroleum resources management system* (pp. 149–160). Sponsored by Society of Petroleum Engineers, American Association of Petroleum Geologists, World Petroleum Council, Society of Petroleum Evaluation Engineers, and Society of Exploration Geophysicists. <http://www.spe.org/

industry/docs/PRMS_Guidelines_Nov2011.pdf> Accessed 16.06.12.

Barker, J. F., & Fritz, P. (1981). The occurrence and origin of methane in some groundwater flow systems. *Canadian Journal of Earth Sciences, 18*, 1802–1816.

Barker, C. E., Lewan, M. D., & Pawlewicz, M. J. (2007). The influence of extractable organic matter on vitrinite reflectance suppression: a survey of kerogen and coal types. *International Journal of Coal Geology, 70*, 67–78.

Barrero, D., Pardo, A., Vargas, C. A., & Martinez, J. F. (2007). *Colombian sedimentary basins: Nomenclature, boundaries, and petroleum geology, a new proposal*. Bogota, Colombia: Agencia Nacional de Hidrocarburos. Calle 99 No. 9A–54 (piso 14).

Barsky, V., Vlasov, G., & Rudnitsky, A. (2011). Composition and structure of coal organic mass, 2. kinetic models of metamorphism. *Chemistry and Chemical Technology, 5*, 139–145. <http://lp.edu.ua/fileadmin/ICCT/journal/Vol. 5/No_2/04.pdf> Accessed 18.03.11.

Bass, F. (1907). *The explosion at Monongah number six and Monongah number eight mines of the Fairmont Coal Company*. p. 39. (Accessed at the MSHA Library, Beaver, West Virginia).

Bassett, R. A., Holtkamp, J. A., & Ryan, R. (2011). U.S. laws and policies regarding capturing methane gas. In *2009 U.S. coal mine methane conference*. Boulder, Colorado, Holland & Hart LLP. Accessed 312344dc30c3/Presentation/PublicationAttachment/8c2a9a52–5b55-446e-8256-79eb4d3eb74f/Capturing_Methane_Gas_White_Paper.pdf.

Basu, A., & Singh, R. N. (1994). *Comparison of Darcy's law and Ficks' law of diffusion to determine the field parameters related to methane gas drainage in coal seams*. International Mine Water Association Proceedings, Nottingham, UK, pp. 59–70. <http://www.imwa.info/docs/imwa_1994/IMWA1994_Basu_059.pdf> Accessed 09.04.12.

Bates, R. L., & Jackson, J. A. (Eds.). (1987). *Glossary of geology* (3rd ed.). (p. 788) Alexandria, Virginia: The American Geological Institute.

Bauder, J. (2006). Environmental considerations in utilizing produced waters for beneficial use. In R. Wickramasignhe (Ed.), *Produced water workshop* (pp. 42–53). Colorado Water Resources Institute, Colorado State University. Information Series No. 102. <http://www.cwi.colostate.edu/publications/is/102.pdf> Accessed 28.11.12.

Beagle, D. (October 13, 2006). Countercurrent ion exchange applied to CBM produced water treatment in the PRB. In *The proceedings of the Rocky Mountain unconventional gas conference*. South Dakota School of Mines and Technology.

Beagle, D., & Dennis, R. (November 5–9, 2007). *Continuous countercurrent ion exchange applied to CBM produced water treatment in the Powder River Basin of Wyoming and Montana*. Houston, Texas: Presented at the 14th International Petroleum Environmental Conference. <http://ipec.utulsa.edu/Conf2007/Papers/Beagle_40.pdf> Accessed 18.12.12.

Beamish, B. B., & Crosdale, P. J. (1998). Instantaneous outbursts in underground coal mines: an overview and association with coal type. In R. M. Flores (Ed.), *International Journal of Coal Geology: Vol. 35. Coalbed methane: From coal-mine outbursts to a gas resource* (pp. 27–55).

Beamish, B. B., & Gamson, P. D. (Eds.), (1992). *Symposium on coalbed methane research and development in Australia* (Vols. 1–5). James Cook University of North Queensland.

Beamish, B. B., & Gamson, P. D. (1993). *Sorption behavior and microstructure of Bowen Basin coals*. Coalseam Gas Research Institute, James Cook University. Technical Report CGRI TR 92/4, February 1993, p. 195.

Beaton, A., Langenberg, W., & Pana, C. (2006). Coalbed methane resources and reservoir characteristics from Alberta Plains, Canada. *International Journal of Coal Geology, 65, 93–113.*

Beaton, A. (2003a). *Coal-bearing formations and coalbed methane potential in the Alberta Plains and Foothills*. CSEG Recorder, 23–29. <http://www.cseg.ca/publications/recorder/2003/11nov/nov03-coal-bearing-formations.pdf> Accessed 15.02.13.

Beaton, A. (2003b). *Production potential of the coalbed methane resources in Alberta*. Alberta Geological Survey. EUB/AGS Earth Sciences Report 2003-03. http://www.ags.gov.ab.ca/publications/ESR/PDF/ESR_2003_03.PDF Accessed 31.03.13.

Beckmann, S., Lueders, T., Kruger, M., Von Netzer, F., Engelen, B., & Cypionka, H. (2011). Acetogens and acetoclastic methanosarcinates govern methane formation in abandoned coalmines. *Applied Environmental Microbiology, 77*, 3749–3756.

Beckwith, C. W., & Baird, A. J. (2001). Effect of biogenic gas bubbles on water flow through poorly decomposed blanket peat. *Water Resources Research, 37*, 551–558.

Beeson, M. (2006). Can coordination of federal agencies with state and local agencies help make produced water "lemons" into lemonade. In *Information Series: Vol. 102. Produce water workshops* (pp. 136–138). Colorado Water Resources Institute, Colorado State University.

Behar, F., Lewan, M. D., Lorant, F., & Vandenbroucke, M. (2003). Comparison of artificial maturation of lignite in hydrous and nonhydrous conditions. *Organic Geochemistry, 34*, 575–600.

Benson, S. M., & Cole, D. R. (2008). CO_2 sequestration in deep sedimentary formations. *Elements, 4*, 325–331.

Bern, C. R., Breit, G. N., Healy, R. W., & Zupaancic, J. W. (2011). *Deep subsurface drip irrigation using coal-bed sodic water: part II. geochemistry*. Agricultural Water Management. <http://www.sciencedirect.com/science/article/pii/S037837741200306X> Accessed 29.12.12.

Bertrand, P. (1984). Geochemical and petrographic characterization of humic coals considered as possible oil source rocks. In P. A. Schenck, J. W. de Leeuw, & G. W. M. Lijmbach (Eds.), *Organic Geochemistry: Vol. 6. Advances in Organic Geochemistry 1983* (pp. 481–488).

Bertard, C., Bruyet, B., & Gunther, J. (1970). Determination of desorbable gas concentration of coal (direct method). *International Journal of Rock Mechanics and Mining Science, 7*, 43–65.

Bestougeff, M. A. (1980). Summary of world coal resources and reserves. In *Twenty Six International Geological Congress, Paris, Colloquium, C-2*. (Vol. 35; pp. 353–366).

BGR. (2010). *Reserves, resources and availability of energy resources, 2010 annual report*. Federal Institute for Geosciences and Natural Resources (BGR), 85.

Bhanja, A. K., & Srivastava, O. P. (2008). *A new approach to estimate CBM gas content from well logs*. Society of Petroleum Engineers. SPE Asia Pacific Oil and Gas Conference and Exhibition, Perth, Australia, October 20–22, Paper No. SPE 115563.

Bhattacharya, D. (2012). An overview of domestic coal resources. *Geological Survey of India*, Accessed ppt on August 30, 2013 at http://www.google.com/search?client=safari&rls=en&q=Geological+Survey+of+India.+Thickness+of+coal+in+Jharia+Coalfield,+India&ie=UTF-8&oe=UTF-8.

Bilgesu, H. I., & Ali, M. W. (2005). *Well-completion strategies for methane gas production from coal seams using stimulation*. Society of Petroleum Engineers. Eastern Regional Meeting, Document ID 98017–6, SBN 978–1–55563–997–6.

Bird, M. I., Fifield, L. K., Chua, S., & Goh, B. (2004). Calculating sediment compaction for radiocarbon dating intertidal sediments. *Radiocarbon, 46*, 421–435.

Bishop, M. G. (2000). *Petroleum systems of the Gippsland Basin, Australia*. U.S. Geological Survey, Open-File Report 99-50-Q, p. 36.

Black, P. M. (1989). Petrographic and coalification variations in the Eastern Southland lignites, New Zealand. In P. C. Lyons, & B. Alpern (Eds.), *Coal: Classification, coalification, mineralogy, trace-element chemistry, and oil and gas potential. International Journal of Coal Geology, 13*, 127–141.

Blasting, T. J. (2013). *Recent greenhouse gas concentrations*. U.S. Department of Energy, Office of Science. <http://dx.doi.org/10.3334/CDIAC/atg.032>. <http://cdiac.ornl.gov/pns/current_ghg.html> Accessed 26.05.13.

Blaut, M. (1994). Metabolism of methanogens. *Antonie Van Leeuwenhoek, 66*, 187–208.

BLM. (2003a). *Environmental consequences*. Bureau of Land Management, PRB Final EIS/Proposed RMP Amendment (Chapter 4). http://www.blm.gov/wy/st/en/info/NEPA/documents/bfo/prb_eis.html Accessed on 05.04.13.

BLM. (2003b). *Final environmental impact statement and proposed plan amendment for the Powder River Basin oil and gas project*. Wyoming Bureau of Land Management. U.S. Department of Interior, WY–070–065, 1–4. <http://www.blm.gov/pgdata/etc/medialib/blm/wy/information/NEPA/prb-feis/vol_1.Par.96397.File.dat/V_1_cvr.pdf> Accessed 11.02.13.

BLM. (2008). *Powder River Basin coal review: Cumulative impact assessment. U.S. Bureau of land management*. Task 3B Report. <http://www.blm.gov/wy/st/en/programs/energy/Coal_Resources/PRB_Coal/prbdocs.html> Accessed 20.01.13.

BLM. (2012a). *Powder River Basin coal review*. Casper, Wyoming: Bureau of Land Management. <http://www.blm.gov/wy/st/en/programs/energy/Coal_Resources/PRB_Coal/prbdocs.html> Accessed 16.12.12.

BLM. (2012b). *Split estate mineral ownership*. United States Department of Interior, Bureau of Land Management. <http://www.blm.gov/wy/st/en/programs/mineral_resources/split-estate.html> Accessed 06.03.13.

Blodau, C., Diems, M., & Beer, J. (2011). Experimental burial inhibits methanogenesis and anaerobic decomposition in water-saturated peats. *Environmental Science Technology, 45*, 9984–9989.

Blum, M. D., & Roberts, H. H. (2009). Drowning of the Mississippi Delta due to insufficient sediment supply and global sea-level rise. *Nature Geoscience, 2*, 488–491.

Bodziony, J., & Lama, R. D. (1996). *Sudden outbursts of gas and coal in underground coalmines*. Australian Coal Association Research Program Project No. C4034, March 3, p. 677.

Bohacs, K., & Suter, J. (1997). Sequence stratigraphic distribution of coaly rocks: fundamental controls and paralic examples. *American Association of Petroleum Geologists Bulletin, 81*, 1612–1639.

Bohor, B. F., & Triplehorn, D. M. (1993). *Tonsteins: Altered volcanic-ash layers in coal-bearing sequences*. Geological Society of America. Special Paper 285, p. 44.

Boivin-Jahns, V., Ruimy, R., Bianchi, A., Daumas, S., & Christin, R. (1996). Bacterial diversity in a deep-subsurface clay environment. *Applied and Environmental Microbiology, 62*, 3405–3412.

Bolger, P. F. (1991). Lithofacies variations as a consequence of late Cainozoic tectonic and palaeoclimatic events in the onshore Gippsland Basin. In M. A. J. Williams, P. DeDeckker, & A. P. Kershaw (Eds.), *The Cainozoic in Australia: A re-appraisal of the evidence* (Vol. 18; pp. 158–180). Geological Society of Australia. Special Publication.

Bonner, T. H., & Wilde, G. R. (2002). Effects of turbidity on prey consumption by prairie stream fishes. *Transactions of the American Fisheries Society, 131*, 1203–1208.

Boral Energy. (2011). Yoalla 2 well proposal T/RL1. NVARW/Y0LLA2PR0P0SAL DOC. 10. p. 17. http://www.mrt.tas.gov.au/mrtdoc/petxplor/download/OR_0439A/OR_439A.pdf Accessed 19.12.11.

Boreck, D. L., & Weaver, J. N. (1984). *Coalbed methane study of the "Anderson" coal deposit, Johnson County, Wyoming — a preliminary report*. U.S. Geological Survey Open-File Report 84—831.

Boreham, C. J., & Powell, T. G. (1993). Petroleum source rock potential of coal and associated sediments: qualitative and quantitative aspects. In B. E. Law, & D. D. Rice (Eds.), *Studies in Geology: Vol. 38. Hydrocarbons from Coal* (pp. 133—157). American Association of Petroleum Geologists.

Boreham, C. J., Blevin, J. E., Radlinski, A. P., & Trigg, K. R. (2003). Coal as a source of oil and gas: a case study from the Bass Basin, Australia. *APPEA Journal, 44*, 117—148. < http://www.ret.gov.au/resources/Documents/acreage_releases/2006/CDcontents/PDF/references/Boreham_2003_Coal_source_Bass.pdf> Accessed 10.12.11.

Boreham, C., Draper, J., & Hope, J. (2004). *Origin of Jurassic coal seam gas, SE Queensland*, 21st Annual Meeting of The Society for Organic Petrologists, Sydney, NSW, Australia, 26 September—1 October, 33—35.

Boreham, C. J., Horsfield, B., & Schenk, H. J. (1999). Predicting the quantities of oil and gas generated from Australian Permian coals, Bowen Basin using pyrolytic methods. *Marine and Petroleum Geology, 16*, 165—188.

Borrego, A. G., Araujo, C. V., Balke, A., Cardott, B., Cook, A. C., David, P., et al. (2006). Influence of particle and surface quality on the vitrinite reflectance of dispersed organic matter. Comparative exercise using data from the qualifying system for reflectance analysis working group of ICCP. *International Journal of Coal Geology, 68*, 151—170.

Bossie-Codreanu, D., Wolf, K.-H. A. A., & Ephraim, R. (2004). A new characterization method for coal bed methane. *Geologica Belgica, 7*, 137—145. <Dan.Bossie-Codreanu@ifp.fr>.

Bostick, N. H. (2011). Measured reflectance suppressed by thin-film interference of crude oil smeared on glass—as on vitrinite in coal or petroliferous rocks: the society of organic petrologists. *Newsletter, 28*, 12—15.

Bostick, N. H., Betterhorn, W. J., Gluskoter, H. J., & Islam, M. N. (1991). Petrology of Permian "Gondwana" coals from boreholes in northwestern Bangladesh, based on semiautomated reflectance scanning. *Organic Geochemistry, 17*, 399—413.

Bouska, V. (1981). *Geochemistry of coal*. Prague: Academia.

Bowler, J., & Tedesco, S. A. (2004). *Bulk density logs estimate coalbed methane adsorbed, desorbed and producible gas volumes*, 2004 Rocky Mountains Section American Association of Petroleum Geologists Meeting.

Boyd, C. L. (1991). *Measurement techniques of coalbed methane gas*. Coalbed Methane Symposium, University of Alabama, Tuscaloosa, Alabama, May, 1991, 34—352.

Boyer, C. M., & Qingzhao, B. (1998). Methodology of coalbed methane resource assessment. *International Journal Coal Geology, 35*, 349—368.

Boyer, C. M., & Reeves, S. R. (1989). Strategy of coalbed methane production development part III: Production operations. In *Proceeding of the 1989 coalbed methane symposium*. Tuscaloosa: The University of Alabama. April 17—20, 1989, pp. 19—24.

BP. (2010). *BP statistical review of world energy, June 2011*. London, England: British Petroleum Company. <http://www.google.com/search?client=safari&rls=en&q=BP+2010+statistical+review&ie=UTF-8&oe=UTF-8> Accessed 19.09.11.

BP. (2011). *BP statistical review of world energy, June 2010*. London, England: British Petroleum Company. < http://www.bp.com/liveassets/bp_internet/globalbp/globalbp_uk_english/reports_and_publications/statistical_energy_review_2008/STAGING/local_assets/2010_downloads/statistical_review_of_world_energy_full_report_2010.pdf> Accessed 19.09.11.

BP. (2012). *BP statistical review of world energy, June, 2012*. London, England: British Petroleum Company. <http://www.bp.com/assets/bp_internet/globalbp/globalbp_uk_english/reports_and_publications/statistical_energy_review_2011/STAGING/local_assets/pdf/statistical_review_of_world_energy_full_report_2012.pdf> Accessed 01.05.13.

Bradfield, J. J. C. (1943). Oil from coal. *Geographical Journal*, 47—52. Queensland.

Bradsher, K. (2009). *China outpaces U.S. in cleaner coal-fired plants*. Asia Pacific: New York Times. <http://www.nytimes.com/2009/05/11/world/asia/11coal.html> Accessed 07.04.13.

Brahic, C. (October 9, 2008). Goldmine bug DNA may be key to alien life. *New Scientist*.

Brantly, J. E. (1971). *History of oil well drilling*. p. 153.

Braun, C. E., Oedekoven, O. O., & Aldridge, C. L. (2002). Oil and gas development in western North America: effects on sagebrush steppe avifauna with particular emphasis on sage grouse. *Transaction of the North American Wildlife and Natural Resources Conference, 67*, 337—349.

Bremer, K. (2000). Early Cretaceous lineages of monocot flowering plants. *The National Academy of Sciences, 97*, 4707—4711.

Briden, J. C., & Irving, E. (1964). Palaeolatitude spectra of sedimentary palaeoclimatic indicators. In A. E. M. Nairn (Ed.), *Problems in Palaeoclimatology* (pp. 199—201). New York: Interscience.

Bridgham, S. D., Pastor, J., Updegraff, K., Malterer, T. J., Johnson, K., Harth, C., et al. (1999). Ecosystem control over temperature and energy flux in northern peatlands. *Ecological Applications, 9*, 1345—1358.

Brinck, E. L., Drever, J. I., & Frost, C. D. (2008). The geochemical evolution of water co-produced with coal bed natural gas in the Powder River Basin, Wyoming. *Environmental Geosciences, 15*, 153–171.

British Columbia Ministry of Energy and Mines. (2011). *Coalbed methane in British Columbia, British Columbia Ministry of Energy and Mines.* <http://www.em.gov.bc.ca/Mining/Geoscience/Coal/CoalBC/CBM/Pages/CBMBrochure.aspx> Accessed 01.10.12.

British Geological Survey. (2012). *European mineral statistics 2006–10: A product of the World Mineral Statistics database.* Keyworth, England: British Geological Survey.

Bromilow, J. G., & Jones, J. M. (1955). Drainage and utilization of firedamp. *Colliery Engineering, 32*, 222–232.

Brooks, J. D., & Smith, J. W. (1967). The diagenesis of plant lipids during the formation of coal, petroleum and natural gas. I. changes in the n-paraffin hydrocarbons. *Geochimica et Cosmochimica Acta, 31*, 2389–2397.

Brown, D. A. (1998). Gas production from an ombrotrophic bog: effect of climate change on microbial ecology. *Climate Change, 40*, 277–284.

Brown, D. A., & Overend, R. P. (1993). Methane metabolism in raised bogs of northern wetlands. *Geomicrobiology Journal, 11*, 35–48.

Brownfield, M. E. (2002). Characterization and Modes of Occurrence of Elements in Feed Coal and Fly Ash—An Integrated Approach, U.S. Geological Survey Fact Sheet FS-038-02, p. 4.

Brownfield, M. E., Cathcart, J. D., Affolter, R. H., Brownfield, I. K., Rice, C. A., O'Connor, J. T., et al. (2004). *Feed coal and coal combustion products from a power plant utilizing low-sulfur coal from the Powder River Basin.* Wyoming: U.S. Geological Survey Scientific Investigations. Report 2004–5271, p. 36.

Brunauer, S., Emmett, P. H., & Teller, E. (1938). Adsorption of gases in multimolecular layers. *Journal of the American Chemical Society, 60*, 309–319.

Bryner, G. (2002). *Coalbed methane development in the intermountain west: Primer.* Boulder: Natural Resource Law Center, University of Colorado. <www.colorado.edu/Law/centers/nrlc/CBM_Primer.pdf> Accessed 23.01.13.

Buillit, N., Lallier-Verges, E., Pradier, B., & Nicolas, G. (2002). Coal petrographic genetic units in deltaic-plain deposits of the Campanian Mesa Verde Group (New Mexico, USA). *International Journal of Coal Geology, 51*, 93–110.

Bullis, K. (2012). *Coal-eating microbes might create vast amounts of natural gas.* MIT Technology Review, Energy News. <http://www.technologyreview.com/news/429682/coal-eating-microbes-might-create-vast-amounts-of-natural-gas/> Accessed 10.04.13.

Bunny, M. (2001). *Biogenic gas discoveries in New South Wales.* Petroleum Exploration Society of Australia. Exploration NSW Supplement.

Burchell, T., & Rogers, M. (2002). *Low pressure storage of natural gas for vehicular application.* Society of Automotive Engineers, Inc., 2000–01-2205, p. 5. <http://www.ornl.gov/~webworks/cpr/pres/107683.pdf> Accessed 31.10.12.

Burnham, A. K., & Sweeney, J. J. (1989). A chemical model of vitrinite maturation and reflectance. *Geochimica et Cosmochimica Acta, 53*, 2649–2657.

Buros, O. K. (1990). *The desalting ABC's.* Riyadh, Saudi Arabia: Saline Water Conversion Corp. International Desalination Association, modified and reproduced by Research Dept.

Busch, A., & Gensterblum, Y. (2011). CBM and CO_2-ECBM related sorption processes in coal: a review. *International Journal of Coal Geology, 87*, 49–71.

Bushong, S. (2006). Who owns the rights to treated produced water. In R. Wickramasignhe (Ed.), *Produced water workshop* (pp. 56–66). Colorado Water Resources Institute, Colorado State University. Information Series No. 102. <http://www.cwi.colostate.edu/publications/is/102.pdf> Accessed 28.11.12.

Bustin, R. M. (2004). *Acid gas sorption by British Columbia coals: Implications for permanent disposal of acid gas in deep coal seams and possible co-production of methane.* Vancouver: University of British Columbia. Final report funding agreement 2000–16: p. 126.

Bustin, A. M. M., & Bustin, R. M. (2008). Coal reservoir saturation: impact of temperature and pressure. *American Association of Petroleum Geologists Bulletin, 92*, 77–86.

Bustin, R. M., Cameron, A. R., Grieve, D. A., & Kalkreuth, W. D. (1983). Coal petrology, its principles, methods, and applications. *Geological Association of Canada.* Short Course Notes, May 1983.

Bustin, R. M., & Clarkson, C. R. (1998). Geological controls on coalbed methane reservoir capacity and gas content. *International Journal of Coal Geology, 38*, 3–26.

Bustin, R. M., Cui, X., & Chikatamarla, L. (2008). Impacts of volumetric strain on CO_2 sequestration in coals and enhanced CH_4 recovery. *American Association of Petroleum Geologists Bulletin, 92*, 15–29.

Bustin, R. M., & England, T. D. J. (1989). Timing of organic maturation (coalification) relative to thrust faulting in the southeastern Canadian Cordillera. *International Journal of Coal Geology, 13*, 327–339.

Bustin, R. M., Ross, J. V., & Rouzaud, J. N. (1995). Mechanisms of graphite formation from kerogen: experimental evidence. *International Journal of Coal Geology, 28*, 1–36.

Butala, S. J., Medina, J. C., Taylor, T. Q., Bartholomew, C. H., & Lee, M. L. (2000). Mechanisms and kinetics of reactions leading to natural gas formation during coal maturation. *Energy & Fuels, 14*, 235–259.

Butland, C. I., & Moore, T. A. (2008). Secondary biogenic coal seam gas reservoirs in New Zealand: a preliminary assessment of gas contents. In R. M. Flores (Ed.), *International Journal of Coal Geology: Vol. 76. Microbes, methanogenesis, and microbial gas in coal* (pp. 151–165).

Butler, J. M. (2011). *The NOAA greenhouse gas index (AGGI)*. Boulder, Colorado: National Oceanic Atmosphere Administration, Earth System Research Laboratory. <http://www.esrl.noaa.gov/gmd/aggi/> Accessed 23.09.13.

Buttler, A. J., Dinel, H., Le vesque, M., & Mathur, S. P. (1991). The relation between movement of subsurface water and gaseous methane in a basin bog with a novel instrument. *Canadian Journal of Soil Science, 71*, 427–438.

Byrne, J. F., & Marsh, H. (1995). Introductory overview. In J. W. Patrick (Ed.), *Porosity in carbons, characterization, and applications* (p. 331). New York: Halsted Press.

Cabrera, L. I., & Saez, A. (1987). Coal deposition in carbonate-rich shallow lacustrine systems: the Calaf and Mequinenza sequences (Oligocene, eastern Ebro Basin, NE Spain). *Journal of the Geological Society, 144*, 451–461.

Cady, G. H. (1942). Modern concepts of physical constituents of coal. *Journal of Geology, 50*, 337–356.

Cameron, C. C. (1973). Peat. In D. A. Brobst, & W. P. Pratt (Eds.), *U.S. Geological Survey Professional Paper: Vol. 820. United States mineral resources* (pp. 505–513). Available at. <http://pubs.er.usgs.gov/usgspubs/pp/pp.820>.

Cameron, C. C. (1975). *Some peat deposits in Washington and southeastern Aroostook Counties, Maine*. U.S. Geological Survey Bulletin 1317–C, pp. 40, for more information see. <http://pubs.er.usgs.gov/usgspubs/b/b1317C>.

Cameron, A. R. (1978). Megascopic description of coal with particular reference to seams in southern Illinois. In R. R. Dutcher (Ed.), *Field description of coal* (pp. 9–32). American Society for Testing Materials, ASTM STP 661.

Cameron, C. C., Esterle, J. S., & Palmer, C. A. (1989). The geology, botany and chemistry of selected peat-forming environments from temperate and tropical latitudes. In P. C. Lyons, & B. Alpern (Eds.), *International Journal of Coal Geology: Vol. 12. Peat and coal: Origin, facies, and depositional models* (pp. 105–156).

Canadian Gas Potential Committee. (2005). *Natural potential in Canada—2005*. <http://www.centreforenergy.com/documents/545.pdf> Accessed 15.02.13.

Cantrill, T. C. (2012). *Coal mining*. Cambridge University Press.

Cao, Y., Davis, A., Liu, R., Liu, X., & Zhang, Y. (2003). The influence of tectonic deformation on some geochemical properties of coals—a possible indicator of outburst potential. *International Journal of Coal Geology, 53*, 69–79.

Cao, D., Li, X., & Deng, J. (2009). Coupling effect between coalification and tectonic-thermal events– geological records of geodynamics of sedimentary basin. *Earth Science Frontiers, 16*, 52–60.

Cao, D., Li, X., & Zhang, S. R. (2007). Influence of tectonic stress on coalification: stress degradation mechanism and stress polycondensation mechanism. *Science in China Series D, Earth Sciences, 50*, 41–54.

CAPP. (2006). *Best management practices for natural gas in coal (NGC)/coalbed methane (CBM)*. Canadian Association of Petroleum Producers, 2005-0040, Calgary, Alberta, p. 79. < http://www.capp.ca/getdoc.aspx?DocID=103407> Accessed 24.10.12.

Cardott, B. J. (2001). *Oklahoma coal bed methane workshop*. Oklahoma Geological Survey. Open-File Report 2–2001.

Cardott, B. J. (2002). Coalbed methane development in Oklahoma. In S. D. Schwochow, & V. F. Nuccio (Eds.), *Coalbed methane of North America, II* (pp. 83–98). Rocky Mountain Association of Geologists.

Cardot, B. J. (March 21, 2006). *Oklahoma coalbed-methane activity, 2006 update. OGC conference on coalbed methane and gas shales in the Midcontinent, Oklahoma city*. <http://www.ogs.ou.edu/fossilfuels/coalpdfs/Activity2006%20Revised.pdf> Accessed 12.11.12.

Carlson, F. M., Hartman, R. C., Obluda, J. C., & Miller, M. T. (2008). *Field validation of new methodology for resource assessment of undersaturated CBNG reservoirs*. International Coalbed & Shale Gas Symposium, University of Alabama, Tuscaloosa, Alabama, Paper 819, unpaginated.

Carlton, D. R. (2006). Discovery and development of a giant coalbed methane resource, Raton Basin, Las Animas County, Southeast Colorado. In S. A. Sonnenberg, & E. D. Dolly (Eds.), *The Mountain Geologist: Vol. 43. Rocky Mountain gas reservoirs revisited* (pp. 231–236).

Carothers, W. W., & Kharaka, Y. K. (1980). Stable carbon isotopes of HCO_3 in oil-field waters: Implications for the origin of CO_2. *Geochemica et Cosmochemica Acta, 44*, 323–332.

Carr, A. D. (2000). Suppression and retardation of vitrinite reflectance, part 1. Formation and significance for hydrocarbon generation. *Journal of Petroleum Geology, 23*, 313–343.

Carr, A. D. (2003). Thermal history model for the south central graben, North Sea, derived using both tectonics and maturation. *International Journal of Coal Geology, 54*, 3–19.

Carroll, R. E., & Pashin, J. C. (2003). Relationship of sorption capacity to coal quality: CO_2 sequestration potential of coalbed methane reservoirs in the Black Warrior Basin. *Geological Survey of Alabama*, 11.

Cartz, L., Diamond, R., & Hirsch, P. B. (1956). New X-ray data on coals. *Nature, 177*, 500.

Casshyap, S. M., & Tewari, R. C. (1984). Fluvial models of the Lower Permian coal measures of Son-Mahanadi and Koel-Damodar Valley basins, India. In R. A. Rahmani, & R. M. Flores (Eds.), *International Association of Sedimentologists, Special Publication: Vol. 7. Sedimentology of coal and coal-bearing sequences* (pp. 121–147). Blackwell Scientific Publications.

Castellan, G. W. (1983). *Physical chemistry* (3rd ed.). Addison-Wesley.

Cattaneo, L. (2013). The enduring technology of coal. *MIT Technology Review, 116*, 15.

Catuneanu, O. (2006). *Principles of sequence stratigraphy* (1st ed.). Amsterdam: Elsevier. p. 375.

Cazzulo-Klepzig, M., Guerra-Sommer, M., Menegat, R., Simas, M. W., & Filho, J. G. M. (2007). Peat-forming environment of Permian coal seams from the Faxinal coalfield (Parana Basin) in southern Brazil, based on palynology and palaeobotany. *Revista Brasileira de Paleontologia, 10*, 117–127.

CDX Gas. (2005). *Unconventional plays: Enhancing performance with new technologies*. Summer NAPE Expo 2005, Doug White.

Cecil, C. B., Delong, F. T., Cobb, J. C., & Supardi. (1993). Allogenic and autogenic controls on sedimentation in the central Sumatra basin as an analogue for Pennsylvanian coal-bearing strata in the Appalachian basin. In J. C. Cobb, & C. B. Cecil (Eds.), *Geological Society of America Special Paper: Vol. 286. Modern and ancient coal-forming environments* (pp. 3–22).

Cedigaz. (2010). *Indonesia: CBM could account for 15% of gas supply*. U-Gas News Report, No. 50 (October 2010).

Cedigaz. (2011). *Indonesia: First coalbed methane to flow in 2011*. U-Gas News Report, No. 45 (May 2010).

Ceglarsk-Stefańska, G., & Zarebska, K. (2002). The competitive sorption of CO_2 and CH_4 with regard to the release of methane from coal. *Fuel Processing Technology, 77–78*, 423–429.

Chalmers, G. R. L., & Bustin, R. M. (2007). On the effect of petrographic composition on coalbed methane sorption. *International Journal of Coal Geology, 69*, 288–304.

Chan, P. (2011). Unconventional resources estimation: Introduction. In *Guidelines for application of the petroleum resources management system* (pp. 141–153). Sponsored by Society of Petroleum Engineers, American Association of Petroleum Geologists, World Petroleum Council, Society of Petroleum Evaluation Engineers, and Society of Exploration Geophysicists. <http://www.spe.org/industry/docs/PRMS_Guidelines_Nov2011.pdf> Accessed 16.06.12.

Chao, E. C. T., Minkin, J. A., & Thompson, C. L. (1982). Application of automated image analysis to coal petrography. *International Journal of Coal Geology, 2*, 113–150.

Chappellaz, J. A., Fung, I. Y., & Thompson, A. M. (1993). The atmospheric CH_4 increase since the Last Glacial Minimum: (1). Source estimates. *TELUS, 45B*, 228–241.

Charpentier, R. R., & Ahlbrandt, T. S. (2003). Petroleum (oil and gas) geology and resources. In B. De Vivo, B. Grasemann, & K. Stüwe (Eds.), *Developed under the Auspices of the UNESCO: Vol. 5. Geology, encyclopedia of life support systems (EOLSS)* (pp. 31–53). Oxford, UK: EOLSS Publishers. < http://www.eolss.net>. Accessed 19.10.11.

Charpentier, R. R., & Klett, T. R. (2000). *Monte Carlo simulation method*. U.S. Geological Survey World Petroleum Assessment 2000–Description and Results. In *U.S. Geological Survey Digital Data Series* (Vol. 60). Chapter MC, MC1–MC13.

Chase, R. W. (1979). A comparison of methods used for determining the natural gas content of coalbeds from exploratory cores. *U.S. Department of Energy*. METC/CR-79/18, p. 25.

Chen, G., & Harpalani, S. (1995). Study of fracture network in coal. In L. R. Myer, N. G. W. Cook, R.,E. Goodman, & C. F. Tsang (Eds.), *Fractured and jointed rock masses* (pp. 431–434). Rotterdam: Balkema.

Chen, X., Barron, K., & Chan, D. (1995). A fee factors influencing outbursts in underground coalmines. In R. D. Lama (Ed.), *"Management and control of high gas emissions and outbursts in underground coal mines"* (pp. 39–47). Kiama, North South Wales, Australia: Westonprint.

Chen, Y., Yang, Y., & Luo, J. (2012). Uncertainty analysis of coalbed methane economic assessment with Monte Carlo method. *Procedia Environmental Sciences, 12*, 640–645.

Cheung, K., Sanei, H., Klassen, P., Mayer, B., & Goodarzi, F. (2009). Produced fluids and shallow groundwater in coalbed methane (CBM) producing regions of Alberta, Canada: trace element and rare earth element geochemistry. *International Journal of Coal Geology, 77*, 338–349.

Chizmeshya, A. V. G., McKelvy, M. J., Marzke, R., Ito, N., & Wolf, G. (2007). Investigating geological sequestration reaction processes under in situ process conditions. In B. Sakkestad (Ed.), *32nd International Technical Conference on Coal Utilization and Fuel Systems: Proceedings, Coal Utilization and Fuel Systems* (pp. 429–440) (CD-ROM).

Choate, R., Johnson, C. A., & McCord, J. P. (1984). Geologic overview, coal deposits, and potential for methane recovery from coal beds – Powder River Basin. In C. T. Rightmire, G. E. Eddy, & J. N. Kirr (Eds.), *American Association of Petroleum Geologists Studies in Geology Series: Vol. 17. Coalbed methane resources of the United States* (pp. 335–351).

Chow, T. L., Rees, H. W., Ghanem, I., & Cormier, R. (1992). Compactibility of cultivated *Sphagnum* peat material and its influence on hydrologic characteristics. *Soil Science, 153*, 300–306.

Chung, Y. H., Yang, S. Y., & Kim, J. W. (2011). *Numerical simulation of deep biogenic gas play northeastern Bay of Bengal, northwestern Myanmar*. American Association of Petroleum Geologists International Conference and Exhibition, Milan, Italy, October 25–26, 2011.

Cienfuegos, O., & Laredo, J. (2010). Coalbed methane resource assessment in Asturias (Spain). *International Journal of Coal Geology, 83*, 366–376.

Clark, L. (2008). *Gas riches from our coal*. Contrafed Publishing. Energy Ez No. 7, Summer 2008. <http://www.contrafedpublishing.co.nz/Energy+NZ/Issue+7+Summer+2008/Gas+riches+from+our+coal.html>.

Clarkson, C. R., & Barker, G. J. (2011). Coalbed methane. In *Guidelines for application of the petroleum resources management system* (pp. 141–154). Sponsored by Society of Petroleum Engineers, American Association of Petroleum Geologists, World Petroleum Council, Society of Petroleum Evaluation Engineers, and Society of Exploration Geophysicists. <http://www.spe.org/industry/docs/PRMS_Guidelines_Nov2011.pdf> Accessed 16.06.12.

Clarkson, C. R., & Bustin, R. M. (1996). Variation in micropore capacity and size distribution with composition in bituminous coal of the Western Canadian Sedimentary Basin. *Fuel, 75*, 1483–1498.

Clarkson, C. R., & Bustin, R. M. (1997). Variation in permeability with lithotype and maceral composition of Cretaceous coals of the Canadian Cordillera. *International Journal of Coal Geology, 33*, 135–151.

Clarkson, C. R., Bustin, R. M., & Levy, H. (1997). Application of the mono/multilayer and adsorption potential theories to coal methane adsorption isotherms at elevated temperature and pressure. *Carbon, 35*, 1689–1705.

Clarkson, C. R., Jordan, C. L., Gierhart, R. R., & Seidle, J. P. (2008). *Production data analysis of coalbed-methane wells*. In *SPE reservoir evaluation & engineering* (Vol. 11). Society of Petroleum Engineers. p. 15.

Clarkson, C. R., & McGovern, J. M. (2001). *Study of the Potential Impact of Matrix Free Gas Storage Upon Coalbed Gas Reserves and Production Using a New Material Balance Equation*. Paper 0113 presented at the International Coalbed Methane Symposium, (pp. 14–18) May. Tuscaloosa, Alabama.

Claypool, G. E., & Kaplan, I. R. (1974). The origin and distribution of methane in marine sediments. In *Natural gases in marine sediments* (pp. 99–139). New York: Plenum Press.

Clayton, J. L. (1993). Composition of crude oils generated from coals and coaly organic matter in shales. In B. E. Law, & D. D. Rice (Eds.), *Studies in Geology: Vol. 38. Hydrocarbons from coal* (pp. 185–201). American Association of Petroleum Geologists.

Clayton, J. L. (1998). Geochemistry of coalbed gas—a review. In R. M. Flores (Ed.), *International Journal of Coal Geology: Vol. 35. Coalbed methane: From coal-mine outbursts to a gas resource* (pp. 159–173).

Clearwater, S. J., Morris, B. A., Meyer, J. S., & J.S. (2002). *A comparison of coalbed methane product water quality versus surface water quality in the Powder River Basin of Wyoming, and an assessment of the use of standard aquatic toxicity testing organisms for evaluating the potential effects of coalbed methane product waters*. Laramie, Wyoming, USA: Department of Zoology and Physiology, University of Wyoming. Report to Wyoming Department of Environmental Quality, p. 131.

Clement, B. G., Ferry, J. G., & Underwood, S. (2012). *Methods to stimulate biogenic methane production from hydrocarbon-bearing formations*. United States Patent Application No. 20120138290. <http://www.faqs.org/patents/app/20120138290#b> Accessed 27.03.13.

Cline, J. (2006). Opportunities and liabilities for produced water. In R. Wickramasighne (Ed.), *Produced water workshop* (pp. 36–41). Colorado Water Resources Institute, Colorado State University. Information Series No. 102. <http://www.cwi.colostate.edu/publications/is/102.pdf> Accessed 28.11.12.

Close, J. C. (1990). *Update on coalbed methane potential of Raton Basin, Colorado and New Mexico Society of Petroleum Engineers*. SPE Annual Technical Conference and Exhibition, September 23–26, 1990, New Orleans, Louisiana.

Close, J. C. (1993). Natural fractures in coal. In B. E. Law, & D. D. Rice (Eds.), *Studies in Geology: Vol. 38. Hydrocarbons from coal* (pp. 119–184). American Association of Petroleum Geologists.

Close, J. C., & Mavor, M. J. (1991). Influence of coal composition and rank on fracture development in Fruitland coal gas reservoirs of San Juan Basin. In S. Schwochow, D. K. Murray, & M. F. Fahy (Eds.), *Coalbed methane of western North America* (pp. 109–121). Rocky Mountain Association of Geologists, Field Conference.

Clymo, R. S. (1978). A model of peat bog growth. In O. W. Heal, & D. F. Perkins (Eds.), *Production ecology of British moors and Montane Grasslands* (pp. 183–223). Berlin, Heidelberg, New York: Springer.

Clymo, R. S. (1983). Peat. InA. J. P. Gore (Ed.), *Ecosystems of the world, Mires: Swamp, bog, fen, and moor* (Vol. 4A; pp. 159–224). Amsterdam: Elsevier.

Clymo, R. S. (1984). The limits to peat bog growth. *Philosophy Transaction Royal Society London,* 303B, 605–654.

Coal Authority. (2008). *Annual report and accounts 2007–2008*. United Kingdom: The Stationary Office Limited. The Coal Authority, London.

Coalfields Geology Council of New South Wales and the Queensland Mining Council. (2001). *Guidelines for the estimation and reporting of Australian black coal resources and reserves* (as referred to in the Joint Ore Reserves Committee Code {'The JORC Code'} 1999 edition), p. 9. <http://www.jorc.org/pdf/coalguidelines2001.pdf> Accessed 23.07.12.

Coates, D. A., & Heffern, E. L. (1999). Origin and geomorphology of clinker in the Powder River Basin,

Wyoming and Montana. In W. R. Miller (Ed.), *Coalbed methane and the Tertiary geology of the Powder River Basin, Wyoming and Montana* (pp. 211–229). Wyoming Geological Association. Fiftieth Field Conference Guidebook.

Coates, D. A., Stricker, G. D., & Landis, E. R. (1990). Coal geology, coal quality, and coal resources in Permian rocks of the Beacon Supergroup, Transarctic Mountains, Antarctica. In J. F. Splettstoesser, & G. A. M. Dreschoff (Eds.), *Antarctic Research Series: Vol. 51. Mineral resources potential of Antarctica* (pp. 133–162).

Cobb, J. C., & Cecil, C. B. (Eds.), (1993). *Geological Society of America Special Paper: 286. Modern and ancient coal-forming environments* (p. 193).

Cohen, A. D., & Bailey, A. M. (1997). Petrographic changes induced by artificial coalification of peat: comparison of two planar facies (*Rhizophora* and *Cladium*) from the Everglades-mangrove complex of Florida and a domed facies (*Cyrilla*) from the Okefenokee Swamp of Georgia. *International Journal of Coal Geology, 34*, 163–194.

Cohen, H. A., Grigsby, F. B., Bessinger, McAteer, J. J., Jr., & Baldassare, F. (2012). *Review of U.S. EPA's December 2011 draft report "investigation of ground water contamination near Pavilion, Wyoming"*. Prepared for the Independent Petroleum Association of America, Washington, D.C. S.S. Papadopulos & Associates, Inc. p. 30. <http://www.velaw.com/UploadedFiles/VEsite/E-comms/Pavillion Report2012.pdf> Accessed 12.10.12.

Cohen, A. D., Raymond, R., Jr., Ramirez, R., Morales, Z., & Ponce, F. (1989). The Changuinola peat deposit of northwestern Panama: a tropical, back-barrier, peat (coal)-forming environment. In P. C. Lyons, & B. Alpern (Eds.), *International Journal of Coal Geology: Vol. 12. Peat and coal: Origin, facies, and depositional models* (pp. 157–192).

Cole, G. A. (1994). *Textbook of limnology* (4th ed.). Waveland Press.

Cole, J. M., & Crittenden, S. (1997). Early Tertiary basin formation and the development of lacustrine and quasi-lacustrine/marine source rocks on the Sunda Shelf of SE Asia. In A. J. Fraser, S. J. Matthews, & R. W. Murphy (Eds.), *Geological Society Special Publication: Vol. 126. Petroleum geology of southeast Asia* (pp. 147–183).

Colmenares, L., & Zoback, M. (2007). Hydraulic fracturing and wellbore completion of coalbed methane wells in the Powder River Basin, Wyoming: implications for water and gas production. *American Association of Petroleum Geologists Bulletin, 91*, 51–67.

Condie, K. C. (1997). *Plate tectonics and crustal evolution* (4th ed.). Oxford, UK: Butterworth-Heinemann.

Condie, K. C. (2011). *Earth as an evolving planetary system* (2nd ed.). Elsevier.

Conrad, J. M., Miller, M. J., Phillips, J., & Ripepi, N. (2006). *Characterization of central Appalachian Basin CBM Development: Potential for carbon sequestration and enhanced CBM recovery*. Proceedings. 2006 International Coalbed Methane Symposium, Preprint 0625. May 22–26, 2006, Tuscaloosa, AL. <http://www.energy.vt.edu/Publications/2006_CBM_Conrad.pdf> Accessed 10.02.13.

Conybeare, W. D., & Phillips, W. (1822). *Outlines of the geology of England and Wales: With an introductory compendium of the general principles of that science, and comparative views of the structure of foreign countries*. George Yard, Lombard Street, London: William Phillips. p. 470.

Cook, T. (2004). Calculation of estimated ultimate recovery (EUR) for wells in assessment units of continuous hydrocarbon accumulations. In, *U.S. Geological Survey Digital Data Series: DDS–69–C. Petroleum system and assessment of coalbed gas in the Powder River Basin province, Wyoming and Montana* (p. 6). Chapter 7.

Cook, R., & Gregg, R. (1997). Overview of New Zealand's petroleum systems and potential. *Oil & Gas Journal*, 55–58.

Cooley, H., & Donnelly, K. (2012). *Hydraulic fracturing and water resources: Separating the frack from the friction*. Pacific Institute, 654 13th Street, Preservation Park, Oakland, California 94612, p. 34. <http://www.pacinst.org/reports/fracking/full_report.pdf> Accessed 18.10.12.

Cooper, J. R. (2006). *Igneous intrusions and thermal evolution in the Raton Basin, CO-NM: Contact metamorphism and coal-bed methane generation*. Masters of Science Thesis. Columbia: University of Missouri. p. 249.

Cooper, J. R., Crelling, J. C., Rimmer, S. M., & Whittington, A. G. (2007). Coal metamorphism by igneous intrusion in the Raton Basin, CO and NM: implications for generation of volatiles. *International Journal of Coal Geology, 71*, 15–27.

Correa, G., Osorio, T. F., & Restrepo, D. P. (2009). *Unconventional gas reservoirs*. Colombia: Energética, Universidad Nacional de Colombia. Num 41, December–July 2009.

Correia, M. (1969). *Contribution a la recherche de zones favorable a la genese du petrole par l'observation microscopique de la matiere organique figuree: Revue de l'Institut Francais du Petrole, 24*, 1417–1454.

Cottingham, K. D., Gribb, N. W., Hays, R. J., & McLaughlin, J. F. (2012). *2009 coalbed natural gas regional groundwater monitoring report: Powder River Basin*. Wyoming: Wyoming State Geological Survey. Technical Memorandum 2012, p. 379. Accessed 04.09.12.

Couwenberg, J., & Fritz, C. (2012). Towards developing IPCC methane 'emission factors' for peatlands (organic soils). *Mires and Peat, 10*, 1–17.

Cowardin, L. M., Carter, V., Golet, F. C., & LaRoe, E. T. (1979). *Classification of wetlands and deepwater habitats of the United States*. Washington, D.C: U.S. Department of the Interior, Fish and Wildlife Service, Office of Biological Services. FWS/OBS-79/31.

Craig, H. (1961). Isotopic variations in meteoric waters. *Science, 133*, 1702–1703.

Creedy, D. P. (1983). The quantity and distribution of gas in coal seams. *Access*, 79–85.

Creedy, D. P. (1988). Geological controls on the formation and distribution of gas in British coal measure strata. *International Journal of Coal Geology, 10*, 1–31.

Creedy, D. P. (1994). Prospects for coalbed methane in Britain. In *Coalbed methane extraction, an analysis of UK and European Resources and Potential for Development*. Cavendish Conference Centre, London January 1994.

Crockett, F., Ellis, M., Stricker, G., Gunther, G., Ochs, A., & Flores, R. (2001). An estimate of recoverable coal gas resources in the Powder River Basin, Wyoming. In D. Stilwell (Ed.), *Fifty Second Field Conference, Wyoming Geological Association Guidebook* (pp. 183–189).

Croft, G. D., & Patzek, T. W. (2009). Potential for coal-to-liquids conversion in the United States-resource base. *Natural Resources Research, 18*, 173–179.

Crosdale, P. J. (1993). Coal maceral ratios as indicators of environment of deposition: do they work for ombrogenous mires? An example from the Miocene of New Zealand. *Organic Geochemistry, 20*, 797–809.

Crosdale, P. J. (1995). Lithotype sequences in the early Miocene Maryville coal measures, New Zealand. *International Journal of Coal Geology, 28*, 37–50.

Crosdale, P. J., & Beamish, B. (1995). Methane diffusivity at South Bulli (NSW) and Central (Old) collieries in relation to coal maceral composition. In R. D. Lama (Ed.), *Management and control of high gas emission and outburst in underground coal mines* (pp. 363–367). Proceedings of International Symposium Cum-Workshop, Wollongong. Kiama, NSW, Australia.

Crosdale, P. J., Beamish, B. B., & Valix, M. (1998). Methane sorption related to coal composition. *International Journal of Coal Geology, 35*, 147–158.

Crosdale, P. J., Moore, T. A., & Mares, T. E. (2008). Influence of moisture content and temperature on methane adsorption isotherm analysis for coals from a low-rank, biogenically-sourced gas reservoir. *International Journal of Coal Geology, 76*, 166–174.

Cross, A. T., & Phillips, T. L. (1990). Coal-forming plants through time in North America. In P. C. Lyons, T. G. Callcott, & B. Alpern (Eds.), *International Journal of Coal Geology: Vol. 16. Peat and coal—origin, facies, and coalification* (pp. 1–46).

Crowley, S. S., Ruppert, L. F., Stanton, R. W., & Belki, H. E. (1994). Geochemical studies of the Anderson-Dietz 1 coal bed, Powder River Basin: origin of inorganic elements and environmental implications. In R. M. Flores, K. T. Mehring, R. W. Jones, & T. L. Beck (Eds.), *Organics and the Rockies field guide* (pp. 173–184). Wyoming: The Society for Organic Petrology Eleventh Annual Meeting, Jackson. Wyoming State Geological Survey Public Information Circular No. 33.

Cui, X., & Bustin, R. M. (2006). Controls of coal fabric on coalbed gas production and comparative shift in both field production and canister tests. *Society of Petroleum Engineers Journal, 11*, 111–119.

Cui, X., Bustin, A. M. M., & Bustin, R. M. (2009). Measurement of gas permeability and diffusivity of tight reservoir rocks: different approaches and their applications. *Geofluids, 9*, 208–223.

Cui, X., Bustin, R. M., & Chikatamarla, L. (2007). Adsorption-induced coal swelling and stress: implications for methane production and acid gas sequestration into coal seams. *Journal of Geophysical Research, 112*, B10202.

Cui, X., Bustin, M., & Dipple, G. (2004). Differential transport of CO_2 and CH_4 in coalbed aquifers: Implications for coalbed gas distribution and composition. *International Journal of Coal Geology, 88*, 1149–1161.

Curiale, J. A., Decker, J., Lin, R., & Morley, R. J. (2006). *Oils and oil-prone coals of the Kutei Basin, Indonesia.* American Association of Petroleum Geologists International Conference and Exhibition, Perth, West Australia. November 5–8, 2006. <http://www.pesa.com.au/aapg/aapgconference/pdfs/abstracts/tues/OilfromCoalSources.pdf> Accessed 10.12.2011.

Curry, D. J., Emmett, J. K., & Hunt, J. W. (1994). Geochemistry of aliphatic-rich coals in the Cooper basin, Australia and Taranaki basin, New Zealand: implications for the occurrence of potentially oil-generative coals. In A. C. Scott, & A. J. Fleet (Eds.), *The Geological Society Special Publication: Vol. 77. Coal and coal-bearing strata as oil-prone source rocks?* (pp. 149–181). London.

Curtis, J. B. (2011). *Potential gas committee reports substantial increase in magnitude of U.S. natural gas resource base.* Colorado School of Mines, Golden, CO 80401–1887: Potential Gas Agency. <http://www.potentialgas.org/> Accessed 14.11.11.

CWRRI. (April 4–5, 2006). *Produced water workshop.* In *Information series* (Vol. 02). Fort Collins, Colorado: Colorado Water Resources Institute, Colorado State University. p. 242.

Daddow, P. B. (1986). *Potentiometric-surface map of the Wyodak–Anderson coal bed.* Wyoming: Powder River structural sasin, 1973–84: U.S. Geological Survey Water-Resources Investigations Report 85-4305, 1 sheet, Scale 1:250,000.

Dahm, K. G., Guerra, K. L., Xu, P., & Drewes, J. E. (2011). Composite geochemical database for coalbed methane produced water quality in the Rocky Mountain region. *Environmental Science Technology, 45*, 7655–7663.

Dallegge, T. A., & Barker, C. E. (2000). *Coal-bed methane gas-in-place resource estimates using sorption isotherms and burial history reconstruction: An example from the Ferron Sandstone member of the Mancos Shale, Utah*. U.S. Geological Survey Professional Paper 1625–B, L1–L26.

Damberger, H. H. (1991). Coalification in North America coal fields. In H. J. Gluskoter, D. D. Rice, & R. B. Taylor (Eds.), *Economic geology, the geology of North America* (Vol. p-2; pp. 503–522). Geological Society of America.

Dang, Y., Zhao, W., Su, A., Zhang, S., Li, M., Guan, Z., et al. (2008). Biogenic gas systems in eastern Qaida Basin. *Marine and Petroleum Geology, 25*, 344–356.

Daniels, L., Fulton, R., Spencer, W., & Orme-Johnson, W. H. (1980). Origin of hydrogen in methane produced by Methanobacterium thermoautotrophicum. *Journal of Bacteriology, 141*, 694–698.

Dapples, E. C., & Hopkins, M. E. (Eds.), (1969). *Geological Society of America Special Paper: Vol. 114. Environments of coal deposition* (p. 204).

Darton, N. H. (1915). Occurrence and utilization of firedamp. *U.S. Bureau of Mines Bulletin, 72*, 101–102.

Davidson, S. C. (1992). Some aspects of the transient testing of Bowen Basin coal seams. In B. B. Beamish, & P. D. Gamson (Eds.), *Symposium on coalbed methane research and development in Australia* (Vol. 2; pp. 153–169). James Cook University of North Queensland.

Davidson, S. C. (1995). Horizontal wells for coalbed methane stimulation. *Energy Research and Development Corporation, 262*, 87.

Davis, R. W. (1976). Hydrologic factors related to coal development in the eastern Powder River Basin. In R. B. Laudon (Ed.), *Geology and energy resources of the Powder River Basin* (pp. 203–207). Wyoming Geological Association, 28th Annual Field Conference Guidebook.

Davis, A. (1978). Compromise in coal seam description. In R. R. Dutcher (Ed.), *Field description of coal* (pp. 23–40). American Society for Testing Materials, ASTM STP 661.

Davis, R. C., Noon, S. W., & Harrington, J. (2007). The petroleum potential of Tertiary coals from western Indonesia: relationship to mire type and sequence stratigraphic setting. *International Journal of Coal Geology, 70*, 35–52.

Davis, J. G., Waskom, R. M., & Bauder, T. A. (2012). *Managing sodic soils*. Colorado State University Extension. No. 0.504. <http://www.ext.colostate.edu/pubs/crops/00504.html> Accessed 30.11.12.

Davis, W. N., Zale, A. V., & Bramblett, R. G. (2008). *Effects of coalbed natural gas development on fish assemblages in tributary streams in the Powder River Basin, Montana and Wyoming*. Final Report to Wild Fish Habitat Initiative. Bozeman, Montana: Montana University System Water Center. p. 125.

Dawson, G. K. W., & Esterle, J. S. (2010). Controls on coal cleat spacing. In S. D. Golding, V. Rudolph, & R. M. Flores (Eds.), *International Journal of Coal Geology: Vol. 82. Asia Pacific coalbed methane symposium: Selected papers from the 2008 Brisbane Symposium on coalbed methane and CO_2-enhanced coalbed methane* (pp. 213–218).

Dawson, F. M., Marchioni, D. L., Anderson, T. C., & McDougall, W. J. (2000). An assessment of coalbed methane exploration projects in Canada. *Geological Survey of Canada, Bulletin, 549*, 218.

Dawson, M., & Sloan, G. (2001). *Geological setting and play concepts for Cretaceous and Tertiary age coalbed methane opportunities in western Canada*. The Canadian Society of Petroleum Geologists. Rock the Foundation Convention June 18–22, 2001.

Dawson, K. S., Strapoc, D., Huizinga, B., Lidstrom, U., Ashby, M., & Macalady, J. L. (2012). Quantitative fluorescence *in situ* hybridization analysis of microbial consortia from a biogenic gas field in Alaska's Cook Inlet Basin. *Applied Environmental Microbiology, 78*, 3599–3605. http://dx.doi.org/10.1128/AEM.07122-11. Published Ahead of Print 16 March 2012.

Day, S., Fry, R., & Sakurovs, R. (2012). Swelling of coal in carbon dioxide, methane and their mixtures. *International Journal of Coal Geology, 93*, 40–48.

Day, J. W., Kemp, G. P., Reed, D. J., Cahoon, D. R., Boumans, R. M., Suhayda, J. M., et al. (2011). Vegetation death and rapid loss of surface elevation in two contrasting Mississippi delta salt marshes: the role of sedimentation, autocompaction and sea-level rise. *Ecological Engineering, 37*, 229–240.

De Berg, M., van Kreveld, M., Overmars, M., & Schwarzkopf, O. (2000). *Computational geometry* (2nd ed.). Springer-Verlag. ISBN 3-540-65620-2, Chapter 7, pp. 147–163.

DECC. (2010). *The unconventional hydrocarbon resources of Britain's onshore basins coalbed methane (CBM)*. Department of Energy and Climate Change.

DeCicco, J., & Fung, F. (2006). *Global warming on the road: The climate impact of America's automobiles*. New York: Environmental Defense. <www.environmentaldefense.org/documents/5301_Globalwarmingontheroad.pdf> Accessed 29.10.11.

DEEDI. (2011). *Gas market review Queensland*. Queensland Government, Department of Employment, Economic Development and Innovation. < http://www.deedi.qld.gov.au/energy/gas-market-rev.html> Accessed 15.01.13.

DEEDI. (2012). *Queensland coal seam gas overview*. Queensland Government, Department of Employment, Economic Development and Innovation. <www.mines.industry.qld.gov.au> Accessed 09.06.13.

Deguchi, G., Yu, B., & Jiao, J. (1995). Japan/China research cooperation on prevention of gas outbursts. In R. D. Lama (Ed.), *Management and control of high gas emissions and outbursts in underground coal mines* (pp. 139–146). Kiama, North South Wales, Australia: Westonprint.

Dehmer, J. (1989). Petrographical and organic geo- chemical investigation of the Oberpfalz brown coal deposit, West Germany. *International Journal of Coal Geology, 11*, 273–290.

Dehmer, J. (1995). Petrological and organic geochemical investigation of recent peats with known environments of deposition. *International Journal of Coal Geology, 28*, 111–138.

Deisman, N., Gentzis, T., & Chalatumyk, R. J. (2008). Unconventional geomechanical testing on coalbed reservoir well design: the Alberta Foothills and Plains. *International Journal of Coal Geology, 75*, 15–26.

Dembicki, H., Jr. (2009). Three common source rock evaluation errors made by geologists during prospect or play appraisals. *American Association of Petroleum Geologists Bulletin, 93*, 341–356.

Demchuk, T. D. (1992). Epigenetic pyrite in a low-sulphur, sub-bituminous coal from the central Alberta Plains. *International Journal of Coal Geology, 21*, 187–196.

Demchuk, T. D., & Moore, T. A. (1993). Palynofloral and organic characteristics of a Miocene bog-forest, Kalimantan, Indonesia. *Organic Geochemistry, 20*, 119–134.

Demirel, I. H., & Karatigit, A. I. (1999). Quality and petrographic characteristics of the lacustrine Ermenek coal (Early Miocene), Turkey. *Energy Sources, 21*, 329–338.

Deppenmeier, U. (2002). The unique biochemistry of methanogenesis. *Progress in Nucleic Acid Research and Molecular Biology, 71*, 223–284.

Deul, M., & Kim, A. G. (1986). Methane control research: summary of results, 1964–80. *U.S. Bureau of Mines Bulletin, 687*, 174.

DeVanney, N. (2001). Impact of sample desiccation on the mean maximum vitrinite reflectance for various ranks of coal. *The Society of Organic Petrologists Newsletter, 18*, 5–20.

Dewing, K., & Obermajer, M. (2009). Lower Paleozoic thermal maturity and hydrocarbon potential of the Canadian Arctic Archipelago. *Bulletin of Canadian Petroleum Geology, 57*, 141–166.

DGH. (2010). *CBM Exploration, Directorate General of Hydrocarbons, Ministry of Petroleum and Natural Gas, New Delhi, India.* <http://www.dghindia.org/Index.aspx.> Accessed 10.02.13.

Diamond, W. P. (1982). *Site-specific and regional geologic considerations for coalbed gas drainage.* Pittsburgh, PA: US Dept. of Interior, US Bureau of Mines. Information Circular No. 8898.

Diamond, W. P. (1993). Methane control for underground coal mines. In B. E. Law, & D. D. Rice (Eds.), *American Association of Petroleum Geologists Studies in Geology: Vol. 38. Hydrocarbons from coals* (pp. 217–267).

Diamond, W. P., LaScola, J. C., & Hyman, D. M. (1986). *Results of direct-method determination of the gas content of U.S. Coalbeds.* United States Bureau of Mines Information Circular 9067. p. 95.

Diamond, W. P., & Levine, J. R. (1981). *Direct method determinations of the gas content of coals, procedures and results.* U.S. Bureau of Mines Report of Investigations 8515, p. 36.

Diamond, W. P., Murrie, G. W., & McCulloch, C. M. (1976). *Methane gas content of the Mary Lee Group of coal beds, Jefferson, Tuscaloosa, and Walker Counties, AL.* US Bureau of Mines. Report of Investigations 8117, p. 9.

Diamond, W. P., & Schatzel, S. J. (1998). Measuring the gas content of coal: a review. In R. M. Flores (Ed.), *International Journal of Coal Geology: Vol. 35. Coalbed methane: From coal mine outbursts to a gas resources* (pp. 311–331).

Diamond, W. P., Schatzel, S. J., Garcia, F., & Ulery, J. P. (2001). *The modified direct method: A solution for obtaining accurate coal desorption measurements.* National Institute for Occupational Safety and Health (NIOSH), Office for Mine Safety and Health Research, Pittsburgh Research Laboratory. <http://www.cdc.gov/niosh/mining/pubs/pdfs/mdmsoa.pdf> Accessed 13.08.2012.

Diemont, W. H., & Supardi. (1987). Accumulation of organic matter and inorganic constituents in a peat domed in Sumatra, Indonesia. In *"International peat society symposium on tropical peats and peatland for development," Yogyakarta, Indonesia.*

Diessel, C. F. K. (1986). *On the correlation between coal facies and depositional environments.* Proceeding 20th Symposium. N.S.W: Department of Geology, University of Newcastle. pp. 19–22.

Diessel, C. F. K. (1992). *Coal-bearing depositional systems.* Berlin: Springer-Verlag.

Diessel, C., Boyd, R., Wadsworth, J., Leckie, D., & Chalmers, G. (2000). On balanced and unbalanced accommodation/peat accumulation ratios in the Cretaceous coals from gates Formation, Western Canada, and their sequence-stratigraphic significance. *International Journal of Coal Geology, 42*, 143–186.

Diessel, C. F. K., Brothers, R. N., & Black, P. M. (1978). Coalification and graphitization on high-pressure schists in New Caledonia. *Contributions to Mineralogy and Petrology, 68*, 63–78.

Diessel, C. F. K., & Gammidge, L. (1998). Isometamorphic variations in the reflectance and fluorescence of vitrinite—a key to depositional environment. *International Journal of Coal Geology, 37*, 179–206.

DiMichele, W. A., & Phillips, T. L. (1996). *Climate change, plant extinctions and vegetational recovery during the Middle-Late Pennsylvanian Transition: The Case of tropical peat-forming environments in North America.* In *Geological Society, London, Special Publication* (Vol. 102; pp. 201–221).

Dinwiddie, R., Lamb, S., & Reynolds, R. (2011). Violent earth. In *Cooperation with Smithsonian Institution* (1st ed.). New York: D.K. Publishing.

Dockins, W. S., Olson, G. J., McFeters, G. A., Turbak, S. C., & Lee, R. W. (1980). *Sulfate reduction in groundwater of southeastern Montana.* U.S. Geological Survey Water-Resources. Investigations Report 80–9, p. 13.

Doherty, M.K. (2007). *Mosquito populations in the Powder River Basin, Wyoming: a comparison of natural, agricultural and effluent coal bed natural gas aquatic habitats* (M.S. Thesis), Bozeman, Montana: Montana State University, p. 94. http://www.voiceforthewild.org/documents/doherty_m.pdf Accessed December 2012.

Dolly, E. D., & Meissner, F. F. (1977). Geology and gas exploration potential, Upper Cretaceous and lower tertiary strata, northern Raton Basin, Colorado. In H. K. Veal (Ed.), *Rocky Mountain Association of Geologists. Exploration frontiers of the central and southern rockies: Denver, Colo* (pp. 247–270).

Dreesen, R., Bossiroy, D., Dusar, M., Flores, R. M., & Verkaeren, P. (1995). Coal-seam discontinuities influenced by synsedimentary tectonics and paleofluvial systems in the Westphalian, C., Campine Basin, Belgium. In M. K. G. Whateley, & D. A. Spears (Eds.), *London, Geological Society Special Publication: Vol. 82. European coal geology* (pp. 215–232).

Drexler, J. Z., de Fontaine, C. S., & Brown, T. A. (2009). Peat accretion histories during the Past 6000 years in marshes of the Sacramento–San Joaquin Delta, CA, USA. In *Journal of the Coastal and Estuarine Research Federation: Vol. 32. Estuaries and coasts* (pp. 871–892).

Driessen, P. M. (1978). *Peat soils. Soils and rice symposium.* Manila: Philippines.

Duan, Y., Sun, T., Qian, Y., He, J., Zhang, X., & Wu, B. (2011). Pyrolysis experiments of forest marsh peat samples with different maturities: an attempt to understand the isotopic fractionation of coalbed methane during staged accumulation. *Fuel, 94*, 480–485.

Duan, Y., Wu, B., & Zheng, C. (2005). Thermal simulation of the formation and evolution of coalbed gas. *Chinese Science Bulletin, 50*, 40–44.

Dubaniewicz, T. H., Jr. (2009). *From Scotia to Brookwood, fatal US underground coalmine explosions ignited in intake air courses.* P.O. Box 18070, Pittsburgh, PA 15236: National Institute for Occupational Safety and Health, Pittsburgh Research Laboratory. <http://www.cdc.gov/niosh/mining/pubs/pdfs/fstbf.pdf>, 12 pages references, 37 figures and 4 tables. Accessed 16.12.11.

Dubinin, M. M. (1966). Porous structure and adsorption properties of active carbons. In P. L. Walker (Ed.), *Chemistry and physics of carbon* (pp. 51–120). N.Y.: Marcel Dekker.

Dunn, B. W. (1995). Vertical well degasification in advance of mining: a potential gas management method for reducing gas emissions and outbursts. In R. D. Lama (Ed.), *Management and control of high gas outbursts in underground coal mines* (pp. 267–275). Kiama, North South Wales, Australia: Westonprint.

Durica, D., & Nemec, J. (1997). *Preliminary results of the coalbed methane exploration in the Czech Republic*, 9th International Mining Conference. < http://actamont.tuke.sk/pdf/1997/n5/27durica.pdf> Accessed 28.01.13.

Durrani, N. A., & Warwick, P. D. (1992). Regional characterization and resource evaluation of Paleocene and Eocene coal-bearing rocks in Pakistan. *University of Peshawar Geological Bulletin, 24*, 229–237.

Durucan, S., Ahsan, M., & Shi, J.-Q. (2009). Matrix shrinkage and swelling characteristics of European coals. *Energy Procedia, 1*, 3055–3062.

Durucan, S., Daltaban, T. S., Shi, J. Q., & Sinka, I. C. (1995). Coalbed methane well stimulation: the effect of structural and geotechnical parameters on connectivity in coal seams. In *Planning for profit coalbed methane in the UK and Europe.* March 30–31, 1995 (Selfridge Hotel, London).

Durucan, S., & Edwards, J. S. (1986). The effects of stress and fracturing on permeability of coal. *Mining Science and Technology, 3*, 205–216.

Dutta, S., Pandey, A., & Mishra, S. S. (2011). *Economics of coalbed methane as potential energy in India.* ICOQM-10, International Conference on Operations and Quantitative Management, 1209–1214. <http://www.icmis.net/infoms/icoqm10/ICOQM10CD/pdf/P448-Final.pdf> Accessed 28.01.13.

Eaton, N. G., Redford, S. R., Segrest, C., & Christensen, E. (1991). *Pressure coring: better cores for better analyses.* pp. 79–94 <http://www.scaweb.org/assets/papers/1991_papers/1-SCA1991-05EURO.pdf> Accessed 28.07.12.

Eavenson, H. N. (1939). *Coal through the years.* New York: American Institute of Mining and Metallurgical Engineers.

Eddens, B., & Dimatteo, I. (2007). *Classification issues for mineral and energy resources.* Eleventh Meeting of the London Group on Environmental Accounting, March 26–30, 2007, p. 8. <http://www.unece.org/ie/se/pdfs/UNFC/UNFCemr.pdf> Accessed 20.10.11.

Ekono. (1981). *Report on energy use of peat* (Contribution to U.N. Conference on New and Renewable Sources of Energy, Nairobi).

Elliott, R. E. (1984). Quantification of peat to coal compaction stages, based especially on phenomena in the East Pennine Coalfield, England. *Proceedings of the Yorkshire Geological Society, 45*, 163–172.

Ellis, D. V., & Singer, J. M. (2010). *Well logging for Earth Scientists* (2nd ed.). Springer.

Ellis, M. S. Gunther, G. L., Flores, R. M., Ochs, A. M., Stricker, G. D., Roberts, S. B., Taber, T. Y., Bader, L. R., and Schuennemeyer, J. H. (1999). Preliminary report on coal resources of the Wyodak-Anderson coal zone, Powder River Basin, Wyoming and Montana: U.S. Geological Survey Open-File Report 98–789A, p. 49, 17 figs., 11 tables.

Ely, J. W. (1989). Fracturing fluids and additives. In J. L. Gidley, S. A. Holditch, D. E. Nierode, & R. W. Veatch, Jr. (Eds.), *Recent advances in hydraulic fracturing* (pp. 130–146). Society of Petroleum Engineers, Monograph 12, Richardson, Texas, U.S.A.

Engelder, T., & Whitaker, A. (2006). Early jointing in coal and black shale: evidence for an Appalachian-wide stress field as a prelude to the Alleghanian orogeny. *Geology, 34*, 581–584.

Engelder, T. (1987). Joints and shear fractures in rock. In B. K. Atkinson (Ed.), *Fracture mechanics of rock* (pp. 27–69). Orlando: Academic Press.

Engle, M. A., Bern, C. R., Healy, R. W., Sams, J. I., & Zuapancic, J. W. (2011). *Tracking solutes and water from subsurface drip irrigation application of coalbed methane-produced water, Powder River Basin, Wyoming*. USGS Staff-Published. Research Paper 508. <http://digitalcommons.unl.edu/usgsstaffpub/508> Accessed 28.12.12.

ERCB. (2011). *Changes to the province-wide framework to well spacing for conventional and unconventional oil and gas reservoirs*. Energy Resources Conservation Board, Bulletin 2011-29. <http://www.ercb.ca/bulletins/Bulletin-2011-29.pdf> Accessed 08.11.12.

Erdmann, M., & Horsfield, B. (2006). Enhanced late gas generation potential of petroleum source rocks via recombination reactions: evidence from the Norwegian North Sea. *Geochimica et Cosmochimica Acta, 70*, 3943–3956.

Erslev, E. A. (1993). Thrusts, back-thrusts, and detachment of Rocky Mountain foreland arches. In C. J. Schmidt, R. B. Chase, & E. A. Erslev (Eds.), *Geological Society of America Special Paper: Vol. 280. Laramide basement deformation in the Rocky Mountain foreland of western United States* (pp. 339–358).

Ersoy, M. (2005). *Overview of earlier case studies – coal, uranium and other solid minerals. Second session of the ad-hoc group of experts on harmonization of reserves and resources terminology*. November 9–11, 2005, Geneva.

ESRI. (2009). *ArcGIS-ArcInfo computer software*. Redwood, California: Environmental Systems Research Institute, Inc.

Estele, J. S., & Ferm, J. C. (1994). Spatial variability in modern tropical peat deposits from Sarawak, Malaysia and Sumatra, Indonesia: analogues for coal. *International Journal of Coal Geology, 26*, 1–41.

Esterle, J. S. (1990). *Trends in petrographic and chemical characteristics of domed peats in Indonesia and Malaysia as analogues for coal*. Unpublished Ph.D. thesis. University of Kentucky, p. 270.

Esterle, J. S., & Ferm, J. C. (1994). Spatial variability in modern tropical peat deposits from Sarawak, Malaysia and Sumatra, Indonesia: analogues for coal. *International Journal of Coal Geology, 26*, 1–41.

Esterle, J. S., Ferm, J. C., Durig, D. T., & Supardi. (1987). *Physical and chemical properties of peat near Jambi, Sumatra, Indonesia*. Directorate Mineral Resources Special Report No. 21.

Esterle, J. S., Ferm, J. C., & Yiu-Lion, T. (1989). A test for the analogy of tropical domed peat deposits to "dulling up" sequences in coal beds–preliminary results. *Organic Geochemistry, 14*, 333–342.

Esterle, J. S., Williams, R., Sliwa, R., & Malone, M. (2006). *Variability in coal seam gas content that impacts on fugitive gas emissions estimations for Australian black coals*, 36th Sydney Basin Symposium 2006: Advances in the Study of the Sydney Basin. November 26–28, 2006. Wollongong, Australia: University of Wollongong. p. 10. <http://www.geogas.com.au/files/u2/SydSymp_Paper_FugitiveGas_final.pdf> Accessed 12.11.11.

Esterle, J. S. (2011). *Introduction to coal geology for gas reservoir characterisation, Lecture Presentation*. Brisbane, Australia: University of Queensland.

Ethridge, F. G. (2011). Interpretation of ancier nt fluvial channel deposits: Review and recommendations. In *From Rivers to Rock Record: The Preservation of Fluvial Sediments and their Subsequent interpretation, Society of Economic Paleontologists and Mineralogists, Special Publication No: 97*. (pp. 9–35).

Ettinger, I., Eremin, I., Zimakov, B., & Yanovskaya, M. (1966). Natural factors influencing coal sorption properties: I. petrographic and sorption properties of coals. *Fuel, 45*, 267–275.

Eychaner, J. (2000). *Effects of mountaintop coal mining on groundwater*. Workshop on Mountaintop Mining Effects on Groundwater. Charleston, W. VA., May 9, 2000. <http://www.epa.gov/region3/mtntop/pdf/appendices/g/groundwater-symposium/Proceedings/Presentations/Eychaner.pdf> Accessed 22.11.11.

Faiz, M. M., Aziz, N. I., Hutton, A. C., & Jones, B. G. (1992). Porosity and gas sorption capacity of some eastern Australian coals in relation to coal rank and composition. In *Proceeding of symposium on coal bed methane resource development* (Vol. 4; pp. 9–20). Australia.

Faiz, M., & Hendry, P. (2006). Significance of microbial activity in Australian coal bed methane reservoirs: a review. *Bulletin of Canadian Petroleum Geology, 54*, 261–272.

Faiz, M., Murphy, A., Lidstrom, U., & Ashby, M. (2012). Geochemical and microbiological evidence for degradation of heavy hydrocarbons in Australian CBM reservoirs. In *Proceedings of the Thirthy Fourth International Geological Congress*. Brisbane, Australia, August 5–10, 2012.

Faiz, M., Saghafi, A., Bareclay, S. A., Stalker, L., Sherwood, N. L., & Whitford, D. J. (2007). Evaluating geological sequestration of CO_2 in bituminous coals: the southern Sydney Basin, Australia as a natural analogue. *International Journal of Greenhouse Gas Control, 1*, 223–235.

Faiz, M., Saghafi, A., Sherwood, N., & Wang, I. (2007). The influence of petrological properties and burial history on coal seam methane reservoir characterisation, Sydney Basin, Australia. *International Journal of Coal Geology, 70*, 193–208.

Faiz, M., Stalker, L., Sherwood, N., Saghafi, A., Wold, M., Barclay, S., et al. (2003). Bio-enhancement of coalbed methane resources in the southern Sydney Basin. *Australian Petroleum Production and Exploration Association Journal, 43*, 595–610.

Falini, F. (1965). On the formation of coal deposits of lacustrine origin. *Geological Society of America Bulletin, 76*, 1317–1346.

Fallgren, P. H., Zeng, C., Ren, Z., Lu, A., & Ren, S. (2013). Feasibility of microbial production of new natural gas from non-gas-producing lignite. *International Journal of Coal Geology, 115*, 79–84.

Fallgren, P. H., Jin, S., Zeng, C., Ren, Z., Lu, A., & Colberg, P. J. S. (2013). Comparison of coal rank for enhanced biogenic natural gas production. *International Journal of Coal Geology, 115*, 92–96.

Fallgren, P. H., Zeng, C., Ren, Z., Lu, A., & Ren, S. (2013). Feasibility of microbial production of new natural gas from non-gas-producing lignite. *International Journal of Coal Geology*. Biogenic Coal Bed Natural Gas International Conference, Laramie, Wyoming, June 20–21, 2012. <http://dx.doi.org/10.1016/j.coal.2013.01.014>.

Farabee, M. J., Taylor, E. L., & Taylor, T. N. (1990). Correlation of Permian and Triassic palynomorph assemblages from the central Transantarctic Mountains, Antarctica. *Review of Palaeobotany and Palynology, 65*, 257–265.

Farag, A. M., & Harper, D. D. (Eds.), (2012). *The potential effects of sodium bicarbonate, a major constituent of produced waters from coalbed natural gas production, on aquatic life*. U.S. Geological Survey Scientific Investigations Report 2012–5008. p. 101.

Faraj, B. S. M., Fielding, C. R., & Mackinnon, I. D. R. (1996). Cleat mineralization of upper Permian Baralaba/Rangal coal measures, Bowen Basin, Australia. *Geological Society, 109*, 151–164. Special Publications, London.

Farnham, R. S., & Finney, H. R. (1965). Classification and properties of organic soils. *Advance Agronomy, 17*, 135–163.

Fassett, J. E., & Rigby, J. K., Jr. (Eds.), (1987). *Geological Society of America, Special Paper: Vol. 209. The Cretaceous–Tertiary boundary in the San Juan and Raton Basins, New Mexico and Colorado* (p. 200).

Fassett, J. E. (1989). Coal-bed methane; A contumacious, free-spirited bride, the geologic handmaiden of coal beds. In J. C. Lorenz, & S. G. Lucas (Eds.), *Energy frontiers in the Rockies* (pp. 131–146). Albuquerque, New Mexico: Albuquerque Geological Society (Transactions/Summary volume, American Association of Petroleum Geologists Rocky Mountain Section Meeting, Albuquerque, New Mexico).

Fassett, J. E. (2000). *Geology and coal resources of the Upper Cretaceous Fruitland Formation, San Juan Basin, New Mexico and Colorado*. U.S. Geological Survey Professional Paper 1625–B, Chapter Q, Version 1.0, Q1–Q132. CD-Rom.

Fassett, J. E. (2010a). *Oil and gas resources of the San Juan Basin, New Mexico and Colorado*. New Mexico Geological Society Guidebook, 61 Field Conference, Four Corners Country.

Fassett, J. E. (2010b). *The San San Juan Basin, a complex giant gas field, New Mexico and Colorado, American Association of Petroleum Geologists Rocky Mountain Section 58th Annual Rocky Mountain Rendezvous*. Durango, Colorado, June 13–16, 2010.

Faure, K., de Wit, M. J., & Willis, J. P. (1995). Late Permian global coal hiatus linked to ^{13}C-depleted CO_2 flux into the atmosphere during final consolidation of Pangea. *Geology, 23*, 507–510.

Fechner-Levy, E. J., & Hemond, H. F. (1996). Trapped methane volume and potential effects on methane ebullition in a northern peatland. *Limnology and Oceanography, 41*, 1375–1383.

Fedor, F., & Vido, M. H. (2003). Statistical analysis of vitrinite reflectance data—a new approach. *International Journal of Coal Geology, 56*, 277–294.

Feng, S., Ye, J., & Zhang, S. (2002). Coalbed methane resources in the Ordos basin and its development potential. *Geological Bulletin of China, 21*(10), 658–662 (in Chinese with English abstract).

Ferm, J. C. (1979). Allegheny deltaic deposits: a model for the coal-bearing strata. In J. C. Ferm, & J. C. Horne (Eds.), *Carboniferous depositional environments in the Appalachian region* (pp. 291–294). Columbia: University of South Carolina.

Ferm, J. C., & Cavaroc, V. V. (1968). A nonmarine sedimentary model for the Allegheny rocks of West Virginia. In G. deVries Klein (Ed.), *Geological Society of America Special Paper: Vol. 108. Late Paleozoic and Mesozoic continental sedimentation, northeastern North America* (pp. 1–19).

Ferm, J. C., & Horne, J. C. (Eds.). (1979). *Carboniferous depositional environments in the Appalachian region* (p. 760). Columbia: University of South Carolina.

Ferm, J. C., & Staub, J. R. (1984). Depositional controls of minable coal bodies. In R. A. Rahmani, & R. M. Flores (Eds.), *International Association of Sedimentologists, Special Publication: Vol. 7. Sedimentology of coal and coal-bearing sequences* (pp. 275–290). Oxford: Black- well Scientific Publications.

Ferm, J. C., Smith, G. C., Weisenfluh, G. A., & DuBois, S. B. (1985). *Cored rocks in the Rocky Mountain and High Plains coal fields*. Lexington: Department of Geology, University of Kentucky.

Ferm, J. C., Staub, J. R., Baganz, B. P., Clark, W. J., Galloway, M. C. C. B., Hohos, E. F., et al. (1979). The shape of coal bodies. In J. C. Ferm, & J. C. Horne (Eds.), *Carboniferous deposition environments in the Appalachian region* (p. 760). Columbia: University of South Carolina.

Ferm, J. C., & Weisenfluch, G. A. (1989). Evolution of some depositional models in Late Carboniferous rocks of the Appalachian coal fields. In P. C. Lyons, & B. Alpern (Eds.), *International Journal of Coal Geology: Vol. 12. Peat and coal: Origin, facies, and depositional models* (pp. 259–292).

Fernandez-Rubio, R. (1986). Water problems in Spanish coal mining. *International Journal of Mine Water, 5*, 13–28.

Ferris, J. D., Knowles, D. B., Brown, R. H., & Stallman, R. W. (1962). *Theory of aquifer test: Groundwater hydraulics*. U.S. Geological Survey Water-Survey. Paper 1536-E, p. 99.

Ferry, J. G. (1993). Fermentation of acetate. In J. Ferry (Ed.), *Methanogenesis* (pp. 304–334). London: Chapman and Hall Inc.

Fielding, C. R. (1984). 'S' or 'Z' shaped coal seam splits in the coal measures of County Durham. *Proceedings of the Yorkshire Geological Society, 45*, 85–89.

Fielding, C. R. (1987). Coal depositional models for deltaic and alluvial sequences. *Geology, 15*, 661–664.

Fielding, C. R. (1992). A review of Cretaceous coal-bearing sequences in Australia. In P. J. McCabe, & J. T. Parrish (Eds.), *Geological Society of America Special Paper: Vol. 267. Controls on the distribution and quality of Cretaceous coals* (pp. 303–324).

Fielding, C. R., & Webb, J. A. (1996). Facies and cyclicity of the late Permian Bainmedart coal measures in the northern Prince Charles mountains, MacRobertson land, Antarctica. *Sedimentology, 43*, 295–322.

Figueroa, J. D., Fout, T., Plasynski, S., McIlvried, H., & Srivastava, R. D. (2008). Advances in CO_2 capture technology – the U.S. department of Energy's carbon sequestration program. *International Journal of Greenhouse Gas Controls, 2*, 9–20.

Finkelman, R. B. (1994). Abundance, source and mode of occurrence of the inorganic constituents in coal. In O. Kural (Ed.), *Coal* (pp. 115–125). Istanbul: Istanbul Technical University.

Finkelman, R. B. (1995). Modes of occurrence of environmentally sensitive trace elements in coal. In D. J. Swaine, & F. Goodarzi (Eds.), *Environmental aspects of trace elements in coal* (pp. 24–50). Dordrecht: Kluwer Academic Publishing.

Finkelman, R. B., & Brown, R. D., Jr. (1991). Coal as a host and as an indicator of mineral resources. In D. C. Peters (Ed.), *Geology in coal resource utilization* (pp. 471–481). Fairfax, VA: Tech Books.

Finkelman, R. B., & Stanton, R. W. (1978). Identification and significance of accessory minerals from a bituminous coal. *Fuel, 57*, 763–768.

Finkelman, R. B., Bostick, N. H., Dulong, F. T., Senftle, F. E., & Thorpe, A. N. (1998). Influence of an igneous intrusion on the inorganic geochemistry of a bituminous coal from Pitkin County, Colorado. *International Journal of Coal Geology, 36*, 223–241.

Finklestein, M., DeBryuyn, R. P., Weber, J. L., & Dodson, J. B. (2005). Buried hydrocarbons: a resource for biogenic methane generation. *World Oil, 226*.

Finklestein, M., Pfeiffer, R. S., & Ulrich, G. A. (2010). *Chemical amendments for the stimulation of biogenic gas generation in deposits of carbonaceous material*. United States Patent No. 7696132 B2, April 13, 2010.

Fisher, J. B. (2001). *Environmental issues and challenges in coal bed methane production*. Tulsa, OK: Exponent, Inc. Available from. < http://ipec.utulsa.edu/Conf2001/fisher_92.pdf>.

Flint, S., Aitken, J., & Hampson, G. (1995). *Application of sequence stratigraphy to coal-bearing coastal plain successions: implications for the UK coal measures*. In *Geological Society, London, Special Publications* (Vol. 82; pp. 1–16).

Florentin, R., Aziz, N., Black, D., & Neighm, L. (2009). Characteristics of coal, particle size, gas type, and time. In N. Aziz (Ed.), *Coal 2009* (pp. 208–216). Coal Operators' Conference, University of Wollongong & the Australasian Institute of Mining and Metallurgy, 2009.

Flores, R. M. (1981). Coal deposition in fluvial paleoenvironments of the Paleocene Tongue river Member of the Fort Union Formation, Powder river area, Powder river basin, Wyoming and Montana. In F. G. Ethridge, & R. M. Flores (Eds.), *Society of Economic Paleontologists and Mineralogists, Special Publication: Vol. 31. Nonmarine depositional environments: Models for exploration* (pp. 169–190).

Flores, R. M. (1983). Basin facies analysis of coal-rich Tertiary fluvial deposits in the northern Powder River Basin, Montana and Wyoming. In J. D. Collinson, & J. Lewin (Eds.), *International Association of Sedimentologists, Special Publication: Vol. 6. Modern and ancient fluvial systems*. 501–515.

Flores, R. M. (1984). Comparative analysis of coal accumulation in Cretaceous alluvial deposits, southern United States Rocky Mountain basins. In D. F. Stott, & D. Glass (Eds.), *Mesozoic of middle North America* (pp. 373–385). Canadian Society of Petroleum Geologists Memoir, 9.

Flores, R. M. (1986). Styles of coal deposition in Tertiary alluvial deposits, Powder River Basin, Montana and Wyoming. In P. C. Lyons, & C. L. Rice (Eds.), *Paleoenvironmental and tectonic controls in coal-forming Basins of the United States* (pp. 79–104). Geological Society of America. Special Paper 210.

Flores, R. M. (1998). Coalbed methane: from a mining hazard to gas resource. In R. M. Flores (Ed.), *International Journal of Coal Geology: Vol. 35. Coalbed methane: From coal-mine outbursts to a gas resource* (pp. 3–26). Special Issue.

Flores, R. M. (2000). Biogenic gas in low rank coal: a viable and economic resource in the United States. In Sukarjo, Herudyanto, E. Djaelani, & Komaruddin

(Eds.), *Southeast Asian coal geology conference proceedings* (pp. 1–7). Published by Directorate of Mineral Resources in association with Indonesian Association of Geologists, ISBN: 979-8126-03-3.

Flores, R. M. (2003a). Paleocene paleogeographic, paleotectonic, and paleoclimatic patterns of the northern Rocky Mountains and Great Plains region. In R. Raynolds, & R. M. Flores (Eds.), *Society of Economic Paleontologists and Mineralogists Rocky Mountain Section, Special Publication Cenozoic paleogeography* (pp. 63–106).

Flores, R. M. (2003b). Coal buildup in tide-influenced coastal plains in the Eocene Kapuni Group, Taranaki Basin, New Zealand. In J. C. Pashin, & R. A. Gastaldo (Eds.), *Studies in geology: Vol. 51. Coal-bearing strata: Sequence stratigraphy, paleoclimate, and tectonics* (pp. 45–70). American Association of Petroleum Geologists.

Flores, R. M. (2004). *Coalbed methane in the Powder River Basin, Wyoming and Montana: An assessment of the Tertiary-upper Cretaceous coalbed methane total petroleum system.* In *U.S. Geological Survey Digital Data Series DDS–69–C, version 1.* Chapter 2, p. 56.

Flores, R. M. (2008a). *Coalbed methane: gas of the past, present, and future. Sci Topic Pages.* Elsevier Publication. <http://topics.scirus.com/Coalbed_Methane_Gas_of_the_Past_Present_and_Future.html.>

Flores, R. M. (Ed.), (2008b). *International Journal of Coal Geology: Vol. 76. Microbes, methanogenesis, and microbial gas in coal* (pp. 1–185).

Flores, R. M. (2008c). Methanogenic pathways of coal-bed gas in the Powder River Basin, United States: The geologic factor. In R. M. Flores (Ed.), *International Journal of Coal Geology: Vol. 76. Microbes, methanogenesis, and microbial gas.* (pp. 52–75).

Flores, R. M., & Bader, L. R. (1999a). A summary of Tertiary coal resources of the Raton Basin, Colorado and New Mexico. In Fort Union Coal Assessment Team (Ed.), *1999 resource assessment of selected Tertiary coal beds and zones in the Northern Rocky Mountains and Great Plains Region.* U.S. Geological Survey Professional Paper 1625–A, Chapter SR, Disc 1, Version 1.0, p. SR–1–SR-35.

Flores, R. M., & Bader, L. R. (1999b). Fort Union coal in the Powder River Basin, Wyoming and Montana: a synthesis. In *Fort Union coal assessment team, 1999 resource assessment of selected tertiary coal beds and zones in the northern Rocky Mountains and Great Plains region.* U.S. Geological Survey Professional Paper 1625-A, Chapter GS, Disc 1, version 1.0, pp. GS-1–GS-36.

Flores, R. M., & Cross, T. A. (1991). Cretaceous and Tertiary coals of the Rocky Mountains and Great Plains regions. In H. Gluskoter (Ed.), *Geological Society of America, economic geology: Vol. P-2. Economic geology, decade of North American geology* (pp. 547–571).

Flores, R. M., & Erpenbeck, M. F. (1981). *Differentiation of delta front-barrier lithofacies of the Upper Cretaceous pictured cliffs sandstone.* Southwest San Juan Basin, New Mexico: The Mountain Geologists. 18, pp. 23–34.

Flores, R. M., & Hadiyanto (2006). Patterns of Sumatran domed peatlands from alluvial to coastal settings: geological Society of America National Meetings Abstracts in Programs. Oct. 22–25, 2006, Philadelphia, p. 233.

Flores, R. M., & McGarry, D. E. (2004). *Coalbed natural gas reservoirs in Fort Union Formation, Powder River Basin: Coalbed stratigraphy and regional coalbed aquifers.* American Association of Petroleum Geologists (Rocky Mountain Section Meeting, Abstracts with Programs).

Flores, R. M., McGarry, D. E., & Stricker, G. D. (2005). CBNG development: Confusing coal stratigraphy and gas production in the Powder River Basin: Canadian Society of Petroleum Geologist Gussow Conference, Canmore, Canada, Abstracts with Programs.

Flores, R. M., & M'Gonigle, J. W. (1991). Oligocene-Miocene lacustrine rudite-dominated alluvial-fan delta, southwest Montana, U.S.A. In C. J. Dabrie (Ed.), *II fan-delta workshop* (pp. 241–278). Universidad Computense de Madrid. Special Issue.

Flores, R. M., Moore, T. A., Stanton, R. W., & Stricker, G. D. (2001). *Textural controls on coalbed methane content in the subbituminous coal of the Powder River Basin.* Geological Society of America. Annual Meeting, Boston. Paper No. 23–0.

Flores, R. M., Roberts, S. B., & Perry, W. J., Jr. (1994). Paleocene paleogeography of the Wind River, Bighorn, and Powder River Basins, Wyoming. In R. M. Flores, K. T. Mehring, R. W. Jones, & T. L. Beck (Eds.), *Organics and the Rockies field guide* (pp. 1–16). Public Information Circular No. 33, Wyoming State Geological Survey.

Flores, R. M., Spear, B. D., Kinney, S. A., Purchase, P. A., & Gallagher, C. M. (2010). *After a century – Revised Paleogene coal stratigraphy, correlation, and deposition, Powder River Basin, Wyoming and Montana.* U.S. Geological Survey Professional Paper 1777, p. 97, CD–ROM in pocket.

Flores, R. M., & Stricker, G. D. (2004). Potential CO_2 sequestration and enhanced recovery of coalbed methane in subbituminous coals in the Powder River Basin, United States. In *Twenty-first annual meeting, September 27–October 1, 2004, Sydney, Australia* (Vol. 21). The Society of Organic Petrology.

Flores, R. M., & Stricker, G. D. (2010). CBM adsorption isotherms of Philippines versus U.S. coals: From tectonic control to resource evaluation: American Association of Petroleum Geologists, International Convention and Exhibition, September 12–15, 2010.

Flores, R. M., & Stricker, G. D. (2012). *Modeling methane adsorption capacity of Tertiary coals from U.S. and other countries*, 34th International Geological Congress, Brisbane, Australia, August 5–10, 2012.

Flores, R. M., Stricker, G. D., & Kinney, S. A. (2003). Alaska coal resources and coalbed methane potential. *U.S. Geological Survey Bulletin, 2198*, 4. <http://pubs.usgs.gov/bul/b2198/> Accessed 13.09.13.

Flores, R. M., Stricker, G. D., & Kinney, S. A. (2004). *Alaska coal geology, resources, and coalbed methane potential.* In *U.S. Geological Survey, Digital Data Series* (Vol. 77). version 1.0 CD-ROM, p. 68, 118 figs. 7 tables (Also available at <http://pubs.usgs.gov/dds/2004/77/>).

Flores, R. M., Stricker, G. D., & Kinney, S. A. (2005). *Alaska coal geology, resources, and coalbed methane potential.* U.S. Geological survey DDS 77, version 1.0.

Flores, R. M., Stricker, G. D., Meyer, J. F., Doll, T., Norton, Ph., Livingston, R. J., et al. (2001). *A field conference on impacts of coalbed methane development in the Powder River Basin, Wyoming: U.S. Geological Survey Open-File Report 01-126.* p. 26, 32 figs, 1 table. <http://greenwood.cr.usgs.gov//enrgy/OF01-126/>.

Flores, R. M., Stricker, G. D., Papasin, R. F., Pendon, R. R., del Rosario, R. A., Malapitan, R. T., et al. (2006). *The Republic of the Philippines coalbed methane assessment: Based on seventeen high pressure methane adsorption isotherms.* U.S. Geological Survey Open File Report 2006–1063, Appendices A1–49.

Flores, R. M., Stricker, G. D., Rice, C. A., Ellis, M. S., Osvald, K. S., & McGarry, D. E. (2006). *U.S. Geological Survey and Bureau of Land Management cooperative coalbed methane project in the Powder River Basin.* U.S. Geological Survey Fact Sheet 2006, 3132, p. 6. < http://energy.cr.usgs.gov/oilgas/cbmethane/> and at < http://www.wy.blm.gov/minerals/og/ogcbm.htm> Accessed 27.07.12.

Flores, R. M., Stricker, G. D., Rice, C. A., Warden, A., & Ellis, M. S. (2008). Methanogenic pathways in the Powder River Basin: the geologic factor. In R. M. Flores (Ed.), *International Journal of Coal Geology: Vol. 76. Microbes, methanogenesis, and microbial gas* (pp. 52–75).

Flores, R. M., & Sykes, R. (1993). Depositional controls on coal distribution and quality in the Eocene Brunner coal measures, Buller Coalfield, South Island, New Zealand. *International Journal of Coal Geology, 29*, 291–336.

Flores, R. M., Toth, J. C., & Moore, T. A. (1982). *Use of geophysical logs in recognizing depositional environments in the Tongue River Member of the Fort Union Formation, Powder River area, Wyoming and Montana.* U.S Geological Survey Open-File Report 82–576, p. 40.

Folger, P. (2013). *FutureGen: A brief history and issues for Congress.* Congressional Research Service, 7-5700,. <www.crs.gov>. R43028, p. 12.

Formolo, M., Martini, A., & Petsch, S. (2008). Biodegradation of sedimentary organic matter associated with coalbed methane in the Powder River and San Juan Basins, U.S.A. In R. M. Flores (Ed.), *International Journal of Coal Geology: Vol. 76. Microbes, methanogenesis, and microbial gas* (pp. 52–75).

Fort Union Coal Assessment Team. (1999). *Resource assessment of selected Tertiary coal beds and zones in the northern Rocky Mountains and Great Plains region.* U.S. Geological Survey Professional Paper 1625-A, Chapter ES, Disc 1, Version 1.0.

Forward Energy. (2009). *Coalbed methane play characterization.* < http://www.forwardenergy.ca/pdf%20files%202009/Coalbed%20Methane.pdf> Accessed 15.02.13.

Frankenberg, D. (1999). Wetlands of coastal plains. In *Carolina environmental diversity explorations.* < http://www.learnnc.org/lp/pages/4455> Accessed 27.0212.

Franklin, P. (2004). *Methane recovery opportunities from abandoned mines: A unique resource.* The Fifth Annual Coalbed & Coalmine Methane Conference. New York: Strategic Research Institute. p. 16.

Franklin, P., Jemelkova, B., & Somers, J. (2008). *Identifying U.S. coalmine methane project opportunities.* United States Environmental Protection Agency. Coalbed Methane Outreach Program, Roundtable for CMM Project Developers and Financiers, p. 34. < http://www.epa.gov/cmop/docs/cmop_ppt_roundtable_072208.pdf> Accessed 24.11.11.

Fraser, C. J. D., Roulet, N. T., & Lafeur, M. (2001). Groundwater flow patterns in a large peatland. *Journal of Hydrology, 246*, 142–154.

Frazier, D. E. (1967). Recent deltaic deposits of the Mississippi River; their development and chronology. Gulf Coast Association of Geological Societies. *Transactions, 17*, 287–315.

Friederich, M. C., Langford, R. P., & Moore, T. A. (1999). The geological setting of Indonesian coal deposits. *PACRIM, 99*, 625–631.

Friederich, M. C., Liu, G., Langford, R. P., Nas, C., & Ratanashien, B. (2000). Coal in Tertiary rift systems in Southeast Asia. In *Proceedings Southeast Asian coal geology conference* (Herudyanto, Sukarjo, Endang Djaelani, and Komaruddin, Compilers.) Directorate of Mineral Resources, pp. 33–43. ISBN: 979-8126-03-3.

Frisch, W., Meschede, M., & Blakey, R. C. (2011). *Continental drift and mountain building* (Vol. VII). Springer.

Frodsham, K., & Gaye, R. A. (1999). The impact of tectonic deformation upon coal seams in the South Wales coalfield, UK. *International Journal of Coal Geology, 38*, 297–332.

Fu, X., Qin, Y., Wang, G. G. X., & Rudolph, V. (2009). Evaluation of gas content of coalbed methane reservoirs with the aid of geophysical logging technology. *Fuel, 88*, 2269–2277.

Galand, P. E., Fritze, H., Conrad, R., & Yrjala, K. (2005). Pathways for methanogenesis and diversity of methanogenic Archaea in three boreal peatland ecosystems. *Applied Environmental Microbiology, 71*, 2195–2198.

Galatowitsch, S. M., Anderson, N. O., & Ascher, P. D. (1999). Invasiveness in wetland plants in temperate North America. *Wetlands, 19*, 733–755.

Gamson, P. D., Beamish, B. B., & Johnson, D. P. (1993). Coal microstructure and micropermeability and their effects on natural gas recovery. *Fuel, 72*, 8–99.

Gan, H., Nandi, S. P., & Walker, P. L. (1972). Nature of the porosity in American coals. *Fuel, 51*, 272–277.

Garcia, J. L., Patel, B. K. C., & Ollivier, B. (2000). Taxonomic, phylogenetic, and ecological diversity of methanogenic Archaea. *Anaerobe, 6*, 205–226.

Garcia–Gonzalez, M. (2010). *Coalbed methane resources in Colombia*. American Association of Petroleum Geologists International Conference and Exhibition, Calgary, Alberta, Canada, September 12–15, 2010.

Gatens, M. (2005). Coalbed methane development: practices and progress in Canada. *Journal of Canadian Petroleum Technology, 44*, 6.

Gazprom. (2010). *Russian forecasted CBM resources*. Gazprom <http://www.gazprom.com/production/extraction/metan/> Accessed 18.01.13.

Gazprom. (2013). *Production of coalbed methane*. <http://www.gazprom.com/about/production/extraction/metan/> Accessed 16.01.13.

Gentzis, T. (2009). Stability analysis of a horizontal coalbed methane well in the Rocky Mountain Front Ranges of southeast British Columbia, Canada. *International Journal of Coal Geology, 77*, 328–337.

Gentzis, T., Deisman, N., & Chalatumyk, R. (2009). Effect of drilling fluids on coal permeability: impact on horizontal wellbore stability. *International Journal of Coal Geology, 78*, 171–191.

Gentzis, T., Goodarzi, F., Cheung, F. K., & Laggoun-Defarge, F. (2008). Coalbed methane producibility from the Manville coals in Alberta, Canada: a comparison of two areas. *International Journal of Coal Geology, 74*, 237–249.

Geological Survey of India. (1977). *Coal resources of India*. In *Memoir Geological Survey of India*. 575.

Geological Survey of India. (2012). Coal, lignite, coal bed methane (CBM) exploration: "Standard operational procedure" (SOP). In *Part II, Geological Survey of India-Mission II – Final SOP on mineral exploration and coal, lignite & CBM exploration* (pp. 87–98). <http://www.portal.gsi.gov.in/portal/page?_pageid=108, 669637&_dad=portal&_schema=PORTAL> Accessed 21.06.12.

Georage, A. M., & Mackay, G. H. (1991). Petrology. In R. A. Durie (Ed.), *The science of Victorian brown coal: Structure, properties and consequences for utilisation* (pp. 45–102). Butterworth: Heinemann.

George, A. M. (1975). *Brown coal lithotypes in the Latrobe Valley deposits*. Victoria State Electricity Commission Planning and Investigation Department. Petrological Report, No. 17, p. 36.

Geoscience Australia. (2005). *2005 coalbed methane*. Australian Government. <http://www.ga.gov.au/oceans/pgga_ogra2005_CoalbedMethane.jsp> Accessed 18.02.13.

Geoscience Australia and BREE. (2012). *Australian gas resource assessment 2012*. Canberra. <http://www.bree.gov.au/documents/publications/_other/gasresource assessment.pdf> Accessed 27.01.13.

Gerami, S., Pooladi-Darvish, M., Morad, K., & Mattar, L. (2007). *Type curves for dry CBM reservoirs with equilibrium desorption*. Paper 2007-011, Petroleum Society's Canadian Petroleum Conference, Calgary, Alberta, Canada, June 12–14 2007.

Gilman, A., & Beckle, R. (2000). Flow of coal-bed methane to a gallery. In *Earth and environmental science: Vol. 41. Transport in Porous Media* (pp. 1–16).

Glaser, P. H., Chanton, J. P., Morin, P., Rosenberry, D. O., Siegel, D. I., Ruud, O., et al. (2004). Surface deformations as indicators of deep ebullition fluxes in a large northern peatland. *Global Biogeochemical Cycles, 18*, GB1003. http://dx.doi.org/10.1029/2003GB002069.

Gleeson, T., Wada, Y., Bierkens, M. F. P., & van Beek, L. P. H. (August 9, 2012). Water balance of global aquifers revealed by groundwater footprint. *Nature, 488*, 197–200. Letter Research. <http://civil-staff.mcgill.ca/gleeson/publications/Gleeson%20et%20al%202012%20Nature_Groundwater%20footprint.pdf> Accessed 23.12.12.

GLG. (May 25, 2010). *China launching the second wave of coal industry consolidation in Henan*. Gerson Lehrman Group. <http://www.glgroup.com/News/China-launching-the-second-wave-of-coal-industry- consolidation-in-Henen-48616.html. > Accessed 23.11.11.

Glossner, A., Schmidt, R., Flores, R. M., & Mandernack, K. (2008). *Phospholipid evidence for methanogenic Archaea and sulfate-reducing bacteria in coalbed methane wells in the Powder River Basin, Wyoming: EOS Transactions*. American Geophysical Union National Meeting, San Francisco, December 8–13, 2008, 89, no. 53.

Glossner, A., Schmidt, R., Flores, R. M., & Mandernack, K. (2009). *Evidence of phospholipids for methanogenic Archaea and sulfate-reducing bacteria in coalbed methane wells in the Powder River Basin, Wyoming*. Denver, Colorado: American Association of Petroleum Geologists National meetings. June 7–10, 2009.

Gloyn, R. W., & Sommer, S. N. (May 31, 1993). Exploration for coalbed methane gains momentum in Uinta Basin. *Utah Geological Survey, Oil & Gas Journal, Exploration*, 73–76.

Gluskoter, H. R. (2009). Coal resources and reserves. In *Meeting projected coal production demands in the U.S.A* (pp. 30–67).

National Commission of Energy Policy, The Virginia Center for Coal and Energy Research, Virginia Polytechnic Institute and State University. <www.energy.vt.edu/NCEP Study/.../Coal_Production_Demands_Chapter2.p. >. . . . File Format: PDF/Adobe Acrobat – View as HTML Accessed 23.09.11.

Gochioco, L. M. (1992). Modeling studies of interference reflections in thin-layered media bounded by coal seams. *Geophysics, 57*, 1209–1216.

Golding, S. D., Rudolph, V., & Flores, R. M. (Eds.), (2010). Asia Pacific coalbed methane symposium: selected papers from the 2008 Brisbane symposium on coalbed methane and CO_2-enhanced coalbed methane. *International Journal of Coal Geology, 82*, 298.

Golitsyn, A., Courel, L., & Debriette, P. (1997). A fault-related coalification anomaly in the Blanzy-Montceau Coal Basin (Massif Central, France). *International Journal of Coal Geology, 33*, 209–228.

van der Gon, H. A. C. D., & Neue, H. U. (1995). Influence of organic matter incorporation on the methane emission from a wetland rice field. *Global Biogeochemical Cycles, 9*, 11–22.

Gopal, B. (1999). Natural and constructed wetlands for wastewater treatment: potentials and problems. *Water Science Technology, 40*, 27–35.

Gore, A. J. P. (Ed.), (1983). *Ecosystems of the world, Mires: Swamp, bog, fen, and moor: Vol. 4A*. Elsevier, Amsterdam.

Gorham, E., Janssens, J. A., & Glaser, P. H. (2003). Rates of peat accumulation during the postglacial period in 32 sites from Alaska to Newfoundland, with special emphasis on northern Minnesota. *Canadian Journal of Botany, 81*, 429–438.

Gorody, A. W. (1999). The origin of natural gas in the Tertiary coal seams on the eastern margin of the Powder River Basin. In W. R. Miller (Ed.), *Coalbed methane and the Tertiary geology of the Powder River Basin: Wyoming* (pp. 89–101). Geological Association, 50th Annual Field Conference Guidebook.

Gossling, J. M. (1994). *Coalbed methane potential of the Hartshorne coals in parts of Haskell, Latimer, LeFlore, McIntosh, and Pittsburg Counties*. Oklahoma: Norman, University of Oklahoma, unpublished M.S. thesis, 155 p.

Gradall, K. S., & Swenson, W. A. (1982). Responses of brook trout and creek chubs to turbidity. *Transactions of the American Fisheries Society, 111*, 392–395.

Granau, H. R. (1984). Natural gas in major basins worldwide attributed to source type, thermal maturity and bacterial origin. *Proceedings of the 11th World Petroleum Congress, 2*, 293–302.

Gray, I. (1987). *Reservoir engineering in coal seams: Part 1. The physical process of gas storage and movement in coal seams* (Vol. 2). Society of Petroleum Engineers, Reservoir Engineering. pp. 28–34.

Greb, S. F., Andrews, W. M., Eble, C. F., DiMichele, W., Cecil, C. B., & Hower, J. C. (2003). Desmoisnesian coal beds of the Eastern Interior and surrounding basins: the largest peat mires in earth history. In M. A. Chan, & A. W. Archer (Eds.), *Geological Society of America Special Paper: Vol. 370. Extreme depositional environments: Mega end members in geologic time* (pp. 127–150).

Greb, S. F., DiMichele, Q. W. A., & Gastaldo, R. A. (2006). Evolution and importance of wetlands in the Earth's history. In S. F. Greb, & W. A. DiMichele (Eds.), *Geological Society of America Special Paper: Vol. 399. Wetlands through time* (pp. 1–40).

Greb, S. F., Eble, C. F., & Hower, J. C. (1999). Depositional history of the fire clay coal bed (Late Duckmantian), eastern Kentucky, USA. *International Journal of Coal Geology, 26*, 255–280.

Greb, S. F., Eble, C. F., Peters, D. C., & Papp, A. R. (2006). Coal and the environment. American Geological Institute. *Environmental Awareness Series, 10*, 64.

Green, M. S., Flanegan, K. C., & Gilcrease, P. C. (2008). Characterization of a methanogenic consortium enriched from a coalbed methane well in the Powder River Basin, U.S.A. In R. M. Flores (Ed.), *International Journal of Coal Geology: Vol. 76. Microbes, methanogenesis, and microbial gas in coal* (pp. 34–45).

Gregg, S. J., & Sing, K. S. W. (1982). *Adsorption, surface area and porosity* (2nd ed.). Academic Press. p. 303.

Gu, F., & Chalaturnyk, R. J. (2006). Numerical simulation of stress and strain due to gas sorption/desorption and their effects on in situ permeability of coalbeds. *Journal of Canadian Petroleum Technology, 45*, 1–11.

Guar Gum. (2006). *Guar gum*, 88/2 G.I.D.C. Estate, Vatva, Ahmedabad – 382 445, Gujarat, India. <http://www.guargum.biz/index.html> Accessed 12.10.12.

Guion, P. D., Fulton, I. M., & Jones, N. S. (1995). Sedimentary facies of the coal-bearing Westphalian A and B north of the Wales-Brabant High. In M. K. G. Whateley, & D. A. Spears (Eds.), *Geological Society, London, Special Publication: Vol. 82. European coal geology* (pp. 45–78).

Gulik, C. (2008). *Natural gas prices in recession. Natural gas & electricity*. Wiley Periodicals, Inc. <http://bateswhite.com/media/pnc/3/media.293.pdf. > Accessed January 2013.

Gunter, W., Wong, S., Law, D., Sanli, F., Jianping, Y., & Zhiqiang, F. (2005). *Enhanced coalbed methane (ECBM) field test at south Qinshui Basin, Shanxi Province, China*. Global Climate and Energy Project, Stanford University. GCEP Workshop, August 23–25, 2005, China. <http://gcep.stanford.edu/pdfs/wR5MezrJ2SJ6NfFl5sb5Jg/15_china_gunter.pdf> Accessed 23.02.13.

Guo, B. (2006). Status and inspiration of CBM exploration and production in Hancheng Block. In J. Ye, & Z. Fan (Eds.), *Progress in exploration, development and utilization techniques of coalbed methane in China* (pp. 64–67). Geological Publishing House (in Chinese).

Guo, H., Liu, R., Yu, Z., Zhang, H., Yun, J., Li, Y., et al. (2012). Pyrosequencing reveals the dominance of methylotrophic methanogenesis in a coalbed methane reservoir associated with Eastern Ordos Basin in China. *International Journal of Coal Geology, 93*, 56–61.

Gupta, R. K., Bhumbla, D. K., & Abrol, I. P. (1984). Effect of sodicity, pH, organic matter, and calcium carbonate on the dispersion behavior of soils. *Soil Science, 137*, 245–251.

Gurba, L. W., Gurba, A., Ward, C. R., Wood, J., Filipowski, A., & Titheridge, D. (2001). The impact of coal properties on gas drainage efficiency. In R. Doyle, & J. Moloney (Eds.), *Geological Hazards—The impact to mining* (pp. 215–220). Newcastle, Australia: Coalfield Geology Council of New South Wales.

Gurba, L. W., & Weber, C. R. (2000). *Coal petrology and coalbed methane occurrence in the Gloucester Basin, NSW, Australia.* The Society of Organic Petrology (TSOP), 17th Annual Meeting Abstracts and Program.

Gurba, L. W., & Weber, C. R. (2001). Effects of igneous intrusions on coalbed methane potential, Gunnedah Basin, Australia. *International Journal of Coal Geology, 46*, 113–131.

Gurdal, G., & Yalcin, N. M. (2001). Pore volume and surface area of Carboniferous coals from the Zonguldak basin (NW Turkey) and their variations with rank and maceral composition. *International Journal of Coal Geology, 48*, 133–144.

Guzman, R. (June 8, 2011). *Potential resources of unconventional hydrocarbons in Colombia.* Bogota, Colombia: ANH Unconventional Hydrocarbons Workshop.

Habitch, J. K. A. (1979). *Paleoclimate, paleomagnetism and continental drift.* In *Studies in Geology* (Vol. 9). American Association of Petroleum Geologists. p. 29, 11 fold out plates.

Hagemann, H. W., & Hollerbach, A. (1980). Relationship between the macropetrographic and organic geochemical composition of lignites. *Physics and Chemistry of the Earth, 2*, 631–638.

Hagmaier, J.L. (1971). *Groundwater flow, hydrogeochemistry, and uranium deposition in the Powder River Basin, Wyoming* (Ph.D. Thesis), Grand Forks: University of North Dakota, p. 166.

Hall, H. (1997). *Cenozoic tectonics of SE Asia and Australia.* Proceedings of the Petroleum Systrems of SE Asia and Australasia Conference, May 1997, 47–32.

Hall, R., & Wilson, M. E. J. (2000). Neogene sutures in eastern Indonesia. *Journal of Asian Earth Sciences, 18*, 781–808.

Halliburton. (2007). Water production and disposal. In *Coalbed Methane: Principles and practices* (pp. 421–460). Halliburton Inc. < http://www.halliburton.com/public/pe/contents/Books_and_Catalogs/web/CBM/H06263_Chap_09.pdf> Accessed 15.12.12.

Halliburton. (2007). *Coalbed methane: Principles and practices.* Halliburton Inc. < http://www.halliburton.com/public/pe/contents/Books_and_Catalogs/web/CBM/CBM_Book_Intro.pdf> Accessed 09.10.12.

Ham, Y., & Kantzas, A. (2008). *Development of coalbed methane in Australia: Unique approaches and tools.* Society of Economic Petroleum Engineers. CIPC/SPE Gas Technology Symposium 2008 Joint Conference, June 16–19, 2008, Calgary, Alberta, Canada.

Hamilton, D. S., & Tadros, N. Z. (1994). Utility of coal seams as genetic stratigraphic sequence boundaries in nonmarine basins: an example from the Gunnendah basin, Australia. *American Association of Petroleum Geologists Bulletin, 78*, 267–286.

Hammond, R. (1981). *The peatlands of Ireland.* Dublin, Ireland: An Foras Talu'ntais.

Hancock, P. J. (1985). Brittle microtectonics: Principles and practices. *Journal of Structural Geology, 7*, 437–458.

Hanson, B., Grattan, S., & Fulton, A. (1999). *Water management handbook series. Agricultural salinity and drainage Publication #93–01alifornia irrigation program.* Davis: University of California. p. 141.

Hardjono, O., & Supardi (1989). Gambut di Sumatra. Direktorat Sumberdaya Mineral, Bandung, Laporan No. P/001/DSM, p. 17.

Hardwicke, R. E. (1935). The rule of capture and its implications as applied to oil and gas. *Texas Law Review, 401*.

Hargraves, A. J. (1983). Instantaneous outbursts of coal and gas— a review. *Proceedings of Australia's Institute of Mining and Metallurgy, 285*, 1–37.

Harpalani, S., & Chen, G. (1995). Estimation of changes in fracture porosity of coal with gas emission. *Fuel, 74*, 1491–1498.

Harpalani, S., & Chen, G. (1997). Influence of gas production induced volumetric strain on permeability of coal. *Geotechnical & Geological Engineering, 15*, 303–325.

Harpalani, S., & Schraugnagel, A. R. (1990). *Influence of matrix shrinkage and compressibility on gas production from coalbed methane reservoirs.* Proceedings of Annual Technical Conference and Exhibition, September 1990. Society of Petroleum Engineers. Paper SPE 20729.

Harris, L. A., & Yust, C. S. (1976). Transmission electron microscopy observations of porosity in coal. *Fuel, 55*, 233–236.

Harris, S. H., Smith, R. L., & Barker, C. E. (2008). Microbial and chemical factors influencing methane production in laboratory incubations of low-rank subsurface coals. In R. M. Flores (Ed.), *International Journal of Coal Geology: Vol. 76. Microbes, methanogenesis, and microbial gas in coal* (pp. 46–51).

Harrison, M. J., Marshak, S., & Onasch, C. M. (2004). Stratigraphic control of hot fluids on anthracitization,

Lackawanna synclinorium, Pennsylvania. *Tectonophysics, 378*, 85–103.

Harrison, S. M., Gentzis, T., & Payne, M. (2006). Hydraulic, water quality, and isotopic characterization of Late Cretaceous–Tertiary Ardley coal waters in a key test-well, Pembina–Warburg exploration area, Alberta, Canada. *Bulletin of Canadian Petroleum Geology, 54*, 261–272.

Hart, G. F., & Coleman, J. (2004). *The world deltas database framework.* Louisiana State University. <www.geol.lsu.edu/WDD>.

Hartman, R. C., Lasswell, P., & Bhatta, N. (2008). *Recent advances in the analytical methods used for shale gas reservoir gas-in-place assessment.* American Association of Petroleum Geologists. Presented at the Annual Convention, San Antonio, Texas, April 20–23, 2008. <http://www.searchanddiscovery.com/documents/2008/08209hartman/hartman-40317.pdf> Accessed 26.08.11.

Hatcher, P. G., Berger, I. A., Szeverenyi, N., & Maciel, G. E. (1982). Nuclear magnetic resonance studies of ancient buried wood, II. observations on the origin of coal from lignites to bituminous coal. *Organic Geochemistry, 4*, 9–18.

Hatcher, P. G., Wenzel, K. A., & Faulon, J. L. (1993). Reactions of wood during early coalification, a clue to the structure of vitrinite. *American Chemical Society, Division of Fuel Chemistry Preprints, 38*, 1270–1272. <http://web.anl.gov/PCS/acsfuel/preprint%20archive/Files/38_4_CHICAGO_08-93_1270.pdf> Accessed 18.03.12.

Hawkins, J. M., Schraufnagel, R. A., & Olszewski, A. J. (1992). *Estimating coalbed gas content and sorption isotherm using well log data.* Society of Petroleum Engineers, 67th Annual SPE Technical Conference and Exhibition. Washington, DC. SPE 24905.

Healy, R. W., Rice, C. A., Bartos, T. T., & McKinley, M. P. (2008). Infiltration from an impoundment for coal-bed natural gas, Powder River Basin, Wyoming: evolution of water and sediment chemistry. *Water Resources, 44*, 16. <http://www.agu.org/pubs/crossref/2008/2007WR006396.shtml> Accessed 24.12.12.

Heath, R. C. (1993). *Basic ground-water hydrology.* U.S. Geological Survey. Water-Supply Paper 2220, p. 86.

Hedberg, H. D. (1968). Significance of high-wax oils with respect to genesis of petroleum. *American Association of Petroleum Geologists Bulletin, 52*, 736–750.

Heffern, E. L., & Coates, D. A. (1999). Hydrogeology and ecology of clinker in the Powder River Basin, Wyoming and Montana. In W. R. Miller (Ed.), *Coalbed methane and the Tertiary geology of the Powder River Basin, Wyoming and Montana* (pp. 231–252). Wyoming Geological Association. Fiftieth Field Conference Guidebook.

Henderson, R. E., & Doiron, R. (1981). Some identification hints for the field classification of peat. In *Proceedings of organic soils mapping workshop, Fredericton, New Brunswick* (pp. 105–110). Ottawa: Agriculture Canada Land Resources Research Institute.

Hendrix, M. S., Brassell, S. C., Carroll, A. R., & Graham, S. A. (1995). Sedimentology, organic geochemistry, and petroleum potential of Jurassic coal measures: Tarim, Junggar, and Turpan Basins, northwest. *American Association of Petroleum Geologists Bulletin, 79*, 929–959.

Hensel, W. M., Jr. (1989). *A perspective look at fracture porosity, 4.* Society of Petroleum Engineers. Formation Evaluation, 4, 531–534.

Herudiyanto. (2006). *Possible explanation of the increasing coal rank at Sangatta.* Internal Report, Geological Resource Centre, Geological Agency of Indonesia, Bandung. p. 16.

Heward, A. P. (1978). Alluvial fan and lacustrine sediments from the Stephanian A and B (La Magdalena, Cinera–Matallana and Sabero) coalfields, northern Spain. *Sedimentology, 25*, 451–488.

Higley, D. K. (2007). *Petroleum systems and assessments of undiscovered oil and gas in the Raton Basin–Sierra Grande Uplift Province, Colorado and New Mexico–USGS Province 41.* U.S Geological Survey (Chapter 2), p. 124, Digital Data Series DDS–69–N. < http://pubs.usgs.gov/dds/dds-069/dds-069-n/REPORTS/69_N_CH_2.pdf. > Accessed 10.11.12.

Hinaman, K. (2005). *Hydrologic framework and estimates of ground-water volumes in Tertiary and Upper Cretaceous hydrogeologic units in the Powder River Basin.* Wyoming: U.S. Geological Survey Scientific Investigations Report 2005–5008, p. 18.

Hobbs, R. G. (1978). *Methane occurrences, hazards, and potential resources: Recluse analysis area, northern Campbell County, Wyoming.* U.S. Geological Survey Open-File Report 78–401, p. 20.

Hoefer, H. (1915). *Schwundspalten (Schlechten, Lassen) (Shrinkage cracks- cleats, separations).* Vienna: Mitteilungen der Geologischen Gesellschaft, 7, Nos. 1 and 2.

Holdgate, G. R. (1984). *Latrobe Valley brown coals, their subsurface distribution; with appendix 1: the brown coal geological resources of the Gippsland Basin and their comparison to previous estimates* (Vol. 29). Victorian State Electricity Commission, Exploration and Geological Division. Fuel Department Report p. 16.

Holdgate, G. R. (2003). Coal, world-class energy reserves without limits. InW. D. Birch (Ed.), *Geology of Victoria* (Vol. 23; pp. 489–516). Geological Society of Australia. Special Publication.

Holdgate, G. R. (2005). Geological processes that control lateral and vertical variability in coal seam moisture contents–Latrobe Valley (Gippsland Basin) Australia. *International Journal of Coal Geology, 63*, 130–155.

Holdgate, G. R., Kershaw, A. P., & Sluiter, I. R. K. (1995). Sequence stratigraphic analysis and the origins of

Tertiary brown coal lithotypes, Latrobe Valley, Gippsland basin, Australia. *International Journal of Coal Geology, 28*, 249–275.

Holdgate, G. R., & Sluiter, I. R. K. (1991). Oligocene–Miocene marine incursions in the Latrobe Valley depression, onshore Gippsland Basin; evidence, facies relationships and chronology. In M. A. J. De Williams, P. Dekker, & A. P. Kershaw (Eds.), *The Cainozoic in Australia. A re-appraisal of the evidence* (Vol. 18; pp. 137–157). Geological Society of Australia. Special Publication.

Holdgate, G. R., Wallace, M. W., Gallagher, S. J., & Taylor, D. (2000). A review of the Traralgon formation in the Gippsland Basin—a world class brown coal resource. *International Journal of Coal Geology, 45*, 55–84.

Holditch, S. A. (2001). *The increasing role of unconventional reservoirs in the future of the oil and gas business.* <http://www.spegcs.org/attachments/studygroups/6/Holditch2001.pdf>

Holditch, S. A. (2007). *Unconventional gas, Topic Paper #29. Working document of The National Council Global Oil and Gas Study.* p. 52.

Holditch, S. A. (2009). *Stimulation of tight gas reservoirs worldwide.* Offshore Technology Conference, Houston, Texas, May 4–7, 2009, OTC 20267, p. 12. <http://www.holditch.com/Portals/66/OTC-20267_Stimulation%20of%20TGRs.pdf> Accessed 11.10.12.

Holditch, S. A. (2009). *The role of OICs and NOCs in developing unconventional oil and gas reservoirs.* THEWAYAHEAD, Pillars of the Industry, 5, p. 2. <http://www.holditch.com/Portals/66/The%20Role%20of%20IOCs%20and%20NOCs%20in%20UGRs_TWA_Sept.pdf> Accessed 29.08.12.

Holditch, S. A., Perry, K., & Lee, J. (2007). *Unconventional gas reservoirs – Tight gas, coal seams, and shales.* National Petroleum Council, Working Document of the NPC Global Oil and Gas Study, Topic Paper #29, Unconventional Gas, p. 52.

Hong, B. D., & Slatick, E. R. (1994). *Carbon dioxide emission factors for coal.* Quarterly Coal Report, DOE/USEIA-0121(94/Q1), Energy Information Administration, pp. 1–8. <http://www.USEIA.doe.gov/cneaf/coal/quarterly/co2_article/co2.html> Accessed 15.11.09.

Honore, A. (2011). *Economic recession and natural gas demand in Europe: What happened in 2008–2010?* Oxford Institute for Energy Studies, University of Oxford. <http://www.oxfordenergy.org/wpcms/wp-content/uploads/2011/01/NG47-EconomicRecessionandNaturalGasDemandinEurope-whathappenedin20082010-AnoukHonore-2011.pdf> Accessed 16.01.13.

Hooijer, A., Silvius, M., Wösten, H., & Page, S. (2006). *Assessment of CO_2 emissions from drained peatlands in South east Asia.* Delft Hydraulics Report Q3943(2006). <http://www.wetlands.org/publication.aspx?ID=51a80e5f-4479-4200-9be0-66f1aa9f9ca9>.

Hook, M. (2007). *Peat: Energy from the bogs.* Uppsala Hydrocarbon Depletion Study Group, Uppsala University. p. 8. <http://www.tsl.uu.se/uhdsg/Popular/Peat.pdf> Accessed 26.01.12.

Hook, M., & Aleklett, K. (2009). Historical trends in American coal production and possible future outlook. *International Journal of Coal Geology, 78*, 201–216.

Hook, M., & Aleklett, K. (2010). A review on coal to liquid fuels and its coal consumption. *International Journal of Energy Research, 34*, 848–864.

Hook, M., & Aleklett, K. (2010). Trends in United States recoverable coal supply estimates and future production outlooks. *Natural Resources Research, 19*, 189–208.

Hook, M., Zittel, W., Schindler, J., & Aleklett, K. (2008). *A supply-driven forecast for the future global coal production.* Contribution to Association for the Study of Peak Oil 2008, International, Klintvagen 42, 75655 Uppsala, Sweden, p. 48.

Hook, M., Zittel, W., Schindler, J., & Aleklett, K. (2010). Global coal production models based on a logistic model. *Fuel, 89*, 3546–3558.

Horgornez, H., Yalcun, M. N., Cramer, B., Gerlbing, P., Faber, E., Schaefer, R. G., et al. (2002). Isotopic and molecular composition of coal-bed gas in Amarva region (Zonguidak basin—western Black Sea. *Organic Geochemistry, 33*, 1429–1439.

Horkel, A. (1989). On the plate-tectonic setting of the coal deposits of Indonesia and the Philippines. *Mitteilungen der Österreichischen Geologischen Gesellschaft, 82*, 119–133.

Horne, J. C., Ferm, J. C., Carrucio, F. T., & Baganz, B. P. (1978). Depositional models in coal exploration and mine planning in Appalachian region. *American Association of Petroleum Geologists Bulletin, 62*, 2379–2411.

Horsfield, B., & Rullkotter, J. (1994). Diagenesis, catagenesis, and metagenesis of organic matter: Chapter 10: Part III. Processes. In *American Association of Petroleum Geologists Memoir: Vol. 60. The petroleum system—from source to trap* (pp. 189–199).

Horsfield, C. F., Yordy, K. L., & Crelling, J. C. (1988). Determining the petroleum-generating potential of coal using organic geochemistry and organic petrology. *Organic Geochemistry, 13*, 121–129.

Hosgormez, H., Yalcin, M. N., Cramer, B., Gerling, P., Faber, E., Schaefer, R. G., et al. (2002). Isotopic and molecular composition of coal-bed gas in the Amasra region (Zonguldak basin—western Black Sea). *Organic Geochemistry, 33*, 1429–1439.

Hotchkiss, W. R., & Livings, J. F. (1986). *Hydrogeology and simulation of water flow in strata above the Bearpaw Shale and equivalents of eastern Montana and northeastern Wyoming.* U.S. Geological Survey Water-Resources Investigations Report 85-4281, p. 72.

Hou, J., Hong, D., & Hao, W. (2002). *Evaluation methods for coalbed methane reservoirs using log data and their application to the Eastern Hollow, Liaohe Basin, China*, 2002 American Association of Petroleum Geologists Annual Meeting, Houston, Texas.

Hower, J. C. (1990). Hardgrove grindability index and petrology used as an enhanced predictor of coal feed rate. *Energeia, 1*(6), 1–2.

Hower, J. C., Calder, J. H., Eble, C. F., Scott, A. C., Robertson, J. D., & Blanchard, L. J. (2000). Metalliferous coals of the Westphalian A Joggins Formation, Cumberland basin, Nova Scotia, Canada: petrology, geochemistry and palynology. *International Journal of Coal Geology, 42*, 185–206.

Hower, J. C., Esterle, J. S., Wild, G. D., & Pollock, J. D. (1990). Perspectives on coal lithotype analysis. *Journal of Coal Quality, 9*, 48–52.

Hower, J. C., & Gayer, R. A. (2002). Mechanisms of coal metamorphism: case studies from Paleozoic coalfields. *International Journal of Coal Geology, 50*, 215–245.

Hower, J. C., Graese, A. M., & Klapheke, J. G. (1987). Influence of microlithotype composition on Hardgrove grindability for selected eastern Kentucky coal. *International Journal of Coal Geology, 7*, 227–244.

Hower, J. C., Greb, S. F., Kuehn, K. W., & Eble, C. F. (2009). Did the Middlesboro, Kentucky, bolide impact event influence coal rank? *International Journal of Coal Geology, 79*, 92–96.

Hower, J. C., Rathbone, R. F., Robl, T. L., Thomas, G. A., Haeberlin, B. O., & Trimble, A. S. (1998). Case study of the conversion of tangential- and wall-fired units to low-$NO_{(x)}$ combustion: impact on fly ash quality. *Waste Management, 17*, 219–229.

Hower, J. C., & Wagner, N. J. (2012). Notes on the methods on the combined maceral/microlithotype determination in coal. *International Journal of Coal Geology, 95*, 47–53.

Hower, J. C., Williams, D. A., Eble, C. F., Sakulpitakphon, T., & Moecher, D. P. (2001). Brecciated and mineralized coals in Union County, Western Kentucky coal field. *International Journal of Coal Geology, 47*, 223–234.

Hsi, H. P., & Li, E. (2010). *Coalbed methane in China*. Equity Research, Asia Pacific Oil & Gas, Standard Chartered, p. 20. <http://www.enviro-energy.com.hk/admin/uploads/files/1292913959Coalbed_combined%20(Standard%20Chartered)(Dec%202010).pdf.> Accessed 25.10.12.

Hsieh, F. S. (2003). Reserves estimation for a coalbed methane well. *Journal of Canadian Petroleum Technology, 42*, 6. <http://www.methanetomarkets.org/expo/docs/postexpo/coal_santillan.pdf>.

Hu, Q., Chen, J., & Xi, L. (2012). *CBM development through surface boreholes in active and gob areas in China*, 34th International Geological Congress, August 5–10, Brisbane, Australia.

Huang, S. (2007). *Progress and project opportunities of the CMM development and utilization in China*. Presented at the Methane to Markets Partnership Expo, Beijing, China, October 30–November 1, 2007. <http://www.google.com/search?client=safari&rls=en&q=Huang,+S.+(2007).+Progress+and+Project+Opportunities+of+the+CMM+Development+and+Utilization+in+China&ie=UTF-8&oe=UTF-8> Accessed 24.11.11.

Huang, S. (2008). *Development and utilization of coalmine methane in China*. Presented at the Ninth International Symposium on CBM/CMM and Carbon Trading in China, Beijing, China, December 4, 2008.

Huang, S. (2010). *Great potential for CBM/CMM recovery and utilization and preferential policies*. Presented at the Methane to Markets Partnership Expo, New Delhi, India, March 3, 2010. <http://www.globalmethane.org/expo_india10/docs/postexpo/coal_shengchu.pdf> Accessed 28.01.13.

Huang, S. (2013). Current situations of CBM?CMM recovery and utilzation & methane emission reduction in China. In *Global Methane Initiative Expo*. Canada: Vancouver. March 12_15, 2013. (Accessed on September 20, 2013 at < https://www.globalmethane.org/expo-docs/canada13/coal_02_Huang_UPDATED.pdf>.

Huang, F. Y. C., & Natrajan, P. (2006). Feasibility of using natural zeolites to remove sodium from coal bed methane-produced water. *Journal of Environmental Engineering, 132*, 1644–1650.

Huang, X., Zhang, L., Jia, G., & Wang, Y. (2009). *Coalbed methane potential of Qinshui Basin, Shanxi, China*. Rio de Janeiro, Brazil: American Association of Petroleum Geologists. International Conference and Exhibition, November 15–18.

Huat, B. B. K., Kazemian, S., Prasad, A., & Barghchi, M. (2011). State of art review of peat: general perspective. *International Journal of the Physical Sciences, 8*, 1988–1996.

Hubbard, B. (1950). Coal as a possible petroleum source rock. *American Association of Petroleum Geologists Bulletin, 34*, 2347–2351.

Hubbert, M. K. (1956). *Nuclear energy and the fossil fuels*. Southern District Meeting March 7–9, 1956 at San Antonio, Texas. American Petroleum Institute. p. 40. <http://www.hubbertpeak.com/hubbert/1956/1956.pdf> Accessed 04.11.11.

Hubbert, M. K. (May 1982). *Techniques of prediction as applied to production of oil and gas*. In *NBS special publication* (Vol. 631). U.S. Department of Commerce.

Hubbert, M. K., & Rubey, W. W. (1959). Mechanics of fluid-filled porous solids and its application to overthrust faulting. *Geological Society of America Bulletin, 70*, 115–166.

Hucka, V. J., & Bodily, D. M. (1989). Methane drainage from gassy western U.S. coalseams. In *Proceedings from the international symposium on today's technology for the mining and metallurgical industries Kyoto, Japan*. Published by The Mining and Material Processing Institute of Japan. pp. 453–460.

Huffman, G. K., & Brister, B. S. (2003). New Mexico's Raton coalbed methane play. *New Mexico Geology, 25*, 95–110. <http://geoinfo.nmt.edu/publications/periodicals/nmg/downloads/25/n4/nmg_v25_n4_p.95.pdf> Accessed 12.11.12.

Hughes, J. D., Klatzel-Mudry, L., & Nikols, D. J. (1988). *A standardized coal resource/reserve reporting system for Canada/Méthode d'évaluation normalisée des ressources et des réserves canadiennes de charbon*. Geological Survey of Canada, Paper 88–21, 1989, p. 17.

Hull, E. (1861). *The coal-fields of Great Britain: Their history, structure, and resources* (2nd ed.). London: Edward Stanford.

Hunt, J. M. (1979). *Petroleum geochemistry and geology* (1st ed.). San Francisco: W. H. Freeman and Co. p. 617.

Hunt, J. W. (1989). Permian coals of eastern Australia: geological control to of petrographic variation. In P. C. Lyons, & B. Alpern (Eds.), *International Journal of Coal Geology: Vol. 12. Peat and coal: Origin, facies, and depositional models* (pp. 589–634).

Hunt, J. M. (1991). Generation of gas and oil from coal and other terrestrial organic matter. *Organic Geochemistry, 17*, 673–680.

Hutton, A. C. (2009). Geological setting of Australasian coal deposits. In R. Kininmonth, & E. Baafi (Eds.), *Australasian coal mining practice* (pp. 40–84). 15-31 Pelham Street, Carlton Victoria 3053: The Australasian Institute of Mining and Metallurgy.

Hunt, A. M., & Steele, D. J. (1991). *Coalbed methane development in the northern and central Appalachian Basins—Past, present and future*. Proceedings Coalbed Methane Symposium, Tuscaloosa, Alabama (May 1991) 127–141.

ICCP. (1963). *International handbook of coal petrology* (2nd ed.). Paris, France: Centre National de la Recherche Scientifique. International Committee for Coal Petrology.

ICCP. (1971). *International handbook on coal petrography*. International Committee for Coal Petrology. First Supplement to Second Edition, Paris, France (Cent. Nat. Rech. Sci.) p. 186.

ICCP. (1993). *International handbook of coal petrography*. International Committee for Coal Petrology. Third supplement to the 2nd Edition.

ICCP. (1998). The new vitrinite classification (ICCP System 1994). *Fuel, 77*, 349–358.

ICCP. (2001). The new inertinite classification (ICCP system 1994). *Fuel, 80*, 459–471.

ICCP. (2011). *Organic petrology, macerals, microlithotypes, lithotypes, minerals, rank*. International Commission of Coal Petrology, Training Course on Dispersed Organic Matter, Chapter 2. Portugal: Universidada do Porto, 70.

ICF Resources. (1990). The United States coalbed methane resource. *Quarterly Review of Methane from Coal Seams Technology, 7*, 10–28.

ICMM. (2010). Fugitive methane emissions in coal mining: climate change. *International Council on Mining and Metals*, 3. < http://www.google.com/search?client=safari&rls=en&q=figitive+coal+mine+emissions&ie=UTF-8&oe=UTF-8> Accessed on 27.11.11.

IEA. (2006a). *Focus on clean coal*. International Energy Agency. p. 8. <http://www.iea.org/publications/freepublications/publication/focus_on_coal.pdf> Accessed 07.04.13.

IEA. (2006b). *Ukraine: Energy policy review 2006*. International Energy Agency. OECD Publishing. <www.iea.org/w/bookshop/pricing.html> Accessed 30.01.13.

IEA. (2007). *Annual energy outlook, China and India Insights*. International Energy Agency. <http://www.iea.org/textbase/nppdf/free/2007/weo_2007.pdf> Accessed 19.09.11.

IEA. (2009a). *Coal information 2009 edition, documentation for beyond 2020 files*. International Energy Agency. p. 18. <http://wds.iea.org/wds/pdf/doc_Coal_2009.pdf> Accessed 20.03.12.

IEA. (2009b). *Coalmine methane in China: A budding asset with the potential to bloom*. Paris, France: International Energy Agency. Information Report, February 2009, p. 35. <http://www.google.com/search?client=safari&rls=en&q=IEA+(2009).+Coal+mine+methane+in+China:+A+budding+asset+wi+the+potential+to+bloom.&ie=UTF-8&oe=UTF-8> Accessed 11.11.11.

IEA. (2011). *World energy outlook 2011*. International Energy Agency. ISBN: 978-92-64-12413-4. <http://www.worldenergyoutlook.org/publications/weo-2011/> Accessed 15.01.13.

IEA. (2012a). *Coal's share of global energy mix to continue rising with coal closing in on oil as world's top energy source by 2017*. International Energy Agency. <http://www.iea.org/newsroomandevents/pressreleases/2012/ december/name, 34441, en.html> Accessed 13.01.13.

IEA. (2012b). *The medium-term coal market report 2012 factsheet*. International Energy Agency. <http://www.iea.org/publications/medium-termreports/> Accessed 06.04.13.

IEA. (2013). *FAQS: Natural gas*. International Energy Agency. <http://www.iea.org/aboutus/faqs/gas/> Accessed 12.01.13.

Ignatchenko, N. A. (1960). Geologicheskoe stroenie i ugol'nye mestorozhdeniia zapadnoi chasti Lenskogo ugol'nogo basseina. Moscow.

IGU. (2011). World LNG report 2011. In *Gas: Sustaining future global growth*. International Gas Union 25th World Gas Conference, Kuala Lumpur, Malaysia, June 4–8, 2012. < http://www.igu.org/igu-publications/LNG%20Report%202011.pdf> Accessed 17.02.13.

Igunnu, E. T., & Chen, G. Z. (2012). Produced water treatment technologies. *International Journal of Low-Carbon Technologies,* 1–23. < http://ijlct.oxfordjournals.org/content/early/2012/07/04/ijlct.cts049.full.pdf+html> Accessed 03.12.12.

Immirzi, P., & Maltby, E. (1992). *The global status of wetlands and their role in the carbon cycle.* United Kingdom: Wetland Ecosystems Research Group, University of Exeter.

India Ministry of Petroleum & Natural Gas. (2009). Basic statistics on Indian petroleum and natural gas, 2008–09, "Table 11: Production of crude oil and natural gas", September 2009, p. 12.

Indonesia Ministry of Energy & Mineral Resources. (2012). *Indonesian standard on coal resources & reserves classification.* "Workshop on UNFC Resources Classification" Bangkok, 9–10 February 2012. <http://www.ccop.or.th/eppm/projects/35/docs/INDO_Coalpercent20Classification.pdf> Accessed 20.06.12.

Ingelson, A. (2010). Clarification of CBM ownership on freehold lands in Alberta. *The University of Calgary Faculty of Law blog on Developments of Alberta Law.* < http://ablawgca/wp-content/uploads/2010/11/blog_ai_bill26_nov2010.pdf> Accessed 26.08.13.

Ingelson, A., McLean, P. K., & Gray, J. (2006). *CBM produced water—The emerging Canadian regulatory framework.* University of Calgary, Institute of Sustainable Energy, Environment, and Economy. Paper No. 4 of the Alberta Energy Futures Project, p. 38. <http://www.iseee.ca/media/uploads/documents/AB%20Energy%20Futures/policy papers/4-CBM%20Produced%20Water.pdf> Accessed 23.11.12.

Ingram, H. A. P. (1978). Soil layers in mires: function and terminology. *Journal Soil Science, 29,* 224–227.

Ingram, H. A. P. (1983). Hydrology. In A. J. P. Gore (Ed.), *Ecosystems of the world, Vol. 4A, Mires: Swamp, bog, fen and moor, general studies* (pp. 67–158). New York: Elsevier.

Ingram, F. T., & Robinson, V. A. (1996). Petroleum prospectivity of the Clarence–Moreton Basin in New South Wales. In R. A. Facer (Ed.), *Petroleum Bulletin. Vol. 3.* Department of Mineral Resources.

Inouye, K. (1913). The coal resources of China. In *The coal resources of the world* (Vol. 1; pp. 169–214). Toronto, Canada: Morang and Co., Ltd.

Institut Pertanian Bogor Team. (1980). *Soil surveys and soil mapping in Mesuji area.* Sub P4S South Sumatra, Final Report. Institut Pertanian Bogor, Bogor Agricultural University. p. 236.

Institute of Materials, Minerals & Mining Working Group. (2001). *Code for reporting of mineral exploration results, mineral resources and mineral reserves (The Reporting Code).* European Federation of Geologists, the Geological Society of London, and Institute of Geologists of Ireland. p. 34. <http://www.crirsco.com/nat_uk_euro.pdf> Accessed 26.10.11.

IPCC. (1996). *Climate change 1996; impacts, adaptations and mitigation of climate change: Scientific technical analysis.* Cambridge, UK: Cambridge University Press. Contribution of working group II to the second assessment report of the International Panel on Climate Change.

IPCC. (1997). *Revised 1996 IPCC guidelines for national greenhouse gas inventories, reference manual, 3, Intergovernmental Panel on Climate Change.* < http://www.ipcc-nggip.iges.or.jp/public/gl/invs1.htm> Accessed 27.11.11.

IPCC. (2007). *Climate change 2007; the physical science basis.* Contribution of working group I to the fourth assessment report of the Intergovernmental Panel of Climate Change. Cambridge UK: Cambridge University Press. p. 996.

Jangbarwala, J. (2007). CBM-produced water: a synopsis of effects and opportunities. *Water Conditioning & Purification,* 4. <http://www.wcponline.com/pdf/0712Jangbarwala.pdf> Accessed 29.11.12.

Jasper, K., Krooss, B. M., Flajs, G., Hartkopf-Fröder, C., & Littke, R. (2009). Characteristics of type III kerogen in coal-bearing strata from the Pennsylvanian (Upper Carboniferous) in the Ruhr Basin, western Germany: comparison of coals, dispersed organic matter, kerogen concentrates and coal-mineral mixtures. *International Journal of Coal Geology, 80,* 1–19.

Jenkins, C. D. (1999). Fruitland coal description and analysis, Ignacio Blanco Field, La Plata County, Colorado: ARCO Technology and Operations Services. *Technical Series Report,* 1999-0025, p. 8 plus figures, plates, and appendices.

Jenkins, C. D. (2008). *Practices and pitfalls in estimating coalbed methane resources and reserves.* Search and Discovery Article #80011 Posted October 29, 2008.

Jenkins, C. D., & Boyer, C. M. (2008). Coalbed- and shale-gas reservoirs. *Journal of Petroleum Technology, 60,* 92–99. Society of Petroleum Engineers, Distinguished Authors Series, Paper SPE 103514, 92–99.

Jensen, D., & Smith, L. K. (1997). *A practical approach to coalbed methane reserve prediction using a modified material balance technique.* Paper 9765, International Coalbed Methane Symposium, Tuscaloosa, Alabama, May 12–16, 1997.

Jervey, M. T. (1988). Quantitative geological modeling of siliciclastic rock sequences and their seismic expression. In C. K. Wilgus, B. S. Hastings, C. G. St. Kendall, H. W. Posamentier, C. A. Ross, & J. C. Van Wagoner (Eds.), *Society of Economic Paleontologists and Mineralogists, Special Publication: Vol. 42. Sea level changes – An integrated approach* (pp. 47–69).

Jevons, W. S. (1856). The coal question. In A. W. Flux (Ed.), *An inquiry concerning the progress of the nation, and the probable exhaustion of our coal-mines* (3rd ed. 1905). New York: Augustus M. Kelley Publications. ISBN 978-0678001073.

<http://www.econlib.org/library/YPDBooks/Jevons/jvnCQCover.html>. Accessed 18.09.11.

Ji, T., & Yang, D. (2007). Evaluation of coalbed methane geology conditions in Qinshui Basin [J]. *Coal Engineering, 10*, 83–86.

Jia, H. (2009). *Coalbed methane efforts dampened by divisional barriers*. Chemistry World, Royal Society of Chemistry. <http://www.rsc.org/chemistryworld/Issues/2009/August/CoalbedMethaneEffortsDampenedByDivisionalBarriers.asp> Accessed 16.01.13.

Jiang, B. (2007). *Inside Gui Zhou: What are the prospects? Newsletter of Gui Zhou coal mine drainage and use promotion project*. Issue No. 1, China.

Jiang, H., Ye, S., Lei, Z., Wang, X., Zhu, G., & Chen, M. (2010). The productivity evaluation model and its application for finite conductivity horizontal wells in fault block reservoirs. *Petroleum Science, 7*, 530–535.

Jin, S. (2007). *Enhancement of biogenic coalbed methane production and back injection of coalbed methane co-produced water*. Final Report to U.S. Department of Energy, WRI 07-R008.

Jin, S., Bland, A. E., & Price, H. K. (November 16, 2010). *Biogenic methane production enhancement systems*. United States Patent No. 7,832,475 B2.

Johnson, R. C., & Finn, T. M. (2001). Potential for a basin-centered gas accumulation in the Raton Basin, Colorado and New Mexico. In V. F. Nuccio, & T. S. Dyman (Eds.), *Geologic studies of basin-centered gas systems* (p. 18). U.S. Geological Survey Bulletin 2184–B.

Johnston, J. H., Collier, R. J., & Maidment, A. I. (1991). Coal as source rocks for hydrocarbon generation in the Taranaki Basin, New Zealand: a geochemical biomarker study. *Journal of Southeast Asian Sciences, 5*, 283–289.

Johnston, C. R., Vance, G. F., & Ganjegunte, G. K. (2008). Irrigation with coalbed natural gas co-produced water. *Agricultural Water Management, 95*, 1243–1252.

Joint Ore Reserves Committee Code (JORC Code) and Guidelines. (2004). *Australasian code for reporting of exploration results, minerals resources and ore reserves*. Prepared by The Joint Ore Reserves Committee of the Australasian Institute of Mining and Metallurgy, Australian Institute of Geoscientists and Minerals Council of Australia, p. 20. <http://www.crirsco.com/jorc2004print.pdf> Accessed 19.09.11.

Jones, N. S. (2005). *A review of the AMM and CMM resources in the Kuznetsk (Kuzbass) coal basin, Russia*. British Geological Survey.

Jones, W. J., Leigh, J. A., Mayer, F., Woese, C. R., & Wolfe, R. S. (1983). *Methanococcus jannaschii* sp. nov., an extremely thermophilic methanogens from a submarine hydrothermal vent. *Archives of Microbiology, 136*, 254–261.

Jones, N. S., Holloway, S., Creedy, D. P., Garner, K., Smith, N. J. P., Browne, M. A. E., et al. (2004). *UK coal resource for new exploitation technologies: Final report*. British Geological Survey. Commissioned Report CR/04/015N. Cleaner Coal Technology Programme, COAL R271.

Jones, D. M., Head, I. M., Gray, N. D., Adams, J. J., Rowan, A. K., Aitken, C. M., et al. (2008). Crude-oil biodegradation via methanogenesis in subsurface petroleum reservoirs. *Nature, 451*, 176–180.

Joosten, H. (2011). Peatlands, policies and markets. Report to IUCN UK Peatland Programme, Edinburgh. <www.iucn-uk-peatlandprogramme.org/scientificreviews>

Juntgen, H., & Karwell, J. (1966). Formation and storage of gases in bituminous coal seams, part 1, gas formation and part 2, Gas storage (English translation). *Erdol und Kohle-Erdgas-Petrochemie, 19*, 251–258, 339–344.

Juntgen, H., & Klein, J. (1975). Entstehung von erdgas aus kohligen sedimenten. *Erdol und Kohle-Erdgas-Petrochemie, 28*, 52–69.

Kaiser, E. (1908). Erlauterungen zur geologische Karte von Preussen. (Explanation of the geological map of Prussia). *Blatt Bruhe*, 44.

Kaiser, W. R., & Swartz, T. E. (1989). *Fruitland Formation hydrology and producibility of coalbed methane in the San Juan Basin, New Mexico and Colorado*. Tuscaloosa, Alabama: Proceedings Coalbed Methane Symposium. April 1989, p. 87.

Kaiser, W. R., Swartz, T. E., & Hawkins, G. J. (1994). Hydrologic framework of the Fruitland formation, San Juan Basin. In *Bureau of Mines and Minerals Bulletin: Vol. 146. Coalbed methane in the upper Cretaceous Fruitland Formation, San Juan Basin New Mexico and Colorado* (pp. 133–164).

Kalkreuth, W., & Holz, M. (2000). The coalbed methane potential of the Santa Terezinha coalfield, Rio Grande do Sul, Brazil. *Revista Brasileira de Geociências, 30*, 342–345. < http://sbgeo.org.br/pub_sbg/rbg/vol30_down/3002/3002342.pdf> Accessed 28.01.13.

Kalkreuth, W., Kotis, T., Papanicolaou, C., & Kokknakis, P. (1991). The geology and coal petrology of a Miocene lignite profile at Meliadi Mine, Katerini, Greece. *International Journal of Coal Geology, 17*, 51–67.

Kanana, Y. F., & Matveyev, A. K. (1989). Temperature and geologic time in regional metamorphism of coal. *International Geology Review, 31*, 258–261.

Kansas Geological Survey. (2002). Public Information Circular (PIC) 19. http://www.kgs.ukans.edu/publications/pic19.html.

Karacan, O., Ruiz, F. A., Cote, M., & Phipps, S. (2011). Coal mine methane: A review of capture and utilization practices with benefits to mining safety and to greenhouse gas reduction. *International Journal of Coal Geology, 86*, 121–156.

Karacan, C. O. (2003). Heterogeneous sorption and swelling in a confined and stressed coal during CO_2 injection. *Energy & Fuels, 17*, 1595–1608.

Karacan, C. Ö. (2007). Swelling-induced volumetric strains internal to a stressed coal associated with CO_2 sorption. *International Journal of Coal Geology, 72*, 209–220.

Karacan, C. O., & Mitchell, G. D. (2003). Behavior and effect of different coal microlithotypes during gas transport for carbon dioxide sequestration in coal seams. *International Journal of Coal Geology, 53*, 201–217.

Karacan, C. O., & Okandan, E. (2000). Fracture/cleat analysis of coals from Zonguldak Basin (northwestern Turkey) relative to the potential of coalbed methane production. *International Journal of Coal Geology, 44*, 109–125.

Karacan, C. O., Ruiz, F. A., Cote, M., & Phipps, S. (2011). Coal mine methane: a review of capture and utilization practices with benefits to mining safety and to greenhouse gas reduction. *International Journal of Coal Geology, 86*, 121–156.

Karayigit, A. I., Spears, D. A., & Booth, C. A. (2000). Antimony and arsenic anomalies in the coal seams from the Gokler coalfield, Gediz, Turkey. *International Journal of Coal Geology, 44*, 1–17.

Karweil, J. (1969). Aktuelle problem der geochemic der kohle. In P. A. Schenk, & I. Havernaar (Eds.), *Advances in Geochemistry* (pp. 59–84). Oxford: Pergamon Press.

Katz, B. J. (2011). Microbial processes and gas accumulations. *The Open Geology Journal, 5*, 75–83.

Kaufmann, R. K., Kauppi, H., Mann, M. L., & Stock, J. H. (2011). *Reconciling anthropogenic climate change with observed temperature 1998–2008*. PNAS. <http://wattsupwiththat.files.wordpress.com/2011/07/pnas-201102467.pdf> Accessed 24.05.10, p. 4.

Kawata, Y., & Fujita, K. (2001). *Some predictions of possible unconventional hydrocarbons availability until 2100*. Society of Petroleum Engineers. Paper SPE 68755 presented at the SPE Asia Pacific Oil and Gas Conference, Jakarta, 17–19 April. <http://dx.doi.org/10.2118/68755-MS>.

Kaye, C. A., & Barghoorn, E. S. (1964). Late Quaternary sea-level change and crustal rise at Boston, Massachusetts, with notes on autocompaction of peat. *Geological Society of America Bulletin, 75*, 63–80.

Keary, P., & Vine, F. J. (1996). *Global tectonics* (2nd ed.). Oxford: Blackwell.

Keith, K., Bauder, J., & Wheaton, J. (2003). *Frequently asked questions: Coal bed methane (CBM)*. < http://waterquality.montana.edu/docs/methane/cbmfaq.shtml> Accessed 12.12.12.

Kelafant, J. R. (2011). Outlook for CMM in Mexico and Colombia 2011 U.S. Coal Mine Methane Conference, U.S. Environmental Protection Agency, Coalbed Methane Outreach Program, October 19–20, 2011, Salt Lake, Utah.

Kellner, E., Price, J. S., & Waddington, J. M. (2004). Pressure variations in peat as a result of gas bubble dynamics. *Hydrological Processes, 18*, 2599–2605. <http://dx.doi.org/10.1002/hyp.5650>.

Kellner, E., Waddington, J. M., & Price, J. S. (2005). Dynamics of biogenic gas bubbles in peat: potential effects on water storage and peat deformation. *Water Resources Research, 41*, W08417, 1029/2004WR003732.

Keltjens, J. T. (1984). Coenzymes of methanogenesis from hydrogen and carbon dioxide. *Biochemistry, 50*, 383–396.

Kendall, P. F., & Briggs, H. (1933). *The formation of rock joints and the cleat of coal* (Vol. 53). Edinburgh: Proceedings of the Royal Society. 164–187.

Kennedy, G. W., & Price, J. S. (2005). A conceptual model of volume-change controls on the hydrology of cutover peats. *Journal of Hydrology, 302*, 13–27.

Khoudin, Y. L. (1995). Prevention of sudden outbursts hazards in coal faces by means of control of state of stress and strain of coal and rock mass. In R. D. Lama (Ed.), *Management and control of high gas emissions and outbursts in underground coal mines* (pp. 509–512). Kiama, North South Wales, Australia: Westonprint.

Kiehl, J. T., & Trenberth, K. E. (1997). Earth's annual global mean energy budget. *Bulletin of the American Meteorological Society, 78*, 197–208.

Killops, S., & Killops, V. (2005). *Introduction to organic geochemistry* (2nd ed.). Oxford, UK: Blackwell Publishing.

Killops, S. D., Mills, N., & Johansen, P. E. (2008). Pyrolytic assessment of oil generation and expulsion from a suite of vitrinite-rich New Zealand coals. *Organic Geochemistry, 39*, 1113–1118.

Killops, S. D., Woolhouse, A. D., Weston, R. J., & Cook, R. A. (1994). A geochemical appraisal of oil generation in the Taranaki Basin, New Zealand. *American Association of Petroleum Geologists Bulletin, 78*, 60–85.

Kim, A. G. (1977). *Estimating methane content of bituminous coal beds from adsorption data*. U.S. Bureau of Mines, RI8245, 1–22.

Kim, A. G. (1978). *Experimental studies on the origin and accumulation of coalbed gas*. U.S. Bureau of Mines Report of Investigations 8317, p. 18.

King, G. (2008). *Origin Energy*. Presented at the UBS Sixth Annual Australian Energy and Resources Conference, 18–19 June 2008. Sydney Australia.

King, G. R. (1990). *Material balance techniques for coal seam and Devonian shale gas reservoirs*. Society of Petroleum Engineers. Paper 20730, SPE Annual Technical Conference and Exhibition, New Orleans, LA, September 23–26, 1990.

King, P. R., Cook, R. A., Field, B. D., Funnell, R. H., Killops, S. D., & Sykes, R. (1998). *Petroleum systems of the Taranaki, East Coast, and Great South Basins: Some key elements*. New Zealand Petroleum Conference Proceedings, pp. 71–80. <http://www.nzpam.govt.nz/cms/pdf-library/petroleum-conferences-1/1998-petroleum-conference-proceedings/king-890-kb-pdf> Accessed 11.12.11.

King, P. R., & Thrasher, G. P. (1996). Cretaceous-Cenozoic geology and petroleum systems of the Taranaki Basin: New Zealand. *Institute of Geological & Nuclear Sciences Monograph, 13*, 243.

Kious, W. J., & Tilling, R. I. (1996). *This dynamic earth: The story of plate tectonics*. U.S. Geological Survey. Portable Document Format Online Version. <http://pubs.usgs.gov/gip/dynamic/dynamic.html> Accessed 20.08.11.

Kirkpatrick, A. D. (2005). *Assessing constructed wetlands for beneficial use of saline-sodic water* (MS Thesis), Bozeman: Montana State University, p. 82 <http://etd.lib.montana.edu/etd/2005/kirkpatrick/KirkpatrickA0505.pdf> Accessed 13.12.12.

Kirwan, M. L., & Guntenspergen, G. (2010). *Threshold sea level rise rates from wetland survival: Limits to eco-geomorphic feedbacks*. Ecological Society of America Conference, Pittsburg, Pennsylvania, August 1–6, 2010 <http://www.usgs.gov/newsroom/article.asp?ID=2561> Accessed 25.02.12.

Kissell, F. N., McCulloch, C. M., & Elder, C. H. (1973). *The direct method of determining methane content of coalbeds for ventilation design*. U.S. Bureau of Mines Report of Investigations 8043, p. 22.

Kivinen, E., & Pakarinen, P. (1980). *Peatland areas and the proportion of virgin peatlands in different countries*. Proceedings of the 6th International Peat Congress, Duluth, pp. 52–54.

Kleesattel. D.R. (1984). Distribution, abundance, and maceral content of the lithotypes in the Beulah-Zap bed of North Dakota. Rocky Mountain Coal Symposium, North Dakota geological Survey, Bismarck, North Dakota.

Kleikemper, J., Pombo, S. A., Schroth, M. H., Sigler, W. V., Pesaro, M., & Zeyer, J. (2005). Activity and diversity of methanogens in a petroleum hydrocarbon-contaminated aquifer. *Applied and Environmental Microbiology, 71*, 149–158.

Klein, D., Flores, R. M., Venot, C., Gabbert, K., Schmidt, R., Stricker, G. D., et al. (2008). Molecular sequences derived from Paleocene Fort Union Formation coals vs associated produced waters: implications for CBM regeneration . In R. M. Flores (Ed.), *International Journal of Coal Geology: Vol. 76. Microbes, methanogenesis, and microbial gas* (pp. 3–13).

Knight, J. A., Burger, K., & Bieg, G. (2000). The pyroclastic tonsteins of the Sabero Coalfield, northwestern Spain, and their relationship to the stratigraphy and structural geology. *International Journal of Coal Geology, 44*, 187–226.

Knutson, C., Dastgheib, S. A., Yang, Y., Ashraf, A., Duckworth, C., Sinata, P., et al. (2012). *Reuse of produced water from CO_2 enhanced oil recovery, coal-bed methane, and mine pool water by coal-based power plants*. Illinois Geological Survey, Prairie Research Institute, DOE. Award number: DE-NT0005343, p. 150. < http://www.netl.doe.gov/technologies/coalpower/ewr/water/pp-mgmt/pubs/05343FSRFG043012.pdf> Accessed 28.11.12.

Koesoemadinata, R. P. (2000). Tectono-stratigraphic framework of Tertiary coal deposits of Indonesia. In Proceedings Southeast Asian coal geology conference (Herudyanto, Sukarjo, Endang Djaelani, and Komaruddin, Compilers), Directorate of Mineral Resources, 8–16. ISBN: 979-8126-03-3.

Koesoemadinata, R. P., Hardjono, U. I., & Sumadirdija, H. (1978). Tertiary coal basins of Indonesia. *United Nations ESCAP, CCOP Technical Bulletin, 12*, 43–86.

Koide, H., & Kuniyasu, M. (2006). *Deep unmineable coalbeds in Japan: CO_2 sink and untapped energy resource*. Proceedings of the 8th International Conference on Greenhouse Gas Technologies. Oxford: Elsevier. CD-ROM. < http://co2.eco.coocan.jp/ref/A6GHGT8.pdf> Accessed 29.01.13.

Koide, H., & Yamazaki, K. (2001). Subsurface CO_2 disposal with enhanced gas recovery and biogeochemical carbon recycling. *Environmental Geosciences, 8*, 218–224.

Kolcon, I., & Sachsenhofer, R. F. (1998). Coal petrography and palynology of the Early Miocene lignite seam from the opencast mine Oberdorf (N Voitsberg, Styria, Austria). In F. F. Steininger (Ed.), *Jahrbuch der Geologischen Bundesanstalt: Vol. 140. The early Miocene lignite deposits of Oberdorf, N Voitsberg (Styria, Austria)* (pp. 433–440).

Kolcon, I., & Sachsenhofer, R. F. (1999). Petrography, palynology and depositional environments of the early Miocene Oberdorf lignite seam (Styrian Basin, Austria). *International Journal of Coal Geology, 41*, 275–308.

Kolesar, J. E. (1986). *Nature of sorption and diffusion phenomena in the micropore structure of coal*. Society of Petroleum Engineers. p. 40.

Kolker, A., & Chou, C.-L. (1994). Cleat-filling calcite in Illinois Basin coals: trace element evidence for meteoric fluid migration in a coal basin. *Journal of Geology, 102*, 111–116.

Kong, C., Chen, K. C., Irawan, S., Sum, C. W., & Tunio, S. Q. (2011). Preliminary study on gas storage capacity and gas-in-place for CBM potential in Balingian coalfield, Sarawak Malaysia. *International Journal of Applied Science and Technology, 1*, 82–94.

Kopp, O. C., Bennett, M. E., & Clark, C. E. (2000). Volatiles lost during coalification. *International Journal of Coal Geology, 44*, 69–84.

Kortenski, J., & Kostova, I. (1996). Occurrence and morphology of pyrite in Bulgarian coals. *International Journal of Coal Geology, 29*, 273–290.

Kotarba, M. J. (1988). *Geochemical criteria for the origin of natural gases from the Upper Cretaceous Zacher coal-bearing formations in Walbrzych coal basin (in Polish and English summary)*. Stanislaw Staszio Academy of Mining and Metallurgy Scientific Bulletin, 1199.

Kotarba, M. J. (1998). Composition and origin of gaseous hydrocarbons in the Miocene strata of the Polish part of the Carpathian foredeep. *Przeglad Geologische, 46*, 751–758.

Kotarba, M. J. (2001). Composition and origin of coalbed gases in the Upper Silesian and Lublin Basins, Poland. *Organic Geochemistry, 32*, 163–180.

Kotarba, M. (2004). Characterizing thermogenic coalbed gas from Polish coals of different ranks by hydrous pyrolysis. *Organic Geochemistry, 35*, 615–646.

Kotarba, M. J., & Lewan, M. D. (2004). Characterizing thermogenic coalbed gas from Polish coals of different ranks by hydrous pyrolysis. *Organic Geochemistry, 35*, 615–646.

Kotarba, M. J., & Pluta, I. (2009). Origin of natural waters and gases within the Upper Carboniferous coal-bearing and autochthonous Miocene strata in the South-Western part of the Upper Silesian Coal Basin, Poland. *Applied Geochemistry, 24*, 876–889.

Kotarba, M., & Rice, D. D. (1995). Composition and origin of of coalbed gases in Lower Silesia coal basin, NW Poland. In *Report from research cooperation within U.S.-Polish Maria Sklodowska-Curie joint fund II, origin and habitat of coal gases in Polish basins, isotopic and geological approach* (pp. 69–78).

Kotelnikova, S. (2002). Microbial production and oxidation of methane in deep subsurface. *Earth-Science Reviews, 58*, 367–395.

Koukouzas, C., & Koukouzas, N. (1995). Coals of Greece: distribution, quality and reserves. In M. K. G. Whateley, & D. A. Spears (Eds.), *Geological Society, London, Special Publication: Vol. 82. European coal geology* (pp. 171–180).

Kremp, G. D. W. (1977). The positions and climatic changes of Pangea and five southeast Asia plates during Permian and Triassic times. *Palynodata,* (Part 1), 108.

Krevor, C. M., & Lackner, K. S. (2011). Enhancing serpentine dissolution kinetics for mineral carbon dioxide sequestration. *International Journal of Greenhouse Gas Control.* <http://dx.doi.org/10.1016/j.ijggc.2011.01.006>.

Kruger, J., & Kjaer, K. H. (1999). A data chart for field description and genetic interpretation of glacial diamicts and associated sediments with examples from Greenland, Iceland, and Denmark. *Boreas, 28*, 386–402.

Krumbein, W. C., & Sloss, L. L. (1963). *Stratigraphy and sedimentation* (2nd ed.). San Francisco, California: W.H. Freeman and Company. p. 660.

Kulander, B. R., & Dean, S. L. (1993). Coal cleat domain and domain boundaries in the Allegheny Plateau of West Virgina. *American Association of Petroleum Geologists Bulletin, 77*, 1374–1388.

Kurbatov, I. M. (1968). The question of the genesis of peat and its humic acids. In R. A. Robertson (Ed.), *Transactions of the 2nd International Peat Congress, Leningrad* (Vol. 1; pp. 133–137). Edinburgh: HMSO.

Kurniawan, Y., Bhatia, S. K., & Rudolph, V. (Mar 2006). Simulation of binary mixture adsorption of methane and CO_2 at supercritical conditions in carbons. *AIChE Journal, 52*, 957–967.

Kuuskraa, V. A., Boyer, C. M. II, & Kelafant, J. A. (1992). Hunt for quality basins goes abroad. *Oil and Gas Journal, 90*, 49–54.

Kuuskraa, V. A., & Boyer, C. M. II (1993). Economic and parametric analysis of coalbed methane. In B. E. Law, & D. D. Rice (Eds.), *American Association of Petroleum Geologists Studies in Geology, Hydrocarbons from Coal.* (Vol. 38; pp. 373–394).

Kuuskraa, V. A., & Brandenburg, C. F. (1989). Coalbed methane sparks a new energy industry. *Oil and Gas Journal, 87*, 49.

Kuuskraa, V. A., & Stevens, S. H. (2009). Worldwide gas shales and unconventional gas: a status report. In *Natural gas, renewables and efficiency: Pathways to a low-carbon economy.* Presented at the United Nations Climate Change Conference, COP15, sponsored by the American Clean Skies Foundation (ACSF), the UN Foundation (UNF) and the Worldwatch Institute, Copenhagen. December 7–18, 2009.

Kuuskraa, V. A., & Van Leeuwen, T. (2011). *Economic and market impacts of abundant shale gas resources.* Advanced Resources International, Inc. <http://www.acus.org/files/EnergyEnvironment/Unconventional25Jan/Vello_Kuuskraa.pdf> Accessed 27.12.11.

Kwiecinska, B. K., Hamburg, G., & Vleeskens, J. M. (1992). Formation temperatures of natural coke in the lower Silesian coal basin, Poland: evidence from pyrite and clays by SEM-EDX. *International Journal of Coal Geology, 21*, 217–235.

Kwiecinska, B., & Petersen, H. I. (2004). Graphite, semi-graphite, natural coke, and natural char classification—ICCP system. *International Journal of Coal Geology, 57*, 99–116.

Kwiecińska, B., & Wagner, M. (1997). *Typizacja cech jakościowych wegla brunatnego z krajowych złóż według kryteriów petrograficznych i chemiczno-technologicznych dla celów dokumentacji geologicznej złóż oraz obsługi kopalń classification of qualitative features of brown coal from Polish deposits according to petrographical, chemical and technological criteria.* Cracow: Polish Academy of Sciences. p. 87.

Lakatos, I., Foldessy, J., Nemedi Varga, Z., Toth, J., Fodor, B., & Csecsei, T. (2006). *The coalbed methane extraction – CO_2 sequestration potential of the Mecsek Mountains, Hungary.* International Coalbed Methane Symposium, Tuscaloosa, Alabama Proceedings, Paper 0652. < http://fold1.ftt.uni-miskolc.hu/pdf/060522.pdf> Accessed 01.02.13.

Lama, R. D. (Ed.), (1995). *Management and control of high gas emissions and outbursts in underground coal mines International symposium-cum-workshop proceedings", Wollongong, New South Wales, Australia* (pp. 618). ISBN 0 646 22810 2.

Lama, R. D., & Bodziony, J. (1998). Management of outbursts in underground coalmines. In R. M. Flores (Ed.), *Coalbed methane: From coal-mine outbursts to gas resource* (pp. 83–116). International Journal of Coal Geology, 35.

Lama, R. D., & Saghafi, A. (2002). *Overview of gas outbursts and unusual emissions.* Underground Coal Operator's Conference Paper 196, pp. 74–88. <http://ro.uow.edu.au/coal/196> Accessed 20.11.11.

Lamarre, R. A. (2003). Hydrodynamic and stratigraphic controls for a large coalbed methane accumulation in Ferron coals of east-central Utah. *International Journal of Coal Geology, 56*, 97–110.

Lamarre, R. A., & Burns, T. D. (1996). *Drunkard's Wash unit: Coalbed methane production from Ferron coals in east-central, Utah.* GSA Abstracts with Programs.

Lamarre, R. A., & Pope, J. (2007). Critical-gas-content technology provides coalbed-methane reservoir data. *Journal of Petroleum Technology, 59*, 108–113.

Lamberson, N., & Bustin, R. M. (1993). Coalbed methane characteristics of Gates Formation coals, Northeastern British Columbia: effect of maceral composition. *American Association of Petroleum Geologists Bulletin, 12*, 2062–2076.

Lambert, S. W., Niederhofer, J. D., & Reeves, F. R. (1990). Multiple-coal-seam completions experience in the Deerlick field. *Journal of Petroleum Technology, 42*, 1360–1363.

Lambert, S. W., & Trevits, M. A. (1978). *Improved methods for monitoring production from vertical degasification wells.* U.S. Bureau of Mines Report of Investigations 8309, p. 14.

LaMonica, M. (2010). *Luca Tech feeds coal-eating microbes to make gas. C/Net.* <http://news.cnet.com/8301-11128_3-20012875-54.html> Accessed 23.03.13.

Landis, E. R., Rohrbacher, T. J., Barker, C. E., Fodor, B., & Gombar, G. (2002). *Coalbed gas in Hungary – a preliminary report.* U.S. Geological Survey Open File Report 01–473, version 1.0.

Landis, E. R., & Weaver, J. N. (1993). Global coal occurrence. In B. E. Law, & D. D. Rice (Eds.), *Studies in Geology: Vol. 38. Hydrocarbons from Coal* (pp. 1–77). American Association of Petroleum Geologists.

Langmuir, I. (1918). The adsorption of gases on plane surfaces of glass, mica and platinum. *Journal of the American Chemical Society, 40*, 1361–1403.

Lansdown, J. M., Quay, E. D., & King, S. L. (1992). CH_4 production via CO_2 reduction in a temperate bog: a source of ^{13}C depleted CH_4. *Geochimica et Cosmochimica Acta, 56*, 3493–3503.

Lapo, A. V., & Drozdova, I. N. (1989). Phyterals of humic coals in the U.S.S.R. In P. C. Lyons, & B. Alpern (Eds.), *International Journal of Coal Geology: Vol. 12. Peat and coal: origin, facies, and depositional models* (pp. 477–510).

Large, D. J., Marshall, C., Meredith, W., Snape, C. E., & Spiro, B. F. (2011). *Potential to generate oil prone coal source rocks in Arctic environments*, 3P Arctic – The Polar Petroleum Potential Conference & Exhibition, American Association of Petroleum Geologists and Allworld Exhibitions, Halifax, Nova Scotia, Canada, August 30–September 2, 2011. Search and Discovery Article #40806. <http://www.searchanddiscovery.com/documents/2011/40806large/ndx_large.pdf> Accessed 03.12.11.

Larsen, J. W. (2004). The effects of dissolved CO_2 on coal structure and properties. *International Journal of Coal Geology, 57*, 63–70.

Laubach, S. E., Marrett, R. A., Olson, J. E., & Scott, A. R. (1998). Characteristics and origins of coal cleat: a review. In R. M. Flores (Ed.), *International Journal of Coal Geology: Vol. 35. Coalbed methane: From coal-mine outbursts to gas resource* (pp. 175–207).

Laubach, S. E., & Tremain, C. M. (1991). Regional coal fracture patterns and coalbed methane development. In J. C. Roegiers (Ed.), *Proceedings of Thirty Second U.S. Symposium of Rock Mechanics.* Rotterdam: Balkema (pp. 851–851).

Law, B. E. (1991). Relationship between coal rank and cleat density–a preliminary report. *American Association of Petroleum Geologists Bulletin, 75*, 1131.

Law, B. E. (1993). *The relation of coal rank and cleat spacing: implications for the prediction of permeability in coal.* Proceedings of the International Coalbed Methane Symposium II, 435–442.

Law, B. E., & Rice, D. D., (Eds.), *Studies in Geology: Vol. 38. Hydrocarbons from Coal* (p. 400). American Association of Petroleum Geologists.

Law, B. E., Rice, D. D., & Flores, R. M. (1991). Coal-bed gas accumulations in the Paleocene Fort Union Formation, Powder River Basin, Wyoming. In S. D. Schwochow, D. K. Murray, & M. F. Fahy (Eds.), *Coalbed methane of western North America* (pp. 179–190). Rocky Mountain Association of Geologists.

Lawall, C. E., & Morris, L. M. (1934). Occurrence and flow of gas in the Pocahontas No. 4 coalbed in southern West Virginia. *Transactions of the American Institute of Mining and Metallurgical Engineers, 108*, 11–30.

Lawson, P. (2002). *An introduction to Russian coal industry.* Vice President, Energy & Mineral Resources Division Marshall Miller & Associates, Charleston, West Virginia, United States, p. 12. < http://www.mma1.com/company/pdf/papers/An%20Introduction%20of%20the%20Russian%20Coal%20Industry.pdf> Accessed 19.09.11.

Laxminarayana, C., & Crosdale, P. J. (2002). Controls on methane sorption capacity of Indian coals. *American Association of Petroleum Geologists Bulletin, 86*, 201–212.

Leahy, J. G., & Colwell, R. R. (1990). Microbial degradation of hydrocarbons in the environment. *Microbiological Reviews, 54*, 305–315.

Leckel, D. (2009). Diesel production from Fischer–Tropsch: the past, the present, and new concepts. *Energy Fuels, 23*, 2342–2358.

Leckie, D. A., & Boyd, R. (2003). *Towards a nonmarine sequence stratigraphic model*. American Association of Petroleum Geologists Annual Convention, Salt Lake City, 11–14 May 2003, Official Program, 12, p. A101.

Lee, R. W. (1981). *Geochemistry of water in the Fort Union Formation of the northern Powder River Basin, southeastern Montana*. U.S. Geological Survey Water-Supply. Paper 2076, p. 17.

Lehman, L. V., Blauch, M. E., & Robert, L. M. (1998). *Desorption enhancement in fracture-stimulated coalbed methane wells*. Society of Petroleum Engineers, SPE Eastern Regional Meeting, November 9–11, 1989, ID 51063–MS, p. 8.

Lennon, R. G., Suttill, R. J., Guthrie, D. A., & Waldron, A. R. (1999). The renewed search for oil and gas in the Bass Basin; results of Yolla–2 and white Ibis–1. *APPEA Journal, 39*, 248–261.

Lesher, C. E. (1916). Coal. In *Mineral resources of the United States, 1914, Part 11-Nonmetals* (pp. 587–746). U.S. Geological Survey.

Levine, J. R. (1987). *Influence of coal composition on the generation and retention of coalbed natural gas*. Proceedings of 1987 Coalbed Methane Symposium, 15–18.

Levine, J. R. (1991a). *Coal petrology with applications to coalbed methane R & D. Short course notes*. Tuscaloosa, Alabama: Coalbed Methane Symposium.

Levine, J. R. (1991b). *The impact of oil formed during coalification on generation and storage of natural gas in coal bed reservoir systems*. Third Coalbed Methane Symposium Proceedings, Tuscaloosa, Alabama, May 13–16, 1991, 307–315.

Levine, J. R. (1992). Oversimplifications can lead to faulty coalbed gas reservoir analysis. *Oil Gas Journal, 23*, 63–69.

Levine, J. R. (1993). Coalification: the evolution of coal as source rock and reservoir rock for oil and gas. In B. E. Law, & D. D. Rice (Eds.), *Studies in Geology: Vol. 38. Hydrocarbons from Coal* (pp. 1–12). American Association of Petroleum Geologists. Chapter 3.

Levine, J. R. (1996). Model study of the influence of matrix shrinkage on absolute permeability of coal bed reservoirs. In R. Gayer, & I. Harris (Eds.), *Coalbed methane and coal geology* (pp. 197–212). London: The Geological Society. Special Publication No. 109.

Levine, D. G., Schlosberg, R. H., & Silbernagel, B. G. (1982). Understanding the chemistry and physics of coal structure (a review). *Proceedings of National Academy of Sciences, 79*, 3365–3370.

Levorsen, A. I. (1967). Geology of petroleum, sections on hydrodynamics and capillary pressure revised. In Fredrick A. F. Berry, W. H. Freeman, & Company (Eds.), (2nd ed.). p. 724.

Lewan, M. D. (1992). Water as a source of hydrogen and oxygen in petroleum formation by hydrous pyrolysis. *American Chemical Society, Division of Fuel Chemistry, 37*, 1643–1649.

Lewan, M. D. (2010). *Lower organic carbon limits and upper thermal maturity limits for methane generation from type-III kerogen in coals and fine-grained rocks*. The Society of Organic Petrology Annual Meeting Abstracts and Programs, September 12–16, Denver, Colorado, U.S., 14.

Lewis, J. V. (1934). The evolution of mineral coals. *Economic Geology, 29*, 157–212.

Li, T. (2009). *Liquefaction of Coal. Coal, Oil Shale, Natural Bitumen, Heavy Oil and Peat, Encyclopedia of Life Support Systems (EOLSS)*. EOLSS Publishers Co Ltd. < http://www.eolss.net/Sample-Chapters/C08/E3-04-03-05.pdf> Accessed 04.01.11.

Li, W., & Gong, S. (2000). A new understanding of south China Permian coal-bearing strata and coal accumulation regularity. *Acta Geologica Sinica, 74*, 711–716.

Li, R. X., Li, Y. Z., & Gao, Y. W. (2007). Catagenesis of organic matter of oil source rocks in Upper Paleozoic coal formation of Bohai Gulf Basin (eastern China). *Russian Geology and Geophysics, 48*, 415–425.

Li, D., Hendry, P., & Faiz, M. (2008). A survey of the microbial populations in some Australian coalbed methane reservoirs. In R. M. Flores (Ed.), *International Journal of Coal Geology: Vol. 76. Microbes, methanogenesis, and microbial gas* (pp. 14–24).

Li, R., Jin, K., & Lehrmann, D. J. (2008). Hydrocarbon potential of Pennsylvanian coal in Bohai Gulf Basin, eastern China, as revealed by hydrous pyrolysis. *International Journal of Coal Geology, 73*, 88–97.

Li, S., Li, B., Yang, S., Huang, J., & Li, Z. (1984). Sedimentation and tectonic evolution of Late Mesozoic faulted coal basins in north-eastern China. In R. A. Rahmani, & R. M. Flores (Eds.), *International Association of Sedimentologists, Special Publication: Vol. 7. Sedimentology of coal and coal-bearing sequences* (pp. 387–406). Oxford: Blackwell Scientific Publications.

Li, J., Lui, D., Yao, Y., Cai, Y., & Qui, Y. (2011). Evaluation of the reservoir permeability of anthracite coals by geophysical logging data. *International Journal of Coal Geology, 87*, 121–127.

Li, S., Mao, B. M., & Lin, C. (1995). Coal resources and coal geology in China. *Episodes, 18*, 26–30.

Li, L., Ryan, A., Nenoff, T. M., Dong, J., & Lee, R. O. (September 26–29, 2004). *Purification of coal-bed methane produced water by zeolite membrane*. SPE Annual Technical Conference and Exhibition held in Houston. Texas, U.S.A: Society of Petroleum Engineers Inc. p. 6. <http://baervan.nmt.edu/research_groups/carbon_sequestration_membrane_technology/pages/Publications/B-10.pdf> Accessed 15.12.12.

Lin, W. (2010). Gas sorption and the consequent volumetric and permeability change of coal (Ph.D. Dissertation). Palo Alto, California: Stanford University. p. 195.

Lin, X., & Su, X. (2007). Resurveying mechanism of coalbed methane in southern Qinshui Basin. *Natural Gas Industry, 27*(7), 8–12.

Lindbergh, K., & Provorse, B. (1977). *Coal, a contemporary energy study*. New York: Scribe Publishing Corporation and Coal Age – E/MJ.

Lindqvist, J. K., & Issac, M. J. (1991). Silicified conifer forests and potential mining problems in seam M2 of the Gore Lignite Measures (Miocene), Southland, New Zealand. *International Journal of Coal Geology, 17*, 149–169.

Lipinski, B.A. (2007). *Integrating geophysics and geochemistry to evaluate coalbed natural gas produced water disposal, Powder River Basin, Wyoming* (Ph.D. Dissertation). University of Pittsburgh, p. 113.

Lis, G. P., Mastalerz, M., Schimmelmann, A., Lewan, M. D., & Stankiewicz, B. A. (2005). FTIR absorption indices for thermal maturity in comparison with vitrinite reflectance R_o in type-II kerogens from Devonian black shales. *Organic Geochemistry, 36*, 1533–1552.

Littke, R., & Haven, H. L. T. (1989). Paleoecologic trends and petroleum potential of upper carboniferous coal seams of western Germany as revealed by their petrographic and organic geochemical characteristics. In P. C. Lyons, & B. Alpern (Eds.), *Coal: Classification, coalification, mineralogy, trace element chemistry, and oil and gas potential* (pp. 529–574). International Journal of Coal Geology, 13.

Littke, R., & Leythaeuser, D. (1993). Gas sorption on coal and measurement of gas content. In B. E. Law, & D. D. Rice (Eds.), *Studies in Geology: Vol. 38. Hydrocarbons in Coal* (pp. 219–236). American Association of Petroleum Geologist.

Liu, G. (1990). Permo-Carboniferous paleogeography and coal accumulation and their tectonic control in the north and south China continental plates. *International Journal of Coal Geology, 11*, 73–117.

Liu, W. (2007). *Case study on CMM/CBM projects in China*. presented at CMM Development in the Asia-Pacific Region: Perspectives and Potential. Brisbane, Australia: China Coal Information Institute. October 4–5, 2006.

Liu, G. (2008). *CBM development in China*, 33rd International Geological Congress, August 6–14, 2008, Oslo, Norway.

Liu, Y., Beer, L.L., & Whitman, W.B. (2012). Methanogens: a window into ancient sulfur metabolism. *Trends in Microbiology, 20*, 251–258.

Liu, S., Song, Y., Gao, X., Zheng, Y., & Hong, F. (2011). *Coalbed methane and production water of Fuxin Basin, northeast China: chemical and isotopic geochemistry*. American Association of Petroleum Geologists Hedberg Conference, May 9–12, 2011, Beijing, China.

Lo, H. B. (1993). Correction criteria for the suppression of vitrinite reflectance in hydrogen-rich kerogens: preliminary guidelines. *Organic Geochemistry, 20*, 653–657.

Lo, H. B., Wilkins, R. W. T., Ellacott, M. V., & Buckingham, C. P. (1997). Assessing the maturity of coals and other rocks from North America using fluorescence alteration of multiple macerals (FAMM) technique. *International Journal of Coal Geology, 33*, 61–71.

Lobmeyer, D. H. (1985). *Freshwater heads and ground-water temperatures in aquifers of the Northern Great Plains in parts of Montana, North Dakota, South Dakota, and Wyoming*. U.S. Geological Survey. Professional Paper 1402–D, D1–D11.

Lode, E. (1999). Wetland restoration: a survey of options for restoring peatlands. *Studia Forestalia Suecica, 205*, 30. <http://pub.epsilon.slu.se/3014/1/SFS205.pdf> Accessed 02.02.12.

Logan, T. L. (1993). Drilling techniques for coalbed methane. In B. E. Law, & D. D. Rice (Eds.), *Studies in Geology: Vol. 38. Hydrocarbons from Coal* (pp. 269–285). American Association of Petroleum Geologists.

Lohila, A., Minkkinen, K., Aurela, M., Tuovinen, J.-P., Penttila, T., Ojanen, P., et al. (2011). Greenhouse gas flux measurements in a forestry-drained peatland indicate a large carbon sink. *Biogeosciences, 8*, 3203–3218.

Lombardi, T. E., & Lambert, S. W. (2001). *Coalbed methane in the Illinois Basin: Development challenges in a frontier area*. Society of Petroleum Engineers, ID72367–MS, SPE Eastern Regional Meeting, October 17–19, 2001, Canton, Ohio, p. 8. <http://www.onepetro.org/mslib/servlet/onepetropreview?id=00072367&soc=SPE> Accessed 01.11.12.

Long, A. J., Waller, M. P., & Stupples, P. (2006). Driving mechanisms of coastal change: peat compaction and the destruction of late Holocene coastal wetlands. *Marine Geology, 225*, 63–84.

Lowry, M. E., & Cummings, T. R. (1966). *Ground-water resources of Sheridan County*. Wyoming: U.S. Geological Survey Water-Supply. Paper 1807, p. 77.

Lowry, M. E., Daddow, P. B., & Rucker, S. J. (1993). *Assessment of the hydrologic system and hydrologic effects of uranium exploration and mining in the southern Powder River Basin uranium districts and adjacent areas, Wyoming*. U.S. Geological Survey Water-Resources Investigations Report 90–4154, p. 42.

Lowry, M. E., & Wilson, J. F. (1986). *Hydrology of area 50, Northern Great Plains and Rocky Mountain coal provinces, Wyoming and Montana*. U.S. Geological Survey Water-Resources. Investigations Open-File Report 83–545, p. 137.

Lu, S.-T., & Kaplan, I. R. (1990). Hydrocarbon-generating potential of humic coals from dry pyrolysis. *American Association of Petroleum Geologists Bulletin, 74*, 163–173.

Lu, Z., Streets, D. G., Zhang, Q., Wang, S., Carmichael, G. R., Cheng, Y. F., et al. (2010). Sulfur dioxide emissions in China and sulfur trends in East Asia since 2000. *Atmospheric Chemistry and Physics Discussions, 10*, 8657–8715. <http://dx.doi.org/10.5194/acpd-10-8657-2010>.

Lucarelli, B. (2010). *The history and future of Indonesian coal industry: Impacts of politics and regulatory framework on industry structure and performance*. In "*Working Paper #93, July 2010*," p. 87, Program on Energy and Sustainable Development, Stanford University, California. <http://iis-db.stanford.edu/pubs/22953/WP_93_Lucarelli_revised_Oct_2010.pdf> Accessed 26.10.11.

Lucas, R. E. (1982). *Organic Soils (Histosols): Formation, distribution, physical and chemical properties and management for crop production*. Michigan State University. Research Report 435 Farm Science.

Luhowy, V. M. (1993). Horizontal wells prove effective in Canadian heavy-oil field. *Oil and Gas Journal, 91*, 47–50.

Luly, J., Sluiter, I. R. K., & Kershaw, A. P. (1980). *Pollen studies of Tertiary brown coals: preliminary analysis of lithotypes within the Latrobe Valley*. Victoria: Monash Publications in Geography, 23.

Lyons, P. C. (1996). *Coalbed methane potential in the Appalachian states of Pennsylvania, West Virginia, Maryland, Ohio, Virginia, Kentucky, and Tennessee – an overview*. U.S. Geological Survey, Open-File Report 96–735.

Lyons, P. C., & Alpern, B. (Eds.), (1989a). *International Journal of Coal Geology: Vol. 12. Peat and coal: origin, facies, and depositional models* (pp. 798).

Lyons, P. C., & Alpern, B. (Eds.), (1989b). *International Journal of Coal Geology: Vol. 13. Coal: Classification, coalification, mineralogy, trace-element chemistry, and oil and gas potential* (p. 626).

Lyons, P. C., & Rice, C. L. (1986). Paleoenvironmental and tectonic controls in coal-forming basins in the United States. *Geological Society of America Special Paper, 210*, 200.

M2M-Australia. (2005). *Methane to markets partnership coal mine methane technical subcommittee country specific profile – Australia*. Australian Government, Department of Industry, Tourism and Resources & Department of the Environment and Heritage. <http://www.methanetomarkets.org/documents/events_coal_20050427_australia_profile.pdf> Accessed 23.11.11.

M2M Projects. (2010a). *Methane to markets international coal mine methane projects database, methane to markets, 2008*. <http://www2.ergweb.com/cmm/index.aspx> Accessed 23.11.11.

M2M Projects. (2010b). *Methane to markets international coal mine methane projects database, 2010*. <http://www2.ergweb.com/cmm/projects/projectlist.aspx?sortField=country&sortOrder=A>.

MacCartney, H. (2011). *Hydraulic fracturing in coalbed methane development, Raton Basin, southern Colorado*. EPA Hydraulic Fracturing Workshop, Arlington, Virginia, March 10–11, 2011. <gov/hfstudy/hfincoalbedmethanedevelopment.pdf> Accessed 10.02.13.

MacCarthy, F. J., Tisdal, R. M., & Ayers, W. B. (1996). Geological controls on coalbed prospectivity in part of the North Staffordshire Coalfield, UK. In R. Gayer, & L. Harris (Eds.), *Coalbed methane and coal geology* (Vol. 109; pp. 27–42). Geological Society Special Publication.

Macgregor, D. S. (1994). Coal-bearing strata as source rocks - a global overview. In A. C. Scott, & A. J. Fleet (Eds.), *Coal and coal-bearing strata as oil-prone source rocks?* (Vol. 77; pp. 107–116). London: The Geological Society Special Publication.

Magoon, L. B., & Anders, D. E. (1990). Oil-source rock correlation using carbon isotope data and biological marker compounds, Cook Inlet, Alaska peninsula, Alaska. *American Association of Petroleum Geologists Bulletin, 63*, 711.

Magoon, L. B., & Dow, W. G. (1994). Petroleum system. In L. B. Magoon, & W. G. Dow (Eds.), *American Association of Petroleum Geologists Memoir: Vol. 60. The petroleum system from source to trap* (pp. 3–24).

Magoon, L. B., Molenaar, C. M., Bruns, T. R., Fisher, M. A., & Valin, Z. C. (1995). *Southern Alaska province (003) U.S. Geological Survey*. p. 23. < http://certmapper.cr.usgs.gov/data/noga95/prov3/text/prov3.pdf> Accessed 31.12.11.

Majewska, Z., Majewski, S., & Zietek, J. (2010). Swelling of coal induced by cyclic sorption/desorption of gas: experimental observations indicating changes in coal structure due to sorption of CO_2 and CH_4. *International Journal of Coal Geology, 83*, 475–483.

Mallet, C. W., Pattison, C., McLennan, T., Balfe, P., & Sullivan, D. (1995). Bowen Basin. In C. R. Ward, H. J. Harrington, C. W. Mallet, & I. W. Beeston (Eds.), *Geology of Australian coal basins* (pp. 299–339). Geological Society of Australia, Coal Geology Group, Special Publication 1.

Malmrose, P., Lozier, J., Mickley, M., Reiss, R., Russell, J., Schaefer, J., et al. (2004). Residual management research committee subcommittee on membrane residual management. In *Journal of the American Water Resources Association: Vol. 96. 2004 committee report: Residuals management for desalting membranes* (pp. 73–87).

Mandal, D., Tewari, D. C., & Rautela, M. S. (2004). *Analysis of macro-fractures in coal for coal bed methane exploitation in Jharia coal field*, 5th Conference and Exposition on Petroleum Geophysics, Hyderabad, India.

Mandernack, K. W., Schmidt, R. A., Flores, R. M., & Klein, D. A. (2007). *Characterization of the bacterial and archaeal communities in coalbed methane aquifers in the Wyodak-Anderson coal zone of the Powder River Basin, WY using polar lipid analysis techniques*. Denver, CO: Geological Society of America with Programs. 39, p. 255.

Manrique, J. F., Poe, B. D., Jr., & England, K. (2001). *Production optimization and practical reservoir management of coal bed methane reservoirs*. Society of Petroleum Engineers, SPE Production and Operations Symposium, March 24–27, 2001, Oklahoma City, Oklahoma, ID67315–MS, p. 13. <http://www.onepetro.org/mslib/servlet/onepetropreview?id=00067315> Accessed 06.10.12.

Maracic, N., Mhoaghegh, S. D., & Artun, E. (2005). *A parametric study on the benefits of drilling horizontal and multilateral wells in coalbed methane reservoirs*. Society of Petroleum Engineers, SPE Annual Technical Conference and Exhibition in Dallas, Texas, October 9–12, 2005, Paper No. 96018, p. 8. <www.pe.wvu.edu/Publications/Pdfs/96018.pdf> Accessed 06.10.12.

Marchioni, D. L. (1980). Petrography and depositional environment of the Liddell seam, upper Hunter Valley, New South Wales. *International Journal of Coal Geology, 1*, 35–61.

Marchioni, D. L. (2003). *Coal gas contents in Western Canada Basin (and some comments on measurement)*. Calgary: Alberta, 2nd Annual Coalbed Methane Symposium.

Mardon, S. M., Takacs, K. G., Eble, C. F., Hower, J. C., & Mastalerz, M. (2007). *Evaluation of coalbed methane potential and gas adsorption capacity in the Western Kentucky coalfield*. Lexington, KY: American Association of Petroleum Geologists, 36th Annual Eastern Section Meeting, September 16–18.

Mares, T. E., & Moore, T. A. (2008). The influence of macroscopic texture on biogenically derived coalbed methane, Huntly coalfield, New Zealand. In R. M. Flores (Ed.), *Microbes Methanogenesis, and Microbial Gas in coal. International Journal of Coal Geology, 76*; pp. 175–185).

Mares, T. E., Radlinski, A. P., Moore, T. A., Cookson, D., Thiyagarajan, P., Ilavsky, J., et al. (2009). Assessing the potential for CO_2 adsorption in a subbituminous coal, Huntly Coalfield, New Zealand, using small angle scattering techniques. *International Journal of Coal Geology, 77*, 54–68.

Markič, M., & Sachsenhofer, R. F. (1997). Petrographic composition and depositional environments of the Pliocene Velenje lignite seam (Slovenia). *International Journal of Coal Geology, 33*, 229–254.

Markic, M., & Sachsenhofer, R. F. (2010). *The Velenje lignite –Its petrology and genesis*. Ljubljana: Geoloski Zavod Slovenije. p. 218.

Markič, M., Zavšek, S., Pezdič, J., Skaberne, D., & Kočevar, M. (2001). Macropetrographic characterization of the Velenje lignite (Slovenia). *Acta Universitatis Carolinae, Geologic, 45*, 81–97.

Marks, A.G. (2006). Investigation of intramolecular bonding in four high-volatile bituminous coals using wide angle X-ray scattering analysis. (Ph.D. Dissertation). The University of Southern Mississippi, p. 144.

Marland, G., Boden, T., & Andres, R. J. (2008). *National CO_2 emissions from fossil-fuel burning, cement manufacture, and gas flaring: 1751–2005*. Carbon Dioxide Information Analysis Center, Oak Ridge National Laboratory. <http://www.cdiac.ornl.gov/ftp/trends/co2_emis/.usa.dat> Accessed 08.04.09.

Marsh, H. (1987). Adsorption methods to study microporosity in coals and carbons-a critique. *Carbon, 25*, 49–58.

Marshall, J. S., Pilcher, R. C., & Bibler, C. J. (1996). Opportunities for the development of coalbed methane in three coal basins in Russia and Ukraine. In R. Gayer, & I. Harris (Eds.), *Coalbed methane and coal geology* (pp. 89–101). Geological Society Special Publication No 109.

Marshall, J. S. (September 1, 2001). Coalbed methane opportunities in Bulgaria. *World Coal*.

Martinez, R. (December 6, 2004). *Presentation to UNECE ad hoc group of experts on coal mine methane*. Instituto Geologico y Minero de Espagne (IGME). <http://www.unece.org/ie/se/pdfs/cmm/ppp09dec/IGME_Martinez_UNECECMM.pdf> Accessed February 23.

Martini, A. M., Walter, L. M., Budai, J. M., Ku, T. C. W., Kaiser, C. J., & Schoell, M. (1998). Genetic and temporal relations between formation waters and biogenic methane: upper Devonian Antrim Shale, Michigan Basin, USA. *Geochimica et Cosmochimica Acta, 62*, 1699–1720.

Masing, V. (1975). Mire typology of the Estonian S.S.R. In L. Laasimer (Ed.), *Some aspects of botanical research in the Estonian S.S.R* (pp. 123–138). Academy of Sciences of the Estonian S.S.R. Tartu.

Massarotto, P., Golding, S. D., Bae, J. S., Iyer, R., & Rudolph, V. (2010). Changes in reservoir properties from injection of supercritical CO_2 into coal seams—a laboratory study. In S. D. Golding, V. Rudolph, & R. M. Flores (Eds.), *International Journal of Coal Geology: Vol. 82. Asia Pacific coalbed methane symposium: Selected papers from the 2008 Brisbane symposium on coalbed methane and CO_2-enhanced coalbed methane* (pp. 269–279).

Mastalerz, M., Gluskoter, H., & Rupp, J. (2004). Carbon dioxide and methane adsorption in high volatile bituminous coals from Indiana. *International Journal of Coal Geology, 60*, 43–57.

Mastalerz, M., Drobniak, A., & Rupp, J. (2008). Meso- and micropore characteristics of coal lithotypes: Implications for CO_2 adsorption. *Energy and Fuels, 22*, 4049–4061.

Mastalerz, M., Goodman, A., & Chirdon, A. (2012). Coal lithotypes before, during, and after exposure to CO_2: insights from Direct Fourier Transform Infrared investigation. *Energy & Fuels, 26*, 3586–3591.

Mastalerz, K. (1995). Deposits of high-density turbidity currents on fan-delta slopes: an example from the upper Visean Szczawno formation, Intrasudetic Basin, Poland. *Sedimentary Geology, 98*, 121–146.

Matthews, M. (2007). *Coalbed methane producers have some new options*. McGraw-Hill. Construction/ENR 258, p. 19.

Matveev, A. K. (1976). *Exploration: First International Symposium*. San Francisco, California: Miller Freeman Publications, 77–88.

Mavor, M. J., Close, J. C., & Pratt, T. J. (1991). *Western Cretaceous coal seam project summary of completion optimization and assessment laboratory (COAL) site*. Gas Research Institute Topical Report No. GRI–91/0377.

Mavor, M. J., Dhir, R., McLennan, J. D., & Close, J. C. (1991). *Evaluation of the hydraulic fracture stimulation of the Colorado 32-7 No. 9 well, San Juan Basin*. In *Rocky Mountain Association of Geologists Guidebook, "Coalbed methane of western North America"*. Fall Conference and Field Trip, 241–249.

Mavor, M. J., & Gunter, W. D. (2004). Secondary porosity and permeability of coal vs gas composition and pressure. Society of Petroleum Engineers. *Reservoir Evaluation and Engineering, 9*, 114–125.

Mavor, M. J., & Nelson, C. (1997). Coalbed gas-in-place methodology and error summary. In *Coalbed reservoir gas-in-place analysis*. Gas Research Institute.

Mavor, M. J., & Pratt, T. J. (1996). *Improved methodology for determining total gas content*. In *Comparative evaluation of the accuracy of gas-in-place estimates and review of lost gas models* (Vol. II). Gas Research Institute. Topical Report GRI-94/0429.

Mavor, M. J., Pratt, T. J., & Britton, R. N. (1994). Improved methodology for determining total gas content. In *Canister gas desorption data summary* (Vol. 1). Gas Research Institute. Topical Report GRI-93/0410.

Mavor, M. J., Pratt, T. J., & Nelson, C. R. (1995). *Quantitative evaluation of coal seam gas content estimate accuracy*. Society of Petroleum Engineers. Joint Rocky Mountain Regional/Low-Permeability Reservoirs Symposium, Denver, Colorado March 20–22, 1995, 327–340, paper Id. 29577.

Mazumder, S., Sosrowidjojo, B., & Ficarra, A. (2010). *The late Miocene coalbed methane system in south Sumatra Basin in Indonesia*. <http://dx.doi.org/10.2118/133488-MS>. SPE Asia Pacific Oil and Gas Conference and Exhibition, October 18–20, Brisbane, Queensland, Australia.

Mazumder, S., Wolf, K.-H. A. A., Elewaut, K., & Ephraim, R. (2006). Application of X-ray computed tomography for analyzing cleat spacing and cleat aperture in coal samples. *International Journal of Coal Geology, 68*, 205–222.

McBeth, I., Reddy, K. J., & Skinner, Q. D. (2003). Chemistry of trace elements in coalbed methane product water. *Water Research, 37*, 884–890. <http://144.206.159.178/ft/1092/78769/1329554.pdf> Accessed 29.11.12.

McBride, E. F. (1977). Short note: secondary porosity-importance in sandstone reservoirs in Texas. *Gulf Coast Association of Geological Societies Transaction, 27*, 121–122.

McCabe, P. J. (1984). Depositional environment of coal and coal-bearing strata. In R. A. Rahmani, & R. M. Flores (Eds.), *Sedimentology of coal and coal-bearing sequences* (pp. 13–42). Oxford: Blackwell Scientific Publications.

McCabe, R., Almasco, J. N., & Yumul, G. (1985). Terranes of the central Philippines. In D. Howell (Ed.), *Circum Pacific Council for Energy and Mineral Resources Series: Vol. 1. Tectonostratigraphic terranes of the circum-Pacific region* (pp. 421–435).

McCabe, P. J., & Parrish, J. T. (Eds.), (1992). *Geological Society of America Special Paper: Vol. 267. Tectonic and climatic controls on the distribution and quality of cretaceous coals* (p. 395).

McCartney, K., Hoffman, A. R., & Tredoux, M. (1990). A paradigm for endogenous causation of mass extinctions. In V. L. Sharpton, & P. D. Ward (Eds.), *Geological Society of America Special Paper, Global catastrophes in earth history* (pp. 125–138).

McCormack, N., Pepper, A., & Adams, M. (November 2006). *Modeling expelled petroleum fluids from coals*. American Association of Petroleum Geologists International Conference and Exhibition, Perth, West Australia 5–8. <http://www.pesa.com.au/aapg/aapgconference/pdfs/abstracts/tues/OilfromCoalSources.pdf> Accessed 10.12.11.

McCulloch, C. M., Deul, M., & Jeran, P. W. (1974). *Cleat in bituminous coal beds*. U.S. Bureau of Mines. Report of Investigations 7910, p. 25.

McCulloch, C. M., Levine, J. R., Kissell, F. N., & Deul, M. (1975). *Measuring the methane content of bituminous coalbeds*. U.S. Bureau of Mines Report of Investigations 8515, p. 22.

McCune, D. (2003). *Coalbed methane development in the Cherokee and Forest City Basins*. Paper 0313 presented at the International Coalbed Methane Symposium, Tuscaloosa, Alabama, May 7–8, 2003.

McCutcheon, S. C., & Schnoor, J. L. (Eds.), (2003). *Phytoremediation—Transformation and control of contaminants*. Wiley-Interscience, A John Wiley & Sons, Inc., Publication. Environmental Science and Technology.

McDaniel, B. W. (1990). *Hydraulic fracturing techniques used for stimulation of coalbed methane*. Society of Petroleum Engineers. SPE Eastern Regional Meeting, October 31–November 2, 1990. Columbus, Ohio SPE 21292. <http://www.onepetro.org/mslib/app/Preview.do?paperNumber=00021292&societyCode=SP> Accessed 14.02.13.

McFall, K. S., Wicks, D. E., and Kuuskraa, V. A., 1986. A geologic assessment of natural gas from coal seams in the warrior basin, Alabama. Gas Research Institute, Topical Report GRI-86/0272.

McFarland, J. R., Herzog, H. J., & Jacoby, H. D. (2004). *The future of coal consumption in a carbon constrained world*. Presented at the Seventh International Conference on Greenhouse Gas Control Technologies, Vancouver,

Canada, September. <http://sequestration.mit.edu/pdf/GHGT7_paper_McFarland_etal.pdf> Accessed 24.09.11.

McGarry, D.E. (2003). Resources in motion. U.S. Bureau of Land Management Resource Management Tools Conference, April 21–25, 2003, Phoenix, Arizona.

McGraw-Hill. (2003). *Dictionary of scientific and technical terms* (6th ed.). Mc-Graw-Hill Companies, Inc.

McIntosh, J., Martini, A., Petsch, S., Huang, R., & Nusslein, K. (2008). Biogeochemistry of the Forest City Basin coalbed methane play. In R. M. Flores (Ed.), *International Journal of Coal Geology: Vol. 76. Microbes, methanogenesis, and microbial gas* (pp. 111–118).

McKee, C. R., Bumb, A. C., & Koenig, R. A. (1988). Stress-dependent permeability and porosity of coal and other geologic formations. *SPEFE, 3*(1), 81–91. <http://dx.doi.org/10.2118/12858-PA SPE-12858-PA>.

McKee, C. R., Bumb, A. C., Way, S. C., Koenig, R. A., & Brandenburg, C. F. (1986). Using permeability vs depth correlations to assess the potential for producing gas from coal seams. *Methane from Coal Seams Technology*, 415–426.

McKelvey, V. E. (1972). Mineral resource estimates and public policy. *American Scientist, 60*, 32–40.

McLennan, J. D., Schafer, P. S., & Pratt, T. J. (1995). *A guide to determining coalbed gas content*. Gas Research Institute. GRI-94/0396.

McPherson, B. J. (2006). *New Mexico sequestration test project overview*. Pittsburgh: National Energy Laboratory Technology. Southwest Regional Partnership on Carbon Sequestration. <http://www.netl.doe.gov/publications/proceedings/06/rcsp/oct%204/33%20Brian%20McPherson%20oct%204%20san%20juan%20basin.pdf> Accessed 11.02.13.

McVay, D. A., Ayers, W. B., Jr., & Jensen, J. L. (2007). *CO_2 sequestration potential of Texas low-rank coals*. Texas A & A University. Final Technical Report. Texas Engineering Experiment Station.

Medek, J., Weishauptova, Z., & Kovar, L. (2006). Combined isotherm of adsorption and absorption on coal and differentiation of both processes. *Microporous and Mesoporous Materials, 89*, 276–283.

van de Meene, E. A. (1984). *Geological aspects of peat formation in the Indonesian—Malaysian lowlands*. Republic of Indonesia: Bulletin of the Geological Research and Development Centre.

Melnichenko, Y. B., He, L., Sakurovs, R., Kholodenko, A. L., Blach, T., Mastalerz, M., et al. (2012). Accessibility of pores in coal to methane and carbon dioxide. *Fuel, 91*, 200–208.

Meneley, R. A., Calverley, A. E., Logan, K. G., & Proctor, R. M. (2003). Resource assessment methodologies: current status and future direction. *American Association of Petroleum Geologists Bulletin, 87*, 535–540.

Meng, Z. P., & Tian, Y. D. (2007). *Theory and method of coalbed methane development geology*. p. 298. <www.sciencep.com> (in Chinese).

Mercier, T. J., Brownfield, M. E., & Johnson, R. C. (2010). *Methodology for calculating oil shale resources for the Uinta Basin, Utah and Colorado*. U. S. Geological Survey DDS-69-BB, p. 54.

Metje, M., & Frenzel, P. (2007). Methanogenesis and methanogenic pathways in a peat from subarctic permafrost. *Environmental Microbiology, 9*, 954–964.

Miall, A. D. (1977). A review of the braided-river depositional environment. *Earth-Science Reviews, 13*, 1–62.

Middleton, M. F., & Hunt, J. W. (1989). Influence of tectonics on Permian coal-rank patterns in Australia. In P. C. Lyons, & B. Alpern (Eds.), *International Journal of Coal Geology: Vol. 13. Coal: Classification, coalification, mineralogy, trace-element chemistry, and oil and gas potential* (pp. 391–411).

Midgley, D. J., Hendry, P., Pinetown, K. L., Fuentes, D., Gong, S., Mitchell, D. L., et al. (2010). Characterisation of a microbial community associated with a deep, coal seam methane reservoir in the Gippsland Basin, Australia. In S. D. Golding, V. Rudolph, & R. M. Flores (Eds.), *International Journal of Coal Geology: Vol. 82. Asia Pacific coalbed methane symposium* (pp. 232–239).

Miki, T. (1994). Tertiary sedimentation and organic maturation in Kyushu, Japan. In *Journal of Southeast Asian Earth Sciences: Vol. 9. Symposium on the dynamics of subduction and its products* (pp. 123–128).

Milici, R. C. (2004). *Assessment of Appalachian oil and gas resources: Carboniferous coal bed gas total petroleum system*. U.S Geological Survey Open-File Report 2004-1272. Online publication only version 1.0. <http://pubs.usgs.gov/of/2004/1272/2004-1272.pdf> Accessed 10.11.12.

Milici, R. C., & Hatch, J. R. (2004). *Assessmet of undiscovered Carboniferous coal-bed gas resources of the Appalachian Basin and Black Warrior Basin Provinces*. U.S. Geological Survey Fact Sheet FS-2004–3092.

Milici, R., & Houseknecht, D. (2009). Akoma Basin coalbed gas. U.S. Geological Survey Presentation. In *Natural Gas Assessment of the Arkoma Basin, Ouachita Thrust Belt, and Reelfoot Rift, Oklahoma Geological Survey*. November 3, 2009.

Milici, R. C., Flores, R. M., & Stricker, G. D. (2009). *Estimation of U.S. peak coal production*. Geological Society of America National Meetings, Portland, Oregon, October 18–21, 2009, Abstracts with Programs for the 2009 Annual Meeting and Exposition, 550.

Milici, R. C., Flores, R. M., & Stricker, G. D. (2012). Coal resources, reserves and peak coal production in the United States. *International Journal of Coal Geology*. < http://dx.doi.org/10.1016/j.coal.2012.10.002> Accessed 04.04.13.

Milkov, A. V. (2011). Worldwide distribution and significance of secondary microbial methane formed during petroleum biodegradation in conventional reservoirs. *Organic Geochemistry, 42*, 184–207.

Mishra, H. K., & Ghosh, R. K. (1996). Geology, petrology and utilization potential of some Tertiary coals in northeastern region of India. *International Journal of Coal Geology, 30*, 65–100.

MIT. (2007). *The future of coal: Options for a carbon-constrained world*. Massachusetts Institute of Technology. ISBN 978-0-615-14092-6, p. 174. <http://web.mit.edu/coal/The_Future_of_Coal.pdf> Accessed 02.12.11.

Mitchell, B. R. (1988). *British historical statistics*. Cambridge: Cambridge University Press.

Mitchell, N. (2005). *Coal seam gas drilling*. Mitchell Drilling Contractors Presentation, June 16, 2005. < http://www.drillsafe.org.au/06-05_pres/DrillSafe_Forum_Jun05_MITCHELL_DRILLING_Coal_Seam.pdf> Accessed 05.10.12.

Miyamoto, K. (Ed.), (1997). *Food Agricultural Organization of the United Nations, Agricultural Services Bulletin: Vol. 128. Renewable biological systems for alternative sustainable energy production*. < http://www.fao.org/docrep/w7241e/w7241e0f.htm#4.1%20microbial%20consortia%20and%20biological%20aspects%20of%20methane%20fermentation>. Accessed 24.02.12.

MOC. (2010). *Inventory of coal resources of India*. Ministry of Coal. <http://coal.nic.in/reserve2.htm> Accessed 23.01.13.

Mohr, S. H., & Evans, G. M. (2009). *Forecasting coal production until 2100*. p. 22. <http://www.theoildrum.com/files/coalmodel.pdf> Accessed 19.09.11.

Monash University (2011). Chapter 5: Thermal maturity. In "Crustal Heat Flow". <http://www.geosci.monash.edu.au/heatflow/chapter5.html> Accessed 07.12.11.

Moniz, E. J. (2007). *The future of coal: Summary*. <http://web.mit.edu/miyei/> Accessed 03.04.13.

Montgomery, S. L., & Barker, C. E. (2003). Coalbed methane, Cook Inlet, south-central Alaska: a potential giant gas resource. *American Association of Petroleum Geologists Bulletin, 87*, 1–13.

Moody, J. D. (1978). The world hydrocarbon resource base and related problems. In G. M. Philip, & K. L. Williams (Eds.), *University of Sydney Earth Resources Foundation Occasional Publication: Vol. 1. Australia's mineral energy resources assessment and potential* (pp. 63–69).

Moore, E. S. (1922). *Coal* (1st ed.). New York: John Wiley and Sons, Inc.

Moore, P. D. (1987). Ecological and hydrological aspects of peat formation. In A. C. Scott (Ed.), *Geological Society Special Publication: Vol. 32. Coal and coal-bearing strata: Recent advances* (pp. 7–15).

Moore, P. D. (1989). The ecology of peat-forming processes – a review, in peat and coal. In P. C. Lyons, & B. Alpern (Eds.), *International Journal of Coal Geology: Vol. 12. Origin, facies, and depositional models* (pp. 89–104).

Moore, T. A. (1994). Organic compositional clues to a stacked mire sequence in the Anderson-Dietz coal bed (Paleocene), Montana. In R. M. Flores, K. T. Mehring, R. W. Jones, & T. L. Beck (Eds.), *Organic and the Rockies field guide* (pp. 163–172). Wyoming State Geological Survey, Public Information Circular No. 33, The Society of Organic Petrology Eleventh Annual Meeting, Jackson Hole, Wyoming.

Moore, T. A. (1995). *Developing models for spatial prediction of mining and utilization potentials in coal seams: An example from the Greymouth coalfield*. Proceedings 6th New Zealand Coal Conference, pp. 385–402.

Moore, T. A. (2012). Coalbed methane: a review. *International Journal of Coal Geology, 101*, 36–81.

Moore, T. A., & Friederich, M. C. (2010). A probabilistic approach to estimation of coalbed methane resources for Kalimantan, Indonesia. In N. I. Basuki, & S. Prihatmoko (Eds.), *Kalimantan Coal and Mineral Resources, MGEI-IAGI Seminar, 29–30 March 2010, Balikpapan, Indonesia* (pp. 61–71).

Moore, P. D., & Bellamy, D. J. (1973). *Peatlands*. Springer-Verlag (Digitize Aug. 8, 2011).

Moore, T. A., Bowe, M., & Nas, C. (2012). *Effects of a shallow-seated heat-source on coalbed methane reservoir character, Kalimantan Timur (Borneo), Indonesia, Proceedings and Abstracts*, 29th Annual Meeting of The Society of Organic Petrology, 20-24 September 2012, Beijing, China, 9, 37–39 (ISSN 1060-7250).

Moore, T. A., Esterle, J. S., & Shearer, J. C. (2000). The role and development of texture in coal beds: examples from Indonesia and New Zealand. In Hadiyanto (Ed.), *Southeast Asian Coal Geology Conference* (pp. 239–252). Bandung: Ministry of Mines and Energy.

Moore, T. A., Gillard, G. R., Boyd, R., Flores, R. M., Stricker, G. D., & Galceran, C. M. (2004). *A mighty wind: determining the methane content of New Zealand coal seams. The Society for Organic Petrology*. Sydney, Australia: 21st Annual Meeting (pp. 114–116).

Moore, T. A., & Ferm, J. C. (1992). Composition and grain-size of an Eocene coal bed in southeastern Kalimantan, Indonesia. *International Journal of Coal Geology, 21*, 1–30.

Moore, T. A., Flores, R. M., Stanton, R. W., & Stricker, G. D. (2001). The role of macroscopic texture in determining coal bed methane variability in the Anderson–Wyodak coal seam, Powder River Basin, Wyoming. In C. R. Robinson (Ed.), *Eighteenth annual meeting* (pp. 85–88). Houston, Texas: The Society for Organic Petrology.

Moore, T. A., & Hilbert, R. E. (1992). Petrographic and anatomical characteristics of plant material from two peat deposits of Holocene and Miocene age, Kalimantan, Indonesia. *Review of Palaeobotany and Palynology, 72*, 174–199.

Moore, T. A., & Nas, C. (2012). The enigma of the Pinang Dome (Kalimantan Timur): A review of its origin, significance and influence on coal rank and coalbed methane properties. Proceedings, Indonesian Petroleum Association, 37th Annual Convention and Exhibition, May 15–17, Jakarta, Indonesia, Paper IPA13-G-119.

Moore, T. A., & Shearer, J. C. (1999). Coal: types and characteristics. In C. P. Marshall, & R. E. Fairbridge (Eds.), *The encyclopedia of geochemistry* (pp. 87–90). Dordrecht: Kluwer Academic Publishers.

Moore, T. A., & Shearer, J. C. (2003). Peat/coal type and depositional environment–are they related? *International Journal of Coal Geology, 56*, 233–252.

Moore, T.A., Mares, T.E., and Moore, C.R. (2009). Assessing uncertainty of coalbed methane resources. Proceedings, Indonesian Petroleum Association, 33rd Annual Convention & Exhibition. Indonesian Petroleum Association, Jakarta, Indonesia. Paper IPA09-G-056, unpaginated.

Moore, T. A., Shearer, J. C., & Esterle, J. S. (1993). Quantitative macroscopic textural analysis. *The Society of Organic Petrology Newsletter, 9*, 13–16.

Moores, E. M., & Twiss, R. J. (1995). *Tectonics* (11th ed.). New York: W.H. Freeman.

Morad, K. (2008). *Application of flowing p/Z* material balance for dry coalbed-methane reservoirs.* Society of Petroleum Engineers. CIPC/SPE Gas Technology Symposium 2008 Joint Conference, 16–19 June 2008, Calgary, Alberta, Canada, p. 8.

Moran, J. E., Fehn, U., & Hanor, J. S. (1995). Determination of source ages and migration of brines from the U.S. Gulf coast basin using ^{129}I. *Geochimica et Cosmochimica Acta, 24*, 5055–5069.

Morgan, J. P. (1970). *Deltaic sedimentation, modern and ancient.* In *Society of Economic Paleontologists and Mineralogists, Special Publication* (Vol. 15).

Morley, R. J. (1981). Development and vegetation dynamics of a lowland ombrogenous peat swamp in Kalimantan Tengah, Indonesia. *Journal of Biogeography, 8*, 383–404.

Morley, R. J. (2000). *Origin and evolution of tropical rain forests.* London: Wiley.

Morozova, D., Mohlmann, D., & Wagner, D. (2006). *Survival of methanogenic Archaea from Siberian permafrost under simulated Martian thermal conditions.* Origins of Life and Evolution of the Biosphere, Springer Science. p. 12. <http://epic.awi.de/14473/1/Mor2006e.pdf> Accessed 01.04.12.

Morris, P. J., & Waddington, J. M. (2011). Groundwater residence time distributions in peatlands: implications for peat decomposition and accumulation. *Water Resources Research, 47*, W02511. <http://dx.doi.org/10.1029/2010WR009492>.

Mosher, K. (2011). *The impact of pore size on methane and CO_2 adsorption in carbon* (MS thesis). Department of Energy Resources and Engineering, Stanford University, p. 63.

Mosle, B., Kukla, P., Stollhofen, H., & Prueße, A. (2009). *Coal bed methane production in the Munsterland Basin, Germany.* European Geosciences Union General Assembly 2009, Vienna, Austria, April 19–24, 2008. <http://adsabs.harvard.edu/abs/2009EGUGA..11.4267M> Accessed 28.01.13.

Mosquera-Corral, A., Gonzalez, F., Campos, J. L., & Mendez, R. (2005). Partial nitrification in a SHARON reactor in the presence of salts and organic carbon compounds. *Process Biochemistry, 40*, 3109–3118.

Mott, R. (1943). The origin and composition of coals. *Fuel, 22*, 20–26.

MSHA. (2011). *Historical data on mine disasters in the United States.* U.S. Mine Safety and Health Administration Fact Sheet 95–8. <http://www.msha.gov/MSHAINFO/FactSheets/MSHAFCT8.HTM> Accessed 18.11.11.

Mudd, M. B. (2012). Montana v. Wyoming: a opportunity to right the course from coalbed methane development and prior appropriation. *Golden Gate University Environmental Law Journal, 5*, 297–341. Pacific Region Edition, Article 6.

Mukhopadhyay, P. K. (1986). Petrography of selected Wilcox and Jackson Group lignites from the Tertiary of Texas. In R. B. Finkelman, & D. J. Casagrande (Eds.), *Geology of Gulf Coast Lignites* (pp. 126–145). Geological Society of America Field Trip Guide Bopok.

Mukhopadhyay, P. K., & Dow, W. G. (Eds.), (1994). *Symposium series 570. Vitrinite reflectance as a maturity parameter, applications and limitations.* Washington, D.C: American Chemical Society, ACS.

Mukhopadhyay, P. K., & Hatcher, P. G. (1993). Composition of coal. In B. E. Law, & D. D. Rice (Eds.), *Studies in Geology: Vol. 38. Hydrocarbons from Coal* (pp. 79–118). American Association of Petroleum Geologists. <http://dx.doi.org/10.1007/s11084-006-9024-7>. Chapter 4.

Mukhopadhyay, G., Mukhopadhyay, S. K., Roychowdhury, M., & Parui, P. K. (2010). Stratigraphic correlation between different Gondwana basins of India. *Journal Geological Society of India, 76*, 251–266.

Mullen, M. J. (1988). *Log evaluation in wells drilled for coal-bed methane.* Rocky Mountain Association of Geologists. pp. 113–124.

Mullen, M. J. (1989). Coalbed methane resource evaluation from wireline logs in the northeastern San Juan basin: A Case Study. *Proceeding the 1989 Coalbed Methane Symposium.* Tuscaloosa, AL: University of Alabama (pp. 167–184).

Mullen, M. J. (1991). *Coalbed methane resource evaluation from wireline logs in northeastern San Juan Basin: A case study.* Society of Petroleum Engineers. Joint Rocky Mountain Regional/Low Permeability Reservoirs Symposium and Exhibition, Denver, Colorado, 6–8 March, 1991, 161–172, paper Id. 18946.

Muramatsu, Y., & Wedepohl, K. H. (1998). The distribution of iodine in the earth's crust. *Chemical Geology, 147*, 201–216.

Murray, D. K. (1996). In *Geological Society London, Special Publications: Vol. 109. Coalbed methane in the USA: Analogues for worldwide development* (pp. 1–12).

Mustafa, B., & Ayar, G. (2004). Turkey's coal reserves, potential trends and pollution problems of Turkey. *Energy Exploration & Exploitation, 22*, 367–376.

Myers, T. (2012). Potential contaminant pathways from hydraulically fractured shale to aquifers. *Ground Water*. <http://dx.doi.org/10.1111/j.1745-6584.2012.00933.x> <http://catskillcitizens.org/learnmore/Fracking-Aquifers.pdf> Accessed 11.10.12.

Nadon, G. C. (1998). Magnitude and timing of peat to coal compaction. *Geological Society of America, Geology, 26*, 727–730.

Naeser, N. D., & McCulloh, T. H. (Eds.), (1989). *Thermal history of sedimentary basins*. New York: Springer-Verlag.

Nambo, H. (2000). The abandoned coalmine gas project in northern Japan. In *Minimize hazards and maximize profits through the recovery of coalbed and coal mine methane* (pp. 17). New York: Strategic Research Institute.

NASA and STSci. (2006). *Mars may be cozy place for hardy microbes*. News Release Number: STSci–2006–48. <http://hubblesite.org/newscenter/archive/releases/2006/48/image/a/> Accessed 02.04.12.

Navale, G. K. B., & Saxena, R. (1989). An appraisal of coal petrographic facies in Lower Gondwana (Permian) coal seams of India. In P. C. Lyons, & B. Alpern (Eds.), *International Journal of Coal Geology: Vol. 12. Peat and coal: Origin, facies, and depositional models* (pp. 553–588).

NC Division of Water Resources. (2012). *Basic hydrology: Ground water, aquifers and confining beds*. North Carolina Department of Environmental and Natural Resources. <http://www.ncwater.org/Education_and_Technical_Assistance/Ground_Water/Hydrogeology/> Accessed 25.10.12.

NCC. (2006). *Americas energy future*. National Coal Council, Coal. <http://nationalcoalcouncil.org/> Accessed 12.12.11.

NEB. (2007). *Coalbed methane – Fact sheet*. National Energy Board of Canada. <http://www.neb.gc.ca/clf-nsi/rnrgynfmtn/nrgyrprt/ntrlgs/hrsshcnynclbdmthn2007/clbdmthnfctsht-eng.html> Accessed 08.11.12.

Negri, M. C., Hinchman, R. R., & Settle, T. L. (2003). Salt tolerant plants to concentrate saline waste streams. In S. C. McCutcheon, & J. L. Schnoor (Eds.), *Environmental science and technology. Phytoremediation–Transformation and control of contaminants* (pp. 753–762). Wiley-Interscience, A John Wiley & Sons, Inc., Publication.

Nelson, C. R. (1999). *Effects of coalbed reservoir property analysis methods on gas-in-place estimates*. Society of Petroleum Engineers. Paper no. SPE 57443.

NETL. (2007). *Cost and performance baseline for fossil energy plants*. National Energy Technology Laboratory. DOE/NETL-2007/1281, May 2007. <http://www.netl.doe.gov/energy- analyses/pubs/Bituminous%20Baseline_Final%20Report.pdf> Accessed 04.11.11.

NETL. (2009). *Tracking new coal-fired power plants*. National Energy Technology Laboratory. <http://www.netldoe.gov/coal/refshelf/ncp.pdf> Accessed 08.04.09.

NETL. (2012a). *Produced water management technology descriptions fact sheet - ion exchange*. Department of Energy, National Energy Technology Laboratory. <http://www.netl.doe.gov/technologies/pwmis/techdesc/ionex/index.html> Accessed 16.12.12.

NETL. (2012b). *Produced water management information system: Introduction to produced water*. U.S. Department of Energy, National Energy Technology Laboratory. <http://www.netl.doe.gov/technologies/pwmis/intropw/index.html> Accessed 28.11.12.

Neuzil, S. G., & Cecil, C. B. (1985). *Classification of peat as a basis for modeling coal quality*. Geological Society of America Annual Meeting, Orlando, FL, U.S.A., p. 676.

Newman, J. (1997). New approaches to detection and correction of suppressed vitrinite reflectance. *APEA Journal, 37*, 524–535.

Newman, J., Eckersley, M., Francis, D. A., & Moore, N. A. (2000). *Application of vitrinite-inertinite reflectance and fluorescence (VIRF) to maturity assessment in the East Coast and Canterbury Basins of New Zealand*. New Zealand Petroleum Conference Proceedings, March 19–22, 2000, p. 19.

Newnham, R. M., de Lange, P. J., & Lowe, D. J. (1995). Holocene vegetation, climate and history of a raised bog complex, northern New Zealand based on palynology, plant macrofossils and tephrochronology. *The Holocene, 5*, 267–282.

Nichols, D. J. (1999). Biostratigraphy, Powder River Basin. In *1999 Resource assessment of selected Tertiary coal beds and zones in the northern Rocky Mountains and Great Plains region* (Fort Union Assessment Team), U.S. Geological Survey Professional Paper 1625A, Chapter PB, PB1–PB12.

Nichols, D. J., Brown, J. L., Attrep, M., Jr., & Orth, C. J. (1992). A new Cretaceous–Tertiary boundary locality in the western Powder River Basin, Wyoming: biological and geological implications. *Cretaceous Research, 13*, I3–I30.

Nichols, D. J., Jarzen, D. M., Orth, C. J., & Oliver, P. Q. (1986). Palynological and iridium anomalies at Cretaceous–Tertiary boundary, south-central Saskatchewan. *Science, 231*, 714–717.

Nicot, J. P., Hebel, A. K., Ritter, S. M., Walden, S., Baier, R., Galusky, O., et al. (2011). *Current project water use in the Texas, Texas mining and oil and gas industry*. Draft Report – February 2011, Prepared for Texas Water Development Board, p. 314, Appendix. <http://www.frackinginsider.com/API_Comments_to_EPA_Fracking_Study.pdf> Accessed 11.10.12.

NIEA. (2010). *Global peatlands*. Northern Ireland Environmental Agency, Department of Environment. <http://www.doeni.gov.uk/wonderfulni/biodiversity/habitats-2/peatlands/about_peatlands/global_peatland.htm> Accessed 11.03.12.

NIOSH. (2010). *Coal mining disasters*. National Institute for Occupational Safety and Health, Mining Safety and Health Research website. <http://www.cdc.gov/niosh/mining/statistics/discoal.htm> Accessed 20.11.11.

NMA. (2009). *Mountaintop mining fact book*. National Mining Association. p. 11. <http://www.nma.org/pdf/fact_sheets/mtm.pdf> Accessed 17.11.11.

NMA. (2011a). *Fast facts about coal*. National Mining Association. <http://www.nma.org/statistics/fast_facts.asp> Accessed 22.11.11.

NMA. (2011b). *Liquid fuels from U.S. coals*. National Mining Association. <http://www.nma.org/pdf/liquid_coal_fuels_100505.pdf> Accessed 05.12.11.

North American Commission on Stratigraphic Nomenclature. (2005). North American stratigraphic code. *American Association of Petroleum Geologists Bulletin, 89*, 1547–1591. <http://ngmdb.usgs.gov/Info/NACSN/05_1547.pdf> Accessed 17.09.11.

NPC. (2007). Facing the Hard Truths about Energy and supplementary material: Topic Paper #18 – Coal to Liquids and Gas. National Petroleum Council. http://www.npchardtruthsreport.org/ Accessed 12.12.11.

NPC. (2011). *Transmission natural gas demand*. National Petroleum Council Working Document of the NPC North American Development Study Paper #3–4 made available September 15, 2011, p. 34. <http://www.npc.org/Prudent_Development-Topic_Papers/3-4_Transmission_Natural_Gas_Demand_Paper.pdf> Accessed 28.10.11.

NRC. (2010). *Management and effects of coalbed methane produced water in the United States*. National Academy of Sciences, The National Academies Press. Committee on Earth Resources; National Research Council, ISBN: 0-309-15433-2, 270 pages, 7 × 10. <http://www.nap.edu/catalog/12915.html> Accessed 28.11.12.

NSF. (2012). *From farm waste to fuel tanks*. National Science Foundation. Press release 07–011. <http://www.nsf.gov/news/news_summ.jsp?cntn_id=108390> Accessed 31.10.12.

Nuccio, V. (2000). *Coal-bed methane: Potential and concerns*. U.S. Geological Survey Fact Sheet FS–123–00, p. 2.

Nummedal, D., Towler, B., Mason, C., & Allen, M. (2003). *Enhanced oil recovery in Wyoming: Prospects and challenges*. University of Wyoming, Department of Geology and Geophysics. p. 22. < http://www.uwyo.edu/acadaffairs/_files/docs/eorfinal.pdf> Accessed 25.12.12.

NYSDEC. (2011). *Revised draft supplemental generic environmental impact statement on the oil, gas and solution mining regulatory program-well permit issuance for horizontal drilling and high-volume hydraulic fracturing to develop the Marcellus Shale and other low permeability gas reservoirs*. New York State Department of Environmental Conservation, Bureau of Oil and Gas Regulation, Division of Mineral Resources.

Ogala, J. E. (2011). Hydrocarbon potential of upper Cretaceous coal and shale units in the Anambra Basin, southeastern Nigeria. *Petroleum & Coal, 53*, 35–44. <www.vurup.sk/pc> Accessed 02.12.11.

OGJ. (April 4, 2011). Sangatta West produces coalbed gas in Indonesia. *Oil and Gas Journal*. <http://www.ogj.com/articles/2011/04/sangatta-west-produces.html> Accessed 02.03.13.

Ojeifo, E., Abaa, K., Orsulak, M., Pamidimukkala, P. K., Sircar, S., Sharma, S., et al. (2010). *Coalbed methane: Recovery 7 utilization in northwestern San Juan Basin, Colorado EME, integrative design*. Pennsylvania State University, Department of Energy and Mineral Engineering. <http://www.ems.psu.edu/~elsworth/courses/egee580/2010/Final%20Reports/CBM_Report.pdf> Accessed 25.11.12.

Ojha, K., Karmakar, B., Mandal, A., & Pathak, A. K. (2010). Coal bed methane in India: difficulties and prospects. *International Journal of Chemical Engineering and Applications, 2*, 256–260. <http://www.ijcea.org/papers/113-A618.pdf> Accessed 31.10.12.

Oldaker, P. R., & Fehn, U. (2005). *Dating of coal bed methane reservoir and surface waters in the Raton Basin, Colorado and the Susitna Basin, Alaska*. Rocky Mountain Section Meeting. Jackson Hole, Wyoming: American Association of Petroleum Geologists. pp. 43–44.

Oldeman, L. R. (1978). *Agro-climatic map of Sumatra, scale 1:3,000,000*. Institute Pertanian Bogor.

Oldham, D. W. (1997). Exploration for shallow compaction-induced gas accumulations in sandstones of the fort union formation, Powder River Basin, Wyoming. *Mountain Geologist, 34*, 25–38.

Oleszczuk, R., Szatylowicz, J., Brandyk, T., & Gnatowski, T. (2000). An analysis of the influence of shrinkage on water retention characteristics of fen peat-moorsh soil. *Suo, 51*, 139–147.

Ollivier, B., & Magot, M. (Eds.), (2002). *Petroleum microbiology*. Washington, D.C: American Society for Microbiology Press.

Olovyanny, A. G. (2005). Mathematical modeling of hydraulic fracturing coal seams. *Journal of Mining Science, 41*, 61–67. <http://web.ebscohost.com/ehost/pdfviewer/pdfviewer?sid=7942ea9c-7f9c-4285-bb28-07bd6c2789aa%40sessionmgr13&vid=2&hid=122> Accessed 10.10.12.

Olsen, T.N., Brenize, G., & Frenzel, T. (2003). Improvement processes for coalbed natural gas completion and stimulation. Society of Petroleum Engineers Annual Technical Conference and Exhibition, October 5–8, 2003, Denver, Colorado, ID 84122–MS, ISBN 978–1–55563–152–9.

Olsen, G. J., & Woese, C. R. (1993). Ribosomal RNS: a key to Phylogeny. *FASEB Journal, 7*, 113–123.

Olson, G. J., Dockins, W. S., McFeters, G. A., & Iverson, W. P. (1981). Sulfate-reducing and methanogenic bacteria from deep aquifers in Montana. *Geomicrobiology Journal, 2*, 327–340.

Olson, J. E., & Pollard, D. D. (1989). Inferring paleostresses from natural fracture patterns: a new method. *Geology, 17*, 345–348.

Olsson-Francis, K., de la Torre, R., Towner, M. C., & Cockell, C. S. (2009). Survival of akinetes (resting-state cells of cyanobacteria) in low earth orbit and simulated extraterrestrial conditions. Origins of life and evolution of the biosphere. *Springer Science, 39*, 565–579.

O'Neil, P. E., Harris, S. C., Drottar, K. R., Mount, D. R., Fillo, J. P., & Mettee, M. F. (1989). Biomonitoring of a produced water discharge from the Cedar Cove degasification field, Alabama. *Alabama Geological Society Bulletin, 135*, 195.

Otton, J. (2005). Estimated volume and quality of produced water associated with projected energy resources in the western U.S. In R. Wickramasighne (Ed.), *Produced Water Workshop* (pp. 26–35). Colorado Water Resources Institute, Colorado State University. Information Series No. 102. <http://www.cwi.colostate.edu/publications/is/102.pdf> Accessed 28.11.12.

Overeem, I., Syvitski, J., Kettner, A., Hutton, E., & Brakenridge, B. (2010). *Sinking deltas due to human activities.* Tulsa Geological Society presentation, Search and Discovery Article #70094(2010), Posted March 14, 2011. <http://www.searchanddiscovery.com/documents/2011/70094overeem/ndx_overeem.pdf> Accessed 24.02.10.

Owen, L. B., & Sharer, J. (1992). Method calculates gas content per foot of coalbed methane pressure core. *Oil Gas J, 2*, 47–49.

Ozdemir, E. (2004). Chemistry of the adsorption of carbon dioxide by Argonne premium coals and a model to simulate CO_2 sequestrations in coal seams (Ph.D. Dissertation). University of Pittsburgh, p. 339.

Ozdemir, E., Morsi, B. I., & Schroeder, K. (2004). CO_2 adsorption capacity of Argonne premium coals. *Fuel, 83*, 1085–1094.

Ozdemir, E., & Schroeder, K. (2009). Effect of moisture on adsorption isotherms and adsorption capacities of CO_2 on coals. *Energy & Fuels, 23*, 2821–2831.

Paetz, R.J., & Maloney, S. (2002). Demonstrated economics of managed irrigation for CBM produced water. Presented at the 2002 Ground Water Protection Council Produced Water Conference, Colorado Springs, CO, October 16–17. http://www.gwpc.org/meetings/special/PW%202002/Papers/Steel_Maloney_PWC2002.pdf Accessed 18.12.12.

Page, S. E., Wust, R. A. J., Weiss, D., Rieley, J. O., Shotyk, W., & Limin, S. H. (2004). A record of Late Pleistocene and Holocene carbon accumulation and climate change from an equatorial peat bog (Kalimantan, Indonesia): implications for past, present and future carbon dynamics. *Journal of Quaternary Science, 19*, 625–635.

Pahari, S., Singh, H., Prasad, I. V. S. V., & Singh, R. R. (2008). Petroleum systems of upper Assam Shelf, India. *Geohorizons, 14*, 14–21.

Pakham, R., Cinar, Y., & Moreby, R. (2009). Application of enhanced gas recovery to coal mine gas drainage systems. In N. Aziz (Ed.), *Coal 2009: Coal Operators' Conference* (pp. 225–235). University of Wollongong & the Australasian Institute of Mining and Metallurgy.

Palmer, I. D. (1992). *Review of coalbed methane well stimulation.* Society of Petroleum Engineers, Proceedings of SPE International Meeting of Petroleum Engineers, Beijing, China, SPE 22395, 679–703.

Palmer, I. D. (2007). *Coalbed methane well completions and production: Field data, best practices, new aspects.* International Coalbed Methane Symposium, Short Course, Tuscaloosa, Alabama, May 22, 2007. Accessed at <http://www.docstoc.com/docs/49845953/INTERNATIONAL-COALBED-SHALE-GAS-SYMPOSIUM>.

Palmer, I. (2010). Coalbed methane completions: a world review. In S. D. Golding, V. Rudolph, & R. M. Flores (Eds.), *International Journal of Coal Geology: Vol. 82. Asia Pacific coalbed methane symposium: Selected papers from the 2008 Brisbane Symposium on coalbed methane and CO_2-enhanced coalbed methane* (pp. 184–195).

Palmer, I. D., Lambert, S. W., & Spitler, J. L. (1993). Coalbed methane well completions and stimulations. In B. E. Law, & D. D. Rice (Eds.), *American Association of Petroleum Association Studies in Geology: Vol. 38. Hydrocarbons from Coal* (pp. 303–339).

Palmer, I., & Mansoori, J. (1998). How permeability depends on stress and pore pressure in coalbeds: a new model. *SPEREE, 1*(6), 539–544. SPE-52607-PA.

Pan, Z., & Connell, L. D. (2011). Modelling of anisotropic coal swelling and its impact on permeability behaviour for primary and enhanced coalbed methane recovery. *International Journal of Coal Geology, 85*, 257–267.

Pan, Z., Connell, D., & Camilleri, M. (2010). Laboratory chracterisation of coal reservoir permeability for primary and enhanced coalbed methane recover. In S. D. Golding, V. Rudolph, & R. M. Flores (Eds.), *International Journal of Coal Geology: Vol. 82. Asia Pacific coalbed methane symposium: Selected papers from the 2008 Brisbane symposium on coalbed methane and CO_2-enhanced coalbed methane* (pp. 252–261).

Pande, S. K. (1996). *Coal-bed methane: Indian scenario.* International Conference on Business and Investment

Opportunities in Mining Industry. New Delhi: Organized by the Mining Geological and Metallurgical Institute of India. December 16–18, 1996.

Papadopulus and Associates, Inc. (2008). *Colorado methane stream depletion assessment study – Raton Basin, Colorado.* S.S. Papadopulos & Associates, Inc in conjunction with Colorado Geological Survey. < http://geosurvey.state.co.us/water/CBM%20Water%20Depletion/Documents/RatonCBMdepletion_FINAL.pdf> Accessed 25.11.12.

Papendick, S. L., Downs, K. R., Vo, K. D., Hamilton, S. K., Dawson, G. K. W., Golding, S. D., et al. (2011). Biogenic methane potential for Surat Basin, Queensland coal seams. *International Journal of Coal Geology, 88*, 123–134.

Parkash, S., & Chakrabartty, S. K. (1986). Microporosity in Alberta Plains coals. *International Journal of Coal Geology, 6*, 55–70.

Parkyns, N. D., & Quinn, D. F. (1995). Natural gas adsorbed on carbon. In J. W. Patrick (Ed.), *Porosity in carbons, characterization and applications* (p. 331). New York: Halsted Press.

Parsegian, V. A. (2006). *Van der Waals forces: A Handbook for biologists, chemists, engineers, and physicists.* Cambridge: Cambridge University Press.

Pashin, J. C. (1998). Stratigraphy and structure of coalbed methane reservoirs in the United States. In R. M. Flores (Ed.), *International Journal of Coal Geology: 35. Coalbed methane: From coal-mine outbursts to gas resource* (pp. 209–240). Special Issue.

Pashin, J. C. (2007a). *Hydrodynamics of coalbed methane reservoirs in the Black Warrior Basin in Alabama: eustatic snapshots of a foreland basin tectonism.* In *American Association of Petroleum Geologists Studies in Geology* (Vol. 51; pp. 99–217).

Pashin, J. C. (2007b). Hydrodynamics of coalbed methane reservoirs in the Black Warrior Basin: key to understanding reservoir performance and environmental issues. *Applied Geochemistry, 22*, 2257–2272.

Pashin, J. C. (2010). Variable gas saturation in coalbed methane reservoirs of the Black Warrior Basin: implications for exploration and production. In S. D. Golding, V. Rudolph, & R. M. Flores (Eds.), *International Journal of Coal Geology: Vol. 82. Asia Pacific coalbed methane symposium: Selected papers from the 2008 Brisbane symposium on coalbed methane and CO_2-enhanced coalbed methane* (pp. 135–146).

Pashin, J. C., & Hinkle, F. (1997). Coalbed methane in Alabama. *Geological Survey of Alabama Circular, 192*, 71.

Pastor, J., Solin, J., Bridgham, S. D., Updegraff, K., Harth, C., Weishampel, P., et al. (2003). Global warming and the export of dissolved organic carbon from boreal peatlands. *Oikos, 100*, 380–386.

Patzek, T. W., & Croft, G. D. (2009). *Potential for coal-to-liquids conversion in the United States—Fischer–Tropsch Synthesis Natural Resources Research*, 18, 181–191. <http://gaia.pge.utexas.edu/papers/CoalConversion2.pdf> Accessed 30.03.13.

Patzek, T. W., & Croft, G. D. (2010). A global coal production forecast with multi-Hubbert cycle analysis. *Energy, 35*, 3109–3122.

Pawlewicz, M. D. (1985). Seam profiling of three coals from upper Cretaceous Menefee formation near Durango, Colorado. *American Association of Petroleum Geologists Bulletin, 69*.

Pearce, F. (November 2002). Indonesian wildfires spark global warming fears. *New Scientist.* see. <http://www.newscientist.com/article.ns?id=dn3024>.

Peatland Ecology Research Group. (2009). *Hydrology.* Quebec, Canada: Universite Laval. <http://www.gret-perg.ulaval.ca/joindre-gret.html> Accessed 20.02.12.

Pedlow, G. W. (1977). A peat island hypothesis for the formation of thick coal deposits. PhD. Thesis, University of South Carolina, Columbia, p. 104.

Peng, C. (2000). Study on integrated classification system for Chinese coal. *Fuel Processing Technology, 62*, 77–87.

Peng, S. S. (2006). *Longwall mining* (2nd ed.). Morgantown, West Virginia: Syd S. Peng.

Penner, T. J., Foght, J. M., & Budwill, K. (2010). Microbial diversity of western Canadian subsurface coal beds and methanogenic coal enrichment cultures. *International Journal of Coal Geology, 82*, 81–93.

Pennsylvania EP. (2012). *Tritium facts and information.* Pennsylvania Department of Environmental Protection. <http://www.dep.state.pa.us/brp/Radiation_Control_Division/Tritium.htm> Accessed 22.11.12.

Penuela, G., Ordonez, A., & Bejarno, A. (1998). *Linearization of a generalized material balance equation for coalbed methane reservoir.* C.T.F. Cienca, Tecnologia Y Futuro, 1, Bucaramanga, Colombia. <http://www.scielo.org.co/scielo.php?pid=S0122-53831998000100002&script=sci_artt> Accessed 17.06.12.

Pependick, S. L., Downs, K. R., Vo, K. D., Hamilton, S. K., Dawson, G. K. W., Golding, S. D., et al. (2011). Biogenic methane potential for Surat Basin, Queensland coal seams. *International Journal of Coal Geology, 88*, 123–134.

Perry, G. J., Allardice, D. J., & Kiss, L. T. (1984). Chemical characteristics of Victorian brown coals. In H. H. Schobert (Ed.), *Chemistry of low-rank coals* (Vol. 264; pp. 3–14). American Chemical Society. Symposium Series.

Peters, V., & Conrad, R. (1995). Methanogenic and other strictly anaerobic bacteria in desert soil and other oxic soils. *Applied and Environmental Microbiology, 61*, 1673–1676.

Peters, J. (2000). *Evaluation of coalbed methane potential of Jharia Basin, India.* Society of Economic Petroleum Engineers, Asia Pacific Oil Conference, Brisbane, Australia, SPE 64457, p. 14.

Petersen, H. I. (2004). Oil generation from coal source rocks: the influence of depositional conditions and stratigraphic age. In *Geological Survey of Denmark and Greenland Bulletin, Nr. 7, review of survey activities 2004* (pp. 9–12). <http://www.geus.dk/publications/bull/nr7/nr7_p.09-12-uk.htm> Accessed 13.12.11.

Petersen, H. I. (2006). The petroleum generation potential and effective oil window of humic coals related to coal composition and age. *International Journal of Coal Geology, 67*, 221–248.

Petersen, H. I., & Brekke, T. (2001). Source rock analysis and petroleum geochemistry of the Trym discovery, Norwegian North Sea: a middle Jurassic coal-sourced petroleum system. *Marine and Petroleum Geology, 18*, 889–908.

Petersen, H. I., Lindström, S., Nytoft, H. P., & Rosenberg, P. (2009). Composition, peat-forming vegetation and kerogen paraffinicity of Cenozoic coals: relationship to variations in the petroleum generation potential (Hydrogen Index). *International Journal of Coal Geology, 78*, 119–134.

Petersen, H. I., Nielsen, L. H., Koppelhus, E. B., & Sorensen, H. S. (2003). Early and Middle Jurassic mires of Bornholm and the Fennoscandian Border Zone: a comparison of depositional environments and vegetation. *Geological Survey of Denmark and Greenland Bulletin, 1*, 631–656.

Petersen, H. I., & Nytoft, H. P. (2007). Are Carboniferous coals from the Danish North Sea oil-prone? In M. Sønderholm, & A. K. Higgins (Eds.), *Geological Survey of Denmark and Greenland Bulletin: Vol. 13. Review of survey activities 2006* (pp. 13–16).

Petersen, H. I., Sherwood, N., Mathiesen, A., Fyhn, M. B. W., Dau, N. T., Russe, N., et al. (2009). Application of integrated vitrinite reflectance and FAMM analyses for thermal maturity assessment of the northeastern Malay Basin, offshore Vietnam: implications for petroleum prospectivity evaluation. *Marine and Petroleum Geology, 26*, 319–332.

Petersen, J. M., Zielinski, F. U., Pape, T., Seifert, R., Moraru, C., Amann, R., et al. (2011). Hydrogen is an energy source for hydrothermal vent symbioses. *Nature, 476*(7359), 176.

Petrel Robertson Consulting. (2010). *Assessment of Canada's natural gas resources*. Prepared for Canadian Society of Unconventional Gas. <http://www.csug.ca/images/news/2010/Petrel%20Gas%20In%20Place%20Report%20April%202010.pdf> Accessed 31.01.13.

Pew Center. (2011). *Coal and climate change facts*. < http://www.pewclimate.org/global-warming-basics/coalfacts.cfm> Accessed 14.09.11.

Pfeiffer, R. S., Ulrich, G., Vanzin, G., Dannar, V., DeBruyn, R. P., & Dodson, J. B. (2008). *Biogenic fuel gas generation in geologic hydrocarbon deposits*. United States Patent No. 7,426,960 B2., September 23, 2008.

PGC. (2013). *Estimates of the United States potential supply of natural gas*. Colorado School of Mines. Press release report of the Potential Gas Committee (2013). <http://potentialgas.org/press-release> Accessed 10.06.13.

Phillips, F. M., Peeters, L. A., Tansey, M. K., & Davis, S. N. (1986). Paleoclimatic inferences from an isotopic investigation of groundwater in the central San Juan Basin, New Mexico. *Quaternary Research, 26*, 179–193.

Phillips, T. L., Peppers, R. A., Avcin, M. J., & Loughman, P. F. (1974). Fossil plants and coal: patterns of change in Pennsylvanian coal swamps of the Illinois Basin. *Science, 184*, 1367–1369.

Phillips, F. M., Tansey, M. K., Peeters, L. A., Cheng, S., & Long, A. (1989). An isotopic investigation of groundwater in the San Juan Basin New Mexico: carbon 14 dating as a basis for numerical flow modeling. *Water Resources Research, 25*, 2259–2273.

Pillmore, C. L., & Flores, R. M. (1987). Stratigraphy and depositional environments of the Cretaceous–Tertiary boundary clay and associated rocks, Raton Basin, New Mexico and Colorado. In J. E. Fassett, & J. K. Rigby, Jr. (Eds.), *Geological Society of America, Special Paper: Vol. 209. The Cretaceous–Tertiary boundary in the San Juan and Raton Basins, New Mexico and Colorado* (pp. 11–130).

Piperi, T., Luzzi-Arbouille, T., Ehlers-Fliege, B., & Zine, M. (2011). *Exploration for coal bed methane in the Greater Nama-Kalahari Basin of Namibia, Botswana, South Africa, and Zimbabwe*. American Association of Petroleum Geologists International Conference and Exhibition, Mila, Italy, October 23–26, 2011, Search and Discovery Article #90135. < http://www.searchanddiscovery.com/abstracts/html/2011/ice/abstracts/abstracts349.html> Accessed 30.01.13.

Pizzuto, J. E., & Schwendt, A. E. (1997). Mathematical modeling of autocompaction of a Holocene transgressive valley-fill deposit, Wolfe Glade, Delaware. *Geology, 25*, 57–60.

Planet Earth. (2013). *Soil – Earth's living skin*. <http://yearofplanetearth.org/content/downloads/Soil.pdf> Accessed 26.03.13.

Polak, B. (1950). *Occurrence and fertility of tropical peat soils in Indonesia*. In *Proceedings 4th International Congress of Soil Science*. (Vol. 2; pp. 183–185).

Posamentier, H. W., Allen, G. P., James, D. P., & Tesson, M. (1992). Forced regression in a stratigraphic sequence framework: concepts, and exploration significance. *American Association of Petroleum Geologists Bulletin, 76*, 1687–1709.

Potential Gas Committee (1995). Potential supply of natural gas in the United States – report of the Potential Gas Committee (December 31, 1994). Colorado School of Mines, Potential Gas Agency, p. 130.

Powell, T. G. (1988). Developments in concepts of hydrocarbon generation from terrestrial organic matter. In

H. C. Wagner, L. C. Wagner, F. F. H. Wang, & F. L. Wong (Eds.), *Houston, Circum-Pacific Council for Energy and Mineral Resources, Earth Science Series: Vol. 10. Petroleum resources of China and related subjects* (pp. 807–824).

Powell, D. (2012). Groundwater dropping globally. *Science News, 181*, 5–6. < http://www.sciencenews.org/view/generic/id/337097/description/Groundwater_dropping_globally> Accessed 23.12.12.

Powell, K. R., Hathcock, R. L., Mullen, M. E., Norman, W. D., & Baycroft, P. D. (1997). *Productivity performance comparisons of high rate water pack and frac-pack completion techniques*. Society of Petroleum Engineers, Proceedings Annual Technical Conference and Exhibition, Drilling and Completion, San Antonio, Oct. 5–8, 1997, SPE 38592, 269–284.

Powell, R. J., McCabe, M. A., Salbaugh, B. F., Terracina, J. M., Yaritz, J. G., & Ferrer, D. (1999). Applications of a new, efficient hydraulic fracturing fluid system. Society of Petroleum Engineers. *Production and Facilities, 14*, 139–143.

Prada, A. G., Rodriquez, S. L., Flores, R. M., Fuentes, J. R., & Guzman, G. N. (2010). *Coalbed methane in Colombia: An overview*. The Society for Organic Petrology Meeting, Denver, Colorado, September 13–14, 2010.

Pravettoni, R. (2009). *Peat distribution in the world*. United Nations Environment Programme (UNEP) GRID-Arendal, Government of Norway as a Norwegian Foundation. < http://www.grida.no/graphicslib/detail/peat-distribution-in-the-world_8660#> Accessed 26.02.12.

Prescott, L. M., Harley, J. P., & Klein, D. A. (2002). *Microbiology* (5th ed.). Mc-Graw Hill. p. 1026.

Price, J. S. (2003). The role and character of seasonal peat soil deformation on the hydrology of undisturbed and cutover peatlands. *Water Resource Research, 39*, 1241. <http://dx.doi.org/10Ð1029/2002WR001302>.

Price, J. S., Cagampan, J., & Kellner, E. (2005). Assessment of peat compressibility: is there an easy way? *Hydrological Processes, 19*, 3469–3475.

Price, P. H., & Headlee, J. W. (1943). Natural coal gas in West Virginia. *American Association of Petroleum Geologists Bulletin, 27*, 529–537.

PRMS. (2011). *Guidelines for application of the petroleum resources management system*. Sponsored by Society of Petroleum Engineers, American Association of Petroleum Geologists, World Petroleum Council, Society of Petroleum Evaluation Engineers, and Society of Exploration Geophysicists. < http://www.spe.org/industry/docs/PRMS_Guidelines_Nov2011.pdf> Accessed 16.06.12.

ProPublica. (2011). *Climate benefits of natural gas may be overstated*. <http://www.propublica.org/article/natural-gas-and-coal-pollution-gap-in-doubt> Accessed 07.01.12.

Puri, R., King, G. E., & Palmer, I. D. (May 1991). *Damage to coal permeability during hydraulic fracturing*. Proceedings Coalbed Methane Symposium, Tuscaloosa, Alabama, 247–255.

Puttmann, W., Wolf, M., & Bujnowska, B. (1991). Chemical characteristics of subbitumious coal lithotypes. *Fuel, 70*, 227–233.

Qian, D., Yan, T., Li, S., Zhou, S., Bai, H., Li, X., et al. (1999). *Classification for resources/reserves of solid fuels and mineral commodities*. General Administration of Quality Supervision, Inspection and Quarantine of the People's Republic of China, p. 11.

Qiu, H. (April 20–23, 2008). *Coalbed methane exploration in China*. San Antonio, Texas: American Association of Petroleum Geologists Annual Convention.

Querol, X., Chinenon, S., & Lopez-Soler, A. (1989). Iron sulphide precipitation sequence in Albian coals from the Maestrazgo Basin, southeastern Iberian Range, northeastern Spain. *International Journal of Coal Geology, 11*, 171–189.

Querol, X., Whately, M. K. G., Fernandez-Turiel, J. L., & Tuncali, E. (1997). Geological controls on the mineralogy and geochemistry of the Beypazari Lingnite, central Anatolia, Turkey. *International Journal of Coal Geology, 33*, 225–271.

Quick, J. C., & Tabet, D. E. (2003). Suppressed vitrinite reflectance in the Ferron coalbed gas fairway, central Utah: possible influence of overpressure. *International Journal of Coal Geology, 56*, 49–67.

Quinty, F., & Rochefort, L. (2003). *Peatland restoration guide* (2nd ed.). Québec, Québec: Canadian Sphagnum Peat Moss Association and New Brunswick Department of Natural Resources and Energy. p. 106.

Radlinski, A. P., Mastalerz, M., Hinde, A. L., Hainbuchner, M., Rauch, H., Baron, M., et al. (2004). Application of SAXS and SANS in evaluation of porosity, pore size distribution and surface area of coal. *International Journal of Coal Geology, 59*, 245–271.

Rahmani, R. A., & Flores, R. M. (Eds.), (1984), *International Association of Sedimentologists, Special Publication: Vol. 7. Sedimentology of coal and coal-bearing sequences* (p. 412). Oxford: Blackwell Scientific Publications.

Raina, A., Pande, H. C., Sharma, P., Singh, H., & Singh, R. R. (2009). Geochemistry of coal bed gases from Gondwana Basins, India. *Journal of Applied Geochemistry, 11*, 300–309.

Ramaswamy, S. (2007). *Selection of best drilling completion, and stimulation methods for coalbed methane reservoirs*. M.S. Thesis, Teas A&M University, p. 135 <http://www.pe.tamu.edu/wattenbarger/public_html/Dissertations%20and%20Theses/2007%20MS%20-%20Ramaswmay.pdf> Accessed 27.09.12.

Ramaswamy, S., Ayers, W. A., & Holditch, S. A. (2008). Best drilling, completion and stimulation methods for CBM reservoirs. *World Oil, 229*, 125–132.

Ramirez, P. (2005). *Assessment of contaminants associated with coal bed methane-produced water and its suitability for wetland creation or enhancement projects.* Cheyenne, Wyoming: U.S. Fish and Wildlife Service. Contaminant Report R6/721C/05.

Rampelotto, P. H. (2010). Resistance of microorganisms to extreme environmental conditions and its contribution to astrobiology. *Sustainability, 2,* 1602–1623. <http://dx.doi.org/10.3390/su2061602>.

Ramurthy, M., Young, G. B. C., Daves, S. B., & Witsell, F. (2003). *Case history: Reservoir analysis of the Fruitland coals results in optimizing coalbed methane completions in the Tiffany area of the San Juan Basin.* Society of Petroleum Engineers. SPE Annual Technical Conference and Exhibition, October 5–8, 2003, Denver, Colorado. <http://www.onepetro.org/mslib/app/Preview.do?paperNumber=00084426&societyCode=SPE> Accessed 14.02.13.

Randall, A. G. (1989). Shallow Tertiary gas production, Powder river basin, Wyoming. In J. L. Eisert (Ed.), *Wyoming Geological Association, 40th annual field Conference guidebook, Gas resources of Wyoming* (pp. 83–96). Casper, Wyoming: September 1989.

Range Resources–Appalachia, LLC. (2010). *Well record and completion report, John Miller Unit #8H.* Commonwealth of Pennsylvania, Department of Environmental Protection, Oil and Gas Management Program, p. 4. < http://www.rangeresources.com/rangeresources/files/8a/8a276f10-ba9c-406a-9b7f-af75f6c3244e.PDF> Accessed 14.10.12.

Ranney, L. (1941 June). *The recovery of natural gas from coal.* Ohio State University, Engineering Experiment Station News, 6–8.

Rao, C. P., & Gluskoter, H. J. (1973). *Occurrence and distribution of minerals in Illinois coals.* Illinois State Geological Survey. Circular 476, p. 56.

Rao, P. D., & Walsh, D. E. (1999). Influence of environments of coal deposition on phosphorus accumulation in a high latitude, northern Alaska, coal seam. *International Journal of Coal Geology, 38,* 261–284.

Rawn-Schatzinger, V., Arthur, D., & Langhus, B. (2004). *Coalbed natural gas resources: Beneficial use alternatives.* GasTIPS. Winter 2004. EPA-HQ-OW-2008-0517, DCN 07229.

Reading, H. G. (1996). *Sedimentary environments and facies.* Blackwell Scientific Publications.

Regele, S., & Stark, J. (2000). Coal-bed methane gas development in Montana, some biological issues. Presented at the interactive forum on surface mining reclamation approaches to bond release: cumulative hydrologic impacts assessment (CHIA) and hydrology topics for the arid and semi-arid west coal-bed methane workshop. Sponsored by USDI office of surface mining, Denver, CO, September 2000, Montana Department of Environmental Quality, Helena, MT and Montana Bureau of Mines and Geology, Butte, MT.

Reiche, M., Gleixner, G., & Kusel, K. (2010). Effect of peat quality on microbial greenhouse gas formation in an acidic fen. *Biogeosciences, 7,* 187–198.

Reid, C. (1966). Principles of reverse osmosis. In U. Merten (Ed.), *Desalination by reverse osmosis* (pp. 1–14). Cambridge, Massachusetts: Massachusetts Institute of Technology Press.

Rendu, J. M. (2000). *International aspects of resource and reserve reporting standards.* MICA, The Codes Forum, International Aspects of Resources and Reserves Reporting, Sydney, pp. 91–100.

Retallack, G. J., Veevers, J. J., & Morante, R. (1996). Global coal gap between the Permian–Triassic extinction and Middle Triassic recovery of peat-forming plants. *Geological Society of America Bulletin, 108,* 195–207.

Rhee, C. W., & Chough, S. K. (1993). The Cretaceous Pyonghae sequence, SE Korea: terminal fan facies. *Palaeogeography, Palaeoclimatology, Palaeoecology, 105,* 139–156.

Rhone, R. D. (2011). *Anthracite coal mines and mining.* eHistory @ The Ohio State University. <http://ehistory.osu.edu/osu/about.cfm> Accessed 14.09.11.

Rice, D. D. (1992). Controls, habitat, and resource potential of ancient bacterial gas. In R. Vialy (Ed.), *Bacterial gas* (pp. 91–118). Editions Technip.

Rice, D. D. (1993). Composition and origins of coalbed gas. In B. E. Law, & D. D. Rice (Eds.), *Studies in Geology: Vol. 38. Hydrocarbons from Coal* (pp. 159–184). American Association of Petroleum Geologists.

Rice, C. A. (2003). Production waters associated with the Ferron coalbed methane fields, central Utah: chemical and isotopic composition and volumes. *International Journal of Coal Geology, 56,* 141–169.

Rice, C. A., Bartos, T. T., & Ellis, M. E. (2002). Chemical and isotopic composition of water in the Fort Union and Wasatch Formations of the Powder River Basin, Wyoming and Montana: implications for coalbed methane development. In S. D. Schwochow, & V. F. Nuccio (Eds.), *Coalbed methane of North America, II* (pp. 53–70). Rocky Mountain Association of Geologists.

Rice, D. D., & Claypool, G. E. (1981). Generations, accumulation, and resource potential of biogenic gas. *American Association of Petroleum Geologists Bulletin, 65,* 5–25.

Rice, D. D., & Flores, R. M. (1990). Coal-bed methane potential of Tertiary coal beds and adjacent sandstone deposits, Powder River Basin, Wyoming and Montana. *American Association of Petroleum Geologists Bulletin, 74,* 1343.

Rice, D. D., Clayton, J. L., & Pawlewicz, M. J. (1989). Characterization of coal-derived hydrocarbons and source-rock potential of coal beds, San Juan Basin, New Mexico and Colorado. In P. C. Lyons, & B. Alpern (Eds.), *International Journal of Coal Geology: Vol. 13. Coal: Classification, coalification, mineralogy, trace-element chemistry, and oil and gas potential* (pp. 597–626).

Rice, C. A., Ellis, M. S., & Bullock, J. H., Jr. (2000). *Water co-produced with coalbed methane in the Powder River Basin, Wyoming – Preliminary compositional data.* U.S. Geological Survey Open-File Report, 00–372. p. 20.

Rice, D. D., & Finn, T. M. (1996). Powder River Basin (033). In D. L. Gautier, G. L. Dolton, K. I. Takahashi, & K. L. Varnes (Eds.), *U.S. Geological Survey Digital Data Series: Vol. DDS-30. 1995 National assessment of United States oil and gas resources—Results, methodology, and supporting data* (pp. 40–46).

Rice, D. D., & Flores, R. M. (1989). Nature of natural gas in anomalously thick coal beds, Powder River Basin, Wyoming. *American Association of Petroleum Geologists Bulletin*, 1172.

Rice, D. D., & Flores, R. M. (1991). Controls on bacterial gas accumulations in thick Tertiary coalbeds and adjacent channel sandstones, Powder River Basin, Wyoming and Montana. *American Association of Petroleum Geologists Bulletin, 75*, 661.

Rice, C. A., Flores, R. M., Stricker, G. D., & Ellis, M. S. (2008). Chemical and stable isotopic evidence for water/rock interaction and the biogenic origin of coalbed methane, Fort Union Formation, Powder River Basin, Wyoming and Montana, U.S.A. In R. M. Flores (Ed.), *International Journal of Coal Geology: Vol. 76. Microbes, Methanogenesis, and Microbial Gas* (pp. 76–85).

Rice, C. A., & Nuccio, V. (2000). *Water produced with coal-bed methane.* U.S. Geological Survey. Fact Sheet FS-156-00, p. 2.

Rice, D. D., & Trelkeld, C. N. (1982). *Occurrence and origin of natural gas in groundwater, southern Weld County, Colorado.* U.S. Geological Survey Open File Report 82–496.

Rice, D. D., Young, G. B., & Paul, G. W. (1995). Methodology for assessment of potential additions to reserves of coalbed gas. In *U.S. Geological Survey Digital Data Series: Vol. 30. Results, methodology, and supporting data for the 1995 national assessment of the United States Oil and Gas Resources.* (CD-ROM).

Rich, F. J. (1980). Brief survey of chemical and petrographic characteristics of Powder River Basin coals. In F. G. Ethridge, T. J. Jackson, R. M. Flores, S. L. Obernyer, F. J. Rich, B. H. Kent, et al. (Eds.), *Guidebook to the coal geology of the Powder River Basin, Wyoming Field Trip No. 5* (pp. 125–149). American Association of Petroleum Geologists.

Richardson, M. (2007). Does China's coal abundance threaten the Middle East? *ICIS Chemical Business*, 2, 19. <http://web.ebscohost.com/ehost/detail?sid=2eff669f-c0db-4001-af32-7942ab26b669%40sessionmgr4&vid=1&hid=24&bdata=JnNpdGU9ZWhvc3QtbGl2ZQ%3d%3d#db=buh&AN=24348691> Accessed 28.12.2011.

Richardson, R. J. (2010). *Alberta's 2 trillion tonnes of 'unrecognized coal'. Alberta – innovates energy and environmental solutions.* CanZealand Geoscience Limited. p. 44. <http://eipa.alberta.ca/media/43006/alberta_2_trillion_tonnes_coal.pdf> Accessed 26.10.11.

Rieke, H. H. (1980). *Preliminary methane resource assessment of coalbeds in Arkoma Basin.* Society of Petroleum Engineers, Paper SPE/DOE 8926 presented at the SPE/DOE Symposium of Unconventional Gas Recovery, Pittsburgh, Pennsylvania, May 18–21, 1980.

Rieke, H. H., & Kirr, J. N. (1984). Geologic overview, coal, and coalbed methane resources of the Arkoma Basin – Arkansas and Oklahoma. In C. T. Rightmire, G. E. Eddy, & J. N. Kirr (Eds.), *American Association of Petroleum Geologists Studies in Geology Series #17. Coaled methane resources of the United States* (pp. 135–161).

Rieley, J. O. (2007). Tropical peatland – the amazing dual ecosystem: co-existence and mutual benefit. In J. O. Rieley, C. J. Banks, & B. Ragjagukguk (Eds.), *Carbon-climate-human interactions on tropical Peatland: Carbon pools, fires, mitigation, restoration and wise use.* Proceedings of the International Symposium and Workshop on Tropical Peatland, Yogyakarta, August 27–29, 2007, p. 15. <http://www.geog.le.ac.uk/carbopeat/media/pdf/yogyapapers/yogyaproceedings.pdf> Accessed 23.02.12.

Rieley, J. O., Ahmad-Shah, A.-A., & Brady, M. A. (1996). The extent and nature of tropical peat swamp. In E. Maltby, C. P. Immirzi, & R. J. Stafford (Eds.), *Tropical lowland peatlands of southeast Asia.* Proceedings of a Workshop on Integrated Planning and Management of Tropical Lowland Peatlands held at Cisarua, Indonesia, July 3–8, 1992. IUCN, Gland, Switzerland, p. 294, ISBN 2-8317-0310-7.

Riese, W. C., Pelzmann, W. L., & Snyder, G. T. (2005). *New insights on the hydrocarbon system of the Fruitland Formation coal beds, northern San Juan Basin, Colorado and New Mexico, USA.* Geological Society of America. Special Papers 387, pp. 73–111.

Rightmire, C. T. (1984). Coalbed methane resources. In C. T. Rightmire, G. E. Eddy, & J. N. Kirr (Eds.), *American Association of Petroleum Geologists Studies in Geology Series: Vol. 17. Coalbed methane resources of the United States* (pp. 1–13).

Rightmire, C. T., Eddy, G. E., & Kirr, J. N. (Eds.), (1984). *American Association of Petroleum Geologists Studies in Geology Series: Vol. 17. Coalbed methane resources of the United States.*

Rippon, J. H., Ellison, R. A., & Gayer, R. A. (2002). A review of joints (cleats) in British Carboniferous coals: indicators of palaeostress orientation. *Proceedings of the Yorkshire Geological Society, 56*, 15–30.

Ritter, U., & Grover, A. (2005). Adsorption of petroleum compounds in vitrinite: implications for petroleum expulsion from coal. *International Journal of Coal Geology, 62*, 183–191.

RMB Earth Science Consultants. (2005). High pressure methane and carbon dioxide adsorption analyses. In *Your isotherms, appendix VII*. 327 Rosehill Wynd Delta, Calgary, Canada: RMB Earth Science Consultants Ltd.

Roberts, S. (2008). Geologic assessment of undiscovered, technically recoverable coalbed-gas resources in Cretaceous and Tertiary rocks, North Slope and adjacent State Waters, Alaska. In *U.S. Geological Survey Digital Data Series* (Vol. DDS–69–S). Chapter 2, p. 102.

Roberts, S. S., Stanton, R. W., & Flores, R. M. (1994). A debris flow deposit in alluvial, coal-bearing facies, Bighorn Basin, Wyoming, USA: Evidence for catastrophic termination of mire. *International Journal of Coal Geology, 25*, 213–241.

Robinson, S. D., & Moore, T. R. (1999). Carbon and peat accumulation over the past 1200 years in a landscape with discontinuous permafrost, northwestern Canada. *Global Biogeochemical Cycles, 13*, 591–601.

Rodrigues, C. F., & Lemos de Sousa, M. J. (2002). The measurement of coal porosity with different gases. *International Journal of Coal Geology, 48*, 245–251.

Roehler, H. W. (1987). *Depositional environments of coal-bearing and associated formations of Cretaceous age in the National Petroleum Reserve in Alaska*. U.S. Geological Survey Professional Paper 1575, p. 16.

Rogner, H. H. (1997). An assessment of world hydrocarbon resources. *Annual Review of Energy and the Environment, 22*, 217–262. Annual Reviews Inc.

Rollins, M. S., Cohen, A. D., Bailey, A. M., & Durig, J. R. (1991). Organic chemical and petrographic changes induced by early-stage artificial coalification of peats. *Organic Geochemistry, 17*, 451–465.

Romanov, V. N., Goodman, A. L., & Larsen, J. W. (2006). Errors in CO_2 adsorption measurements caused by swelling. *Energy & Fuels, 20*, 415–416.

Romanowicz, E. A., Siegel, D. I., Chanton, J. P., & Glaser, P. H. (1995). Temporal variations in dissolved methane deep in the Lake Agassiz peatlands, Minnesota. *Global Biogeochemical Cycles, 9*, 197–212.

Romanowicz, E. A., Siegel, D. I., & Glaser, P. H. (1993). Hydraulic reversals and episodic emissions during drought cycles in mires. *Geology, 21*, 231–234.

Rooney, M. A., Claypool, G. E., & Chung, H. M. (1995). Modeling thermogenic gas generation using carbon isotope ratios of natural gas hydrocarbons. *Chemical Geology, 126*, 451–465.

Rose, P. R. (2001). Risk analysis and management of petroleum exploration ventures. In *Methods in exploration series* (Vol. 12; p. 164). Tulsa, Oklahoma: American Association of Petroleum Geologists.

Ross, J. G. (2011). Petroleum resources definitions, classification, and categorization guidelines. In *Guidelines for application of the petroleum resources management system*. Sponsored by Society of Petroleum Engineers, American Association of Petroleum Geologists, World Petroleum Council, Society of Petroleum Evaluation Engineers, and Society of Exploration Geophysicists, 7–22. <http://www.spe.org/industry/docs/PRMS_Guidelines_Nov2011.pdf> Accessed 16.06.12.

Rothfuss, F., & Conrad, R. (1998). Effect of gas bubbles on the diffusive flux of methane in anoxic paddy soil. *Limnology Oceanography, 43*, 1511–1518.

Roulet, N. T. (1991). Surface level and water table fluctuations in a subarctic fen. *Arctic and Alpine Research, 23*, 303–310.

Rovenskaya, A. S., & Nemchenko, N. N. (1992). Prediction of hydrocarbons in the West Siberian Basin. *Bulletin Centre de Recherche Exploration-Production Elf Aquitaine, 16*, 285–318.

Royer, D. L., Berner, R. A., Montañez, I. P., Tabor, N. J., & Beerling, D. J. (2004). CO_2 as a primary driver of Phanerozoic climate. *Geological Society of America Today, 4*, 4–10.

RPS Australia East Pty Ltd. (2011). *Onshore co-produced water: Extent and management*. Waterlines Report Series No.54, September 2011, National Water Commission, p. 38. <http://nwc.gov.au/__data/assets/pdf_file/0007/18619/Onshore-co-produced-water-extent-and-management_final-for-web.pdf> Accessed 12.11.12.

RPSEA. (2009). *An integrated framework for treatment and management of produced water: Technical assessment of produced water treatment technologies*. Research Partnership to Secure Energy for America. RPSEA Project 07122-12. Golden, Colorado School of Mines, p. 157. < http://aqwatec.mines.edu/produced_water/treat/docs/Tech_Assessment_PW_Treatment_Tech.pdf> Accessed 18.12.12.

Ruppert, L. F., Stanton, R. W., Cecil, C. B., Eble, C. F., & Dulong, F. T. (1991). Effects of detrital influx in the Pennsylvanian Upper Freeport peat swamp. *International Journal of Coal Geology, 17*, 95–11.

Ruppert, L. F., Kirschbaum, M. A., Warwick, P. D., Flores, R. M., Affolter, R. H., & Hatch, J. R. (2002). The U.S. Geological Survey's national coal resource assessment: the results. *International Journal of Coal Geology, 50*, 247–274.

Ruppert, L. F., Hower, J. C., Ryder, R. T., Levine, J. R., Trippi, M. H., & Grady, W. C. (2010). Geologic controls on thermal maturity patterns in Pennsylvanian coal-bearing rocks in the Appalachian Basin. *International Journal of Coal Geology, 81*, 169–181.

Rushing, J. A., Perego, A. D., & Blasingame, T. A. (2008). *Applicability of the Arps rate-time relationships for evaluating decline behavior and ultimate gas recovery of coalbed methane wells*. Society of Petroleum Engineers. SPE 114514, Presentation at the CIPC/SPE Gas Technology Symposium 2008 Joint Conference in Calgary, Alberta, Canada, 16–19 June 2008. p. 18. <http://www.pe.tamu.edu/blasingame/data/0_TAB_Public/TAB_Publications/

SPE_114514_(Rushing)_Arps_Rate_Time_Relationships_Coalbed_Methane_Wells.pdf> Accessed 26.08.12.

Russell, N. J., & Barron, P. F. (1984). Gelification of Victorian Tertiary soft brown coal wood, II. Changes in chemical structure associated with variation in the degree of gelification. *International Journal of Coal Geology, 4*, 119–142.

Rutledge, D. (2009). *Projections for ultimate coal production from production histories*. Geological Society of America, Annual Meeting, Portland, Oregon, Oct 18–21, 2009, Paper No. 215–1.

Rutledge, D. (2011). Estimating long-term world coal production with logit and probit transforms. *International Journal of Coal Geology, 85*, 23–33.

Rutledge, D. B. (2013). *Projections for ultimate coal production from production histories through 2012*. Geological Society of America, Paper No. 160-7, Annual Meeting and Expo, Denver, Colorado, October 27-30, 2013.

Ryan, B. D. (2003). *Cleat development in some British Columbia coals*. British Columbia Geological Survey. Geological Fieldwork 2002, Paper 2003–1. 237–255.

Ryan, B. (2006). *A discussion on moisture in coal implications for coalbed gas and coal utilization*. British Columbia Ministry of Energy, Mines and Petroleum Resources, 139–149.< http://www.empr.gov.bc.ca/Mining/Geoscience/Publications Catalogue/OilGas/OGReports/Documents/2006/OG_Rpt2006Ryan%20CBG%202.pdf> Accessed 30.11.12.

Ryan, M. (2012). *Coalbed methane potential of the anthracite Groundhog/Klappan coalfield, northern Bowser Basin*. Ministry of Energy and Mines British Columbia, Resources Development and Geosceiences Branch. <http://www.empr.gov.bc.ca/OG/oilandgas/petroleumgeology/CoalbedGas/TechandRegInfo/TechnicalInformation/Documents/Northern_Bowser_Basin.pdf> Accessed 04.04.12.

Ryan, B. D., & Dawson, F. M. (1993). *Coalbed methane canister desorption techniques; in geological fieldwork 1993*. British Colombia Ministry of Energy, Mines and Petroleum Resources. Paper 1994–1, pp. 245–256.

Rydin, H., & Jeglum, J. (2006). *The biology of peatlands*. New York: Oxford University Press Inc.

Ryer, T. A. (1984). Deltaic coals of Ferron Sandstone Member of the Mancos Shale—predictive model for Cretaceous coal-bearing strata of Western Interior. *American Association of Petroleum Geologists Bulletin, 65*, 2323–2340.

Ryer, T. A., & Langer, A. W. (1980). Thickness change involved in the peat-to-coal transformation for a bituminous coal of Cretaceous age in Central Utah. *Journal of Sedimentary Petrology, 50*, 987–992.

Saavedra, N. F., & Joshi, S. D. (2002). Application of horizontal well technology in Colombia. *Journal of Canadian Petroleum Technology, 41*, 33–39.

Sabiham, S. (2010). Properties of Indonesian peat in relation to the chemistry of carbon emission. In *Proceedings of international workshop on evaluation and sustainable management of soil carbon sequestration in Asian countries* (pp. 205–216). Bogor, Indonesia Sept. 28–29, 2010.

Sachsenhofer, R. F., Jelenb, B., Hasenhuttl, C., Dunkl, I., & Rainer, T. (2001). Thermal history of Tertiary basins in Slovenia. *Tectonophysics, 334*, 77–99.

Saghafi, A., Day, S. J., Fry, R., Quintanar, A., Roberts, D., Williams, D. J., et al. (2005). *Development of an improved methodology for estimation of fugitive seam gas emissions from open cut mining*. Final Report for ACARP Project C12072.

Saghafi, A., & Hadiyanto. (2000). Methane storage properties of Indonesian Tertiary coals. In Herudyanto, et al. (Eds.), *Proceedings of Southeast Asian Coal Geology Conference* (pp. 121–124). Bandung: Ministry of Mines and Energy of The Republic of Indonesia.

Saghafi, A., Hatherly, P., & Pinetown, K. (2011). *Towards an optimal gas sampling and estimation guideline for GHG emissions from open cut coal mines* (ACARP project C19005). CSIRO ePublish number EP113374-20 July 2011, p. 72. <https://publications.csiro.au/rpr/download?pid=csiro:EP113374&dsid=DS4> Accessed 11.08.12.

Saghafi, A., & Williams, D. J. (1998). *Factors influencing the accuracy of measurement of gas content of coal: Inter-comparison of the quick crush technique*. ACARP Project C6023, p. 70.

Saghafi, A., Williams, D. J., & Battino, S. (1998). *Accuracy of measurement of gas content of coal using rapid crushing techniques*. Proceedings of the 1st Australian Coal Operators Conference COAL'98, Wollongong 18–20 February 1998, Australia, 551–559, also available from <http://ro.uow.edu.au/coal/273/>

Samingan, M. T. (1980). Notes on the vegetation of the tidal areas of South Sumatra, Indonesia, with special reference to Karang Agung. In *Proceedings of the 5th international symposium, tropical ecology and development. Kuala Lumpur* (pp. 1107–1112).

SAMREC Working Group. (2009). *The South African code for the reporting of exploration results, mineral resources and mineral reserves the SAMREC Code*. p. 61. <http://www.samcode.co.za/downloads/SAMREC2009.pdf> Accessed 26.10.11.

Samuelsson, J., & Middleton, M. F. (1998). A thermal maturation study of the Carnarvon Basin, Australia and the northern North Sea, Europe. *Exploration Geophysics, 29*, 597–604.

Sanstrom, B., & Longorio, P. (2002). Innovative 3D visualization tool promotes development-drilling efficiency. *Oil and Gas Journal, 100*, 79–84.

Santos. (2010). *Gunnedah Basin – geology and pilot testing*. Santos. <http://www.santos.com/library/Gunnedah_presentation_geology_pilot_Mar2010.pdf> Accessed 07.06.13.

Sarate, O. S. (2010). Study of petrographic composition and depositional environment of the coals from Queen seam of Yellundu area, Godavari Valley coalfield, Andhra Pradesh. *Society of Journal of Geological India, 75*, 488–494.

Sarmiento, Z. F., & Steingrimsson, B. (2007). *Computer programme for resource assessment and evaluation using Monte Carlo simulation*. United Nations University. Geothermal Training Programme, Presented at Short Course on Geothermal Development in Central America – Resource Assessment and Environmental Management, organized by UNU-GTP and LaGeo, in San Salvador, El Salvador, 25 November – 1 December, 2007, p. 13 <http://www.os.is/gogn/unu-gtp-sc/UNU-GTP-SC-04-13.pdf> Accessed 16.06.12.

Sasol. (2011). *Unlocking the potential wealth of coal, information brochure*. <http://www.sasol.com/sasol_internet/downloads/CTL_Brochure_1125921891488.pdf> Accessed 06.12.11.

Sawhney, P. (2010). *CBM field development and challenges*. Oil and Gas Conference, India Drilling and Exploration Conference (IDEC) 2010, May 20–21, 2010. <http://www.oilnmaritime.com/21st%20may%202010%20%20nd%20Day2/no%206%20CBM%20Exploration/Prem_Sawhne_Essar.pdf> Accessed 08.11.12.

Sawyer, W. K., Zuber, M. D., Kuuskraa, V. A., & Horner, D. M. (1987). *A field derived inflow performance relationship for coalbed gas wells in the Black Warrior Basin* Proceedings of the 1987 International Coalbed Methane Symposium (pp. 207–215), Tuscaloosa, Alabama.

Schakel, S. J., Hyman, D. M., Sainato, A., & LaScola, J. C. (1987). *Methane contents of oil shale from the Piceance Basin, CO*. U.S. Bureau of Mines, Report of Investigation 9063, p. 32.

Schimmelmann, A., Sessions, A. L., Boreham, C. J., Edwards, D. S., Logan, G. A., & Summons, R. E. (2004). D/H ratios in terrestrially sourced petroleum systems. *Organic Geochemistry, 35*, 1169–1195.

Schimmelmann, A., Sessions, A. L., & Mastalerz, M. (2006). Hydrogen isotopic (D/H) composition of organic matter during diagenesis and thermal maturation. *The Annual Review of Earth and Planetary Science, 34*, 501–533.

Schlachter, G. (2007). *Using wireline formation evaluation tools to characterize coalbed methane formations*. Society of Petroleum Engineers, Eastern Regional Meeting, October 17–19, 2007, ID 111213–MS, p. 7. <http://www.onepetro.org/mslib/servlet/onepetropreview?id=SPE-111213-MS> Accessed 16.10.12.

Schlumberger. (1991). *Log interpretation principles/applications*. Schlumberger Ltd. <http://www.slb.com/resources/publications/books/lipa.aspx> Accessed 16.10.12.

Schlumberger. (2009). *Logging solutions for optimizing field development—Well evaluation for coalbed methane*. p. 8. < http://www.slb.com/~/media/Files/industry_challenges/unconventional_gas/brochures/well_evaluation_coalbed_methane_08os141.pdf> Accessed 12.10.12.

Schlumberger. (2012a). *Hydrostatic head and hydrostatic pressure*. Schlumberger Oilfield Glossary. <http://www.glossary.oilfield.slb.com/Display.cfm?Term=hydrostatic%20head> Accessed 17.04.12.

Schlumberger. (2012b). *Oilfield glossary: Caliper log*. Schlumberger Ltd. <http://www.glossary.oilfield.slb.com/Display.cfm?Term=caliper%20log> Accessed 16.10.12.

Schlumberger. (2012c). *Schematic diagram of depositional environments*. Oilfield Glossary, Schlumberger. <http://www.glossary.oilfield.slb.com/DisplayImage.cfm?ID=56> Accessed 25.02.12.

Schmidt, S., Gerling, P., Thielemann, T., & Littke, R. (2007). *Comparability of hard coal reserves and resources in Europe*. Germany: BGR.

Schmidt, R. A. (2007). *Characterization of the bacterial and archaeal communities in coalbed methane aquifers in the Wyodak-Anderson coal zone of the Powder River Basin, WY using polar lipid analysis techniques*. MS Thesis. Colorado School of Mines. p. 130.

Schmoker, J. W. (2002). Resource-assessment perspectives for unconventional gas systems. *American Association of Petroleum Geologists Bulletin, 86*, 1993–1999.

Schmoker, J. W. (2004). U.S. Geological Survey assessment concepts and model for continuous petroleum accumulations. In *Total petroleum system and assessment of coalbed gas in the Powder River Basin province, Wyoming and Montana* (USGS Powder River Basin Province Assessment Team), U.S. Geological Survey Digital Data Series-69-C, Chapter 4, CD-ROM version 1, p. 7.

Schmoker, J. W., & Dyman, T. S. (1996). *"New" unconventional, technically recoverable, United States gas resources for the short to intermediate term*. U.S. Geological Survey Open-File-Report 96–066, p. 8.

Schmoker, J. W., & Klett, T. R. (2005). U.S. Geological Survey assessment concepts for conventional petroleum accumulations. In *USGS southwestern Wyoming province assessment team, compilers, petroleum systems and geologic assessment of oil and gas in the southwestern Wyoming province, Wyoming, Colorado and Utah*. U.S. Geological survey Digital data Series DDS-69-D, (Chapter 19), p. 7, CD-ROM. <http://certmapper.cr.usgs.gov/data/noga00/natl/text/CH_19.pdf> Accessed 01.09.11.

Schobert, H. H., Karner, F. R., Kleesattel, D. R., & Olso, E. S. (1985). *Characterization of the components of lithologic layers of North Dakota lignites*. Sydeny, Australia: Conference on Coal Science.

Schoell, M. (1980). The hydrogen and carbon isotopic composition of methane from natural gases of various origins. *Geochimica et Cosmochimica Acta, 44*, 649–661.

Scholes, P. L., & Johnston, D. (1993). Coalbed methane applications of wireline logs. In B. E. Law, & D. D. Rice (Eds.), *American Association of Petroleum Geologists*

Studies in Geology: Vol. 38. Hydrocarbons from Coal (pp. 287–302).

Schopf, J. M. (1956). A definition of coal. *Economic Geology, 51*, 621–627.

Schopf, J. M. (1960). *Field description and sampling of coal beds.* U.S. Geological Survey Bulletin 111B, 25–67.

Schraufnagel, R. A. (1993). Coalbed methane production. In B. E. Law, & D. D. Rice (Eds.), *American Association of Petroleum Geologists Studies in Geology: Hydrocarbons from Coal.*, p. 38.

Schultz, K. (2000). *Environmental and economic reasons to reduce methane emissions from coalmines. Minimize hazards and maximize profits through the recovery of coalbed and coal mine methane.* New York: Strategic Research Institute.

Schulze-Makuch, D., Goodell, P., Kretzchmar, T., & Kennedy, J. F. (2003). Microbial and chemical characterization of a groundwater flow system in an intermontane basin of southern New Mexico, USA. *Hydrogeology Journal, 11*, 401–412.

Schumm, S. A. (1993). River responses to base level changes: Implications for sequence stratigraphy. *Journal of Geology, 101*, 279–294.

Schurr, G. W., & Ridgley, J. L. (2002). Unconventional shallow biogenic gas systems. *American Association of Petroleum Geologists Bulletin, 86*, 1939–1969.

Schweinfurth, S. P. (2002). *Coal: A complex natural resource.* U.S. Geological Survey Circular 1143. p. 39.

Schwerdtfeger, G. (1980). *Comparison of peatland-classification in different national systems of soil science.* Proceedings of the 6th International Peat Congress, Duluth. (pp. 93–95).

Schwochow, S. (1997). The international coal seam gas report. In S. Schwochow (Ed.), *Cairn Point Publishing* (p. 218).

ScienceDaily. (2009). Peat fires drive temperature up: burning rain forest release huge amounts of greenhouse gases. *Science News.* <http://www.sciencedaily.com/releases/2009/11/091127132838.htm> Accessed 08.02.12.

Scientific American. (2011). *Just hot air?: are natural gas environmental benefits overstated?* <http://www.scientificamerican.com/article.cfm?id=are-natural-gas-eco-benefits-overstated> Accessed 07.01.11.

Scientific American. (2012). What makes us human. *Scientific American, Special Collectors Edition, 22*, 111. ISSN 1936–1513.

Scotese, C. R. (2013). *Paleogeographic maps. PALEOMAP project, Arlington, Texas.* p. 52. <www.scotese.com> Accessed 20.03.13.

Scott, A. D., Kaiser, W. R., & Ayers, W. B., Jr. (1991). Composition, distribution, and origin of Fruitland Formation and Pictured Cliffs Sandstone Gases, San Juan Basin, Colorado and New Mexico. In S. D. Schwochow, D. K. Murray, & M. F. Fahy (Eds.), *Coalbed methane in western North America* (pp. 93–109). Rocky Mountain Association of Geologists.

Scott, A. R., Kaiser, W. R., & Ayers, W. B., Jr. (1994). Thermogenic and secondary biogenic gases, San Juan Basin, Colorado and new Mexico—Implications for coalbed gas producibility. *American Association of Petroleum Geologists Bulletin, 78*, 1186–1209.

Scott, A. C., Mattey, D. P., & Howard, R. (1996). New data on the formation of Carboniferous coal balls. *Review of Palaeobotany and Palynology, 93*, 317–331.

Scott, S., Anderson, B., Crosdale, P., Dingwall, J., & Leblang, G. (2007). Coal petrology and coal seam gas contents of the Walloon Subgroup- Surat Basin, Queensland, Australia. *International Journal of Coal Geology, 70*, 209–222.

Scott, A. R. (1997). *Timing of cleat development in coal beds.* Dallas, Texas: American Association of Petroleum Geologists Annual Convention.

Scott, A. R. (2004). *Historical perspectives and future opportunities of coalbed methane.* American Association of Petroleum Geologists, Hedberg Research Conference, September 12–16, 2004, Vancouver, Canada. <http://www.searchanddiscovery.com/documents/abstracts/2004hedberg_vancouver/extended/scott/scott.htm9> Accessed 15.09.11.

Scott, A. R., Zhou, N., & Levine, J. R. (1995). A modified approach to estimating coal and coal gas resources: example from the Sand Wash Basin, Colorado. *American Association of Petroleum Geologists Bulletin, 79*, 1320–1336.

SeaCrest Group. (2003). *Water quality data collected from water wells in the Raton Basin, Colorado.* Prepared for the Colorado Oil and Gas Conservation Commission, p. 34. < http://cogcc.state.co.us/Library/RatonBasin/Seacrest_Final_Report.pdf> Accessed 28.11.12.

Secor, D. T. (1965). Role of fluid pressure in jointing. *American Journal of Science, 263*, 633–646.

Seidle, J.P. 1999. A Modified p/Z Method for Coal Wells. Paper SPE 55605 presented at the SPE Rocky Mountain Regional Meeting, Gillette, Wyoming, USA, 15–18 May. DOI: 10.2118/55605-MS.

Seidle, J. P., & Huitt, L. G. (1995). *Experimental measurement of coal matrix shrinkage due to gas desorption and implications for cleat permeability increase.* Beijing, China: SPE International Meeting on Petroleum Engineering, 14–17 November. SPE 30010.

Selley, R. C. (2005). *UK shale-gas resources.* In *Geological Society, London Petroleum Geology Conference Series* (Vol. 6). 707–714.

Senel, I. G., Gürüz, A. G., & Yücel, H. (2001). Characterization of pore structure of Turkish coals. *Energy & Fuels, 15*, 331–338.

Senturk, Y. (2011). Assessment of petroleum resources using deterministic procedures. In *Guidelines for application of the petroleum resources management system.* Sponsored by Society of Petroleum Engineers, American Association of

Petroleum Geologists, World Petroleum Council, Society of Petroleum Evaluation Engineers, and Society of Exploration Geophysicists, Chapter 4, pp. 35–77. <http://www.spe.org/industry/docs/PRMS_Guidelines_Nov2011.pdf> Accessed 16.06.12.

Setiawan, B. (March 27, 2009). *Indonesia's coal policy*. Directorate General of Mineral, Coal, and Geothermal, Ministry of Energy and mineral Resources of Indonesia, Japan Coal Seminar.

Settle, T., Mollock, G. N., Hinchman, R. R., & Negri, M. C. (1998). *Engineering the use of green plants to reduce produced water disposal volume*. Society of Petroleum Engineers. SPE Permian Basin Oil and Gas Recovery Conference, 23–26 March 1998, Midland, Texas, p. 7.

Sewandono, M. (1937). Inventarisatie en inrichting van de veenmoerasbosschen in het Panglonggebied van Sumatra's Oostkust. *Tectona, 30*, 660–679 (Inventory and distribution of peat swamp forests in the Panglong area along Sumatra's east coast. In Dutch.).

Seyler, C. A. (1924). Chemical classification of coal. *Fuel in Science and Practice, 3*, 15, 41, 79.

Shanley, K. W., & McCabe, P. J. (Eds.), (1998). Relative role of eustasy, climates, tectonism in continental rocks. *Society of Sedimentary Geology Special Publication* (Vol. 59; p. 234).

Shanmugam, G. (1985). Significance of coniferous rain forests and related organic matter in generating commercial quantities of oil, Gippsland Basin, Australia. *American Association of Petroleum Geologists Bulletin, 69*, 1241–254.

Sharma, S., & Frost, C. D. (2008). Tracing coal bed natural gas co-produced water using stable isotopes of carbon. *Ground Water, 4*, 329–334.

Sharma, S., Kyotani, T., & Tomita, A. (2000). Direct observation of layered structure of coals by a transmission electron microscope. *Energy & Fuels, 14*, 515–516.

Shearer, J. C., & Moore, T. A. (1994). Botanical control on banding character in two New Zealand coalbeds. *Palaeogeography, Palaeoclimatology, Palaeoecology, 110*, 11–27.

Shearer, J. C., & Moore, T. A. (1996). Effects of experimental coalification on texture, composition and compaction in Indonesian peat and wood. *Organic Geochemistry, 24*, 127–140.

Shearer, J. C., Moore, T. A., & Demchuk, T. D. (1995). Delineation of the distinctive nature of Tertiary coal beds. *International Journal of Coal Geology, 28*, 71–98.

Shepard, R. B. (2006). *Wetlands: An introduction for the non-scientist*. Troutdale, Oregon: Applied Ecosystem Services, Inc. <http://www.appl-ecosys.com/publications/tech-series-2.pdf> Accessed 25.02.12.

Sherwood, K. W., Larson, J., Comer, C. D., Craig, J. D., & Reitmeier, C. (2006). *North Aleutian Basin planning area: Assessment of undiscovered technically-recoverable oil and gas*. Anchorage, Alaska: U.S. Department of Interior, Minerals Management Service. <http://www.alaska.boemre.gov/re/reports/NAB06/North%20Aleutian%20Basin%20Assessment%20Report.pdf> Accessed 31.12.11.

Shi, J. Q., & Durucan, S. (2005). A model for changes in coalbed permeability during primary and enhanced methane recovery. *SPEREE, 8*, 291–299. SPE-87230-PA.

Shi, J. Q., & Durucan, S. (2005). Gas storage and flow in coalbed reservoirs: Implementation of a bidisperse pore model for gas diffusion in a coal matrix. *Society of Petroleum Engineers Reservoir Evaluation & Engineering, 8*, 169–175.

Shimizu, S., Akiyama, M., Ishijima, Y., Hama, K., Kunimaru, T., & Naganuma, T. (2006). Molecular characterization of microbial communities in fault-bordered aquifers in the Miocene formation of northernmost Japan. *Geobiology, 4*, 147–223.

Shimizu, S., Akiyama, M., Naganuma, T., Fujioka, M., Nako, M., & Ishijima, Y. (2007). Molecular characterization of microbial communities in deep coal seam groundwater of northern Japan. *Geobiology, 5*, 423–433.

Shimoyama, T., & Iijima, A. (1974). Influence of temperature on coalification of Tertiary coal in Japan. *American Association of Petroleum Geologists Bulletin, 58*, 1458.

Shoup, R. C., Lambiase, J., Cullen, A. B., & Caughery, C. A. (2010). Variations in fluvial-deltaic and coastal reservoirs deposited in tropical environments. *American Association of Petroleum Geologists Bulletin, 94*, 1477–1484.

Shtepani, E., Noll, L. A., Elrod, L. W., & Jacobs, P. M. (2010). A new regression-based method for accurate measurement of coal and shale gas content. *Society of Petroleum Engineers Reservoir Evaluation & Engineering, 13*, 359–364. ISSN 1094–6470.

Sidi, F. H., & Darman, H. (1995). Introduction: an overview of tectonics and sedimentation relationship. In H. Darman, & F. H. Sidi (Eds.), *Tectonics and sedimentation of Indonesia*. Indonesian Sedimentologists Forum Regional Seminar to commemorate 50th anniversary of the Geology of Indonesia Proceeding, p. 2.

Siegel, D. I., Reeve, A. S., Glaser, P. H., & Romanowicz, E. A. (1995). Climate-driven flushing of pore water in peatlands. *Nature, 374*, 531–533.

Sielfermann, G., Rieley, J. O., & Fournier, M. (1992). *The lowland peat swamps of central Kalimantan (Borneo): a complex and vulnerable ecosystem*. In *International conference on geography in the Asian region, Fakultas Geografica* (Vol. 1; pp. 1–26). University of Mada.

Sierra Club. (2009). *Stopping the coal rush*. <http://www.sierraclub.org/environmentallaw/coal> Accessed 08.04.09.

da Silva, Z. C. C. (1989). The rank classification of South Brazilian Gondwana coals on the basis of different chemical and physical properties. In P. C. Lyons, & B. Alpern (Eds.), *International Journal of Coal Geology: Vol. 13. Coal: Classification, coalification, mineralogy, trace-element chemistry, and oil and gas potential* (pp. 21–39).

Simpson, D. A. (2003). *Producing coalbed methane at high rates at low pressures*. Society of Petroleum Engineers, SPE Annual Technical Conference and Exhibition, Denver, Colorado, October 5–8, 2003, SPE84509, p. 8. <http://www.muleshoe-eng.com/sitebuildercontent/sitebuilderfiles/spe84509.pdf> Accessed 26.10.12.

Simpson, D. A. (2008). Coal bed methane. In J. F. Lea, H. V. Nickens, & M. R. Wells (Eds.), *Gas well deliquification* (2nd ed.). (pp. 405–422) Gulf Publishing. < http://www.sgtk.ch/rkuendig/dokumente/Coal_bed_methane.pdf>. Accessed 26.10.12.

Simpson, D. A., Lea, J. F., & Cox, J. C. (2003). *Coal bed methane production*. Society of Petroleum Engineers, SPE Production and Operations Symposium, March 23–25, 2003, SPE 80900, p. 10. <http://www.muleshoe-eng.com/sitebuildercontent/sitebuilderfiles/spe80900.pdf> Accessed 29.10.12.

Sing, K. S. W. (1982). Reporting physisorption data for gas solid systems - with special reference to the determination of surface-area and porosity. *Pure Applied Chemistry, 54*, 2201–2218.

Singh, S. (1997). *Coal – its origin and occurrence*. pp.1 Copyright © 1999–2011 by The Durham Mining Museum and its contributors Registered Charity No: 1110608, Page last updated: 03 Nov 2009. <www.dmm.org.uk/educate/cioao1.htm> Accessed 14.09.11.

Singh, R. D. (2004). *Principles and practices of modern coal mining*. New Age International (P) Limited. p. 688.

Singh, A. K., Kispotta, J., Singh, H., & Mendhe, V. A. (1999). Indian scenario of coal bed methane. In T. N. Singh, TN, & M. L. Gupta (Eds.), *Proceedings of International Symposium. On Clean Coal Initiatives* (pp. 729–737). New Delhi, India.

Singh, D. N., Kumar, A., Sarbhai, M. P., & Tripathi, A. K. (2012). Cultivation-independent analysis or archael and bacterial communities of formation water in an Indian coal bed to enhance biotransformation of coal into methane. *Applied Microbiology and Biotechnology, 93*, 1337–1350.

Singh, M. P., & Rakesh, S. R. (2004). Petrographic characteristics and depositional conditions of Permian coals of Pench, Kanhan, and Tawa Valley coalfields of Satpura Basin, Madhya Pradesh, India. *International Journal of Coal Geology, 59*, 209–243.

Singh, M. P., & Singh, G. P. (1995). Petrological evolution of Paleogene coal deposits of Jammu, Jammu and Kashmir, India. *International Journal of Coal Geology, 27*, 171–199.

Singh, P., Singh, M. P., Prachiti, P. K., Kalpana, M. S., Manikyamba, C., Lakshminarayana, G., et al. (2012). Petrographic characteristics and carbon isotopic composition of Permian coal: implications on depositional environment of Sattupalli coalfield, Godavari Valley, India. *International Journal of Coal Geology, 90–91*, 34–42.

Singh, P. K., Singh, G. P., & Singh, M. P. (2011). Characterization of coal seams II, III, and IIIA from Ramagundan coalfield, Godavari Valley, Andhra Pradesh, India. *Energy Sources, Part A, 33*, 1863–1870.

Singh, P. K., Singh, M. P., Singh, A. K., & Arora, M. (2010). Petrographic characteristics of coal from Lati formation, Tarakan Basin, east Kalimantan, Indonesia. *International Journal of Coal Geology, 81*, 109–116.

Skiba, U. M. (January 2009). *Methane to markets coal subcommittee report Poland, Jacek Skiba and Rafał Wojciechowski*. < http://www.methanetomarkets.org/documents/events_coal_20090127_subcom_poland.pdf> Accessed 23.12.11.

Sluiter, I. R. K., Kershaw, A. P., Holdgate, G. R., & Bulman, D. (1995). Biogeographic. ecological and stratigraphic relationships of the Miocene brown coal floras, Latrobe Valley, Victoria, Australia. *International Journal of Coal Geology, 28*, 277–302.

Smakowski, T., & Paszcza, H. (2010). *Hard coal reserves and resources in Poland and relationship to UNFC-2009*. International workshop on United Nations Framework Classification for Fossil Energy and Mineral Reserves and Resources 2009 (UNFC 2009) – Theory and Practice. Warsaw, 21–22 June 2010.

Smith, G. G. (1989). *Coal resources of Canada*. Geological Survey of Canada, Paper 89-4.

Smith, T. N. (1995). *Coalbed methane potential for Alaska and drilling results for the upper Cook Inlet Basin: Intergas*. Tuscaloosa: University of Alabama. May 15–19, pp. 1–21, 1995.

Smith, G. G., Cameron, A. R., & Bustin, R. M. (1994). *Coalfields of the western Canada sedimentary basin – Alberta Geological Survey*. < http://www.ags.gov.ab.ca/publications/DIG/ZIP/DIG_2008_0349.zip> Accessed 16.01.13.

Smith, G. G., Cameron, A. R., & Bustin, R. M. (2004). Coal resources in the eastern Canada sedimentary basin. In G. D. Mossop, & I. Shetsen, Comp. (Eds.), *Geological atlas of the western Canada sedimentary basin*. Canadian Society of Petroleum Geologists and Alberta Research Council. <http://www.ags.gov.ab.ca/publications/wcsb_atlas/atlas.html> Accessed 14.02.13.

Smith, J. W., & Pallasser, R. J. (1996). Microbial origin of Australian coalbed methane. *American Association of Petroleum Geologists Bulletin, 80*, 891–897.

Smith, D. M., & Williams, F. L. (1981). A new method for determining methane content of coal. In *Proceedings, 16th Intersociety Energy Conversion Engineering Conference, Atlanta, Ga., August 9–14, 1981* (pp. 1267–1272). American Society of Mechanical Engineers.

Smyth, M., & Buckley, M. J. (1993). Statistical analysis of the microlithotype sequences in the Bulli seam, Australia, and relevance to permeability of coal gas. *International Journal of Coal Geology, 22*, 167–187.

Smyth, M. (1983). Nature of source material for hydrocarbons in Cooper Basin, Australia. *American Association of Petroleum Geologists Bulletin, 67*, 1422–1426.

Smyth, M. (1989). Organic petrology and clastic depositional environments with special reference to Australian coal basins. *International Journal of Coal Geology, 12*, 635–656.

Snowdon, L. R. (1991). Oil from Type III organic matter: resinite revisited. *Organic Geochemistry, 17*, 743–747.

Snyder, G. T., Riese, W. C., Franks, S., Fehn, U., Pelzmann, W. L., Gorody, A. W., et al. (2003). Origin and history of waters associated with coalbed methane: 129I, 13 Cl, and stable isotope results from the Fruitland formation, CO and NM. *Geochimica et Cosmochimica Acta, 67*, 4529–4544. <http://www.geo.umass.edu/faculty/boutt/Courses/Fluids/Snyder%20et%20al%202003%20 Origin%20of%20water%20Fruitland%20Formation.pdf> Accessed 23.11.12.

Snyman, C. P., & Barclay, J. (1989). The coalification of South African coal. In P. C. Lyons, & B. Alpern (Eds.), *International Journal of Coal Geology: Vol. 13. Coal: Classification, coalification, mineralogy, trace-element chemistry, and oil and gas potential* (pp. 375–390).

Soil Survey Staff. (1975). *Soil taxonomy – a comprehensive system*. United States Department of Agriculture.

Solano-Acosta, W., Mastalerz, M., & Schimmelmann, A. (2007a). Cleats and their relation to geologic lineaments and coalbed methane potential in Pennsylvanian coals in Indiana. *International Journal of Coal Geology, 72*, 187–208.

Solano-Acosta, W., Schimmelmann, A., Mastalerz, M., & Arango, I. (2007b). Diagenetic mineralization in Pennsylvanian coals from Indiana, USA: 13C/12C and 18/16O implications for cleat origin and coalbed methane generation. *International Journal of Coal Geology, 73*, 219–236.

Somers, J. (2012). *Coalmine methane recovery and utilization*. U.S. Coal Mine Methane Conference, September 24, 2012. Las Vegas Nevada. <http://www.epa.gov/cmop/docs/cmm_conference_sep12/01_Somers.pdf> Accessed 16.01.13.

Soot, P. M. (1988). Non-conventional fuel tax. In J. E. Fassett (Ed.), *Geology and coal-bed methane resources of southern San Juan Basin, Colorado and New Mexico* (pp. 253–255). Rocky Mountains Association of Geologists.

Sorkhabi, R. (2012). Borneo's petroleum plays. GEOExPro Magazine. *South East Asia, 9*, 20–25.

Sowers, K. R., & Ferry, J. G. (2002). Methanogenesis in the marine environment. In G. Bitton (Ed.), *Encyclopedia of Environmental Microbiology* (pp. 1913–1923). New York: John Wiley & Sons, Inc.

Spackman, W. (1958). The maceral and the study of modern environments as a means of understanding the nature of coal. *Transaction, New York, Academy of Science Series II, 20*, 411–423.

Spackman, W., Dolsen, C. P., & Riegel, W. (1966). Phytogenic organic sediments and sedimentary environments in the Everglades-mangrove complex. *Palaeontographica, B111*, 135–152.

Spafford, S. D., & Schraufnagel, R. A. (1992). Multiple coal seam project. *Quarterly Review of Methane from Coal Seams Technology, 10*, 15–18.

Spath, P. L., & Mann, M. K. (2001). *Capturing and sequestering CO_2 from a coal-fired – assessing the net energy and greenhouse gas emissions*. U.S. Department of Energy, National Renewable Energy Laboratory. <http://www.netl.doe.gov/publications/proceedings/01/carbon_seq/p.4.pdf> Accessed 14.04.13.

Spears, D. A. (1987). Mineral matter in coal with special reference to the Pennine coalfields. In A. C. Scott (Ed.), *Coal and coal- bearing strata—Recent advances* (Vol. 32; pp. 171–185). Geological Society Special Publication.

Spears, D. A., & Caswell, S. A. (1986). Mineral matter in coals: cleat minerals and their origin in some coals from the English midlands. *International Journal of Coal Geology, 6*, 107–125.

Speight, J. G. (2005). *Handbook of coal analysis* (Vol. 166). Hoboken, New Jersey, U.S.: A John Wiley & Sons, Inc. p. 222.

Spindler, M. L., & Poundstone, W. N. (1960). *Experimental work on degasification of the Pittsburgh coal seam by horizontal and vertical drilling*. American Institute of Mining and Metallurgical Engineers Annual Meeting, Preprint No.60F106, New York, p. 28.

Stach, E., Mackowsky, M. Th., Teichmuller, M., Taylor, G. H., Chandra, D., & Teichmuller, R. (1982). *Stach's textbook of coal petrology* (3rd ed.). Stuttgart: Borntraeger.

Stammers, J. (2012). *Coal seam gas: Issues fro consideration in the Illawara region, NSW, Australia*. Ph.D. Dissertation. University of Wollongong. Research Online.

Stanford, J. A., & Hauer, F. R. (2003). *Coalbed methane (CBM) in Montana: Problems and solutions*. A White Paper. Flathead Lake Biological Station Division of Biological Sciences. University of Montana. p. 13. <http://www2.umt.edu/flbs/Research/_ResearchAssets/CBMFinal2-5-03.pdf> Accessed 12.12.12.

Stanley, R. G., Charpentier, R. R., Cook, T. A., Houseknecht, D. W., Klett, T. R., Lewis, K. A., et al. (2011). *Assessment of undiscovered oil and gas resources of the Cook Inlet region, south-central Alaska, 2011*. U.S. Geological Survey Fact Sheet fs2011–3068.

Stanton, R. W., Flores, R. M., Warwick, P. D., & Gluskoter, H. G. (2001). *Coal bed sequestration of carbon dioxide*. National Energy Technical Laboratory-Department of Energy. NETL/114, p. 12.

Stanton, R. W., Moore, T. A., Warwick, P. D., Crowley, S. S., & Flores, R. M. (1989). Comparative facies formation in selected beds of the Powder River Basin. In R. M. Flores, P. D. Warwick, & T. A. Moore

(Eds.), *Tertiary and Cretaceous coals in the Rocky Mountains Region* (pp. 19–27). American Geophysical Union, 28th International Geological Congress, Field Trip Guidebook T132.

Stark, T. J., & Cook, C. W. (2009). *Factors controlling coalbed methane production from Helper, Drunkards Wash and Buzzard Bench Fields, carbon and Emery Counties, Utah.* American Association of Petroleum Geologists Annual Convention and Exhibition, Denver, Colorado, June 7–10, 2009. <http://www.searchanddiscovery.net/abstracts/html/2009/annual/abstracts/stark.htm> Accessed 27.03.13.

Stasiulaitiene, I., Denafas, G., & Sliaupa, S. (2007). Lithuanian serpentines as potential carbon dioxide binders. In B. Sakkestad (Ed.), *32nd International Technical Conference on Coal Utilization and Fuel Systems: Proceedings, Coal Utilization and Fuel Systems* (pp. 429–440) (CD-ROM).

Staub, J. R., Among, H. L., & Gastaldo, R. A. (2000). Seasonal sediment transport and deposition in the Rajang River delta, Sarawak, East Malaysia. *Sedimentary Geology, 133*, 249–264.

Staub, J. R., & Esterle, J. S. (1992). Evidence for a tidally influenced upper carboniferous ombrogenous mire system; Upper Bench, Beckley Bed (Westphalian A), southern West Virginia. *Journal of Sedimentary Research, 62*, 411–428.

Staub, J. R., & Esterle, J. S. (1993). Provenance and sediment dispersal in the Rajang River delta/coastal plain setting, Sarawak, Easy Malaysia. In C. R. Fielding (Ed.), *Sedimentary Geology: Vol. 85. Current research in fluvial sedimentology* (pp. 191–201).

Staub, J. R., & Esterle, J. S. (1994). Peat-accumulating depositional systems of Sarawak, East Malaysia. *Sedimentary Geology, 80*, 91–106.

Staub, J. R., & Richards, B. K. (1993). Development of low-ash, planar peat swamps in an alluvial plain setting; the No. 5 block beds (Westphalian D) of southern West Virginia. *Journal of Sedimentary Research, 63*, 714–726.

Steidl, P. F. (1991). *Observations of induced fractures intercepted by mining in the Warrior Basin, Alabama.* Gas Research Institute. Topical Report 91/0327, p. 39.

Stevens, M. G. (2004). *Current perspectives on coal reporting guidelines.* Pincock Perspectives, Pincock, Allen & Holt. Issue 62, p. 4. < http://www.pincock.com/perspectives/Issue61-Coal-Reporting.pdf> Accessed 26.10.11.

Stevens, S. H., & Hadiyanto. (2004). *Indonesia: Coalbed methane indicators and basin evaluation.* Society of Petroleum Engineers, SPE Asia Pacific Oil and Gas Conference and Exhibition, Perth, Australia, October 18–20, 2004, SPE 88630, p. 9. <http://www.ijcea.org/papers/113-A618.pdf> Accessed 31.10.12.

Stewart, R., & Alder, D. (Eds.), (1995). *New South Wales petroleum potential* (p. 188). Sydney: New South Wales Department of Mineral Resources.

Stewart, W. N., & Rothwell, G. W. (1993). *Paleobotany and the evolution of plants* (2nd ed.). Cambridge: Cambridge University Press.

Stoeckli, H. F. (1995). Characterization of microporous carbons by adsorption and immersion techniques. In J. W. Patrick (Ed.), *Porosity in carbons, characterization, and applications* (p. 331). New York: Halsted Press.

Stoker, P. (2009). *Progress on the revision of the Chinese mineral resources and mineral reserves reporting standard.* < http://www.jorc.org/pdf/progress_revision_chinese_mineral1.pdf> Accessed 26.10.11.

Stoker, P. (2009). *The JORC code and coal.* Hunter Regional Branch Aus IMM Talk.

Stoltz, L. R., & Jones, M. S. (1998). *Probabilistic reserves assessment using a filtered Monte Carlo method in a fractured limestone reservoir.* Society of Petroleum Engineers. ID 39714-MS, ISBN 978-1-55563-390-5.

Stopes, M. (1919). On the four visible ingredients in banded bituminous coal. *Proceeding Royal Society of London, B9*, 470–487.

Stopes, M. C. (1935). On the petrology of banded bituminous coal. *Fuel, 14*, 4–13.

Stout, S. A., Boon, J. J., & Spackman, W. (1988). Molecular aspects of the peatification and early coalification of angiosperm and gymnosperm woods. *Geochimica Cosmochimica Acta, 52*, 405–414.

Stout, S. A., & Spackman, W. (1988). Notes on the compaction of a Florida peat and the Brandon lignite as deduced from the study of compressed wood. *International Journal of Coal Geology, 11*, 247–256.

Stracher, G. B., Prakash, A., & Sokol, E. V. (2010). *Coal and peat fires: A global perspective* (1st ed.). In *Coal – Geology and combustion* (Vol. 1) Elsevier Science. p. 380.

Strack, M., Kellner, E., & Waddington, J. M. (2005). Dynamics of biogenic gas bubbles in peat and their effects on peatland biogeochemistry. Global Biogeochemical Cycles. *American Geophysical Union, 19*, 1–9.

Strapoc, D., Maastalerz, M., Eble, C., & Schimmelmann, A. (2007). Characterization of the origin of coalbed gases in southeastern Illinois Basin by compound-specific carbon and hydrogen stable isotope ratio. *Organic Geochemistry, 38*, 267–287.

Strapoc, D., Mastalerz, M., Dawson, K., Macalady, J., Callaghan, A. V., Wawrik, B., et al. (2011). Biogeochemistry of microbial coal-bed methane. *Annual Review of Earth and Planetary Science, 39*, 617–656.

Strapoc, D., Mastalerrz, M., Schimmelmann, A., Drobniak, A., & Hedges, S. (2008). Variability of geochemical properties in a microbially dominated coal-bed gas system from the eastern margin of the Illinois

Basin, USA. In R. M. Flores (Ed.), *International Journal of Coal Geology: Vol. 76. Microbes, methanogenesis, and microbial gas in coal* (pp. 98–110).

Strapoć, D., Picardal, F. W., Turich, C., Schaperdoth, I., Macalady, J. L., Lipp, J. S., et al. (2008). Methane-producing microbial community in a coal bed of the Illinois Basin. *Applied and Environmental Microbiology, 74*, 2424–2432.

Stricker, G. D., & Flores, R. M. (2002). *Coalbed methane content in the Powder River Basin, Wyoming: Saturation by coal rank and depth*. Nineteenth Annual International Pittsburgh Coal Conference Proceedings, Pittsburgh, Pennsylvania, September 22–27, 2002, Session 6, p. 15. ISBN 1-890977-19-5, CD-ROM, unpaginated.

Stricker, G. D., & Flores, R. M. (2009). Depletion of coal reserves and its effect on carbon dioxide emissions. In B. Sakkestad (Ed.), *34th International Technical Conference on Coal Utilization and Fuel Systems: Proceedings, Coal Utilization and Fuel Systems* (p. 12).

Stricker, G. D., & Flores, R. M. (2009). *Factors controlling trace elements of environmental concern for Powder River Basin coal, Wyoming*. Portland, Oregon: Geological Society of America Annual Meeting. October.

Stricker, G. D., Flores, R. M., Ellis, M. E., & Klein, D. A. (2008). Post-combustion CO_2 capture: let the microbes ruminate! In B. Sakkestad (Ed.), *33rd International Technical Conference on Coal Utilization and Fuel Systems: Proceedings, Coal Utilization and Fuel Systems* (p. 11) (CD-ROM).

Stricker, G. D., Flores, R. M., McGarry, D. E., Stilwell, D. P., Hoppe, D. J., Stilwell, K. R., et al. (2006). *Gas desorption isotherm studies in coals in the Powder River Basin and adjoining basins in Wyoming and North Dakota*. U.S. Geological Survey Open File Report 2006–1174, p. 273.

Stricker, G. D., Flores, R. M., & Trippi, M. H. (September 12–15, 2010). *Coalbed reservoir characterization of coal lithotypes and cleat spacing in the Powder River Basin*. Wyoming: American Association of Petroleum Geologists. International Convention and Exhibition.

Styan, W. B., & Bustin, R. M. (1983). Petrography of some Fraser River delta peat deposits: coal maceral and microlithotype precursor in temperate-climate peats. *International Journal of Coal Geology, 2*, 321–370.

St. George, J. D., & Barakat, M. A. (2001). The change in effective stress associated with shrinkage from gas desorption in coal. *International Journal of Coal Geology, 45*, 105–113.

Su, X., Feng, Y., Chen, J., & Pan, J. (2001). The characteristics and origins of cleat in coal from Western North China. *International Journal of Coal Geology, 47*, 51–62.

Su, K. H., Shen, J. C., Chang, Y. J., & Huang, W. L. (2006). Generation of hydrocarbon gases and CO_2 from a humic coal: experimental study on the effect of water, minerals and transition metals. *Organic Geochemistry, 37*, 437–453.

Suarez-Ruiz, I., & Crelling, J. C. (Eds.), (2008). *Applied coal petrology: The role of petrology in coal utilization*. Burlington, MA: Academic Press, Elsevier Ltd. p. 381.

Sudarmono, S. T., & Eza, B. (1997). Paleogene basin development in Sundaland and its role to the petroleum systems in western Indonesia. In *Proceedings Indonesian Petroleum Association, 26th Annual Convention* (pp. 545–560).

Suggate, P. (1959). New Zealand coals: their geological setting and its influence on their properties. *New Zealand Department of Scientific and Industrial Research Bulletin, 134*, 113.

Suggate, R. P. (2002). Application of rank (Sr), a maturity index based on chemical analyses of coals. *Marine and Petroleum Geology, 19*, 929–950.

Suggate, R. P. (2006). Coal rank, coal type, and marine influence in the north Taranaki coalfields, New Zealand. *New Zealand Journal of Geology and Geophysics, 49*, 255–268.

Sumi, L. (2005). *Produced water from oil and gas production*. Oil and Gas Accountability Project, Presentation at the 2005 People's Oil and Gas Summit, Farmington, New Mexico, <http://www.earthworksaction.org/files/publications/Sumi2.pdf> Accessed 28.01.12.

Suping, P., & Flores, R. M. (1996). Modern Pearl River delta and Permian Huainan coalfield, China: A comparative sedimentary facies study. *Organic Geochemistry, 24*, 159–179.

Supreme Court of the United States. (1999). *AMOCO Production Company, on behalf of itself and the class it represents v. Southern Ute Indian Tribe, et al.* On Writ of Certiorari of the United States Court of Appeals for the Tenth Circuit, Justice Kennedy delivered the opinion of the court, No. 98–830, p. 14. <http://www.law.cornell.edu/supct/html/98-830.ZS.html> Accessed 24.09.12.

Susilawati, R., Papendick, S., Gilcrease, P., Esterle, J., & Golding, S. (2012). *Biogenic origins of coal seam gas in Indonesia – a natural analogue for bio-renewable energy in tandem with CO_2 sequestration*, 34th International Geological Congress Proceedings, Brisbane, Australia, August 5–10, 2012.

Suwarna, N., & Hermanto, B. (2007). Berau coal in East Kalimantan; its petrographics, characteristics, and depositional environment. *Jurnal Geologi Indonesia, 2*, 196–201.

Swinkels, W. J. A. M. (2010). Probabilistic reserves estimations. In *Guidelines for application of the petroleum resources management system* (pp. 78–91). Sponsored by Society of Petroleum Engineers, American Association of Petroleum Geologists, World Petroleum Council, Society of Petroleum Evaluation Engineers, and Society of Exploration Geophysicists. <http://www.spe.org/industry/docs/PRMS_Guidelines_Nov2011.pdf> Accessed 16.06.12.

Swinkels, W. J. A. M. (2011). Probabilistic reserves estimation. Chapter 5. In *Guidelines for Application of the Petroleum Resources Management System* (pp. 78–91).

Sykes, R., & Johansen, P. E. (November 2006). *Facies model for recognition of planar and raised mire coals and why the former are more oil-prone.* American Association of Petroleum Geologists International Conference and Exhibition, Perth, West Australia 5–8. < http://www.pesa.com.au/aapg/aapgconference/pdfs/abstracts/tues/OilfromCoalSources.pdf> Accessed 10.12.11.

Sykes, R., & Lindqvist, J. K. (1993). Diagenetic quartz and amorphous silica in New Zealand coals. *Organic Geochemistry, 20*(6), 855–866.

Sykes, R., Suggate, R.P., & King, P.R. (1991). Timing and depth of maturation in southern Taranaki Basin from reflectance and rank(s), New Zealand Petroleum Conference Proceedings (373–389). <http://www.nzpam.govt.nz/cms/pdf-library/petroleum-conferences-1/1991-petroleum-conference-proceedings/r-sykes-1-3-mb-pdf> Accessed 03.12.11.

Sykorova, I., Pickel, W., Christianis, K., Wolf, M., Taylor, G. H., & Flores, D. (2005). Classification of huminite—ICCP system 1994. *International Journal of Coal Geology, 62*, 85–106.

Szadecky-Kardoss, E. (1946). Uj elegyresek a neogenkoru barnaszeneinkbol (Neue Gememgteile aus den neogenen Braunkohlen Ungarns). *Banyász. Kohász. L., 79*, 25–30.

Tadros, N. Z. (Ed.), (1993), *Geological Survey of New South Wales, Memoir Geology: 12. The Gunnedah Basin, New South Wales* (p. 649).

Tanaka, Y., & Hamada, Y. (2005). *Summary of source rock potential of SC46 central Tanon Strait, Visayan Basin.* The Geological Society of the Philippines. Geocon 2005.

Tang, Y., Jenden, P. D., Nigrini, A., & Teerman, S. C. (1996). Modeling early methane generation in coal. *Energy & Fuels, 10*, 659–671.

Tao, M., Wang, W., Xie, G., Li, J., Wang, Y., Zhang, X., et al. (2005). Secondary biogenic coalbed gas in some coal fields of China. *Chinese Science Bulletin, 50*, 24–29.

Tasch, K. H. (1960). Die moglichkeiten der flozgleichstellung unter zuhhilfenahme von flozbildungsdiagrammen. *Berbau–Rundschau, 12*, 153–157.

Taylor, M., Bustin, M., Hancock, B., & Solinger, R. (2008). *Coalbed methane development in Canada-challenges and opportunities*, 33rd International Geological Congress, Oslo, Norway, August 6–14, 2008. < http://www.cprm.gov.br/33IGC/1324175.html> Accessed 13.02.13.

Taylor, G. H., Liu, S., & Shibaoka, M. (1983). Huminite and vitrinite macerals at high magnification. In *Proceedings International Conference in Coal Science.* Pittsburgh: International Energy Agency (pp. 397–400).

Taylor, C. D., & Ozgen, K. (2011). Historical development of technologies for controlling methane in underground coalmines. *International Journal of Coal Geology, 72*, 221–231.

Taylor, G. H., Teichmuller, M., Davis, A., Diessel, C. F. K., Littke, R., & Robert, P. (1998). *Organic petrology.* Borntraeger, Berlin: Gebruder. p. 704.

Teichmuller, M. (1989). The genesis of coal from the viewpoint of coal petrology. In P. C. Lyons, & B. Alpern (Eds.), *International Journal of Coal Geology: Vol. 12. Peat and coal: Origin, facies, and depositional models* (pp. 1–87).

Teichmuller, M., & Teichmuller, R. (1966). Geological causes of coalification. In R. F. Goyld (Ed.), *Coal science.* American Chemical Society Series 55, 133–155.

Tejoyuwono, N. R. M. (1979). Peat deposition, an idle stage in the natural cycling of nitrogen and its possible activation for agriculture. In *Nitrogen cycling in south-east Asian wet monsoonal ecosystems* (pp. 139–147). Australian Academy of Science, 1981.

Testa, S. M., & Pratt, T. J. (2003). Sample preparation for coal and shale gas resource assessment. In *Proceedings, International Coalbed Methane Symposium, Tuscaloosa, Alabama, May 5–9, 2003.* University of Alabama. CD-ROM, Paper 356, p. 12.

Thakur, P. C. (2006). *Handbook for methane control in mining: Chapter 6 — Coal seam degasification.* Pittsburg, PA: U.S. Department of Health and Human Services, Centers for Disease Control and Prevention, National Institute for Occupational Safety and Health. Information Circular No. 9486.

Theissen, R. (1920). Compilation and composition of bituminous coals. *Journal of Geology, 28*, 185–209.

Thielemann, T., Cramer, B., & Schippers, A. (2004). Coalbed methane in the Ruhr Basin, Germany: a renewable energy source? *Organic Geochemistry, 34*, 1537–1549.

Thiessen, R. (1920). *Structure in Paleozoic bituminous coal.* U.S. Bureau of Mines Bulletin Number 117, p. 296.

Thomas, J., Jr., & Damberger, H. H. (1976). *Internal surface area, moisture content, and porosity of Illinois coals: Variations with coal rank.* Illinois State Geological Survey Circular 493, p. 38.

Thomas, L. (2002). *Coal geology* (1st ed.). San Francisco: John Wiley & Sons.

Thomas, L. (2012). *Coal geology.* John Wiley & Ssons, Ltd. p. 431.

Thorez, J., Flores, R. M., & Bossiroy, D. (1997). *Tectonic control on Paleocene Rocks in the Northern Rocky Mountains, USA: Supported by Smectite Composition and Spatial-temporal Distribution.* International Clay Conference, Abstracts with Programs, Ottawa, Canada.

Thrasher, G. P. (1991). *Late Cretaceous source rocks of Taranaki Basin.* New Zealand Oil Exploration Conference. Proceedings, pp. 147–154.

Ticleanu, N., Scradeanu, D., Popa, M., Milutinovici, S., Popa, R., Preda, I., et al. (1999). The relation between the lithotypes of Pliocene coals from Oltenia and their main quality characteristics. *Bulletin of the Czech Geological Survey, 74*, 169–174.

Ting, F. T. C. (1977). Microscopical investigation of the transformation (diagenesis) from peat to lignite. *Journal of Microscopy, 109*, 75–83.

Ting, F. T. C. (1977). Origin and spacing of cleats in coal beds. *Journal of Pressure Vessel Technology, 99*, 624–626.

Tissot, B. P., Durand, B., Espitalie, J., & Combaz, A. (1974). Influence of the nature and diagenesis of organic matter in the formation of petroleum. *American Association of Petroleum Geologists Bulletin, 58*, 499–506.

Tissot, B. P., & Welte, D. H. (1984). *Petroleum formation and occurrence* (2nd ed.). Springer & Verlag. p. 699.

Tonnsen, R. R., & Miskimins, J. L. (2010). *A conventional look at an unconventional reservoir: Coalbed methane production potential in deep environments*. American Association of Petroleum Geologists Annual Convention and Exhibition, New Orleans, Louisiana, April 11–14, 2010. <http://www.ebookpp.com/to/tonnsen-pdf.html> Accessed 12.11.11.

Tornqvist, T. E., Wallace, D. J., Storms, J. A., Wallinga, J., van Dam, R. L., Blaauw, M., et al. (2008). Mississippi delta subsidence primarily caused by compaction of Holocene strata. *Letters, Nature Geoscience, 1*, 173–176. < http://www.tulane.edu/~tor/documents/NG2008.pdf> Accessed 24.02.12.

Treese, K. L., & Wilkinson, B. H. (1982). Peat-marl deposition in a Holocene paludal-lacustrine basin—Sucker Lake, Michigan. *Sedimentology, 29*, 375–390. < http://hdl.handle.net/2027.42/72052>.

Tremain, C. M., Laubach, S. E., & Whitehead, N. H. (1991). Coal fracture (cleat) patterns in upper Cretaceous Fruitland Formation, San Juan Basin, Colorado and New Mexico: implications for exploration and development. In S. Schowchow, D. K. Murray, & M. F. Fahy (Eds.), *Coalbed methane of western North America* (pp. 49–59). Rocky Mountain Association of Geologists.

Tretkoff, E. (2011). Research Spotlight: analyzing the ability of peat to trap gas bubbles. *EOS Transactions, American Geophysical Union, 92*, 168. <http://dx.doi.org/10.1029/2011EO190011>.

Trevelyan, M. (1909). *Folk-lore and folk-stories of wales*. London. Republished by EP Publishing, 1973, East Adsley, Wakefield, Yorkshire, England, p. 351. < http://books.google.com/books?id=zmYHrsC6cYIC&printsec=frontcover#v=onepage&q&f=false> Accessed 08.02.12.

Trigg, K. R., & Blevin, J. E. (2004). *The Bass Basin in a northwest-trending, intracratonic rift basin located in offshore Tasmania*. Provexplorer, Geoscience Australia, Australian Government, Entity ID: 22295. <http://www.ga.gov.au/provexplorer/provinceDetails.do?eno=22295> Accessed 09.12.11.

Trippi, M. H., Stricker, G. D., Flores, R. M., Stanton, R. W., Chiehowsky, L. A., & Moore, T. A. (2010). *Megascopic lithologic studies of coals in the Powder River Basin, Wyoming and adjacent basins in Wyoming and North Dakota*. U.S. Geological Survey Open-File Report 2010-1114, p. 17 (includes Appendices p. 195).

Tu, J. J. (2007). *Coal mining safety: China's Achilles' heel*. In *China security, 3*. World Security Institute. 36–53. <http://www.wsichina.org/cs6_3.pdf> Accessed 23.11.11.

Tully, J. (1996). Coal fields of the conterminous United States. *U.S. Geological Survey Open-File Report OF*, 96-92.

Tyler, R., Kaiser, W. R., Scott, A. R., Hamilton, D. S., & Ambrose, W. A. (1995). *Geologic and hydrologic assessment of natural from coal: Greater Green River, Piceance, Powder River, and Raton Basins, western United States*. Texas: Bureau of Economic Geology. Report of Investigations 228.

Ugle Metan. (2005). Methane in the atmosphere over Russia: TROICA experiments. In *Workshop on Modern Technologies of Detection and Elimination of Methane Leakages from Natural Gas Systems*. Tomsk, Russia, September 14—16-2005.

Ugle Metan. (2010). *Russia - removing barriers to coalmine methane recovery and utilization*. International Coal & Methane Research Center. <http://www.uglemetan.ru/undp.htm> Accessed 23.12.11.

Uglemetan. (2002). *Status of coalbed methane recovery and utilization in former Soviet Union*. Online report at <http://www.uglemetan.ru/HTML/WhitePapers3.htm>.

Ulery, J. P. (2008). *Explosion hazards from methane emissions related to geologic features in coalmines*. Pittsburg, PA: U.S. Department of Health and Human Services, Centers for Disease Control and Prevention, National Institute for Occupational Safety and Health. Information Circular No. 9503.

Ulery, J. P., & Hyman, D. M. (1991). The modified direct method of gas content determination—Applications and results. In *"The 1991 Coalbed Methane Symposium Proceedings," Tuscaloosa, Alabama, May 13–17, 1991* (pp. 489–500). Gas Research Institute, University of Alabama, U.S. Mine Safety and Health Administration, and Geological Survey of Alabama. Paper 9163.

Ulrich, G., & Bower, S. (2008). Active methanogenesis and acetate utilization in Powder River Basin coals, United States. In R. M. Flores (Ed.), *International Journal of Coal Geology: Vol. 76. Microbes, methanogenesis, and microbial gas in coal* (pp. 25–33).

UMEC. (2009). *Coahuila presentation to M2M meeting, Monterrey, Mexico, 2009*. Coal resource figures sourced from the Undersecretariat of Mining and Energetics of Coahuila. <http://www.methanetomarkets.org/documents/events_coal_20090127_techtrans_cabrera.pdf>.

Underschultz, J., Jeffrey, Connell L., & Sherwood, N. (September 12–15, 2010). *Coal seam gas in Australia: Resource potential and production issues*. Calgary, Alberta, Canada: American Association of Petroleum Geologists. International Conference and Exhibition.

UNDP. (2006). *Malaysia's peat swamp forests: conservation and sustainable use.* United Nation Development Programme, Malaysia, p. 33. <http://www.undp.org.my/uploads/Malaysia%20Peat%20Swamp%20Forest.pdf> Accessed 15.01.15.

UNDP. (2009). *Coal bed methane recovery and commercial utilization – Factsheet, United Nations development programme – India, New Delhi, India.* <http://data.undp.org.in/factsheets/ene/Coal-Bed-Methane-Recovery-Commercial-Utilization.pdf> Accessed 15.02.13.

UNEP. (2012a). *Alternative technologies for freshwater augmentation in small island developing states: Technologies applicable to small islands with specific problems or circumstances.* United Nations Environment Programme. <http://www.unep.or.jp/ietc/Publications/techpublications/TechPub-8d/desalination.asp> Accessed 28.12.12.

UNEP. (2012b). *Phytoremediation: An environmentally sound technology for pollution prevention, control, and remediation.* United Nations Environment Programme, Division of Technology, Industry and Economics, Newsletter and Technical Publications. Freshwater Management Series No. 2. <http://www.unep.or.jp/ietc/publications/freshwater/fms2/4.asp> Accessed 22.12.12.

UNESCO. (2003). *Water for people, water for life.* United Nations Educational, Scientific and Cultural Organization (United Nations World Water Development Report).

UNFC (2010). *United Nations framework classification for fossil energy and mineral reserves and resources 2009.* Economic Commission for Europe Energy Series No. 39, United Nations Publication ISBN, ISSN 1014–7225, p. 20, New York and Geneva. <http://www.unece.org/fileadmin/DAM/energy/se/pdfs/UNFC/unfc2009/UNFC2009_ES39_e.pdf> Accessed 27.10.11.

UNFCCC. (2009). *Russian Federation National Inventory Report 2009.* United Nations Framework Convention on Climate Change, Annex I Party GHG Inventory Submission, September 11, 2009. <http://unfccc.int/national_reports/annex_i_ghg_inventories/national_inventories_submissions/items/4771.php> Accessed 23.12.11.

Uniservices. (2002). *Report on laboratory coal permeability measurements – Bullwacker Cr samples.* Auckland Uniservices Limited, The University of Auckland. p. 13.

United Nations. (1988a). *International codification system for medium and high rank coals: World coal trade, electric power stations, coke-oven plants, and gasification plants.* United Nations, Geneva: Economic Commission for Europe. GE.88–31401/4178G Sales No. E.88.II.E.15, ISBN 92–1–116420–6, p. 30.

United Nations. (1998b). *International classification of in seam coals. Economic Commission for Europe.* United Nations, Geneva: Committee on Sustainable Energy. GE 98–31816, p. 42.

United Nations. (2004). Lignite (low rank) coal. In *International classification of in-seam coals* (Economic Commission for Europe. United Nations).

United Nations. (2008). *UNECE Ad Hoc group of experts harmonization of fossil energy and mineral resources terminology.* "Report of the Task Force on Mapping of the United Nations Framework Classification for Fossil Energy and Mineral Resources", p. 110 Accessed 15.07.12.

Unsworth, J. F., Fowler, C. S., & Jones, L. F. (1989). Moisture in coal: 2. maceral effects on pore structure. *Fuel, 69*, 18–26.

USDOE. (1993). *Drilling sideways – A review of horizontal well technology and its domestic application.* U.S. Department of Energy, Energy Information Administration, Office of Oil and Gas. p. 24. Washington, D.C., DOE/EIA–TR–0565.

USDOE. (2002). *Powder River Basin coalbed methane development and produced water management study.* U.S. Department of Energy, National Energy Technology Laboratory. p. 111. <http://www.fe.doe.gov/programs/oilgas/publications/coalbed_methane/PowderRiverBasin2.pdf> Accessed 21.12.11.

USDOE. (2003). *Multi-seam well completion technology: Implications for Powder River Basin coalbed methane production.* U.S. Department of Energy, Office of Fossil Energy and National Energy Technology Laboratory. DOE/NETL–2003/1193, p. 101, plus Appendix. <http://www.adv-res.com/pdf/Multi-Seam%20Well%20Completion%20Technology%20-%20Powder%20River%20Basin.pdf> Accessed 01.10.12.

USDOE. (2005). *Liquefied natural gas: Understanding the basic facts.* U.S. Department of Energy, Office of Fossil Energy. DOE/FE-0489, p. 20. <http://www.fe.doe.gov/programs/oilgas/publications/lng/LNG_primerupd.pdf> Accessed 06.11.12.

USDOE. (2008). *Coal-to-liquids technology: Clean liquid fuels from coal.* United States Department of Energy. < http://fossil.energy.gov/programs/fuels/publications/2-28-08_CTL_Brochure.pdf> Accessed 07.12.11.

USDOE. (2010a). *Carbon dioxide enhanced oil recovery.* U.S. Department of Energy, National Energy Technology Laboratory. < http://www.netl.doe.gov/technologies/oil-gas/publications/EP/CO2_EOR_Primer.pdf> Accessed 13.04.13.

USDOE. (2010b). *FutureGen 2.0. U.S. Department of Energy.* <http://www.fossil.energy.gov/programs/powersystems/futuregen/> Accessed 28.08.11.

USDOE. (2011). *Innovations for existing power plants.* U.S Department of Energy. <http://www.fossil.energy.gov/programs/powersystems/pollutioncontrols/> Accessed 03.09.11.

USDOE. (2012). *Liquid natural gas.* U.S. Department of Energy, Office of Fossil Energy. < http://www.fossil.energy.gov/

programs/oilgas/storage/lng/feature/whoproducesit.html> Accessed 30.10.12.

USDOI. (2005). *Assessment of contaminants associated with coal bed methane-produced water and its suitability for wetland creation or enhancement projects*. U.S. Fish & Wildlife. Service Region 6, Contaminant Report Number: R6/721C/05. EPA–HQ–OW–2004–0032–2422.

USEIA. (1991). *Annual energy review*. U.S. Energy Information Administration, Department of Energy. no. 0384–90.

USEIA. (1995). *Longwall mining*. Department of Energy, Energy Information Administration. DOE/EIA-TR-0588 Distribution Category UC-950, p. 63. < ftp://ftp.eia.doe.gov/coal/tr0588.pdf> Accessed 24.11.11.

USEIA. (1996). *U.S. coal reserves: A review and update*. U.S Energy Information Administration, Department of Energy/Energy Information Administration–0529(95) Distribution Category UC–950 p. 105. < ftp://ftp.USEIA.doe.gov/coal/052995.pdf> Accessed 19.09.11.

USEIA. (1998). *The natural gas 1998: Issues and trends*. U.S. Energy Information Administration. <http://www.USEIA.gov/oil_gas/natural_gas/analysis_publications/natural_gas_1998_issues_and_trends/it98.html> Accessed 03.11.11.

USEIA. (2003). *The global liquefied natural gas market and outlook*. U.S. Energy Information Administration. DOE/EIA <http://www.eia.gov/oiaf/analysispaper/global/exporters.html>.

USEIA. (2007). *US coalbed methane – past, present and future*. U.S. Energy Information Administration, 2 Parts. <http://www.eia.gov/oil_gas/rpd/cbmusa2.pdf> Accessed 05.02.13.

USEIA. (2008). *About natural gas pipelines: Intrastate natural gas pipeline segment*. United States Energy Information Administration. <http://www.eia.gov/pub/oil_gas/natural_gas/analysis_publications/ngpipeline/intrastate.html> Accessed 27.10.12.

USEIA. (2010a). *Country analysis briefs – Colombia*. Energy Information Administration. Washington, DC. <http://www.eia.doe.gov/cabs/Colombia/Coal.html> Accessed March, 2010.

USEIA. (2010b). *International energy outlook, 2010*. U.S. Energy Information Administration, 2010, DOE/USEIA-0484(2010), pp. 61–76. <http://205.254.135.24/oiaf/ieo/pdf/0484(2010).pdf> Accessed 15.09.11.

USEIA. (2010c). *Energy-related carbon dioxide emission*. U.S. Energy Information Administration. Report #: DOE/USEIA-0383(2010), release date: May 11, 2010. < http://205.254.135.24/forecasts/ieo/more_highlights.cfm#world> Accessed 19.09.11.

USEIA. (2011a). *International energy outlook 2011, coal, natural gas, electricity*. Energy Information Administration, Independent Statistics and Analysis, <http://www.eia.gov/forecasts/ieo/pdf/0484(2011).pdf> Accessed 12.11.11.

USEIA. (2011b). *Natural gas: International energy outlook 2011*. U.S. Energy Information Administration. <http://www.eia.gov/forecasts/ieo/nat_gas.cfm> Accessed 14.01.13.

USEIA. (2012a). *U.S. coalbed methane production*. United States Energy Information Administration. <http://www.eia.gov/dnav/ng/hist/rngr52nus_1a.htm> Accessed 28.03.13.

USEIA. (2012b). Early declines in coal production are more than offset by growth after 2015. In *Market trends – coal*. Energy Outlook 2012. United States Energy Information Administration. <http://www.eia.gov/forecasts/aeo/MT_coal.cfm> Accessed 02.04.13.

USEIA. (2012c). *U.S. crude oil, natural gas, and natural gas liquids proved reserves, 2010*. U.S. Energy Information Administration. <http://www.eia.gov/naturalgas/crudeoilreserves/pdf/uscrudeoil.pdf> Accessed 28.01.13.

USEIA. (2013a). *Natural gas weekly update*. U.S. Energy Information Administration. < http://www.webcrawler.com/info.wbcrwl.300.7/search/web?qsi=11&q=natural%20gas%20chart%20price&fcoid=4&fcop=results-bottom&fpid=2> Accessed 16.01.13.

USEIA. (2013b). *Coalbed methane data series*. U.S. Energy Information Administration; <http://www.eia.gov/dnav/ng/NG_ENR_COALBED_A_EPG0_R52_BCF_A.htm>. <http://www.eia.gov/dnav/ng/ng_prod_coalbed_s1_a.htm> Accessed 04.02.13.

USEIA. (2013c). *Coal regains some electric generation market share from natural gas*. U.S. Energy Information Administration. < http://www.eia.gov/todayinenergy/detail.cfm?id=11391> Accessed 01.05.13.

USEIA. (2013d). *Short-term energy outlook–coal*. United States Energy Information Administration. < http://www.eia.gov/forecasts/steo/report/coal.cfm> Accessed 06.04.13.

USEIA–DOE. (2008). *International energy outlook 2008, Coal, Chapter 4*. U.S. Energy Information Administration. <http://www.tulane.edu/~bfleury/envirobio/readings/International%20Energy%20Outlook%2008.pdf> Accessed 02.11.11.

USEPA. (1996). *Reducing methane emissions from coal mines in Russia: A handbook for expanding coalbed methane recovery and use in the Kuznetsk Coal Basin*. U.S. Environmental Protection Agency, Coalbed Methane Outreach Program, September 1996 <http://www.epa.gov/cmop/docs/int005.pdf>.

USEPA. (1999). *White paper: Guidebook on coalbed methane drainage for underground coalmines*. U.S. Environmental Protection Agency. p. 45. <http://www.epa.gov/cmop/docs/red001.pdf> Accessed 23.11.11.

USEPA. (2000). *Benefits of an enclosed gob well flare design for underground coalmines: Addendum to conceptual design for a coalmine gob well flare*. U.S. Environmental Agency EPA430–R–99–012, p. 8. <http://www.epa.gov/cmop/docs/022red.pdf> Accessed 25.11.11.

USEPA. (2001). Economic impact analysis of disposal options for produced waters from coalbed methane operations in EPA Region 8. Presented at a public meeting held by EPA, Billings, MT, September 25. Presentation is available at <http://www.epa.gov/region8/water/wastewater/npdeshome/cbm/PPT010925.pdf>.

USEPA. (2004). *Evaluation of impacts to underground sources of drinking water by hydraulic fracturing of coalbed methane reservoirs*. U.S. Environmental Protection Agency. EPA 816-R-04–003, A1–33. <http://www.epa.gov/ogwdw/uic/pdfs/cbmstudy_attach_uic_attach01_sanjuan.pdf> Accessed 18.11.12.

USEPA. (2005a). *Mountaintop mining/valley fills in Appalachia. Final programmatic environmental impact statement.* U.S. Environmental Protection Agency. Summary, p. 10. <http://www.epa.gov/region03/mtntop/index.htm> Accessed 17.11.11.

USEPA. (2005b). *Identifying opportunities for methane recovery at U.S. Coal Mines: Profiles of selected gassy underground coal mines 1999–2003.* U.S. Environmental Protection Agency. EPA 430-K-04-003. <http://www.epa.gov/cmop/docs/profiles_2003_final.pdf> Accessed 12.02.13.

USEPA. (2005c). *Agency focus on desalination: The underground injection control (UIC) program outlook.* U.S. Environmental Protection Agency.

USEPA. (2006). *Global mitigation of non-CO2 greenhouse gases, Executive summary*. United States Environmental Protection Agency, EPA 430–R–06–005. <http://www.epa.gov/climatechange/Downloads/EPAactivities/GlobalMitigationFullReport.pdf> Accessed 20.08.13.

USEPA. (2006a). Power plant to be largest run on coalmine methane. In *Coalbed methane notes.* U.S. Environmental Protection Agency. Coalbed Methane Outreach Program, May, 18, 2006. <http://yosemite.epa.gov/opa/admpress.nsf/4d84d5d9a719de8c85257018005467c2/8ec89e33e48a863f852571720063e8d7!OpenDocument> Accessed 23.11.11.

USEPA. (2006b). *Global anthropogenic non-CO_2 greenhouse gas emissions: 1990–2020.* Draft, U.S. Environmental Protection Agency, Office of Atmospheric Programs, Climate Change Division, June, 2006. <http://www.epa.gov/nonco2/econ-inv/international.html> Accessed 21.12.11.

USEPA. (2007). *Natural gas: Electricity from natural gas.* U.S. Environmental Protection Agency. <http://www.epa.gov/cleanenergy/energy-and-you/affect/natural-gas.html> Accessed 7.01.11.

USEPA. (2008). Global overview of CMM opportunities. In *Coalbed Methane Outreach Program.* United States Environmental Protection Agency, September 2008. < http://www.epa.gov/cmop/> Accessed 25.08.13.

USEPA. (September 2008). Global overview of CMM opportunities. In *Coalbed methane outreach program.* <http://www.methanetomarkets.org/resources/coalmines/docs/overviewfull.pdf>.

USEPA. (2008a). *Coalbed methane outreach program: Promoting coalmine recovery and use.* U.S. Environmental Protection Agency. p. 4. <http://www.epa.gov/cmop/docs/cmop_outreach_brochure.pdf> Accessed 24.12.11.

USEPA. (2008b). *Identifying Opportunities for Methane Recovery at U.S. Coal Mines: profile of selected gassy underground mines 2002-2006.* report No. EPA 430-K-04-003, U.S. Environmental Protection Agency, Coalbed Methane Outreach Program, Washington DC, September 2008 (revised January 2009). Accessed 12.12.11.

USEPA. (2008c). *Upgrading drained coal mine methane to pipeline quality: A report on the commercial status of systems suppliers.* United States Environmental Protection Agency. EPA-430–R08–004, p. 20. <http://www.epa.gov/cmop/docs/red24.pdf> Accessed 08.11.12.

USEPA. (2008d). *U.S. abandoned coal mine methane recovery project opportunities.* U.S. Environmental Protection Agency. EPA 430–R–08–002. <http://www.epa.gov/cmop/docs/cmm_recovery_opps.pdf> Accessed 12.02.13.

USEPA. (2009). *Coal mine methane recovery: A primer.* U.S Environmental Protection Agency EPA–430–R–09–013, p. 63. <http://www.epa.gov/cmop/docs/cmm_primer.pdf> Accessed 24.11.12.

USEPA. (2010a). *Coal mine methane country profiles.* U.S. Environmental Protection Agency. Coalbed methane outreach program. Global Methane Initiative. p. 320. <http://www.epa.gov/cmop/docs/spring_2011.pdf> Accessed 20.01.13.

USEPA. (2010b). *Detailed methane extraction: Detailed study report.* United States Environmental Protection Agency. Office of Water, EPA–820–R–10–002, p. 91. <http://water.epa.gov/lawsregs/lawsguidance/cwa/304m/upload/cbm_report_2011.pdf> Accessed 27.11.12.

USEPA. (2011a). *Coalbed methane outreach program (CMOP).* U.S. Environmental Protection Agency. <http://www.epa.gov/cmop/basic.html> Accessed 30.11.11.

USEPA. (2011b). *Green power equivalency calculator methodologies*. United States Environmental Protection Agency. p. 7. <http://www.USEPA.gov/greenpower/pubs/calcmeth.htm> Accessed 29.10.11.

USEPA. (2011c). *India pushes ahead with CMM projects*. U.S. Environmental Protection Agency. EPA-430-N-00–004.

USEPA. (2011d). *Investigation of ground water contamination in near Pavilion, Wyoming draft*. United States Environmental Protection Agency, Office of Research and Development. EPA 600–00–000, Ada, Oklahoma, p. 42 and Appendix.

USEPA. (2012a). *Drinking water contaminants.* United States Environmental Protection Agency. <http://water.epa.gov/drink/contaminants/index.cfm> Accessed 30.11.12.

USEPA. (2012b). *Basic information of injection wells.* United States Environmental Protection Agency. <http://water.epa.gov/type/groundwater/uic/basicinformation.cfm#how_do> Accessed 14.12.12.

USEPA. (2012c). *Greenhouse gas emissions.* United States Environmental Protection Agency. <http://epa.gov/climatechange/ghgemissions/gases/ch4.html> Accessed 07.04.13.

USEPA. (2012d). *Hydraulic fracturing background information.* United States Environmental Protection Agency. <http://water.epa.gov/type/groundwater/uic/class2/hydraulicfracturing/wells_hydrowhat.cfm> Accessed 10.10.12.

USGS. (1976). Coal resource classification system of the U.S. Bureau of Mines and U.S. Geological Survey. *U.S. Geological Survey Bulletin,* 7, 1450–B.

USGS. (1980). Principles of a resource/reserve classification for minerals. *U.S. Geological Survey, 831,* 5.

USGS. (1982). *Geological Survey research 1981.* U.S. Geological Survey Professional Paper 1275, p. 395.

USGS. (1998). *Estimated use of water in the United States in 1995.* U.S. Geological Survey. Circular 1200.

USGS. (2002). *Coal bed gas resources of the Rocky Mountain region.* United States Geological Survey Fact Sheet FS-158-02.

USGS. (2004a). *New Zealand coal resources.* U.S. Geological Survey Fact Sheet 2004–3089, September 2004, Online version 1.0.

USGS. (2004b). *Resources on isotopes, periodic table-oxygen.* U.S. Geological Survey Isotope Tracers Project. <http://wwwrcamnl.wr.usgs.gov/isoig/period/o_iig.html> Accessed 22.11.12.

USGS. (2006). *Toxicity of sodium bicarbonate to fish from coal-bed natural gas production in the Tongue and Powder River drainages, Montana and Wyoming.* U.S. Geological Survey Fact Sheet 2006–3092.

USGS. (2007). *Assessment of undiscovered oil and gas resources of the Illinois Basin, 2007.* U.S. Geological Survey Fact Sheet 2007–3058.

U.S. Geological Survey. (2011). *Assessment of undiscovered oil and gas resources of the Cook Inlet region, south-central Alaska, 2011.* U.S. Geological Survey, National Assessment of Oil and Gas Fact Sheet 2011–3068, p. 2. <http://pubs.usgs.gov/fs/2011/3068/fs2011-3068.pdf> Accessed 31.12.11.

USLegal, Inc. (2012). *Interest of landowner with full gas and oil rights.* <USLEGAL.Com.>. <http://oilandgas.uslegal.com/interest-of-landowner-with-full-gas-and-oil-rights/> Accessed 26.10.12.

USMRA. (2011). *Mine disasters in the United States.* <http://www.usmra.com/Mine_Disasters/> Accessed 16.12.11.

Valentine, D. L., Chidthaisong, A., Rice, A., Reeburgh, W. S., & Tyler, S. C. (2004). Carbon and hydrogen isotope fractionation by moderately thermophilic methanogens. *Geochimica et Cosmochimica Acta, 68,* 1571–1590.

Van Hamme, J. D., Singh, A., & Ward, O. P. (2003). Recent advances in petroleum microbiology. *Microbiology and Molecular Biology Reviews, 67,* 503–549.

Van Heek, K. H., Juntgen, H., Luft, K. F., & Teichmuller, M. (1971). Aussagen zur gasbilding in fruhen inkohlungsstadied auf grund von pyroyseversuchen. *Erdol und Kohle, 24,* 566–572.

Van Krevelen, D. W. (1961). *Coal: Typology, chemistry, physics, constitution.* Amsterdam: Elsevier.

Van Krevelen, D. W. (1981a). *Coal-typology, chemistry, physics, constitution.* Amsterdam: Elsevier. p. 514.

Van Krevelen, D. W. (1981b). *Coal: Coal science and technology* (Vol. 3). New York: Elsevier Scientific Publishing Co.. p. 407.

Van Krevelen, D. W. (1993). *Coal: Typology–Physics–Chemistry–Constitution* (3rd ed.). Amsterdam: Elsevier. p. 979.

Van Vliet, O. P. R., Faaij, A. P. C., & Turkenberg, W. C. (2009). Fischer–Tropsch diesel production in a well-to-wheel perspective: a carbon, energy flow and cost analysis. *Energy Conversion and Management, 50,* 855–876.

Van Voast, W. A. (1991). Hydrologic aspects of coal-bed methane occurrence, Powder River Basin. *American Association of Petroleum Geologists Bulletin, 75,* 1142–1143.

Van Voast, W. A. (2003). Geochemical signatures of formation waters associated with coalbed methane. *American Association of Petroleum Geologists Bulletin, 87,* 667–676.

VandrÈ, C., Cramer, B., Gerling, P., & Winsemann, J. (2007). Natural gas formation in the western Nile delta (Eastern Mediterranean): thermogenic versus microbial. *Organic Geochemistry, 38,* 523–539.

Vasander, H. (2005). Carbon fluxes from tropical peat swamps. In H. Wosten (Ed.), *Strategies for implementing sustainable management of peatlands in Borneo.* European Commission Inco-Dev. Contract No: ICA4-CT-2001-10098, pp. 103–122. <http://www.strapeat.alterra.nl/download/Final%20report%20STRAPEAT.pdf> Accessed 24.02.12.

Veil, J. A., Pruder, M. G., Elcock, D., & Redweik, R. J., Jr. (2004). *A white paper describing produced water from production of crude oil, natural gas, and coal bed methane.* Argonne National Laboratory Prepared for U.S. Department of Energy, National Energy Technology Laboratory, Contract W–31–109–Eng–38, p. 79. <http://www.circleofblue.org/waternews/wp-content/uploads/2010/08/prodwaterpaper1.pdf> Accessed 23.11.12.

Venter, J., & Stassen, P. (1953). *Drainage and utilization of firedamp.* U.S. Bureau of Mines Information Circular 7670, p. 22.

Verhoeven, J. T. A. (1986). Nutrient dynamics in minerotrophic peat mires. *Aquatic Botany, 25,* 117–137.

Verkade, J. G., Oshel, R. E., & Downey, R. A. (2009). *Pretreatment of coal.* United States Patent Application 20090193712.

Verma, A. (2005). *Experience and opportunities for methane projects in India.* New Delhi, India: Directorate General of Hydrocarbons.

Veto, I., Futo, I., Horvath, I., & Szanto, Z. (2004). Late and deep fermentative methanogenesis as reflected in the H–C–O–S isotopy of the methane-water system in deep aquifers of the Pannonioan Basin (SE Hungary). *Organic Geochemistry, 35*, 713–723.

Vidas, H., & Hugman, R. (2008). *Availability, economics, and production potential of North American unconventional gas supplies.* INGAA Foundation, Inc.

Virginia Department of Conservation and Recreation. (2012). *Palustrine system.* The Natural Communities of Virginia Classification of Ecological Community Groups, Virginia Department of Conservation and Recreation, Second Approximation (Version 2.5). <http://www.dcr.virginia.gov/natural_heritage/ncpalustrine.shtml> Accessed 20.02.12.

Vogler, L. (2006). *Canadian coal deposits.* Technische Universitat Bergakademie Freiberg. Germany. <http://www.geo.tu-freiberg.de/oberseminar/os06_07/Canadian%20Coal%20Deposits%20-%20Luise%20Vogler.pdf> Accessed 14.02.13.

Volkov, V. N. (1965). *On possible thickness decrease of layers in the interval peat–anthracite.* Soviet Geologiya (Soviet Geology), No. 5. p. 85–97.

Volkwein, J. C. (1995). *Method for in situ biological conversion of coal to methane.* United States Patent No. 5,424,195.

Vorobyov, A. E., Chekushina, T. V., Balykhin, G. A., & Gladush, A. D. (2004). Exploitation of technologically generated methane deposits by means of surface wells. In J. Kicki, & E. J. Sobcyzk (Eds.), *New technologies in underground mining safety in mines* (pp. 79–92). London, UK: International Mining Forum, 2004, Taylor & Francis Group plc.

Waechter, N. B., Hampton, G. L., & Shipps, J. C. (2004). *Overview of coal and shale gas measurement: field and laboratory procedures.* Proceedings of the 2004 Inte, 1–17. International Coalbed Methane Symposium, May 2004, The University of Alabama, Tuscaloosa, Alabama.

Wahyunto, S. R., & Subagjo, H. (2003). *Maps of peatland distribution and carbon content in Sumatra, 1990 – 2002, Wetlands International, Bogor, Indonesia. Reproduced within interactive Atlas of Indonesia's forests (CD-ROM).* Washington, D.C.: World Resources Institute.

Walker, P. L., Jr. (1981). Microporosity in coal: its characterization and its implication for coal utilization. *Philosophical Transaction Royal Society London, A, 300*, 65–81.

Walker. (2000). *Major coalfields of the world.* London, UK: IEA Coal Research. CCC/32.

Walsh, M. P. (1994). The new generalized material balance as an equation of a straight line: Part 2. In *Applications to saturated and non-volumetric reservoirs.* SPE 27728 presented at the 1994 Society of Petroleum Engineers, Permian Basin Oil and Gas Recovery Conference, Midland, TX. March 16–18, 1994.

Walsh, M. P. (1995). A generalized approach to reservoir material balance calculations. Presented at the International Technical Conference of Petroleum Society of CIM, Calgary, Canada, May 9–13. *Journal of Canadian Petroleum Technology, 34.* ID 956–01–07. <http://www.onepetro.org/mslib/servlet/onepetropreview?id=PETSOC-95-01-07> Accessed 17.06.12.

Walsh, M. P., Ansah, J., Raghavan, R., & R. (1994). The new generalized material balance as an equation of a straight line: Part 1. In *Applications to undersaturated and volumetric reservoirs.* Paper SPE 27684 presented at the 1994 Society of Petroleum Engineers Permian Basin Oil and Gas recovery Conference, Midland, TX. March 16–18, 1994.

Wandless, A. M., & Macrae, J. C. (1954). The banded constituents of coal. *Fuel, 13*, 4–15.

Wang, X. (2006). Influence of coal quality factors on seam permeability associated with coalbed methane production (Ph.D. thesis). Australia, Sydney: University of New South Wales. p. 338.

Wang, K. (2011). Adsorption characteristics of lignite in China. *Journal of Earth Science, 22*, 371–376.

Wang, Y. (2011). *Main technologies for CBM exploration and development in China.* International Gas Union Research Conference (IGRC), October 19–21, 2011, Seoul, Republic of Korea, Research Institute of Petroleum Exploration and Development, Langfang, China, October 13, 2010. <http://www.google.com/search?client=safari&rls=en&q=CBM+in+Qinshui+Basin&ie=UTF-8&oe=UTF-8#q=CBM+in+Qinshui+Basin&hl=en&client=safari&rls=en&psj=1&ei=5KspUbX8Hsj7qwG26YDoDQ&start=190&sa=N&bav=on.2, or.r_gc.r_pw.r_qf.&fp=9386443df6a5d2f6&biw=887&bih=627> Accessed 23.02.13.

Wang, Z. P., Delaune, R. D., Masscheleyn, P. H., & Patrick, W. H. (1994). Soil redox and pH effects on methane production in a flooded rice soil. *Soil Science Society American Journal, 57*, 382–385.

Wang, G. X., Massarotto, P., & Rudolph, V. (2009). An improved permeability model of coal for coalbed methane recovery and CO_2 geosequestration. *International Journal of Coal Geology, 77*, 127–136.

Wang, X., & Ward, C. (2009). *Experimental investigation of permeability changes with pressure depletion in relation to coal quality.* Tuscaloosa, Alabama: University of Alabama. International Coalbed Methane Symposium.

Wanless, H. R., & Weller, J. R. (1932). Correlations and extent of Pennsylvanian cyclothems. *Geological Society of America Bulletin, 42*, 1003–1026.

Ward, C. R. (Ed.), (1984). *Coal geology and coal technology* (pp. 345). Oxford: Blackwell.

Ward, C. R. (2002). Analysis and significance of mineral matter in coal seams. *International Journal of Coal Geology, 50*, 135–168.

Ward, R. G. (2005). *Coal bed methane well automation*. International Coalbed Methane Symposium. University of Alabama. Tuscaloosa Paper 0524, p. 5. <http://www.controlmicrosystems.com/media/page-body-files/white-papers/WP_6.pdf>. Accessed 26.10.12.

Ward, C. R., Crouch, A., & Cohen, D. R. (1997). Identification of frictional ignition potential for rocks in Australian coal mines. In R. Doyle, J. Moloney, J. Rogis, & M. Sheldon (Eds.), *Safety in mines: The role of geology* (pp. 169–175). Newcastle: Coalfield Geology Council of New South Wales.

Warpinski, N. R., Branagan, P., & Wilmer, R. (1985). In-situ stress measurements at U.S. DOE's multiwell experiment site, Mesaverde Group, Rifle, Colorado. *Journal of Petroleum Technology, 37*, 527–536.

Warr, L. N. (2002). The Variscan orogeny: the welding of Pangea. In N. Woodcock, & R. Strachan (Eds.), *Geological history of Britain and Ireland* (pp. 271–294). Oxford: Blackwell.

Warwick, P. D. (2005). *Coal systems analysis*. Geological Society of America Issue. 378, p. 111.

Warwick, P. D. (2005). *Cretaceous and Tertiary coal-bed gas resources in the Gulf of Mexico coastal plain*. <http://www.searchanddiscovery.com/documents/abstracts/2005eastern/abstracts/warwick.htm> Accessed 10.11.11.

Warwick, P. D., Breland, F. C., Jr., & Hackley, P. C. (2008). Biogenic origin of coalbed gas in the northern Gulf Coast of Mexico coastal plain, U. S. A. In R. M. Flores (Ed.), *International Journal of Coal Geology: Vol. 76. Microbes, methanogenesis, and microbial gas in coal* (pp. 119–137).

Warwick, P. D., Charpentier, R. R., Cook, T. A., Klett, T. R., Pollastro, R. M., & Schenk, C. J. (2007). *Assessment of undiscovered oil and gas resources in Cretaceous-Tertiary coal beds of the Gulf Coast region, 2007*. U.S. Geological Survey Fact Sheet 2007-3039, p. 2.

Warwick, P. D., & Flores, R. M. (2008). *Geologic and hydrologic characteristics of North American coal-bearing basins with biogenic gas accumulations*, 33rd International Geological Congress Proceedings, Oslo, Norway, August 6–15, 2008, CD-Rom.

Warwick, P. D., & Stanton. (1988). Depositional models for two Tertiary coal-bearing sequences in the Powder River Basin, Wyoming, USA. *Journal of the Geological Society, 145*, 613–620.

Waters, C. N. (2009). Carboniferous of northern England. *Journal Open University Geological Society, 30*, 5–16. <http://nora.nerc.ac.uk/10713/1/Waters_Final.pdf> Accessed 19.02.13.

WCA. (2013). *Coal bed methane*. World Coal Association. <http://www.worldcoal.org/coal/coal-seam-methane/coal-bed-methane/> Accessed 17.01.13.

WDEQ. (2006). *Potential groundwater impacts from coalbed methane impoundments*. Wyoming Department of Environmental of Quality. <http://deq.state.wy.us/wqd/wypdes_permitting/WYPDES_cbm/Pages/CBM_Watershed_Permitting/Tongue_PrairieDog_HangingWoman_Badger_Creek]/Tongue%20PHB%20Downloads/Meeting%202%206-8-06/Water%20Quality%20Div%20Impoundments%20Presenation.pdf> Accessed 11.12.12.

WDEQ. (2007). *CBM impoundment groundwater quality data update 2007*. Wyoming Department of Environmental Quality Groundwater Program. <http://www.wy.blm.gov/prbgroup/07minutes/05_fisher.pdf> Accessed 24.12.12.

WDEQ. (2011). *CBM permitting guideline for discharges to irrigated drainages of the Powder River Basin*. Wyoming Department of Environmental. p. 15. <http://deq.state.wy.us/wqd/wypdes_permitting/WYPDES_cbm/downloads/WDEQ_CBM_Permitting_Guideline_5-2011.pdf> Accessed 28.12.12.

Weaver, J. N. (1992). Coal in Latin America: 1992, Uruguay, Argentina, Chile, Peru, Ecuador, Colombia, Venezuela, Brazil, and Bolivia. U.S. Geological Survey Open-File Report 93–239, p. 60.

Webb, P. A. (2001). *Volume and density determinations for particle technologists*. Micrometrics Instrument Corporation. p. 16.

WEC. (2007a). *2007 survey of energy resources*. World Energy Council.

WEC. (2007b). *Survey of energy resources 2007*. <http://www.worldenergy.org/publications/survey_of_energy_resources_2007/default.asp> Accessed 27.01.13.

Wegener, A. (1912). Die Herausbildung der Grossformen der Erdrinde (Kontinente und Ozeane), auf geophysikalischer Grundlage (in German). *Petermanns Geographische Mitteilungen, 63*, 185–195, 253–256, 305–309. Presented at the annual meeting of the German Geological Society, Frankfurt am Main (January 6, 1912).

Welch, J. P. (2009). Reverse osmosis treatment of CBM produced water continues to evolve. *Oil and Gas Journal* 6. <http://www.water.siemens.com/SiteCollectionDocuments/Industries/Oil_and_Gas/Brochures/RO_Coalbd_Methane-OGJ.pdf> Accessed 16.12.12.

Weller, J. M. (1930). Cyclical sedimentation of the Pennsylvanian period and its significance. *Journal of Geology, 38*, 97–135.

Wetlands International. (2007). *Peatlands. Wetlands International Malaysia, Petaling Jaya, Selangor, Malaysia*. < http://malaysia.wetlands.org/OurWetlands/Peatlands/tabid/1641/Default.aspx> Accessed 27.02.12.

Whateley, M. K. G., & Spears, D. A. (Eds.). (1995), *London, Geological Society Special Publication: Vol. 82. European coal geology* (p. 324).

Whateley, M. K. G., & Tuncali, E. (1995). Origin and distribution of sulphur in the Neogene Beypazari lignite basin, central Anatolia, Turkey. In M. K. G. Whateley, & D. A. Spears (Eds.), *Geological Society, London, Special Publication: Vol. 82. European coal geology.* (pp. 307–303).

Whately, M. K. G., & Jordan, G. R. (1989). *Fan-delta-lacustrine sedimentation and coal development in the Tertiary Ombilin Basin, W Sumatra, Indonesia*. Geological Society, London. Special Publications, (Vol. 41; pp.317-332).

Wheaton, J., & Meredith, E. (March 30, 2009). *Montana regional coalbed methane ground-water monitoring program. Montana bureau of mines and geology*. Denver, CO: Montana Tech of the University of Montana. Presentation to the National Research Council Committee on Management and Effects of Coalbed Methane Development and Produced Water in the Western United States.

Wheaton, J., & Metesh, J. (2002). *Potential ground-water drawdown and recovery from coalbed methane development in the Powder River Basin, Montana*. Montana Bureau of Mines and Geology. Open-File Report MBMG 458, p. 53.

Wheeler, B. D. (1999). Water and plants in freshwater wetlands. In A. J. Baird, & R. L. Wilby (Eds.), *Eco-hydrology* (pp. 119–169). Routledge.

Whipkey, C. W., Cavaroc, V. V., & Flores, R. M. (1991). Uplift of the Bighorn Mountains, Wyoming and Montana—a sandstone provenance study. In *Evolution of sedimentary basins—Powder River Basin* (Vol. 1917; pp. D1–D20). U.S. Geological Survey Bulletin.

White, J. M. (1986). Compaction of Wyodak coal, Powder River Basin, Wyoming, U.S.A. *International Journal of Coal Geology, 6*, 139–147. This edition published in the Taylor*Francis e-Library, 2005. <http://books.google.com/books?id=S5OoOm3ZMXUC&pg=PA122&lpg=PA122&dq=topogenous+versus+ombrogenous&source=bl&ots=gJhly1IWAL&sig=sC8mUQGZB-WxYqn0pu__AzbPToA&hl=en&sa=X&ei=3f4qT_z-MYGC2AXWuN3fDg&ved=0CEEQ6AEwBQ#v=onepage&q=topogenous%20versus%20ombrogenous&f=false> Accessed 02.02.12.

Whitehead, R. L. (1996). *Ground water atlas of the United States*. U.S. Geological Survey Hydrologic. Investigations Atlas HA–0730–1, p. 24.

Whiticar, M. J. (1999). Carbon and hydrogen isotope systematics of microbial formation and oxidation of methane. *Chemical Geology, 161*, 291–314.

Whiticar, M. J., Faber, E., & Schoell, M. (1986). Biogenic methane formation in marine and freshwater environments: CO_2 reduction vs acetate fermentation – isotope evidence. *Geochimica et Cosmochimica Acta, 50*, 693–709.

Whitmore, T. C. (1975). *Tropical rain forests of the Far East*. Oxford: Clarendon Press.

Wicks, D. E., Schewerer, F. C., Militzer, M. R., & Zuber, M. D. (1986). *Effective production strategies for coalbed methane in the Warrior Basin*. Society of Petroleum Engineers, Paper SPE 15234, Unconventional Gas Technology Symposium, Louisville, Kentucky, 18–21 May 1986.

Widera, M. (2012). Macroscopic lithotype characterisation of the first polish (first Lusatian) lignite seam in the Miocene of central Poland. *Geologos, 18*, 1–11.

Wildman, J., & Derbyshire, F. (1991). Origins and functions of macroporosity in activated carbons from coal and wood precursors. *Fuel, 70*, 655–661.

Wilford, G. E. (1960). *Radiocarbon age determinations of Quaternary sediments in Brunei and northeast Sarawak*. British Borneo Geological Survey Annual Report, 1959.

Wilkins, R. W. T., & George, S. C. (2002). Coal as a source rock for oil: a review. *International Journal of Coal Geology, 50*, 317–361.

Wilkins, R. W. T., Wilmshurst, J. R., Hladky, G., Ellacott, M. V., & Buckingham, C. P. (1995). Should fluorescence replace vitrinite reflectance as a major tool for thermal maturity determination in oil exploration? *Organic Geochemistry, 22*, 191–209.

Williams, H. (2003). Modeling shallow coastal marshes using Cesium-137 fallout: preliminary results from the Trinity River, Texas. *Journal of Coastal Research, 19*, 180–188.

Williams, C. (2011). Sense and sense ability. *New Scientist, 201*, 20–29.

Windley, B. F. (1995). *The evolving continents* (3rd ed.). New York: John Wiley & Sons.

Winston, R. B. (1986). Characteristics features and compaction of plant tissues traced from permineralized peat to coal in Pennsylvanian coals (Desmoinesian) from the Illinois basin. *International Journal of Coal Geology, 6*, 21–41.

Winston, R. B. (1994). Models of the geomorphology, hydrology, and development of domes peat bodies. *Geological Society of America Bulletin, 106*, 1594–1604.

WOGCC. (2001). *Docket #226-2001*. Pennaco Energy Inc. Exhibit E–2. Wyoming Oil and Gas Conservation Commission, 2211 King Blvd., Casper Wyoming, p. 25.

WOGCC. (2013). *Coalbed methane database*. <http://wogcc.state.wy.us/> Accessed 05.04.13.

Wold, M. B., Davidson, S. C., Wu, B., Choi, S. K., & Koenig, R. A. (1995). *Cavity completion for coalbed methane stimulation – An integrated investigation and trial in the Bowen Basin, Queensland*. Society of Petroleum Engineers, SPE Annual Technical Conference and Exhibition, October 22–25, 1995, ID 30733–MS, p. 13.

Wold, M. B., & Jeffrey, R. G. (May 15–18, 1999). *A comparison of coal seam directional permeability as measured in laboratory core tests and in well interference tests*. Rocky Mountain Regional Meeting. Gillette, Wyoming: Society of Petroleum Engineers. ISBN 978–1–55563–364–6.

Wolf, M. (1988). Torf und Kohle. In H. Füchtbauer (Ed.), *Sedimente und sedimentgesteine – Teil II* (pp. 683–730). Stuttgart: Schweizerbart.

Wood, G. H., Jr., Kehn, T. M., Carter, M. D., & Culbertson, W. C. (1982). *Coal resource classification system of the United States Geological Survey*. U.S. Geological Survey. Circular 891, p. 65.

World Coal Association. (2010). *Coal statistics*. <http://www.worldcoal.org/resources/coal-statistics/> Accessed 05.09.11.

World Coal Association. (2011). *Coal mining*. < http://www.worldcoal.org/coal/coal-mining/>.

World Coal Association. (2012). *What is coal?* World Coal Association. <http://www.worldcoal.org/coal/what-is-coal/> Accessed 20.03.12.

World Energy Conference. (1978). *Survey of energy resources*. London: World Energy Conference.

World Energy Council. (2007). *2007 survey of energy resources, executive summary*. World Energy Council. p. 20. < http://www.worldenergy.org/documents/ser2007_executive_summary.pdf> Accessed 24.01.12.

World Energy Council. (2010). *2010 Survey of energy resources – Coal* (Vol. 1, pp. 1–40).

World Nuclear Association. (2011). *Safety of nuclear power reactors: Appendix 1. The hazards of using energy*. < http://www.world-nuclear.org/info/inf06app.html> Accessed 16.12.11.

Worsely, T. R., Moore, T. L., Fraticelli, C. M., & Scotese, C. R. (1994). Phanerozoic CO_2 levels and Global Temperatures inferred from changing paleogeography. In G. deV Klein (Ed.), *Geological Society of America, Special Paper: Vol. 288. Pangea: Paleoclimate, tectonics and sedimentation during accretion, zenith and breakup of a supercontinents* (pp. 57–74).

WQA. (2012). *Reverse osmosis*. Lisle, Illinois: The Water Quality Association. <http://www.wqa.org/sitelogic.cfm?ID=872> Accessed 16.12.12.

Xamplified. (2010). *Langmuir adsorption isotherm*. < http://www.chemistrylearning.com/langmuir-adsorption-isotherm/> Accessed 11.04.11.

Xue, S., Xing, S., Li, S., & Xie, J. (2011). A new sampling system for measurement of gas content in soft coal seams. *Applied Mechanics and Materials, 121–126*, 2459–2464.

Yakutseni, V. P., Petrova, Y. E., Law, B. E., & Ulmishek, G. F. (1996). Coalbed methane potential of the Pechora coalfield, Timan-Pechora Basin, Russia. *American Association of Petroleum Geologists Bulletin, 5*. Annual Convention, San Diego, CA, May 19–22, 1966.

Yalçin, M. N., Inan, S., Gürdal, G., Mann, U., & Schaefer, R. G. (2002). Carboniferous coals of the Zonguldak Nason (northwest Turkey): implications for coalbed methane potential. *American Association of Petroleum Geologists Bulletin, 86*, 1305–1328.

Yale. (2009). In ocean's depths, heat-loving "Extremophile" evolves a strange molecular trick. *Yale News*.

Yang, M., Ju, Y. W., Li, T., & Xu, G. (2011). Characteristics of coalbed produced water in the process of coalbed methane development. *Environmental Engineering and Management Journal, 10*, 985–993. <http://gdlab-en.gucas.ac.cn/SiteCollectionDocuments/Pub/017_Yang_11.pdf> Accessed 15.11.12.

Yao, S. P., Jiao, K., Zhang, K., Hu, W. X., Ding, H., Li, M. C., et al. (2011). An atomic force microscopy study of coal nanopore structure. *Chinese Science Bulletin, 56*, 2706–2712.

Yao, Y., Liu, D., Tang, D., Tang, S., Che, Y., & Huang, W. (2009). Preliminary evaluation of the coalbed methane production potential and its geological controls in the Weibei coalfield, southeastern Ordos Basin, China. *International Journal of Coal Geology, 78*, 1–15.

Yao, Y., Liu, D., Tang, D., Tang, S., & Huang, W. (2008). Fractal characterisation of adsorption pores of coals from North China: an investigation on CH_4 adsorption capacity of coals. *International Journal of Coal Geology, 73*, 27–42.

Yao, Y., Liu, D., & Qiu, Y. (2013). *Variable gas content, saturation and accumulation characteristics of Weibei coalbed methane pilot-production field in southeast Ordos Basin, China. American Association of Petroleum Geologists Bulletin*, 97, pp. 1371–1393. <http://dx.doi.org/10.1306/02131312123>.

Yee, D., Seidle, J. P., & Hanson, W. B. (1993). Gas sorption on coal and measurement of gas content. In B. E. Law, & D. D. Rice (Eds.), *American Association of Petroleum Geologists Studies in Geology: Vol. 38. Hydrocarbons from Coal* (pp. 203–218). Chapter 9.

Yerima, P. K., & Van Ranst, E. (2005). *Major soil classification systems used in the tropics: soils of Cameroon*. Canada, U.S., and UK.: Trafford Publishing. p. 298. <http://books.google.com/books?id=XXkAhSzJw-wC&pg=PA203&lpg=PA203&dq=topogenous+versus+ombrogenous&source=bl&ots=3LHCvECu4v&sig=mIue2S8b1r6aSdb_SJBpB5bp9C0&hl=en&sa=X&ei=tBIrT8f6KLC-2AW43_yVDw&ved=0CEUQ6AEwBg#v=onepage&q=topogenous%20versus%20ombrogenous&f=false> Accessed 02.02.12.

Yermakov, V. I., Lebedev, V. S., Nemchenko, N. N., Rovenskaya, A. S., & Grachev, A. V. (1970). Isotopic composition of carbon in natural gases in the northern part of the West Siberian Plain in relation to their origin. *Akademiya Nauk SSSR Doklady, 190*, 683–686.

Yi, J., Akkutlu, I., & Deutsch, C. (2008). Gas transport in bidisperse coal particles: Investigation for an effective diffusion coefficient in coalbeds. *Journal of Canadian Petroleum Technology, 47*, 20–26.

Yi, J., Akkutlu, I. Y., Karacan, C. O., & Clarkson, C. R. (2009). Gas sorption and transport in coals: a poroelastic medium approach. *International Journal of Coal Geology, 77*, 137–144.

Yongguo, Y., Yuhua, C., Yong, Q., & Qiuming, C. (2008). Monte-Carlo method for coalbed methane resource assessment in key coal mining areas of China. *Journal of China University of Geosciences, 19*, 429–435.

Yu, Z., Vitt, D. H., Campbell, I. D., & Apps, M. J. (2003). Understanding Holocene peat accumulation pattern of continental fens in western Canada. *Canadian Journal of Botany, 81*, 267–282.

Yumul, G. P., Jr., Dimalanta, C. B., Maglambayan, V. B., & Marquez, E. J. (2008). Tectonic setting of a composite terrane: a review of the Philippine island arc system. *Geosciences Journal, 12*, 7–17.

Zahnle, K. J. (1990). Atmospheric chemistry by large impacts. In V. L. Sharpton, & P. D. Ward (Eds.), *Geological Society of America, Special Paper. Global catastrophes in earth history* (pp. 271–300).

Zaritsky, P. V. (1975). On thickness decrease of parent substance of coal. International congress stratigraphy Carboniferous Geology, 7th Prefeld. *Compte Rendu*, 393–396.

Zarrouk, S. J. (2008). *Reacting flows in porous media: Complex multi-phase, multi-component simulation.* VDM Verlag Dr. Muller (Chapter 3), 66–85.

Zedler, J. B., & Kercher, S. (2004). Causes and consequences of invasive plants in wetlands: opportunities, opportunists, and outcomes. *Critical Reviews in Plant Sciences, 23*, 431–452. <http://www.globalrestorationnetwork.org/uploads/files/LiteratureAttachments/60_causes-and-consequences-of-invasive-plants-in-wetlands.p> Accessed 11.12.12.

Zeng, F. (2005). Coal, oil shale, natural bitumen, heavy bitumen and peat. In *Organic and inorganic geochemistry of coal* (Vol. 1). < http://www.eolss.net/Sample-Chapters/C08/E3-04-01-05.pdf>. Accessed 12.02.12.

Zhang, H. (2001). Genetical type of pores in coal reservoir and its research significance. *Journal of China Coal Society, 26*, 40–44 (in Chinese with English abstract).

Zhang, X., Cao, Z., Tao, M., Wang, W., & Ma, J. (2012). Geological and geochemical characteristics if secondary biogenic gas in coalbed gases. *Huainan Coalfield.* http://www.worldenergy Accessed 29.03.12.

Zhang, Y. G., & Chen, H. J. (1985). Concepts on the generation and accumulation of biogenic gas. *Journal of Petroleum Geology, 8*, 405–422.

Zhang, H., Cui, Y., Tao, M., Peng, G., Jin, X., & Li, G. (2005). Evolution of the CBM reservoir-forming dynamic system with mixed secondary biogenic and thermogenic gases in the Huainan coalfield, China. *Chinese Science Bulletin, 50*, 30–39.

Zhang, E., Hill, R. J., Katz, B. J., & Tang, Y. (2008). Modeling of gas generation from the Cameo coal zone in the Piceance Basin, Colorado. *American Association of Petroleum Geologists Bulletin, 92*, 1077–1106.

Zhang, H., Li, H., & Xiong, C. (1998). *Jurassic coal-bearing strata and coal accumulation in northwest China.* China Scientific Book Service.

Zhang, T., Liu, Q., Dai, J., & Tang, Y. (2007). *Natural gas chemistry in the Tarim Basin, China and its indication of gas filling history.* American Association of Petroleum Geologists. Annual Convention, Long Beach, California, April 1–4, 2007. <http://www.searchanddiscovery.com/documents/2007/07082zhang/> Accessed 20.04.12.

Zhang, S., Tang, S., Tang, D., Pan, Z., & Yang, F. (2010). The characteristics of coal reservoir pores and coal facies in Liulin district, Hedong coal field of China. *International Journal of Coal Geology, 81*, 117–127.

Zhao, X., & Jiang, X. (2004). Coal mining: most deadly job in China. *China Daily News.* <http://www.chinadaily.com.cn/english/doc/2004-11/13/content_391242.htm> Accessed 20.11.11.

Zhao, H., Vance, G. F., Ganjegunte, G. K., & Urynowicz, M. A. (2008). Use of zeolites for treating natural gas co-produced waters in Wyoming, USA. *Desalinization, 228*, 263–276.

Zhao, H., Vance, G. F., Urynowicz, M. A., & Gregory, R. W. (2009). Integrated treatment process using a natural clinoptilolite for remediating produced waters from coalbed natural gas operations. *Applied Clay Science, 42*, 279–385.

Zhong, N., Sherwood, N., & Wilkins, R. W. T. (2000). Laser-induced fluorescence of macerals in relations to its hydrogen richness. *Chinese Science Bulletin, 45*, 946–951.

Zimpfer, G. L., Harmon, E. J., & Boyce, B. C. (1988). Disposal of production waters from oil and gas wells in the Northern San Juan Basin, Colorado. In J. E. Fassett (Ed.), *1988 CBM Symposium. Geology and coal-bed methane resources of the northern San Juan Basin, New Mexico and Colorado* (pp. 183–198). Denver, CO: Rocky Mountain Association of Geologists. DCN 01190.

Zinder, S. H. (1993). Physiological ecology of methanogens. In J. G. Ferry (Ed.), *Methanogenesis: Ecology, physiology, biochemistry & genetics* (pp. 128–206). Chapman and Hall.

Zuber, M. D. (1996). *Basic reservoir engineering for coal.* Chicago, Illinois: Gas Research Institute. Reservoir Engineering, report GRI–94/0397.

Zuber, M. D., & Boyer, C. M., II (2002). Coalbed-methane evaluation techniques—the current state of the art. *Journal of Petroleum Technology, 54*, 66–68.

Zuber, M. D., & Boyer, C. M., II (2002). *Coalbed-methane evaluation techniques—The current state of the art.* Society of Petroleum Engineers, Technology today series, SPE72274, 66–68. <http://www.spe.org/jpt/print/archives/2002/02/JPT2002_02_techtoday_series.pdf> Accessed 16.10.16.

Index

Page numbers followed by "b", "f," and "t" denote boxes, figures, and tables, respectively.

D

E

F

G

H

I

J

K

L

M

Q

R

X

Y

Z

Printed and bound by CPI Group (UK) Ltd, Croydon, CR0 4YY

08/05/2025

01864848-0001